MW01480536

Biomechanics and Biomaterials in Orthopedics

Springer
London
Berlin
Heidelberg
New York
Hong Kong
Milan
Paris
Tokyo

Dominique G. Poitout (Ed.)

Biomechanics and Biomaterials in Orthopedics

With Forewords by Professor Reinat Kotz and Professor Karl-Göran Thorngren

Springer

Dominique Poitout, MD
Professeur à l'Université de la Mediterranée
Chef de Service
Service de Chirurgie Orthopédique et de Traumatologie
Centre Hospitalo-Universitaire Nord
Marseille Cedex 20
France

British Library Cataloguing in Publication Data
Biomechanics and biomaterials in orthopedics
1. Orthopedics 2. Biomechanics 3. Human mechanics
4. Musculoskeletal system – Mechanical properties
5. Biomaterials
I. Poitout, Dominique G.
616.7
ISBN 1852334819

Library of Congress Cataloging-in-Publication Data
Biomechanics and biomaterials in orthopedics / [edited by] Dominique G. Poitout ; with forewords by R. Kotz and Karl-Göran Thorngren.
p. ; cm.
Includes bibliographical references and index.
ISBN 1-85233-481-9 (alk. paper)
1. Orthopedic implants. 2. Human mechanics. 3. Biomedical materials. I. Poitout, Dominique G.
[DNLM: 1. Biocompatible Materials. 2. Orthopedic Equipment. 3. Biomechanics. 4. Bone Cements. 5. Prostheses and Implants. 6. Tissue Adhesives. QT 37 B6146 2004]
RD755.5.B575 2004
617'.9 – dc21 2003050534

ISBN 1-85233-481-9 Springer-Verlag London Berlin Heidelberg
Springer–Verlag is part of BertelsmannSpringer Science+Business Media GmbH
springeronline.com

Printed in Singapore

Typeset by SNP Best-set Typesetter Ltd., Hong Kong
28/3830-543210 Printed on acid-free paper SPIN 10834401

Foreword

Reinat Kotz

The book *Biomechanics and Biomaterials in Orthopedics and Traumatology* gives a completely novel description of the problems. For the first time a comprehensive overview on all materials used for implants and their biologic compatibility is given. Furthermore, it contains interesting tissue biomechanics and histomorphometry on fractures and influences by non-biologic and biologic material and especially deals with the problems arising in cases of growing bones. Moreover, the principles of biomechanics are explained by several forms of application in the normal and pathologic skeleton. Materials and techniques are described by specialists from all over the world; this edition, therefore, offers an excellent contemporary overview of biocompatible materials and the biomechanics of the locomotive apparatus. Each author gives his own view of the matter thereby rendering the book a very individual and diverse spectrum of the problems under discussion. The truly international character of the book is reflected authentically by the cooperation with SICOT and SIROT.

Foreword

Karl-Göran Thorngren

The broad science field of the locomotor system covers research and knowledge with a wide span from molecular biology and bioengineering to application of new methods to patients and evaluation of the outcome. The biomechanics and biomaterials of orthopedics and traumatology have become increasingly important as the possibilities have increased to treat patients with foreign material introduced both as optimized osteosynthesis after trauma and as arthroplasties for joint diseases, sequelae of trauma, or in tumor treatment. Furthermore, substitutes for lacking tissues are emerging.

The present book, *Biomechanics and Biomaterials in Orthopedics and Traumatology*, provides an important update within this highly important field. The authors have been chosen among renowned researchers and clinicians from all over the world. The Coordinating Editor, Professor Dominique Poitout, has succeeded in composing a high-quality spectrum of chapters by authors from the international background of SICOT and SIROT. The international orthopedic research organization, SIROT, together with the international clinically founded major organization, SICOT, have both as their goal to provide exchange of knowledge and to present new achievements. Many of the authors have during recent years contributed to international knowledge at the meetings of SIROT and SICOT and this book now provides a unique possibility for permanent access to this gathered international knowledge in the field of locomotor system trauma and disease.

This book covers both basic concepts concerning biomaterials and biomechanics as well as their clinical application and experience from everyday practical use. The presentations here span from laboratory trials to patient satisfaction as well as from basic principles and ideas to long-term experience. It is the continuous feedback between laboratory experiments and clinical practice that has resulted in the high standard that today can be offered to patients with locomotor system trauma and disease. This book will provide an important basis for graduate and postgraduate learning by international specialists of orthopedics and traumatology.

Preface – Biomechanics and Biomaterials

J. C. Y. Leong and W. W. Lu

What is Biomechanics?

Biomechanics seeks to understand the mechanical structure and function of living systems, and can therefore be thought of as the study of biological systems from a mechanical point of view. In this book we concentrate on orthopedic biomechanics and related medical applications, which constitute the majority of recent work in the field of biomechanics. Orthopedic biomechanics has focused on the forces and moments acting on the tissues of the musculoskeletal system, such as bone, cartilage, growth plate, ligament, meniscus, synovial fluid, and tendon, in order to describe their function and behavior. The musculoskeletal system, although complex, serves the basic mechanical function of supporting the body while providing articulation and motion, and as such obeys Newton's basic laws of mechanics. The study of biomechanics has also been important in the development and design of many of the joint replacement and fracture fixation devices commonly used in orthopedic surgery today. Biomechanics helps us to understand the normal function of tissues and joints, predict changes due to aging or pathologic changes, and propose methods of intervention and replacement. Thus, diagnosis, surgery, and prosthetic design and implantation are closely associated with biomechanics.

Orthopedic biomechanics seeks to examine specific pathologic conditions through the study of areas such as joint instability, spinal deformities, gait pathologies, and fracture healing. Surgical procedures designed to restore normal mechanics may be critically evaluated, using techniques such as analysis of tendon force transfer, kinematic studies of ligament repair, or finite element analysis of joint replacements.

The discipline of biomechanics therefore incorporates a broad range of subjects and array of experimental techniques. These include, but are not limited to, the structural and geometric properties of the human body and its individual components; mechanics (elastic, creep, fatigue, failure and dislocation mechanics), materials science (metallurgy, ceramics, composite materials, and biomaterials); kinesiology, the science of human motion and locomotion, and the effect of environment, such as vibrations, etc. It is difficult to find any biological system that does not involve some of these areas.

What Contributions has Biomechanics made to Orthopedics and Traumatology?

Biomechanics has participated in many advances of medical science and technology. Molecular biology may appear far removed from biomechanics, but in its deeper reaches one has to understand the mechanics of the formation, design, function, and production of the molecules. Surgery seems to be an activity unrelated to mechanics, yet healing and rehabilitation are intimately related to the stress and strain in the tissues.

The achievements of biomechanics in the area of orthopedics and traumatology have included an increased understanding of the function of bones, muscles, ligaments, and tendons; including the relationship between stress and bone turnover, and the interaction between stress and piezoelectric behavior in bones. These efforts have led to a reduction in the non-union rate of fusions and fractures, and have also resulted in the production of substantially more effective prosthetic and orthotic devices, including total hip replacements, and internal and external fixators. Restoration of normal joint lubrication and other corrective processes have also been achieved. A special challenge has been the development of biocompatible and biodegradable implantable materials and the satisfaction of the necessary interface conditions. Biomechanics has also contributed substantially to the development of advanced surgical procedures such as novel methods of disk repair using suction, arthroscopy, and microsurgery reducing the risk of infection and allowing the patient to return to normal function in the shortest possible time. Today, biomechanics has become an everyday clinical tool in orthopedics. Fundamental research now covers not only surgery, prosthetics, implantable materials, and orthotics, but also cellular and molecular aspects of healing in relation to stress and strain, and tissue engineering of cartilage, tendon, and bone. In the long run, the most important contribution of modern biomechanics to medicine probably lies in its promotion of a better understanding of normal and pathologic human physiology.

Biomechanics Approach and Methodology

Theoretical Approach

Theoretical approaches to orthopedic biomechanics have typically involved the use of Newtonian mechanics in predicting the stresses and strains generated in tissues and joints by lifting, twisting, bending motions, etc. These, in combination with in vitro studies of the tissue failure properties, has allowed the prediction of safe levels of occupational exposure, and strategies to avoid overstraining the tissues involved. More recently, advanced theoretical techniques such as finite element model (FEM) analysis has been used to study loads within the tissues themselves, such as load sharing in fracture fixation implants, and is now a vital tool in implant design. FEM uses basic mechanical concepts combined with a knowledge of the materials stress–strain behavior, and then analyzes the forces in the body by considering it as an assemblage of small, brick-like elements. Depending upon the detail and precision required from the analysis, several thousand elements constituting a fine mesh may be used, or a coarse mesh of as little as a few dozen elements. No matter what the mesh size, each element obeys the rules of mechanics and is assigned appropriate elastic properties (elastic modulus, shear modulus, and Poisson's ratio), as well as appropriate failure properties (yield stress and ultimate stress). All forces crossing the boundaries of the elements are considered, and once the model and parameters have been established, the response of the model to any theoretical external load can be examined by running the model through standard computational software. In this way, theoretical experiments can be performed that would be difficult or impossible to carry out in practice. For example, a CT scan-based FE model, containing 11,604 cube elements, has been used to estimate femoral fracture load under very different loading conditions including simulating impact from a fall and loading during normal gait [1].

Empirical Approach

Biomechanical Study at the Cellular Level

Cell biomechanics may be defined as the application of principles and methods of both engineering and life sciences to understand the structure and function of normal and pathological cells. Cells of the body are exposed to mechanical stresses and strains throughout life. For years it has been recognized that the interaction between cells and mechanical factors is critical to the health and function of various tissues and organs of the body. Because of the staggering complexity of the in vivo environment, systematic study of phenomena of cellular responses to mechanical stimulation has relied heavily on the use of in vitro preparations. Such work frequently has involved cell culture systems with controlled delivery of a mechanical input such as hydrostatic pressure, fluid shear stress, or substrate strain. Because of no direct contact compression or tension of cells, the hydrostatic pressurization method has become popular as it has the advantages of simplicity, spatial homogeneity of the stimulus, ease of configuring multiple loading replicates, and ease of delivering and transducing either static and transient loading inputs.

In order to simulate "physiological" loads experienced by bone cells in vivo, a four-point bending substrate flexure provides an alternative means for delivering longitudinal strains to a culture surface. The low strain levels, typically several hundred to about a few thousand microstrain is similar to the range encountered by bone in vivo. The four-point bending system with rectangular culture plates is suspended within a culture well and permits study of either tensile or compressive strains depending on the sides of the substrate plate being loaded.

Another broad approach to mechanical stimulus of cell cultures has been by applying fluid shear stress. A wide range of in vivo cellular phenomena are recognized as being influenced by fluid shear, including both mechanoreception (e.g. plasma membrane receptors, protein kinase signaling, ion channels, etc.) and response (e.g., intracellular calcium, cytoskeletal remodeling, etc.). A few principal apparatus configurations have been used for fluid shear stress input. One is the cone-and-plate system, in which rotation is imposed about a cone axis oriented perpendicular to the surface of a flat plate. Depending on the conical taper and the imposed angular velocity, a wide range of shear stresses can be achieved, extending even into the turbulent flow regime. Another configuration is the parallel plate flow chamber in which a pressure differential is created between two slit openings at either end of a rectangular chamber, causing uniform laminar flow to develop across the culture surface. The parallel plate approach holds many practical attractions, including homogeneity of the stress stimulus, simplicity of the equipment, ease of medium sampling, and small volumetric fluid requirement. Laboratory apparatus devised for study of cellular response to mechanical stimulation generally feature an appreciable range of complexity and sophistication to replicate the staggering complexity of the in vivo environment.

Biomechanical Study at Tissue and Organ Level

The skeleton is first and foremost a mechanical structure. Its primary functions are to transmit forces from one part of the body to another and to protect certain other organs (e.g., the brain) from mechanical forces that could damage them. Therefore, the main purpose of orthopedic biomechanics is to define the mechanical properties and functions of the musculoskeletal system.

Forces in joints or ligaments and forces acting on bones in vivo have been measured by using transducers (e.g., Buckle transducer), and strain gauges with telemetry systems for signal transfer. For example, transducers have been implanted in the antero-medial portion of the anterior cruciate ligament to evaluate the effects of different types of rehabilitation exercises on strain magnitudes. Based on force-sensitive resistor technology, the three-dimensional plantar pressure distribution can also be measured in vivo. Information from such force sensors is being increasingly used by orthopedic surgeons both in clinical practice and research.

Recently, a robot with a universal force–moment sensor has been used to record the in situ force and path of motions of all the tissues in the knee joint. The robot is further able to define the distribution of the forces of different bundles in a ligament and to repeat the recorded initial knee positions or path of motions. This approach does not require mechanical contact with the tissues, so in situ forces in intact ligaments or tendons can be accurately determined. The data obtained can provide useful information for studying the mechanism of ligament or tendon injury, improvement of surgical reconstruction procedures, and rehabilitation protocols.

In principle, the approach to the study of problems in biomechanics consists of several steps:

Formulation of a hypothesis. A statement of an unknown issue in biomechanics that the investigator wishes to address is expressed as a research question or hypothesis.

Understanding of the anatomy and physiology of the subject, especially the musculoskeletal system. Detailed understanding of the morphology and histology of the tissues, and the structure and ultastructure of the materials, is required in order to know the geometric configuration of the object we are dealing with.

Understanding of the mechanical properties of the musculoskeletal system.

Determination of the mechanical properties of the materials or tissues in question. The material composition and the mechanical properties may change as a natural function of aging.

Application of the basic laws of physics (Newtonian and Maxwellian mechanics, conservation of energy, etc.) to the problem in hand. An accurate understanding of the environment in which the tissues function is also required to establish meaningful biomechanical boundary conditions.

Perform physiological experiments within the boundary conditions to obtain the required data to prove or refute the hypothesis.

What are Biomaterials?

In the field of orthopedics, the term biomaterials frequently refers to man-made materials used to construct prosthetic or other medical devices to replace or augment parts of the human body. However, a general definition of biomaterials includes not only materials designed for use in a biological system, but also those produced by living organisms, from animals to plants. No matter what the source, biomaterials must meet several criteria to perform successfully in orthopedic applications. They must be biocompatible, or able to function in vivo without eliciting an intolerable response in the body either locally or systemically. Appropriate biomaterials must be able to withstand the often hostile environment of the body, and show properties such as resistance to corrosion and degradation, such that the body environment does not adversely affect material performance over the intended performance lifetime of the implant. Adequate mechanical properties are also an important criterion for biomaterials, especially those used in devices intended to replace or reinforce load-bearing skeletal structures.

Implants or prostheses also place demands on the biomaterials from which they are made, depending on the function of the implant. For example, orthopedic biomaterials intended for total joint replacement must possess adequate wear resistance to maintain proper joint function and to minimize biocompatibility problems caused by biological reactions to particulate debris. They must be capable of reproducible fabrication to the highest standards of quality control and, of course, at a reasonable cost. Biomaterials that meet these criteria are fundamental to the practice of orthopedic surgery. Today, many types of biomaterials have been used successfully in the development of devices for internal fixation of fractures, osteotomies and arthrodeses, wound closure, soft-tissue reconstruction, and total joint arthroplasty that have advanced significantly the treatment of musculoskeletal diseases.

As a materials scientist or a clinician, one should disregard any notion that modern technology has the ability to replace any part of a living organism with an artificial organ which will be superior to the original structure. While it is possible to imagine situations in which this might be true in some limited sense, one always finds that the organism as a whole will never work better than when the original organ was in place. The reason for this barrier to improvement is that all living organisms are the result of millions of years of evolution – a process of "cut-and-try" engineering involving trials, thus exceeding by orders of magnitude anything which human engineers can manage. In the new millennium, the ultimate solution to most problems involving internal implants will probably come when we are able to control cell function well enough so that organs can be replaced biologically. With the rapid development of molecular biology and cellular/tissue engineering in the 1980s, this has become a plausible goal.

Biomaterials Commonly Used in Orthopedics

Biomaterials in orthopedic applications can be broadly categorized into metals, polymers, ceramics, and composites. The history of biomaterials in medicine can be traced back to the late 1800s and was precipitated by the rapid development of surgery as something other than an emergency procedure. In 1892 Levert experimented with lead, gold, silver, and platinum wire in dogs, but these metals clearly did not have the desired mechanical attributes, and without anesthesia, human patients could not endure long surgeries in order to implant meaningful prostheses or fixation devices. The advent of anesthesia just prior to the middle of the nineteenth century made surgery infinitely more tolerable to the patient, and the discovery of X-rays by Roentgen in the late 1800s revealed for the first time the true nature of many skeletal problems which had previously been misunderstood.

Following the advances in surgical technique, important developments in materials science led to the widespread use of man-made materials in orthopedic applications. In the 1930s, implants made from cobalt–chrome and stainless steel alloys were developed, and later in the 1950s, the development of polymer chemistry and plastics led to the economical production of high-quality polymeric-bearing materials such as ultra-high-molecular-weight polyethylene (UHMWPE). Further developments in the 1970s and 1980s included the production of resorbable polymers such as polylactic acid (PLA) and polyglycolic acid (PGA). The ability to further develop and fabricate many biomaterial-related devices compatible with biological tissues significantly advanced the ability of orthopedic surgeons to treat a great variety of musculoskeletal problems. It should be noted that most of the implant materials commonly in use today were developed more than 25 years ago, and the intervening years have been ones of gradual refinement. Today, almost all orthopedic implants involve some combination of the following metals, polymers, and ceramics.

Cobalt–chromium alloys were the first corrosion-resistant alloys to be developed, and have proven very effective in surgical implants, beginning in 1938 when Venable reported the use of such an alloy in orthopedics. A modification of the Co–Cr alloy was introduced in 1972, containing 35% nickel, and is thus known as MP 35N. It can be forged and heat-treated to obtain tensile strengths of 1,800 MPa, significantly above those of stainless steel and Co–Cr alloy.

Stainless steel (usually 316 L or 317 L) is a workhorse industrial alloy, which has been very successful as a surgical implant material. Cast stainless steels are unsuitable for orthopedic applications because of their large grain sizes and low fatigue strengths. Therefore, type 316LVM, a low-carbon, vacuum-melt stainless steel is preferred. Both stainless steel and cobalt–chrome alloys owe their corrosion resistance to the formation of a ceramic-like oxidation layer coating the surface, and it is important that this coating not be scratched during implantation.

Titanium alloys are primarily type Ti-6A1-4V, which contain 6% aluminum, 4% vanadium, and 90% titanium. This metal is becoming increasingly popular because its strength is as good as that of stainless steel and cobalt–chrome, but it is only half as stiff. This is potentially important because a large elastic modulus mismatch between implant and bone causes stress concentrations in some places and tends to unload the bone in others, even though the modulus of Ti-6A1-4V is till several times greater than bone.

Polymethyl methacrylate (PMMA) is an extremely common acrylic plastic otherwise known as Lucite, and is commonly used in orthopedics as bone cement, which was introduced by Charnley in the 1970s. Barium and antibiotics are frequently added to this polymer to increase its radiographic visualization and to prevent infection following surgery, respectively. In comparison to the metallic alloys, the mechanical properties of PMMA are similar to those of human bone, and barium and antibiotic additions do not substantially affect these properties.

Ultra-high-molecular-weight polyethylene (UHMWPE) has a very simple chemical structure. While a variety of other polymers have been tried, UHMWPE remains the best one for use as a bearing surface biomaterial because of its relatively low wear against metal. There have been many efforts to develop other polymers in the hopes of finding a material with better wear properties or lower cost than UHMWPE. However, while some slightly better materials have been produced, to date, very few of these have made it into actual clinical practice.

Ceramic materials are most commonly solid, inorganic compounds consisting of metallic and non-metallic elements held together by ionic or covalent bonding. They are very biocompatible and show exceptional wear resistance, but are stiffer and more brittle than other biomaterials. In recent orthopedic applications, ceramics have gained favor as biomaterials in two quite different aspects. The first involves their use in total joint replacement components as fully dense ceramics, such as alumina (Al_2O_3) and zirconia (polycrystalline ZrO_2), that possess inertness and high wear resistance superior to those of metallic alloys or polymers. The second involves the use of less-dense, even porous ceramics, such as hydroxyapatite (HA) and bioglass (Na_2O-CaO-P_2O_5-SiO_2), as bone graft substitutes and as coatings for metallic implants. These bioceramics are osteoconductive, providing surfaces to which bone will bond.

Biomaterials Produced by Human Cellular and Tissue Engineering

Tissue engineering is a multidisciplinary field that enlists the knowledge and experience of scientists involved in materials science, biomedical engineering, cell and molecular biology, and clinical medicine to produce constructs that can replace ill-functioning or missing tissues or organs. These constructs can be composed of biomaterials which, when implanted into the living host, will evoke a cellular response that results in the building of a structure that has the biochemical and structural properties to carry out the required physiological function. Alternatively, the construct may be composed of an artificial scaffold, usually a biodegradable biomaterial seeded with cells and sometimes mixed with growth factors (FGF, EGF, bHLH, etc.), which have been assembled and grown in vitro for a prescribed period of time before implantation into the living host. While some promising results have been found, it should be noted that in common with all implants or replacements, the success of tissue-engineered structures is partly dependent on the skill of the surgeon for proper implantation or grafting.

The first engineered tissues to hit the market have been skin and cartilage products. In 1997, the US Food and Drug Administration approved an engineered skin replacement made by Advanced Tissue Sciences Inc. of La Jolla, California. Consisting of cells from the inner, or dermal, skin layer grown on a biodegradable polymer, the skin can serve as a temporary wound cover for patients with second- and third-degree burns. A cartilage product, Carticel, has also recently won regulatory approval and has been used to replace damaged knee cartilage. This uses cartilage-forming cells (chondrocytes) from cartilage removed from the patient and grown in a degradable matrix. The orthopedic surgeon can then remove the damaged cartilage and replace it with this new tissue.

There are many tissue-engineered, orthopedic-related biomaterials currently under preclinical or clinical trials. For example, Antonios Mikos of Rice University and his colleagues have recently developed an injectable polypropylene fumarate copolymer that hardens quickly in the body and provides a surface that guides regeneration of many of the long bones in rats and goats. Tissue-engineered biomaterials are clearly an important development, and we are only beginning to realize their full potential. Imagine being able to reach into the freezer, take out a cell culture, treat it with growth factors on a scaffold matrix, and produce almost any tissue in the human body. This would have sounded like science fiction in last decade, but may be common clinical practice in the next.

Scope of the Book

The material in this book is divided into basic biomechanical concepts, which presents the fundamentals of tissue biomechanics and biomechanics of bone growth, and the applications of biomechanical principles to orthopedics and traumatology, including anchoring of implants, principles of fixator use, and biomechanics of oncology. More comprehensive and advanced descriptions of articular biomechanics can also be found in the later part of the text. The level of the subject matter is designed for medical students in their senior year or at the graduate level, and young orthopedic surgeons with some foundation in basic physics and mathematics.

Reference

1. Keyak et al., J. Biomech. 1998;125–33.

Selected Bibliography for Biomechanics

1. Berger SA, Goldsmith W, Lewis ER (Eds.). Introduction to Bioengineering. Oxford University Press, 1996.
2. Journal of Biomechanics. Special Issue: Cell Mechanics. J Biomech 2000;33(1).
3. Buckwalter JA, Einhorn TA, Simon SR (Eds.). Orthopedic Basic Science, 2nd ed. American Academy of Orthopedic Surgeons, 1999.
4. Cowin SC (Ed.). Bone Mechanics. Boca Raton, FL: CRC Press, 1989.
5. Fung YC (Ed.). Biomechanics: Mechanical Properties of Living Tissues. New York: Springer-Verlag, 1981.
6. Hertzberg RW (Ed.). Deformation and Fracture Mechanics of Engineering Materials, 3rd ed. New York: John Wiley & Sons, 1989.
7. Hibbeler RC. Engineering Mechanics: Combined Statics and Dynamics, 8th ed. Upper Saddle River, NJ: Prentice Hall, 1998.
8. Hirasawa Y, Sledge CB, Woo SLY (Eds.). Clinical Biomechanics and Related Research. Springer-Verlag, 1994.
9. Johnson KL (Ed.). Contact Mechanics. Cambridge, England: Cambridge University Press, 1985.
10. Lai WM, Rubin D, Krempl E. Introduction to Continuum Mechanics. Oxford, UK: Pergamon Press, 1993.
11. Mow VC, Hayes WC (Eds.). Basic Orthopedic Biomechanics, 2nd ed. Philadelphia, PA: Lippincott-Raven Press, 1997.
12. Mow VC, Ratcliffe A, Woo SL-Y (Eds.). Biomechanics of Diarthrodial Joints. New York: Springer-Verlag, 1990.
13. Ozkaya N, Nordin M (Eds.). Fundamentals of Biomechanics: Equilibrium. Motion, and Deformation, New York: Van Nostrand Reinhold, 1991.
14. Riley WF, Sturges L, Morris D. Mechanics of Materials, 5th ed. New York: John Wiley, 1996.
15. Skalak R, Chien S (Eds.). Handbook of Bioengineering. New York: McGraw Hill, 1987.
16. Woo SL-Y, Buckwalter JA (Eds.). Injury and Repair of the Musculoskeletal Soft Tissues. Park Ridge, IL: American Academy of Orthopedic Surgeons, 1988.

Selected Bibliography for Biomaterials

1. American Society for Testing and Materials. 1998 ASTM Book of Standards, Volume 13.01: Medical Devices and Services. West Conshohocken, PA: American Society for Testing and Materials, 1998.
2. Black J (Ed.). Orthopedic Biomaterials in Research and Practice. New York: Churchill Livingstone, 1988.
3. Buckwalter JA, Einhorn TA, Simon SR (Eds.). Orthopedic Basic Science, 2nd ed. American Academy of Orthopedic Surgerons, 1999.
4. Burstein AH, Wright TM (Eds.). Fundamentals of Orthopedic Biomechanics. Baltimore, MD: Williams & Wilkins, 1994.
5. Ferber D. From the lab to the Clinic. Science 1999; 284(5413):423.
6. Gilges D, Vinit MA, Callebaut I, Coulombel L, Cacheux V. Polydom: a secreted protein with pentraxin, complement control protein, epidermal growth factor, and von Willebrand factor A domains. Biochem J 2000 Nov 15;352(Pt 1):49–59.
7. Horowitz E, Parr JE (Eds.). Characterization and performance of calcium phosphate coatings for implants. West Conshohocken, PA: American Society for Testing and Materials, 1994, Series: ASTM STP 1196.
8. Li YW, Leong JCY, Lu WW, Luk KDK, Cheung KMC, Chiu KY, Chow SP. A novel injectable bioactive bone cement for spinal surgery. J Biomed Mater Res 2000; 52:164–70.
9. Oonishi H, Ishimaru H, Kato A. Effect of cross-linkage by gamma radiation in heavy doses to low-wear polyethylene in total hip prostheses. J Mater Sci 1996;7: 753–63.
10. Oonishi H, Kotani T, Shikita T. Proceedings of the 12 SICOT. Amsterdam.
11. Premnath V, Merrill EW, Jasty M, Harris WH. Melt irradiated UHMWPE for total hip replacements: Synthesis and properties. Trans Orthop Res Soc 1997;22:91.
12. Ratner BD, Hoffman AS, Schoen FJ, Lemons JE (Eds.). Biomaterials Science: An Introduction to Materials in Medicine. San Diego, CA: Academic Press, 1996.
13. Streicher RM. Ionizing radiation for sterilization and modification of high-molecular-weight polyethylenes. Plast Rubb Proc Appl 1988;10:221–9.
14. Stupp SI, Braun PV. Molecular Manipulation of Microstructures: Biomaterials, Ceramics, and Semiconductors. Science 1997;277:1242–8.
15. Von Recum A, Jacobi JE (Eds.). Handbook of Biomaterials Evaluation: Scientific, Technical, and Clinical Testing of Implant Materials, 2nd ed. Philadelphia, PA: Taylor & Francis, 1999.
16. Willmann G. Ceramics for total hip replacement: What a surgeon should know. Orthopedics 1998;21:173–7.

Contents

Principle Contributors

W. H. Akeson
University of California San Diego
San Diego
USA

B. T. Allende
Sanatorio Allende
Cordoba
Argentina

A. A. Amis
Imperial College
London
UK

J.-N. Argenson
Hôpital Sainte Marguerite
Marseille
France

E. Berthonnaud
Group of Applied Research in Orthopaedics
Lyon
France

G. Bollini
Hôpital Timone Enfants
Marseille
France

P. Bonnevialle
Centre Hospitalo-Universitaire Purpan
Toulouse
France

C. Bouvier
Centre Hospitalo-Universitaire Timone
Marseille
France

E. M. Brach del Prever
University of Turn
Torino
Italy

B. Clouet D'Orval
Centre Hospitalo-Univeristaire Nord
Marseille
France

K. R. Dai
Shanghai Second Medical University
Shanghai
PR China

Y. Debacker
BioConsulting
Luxembourg

J. Dubousset
Hôpital St Vincent de Paul
Paris
France

P. Frayssinet
Bioland
St Lys
France

E. Gautier
Hôpital Cantonal
Fribourg
Switzerland

P. Grosbras
Université de Poitiers
Futuroscope Chasseneuil
France

N. Hagemeister
Centre de recherche du Centre hospitalier de l'Université de Montreal
Montreal
Canada

J. A. L. Hart
Alfred Hospital
Melbourne
Australia

P. Hernigou
Hôpital Henri Mondor
Creteil
France

Y. Hu
Xijing Hospital
Xi'an
China

R. Kotz
University of Vienna
Vienna
Austria

A. Kusaba
Fujigaoka University
Yokahama
Japan

J.-Y. Lazennec
Pitié-Salpêtrière Teaching Hospital
Paris
France

J. C. Y. Leong
University of Hong Kong
Pokfulum
Hong Kong

T. S. Lindholm
Universities of Tampere and Oulu
Espoo
Finland

J. X. Lu
Biocitis
Berck Sur Mer
France

P. Mainil-Varlet
University of Bern
Bern
Switzerland

M. M. Mansat
Hôpital Purpan
Toulouse
France

B. Masson
CeramTec
Toulouse
France

Y. Mochizuki
Hiroshima University School of Medicine
Hiroshima
Japan

A. Nehme
Rangueil University Hospital
Toulouse
France

M. Peoc'h
Centre Hospitalier-Universitaire
Saint Etienne
France

D. G. Poitout
Centre Hospitalo-Universitaire Marseille Nord
Marseille
France

S. Saito
Fujigaoka University
Yokahama
Japan

C. Sorbie
Human Mobility Research Centre (Kingston General Hospital and Queen's University)
Kingston
Canada

H. Stein
Rambam Medical Center
Haifa
Israel

S.-I. Suk
Seoul Spine Institute
Seoul
Korea

J. Tamura
Kyoto University
Kyoto
Japan

K.-G. Thorngren
Lund University Hospital
Lund
Sweden

P. Tropiano
Centre Hospitalo-Univeristaire Marseille Nord
Marseille
France

C. T. Wang
Shanghai Jiao Tong University
Shanghai
PR China

T. Yamamuro
Research Institute for Production Development
Sakyo-Ku-Kyoto
Japan

Introduction

1 Bone as Biomaterial

D. G. Poitout

In recent years surgery has seen striking developments in the area of biomaterials and it is becoming increasingly necessary for surgeons from various specialisms to have an in-depth knowledge of the biomechanical properties of and what happens to foreign bodies implanted in the body, whether metallic or biological such as bone. Industrial researchers have to identify and then resolve the mechanical problems which arise when using inert (metallic or plastic) or biological materials to replace joints, ligaments, or even whole bones.

Using human or animal grafts (bone, cartilage, or ligament) in certain surgical, traumatological, or oncological indications requires a combination of various types of knowledge in the areas of immunology, biology, and biomechanics which are necessary for these allografts or these xenografts to be incorporated into the body.

Human bone, whether autologous and therefore bone-forming, allogenic, and simply bone-conducting or even animal bone (xenograft), behave biomechanically in a progressive fashion depending on the extent of the demands placed on it, the rate and degree of its revascularization, and of the procedures used to preserve and sterilize it. Bone substitutes are also currently being studied, whether in the area of hydroxyapatites, vitroceramics, tricalcium phosphates, corals, or even ceramized or heated allografts or xenografts. Mixed compounds combining a massive metallic prosthesis with bone from a bone bank surrounding it are composite biomaterials, the constituents of which each have their own advantages and disadvantages.

Introduction

Biomaterials can be defined as being "natural or synthetic substances, capable of being tolerated permanently or temporarily by the human body".

Indeed, although initially doctors chose mainly precious materials, as dentists still do, the development of new materials such as ceramics, polyethylene, carbon–carbon composites, or titanium have enabled the field of application which used to be limited to joint or dental prostheses to be extended to other areas such as ophthalmology and cardiology.

The use of allografts or xenografts is not recent but progress now being made in the areas of the sterilization and preservation of these products of human or animal origin mean that there is fresh interest in the surgical techniques which use them.

Research in these areas focuses on three aspects:

First, the study of the mechanical, physical, and chemical behavior of the material in its biological environment, i.e., its resistance to fatigue, wear, its elasticity, its resistance to corrosion, its biomechanical behavior, and its possible incorporation into the structures of the human body.

Then the study of its biocompatibility, in particular the analysis and identification of the reactions which occur at the interface between the material and the live tissue (for example, at the interface between the receiving bone and the prosthesis or the graft which has been introduced).

The biochemical growth factors, the role of certain enzymes in the breakdown of the materials used, the problems inherent to rejection or even immunological phenomena in relation to the destruction of an implanted graft are currently the subjects of a great deal of research.

Finally, it is necessary to choose a method which makes it possible to decide on a product which can be implanted in the body and which is also relatively easy to manufacture industrially or, where bone is concerned, preserved and distributed under ideal sterile conditions and the biomechanical behavior of which is compatible with restoring satisfactory and long-lasting joint function.

The Materials Used in Orthopedics

In the field of biomaterials, research has to follow two different but complementary paths:

On the one hand the characteristics and performance alone of the material have to be studied in accordance with its role in the body,

On the other, its biocompatibility has to be studied.

The biomaterials used in orthopedic surgery have developed a great deal in recent years. We now have a better understanding of the advantages they bring and their limitations. We know that steels corrode (vitallium) and that cobalt-chromium alloys wear. The complications connected with intolerance to the debris of metallic wear have meant that metal–metal prostheses are no longer used. The combination of metal and polyethylene also produces wear debris which plays a decisive role in the physiopathology of the loosening of prostheses, and the ceramic–ceramic joint may become blocked if the slightest particle enters the interface.

Plastics, such as polyethylene, which cover the sliding surfaces of many joint prostheses, become deformed, creep, and break down, tending to limit the life of these prostheses.

Cements, made of methyl methacrylate, which are used to fix some joint prostheses in the bone, have a high polymerization temperature if they are used in large quantities (over 70 °C), and for this reason cause bone necrosis (proteins congeal at 54 °C). The salting-out product may be toxic to the heart and when first used caused peroperative cardiac arrest from which the patients did not recover.

In 10% of cases allografts produce considerable immune reactions and are only slowly and incompletely assimilated by the skeleton. Bone substitutes are not necessarily successful in mechanical terms and at present can only be used to a limited extent.

Many materials have disappeared completely from our arsenal of therapeutic options and we may well ask ourselves what can be used in future to replace the biomaterials used at present.

Biodegradable Materials

The need to remove an osteosynthesis product which was implanted a few months or years earlier is inconvenient; it means that the patient has to be hospitalized and operated on again and leads to a search for products based on amino acid-based polymers which would break down and disappear spontaneously in the body within a few years.

Compounds made of polyglycolic or polylactic acid are currently used in the form of suture materials or parietal reinforcing plates and produce reasonable results. Their mechanical strength and life have to be improved and the way they are implanted into the body has to be specified. However, as from now, there is hope that in future they will replace the metallic materials currently used for osteosynthesis.

Bone Replacement Materials

Bone grafts currently have a major role.

Autografts

Autografts (bone graft taken directly from the patient) cannot be used to replace large segments of bone or an osteocartilaginous segment forming part of a joint. Being bone-forming, they alone can induce the formation of new bone and help in the healing of a fracture or the assimilation of an allograft.

Allografts

Since 1979 we have turned our attention to Marseilles, to preservation in tissue banks of allogenic bone fragments (bone graft taken from another person) stored in liquid nitrogen at −196 °C with cryopreservatives.

Currently used in traumatology or in oncology, these allografts make it possible to reconstruct a bone segment which has been destroyed by a tumor or an accident. These allografts are well tolerated by the body and only in exceptional cases (10% of cases) do immunological rejection phenomena occur. They can therefore be used easily in anybody requiring this type of operation.

Xenografts

Xenografts were used several decades ago by French teams (Judet-Sichard). The large number of rejection phenomena experienced with them (more than 50%) led to people refusing to use them. Because of the current shortage of human grafts, new attempts using different sterilization, preparation, or treatment techniques (lyophilization, ceramization, irradiation, heating) try to mitigate the inadequacies of this type of graft.

Bone Substitutes

Derivatives of artificial hydroxyapatite (a combination of hydroxyapatite-collagen, hydroxyapatite cement, corals or madrepores, vitroceramics or bioglasses) are undergoing in-depth mechanical and experimental studies to see how well they are tolerated in-situ and how they can be used. Even if some bone substitutes really are "colonized" by the bone of the host, their mechanical properties are still inadequate and mean that large fragments cannot be used in human clinical medicine. Furthermore, these structures, which are uniquely bone-conducting, do not form new bone, and tend to break down rapidly.

Joint Replacement Materials

There are a great number of plastics including polyethylenes with mechanical properties which allow them to be used in human clinical medicine. Various treatments (irradiation of the grafts or the addition of other compounds, for example) are being used in an attempt to improve their properties and to prolong their life in the body.

Alumina ceramics have been used for more than 15 years and their mechanical properties are well known. As the manufacturing processes are now very well established, it is possible that this material has the best coefficient of friction and produces the least wear debris in the body.

Zirconia ceramics are currently being investigated. They are less hard than alumina ceramics, they are easier to shape, are extremely strong but in some cases can break. Biological tolerance studies are currently being carried out and their biomechanical behavior in use is being characterized.

Silicon carbides could be used as friction surfaces for joint prostheses because they seem to be well tolerated, as the experimental implants have shown, but their long-term fate is not yet completely understood.

The use of massive cartilaginous allografts is being proposed more and more frequently by some international teams producing surprisingly good clinical results. The assimilation of these cartilaginous allografts is excellent as cartilage cells do not need vascularization to survive. They are sustained only by the components of synovial fluid. However, in order for the mechanical behavior of the graft to be adequate for the purpose, it is necessary for the cells contained in the cartilage, which ensure its trophicity in relation to the hydrophilia of the proteoglycans, to be protected during the freezing phase. Hence the advantages of using a cryopreservative when the temperature drops and the option of using secondary sterilization by heat, gas, or irradiation is absent. This has to be particularly rigorous when grafts are being taken and osteocartilaginous fragments are being stored so that the graft is definitely entirely sterile.

Capsuloligament and Joint Replacement Materials

The frequency with which tendons and ligaments tear directs world research towards these areas. Artificial ligaments are used more and more frequently in clinical practice but their long-term fate is unclear.

Carbon fibers sheathed in polylactic or polyglycolic acid, polyamide fibers, or high-density polyethylene threads are currently being tested for fatigue but they are already used in human surgery. Dacron or Teflon ligaments have not given good mechanical results in the medium term and have led to inflammation.

Preserving human ligaments in tissue banks is also an avenue of research which appears to be promising but comes up against the problem of how tissue banks obtain their supplies and of the mechanical behavior of the grafted ligaments while they are being revascularized.

Mineral Structure of Bone

Approximately 70% of mature bone is made up of an inorganic substance: calcium phosphate, and 30% of an organic matrix, the main component of which is a fibrous protein: collagen.

The exact nature of this mineral phase, which has been studied mainly by X-ray diffraction, remains unclear. Furthermore, it appears to be an established fact that the nature of this phase varies as the bone ages.

Several main components are frequently suggested:

brushite: $CaHPO_4 \cdot 2H_2O$

octacalcium phosphate: $Ca_8H_2(PO_4)_6 \cdot 5H_2O$

amorphous tricalcium phosphate: $Ca_3(PO_4)_2$

apatite, classically hydroxyapatite: $Ca_{10}(PO_4)_6(OH)_2$

The crystallites of bone apatite are small and often carry impurities. PO_4^{3-}, Ca^{2+}, and hydroxyapatite hydroxide are replaced by carbonate, Mg^2, and fluoride respectively. Compared with mineral hydroxyapatite, these imperfect crystals are more soluble and easily dissolved during resorption in the acid environment of the brush border of the osteoclastic cells.

The smallest unit of crystalline structure of the apatites contains 18 ions and it appears probable that such a complex structure is formed de novo from ions in solution. Progression through simpler forms has been demonstrated in vitro. However, these forms are unstable and difficult to demonstrate in vivo. The fluid environments of the body are said to be metastable in terms of their calcium and inorganic phosphate concentration. More precisely, that this concentration is below that of the concentration necessary for spontaneous precipitation but well above the concentration needed for the growth of the crystal if apatite crystals are present in the solution.

This therefore leads us to consider two very different phenomena:

the initiation of mineralization or “nucleation”,

the growth of the first crystals formed.

Progression of Mineralization

It has been demonstrated in vivo that more than 90% of mineralization takes place normally by the growth of pre-existing crystals. As far as the growth of the mineral phase is concerned, the problem here is how to control it. Indeed, once mineralization has started in a metastable environment, it should continue until all the ions are used up. If this were the case, we would all be turned into a pillar of salt like Lot’s wife. Mineral growth is therefore tightly controlled and regulated. Three factors play an important role: collagen, certain non-collagenic proteins, and proteoglycan.

Collagen

Initially considered to assist in nucleation, bone collagen essentially of type I helps in the formation of apatite in vitro and in particular organizes crystallization. The crystals are deposited parallel to the axis of the collagen fibrils and denaturing of the collagen disturbs this precipitation. Therefore, although in vivo studies tend to call into question the role of collagen in nucleation, it has an essential organizing role during the growth of the crystals.

Non-collagenic Proteins

Several non-collagenic proteins have been extracted from different calcified matrices. Two large groups have to be distinguished; the phosphoproteins and the GLA proteins (or proteins carrying gammacarboxyglutamic acid). The phosphoproteins have been isolated from bone, dentine, enamel, and calcified cartilage. Some phosphoproteins are more closely bound to collagen. Various roles have been suggested: orientation of the crystals, the control of their shape and size, or even a support role in particular in tissues which do not contain collagen, such as enamel. Osteonectin, a phosphorylated glycoprotein specific to bone tissue, is thought to help in binding calcium to collagen.

GLA proteins have been suggested as being the agent which regulates mineral growth but their role is still unclear and controversial. Their interest lies particularly in the possibility that a radioimmunological assay could be carried out on the serum, which would be a reliable and sensitive marker of bone remodeling activity.

Proteoglycans

These consist of a central protein of hyaluronic acid and of carbohydrate chains formed from the repetition of sulfated disaccharide units. Essential components of cartilage, proteoglycans have also been isolated from mineralized tissues.

Proteoglycans of bone are thought to be smaller and immunologically specific. It has been suggested that they play a role in calcification on account of the fact that there is a lower level of these in calcified tissues than in noncalcified tissues. Furthermore, in epiphyseal cartilage, the proteoglycans are thought to become smaller and fewer in number close to the calcification front. Moreover, proteoglycan aggregates inhibit the formation of apatite. The idea that proteoglycans indispensable to nucleation are transformed has therefore also been suggested. However Blumenthal has shown that the subunits, like the aggregates, inhibit mineralization. Poole et al., using immunofluorescence techniques, challenge the classical ideas of proteoglycans being reduced during endochondral ossification. In their view proteoglycans continue unchanged when mineralization starts and are only modified during immature primary bone modeling.

Bone Remodeling

Bone resorption and formation take place in a perfectly organized manner. The phenomena are most stereotypical in cortical bone. In old bone, and under influences which are currently little understood but which are certainly biochemical in nature, a population of osteoclasts appears which hollows out a resorption cavity which grows 7 to 9 microns a day up to a diameter comparable to that of a haversian osteon, and in particular advances into the bone, in a direction determined in particular by the mechanical constraints at a rate of 40 to 60 microns per day, thus producing a tunnel-like structure. After an intermediate phase (reversal phase), the osteoblasts appear on the walls of the cavity which initially deposit 8 to 10 lamellae of osteoid tissue and then, owing in particular to the osteoblastic alkaline phosphatases, cause the mineralization of this osteoid. Approximately 10% of the osteoblasts remain in the bone tissue formed in this way and, when they mature they become osteocytes, reunited with each other and communicating with the cells remaining on the surface of the residual canal by prolongations using a rich and anastomotic canalicular system. The end structure created in this way is the haversian osteon.

The resorption phase lasts approximately three weeks, the formation phenomena are spread over three months. In the trabeculae of the spongy bone the phenomena are the same but their spatial layout is different. Osteoclastic resorption takes places and advances on the surface of the bony trabeculae, forming Howship's lacuna, subsequently covered, there too, with osteoblasts transforming and then mineralizing the osteoid tissue. In this system, described by Frost, the site being remodeled is called the "basic multicellar unit" (BMU) and the cells which form it are called the "basic structural unit" (BSU), the end result of this remodeling is the haversian osteon.

Any pathological condition of the bone, and in particular diffuse conditions affecting the skeleton, is the result of an anomaly, varying in nature, of remodeling and of its elementary phenomena, with resorption always preceding its formation except in very specific cases (early stages of bony callus or ossifications of the soft tissues for example).

Morphology and Bone Mechanics in Hypodynamia

During its development each bone acquires a shape and a mass which is determined genetically in such a way that it has sufficient mechanical competence to perform the usual human activities. This acquisition requires the bone to be put into control, which allows it to be modeled during growth, followed by permanent remodeling throughout life. Physical activity therefore has a vital role to play in obtaining and then maintaining sufficient bone mass. A sedentary person will have a weaker bone mass and will be more likely to suffer fractures when making unaccustomed efforts. On the other hand, people who have been practicing a sport or an intense physical activity for a long time will have a higher bone mass or bone density than average (weight lifters, ballet dancers, tennis players) and may even thus be able to compensate for a diet which is extremely low in calcium, as is the case in some Equatorial areas.

The osteogenic stimulus therefore has a permanent effect on the bone, which continually adapts to this stimulus. Trabeculae of bone in children organize themselves in line with increasing functional activity, adopting an orthogonal arrangement according to the main force lines. This arrangement gives the system maximum strength with minimum bone tissue. On the other hand, cortical bone does not have the same mechanical requirements and its structural objectives are also different. There does not appear to be any clear relationship between the usual structure of the compact bones and the forces to which they are regularly subjected, but the ability of the bone cortices to react to a high local force is still possible (the end of a hip prosthesis, for example). Functional adaptation therefore affects the shape and mass of the bone from a basic level determined genetically, to a structurally adequate level. Nevertheless, each bone adapts itself independently; it is therefore the bone overall which adapts itself to the mechanical forces rather than specific tissue structures. The cell population of a bone is therefore able to assess the forces exerted on this bone.

Not only is the adaptation of the bone sensitive to the intensity and distribution of the force exerted, but in particular to the variations in this force. Static forces therefore appear only to have a moderate effect on bone remodeling and if they increase excessively, this can have a paradoxically negative effect.

It also seems that four daily compression cycles are sufficient to counterbalance the effect of immobilization, and that 36 daily cycles allow the maximum effect to be obtained.

Hypodynamia has a rapid and negative effect on the bone formed: the absence of forces exerted no longer allows the bone to adapt itself permanently, and opens the field to various biochemical and hormonal influences, of which adequate physical activity is the necessary counterpart. It has an identical effect on the growing bone, which without adequate stimulation does not acquire the architecture or reach the bone mass critical for it to be compatible with normal functional activity (the sequelae of poliomyelitis, for example).

Epiphyseal Cartilage

Continuous axial compression slows down the growth of connecting cartilage. The clinical applications (epiphyseal agraffing when the length of the lower limbs is unequal) are evidence of this.

Increased axial compression leads not only to a resumption of the activity of the epiphyseal cartilage but to an even more rapid rate of growth than normal. (Bonnel's experience, growth spurts observed in children confined to bed). This hypothesis explains the apparently contradictory results for stresses on flexion. During the day, when under pressure, the part of the epiphyseal cartilage subjected to compression in the resolution of a stress on flexion grows at a reduced rate. At night, or when not under pressure, the growth rate of this same part is accelerated. The sum of these two phenomena is thought to have a positive effect on growth with, in all, a more rapid rate of growth than for the part of the cartilage subjected to traction, still in the context of flexion.

These considerations apply, of course, to stresses greater than those physiologically endured by epiphyseal cartilage but less than the pathological stresses for maintaining the biological competence of this cartilage. The effects observed combine to produce a biologically healthy epiphyseal cartilage.

Articular Surfaces and Friction

The types of friction of the articular surfaces can be of the limited type (or Coulomb's type) or of the viscous type. In the limited type, for a light load and a slow rate, friction occurs via a substance with remarkable sliding properties, absorbed in the articular surfaces.

In the viscous type, for a heavy load and a rapid rate, a continuous liquid film permanently separates the two articular surfaces. The thickness of this film depends on the stresses which are exerted normally on the surfaces and the rate at which they move in relation to each other.

These two types of friction occur in human joints. They were demonstrated experimentally by studying the way in which the oscillations of a pendulum decrease when attached to a joint: a linear decrease in the case of limited friction, an exponential decrease in the case of viscous friction.

Lubrication and Pathology

The synovial fluid of joints affected by rheumatoid arthritis has proved to be a slightly less-effective lubricant than normal fluid. The fluid taken from arthrosed joints is thought to be better, almost as good as normal fluid. In the opinion of Little et al. (1969), there is no significant difference between the coefficients of friction of normal hips and those of joints manifesting fibrillation phenomena. There is no evidence to date to suggest that a lubrication disorder is at the root of degenerative phenomena observed in clinical practice.

Finally: Tomorrow, Will Man be Artificial?

If advances in technology continue at the current rate, it may be that many materials used today will be abandoned in years to come, but that, on the other hand, new products will appear on which the arthroplasties of the year 2000 will be based.

The reconstitution of joint cartilage by collagen, osteocartilaginous allografts, or artificial substances will allow huge strides to be made in the treatment of arthroses, the number of which increases as people live longer.

Methods of fixation for joint prostheses – biological fixation, new cements, so-called "intelligent" materials (nitinol and monocrystalline aluminas), or even bone grafts sheathing a metallic prosthesis – will enable the prosthesis to be better tolerated by the body. However, no-one can predict how this area will develop as chemists and metallurgists will without a doubt discover some new materials which will turn the future of the science upside down.

Artificial organs are now part of the usual arsenal of medical solutions. But can we expect to see an artificial man tomorrow? The list of artificial organs which are currently available or are being created is so long that it is becoming increasingly difficult to draw up a comprehensive list of them. Artificial skin is currently being developed for very severe burns. Cell cultures of osteoblasts or chondrocytes could, in the near future, cover bone substitutes or recolonize them.

However, all these artificial organs are expensive. The cost of the worldwide use of artificial kidneys or renal dialysis, for example, is several billion dollars (and in the case of France alone, 1% of the social security budget). It can well be imagined that the cost of creating very complex prostheses which can be used by only a small number of people could well be prohibitive, particularly for the most severely affected patients or the elderly who have relatively limited life expectancy.

Is it preferable to use grafts or artificial organs? In some cases it would be preferable to use prostheses and in others grafts. It would seem that the graft is the final element which would make it possible to save the patient, the prosthesis only allows him to wait until his graft can be implanted.

Combinations of prosthetic materials and biological materials are now used more and more frequently, whether it is a bone graft sheathing a prosthesis, or artificial skin made of human cells and cultured, or even live pancreatic cells developing within a synthetic structure.

In truth, it is worrying to think how far it could go, and whether one day it would be possible to create a wholly artificial man or carry out a succession of grafts aiming to replace the various components of the human body. For the moment it is still impossible to replace live organs with artificial organs which are as reliable, and in particular have the same capacity of self-repair as scar formation. Furthermore, their incorporation will without a doubt pose problems in the long term.

Nevertheless, the progress we are constantly making in the development of biocompatible implantable products – ever smaller circuits, ever more powerful software, and in particular live grafts assimilating perfectly into the body in which they are placed – give us real hope.

Biocompatible Materials

I – Biocompatible Materials

I A – Biomaterials Used in Orthopedics

2 Biomaterials Used in Orthopedics

D. G. Poitout

The great advances in orthopedic surgery over the past few decades and the fact that it constantly out-performs itself are the result of a policy of rigor in various areas.

Rigor in the training of the surgeons in this discipline, which demands a long period of training in specialist departments.

Rigor in performing operating techniques as a result of which hazardous improvisation is excluded.

Rigor in the choice of materials, the use of which has opened up the way to progress but the quality of which determines the results.

Precision and reliability are therefore the key words of the orthopedic surgeon who is preparing and executing an osteotomy in the same way as an engineer approaches the bridges and road surfaces for the arch of a bridge. He needs a good knowledge of the laws of physics and of the rules of mechanics, but he also has to be able to apply this knowledge to living matter.

I also believe it to be important to stress that orthopedists are clinicians and care for patients and that, if clinical practices develop in a direction which is not in line with their wishes, even though the theory and the calculations are accurate, we should not try to understand how this should work but why it does not work. Indeed, there are so many parameters involved in human clinical medicine that it is often difficult, when trying to describe a movement or define the stresses on a particular material, to take all the normal physiological parameters into account.

Behavior of Biomaterials in Situ

Although the functional aspects of implanted materials can be anticipated fairly reliably, it is very often difficult to anticipate how well they will be tolerated clinically. For materials of any kind there are two aspects which have to be taken into account. They are:

on the one hand the *adhesion* between a biomaterial and the part of the human body with which it will be in contact,

on the other, the *aging* of the product implanted.

Adhesion involves all the problems of using cements and adhesives, the role of which is to transmit and distribute the stresses over the largest area of contact possible. This adhesion problem is far from being resolved satisfactorily from the practical point of view and there is still plenty of scope for the researchers to investigate. Should a prosthesis be cemented, screwed, or introduced with force, hoping that its irregular surface will allow the bone to grow again and for the prosthesis to be fixed into the bone? More and more surgeons are currently abandoning these latter methods because of the frequency of painful failed fixations requiring surgery to be repeated (6–8% on average after 12 months). Cement has its drawbacks but according to the current state of knowledge seems to be the best compromise for fixing material into bone.

Aging. As soon as it has been implanted in the body, the biomaterial finds itself in an environment which is more aggressive than sea water, not least on account of its higher temperature and its sodium chloride content. Furthermore, there are also the variations in pH which may lead to a rapid breakdown of plastics and may accelerate_metal corrosion.

I would like to dwell on this problem of metal corrosion for a few moments. Some metallic materials are very resistant to generalized corrosion. This is the case for Vitallium, stainless steels, or alloys based on titanium, but they are still vulnerable to corrosion if pitted, the risk of which increases with contact friction which leads to breaks in the protective passive layer. It is also necessary to take into account the simultaneous action of the corrosive environment on the prostheses and the mechanical stresses to which they are subjected. This results in the risk of corrosion under stress, and corrosion due to fatigue which can lead to the appearance of weak points with the risk of breakage. Another well-known case of corrosion is galvanic corrosion caused by placing two different metals in contact with each other in a conducting liquid which then behave like an electric battery.

When there is corrosion, metal ions pass into the body. Therefore, some studies have shown that for austenitic stainless steel osteosynthesis plates, 9.1 mg of the alloy passed into the body two years after having been implanted. That is to say that there is a release of iron, nickel, and chromium in an equal proportion to that of the composition of the alloy. For example, in an individual who had had intramedullary pinning of the tibia, after 18 years he was found to have a nickel concentration in his serum, urine, hair, and nails which was up to 18 times the normal concentration, almost the same level as is found in workers in the nickel industry.

More generally, the implantation of foreign material, and particularly a metallic material, always has consequences for the surrounding biological environment. It was even possible to demonstrate a transformation of the proteins left in contact with nickel, in particular by electron transfer at the metal–electrolyte interface.

The problems listed above therefore require the practitioner to know the mechanical and chemical properties of the materials to be implanted without, of course, forgetting the sterilization conditions which can alter certain materials (such as gamma rays on plastics, ethylene dioxide absorbed by certain materials then released producing toxic reactions).

If the surgeon cannot check all the properties of the material he uses by appropriate tests, he has to rely on the manufacturer's literature to make his choice. But if he knows the properties that he can expect for a given application, the dialog will be more to the point.

That is the current direction in the area of French orthopedics.

Biomaterials Used in Orthopedics

As it would be excessive to give an exhaustive list of all the biomaterials used in orthopedics, we will only take a few examples from each of the five main classes of orthopedic biomaterials;

metals and metal alloys,

ceramics and ceramo-metallic materials,

bone replacement materials and allografts

carbon materials and composites, polymers.

Metal Alloys and Metals

First, where steels are concerned, the introduction of alloys leads to a spectacular improvement in oxidation. Molybdenum plays an essential role in resistance to corrosion caused by pitting.

Chromium also plays an essential role from the point of view of corrosion. Indeed, exposed to the air or to an oxidizing environment, chromium allows a very thin, invisible film of chromium oxide to form – this is called the passivation phenomenon. A minimum chromium content of 12% is necessary to give steel its stainless properties.

Other elements can be added; this is true for nickel which, when in a proportion of 10–14%, makes it possible to obtain an improvement in mechanical performance without leading to brittleness.

Steel with a high carbon content is therefore suitable for temporary surgical implants (osteosynthesis plates, intramedullary nails) because of its malleability and its stainless properties. But its poor prolonged resistance to corrosion means that it has to be removed after a few years.

Alloys based on cobalt–chromium are shaped by microfusion or casting, which is less good mechanically, and only very rarely has it been possible to make forgeable alloys, owing to considerable additions of molybdenum, tungsten, and nickel.

Although these materials have a resistance to corrosion and a breaking load which is better than stainless steel, their elastic limit is very close to the breaking load, which prevents any possibility of permanent deformation. And, as their resistance to fatigue is low, a significant breakage rate has been seen for femoral implants.

Their modulus of elasticity is high, at around 200,000 MPa, which poses the same problems as when using stainless steels (the modulus of elasticity of a bone being less than 20,000 MPAI. Due to their great hardness, alloys based on chromium and cobalt are the best compromise to date for making prosthetic femoral heads.

Titanium alloys have high resistance to all forms of corrosion and have good mechanical properties. Their modulus of elasticity is low, 110,000 MPa, which is half that of other alloys such as stainless steels. They have excellent biocompatibility, a high breaking load, and an elastic limit close to that of the breaking load, which eliminates any problems of permanent deformation in the case of high stresses, but also limits their use as a material in osteosynthesis. Owing to the passivation phenomenon, titanium covers itself spontaneously with a protective film of titanium oxide which renders it remarkably resistant to corrosion. This can be increased even further by the chemical process of anodization. There is one negative element that should be emphasized which is that titanium alloys have poor friction properties in that it is not possible to use them as prosthetic femoral heads or in the axis of a hinged prosthesis. Current trials, aiming to improve the friction characteristics by laying down deposits of titanium nitride or carbide, have not been very successful because these deposits are irregular and thin so that the layers abrade after a few thousand cycles.

Hydrogen or nitrogen ion inclusion techniques are still at the experimental stage.

Finally, the alloy most frequently used currently is an alloy containing a combination of aluminum and vanadium; Ti_6AP_4V, which has properties clearly superior to those of nickel–chromium–cobalt alloys. This is certainly the best solution today for all diaphyseal implants, particularly femoral hip implant which is subjected to high mechanical stresses.

Other metallic biomaterials could, in future, be useful in orthopedics; more specifically zirconium, tantalem, and nobium, all three of which display excellent biotolerance. However, progress still has to be made with alloys before they can rival titanium alloys.

Ceramics and Ceramic–Metal Compounds

Ever since man discovered that fire can modify the properties of clay (hydrated aluminum silicate), ceramics have never stopped developing. New ceramics have been developed and these materials take various forms:

oxides: aluminum oxide (Al_2O_3), zirconium oxide (ZrO_2),

carbides: silicon carbide (SiC),

nitrides, bromides, and fluorides.

The science of ceramics has also meant that new textures can be created such as ceramic composites with various fibers combining metals and ceramics, which are called ceramic–metals or even cermets. There are also controlled crystallization glasses called vitroceramics.

The New Ceramics

Sintered oxides are either pure oxides such as alumina or mixtures of oxides. When high-purity alumina is used in the medical field, the specification is extremely precise. Alumina is a hydrophilic material (unlike polyethylene which is hydrophobic), it is very hard, slightly less so than diamond (which is, moreover, used to grind and polish it), and its modulus of elasticity is 380,088 MPa, which is practically twice that of the metal alloys. Its resistance to flexion, however, is low, which limits the indications in which it can be used as an osteosynthesis rod or plate. When alumina was first used as a prosthetic hip compound, there were many failures of the femoral head when used with an acetabulum also made of alumina.

The two pieces machined for each other:

tended to jam if the slightest particle of wear debris came between them.

produced very little wear debris, certainly, but as these were crystals they led to synovial reactions comparable to microcrystalline arthritis.

prevented any isolated change in one of the pieces of the prosthesis if only one became damaged.

The existence of a high modulus of elasticity, far higher than that of methyl methacrylate and that of cortical bone, led to problems when sealing an alumina acetabulum with methyl methacrylate because unsealing occurred more frequently and usually occurred between the cement and the acetabulum and not between the bone and cement, as is normally the case. On the other hand, if the alumina acetabulum is directly screwed into the bone, the quality of the fixation is exceptional and the mobility of the implant normal because of the almost inevitable appearance of a film of fibrous tissue between the implant and the bone. The use of alumina currently, therefore, seems to be restricted to femoral heads and sliding surfaces in contact with polyethylene.

Zirconia (ZrO_2) also has excellent mechanical properties, in particular flexion, together with satisfactory resistance to wear and friction, but in some cases it breaks! We hope that zirconias stabilized by yttrium oxide (Y_2O_3) and by alumina ($R_{12}O_3$) will be used routinely as friction components in total prostheses of the hip.

Carbides and Nitrides: These new materials include silicon carbide, which appears to have greater resistance to flexion than alumina as well as a higher modulus of elasticity, but its coefficient of friction is lower than that of alumina.

Ceramic–Ceramic and Ceramic–Metal Compounds

Fiber composites are a compromise between a deformable solid (for example, carbon fibers or alumina fibers) and a matrix which resists deformation (such as alumina or silicon carbide). To date, the first experiments with mixtures of aluminum oxide and iron have not produced useful results for improving the properties of the material. On the other hand, other combinations with molybdenum and its carbide, with tungsten and its carbide, or with titanium combined with zirconium oxide, seem to improve the resilience and toughness of the material considerably.

Glass and Vitroceramics

The mechanical strength of some glasses can be greatly improved by being transformed into vitroceramics. Direct anchoring, as for conventional ceramics, can, together with glasses and the vitroceramics, be performed by mechanical or chemical processes. In the case of vitroceramics anchored mechanically the dimensions of the interconnections between the pores are sufficiently large to allow colonization by bone tissue. Unfortunately, the mechanical properties of these vitroceramics are relatively poor. Resistance to breakage on flexion remains around 20 MPa, which is far too low for use in internal prostheses.

It seems that glasses and vitroceramics anchored chemically give better results. These materials initially have better mechanical strength than those of porous materials and are better than those of bone, but these criteria do not last. On the other hand, adhesion only seems to occur if the implant is immediately placed into intimate contact with the bone tissue, which is not always easy to do in practice, because, as in the case of bio-inert materials, a fibrous capsule forms which isolates the material from the bone.

Natural, Biological, or Synthetic Bone Replacement Materials

Bone loss can be remedied today either by natural autologous or homologous bone grafts or with ceramic-like materials. This is particularly true for madreporic coral or synthetic coral which consist of calcium phosphates and fluoroapatites and are comparable to the vitroceramics we have been discussing.

Natural calcium carbonates are skeletons of madreporic corals with their organic part removed. They consist of virtually pure aragonite ($CaCO_3$). Used experimentally to replace bone substance losses or to fill cavities, it seems that the tendency is for the fragment of natural calcium carbonate to be resorbed, then for the carbonated skeleton to be replaced centripetally and gradually by bone. The structure of coral skeleton makes it possible to re-establish the intra-medullary circulation and its resorption releases calcium ions reused by the body for the precipitation of phosphocalcium apatite. However, the mechanical properties of the corals, which have a strength under flexion of the order of 3 MPa, and under compression of 16 MPa, are much inferior to those of bone and the clinical applications are comparable to bone autografts and allografts.

Materials Obtained by Synthesis

With comparable porosity, the mechanical properties of synthetic materials are generally superior to those of natural materials. Only the compressive strength of tricalcium phosphate, which is between 7 and 21 MPa, is of the order of magnitude of that of coral. As for the latter, there are ultimately extremely few clinical applications.

Allografts

Bone is a living tissue consisting of cells as well as of a prosthetic structure on which calcium and phosphorus have been precipitated. The introduction into the body of a bone graft of any kind will lead to the progressive destruction of its cells without modifying the supporting protein lattice. Indeed, although the cells are antigenically specific to any particular individual (various HLR groups), the collagen which forms the architecture of the bone is the same throughout the human race and will not give rise to rejection phenomena. Whether we use an autograft or an allograft, the clinical development of this tissue is approximately comparable and the cells will die. The protein structure on which the phosphocalcium raster is fixed will no longer exist and the bone cells of the host will recolonize the bone which serves as a mold. After several years, new bone will be reformed from the cells of the host.

As massive samples cannot be taken from the same person without running the risk of causing problems at the donor site, we turned to preservation by cryopreservation of the bone homografts in bone banks. In order for it to be preserved "indefinitely", it is necessary for the bone to be stored in very cold conditions below −80 °C. For these technical reasons, we chose to store the cryopreserved bone – preserved in 10% DMSO in liquid nitrogen at −196 °C; which, subject to certain precautions, gives the most reliable results. Cryopreserved bone makes it possible to reconstruct a bone segment which had to be resected due to the existence of a bone tumor at that site and also to reconstruct the locomotor architecture after a considerable loss of bone substance due to trauma.

Massive osteocartilaginous fragments are used ever more frequently to reconstruct articular surfaces which have been damaged or removed as part of the excision of a tumor. Smaller, spongy fragments can also be used in addition to osteosynthesis to fill a bone cavity or to complete the fixation of an arthroplasty. The results we are obtaining currently are wholly encouraging and in many cases have made it possible to avoid amputation or the use of massive prostheses, the long-term mechanical future of which is not guaranteed. Between 1978 and 2000, the Marseilles Bone Bank has supplied 1744 massive bone parts used for grafts.

Carbon Compounds

Since 1967, numerous procedures have been used to create biomedical carbon but so far none have given absolute biological stability. It cannot, therefore, yet be considered for use routinely in human biology, in spite of the very many suggestions which have been made (osteosynthesis plates, nails, joint prostheses) and in spite of its unrivalled endurance to fatigue (easily able to exceed 10 million cycles). The natural communicating porosity of its structure allows colonization into the mass of the prosthesis by the surrounding biological tissues, and the structural flexibility of the composites harmonize with the elasticity of the host bone. The fact that they cannot be deformed means that they cannot be used as osteosynthesis plates and as the carbon fibers cannot tolerate lengthening, even to a very small extent, nor can they be bent to more than 30° without breaking. They cannot be used as a prosthetic ligament because fixing this ligament into bone is very difficult.

Finally, the many particles from wear found in the ganglions, and even in the spleen, mean that we have to be careful when using these composites. Owing to the hardness of the surfaces obtained by the ceramization treatment, it may be possible to consider using carbon as an articular surface, placing polyethylene in between the opposing surfaces.

Polymers

Numerous products have been suggested but, of course, they cannot all be considered.

As far as their common properties are concerned, it is important to stress the fact that they age physically and chemically.

The following will be discussed:

1. Silicones, which are chemically inert, have good biotolerance and a high hydrophobic capacity. They are used in plastic surgery or in orthopedics in the form of elastomer rubbers for joint prostheses of the fingers, for example.
2. Polyacrylics, and more specifically, methyl polymethacrylate, are well known in the area of orthopedics as they are used as a cement for fixing prostheses. The time that cements take to grip varies considerably depending on the type of used; also the polymerization reaction, which is very exothermic. If none of the heat were to be dissipated to the exterior while polymerizing, the mass of cement could reach more than 70 °C. It is thought that the maximum temperature should generally be no more than 40–50 °C in vitro, which is relatively close to the coagulation point for proteins (56 °C) and that of bone collagen (70 °C). It would therefore be desirable to find a new, weakly exothermic cement, which sets relatively slowly, but this is not yet available.

Currently, the cement penetrates the interstices of the bone more effectively and leads to even more secure anchorage if it is more fluid or less viscous. It is therefore preferable to use a cement with a viscosity of less than 100 Newton/s/m^2 after mixing.

Similarly, the porosity is a decisive factor in the mechanical behavior of the cement. For a particular cement, the size of the pores does not depend on the maximum temperature, but on the mixing and usage conditions. On the other hand, the number of bubbles per unit volume, for any particular cement, depends on the maximum polymerization temperature.

Finally, all acrylic cements show volume changes between the beginning of the mixing and the end of hardening. Currently, it appears that cement starts by contracting approximately 2.5 to 6.5 microns per 2 mm thickness. As far as the mechanical properties of cement are concerned, the Young's modulus is low (of the order of 3,000 MPa) and traction strength and compressive strength are approximately a quarter of the strength of normal bone. It is therefore important to emphasize the preparation of the cement, the frequency of the movements, and the role of the additives. In this area, the addition of powders only very slightly changes their mechanical properties. On the other hand, when a liquid, such as an antibiotic, is to be added, this leads to serious weak points appearing and causes fractures to start which will only spread under stress. Finally, irradiation does not cause any significant changes in the mechanical behavior of the cement.

3. Saturated polyesters, which are condensation polymers, are essentially represented by polyethylene terephthalate. This polymer has good resistance to chemical agents, good tolerance in solid form and good mechanical properties. However, its behavior in a humid

environment is poor, with a sharp reduction of its mechanical properties. It is used in orthopedics in the form of plaited threads to make prosthetic ligaments (Dacron or Rodergon, for example). The poor elastic elongation properties (1.25 Y approximately) seem to be a very worrying factor for how this prosthesis behaves over time because the relative physiological elongation of the cruciate ligaments of the knee, for example, is 26 to 25 Y.

4. Polyolefins. In this group it is UHMW (Ultra-high-molecular-weight) polyethylene which is used for making friction components for prostheses of the hip, knee, and elbow because of its mechanical properties. A great deal of research is currently being carried out to improve its properties, and in particular its resistance to creep with, for example, the incorporation of carbon fibers. Polyethylene reticulated by ionizing radiation with grafting of polytetrafluoroethylene should also improve the resistance to wear and creep. The use of a metal backing for prosthetic cupulae also seems to limit the extent of creep. Polypropylene can be used for ligament use but here, too, its elastic elongation risks breaks in or detachment of the implant.

To conclude, how do these biomaterials behave in use? It should be borne in mind that the main reasons why these materials fail are due to an as yet inadequate understanding of the properties of the materials used. Detachment is due to a breakdown in the cements and requires research to be carried out into their properties together with research into the mechanics of the transfer of loads between the implant and the bone. The extent of wear on the polyethylene parts will mean that the properties of these products will have to be changed, while amending the design of the parts. The introduction of ceramics to reduce the extent of wear has not managed to stop it, and until these materials are made less brittle, there will still be the risk of accidents.

There is still insufficient experience with carbon composite materials and only rigorously controlled experiments will enable us to say whether the hoped for advantages of these new materials are accompanied by serious disadvantages linked to a possible fragmentation of the fibers.

Finally, in the case of metal alloys, an analysis of the behavior of the parts in use shows that the resistance to fatigue corrosion should be studied in experimental conditions to enable easier comparison of the advantages and disadvantages of the various alloys proposed.

Care should be taken not to reach too hasty a conclusion as to the risks of certain techniques and, perhaps even more importantly, the wholly beneficial effect of the new techniques where it is not possible to be entirely sure of the scientific objectivity of the measures. In practical terms, all the phenomena involved in the behavior of implantable materials start at the surface of the implants. It is therefore by studying the surfaces and their changes by physicochemical or mechanical treatment that advances can be made in the current techniques for manufacturing surgical implants. Reconstruction of joint cartilage with collagen, osteocartilaginous allografts, or artificial substances will allow enormous advances to be made in the treatment of arthroses, the number of cases of which rise as life expectancy increases.

Finally, many materials used today will probably be abandoned in the years to come. On the other hand, new products will appear which will be based on the arthroplasties of the year 2000. Today we are probably only aware of one third of the materials we will be using in 20 years time.

3 Bioceramics

T. Yamamuro

Definition and Classification

Ceramic is defined as "synthesized inorganic, solid, crystalline materials, excluding metals". Ceramics, used as biomaterials to fill defects in tooth and bone, to fix bone grafts, fractures, or prostheses to bone, and to replace diseased tissue, are called bioceramics. They must be highly biocompatible and antithrombogenic, and should not be toxic, allergenic, carcinogenic, or teratogenic. Bioceramics can be classified into three groups; (1) **bioinert ceramics**, (2) **bioactive ceramics**, and (3) **bioresorbable ceramics.** Bioinert ceramics have a high chemical stability in vivo as well as high mechanical strength as a rule, and when they are implanted in living bone, they are incorporated into the bone tissue in accordance with the pattern of "**contact osteogenesis**". On the other hand, bioactive ceramics have the character of osteoconduction and the capability of chemical bonding with living bone tissue. In other words, when bioactive ceramics are implanted in living bone, they are incorporated into the bone tissue in accordance with the pattern of "**bonding osteogenesis**". The mechanical strength of bioactive ceramics is generally lower than that of bioinert ceramics. Bioresorbable ceramics are gradually absorbed in vivo and replaced by bone in the bone tissue. The pattern of their incorporation into the bone tissue is considered similar to contact osteogenesis, although the interface between bioresorbable ceramics and bone is not stable as that observed with bioinert ceramics.

Bioinert Ceramics

In 1969, Benson [1] predicted that carbon ceramic will be brought into clinical application as a biomaterial in the near future, as it has excellent biocompatibility, a high compressive strength, and a reasonable elastic modulus. When carbon fiber was used as artificial ligaments, however, it tended to undergo fragmentation. Recently, such mechanically stronger carbons as low-temperature isotropic carbon (LTI carbon) and carbon-fiber-reinforced carbon (CFRC) have been developed, but their clinical application has not yet been brought to realization.

Bioinert ceramics such as alumina ceramic (Al_2O_3) and zirconia ceramic (ZrO_2) have a higher compressive and bending strength and better biocompatibility than stainless steel (SUS 316L) or Co–Cr alloy. Alumina ceramic particularly, therefore, was used for osteosynthetic devices (alumina monocrystal) or to fabricate bone and joint prostheses (alumina monocrystal + polycrystal) in the 1980s [2]. Recently, however, due to their brittleness and too high elastic modulus as compared to those of human bone, they are very little used for these purposes. On the other hand, it has been known that a ball made of alumina or zirconia exhibits a wear-resistant character when its surface is polished to an average surface roughness of 0.02 μm. At present, therefore, the clinical application of alumina and zirconia is almost solely limited to the bearing surface of joint prostheses.

It is well known that one of the important factors causing loosening of joint prostheses is periprosthetic osteolysis, which is due mainly to excessive macrophage activity against wear debris, particularly of polyethylene, around the prosthesis. There have been various attempts, therefore, to reduce the amount of wear debris by changing the bearing surfaces of prostheses from a metal-on-polyethylene combination to a metal-on-metal, ceramic-on-ceramic, or ceramic-on-polyethylene combination. In 1970, Boutin [3] started to use an alumina-on-alumina combination for the bearing surface of hip prostheses. According to Sedel [4], alumina ceramic used for hip prostheses between 1970 and 1979 had a mean grain size of 7 μm and the linear wear of its bearing surface was 5 to 9 μm per year, while alumina ceramic currently used has a grain size of 2 μm and the linear wear is in the order of 3 μm per year. Sedel further described that the overall wear of the currently used alumina-on-alumina hip prostheses, calculated by the weight of debris generated, was approximately 1,000 times less than for metal-on-polyethylene and 40 times less than for metal-on-metal joints, if all requirements for alumina quality, sphericity, circularity, and clearance of the bearing components are met. Therefore, in spite of the fact that the alumina-on-alumina hip prostheses used in the 1970s did not show significantly better 10–15 year results as compared to those of the Charnley hip prostheses, the current alumina-on-alumina hip prostheses are expected to bring about much better long-term results than those used in the 1970s.

On the other hand, it is a well-known rule for hip prostheses that the smaller the head size the less the volumetric wear of the bearing surface. There has been much work on ceramic-on-polyethylene hip prostheses in attempts to reduce the volumetric wear of polyethylene. The diameter of most alumina femoral heads of hip prostheses has been limited to 26–32 mm even with new alumina ceramic, because it exhibits only moderate bending strength and toughness. In an attempt to reduce the head size to 22 mm while still guarding against breakage, zirconia ceramic has been considered as a constituent material, as it has the advantages over alumina of higher bending, compressive, and impact strength, higher fracture toughness, and lower elastic modulus. Zirconia ceramic was not brought into clinical application until recently because zirconia synthesized in the 1980s was abnormally radioactive [5] and tended to biodegrade in vivo. Modern technology, however, made it possible to synthesize a new zirconia which is not abnormally radioactive and is stable in vivo. This has been accomplished mainly by developing a refining technique to obtain pure zirconium from a raw ore and by adding chemical stabilizers such as yttrium oxide or cerium oxide during the sintering process. The estimated amount of radioactivity exhibited from zirconia, prepared by the Kobe Steel Company since 1993 is 1.152 μR, while the normal background radioactivity is approximately 100 mR [6]. Thus, the radioactivity of new zirconia is considered negligible.

Concerning crystallographic stability, alumina ceramic (usually α-alumina) consists entirely of hexagonal crystals and is hence chemically very stable in vivo. On the other hand, zirconia ceramic usually consists of three crystallographical phases; cubic, tetragonal, and monoclinic, and transformation of the phase takes place under various conditions such as change of temperature, mechanical stress, and humidity. The phase transformation often results in self destruction of the ceramic. Until the 1980s, this crystallographical instability was one reason for not using zirconia as a constituent material for the bearing component of joint prostheses in which high stress concentrations may be created on the ceramic surface by repeated loading under wet conditions. In the 1990s, however, new sintering methods have been introduced to prepare crystallographically stable zirconia ceramics by adding such chemical stabilizers as Y_2O_3, CeO_2, and Al_2O_3 in the sintering process. These are called partially stabilized zirconia (PSZ). As an example, to prepare a zirconia femoral head of 22 mm in outer diameter, zirconia powder with a grain size of less than 1 μm is mixed with chemical stabilizers (Table 3.1), and is molded into a ball using rubber at room temperature. The ball is then sintered for two hours at 1,500 °C. The sintered zirconia ball undergoes machining to shape a precise spherical ball with an outer diameter of 22 mm and a tapering fit (Figure 3.1). The ball is finally polished to obtain an average surface roughness of less than 0.02 μm.

A zirconia femoral head made in this way consists mainly of the tetragonal phase with 1–2% monoclinic and cubic phases.

The mechanical properties of PSZs are comparable with those of new alumina ceramics with a grain size of less than 2 μm (Table 3.2). PSZs have significantly higher bending strength, compressive strength, fracture toughness, and impact strength, but have a lower Vickers hardness and elastic modulus than alumina ceramic, although they are slightly different depending on the grain size and kind of chemical stabilizers used. Among them, yttrium oxide PSZ (Y-PSZ) has the highest bending strength and fracture toughness followed by cerium oxide PSZ (Ce-PSZ). Breaking tests for Y-PSZ and alumina femoral heads, 22 mm in outer diameter, were performed by static loading over a polyethylene liner which was set against the ceramic head. The alumina heads were broken by loads of 2,400–3,400 kg (average 2,800 kg), while Y-PSZ heads were broken by loads of 2,770–4,480 kg (average 3,700 kg). Thus, Y-PSZ heads were significantly stronger than the alumina heads against breakage [7]. Fatigue testing was performed on eight Y-PSZ femoral heads on a hip simulator in physiological saline at 37 °C, by applying 10^7 cycles of repeated loading with 450 kg. This loading is considered to correspond approximately to 20 years of a person walking. After the test, no breakage was observed in all eight Y-PSZ heads.

Wear tests for the polyethylene liner against the Y-PSZ, alumina, and stainless steel head, all

Table 3.1. Chemical composition of zirconia ceramic

	Weight %
$SiO_2 + M_2O$ (M_2O: Na_2O, K_2O etc)	0.1
Fe_2O_3	0.1
Al_2O_3	0.5
Y_2O_3	4.8 ± 0.7
ZrO_2	remainder

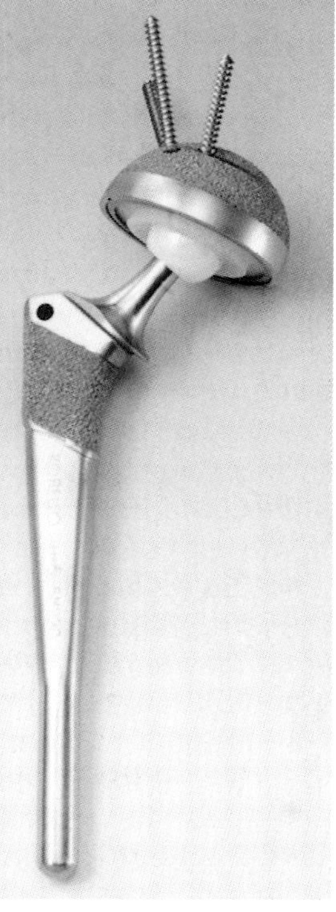

Figure 3.1. **a** A zirconia femoral head with an outer diameter of 22 mm. **b** A cementless hip prosthesis made of titanium alloy with a combination of zirconia head and polyethylene socket for the bearing component.

Table 3.2. Mechanical properties of bioinert ceramics

		Zirconia	Alumina
Bending strength	(kgf/mm^2)	170	>40.8
Compressive strength	(kgf/mm^2)	500	408
Fracture toughness	(MPa, m$^{1/2}$)	5.2	3.4
Impact strength	(kgf/mm^2 or kJ/m^2)	14	4
Vickers hardness	(HV kgf/mm^2)	1270	2300
Elastic modulus	(kgf/mm^2)	20500	>38800
Density	(g/cm^3)	6.05	>3.9
Crystal size	(μm)	0.2	<7

22 mm in outer diameter, were performed using a hip simulator in physiological saline at 37 °C by applying a load of 450 kg at 1 Hz. After 5×10^5 cycles of loading, the polyethylene liner against the stainless steel head showed significant wear, while those against the Y-PSZ head and alumina head did not show any measurable wear, even after 2×10^6 loading [7].

Thus, alumina is chemically more stable than PSZ in vivo, while PSZ is mechanically stronger than alumina, and both of them exhibit much better wear-resistant characteristics compared to stainless steel or Co–Cr alloy when assessed in the form of bearing components for hip prostheses. For these reasons, alumina is used to fabricate ceramic-on-ceramic hip prostheses where head size is not a key issue, while PSZ is used to fabricate ceramic-on-polyethylene hip prostheses where the head size must be made reasonably small. One reason why zirconia-on-zirconia or alumina-on-zirconia hip prostheses have not yet been brought to the market is that, even with PSZ, its long-term crystallographical stability in vivo has not been confirmed.

Bioactive Ceramics

Bioactive ceramics include glasses, glass–ceramics, and ceramics that elicit a specific biological response at the interface between the material and the bone tissue which results in the formation of a bond between them. The first evidence of direct bone bonding to a glass implant was discovered by Hench et al. in 1970 [8]. Since then, some other glasses, glass–ceramics, and ceramics had been proved to have bone-bonding capability. Among them, Bioglass®, apatite- and wollastonite-containing glass–ceramics (AW-GC) and synthetic hydroxyapatite (HA) are representative materials currently used for clinical applications.

In 1970, Hench et al. [8] synthesized a bioactive glass with a chemical composition of SiO_2 45, CaO 24.5, P_2O_5 6, Na_2O 24.5 (wt%). This glass is called 45S5 Bioglass® and is known to exhibit the strongest bioactivity among hitherto developed bioactive ceramics. Wilson et al. [9] proved that when the implant–tissue interface was immobilized, collagen fibers of soft tissue became embedded and bonded within the growing silica-rich and hydroxy-carbonate apatite layer on the 45S5 Bioglass®. Such soft-tissue bonding has never been observed with other bioactive ceramics or glass–ceramics. However, as Bioglass® is mechanically much weaker than human cortical bone, it can not be used as a weight-bearing bone prosthesis. Instead, it has been used as a bone void filler in the form of granules, coating materials on metallic prostheses, and to fabricate a middle-ear prosthesis.

Aoki et al. [10] in 1966 and Jarcho et al. [11] in 1976 independently developed a process for producing dense hydroxyapatite implants with considerably high mechanical strength. Synthetic hydroxyapatite ($Ca_{10}(PO_4)_6(OH)_2$) has the capability of chemical bonding with living bone tissue, but it takes much longer than Bioglass® for bone bonding. Its mechanical properties are shown in Table 3.3 in comparison with that of natural bone and AW-GC. The bending strength of HA is lower than that of natural cortical bone, and hence HA can not be used to fabricate weight-bearing bone prostheses with absolute safety against breakage in vivo. It has been used as a bone void filler in the form of granules with various particle sizes (Figure 3.2), as a coating

Table 3.3. Mechanical property of natural bone and bioactive ceramics

	Bending strength (Mpa)	Compressive strength (Mpa)	Elastic modulus (Gpa)
Natural bone	30–190	90–230	3.8–17
Synthesized Hydroxyapatite	110–170	500–900	35–120
A-W Glass-Ceramic	220	1000	120

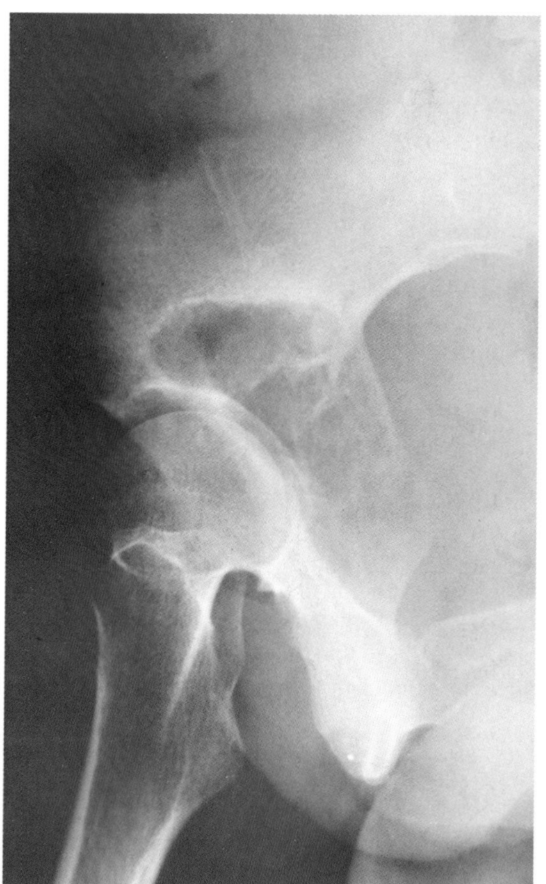

a

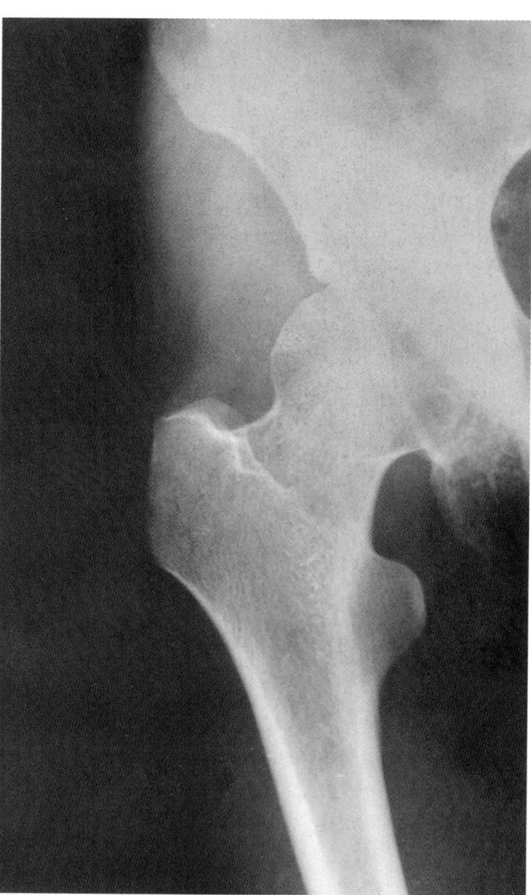

b

Figure 3.2. **a** A giant cell tumor developed in the right ilium and ischium of a 27-year-old female. The tumor was excised and the remaining large bone defect was filled with a mixture of autogenous bone chips, HA granules, and fibrin glue. **b** 20 years postoperatively, HA has been well incorporated into the surrouding bone and the patient has no symptoms.

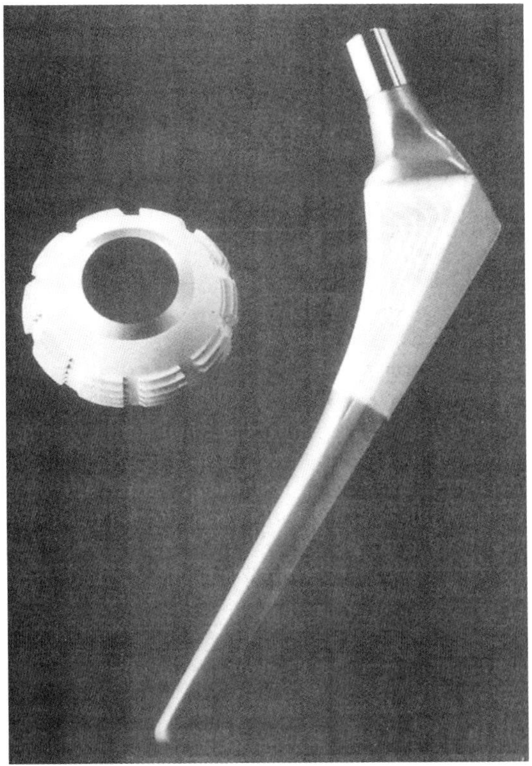

Figure 3.3. HA-coated hip prosthesis developed by Geesink et al.

material on metallic prostheses, and to fabricate an iliac crest prosthesis and a laminoplasty spacer in which high mechanical strength is not required. In 1987, Geesink et al. [12] developed a HA-coated hip prosthesis and recently reported an excellent ten-year clinical result in a large number of patient (Figure 3.3). At present, HA is the bioactive ceramic most widely used for clinical applications as a bone void filler and a coating material for hip prostheses which are employed in cementless hip replacement.

Aiming at producing a mechanically stronger bioactive material, Kokubo et al. [13] in 1982 developed apatite- and wollastonite-containing glass–ceramic (AW-GC or Cerabone AW®) with the chemical composition SiO_2 34.0, CaO 44.7, P_2O_5 16.2, MgO 4.6, CaF_2 0.5 (wt%). As shown in Table 3.3, AW-GC has a significantly greater bending and compressive strength than human cortical bone and dense HA. Bioacitivity was compared among Bioglass®, HA, and AW-GC by implanting them into living bone tissue and carrying out detaching tests in different post-implantation periods. It was demonstrated that bone bonding occurred earliest with Bioglass® followed by AW-GC and then HA. The essential mechanism of bone bonding for Bioglass® and AW-GC is considered similar; i.e., the formation of an apatite layer on the implant surface in the body environment. This surface apatite formation takes place by the chemical reaction of Ca^{2+} and $HSiO^{3-}$ ions dissolved from the implant surface. This apatite is called chemical apatite (Figure 3.4). At the same time, on the cut surface of bone, an apatite layer accompanied by collagen fibers is formed by the activity of osteoblasts. This apatite layer is called biological apatite. Neo et al. [14] observed under transmission electron microscopy that the chemical apatite and the biological apatite were intermingled at the bone bonding interface (Figure 3.5). The HA implant also showed similar bone-bonding morphology under transmission electron microscopy, but HA took longer than Bioglass® or AW-GC for bone bonding. This is presumably due to the fact that HA consists solely of crystals, while others contain the glass phase which is dissolved faster than the crystal; also, dissolved $HSiO_3^-$ ions might provide favorable sites for nucleation of the apatite [15].

Yamamuro et al. [16] replaced vertebral bodies of sheep with a vertebral prosthesis made of AW-GC and found that the prosthesis bonded directly to the adjacent vertebrae within about one year (Figure 3.6). Then, using AW-GC, various bone prostheses were fabricated such as iliac crest prostheses, vertebral prostheses, intervertebral spacers, and laminoplasty spacers (Figure 3.7) [17]. The iliac crest spacer is used to substitute a bone defect remaining after harvesting a large bone graft from the iliac crest in various orthopedic operations. The vertebral prosthesis is used as a substitute for vertebral bodies suffering from benign and malignant tumors, compression fractures, and burst fractures (Figure 3.8). The intervertebral spacer is used for interbody fusion through either an anterior or posterior approach (Figure 3.9). The laminoplasty spacer is used to maintain bilateral laminae opened after surgical enlargement

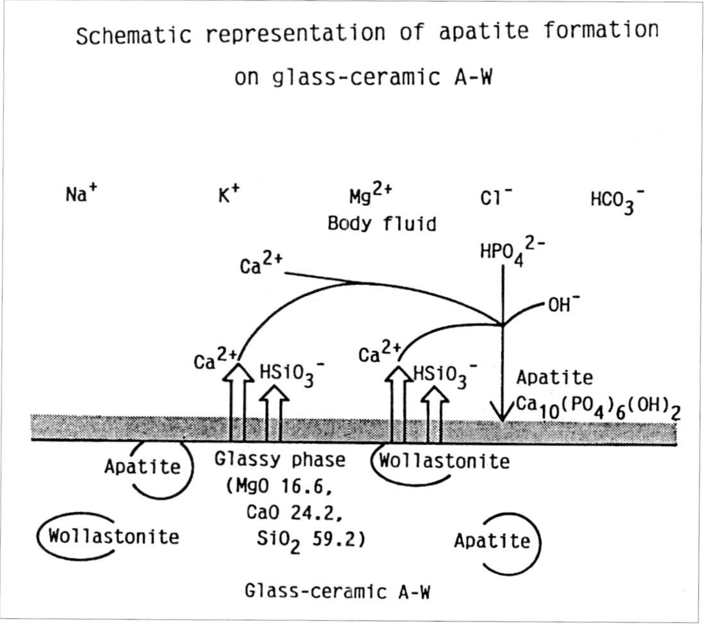

Figure 3.4. Schematic representation of apatite formation on AW-GC.

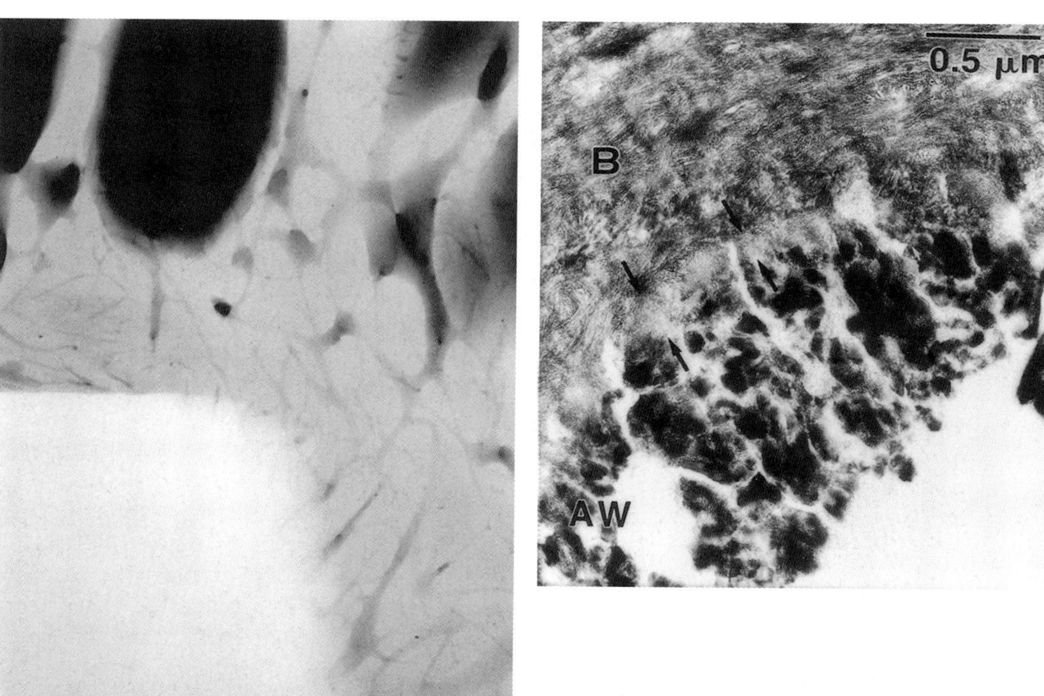

Figure 3.5. **a** A contact micro-radiograph showing the bone bonding between AW-GC implant and newly formed bone. **b** A transmission electron micrograph showing the bonding interface between AW-GC and bone. AW: AW-GC crystals, B: bone tissue.

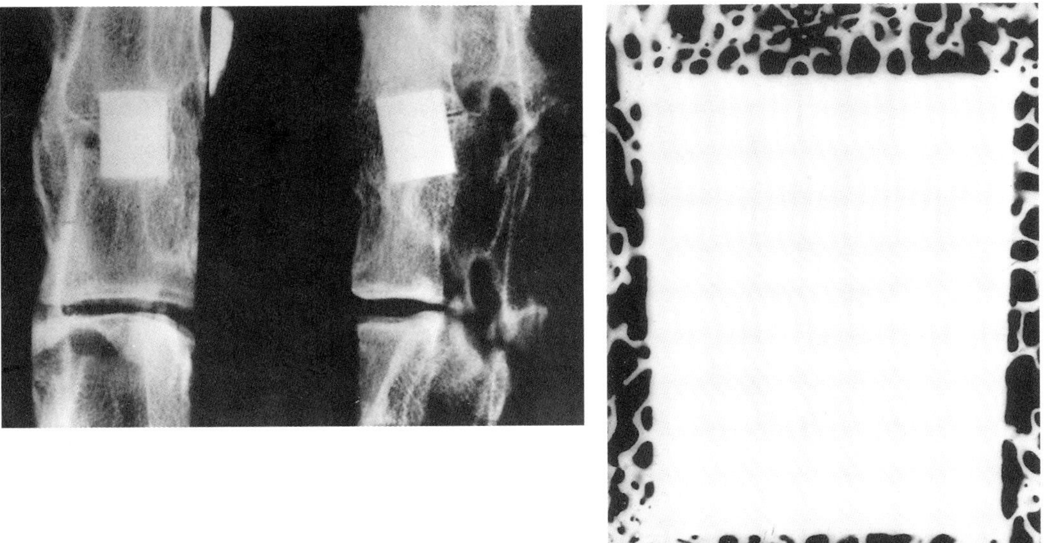

Figure 3.6. **a** X-ray showing an AW-GC implant used for interbody fusion of the lumbar vertebrae of a sheep. **b** A contact micro-radiograph demonstrating direct bonding between the AW-GC implant and bone trabeculae of the lumbar vertebrae, one year post-implantation.

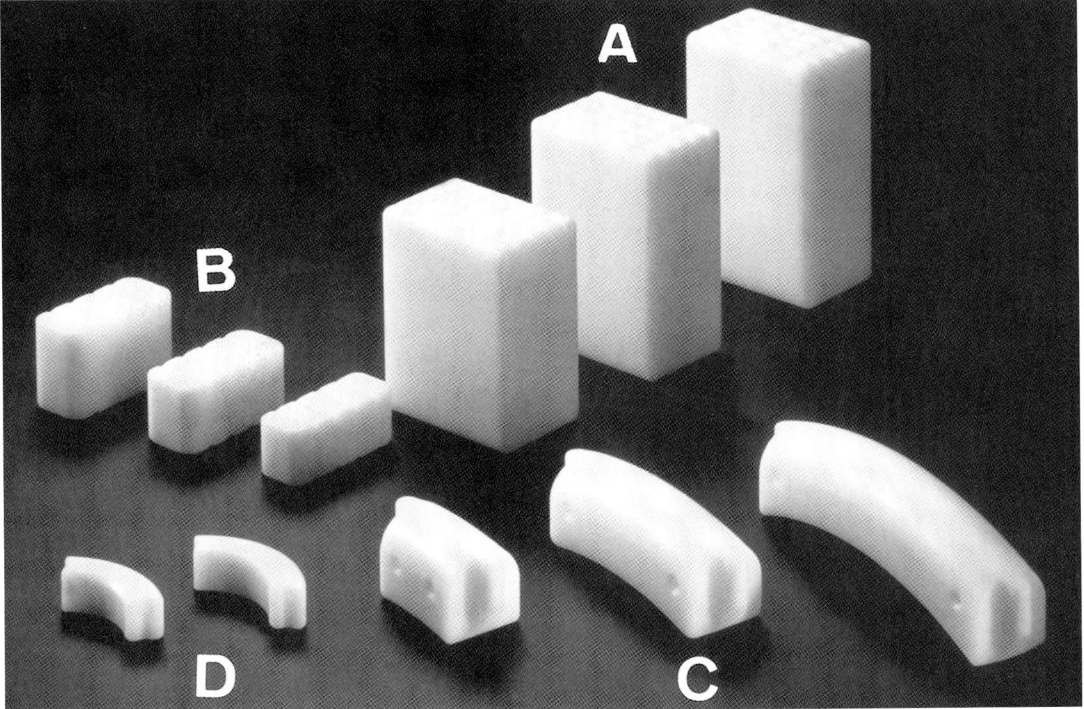

Figure 3.7. Various bone prostheses. A: vertebral prosthesis, B: intervertebral spacer, C: iliac crest prosthesis, D: laminoplasty spacer.

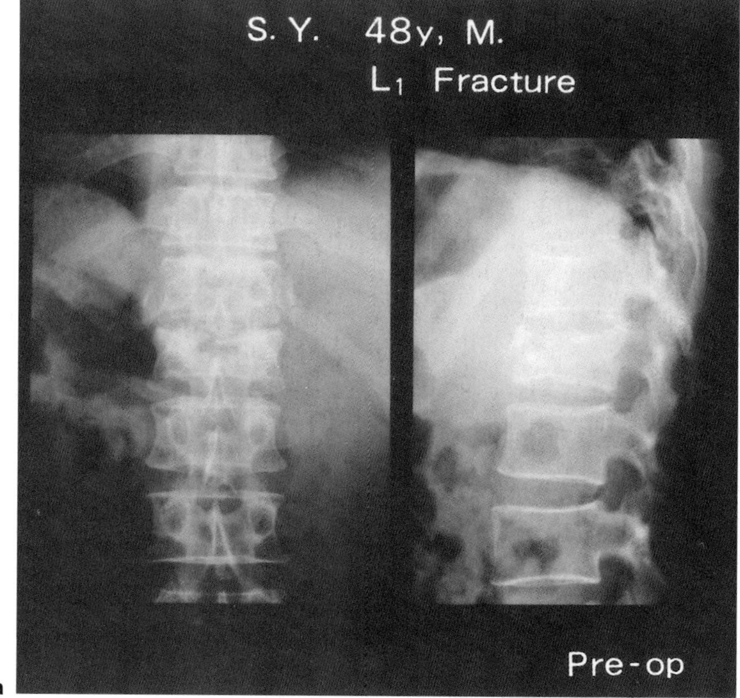

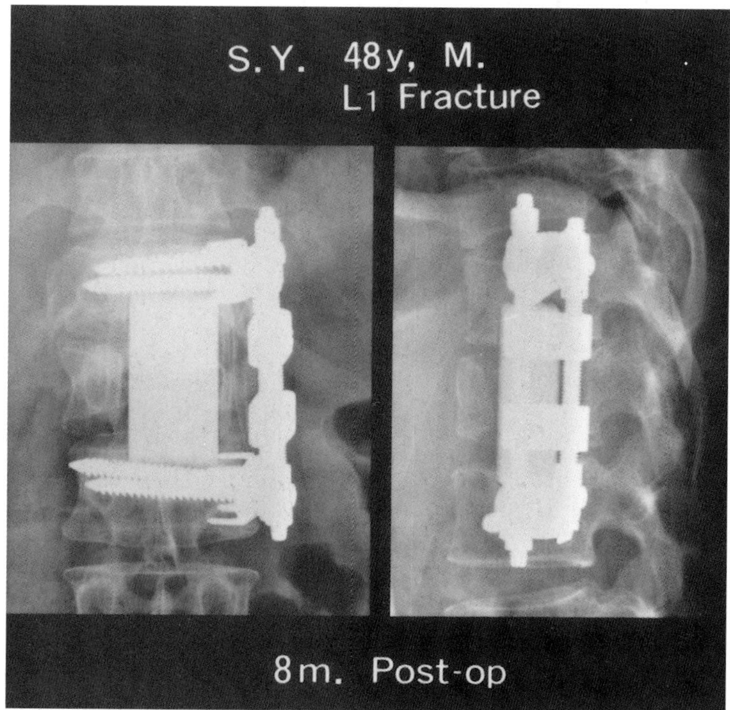

Figure 3.8. **a** X-ray demonstrating L_1 burst fracture associated with paraplegia developed in a 48-year-old male. **b** Postoperative X-ray showing a vertebral prosthesis used for the reconstruction of the lumbar spine.

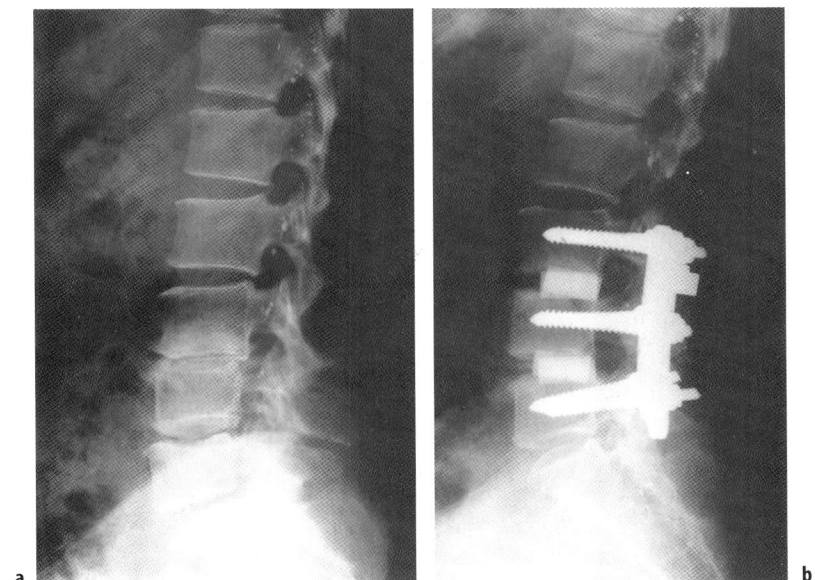

Figure 3.9. **a** A case of multiple degenerative spondylosis of the lumbar spine developed in a 58-year-old female. **b** Postoperative X-ray showing the results of postero-lateral interbody fusion using intervertebral spacers made of AW-GC.

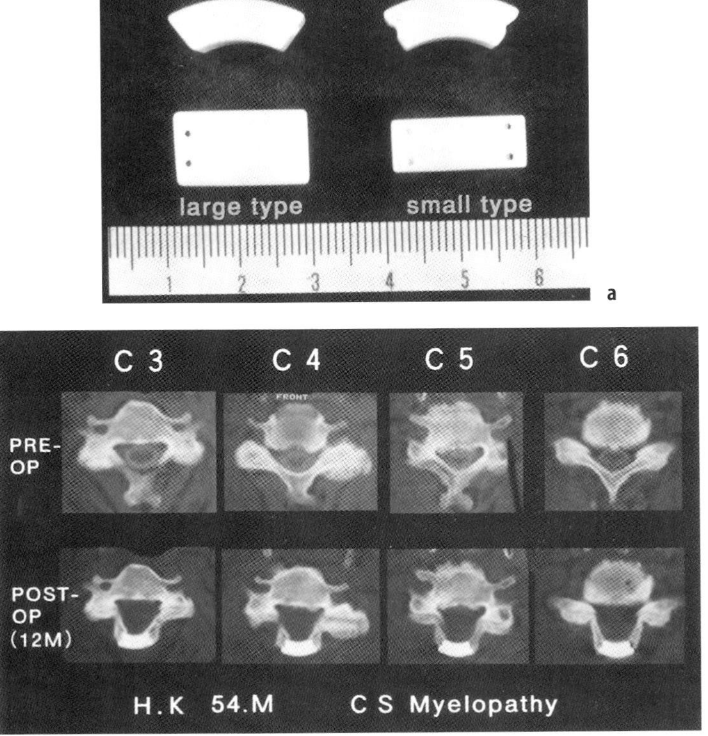

Figure 3.10. **a** Laminoplasty spacers made of AW-GC. **b** CT images of a 54-year-old male suffering from cervical myelopathy due to spondylosis (upper row). Enlargement and reconstruction of the spinal canal was performed by the use of laminoplasty spacers in four levels (lower row).

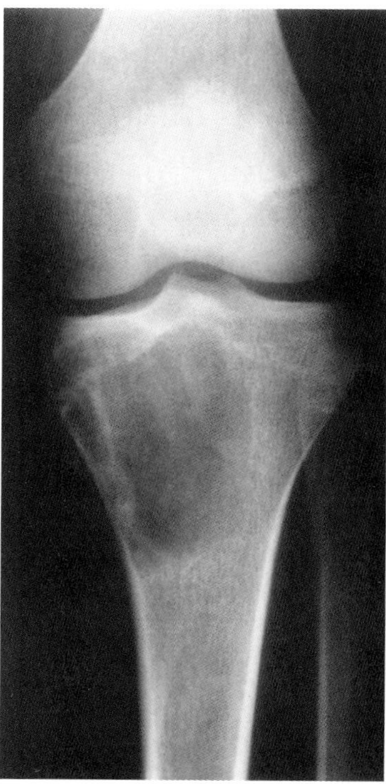

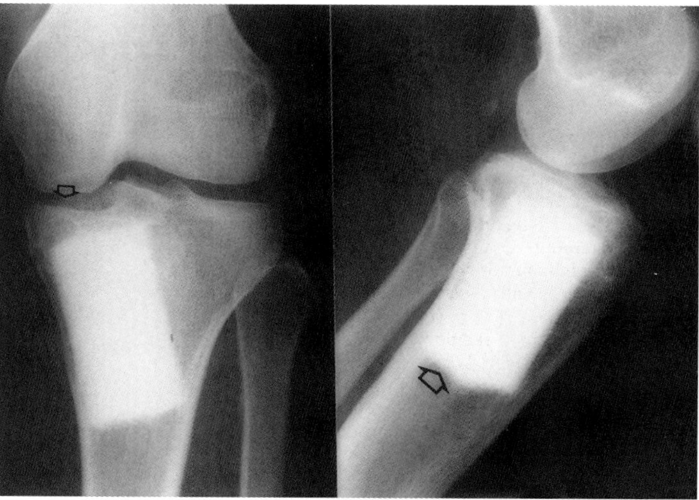

Figure 3.11. **a** A giant cell tumor developed in the proximal tibia of a 15-year-old female. **b** Ten years post-implantation of an AW-GC block that was used in combination with autogenous bone chips to replace the tumor. No clinical symptoms postoperatively.

of the cervical spinal canal in degenerative spondylosis and ossification of the posterior longitudinal ligament (Figure 3.10). AW-GC has also been used as a bone substitute in the form of either blocks (Figure 3.11) or granules. When bioactive ceramic granules are used as a bone void filler together with fibrin glue, osteoconduction and bone bonding are accelerated. AW-GC is also widely used for bioactive coating of hip prostheses. Its details are described in the chapter on Ceramic Coating.

References

1. Benson J. Presurvey on biomaterial application of carbons. North American Rockwell, Rocketdyne Report R-7855, 1969.
2. Yamamuro T, Kotoura Y, Kasahara K, Takahashi M, Abe M. Intraoperative radiotherapy and ceramic prosthesis replacement for osteosarcoma. In: Yamamuro T, editor. New development for limb salvage in musculoskeletal tumors. Tokyo: Springer Verlag, 1989;327–36.
3. Boutin P, Blanquaert D. Le frottement alumine–alumine en chirurgie de la hanche 1,205 arthroplasties totales: avril 1970–juin 1980. Rev Chir Orthop 1981;67: 279–87.
4. Sedel L. Evolution of alumina-on-alumina implants. Clin Orthop Rel Res 2000;379:48–54.
5. Cales B, Peille CN. Radioactive properties of ceramic hip implants. Bioceramics 1988;1:152–5.
6. Yamamuro T. Zirconia ceramic for the femoral head of a hip prosthesis. In: Sedel L, Cabanela ME, editors. Hip surgery, materials and development. London: Martin Duntz, 1998;41–4.
7. Yamamuro T. A new model of bone-conserving cementless hip prosthesis made of high-tech materials: Kobelco H-5. In: Imura S, Wada M, Omori H, editors. Joint arthroplasty. Tokyo: Springer Verlag, 1990;213–24.
8. Hench LL, Greenlee TK Jr, Allen WC, Piotrowski G. U.S. Army Research and Development Command, Contract No. DADA 17-70-C-0001, University of Florida, Gainesville, 1970.
9. Wilson J, Pigott GH, Schoen FJ, Hench LL. Toxicology and biocompatibility of bioglasses. J Biomed Mater Res 1981;15:805–17.
10. Aoki H, Shin Y, Akao M, Tsuji T, Togawa T, Ukegawa Y, Kikuchi R. Sintered hydroxyapatite for a percutaneous device. In: Christel P, Meunier A, Lee AJC, editors. Biological and biomechanical performances of biomaterials. Amsterdam: Elsevier, 1966;1–3.

11. Jarcho M, Bolen CH, Thomas MB, Nobick J, Kay JF, Doremus RH. Hydroxyapatite synthesis and characterization in dense polycrystalline form. J Mater Sci 1976; 11:2027–34.
12. Geesink RGT, de Groot K, Klein CPAT. Chemical implant fixation using hydroxyl-apatite coatings: the development of a human total hip prosthesis for chemical fixation to bone using hydroxyl-apatite coating on titanium substrates. Clin Orthop Rel Res 1987;225:147–70.
13. Kokubo T, Shigematsu M, Nagashima Y, Tashiro M, Nakamura T, Yamamuro T, et al. Apatite- and wollastonite-containing glass-ceramic for prosthetic application. Bull Inst Chem Res Kyoto Univ 1982;60:260–8.
14. Neo M, Kotani S, Nakamura T, Yamamuro T, Ohtsuki C, Kokubo T, Bando Y. A comparative study of ultrastrctures of the interface between four kinds of surface-active ceramic and bone. J Biomed Mater Res 1992;26:1419–32.
15. Kokubo T. Bonding mechanism of bioactive glass-ceramic A–W to living bone. In: Yamamuro T, Hench LL, Wilson J, editors. Handbook of bioactive ceramics, Vol. 1: Bioactive glasses and glass-ceramics, Boca Raton: CRC Press, 1990;41–9.
16. Yamamuro T, Shikata J, Okumura H, Kitsugi T, Kakutani Y, Matsui T, Kokubo T. Replacement of the lumbar vertebrae of sheep with ceramic prosthesis. J Bone Joint Surg 1990;72-B:889–93.
17. Yamamuro T. A/W glass-ceramic: Clinical applications. In: Hench LL, Wilson J, editors. An introduction to bioceramics, Singapore: World Scientific, 1993;89–103.

4 Tissue Adhesives in Orthopedic Surgery

P. Mainil-Varlet, X. Wang, and R. P. Jakob

In times of change, learners inherit the earth, while the learned find themselves beautifully equipped to deal with a world that no longer exists.

(Eric Hoffer)

Introduction

Tissue adhesives have important applications in clinical practice. These biological glues, most especially fibrin, have now been used in European countries for several decades for tissue welding, as hemostatic agents in the control of bleeding, as fluid and gas barriers in fistulas and vascular grafts, as well as in meningeal- and pulmonary-puncture coatings, and, not least, as implants, in which situation they have served as scaffolds and/or as drug-delivery vehicles in wound healing. Fibrin has recently been approved for clinical use in the USA, albeit solely in the capacity of a surgical hemostatic homeostasis agent.

Tissue adhesives possess properties which render them of particular value in wound care. Unlike most dressings they adhere to the lesion and conform to its contours. They can also bond two pieces of tissue together during the course of healing without engendering the uneven distribution of stress evoked by suturing, which can lead to localized trauma, scarring, dehiscence, or fistularization. In most instances they can also be applied more rapidly and with greater ease than is possible with suturing. Furthermore, their applications are more diversified in that they can be used for attaching implants to host tissue, for sealing tissue, and for inducing hemostasis. The ability to undergo degradation at an appropriate rate is a critical feature for tissue adhesives. If their breakdown is too rapid, then they will be deficient in bonding strength, which will result in premature tissue separation. On the other hand, if they degrade too slowly, then their physical presence may interfere with the process of regeneration. The ideal system would be one in which the regeneration process itself controls the breakdown, which could be termed biofeedback-controlled degradation. A system of this kind would automatically take into account patient-to-patient variability in healing rate, would adjust the healing rate, and would attune to temporal fluctuations in this parameter; as such, it would have a broad spectrum of applications in many tissues with different healing rates.

These tissue adhesives are often applied to injured areas rendering them attractive vehicles for the direct delivery of healing substances thereto. Antibiotics, for example, can help reduce the risk of infection, and angiogenic agents can accelerate the process of wound healing by stimulating capillary ingrowth. Adhesive agents may also be used for the direct delivery of chemotherapeutic drugs to tumors. Such topical application renders possible the use of lower drug doses, thereby minimizing the risks associated with the systemic administration of high doses. Furthermore, a tissue adhesive can be designed to release a medication over a sustained period of time, thus obviating the need for repeated doses.

Types of Tissue Sealant

Fibrin

Chemistry and History of Development

The hemostatic and sealing effects of fibrinogen have been attracting the attention of physicians

Table 4.1. Compositions and mechanical properties of tissue adhesives [9–13]

Tissue adhesive	Ultimate tensile strength (Mpa)	Young's modulus	Clotting time (min)	Incisional strength (kPa) Days after surgery			Skin graft adhesion (kPa)	
				0	4–7	14	Tensile	Shear
Methyl-a-CA	28–55	210–340	1	16	9	32		
Propyl-2-CA			1	30	13	30		
Isobutyl-2-CA	65		1					
Heptyl-2-CA			1	32	19	22		
Fibrin	0.1–0.2	0.15	1–120		8	63	20	6–30
Albumin (25)/PEG (15)			2		36	90		37
Albumin (25)/PEG (10)			2		7	28		
Albumin (30)/PEG (10)			2		13	27		
MAP			2–20					
GFR				62				
Sutures	0.7		NA	8	16	53		

Micaceous Bouquique apparently has less weight loss. This is the first series and samples were very small. Although results are interesting as a first series and perhaps indicative, further analyses are necessary.

for more than a century now. Indeed, the use of blood preparations with sealing and hemostatic properties in wound treatment dates back to the end of the eighteenth century [1,2]. In 1909, Bergel used fibrin fleece as a degradable hemostatic agent [3]. The first recorded use of fibrinogen as a tissue adhesive was in 1940, when Young and Medawar employed it for bonding first peripheral nerves, and then skin [4]. After this time, interest in fibrin as an adhesive languished, owing to its inadequate strength and premature repair failures. It took a further three decades before the concept of a "blood glue" was successfully reintroduced by Matras in 1972 [5]. He used a cryoprecipitate containing elevated concentrations of fibrinogen, factor XIII, and fibronectin, as well as of other substances. Later, a plasma product employed in factor VIII-deficiency therapy was used in conjunction with a solution of bovine thrombin. Commercial versions of this technology, marketed under the trade names Tisseel® and Tissucol®, were launched in Europe in the early 1980s by Immuno AG. Each product is sold as a kit consisting of two components: (1) a lyophilized concentrate of pooled human fibrinogen/factor XIII, which is reconstituted with an antifibrinolytic solution; and (2) bovine thrombin, which is reconstituted with a solution of calcium chloride. A competing fibrin sealant product, Beriplast® (Behringwerke, Marburg, Germany), utilizes a different fibrinogen/factor XIII purification method [6,7]. Another, which is prepared by Cohn fractionation (Blood Transfusion Centre, Lille, France), is marketed as Biocoll® [8]. Table 4.1 details the compositions of some commercially available fibrin sealants.

Fibrinogen, the structural blood protein that is instrumental in clot formation, is converted by thrombin into fibrin monomer. These assemble into fibrils, which eventually aggregate to form a three-dimensional fibrous matrix or gel that effectively prevents further loss of blood. Factor XIII is activated by thrombin in the presence of Ca^{2+}, in which state it induces covalent bond formation between the assembled fibrin monomers. This renders the fibrin gel less susceptible to proteolytic digestion by plasmin and increases its overall strength and stiffness. The fibrin gel adheres to a variety of molecules such as collagen and cell-surface receptors, most notably integrins on platelets and other cells. Fibrinopeptide A, formed during the fibrinogen-to-fibrin conversion, acts as a chemotactic agent for polymorphonuclear leukocytes, whilst fibronectin, incorporated covalently into the gel structure, serves as an attachment site for migrating fibroblasts. Fibrin degradation products, formed during proteolytic digestion of the matrix, stimulate the migration of monocytes, which then transform into macrophages. These cells, in turn, phagocytoze remnants of the degrading fibrin network. Neovascularization follows shortly thereafter. This composite of fibrin, macrophages, fibroblasts, collagen, and vascular buds comprises the granulation tissue (Figure 4.1). In its unaggregated configuration, is a "sticky" liquid which readily adheres to wet surfaces, but once polymerized it forms a semi-rigid, hemostatic, fluid-tight adhesive mass, which is capable of holding tissue or materials in a desired shape.

Source of Fibrin

The fibrinogen and factor XIII components of fibrin sealants are prepared from blood plasma. Variations between products arise from differences in the source of plasma, in the mode of precipitation, and in the subsequent purification steps performed (if any). The basic methods adopted can be broadly categorized according to the plasma source: pooled donor (commercial) and single or autologous donor.

Pooled Donor (Commercial): The initial step in the preparation of Tissucol® (Immuno AG, Vienna, Austria) [14,15] involves the generation of a cryoprecipitate, which is then washed with a citrate buffer to extract the cold-soluble proteins. The extract is lyophilized in a citrate–glycine buffer and then pasteurized by steam [16]. Behringwerke AG, (Marburg/Lahn, Germany) [6] also utilizes cryoprecipitation in the preparation of Beriplast®, this step being followed by glycine precipitation to further purify the fibrinogen. Biocoll® (Blood Transfusion Centre, Lille, France) is prepared according to Cohn; ethanol fractionation technique [8,17].

The only reservations expressed concerning the use of pooled-source fibrinogen products – and these are, admittedly, major ones – relate to the risk of transmitting AIDS and hepatitis [18–20]. However, according to Hilfenhaus, current methods of sterilization render fibrin sealants safe with respect to virus infections [21], and on the basis of their long-term investigations with many patients, Koveker [22], Borst, and Haverich claim infection via fibrin components to be either very rare or non-existent. Despite a large body of supporting evidence and a patent for the viral deactivation process, Immuno AG has only recently (1998) gained FDA approval for the restricted use (topical applications to help control bleeding) of their fibrin sealant [23,24].

Single or Autologous Donor: Cryoprecipitation is the most commonly used method for fibrin sealant production in the USA [25,26]. The use of single-donor plasma as a source of fibrin sealant is popular owing to the simplicity of the procedure involved and its conformity to standard blood centre practice.

Gel Structure

The three-dimensional structure of a fibrin gel may be modified by changing any one of several parameters: the concentrations of fibrinogen, thrombin, or Ca^{2+}; ionic strength; pH; and, to a lesser extent, temperature [27,28]. Varying the ionic strength or pH will produce "fine" gels (pore size $<2\,\mu m$) whereas these parameters will yield "coarse" ones (pore size $>2\,\mu m$). Polymerization rate increases with increasing temperature, reaching a maximum at 37 °C, above which degradation (denaturation) begins. The presence of fibronectin does not appear to influence pore size or gelation rate, although it decreases fibril diameter [29]. Factor XIII has no affect either on gel porosity or on other geometrical attributes [30].

Mechanical and Adhesive Characteristics

Since fibrin sealants are used not only as tissue adhesives but also as physical barriers and fillers, both their material mechanics and gluing properties must be considered.

Shear loading of fibrin gels has been extensively studied under a broad range of conditions. "Fine" gels, irrespective of the degree of factor-XIII crosslinkage, have a lower shear compliance than do "coarse" ones, although they are more prone to permanent deformation, i.e., they are characterized by a more viscous behavior.

The tensile mechanical properties of fibrin sealants have also been evaluated [31]. Increasing the thrombin concentration of Tissucol® up to 500 units/ml results in an increase in its ultimate tensile strength and Young's modulus.

An in vivo model has been developed to test the adhesive strength of Tissucol® in the closure of full-thickness incisional wounds within the

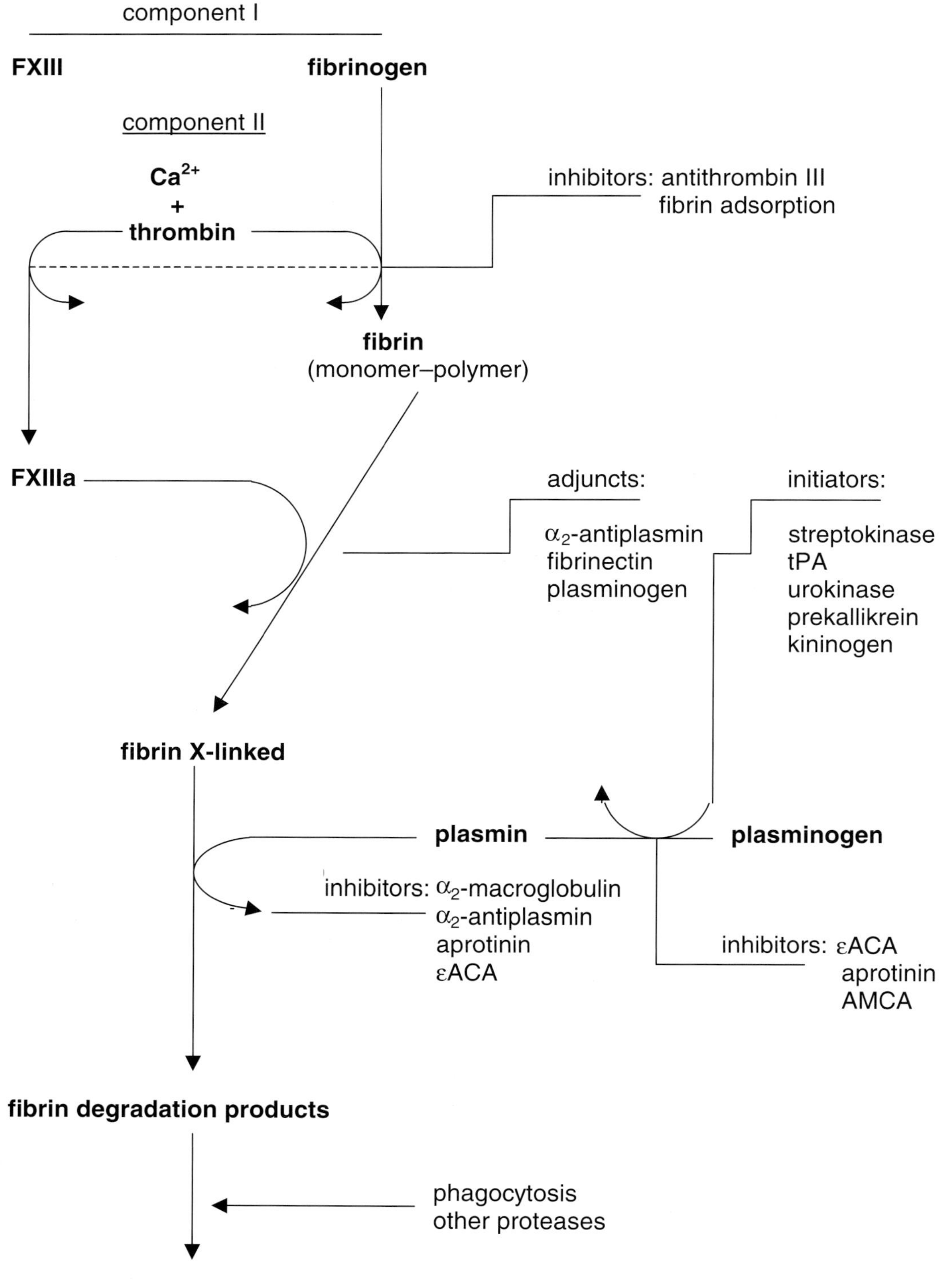

Figure 4.1. Coagulation and degradation cascades of fibrin tissue sealant.

Table 4.2. Composition of various fibrin adhesives

Fibrin glue Pooled-donor source	Fibrinogen (g/l)	Thrombin (units/ml)	Factor XIII (units/ml)
Tisseel® (Immuno AG, Vienna, Austria) [14,15]	≥75	400–500	≥10
Beriplast P (Behringwerke AG, Marburg, Germany) [6]	65–115	400–600	40–80
Lille (CNTS Fractionation Centre, Lille, France) [8]	90–95	?	25
Transglutine (CNTS Fractionation Centre, Strasbourg, France) [33]	≥70	200	?
Single autologous donor			
Durham et al. [34]	18–27	5–50	?
Dresdale et al. [35]	21–6	500–1,000	?

rodent dorsum [32]. The fibrin sealant repairs had a higher bonding strength than did suture controls up to the fourth postoperative day, after which time no difference between the two groups was observed. It was noted, however, that in the fibrin sealant group, maximal mechanical strength was not attained until 20–40 days following surgery, whereas in suture controls, it was achieved immediately [8]. Adhesive strength was shown to increase with time from about 50 g/cm^2 at 1 minute to 200 g/cm^2 at 2 hours. The breaking strength of 1 cm^2 grafts of human split-thickness skin samples glued with a fibrin glue preparation obtained from pooled human plasma for 3.5 minutes was demonstrated by the same authors to be greater than that of sutured ones when separated with a dynamometer [8]. Table 4.2 details the composition of various fibrin glues.

Methods of Application

Commercially available fibrin sealants are sold in powdered form as part of a kit, which also includes thrombin, calcium chloride, and aprotonin. As already mentioned, almost all marketed formulations of fibrin sealant are purchased as two-component systems, analogous to epoxy resin catalyst adhesives. Fibrinogen and thrombin components are mixed either immediately before their contact with tissue or upon the tissue surface itself. Corus Medical Corp. has developed a dispenser for the delivery of FS in aerosol form [36].

Composites

Fibrin composites combine two or more materials with a view to improving mechanical or other properties. Glues composed exclusively of fibrin are not very strong, and they become less so if their porosity is increased to promote vascular ingrowth. Calcium phosphate salts are used in conjunction with fibrin to form a mineralized composite similar to bone. Extracellular matrix components of cartilage and bone, such as glucosaminoglycans, hyaluronic acid, and chondroitin sulfate, would be expected to confer viscoelastic properties upon a composite.

Hyaluronic Acid: When mixed with a fibrin adhesive, hyaluronan dose indeed increases the viscosity of the glue; it has also been shown to enhance the quantity of fibrin entering blood vessels after reanastomosis [37,38].

Fibrin–Collagen: Fibrin–collagen composites, prepared by adding fibrillar type I collagen to fibrinogen–factor XIII solutions, have a higher binding strength than purely fibrinous glues, owing to the interaction between collagenous and fibrinous fibers [9]. Collagen has also been combined with fibroblast growth factor (FGF), and heparin complexed to fibrin [39]. In vitro, fibroblast growth factor bound to this matrix is released rapidly, but only partially, in the presence of heparin. Subcutaneous implantation of collagen sponges impregnated with the FGF/heparin/fibrin mixture promotes tissue growth into the sponges, the resulting fibroblast-infiltrated tissue resembling a normal dense connective tissue.

Other Composites: In orthopedic surgery, composites are effectively generated by mixing fragments of bone or osteochondral tissue with fibrin adhesives, these being used for the filling of defects [40,41]. A mixture of Triosite® and fibrin adhesive has been successfully used to narrow the nasal fossae in patients with atrophic rhinitis.

Albumin

Serum albumin represents the largest protein component of human blood, normally constituting 50–60% of its dry mass. It plays an important role in plasma-volume regulation and in tissue-fluid balance by virtue of its contribution to the coloid oncotic pressure of plasma. Albumin has a molecular weight of 66,300–69,000 and a melting temperature of 63 °C [42].

Types of Crosslinked Systems

Crosslinkage of albumin results in polymerization, when its mechanical properties undergo a marked change. Not least, this process confers upon albumin excellent adhesiveness. Being a naturally occurring protein, externally applied albumin will of course be degraded and removed by physiological mechanisms during the natural course of healing. Human albumin has also been used as a vehicle for site-specific delivery of growth factors to accelerate tissue repair; for example, after wound closure by laser welding [43–48]. For such purposes, albumin is generally utilized in the form of microspheres. Various agents have been employed for the crosslinkage of such microbeads, including glutaraldehyde, dexamethasone [49], and radiation [50,51]. In the latter instance, however, special facilities are required. Glutaraldehyde–albumin microsphere complexes have been not infrequently used as vehicles for the delivery of proteins and polymers, including progesterone, insulin, heparin, and polyamidoamine polythylene glycol and polythioetheramido polythylene glycol [52–55]. The glutaraldehyde component of such systems does, however, have several undesired effects; not least, it inhibits cell migration and induces a tissue inflammatory response [56,57]. The fibrous capsules formed in consequence around the microspheres have been shown to retard the release of insulin carried therein [52,58,59]. Crosslinkage of albumin with activated polythylene glycol yields a hydrogel which has proved to be useful for the delivery of drugs such as acetominophen, theophyline, hydrocortisone, and gentamycin, as well as enzymes such as alkaline phosphatase and lysozyme [60,61]. When crosslinked with certain activated derivatives of polythylene glycol, such as polyethylene disuccinimidylsuccinate, the adhesive properties of albumin are markedly enhanced. In the aforementioned instance, the peel force (a measure of adhesive strength) of albumin has been shown to be similar to that of cyanoacrylate (see below) and much greater than that of fibrin [62,63].

Device Coating

Prostheses become rapidly coated with a natural layer of proteins upon implantation, albumin constituting a major component of such films. Albumin has, indeed, been used as a substitute for blood or fibrin in the coating Dacron® vascular grafts. Although albumin has been utilized in both its unpolymerized [64,65] and polymerized [66–68] forms, crosslinkage is now considered to be desirable, in that it renders this protein more resistant to degradation and tissue ingrowth. Titanium implants coated with carbodiimide-crosslinked albumin inhibit the adhesion of both *Staphylococcus aureus* and *Staphylococcus epidermis* [67]. The coating undergoes little degradation during the first 20 days following implantation, the adherence-inhibition of bacteria being sustained at a level of 85% throughout this period. Furthermore, coating of hydroxyapatite beads or human teeth with crosslinked albumin has been observed to suppress significantly the adhesion of oral bacteria.

Cyanoacrylates

The cyanoacrylates were first synthesized in 1949, although their adhesive properties were not discovered until a decade later [69,70]. The methyl-2-cyanoacrylates were the systems initially used as surgical adhesives, but problems with wettability and histotoxicity led to the development of longer-chain homopolymers, such as ethyl-2-cyanoacrylate (Krazy Glue®), isobutyl-2-cyanoacrylate (Bucrylate®), and butyl-2-cyanoacrylate (Hexacryl®) [71,72]. Longer side-chains increase the hardness and tensile strength of the adhesive, but at the cost of flexibility and degradation rate. During the 1970s, reports on the histotoxity of Bucrylate® and other cyanoacrylates began to appear in the literature [73–82]. Since then, Hexacryl® has been the system of choice, although this substance, too, has been shown to be histotoxic when applied subcutaneously [82]. Cyanoacrylates have a thrombotic effect, which is generally deemed to be undesirable, but can be useful for embolizing arteriovenous malformations [83–85]. Owing to their being considered less toxic than other classes of cyanoacrylates, n-butyl derivatives (Nexacryl, Avacryl, Occyldent, and TPS) have been developed for various applications, namely, tissue adhesion, tissue sealing in periodontal situations, and as drug-delivery systems. In general, the n-butyl cyanoacrylates have proved to be promising candidates for tissue adhesion, owing to their ease of application, reliability, and good bond strength. They afford a stronger attachment than do sutures, are less expensive, and are more readily administered. Doubts have been raised, however, about the suitability of these thrombogenic agents for well-vascularized tissue. Since cyanoacrylates undergo an exothermic reaction during polymerization, these adhesives must be carefully designed and applied so as to avoid thermal injury to tissues. An additional concern is that many therapeutic proteins and medications denature or degrade at higher temperatures. Although materials can be added to cyanoacrylates to act as heat sinks, this can adversely affect the desired mechanical properties and the rate of bonding. It is also of significance that healing tissue cannot grow through some cyanoacrylate adhesives, which can remain unresorbed for 12 months or longer. But since they form a barrier to vascular ingrowth in wounds, they can be used to embolize arteries, as mentioned above.

Mucopolysaccharides

The blue mussel, *Mytilus edulis*, synthesizes a 3,4-dihydroxyphenylalanine (DOPA)-rich polyphenolic protein which mediates attachment together with other proteins found in the byssus via collagenous threads or tethers [86–90]. This protein, referred to as the mussel adhesive protein (MAP), serves to affix the mollusk to rocks or other surfaces in turbulent tidal zones. MAP appears to have an affinity for both soluble and insoluble collagen. In vivo studies have demonstrated that MAP provides sufficient holding strength within the first minute of secretion, the firmness of the attachment increasing continually thereafter over the course of several hours [88,91,92]. MAP has been used as an artificial basement membrane, in which situation it permits the diffusion of insulin and dextran, two large non-electrolyte molecules [90] and because of its adhesiveness for cells, it has been employed for the fixation of osteoblasts and chondrocytes. Both MAP and fibrin glue have been tested in vivo for their ability to fix internal chondrocyte allografts. Whilst results for fibrin were at best inconclusive, those pertaining to MAP were considered to be highly promising [93]. This natural glue may be of clinical use in situations requiring permanent adhesion, such as in implant fixation in hard tissues. And looking torwards the future, if the native strength of MAP could be reproduced under high-share clinical conditions, then it could prove to be indispensable for vascular procedures.

Other Adhesives

A gelatine–resorcinol–formaldehyde system has been developed as a tissue adhesive, its principal advantages being that it is readily applied and maintains its properties in a moist environment [10,94]. Gelatine is denatured collagen, the most common structural protein within the body. Formaldehyde increases the cohesive strength of the adhesive and decreases its solubility by crosslinking the gelatine. Resorcinol reacts with the formaldehyde to further stabilize the system and reduce its overall viscosity. As yet, the gelatine–resorcinol–formaldehyde system has been tested almost exclusively in animal models [95–98]. Only a few clinical trials have been undertaken, this circumstance reflecting concerns for the cytotoxicity of formaldehyde and glutaraldehyde. With a view to improving biocompatibility, investigators have also tested the effects of incorporating two less-toxic aldehydes, namely, pentanedial and ethanedial, but no data relating to the use of such systems in animal models are available at present.

Orthopedic Applications of Tissue Adhesives

The Influence of Sealants on Bone Healing

Adhesives may offer a number of advantages over conventional metal osteosynthesis in the treatment of fractures: They improve the ease and speed of fixation; and they have the plasticity to anatomically coapt small fragments, thereby obviating the need for hardware removal. The potential benefits to be derived from using a fibrin sealant in osteo-regeneration are debatable [99,100], although factor XIII itself is known to accelerate bone healing when administered systemically, as revealed by testing the tensile strength of healed osteotomies [101–103]. To study the influence of fibrin on bone healing, Bösch et al. [104] created standardized cortical defects within rabbit tibia and then closed them with homologous plugs of this substance [104]. Fibrin accelerated the sprouting of capillary vessels and the ingrowth of connective tissue cells, thereby facilitating the rapid neoformation of bone. The effects of fibrin were first manifested at the edges of each defect where intensive osseogenesis was observed within 14 days. In control lesions, a homogenous connection developed between the callus and cortex within 31 days. In this latter group, a sclerosing edge was observed, which corresponded to an inactive marginal zone. Use of fibrin was also associated with an accelerated rate of remodeling and an absence of osteocyte regression from the center of the autologous cancellous bone reimplants [105–108], which indicates an earlier onset of nutrition than in the control group. Some authors have also investigated the influence of fibrin on heterologous cancellous transplants (Kiel spongy-bone grafts) [106,107] which were osseously connected to the recipient bone. The trabeculae became surrounded by newly formed lamellar bone and were subsequently resorbed. As a xenogenic replacement the Kiel graft possesses no osteogenic potency. In the control group, such transplants were almost completely resorbed and replaced by connective tissue with poor metaplastic bone formation. Filling the spongy medullary cavities of the Kiel grafts with fibrin sealant improved the proliferation of invading vessels – a critical event in osteogenesis. Pflüger et al. [109] have improved the osseointegration of stainless steel cylindrical implants by increasing their pore sizes. The ingrowing bone tissue, which is responsible for anchoring the cylinders, was incorporated more rapidly in the presence of a fibrin sealant. The force needed to remove fibrin-fixed implants was greater than that required to dislodge control ones. Fibrin sealant appears to accelerate the incorporation of implants into bone. The reason for this presumably resides in an improvement in implant-immobilization and possibly also in osteogenic potency. Schumacher et al. [110–112] have likewise reported the incorporation of alloplastic implants into bone to be accelerated by the applications of a fibrin sealant [113]. The fibrin-sealant system has also been tested in a double osteotomy rabbit tibia model, the tibia being stabilized by means of DC6 six-hole metal plates. In the test group, a 5-mm-long bone cylinder was removed, reinserted, and then sealed with homologous fibrin. In the control group, the cylinder was introduced into the surrounding soft tissue and fixed by sutures. The fibrin-sealant group revealed better osseous integration than did the control one 5–7 weeks after surgery. The bending strain quotient (calculated from the bending strain of the operated and of the healthy contralateral tibia) was significantly higher in the fibrin-sealant group after the fifth week ($p < 0.05$). But after seven weeks, no difference was apparent between the two categories. The fibrin sealant appeared to have accelerated the healing of the reimplanted bone cylinders not by virtue of its osteogenic potency but by preventing the formation of large hematomas and by improving the anchorage of the said bone cylinders. Albrektsson et al. [114,115] were unable to demonstrate any positive effects of fibrin sealant on initial osteogenesis using a newly developed bone-growth chamber. Early bone growth appeared rather to be delayed and the results did not support those reported by Pflüger et al. [109,116,117]. The authors believed that their failure to demonstrate the osteogenic potency of fibrin lay in this agent's coming to lie within a titanium tube that formed a part of their system, which ruled out biological degradation. Unless fibrin undergoes degradation by tissue plasminogen activators and by cells such as polymorphonuclear neutrophils and macrophages, it may persist locally for days and form a barrier to natural tissue growth [40,118]. Zilch and Noffke [111,112,119] have performed animal experiments similar to those conducted by Böhler et al. They observed no significant difference in the quantity of newly formed bone laid down between control and fibrin-treated groups after the third week. But after four weeks, microradiographic examination revealed the amounts in each case to be 46.5% and 52.3%, respectively. Albeit so, fibrin had exerted no influence on remodeling into lamellar bone at this stage. Lucht et al. [120,121] have studied the effects of fibrin on bone formation and regional blood flow in autologous cancellous bone transplantation. They observed no significant influence of this agent on either of these parameters, although it had a tendency to diminish new bone formation in some grafts. These blood-flow findings are in contrast to those reported by other investigators, who have demonstrated that fibrin sealing accelerates vascularization [111,112]. In a study performed on Sprague–Dawley rats, using the nasal critical-size defect model, Tholpady et al. [122] have shown that osteoprogenitor cells contained within the fibrin sealant augment bone regeneration. The potential of fibrin glue to improve the fusion of bone grafts has been investigated in various animal models. In one such study involving cats, cortico-cancellous bone grafts (derived from the disc space in the anterior cervical region) were fused together with a fibrin adhesive (Tisseel®). The authors observed that the allograft fusion mass was more voluminous in control (untreated) animals than in fibrin-treated ones, as revealed by CT measurements. In another study using cats, Turgut et al. [123] reported local sealing with fibrin to significantly retard osteogenic fusion in a cortico-cancellous bone grafting model, this finding having been attributed to reduced vascularization of the graft as well as to diminished new-bone formation. These authors concluded that fibrin sealant was not suitable for the fixation of bone fragments in anterior cervical fusion. In a similar study conducted with mongrel dogs (spinal-fusion model), Jarzem et al. [124] demonstrated that Tisseel®-treated allografts had a significantly smaller fusion volume than did controls (as revealed by computed tomographic volumetric analysis). They, too, were of the opinion that Tisseel® is not an ideal material for achieving or augmenting intervertebral arthrodesis [124]. Plaga et al. [125] have compared the results achieved using a fibrin sealant, polydioxanone pins, and Kirschner wires in the fixation of standardized rabbit-knee osteochondral fractures within the medial femoral condyle. In the fibrin-treated group, fracture healing occurred in only 50% of cases as compared with 100% in the Kirschner-wire one.

Sealants in Bone Replacement

Using a femur of iliac-crest defect model in rabbits, Palacios-Carvajal and Moina [126] have compared the results achieved by implantating either tricalcium phosphate (TCP) mixed with fibrin sealant or TCP alone. In the fibrin sealant/TCP group, a dense network of osteoid tissue surrounded each ceramic grain, whereas in the TCP group (no fibrin sealant), only a narrow and a discontinuous layer of osteoid material mantled each particle. Histomorphometry revealed a difference in osteoid thickness of about 15% between the two groups. A similar study has been carried out by Siebert et al. [127] using a rat femur defect model. The plasticity of the implant was considerably improved by adding the fibrin sealant to TCP, but after 40 days, all defects in each of the groups [TCP/fibrin sealant and controls (open defects)] were alike filled with spongiosa. An experimental study has also been performed to evaluate the influence of fibrin on bone morphogenic protein (BMP)-dependent osteoinduction in rats [128]. Autolyzed antigen-extracted allogeneic bone or bone gelatine implanted within a muscle bed invariably led to new bone formation. Addition of allogeneic fibrin to the implants had no influence on the histological result achieved after 21 days. Allogeneic bone gelatine, both with and without allogeneic fibrin, has also been implanted within 7 mm diameter in trepanation defects in given animal [129]. After 26 days, histomorphometric analyses revealed identical findings in each group, most of the defects having been bridged with lamellar bone and marrow being apparent between the trabeculae. No influence of fibrin has been demonstrated on osteoinduction either in rats [99,100] or in adult rabbits (five years after surgery) [129]. On the other hand, a positive influence of this sealant on the repair of cancellous bone cavities filled with a porous, resorbable ceramic has been reported by Kania et al. [130]. These authors tested the effects of two fibrin sealants: Autocolle® and Tissucol®, the former having been enriched with platelet factors during its preparation. Cavities 10 mm deep and 5 mm in diameter were drilled in the lateral condyles of 45 New Zealand rabbits. These were then filled with either coral granules, a mixture of fibrin sealant (Autocolle® or Tissucol®) and coral granules, or left empty. One month later, the fibrin (Autocolle® or Tissucol®)/coral mixture had elicited a significant increase in bone formation compared to the result achieved using coral alone. After two months, a marked enhancement in fibrin-mediated bone repair was observed only with Autocolle. By the six-month stage, the degree of bone formation was similar in each group and comparable to the physiological picture in normal, untreated animals. Control cavities, on the other hand, were filled exclusively with fibrous connective tissue at this latter juncture. Fibrin scaffolds have been shown to serve as suitable carriers for pluripotent mesenchymal cells in the reconstruction of critical-size bone defects, as demonstrated by Perka et al. [131] using New Zealand white rabbits.

Matsumoto et al. [132] have used fibrin as a carrier for analogous bone dust in the repair of craniotomy defects. This mixture conformed readily to the defect contours and yielded favorable cosmetic results 1–5 years after surgery.

Fibrin Sealant in Cartilage Repair

In cartilage surgery, fibrin adhesives have been employed in various capacities, but with the common aim of boosting the repair process. As in other fields, these substance have been utilized as scaffolds to support the ingrowth of host tissue, as carriers for transplanted cells, as vehicles for the site-specific delivery of drugs, and as glues for implant anchorage.

In vitro studies have indicated that fibrin glues release platelet-derived growth factors and transforming growth factor beta which promote the migration of chondrogenic cells into the defect volume, their proliferation therein and subsequent differentiation into chondrocytes, this phenotypic expression being accompanied by the synthesis of a cartilaginous matrix [133]. In vivo, treatment of rabbit-knee osteochondral defects with the fibrin adhesive Tisseel® did not merely have an effect on the natural repair process, but actually inhibited it [134]. Inclusion of growth hormone did not improve the situation. Furthermore, the same authors found that whilst cultured chondrocytes migrated readily into both human and rabbit clots under in vitro conditions, they failed to penetrate Tisseel® gels [134].

Although Tisseel® apparently inhibits the migration of chondrocytes into its substance, cells artificially lodged within such gels during their polymerizations do not appear to be deleteriously affected by this milieu. Homminga et al. [135] demonstrated that chondrocytes not only retained their phenotype within fibrin glues (Tissucol®) maintained in culture but also underwent mitotic division. However, as soon as the fibrin clots started to disintegrate (after seven days), the chondrocytes began to dedifferentiate.

Instead of chondrocytes, chips of devitalized allogenetic cartilage have also been embedded with fibrin glues during their polymerization [136]. Twelve weeks after implantation on the backs of nude mice, not only was the initial mass of such composites retained but also the chondrocyte lodged therein had preserved their native phenotype.

Fibrin products have been widely used simply as glues for fixing homologous cartilage implants [137], perichondral grafts [138–146], and periosteal tissue to articular cartilage [147–151], as well as for securing periosteal flaps to host cartilage surrounding defects filled with autologous chondrocyte suspensions [152] and for plugging lesions containing chondrocyte allografts [153–155]. In this latter instance, the resurfacing results achieved using fibrin were found to be superior to those realized using non-grafting controls, and the tissue laid down during its remodeling contained a higher concentration of aggrecans and type II collagen molecules. However, a general survey of the findings pertaining to fibrin's efficacy as an adhesive or sealant do not yield an overriding good or bad impression, the data and experimental conditions employed being too variable to warrant the drawing of a valid conclusion. One feature that may have contributed to the unpredictable results is the intrinsic anti-adhesiveness of cut cartilage tissue surfaces (which may vary according to the experimental conditions operative), this property being attributable to extracellular matrix proteoglycans, especially decorin and biglycan [156,157]. Superficial removal of such matrix macromolecules, achieved by treating defect surfaces with chondroitinase ABC or trypsin [158], has been shown to improve the adhesiveness of fibrin matrices to the floor and walls of purely cartilaginous lesions. Apropos of this, tissue-engineered cartilage produced using a fibrin-based polymer has been demonstrated to adhere well to native cartilage. When this product was used to join two pieces of native cartilage in nude mice, the strength of the bonding achieved significantly exceeded that realized with a commercial fibrin glue [159].

Sealants in Trauma/Reconstructive Surgery and Soft-tissue Co-optation

Fibrin matrices have been applied in diverse clinical situations, including cardiovascular surgery, neurosurgery, ophthalmic surgery, general/trauma surgery, plastic surgery, middle ear, [160] and orthopedic surgery. In cardiovascular surgery, fibrin has been used as a hemostatic sealant in the attachment of vascular grafts, cardiovascular patches, and heart valves, to preclot porous vascular grafts, and to seal ventricular septal defects. In neurosurgery, it has been utilized for peripheral nerve reattachment and to prevent cerebrospinal fluid leakage during dural sealing [161]. Cyanoacrylate adhesives have likewise been successfully employed to seal experimental dural leaks, but they evoke a stronger inflammatory response than fibrin. In ophthalmic surgery, fibrin adhesives have been used to reattach retinas and to seal perforated corneas. In trauma surgery, fibrin has served as both a sealant and a hemostatic agent in the repair of spleen, kidney, and liver ruptures [162–168]. Topical application has proved to be effective for less serve injuries, as has intraparenchymal injection for more severe ones involving larger vessels [166]. In patients with non-suturable hemorrhages, the spraying of fibrin glue in aerosol form has been shown to be efficacious in staunching blood flow, especially from parenchymal organs, skin-graft donor sites, and retroperitoneal and pleural surfaces [164]. Fibrin glues have also been successfully used to repair esophageal perforations, pulmonary air leaks, and fistulas in both animals and humans [169]. In general surgery, fibrin matrices have been widely employed for packaging and hemostasis in plastic surgery, and have been utilized for blepharoplasty, face lifts, and rhinoplasty [170,171]. Skin grafts have been successfully affixed using fibrin adhesives, even in regions of complex anatomical contouring such as the hands. In such instances, they ensure a firmer attachment than do sutures when the source is autologous [172,173]. Fibrin has proved to be especially useful for the fixation of skin grafts in topographic locations that do not lend themselves readily to bandaging [174].

In orthopedic surgery, fibrin has been found to be a useful alternative to suturing for ruptured Achilles' tendons. In a prospective study conducted at the trauma centre in Zurich, they were either treated with a fibrin sealant or received sutures. Of the 25 individuals (56%) who were sutured, rerupturing occurred in three cases (12%) after 8–10 weeks, whereas only one (5%) of the 20 (44%) who were treated with fibrin suffered this occurrence. The period of time elapsing before patients were deemed fit to resume social and recreational activities was nine months in the former group of patients and seven months in the latter. The long-term functional and cosmetic results achieved were significantly better after treatment with fibrin than after suturing [175].

In an experimental study using rabbits, however, no significant difference in the histological appearance and tensile strength of sutured and fibrin-treated Achilles tendon ruptures were observed up to 27 days after surgery [176].

The effect of fibrin sealants on adhesion in flexor tendon surgery has been investigated in both immobilized and active weight-bearing rabbits' paws [177]. Partial lacerations were produced and either sutured or sutured and glued. In immobilized tendons, no difference in the strength of the adhesion achieved was observed between the two groups. But in mobile ones, the gross and microscopic appearance, as well as the strength of the bonding attained, were vastly superior in the fibrin group.

Several investigations have been performed to ascertain whether the strength of bone–tendon junctions can be improved by use of a fibrin sealant, employed either alone or in conjunction with bone morphogenetic protein (BMP) [178,179]. Fibroblasts and undifferential mesenchymal stem cells were found to migrate into the critical zone, wherein they led to an early and accelerated restoration of the tissue surrounding the tendon. Early union of the tendon with the bone was achieved in both groups, but neither fibrin alone nor the fibrin/BMP combination contributed to enthesis formation.

In meniscal surgery, it is now becoming clear that a combined approach, involving the insertion of a structural fibrin matrix reinforced with collagenous material, is probably advisable. Using such a composite, the stability of longitudinal meniscal tears is enhanced, due both to improved tissue adhesion at the onset and to the induction of an earlier and more exuberant fibrous response. But for more complex rents, healing poses more of a problem. Although various measures have been adopted to enhance repair (e.g., the use of growth factors, synthetic matrices, or stem cells), the results are as yet speculation concerning their potential for clinical application. It is worthy of note, however, that in various animal models [180–183], use of fibrin sealant together with endothelial cell growth factor enhanced neovascularization and the formation of granulation tissue, which accounted for the improved repair results achieved in a vascular portion of the menisci. The repair of white–white meniscal tears still represents something of an enigma. Although various techniques such as fibrin clot implantation and synovial abrasion have been shown to improve the healing rates in certain red–white meniscal lesions [184–188], the results obtained for white–white ones have not been encouraging.

Tissue Adhesives as Drug Delivery Systems

One of the most promising attributes of tissue adhesives is their potential to serve as scaffolds for the ingrowth of cells and the deposition of repair tissue, and as vehicles for the delivery of biological-response modifiers, such as cytokines, growth factors, or receptors. Angiogenic substances are particularly promising candidates for such situations, because wound healing depends critically upon adequate tissue perfusion or blood flow. Repair cells such as fibroblasts require oxygen for their many functions, including proliferation, migration, attachment, and protein production. Hence, their proximity to a capillary bed will have a direct influence on their activity level as well as on their survival. The capacity of fibroblasts to migrate into a provisional repair matrix will thus depend on the vascularity of the tissue bed. But blood vessels require a scaffold to support their ingrowth, in which capacity either the implanted matrix or the collagenous network laid down by the fibroblasts themselves can function. Fibroblasts, together with the collagenous meshwork they produce, therefore grow into the implant just ahead of the blood vessels. The oxygen gradient will be such as to provide the necessary high tension for fibroblasts at the edge of the wound, but a low one at its center, which will conduce macrophages to release the cytokines that will stimulate chemotaxis, mitosis, and other events constituting an integral part of the healing process. The use of a degradable matrix to deliver biological-response modifiers can protect these agents until they are released, this being expedient owing to their short half-lives in vivo. This circumstance has hitherto posed a stumbling block in clinical studies, having led to reduced efficacy and/or the increased expense associated with daily administration.

Applications in Wound Healing

Fibrin would appear to be a good choice for many wound-healing applications owing to its hemostatic and angiogenic effects. The addition of growth factors or other substances to such a system would further enhance its potential for healing. For agents that diffuse readily out of the matrix, simple diffusion models can be used to estimate their rate of release, but in other cases (such as when growth factors or other therapeutic peptides are employed), binding to fibrin may occur, in which instance release would be tied to the degradation of the matrix. Since tissue enzymes control the breakdown of fibrin, the release of bound drugs is controlled by a biofeedback mechanism. Fibroblast growth factor 1 (FGF-1) is commonly selected as the angiogenic agent of choice for inclusion in fibrin scaffold. Both in vitro and in vivo studies have indicated that the tissue repair response elicited is dose dependent, maximal effects being attained using intermediate doses. Interestingly, the strength of the fibrin clot has been found to increase, and its degradation rate to decrease, as a function of increasing FGF-1 concentration. When used in open skin wounds, the fibrin/FGF-1 system has yielded better overall healing results than any other treatment, complete re-epithelialization being achieved with minimal tissue contraction. Albumin matrices have also proved to be promising vehicles for drug delivery in wound-healing applications. Albumin, like fibrin, is degraded by physiological processes, and, possessing drug-binding regions, can serve as a controlled-release system for therapeutic agents. The local concentration of a particular drug will be determined by its albumin-binding affinity.

Design of Drug-delivery System

Apart from the aforementioned applications in wound healing, tissue adhesives can also be used to delivery medications to specific areas of the body, thereby localizing the release of the therapeutic agent to the intended site of action [9,189]. Consequently, less toxic doses can be employed than when the drug is administered systemically, non-specific side-effects being thereby largely avoided. For example, chemotherapeutic agents used in the handling of cancer have been incorporated into tissue adhesives and applied topically to the area of surgical tumor resection to destroy any remaining cancerous cells [190]. Likewise, fibrin adhesives containing cefotaxime mixtures have been utilized locally to treat osteitis in human subjects [191]. With natural biomaterials such as fibrin and albumin, biological-response modifiers can be carried in several ways. They can be entrapped within the polymerized matrix by attachment to its polymer chains, or incorporated during the matrix polymerization process itself. If in this latter case the active agent is larger than the intrafibrillar pores, its release will be coupled to the degradation of the material, i.e., under biofeedback control. Moreover, matrix degradation begins at the wound edge, which will provide the appropriate drug gradient to further stimulate angiogenesis and tissue healing. Overall, these studies point to the feasibility of using natural biomaterials not only as tissue adhesives, but also as angiogenic tissue scaffolds. Such systems can protect the biological-response modifier up to the time of its release, serve as biofeedback-controlled drug delivery systems, and provide an adherent tissue scaffold during healing. In the case of fibrin, its adhesive properties do not suffice for a number of applications; and, unfortunately, manipulations to increase its porosity lead to a further depreciation of this quality. Current efforts to improve the adhesive properties of fibrin via chemical modification or by combining it with other materials have met with encouraging results, and these avenues are thus deemed worthy of further exploration.

Conclusions

Purified sealants of human origin have undergone considerable scientific development, with their viral safety having been demonstrated and their hemostatic effectiveness established beyond dispute. Since different applications (e.g., hemostatic agent, sealant, drug-delivery system, and tissue scaffold) require particular attributes, the ideal tissue adhesive is defined in terms of its intended use. But certain features are called for in all systems, namely, ease of handling and biocompatibility. To satisfy the former requisite, the adhesive should be moderately viscous so that once applied it remains in the field, but yet spreads readily and conforms to the wound contours. For most applications, it also needs to set rapidly. For biocompatibility, both the host response and the degradation of the implant need to be optimized for each application. Tissue compatibility is a major concern with both cyanoacrylates and gelatine–resorcinol–formaldehyde composites, since both systems deliver formaldehyde and other cytotoxic products, which provoke inflammation. Although these systems have the greatest initial strength, the nature of the host response limits their utility. Natural tissue adhesives such as fibrin and albumin not only possess an inherent bioactivity, but are also amenable to structure modification to improve their porosity, and can serve as vehicles for the delivery of agents to augment their own healing properties. Efforts to develop a clinically approved cyanoacrylate that will evoke a more muted host response are nevertheless deemed merited. The mechanical properties of a tissue adhesive at the wound edge are also of importance for many applications, and should be comparable to those of the adjacent native tissue. Although natural adhesives tend to be weaker than artificial ones, the sacrifice in initial strength is often compensated for by a swifter repair response and more complete healing. Not all features desired from a biological point of view are mutually compatible from a structural one within a single matrix construct, and a working compromise between initial strength, degradation rate, and healing rate is thereby necessitated.

References

1. Hunter J. A treatise on the blood, inflammation and gunshot wounds, London, 1774. In: Brunner C, editor. Handbuch der Wundbehandlung: (Neue deutsche Chirurgie, Bd. 20). Stuttgart, 1916.
2. Haeberlein C. Welche geschnittene oder gehauene Wunden sollen durch die Vereinigung und welche sollen durch die Eiterung geheilt werden? In: Rather PW, editor. Gewe-beklebstoffe in der Medizih. Munich: Goldmann, 1972.
3. Bergel S. Uber wirkungen des fibrins. Deutsch Med Wochenschr 1909;35:633–65.
4. Young JZ, Medawar RB. Fibrin Suture of Peripheral Nerves. Lancet 1940;2:126–32.
5. Matras H. The use of fibrin sealant in oral and maxillofacial surgery. J Oral Maxillofac Surg 1982;40(10): 617–22.
6. Fuhge P, Heimburger N, Stohr HA, Burk W. 1987. United States Patent #4,650,678.
7. Heimburger N, Fuhge P, Ronneberge H. 1987. European Patent #87,109,374.
8. Burnouf-Radosevich M, Burnouf T, Huart JJ. Biochemical and physical properties of a solvent-detergent-treated fibrin glue. Vox Sang 1990;58(2):77–84.
9. Sierra DH. Fibrin sealant adhesive systems: a review of their chemistry, material properties and clinical applications. J Biomater Appl 1993;7(4):309–52.
10. Bachet J, Gigou F, Laurian C, Bical O, Goudot B, Guilmet D. Four-year clinical experience with the gelatin–resorcine–formol biological glue in acute aortic dissection. J Thorac Cardiovasc Surg 1982;83(2):212–17.
11. Huang ST, Kilpadi DV, Feldman DS. A comparison of the shear strength of fibrin and albumin glues. Transactions of the Wound Healing Society 1997;7:63.
12. Leonard F. The n-alkylalphacyanoacrylate tissue adhesives. Ann N Y Acad Sci 1968;146:203–13.
13. Feldman D, Sierra D. Tissue adhesives in wound healing. In: Dekker M, editor. Encyclopedic Handbook of Biomaterials and Bioengineering. New York, 1995.
14. Linder A, Linnau Y. 1986. United States Patent #4,600,574.
15. Schwartz, Linnau Y, Loblich F, Seelich T. 1980. United Kingdom Patent Application GB 2 041942.
16. Seelich T. European Patent #85,890,227.
17. Dahlstrom KK, Weis-Fogh US, Medgyesi S, Rostgaard J, Sorensen H. The use of autologous fibrin adhesive in skin transplantation. Plast Reconstr Surg 1992;89(5): 968–72; discussion 973–6.
18. Conte JE, Jr., Hadley WK, Sande M. Infection-control guidelines for patients with the acquired immunodeficiency syndrome (AIDS). N Engl J Med 1983;309(12): 740–4.
19. Sedlarik KM, Ursinus W, Lichey C, Reichert A, Schilling B. The sealing of vascular prostheses using autologous electrically activated blood. Thorac Cardiovasc Surg 1984;32(5):329–30.
20. Dresdale A, Bowman FO, Jr., Malm JR, Reemtsma K, Smith CR, Spotnitz HM, et al. Hemostatic effectiveness

of fibrin glue derived from single-donor fresh frozen plasma. Ann Thorac Surg 1985;40(4):385–7.
21. Mauler R, Hilfenhaus J. Inactivation of viruses in Factor XIII concentrate by pasteurization. Artzneimittelforschung 1984;34(11):1524–7.
22. Koveker G. Clinical application of fibrin glue in cardiovascular surgery. Thorac Cardiovasc Surg 1982; 30(4):228–9.
23. FDA. New fibrin sealant approved to help control bleeding in surgery, 1998 May 1.
24. FDA. Revocation of Fibrinogen Licences: FDA, Drug Bull, 1978;8:15.
25. Jackson MR, Alving BM. Fibrin sealant in preclinical and clinical studies. Curr Opin Hematol 1999;6(6): 415–19.
26. Jackson MR, MacPhee MJ, Drohan WN, Alving BM. Fibrin sealant: current and potential clinical applications. Blood Coagul Fibrinolysis 1996;7(8): 737–46.
27. Ferry JD, Morrison PR. Preparation and Properties of Serum and Plasma Proteins. VIII. The Conversion of Human Fibrinogen to Fibrin under Various Conditions. J Amer Chem Soc 1947;69:388–400.
28. Ferry JD. The conversion of fibrinogen to fibrin: events and recollections from 1942 to 1982. Ann N Y Acad Sci 1983;408:1–10.
29. Okada M, Blomback B, Chang MD, Horowitz B. Fibronectin and fibrin gel structure. J Biol Chem 1985;260(3):1811–20.
30. Blomback B, Okada M. Fibrin gels and their possible implication for surface hemorheology in health and disease. Ann N Y Acad Sci 1983;416:397–409.
31. Nowotony R, Chalupka A, Nowotony C, Bosch P. Mechanical Properties of Fibrinogen Adhesive Material. In: Winter, GD, Gibbons GF, Plenk H, editors. Biomaterials. London: John Wiley and Sons, 1982.
32. Jorgensen PH, Jensen KH, Andreassen TT. Mechanical strength in rat skin incisional wounds treated with fibrin sealant. J Surg Res 1987;42(3):237–41.
33. Byrne DJ, Hardy J, Wood RA, McIntosh R, Cuschieri A. Effect of fibrin glues on the mechanical properties of healing wounds. Br J Surg 1991;78(7):841–3.
34. Durham LH, Willatt DJ, Yung MW, Jones I, Stevenson PA, Ramadan MF. A method for preparation of fibrin glue. J Laryngol Otol 1987;101(11):1182–6.
35. Dresdale A, Rose EA, Jeevanandam V, Reemtsma K, Bowman FO, Malm JR. Preparation of fibrin glue from single-donor fresh-frozen plasma. Surgery 1985;97(6): 750–5.
36. Avoy DR. 1990. United States Patent #4,902,281.
37. Wadstrom J, Wik O. Fibrin glue (Tisseel) added with sodium hyaluronate in microvascular anastomosing. Scand J Plast Reconstr Surg Hand Surg 1993;27(4): 257–61.
38. Wadstrom J, Tengblad A. Fibrin glue reduces the dissolution rate of sodium hyaluronate. Ups J Med Sci 1993;98(2):159–67.
39. DeBlois C, Cote MF, Doillon CJ. Heparin-fibroblast growth factor–fibrin complex: in vitro and in vivo applications to collagen-based materials. Biomaterials 1994;15(9):665–72.
40. Schlag G, Redl H. Fibrin sealant in orthopedic surgery. Clin Orthop 1988;227:269–85.
41. Ono K, Shikata J, Shimizu K, Yamamuro T. Bone–fibrin mixture in spinal surgery. Clin Orthop 1992(275): 133–9.
42. Pico G. Thermodynamic aspects of thermal stability of human serum albumin. Biochem Mol Biol Int 1995;36:1017–23.
43. Bleustein CB, Walker CN, Felsen D, Poppas DP. Semisolid albumin solder improved mechanical properties for laser tissue welding. Lasers Surg Med 2000;27(2): 140–6.
44. Phillips AB, Ginsburg BY, Shin SJ, Soslow R, Ko W, Poppas DP. Laser welding for vascular anastomosis using albumin solder: an approach for MID-CAB. Lasers Surg Med 1999;24(4):264–8.
45. Lauto A, Poppas DP, Murrell GA. Solubility study of albumin solders for laser tissue welding. Lasers Surg Med 1998;23(5):258–62.
46. Massicotte JM, Stewart RB, Poppas DP. Effects of endogenous absorption in human albumin solder for acute laser wound closure. Lasers Surg Med 1998;23(1):18–24.
47. Scherr DS, Poppas DP. Laser tissue welding. Urol Clin North Am 1998;25(1):123–35.
48. Poppas DP, Wright EJ, Guthrie PD, Shlahet LT, Retik AB. Human albumin solders for clinical application during laser tissue welding. Lasers Surg Med 1996;19(1):2–8.
49. Pavanetto F, Genta I, Giunchedi P, Conti B, Conte U. Spray-dried albumin microspheres for the intra-articular delivery of dexamethasone. J Microencapsul 1994;11(4):445–54.
50. Luft JH. Fixation for biological ultrastructure. II. Cross-linking of bovine serum albumin by nanosecond pulses of ionizing radiation. J Microsc 1992;167(Pt 3):259–72.
51. Luft JH. Fixation for biological ultrastructure. I. A viscometric analysis of the interaction between glutaraldehyde and bovine serum albumin. J Microsc 1992;167(Pt 3):247–58.
52. Goosen MF, Leung YF, Chou S, Sun AM. Insulin–albumin microbeads: an implantable, biodegradable system. Biomater Med Devices Artif Organs 1982; 10(3):205–18.
53. Lee TK, Sokoloski TD, Royer GP. Serum albumin beads: an injectable, biodegradable system for the sustained release of drugs. Science 1981;213(4504):233–5.
54. Cremers HF, Wolf RF, Blaauw EH, Schakenraad JM, Lam KH, Nieuwenhuis P, et al. Degradation and intrahepatic compatibility of albumin–heparin conjugate microspheres. Biomaterials 1994;15(8):577–85.
55. Lin W, Garnett MC, Davies MC, Bignotti F, Ferruti P, Davis SS, et al. Preparation of surface-modified albumin nanospheres. Biomaterials 1997;18(7):559–65.
56. Ben Slimane S, Guidoin R, Merhi Y, King MW, Domurado D, Sigot-Luizard MF. In vivo evaluation of polyester arterial grafts coated with albumin: the role and importance of cross-linking agents. European Surgical Reseach 1988;20:66–74.
57. Chafke N, Gasser B, Lindner V, Rouyer N, Rooke R, Kretz JG, et al. Albumin as a sealant for a polyester vascular prosthesis: its impact on the healing sequence in humans. J Cardiovasc Surg (Torino) 1996;37(5): 431–40.
58. Leung YF, O'Shea GM, Goosen MF, Sun AM. Microencapsulation of crystalline insulin or islets of Langerhans: an insulin diffusion study. Artif Organs 1983; 7(2):208–12.
59. Goosen MF, Leung YF, O'Shea GM, Chou S, Sun AM. Slow release of insulin from a biodegradable matrix implanted in diabetic rats. Diabetes 1983;32(5):478–81.

60. D'Urso EM, Jean-Francois J, Doillon CJ, Fortier G. Poly(ethylene glycol)-serum albumin hydrogel as matrix for enzyme immobilization: biomedical applications. Artif Cells Blood Substit Immobil Biotechnol 1995;23(5):587–95.
61. Gayet JC, Fortier G. Drug release from new bioartificial hydrogel. Artif Cells Blood Substit Immobil Biotechnol 1995;23(5):605–11.
62. Evaluation of a new tissue sealant material: Serum albumin crosslinked in vivo with poolyethylene glycol. Fifth World Bioniaterials Conference; 1996 May 29–June 2; Toronto.
63. In vitro analysis of mechanical properties of a new tissue sealant material: polyethylene glycol crosslinked serum albumin; 1996 May 29–June 2; Toronto.
64. Warkentin P, Walivaara B, Lundstrom I, Tengvall P. Differential surface binding of albumin, immunoglobulin G and fibrinogen. Biomaterials 1994;15(10):786–95.
65. Klinger A, Steinberg D, Kohavi D, Sela MN. Mechanism of adsorption of human albumin to titanium in vitro. J Biomed Mater Res 1997;36(3):387–92.
66. An YH, Stuart GW, McDowell SJ, McDaniel SE, Kang Q, Friedman RJ. Prevention of bacterial adherence to implant surfaces with a crosslinked albumin coating in vitro. J Orthop Res 1996;14(5):846–9.
67. McDowell SG, An YH, Draughn RA, Friedman RJ. Application of a fluorescent redox dye for enumeration of metabolically active bacteria on albumin-coated titanium surfaces. Lett Appl Microbiol 1995;21(1): 1–4.
68. An YH, Bradley J, Powers DL, Friedman RJ. The prevention of prosthetic infection using a cross-linked albumin coating in a rabbit model. J Bone Joint Surg Br 1997;79(5):816–19.
69. Ardis AE. 1949. US patent 2,467,927.
70. Coover HW, Joyner FB, Shearer NH, Wicker TH. Chemistry and performance of cyanoacrylate adhesives. J Soc Plastic Engineering 1959;15:413–17.
71. Coover HW, McIntire JM. The chemistry of cyanoacrylate adhesives. In: T. M-m, editor. Tissue Adhesives in Surgery. New York: Medical Examination, 1972.
72. Smith DC. Lutes, glues, cements and adhesives in medicine and dentistry. BioMedical Engineering 1973;8:108–15.
73. Kaufman RS. The use of tissue adhesive (isobutyl cyanoacrylate) and topical steroid (0.1 percent dexamethasone) in experimental tympanoplasty. Laryngoscope 1974;84:793–804.
74. Greer RO. Studies conccrning the histotoxicity of isobutyl-2-cyanoacrylate tissue adhesive when employed as an oral hemostat. Oral Surgery, Oral Medicine, Oral Pathologv 1975;40:659–69.
75. Hunter KM. Cyanoacrylate tissue adhesive in osseous repair. Br J Oral Surg 1976;14(1):80–6.
76. Diaz FG, Mastri AR, Chou SN. Neural and vascular tissue reaction to aneurysm-coating adhesive (ethyl 2-cyanoacrylate). Neurosurgery 1978;3(1):45–9.
77. Hood TW, Mastri AR, Chou SN. Neural and vascular tissue reaction of cyanoacrylate adhesives: a further report. Neurosurgery 1982;11(3):363–6.
78. Zumpano BJ, Jacobs LR, Hall JB, Margolis G, Sachs E, Jr. Bioadhesive and histotoxic properties of ethyl-2-cyanoacrylate. Surg Neurol 1982;18(6):452–7.
79. Vinters HV, Galil KA, Lundie MJ, Kaufmann JC. The histotoxicity of cyanoacrylates. A selective review. Neuroradiology 1985;27(4):279–91.
80. Vinters HV, Lundie MJ, Kaufmann JC. Long-term pathological follow-up of cerebral arteriovenous malformations treated by embolization with bucrylate. N Engl J Med 1986;314(8):477–83.
81. Kerr AG, Smyth GD. Bucrylate (isobutyl cyanoacrylate) as an ossicular adhesive. Arch Otolaryngol 1971;94(2):129–31.
82. Toriumi DM, Raslan WF, Friedman M, Tardy ME, Jr. Variable histotoxicity of histoacryl when used in a subcutaneous site: an experimental study. Laryngoscope 1991;101:339–43.
83. Purdy PD, Batjer HH, Risser RC, Samson D. Arteriovenous malformations of the brain: choosing embolic materials to enhance safety and ease of excision. J Neurosurg 1992;77(2):217–22.
84. Cognard C, Miaux Y, Pierot L, Weill A, Martin N, Chiras J. The role of CT in evaluation of the effectiveness of embolisation of spinal dural arteriovenous fistulae with N-butyl cyanoacrylate. Neuroradiology 1996; 38(7):603–8.
85. DeMeritt JS, Pile-Spellman J, Mast H, Moohan N, Lu DC, Young WL, et al. Outcome analysis of preoperative embolization with N-butyl cyanoacrylate in cerebral arteriovenous malformations. AJNR Am J Neuroradiol 1995;16(9):1801–7.
86. Tamarin A, Lewis P, Askey J. The structure and formation of the byssus attachment plaque in Mytilus. J Morphology 1976;149:199–221.
87. Strausberg RL, Link RP. Trends in Biotechnology. Protein-based medical adhesives, 1990:53–7.
88. Strausberg RL, Link RP. Protein-Based Medical Adhesives. London: Elsevier Science, 1990.
89. Filpula DR, Lee SM, Link RP, Strausberg SL, Strausberg RL. Structural and functional repetition in a marine mussel adhesive protein. Biotechnology Progress 1990;6:171–7.
90. Green K. Mussel adhesive protein. In: Sierra D. H. SR, editor. Surgical Adhesives and Sealants – Current Technology and Applications. Lancaster: Technornic, 1996:19–27.
91. Robin JB, Lee CF, Riley JM. Preliminary evaluation of two experimental surgical adhesives in the rabbit cornea. Refractive & Corneal Surgery 1989;5:302–6.
92. Liggett PE, Cano M, Robin JB, Green RL, Lean JS. Intravitreal biocompatibility of mussel adhesive protein. A preliminary study. Retina 1990;10:144–7.
93. Pitman MI, Menche D, Song EK, Ben-Yishay A, Gilbert D, Grande DA. The use of adhesives in chondrocyte transplantation surgery: in vivo studies. Bulletin of the Hospital for Joint Diseases Orthopedic Institute, 1989;49:213–21.
94. Bachet J, Guilmet D. The use of biological glue in aortic surgery. Cardiol Clin 1999;17(4):779-96, ix–x.
95. Braunwald NS. A clinical evaluation of methyl-2-cyanoacrylate monomer as a hemostatic agent on the aorta. Ann Surg 1966;164(6):967–72.
96. Tatooles CJ, Braunwald NS. The use of crosslinked gelatin as a tissue adhesive to control hemorrhage from liver and kidney. Surgery 1966;60(4):857–61.
97. Braunwald NS, Gay W, Tatooles CJ. Evaluation of crosslinked gelatin as a tissue adhesive and hemostatic agent: an experimental study. Surgery 1966;59(6): 1024–30.
98. Braunwald NS, Tatooles CJ. Use of a crosslinked gelatin tissue adhesive to control hemorrhage from liver and kidney. Surg Forum 1965;16:345–6.

99. Schwarz N, Redl H, Zeng L, Schlag G, Dinges HP, Eschberger J. Early osteoinduction in rats is not altered by fibrin sealant. Clin Orthop 1993(293):353–9.
100. Schwarz N. The role of fibrin sealant in osteoinduction. Ann Chir Gynaecol Suppl 1993;207:63–8.
101. Gerngross H, Burri C, Claes L. Experimental studies on the influence of fibrin adhesive, factor XIII, and calcitonin on the incorporation and remodeling of autologous bone grafts. Arch Orthop Trauma Surg 1986;106(1):23–31.
102. Claes L, Burri C, Gerngross H, Mutschler W. Bone healing stimulated by plasma factor XIII. Osteotomy experiments in sheep. Acta Orthop Scand 1985;56(1): 57–62.
103. Claes L, Burri C, Gerngross H, Mutschler W. Acceleration of fracture healing with factor XIII. Helv Chir Acta 1984;51(2):209–12.
104. Bosch P, Braun F, Eschberger J, Kovac W, Spangler HP. The action of high-concentrated fibria on bone healing. Arch Orthop Unfallchir 1977;89(3):259–73.
105. Bosch P. Bone grafting with fibrin glue. Wien Klin Wochenschr Suppl 1981;124:1–26.
106. Arbes H, Bosch P, Lintner F, Salzer M. First clinical experience with heterologous cancellous bone grafting combined with the fibrin adhesive system (FAS). Arch Orthop Trauma Surg 1981;98(3):183–8.
107. Bosch P, Lintner F, Arbes H, Brand G. Experimental investigations of the effect of the fibrin adhesive on the Kiel heterologous bone graft. Arch Orthop Trauma Surg 1980;96(3):177–85.
108. Bosch P, Braun F, Spangler HP. The technic of fibrin glue in cancellous bone transplants. Arch Orthop Unfallchir 1977;90(1):63–75.
109. Pfluger G, Bosch P, Grundschober F, Kristen H, Plenk H, Jr., Schider S. Investigation of bone growth into porous metal implants. Wien Klin Wochenschr 1979;91(14):482–7.
110. Bernett P, Pfister A, Paar O, Deigentesch N. Klebung von Knorpelfrakturen am Knie- und Sprunggelenk mit Hilfe der Fibrinklebung. In: Chapchal G, editor. Sportverletzungen und Sportschäden. Stuttgart: Thieme-Verlag, 1983:159.
111. Böhler N, Bösch P, Sandbach G, Eschberger J, Schmid L. Experimentelle Erfahrungen mit der Einklebung von Kortikaliszylindern. In: Cotta H, Braun A, editors. Fibrinkleber in Orthopädie und Traumatologie. Stuttgart: Thieme-Verlag, 1982:68.
112. Böhler N, Bösch P, Sandbach G, Schlag g, Eschberger J, Schmid L. Der Einfluss von homologen Fibrinogen auf die Osteomieheilung beim Kaninchen. Unfallheilkunde 1977;80:501.
113. Schumacher G, Braun A, Heine W. Das Alloimplantat am Knochen unter Verwendung des fibrinklebesystems Tierexperimentelle Ergebenisse. In: H C, A B, editors. Fibrinkleber in Orthopädie und Traumatologie. Stuttgart: Thieme-Verlag, 1982:71.
114. Kalebo P, Buch F, Albrektsson T. Bone formation rate in osseointegrated titanium implants. Influence of locally applied haemostasis, peripheral blood, autologous bone marrow and fibrin adhesive system (FAS). Scand J Plast Reconstr Surg Hand Surg 1988;22(1):53–60.
115. Albrektsson T, Bach A, Edshage S, Jonsson A. Fibrin adhesive system (FAS) influence on bone healing rate: a microradiographical evaluation using the bone growth chamber. Acta Orthop Scand 1982;53(5): 757–63.
116. Pfluger H, Kirchheimer J, Ritschl P, Koller A, Hienert G, Binder BR. Tissue plasminogen activator activity in early prostatic cancer and in bone metastases of prostatic cancer. Wien Klin Wochenschr 1984;96(17): 658–61.
117. Pfluger H, Redl H. In vivo and in vitro degradation of fibrin adhesives (studies in rats). Z Urol Nephrol 1982;75(1):25–30.
118. Schlag G, Redl H. Fibrin adhesive system in bone healing [letter]. Acta Orthop Scand 1983;54(4): 655–8.
119. Zilch H, Noffke B. The influence of the fibrinogen adhesive system on bone healing. Unfallheilkunde 1981;84(9):363–72.
120. Lucht U, Bunger C, Moller JT, Joyce F, Plenk H, Jr. Fibrin sealant in bone transplantation. No effects on blood flow and bone formation in dogs. Acta Orthop Scand 1986;57(1):19–24.
121. Keller J, Andreassen TT, Joyce F, Knudsen VE, Jorgensen PH, Lucht U. Fixation of osteochondral fractures. Fibrin sealant tested in dogs. Acta Orthop Scand 1985;56(4):323–6.
122. Tholpady SS, Schlosser R, Spotnitz W, Ogle RC, Lindsey WH. Repair of an osseous facial critical-size defect using augmented fibrin sealant. Laryngoscope 1999;109(10):1585–8.
123. Turgut M, Erkus M, Tavus N. The effect of fibrin adhesive (Tisseel) on interbody allograft fusion: an experimental study with cats. Acta Neurochir (Wien) 1999;141(3):273–8.
124. Jarzem P, Harvey EJ, Shenker R, Hajipavlou A. The effect of fibrin sealant on spinal fusions using allograft in dogs. Spine 1996;21(11):1307–12.
125. Plaga BR, Royster RM, Donigian AM, Wright GB, Caskey PM. Fixation of osteochondral fractures in rabbit knees. A comparison of Kirschner wires, fibrin sealant, and polydioxanone pins. J Bone Joint Surg [Br] 1992;74(2):292–6.
126. Palacios-Carvajal J, Moina E. The mixture of fibrin sealant and a porous ceramic as osteoconductor: An experimental study. In: Schalg G, Redl H, editors. Fibrin sealant Operative Medicine. Berlin: Springer-Verlag, 1986.
127. Siebert HR, Rueggr J, Weidner R, Pannike A. Histomophologische Verlaufsbeobachtungen der einheilung eines neuantigen Knochenersatzmittels in Knochendefecten am femur der Ratte. Langenbecks Arch. Chir. 1982:147.
128. Urist MR. Bone transplants and implants. In: Urist MR, editor. Fundamental and Clinical Bone Physiology. Phildelphia: Lippincott, 1980:331.
129. Oberg S, Rosenquist JB. Bone healing after implantation of hydroxyapatite granules and blocks (Interpore 200) combined with autolyzed antigen-extracted allogeneic bone and fibrin glue. Experimental studies on adult rabbits. Int J Oral Maxillofac Surg 1994;23(2): 110–4.
130. Kania RE, Meunier A, Hamadouche M, Sedel L, Petite H. Addition of fibrin sealant to ceramic promotes bone repair: long-term study in rabbit femoral defect model. J Biomed Mater Res 1998;43(1):38–45.
131. Perka C, Schultz O, Spitzer RS, Lindenhayn K, Burmester GR, Sittinger M. Segmental bone repair by tissue-engineered periosteal cell transplants with bioresorbable fleece and fibrin scaffolds in rabbits. Biomaterials 2000;21(11):1145–53.

132. Matsumoto K, Kohmura E, Kato A, Hayakawa T. Restoration of small bone defects at craniotomy using autologous bone dust and fibrin glue [see comments]. Surg Neurol 1998;50(4):344–6.
133. Sierra DH, Saltz R. Surgical adhesives and sealants: Current technology and Application: Techomic Publishing Company, 1996.
134. Brittberg M, Sjogren-Jansson E, Lindahl A, Peterson L. Influence of fibrin sealant (Tisseel) on osteochondral defect repair in the rabbit knee. Biomaterials 1997; 18(3):235–42.
135. Homminga GN, Buma P, Koot HW, van der Kraan PM, van den Berg WB. Chondrocyte behavior in fibrin glue in vitro. Acta Orthop Scand 1993;64(4):441–5.
136. Peretti GM, Randolph MA, Villa MT, Buragas MS, Yaremchuk MJ. Cell-based tissue-engineered allogeneic implant for cartilage repair [In Process Citation]. Tissue Eng 2000;6(5):567–76.
137. Passl R, Plenk H, Jr., Sauer G, Spangler HP, Radaszkiewicz T, Holle J. Homologous articular cartilage transplantation in animal experiments. Preliminary studies on sheep. Arch Orthop Unfallchir 1976;86(2):243–56.
138. Ohlsen L, Widenfalk B. The early development of articular cartilage after perichondrial grafting. Scand J Plast Reconstr Surg 1983;17(3):163–77.
139. Homminga GN, van der Linden TJ, Terwindt-Rouwenhorst EA, Drukker J. Repair of articular defects by perichondrial grafts. Experiments in the rabbit. Acta Orthop Scand 1989;60(3):326–9.
140. Widenfalk B, Engkvist O, Ohlsen L, Segerstrom K. Perichondrial arthroplasty using fibrin glue and early mobilization. An experimental study. Scand J Plast Reconstr Surg 1986;20(3):251–8.
141. Bouwmeester SJ, Beckers JM, Kuijer R, van der Linden AJ, Bulstra SK. Long-term results of rib perichondrial grafts for repair of cartilage defects in the human knee. Int Orthop 1997;21(5):313–17.
142. Bruns J, Kersten P, Silbermann M, Lierse W. Cartilage-flow phenomenon and evidence for it in perichondrial grafting. Arch Orthop Trauma Surg 1997;116(1–2): 66–73.
143. Bruns J, Kersten P, Lierse W, Silbermann M. Autologous rib perichondrial grafts in experimentally induced osteochondral lesions in the sheep-knee joint: morphological results. Virchows Arch A Pathol Anat Histopathol 1992;421(1):1–8.
144. Bruns J, Kersten P, Lierse W, Silbermann M. Autologous transplantation of rib perichondrium in treatment of deep cartilage defects of the knee joint of sheep. Morphologic comparison of two resorbable fixation methods. Unfallchirurg 1993;96(9):462–7.
145. Homminga GN, Bulstra SK, Bouwmeester PS, van der Linden AJ. Perichondral grafting for cartilage lesions of the knee. J Bone Joint Surg [Br] 1990;72(6):1003–7.
146. Puzas J, editor. Ten-year follow-up results of a prospective study of human perichondrial grafting versus debridement of cartilage defects in the knee. Orthopaedic Research Society; 2000; Orlondo, Florida. Orthopaedic Research Society.
147. Orr TE, Patel AM, Wong B, Hatzigiannis GP, Minas T, Spector M. Attachment of periosteal grafts to articular cartilage with fibrin sealant. J Biomed Mater Res 1999;44(3):308–13.
148. Niedermann B, Boe S, Lauritzen J, Rubak JM. Glued periosteal grafts in the knee. Acta Orthop Scand 1985;56(6):457–60.
149. Bruns J, Steinhagen J. Transplantation of chondrogenic tissue in the treatment of lesions of of the articular cartilage. Orthopade 1999;28(1):52–60.
150. Kreder HJ, Moran M, Keeley FW, Salter RB. Biologic resurfacing of a major joint defect with cryopreserved allogeneic periosteum under the influence of continuous passive motion in a rabbit model. Clin Orthop 1994(300):288–96.
151. Tsai CL, Liu TK, Fu SL, Perng JH, Lin AC. Preliminary study of cartilage repair with autologous periosteum and fibrin adhesive system. J Formos Med Assoc 1992;91 Suppl 3:S239–45.
152. Gillogly SD, Voight M, Blackburn T. Treatment of articular cartilage defects of the knee with autologous chondrocyte implantation. J Orthop Sports Phys Ther 1998;28(4):241–51.
153. Hendrickson DA, Nixon AJ, Grande DA, Todhunter RJ, Minor RM, Erb H, et al. Chondrocyte–fibrin matrix transplants for resurfacing extensive articular cartilage defects. J Orthop Res 1994;12(4):485–97.
154. van Susante JL, Buma P, Schuman L, Homminga GN, van den Berg WB, Veth RP. Resurfacing potential of heterologous chondrocytes suspended in fibrin glue in large full-thickness defects of femoral articular cartilage: an experimental study in the goat. Biomaterials 1999;20(13):1167–75.
155. van Susante JL, Buma P, Homminga GN, van den Berg WB, Veth RP. Chondrocyte-seeded hydroxyapatite for repair of large articular cartilage defects. A pilot study in the goat. Biomaterials 1998;19(24):2367–74.
156. Lewandowska K, Choi HU, Rosenberg LC, Zardi L, Culp LA. Fibronectin-mediated adhesion of fibroblasts: inhibition by dermatan sulfate proteoglycan and evidence for a cryptic glycosaminoglycan-binding domain. J Cell Biol 1987;105(3):1443–54.
157. Loncar D. Ultrastructural analysis of differentiation of rat endoderm in vitro. Adipose vascular-stromal cells induce endoderm differentiation, which in turn induces differentiation of the vascular-stromal cells into chondrocytes. J Submicrosc Cytol Pathol 1992;24(4):509–19.
158. Hunziker EB, Rosenberg LC. Repair of partial-thickness defects in articular cartilage: cell recruitment from the synovial membrane. J Bone Joint Surg Am 1996;78(5):721–33.
159. Silverman RP, Bonasser L, Passaretti D, Randolph MA, Yaremchuk MJ. Adhesion of tissue-engineered cartilate to native cartilage. Plast Reconstr Surg 2000;105(4): 1393–8.
160. Katzke D, Pusalkar A, Steinbach E. The effects of fibrin tissue adhesive on the middle ear. J Laryngol Otol 1983;97(2):141–7.
161. Kudema H, Matras H. Die Klinische Anwendung der Klebung von Nerveanastomosen bei der Rekonstruktion Verletzer Peripherer Nerven. Wein Klin. Wochenschr 1975;87:495–501.
162. Chen RJ, Fang JF, Lin BC, Hsu YB, Kao JL, Kao YC, et al. Selective application of laparoscopy and fibrin glue in the failure of nonoperative management of blunt hepatic trauma. J Trauma 1998;44(4):691–5.
163. Berguer R, Staerkel RL, Moore EE, Moore FA, Galloway WB, Mockus MB. Warning: fatal reaction to the use of fibrin glue in deep hepatic wounds. Case reports. J Trauma 1991;31(3):408–11.
164. Kram HB, Evan T, Clark T. Techniques of hepatic hemostasis using fibrin glue. Contemp Surg. 1990; 37:11.

165. Kram HB, Reuben BI, Fleming AW, Shoemaker WC. Use of fibrin glue in hepatic trauma. J Trauma 1988;28(8): 1195–201.
166. Ochsner MG, Maniscalco-Theberge ME, Champion HR. Fibrin glue as a hemostatic agent in hepatic and splenic trauma. J Trauma 1990;30(7):884–7.
167. Uranus S, Mischinger HJ, Pfeifer J, Kronberger L, Jr., Rabl H, Werkgartner G, et al. Hemostatic methods for the management of spleen and liver injuries. World J Surg 1996;20(8):1107–11; discussion 1111–12.
168. Hauser CJ. Hemostasis of solid viscus trauma by intraparenchymal injection of fibrin glue. Arch Surg 1989;124(3):291–3.
169. McCarthy PM, Trastek VF, Schaff HV, Weiland LH, Bernatz PE, Payne WS, et al. Esophagogastric anastomoses: the value of fibrin glue in preventing leakage. J Thorac Cardiovasc Surg 1987;93(2):234–9.
170. Siedentop KH, Harris DM, Sanchez B. Autologous fibrin tissue adhesive. Laryngoscope 1985;95(9 Pt 1): 1074–6.
171. Staindl O. Tissue adhesion with highly concentrated human fibrinogen in otolaryngology. Ann Otol Rhinol Laryngol 1979;88(3 Pt 1):413–8.
172. Salasche SJ, Feldman BD. Skin grafting: perioperative technique and management. J Dermatol Surg Oncol 1987;13(8):863–9.
173. Chakravorty RC, Sosnowski KM. Autologous fibrin glue in full-thickness skin grafting. Ann Plast Surg 1989;23(6):488–91.
174. Kjaergard D, Weis-Fogh U, Medgyesi S. The use of autologous fibrin adhesive in skin transplantation. Skin Transplantation 1992;89:968–975.
175. Redaelli C, Niederhauser U, Carrel T, Meier U, Trentz O. Rupture of the Achilles tendon – fibrin gluing or suture? Chirurg 1992;63(7):572–6.
176. Lusardi DA, Cain JE, Jr. The effect of fibrin sealant on the strength of tendon repair of full thickness tendon lacerations in the rabbit Achilles tendon. J Foot Ankle Surg 1994;33(5):443–7.
177. Frykman E, Jacobsson S, Widenfalk B. Fibrin sealant in prevention of flexor tendon adhesions: an experimental study in the rabbit. J Hand Surg [Am] 1993;18(1):68–75.
178. Itoh O. An experimental study on effect of bone morphogenetic protein and fibrin sealant in tendon implantation into bone. Nippon Seikeigeka Gakkai Zasshi 1991;65(8):580–90.
179. Shoemaker SC, Rechl H, Campbell P, Kram HB, Sanchez M. Effects of fibrin sealant on incorporation of autograft and xenograft tendons within bone tunnels. A preliminary study. Am J Sports Med 1989;17(3):318–24.
180. Hashimoto J, Kurosaka M, Yoshiya S, Hirohata K. Meniscal repair using fibrin sealant and endothelial cell growth factor. An experimental study in dogs. Am J Sports Med 1992;20(5):537–41.
181. Roddecker K, Munnich U, Jochims J, Nagelschmidt M. Measurement of the biomechanical stability of the healing menisci in animals: fibrin gluing, an alternative to traditional therapy methods? Z Orthop Ihre Grenzgeb 1991;129(4):350–4.
182. Nabeshima Y, Kurosaka M, Yoshiya S, Mizuno K. Effect of fibrin glue and endothelial cell growth factor on the early healing response of the transplanted allogenic meniscus: a pilot study. Knee Surg Sports Traumatol Arthrosc 1995;3(1):34–8.
183. Ritchie JR, Miller MD, Bents RT, Smith DK. Meniscal repair in the goat model. The use of healing adjuncts on central tears and the role of magnetic resonance arthrography in repair evaluation. Am J Sports Med 1998;26(2):278–84.
184. Ishimura M, Tamai S, Fujisawa Y. Arthroscopic meniscal repair with fibrin glue. Arthroscopy 1991;7(2): 177–81.
185. Kollias SL, Fox JM. Meniscal repair. Where do we go from here? Clin Sports Med 1996;15(3):621–30.
186. McAndrews PT, Arnoczky SP. Meniscal repair enhancement techniques. Clin Sports Med 1996;15(3):499–510.
187. Arnoczky SP, Warren RF, Spivak JM. Meniscal repair using an exogenous fibrin clot. An experimental study in dogs. J Bone Joint Surg [Am] 1988;70(8):1209–17.
188. Henning CE, Lynch MA, Yearout KM, Vequist SW, Stallbaumer RJ, Decker KA. Arthroscopic meniscal repair using an exogenous fibrin clot. Clin Orthop 1990(252):64–72.
189. Lasa C, Jr., Hollinger J, Drohan W, MacPhee M. Delivery of demineralized bone powder by fibrin sealant. Plast Reconstr Surg 1995;96(6):1409–17; discussion 1418.
190. Miura S, Mii Y, Miyauchi Y, Ohgushi H, Morishita T, Hohnoki K, et al. Efficacy of slow-releasing anticancer drug delivery systems on transplantable osteosarcomas in rats. Jpn J Clin Oncol 1995;25(3):61–71.
191. Zilch H, Lambiris E. The sustained release of cefotaxim from a fibrin-cefotaxim compound in treatment of osteitis. Pharmacokinetic study and clinical results. Arch Orthop Trauma Surg 1986;106(1):36–41.

5 Bone Materials and Tissue Banks

D. G. Poitout

Introduction

When there is a considerable loss of bone substance after the excision of a large tumor in the pelvis, it is always difficult to reconstruct the cotyloid cavity. A massive metal prosthesis fixed on the remaining bone with a plate, screws, or cement is often unstable. Fixation of the greater trochanter on the remaining bone, whether the sacrum or the wing of the ilium or, as some authors suggest, the fact of leaving a swinging hip, makes it difficult for the limb operated on to bear weight. We are of the opinion that the best way to return relatively satisfactory function to the limb is to replace the head of the femur, using a massive allograft. Autografts are not sufficiently large to replace the loss of substance, even though they alone are osteoforming and when incorporated into the skeleton have undeniable mechanical properties. Other types of grafts have to be used. Our decision to use a massive allograft is essentially linked to the rapid muscular refixation which it allows and the anatomical reconstruction of the pelvic ring on which a normal hip prosthesis has to be supported.

Since 1982, we have chosen to use massive cryopreserved and non-irradiated allografts because the experimental and clinical results seem to show that secondary irradiation of the allografts significantly reduces their mechanical strength in the short and medium term (and does not give complete safety from viruses). We prefer to use a bone removed under sterile and safe conditions at the fourth month rather than a bone removed under non-sterile conditions and irradiated later. The irradiation of tissue in the paste phase (as frozen bone is) requires doses of irradiation which are far higher than those usually used, or for the graft to be thawed. Freezing it again subsequently is not biologically satisfactory in our opinion. If the intention is to irradiate it before freezing, the waiting times are harmful to the mechanical quality of the bone, cartilage, or ligament allograft.

History

In 185 AD, Cosmas and Damian, the patron saints of surgeons, were canonized because they performed a posthumous miracle by grafting the limb of a person who had died onto the sacristan of their basilica in Rome, who had a tumor of the tibia. It was the first massive allograft reported in history (grafting a human bone onto another human being).

Larrey tells the story of a Polish nobleman who had been struck by a Tartar's saber and had lost part of the top of his cranium. Seeing a large dog passing close by, he decided to graft the dog's cranium onto his head. He then noted that it bonded very well. This was the first xenograft (graft from an animal onto man).

As it is not possible to graft a large part of a bone or joint from one and the same subject onto a different site in the same person (autograft), it was necessary to develop bone preservation procedures and to create a tissue bank. Ollier (1887) [1] is the nineteenth-century author who most studied the various types of graft, whether autologous grafts, allografts, or xenografts. Many other cases where massive grafts were used have also been published by Albee (1930) [2], Abbot (1947) [3], Sicard (1949, 1951, 1952, 1953) [4,5,6,7], Judet (1949, 1952) [9,10,11], Ottolenghi (1972) [12], and Parrish

(1971, 1972) [13,14], but although in more than 50% of cases these grafts gave results considered to be satisfactory, fifteen to twenty years later, the problems posed by removing, sterilizing, and storing them deterred some surgeons, who turned towards other procedures for reconstructing the skeleton.

The progress made in cryopreservation, as well as in understanding the immunological and biological information provided by grafts, has meant that for 15 years now the use of preserved allografts has returned to popularity. In 1979 we saw the possibility of preserving large bone and cartilage fragments and set up a bone bank in Marseilles Poitout (1985, 1986) [15,16,17]. In 1978 we became interested in the problems posed by the removal, preservation, and use of bone from bone banks. Several preservation methods have been suggested since the use of allografts was considered. The techniques employed used:

Liquid preservatives: alcohol, phenol, ether, hydrochloric acid, or the sublimate among other things adversely alter the bone architecture more or less rapidly and destroy the cells inside. Furthermore they often have an inhibiting effect on osteogenesis, which in general led to their being abandoned (Herbert, 1949) [18].

Sterilization by boiling: studied by Gallie in 1912, was heavily used until 1920, in particular by Rouvillois' team. But problems with preparing the grafts and the many cases of resorption reported meant that this type of sterilization was abandoned; it appears to be becoming popular again.

Drying in a vacuum or lyophilization: makes it easier to store products prepared by this method. However, as it destroys the rigidity of the bone structure, it makes them brittle. Bone crumbles under cutting forceps and is not mechanically adequate to withstand the usual mechanical stresses. Therefore, they cannot be used to replace massive diaphyseal bone segments or articular surfaces (Fasquelle, 1950) [19].

Ethylene oxide would certainly be an excellent method for sterilizing bone grafts if the gas stayed inside the tissues during storage and if it did not lead to the formation of toxic and even carcinogenic products.

Irradiation has disadvantages as well as advantages. The legal dose is 2.5 megarads. At this dose we are going to destroy the bacteria as well as all the cells but it is not certain that all the viruses will be inactivated and destroyed. The irradiation dose needed to completely disorganize their DNA or RNA molecules would be very high and would at the same time lead to the destruction of the chains of protein molecules which form the architecture of the bone (Roy-Camille, 1981; Hernigou, 1986; Loty, 1988). As far as osteocartilaginous parts are concerned, this method of preservation cannot be considered because it alters the cartilaginous structure of the graft profoundly. Therefore, in spite of certain undeniable advantages, such as that of allowing grafts to be removed in a non-sterile environment (which seems to us to be debatable), we have not used it (Kouvalchouk, 1986).

Bone, cartilage, or ligament grafts are used more and more frequently for treating bone tumors, performing reconstructive surgery after multiple operations such as, for example, on the hip and pelvis, or following traffic accidents. For the first time we are reconstructing the skeleton by bringing in new bone which will assimilate to the skeleton in a few months or years, whereas, to date, bone fragments were removed to be replaced by prostheses, which have a limited life. Currently 15–20% of rejections can be solved by using anti-rejection drugs (Sandimmum®).

Cryopreservation: Since 1876, Ollier recommended cryopreservation and had even published an experimental study on processes for preserving bone and periosteum grafts at −2 °C. In 1948 Jean and Robert Judet defined the broad lines of the cryopreservation processes. In Aix-les-Bains, Herbert organized the first reserve of frozen tissue and in 1951 at the Hôpital Beaujon in Paris, Sicard created the first bank of frozen bone tissue. Since 1979 we have been preserving massive osteocartilaginous grafts in liquid nitrogen at −196 °C [25,26]. This method allows whole bones to be preserved indefinitely and preserves the viability of the cartilage cells. To prevent ice macrocrystals forming, it is necessary to impregnate the bone or cartilage tissues with a suitable cryoprotector (10% DMSO). If we use relatively conservative storage temperatures (higher than –20 °C), the enzymes present in the tissues are not inactivated and will destroy the graft in a few weeks. At –80 °C (the limit for electric freezers), enzyme action, although being clearly reduced, is not totally prevented, because at this temperature only the collagenase is inactivated. On the other hand, DMSO has a prolonged efficacy limit of around –60 °C (eutectic point). Above this, the ice macrocrystals can combine again into macrocrystals and burst the cells and disrupt the architecture. At –196 °C, all enzyme activity is stopped and tissue preservation is unlimited. The temperature of the graft must not be lowered too wuickly and it will be necessary to vary the rate of the lowering in accordance with the temperature obtained. After studying the percentage of live cartilage cells after thawing, it seems to us that the optimum curve is 2 °C per minute down to –40 °C, then 5 °C per minute down to –140 °C, the temperature at which the graft is then plunged directly into a tank of liquid nitrogen where it will be stored.

Thawing, on the other hand, has to be quick so that the largest number of cells stay alive. Plasma or Ringer's lactate solution at 40/41 ° will be used to thaw out and wash the grafts, which will also eliminate the DMSO. Bone fragments are usable for a period of approximately 24 to 36 hours approximately two hours after having been removed from the liquid nitrogen tank, which allows them to be taken to any point in Europe. They cannot, of course, be refrozen after thawing.

From 1978 to 1983 studies were carried out in the Blood Transfusion Center where a tissue bank has been created in order to prepare bone for preservation. After various attempts, it was decided to use a programmed drop in temperature up to –140 °C, with preservation being in nitrogen vapor at –150 °C or in liquid nitrogen at –196 °C. It was then necessary to develop ways of removing the tissue. These discussions took place in the context of the national "France-Tissues" association, then with the collaboration of the French Graft Institute. Ever more stringent controls make it possible to guarantee the sterility of the fragments supplied and currently two successive controls, with an interval of four months between them, make it possible to deliver parts of various sizes four months after removal under wholly sterile conditions. In order to allow surgical teams from the private sector to benefit from the same safety when using tissue fragments, we gave them access to the bank, where they sent the femoral heads removed. The fragments they need are supplied to them after the usual sterility checks. In view of their number, femoral heads are sent by the Tissue Bank on a simple request, but where the massive removal of tissue is concerned (femur, tibia, pelvis, etc.) performed at the AP-HM (hospital), the consent of the referrer is required. These fragments, which are few in number, are delivered according to the instructions given by the surgeons. Grafts are sent throughout France. Owing to this chain of goodwill, we have been able to carry out more than 5,000 grafts.

A surgical world first was performed in 1985 by Professors Poitout and Trifaud, which consisted of replacing a complete femur with a bone from the bone bank sheathing a hip prosthesis and a knee prosthesis. However, in order to be able to perform these grafts, we should remember that they had to be removed and preserved. Informing the public and carers is therefore of prime importance if we are to collect as many grafts as possible which are going to give appreciable relief to patients for whom other surgical techniques would be far too incapacitating.

Bone Replacement Materials

Pride of place is currently reserved for bone grafts.

Autografts

Being osteo-forming, autografts alone can induce the formation of new bone and promote the healing of a fracture or the assimilation of an allograft.

Allografts

Immunology of Bone Allografts

The immune response is said to be *matrical* when there is a specific response to molecules isolated from collagen or from proteoglycans. It appears that this response is practically non-existent in the case of massive bone allografts. On the other hand, the *cellular* immune response caused by cells contained in the bone marrow (osteoblasts, osteocytes, fibroblasts, fat, vascular, nerve, or hematopoietic cells) is important provided, however, that all the cells are alive [27,28,29,30]. Clinically, rejection reactions occur infrequently (10% delay in healing and 10% true rejection with the graft having a lytic appearance), in spite of the use of massive diaphyso-metaphyso-epiphyseal grafts. When the osteosynthesis material is removed and a few years later a biopsy is carried out at the site of the graft, the HLA haplotype is always that of the host. And if a systematic study is carried out of the HLA groups of donors and hosts every month, after an allograft, if a monoclonal antibody specific to the HLA group of the donor is used, we can see that in 80% of cases no immune HLA antibodies specific to the donor appear. The immune HLA groups which may appear after the graft are connected with the introduction into the body of leukocytes from the blood transfusions usually accompanying grafts. The preservation methods (lyophilization or freezing) do not significantly change the extent of the immunological responses:

Fate of the Grafted Tissue

The allograft does not assimilate with the host bone as it is; it undergoes resorption and then reconstruction phenomena which end in the formation of new bone which, in time, will replace the graft [31,32,33]. The mechanisms governing the assimilation of the graft are now well known and were already being mentioned by Sicard more than 30 years ago.

It is necessary to stress the importance of:

Age: the younger the graft, the better the assimilation of the allograft.

The fixing ability of the graft, which is essential if it is to be rehabilitated. Indeed, the precarious nature of the vascularization of the graft, particularly at the beginning of its recolonization, means that it must be particularly firmly fixed and attempts must be made to achieve optimum contact between the bone surfaces.

The site of the graft, which is decisive for its assimilation (muscular environment). If a cortical allograft is placed outside its usual location, far from a bone bed and an adequate muscular environment, it will most frequently be resorbed, which seems to indicate that there is a local stimulus promoting its incorporation. This promoting substance may be Urist's PBH (or even "osteogenin" already mentioned by Lacroix in 1950) which is thought to be a substance produced by live undifferentiated mesenchymatous cells (Urist, 1942, 1967, 1963).

The size of the graft.

Initially, the contours of the graft grow blurred, the bone becomes rarified and then fuses with the adjacent bone. Secondly (around 12–18 months), there is densification of the allograft which indicates the new bone formation which surrounds the graft and consolidates it (Buchardt).

Biomechanics

The revascularization of the graft will only be superficial and only exceptionally and at a very late stage will it be possible to see deeper assimilation of the latter. It also appears to be desirable for this bone lysis to only appear at as late a stage as possible because during the period of revascularization, the mechanical strength of the graft will fall by around 50% between the 12th and the 18th month and it is therefore necessary to have attached it sufficiently firmly to be able to bear the stresses acting on the graft during this period. The mechanical properties of the allografts can be adversely changed by the preservation and storage processes. Low-dose irradiation (less than 2 megarads) only results in minor effects on the graft's strength. Cryopreservation seems to improve the mechanical properties of the allografts, the strength of which is 110–120% of that of fresh bone, but this method of preservation of the graft makes the diaphyseal cortical bone more brittle and liable to breakage. On the other hand, lyophilization or massive irradiation procedures on the bone parts (over 3 megarads) result in a clear reduction in the mechanical strength of the grafts (55% in the case of lyophilization and 65–70% in the case of massive irradiation of the strength of a fresh bone). When a further operation is performed to remove material a few years after a graft, the muscles are intimately fixed to the bone, through a tissue resembling periosteum, and the bone bleeds when this tissue is detached (Hernigou, 1988).

Xenografts

Xenografts were used several decades ago by French teams (Judet-Sicard, Evre, Guilleminet). The large number of rejection phenomena they caused (more than 50%) resulted in their no longer being used, hence the current shortage of human grafts. New attempts using different sterilization, preparation, or treatment techniques (lyophilization, ceramization, irradiation, heating, etc.) are being used in attempts to reduce the shortcomings of this type of graft.

Bone Substitutes

Artificial hydroxyapatite derivatives (hydroxyapatite collagen, hydroxyapatite cements, corals or madrepores, vitroceramics or bioglasses) are the subject of mechanical and experimental studies in order to define their tolerance in situ as well as the ways in which they are used. Even if some bone substitutes really are colonized by the host bone, their mechanical properties are currently still inadequate and do not allow the use of large fragments in human clinical medicine. Furthermore, these structures, which are often only osteo-conducting and sometimes osteo-inducing, do not form new bone (are not osteo-forming), and have a tendency to lyse rapidly or behave like a sequestrum.

Reconstructions with Massive Allografts

Bone reconstruction after surgery for excision of a tumor poses further problems, not all of which have been resolved. In the context of massive grafts we will study those used to replace diaphyseal fragments or articular surfaces. The use of prostheses sheathed by bone from a bone bank is an interim solution allowing joint solidity and stability to be combined, and muscular refixation on the periprosthetic bone. This technique should be used each time the loss of bone substance and soft tissue make revascularization of the graft hazardous.

Going beyond the indications of bone cavities being filled by spongy grafts, and beyond the reconstruction of the acetabulum by cortico-spongy allografts, it may be useful to have larger fragments in the bone bank (such as half-pelvises, for example) so that when the acetabulum is too severely damaged or when there is a bone tumor, the half-pelvis can be replaced by an allograft into which a total hip prosthesis will be fixed.

Massive diaphyseal or epiphyso-metaphyseal grafts will allow the locomotor apparatus to be reconstructed after the excision of a tumor or if substance is lost after trauma. If the intention is to replace a diaphyseal fragment, we believe that it is preferable to use centro-medullary nailing, leaving the muscle masses in close contact with the graft allowing maximum peripheral vascularization. This nail may have to be locked if the upper and lower part of the bone segment of the host is small.

Osteocartilaginous grafts of different sizes may be used to replace articular surfaces destroyed by a tumor or trauma process.

Joint Replacement Options

The use of massive cartilaginous allografts is proposed more and more frequently. These cartilaginous allografts often become very well assimilated. As cartilage cells do not need to be vascularized to survive, they only obtain their nutrition from the constituents of the synovial fluid. However, in order for the mechanical behavior of the graft to be adequate, the cells – contained in the cartilage and ensuring that it is adequately nourished in relation to the hydrophilia of the proteoglycans – are protected during the freezing phase. Hence the advantages of using a cryopreservative when the temperature drops, and the absence of the option of using secondary sterilization by heat, gas, or irradiation. Particular rigor is required when taking and preserving osteocartilaginous parts so as to be certain that the graft is entirely sterile.

Immunogeneity of the Various Constituents of Cartilage

Cartilage is considered to be an immunologically favored tissue because the intact cartilaginous matrix constitutes a real barrier between the chondrocytes and the immunologically competent cells.

Chondrocytes have a histocompatibility system which is comparable to that of the autologous lymphocytes. Langer and Gross have, however, shown the development of cell-type immunity in the rat.

Collagen of human joint cartilage is a type II collagen, synthesized by the chondrocytes. All the collagens can induce an immunological, humoral, and cellular response but collagen II is the most immunogenic.

According to Hermann and Friedlander, the proteoglycans of cartilage are antigenic. An immune response to proteoglycans is possible in the case of cartilaginous lesions, whether they are inflammatory, infectious, or arthrotic. A response of this kind can play a role in the induction or maintenance of the inflammation and destruction of the cartilage.

In summary, the constituents of joint cartilage studied separately are antigenic and immunogenic, but immunogeneity does not manifest itself normally when the various cartilage constituents are intact. Healthy cartilage forms a barrier to their penetration (Davies, 1952; Tomford, 1990) (see Table 5.1).

Table 5.1. Massive bone and osteochondral allografts: 1978–2000

Type	No. of cases
Spongy, cortical and osteocartilaginous allografts	871
Massive diaphyseal and cortico-spongy grafts	303
Hip reconstruction	356
Spongy (femoral heads)	286
Of which	
Acetabula	49
5 with the acetabular joint surface	
Half pelvis	21
Hip prostheses sheathed with associated bone from banks	29
Massive osteocartilaginous grafts	185
Total	1,744

As the joint graft requires capsuloligamentary coaptation, the collar of the capsule taken with the graft can be used, or the capsule and the ligaments of the host can be fixed directly onto the donor bone by trans-bone sutures. This latter method gives better stability to the limb. If articular necrosis were to occur, this would be painless because of the absence of innervation of the bone fragment and would only require partial replacement of the articular surface using a small prosthesis a few years later. The same problem also exists with the possible appearance of incapacitating ligament laxity, which could in the long-term justify stabilization with a prosthetic ligament or a preserved human ligament graft. Our current approach is the systematic insertion of a ligament support by doubling the ligaments grafted, in order to avoid excessive tension on those which are the source of the rupture or elongation.

The risks of secondary deterioration of the articular surface connected with ligament stability problems has led us to suggest the use of prostheses sheathed with bone from a bone bank which has the advantage of removing the cartilage viability problems with its risks of necrosis, and secondary ligament laxity with joint instability. The bone allograft which sheathes the metal core of a prosthesis allows the adjacent muscles to reattach themselves rapidly to the grafted area, which gives far less variable results in terms of function. Twenty-nine reconstructions using hip prostheses sheathed with bone from the bone bank have been used in association with acetabular grafts during the period 1988–2000.

Removal and Preservation of Grafts

All the allografts were removed during multiorgan removals. The grafts were repeatedly checked, i.e., checked when they were removed and at the fourth month, in the host after three successive bacteriological, urological, and immunological examinations had been carried out. The first on the donor, the second on the patients who had been given organs coming from the same donor. Any infection – general, viral, parasitic, or tumoral – or a systemic illness immediately leads to the destruction of all the grafts removed. The grafts are removed in an operating theater under surgically sterile conditions and preserved at −196 °C (in liquid nitrogen, after a progressive lowering of the temperature by 2 °C/min down to −40 °C, then by 5 °C/min down to −140 °C. The DMSO is used systematically at a dosage of 10% and put into contact with the allograft after the latter has been refrigerated when its temperature reaches +4 °C. The allograft and the DMSO are cooled separately and are placed in contact with each other at this temperature (above +10 °C, DMSO is toxic to cells). Thawing has to be rapid, i.e., 1–2 hours, with lavage of the DMSO in Ringer's lactate solution at +40 °C.

Clinical Experience

The first experimental joint transplants date from the beginning of the century, with the work of Henri Judet in 1908. The autologous and homologous osteochondral grafts performed then were complicated by progressive deterioration of the cartilaginous tissue with crumbling of the subchondral necrotic bone. The first massive osteochondral auto-transplants were performed by Reeves and Solmes, (1966), Judet and Padovani (1973), and Goldberg et al. (1973 and 1980). Although the articular auto-transplants demonstrated excellent viability and are maintained in the long term, the allo-transplants produced acute rejection which took two forms:

on the one hand, massive necrosis of the transplant,

on the other, thrombosis of the supply vessels [43].

This was immunological rejection of the supply vessels. The transplant was invaded with lymphocytic cells. Necrosis progressed more rapidly when a homograft of skin or marrow preceded the transplant. Judet and Padovani are of the opinion that a purely immunological mechanism is involved due to the immediate revascularization which created the conditions for accelerated rejection by offering immediate contact between the antigen and the antibody. The delays in assimilation of the graft and its quality only appeared to be slightly altered by the use of cyclosporine A.

Immunosuppressant drugs, such as azathioprine, do not significantly increase the survival time of the transplants. According to Halloran, the anti-lymphocytic serum gives a survival time for the transplant of up to 5 months, but only in the semi-allogenic grafts, and retains identical growth potential to that of the contralateral limb. In our experience, which between 1978 and 2000 includes 185 massive osteocartilaginous transplants, it seems that the clinical result is all the more successful if satisfactory biomechanical conditions are restored to the best extent possible. Histological studies performed by drilling under arthroscopy, even eight years after the graft, showed that the chondrocytes were alive in most cases and that although the superficial layers of the cartilage were sometimes changed and fissured, the deep layers were generally normal. The absence of painful symptoms in patients is a bonus and X-rays do not show any necrotic phenomena or crushing of the grafts. Some people fear the appearance of a tabes-like syndrome when the two parts of the joint are grafted jointly. It is true that the bone itself is no longer innervated and it is unlikely that a new proprioception will reappear during revascularization of the bone tissue. But the ligaments, muscles, and peripheral vasculonervous elements which have not been removed provide the medullary nerve centers with information and stabilize the bone and cartilage, which in fact behave like a metal prosthesis (also not innervated) [44,45]. Clinical experience confirms this interpretation, as no tabes-like symptoms have been reported, even 15 years after these types of massive osteocartilaginous grafts were implanted [46,47].

Capsuloligament and Articular Replacements by Massive Allografts

Preserving human ligaments in tissue banks is also an avenue of research which appears promising but which comes up against the problem of supplies from tissue banks and of the mechanical behavior of ligaments grafted during the period of their revascularization.

Ligament Allografts

These have to be removed from young donors, so that the force and stress values on breakage are close to those of a normal ligament (1,725 Newtons). The allograft which would best meet the morphological and structural criteria would, of course, be an anterior cruciate ligament allograft, however, choosing such an allograft would lead to considerable technical disadvantages and the revascularization of this ligament is risky. The patellar tendon can easily be removed with its two insertions on the patella and the tibia. It is sufficiently long and has mechanical properties which are clearly superior to those of the anterior cruciate ligament even if it is reduced to its central third.

Mechanical Studies

A graft consisting of the central third of the patellar tendon connected to its insertions was studied from a mechanical point of view. Creep tests as well as traction tests right up to rupture, show that:

freezing does not alter the appearance, color, or mechanical properties of grafts;

an irradiated tendon acquires a cardboard-like appearance;

the fibers of an irradiated and lyophilized tendon come apart and they acquire a fibrillary appearance.

The long-term mechanical behavior of these grafts in situ can be problematic when being revascularized. According to the first clinical results, it appears that a considerable percentage of residual articular laxity can be seen. The solution may be to combine a preserved allograft and a reinforcing ligament prostheses which would prevent the stresses being exerted directly on the allograft during its period of rehabilitation (2–3 years). The ligament prosthesis will rupture when the allograft will have regained satisfactory mechanical behavior. This is the technique we use currently but there is insufficient experience of it to be able to publish. Only the test of time will confirm whether or not we were right to make this choice.

Surgical Technique for Reconstructing the Acetabulum and the Pelvis

The reconstruction of a half-pelvis requires a broad approach following in its middle part the iliac crest, in its anterior part the crural arcade curving in from the pubis along the pubioischiatic line, and in its posterior part, continuing horizontally up to the level of the spinous processes of the lumbar vertebrae. An extension upwards or downwards may make it possible to approach the last lumbar vertebrae or the sacrum. The initial locating of the external iliac vessels, the femoral vessels, and of the crural nerve makes it possible to perform a subperitoneal dissection of the tumor which generally pushes back the iliac muscle and only exceeds it at a late stage. This extension then often contraindicates a carcinological surgical maneuver. The prosthesis is inserted and cemented in the half-pelvis allograft after fixing the latter by anterior and posterior plates screwed into the allograft on the remaining bone. When a whole half-pelvis is to be reconstructed, the gluteal muscles on the one hand and the adductors on the other have to be refixed to the bone by trans-bone sutures. In order to avoid the occurrence of an obturator hernia, the obturator is obstructed by a strip of silastic pressed on trans-bone sutures at the edge of the obturator.

The implantation of an acetabular graft may justify a simple transgluteal approach maintaining the continuity of the gluteus medius and of the external vastus muscle in a digastric form and retaining the continuity between the fibrous tendon and the bone in the greater trochanter. After having cut the graft as accurately as possible, filling precisely the resected bone, the anterior and posterior plates are screwed along the whole length of the allograft. The posterior plate rests on the ischium and the remaining iliac wing. The anterior plate rests on the pubis (if necessary the centrolateral pubis) and the remaining iliac wing. The prosthetic acetabulum is implanted after having synthesized the bony allograft. The anchorage holes are drilled right into the host bone, therefore allowing the cement to bridge the area of the allograft. We only insert metal pericotyloidian rings in very exceptional cases. Although in our first cases we did not routinely insert anti-dislocation rings, because of the occurrence of dislocations, we have now chosen to insert them routinely. One or two rings screwed into the rim of the acetabulum make it possible to obtain adequate mechanical setting of the prosthetic femoral head. In the case of operations involving mainly the acetabulum, it is generally useless to refix the gluteal muscles by trans-bone sutures. Only the digastric muscle is closed, and trans-bone sutures fix the part of this digastric muscle on the greater trochanter.

If there is infection, surgery should be carried out in two stages. The first stage is the ablation of all the foreign bodies, prostheses, material, cement, and bone as well as all the fistular courses and necrotic tissue. The limb is placed in traction (trans-femoral traction 1/10 of the body weight). A cement spacer containing antibiotics can be used to replace the part removed. In theory, this method allows the infected area to be sterilized locally and keeps the space free for the future graft. However, the local delivery of antibiotics by the cement spacer only reaches the area from its surface layers. The antibiotic contained inside the cement does not diffuse through this and only antibiotic particles at the surface of the cement enable antibiotics to be delivered at a bactericidal dose in situ. This cement is rapidly surrounded by a fibrous membrane which is impermeable to antibiotics, and in a few days the quantity of antibiotics in the bloodstream can be seen to drop. This technique is little used in the department. Irrigation and drainage may accompany this maneuver, delivering antibiotics in situ and carrying out mechanical cleansing of the infected area. This also makes it possible to carry out sampling on the exit drains, in the search for residual bacteria, making it easier to adapt general antibiotic therapy.

The second stage is that of reconstruction which may be carried out within two months of the removal of the infected tissues, and if the sedimentation rate is less than 20 in the first hour and stable on several successive examinations, as should also be the case with the C-reactive protein level.

Results

Complications associated with the use of massive allografts are what currently determine the limits of this type of surgery, which should only be practiced if the excision of the tumor has the same carcinological value as an amputation, and if draconian precautions are taken when the grafts are removed to avoid these transmitting an iatrogenic pathology. Premature or secondary post-operative deaths were only seen in cases of major surgery of the pelvis for advanced tumoral lesions. The risks of sepsis are comparable in the various groups and are essentially linked to the quality of the cutaneous scarring under chemotherapy (approximately 6%). Fractures of the graft occur when the osteosynthesis is inadequate or physiotherapy is too aggressive (areas unprotected by the osteosynthesis material). Non-healing of the ends of the bones is very much the exception if the junction between the allograft and the host bone is surrounded by autologous spongy tissue right from the first operation. Articular instability and arthrotic lesions depend on the stability of the reconstruction of the ligaments. The cartilage cells are present and, although reduced, the thickness of the cartilage guarantees correct joint function. No Charcot-type arthropathy has been demonstrated, periarticular innervation no doubt maintains articular trophicity.

Discussion

The revascularization of the graft, its assimilation to the skeleton, the fate of the grafted cartilage, and that of the ligament formations refixed to the graft or used as allografts still pose problems which have not been completely resolved. These massive grafts have to be studied over a longer period but the first results after 15 years are promising. The interim solution of the implantation of a prostheses sheathed with bone from a bone bank seems to be indicated in cases where resection of a tumor is large, removing bone, cartilage, ligaments, and muscles. By allowing the muscles to be refixed onto the graft, these sheathed prostheses limit the risks of the stem breaking or becoming detached. The use of a tibial graft including the patellar tendon makes reconstruction of the extensor apparatus easier. However, if physiotherapy is not started rapidly and continued regularly for several months, the graft will develop muscular adhesion, which sometimes considerably limits the freedom of movement of the joints.

Conclusion

Massive grafts are used more and more frequently in current surgical practice. The sterilization and preservation processes make it possible to make these bone and cartilage grafts easily usable and reliable [48].

In biological and clinical terms allografts assimilate perfectly to the skeleton but the complications associated with their use are what currently determine the limits of this type of surgery, which should only be practiced if the excision of the tumor has the same carcinological value as an amputation, and if draconian precautions are taken when the grafts are removed to avoid these transmitting an iatrogenic pathology [49].

Although the legislative problems have been alleviated since the Cavaillet law of 1976, there is still a great deal of opposition of an administrative, personal, and psychological nature from the public and also from some doctors, particularly if graft removals are to be performed on donors who are brain-dead.

On 11 July, 1950, during a meeting of the Academy of Medicine, Professor Moulonguet expressed the wish to see an organization set up to remove, preserve, and deliver human organs recognized as being necessary for therapy under the best possible moral, legal, and scientific conditions (Moulonguet 1950). [50]

This discussion is entirely topical, as the Institut Français des Greffes (French Graft Institute) is now being set up (35 years later, Decree of 10 October, 1994), and although the legal problems have been alleviated, there still remains a great deal of work to be done in this area because the use of human organs currently plays and will continue to play an ever greater role in our arsenal of therapy options. The use of cryopreserved massive allografts is an undeniable advance in our arsenal of therapies.

It is not reasonable, however, to ask of these grafts a result which could not be hoped for in the treatment of a simple fracture. They are often used in extreme conditions and often allow results to be obtained in otherwise hopeless cases. The last problem which has to be resolved is the one concerning the supplying of tissue banks with human grafts, because a great deal of opposition has to be overcome. Even if the legislative problems have been overcome, doctors and the public still do not have sufficient information, and efforts are required from everyone in order for the use of human grafts to be able to play an ever greater role in the future in our therapy.

References

1. Ollier L. Traité expérimental et clinique de la régénération des os et de la production artificielle du tissu osseux. Paris, Masson 1887:2 vol.
2. Albee FH. Principes du traitement des fractures non consolidées. SGO 1930;51:289.
3. Abbot LC, et coll. La valeur de l'os compact et de l'os spongieux comme matériel de greffe. JBJS 1947;29:381.
4. Sicard A, Binet JP. L'utilisation des greffons homogènes conservés. Congrès Fr de Chir Paris 1949.
5. Sicard A, Brièse EG. Les applications chirurgicales de la réfrigération. Sem des Hôp de Paris 1949;25:76,3136.
6. Sicard A. Ce que l'on peut attendre d'une banque d'os. Le Progrès Méd 1951;65:17387–9.
7. Sicard A, Gaudart d'Allaines C. La fracture des greffes rachidiennes. Journ de Chir 1952;68:N° 10.
8. Sicard A, Mouly R. Etude expérimentale des greffes osseuses conservées par le froid. Presse méd 1953;61:N° 44, 905–8.
9. Judet R. Hétérogreffes osseuses. Rev d'Orthop et de Chir de l'appar moteur, Oct Déc 1949;35:N° 6, 532–4.
10. Judet R, Arviset A. Banque d'os et hétérogreffe. Presse méd 1949;57:N° 68.
11. Judet R, Lagrange J, Dunoyer. Hétérotransplants osseux congelés. Acad de Chir, 3 Décembre 1952.
12. Ottolenghi CE. Massive osteo and osteo-articular bone grafts: technic and results of 62 cases. Clin Orthop 1972;87:156–64.
13. Parrish FF. Homografts of bone. Clin Orthop 1972;87: 36–42.

14. Parrish FF. Allograft replacement of all or part of the end of a long bone following excision of a tumor: report of twenty-one cases. J Bone and Joint Surg (Am) 1973; 55-A:1–22.
15. Poitout D. Conservation et utilisation de l'os de banque Cahier d'enseignement de la SOFCOT, 1985, N° 23, 157–177. EXPANSION SCIENTIFIQUE – PARIS.
16. Poitout D. Greffes utilisées pour reconstruire l'appareil locomoteur 1986, Paris, MASSON.
17. Poitout D, Novakovitch G. Allogreffes et banque d'os, Encyclopédie médico-chirurgicale, Paris. Appareil Locomoteur 14015 A10–5, 1986, p. 6.
18. Herbert JJ. De l'utilisation des os conservés comme greffes. La banque d'os. Mém de l'Acad de Chir 19 Janv 1949;75:60–8.
19. Fasquelle, Barbier. A propos de la dessication sous congélation. Sem des Hôp de Paris, 14 Déc 1950;N° 92.
20. Roy-Camille R, Laugier A, Ruyssen S, Chenal C, Bisserie M, Pene F, Saillant G. Evolution des greffes osseuses cortico-spongieuses et radiothérapie. Revue Chirurgicale Orthopédique, 1981;67:599–608.
21. Hernigou P. Allogreffes massives cryopréservées et stérilisées par irradiation. Rev Chir Orthop 1986;72: 267–76.
22. Loty B. Irradiation des allogreffes osseuses. Rev Chir Orthop 1988;74:116–17.
23. Loty B. Allogreffes osseuses massives. Rev Chir Orthop 1988;74:127–31.
24. Kouvalchouk JF, Paszkowski A. Irradiation des homogreffes osseuses. Rev Chir Orthop 1986;72:393–401.
25. Poitout D, Novakovitch G. Utilisation des allogreffes en oncologie et en traumatologie. International Orthopaedics, 11, 169–178, SICOT, 1987.
26. Poitout D. Indications classiques des allogreffes osseuses. Rev Chir Orthop 1988;74:118–19.
27. Friedlaender GE. Bone Allografts: The Biological Consequences of Immunological Events. J Bone and Joint Surg (Am) 1991;8,73-A:1119–20.
28. Horowitz MC, Friedlaender GE. Induction of Specific T-Cell Responsiveness to Allogeneic Bone. J Bone and Joint Surg 1991;73-A:1157–68.
29. Stevenson S, Li XQ, Martin B. The Fate of Cancellous and Cortical Bone After Transplantation of Fresh and Frozen Tissue-Antigen-Matched and Mismatched Osteochondral Alografts in dogs. J Bone and Joint Surg (Am) 1991;73-A:1143–56.
30. Stevenson S, Horowitz M. Current Concepts Review. The Response to Bone Allografts. J Bone and Joint Surg (Am) 1992;74-A:939–50.
31. Leriche P, Policard A. Le périoste et son rôle dans la formation de l'os. Presse méd 1918;26:143–6.
32. Leriche R. De la valeur des signes tenus pour caractéristiques de la vie des greffons. Bull Soc Chir 4 mai 1919.
33. Policard A. Les phénomènes de la réparation des fractures étudiées par la méthode des cultures de tissu. C Acad Sc 1927;184:117.
34. Urist MR. Calcification et ossification. JBJS 1942;24:47.
35. Urist MR, Silverman BF, Büring K, Dubuc FL, Rosenberg JM. The bone induction principle. Clin Orthop 1967;53: 243–83.
36. Urist MR, Delange R, Flnerman G. Bone cell differenciation and growth factors. Science 1983;220:680–6.
37. Burchardt H. The biology of bone graft repair. Clin Orthop 1983;174:28–42.
38. Hernigou P. Conservation des allogreffes osseuses. Rev Chir Orthop 1988;74:114–16.
39. Fèvre, Judet J, Arviset A. Greffes osseuses hétérogènes. MAC 19 Janv 1949.
40. Guilleminet, Stagnara, Dubost-Perret. Transplantations osseuses. Documentation expérimentale sur les hétérotransplants. Mém Acad de Chir 30, Avril 1952.
41. Davis GB, Taylor AN. Greffes osseuses à pédicules musculaires. Arch of Surg Août 1952;65:N° 2, 330–6.
42. Tomford WW, Thongphasuk J, Mankin HJ, Ferraro MJ. Frozen musculoskeletal allografts: a study of the clinical incidence and causes of infection associated with their use. J Bone and Joint Surg (Am) 1990;72-A:1137–43.
43. Teot L. Les transferts osseux libres vascularisés avec cartilage de croissance. Rev Chir Orthop 1982;68(Suppl. II):40–2.
44. Poitout D, Gaujoux G, Lempidakis M. Reconstructions iliaques totales ou partielles à l'aide d'allogreffes de banque. Int Orthop 1990;14:111–19.
45. Poitout D, Lempidakis M, Loncle X. Allogreffes otéo-cartilagineuses massives dans le traitement des nécroses ou des pertes de substance articulaires. Chirurgie 1991;117:193–8.
46. Czitrom AA, Keating S, Gross AE. The viability of articular cartilage in fresh osteochondral allografts after clinical transplantation. J Bone and Joint Surg (Am) 1990;72-A:574–81.
47. Enneking WF, Mindell ER. Observations on Massive Retrieved Human Allografts. J Bone and Joint Surg (Am) 1991;73-A:1123–42.
48. Mankin HJ. Allograft transplantation in the management of bone tumors. Edit. H.K. UHTHOFF – Springerverlag 1984;147–62.
49. Gerard Y. Banque d'Os. Rev Chir Orthop 1988;74; 110–11.
50. Moulonguet P. De la nécessité d'un règlement touchant le prélèvement, la conservation et la délivrance des organes humains utilisés en thérapeutique. Bull de l'Acad Nat de Méd séance du 11 Juillet 1950, pp. 494–6.

6 Bone Banks: Technical Aspects of the Preparation and Preservation of Articular Allografts

D. G. Poitout and Y. Nouaille-de-Gorce

For the technical aspects of this subject the Marseilles team collaborated with the Etablissement de Transfusion Sanguine Alpes-Provence (Alpes-Provence Blood Transfusion Service) to set up a bone bank on their premises because it has a competent cryobiology department equipped with storage tanks containing liquid nitrogen and a temperature-lowering programmer. This laboratory, which for a long time has been storing bone marrow, platelets, and various cryopreserved tissues, has the virology, bacteriology, quality control, and quality assurance laboratories of the Blood Transfusion Service and is accustomed to applying the transfusion safety standards. It was also one of the first in France to obtain the approval of the Microbiological Safety Committee of the Directorate General of Health in April 1996. Banks which were developed nationally have followed the same principles and currently more a hundred have been registered and some are awaiting authorization.

The Removal of Articular and Osteocartilaginous Grafts

Because the bone and cartilage fragments which are removed are not subjected to secondary sterilization, it is imperative that therapeutic maneuvers are performed in wholly sterile conditions.

Selection of the Donors

Selection has to be rigorous so that there is no risk of the transmission of iatrogenic pathology to the host through the graft. This requires adequate knowledge of the history of the illness and the circumstances of the accident, as well as the medical history of the donor. There are many absolute contra-indications. Subjects with a cancerous condition, a systemic illness, collagenosis, an auto-immune disease, or bone dystrophy may not have tissue removed for grafts. We systematically eliminate from the list of donors those suffering from a viral, bacteriological, or parasitic infection, mycosis or tuberculosis as well as those with risk factors and those who have been on artificial ventilation for more than 72 hours in intensive care, as they are potentially infected. As far as the validation of the grafting material is concerned, the banks have to conform to the legislation in force decreed by the French Graft Institute.

Bacterial decontamination is performed using a solution of antibiotics consisting of rifocine and chloramphenicol.

Samples are taken from each graft before and after decontamination and after thawing. Positive bacteriological results will mean that the tissues will have to be destroyed.

Blood samples have to be taken from each donor (live or deceased) in order to perform the obligatory virological examinations.

In accordance with the decrees of 25 February, 1992, of 24 May, 1994, and of 24, July 1996, medical biological analyses are performed to test for infection:

by the hepatitis B virus (HBs antigen and anti-HBc antibodies),

by HIV (antigen P24 and anti-HIV 1 and 2 antibodies),

by HTLV (anti-HTLV 1 and 2 antibodies),

by hepatitis C (anti-HCV antibodies),

for detecting syphilis in two different tests (VDRL and TPHA),

for assaying for transaminases for the live donors.

Decree No. 97-928 of 9 October, 1997 removes the obligation to carry out a search for the agent responsible for toxoplasmosis, for infection by cytomegalovirus, and by the Epstein-Barr virus. The Decree of 1 April, 1997 requires that the results of the examinations performed are examined before the patients have been transfused, as any hemodilution could falsify the tests. Finally, the French Blood Agency and the French Graft Institute recommend that the tissues are placed into quarantine and that the virological tests are repeated 4–6 months after the tissue has been removed. These samplings are performed either on the live donor or on the host of organs coming from the same donor and are performed to reduce the risk in the event of the tissue having been removed during a seroconversion phase. However, placing these products into quarantine is only one of the possible ways of ensuring safety. A decree will specify the conditions under which it is to be performed in the various situations according to the other methods which could be used, in particular directly testing for the viruses by molecular biology techniques (PCR).

Removal Techniques

As the risk of infection is the main concern in this surgery, the various stages of surgery must be performed under the strictest aseptic conditions possible. The tissue therefore has to be removed in an operating theatre, according to the same principles as regulated orthopedic surgery, and it is considered that a maximum time lapse of six hours from the circulation stopping can be reasonably accepted. For joint removals, the whole of the joint capsule is preserved as well as the intra-articular ligaments and the menisces or labra. In the case of the knee, and in order to keep the extensor apparatus intact, we retain the whole of the patellar tendon continuously with the posterior half of the patellar joint. This is also continuous with the quadricipital tendon, the upper part of which is cut into an inverted V. The muscular insertions are scraped and the surgeon removing the tissue cleans all the bone attachments of the ligaments allowing the capsule to be refixed firmly. The part is then placed in a bag which is resistant to very low temperatures (capton-teflon bags). The reconstruction of the skeleton is one of the important stages of the removal. It is a legal obligation and it has to be as perfect as possible.

Coding and Measuring of the Parts

In order to find the desired bone fragment again easily in the bank, it is necessary to fill in the data sheet carefully and to perform X-rays without enlargement or with an enlargement control placed side-by-side with the bone part.

Quality Controls

All the bone banks have to be inspected periodically, and samplings are performed on a very regular basis (20% of the grafts are rejected annually, either immediately after having been removed on account of positive results being found in tests or subsequently at the six-month checks).

Preservation Techniques

Preservation Methods

Many preservation methods have been suggested since the use of allografts was first considered. The techniques of irradiation and sterilization by moist heat will be described in detail subsequently. Cryopreservation is the only current procedure which makes it possible to preserve bone fragments and in particular cartilage cells safely.

Preserving fluids (antiseptics, 1% sodium methyolate, β propriolactoses), as well as ethylene oxide are cytotoxic and the problems involved in handling them have meant that they have been abandoned. The same applies for the methods involving sterilization by boiling.

Drying under vacuum or lyophilization uses plasma preservation processes. These grafts, which are theoretically usable indefinitely at ordinary temperatures, are fragile and are not sufficiently strong in mechanical terms to withstand the usual mechanical stresses and, in particular, all the cells are destroyed preventing the use of friction surfaces.

Irradiation of bone fragments taken under conditions which are not sterile is recommended by some teams who see a practical advantage for taking grafts. However, not only does this irradiation, conventionally performed at 25 K Gray, not guarantee perfect viral sterility but it also causes the destruction of all the cells.

Since 1981 we have been using cryopreservation of massive osteocartilaginous grafts in liquid nitrogen at −196 °C. This method makes it possible to preserve whole bones and complete joints over an extended period as it preserves the viability of the cartilaginous cells, the fibers, and fibroblasts contained in the ligaments and capsules.

Without going into the technical details of cryopreservation and storage, we would like to underline the fact that tissue preservation has to obey two essential rules:

uppression of cadaveric disintegration phenomena,

preservation for an extended period of the architecture of the bone and preservation of the viability of the cartilaginous cells.

To avoid the formation of ice macrocrystals, it is necessary to impregnate the bone, cartilage, and ligament tissues with a suitable cryoprotector. At Marseilles, we use a mixture of macromolecules (4% human albumin which is to be replaced by Héloes) and 10% final DMSO. This solution is maintained at a temperature of 4 °C because DMSO is toxic. In theory, the "deep cold" should enable these objectives to be reached by stopping the action of the tissue enzymes. If relatively moderate freezing temperatures are used (higher than −20 °C), the enzymes present in the tissue are not inactivated and will destroy the architecture of the graft in a few weeks. At −80 °C (limit of electric freezers), although the enzymatic activity is clearly reduced, it is not totally stopped, only collagenase is inactivated at this temperature. At −196 °C all the enzymes are inactivated and the proteins can be preserved over an extended period. On the other hand, DMSO has a eutectic point around −60 °C, at this temperature, the microcrystals of ice can recombine into macrocrystals and make the cells burst. For the same reasons, the temperature of the graft cannot be lowered in an haphazard fashion, and it is necessary to vary the rate at which this takes place in accordance with the temperature obtained. The optimum curve seems to us to be 2 °C per minute down to −40 °C then 5 °C per minute down to −140 °C, the temperature at which the graft is then placed in nitrogen vapor, directly in the tank where it is stored.

Thawing, on the other hand, has to be rapid so that the largest number of cells remain alive. Physiological serum or Ringer's lactate solution at 40–41 °C will be used to thaw out and wash the grafts in order to eliminate the DMSO. Approximately two hours after having removed the bone fragments from the liquid nitrogen tank, they are usable for a period of approximately 24–26 hours, which means that they can be taken to any part of France and Europe. However, donors are becoming rarer and rarer and it is becoming increasingly difficult to obtain tissue.

Biomechanics and Immunology

The mechanical strength of the cortical allograft is only 50–60% of the strength of normal bone during a period ranging from the eighth to the eighteenth month after the graft has been implanted (on account of the revascularization of the bone). Maximum fragility is at around the twelfth month and it is only after 2 to 3 years after the graft has been implanted that the bone regains normal density and biomechanical strength. The fixation of the graft therefore has to be complete in order for this period of fragility to come to an end. The mechanical properties of the allografts can be changed by the preservation and storage processes. Lyophilization, massive irradiation of the grafts (in excess of 3 megarads), or moist heat used for more than 60 minutes at 120 °C adversely affect the mechanical behavior of the grafts considerably. Cryopreservation, on the other hand, seems to improve the mechanical properties of the allograft, the strength of which is 110–120% that of fresh bone but the graft, although it is stronger, does in fact become more brittle (in the mechanical sense of the term) and therefore breakable. In the case of articular allografts, the crucial point is preservation of the ligament structures, of the synovial fluid, and of the menisces. The cells in these formations are preserved, as are their architectural and fundamental structure.

After having been reinstated, the vascularization is linked to the revascularization of the bone insertion area. It is inadequate for several months (or years) which means that they have to be doubled with an artificial ligament during this period to avoid excess stresses which would lead to their being stretched or even ruptured. The extent of the immunologically competent tissue (synovial, capsule, etc.) also risks leading to the occurrence of immunological rejection phenomena justifying the use of immunosuppressants (such as Sandimmum) in the event of the hyperproduction of fluid indicating an immune response.

Transport

Transporting bone parts over long distances will require the use of special containers, in which grafts will be preserved at low temperatures (liquid nitrogen or dry ice). When it is anticipated that these grafts will be used within 24 hours following removal from the Bank, it is preferable to thaw the bone fragment and to dispatch it only after thawing. Once removed from liquid nitrogen, the graft has to be used within 24 hours and it is not possible to refreeze it again if it is not used.

Use

The best indications for using these osteocartilaginous or complete joint grafts are in the knee, ankle, shoulder, elbow, and wrist.

Isolated Osteocartilaginous Grafts

Partial or total graft of the femoral condyle,

graft of the tibial plateau,

graft of the patella,

graft of the tibial pylon,

partial graft of the humeral head,

partial graft of the elbow,

graft of the inferior radius.

The graft is fixed by osteosynthesis material and the ligaments of the host are refixed onto the graft. The functional results are generally excellent.

Osteocartilaginous Graft plus Ligaments from the Donor

The excision is larger and the donor ligaments remain attached to the graft and are refixed onto the host bone or over the ligaments of the host. They have to be protected during the period of revascularization by artificial ligaments.

This applies to:

massive grafts of the inferior extremity of the femur,

massive grafts of the superior or inferior extremity of the tibia,

grafts of the humeral head,

partial grafts of the elbow.

Complete Articular Grafts

A total joint graft with its capsule, its synovial membrane, its ligaments, and the fibrocartilages it contains (meniscus or labrum) is indicated in the case of an extensive tumoral lesion in the joint cavity justifying extensive excision in a single piece.

The problems posed are twofold:

Immunological: connected with the size of the immunologically competent material implanted with apparently an inflammatory reaction which could result in a cutaneous fistula giving rise to an infection.

Mechanical: the ligaments must not be strained during the period of their revascularization and have to be doubled up by artificial ligaments.

Reconstruction Prostheses Sheathed with Bone from a Bone Bank

The indications for using a prosthesis sheathed with bone from a bone bank are often present and have to be analyzed according to the extent of the loss of musculo-ligament and cutaneous substance which requires tumoral excision or which has been produced by the trauma.

Conclusion

It seems to us to be important to emphasize the fact that grafts have to be preserved without breaking the cold chain, without damaging the bags containing the graft, and under strictly technological conditions. The importance of the sterile environment for removing the graft has to be emphasized; for example, out of more than 5,000 parts preserved in the Blood Transfusion Center in Marseilles since 1981, we have had to destroy barely 2% due to a superinfection being found. These encouraging results are due to extreme rigor in the ways in which the grafts are removed and the patients are selected. In our opinion, these preservation methods are the only ones which allow the bone, and particularly the cartilage, to still be guaranteed a normal structure and also good mechanical properties.

Appendix

Recommendations for Setting Up a Tissue Bank for the Locomotor Apparatus

These recommendations are those used by the Blood Transfusion Center of Marseilles to, among other things, preserve bone, cartilage, and ligament grafts.

The Organization of a Tissue Bank

General

The need for:

a tissue removal team approved by the Ministry,

a geographical location for treating, storing, and making the tissue available

Equipment

a preservation department equipped with storage tanks or apparatus, a temperature-lowering programmer indispensable for preserving bone, cartilage, and ligament tissue.

laboratories experienced in the following quality controls:

donor control,

tissue control,

validation of the preservation techniques.

Personnel

At the hospital:

the person in charge of tissue removal checks that removal from a subject in a state of brain death is in line with the regulations,

the surgical teams remove and treat the tissue.

At the tissue bank:

the bank receives the tissue,

the technical staff at the bank treats the tissue,

the laboratories perform viral serological tests on the donor and a bacteriological examination of the tissue removed,

the bank's medical supervisor oversees everything.

Techniques

For each tissue, all the technical aspects of the removal, treatment, storage, quality control, distribution, and results of examinations are recorded in a manual updated regularly.

Information

Information on the donor:

Identity, sex, age;

Radiography of the tissue if necessary;

Cause of death;

Medical history;

ABO blood group and HLA if known;

Operating protocol of removal, therapies used;

Results of any laboratory examinations;

Results of the control examinations;

Tissue removal center and department.

Information on the host:

Identity, sex, age;

Origin of the graft;

Attribution criteria;

Identification of the use of the graft, site and date;

Possible response to implantation of the graft;

ABO blood group and HLA if known;

Results of the culture at the time of the graft;

Note any departure from the guidelines for handling and reconstitution;

An estimate of the clinical results.

Quality Control

Each tissue preservation department has to take part in the development of the methods which make it possible to evaluate the indications of the tissue grafts preserved. Periodical monitoring of the bacteriological status has to be practiced, checking, before they are dispatched, at least 5% of the grafts every six months and more if problems with bacterial contamination are suspected.

Tissue Removal

General Ethical and Legal Considerations

In general, acceptable sources of tissues are cadavers less than six hours after circulation has stopped, patients in a state of brain death, and patients who have had part of their tissue removed for therapeutic purposes (femoral head).

The Caillavet law considers a donor to be any person who, during his lifetime, did not express any opposition to removal of his or her tissue. It is difficult, in practice, not to consider the pain suffered by those close to the person they have lost.

Selection Criteria

These vary depending on the tissue removed. Age may be a limiting factor following the use which will be made of the tissue taken. For cartilage in particular, it seems to be necessary to graft only normal joint surfaces which have been taken from healthy subjects.

Tissue may be removed from the cadaver for five hours after his or her death if it is kept at ambient temperature and for approximately 12 hours if the cadaver is stored at 4 °C immediately after death. The tissue removed from a live patient can be placed in a container, closed immediately and refrigerated at 4 °C. A valid preservation technique can be considered to be up to 12 hours after the tissue has been removed and stored at the preservation temperature.

A medical history of the donor has to be sent. Potential donors will be excluded if their current medical history mentions:

developing septicemia,

a localized infection in the tissue to be removed,

a slowly developing viral episode

malignant neoplasia except for most of the cerebral tumors,

the existence of active hepatitis or unexplained jaundice,

systemic disease,

a patient belonging to the risk groups,

heavy irradiation on risk groups,

treatment with drugs which are toxic to the tissue to be removed.

Laboratory tests have to be performed on the blood of the cadaver or on the live donor:

a test for the hepatitis B virus,

a test for syphilis,

a test for HIV antibodies,

the transaminase levels,

a test for anti-HBc antibodies,

a test for anti-HCV antibodies,

a test for anti-HTL V1 and anti-HTL V2 antibodies

a test for anti-CMV antibodies.

The erythrocyte blood groups and tissue groups should be used and a serum bank should be set up.

Wherever possible, the tissue should be removed under sterile conditions in an operating theatre. If the allografts are removed in a non-sterile manner, it should be ensured that effective sterilization techniques can be used without damaging the tissue structure.

If a collecting medium is used, it has to be sterile and physiological.

If antibiotics are used, the bacterial cultures have to be grown before they are added and the type of antibiotics has to be clearly recorded. A final bacteriological check is recommended before the tissues are dispatched.

Fragments of tissues to be grafted have to undergo bacteriological and fungal studies using current methods and media. Cultures of the donor blood have to be carried out when the tissue is removed as well as a urine culture and possibly also a culture of a pleural effusion.

Secondary sterilization. If it is carried out, the biological and biochemical integrity of the graft has to be maintained. The methods used for decontaminating surfaces are acceptable if only the surface can be contaminated.

Preservation and Storage

The methods used to preserve and store tissue allografts vary according to the type of tissue and the clinical application in which they are included. Although the optimum methods have not been defined, the best for long-term preservation would appear to be preservation at very low temperatures (–80 °C). Continuous monitoring of the temperature may be necessary. Storage for 12 hours at at the most at 4 °C may be practiced from the time the tissue was removed. Only materials which are resistant to low temperatures are suitable for this type of preservation. They have to be sterile. The culture media may vary and have to be defined for each type of cell. Precautions have to be taken to check the persistence of the activity of the cells being cultured and for the absence of contamination.

Bibliography

Babin SR, Katzner M, Vidal PH, Simon P, Kempf JF, Keiling R, Schvingt E. Résection – reconstruction diaphysaire fémorale par allogreffe massive fixée par clou médullaire vérouillé. RCO 1987;73:25–9.

Beaver RJ, Mahomed M, Bachstein D, Davis A, Zukor DJ, Gross AE. Fresh osteochondral allografts for post-traumatic defect in the knee. J Bone Joint Surg (BR) 1992;74-B:105–10.

Brooks DB, Heiple KC, Herdon CH, Powell AE. Immunological factors in homogenous bone transplantation. IV. The effect of various methods of preparation and irradiation on antigenicity. J Bone Joint Surg 1969;45A:1617.

Brown K, Crues SR. Bone and cartilage transplantation, in orthopeadic surgery. J Bone Joint Surg (Am) 1982; 64:270–9.

Burchardt H, Enneking WF. Transplantation of bone. Surg Clin North Am 1978;58:409.

Burchardt H. The biology of bone graft repair. Clin Orthop 1983;174:121–35.

Burchardt H, Jones H, Gloweczewskie F, Rudner C, Enneking WF. Freeze-dried allogeneic segmental cortical bone grrafts in dogs. J Bone Joint Surg 1978;60A:1082.

Burwell RG, Gowland G. Studies in the transplantation of bone. III. The immune responses of lymphnodes draining components of fresh homologous cancellous bone and homologous bone treated by different methods. J Bone Joint Surg 1962;44B:131.

Burwell RG. Studies in the transplantation of bone. V. The capacity of fresh and treated homograftsof bone to evoke transplantation immunity. J Bone Joint Surg 1963;45B:386.

Burwell RG. The fate of bone grafts. In: Apley AG, editor. Recent advances in orthopeadic, Baltimore, the Williams & wilkins Co. 1969;115.

Burwell RG. The fate of freeze-dried bone allografts, transplant. Proc 1976;8(Suppl.1):95.

Burwell RG, Gowland G, Dexter F. Studies in the transplantation of bone. VI. Further observations concerning the antigenicity of homologous cortical and cancellous bone. J Bone Joint Surg 1963;41B:597.

Carr CR, Hyatt GW. Clinical evaluation of freeze-dried bone grafts. J Bone Joint Surg 1955;37A:549.

Carrel A. La conservation des tissus et ses applications en chirurgie. J Am Med – Techniques chirurgicales orthopédique, Paris. 1984;44090:4–7–10.

Chalmers J. Transplantation immunity in bone grafting. J Bone Joint Surg 1959;41B:160.

Charpentier B. Mécanisme du rejet des allogreffes. Presse Méd 1984;13:2697–700.

Chrisman OD, Fessel JM, Southwick WO. Experimental production of synovitis and marginal articular exostose in the knee joint of dogs. Yale J Biol Med 1964;37:409.

Coutelier L, Delloye CH, De Nayer P, et Vincent A. Aspects microradiographiques des allogreffes osseuses chez l'homme. Rev Chir Orthop 1984;70:581–8.

Cracchiolo A, III, Michaeli D, Goldberg LS, Fudenberg HH. The occurrence of antibodies to collagen in synovial fluids. Clin Immonol Immunopathol 1975;3:567.

Curran WJ. The Uniform Anatomical Gift Act. Engl J Med 1969;280:36.

Darcy DA. Reaction of rabbits to frozen homografts. Pathol Bacteriol 1955;70:143.

Devries PH, Badgley CE, Hartman JT. Radiation sterilization of homogenous-bone transplants utilizing radioactiv cobalt. Preliminary report. J Bone Joint Surg (Am) 1958;40:187–203.

Duffy P, Wolf J, Collins G, De Voe AG, Streeten B, Cowen D. Possible person-to-person transmission of Creutzfeld-Jacob disease. N Engl J Med 1974;290:692.

Duparc J, Nordin JY, Olivier H, Augereau B. Les résections-reconstructions dans les tumeurs osseuses des membres et du bassin. Encycl Méd Chir – Techniques Chirurgicales Orthopédie 1984;44090:4–7–10.

Elves MW. Humoralimmune response to allografts of bone. Int Arch Allergy Appl Immunol 1974;47:708.

Elves MW. Newer knowledge of immunology of bone and cartilage. Clin Orthop 1976;120:232.

Elves MW, Ford CHJ. A study of the humoral immunce response to osteoarticular allografts in the sheep. Clin Exp Immunol 1974;17:497.

Friedlaender GE, Mankin HJ. Guidelines for the banking of muqculoskeletal tissues. Am Assoc Tissues Banks Newsletter 1980;4(suppl.):30.

Friedlaender GE. Current concepts. Review: Bone-banking. J Bone Joint Surg (Am) 1982;64:307–11.

Friedlaender GE. Immune response to osteochondral allografts. Clin Orthop 1983;174:58–67.

Friedlaender GE, Ladenbauer-Bellis I, Chrisman OD. Cartilage matrix components as antigenic agents in a osteoarthritis model. Trans Orthop Res Soc 1980;5:170.

Friedlaender GE, Mankin HJ, Kenneth W. Osteochondral allografts (biology, banking and clinical applications). Little Brown, Boston Toronto, 1982.

Friedlaender GE, Strong DM, Sell KW. Donor graft specific anti-HL-A antibodies following freeze-dried bone allografts. Trans Orthop Res Soc 1977;2:87.

Friedlaender GE, Strong DM, SELL KW. Studies on the antigenicity of bone. I. Freeze-dried and deep-frozen bone allografts in rabbits. J Bone Joint Surg 1976;58A:854.

George CR, Chrisman OD. The role of cartilage polysaccharides in osteoarthritis. Clin Orthop 1968;57:259.

Glant T, Hadas E, Nagry TJ. Cell-mediated and humoral immune responses to cartilage antigenic components. Scand J Immunol 1979;9:29.

Golberg V, Heiple K. Experimental hemijoint and whole transplantation. Clin Orthop 1983;174:43–53.

Goldberg V, Bos G, Heiple K, Zita J, Powell A. Improved acceptance of frozen bone allografts in genetically mismatched dogs by immunosuppression. J Bone Joint Surg (Am) 1984;66:937–50.

Greiff D, Milson TJ. Functional activities of isolated lymphocytess following drying by sublimation of ice *in vacus*. I. Rosette formation, stimulation by plant lectins (mitogens) and the mixed lymphocyte reaction. Cryobiology 1980;17:319.

Gresham RB. The freeze-dried cortical bone homograft: a roentgenographic and histologic evaluation. Clin Orthop 1964;37:194.

Gross A, Mc Kee N, Pritzker K, Langer F. Reconstruction of skeletal deficit at the knee. Clin Orthop 1983;174: 96–106.

Gross AE, Langer F, Houpt J, Pritzker K, Friedlander CE. The allotransplantation of the partial joints in the treatment of osteoarthritis of the knee. Transplant Proc 1976;8(suppl.I):129.

Gross AE, Langer F, Silverstein EA, Falk R, Falk J. The allotransplantation of the partial joints in the treatment of osteoarthritis of the knee. Clin Orthop 1975;108:7–14.

Guilleminet, Stagnara, Dubost-Perret, Jarret, Audry. Utilisation d'os hétérogènes réfrigérés en chirurgie humaine. Lyon Chir 47 N°1, Janvier 1952.

Hedde C, Postel M, Kerboul M, Courpied JP. La réparation du cotyle par homogreffe osseuse conservée au cours des révisions de prothèse totale de hanche. Rev Chir Orthop 1986;72:267–76.

Heiple KG, Chase SW, Herndon CH. A comparativ study of the healing process following different types of bone transplantation. J Bone Joint Surg 1963;45A:1593.

Hiky V, Mankin HJ. Radical resection and allograft replacement in the treatment of bone tumors. J Jpn Orthop Assoc 1980;54:475.

Houff SA, Burton RC, Wilson RW, •• AL. Human-to-human transmission of rabies virus by corneal transplant. N Engl J Med 1979;300:603.

Huten D. Utilisation des allogreffes osseuses dans les reconstructions fémorales au cours des reprises de prothèse totale de hanche. Rev Chir Orthop 1988; 74:122–4.

Hyatt GW, Butler MC. Bone grafting. The procurement storage and clinical use of bone homograft. In America Association of Orthopeadic Sugeons: Instructional courses lectures. Ann Arbor, Mich., J.W. Edwards Co. 1957;14:343.

Inclan A. L'emploi des greffes osseuses conservées en Orthopédie. J Bone Joint Surg 1942;26:81–96, Janvier.

Inclan A. Use of preserved bone graft in orthopeadic surgery. J Bone Joint Surg 1942;26:81.

James JIP. Tuberculosis transmitted by banked bone. J Bone Joint Surg 1953;35B:578.

Judet H, et Padovani JP. Transplantation d'articulation complète. Rev Chir Orthop 1983;67:359–60.

Judet J, Aviset A. Homogreffes provenants de la banque d'os. Mem Acad Chir N°, 1948;27–8:671.

Koskinen EV, Salenius P, Alho A. Allogeneic transplantation inlow-grade malignant bone tumors. Acta Orthop Scand 1979;50:129.

Kossowska-Paul B. Studies on the regional lymph node plastic reaction evoked by allogeneic grafts of fresh and preserved bone tissue. Bull Acad Polon Sci 1966;14:651.

Kruez FP, Hyatt GW, Turner TC, Basset AL. The preservation and clinical use of freeze-dried bone. J Bone Joint Surg 1951;33AA:863.

Langer F, Czitrom A, Pritzker KP, Gross AE. The immunogenicity of fresh and frozen allogeneic bone. J Bone Joint Surg 1975;57A:216.

Lee EH, Langer F, Halloran P, Gross AE, Ziv I. The effect of major and minor histocompatibility differences on bone transplant bealing in inbred mice. Trans Orthop Res Soc 1979;4:60.

Lee EH, Langer F, Halloran P, Gross AE, Ziv I. The immunology of osteochondral and massive allografts. Trans Orthop Res Soc 1979;4:61.

Lexter E. Die verwendung der freien knochenplastik nebst versuchen über gelenkversteifung und gelenktransplantation. Arch Klin Chir 1908;86:939.

Lexter E. Joint transplantation and arthrosplasty. Sug Gynecol Obstet 1925;40:782.

Locht R, Gross A, Langer F. Late osteochondral allograft resurfacing for tibia plateau fractures. J Bone Joint Surg (Am) 1984;66:328–35.

Macewen W. Observations concerning trannsplantation of bone. Illustrated by case of inter-human osseous transplantation, where by ovre two-thirds of shaft of a humerus was restored. Pro R Soc Lond 1881;32:232.

Mankin HJ, Doppelt SH, Tomford WW. Clinical experience with allograft implantation. Clin Orthop 1983;174:69–86.

Mankin HJ, Doppelt SH, Sullivan TR, Tomford WW. Osteoarticular and intercalary allograft transplantation in the management of malignant tumors of bone. Cancer 1983;50:613.

Mankin HJ, Fogelson FS, Trasher AZ, Jaffer F. Massive resection and allograft transplantation in the treament of malignant bone tumors. N Engl J Med 1976;294:1247.

Marsh B, Flynn L, Enneking W. Immunologic aspects of osteosarcoma and their application to therapy, a preliminary report. J Bone Joint Surg 1972;54A:1367.

Merle D'Aubigne R. A propos de la résection pour tumeurs du genou. Rev Chir Orthop 1963;67:359–60.

Meyers MH, Chatterjee SN. Osteochondral transplantation. Surg Clin North Am 1978;58:429.

Mnaymneh W, Emerson RH, Brajao F, Head WC, Malinin TI. Massive allografts in slavage revision of failed total knee arthroplasties. Clin Orthop 1990;269:144–53.

Musculo DL, Kawai S, Ray RD. Cellular and humoral immune response analysis of bone-allografted rats. J Bone Joint Surg 1976;58A:826.

Nimelstein SH, Hotti AR, Homan HR. Transformation of a histocompatibility immunogen into a tolerogen. J Exp Med 1973;128:723.

Ollier L. Traité expérimental et clinique de la régénération des os. Victor Masson & fils, Paris 1867.

Ottolenghi CE. Massive osteo and osteo-articular bone grafts: technic and results of 62 cases. Clin Orthop 1972;87:156.

Parrish FF. Allograft replacement of all part of the end of a long bone following excision of a tumor: report of twenty-one cases. J Bone Joint Surg 1973;55A:1.

Penn I. The incidence of malignancies in transplant recipients. Transplant Proc 1975;7(2):323.

Pelkers R, Friedlander G, Markham T. Biomechanical properties of bone allografts. Clin Orthop 1983;174: 54–7.

Poitout D, et Tropiano P. Les reconstitutions du cotyle après chirurgie iteractive de la hanche – A propos de 37 cas. Bulletin de l'Académie Nationale de Médecine. 180-N°3, 1996;515–31.

Poitout D. Allografts of the patella and extensor apparatus. Atlas of open knee surgery. Edition CHAPMAN & HALL, 1996;42–9.

Poitout D. Allotraplanto della rotula e dell'apparato estensor. Atlante di technica chirurgica del ginocchio – Edition Masson – Milan, 1995;42–9.

Poitout D. Banche d'osso: aspetti tecnici criopreservazione di allotraplanti osteocartilagineo. Atlante di technica chirurgica del ginocchio – Edition Masson – Milan, 1995;178–9.

Poitout D. Bone bank: technical aspects of cryopreservation osteocartilaginous grafts. Atlas of open knee surgery. Edition CHAPMAN & HALL, 1996;178–9.

Poitout D. Conservation et utilisation de l'os de banque. Cahier d'enseignement de la S.O.F.C.O.T N°23. Expansion scientifique. Conférence 1985;157–77.

Poitout D. Future of bone allografts in massive bone resection for tumor. Presse Médicale 1996;30;25(11):570–30.

Poitout D. Greffes utilisées pour reconstruire l'appareil locomoteur. Masson, Paris 1986.

Poitout D. Knee reconstruction prosthesis incorporating a large allograft. Atlas of open knee surgery. Edition CHAPMAN & HALL, 1995;145–51.

Poitout D. L'os biomateriaux. « Bulletin de l'Académie de Nationale de médecine ». Communication à l'Académie Nationale de Médecine. Paris, 14 mars 1995. 1995;179;3:517–36.

Poitout D. Les greffes de l'appareil locomoteur. Académie de Chirurgie (Paris, 10 avril 1996). « Mémoire de l'académie de Chirurgie », 1996.

Poitout D. Les reconstructions de cotyle après chirurgie iterative de la hanche. (A propos de 37 cas). Académie Nationale de Médecine. Paris, 5 mars 1996. « Bulletin de l'Académie de Nationale de médecine », 1996;180;3:515–31.

Poitout D. Protesi di ricostruzione di ginocchio con allotraplanto massivo. Atlante di technica chirurgica del ginocchio – Edition Masson – Milan, 1995;145–51.

Poitout D. Reconstruction du cotyle et de l'hémibassin par allogreffe ou prothèse méttalique massive sur mesure. 73ème Réunion Annuelle de la S.O.F.C.O.T. – Paris les 10–13 novembre 1998. Revue de chirurgie Orthopédique, 84, N°218, 1998;120–1.

Poitout D, Bernat M, Martin G, Tropiano P. Indications des greffes osteocartilagineuses massives en traumatologie du genou. Acta Orthopedica Belgica. Vol. 62, N°6, 1996;59–65.

Poitout D, Bernat M, Martin G, Tropiano P. Indications des greffes osteo-cartilagineuses massives en traumatologie du genou. « Acta Orthopaedica Scandinavica », Edition Scandinavian University Press, Vol. 62, 1997; suppl.1.

Poitout D, Bernat M, Moulene JF, Tropiano P. Allogreffes de cotyle et d'hémi-bassin. A propos de 37 cas. 71ème Réunion Annuelle de la S.O.F.C.O.T. – Paris les 12–15 novembre 1996. Revue de chirurgie Orthopédique, 1996;87;152:100.

Poitout D, Bernat M, Moulene JF, Tropiano P. Allogreffes osteochondrales ou prothèses articulaires en chirurgie traumatologique ou oncologique du genou. 71ème Réunion Annuelle de la S.O.F.C.O.T. – Paris les 12–15 novembre 1996. Revue de chirurgie Orthopédique, 1996;69–70,87–99.

Poitout D, Bernat M, Moulene JF, Tropiano P. Devenir des fractures du col du fémur. 71ème Réunion Annuelle de la S.O.F.C.O.T. – Paris les 12–15 novembre 1996. Revue de chirurgie Orthopédique, 1996;87;153:100–1.

Poitout D, Bernat M, Moulene JF, Tropiano P. Indications for cryopreserved allografts in tumoral pathology. European Journal of Orthopeadic Surgery and traumatology. Vol. 7, 1997;100–4.

Poitout D, Bernat M, Moulene JF, Tropiano P. Massive HIP prothese ensheated by allografts. European Journal of Orthopeadic Surgery and traumatology. Vol. 7, 1997; 123–6.

Poitout D, Dubousset JF, Tomeno B. Arthrectomie monobloc du genou. « Revue de Chirurgie Orthopédique », 1995;81–6:565, Paris.

Poitout D, Lempadakis M, Bernat M, Lecoq C, Martin G, Aswad R. Secondary internal osteosynthesis after fixation for recent or lower limb. « Revue de chirurgie orthopédique Réparatrice », 1996;82(2):137–44.

Poitout D, Lempidakis M, et Loncle X. Osteocartilaginous graft of the upper tibia. Atlas of open knee surgery. Edition CHAPMAN & HALL, 1996;170–7.

Poitout D, Lempidakis M, et Loncle X. Allotraplanto osteocartilagineo massivo dell'estremita superiore della tibia. Atlante di technica chirurgica del ginocchio – Edition Masson – Milan, 1995;170–7.

Poitout D, Lempidakis M, et Loncle X. Allotraplanto osteo-cartilagineo massive dell'estremita inferiore del femore. Atlante di technica chirurgica del ginocchio – Edition Masson – Milan, 1995;164–9.

Poitout D, Lempidakis M, et Loncle X. Osteocartilaginous graft of the lower extremity of the femur. Atlas of open knee surgery. Edition CHAPMAN & HALL, 1996;164–9.

Poitout D, Lempidakis M, et Loncle X, Bernat M. Les sarcomes ostéogeniques de l'extrêmité inférieur du fémur. Techniques de reconstruction – L'avenir. « Revue de Chirurgie Orthopédique », Communication à l'Académie Nationale de Chirurgie, Paris le 29 mars 1995. 1995;566:81-6.

Poitout D, Lempidakis M, et Loncle X, Bernat M. Reconstructions massives du cotyle et du fémur proximal. Académie de Chirurgie (Paris, 1994). « Mémoire de l'académie de Chirurgie » Tome N°5, 1994–5;120.

Poitout D, Lempidakis M. Artificial ligament repairs. Atlas of open knee surgery. Edition CHAPMAN & HALL, 1996; 76–85.

Poitout D, Lempidakis M. Legamentoplastiche artificiali. Atlante di technica chirurgica del ginocchio – Edition Masson – Milan, 1995;76–85.

Poitout D, Loncle X. Legamento crociato posteriore. Atlante di technica chirurgica del ginocchio – Edition Masson – Milan, 1995;86–8.

Poitout D, Loncle X. Posterior cruciate ligament. Atlas of open knee surgery. Edition CHAPMAN & HALL, 1996;86–8.

Poitout D, Lu J, Huang ZW, Tropiano P, Clouet D'Orval B, Remusant M, Dejou J, Proust JP. Human biological reactions at the interface between bone tissues and polymethylmethacryalate cement. J Mater Sci Mater Med 2000;803–9.

Poitout D, Nandiegou Y, Lempidakis M. Legamentoplatica artificiale. Atlante di technica chirurgica del ginocchio – Edition Masson – Milan, 1995;89–93.

Poitout D, Nandiegou Y, Lempidakis M. Artificial ligament repair. Atlas of open knee surgery. Edition CHAPMAN & HALL, 1996;89–93.

Poitout D, Novakovich G. Allogreffes et banque d'os. Encyclopédie Médico-chirurgicale (Paris-France), appareil locomoteur 14015AIO, 1986;5–6.

Poitout D, Ozoux P. Intra-articular repair of the posterior cruciate ligament using either semitendinosus or gracilis tendons (Lindenmann's operation). Atlas of open knee surgery. Edition CHAPMAN & HALL, 1996;94–5.

Poitout D, Tropiano P, Bernat M, Loncle X, Martin G. Greffes articulaires, mythe ou réalité? Entretiens de Bichat 1995. Paris 25–30 septembre 1995. « Chirurgie-Spécialité », 1995;43–6.

Poitout D, Tropiano P, Bernat M, Moulene JF. Reconstruction massive de cotyle et du fémur proximal. Les Arcs, 20–25 janvier 1996. « European Journal of Orthopaedic surgery and traumatology. » 1996;6:271–7.

Pool AR, Reiner A, Choi H, Rosenberg LC. Immunological studies of proteoglycan subunit from bovine and human cartilage. Trans Orthop Res Soc 1979;4:55.

Pujet J, Utheza G. Reconstructionde l'os iliaque à l'aide du fémur homoilatéréal après résection pour tumeur pelvienne. Rev Chir Orthop 1986;72:151–5.

Rodrigo JJ. Distal rat femur allografts: a surgical model for induction of humoral cytotoxic antibodies. Trans Orthop Res Soc 1977;2:265.

Rodrigo JJ, Fuller TC, Mankin HJ. Cytotoxic HL-A, antibodies in patient with bone and cartilage allografts. Trans Orthop Res Soc 1976;1:131.

Roy-Camille R, Laugier A, Ruyssen S, Chenal C, Bisserie M, Pene F, Saillant G. Evolution des greffes osseusescortico-spongieuses et radiothérapie. Rev Chir Orthop 1981;67: 599–608.

Sadler AM, •• JR, Sadler BL. Providing cadaver organs: three legal alternatives. Hastings Center Studies I, 1973;14.

Sadler AM, JR, Sadler BL, Stason EB, Stickel DL. Transplantation – a case for consent. N Engl J Med 1969;280:862.

Sagi S, Turianskyj FH, Gyenes L. Immunogenicity of soluble murine histocomptabilicity antigens. Immunol Commun 1974;3:85.

Salama R. Xenogeneic bone grafting in humans. Clin Orthop 1983;174:113–21.

Schachar NS, Friedlander GE, Mankin HJ. Bone transplantation. In Slavin S (ed) Organ transplantation: presentstate, future goals, Amsterdam, Elsevier/North-Holland biomedical Press B.V. (in press).

Schachar NS, Fuller TC, Wadsworth PL, Henry WB, Mankin HJ. A feline model for the study of frozen osteoarticular allografts. Development of lymphocytotoxic antibodies in allograft recipients. Trans Orthop Res Soc 1978;3:131.

Schachar NS, Mankin HJ, Wadsworth PL, Henry WB, Castronovo FP. A feline model for the study of frozen osteoarticular allografts. I. Quantitative assessment of cartilage viability and bone healing. Trans Orthop Res Soc 1978;3:130.

Sell KW, Friedlander GE, Editors. Tissues banking for transplantation, New-York, Grune & Stratton, Inc, 1976.

Sell KW, Friedlander GE, Strong DM. Immunogenicity and freeze-drying, Cryoimmunology 1976;17:187.

Shneider JR, Bright RW. Anterior cervical fusion using preserved bone allografts. Trans Proc 1976;8(suppl.):73.

Shutkin NM. Homologous-serum hepatitis following use of refrigerated bone-bank bone: report of case. J Bone Joint Surg 1954;36A:160.

Solomon L. Bone Grafts. J Bone Joint Surg (Am) 1991;73-B:706–7.

Spence KF, Sell KW, Brown RH. Solitary bone cyst: treatment with freeze-dried cancellous bone allograft. J Bone Joint Surg 1969;51A:87.

Stockley I, Mc Auley JP, Gross AE. Allograft reconstruction in total knee arthroplasty. J Bone Joint Surg (Am) 1992;74-B:393–7.

Syftestad G, Urist M. Bone aging. Clin Orthop 1982;162: 288–97.

Takagi K, Urist M. The reaction of the dura to bone morphogenetic protein (BMP) in repair of skull defects. Ann Surg 1982;196:100–9.

Takagi K, Urist M. The role of bone marrow in bone morphogenetic protein-induced repair of femoral massive diaphyseal defects. Clin Orthop 1982;171:224–30.

Takami H, Doi T, Ninomiya S. Reconstruction of a large tibial defect with a free vascularized fibular graft. Arch Orthop Surg 1984;102:203–5.

Tavernier. Sur les greffes d'os tué. Lyon Chir., 1922, séance du 17 novembre 1921. Utilisation d'os purum. Lyon Chir 1930.

Tomford WW, Fredricks GR, Mankin HJ. Cryopreservation of intact articular cartilage. Trans Orthop Res Soc 1982;7:176.

Tomford WW, Fredricks GR, Mankin HJ. Cryopreservation of isolated chondrocytes. Trans Orthop Res Soc 1982;6:100.

Tomford WW, Mankin HJ, Doppelt S. Bone bank procedures. Clin Orthop 1983;174:15–21.

Tomford WW, Starkweather RJ, Golman MH. A study of the clinical incidence of infection in the use of banked allograft bone. J Bone Joint Surg 1981;63A:244.

Trentham DE, Townes AS, Kang AH, David JR. Humoral and cellular sensitivity to collagen in type II. Collagen induced arthritis in rats. J Clin Invest 1978;61:89.

Tuffier. Des greffes chirurgicales chez l'homme. Bull et Mém Soc Chir 36, 1983, Paris 1910.

Urist MR. Human bone morphogenetic protein. Proc Soc Exp Biol Med 173:194–9.

Urist MR. Practical applications of basic reseach on bone graft physiology. In AAOS: Instructional Course Lectures, 25:1, St Louis, The C.V. Mosby Co, 1976.

Urist MR, Delange R, Finermann G. Bone cell differentiation and growth factors. Science 1983;220:680–6.

Urist MR, Mikulski A, Boyd SD. A chemosterilized antigen-extracted autodigested alloimplant for bone banks. Arch Surg 1975;110:416.

Volkov M, Imamaliyev AS. Use of allogenous articular bone implants as substitutes for autotransplants in adult patients, Clin Orthop 1976;114:192.

Volkov M. Allotransplantation of joints. J Bone and Joint Surg (Br) 1970;52B:49–53.

Weiland A, Moore R, Daniel R. Vascularized bone autografts. Clin Orthop 1983;174.

Weislander J, Heinegard D. Immunochemical analysis of cartilage proteoglycans: antigenic determinants of substructures. Biochem J 1979;179:35.

Wilson PD. Follow-up study of the use of refrigerated homogenous bone transplants in orthopaedic operations. J Bone and Joint Surg 1951;33A:307.

Wilson RE, Penn I. Fate of tumors transplanted with a renal allograft. Transplant Proc 1975;7(2):327.

Wittbjer J, Palmer B, Rohlin M, Thorngren K. Osteogenetic activity in composite grafts of demineralized compact bone and marrow. Clin Orthop 1983;173:229–38.

Yablon I, Brandt KD, Delellis RA. The antigenic determinants of articular cartilage: their role in the homograft rejection. Trans Orthop Res Soc 1977;2:90.

Yablon I, Copperband S, Covall D. Matrix antigens in allografts. Clin Orthop 1982;168:243–51.

Yamane K, Nathenson SG. Biochemical similarity of papain-solubilized H-2d, alloantigens from tumor cells and from normal cells. Biochemistry 1970;9:4743.

Zaleske D, Ehrlich M, Piliero C, May J, Mankin HJ. Growth plate behavior in whole joint replantation in the rabbit. J Bone Joint Surg (Am) 1982;64:249–57.

7 Clinical Applications of Bone Substitutes

P. Bonnevialle and D. Clément

For decades, the use of autografts has been considered as the "gold standard" procedure for the surgical filling of bone defects. But in addition to increasing operative time and morbidity at the harvesting site, autografts are only available in limited volumes. Over time, various substitutes to autografts have been made available for bone grafting. These substitutes can be differentiated from their origin. Besides allografts and xenografts, respectively obtained by treatment of human or animal bone, synthetic bone substitutes present a more recent alternative. Bone substitutes are biocompatible, show osteoconductive properties when implanted in living bone, and can possibly be resorbed, depending on their chemical composition, as evidenced by numerous fundamental or experimental works [1–4]. Clinical applications of bone substitutes may concern general orthopedic surgery as well as spinal, cranial, or maxillofacial surgery. Nevertheless, objective evaluation of these implants is often confronted by methodological problems, as most clinical studies available only consider retrospective results. In France, a list of marketed implants intended for bone grafting is edited and updated by the GESTO ("Association pour l'étude des greffes et des substituts tissulaires en orthopédie"), a medical committee under the aegis of the French Society of Orthopaedic Surgery (SOFCOT). This committee has its motivations in a strictly scientific evaluation of every implant available and publishes its expertise regularly [5].

General Overview of Bone Substitutes

According to the GESTO, three general categories of bone substitutes can nowadays be differentiated, considering the origin of the material constitutive of the implant: calcium phosphate ceramics, xenografts, and a third group of implants of various compositions, such as calcium sulfate or coral. All these implants attempt to mimic alveolar bone structure or the chemical composition of bone minerals.

Calcium Phosphate Bone Substitutes

Biological Properties

Calcium phosphates are obtained by the sintering, generally over 1,000 °C, of a raw material. These materials can either be synthesized by chemical reaction between calcium and phosphate ions or be obtained by thermal treatment of natural animal skeletons. The rate of bone ingrowth and resorption can be varied and controlled depending on the porosity and the chemical composition (Ca/P ratio) of the material. Synthetic calcium phosphate ceramics are produced by molding and/or machining a raw material into various shapes and porosities prior to sintering. In the case of biological ceramics, manufacturing processes mostly consist of thermal treatments to eliminate organic compounds, as in the case of natural bone, but can as well include a chemical reaction to modify the composition of the starting material, as in the case of coral, the calcium carbonate of which can be converted into calcium phosphate by hydrothermal reaction.

Free of organic phases (fats, proteins, etc), calcium phosphate implants are biocompatible and present a negligible biological risk of disease transmission to the patient. When placed in close contact with healthy bone tissue, autologous osteogenic cells grow along the implant and produce extracellular bone matrix directly in contact with the surface, without fibrous interlayering. It has been evidenced that undifferentiated cells from bone marrow can differentiate into specific bone cells on the surface of the ceramics, where they exhibit osteoblastic phenotyping. Secondarily, depending on the material solubility, the implant can be resorbed. For instance, pure crystallized hydroxyapatite ($Ca_{10}(PO_4)_6(OH)_2$) is known to be poorly soluble and can remain for years without significant signs of resorption [6,7], as tricalcium phosphate ($Ca_3(PO_4)_2$) ceramics can be resorbed in a few years [2,8]. The resorption process is controlled by resorbing, osteoclast-like, cells but can also take place by dissolution of the material in extracellular fluids. These properties are enhanced by the presence of cavities (pores) in the material. When the pore size exceeds 80–100 μm (macro-pores), bone cells can penetrate the cavities and form new bone internally. Smaller pore sizes (micropores, 1–10 μm) are thought to favor protein fixation by capillarity and thus facilitate cell differentiation. But if the presence of pores significantly helps bone healing, it also has a negative influence on the mechanical strength of the material. Pure hydroxyapatite or tricalcium phosphate ceramics can be as resistant as cortical bone (approx 150 MPa), should the total porous volume not exceed 30% of the external volume [3]. Increasing the porosity produces a decrease in the mechanical strength, which can be as low as 1–2 MPa when the porous volume reaches 60–70%. Thus, highly porous ceramics cannot be used in load-bearing situations without additional stabilization devices.

Classification

Hydroxyapatite ceramics (HAC)

As bone mineral is a poorly crystallized hydroxyapatite, it is possible to produce HAC from either synthetic (Cérapatite®, Synatite®) or natural (Endobon®, Pro-osteon®) materials. As already discussed, HAC show only limited resorption. Synthetic HAC consists of pure crystallized HA (>95%) as biological HAC can contain trace elements (Mg^{2+}, CO_3^{2-}, etc.). Most products are available in various shapes (granules, blocks, cylinders) for general bone void filling or in the form of anatomical implants designed for specific indications (osteotomy, spinal arthrodesis).

Tricalcium Phosphate Ceramics (TCPC)

TCPC (Biosorb®, Calciresorb®, Vitoss®) are elaborated from purely synthetic materials. Resorption rate and extent vary as a function of the porous volume and pore size, but also depend on the volume of the graft. Total resorption can be radiologically observed after a few months in the case of highly porous implants, as 30% porosity implants are generally still visible two years after implantation. Depending on the manufacturer, product ranges can include various porous volumes (20–65%) in different pore sizes, generally ranging from 10 μm to 400–500 μm. As in the case of HAC, TCPC are made available in different shapes and volumes for bone void filling, but can also have anatomical designs for specific applications.

Biphasic Calcium Phosphate Ceramics (BCPC)

Composed of a mixture of synthetic HA and TCP, BCPC exhibit intermediate integration and resorption rates, depending of the amount of TCP. The higher the amount of TCP, the higher the integration and resorption rate. BCPC ceramics are available in different HA/TCP ratios, ranging from 1:3 to 2:3 (Triosite®, Biosel®, Eurocer®, Physio-6®) and different shapes and packaging. At the present time, these implants are proposed only for general bone void filling and present high porosity factors (45–70%).

Calcium Phosphate Cements

Developed several years ago, this new class of bone substitute (Norian®, Cementek®, Bone-Source® Biocement® . . .) consists of a paste-like structure that can harden in physiological media to form calcium phosphate compounds. During hardening, a hydrolysis or acido-basic chemical reaction takes place between the initial components, mixed together immediately before implantation. In their malleable phase, calcium phosphate cements can possibly be injected or, at the least, be fitted to complex bone defects. Once the hardening reaction is completed (6–72 hours), the porosity of the graft remains low and is composed of low-diameter pores (1–10 μm), thus allowing only limited bone ingrowth. In-vitro mechanical strength ranges from 10 to 50 MPa, but no reliable data is available concerning graft strength after in-vivo hardening in a biological environment. As in the case of pure HAC, resorption rate is also quite poor.

Xenografts

Commercially available xenografts are today from bovine or porcine origin. Cancellous, cortico-cancellous, or cortical specimen are treated by means of different chemical or physical processes to eliminate micro-organisms or proteins that could initiate antigenic of inflammatory reactions. These processes attempt to preserve alveolar structure as well as the mechanical properties of natural bone. Once the treatments are completed, xenografts are simply composed of extracellular bone matrix but contain no osteoinductive proteins. Manufacturing processes are ruled by strict regulations concerning animal selection and safety towards bovine spongiform encephalopathy. Nevertheless, xenografts suffer from considerable suspicion due to the risk of prion-disease transmission.

Others

Calcium Sulfate

Plaster of Paris is among the very first bone substitutes ever implanted in humans. This mineral, $CaSO_4$, 0.5 H_2O forms Gypsum ($Ca(SO_4)$, $2H_2O$) in the presence of water and hardens during transformation. The biocompatibility of Gypsum has been demonstrated in animals and humans. The resorption rate is high (4–8 weeks). It is commercially available in the form of dense pellets, either pure (Osteoset®) or in combination with antibiotics (Osteoset®) delivered during resorption. The initial mechanical strength ranges from 30 to 50 MPa but decreases significantly during the first weeks of implantation.

Coral

Madreporic corals have been experimented with as bone substitutes. Specimens are first chemically treated to eliminate every organic phase, then shaped into various types of implants. Animal experiments have evidenced satisfying biocompatibility and resorption governed over time by carbonic anhydrase dissolution. As in the case of HAC or TCPC, the resorption rate and mechanical strength can be varied by varying the porosity, in this case by using different species of corals. Composed of pure Aragonite (calcium carbonate, $CaCO_3$), a whole range of implants is proposed (Biocoral®) including different shapes, sizes, and porosity.

Clinical Applications

A careful analysis of existing clinical data from the literature allows the objective evaluation of the efficiency of bone substitutes in bone reconstruction. Most studies are retrospective and the results are assessed by means of radiological and clinical evaluation, without histological results.

Bone Tumors

Calcium phosphate ceramics have been used in the filling of bone defects consecutive to benign tumor resection. Inoue and Uchida have used porous blocks and granules of HAC, while Gouin or Yamamoto implanted BCPC. Three main indications were evaluated:

fibrous dysplasia, sometimes leading to voluminous tumoral cavities,

chondromas, more particularly with a metacarpal or phalangeal location (Figures 7.1a, b),

essential bone cysts of children or teenagers.

The clinical results were evaluated as favorable by the authors as bone healing was observed in

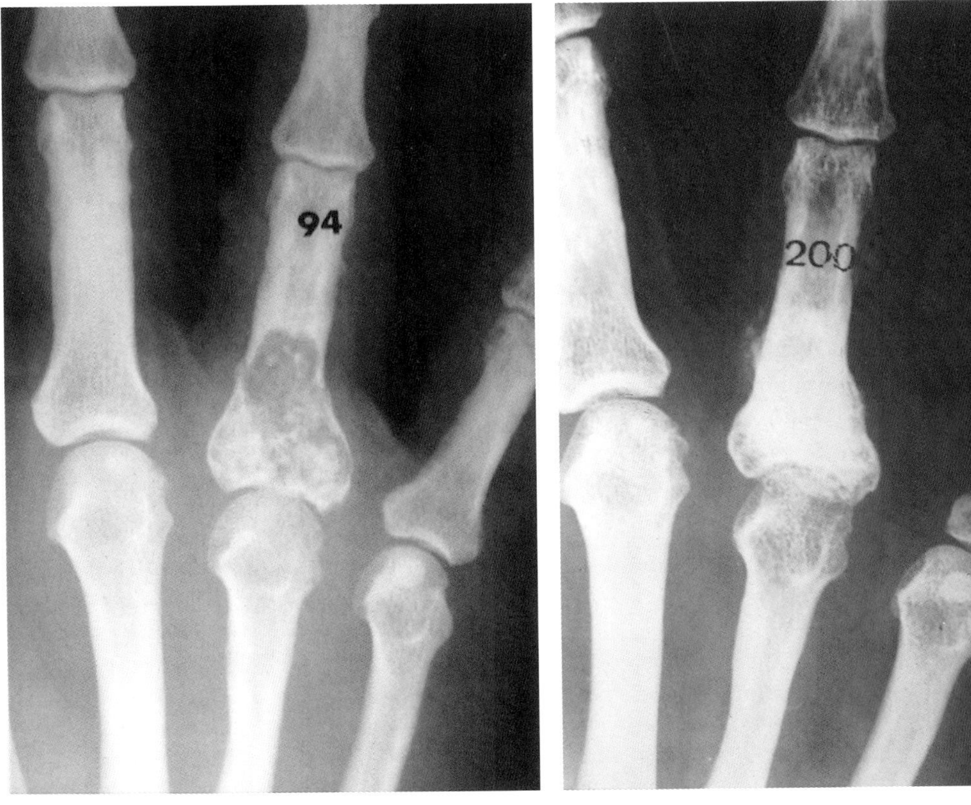

Figure 7.1. **a** Chondroma of the base of the first phalange. Curetage and filling by biphasic calcium phosphate ceramic. **b** Radiological result at six years.

most patients, with only few tumoral recurrences. The exact influence of calcium phosphate ceramics on the clinical results and bone healing rate and extent can nevertheless be discussed. As a matter of fact, few histological analyses have been performed and the necessity of grafting tumoral cavities is still controversial in the treatment of bone cysts and chondromas. Nevertheless, one can suppose that once the tumor has been carefully resected, the osteoconductive potential of calcium phosphate ceramics may annihilate recurrence of the tumor from remaining fragments.

Post-traumatic Reconstruction

Several authors have reported their experience with bone substitutes in the reconstruction of traumatic bone loss, pseudarthrosis, or bone resection consecutive to chronic osteitis. The implants consisted of ceramic granules or cubes or of calcium phosphate cements.

In a randomized study, Bucholz has compared the results of autografts and HAC or TCPC in the treatment of tibial plateau fracture, considering a series of 40 patients operated on with the same osteosynthesis device. There was no significant difference in bone healing rate between the grafts. Seven biopsies where performed during osteosynthesis removal at the ceramic implantation sites one year after initial surgery. Histological results have evidenced that 25–40% of the porous volume was occupied by healthy, newly formed bone.

De Peretti's conclusions about the use of coral in the filling of peri-articular traumatic bone defects (tibial plateau, thalamus, pilon) stabilized by means of osteosynthesis are qualified. Under the tibial plateau, coral implants were successfully incorporated into surrounding bone while maintaining adequate correction. But serous aseptic discharge has been observed at the two other locations with a possible allergic origin. In a methodologically more interesting study, De La Caffinière has compared the radiological healing of two different bone substitutes: natural coral and coraline hydroxyapatite, implanted in tibial or calcanean sites for the correction of depressed cancellous bone defects. Healing and resorption were significantly greater in the cancellous bone of the proximal tibial metaphysis as compared to calcanean situations. Furthermore, aseptic inflammatory reaction was frequently observed in the case of natural coral. Finally, limited signs of resorption have been observed with hydroxyapatite implants. These results confirm the influence of host bone quality on resorption and healing rate of bone substitutes.

A displaced distal radius fracture was the first proposed indication for injectable calcium phosphate cements. Jupiter [9] and Kopylov [10] concluded that, in this indication, calcium phosphate cements are biocompatible and provide satisfying fusion and stabilization but evidence limited signs of bone ingrowth and remodeling. These results have been confirmed by randomized studies comparing calcium phosphate cements versus external fixation [11] or plaster [12].

In a randomized multicentric study, Cornell has compared the fusion rate of diaphyseal pseudarthrosis stabilized by an appropriate osteosynthesis and treated with autograft from the iliac crest or a bone substitute composed of a mixture of hydroxyapatite and collagen. The results show no statistical difference between the patients who received an autograft and those treated with bone substitute. One can discuss in this case the necessity of grafting as a sole mechanically stable fixation might have provided comparable results as well. Finally, bone marrow from the iliac crest was systematically mixed with the bone substitute prior to implantation (Figures 7.2a–f).

In chronic osteitis and osteomyelitis, both infection and loss of bone substance have to be treated. This double problem could hopefully be solved by associating bone substitutes and such antibiotics as gentamicine, trobamicine, or vancomicine, active at a local level against *Staphylococcus aureus*. Sulo [13] has used calcium sulfate in these indications and concluded that this technique is efficient, locally providing a high concentration of antibiotics and osteoconduction by way of calcium sulfate.

Prosthetic Revision

Hip, and to a lesser extent knee, revisions often confront the surgeon with massive bone defects that have to be treated to increase the primary stability of the prosthesis and hopefully provide bone healing to prevent further loosening. If there is a real need for efficient bone grafts, the use of bone substitutes has long been doubted because of the poor quality of surrounding bone, thought to provide only limited bone-healing potential. Compacted allografts are thus often preferred to any other graft. Levaï [14] recently used a bovine xenograft in the filling of acetabular bone defects consecutive to hip prosthesis loosening. Bone-substitute blocks have been used to fill the cavities between the acetabulum and a polyethylene cup cemented into a prosthetic metallic ring. Radiological evaluation showed that 27 of the 30 patients did not evidence motion of the cup or ring after three years and that trabecular bone was present at the xenograft/ring interface in 24 patients.

Oonishi [15] reports the results of 40 acetabular revisions grafted with HAC granules, sometimes associated with a cortical allograft in segmentary defects. In this technique, HAC granules have been tightly packed to reconstruct part of the acetabulum, then stabilized by a thin layer of bone cement prior to fixation of the revision prosthesis in anatomical position. Radiological results are encouraging as only one re-operation was undertaken. The ceramic granules remain visible but no migration of any implant was noted. X-rays evidence in most cases that initial bone defects have disappeared at the HAC/bone interface. No histological results have confirmed the radiological findings. Moreover, the latest follow-up in this series is only three years and results should be confirmed at a significantly greater follow-up to confirm the reliability of this original technique.

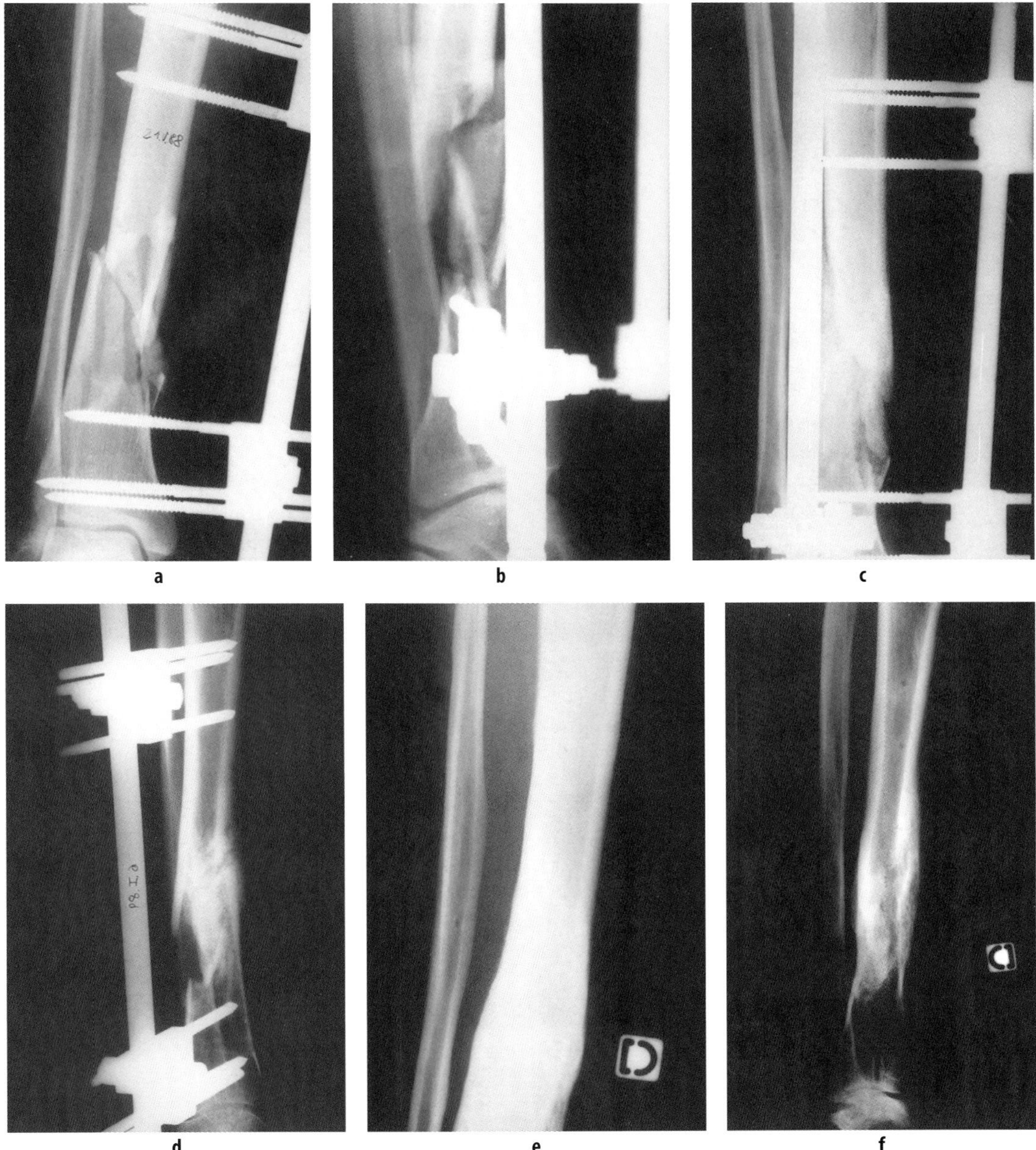

Figure 7.2. **a** Open tibial fracture (Gustilo type III A) fixed by external fixator. **b** idem. **c** after wornd healing filling the bone defect by collagen and calcium phosphate. **d** idem. **e** radiological result at seven years. **f** idem.

We found no comparative study concerning the results of allograft (or autografts) and bone substitutes in hip revision surgery. From our experience, reliable evaluation of the efficiency of bone substitutes in this indication is a major difficulty, as radiological assessment of bone healing is made difficult by the presence of a voluminous metallic device that may alter X-ray imaging (Figures 7.3a–c). Furthermore, the results of allografts evidence that long term follow-up is necessary to evaluate stability and integration of the graft.

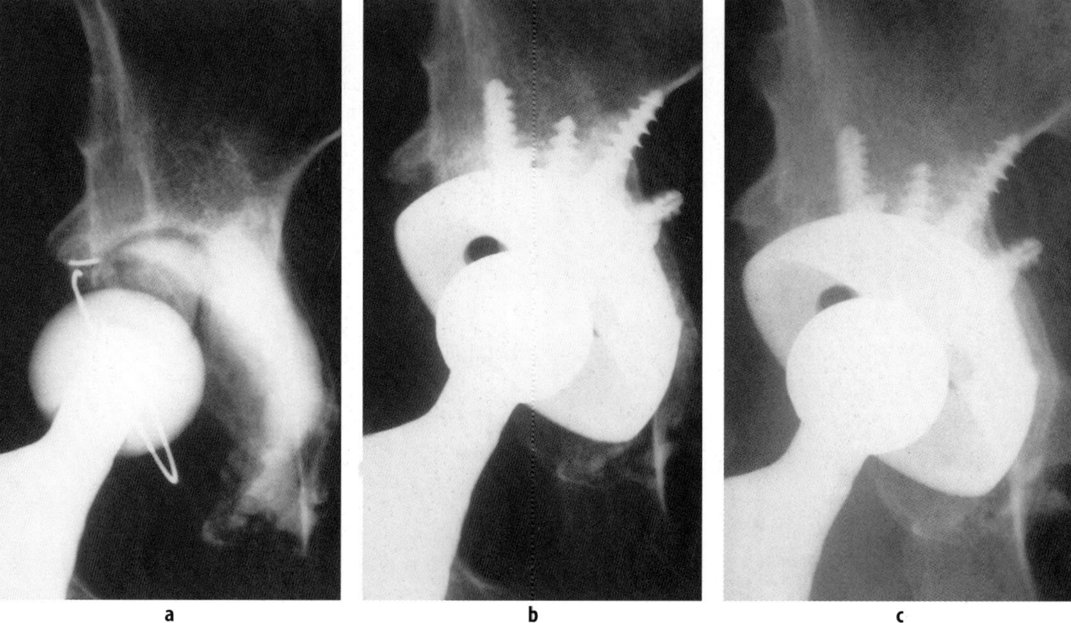

Figure 7.3. **a** acetabular loosening of cemented cup. **b** early radiological result: impacted and screwed component with filling cavitary defect by biphasic calcium phosphate. **c** radiological result at two years.

Spinal Surgery

In spinal surgery, interbody or postero-lateral fusions are the most common indication of bone graft, generally harvested on the anterior iliac crest. Bone substitutes were probably one of its earliest clinical applications. Passuti [16] reports that every 11 patients operated for severe scoliosis with posterior osteosynthesis device and grafted with BCPC blocks showed signs of stable fusion. Three biopsies evidenced bone ingrowth in the pores of the ceramic. Further to this publication, several authors confirmed these findings. In a comparative study, Le Huec [8] showed that in instrumented postero-lateral fusions, porous TCPC achieved comparable fusion rates and curve stabilization than allografts, based on radiological findings. In this study, bone ingrowth has been observed in every two biopsies and complete resorption of the TCPC implants was observed after two years over a four-years follow-up period. More recently, Ransford's randomized study concluded that no significant difference was observed, considering radiological signs of bone fusion, between iliac crest autograft and BCPC. Of 170 patients operated, six biopsies showed similar bone ingrowth and resorption of the implants.

In cervical interbody fusion, Kim [6] indicated that HAC are able to provide early satisfying radiological fusion rates, as compared to the results of autograft, considering a retrospective study including 70 patients. Calcium phosphate cement had been used for filling bone spinal corporeal ostrolysis [17,18].

Deviation Osteotomies

Femoral or tibial osteotomies probably deserve to be differentiated from other indications, as they represent a specific application for bone substitutes [19]. In such cases, the metaphyseal opening plane is filled with a mechanically resistant ceramic implant specifically designed to fit the anatomy of the implantation site (Figures 7.4a, b). An osteosynthesis device is

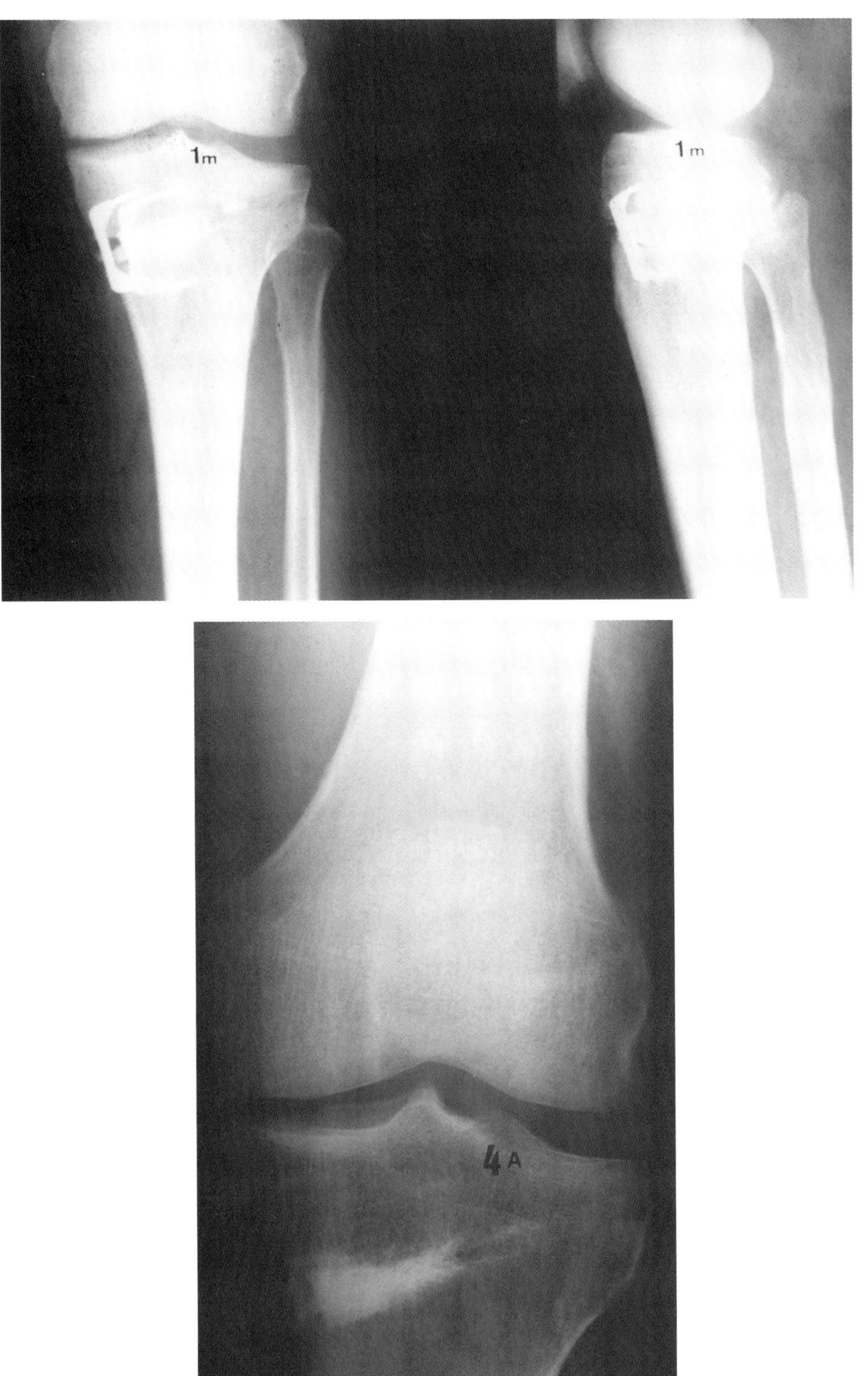

Figure 7.4. **a** Tibial valgisation by proximal metaphyseal opening. Fixation by staple and tricalcium phosphate wedge. Result at two months. **b** radiological result at four years.

necessary in every case for stabilization purposes.

Conclusion

Numerous experimental data have demonstrated the biological properties of calcium phosphate ceramics: biocompatibility, osteoconduction, possible resorption, and adapted mechanical strength. These properties are confirmed by a growing number of clinical studies. The success of the graft depends on two rules which should be followed: tight contact with the living host bone, and mechanical stability of the grafted bone to prevent mechanical overload and avoid micro-movements at the bone/implant interface. Several research programs are in progress to confer on bone substitutes an osteoinductive potential. Among the different hypotheses, association of calcium phosphate ceramics with bone morphogenetic proteins or osteogenic cells look promising in animal models.

Bibliography

Bucholz RW, Carlton A, Holmes R. Interporous hydroxyapatite as a bone graft substitute in tibial plateau fractures. Clin Orthop 1989;240:53–62.

Cornell CN, Lane JM, Chapman M, Merkow R, Seligson D, Henry S et al. Multicenter trial of collagraft as bone graft substitute. J Orthop Trauma 1991;5:1–8.

De La Caffiniere JY, Viehweger E, Worcel A. Evolution radiologique à long terme du corail implanté en os spongieux au membre inférieur corail madréporique versus hydroxyapatite de corail. Rev Chir Orthop 1998;84:501–7.

De Pereti F, Trojani C, Cambas PM, Loubiere R, Argenson C. Le corail comme soutien d'un enfoncement articulaire traumatique. Etude prospective au membre inférieur de 23 cas. Rev Chir Orthop 1996;82:234–40.

Frankenburg EP, Goldstein SA, Bauer TW, Harris SA, Poser RD. Biomechanical and histological evaluation of a calcium phosphate cement. J Bone Joint Surg 1998; 80A:1112–24.

Gouin F, Delecrin J, Passuti N, Touchais S, Pourier P, Bainvel JV. Comblement osseux par céramique phosphocalcique biphasée macroporeuse. A propos de 23 cas. Rev Chir Orthop 1995;81:59–65.

Guillemin G, Patat JL. The use of coral as a bone graft substitute. J Biomed Mater Research 1987;21:557–67.

Heymann D, Passuti N. Bone substitutes: new concepts. Eur J Orthop Surg Traumatol 1999;9:179–84.

Inoue O, Ibaraki K, Shimabukoro OH, Shingaki Y. Packing with high-porosity hydroxyapatite cubes alone for the treatment of simple bone cyst. Clin Orthop 1993;293:287–92.

Matuo MJ, Brunon J, Duthel R, Beauchesne P, Tellawi H. Résultats des xénogreffes intersomatiques cervicales par un substitut osseux. Rachis 1993;5:241–6.

Peltier LF, Jones RH. Treatment of unicameral bone cysts by curettage and packing with plaster of Paris pellets. J Bone Joint Surg 1978;60A:820–22.

Pouliquen JC, Noat M, Verneret C, Guillemin G, Patat JL. Le corail substitué à l'apport osseux dans l'arthrodèse vertébrale postérieure chez l'enfant. Rev Chir Orthop 1989;75:360–9.

Ransford A, Morley T, Edgar MA, Weeb P, Passuti N, Chopin D et al. Synthetic porous ceramic compared with autograft in scoliosis surgery. A prospective randomized study of 341 patients. J Bone Joint Surg 1998;80B:13–18.

Savolainen S, Usenius JP, Hernesniemi J. Iliac crest versus artifical bone grafts in 250 cervical fusions. Acta Neurochir 1994;129:54–7.

Serter HJ, Kortyna R, Kempw R. Anterior cervical discectomie with hydroxyapatite fusion. Neuro Surg 1989; 25:39–43.

Yamamoto T, Onga T, Marui T, Mizuno K. Use of hydroxyapatite to fill cavities after excision of benign bone tumors. Clinical results. J Bone Joint Surg 2000;82B:1117–20.

Yetkinler DN, Ladd AL, Poser RD, Constantz BR, Carter D. Biomechamical evaluation of fixation of intra-articular fractures of the distal radius in cadavers: kirschner wires compared with calcium phosphate bone cement. J Bone Joint Surg 1999;81A:391–9.

References

1. Daculsi G, Passuti N, Martin S, Deudon C, Legeros R, Raher S. Macroporous calcium phosphate ceramics for long bone surgery in humans and dogs. Clinical and histological study. J Biomed Mater Res 1990;24:379–96.
2. Frayssinet P, Trouillet JL, Rouquet N, Azimus E, Autefage A. Osteointegration of macroporous phosphate ceramics having a different chemical composition. Biomaterials 1993;14:423–9.
3. Le Huec JC, Clement D, Lesprit E, Faber J. The use of calcium phosphates, their biological properties. Eur J Orthop Surg 2000;10:223–9.
4. Schwartz C, Lecestre P, Frayssinet P, Liss P. Bone substitutes. Eur J Orthop Surg Traumatol 1999;9:161–5.
5. Les substituts osseux en 1999. Monographie du GESTO.
6. Kim P, Wakai S, Matsuo S, Moriyama T, Takaaki K. Bisegmental cervical interbody fusion using hydroxyapatite implants: surgical result and long-term observation in 70 cases. J Neusurg 1998;88:21–7.
7. Uchida A, Araki N, Shirto Y. The use of calcium hydroxyapatite ceramic in bone tumor surgery. J Bone Joint Surg 1990;72B:298–302.
8. Le Huec JC, Lesprit E, Delavigne C, Clament D, Chauveaux D, Le Rebeller A. Tri-calcium phosphate ceramics and allo-grafts as bone substitutes for spinal fusion in idiopathic scoliosis: comparative clinical results at four years. Acta Orthop Belgica 1997;63:202–11.
9. Jupiter OB, Winters S, Sigman S. Repair of five distal radius fractures with an investigational concellous bone cement: a preliminary report. J Orthop Trauma 1997; 11:110–16.

10. Kopylov P, Jonsson K, Thorngren KG, Aspenberg P. Injectable calcium phosphate in the treatment of distal radial fractures. J Hand Surg 1996;21B:768–71.
11. Kopylov P, Runnqvist K, Jonsson K, Aspenberg P. Norian SRS versus external fixation in redisplaced distal radial fractures: a randomized study in 40 patients. Acta Orthop Scand 1990;70:1–5.
12. Sanchez-Sotelo J, Munuera L, Madero R. Treatment of fractures of the distal radius with a remodelable bone cement. A prospective randomized study using Norian SRS. J Bone Joint Surg 2000;82B:856–63.
13. Sulo I. Granules de plâtre à la gentalline dans le traitement de l'infection osseuse. Rev Chir Orthop 1993;79: 299–305.
14. Levai JB, Boisgard S. Acetabular reconstruction in total hip revision using a bone graft substitute. Clin Orthop 1996;330:108–14.
15. Oonishi H, IwakiI Y, Kin N. Hydroxyapatite in revision of total hip replacements with massive acetabular defect: 4 to 10 year clinical results. J Bone Joint Surg 1997;79B:87–92.
16. Passuti N, Daculsi G, Rogez JM, Martin S, Bainvel JV. Macroporous calcium phosphate ceramics performance in human spine fusion. Clin Orthop 1989;248:169–76.
17. Pasquier G, Flautre B, Blary MC, Anselme K, Hardouin P. Injectable percutaneous bone biomaterial: an experimental study in a rabbit model. J Mater Sci Mater Med 1996;7:683–90.
18. Hardouin P, Lemaitre J. New injectable composites for bone cement. Semin Musculoskeletal Rad 1997;1: 319–23.
19. Bonnevialle P, Abid A, Clement D, Verhaeghe L, Cariven P, Mansat M. Ostéotomie tibiale de valgisation par addition médiale d'un coin de phosphate tricalcique. Technique dite mini invasive. Rev Chir Orthop 2001; Submitted for publication.

8 Orthopedic Bone Cements

J. Lu

The word "cement" comes from the domain of architecture construction. It consists of a system of powder/liquid materials which, when mixed to a paste, set to a hard mass. "Bone cement" uses this system for application in medicine, for example: filling of bone defects and fixation of surgical prostheses, etc.

The history of the application of bone cement dates back more than 100 years. In 1890, Dr. Gluck described the use of the ivory ball-and-socket joints which were especially useful in the treatment of diseases of the hip joint. These joints were stabilized in the bone with a cement composed of colophony, pumice powder, and plaster. He stated that the cement remained walled off in the marrow cavity in the same way as a bullet, the marrow cavity appearing to have almost unlimited tolerance to aseptic implantation [1]. In 1951, Dr. Haboush used self-curing acrylic dental cement to secure a total hip replacement [2]. Also at this time similar resins were being used to repair defects in the skull after brain surgery. Polymethylmethacrylate (PMMA) cement was used primarily in dentistry to fabricate partial dentures, orthodontic retainers, artificial teeth, denture repair resins, and an all-acrylic dental restorative. Dr. Charnely had used a cold-cured acrylic as a possible luting cement to retain the femoral shaft in total hip arthroplasty [3].

From the 1950s to the 1970s numerous studies and long-term clinical trials exposed the biological disadvantages of PMMA cement: (1) the release of monomer toxicity; (2) the high temperature of the cement polymerization; (3) osteonecrosis mediated by inflammatory reaction; (4) osteolysis caused by wear debris formation, or (5) impairment of blood circulation in the bone caused by reaming, then the cement plug [4]. Moreover, this cement is neither biodegradable nor colonizable by bone tissue. Therefore, surgeons sought to ameliorate the PMMA cement looking for new material to replace it. Brown and Chow [5] were the first to develop and patent a calcium orthophosphate cement. Different formulations of the calcium phosphate cement have since been developed by various research groups [5–10]. In vitro and in vivo studies have shown that the calcium phosphate cement (CPC) had excellent biocompatibility, a good bioresorption, and less exothermic, but weaker mechanical properties than PMMA cement.

This paper provides a general regulatory background, chemical composition information, mechanical and biological properties as well as a discussion of the mechanisms of the risks and failures of bone cements. We present principally two bone cements: polymethylmethacrylate cement (PMMA) and calcium phosphate cement (CPC).

Polymethylmethacrylate Cement (PMMA)

Chemical Composition and Polymerization of PMMA

PMMA bone cement, consisting of preformed PMMA beads mixed with methylmethacrylate (MMA) monomers, has remained largely unchanged over the years. PMMA cements include: (1) the weight ratio of powder to the liquid monomer; (2) the use of PMMA or copolymers thereof; (3) the use of benzoyl peroxide as initiator in the powder and MMA; (4)

Table 8.1. General chemical compositions of various commercially available bone cements

	Brand 1	Brand 2	Brand 3	Brand 4
Powder components	***40 g***	***40 g***	***40 g***	***40 g***
PMMA (polymer)	88.85% (w/w)		15.00% (w/w)	89.25% (w/w)
Polystyrene/MMA copolymer			75.00% (w/w)	
MMA/PMMA copolymer		83.55% (w/w)		
Benzoyl peroxide (initiator)	2.00% (w/w)	0.5–1.6% (w/w)		0.75% (w/w)
Sulfate barium (radio-opacifier)	9.10% (w/w)		10.00% (w/w)	10.0% (w/w)
Zirconium dioxide (radio-opacifier)		15.00% (w/w)		
Liquid components	***18.37 g***	***20 ml***	***20 ml***	***20 ml***
MMA (monomer)	98.215% (w/w)	99.26% (w/w)	97.40% (v/v)	97.25% (v/v)
N,N-dimethyl-*p*-toluidine	0.816% (w/w)	1.96% (w/w)	2.62% (v/v)	2.75% (v/v)
(accelerator)	0.002% (w/w)		75 ± 15 ppm	75 ± 10 ppm
Hydroquinone (stabilizer)	15–20 ppm			
Other monomeric additives				
Ethyl alcohol	0.945% (w/w)			
Ascorbic acid	0.022% (w/w)			
Chlorophyll (color additive)		0.002% (w/w)		

the use of MMA as the monomer in the liquid component; and (5) the use of a radio-opaque filler (e.g., barium sulfate or zirconium dioxide). Differences include: (1) the amount of the initiator (benzoyl peroxide) in the powder; (2) the amount of accelerator (*N.N*-dimethyl-*p*-toluidine) in the liquid component; (3) the amount and type of stabilizers (e.g., hydroquinone) in the liquid component; and (4) the addition of chlorophyll used to color the cement green. The chemical composition of the commercially available bone cements is similar, with the minor differences described in Table 8.1.

The polymerizing process of the cement occurs as a result of the reaction between the initiator in the polymer powder and the accelerator in the monomer. These act together to form a complex which produces benzoate and amine radicals. These two radicals then initiate polymerization of the monomer [11]. A radio-opacifier, added to the powder component, enables the surgeon to view the cement in vivo. This process transforms the initial thick liquid to a soft deformable material and finally to a rapidly hardening cement with an associated increase in temperature due to the exothermic polymerization, which can exceed 80 °C. The cement sets through the polymerization of the monomer, which concurrently dissolves and softens the polymer particles. The set mass consists of the polymer matrix uniting the undissolved but swollen original polymer granules. The degree of polymerization is affected by the following: (1) the amount of accelerator and initiator in the powder and liquid monomer; (2) wetting caused by the monomer mixing with the powder; (3) the type of mixing used, and (4) the pro-chilling of the monomer; and the presence of oxygen.

Table 8.2. Physical and mechanical properties

	ISO-5833
Dough time	5 ± 1.5 min
Setting time	3 to 15 min
Exothermic temperature	<90 °C
Compression strength	70 MPa
Tensile strength	50 MPa
Tensile modulus	1.8 GPa

Physical and Mechanical Properties of PMMA

The physical and mechanical characteristics of acrylic bone cement were determined by the ISO 5833-1992 standard [12] (Table 8.2). The liquid and powder mixing procedure should be influenced by various factors in order to modify the properties. These factors include the amount of the ingredient, the temperature and humidity of the mixing environment, the type of sterilization used, and the type of mixing (hand, centrifugation, or vacuum mixing) used to prepare the cement, as well as the surgical installation used in the mixing process.

Physical Properties

Cement viscosity is increased by the addition of fibers, greater molecular weight of the polymer, solubility of the polymer in the monomer, variation in the powder composition or bead size distribution and the temperature of the cement components. Pre-chilling the cement components increases the setting time and reduces the viscosity of the cement, as compared to cement components which are stored at room temperature prior to mixing. In contrast, mixing of bone cement under vacuum generally decreases the setting time. At the time of mixing, the components are usually hand mixed in a bowl. However, with the use of vacuum mixing or centrifugation after mixing, the cement porosity and pore size can be reduced to improve the mechanical properties of the cured cement [13–16]. Greater monomer evaporation may occur if the applied vacuum is too great during vacuum mixing.

Poor monomer wetting in the powder can occur if: (1) the powder is insoluble or only partially soluble in the liquid MMA monomer; (2) an inadequate amount of MMA monomer is mixed into the powder; or (3) the free volume is lowered due to tighter packing of powder. On the other hand, styrene copolymers may have better wetting properties due to a higher free volume which allows for faster monomer diffusion rates [17].

High temperatures of the polymerization process can cause evaporation of the monomer leading to a microporosity in the curing cement. There are a number of factors that affect the maximum exothermic temperature. The following may contribute to a higher cement polymerization temperature: (1) a large cement mass; (2) a cemented device with a low conduction heat; (3) a cemented device that is not cooled before implantation; (4) a lack of irrigation at the implant site; (5) a greater amount of monomer mixed into the powder; or (6) increased levels of accelerators or initiators which may form radicals initiating rapid polymerization [18].

Microporosity in the bulk cement may result from the following: (1) monomer evaporation during the exothermic reaction and/or leaching of the unreacted monomer [19]; (2) flow and wetting during mixing with the beads leading to air entrapment; (3) CO_2 formation due to a benzoyl peroxide reaction with the accelerator; (4) turbulent cement flow during the insertion of the implant into the cement; or (5) the mixing method used to assemble the bone cement components.

Mechanical Properties

The implant–cement–bone interfacial strengths are also considered risk factors [20]. Implant–cement interfacial loosening may result from: (1) cement fracture or poor implant–cement bonding due to foreign matter; (2) inadequate coverage at the implant–cement interface [21]; (3) amount of mechanical interlocking; or (4) a lack of chemical bonding at this interface. More specifically, inadequate cement coverage at the interface may be caused by: (1) shrinkage of the cement due to polymerization; (2) poor mechanical interlocking strength between the implant and cured bone cement [22]; or (3) an increase in the bone cement viscosity over time leading to poor contact between the cement and the implant.

Cement–bone loosening may result from: (1) cement fracture; (2) formation of gaps at the interface; or (3) tissue failure. More specifically, the gaps may form due to: (1) bone resorption; (2) foreign material at the cement–bone interface such as bone particles or blood [23]; (3) shrinkage of the cement after implantation; (4) low cement pressurization during implantation [24]; or (5) movement of the implant before hardening of the cement. For gaps caused by shrinkage, the shrinkage is greater as the porosity decreases.

A fatigue fracture of the cement is a result of a cement stress which exceeds the fatigue limit of the bone cement. High cement stress may be due to an applied stress or a residual stress, cement modulus, implant loosening, or poor cement bonding with the implant or with the bone. Other causes of cement fracture include: fibrous membrane formation between bone and cement, improper cement mantle thickness (a layer too thin or too thick), lamination of the cement due to the presence of blood or other body fluids, or poor canal preparation or areas of increased stress (e.g., the presence of a pore or a sharp corner of an implant). High stress applied to the bone cement may be caused by: (1) increased patient weight or activity; (2) lack of constraint; (3) adverse implant size and orientation; or (4) inadequate bone cement mantle. The latter three are related to the quality of the tissue and the applied surgical technique. A weakness may also be caused by bone resorption or by disease.

Cement mechanical properties are affected by the level of stress at a specific site. This is influenced by: (1) irregular trabecular bone; (2) porous implant coatings; (3) sharp edges on the implant; or (4) a defect in the bulk cement such as a pore or additives. More specifically, localized stress caused by porosity and inclusions (e.g., additive agglomeration, radio-opacifiers and antibiotics) are perhaps the greatest factor affecting cement fatigue properties. The addition of agglomerates (e.g., radio-opacifiers) may also play a similar and significant role in cement fracture.

Biological Properties of PMMA

All biomaterials must be biocompatible. PMMA cements are considered biocompatible despite the toxic potential of the bone cement monomer and the heat generated during the exothermic polymerization.

Cellular Reactions

Initially, the major problems of PMMA bone cement are related to the temperature increase during the polymerization and the release of residual monomer after polymerization. PMMA is non-toxic, but the residual monomer (MMA) can cause an irreversible deterioration of the cells [25,26]. After fifteen minutes of polymerization, there is a residual monomer of approximately 3–5%. This percentage may decrease by up to 1–2% with time [25]. Haas et al. [19] measured the residual MMA content to be 3.3% after 1 h, 2.7% after 24 h, and 2.4% after 215 days under storage in an ambient air environment. According to Schoenfeld et al. [27] the toxicity of the monomer disappears after 4 hours. In our study of cement fragments which were harvested at the time of prosthetic revisions 48 to 78 months after implantation, there was no apparent toxic effect of the cement on the fibroblasts (L929) and human osteoblasts [28]. However, there may be variable reactions to PMMA depending of the cells involved.

PMMA is not cytotoxic with regard to human fibroblasts in vitro. However, it can stimulate proliferation and protein synthetic activity [29]. The increased proliferation of fibroblasts in response to PMMA exposure can be associated with an increased production of collagen and chemical mediators at the bone–cement interface [30,31]. Chemical mediators, such as prostaglandin E2 (PGE2) and other cytokines (interleukine-1), have been shown to mediate inflammation, as well as induce cell division and differentiation [32,33]. Fibroblasts have previously been implicated in the inflammatory response, therefore, it is possible that they are responsible for the recruitment of inflammatory cells at the bone–cement interface via release of chemical mediators such as PGE2.

Monocytes and macrophages are significant agents of the inflammatory reaction. The principal function of the tissue macrophage is phagocytosis and the secretion of cytokines and growth promoters. PMMA particles induce macrophages to secrete protein and to express mRNA of the proinflammatory cytokines, interleukin-1β (IL-1β), interleukin-6 (IL-6), tumor necrosis factor alpha (TNF-α), PGE2, proteinases, collagenases, and oxygen metabolites. Other factors expressed include chemokines such as macrophage-activating and chemotactic protein 1 (MCP-1) as well as macrophage inflammatory protein (MIP) which may be linked to osteolysis [34–43]. Horowitz et al. described a dose-dependent release of arachidonic acid metabolites by murine macrophages induced by PMMA particles [44].

The osteoclast is a multinucleated cell which carries out the unique and highly specialized function of lacunar bone resorption. The osteoclast belongs to the mononuclear phagocyte system which consists of various cell types including monocytes, macrophages, Kupffer cells, and microglia. A common feature of all these cells is their avid and efficient ability to carry out phagocytosis. Most studies have focused on the effect of biomaterial particle phagocytosis on the function of these cells and the observation that specific types of particle enhance the release of mediators thus stimulating osteoclastic bone resorption [45–47]. In addition, it has been shown that macrophages, after having phagocytosed these particulates, are capable of osteoclast differentiation [48]. Wang et al. [49] found that osteoclasts having phagocytosed PMMA wear particles exhibit normal lacunar bone resorption. Also, the phagocytosis of PMMA particles does not appear to compromise the response of osteoclasts to calcitonin or to the ability to carry out lacunar resorption, observation that remains controversial.

PMMA particles can inhibit osteoblast activities causing a decrease in cellular proliferation and collagen synthesis.

Local Tissue Reactions

PMMA bone cement is generally well tolerated and bony tissue generally flourishes on its surface [29,43]. However, there is evidence of the inflammatory potential of bone cement [50]. The tissue reaction around bone cement has several phases. Initially, there is necrosis of the bone tissue and marrow to a depth of 5 mm related to the surgical wound and polymerization. Next, there is a phase of cicatrization lasting up to six months, followed by tissue granulation which develops over a period of two years. This tissue granulation is a characteristic of the chronic inflammation. The cement is then surrounded by a layer of fibrous tissue [51–54] and occasionally by a varying thickness of fibrocartilage [55,56]. Albrektsson [57] reported a 59% reduction in the in growth of cortical bone into titanium bone chambers one month after cement application. Morberg et al. [59,59] reported also decreased bone formation around cemented tibias, being 21% and 31% lower than the non-cemented contralateral tibias after 3 to 11 and 32 to 55 weeks, respectively.

Osteonecrosis

Cell necrosis may occur because of the following: (1) monomer toxicity; (2) the high temperature of cement polymerization; (3) pressure necrosis; (4) osteolysis caused by wear debris generation; or (5) the impairment of blood circulation in the bone caused by reaming and by the presence of cement [4]. Bone cement has been shown to decrease bone metabolism possibly causing a lower revascularisation [60,61].

The production of heat at the bone–cement interface during cement polymerization in vitro is between 60 and 90 °C [62–64] and in vivo between 40 and 50 °C [65,66], both depending on the thickness of the cement. The effect of this heat generation on bone was studied by Lundskog [67] who concluded that the exothermic polymerization did not add to the surgical trauma and had no influence on bone generation. Lee et al. [68] found that the leakage of monomer was very low after the curing. Likewise, Sund and Rosensuist [69] stated "the effect of polymerization heat and monomer toxicity are probably much less important than the trauma effected by blocking of the normal medullar blood supply". Rhinelander et al. [66], who noted a maximal temperature of 55 °C with the placement of thermometers at the bone–cement interface, concluded that thermal necrosis from cement polymerization is not a significant factor. Furthermore, after direct contact with acrylic cement, the delicate trabeculae of cancellous bone in the metaphysis contained healthy-appearing osteocytes after six weeks.

Osteolysis

Bone surrounding an implant may undergo osteolysis leading to loosening and decrease in cancellous bone strength. This may result in a weakening of the cement fixation or the formation of a gap between the cement and bone. Acrylic cement fragments are engulfed by eosinophilic histiocytes which stimulate enzymatic release leading to bone resorption [70,71]. In addition, bone cement particles could accelerate foreign-body deterioration of articulating polyethylene inserts [72,73]. The initial event can be either disintegration of bone cement or deterioration of the articulating surface. Phagocytosis and development of foreign-body granulomas lead to osteolysis of the anchoring bone; thus, disintegration or deterioration areenhanced, accelerating the progress of osteolysis [51].

Formation of Fibrous Membrane

The formation of fibrous tissue is caused by the toxicity of the monomer release as well as the heat production of the polymerization causing chronic inflammation and eventual osteonecrosis and osteolysis. It is a significant factor which induces micromovement and the loosening of surgical implants. The thickness of the fibrous membrane around PMMA cement was 40 μm and 60~70 μm after one and four weeks respectively in the tibiae diaphysis [53,74]. In human femur, the thickness was measured at 20~300 μm at 11 months to 7 years [55] and at 3~5 μm long-term [54].

Implant Loosening

Revision of cemented orthopedic prosthesis may be necessary when pain occurs due to either movement of the prosthesis, bone fracture, bone cement fracture, or prosthesis fracture. More specifically, these complications may result from prosthesis–cement, or cement–bone interfacial loosening or micromotion due to cement fracture or cement creep. Loosening of the prosthesis and fracture of the cement may lead to increased wear and bone cement particle formation. Those particles, less than approximately 5 μm in size, are phagocytosed by macrophages which become activated and directly or indirectly cause bone remodeling and osteolysis [14,75–78]. However, PMMA particles ingested by macrophages cannot be degraded by lysosomal enzymes [45]. The final result is cell death leading to tissue necrosis and chronic inflammation [79]. For the femoral stem, lower-viscosity bone cement had a revision rate 2.5 times greater when compared to the use of higher-viscosity cements. Additionally, a lower-modulus cement had a revision rate that was 8.7 times greater than the higher-viscosity cements [80]. In general, revisions are required between 3.6 and 22.8 years following a total hip prosthesis. The most frequent periods of revision are either during the first three years or after eight years postoperatively [81]. Aseptic loosening of the prosthesis is the principal cause of revision, implicated in 73~74% of cases [82,83]. Subcritical debonding associated with the mechanisms of cyclic fatigue crack growth are particularly relevant considering that these systems will experience over 1,000,000 physiological loading cycles per year, and are expected to survive a minimum of 10–15 years. In these terms, it is critical to understand the progressive debonding of the prosthesis–PMMA cement interface [84].

Secondary Reactions

Systemic and Cardiovascular Reactions

Methylmethacrylate (MMA) is very volatile and is rapidly cleared from body through the lungs resulting in a local concentration that remains very low [85,86]. MMA monomers escaping from the implanted polymerizing cement have been associated with a decrease in both systolic blood pressure and arterial oxygen tension [87] and possibly cardiac arrest [88]. However, many studies have not confirmed this direct correlation between concentration of MMA and blood pressure, heart depression, or vasodilatation [86,89,90]. Circulatory disturbance during hip implantation may be primarily due to either the "implantation syndrome" or to the blockage of pulmonary circulation by fat, bone marrow, and entrapped air rather than MMA monomer. Release of MMA could cause a drop in the partial pressure of arterial oxygen leading to an increased heart rate [91,92]. The possible metabolic pathway of MMA monomer is that the residual monomer is converted to methylacrylic acid rather than methylester. The methylacrylic acid, as a coenzyme A ester, is a normal intermediate in the catabolism of valine, and the existence of an enzymic system would permit methylacrylic acid to enter a normal pathway, leading to carbon dioxide formation. Over 80% of an administered dose of MMA is expired as carbon dioxide within 5–6 hours [90].

Sensitizing

While MMA is considered to be relatively immunologically inert, it can induce phagocytosis, the activation of macrophages and giant cells, as well as the migration of inflammatory mononuclear cells [13,14,52,74,78]. Jensen et al. [52,74] showed that MMA is extremely active in a guinea-pig maximization test. The hospital personnel who repeatedly handle coring acrylic bone cement are potentially at risk of developing a delayed sensitivity [93]. Bengston et al. [94] reported that patients having received a cemented hip prosthesis had increased levels of anaphylatoxines which can contribute to circulatory and respiratory disturbances. In contrast, Kanerva et al. [95] have found allergies to MMA to be rare in a study of patients between 1974 and 1992 (four patients: one orthodontist, three dental technicians).

Improvement of PMMA

The objective of the development of PMMA bone cement is to improve the biocompatibility, to diminish the temperature of polymerization, to eliminate the generation of wear debris and fatigue fractures, as well as to increase the elastic modulus. Therefore, efforts to improve PMMA bone cement have proceeded in two main directions: (1) to change the composition and (2) to improve preparative techniques.

Tertiary aromatic amines are used as accelerators for the benzoyl peroxide (BPO)-initiated MMA polymerization. A complex series of reactions occurs between BPO and the amine, and free radicals are produced that initiate the polymerization process [96,97]. Several types of amine accelerators, such as dimethyl aniline and its derivatives, have been used in the polymerization of MMA by the amines/BPO initiator system. Their relative efficiency as accelerators and their activating effects on the rate of polymerization have been reported [96,97]. Several workers have studied bone cement properties using a number of N,N-dimethyl-p-toluidine derivatives, such as 4-dimethylaminobenzyl methacrylate, and 4-dimethylamino phenethyl alcohol [96–99]. Bone cement products containing residual monomers and amines have been reported in preparation where the amines/BPO molar ratio is outside the equimolar range [11]. Other studies showed that MMA polymerization in the presence of tir-*n*-butylborane used as the cure initiator does not occur too rapidly, and the high temperature during polymerization is lower than that of conventional bone cement. The application time is short enough for clinical use, namely, within 10 min. As for the physical properties, it has a 3% lower elastic modulus and greater ductility than conventional cement [100].

Several workers have added particles or fibers to PMMA bone cement to improve the biocompatibility and the mechanical properties. Fiber-reinforced bone cement possesses significantly greater stiffness and displays poor intrusion characteristics [101–103]. A number of attempts have been made at filling a PMMA matrix with hydroxyapatite and tricalcium phosphate particles, and with bioactive glass [104–106]. The short-term results obtained are encouraging and suggest that the chemical nature of the bone/bioactive materials interface is very important relative to osteoconductivity [107]. For instance, PMMA bone cement can be used only to effectuate mechanical fixation for prostheses or for physically filling bone defects. However, it does not exhibit the functions of osteointegration, biofixation, or bioresorption.

Calcium Phosphate Cement (CPC)

Chemical Compositions and Crystallization of CPC

Calcium phosphate cements can be handled in paste form and set in a wet medium after precipitation of calcium phosphate crystals in the implantation site. Depending on the products involved in the chemical reaction leading to the precipitation of calcium phosphate, different phases can be obtained with different mechanical properties, setting times, and injectability. Numerous components can enter the chemical reaction leading to calcium phosphate precipitation. More than 100 different calcium orthophosphate cements were used to determine the compressive strength and the diametric tensile strength after storage. The setting was carried out on more than 15 formulations. These cements can be divided into four classes: dical-

Table 8.3. Calcium and phosphate compounds

Name	Abbreviation	Formula	Ca/P	Solubility	Acidity	Stability
Monocalcium phosphate monohydrate	MCPM	$Ca(H_2PO_4)_2{\cdot}H_2O$	0.5			
Dicalcium phosphate anhydrous	DCPA	$CaHPO_4$	1.0	+++++	+++++	+
Dicalcium phosphate dihydrate	DCPD	$CaHPO_4{\cdot}2H_2O$	1.0	+++++	+++++	+
Octacalcium phosphate	OCP	$Ca_4H(PO_4)_3$	1.33	++++	++++	++
Amorphous calcium phosphate	ACP	$Ca_9H(PO_4)_6$	1.3–1.5	+++	+++	+++
Tricalcium phosphate	TCP	$Ca_3(PO_4)_2$	1.5	++	++	++++
Hydroxyapatite	HAP	$Ca_{10}(PO_4)_6(OH)_2$	1.67	+	+	+++++
Tetracalcium phosphate	TTCP	$Ca_4(PO_4)_2O$	2.0			

Table 8.4. Properties of CPC

Authors	Powders	Liquid	Setting	Strength	Resorption
Brown & Chow [5]	DCPD/DCP/TTCP/HA	H_2O	30–60 min	10 MPA	minimally
Lemaitre et al. [6]	α-TCP/MCPM	H_3PO_4	10 min	25–35 MPa	completely
Bone Source [110]	TTCP/DCPD	H_2O	10–15 min	36 MPa	minimally
Norian [110]	MCPM/α-TCP/$CaCO_3$	$CaHPO_4$	10 min	55 MPa	completely
Fernandez et al. [10]	DCPA/α-TCP	H_2O		30–40 MPa	yes
Kurashina et al. [8]	α-TCP/DCPD/TTCP	sodium succinate sodium chondroitin sulfate			yes
Liu et al. [9]	TTCP/DCPD/DCPA	H_2O	11 min	70 MPa	yes
Ginebra et al. [7]	α-TCP/β-TCP	Na_2HPO_4	5–12 min	40 MPa	yes

cium phosphate dihydrate, calcium and magnesium phosphates, octocalcium phosphate, and non-stoichiometric apatite cements [108,109]. The calcium and phosphate compounds in Table 8.3 are often used to make CPC. Moreover, adjuvants such as chitosan, lactic acid, and glycerol are added to improve the injectability of the cement, and accelerators such as Na_2HPO_4, sodium phosphate, sodium succinate, and sodium chondroitin sulfate to accelerate its setting time.

The hardening process of CPC is complex and involves the dissolution of solid particles in the liquid, precipitation of HAP from the solution, and the reaction and diffusion on the particle surface. Under ideal conditions, continuing dissolution of the reactions supplies calcium and phosphate ions to the solution, while HAP formation depletes these ions. This process drives the solution composition to an invariant point, which is the intersection of the solubility curves for these two reactants. The pH is about 7.8, but this process is affected by many parameters, such as the component and the particle sizes of the solid phase, presence of HAP seed and properties, aqueous liquid, etc.

Physical and Mechanical Properties of CPC

All CPC are formulated as solid and liquid components that, when mixed in predetermined proportions, react to form HAP. This final reactant is important because it determines whether the end product will be nonresorbable, minimally resorbable, or completely resorbable. The powder component usually consists of two or more calcium phosphate compounds, whereas the liquid component is either water, saline, or sodium phosphate (Table 8.4). Some of the calcium and phosphate compounds involved in bone and mineral formation, or as implants, are listed in Table 8.3. These materials have been well characterized chemically and have not been reported to cause foreign-body reactions or other forms of chronic inflammatory response [110].

The physicochemical reaction that occurs during mixing solid and liquid compounds of CPC is complex. Briefly, when different calcium phosphate salts are mixed in an aqueous environment, dissolution of the solid compounds, then a precipitation or a nucleation, and finally a phase transformation occur. The process leading to final phase transformation of the different forms of calcium phosphate salts is dependent on their solubility, product constant, and pH. It is important to realize that water is not a reactant in the setting reaction of the cement, but it allows dissolution of the solids and precipitation of the products. The nature of the apatite makes the final form biocompatible and promotes a chemical bond to the host bone.

It is possible to transform the cement into an injectable paste by addition of adjuvants without fundamentally modifying the chemical reactions occurring during setting and hardening of the CPC. Leroux et al. [111] found that glycerol greatly improved the injectability and increased the setting time, but decreased the mechanical properties. Lactic acid reduced the setting time, increased the material toughness, but limited the dissolution rate. After injection, the cement did not present any disintegration. The effects of lactic acid were correlated with the formation of calcium complexes. Its association with sodium glycerophosphate is particularly important. Chitosan alone improved the injectability, increased the setting time, and limited the evolution of the cement by maintaining the CPC phase.

CPC have an inherent compressive strength at the final set that can govern their utility. Varying the crystallinity of the HAP or the particle size of materials used in the solid phase may alter the compressive strength. Because CPC are relatively insoluble at neutral and alkaline pHs, their porosity is related to the ratios of powder to liquid used in the starting mixture. Obviously, a cement with a high porosity would be expected to be of low compressive strength. Cements with a high compressive strength would be expected to find utility where they would stabilize non-displaced bone fractures, or repair large bone defects, or fix surgical prostheses. Cements with a low compressive strength would limit their utility and only fill small bone defects.

Biological Properties of CPC

Calcium and phosphate compounds of CPC have attracted considerable attention because they set like a dental cement and form hydroxyapatite as end product, which is the major mineral components of teeth and bone. A number of studies in vitro and in vivo have shown that CPC had no toxicity, negative mutagenicity, and no potential carcinogenicity [112,113], no or slight inflammatory reactions, good osteoconductivity and bioresorption [114] as well as light exothermic temperature (<40 °C) during CPC hardening. However, CPC particles could be harmful for osteoblasts with a decrease of viability, proliferation and production of extracellular matrix, especially when their size was smaller than 10 μm. A dose effect was present, a ratio of 50 CPC particles per osteoblast could be considered as the maximum an osteoblast supported. The acidification of the medium due to the dissolution of CPC could not be responsible for the decrease of osteoblast functions because the control of the pH value of the medium showed no change. It was, then, direct interaction of osteoblasts with particles that was involved in the decrease of osteoblast functions [115]. Some adjuvants of CPC can induce acidification and release some elements to modify the biological properties. We have cultivated osteoblasts on the CPC surface; cell proliferation increased after the first 7 days followed by a decrease and finally an absence of cells at the 21st day. This result indicated that the acidification of the medium and disaggregation of the CPC are the two important factors, directly influencing cellular attachment and proliferation in vitro at cement surface. But, these cements are generally the product of an acid-base reaction which did not seem to induce any necrosis as in vivo, no visible zone of dead tissue was seen because of acid-base balance in the organism.

Tissue reactions to CPC are different in different tissues. When CPC was implanted in cutaneous tissue, a slight inflammatory reaction with numerous macrophages and few foreign-body giant cells were observed in the connective tissue adjacent to the cement implant. However, when CPC was implanted in bone tissue, new bone was formed around the implant within 1 to 2 weeks, cements were resorbed and replaced by bone tissue from 4 to 8 weeks, then bone remodeling occurred in the implanted zone, without inflammatory reaction nor osteonecrosis in all phases [114,116,117]. From this difference, it could be hypothesized that micromovements persist in the materials implanted in the soft tissue which stimulate the tissue around the implant to cause inflammatory reactions. On the other hand, the materials implanted in the bone tissue are immobilized by bone tissue which may explain the absence of this reaction.

There is controversy as to the resorption and replacement of CPC by bone tissue. Ikenaga et al. [118] reported that CPC resorption was about 8% at 2 weeks and 92% at 12 weeks, and new bone formation was about 1% at 2 weeks and 35% at 12 weeks in the femoral condyle of rabbit. When CPC was implanted in same site, Frayssinet et al. [119] found a resorption of 54%, 68%, and 89%, and new bone formation of 25%, 32%, and 23% at 2, 6, and 18 weeks, respectively. Our study has shown that new bone formation increased from 2 to 24 weeks, and the material resorption was about 10%, 15%, 30%, and 60% at 2, 4, 12, and 24 weeks, respectively, in the tibiae condyle of rabbit [114]. In contrast, Costantino et al. [120] made 2.5 cm-diameter, full-thickness parietal skull defects in cats and reconstructed them with CPC. By 6 months, the CPC was replaced by new bone and soft tissue 7.2 mm in depth from the cement surface. Of the replacement tissue, 77.3% was new bone and the remaining portion was soft tissue. Friedman et al. [121] found very little resorption of CPC or new bone deposition when the frontal sinus in the cat was obliterated and reconstructed with CPC. These differences are thought to be caused by many factors, including differences in species and age among the experimental animals, anatomical site, method and duration of implantation, composition of the CPC, etc.

CPC is only an osteoconductor without osteoinduction, and is in direct contact with osteoid or/and bone, but osteoblasts are rarely in direct contact with the CPC surface. This may be due to the fact that a space rapidly is formed by the material degradation at bone/cement interface, and/or that products of the dissolution influence cellular adhesion. The biodegradation of CPC involves the mechanisms of biomaterials, which are resorption by phagocytic cells and dissolution by a physicochemical process. However, the degradation at the beginning is performed by the dissolution with the weak cellular process because of the presence of few osteoclasts, macrophages and foreign-body giant cells. From the second week, numerous macrophages, few foreign-body giant cells, and rare osteoclasts are found around the cement, and CPC particles form at the interface and inside the cells are visible. We consider that the process of biodegradation is directly influenced by the type of crystallization of the calcium phosphate material. For example, sintered calcium phosphate bioceramics processed at high temperature exhibit good crystallization and are primarily degraded by a process dependent on interstitial liquids. However, phosphocalcic bone cement is formed by physicochemical crystallization and is primarily degraded through a cellular process.

The mechanical properties of CPC (compressive strength from 20 to 60 MPa) are less strong than those of PMMA cement (>70 MPa). Biodegradation and new bone formation during implantation modify their properties. Yamamoto et al. [122] tested CPC and showed that compressive strength increased at 3 days and 1 week, and decreased at 4 weeks in vitro; in vivo, it increased at 3 days; 1 and 2 weeks, and decreased at 4 weeks. The values of these results in vivo were only 50–70% of those in vitro. Our study revealed a strong decrease of compressive strength after 2 weeks due to biodegradation, followed by a slight increase from 4 weeks due to new bone formation. There was a general decrease in the elastic modulus with time [114]. This change of the mechanical strength is supposed to be related to the kinetics of recrystallization where the mechanical strength increased according to the progress of recrystallization, but degradation of dissolution subsequently starts after crystallization. This change also suggests that calcium phosphate cement would be remodeled or resorbed in the long term. This is like hydroxyapatite, used as a bone substitute material, and is also the expected characteristic of calcium phosphate cement, used for enhancing the initial fixation of implants and promote biological fixation in the long term.

Clinical Applications of CPC

CPC are resorbable materials with osteoconduction, which are not toxic, not exothermic, and are excellently biocompatible, but their mechanical properties are not ideal, limiting their clinical utilization. They are only used to fill small bone defects or to increase bone volume as bone substitute. Shindo et al. [123] reported that CPC has been used to increase the supraorbital ridge in dogs, as well as in a variety of skull base defects. It was also used in 24 patients to increase or obliterate the frontal and ethmoid sinus regions and mastoid cavities. When these patients were observed for 2 years, it was necessary to remove the material in only one patient. Kveton et al. [124,125] reported on the 2-year follow-up of 15 patients who underwent CPC reconstruction for translabyrinthine, middle cranial fosse, and suboccipital craniectomy; no complication were shown. Stankewich et al. [126] and Goodman et al. [127] showed augmentation of femoral neck fracture with CPC, which significantly improved the initial stability and failure strength of the fractures. The cement has also been used to stabilize distal radius fractures in 6 patients and appeared to promote healing and permit early mobilization of the wrist [128]. Kopylov et al. [129] used an injectable calcium phosphate bone cement with external fixation in the treatment of redisplaced distal radial fractures by a prospective randomized study in 40 patients. The chosen primary effect variable was grip strength at 7 weeks. Patients treated by injection of CPC had better grip strength, wrist extension, and forearm supination at 7 weeks. There was no difference in functional parameters at 3 months or later. None of the methods could fully stabilize the fracture: radiographs showed a progressive redislocation over time.

Development of CPC

The rationale for using CPC is that this material will be completely resorbed and replaced by new bone. Two processes are simultaneously involved: (1) the degradation of CPC performed by osteoclasts and macrophages, and (2) the creation of new bone performed by osteoblasts. The presence of CPC particles could disturb the osteoblast ability to make new bone. An unstable mechanical situation could result if bone formation is delayed by particles resulting from CPC degradation. It would then be important for future CPC development to minimize the generation of particles smaller than $10\,\mu m$.

Since the mechanical properties limit the clinical utilization of CPC, its composition or the adjuvants may be modified to maximize crystallization to improve the mechanical properties. When CPC is rapidly resorbed during implantation and new bone formation is insufficient in the implanted site, or slowly resorbed to prevent new bone formation and CPC looses its initial properties, the mechanical properties are decreased.

In orthopedic surgery, PMMA cements are frequently used to fix prostheses due to its strong mechanical fixation, but this fixation presents loosening, especially in long term, because of the absence of biological fixation by bone tissue. For the fixation of surgical prostheses, it is necessary to obtain mechanical fixation in a short time (1–3 months) and biological fixation in the longer term (beginning after 1 month). The mechanism of this fixation supposes that prostheses are fixed by bone cement or by bone tissue, then the cement is resorbed and conducts new bone formation into the surface of the prosthesis with excellent osteointegration to obtain biological fixation. We think that when noncemented prostheses are combined with CPC, there is: 1) a mechanical fixation due to noncemented prosthesis with a blockage between the prosthesis and bone tissue, and 2) the cement can fill the residual cavity around the prosthesis. Ostoegenesis and osteoconduction will lead to fixation of the prosthesis by new bone formation.

References

1. Gluck T. Referat uber die durch das moderne chirurgische Experiment gewonnenen positiven Resultate, betreffend die Naht und den Ersatz von Defecten hoherer Gewebe, sowie uber die Verwerthung resorbirbarer und lebendiger Tampons in der Chirurgie. Archiv fur klinische Chirurgie 1891;41:187.

2. Haboush EJ. A new operation for arthroplasty of the hip based on biomechanics, photoelasticity, fast-setting dental acrylic, and other considerations. Bull Hosp Joint Dis 1953;14:242.
3. Charnely J. The bonding of prostheses to bone by cement. J Bone Joint Surg [Br] 1964;46:518.
4. Sorensen WG, Bloom JD, Kelly PJ. The effects of intramedullary methylmethacrylate and reaming on the circulation of the tibia after osteotomy and plate fixation in dogs. J Bone Joint Surg [Am] 1079; 61(3):417–24.
5. Brown WE, Chow LC. A new calcium phosphate, water-setting cement. Westerville, OH: American Ceramic Society, 1986.
6. Mirtchi AA, Lemaitre J, Munting E. Calcium phosphate cement: Action of setting regulators on the properties of tricalcium phosphate-monocalcium phosphate cement. Biomaterials 1989;10:634–8.
7. Ginebra MP, Fernandez E, Boltong MG, Bermudez O, Planell JA, Driessens FC. Compliance of an apatitic calcium phosphate cement with the short-term clinical requirements in bone surgery, orthopaedics and dentistry. Clin Mater 1994;17(2):99–104.
8. Kurashina K, Kurita H, Kotani A, Kobayashi S, Kyoshima K, Hirano M. Experimental cranioplasty and skeletal augmentation using an alpha-tricalcium phosphate/dicalcium phosphate dibasic/tetracalcium phosphate monoxide cement: a preliminary short-term experiment in rabbits. Biomaterials 1998;19:701–6.
9. Liu CS, Shen W, Gu YF, Hu LM. Mechanism of the hardening process for a hydroxyapatite cement. J Biomed Mater Res 1997;35:75–80.
10. Fernandez E, Gil FJ, Best SM, Ginebra MP, Driessens FC, Planell JA. Improvement of the mechanical properties of new calcium phosphate bone cements in the $CaHPO_4$-alpha-$Ca_3(PO_4)_2$ system: compressive strength and microstructural development. J Biomed Mater Res 1998;41(4):560–7.
11. Trap B, Wolff P, Jensen JS. Acrylic bone cement: residuals and extractability of methacrylate monomers and aromatic amines. J Appl Biomater 1992;3:51–7.
12. ISO 5833: International standards for implants for surgery acrylic resin cement. International Standards Organisation, 1992.
13. Jasty M, Davies JP, O'Connor DO, Burke DW, Harrigan TP, Harris WH. Porosity of various preparations of acrylic bone cements. Clin Orthop 1990;259:122–9.
14. Jasty M, Maloney WJ, Bragdon CR, Haire T, Harris WH. Histomorphological studies of the long-term skeletal responses to well-fixed cemented femoral components. J Bone Joint Surg [Am] 1990;72(8):1220–9.
15. Davies JP, O'Connor DO, Burke DW, Jasty M, Harris WH. The effect of centrifugation on the fatigue life of bone cement in the presence of surface irregularities. Clin Orthop 1988;229:156–61.
16. Davies JP, Harris WH. Optimization and comparison of three vacuum mixing systems for porosity reduction of Simplex P cement. Clin Orthop 1990;254:261–9.
17. Lautenschlager EP, Stupp SI, Keller JC. Structure and properties of acrylic bone cement. In: Ducheyne P, Hastings GW, editors. Functional behavior of orthopaedic biomaterials, vol II. Application. Florida: CPC Press, 1984;87–119.
18. Ishidara K. Hard tissue compatible polymers. In: Tsuruta et al., editors. Biomedical application of polymeric materials. CPC Press, 1993;143.
19. Haas SS, Brauer GM, Dickson G. A characterization of polymethylmethacrylate bone cement. J Bone Joint Surg [Am] 1975;57:380–91.
20. Harrigan TP, Kareh JA, O'Connor DO, Burke DW, Harris WH. A finite element study of the initiation of failure of fixation in cemented femoral total hip components. J Orthop Res 1992;10(1):134–44.
21. James SP, Schmalzried TP, McGarry FJ, Harris WH. Extensive porosity at the cement–femoral prosthesis interface: a preliminary study. J Biomed Mater Res 1993;27(1):71–8.
22. Keller JC, Lautenschlager EP, Marshall GW Jr, Meyer PR Jr. Factors affecting surgical alloy/bone cement interface adhesion. J Biomed Mater Res 1980;14(5):639–51.
23. Bannister GC, Miles AW, May PC. Properties of bone cement prepared under operating theatre conditions. Clin Mater 1989;4:343–7.
24. Bean DJ, Hollis JM, Woo SL, Convery FR. Sustained pressurization of polymethylmethacrylate: a comparison of low- and moderate-viscosity bone cements. J Orthop Res 1988;6(4):580–4.
25. Linder L. Reactions to bone cement. In: Williams DF, editor. Biocompatibility of Orthopedic implants. Vol. II. Boca Raton, FL: CRC Press, 1982;1–23.
26. De Waal Malefijt J, Sloof TJ, Huiskes R. The actual status of acrylic bone cement in total hip replacement. A review. Acta Orthop Belg 1987;53(1):52–8.
27. Schoenfeld CM, Conard GJ, Lautenschlager EP. Monomer release from methacrylate bone cements during simulated in vivo polymerization. J Biomed Mater Res 1979;13(1):135–47.
28. Lu JX, Huang ZW, Tropiano P, Clouet d'Orval B, Remusat M, Déjou J et al. Human biological reactions at the interface between bone tissue and polymethylmethacrylate cement. J Mater Sci: Mater Med 2000 (accepted).
29. Frondoza CG, Tanner KT, Jones LC, Hungerford DS. Polymethylmethacrylate particles enhance DNA and protein synthesis of human fibroblasts in vitro. J Biomed Mater Res 1993;27(5):611–7.
30. Golds EE, Mason P, Nyirkos P. Inflammatory cytokines induce synthesis and secretion of gro protein and a neutrophil chemotactic factor but not beta 2-microglobulin in human synovial cells and fibroblasts. Biochem J 1989;259:585–8.
31. Dayer JM, Beutler B, Cerami A. Cachectin/tumor necrosis factor stimulates collagenase and prostaglandin E2 production by human synovial cells and dermal fibroblasts. J Exp Med 1985;162:2163–8.
32. Akira S, Hirano T, Taga T, Kishimoto T. Biology of multifunctional cytokines: IL 6 and related molecules (IL 1 and TNF). FASEB J 1990;4(11):2860–7.
33. Swan A, Dularay B, Dieppe P. A comparison of the effects of urate, hydroxyapatite and diamond crystals on polymorphonuclear cells: relationship of mediator release to the surface area and adsorptive capacity of different particles. J Rheumatol 1990;17(10):1346–52.
34. Goldring MB, Goldring SR. Skeletal tissue response to cytokines. Clin Orthop 1990;258:245–78.
35. Murray DW, Rushton N. Macrophages stimulate bone resorption when they phagocytose particles. J Bone Joint Surg [Br] 1990;72(6):988–92.
36. Davis RG, Goodman SB, Smith RL, Lerman JA, Williams RJ III. The effects of bone cement powder on human adherent monocytes/macrophages in vitro. J Biomed Mater Res 1993;27(8):1039–46.

37. Glant TT, Jacobs JJ. Response of three murine macrophage populations to particulate debris: bone resorption in organ cultures. J Orthop Res 1994; 12(5):720–31.
38. Horowitz SM, Rapuano BP, Lane JM, Burstein AH. The interaction of the macrophage and the osteoblast in the pathophysiology of aseptic loosening of joint replacements. Calcif Tissue Int 1994;54(4):320–4.
39. Herman JH, Sowder WG, Anderson D, Appel AM, Hopson CN. Polymethylmethacrylate-induced release of bone-resorbing factors. J Bone Joint Surg [Am] 1989;71(10):1530–41.
40. Takagi M, Konttinen YT, Santavirta S, Sorsa T, Eisen AZ, Nordsletten L et al. Extracellular matrix metalloproteinases around loose total hip prostheses. Acta Orthop Scand 1994;65(3):281–6.
41. Horowitz SM, Gonzales JB. Effects of polyethylene on macrophages. J Orthop Res 1997;15(1):50–6.
42. Lee SH, Brennan FR, Jacobs JJ, Urban RM, Ragasa DR, Glant TT. Human monocyte/macrophage response to cobalt-chromium corrosion products and titanium particles in patients with total joint replacements. J Orthop Res 1997;15(1):40–9.
43. Shanbhag AS, Jacobs JJ, Black J, Galante JO, Glant TT. Human monocyte response to particulate biomaterials generated in vivo and in vitro. J Orthop Res 1995; 13(5):792–801.
44. Horowitz SM, Gautsch TL, Frondoza CG, Riley L Jr. Macrophage exposure to polymethyl methacrylate leads to mediator release and injury. J Orthop Res 1991;9(3):406–13.
45. Amstutz HC, Campbell P, Kossovsky N, Clarke IC. Mechanism and clinical significance of wear debris-induced osteolysis. Clin Orthop 1992;276:7–18.
46. Harris WH. Osteolysis and particle disease in hip replacement. A review. Acta Orthop Scand 1994; 65(1):113–23.
47. Maloney WJ, Smith RL. Periprosthetic osteolysis in total hip arthroplasty: the role of particulate wear debris. Instr Course Lect 1996;45:171–82.
48. Sabokbar A, Fujikawa Y, Murray DW, Athanasou NA. Radio-opaque agents in bone cement increase bone resorption. J Bone Joint Surg [Br] 1997;79(1):129–34.
49. Wang W, Ferguson DJ, Quinn JM, Simpson AH, Athanasou NA. Biomaterial particle phagocytosis by bone-resorbing osteoclasts. J Bone Joint Surg [Br] 1997; 79(5):849–56.
50. Thomson LA, Law FC, James KH, Matthew CA, Rushton N. Biocompatibility of particulate polymethylmethacrylate bone cements: a comparative study in vitro and in vivo. Biomaterials 1992;13(12):811–8.
51. Willert HG, Bertram H, Buchhorn GH. Osteolysis in alloarthroplasty of the hip. The role of bone cement fragmentation. Clin Orthop 1990;258:108–21.
52. Jensen LN, Sturup J, Kramhoft M, Jensen JS. Histological evaluation of cortical bone reaction to PMMA cement. Acta Orthop Belg 1991;57(3):254–9.
53. Nimb L, Sturup J, Jensen JS. Improved cortical histology after cementation with a new MMA-DMA-IBMA bone cement: an animal study. J Biomed Mater Res 1993;27(5):565–74.
54. Revell PA, Braden M, Freeman MA. Review of the biological response to a novel bone cement containing poly(ethyl methacrylate) and n-butyl methacrylate. Biomaterials 1998;19(17):1579–86.
55. Charnley J. The reaction of bone to self-curing acrylic cement. A long-term histological study in man. J Bone Joint Surg [Br] 1970;52(2):340–53.
56. Pizzoferrato A, Ciapetti G, Stea S, Toni A. Cellular events in the mechanisms of prosthesis loosening. Clin Mater 1991;7(1):51–81.
57. Albrektsson T. Osseous penetration rate into implants pretreated with bone cement. Arch Orthop Trauma Surg 1981;102:141.
58. Morberg P, Albrektsson T. Bone reactions to intramedullary insertion of methyl methacrylate. Eur J Muskuloskel Res 1992;1:11.
59. Morberg P et al. Impaired cortical bone formation after intramedullary insertion of bone cement. Clin Mater 1992;10:139.
60. Christensen SB. Osteoarthrosis: Changes of bone, cartilage and synovial membrane in relation to bone scintigraphy. Acta Orthop Scand Suppl 1985;214:1–43.
61. Sturup J, Madsen J, Tondevold E, Jensen JS. Decreased blood perfusion in canine tibial diaphysis after filling with acrylic bone cement compared with inert bone wax. Acta Orthop Scand 1990;61(2):143–7.
62. Mongiorgi R, Valdre G, Giardino R, Maggi G, Prati C, Bertocchi G. Thermodynamical aspects of the polymerization reaction of PMMA cement mixed with phosphatic mineral phases. Boll Soc Ital Biol Sper 1993;69(6):365–72.
63. Berman AT, Reid JS, Yanicko DR Jr, Sih GC, Zimmerman MR. Thermally induced bone necrosis in rabbits. Relation to implant failure in humans. Clin Orthop 1984;186:284–92.
64. Park JB, Turner RC, Atkins PE. EPR study of free radicals in PMMA bone cement: a feasibility study. Biomater Med Devices Artif Organs 1980;8(1):23–33.
65. Biehl G, Harms J, Hanser U. Experimental studies on heat development in bone during polymerization of bone cement. Intraoperative measurement of temperature in normal blood circulation and in bloodlessness. Arch Orthop Unfallchir 1974;78(1):62–9.
66. Rhinelander FW, Nelson CL, Stewart RD, Stewart CL. Experimental reaming of the proximal femur and acrylic cement implantation: vascular and histologic effects. Clin Orthop 1979;141:74–89.
67. Lundskog J. Heat and bone tissue. An experimental investigation of the thermal properties of bone and threshold levels for thermal injury. Scand J Plast Reconstr Surg 1972;9:1–80.
68. Lee AJ, Wrighton JD. Some properties of polymethylmethacrylate with reference to its use in orthopedic surgery. Clin Orthop 1973;95:281–7.
69. Sund G, Rosenquist J. Morphological changes in bone following intramedullary implantation of methyl methacrylate. Effects of medullary occlusion: a morphometrical study. Acta Orthop Scand 1983;54(2): 148–56.
70. Williams RP, McQueen DA. A histopathologic study of late aseptic loosening of cemented total hip prostheses. Clin Orthop 1992;275:174–9.
71. Harris WH. The problem is osteolysis. Clin Orthop 1995;311:46–53.
72. Jasty M, Jiranek W, Harris WH. Acrylic fragmentation in total hip replacements and its biological consequences. Clin Orthop 1992;285:116–28.
73. Jasty M, Bragdon CR, Lee K, Hanson A, Harris WH. Surface damage to cobalt-chrome femoral head prostheses. J Bone Joint Surg [Br] 1994;76(1):73–7.

74. Jensen LN, Jensen JS, Gotfredsen K. A method for histological preparation of undecalcified bone sections containing acrylic bone cement. Biotech Histochem 1991;1(2):82–6.
75. Goodman SB, Fornasier VL, Kei J. The effects of bulk versus particulate polymethylmethacrylate on bone. Clin Orthop 1988;232:255–62.
76. Renvall S. Bone cement and wound healing. An experimental study in the rat. Ann Chir Gynaecol 1991; 80(3):285–8.
77. Quinn J, Joyner C, Triffitt JT, Athanasou NA. Polymethylmethacrylate-induced inflammatory macrophages resorb bone. J Bone Joint Surg [Br] 1992;74(5): 652–8.
78. Santavirta S, Gristina A, Konttinen YT. Cemented versus cementless hip arthroplasty. A review of prosthetic biocompatibility. Acta Orthop Scand 1992; 63(2):225–32.
79. Horowitz SM, Frondoza CG, Lennox DW. Effects of polymethylmethacrylate exposure upon macrophages. J Orthop Res 1988;6(6):827–32.
80. Havelin LI, Espehaug B, Vollset SE, Engesaeter LB. The effect of the type of cement on early revision of Charnley total hip prostheses. A review of 8,579 primary arthroplasties from the Norwegian Arthroplasty Register. J Bone Joint Surg [Am] 1995;77(10):1543–50.
81. Hungerford DS, Jones LC. The rationale of cementless revision of cemented arthroplasty failures. Clin Orthop 1988;235:12–24.
82. Herberts P, Ahnfelt L, Malchau H, Stromberg C, Andersson GB. Multicenter clinical trials and their value in assessing total joint arthroplasty. Clin Orthop 1989;249:48–55.
83. Herberts P, Malchau H. How outcome studies have changed total hip arthroplasty practices in Sweden. Clin Orthop 1997;344:44–60.
84. Ohashi KL, Dauskardt RH. Effects of fatigue loading and PMMA precoating on the adhesion and subcritical debonding of prosthetic–PMMA interfaces. J Biomed Mater Res 2000;51(2):172–83.
85. Dahl OE, Johnsen H, Kierulf P, Molnar I, Ro JS, Vinje A et al. Intrapulmonary thrombin generation and its relation to monomethylmethacrylate plasma levels during hip arthroplasty. Acta Anaesthesiol Scand 1992;36(4):331–5.
86. Gentil B, Paugam C, Wolf C, Lienhart A, Augereau B. Methylmethacrylate plasma levels during total hip arthroplasty. Clin Orthop 1993;287:112–6.
87. Svartling N, Pfaffli P, Tarkkanen L. Methylmethacrylate blood levels in patients with femoral neck fracture. Arch Orthop Trauma Surg 1985;104(4):242–6.
88. Patterson BM, Healey JH, Cornell CN, Sharrock NE. Cardiac arrest during hip arthroplasty with a cemented long-stem component. A report of seven cases. J Bone Joint Surg [Am] 1991;73(2):271–7.
89. Wenda K, Scheuermann H, Weitzel E, Rudigier J. Pharmacokinetics of methylmethacrylate monomer during total hip replacement in man. Arch Orthop Trauma Surg 1988;107(5):316–21.
90. Crout DH, Corkill JA, James ML, Ling RS. Methylmethacrylate metabolism in man. The hydrolysis of methylmethacrylate to methacrylic acid during total hip replacement. Clin Orthop 1979;141:90–5.
91. Orsini EC, Byrick RJ, Mullen JB, Kay JC, Waddell JP. Cardiopulmonary function and pulmonary microemboli during arthroplasty using cemented or non-cemented components. The role of intramedullary pressure. J Bone Joint Surg [Am] 1987;69(6):822–32.
92. Hofmann AA, Wyatt RW, Gilbertson AA, DeKoss L, Miller J. The effect of air embolization from the femoral canal on hemodynamic parameters during hip arthroplasty. Clin Orthop 1987;218:290–6.
93. Fregert S. Occupational hazards of acrylate bone cement in orthopaedic surgery. Acta Orthop Scand 1983;54(6):787–9.
94. Bengtson A, Larsson M, Gammer W, Heideman M. Anaphylatoxin release in association with methylmethacrylate fixation of hip prostheses. J Bone Joint Surg [Am] 1987;69(1):46–9.
95. Kanerva L, Estlander T, Jolanki R, Tarvainen K. Occupational allergic contact dermatitis and contact urticaria caused by polyfunctional aziridine hardener. Contact Dermatitis 1995;33(5):304–9.
96. Brauer GM, Davenport RM, Hansen WC. Accelerating effect of amines on polymerization of methyl methacrylate. Mod Plastic 1956;34(11):154–256.
97. Lal J, Green RJ. Effect of amine accelerators on the polymerization of methacrylate with benzoyl peroxide. J Polym Sci 1955;18:403.
98. Vazquez B, Elvira C, Levenfeld B, Pascual B, Goni I, Gurruchaga M et al. Application of tertiary amines with reduced toxicity to the curing process of acrylic bone cements. J Biomed Mater Res 1997;34(1):129–36.
99. Fritsch EW. Static and fatigue properties of two new low-viscosity PMMA bone cements improved by vacuum mixing. J Biomed Mater Res 1996;31(4):451–6.
100. Morita S, Kawachi S, Yamamoto H, Shinomiya K, Nakabayashi N, Ishihara K. Total hip arthroplasty using bone cement containing tri-n-butylborane as the initiator. J Biomed Mater Res 1999;48(5):759–63.
101. Pourdeyhimi B, Wagner HD. Elastic and ultimate properties of acrylic bone cement reinforced with ultra-high-molecular-weight polyethylene fibers. J Biomed Mater Res 1989;23(1):63–80.
102. Saha S, Pal S. Improvement of mechanical properties of acrylic bone cement by fiber reinforcement. J Biomech 1984;17(7):467–78.
103. Topoleski LD, Ducheyne P, Cuckler JM. The fracture toughness of titanium-fiber-reinforced bone cement. J Biomed Mater Res 1992;26(12):1599–617.
104. Ishihara K, Arai H, Nakabayashi N, Morita S, Furuya K. Adhesive bone cement containing hydroxyapatite particle as bone compatible filler. J Biomed Mater Res 1992;26(7):937–45.
105. Beruto DT, Mezzasalma SA, Capurro M, Botter R, Cirillo P. Use of alpha-tricalcium phosphate (TCP) as powders and as an aqueous dispersion to modify processing, microstructure, and mechanical properties of polymethylmethacrylate (PMMA) bone cements and to produce bone-substitute compounds. J Biomed Mater Res 2000;49(4):498–505.
106. Heikkila JT, Aho AJ, Kangasniemi I, Yli-Urpo A. Polymethylmethacrylate composites: disturbed bone formation at the surface of bioactive glass and hydroxyapatite. Biomaterials 1996;17(18):1755–60.
107. Edwards JT, Brunski JB, Higuchi HW. Mechanical and morphologic investigation of the tensile strength of a bone–hydroxyapatite interface. J Biomed Mater Res 1997;36(4):454–68.
108. Driessens FCM, Boltong MG, Bermudez O, Planell JA. Formulation and setting times of some calcium

orthophosphate cements: a pilot study. J Mater Sci: Mater Med 1993;4:503–8.

109. Driessens FCM, Boltong MG, Bermudez O, Planell JA, Ginebra MP, Fernandez E. Effective formulations for the preparation of calcium phosphate bone cements. J Mater Sci: Mater Med 1994;5:164–70.

110. Schmitz JP, Hollinger JO, Milam SB. Reconstruction of bone using calcium phosphate bone cements: a critical review. J Oral Maxillofac Surg 1999;57(9):1122–6.

111. Leroux L, Hatim Z, Freche M, Lacout JL. Effects of various adjuvants (lactic acid, glycerol, and chitosan) on the injectability of a calcium phosphate cement. Bone 1999;25(2 Suppl):31S–4S.

112. Liu CS, Wang W, Shen W, Chen TY, Hu LM, Chen ZW. Evaluation of the biocompatibility if a nonceramic hydroxyapatite. J Endodont 1997;23(8):490–3.

113. Higashi S, Ohsumi T, Ozumi K, Kuroki K, Inokuchi Y, Terashita M. Evaluation of cytotoxicity of calcium phosphate cement consisting of alpha-tricalcium phosphate and dicalcium phosphate dihydrate. Dent Mater J 1998;17(3):186–94.

114. Lu JX, About I, Stephan G, Van Landuyt P, Dejou J, Fiocchi M et al. Histological and biomechanical studies of two bone colonizable cements in rabbits. Bone 1999;25(2 Suppl):41S–5S.

115. Pioletti DP, Takei H, Lin T, Van Landuyt P, Ma QJ, Kwon SY et al. The effects of calcium phosphate cement particles on osteoblast functions. Biomaterials 2000; 21(11):1103–14.

116. Ohura K, Bohner M, Hardouin P, Lemaitre J, Pasquier G, Flautre B. Resorption of, and bone formation from, new beta-tricalcium phosphate-monocalcium phosphate cements: an in vivo study. J Biomed Mater Res 1996;30(2):193–200.

117. Miyamoto Y, Ishikawa K, Takechi M, Toh T, Yoshida Y, Nagayama M et al. Tissue response to fast-setting calcium phosphate cement in bone. J Biomed Mater Res 1997;37(4):457–64.

118. Ikenaga M, Hardouin P, Lemaitre J, Andrianjatovo H, Flautre B. Biomechanical characterization of a biodegradable calcium phosphate hydraulic cement: a comparison with porous biphasic calcium phosphate ceramics. J Biomed Mater Res 1998;40(1):139–44.

119. Frayssinet P, Gineste L, Conte P, Fages J, Rouquet N. Short-term implantation effects of a DCPD-based calcium phosphate cement. Biomaterials 1998;19 (11–12):971–7.

120. Costantino PD, Friedman CD, Jones K, Chow LC, Sisson GA. Experimental hydroxyapatite cement cranioplasty. Plast Reconstr Surg 1992;90(2):174–85.

121. Friedman CD, Costantino PD, Jones K, Chow LC, Pelzer HJ, Sisson GA Sr. Hydroxyapatite cement. II. Obliteration and reconstruction of the cat frontal sinus. Arch Otolaryngol Head Neck Surg 1991;117(4): 385–9.

122. Yamamoto H, Niwa S, Hori M, Hattori T, Sawai K, Aoki S et al. Mechanical strength of calcium phosphate cement in vivo and in vitro. Biomaterials 1998;19(17): 1587–91.

123. Shindo ML, Costantino PD, Friedman CD, Chow LC. Facial skeletal augmentation using hydroxyapatite cement. Arch Otolaryngol Head Neck Surg 1993; 119(2):185–90.

124. Kveton JF, Friedman CD, Piepmeier JM, Costantino PD. Reconstruction of suboccipital craniectomy defects with hydroxyapatite cement: a preliminary report. Laryngoscope 1995;105(2):156–9.

125. Kveton JF, Friedman CD, Costantino PD. Indications for hydroxyapatite cement reconstruction in lateral skull base surgery. Am J Otol 1995;16(4):465–9.

126. Stankewich CJ, Swiontkowski MF, Tencer AF, Yetkinler DN, Poser RD. Augmentation of femoral neck fracture fixation with an injectable calcium-phosphate bone mineral cement. J Orthop Res 1996;14(5):786–93.

127. Goodman SB, Bauer TW, Carter D, Casteleyn PP, Goldstein SA, Kyle RF et al. Norian SRS cement augmentation in hip fracture treatment. Laboratory and initial clinical results. Clin Orthop 1998;348: 42–50.

128. Kopylov P, Jonsson K, Thorngren KG, Aspenberg P. Injectable calcium phosphate in the treatment of distal radial fractures. J Hand Surg [Br] 1996;21(6):768–71.

129. Kopylov P, Runnqvist K, Jonsson K, Aspenberg P. Norian SRS versus external fixation in redisplaced distal radial fractures. A randomized study in 40 patients. Acta Orthop Scand 1999;70(1):1–5.

9 Calcium Phosphates in Orthopedic Surgery

P. Frayssinet

The first-generation calcium phosphate ceramics have been in daily use in human surgery for more than a decade. They were first used as bone substitutes in devices for guided tissue regeneration then as thin coatings to improve metal biocompatibility. Calcium phosphate ceramics were the first materials to be synthesized almost exclusively for orthopedic use. They were initially used because of their chemical similarity to the bone mineral matrix which consists of A or AB carbonated apatite [1,2,3].

Three types of calcium phosphate materials are commercially available for orthopedic surgery:

Calcium phosphate ceramics for use as bone substitutes or spacers

Calcium phosphate cements for injection in paste form

Calcium phosphate coatings on metal

These three types of material have different characteristics which means that their subsequent behavior in bone or connective tissue is also different.

Calcium Phosphates Used in Orthopedic Surgery

Different salts of orthophosphoric acid (Table 9.1) can be used as orthopedic materials. Their solubility increases as the Ca/P ratio decreases. The solubility of these compounds is greatly affected by pH and they become more soluble as the pH decreases [4].

The chemical formula of the synthetic compound HA is closest to that of the mineral phase of bone. This is the main reason why it has been intensively investigated for bone biocompatibility. HA can also serve as a reservoir for various ions in the human body because of the numerous substitutions that can occur in its lattice.

Calcium Phosphate Ceramics

Ceramics have been processed in the same way for thousands of years. The calcium phosphate powder is suspended in a slurry, which is liquid, then shaped before being sintered. The slurry is heated almost to fusion temperature which favors the migration of matter between the grains and the formation of bridges. As the surface energy is smaller for large than for small grains, their size increases and the distance between the grain centers decreases. The sintering process also makes the surface area of the ceramic smaller [5].

Calcium phosphate ceramics thus consist of calcium phosphate grains with various characteristics depending on the manufacturing process and the properties of the raw powder. The characteristics of calcium phosphate bioceramics affect their behavior once implanted in a biological medium. Calcium phosphate ceramics have been described as bioactive materials [6]. This bioactivity has been attributed to the epitaxial nucleation of carbonated apatite at the surface of the grains [7]. This characteristic formation of a carbonate apatite layer at the material surface is shared by all bioactive materials, e.g., calcium phosphate ceramics and bioactive glasses.

The microstructure of the calcium phosphate ceramic also affects the behavior of the material.

Degradation of the material occurs preferentially at the grain boundaries (Figure 9.1). This leads to a release of individual ceramic grains or agglomerates (Figure 9.2). Once released, these grains are then phagocytosed by macrophages or giant cells. They are internalized into lysozomes which are low-pH cell compartments [8]. The calcium phosphate particles are dissolved at this pH but the released calcium and phosphate are not used locally as has been shown by historadiology.

Table 9.1. Main salts of orthophosphoric acid that can be used in orthopedic devices

Symbol	Name	Formula	Ca/P
DCPD	Brushite	$CaHPO_4,2H_2O$	1
OCP	Octocalcium phosphate	$Ca_8H_2(PO_4)_65H_2O$	1.33
TCP	Tricalcium phosphate	$Ca_3(PO_4)_2$	1.5
HA	Hydroxyapatite	$Ca_{10}(PO_4)_6(OH)_2$	1.67
TCPM	Tetracalcium phosphate	$Ca_4(PO_4)_2O$	2

The Influence of Particle Release on Calcium Phosphate Ceramics Biocompatibility

The characteristics of the released particles are of importance in the biocompatibility of the calcium phosphate device (Figure 9.3). It has been shown in vitro that, when macrophages are grown in the presence of calcium phosphate particles, the amount of cytokines (Il-1 and Il-6) and growth factors (TNF-α) synthesized differ, depending on the particle characteristics. It was also shown that for certain characteristics such as shape, the increase in cytokine synthesis was transitory. Cytotoxicity based on the Na/K of the cytoplasm also differed depending on the particle characteristics. It was also evidenced that cytokine synthesis did not increase linearly with cytotoxicity. This study clearly demonstrated that the biocompatibility of calcium ceramics is largely conditioned by the

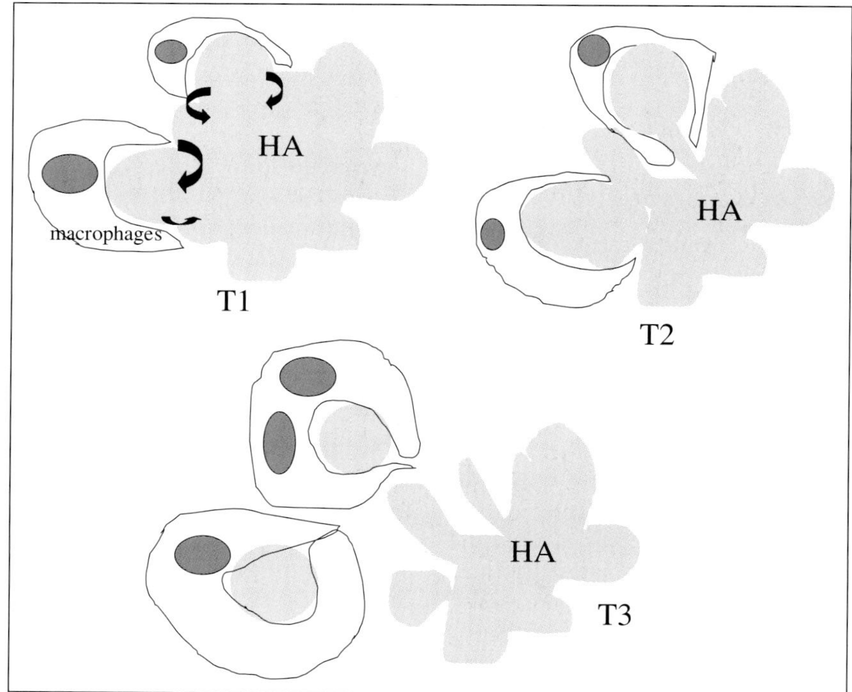

Figure 9.1. Dissolution of the ceramic at the grain boundaries (arrows). The liberated particles are then phagocytosed into macrophages present at the ceramic surface.

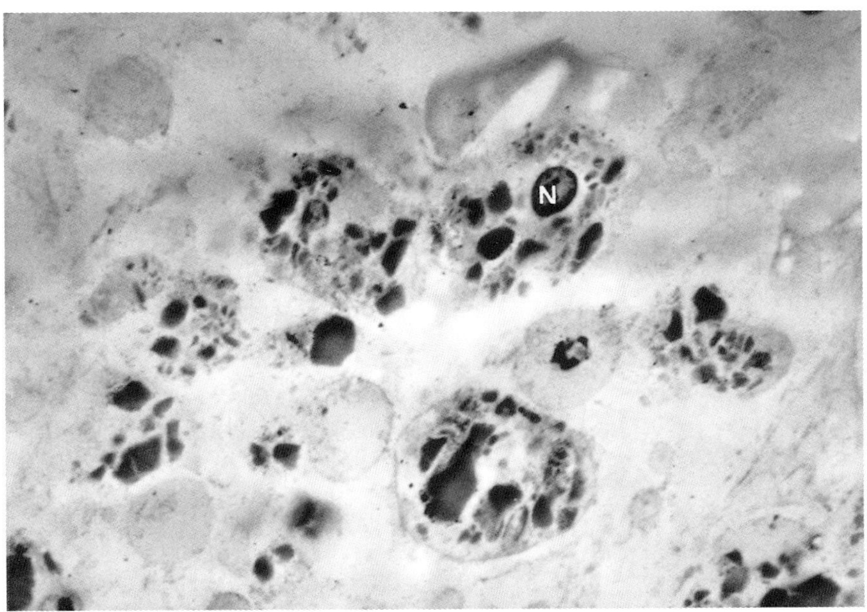

Figure 9.2. Calcium phosphate grains (black) released from a ceramic and phagocytosed into mononuclear cells (N: nucleus). Backscattered MEB ×1,000.

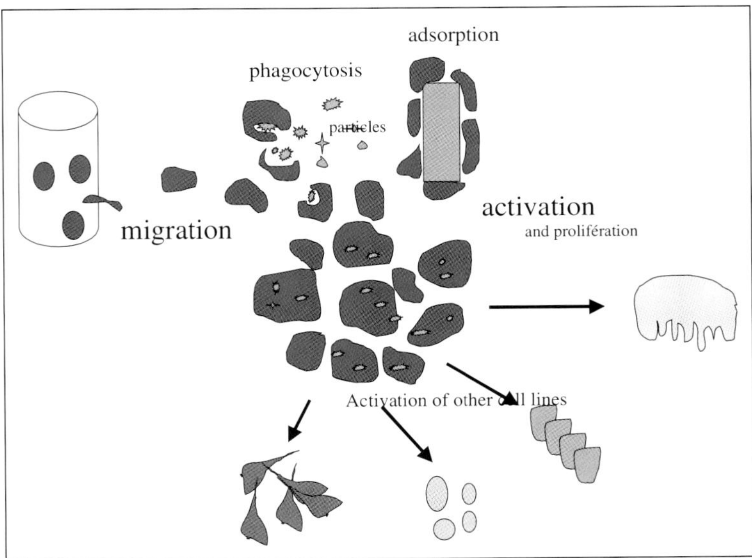

Figure 9.3. Once phagocytosed within macrophages, the particles trigger a synthesis of cytokines and growth factors which in turn activate other cells such as osteoclasts.

released particles which may activate the macrophages phagocytosing them and lead to the release by these cells of cytokines which in turn activate the osteoclasts or the osteoblasts (Figure 9.3).

Integration of Calcium Phosphate Ceramics

The first cells to be in contact with the ceramic are circulating cells of monocyte origin which come into contact with the ceramic even if no connective tissue or blood vessels are present in the pores of the ceramic. These cells are located exclusively at the material surface and there is no doubt that it is their role to synthesize information factors aimed at the cells of the healing tissue.

These cells are followed by a connective tissue which produces a densified collagen network in contact with the ceramic surface. Osteoblasts then differentiate at the contact with the ceramic surface. They do not differentiate anywhere in the volume delimited by the ceramic pore, but differentiate almost exclusively at the ceramic surface where they synthesize an immature bone on the pore walls. Thus, the stem cells are located in the pore center whereas the differentiated cells are in the outer layer (Figure 9.4). The osteogenic stem cells are characterized by their fibroblast-like aspect and an alkaline phosphatase activity of the membrane.

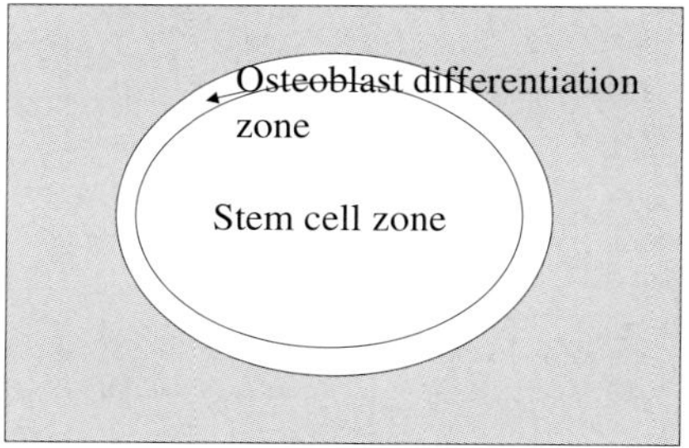

Figure 9.4. The different zones on a section of a calcium phosphate ceramic (gray) pore regarding the osteogenic cell differentiation.

The final stage is one of remodeling during which the immature bone, synthesized by the osteoblasts at the pore surface, is replaced by a layered bone and the ceramic degraded in the biological fluids is fragmented and progressively replaced by bone (Figure 9.5).

HA Coatings

HA coatings are made of calcium phosphate grains introduced into a plasma gas formed between two electrodes subjected to a high

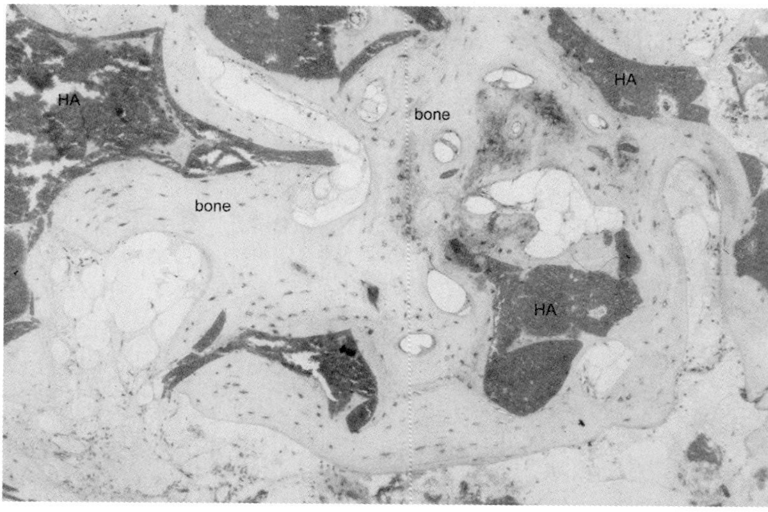

Figure 9.5. Fragmentation of a calcium phosphate ceramic (HA) after one year of implantation in human bone. Newly formed bone (bone) has integrated the fragments.

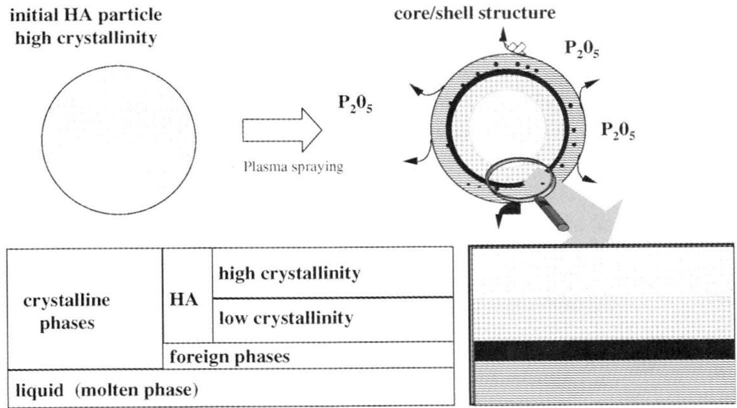

Figure 9.6. Structure of HA particle during plasma spraying showing the different phases appearing in the HA particles.

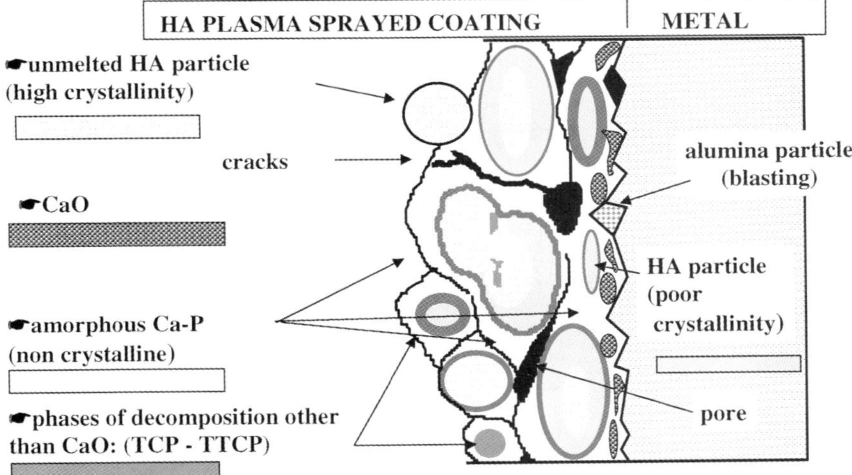

Figure 9.7. Structure of a plasma coating showing grains of crystalline HA dispersed in a continuum phase made of amorphous calcium phosphate, and β-TCP.

potential difference. The recombination of electrons with the ions on the electrodes produces great energy causing the gas to expand and project the particles towards a target.

The HA particle is transformed when introduced into the plasma. While the crystalline core of the particle remains unchanged, the superficial layer melts and, depending on the cooling speed of the particle at the target surface, certain contaminating phases may appear in the periphery of the crystalline phase (Figures 9.6, 9.7). With plasma spraying under atmosphere, the coating thus consists of a dispersed crystalline phase within a continuous phase composed of amorphous calcium phosphate, tricalcium phosphate, or even CaO. In general, any CaO is located preferentially at the junction between the coating and the sandblasted metal. This continuous phase is difficult to characterize due to the presence of the amorphous phase. The fate of calcium phosphate coatings once implanted depends on their microstructure. Not all the different calcium phosphate phases present in the coatings have the same solubility. TCP, amorphous calcium phosphates, and CaO are much more soluble at

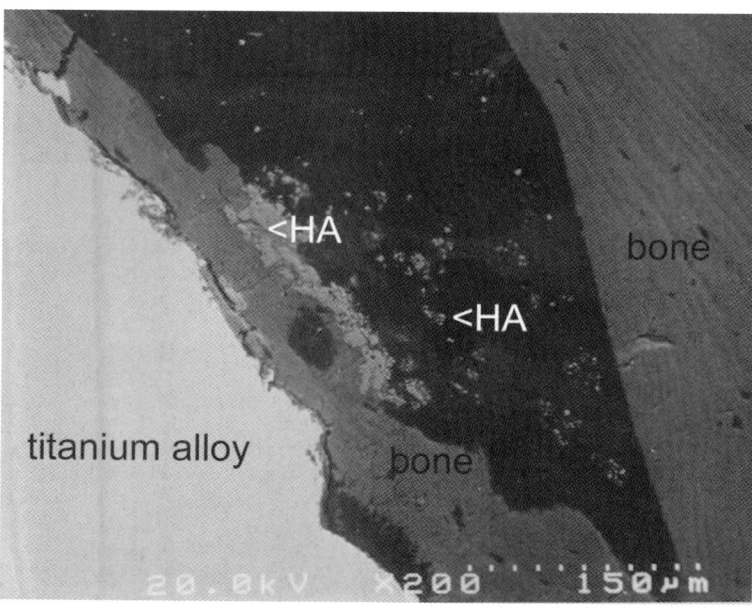

Figure 9.8. SEM of a section of a HA-coated prosthesis implanted two years in a human showing that the coating has been almost totally destroyed. The big fragments of HA are integrated in the newly formed bone at the metal surface (titanium alloy) while small particles are phagocytosed by macrophages and are located in the bone marrow cavity.

neutral pH than the HA phase. Thus, the crystalline phase in the form of grains appears progressively and is released from the coating in the extracellular fluids where the behavior of these particles will differ depending on their size. Particles smaller than a few dozen microns are phagocytosed in macrophages or giant cells. Large particles can be inserted in the newly formed bone or stay in the connective tissue (Figure 9.8).

The increased solubility due to a contaminating phase at the interface between the ceramic and the metal can be visualized at the histological level by the preferential progression of bone trabeculae between the coating and the sandblasted metal surface or even delamination by connective tissue (Figure 9.9).

The process of integration is identical with that of bone substitute. Furthermore, the integration is highly reproducible as the stem is implanted in an osteogenic medium. It is well known that surgical injury to the bone marrow triggers a synthesis of osteogenic factors [9]. The presence of the coating is responsible for perfect integration of the sandblasted metal surface into bone even several years after the calcium phosphate coating has disappeared.

It should be noted that the degradation rate of the coating is not the same at all levels of the stem. It is much higher in the proximal zone than in the distal one [10] for prostheses having a total coated surface. Several explanations can be given. The smaller amount of bone around the proximal coating makes a higher surface of material available for resorbing cells migrating from the bone marrow. The mechanical status of the different levels may also be of importance. The stress shielding occurring at the proximal level could also be responsible for a deficit in bone and higher remodeling of the coating.

Calcium Phosphate Cements

Calcium phosphate cements are materials which precipitate under a phase different from that in suspension in the paste. They are obtained by an acid-base or hydrolysis reaction. The resulting material is unlike a ceramic. Its mechanical cohesion is obtained by entanglement of the

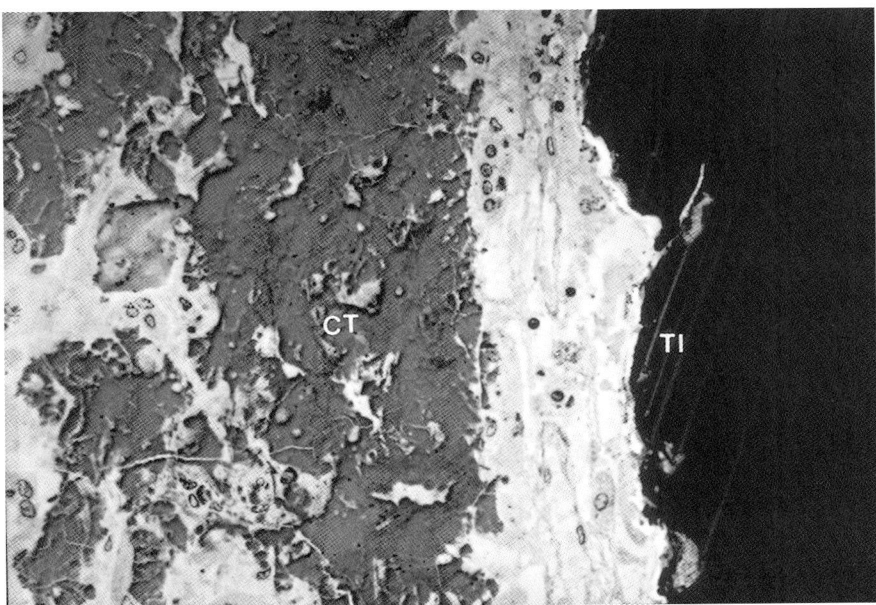

Figure 9.9. Backscattered SEM of HA coating (CT) from a Coral® stem (DePuy-France) implanted in humans for two years. The coating is separated from the sandblasted metal (TI) and shows multinuclear cells at its surface resorbing it. Backscattered MEB ×250.

precipitated crystals and not by grain fusion. This microstructure leads to a very high surface area and a higher dissolution rate than ceramics even with the same chemical composition. Four kinds of cements have been described: apatite, brushite, octocalcium phosphate, or amorphous calcium phosphate cements [11,12].

The precipitation reaction is a stoichiometric reaction with water and other ions or molecules. This means that setting of the material in a biological medium is affected by the amount of water and ions in that medium. The mechanical properties of a cement set in wet or dry medium are not at all similar. A cement set in a wet medium exhibits much poorer compressive properties than the same cement set in a dry medium.

The biological properties of some cements are also affected by the setting of the cement in situ [13]. In certain cases, cements implanted already set show better osseointegration than the same cement implanted in paste form.

Histological analysis of some cements implanted in humans also demonstrated that when integrated into bone, many cracks are present in the material and are enlarged by the progression of bone trabeculae.

Calcium phosphate cements have been proposed for the stabilization of unstable fractures by injection in the fracture zone [14]. It is obvious that the mechanical properties of a cement injected into the fracture gap cannot be predicted as the shape is complex and never the same, and the setting irregular due to the various amounts and composition of the extracellular medium in which the cement is injected. Other applications for these materials should appear such as drug carriers for bone infection or for metastasis or as materials for the incorporation of growth factors.

Conclusions

Calcium phosphate ceramics and/or cements have been specifically developed as orthopedic materials. They show very good biocompatibility and bioactivity when implanted into bones. They can also be remodeled and progressively replaced by newly formed bone. Their degradation products can be taken into the calcium and phosphate turnover of the organism while any large fragments are or can be incorporated into the bone. The first generation of calcium phosphate materials was osteoconductive. A new generation is being developed which will be osteogenic by association of osteogenic cells or osteoinductive by the association of growth and/or morphogenetic factors.

References

1. Paschalis EP, DiCarlo E, Betts F, Sherman P, Mendelsohn R, Boskey AL. FTIR microspectroscopic analysis of human osteonal bone. Calcif Tissue Int 1996;59:480–7.
2. Kuhn LT, Wu Y, Rey C, Gerstenfeld LC, Grynpas MD, Ackerman J et al. Structure, composition, and maturation of newly deposited calcium-phosphate crystals in chicken osteoblast cell cultures. J Bone Miner Res 2000; 15:1301–9.
3. Legros R. Apport de la physico-chimie à l'étude de la phase minérale des tissus calcifiés. Thèse. Institut National Polytechnique de Toulouse 1984.
4. Nancollas GH. In vitro studies of calcium phosphate crystallization. In: Mann S, Webb J, Williams RJP, editors. Biomineralization. Chemical and Biochemical Perspectives. Weinheim, Germany: VCH, 1989;157–87.
5. Frayssinet P, Rouquet N, Fages J, Durand M, Vidalain PO, Bonel G. Influence of HA-ceramic sintering temperature on the proliferation of cell grown at their contact. J Biomed Mater Res 1997;35:337–47.
6. Hench LL. Bioceramics. J Am Ceram Soc 1998;81: 1705–28.
7. Nakamura T, Neo M, Kokubo T. Bone bonding of biomaterials and apatite formation on biomaterials. In: Li P, Calvert P, Kokubo T, Levy R, Scheid C, editors. Mineralization in Natural and Synthetic Biomaterials. Warrendale, PA: Materials Research Society, 2000;15–27.
8. Frayssinet P, Rouquet N, Tourenne F, Fages J, Bonel G. In vivo degradation of calcium phosphate ceramics. Cells and Materials 1994;4:383–94.
9. Amsel S, Maniatis A, Tavassoli M, Crosby WH. The significance of intramedullary cancellous bone formation in the repair of bone marrow tissue. Anat Rec 1969; 164:101–12.
10. Frayssinet P, Hardy D, Hanker J, Giammara B. Natural history of HA-coated hip prosthesis from a series of 15 cases implanted in humans. Cells and Materials 1995; 5(2):125–38.
11. Driessens FCM, Boltong MG, Bermudez O, Planell JA, Ginebra MP, Fernandez E. Effective formulations for the preparation of calcium phosphate bone cements. J Mater Sci: Mater Med 1994;5:164–70.
12. Chow LC. Calcium phosphate cements: chemistry, properties and applications. In: Li P, Calvert P, Kokubo T, Levy R, Scheid C, editors. Mineralization in Natural and Synthetic Biomaterials. Warrendale, PA: Materials Research Society, 2000;27–39.
13. Frayssinet P, Roudier M, Lerch A, Ceolin JM, Deprès E, Rouquet N. Tissue reaction against a self-setting calcium phosphate cement set in bone or outside the bone. J Mater. Materials in Medecine (sous presse).
14. Frankenburg EP, Goldstein SA, Bauer TW, Harris SA, Poser RD. Biomechanical and histological evaluation of a calcium phosphate cement. J Bone Joint Surg 1998; 80A:1112–24.

I B – Biomaterials with Pharmacologic Activities

10 Cement with Antimitotics

P. Hernigou

There are two causes of failure in the surgical treatment of metastatic tumors: first, local recurrence of the tumor is not always prevented, even after extra-tumoral surgical exeresis and systemic chemotherapy, and secondly, failure of osteosynthesis after surgery. For these reasons, we thought that it would be helpful to provide local chemotherapy during and immediately after surgery, for instance, by adding an antimitotic to the acrylic cement used to replace the bone loss or to seal reconstruction prostheses. It was thought that the antimitotic would be likely to be released into the surrounding tissues in the same way as many antibiotics. Diffusion into the surrounding tissues is well established for numerous antibiotics [1–6].

We performed a number of experiments [7–9] to assess acrylic cement as a vehicle for local chemotherapy: 1) Diffusion of antimitotic drugs from acrylic cement was studied in vitro to determine that these drugs were released and were still biologically active after exposure to highly reactive monomer and the exothermic curing reaction. 2) Experiments in vivo were performed on two groups of animals. We tested the effect of such local chemotherapy on experimental osteosarcoma of the rat and on dogs with spontaneous osteosarcoma. General and local tolerance of the antimitotic-loaded cement was assessed.

Finally, we report our preliminary clinical investigations [10–12] with pharmacological data from patients. It was possible to envisage using cement/drug mixtures to treat orthopedic complaints calling simultaneously for mechanical consolidation of the bone [13] and in-situ release of a drug: one example would be the strengthening of bone with cement after resection of a bone tumor [14] plus the local release of antimitotic drugs from the implant.

Cement was the first vehicle to be studied for the purpose of releasing local chemotherapy. Methyl polymethacrylate (PMMA) fulfils the two following criteria: it has good biocompatibility, since the system has to remain in situ throughout the rest of the patient's life; it is not biodegradable, so that it provides mechanical support for bone which has been weakened by the surgical exeresis of a neoplastic site.

Many antimitotic drugs are available; for our first investigations we used methotrexate and cisplatine. Methotrexate was chosen because its concentration is easy to determine by spectrophotometry, and because there is an antidote (citrovorum rescue) for adverse effects. We used the acrylic bone cement currently employed by the authors for clinical arthroplasty.

Study of the Release of Methotrexate (MTX)

The first study investigated the in-vitro release of antimitotics included in acrylic cement. After confirming that this release does actually occur, starts rapidly, and is maintained over a prolonged period, two further studies were then carried out in vivo: one in dogs suffering from spontaneous osteosarcoma, in order to investigate the release of the antimitotic from the cement into the plasma, the systemic safety, and the local activity of the antimitotic-loaded cement following exeresis of the neoplasm.

The second study was conducted in laboratory rats with implanted osteosarcomas. This type of tumor was used so that a large number of animals with tumors could be studied and divided into uniform groups. Under these experimental conditions, it was possible to monitor the progress of the tumors left in situ as well as the histopathological changes brought about by the local action of antimitotics released from implants.

Kinetic Profile of the Release of MTX from Implants

To investigate the release of MTX from a block of acrylic cement implanted into the tissues, cubic test pieces were placed in 32 ml of physiological saline, which was changed every day. The concentration in the elution fluid was measured before each change. These test pieces were made from a mixture of 500 mg methotrexate powder, 46.5 g of polymer, and 20 ml of monomer, poured into 2 cm cubic moulds. Each cube weighed about 13 g and contained approximately 100 mg of methotrexate. Methotrexate elution was evaluated daily for 15 days and then weekly for six months for six specimens, the results being given as an average of the six.

The release profiles from implants containing 1% w/w have shown that methotrexate is released more rapidly during the first two hours and 10% of the load is released within the first 18 hours. The rate of release then slows. Implants immersed in an extraction medium which is changed regularly continue to release methotrexate for six months; the quantities released initially being greater the greater the initial load.

The Release of MTX from Acrylic Cement

This has been investigated in vivo in dogs with spontaneous sarcoma. We therefore chose an animal with a weight close to that of man, and a spontaneous tumor with an evolution like that of human osteosarcoma, similarly hypervascular because this may influence the diffusion of cisplatine. In experiments at the National Veterinary School of Maisons-Alfort, we used dogs with spontaneous osteosarcoma. This is a malignant tumor [15,16] with the same aggressive properties as the human type. It affects the very large breeds of dog such as the Saint Bernard (mean weight 70 kg), the mastiff (55 kg), and the boxer (30 kg). It progresses rapidly in the absence of treatment, and death is the rule in a few months [17,18]. Simple resection of the tumor rapidly leads to local relapse, and even after amputation 85% of dogs die within seven months of diagnosis [15,19,20]. The loss of substance resulting from the exeresis of the tumor was compensated for using freshly prepared methotrexate-loaded cement. The dose of methotrexate received ranged from 1.6 to 16 mg/kg. Two hours after being implanted, plasma levels of methotrexate ranged from 0.08 to 0.02 micromoles/liter (1 micromole of methotrexate = 0.455 mg). After 24 hours, the plasma levels were between 0.1 and 0.02 micromoles/liter and by the third day were no longer detectable. Toxic effects were observed on day 4 in the three animals which had received a dose of more than 200 mg of methotrexate. The other animals, which had received a dose of between 100 and 150 mg, did not display any signs of toxicity. The survival curve of the animals in this group seemed to be better than that of the animals which underwent surgery without adjuvant treatment, where 85% of the animals had died within seven months.

The Efficacy of MTX-loaded Implants

This was investigated using the experimental model of osteosarcoma in the rat [21,22]. Using implants equivalent to 1.5 mg of active constituent, tumor growth was temporarily slowed and the survival time of the animals significantly prolonged.

These experiments have shown that the rise in temperature which accompanies the polymerization of the cement does not destroy MTX and, like antibiotics, MTX can be released from the cement. Migration probably occurs as a result of diffusion; the cement constitutes a network of pores and micro-fissures which makes it accessible to the liquid medium in which it is immersed. This liquid penetrates into the system and dissolves the crystals of MTX which then diffuse into the surrounding medium. This mechanism is certainly the dominant one at work during the early stages of MTX release. It probably accounts for the initial peak which characterizes the kinetics of MTX release. It is logical to suppose that the outer layers of the cement are more accessible to the liquid medium than the inner layers.

Study of the Release of Cisplatin

Cisplatin is one of the antimitotics which would be suitable for mixing with cement and it has the following characteristics: it is often used to treat primary bone tumors; in the context of bone metastases from visceral tumors, it is generally used in multiple-drug therapy of tumors which are characterized particularly by being radio-resistant and resistant to other antimitotics [23–27]: hence the appeal of a local cisplatin-based therapy, which has the advantage of being radio-sensitizing [28].

Cisplatin takes the form of a whitish-yellow crystalline powder. It has no melting point as it decomposes without melting at 270 °C. Cisplatin has a solubility in water at room temperature of 1 mg/ml.

In the solid state, cisplatin is relatively stable. In contrast, in solution it forms *mono-aquo* and *di-aquo* derivatives by the successive shedding of chloride ions. Cisplatin is most stable in solution at an acid pH and in the presence of chloride ions, which prevent a shift in the reaction equilibrium towards the formation of degradation products. It should also be noted that cisplatin has a chemically inert structure with few reactive groups. Differential thermal analysis did not provide further information in this regard, as cisplatin decomposes without melting at around 270 °C and parasite peaks from PMMA superimpose on the cisplatin peak at these temperatures. X-ray diffraction did, however, allow us to demonstrate that there is no difference between the spectrum of the physical mixture and that of the cisplatin implants. Moreover, this hypothesis was supported by a very simple experiment in which an accurately weighed 5% cisplatin implant was dissolved in methylene chloride, a solvent for the polymer but not the cisplatin. The cisplatin crystals were sedimented and extracted by a 9 p. 1000 solution of sodium chloride in a separating funnel. Assay of this solution revealed that it contained all of the active ingredient present in the implant. The very slow release of cisplatin does not therefore appear to be due to a chemical bond with the polymer, but rather to the fact that a large proportion of cisplatin is trapped in the matrix.

The mixture was prepared as follows: during the first step, the active constituent, cisplatin, was mixed with the polymer. A predetermined weight of polymer was placed in a porcelain mortar. A known quantity of cisplatin was then added in small fractions. In the second step, the monomer was added, depending on the quantity of polymer taken, the volume of polymer being that recommended by the manufacturer. The constituents were then thoroughly mixed for four minutes to form a homogeneous paste.

This paste was then poured into the barrel of a stoppered syringe. The mixture was then expelled by the pressure of the piston into polyethylene moulds measuring 6.7 mm (inside diameter) by 10.3 mm in height (cylindrical mould, Prolabo, Paris, France). The implants were left in the moulds for 24 hours to allow complete polymerization to take place, and then tipped out and kept in darkness and at room temperature. The in-vitro release of cisplatin was investigated by placing the implants in a release medium with the following composition: sodium chloride: 9 g, distilled water, qsp 1000 ml, 1 N hydrochloric acid, qsp, pH = 4. After weighing, the implants were placed in the release medium at 37 °C and stirred in darkness. Samples were taken at regular intervals and an equal volume of fresh medium added to replace the reaction mixture removed. "Sink" conditions were maintained, i.e., the concentration of cisplatin in the release medium was never more than one tenth of the saturation concentration (i.e., 100 mg of cisplatin per liter).

The in-vitro release data obtained from implants containing various loads of cisplatin (from 1 to 20% w/w) are shown as a function of time: the quantities released were related to the initial concentration of cisplatin in the implants. For instance, after 90 days, the implants with the highest load had released about 12% of cisplatin, whereas implants containing 1% had released only 3% under these experimental conditions. It should also be noted that the release was incomplete from all the implants and never reached 100% of the initial load.

In the case of pure PMMA films, diffusion experiments have shown that cisplatin in solution had great difficulty in crossing even a thin membrane. Cisplatin therefore appears to be unable to cross pure PMMA. In the case of PMMA/MMA films, we found that up to a certain thickness, cisplatin was readily able to diffuse across the membrane. This diffusion can be accounted for by the structure of the polymer, which is not a uniform matrix, but a layer of spheres of PMMA linked to one another by the polymerized monomer. In solution, cisplatin must be able to diffuse into the relatively less compact zone between the spheres, which may have defects of structure and cohesion. However, at thicknesses from 133 microns, diffusion is slower, as if a thicker layer of spheres impedes the diffusion of the active constituent. These findings should be interpreted in the light of the structure of the cement viewed under electron microscopy, which reveals areas of regular polymer and defects, fissures which doubtless permit the diffusion of cisplatin.

Clinical Experience

The clinical research was done at the Henri Mondor hospital and has confirmed the experimental data obtained in animal studies. It provided the basis of the protocol for clinical use.

During surgery, a dose of 100 mg of MTX mixed with a complete dose of cement (46 g of polymer and 20 ml of monomer) was administered, followed by an intramuscular administration of folinic acid between 72 and 86 hours later. This was well tolerated by the patient. The local concentration of MTX found in the drains within the first few hours may reach levels 10,000 times greater than the plasma concentration and remained 100 times greater than the plasma concentration for the next three days if the drain was kept in place. Systemic distribution of this local chemotherapy was observed, as can be seen from the blood levels of MTX. The release and diffusion of MTX from the cement was continued well beyond 10 days (when MTX could still be assayed in one patient), since urinary excretion continued for at least three weeks.

In the case of cisplatin, which has a lower rate of diffusion from the cement than methotrexate, a dose of 200 mg of cisplatin mixed with one packet of cement was used without any postoperative adverse hematological or renal effects being observed. If further developments in the investigations we have initiated [7–9] confirm these early findings, this method of local neoplastic chemotherapy could offer an adjuvant therapy which is likely to be easier to handle. There is, of course, no question that this therapy could offer a substitute for systemic chemotherapy or radiotherapy when these therapies are indicated. Several other studies [29–32] have confirmed the experimental data of the diffusion of antimitotics from cement.

References

1. Buchholz HW, Engelbrecht H. Uber die Depotwirkung einiger Antibiotica bei Vermischung mit dem Kunstharz Palacos Chirurg 1970;41:511–5.
2. Carlsson AS, Josefsson G, Lindberg L. Revision with gentamicin-impregnated cement for deep infections in total hip arthroplasties. J Bone Joint Surg [Am] 1978; 60-A:1059–64.
3. Elson RA, McGechie DB. Antibiotics and acrylic bone cement. J Bone Joint Surg [Br] 1976;58-B:134.
4. Elson RA, Jephcott AE, McGechie DB, Verettes D. Antibiotic-loaded acrylic cement. J Bone Joint Surg [Br] 1977;59-B:200–5.
5. Fischer L-P, Gonon G-P, Carret J-P, Vulliez Y, de Mourgues G. Association methacrylate de methyle (Ciment acrylique) et antibiotique: etude bacteriologique et mecanique. Rev Chir Orthop 1977;63:361–72 (Eng. abstr.).
6. Marks KE, Nelson CL, Lautenschlager EP. Antibiotic-impregnated acrylic bone cement. J Bone Joint Surg [Am] 1976;58-A:358–64.
7. Hernigou P, Thiery JP, Benoit J et al. Release of antimitotic drugs from acrylic cement and plaster. Eur Surg Res 1987;19(Suppl 1):25.
8. Hernigou P, Thiery JP, Benoist M et al. Etude experimentale sur l'osteosarcome d'une chimiotherapie locale diffusant a partir de ciment acrylique chirurgical et de platre. Rev Chir Orthop 1987;73:517–25.
9. Hernigou P, Thiery JP, Benoit J et al. Methotrexate diffusion from acrylic cement. Bone Joint Surg 1989;71-B:804–11.
10. Hernigou P, Brun P, Thiery JP et al. Antimitotic loaded acrylic cement. In: Langlais F, editor. Berlin: Springer Verlag, 1991.

11. Hernigou Ph, Brun B, Autier A et al. Osteosarcoma in Adolescent and Young Adults. Boston: Kluwer Academic, 1993.
12. Hernigou Ph, Brun B, Astier A et al. Diffusion of methotrexate from acrylic surgical cement. Cancer Treat Res 1993;62:231–5.
13. Harrington KD, Sim FH, Enis JE et al. Methylmethacrylate as an adjunct in internal fixation of pathological fractures: experience with 375 cases. J Bone Joint Surg (Am) 1976;58-A:1047–55.
14. Strube HD, Komitowski D. Experimental studies of the treatment of malignant tumors with bone cement. In: Enneking WF, editor. Limb Salvage in Musculoskeletal Oncology. London: Churchill Livingstone, 1987; 459–69.
15. Brodey RS, Abt DA. Results of surgical treatment in 65 dogs with osteosarcoma. J Am Vet Med Assoc 1976; 168:1032–5.
16. Ling GV, Morgan JP, Pool RR. Primary bone tumors in the dog: a combinated clinical, radiographic and histologic approach to early diagnosis. J Am Vet Med Assoc 1974;165:55–67.
17. Misdorp W, Van der Heul RO. Tumours of bones and joints. Bu U WHO 1976;53:265–82.
18. Owen LN. Cancer chemotherapy. Vet Ann 1979;19: 204–11.
19. Parodi AL. L'osteosarcome (sarcome osteogenigne) chez le chien. Chirurgie1970;96:75–80.
20. Pool RR. Bone and cartilage. In: Moulton JE, editor. Tumors in Domestic Animals. 2nd ed. Berkeley: CA, University of California Press, 1978;89–149.
21. Klein B, Pals S, Masse R et al. Studies of bone and soft-tissue tumours induced in rats with radioactive cerium chloride. Int J Cancer 1977;10(1):12–19.
22. Thiery JP, Perdereau B, Gongora R, Gongora G, Mazabraud A. Un modele experimental d'osteosarcome chez le rat: 11. L'osteosarcome greffable du rat. Sem Hop Paris 1982;58:1686–9.
23. Baum ES, Gaynon P, Greenberg L, Krivit W, Hammond D. Phase II study of cis-dichlorodiammineplatinum(II) in childhood osteosarcoma: Children's cancer study group report. Cancer treatment Reports 1979;63(9–10): 1621–7.
24. Hiroyuki Tsuchiya, Katsuro Tomita, Hidetoshi Yasutake, Yasutaka Takagi, Shinichi Katsuo et al. Intra-arterial cisplatin and caffeine with/without doxorubicin for musculoskeletal high-grade spindle cell sarcoma. Oncol Rep 1994;1:27–36.
25. Jaffe N, Keifer R, Robertson R, Cangir A, Wang A. Renal toxicity with cumulative doses of Cis-diamminedichloroplatinum-II in pediatric patients with osteosarcoma. Effect on creatinine clearance and methotrexate excretion. Cancer 1987;59:1577–81.
26. Litterst CL, LeRoy AF, Guarino AM. Disposition and distribution of platinum following parenteral administration of cis-Dichlorodiammineplatinum(II) to animals. Cancer Treat Rep 1979;63(9–10):1485–92.
27. Straw RC, Withrow SJ, Douple EB, Brekke JH, Cooper MF et al. Effects of Cis-Diamminedichloroplatinum II released from D,L-Polylactic acid implanted adjacent to cortical allografts in dogs. J Orthop Research 1994;12: 871–7.
28. Withrow SJ, Thrall DE, Straw RC, Powers BE, Wrigley RH et al. Intra-arterial cisplatin with or without radiation in limb-sparing for canine osteosarcoma. Cancer 1993;71: 2484–90.
29. Janmin L. Experimental observations on acrylic bone cement containing antitumor drugs. Natl Med J China 1989;69:143.
30. Marshall GJ, Kirchen ME, Lee JH, Menendez LA. The effect of methotrexate eluted from bone cement on giant cell tumour lines in vitro. Trans Orthop Res Soc 1990;15:537.
31. Wang HM, Galasko CS, Crank S, Oliver G, Ward CA. Methotrexate-loaded acrylic cement in the management of skeletal metastases. Clin Orthop 1995;312:173–86.
32. Wu Yang Guan, Wang Tai Yi, Ma Yun Zhi, Sun Shu Zhen. Experimental research on the use of an antineoplastic drug with a bone implant. Int Orthop 1990;14:387–91.

I – Biocompatible Materials

I C – Mechanical and Physicochemical Aspects

11 Striated Muscles, an Underestimated Natural Biomaterial: Their Essential Contribution to Healing and Reconstruction of Bone Defects

H. Stein and M. Solomonow

Mechanical and Biological Factors

Introduction

Surgery of the musculoskeletal system is the most vibrant, quickly developing and enlarging reconstructive surgical specialty of this century. The current ten years are dedicated to this subject, and entitled "The bone and joint decade" which is a tribute to this subject's significance.

The past 35 years have been the stage for more significant developments and advances in reconstructive surgery of the musculoskeletal system then all the previous decades from the time the name of "Ortho-Paeis" was cornered during the industrial revolution.

Studies into the mechanical properties of the thin-wire hybrid three-plane, circular external fixator have opened and enlarged the understanding of biological processes stimulated and supported by the above-mentioned mechanical environment. Thus, with time, the crucial role played by striated muscles in the physiology of bone growth and repair is attaining broader recognition and understanding.

Mechanical Properties

Orthopedic surgeons need to be updated on mechanical terminology.

Compliance is the inverse of stiffness.

Higher compliance allows higher deformation.

Stiffness is the ratio between load and deformation. If measured in bending, it is expressed as the ration between the bending moment and the angular displacement it has caused. It is expressed in newtonmeters by degree (Nm/degree). This modality is well accepted and used by engineers but less so by medical graduates. The latter prefer the use of compliance.

Stiffness relates inversely to the angle of deformation. In vivo, bones are surrounded by soft tissues. The latter create *ligamentotaxis*, which opposes bone displacement, *provided bone integrity* is preserved.

The ring fixation frame, introduced by Ilizarov into clinical practice, is minimally invasive. It is connected to bone by thin K-wires which create minimal damage to both muscle and bone. These wires are the only means of osseous fixation for the frame, and the varying degrees of tension applied to them determine their frame's stiffness.

The ring configuration of the frame resists torsion, shear, and bending.

The factors which determine a ring fixator's stiffness are:

- the K-wire tension,
- the ring diameter,
- the centric or eccentric location of the bone within the ring.

In clinical practice, the ring diameter is often dependent on the circumference of the soft tissues present at the site. The larger the ring diameter, the lower the stiffness. The anatomical tissue topography further dictates the location of the bone proper in the ring. The more eccentric the bone, the higher the stiffness. Hence, the only control over frame stiffness left to the surgeon is the control of the tension applied to the transfixing K-wires.

The Biological Result

The mechanical properties of external fixation frames have a direct influence on the rate of fracture healing [1–27].

In other words, the mechanical properties of external fixation frames have an unequivocal and direct influence on the biological conditions needed for successful (bone gap) fracture healing. This issue has been discussed also by Chao in his presentation to SICOT in Amsterdam [28].

Ring fixation frames harbor a "trampoline" effect which allows both axial compression and distraction during gait. Intermittent distraction appears to be of crucial importance for physiological bone growth. The proof of this statement are tubercles, tuberosities, and bone outgrowths wherever muscles are inserted into bone in the skeleton. Muscle tonus consists of intermittent contractions of muscle fibers. At the site of muscle tendon insertion, these physiological muscle contractions create intermittent distraction. The latter induce local bone growth, the end result being a tubercle, a tuberosity, or an outgrowth.

Frame stiffness of any ring fixation frame is one of its more important mechanical properties. This stiffness depends upon the frame's K-wire tension. For how long does this tension persist in vivo? Under continuous axial loading, it has been shown that the loading causes a decline in the tension of transverse positioned K-wires [29]. *Continuous* – in vivo – measurements of the residual tension present in such K-wires in experimental animals has shown a rapid decline within two weeks of mounting the frame, the measured tension values disappearing altogether within five weeks of tensioning them in the rings. Nevertheless, the first signs of intramembranous ossification in the fracture gap were radiographically detectable 14 days after fracture stabilization by the ring frame, the whole fracture gap was radiographically detectable 14 days after fracture stabilization by the ring frame, and the whole fracture gap was filled by woven bone on day 35.

The bridging callus starts and proliferates from the muscle bed of the fracture. In the tibia, where the significant muscle bed is always postero-medial, this is easy to observe and follow (Figure 11.1). In both clinical cases and in experimental fractures in the laboratory, the callus has been shown to start always in the posterior plane, next to the muscle bed, and progress from there anteriorly [30–32]. In these publications, the intimate relationship between the developing callus and the muscle bed has been described.

Most of the load on weight bearing is transmitted through the fixator [33]. In clinical practice, the choice of the fixator frame means choosing the mechanical configuration which will induce the optimal enhancement in callus formation and fracture healing [33]. A decline in frame stiffness, as initiated by a spontaneous decline in K-wire tension, appears to be an important stimulus for bone formation and healing [7]. Thus, it is very likely that bone mass is regulated by mechanical strain [5]. In

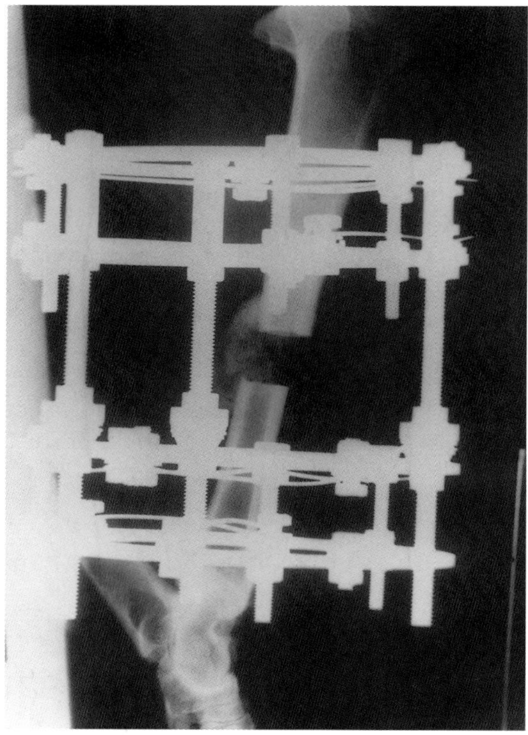

Figure 11.1. Experimental fracture in a sheep tibia (lateral view). Note the healing callus, which starts on the posterior plane next to the muscle bed.

fractures, or in the presence of bone gaps, *this mechanical strain needs to be induced and maintained in the muscle tissue* present in the fracture bed. Micromovements in the axial plane of the fracture stabilized by a ring fixator induce efficient medullary and periosted callus formation. This callus undergoes early metaplasia into membranous bone, and thus stabilizes the fracture gap [34]. Similar observations have been reported by Wu [27] and by Younger [35]. The mechanical efficiency of the stabilizing fixation frame can be evaluated by the amount of fracture callus formation, the stiffness of the latter being capable of early replacing the spontaneously declining stiffness of the ring fixation frame.

In conclusion, therefore, the preservation of the physiological function of striated muscles in the fracture bed appears to be the key biological factor for the efficient repair of a fracture with or without bone loss.

Muscle Assessments

The tools to assess the physiological function of striated muscles are either muscle charting or EMG.

Muscle Charting

This is a time-honored manual semi-quantitative measurement, practiced by physiotherapists. It is a time consuming, not strictly scientific method, the human factor being responsible for all variations in it.

During the poliomyelitis epidemics, there was a whole generation of highly experienced physical therapists who, by repeated muscle charting of the same patient, could provide reliable proof of the progress of the disease, or improvement of the patients due to active aggressive physiotherapy.

At present, reliable, quantitative computerized muscle charting has become available with the aid of ARCON (http://www.arcon-rehab.com/dynamic.asp). This computerized assessment is particularly reliable in measuring the motor power of various muscle groups.

Electromyography as an Electrophysiological Tool

Its usefulness and pitfalls are described in detail in the following section by Professor M. Solomonow.

Electrophysiology and Biomechanics

The EMG is an important tool in the assessment of muscle activities in various occupational tasks, only if used properly. The recording, processing, and interpretation of the EMG should be done while considering the proper electrodes size, interelectrodes distance, muscle architecture, contraction rate, motor unit recruitment pattern, muscle length, cross-talk from nearby muscles, sampling rate, filter bandwidths, and smoothing time constants. Each of the above factors, if not considered, can introduce substantial error and result in false or misleading interpretation.

The EMG is a "by-product" of muscle activity and is therefore a dependent signal, influenced by various physiological and anatomical factors as well as the technical aspects of the recording/processing protocol.

The anatomical factors that have profound impact on the EMG are muscle fiber penation and predominant fiber composition (fast or slow twitch types). Physiological factors that strongly impact the EMG are motor unit firing rate and recruitment pattern, force generation rate (isometric), and muscle length changes (shortening or lengthening). Secondary anatomical/physiological issues are the size of the muscle, its depth (below the skin and other muscles), the presence of adipose tissue, nearby muscles, and proximity to large blood vessels and bones.

The recording/processing hardware and software that influence the surface EMG are electrode size, material, interelectrode distance, electrode/skin interface, electrode location relative to muscle shape and axis, differential recordings, recording frequency bandwidth, sampling rates, and smoothing time constant as the most important.

It becomes clear that the EMG is a complex dependent signal, and its recording and interpretation should not be taken lightly. In this paper, several of the most important factors affecting the EMG will be reviewed in the context of the EMG vs. force relationship and finally, insights into the controversial issue of EMG cross-talk will be highlighted.

EMG vs. Force Relations and Predominant Fiber Type

Different muscles are composed of different proportions of fast and slow twitch fibers. The soleus, for example is composed of 80% slow twitch, fatigue-resistant fibers of oxidative metabolism whereas its companion gastrocnemius muscle is composed of nearly 80% fast twitch, fast-to-fatigue fibers activated by limited stored energy. In highly controlled studies in which both soleus and gastrocnemius were activated with the same contraction control, the EMG vs. force relations of the two muscles were significantly different. The relations were linear for gastrocnemius but distinctly non-linear for the soleus [36].

While such findings explain, in part, the controversy of the linearity (or non-linearity) of the EMG–force relations, it issues a warning against the use of the relations without considering the predominant fiber composition of the tested muscle. Such data is available in the literature [37]. Investigators should, however, also consider that predominant the fiber type of a muscle can change in individuals who acquired skill in a certain task.

EMG vs. Force and Contraction Rate

It is known that the rate at which the muscle contracts and generates force may significantly influence the maximal force obtainable. How would such phenomena influence the EMG vs. force relations? In highly controlled studies, Solomonow et al. [38] recorded the EMG for a muscle increasing its force from 0 to 100% (maximal force) at rates which varied from 36% to 360% maximal force/sec. Calculations of the normalized EMG vs. force curves did not yield any statistical difference. The maximal force obtained in the fast contraction, however, was 25% larger than that of the slowest contraction.

Investigators are cautioned to compensate for the contraction rate when EMG is recorded as a measure of force, or risk up to 25% error that can be associated with highly misleading conclusions. A proper approach is to define the EMG vs. force curves at the contraction rate that is to be tested in the field.

EMG vs. Force and Motor Unit Recruitment

It is known that skeletal muscles utilize the size principle when recruiting motor units in order to increase the force. Small motor units are recruited initially, with gradually larger motor units as more force is required [39]. It is also known that motor unit recruitment is completed by the time the initial 50–80% MVC is generated. The remaining force segment up to 100% MVC is accomplished by increasing the firing rate of the motor units. It is also known that a given muscle recruits all its motor unit pool in the initial 50% MVC during fast contractions, and in the initial 80% MVC in a slower, more accurate contraction.

Solomonow et al. [40] demonstrated that changing motor unit recruitment strategy has pronounced effect on the EMG vs. force relations. When all the motor units of gastrocnemius muscle were recruited in the initial 50% maximal force segment, the EMG vs. force was linear, however, when all the motor units were recruited over the initial 60%, 70%, and 80% of the maximal force, the relations were non-linear. Furthermore, the non-linearity was different at each of the three contractions, with the full recruitment completed at 80% maximal force being the "most" non-linear.

The advice for proper use of the EMG vs. force relations in ergonomics research is to assess the recruitment pattern of the muscle in question [40,41] before attempting to relate EMG recording to force.

EMG Cross-talk and Adipose Tissue

Many studies of static and dynamic movements of the extremities and spine demonstrate that co-activation of agonist and antagonist muscles takes place. However, the co-activation principle became controversial, as the low-level EMG recorded from the antagonist muscles was considered by many to be cross-talk from the agonist muscles transmitted by volume conduction in the tissues. The sources of cross-talk could be numerous, including; electrodes too large for the muscle under investigation, interelectrode distance larger than necessary for the given muscles, proximity to large blood vessels, proximity to major bones, etc. If one utilizes the correct electrodes size, interelectrode distance, and placement as delineated by Fugelvand et al. [42], such cross-talk can be fully eliminated.

Cross-talk EMG was reported despite proper utilization of electrodes, re-initiating the controversy. A more detailed study by Solomonow et al. [43], in a highly controlled experimental animal using surface and intramuscular wire electrodes, gave a clear insight into the problem. Surface EMG recorded from muscles covered by adipose tissue gave as much as 36% (of 100% EMG from nearby muscles) cross-talk whereas wire electrodes from the same muscle yielded only 1–2% cross-talk. Cutting the motor nerve of all the muscles in the leg confirmed that adipose tissue was indeed the source/cause of the cross-talk. A warning is issued against the recording of surface EMG from muscles covered by adipose tissues, such as abdominal and buttock muscles. Wire recordings should be employed in such cases. Otherwise, with the absence of adipose tissue, the cross-talk is at the noise level and should be of no concern for the investigator.

EMG Processing Issues

Several factors associated with the recording and processing of the EMG have pronounced impact on the validity of the data and its interpretation. The raw EMG signal is commonly sampled into personal computers for later processing. One of the important aspects of such a procedure is the sampling rate used. The power spectrum of the surface EMG contains frequencies in the range 0–350 Hz. Less than 5% of the power is in frequencies above 350 Hz. Furthermore, movement artifacts associated with isometric or most free movements are associated with frequencies below 10 Hz. Therefore, in order to eliminate movement artifacts, a high-pass filter with a frequency cut-off of 10 Hz is necessary. Also, in order to prevent any high frequencies (above 350 Hz) from contaminating the signal and to avoid aliasing, a low-pass filter with a cut-off frequency of 350 Hz is required. Overall, a bandpass filter of 10–350 Hz will preserve all the important features of the surface EMG.

Sampling into the computer requires a rate of at least twice the highest frequency in the signal. In this case 350 Hz × 2 = 700 Hz. For better accuracy, sampling rates higher than 700 Hz should be used. Sampling rates lower than 700 Hz will deform the EMG, cause loss of important features, and will lead to wrong or misleading conclusions. Furthermore, utilization of a bandpass filter with a frequency cut-off above 350 Hz has very marginal improvement in signal content, but also requires higher sampling rates, which in turn limit the number of channels that could be used, limit A/D capabilities, and heavily tax storage space on a PC and time required to process the data.

Another important issue is the processing of the EMG into a smoothed curve that represents the changes in the level of activity of the muscle. Commonly, a low-pass filter of a time constant ranging between 50 and 300 ms is used. If such a filter is constructed from hardware, one should assess the time delay imposed by the filter and compensate for such a delay when comparing EMG and force on the same time axis. Without applying the time compensation, and especially for fast movements, a discrepancy between force and EMG may lead to false conclusions.

EMG and Muscle Fatigue

When muscle fatigue is being studied via its EMG, the power spectra frequencies are commonly used. The Median Frequency (MF) of the power spectra represents the average conduction velocity of action potentials in the muscle. It is well known that the average conduction velocity decreases with muscle fatigue, giving rise to a decrease in the MF. The assessment of muscle fatigue could be useful in ergonomics, especially in optimizing task durations, work – rest time ratios, etc. Common pitfalls in using the MF of the surface EMG to study fatigue are especially in low-level, prolonged contractions in the presence of noise. Inherently, contractions producing low levels of force are producing low-amplitude EMG. The assessment of fatigue under such circumstances can become a grossly misleading procedure if the noise is not eliminated.

A method to eliminate such a problem was developed by Baratta et al. [44]. It consists of a noise-subtraction step before the power spectra estimate is calculated. The results are much improved accuracy in the MF calculations and its proper and reliable interpretation.

MVC Determination

Determination of the Maximal Voluntary Contraction can introduce errors in the interpretation of EMG studies if not obtained correctly. In general, if an individual is asked to provide a maximal elbow flexion, for example, he will provide a strong contraction which is perceived by him as maximal. If the force output is displayed on a screen as a moving line, and the individual is asked to exceed the line location obtained from his perceived maximal contraction by 10%, it is easily obtainable. Such feedback attempted normally results in final maximal force which is 25–35% higher than the maximal force obtained in the first trial.

Such errors may have an impact on the interpretation of the EMG, and EMG vs. force relations, as well as MF calculations across the full force range. The error, however, is significantly amplified in the low force ranges, and especially in low-level, static contractions where the casually obtained MVC and the true MVC obtained by training could be miscalculated by 300% at a 10% MVC level.

It is difficult to give any advice to avoid or compensate for this type of error. The true (e.g., trained) MVC is the most reliable baseline for all types of EMG work where relationships to force are made.

Bibliography

Acierno S et al. Manual For Understanding and Using EMG. Bioengineering Laboratory, LSU Health Sciences Center, 1996.

Solomonow M et al. EMG – force model of a single muscle acting across the joint: dependence on joint angle. J EMG & Kinesiology 1991;1:58–67.

Solomonow M, Guzzi A et al. Antagonistic muscles: gravity, joint geometry and recruitment. Am J Phys Med 1986;65:223–42.

Standards for Reporting EMG Data. J EMG & Kinesiology (printed in every issue since 1996;6(1)).

References

1. Calhoun JH, Li F, Bauford WL, Lehman T, Ledbetter FR, Lowery R. Rigidity of half-pins for the Ilizarov external fixator. Bull Hosp Jt Dis Orthop Inst 1992;52(1):21–6.
2. Ilizarov GA. Clinical application of the tension stress effect for limb lengthening. Clin Orthop 1990;250:8–26.
3. Ilizarov GA. Transosseous osteosynthesis. Heidelberg: Springer, 1991;3–279.
4. Jorgens C, Schmidt HG, Schumann U, Fink B. Ilizarov ring fixation and its technical application. Unfallchirurg 1992;95(11):529–33.
5. Paley D, Catangi M, Argnani F, Villa A, Benedetti GB, Cattaneo R. Ilizarov treatment of tibial non-unions with bone loss. Clin Ortho 1989;141:146.
6. Gasser B, Bowman B, Wyder D, Schneider E. Stiffness characteristics of the circular Ilizarov device as opposed to conventional external fixator. J Biomech Eng 1990; 112:15.
7. Aronson IA, Harp JH. Mechanical considerations in using tensioned wires in a transosseous external fixation system. Clin Ortho 1992;280:23–9.
8. Monticelli G, Spinelli R. Limb lengthening by closed metaphyseal corticotomy. Ital J Ortho Traumatol 1983;4: 139–50.
9. Hardy JM. Le fixateur externe monolateral "CAPUCINE". Presented at the 18th SICOT meeting. September 1990. Montreal, Canada. Poster No. 94, p. 492.
10. Wasserstein I, Correl J, Niethard FU. Closed distraction epiphysiolysis for leg lengthening and axis correction of the leg in children. Z Orthop 1986;124(B):743–50.
11. Wagner R. Operative lengthening of femur. Clin Orthop 1978;136:125–42.
12. Green SA, Harris NL, Wall DM, Iskanian J, Marinow H. The Rancho mounting technique for Ilizarov method. A preliminary report. Clin Orthop 1992;280:104–16.
13. DeBastiani G, Aldergheri R, Renzi-Brivio L, Trivella G. Limb lengthening by callus distraction (Callotasis). J Pediatr Orthop 1987;7:129–34.
14. Kenwright J. The influence of cyclic loading upon fracture healing. J R Coll Surg Ed 1989;34(3):160.

15. Fleming B, Paley D, Kristiansen T, Pope M. A biomechanical analysis of the Ilizarov external fixator. Clin Orthop 1989;241:95–105.
16. Green SA. The use of wires and pins. Techn Orthop 1990;5:19–25.
17. Alonso JE. Regazzoni P. The use of Ilizarov concept with the AO/ASIF tubular fixator in the treatment of segmental defects. Orthop Clin North Am 1990;21(4): 655–65.
18. Uhli RL, Goldstock L, Carter AT, Lozman J. Hybrid external fixation for bicondylar tibial plateau fractures. Presented at the 61st American Academy of Orthopaedic Surgeons Meeting. 26 February 1994, New Orleans, LA 278, p. 192.
19. Weiner L. Fixation for complex tibial plateau fractures hybrid fixator. Presented at the Orthopaedic Trauma Association Specialty Day Symposium, 61st American Academy of Orthopaedic Surgeons Meeting, 26 February 1994, New Orleans, LA
20. Chamay A, Tschentz P. Mechanical influence in bone remodeling. Experimental research on Wolffs law. J Biomech 1972;5:173.
21. Goodship AE, Kenwright J. The influence of induced micro-motion upon the healing of experimental tibia fractures. J Bone Joint Surg 1985;67(b):650.
22. Kempson GE. Campbell D. The comparative stiffness of external fixation frames. Injury 1981;12:297.
23. Kristiansen T, Fleming B, Neal G, Reinecke S, Pope MH. Comparative study of fracture gap motion in external fixation. Clin Biomech 1987;2:191.
24. Panjoli MM, White AA, Wolf JW. A biomechanical cyclic compression of fracture healing in long bones. Acta Orthop Scand 1979;50:653.
25. Rubin CT, Lonjon LE. Regulation of bone formation by applied dynamic loads. J Bone Joint Surg 1987;66(A): 397.
26. Sarmiento A, Schaeffer JF, Beckerman L, Latta L, Emis JE. Fracture healing in rat femur is affected by functional weight bearing. J Bone Joint Surg 1977;59(A):367.
27. Wu JJ, Shyr HS, Chao EYS, Kelly PJ. Comparison of osteotomy healing under external fixation devices with different stiffness characteristics. J Bone Joint Surg 1984;66(A):1258.
28. Chao EYS. Orthopaedic biomechanics. The past, present and future. Int Orthop 1996;20:239–43.
29. Stein H, Perren SM, Moscheiff R, Baumgart F, Cordey J. The spontaneous decline in the transfixing K-wire's tension of the circular external fixator. Orthopedics 2001 (in press).
30. Stein H, Cordey J, Perren SM. Segment transport for biological reconstruction of bone defects. Injury 1993;Suppl 24(2):20-4.
31. Stein H, Coleman R, Mosheiff R, Cordey J, Rahn BA, Reznick A. Changes induced in limb muscles by distraction osteogenesis. Trans 43rd ORS Meeting, San Francisco, CA 1997, p. 703.
32. Mosheiff R, Cordey J, Rahn BA, Perren SM, Stein H. The vascular supply to bone formed by distraction osteogenesis. An Experimental Study. J Bone Jt Surg 1996;78-B:497–8.
33. Delprete C, Golo MM. Mechanical performance of external fixator with wires for the treatment of bone fractures. Part 1. Load displacement behavior. J Biomech Eng 1993;115:29–36.
34. Stein H, Cordey J, Mosheiff R, Perren SM. Observation on the stiffness of neogenetic bone produced by distraction or segment transport and its relationship to bone density. In: Wolter D, Hansis M, Havemann D, editors. 150 years Fixateursysteme. Berlin, Heidelberg, New York: Springer Verlag, 1995;47–9.
35. Younger ASE, Mackenzie WG, Morrison JB. Femoral forces during limb lengthening in children. Clin Orthop 1994;301:55–63.
36. Solomonow M et al. EMG – force model: dependence on control strategy and fiber composition. IEEE Trans Biomed Eng 1987;34:692–702.
37. Johnson M et al. Data on the distribution of fiber types in 36 human muscles. J Neurophysiol 1965;28:85–99.
38. Solomonow M et al. EMG – force of skeletal muscle: contraction rate and motor units control strategy. EMG & Clin Neurophysiol 1990;30:141–52.
39. Henneman E et al. Functional significance of cell size in spinal motor neurons. J Neurophysiol 1965;28:560–80.
40. Solomonow M et al. EMG power spectra associated with recruitment strategies. J Appl Physiol 1990;68:1177–85.
41. Bernardi M et al. Motor unit recruitment strategy changes with skill acquisition. Eur J Appl Physiol 1996; 74:52–9.
42. Fugelvand A et al. Detection of motor unit action potentials with surface electrodes: electrodes size and spacing. Biol Cybernetics 1992;67:143–53.
43. Solomonow M et al. Surface and wire EMG cross-talk in neighbouring muscles. J EMG & Kinesiology 1994;4: 131–42.
44. Baratta RV et al. Methods to reduce the variability of EMG power spectrum estimates. J EMG & Kinesiology 1998;8:279–85.

12 Wound Healing: Potential Therapeutic Modulation

W. H. Akeson and A. Giurea

Introduction

Injury invokes a vigorous healing response in soft tissue as it does in bone. The needs of individual survival undoubtedly required such an evolutionary response as a survival mechanism. Inevitably the control mechanisms of the healing response will not infrequently extend beyond the range of the ideal. When the response to soft tissue injury is excessively exuberant, complications are encountered such as keloid formation, peritoneal adhesion, intestinal stricture, tendon adhesion, epidural fibrosis, and arthrofibrosis with attendant joint contractures, to name just a few. An important case in point is the loss of range of motion which occurs occasionally after knee injuries or knee surgery in spite of seemingly appropriate initial management and subsequent rehabilitation. It follows that an understanding of the healing process is necessary to explore potential therapeutic remedies to minimize these complications. The purpose of this chapter is to review the biological processes leading to wound healing and to outline potential pharmacological interventions to impede or accelerate those processes.

Overview of Normal Wound Healing

The healing of soft tissues is a continuum of five general phases [1].

hemorrhage and formation of a blood clot followed by platelet degranulation

leukocyte trafficking into the wound

inflammation, during which neutrophils phagocytose bacteria and debris, degranulate releasing proteolytic enzymes, and the monocytic cells become transformed into macrophages

repair, during which macrophages stimulate fibroblast proliferation and the resulting extracellular matrix synthesis

remodeling of the initial scar

Initial bleeding is followed promptly by platelet aggregation and degranulation, and vasoconstriction of neighboring vessels. Extravasation of blood proteins occurs with the initial hemorrhage which includes fibronectin, vitronectin, and fibrinogen. Fibrin formation and clotting ensue. An inflammatory process occurs next triggered by growth factors and cytokines initiating a cascade of events starting with activation of selectins and cell adhesion molecules (CAMs) which modulate the initial leukocyte trafficking across the endothelium of post-capillary venules. Platelets will have released TGF-beta and TNF-alpha among other factors immediately after wounding, causing leukocyte margination, or "rolling". This results in slowing of the passage of leukocytes through the venule which allows the leukocytes to monitor the environment for chemoattractants. The CAMs and chemoattractants are programmed in a sequential manner to form a leukocyte-endothelial cell adhesion cascade [2–5]. Chemoattractants bind to serpentine receptors on the leukocytes, which then activate G proteins which signal up-regulation of integrins on the leukocyte cell surface. The arrested cells subsequently undergo transmigration between the endothelial cells under the influence of the adhesion proteins plus PECAM-1 (platelet endothelial cell adhesion molecule-1), chemoattractant gradients aided by chemoattractant–proteoglycan binding [6–8], and chemotaxins in the wound environment. After entry into the wound, these cells are free to react with matrix proteins, and to stimulate mitogenesis and additional chemotaxis through autocrine and paracrine functions.

The sequence of leukocyte classes recruited into the wound environment is observed to include neutrophils in the first few hours followed by monocytes at 18 hours and T cells and macrophages at 36 to 48 hours. Sequential changes in the adhesion molecule system is presumably the mechanism of selection and control of specific leukocyte transendothelial migration. Of note is the observation that there are clear differences in adhesion requirements for particular types of inflammation [2,9]. The polymorphonuclear leukocytes which initially populate the wound sterilize the wound by phyagocytosis and the release of oxygen radicals. They bind to complement fragments on bacterial surfaces. Neutrophils are quickly followed by lymphocytes and monocytes. Between days 3 and 4 monocytes proliferate rapidly. The infiltrating cells release a plethora of cytokines, growth factors, and chemokines which collectively orchestrate the local cellular response including the transformation of monocytes into macrophages. The macrophages phagocytose cellular, bacterial, and matrix debris. Macrophages assume a crucial role in further modulating the healing process by stimulating the proliferation of reparative cells, including fibroblasts, epithelial, and capillary endothelial cells [10]. This step is followed by the synthesis of complex components which form the extracellular matrix. The breadth of activities of TGF beta have given it the appellation "the conductor of the symphony" with regard to the healing process [11]. Initially, the extracellular matrix is high in proteoglycan content, which soon gives way to increased collagen synthesis and scar formation. The initial scar presents as a random display of collagen fibers in what is commonly referred to as a haystack pattern [12]. The final chronic remodeling phase occurs over many months. Mechanical factors, particularly tensile stresses, are perceived by adhesion receptors and mechano-receptor of the fibroblasts, and the matrix is gradually remodeled by an iterative process of matrix resorption and new matrix synthesis. In this complex process the collagen molecules become aggregated by intra and inter-molecular crosslinks modulated in part by other matrix components such as the minor collagens and the small proteoglycans described in earlier chapters [12].

Definitive candidates for post-surgical scar control have recently emerged following an explosion of progress in understanding and prevention of a variety of chronic fibrotic conditions of skin, lung, heart, liver, kidney, and vasculature. Extensive studies on the processes of leukocyte trafficking in wounds has broadened the understanding of the underlying processes at work in the early wound. These concepts have been clarified and phenomenologically synthesized by Butcher [13] and others [3,14,15]. Progress in this field now provides an exciting opportunity to approach the problem of control of fibrosis with a plethora of potentially important scar-neutralizing agents. This discussion will focus on some of the new concepts which are of interest as potential therapeutic approaches to the problem of arthrofibrosis and related problems of interest to the orthopedic surgeon.

Fetal Healing

Fetal healing – the healing which occurs in the early fetus without scar – was observed in 1971 by Burrington in the fetal lamb [16]. Subsequently it was noted that the effect is not seen uniformly during development. Fetal healing is dependent on stage of gestation and on wound characteristics [17]. Incisional, scarless healing is most evident in the early fetal gestation periods, generally through mid-gestation. Later in gestation the healing of incisional wounds tends with time gradually to resemble the healing of adult wounds. In mice, for example, fetal wounds heal without scar at gd (gestational day) 14, but at gd 18 heal with scar (mouse term = 20 days) [18]. Excisional wounds heal without scar in the very early fetus, but lose that capacity sooner than in the case of the incisional wound healing with respect to time of gestation [17].

Fetal healing occurs in a unique extracellular matrix environment. Key conditions include a high concentration of high mw hyaluronan [18–24] and very low levels of TGF beta 1 and 2 [25,26]. Other differences are gradually being reported such as the presence of low levels of hyaluronidase [23,26], increased amounts of tenacin [27], and differences in content of the CD44 HA receptor noted to be 4x the level in the fetal compared to the adult wound [24].

If the early fetal environment could be reproduced therapeutically, excessive scar formation could potentially be prevented in strategic sites post-surgically. Indeed, several investigators have provided evidence that conditions which partially mimic the environment of fetal connective tissue matrix can, in fact, result in modified healing in certain animal models which demonstrate reduced scar formation, with more orderly collagen patterning, and with tensile strength of postoperative incisional wounds equivalent to that of normal adult wounds [18,20,26–28]. Potential applications of fetal healing concepts in orthopedic surgery include the modulation of scar formation after tenolysis performed to correct tendon adhesions, prevention of arthrofibrosis, and prevention of epidural adhesions, to name just a few examples. And strikingly, a new branch of surgery is on the doorstep by which the possibility of correcting certain fetal deformities in utero is becoming a reality [28].

Ferguson was the first to show that TGF beta 1 was missing in the connective tissue matrix of the early fetus [25,26]. He demonstrated that blockade of TGF beta 1 and 2 activity encouraged scarless healing in the skin of adult animal models [27]. Effective agents Ferguson used to inhibit TGF beta 1 and 2 included mannose-6-phosphate and decorin as well as antibodies against TGF beta 1 and 2.

A profusion of reports in recent years have described additional differences between the fetal and adult wound. An abbreviated listing of relevant observations includes the following: Adult wound fluids have a higher concentration of hyaluronidase than fetal wound fluids [23]; Fetal healing response to injury can be converted toward the adult form of healing by the injection of TGF beta in the ICR mouse fetus [29]; Hyaluronan injected repeatedly into sponge implants in mice (q 3 days) over one and two weeks altered the adult healing to resemble fetal healing in respect to the histological amount and character of collagen fibers. Injection of hyaluronidase into the sponges reversed that effect [30]; fetal articular cartilage heals/adult cartilage does not [31], (a possible focus of importance in the field of cartilage healing research); the fetal immune response differs – the cellular infiltrate is mainly a small number of macrophages, with few polymorphonuclear leucocytes [32,33]; fetal cytokine profile differences exist [33]; fetal fibroblasts have different collagen gel contraction ability [34,35]; metalloproteinases differ between fetal and adult wounds [36]; COL1A1 gene expression is absent in fetal fibroblasts but is up-regulated by the addition of TGF-beta [37]; Type V collagen alpha1(V)/alpha2(V) chains differ in fetal vs. adult sheep sponges, Type I collagen cross links increase during gestational development as scar formation begins to appear [38]; IL-8 is absent in fetal fibroblasts [39]; HA inhibits fetal platelet aggregation and impairs the release of PDGF-AB [40]; there is more rapid up-regulation of integrins (alpha 2, 3, 5, 6 and b; beta 4, and 6) in the fetal wound and more rapid re-epithelialization by keratinocytes [41].

This necessarily sketchy account of the recent literature suggests the breadth of interest in the fetal healing phenomenon. It is not yet possible to synthesize all these findings meaningfully, but many if not all may be derivative of the two central observations in the fetal wound: the high concentration of HA and the low concentration of TGF beta 1 and 2 which inhibit inflammation.

Hyaluronan

Hyaluronan is a ubiquitous component of the ECM and occurs transiently in both the cell nucleus and cytoplasm. It promotes cell motility, adhesion, and proliferation and has an important role in morphogenesis, wound repair, and tumor metastasis [42,43]. Cell motility is central to the effect of each of these processes. HA is actively synthesized during wound healing and is an important substrate for leukocyte migration during inflammation [24,42,44–47]. Disturbance of regulation of these processes has been alleged to be responsible for errors in morphogenesis, aberrant repair, exaggerated inflammatory responses, and tumorigenesis. HA binding, ligand specificity, and stimulation of signal pathways can be modulated by soluble forms of the receptors, by alternatively spliced cell surface isoforms, and by glycosylation variants of the receptors [42].

Scarless healing has been noted in the sheep embryo up through the 130th day of gestation during which time high levels of HA are found. After 130 days HA was observed to decline to a trough [21]. In an adult rat model it was possible to inhibit scar formation in skin wounds by treating the wound with a HA–protein–collagen form of HA [48]. In this experiment the treated wounds had more organized collagen fibers, less TGF beta 1 and 2, and increased TGF beta 3 (an inhibitor of TGF beta 1 and 2). An in vitro model has also been studied in which HA concentration was varied in fibroblast cell culture [49]. In this experiment streak "wounds" were created in confluent fibroblast cultures under the influence of exogenous HA (5 mg/ml). These wounds closed more rapidly than untreated wounds. Studies on incisional wound healing in fetal limb organ cultures to which exogenous HA was added with each change of media showed results similar to the scarless healing seen in other models mimicking fetal healing. Repair site collagen fibers in the treatment group had the typical "basket weave" pattern of normal dermal collagen in contrast to the less well organized collagen in the control groups [30].

Crucial to the work in the proposed project is the discovery of the bimodal action of HA in the extracellular matrix (ECM) with respect to reaction of HA with various receptors. A collection of inflammatory genes is induced in macrophages by HA oligosaccharides, but not by native high MW HA [50]. High MW HA is anti-angiogenic, whereas oligosaccharide degradation products of HA actively stimulate endothelial cell proliferation and migration and induce angiogenesis in vivo [51].

Clearly, it is the high MW form of HA which is the requisite molecular form of HA in the fetal scarless healing environment. A fascinating complement to this requirement is the observation that low hyaluronidase levels exist in the fetal connective tissue matrix [23]. A potential mechanism for control of this process has been found by the discovery that anti-CD44 mAb, which blocks the fibroblast cell receptor for low mw HA binding, also inhibits the low mw HA induced release of IL-12 by elicited macrophages [52]. Furthermore, native high mw HA has a dose-dependent inhibitory effect on induced gene expression by HA oligosaccharides [53]. Three cell surface receptors have been identified which are important for the cell interaction with HA of the matrix: CD44, RHAMM (Receptor for HA-Mediated Motility – the HA motogenic receptor), and ICAM-1 [42]. As with other adhesion receptors, binding of the cell via the HA-mediated receptors triggers signal transduction events. These events are central to the control of cytoskeletal structure and cell trafficking.

Also central to arthrofibrosis research is the recent discovery that TGF-beta is involved in RHAMM message regulation [54]. This finding presents an interesting juxtaposition of a counter-mechanism between TGFβ and HA. The half life of RHAMM mRNA has been shown to be increased threefold in cells treated with TGF-beta [55]. Possibly, by blocking TGF-beta, RHAMM synthesis may be inhibited a step which could thereby inhibit leukocyte trafficking into the area of concern [50].

The author's laboratory reported in the 1970s that a single injection of HA inhibited contracture formation by about 50% in an experimental animal model [56]. More recently, we have shown the efficacy in reducing epidural scar postoperatively with topical HA in a standard rat model [57]. The observation that HA injection into arthritic joints may be beneficial symptomatically is likely secondary to its anti-inflammatory action. That action would also hold promise as a neutralizing agent against the inflammatory response post knee-joint manipulation in the management complications of knee-joint contracture occurring postoperatively.

TGF-beta

The production of scar through the stimulation of collagen synthesis by CTGF (connective tissue growth factor) in the injured tissue is believed to be coordinated by TGF-beta. TGF-beta has been called the "conductor of the symphony" of the healing response by Grotendorst [11]. TGF-beta stimulates connective tissue cell growth, stimulates extracellular matrix synthesis, and modulates the immune response. It acts on both fibroblasts and on smooth muscle cells. It increases mRNA for CTGF. CTGF is chemotactic and mitogenic for connective tissue cells and stimulates extracellular matrix production. TGF-beta exists in tissue in a latent form, bound by LAP (latency-associated peptide), LTBP (latent TGF-beta-binding protein), decorin, and biglycan [58]. Recombinant LAP is a potent inhibitor of TGF-beta in vivo and in vitro [59]. TGF-beta exists in several isoforms including beta 1, 2, and 3. Neutralization of TGF-beta 1 and 2 or the addition of TGF-beta 3 reduces scarring in several models [27]. Different fibroblasts react differently to TGF betas, emphasizing the fact that all fibroblasts are not the same [58]. Work in our laboratory characterizing differences between ligaments of the knee has clearly shown that ligaments which heal well (medial collateral ligaments) have quite different fibroblastic characteristics than ligaments which fail to heal (anterior cruciate ligaments) [61–64].

TGF-beta is overproduced in fibrotic lesions and is a key factor in the pathogenesis of organ fibrogenesis [9,27,65,66]. Fibrosis resembles normal wound healing, but fails to terminate, leading to replacement of normal tissue with scar. Most fibrotic reactions (lung, heart, vascular system, kidney, liver, skin, brain, GI tract, synovial joints) appear secondary to trauma, infection, or inflammation. Studies using immunohistochemical techniques have demonstrated that TGF-beta is over-produced in areas of chronic fibrosis. The reason for the continued scar proliferation is not always clear, but it is possible that some fibroblasts become permanently altered and do not respond appropriately to the usual regulatory controls [60]. Both TGF-beta 3 and decorin, which inhibit the actions of TGF-beta 1 and 2, may represent local regulatory control elements [27,59,67]. The exaggerated behavior of TGF-beta in the fibrotic syndromes has been termed "The Dark Side of TGF-beta" in a paper by Border and Roushlati [3] and termed "The Good, the Bad and the Ugly" by Wahl [68] A hemarthrosis post trauma, postoperatively, or post manipulation will inevitably contain platelets which release cytokines into the joint. These cytokines, including TGF-beta, provoke an inflammatory response which can result in arthrofibrosis and joint contracture. The use of neutralizing agents against TGF-beta hold promise for minimizing this troublesome complication of knee injuries. Controlled clinical studies on their use should be forthcoming in the near future.

Decorin

Decorin is a powerful molecule with wide-ranging regulatory effects. Decorin is characterized as a member of the class of small proteoglycan molecules. It contains a protein core with leucine-rich motifs and a single side chain of dermatan/chondroitin sulfate. It is ubiquitously distributed in the extracellular matrix of mammals. It binds to fibrillar collagens, including minor collagens, influencing kinetics of fibril formation and final diameter of fibrils [69–70]. It has profound effects on matrix assembly and cellular growth including cytostatic effects on transformed cells with diverse histogenic backgrounds [71]. It influences embryogenesis, inflammation, wound healing, and neoplastic growth [71–75]. Decorin antagonistically regulates the action of TGF-beta [76], and binds TGF-beta 1, co-localizing in many tissues [77,78]. It both inhibits some actions of TGF-beta and its synthesis is stimulated by TGF-beta, suggesting that it provides a negative feedback control for TGF-beta activity [77–79]. The bound form may serve as a tissue reservoir of TGF-beta in a manner similar to the binding interaction between TGF-beta and either latency-associated peptide or betaglycan [80]. Release of TGF-beta from its bound form with decorin is regulated by proteases including MMP-2, 3, and 7 [81]. Decorin also reacts with

HA [82] and has been reputed to possess close similarities to the CD44 family [83].

Decorin also interacts with other growth factors. TNF-alpha has the ability to transcriptionally inhibit decorin gene expression in growth-arrested cells and may be a key modulator of decorin [84]. Decorin is proposed as a novel ligand for EGF and may in this role regulate cell growth in tissue remodeling and cancer [71]. Decorin activates the EGF receptor, triggering a signaling cascade which leads to phosphorylation of MAP kinase, induction of p21, and growth suppression. IL-1 and IL-4 inhibit decorin expression [85,86], Over-expression of v-src selectively abolishes the expression of decorin [87]. Because of its interaction with growth factors, decorin expression can substantially alter the cellular response to injury [88].

Decorin synthesis has been inhibited by antisense nucleotides in ligament-healing studies using gene-delivery methods and this has been proposed as a technique to accelerate scar production [89]. However, most potential pharmacological applications of decorin have addressed application as a TGF-beta inhibitor for decreasing inflammation and scar production and for increasing the tissue immune response [72,75]. The scar-inhibitory activity of decorin has been proposed for treatment of inflammatory kidney diseases [90], pulmonary fibrosis [91], tuberculosis [92], diabetic vascular disease [93] among just a few of the proposed applications. The exciting breadth of antifibrotic agents becoming available for research studies on fibrosis inhibition can be illustrated by the following examples of neutralizing agents to consider:

Other Antagonists to TGF-beta: Blocking peptides to TGF-beta 1 and 2, LAP (latency-associated peptide), TGF-beta 3 (which neutralizes isoforms 1 and 2), decorin, mannose-6-phosphate, TGF-beta antisense nucleotides, non-viral gene therapy, soluble receptors to TGF-beta 1 and 2, and peptidomimetics which react with and block the TGF-beta receptors.

Other classes of anti-fibrotics: These include the interferons, relaxin, certain peptidomimetics such as RGD peptides or Fn fragments against integrins, certain glycomimetics, mAbs against selectins, lysyl oxidase inhibitors of collagen cross linking, collagen prolyl hydroxylase inhibitors, to name just some of the potential candidates. Recently, anti-cd44 antibody has been shown to induce cultured fibroblast detachment from substratum and morphological change compatible with apoptosis [94].

Enhancement of Wound Healing in the Treatment of Chronic Diabetic Ulcers

The very large literature on proposed treatment for chronic diabetic foot ulcers testifies to the intractability of the problem. It is estimated that 54,000 extremity amputations of diabetic patients occur annually in the United States secondary to failed treatment of chronic ischemic foot ulcers [95]. Foot infection is the most common diagnosis of hospital admissions of diabetic patients [96]. This outcome is observed in spite of some reports of as high as a 90% success rate of healing such ulcers in compliant patients [97]. However, lack of compliance resulted in as high as 54 times more likely occurrence of ischemic foot ulcerations in diabetics in one prospective study [98]. A meta-analysis of the literature of diabetic ulcer treatment indicated that about one third of patients heal their ulcers after 20 weeks of standard treatment [99]. However, one study reported 90% healing with "proper" conservative treatment [97]. It is clear from the foregoing that prospective patient educational efforts about foot care and footwear in compliant patients are essential for ulcer prevention and for successful treatment of ulcers which develop. Prediction of successful healing of chronic diabetic ulcers in ischemic lower extremities has been attempted with TcPO2 measurement (transcutaneous oxygen pressure),TBP (toe blood pressure determination), or vibration perception threshold with a few promising reports, but with generally disappointing results [95,100,101].

The range of treatment modalities for chronic neuropathic and ischemic lower-extremity diabetic ulcers is quite large, but there is inadequate evidence for efficacy of new treatment recommendations in most of the examples cited [102]. It is important that baseline conservative treatment be meticulously applied and that surgical and antibiotic management of underlying osteomyelitis be provided if other modalities are to be useful complementary adjuncts. Treatment modalities based upon physical methods include hyperbaric oxygen [103], electrical stimulation [95,104], and low-intensity laser treatment [105]. Treatment with cultured skin cells such as Dermagraft® [106], with cultured algeneic keratinocyte epithelial cell sheets [107], fibroblasts in gelatin sponge [107], or with biologically derived materials such as collagen alginate wound dressings [108] have advocates. The utility of cytokines to enhance the wound healing of diabetic foot ulcers has been explored extensively. Recombinant human platelet-derived growth factor-BB (becaplermin gel) has received considerable attention in several reports indicating efficacy [109–115]. Some studies report as much as twice the rate of improvement of wound healing with the PDGF-BB product [115]. Other cytokines of potential utility for this purpose include fibroblast growth factor-beta (bFGF) [116], vascular endothelial growth factor (VEGF) [117], and granulocyte-macrophage colony stimulating factor (GM-CSF) [118]. Biopsies of chronic wounds of diabetic lower extremities have not demonstrated deficiencies of PDGF, EGF, bFGF, or TGF-beta cytokine content [119], but the persistent increase of the proinflamatory cytokines interleukin-1 and 6 and tumor necrosis factor alpha are thought to be adverse to the wound-healing process [120]. Unfortunately, in vitro lattice models mimicking wound healing have not been found to be useful in studying this process [121] and animal models have offered only limited promise [122].

In the near future, carefully performed prospective studies will clarify many of the confusing aspects concerning treatment of chronic ulcers in diabetic patients with ischemic and neuropathic lower extremities. In the end, however, it is clear that the most important elements of treatment of the diabetic with ischemic and neuropathic lower extremities remains preventive care, including detailed patient education and careful long-term follow-up. But, for any program to be successful, a compliant patient is an essential ingredient.

References

1. Schall T, Bacon K. Chemokines, leukocyte trafficking and inflammation. Curr Op Im 1994;6:865–73.
2. von Andrian UJ. Chambers L, McEvoy R, Bargatze K, Arfors, Butcher E. Two-step model of leukocyte-endothelial cell interaction in inflammation. Proc Nat Acad Sci, USA 1991;88:7538–42.
3. Border WN, Noble T, Yamamoto S, Tomooka, Kagami S. Antagonists to TGF-beta: Treatment of glomerulonephrities and preventionof glomerulosclerosis. Kidney Int 1992;41:566–70.
4. Pober J, Cotran R. The role of endothelial cells in inflammation. Transplantation 1990;50:537–44.
5. Smith C. Endothelial adhesion molecules and their role in inflammation. Can J Physiol Pharmacol 1993;71: 76–87.
6. Muller WS, Weigl X, Deng, Phillips D. PECAM-1 is required for transendothelial migration of leukocytes. Jf Exp Med 1993;178:449–60.
7. Webb LM, Ehrengruber I, Clark-Lewis M, Baggiolini, Rot A. Binding to heparan sulfate or heparin enhances neutrophil responses to interleukin 8. Proc Natl Acad Sci USA 1993;90:7158–62.
8. Witt D, Lander A. Differential vinding of chemokines to glycosaminoglycan subpopulations. Curr Biol 1994;4:394–400.
9. Weyrich AX, Ma D, Lefer K, Albertine, Lefer A. In vivo neutralization of P-selectin protects feline heart and endothelium in myocardial ischemia and reperfusion injury. J Clin Invest 1993;91:2610–29.
10. Hernandez-Pando R, Orozco H, Arriaga K, Sampieri A, Larriva-Sahd J, Madrid-Marina V. Analysis of the local kinetics and localization of interleukin-1 alpha, tumour necrosis factor-alpha and transforming growth factor-beta, during the course of experimental pulmonary tuberculosis. Immunology 1997;90:607–17.
11. IBC Int Conf on therapeutic advances in fibrosis. 1996, Washington, DC.
12. Akeson WHD, Amiel GL, Mechanic SL, Woo FL, Harwood, Hamer ML. Collagen cross-linking alterations in joint contractures: changes in the reducible cross-links in periarticular connective tissue collagen after nine weeks of immobilization. Connect Tissue Res 1997;5:15–19.
13. Butcher E. Leukocyte-endothelial cell recognition: Three (or more) steps to specificity and diversity. Cell 1991;67:1033–6.
14. Collins T. Adhesion molecules in leukocyte emigration. Sci Am Sci Med 1995;Nov/Dec:28–37.
15. Schall T, Bacon K. Chemokines, leukocyte trafficking and inflammation. Curr Op Im 1994;6:865–73.
16. Burrington J. Wound healing in the fetal lamb. Jl Pediatr Surg 1971;6:523–8.
17. Cass DL, Bullard KM, Sylvester KG, Yang, EY Longaker MT, Adzick NS. Wound size and gestational age modulate scar formation in fetal wound repair. J Pediatr Surg 1997;32:411–5.
18. Ioncono JH, Ehrlich K, Keefer, Krummel T. Hyalluronan induces scarless repair in mouse limb organ culture. J Pediatr Surg 1998;33:546–7.
19. DePalma R, Krummel T, Durham L, Michna B, Thomas B, Nelson J et al. Characterization and quantitation of wound matrix in the fetal rabbit. Matrix 1989;9:224–31.

20. Cabrera R, Siebert J, Eidelman Y, Gold L, Langaker M, Garg H. The in vivo effect of hyaluronan associated protein-collagen complex on wound repair. Biochem Molec Biol Int 1995;37:151–8.
21. Freund R, Siebert J, Carera R, Longaker M, Eidelman Y, Adzick N et al. Serial quantitation of hyaluronan and sulfated glycosaminoglycans in fetal sheep skin. Bioch Molec Biol Int 1993;29:773–83.
22. Shepard S, Becker H, Hartman L. Using hyaluronic acid to create a fetal-like environment in vitro. Ann Plastic Surg 1996;36:65–9.
23. West DC, Shaw DM, Lorenz P, Adzick NS, Longaker MT. Fibrotic healing of adult and late gestation fetal wounds correlates with increased hyaluronidase activity and removal of hyaluronan. Int J Biochem Cell Biol 1997;29:201–10.
24. Alaish S, Yager D, Diegelmann R, Cohen I. Biology of fetal wound healing: hyaluronate receptor expression in fetal fibroblasts. J Pediatr Surg 1994;29:1040–3.
25. Whitby DJ. The extracellular matrix of lip wounds in fetal neonatal and adult mice. Development 1991;12: 651–68.
26. Whitby DJ. Immunohistochemical localization of growth factors in fetal wound healing. Devel Biol 1991;147:207–15.
27. Shah M, Foreman D, Ferguson M. Neutralisation of TGF-beta 1 and TGF-beta 2 or exogenous addition of TGF-beta 3 to cutaneous rat wounds reduces scarring. J Cell Sci 1995;108:985–1002.
28. Adzick NS. Fetal Wound Healing. New York: Elsevier, 1991.
29. Stelnicki EJ, Bullard KM, Harrison MR, Cass DL, Adzick NS. A new in vivo model for the study of fetal wound healing. Ann Plast Surg 1997;39:374–80.
30. Ioncono J, Krummel T, Keefer K, Allison G, Paul H. Repeated additions of hyaluronan alters granulation tissue deposition in sponge implants in mice. Wound Repair Regen 1998;6:442–8.
31. Namba RS, Meuli M, Sullivan KM, Le AX, Adzick NS. Spontaneous repair of superficial defects in articular cartilage in a fetal lamb model. J Bone Joint Surg Am 1998;80:4–10.
32. Mackool RJ, Gittes GK, Longaker MT. Scarless healing. The fetal wound. Clin Plast Surg 1998;25:357–65.
33. Cowin AJ, Brosnan MP, Holmes TM, Ferguson MW. Endogenous inflammatory response to dermal wound healing in the fetal and adult mouse. Dev Dyn 1998;212:385–93.
34. Coleman C, Tuan TL, Buckley S, Anderson KD, Warburton D. Contractility, transforming growth factor-beta, and plasmin in fetal skin fibroblasts: role in scarless wound healing. Pediatr Res 1998;43: 403–9.
35. Irwin CR, Myrillas T, Smyth M, Doogan J, Rice C, Schor SL. Regulation of fibroblast-induced collagen gel contraction by interleukin-1beta. J Oral Pathol Med 1998;27:255–9.
36. Bullard KM, Banda MJ, Adzick NS. Transforming gowth factor beta-1 decreases interstitial collagenase in healing human fetal skin. J Ped Surg 1997;32: 1023–7.
37. Gallivan BA, Moriarty KP, Pajerski ME, O'Donnell C, Crombleholme TM. Differential collagen1 gene expression in fetal fibroblasts. J Ped Surg 1997;32:1033–6.
38. Lovvorn HN 3rd, Cheung DT, Nimni ME, Perelman N, Estes JM, Adzick NS. Relative distribution and crosslinking of collagen distinguish fetal from adult sheep wound repair. J Pediatr Surg 1999;34:218–23.
39. Liechty KW, Crombleholme TM, Cass DL, Martin B, Adzick NS. Diminished interleukin-8 (IL-8) production in the fetal wound healing response. J Surg Res 1998;77:80–4.
40. Olutoye OO, Barone EJ, Yager DR, Uchida T, Cohen IK, Diegelmann RF. Hyaluronic acid inhibits fetal platelet function: implications in scarless healing. J Pediatr Surg 1997;32:1037–40.
41. Cass DL, Bullard KM, Sylvester KG, Yang EY, Sheppard D, Herlyn M et al. Epidermal integrin expression is upregulated rapidly in human fetal wound repair. J Pediatr Surg 1998;33:312–6.
42. Entwistle J, Hall CL, Turley EA. HA receptors: regulators of signalling to the cytoskeleton. J Cell Biochem 1996;61:569–77.
43. Naot D, Sionov RV, Ish-Shalom D. CD44: structure, function, and association with the malignant process. Adv Cancer Res 1997;71:241–319.
44. Asari A, Morita M, Sekiguchi T, Okamura K, Horie K, Miyauchi S. Hyaluronan, CD44 and fibronectin in rabbit corneal epithelial wound healing. Jpn J Ophthalmol 1996;40:18–25.
45. Rudzki Z, Jothy S. CD44 and the adhesion of neoplastic cells. Mol Pathol 1997;50:57–71.
46. Shyjan AM, Heldin P, Butcher EC, Yoshino T, Briskin MJ. Functional cloning of the cDNA for a human hyaluronan synthase. J Biol Chem 1996;271:23395–9.
47. Wang C, Entwistle J, Hou G, Li Q, Turley EA. The characterization of a human RHAMM cDNA: conservation of the hyaluronan-binding domains. Gene 1996; 174:299–306.
48. Greco RM, Iocono JA, Ehrlich HP. Hyaluronic acid stimulates human fibroblast proliferation within a collagen matrix. J Cell Physiol 1998;177:465–73.
49. Shepard S, Becker H, Hartmann JX. Using hyaluronic acid to create a fetal-like environment in vitro. Ann Plast Surg 1996;36:65–9.
50. McKee, CM, Penno MB, Cowman M, Burdick MD, Strieter RM, Bao C et al. Hyaluronan (HA) fragments induce chemokine gene expression in alveolar macrophages. The role of HA size and CD44. J Clin Invest 1996;98:2403–13.
51. Lees, VC, Fan TP, West DC. Angiogenesis in a delayed revascularization model is accelerated by angiogenic oligosaccharides of hyaluronan. Lab Invest 1995;73: 259–66.
52. Hodge-Dufour J, Noble PW, Horton MR, Bao C, Wysoka M, Burdick MD et al. Induction of IL-12 and chemokines by hyaluronan requires adhesion-dependent priming of resident but not elicited macrophages. J Immunol 1997;159:2492–500.
53. Deed R, Rooney P, Kumar P, Norton JD, Smith J, Freemont AJ et al. Early-response gene signalling is induced by angiogenic oligosaccharides of hyaluronan in endothelial cells. Inhibition by non-angiogenic, high-molecular-weight hyaluronan. Int J Cancer 1997; 71:251–6.
54. Hall C, Laange L, Prober D, Zhang S, Turley E. pp60 (c-src) is required for cell locomotion regulated by the hyaluronan receptor RHAMM. Oncogene 1996;13: 2213–24.
55. Amara FM, Entwistle J, Kuschak TI, Turley EA, Wright JA. Transforming growth factor-beta1 stimulates multiple protein interactions at a unique cis-element in the

3'-untranslated region of the hyaluronan receptor RHAMM mRNA. J Biol Chem 1996;271:15279–84.
56. Amiel D, Frey C, Woo SL, Harwood F, Akeson W. Value of hyaluronic acid in the prevention of contracture formation. Clin Orthop Related Res 1985;196:306–11.
57. Waters SN, Massie JB, Amiel D, Akeson WH. A role for anti-fibrotics in the prevention of epidural fibrosis. Proc Orthop Res Soc 2000;0071, Mar 12–15.
58. Tamaki K, Okuda S, Miyazono K, Nakayama M, Fujishima M. Matrix-associated latent TGF-beta with latent TGF-beta binding protein in the progressive process in adriamycin-induced nephropathy. Lab Invest 1995;73:81–9.
59. Bottinger EP, Factor VM, Tsang ML, Weatherbee JA, Kopp JB, Qian SW et al. The recombinant proregion of transforming growth factor beta1 (latency-associated peptide) inhibits active transforming growth factor beta1 in transgenic mice. Proc Natl Acad Sci USA 1996;93:5877–82.
60. Davidson J. Cell biology of tissue repair and fibrosis. In IBC Int Conf on Therapeutic Advances in Fibrosis, Washington DC, 1996.
61. Nagineni CN, Amiel D, Green MH, Berchuck M, Akeson WH. Characterization of the intrinsic properties of the anterior cruciate and medial collateral ligament cells: an in vitro cell culture study. J Orthop Res 1992;10: 465–75.
62. Schreck PJ, Kitabayashi LR, Amiel D, Akeson WH, Woods VL Jr. Integrin display increases in the wounded rabbit medial collateral ligament but not the wounded anterior cruciate ligament. J Orthop Res 1995;13: 174–83.
63. Geiger MH, Green MH, Monosov A, Akeson WH, Amiel D. An in vitro assay of anterior cruciate ligament (ACL) and medial collateral ligament (MCL) cell migration. Connect Tissue Res 1994;30:215–24.
64. Amiel D, Kuiper SD, Wallace CD, Harwood FL, VandeBerg JS. Age-related properties of medial collateral ligament and anterior cruciate ligament: a morphologic and collagen maturation study in the rabbit. J Gerontol 1991;46:B159–65.
65. Albelda S, Smith C, Ward P. Adhesion molecules and inflammation Injury. The FASEB Journal 1994;8: 504–12.
66. Kavanaugh A, Heudebert G, Cush J, Jain R. Cost evaluation of novel therapeutics in rheumatoid arthritis (CENTRA). Seminars in Arthritis and Rheumatism 1995;25:1–12.
67. Shah M, Whitby D, Ferguson M. Fetal wound healing and scarless surgery. In: Jackson D, Sommerlad B, editors. Recent Advances in Plastic Surgery. Vol. 5. Edinburgh: Churchill Livingstone, 1996.
68. Wahl SM. Transforming growth factor beta: the good, the bad, and the ugly. J Exp Med 1994;180: 1587–90.
69. Weber IT, Harrison RW, Iozzo RV. Model structure of decorin and implications for collagen fibrillogenesis. J Biol Chem 1996;271:31767–70.
70. Kresse H, Liszio C, Schonherr E, Fisher LW. Critical role of glutamate in a central leucine-rich repeat of decorin for interaction with type I collagen. J Biol Chem 1997;272:18404–10.
71. Iozzo RV, Moscatello DK, McQuillan DJ, Eichstetter I. Decorin is a biological ligand for the epidermal growth factor receptor. J Biol Chem 1999;274:4489–92.
72. Munz C, Naumann U, Grimmel C, Rammensee HG, Weller M. TGF-beta-independent induction of immunogenicity by decorin gene transfer in human malignant glioma cells. Eur J Immunol 1999;29: 1032–40.
73. Olsson U, Bondjers G, Camejo G. Fatty acids modulate the composition of extracellular matrix in cultured human arterial smooth muscle cells by altering the expression of genes for proteoglycan core proteins. Diabetes 1999;48:616–22.
74. Okamoto O, Fujiwara S, Abe M, Sato Y. Dermatopontin interacts with transforming growth factor beta and enhances its biological activity. Biochem J 1999;337: 537–41.
75. Stander M, Naumann U, Dumitrescu L, Heneka M, Loschmann P, Gulbins E et al. Decorin gene transfer-mediated suppression of TGF-beta synthesis abrogates experimental malignant glioma growth in vivo. Gene Ther 1998;5:1187–94.
76. Khanna A, Li B, Li P, Suthanthiran M. Transforming growth factor-beta 1: regulation with a TGF-beta 1 antisense oligomer. Kidney Int Suppl 1996;53:S2–6.
77. Redington AE, Roche WR, Holgate ST, Howarth PH. Co-localization of immunoreactive transforming growth factor-beta 1 and decorin in bronchial biopsies from asthmatic and normal subjects. J Pathol 1998;186: 410–5.
78. Schonherr E, Broszat M, Brandan E, Bruckner P, Kresse H. Decorin core protein fragment Leu155-Val260 interacts with TGF-beta but does not compete for decorin binding to type I collagen. Arch Biochem Biophys 1998;355:241–8.
79. Asakura S, Kato H, Fujino S, Konishi T, Tezuka N, Mori A. Role of transforming growth factor-beta 1 and decorin in development of central fibrosis in pulmonary adenocarcinoma. Hum Pathol 1999;30: 195–8.
80. Mogyorosi A, Ziyadeh FN. Increased decorin mRNA in diabetic mouse kidney and in mesangial and tubular cells cultured in high glucose. Am J Physiol 1998;275: F827–32.
81. Imai K, Hiramatsu A, Fukushima D, Pierschbacher MD, Okada Y. Degradation of decorin by matrix metalloproteinases: identification of the cleavage sites, kinetic analyses and transforming growth factor-beta 1 release. Biochem J 1997;322:809–14.
82. Roughley PJ, White RJ, Mort JS. Presence of pro-forms of decorin and biglycan in human articular cartilage. Biochem J 1996;318:779–84.
83. Ehnis T, Dieterich W, Bauer M, Lampe B, Schuppan D. A chondroitin/dermatan sulfate form of CD44 is a receptor for collagen XIV (undulin). Exp Cell Res 1996;229:388–97.
84. Mauviel A, Santra M, Chen YQ, Uitto J, Iozzo RV. Transcriptional regulation of decorin gene expression. Induction by quiescence and repression by tumor necrosis factor-alpha. J Biol Chem 1995;270:11692–700.
85. Kuroda K, Shinkai H. Decorin and glycosaminoglycan synthesis in skin fibroblasts from patients with systemic sclerosis. Arch Dermatol Res 1997;289:481–5.
86. Demoor-Fossard M, Redini F, Boittin M, Pujol JP. Expression of decorin and biglycan by rabbit articular chondrocytes. Effects of cytokines and phenotypic modulation. Biochim Biophys Acta 1998;1398: 179–91.

87. Kolettas E, Rosenberger RF. Suppression of decorin expression and partial induction of anchorage-independent growth by the v-src oncogene in human fibroblasts. Eur J Biochem 1998;254:266–74.
88. Brown C, Nugent M, Lau F, Trinkaus-Randall V. Characterizationof proteoglycans synthesized by cultured coneal fibroblasts in response to transfoming growth factor-beta and fetal calf serum. Journal of Biological Chemistry 1999;274:7111–9.
89. Nakamura N, Timmermann SA, Hart DA, Kaneda Y, Shrive NG, Shino K et al. A comparison of in vivo gene delivery methods for antisense therapy in ligament healing. Gene Ther 1998;5:1455–61.
90. Peters H, Noble NA, Border WA. Transforming growth factor-beta in human glomerular injury. Curr Opin Nephrol Hypertens 1997;6:389–93.
91. Giri SN, Hyde DM, Braun RK, Gaarde W, Harper JR, Pierschbacher MD. Antifibrotic effect of decorin in a bleomycin hamster model of lung fibrosis. Biochem Pharmacol 1997;54:1205–16.
92. Hirsch CS, Ellner JJ, Blinkhorn B, Toossi Z. In vitro restoration of T cell responses in tuberculosis and augmentation of monocyte effector function against Mycobacterium tuberculosis by natural inhibitors of transforming growth factor beta. Proc Natl Acad Sci USA 1997;94:3926–31.
93. Yokoyama H, Deckert T. Central role of TGF-beta in the pathogenesis of diabetic nephropathy and macrovascular complications: a hypothesis. Diabet Med 1996; 13:313–20.
94. Henke C, Bitterman P, Roongta U, Ingbar D, Polunovsky V. Induction of fibroblast apoptosis by anti-CD44 antibody: implications for the treatment of fibroproliferative lung disease. Am J Pathol 1996;149: 1639–50.
95. Gilcreast, DM, Stotts NA, Froelicher ES, Baker LL, Moss KM. Effect of electrical stimulation on foot skin perfusion in persons with or at risk for diabetic foot ulcers. Wound Repair Regen 1998;6:434–41.
96. Pinzur MS, Slovenkai MP, Trepman E. Guidelines for diabetic foot care. The Diabetes Committee of the American Orthopaedic Foot and Ankle Society. Foot Ankle Int 1999;20:695–702.
97. Daniels TR Diabetic foot ulcerations: an overview. Ostomy Wound Manage 1998;44:76–80, 82, 84; quiz 85–6 passim.
98. Armstrong DG, Harkless LB. Outcomes of preventative care in a diabetic foot specialty clinic. J Foot Ankle Surg 1998;37:460–6.
99. Margolis DJ, Kantor J, Berlin JA. Healing of diabetic neuropathic foot ulcers receiving standard treatment. A meta-analysis. Diabetes Care 1999;22:692–5.
100. Kalani M, Brismar K, Fagrell B, Ostergren J, Jorneskog G. Transcutaneous oxygen tension and toe blood pressure as predictors for outcome of diabetic foot ulcers. Diabetes Care 1999;22:147–51.
101. Abbott CA, Vileikyte L, Williamson S, Carrington AL, Boulton AJ. Multicenter study of the incidence and predictive risk factors for diabetic neuropathic foot ulceration. Diabetes Care 1998;21:1071–5.
102. Mason J, O'Keeffe C, McIntosh A, Hutchinson A, Booth A, Young RJ. A systematic review of foot ulcer in patients with Type 2 diabetes mellitus. I: prevention. Diabet Med 1999;16:801–12.
103. Mitton C, Hailey. D Health technology assessment and policy implicatiions of hyperbaric oxygen treatment [abstract]. Annual Meeting of International Society of Technology Assessment in Health Care 1999; 15:140.
104. Kloth LC, McCulloch JM. Promotion of wound healing with electrical stimulation. Adv Wound Care 1996;9: 42–5.
105. Schindl A, Schindl M, Pernerstorfer-Schon H, Kerschan K, Knobler R, Schindl L. Diabetic neuropathic foot ulcer: successful treatment by low-intensity laser therapy. Dermatology 1999;198:314–6.
106. Allenet B, Paree F, Lebrun T, Carr L, Posnett J, Martini J et al. Cost-effectiveness modeling of Dermagraft for the treatment of diabetic foot ulcers in the French context. Diabetes Metab 2000;26:125–32.
107. Harvima IT, Virnes S, Kauppinen L, Huttunen M, Kivinen P, Niskanen L et al. Cultured allogeneic skin cells are effective in the treatment of chronic diabetic leg and foot ulcers. Acta Derm Venereol 1999;79: 217–20.
108. Donaghue VM, Chrzan JS, Rosenblum BI, Giurini JM, Habershaw GM, Veves A. Evaluation of a collagen-alginate wound dressing in the management of diabetic foot ulcers. Adv Wound Care 1998;11:114–9.
109. Smiell JM, Wieman TJ, Steed DL, Perry BH, Sampson AR, Schwab BH. Efficacy and safety of becaplermin (recombinant human platelet-derived growth factor-BB) in patients with nonhealing, lower-extremity diabetic ulcers: a combined analysis of four randomized studies. Wound Repair Regen 1999;7:335–46.
110. Embil JM, Papp K, Sibbald G, Tousignant J, Smiell JM, Wong B et al. Recombinant human platelet-derived growth factor-BB (becaplermin) for healing chronic lower-extremity diabetic ulcers: an open-label clinical evaluation of efficacy. Wound Repair Regen 2000;8: 162–8.
111. Miller MS. Use of topical recombinant human platelet-derived growth factor-BB (becaplermin) in healing of chronic mixed arteriovenous lower-extremity diabetic ulcers. J Foot Ankle Surg 1999;38:227–31.
112. Rees RS, Robson MC, Smiell JM, Perry BH. Becaplermin gel in the treatment of pressure ulcers: a phase II randomized, double-blind, placebo-controlled study. Wound Repair Regen 1999;7:141–7.
113. Castronuovo JJ Jr, Ghobrial I, Giusti AM, Rudolph S, Smiell JM. Effects of chronic wound fluid on the structure and biological activity of becaplermin (rhPDGF-BB) and becaplermin gel. Am J Surg 1998; 176:61S–67S.
114. Smiell JM Clinical safety of becaplermin (rhPDGF-BB) gel. Becaplermin Studies Group. Am J Surg 1998;176: 68S–73S.
115. Wieman TJ. Clinical efficacy of becaplermin (rhPDGF-BB) gel. Becaplermin Gel Studies Group. Am J Surg 1998;176:74S–79S.
116. Robson MC, Hill DP, Smith PD, Wang X, Meyer-Siegler K, Ko F et al. Sequential cytokine therapy for pressure ulcers: clinical and mechanistic response. Ann Surg 2000;231:600–11.
117. Dvorak HF, Detmar M, Claffey KP, Nagy JA, van de Water L, Senger DR. Vascular permeability factor/vascular endothelial growth factor: an important mediator of angiogenesis in malignancy and inflammation. Int Arch Allergy Immunol 1995;107:233–5.
118. Remes K, Ronnemaa T. Healing of chronic leg ulcers in diabetic necrobiosis lipoidica with local granulocyte-

macrophage colony stimulating factor treatment. J Diabetes Complications 1999;13:115–8.
119. Harris IR, Yee KC, Walters CE, Cunliffe WJ, Kearney JN, Wood EJ. Cytokine and protease levels in healing and non-healing chronic venous leg ulcers. Exp Dermatol 1995;4:342–9.
120. Trengove NJ, Bielefeldt-Ohmann H, Stacey MC. Mitogenic activity and cytokine levels in non-healing and healing chronic leg ulcers. Wound Repair Regen 2000;8:13–25.
121. Kuhn MA, Smith PD, Hill DP, Ko F, Meltzer DD, Vande Berg JS et al. In vitro fibroblast populated collagen lattices are not good models of in vivo clinical wound healing. Wound Repair Regen 2000;8:270–6.
122. Chen C, Schultz GS, Bloch M, Edwards PD, Tebes S, Mast BA. Molecular and mechanistic validation of delayed healing rat wounds as a model for human chronic wounds. Wound Repair Regen 1999;7:486–94.

13 Ceramic Coating

J. Tamura, K. Kawanabe, and T. Nakamura

Introduction

To overcome the problems in cemented arthroplasty and obtain reliable permanent fixation in the absence of cement, several types of cementless prosthesis have been developed. The uncoated, press-fit, cementless stem for total hip arthroplasty (THA) used in young age groups frequently result in marked subsidence and loosening around ten years postoperatively, even with the most precise surgical techniques. The porous, full-coated, cementless stem was then developed in an attempt to encourage bony ingrowth into the pores, thus aiming at firm mechanical anchoring between the stem and the femur. This prosthesis model, however, provoked bone absorption along the stem presumably due to the stress-shielding effect on the femur.

Next, in an attempt to avoid stress shielding, prosthesis models with a partial, porous coating over only the proximal part of the stem were developed in the 1980s. In these porous-coated models, however, bone ingrowth into the porous surface did not take place sufficiently to prevent loosening of the stem for longer than 15 years. The majority of the human implant retrieval studies have demonstrated that skeletal attachment by bony ingrowth is not always reliably achieved with porous-coated surfaces. Failure of bony ingrowth with an intervening layer of fibrous tissue between the bone and the implant has often been observed. It was thought that the wear debris of the ultrahigh-molecular-weight polyethylene (UHMWPE) socket migrated through the gaps and caused osteolysis around the implants. A porous coating is biologically passive, so that bone may not grow into the pores, depending on the local environment and stability of the implants.

In an attempt to achieve reproducible bone growth onto the implant, trials of bioactive and osteoconductive implant coatings have been carried out with calcium phosphate ceramics or bioactive glass–ceramics.

Hydroxyapatite Coating

Hydroxyapatite as an implant material was initially used for dental reconstructive surgery and orthopedic surgery as material filler providing direct bonding with living bone. Its excellent early results in this area led to its trial in the field of total joint arthroplasty as a coating material for cementless prostheses.

Geesink et al. performed the study using hydroxyapatite-coated titanium plugs implanted in dog femora, and demonstrated excellent bonding and osteoconductive capacities under non-weight-bearing conditions [1,2]. The coated plugs showed a bone-bonding shear strength of 64 MPa, which was comparable to the strength of cortical bone. Histologic examination showed good filling of the defect around the implant without any sign of fragmentation, whereas the uncoated controls were surrounded by a fibrous layer and were easily extracted from the femur at any postoperative time. The coating thickness, however, had decreased in most places. Another canine study by Stephenson et al. with hydroxyapatite-coated titanium plate also demonstrated osteoconductive effects [3]. Grooved titanium plates, hydroxyapatite-coated on one surface and simply shot-blasted on the other surface, were implanted. The presence of hydroxyapatite was demonstrated to encourage bone to fill more rapidly. With these experimental results it was thought that hydroxyapatite coating had a capacity to bond to bone and to induce bone formation on its surface.

The study under loaded conditions also showed good results on hydroxyapatite-coated implants. Hydroxyapatite-coated and non-coated canine total hip replacements were performed and were examined up to two years postoperatively [1]. Radiographically, the hydroxyapatite-coated implants showed a complete lack of a radiolucent line around the implant. Under histological examination, a fibrous tissue membrane surrounded the non-coated prostheses, whereas the coated prostheses consistently showed bone formation on its surface. There was no microscopic evidence of fragmentation or coating delamination. Oonishi et al. performed a study with hydroxyapatite-coated and non-coated implants in rabbits and goats [4]. With regard to the bonding strength, the coated implants showed a bonding strength four times greater than that of noncoated implants at two weeks and two times greater at six weeks.

Hydroxyapatite coatings applied by plasma spray to orthopedic implants were first used clinically in 1985. The medium-term clinical results of coated femoral components are encouraging, demonstrating satisfactory survival rates and overcoming the problems of non-coated cementless prostheses [5,6,7]. However,

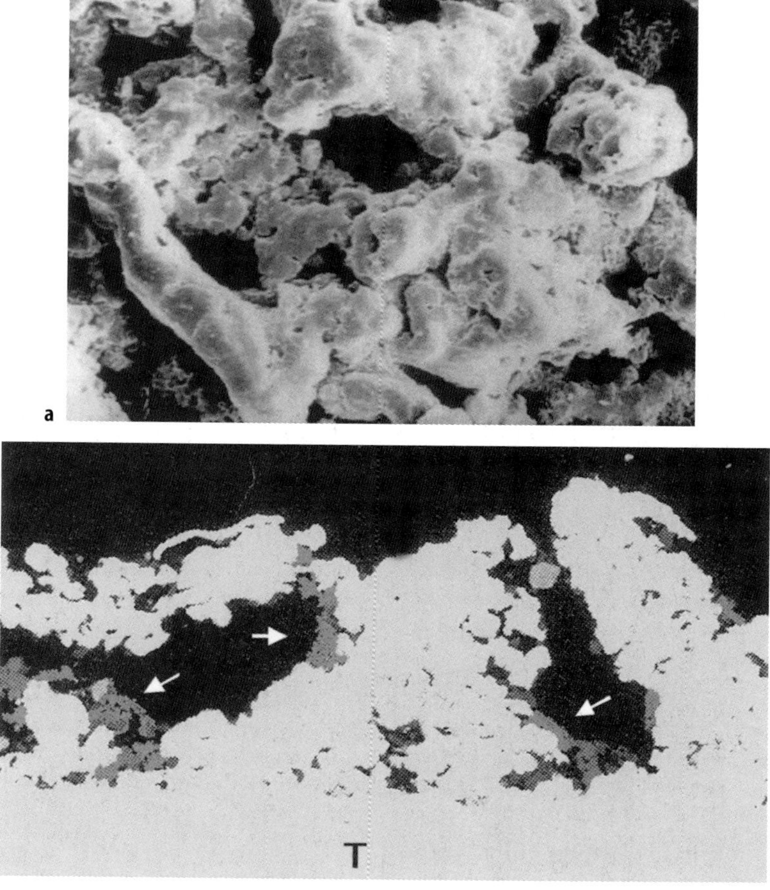

Figure 13.1. AW-GC bottom coating. **a** Surface structure of the plasma-sprayed titanium alloy with the AW-GC bottom coating (SEM image, ×50). **b** Cross-section of the bottom coating. Arrows indicate coated glass–ceramic (×50). T: titanium.

several authors reported possible dangers of the hydroxyapatite coating. In the retrieval study, Bloebaum et al. reported hydroxyapatite particles, possibly migrating from the coated surface, in the osteolytic regions of the periprosthetic tissue with an inflammatory reaction [8]. They also reported hydroxyapatite granules embedded in the polyethylene liner, suggesting the presence of third-body wear caused by the hydroxyapatite fragments. Some retrieval studies demonstrated the absorption of the hydroxyapatite layer from the implant surface [9,10].

Today, it is thought that once the fixation between the bone and the coating was achieved, the hydroxyapatite is gradually replaced by bone. The final fixation depends on the surface structure of the metal implant, not on the surface coating.

Bioactive Glass–Ceramic Coating

Bioactive glass or glass-ceramic was first developed as artificial bone and applied as an

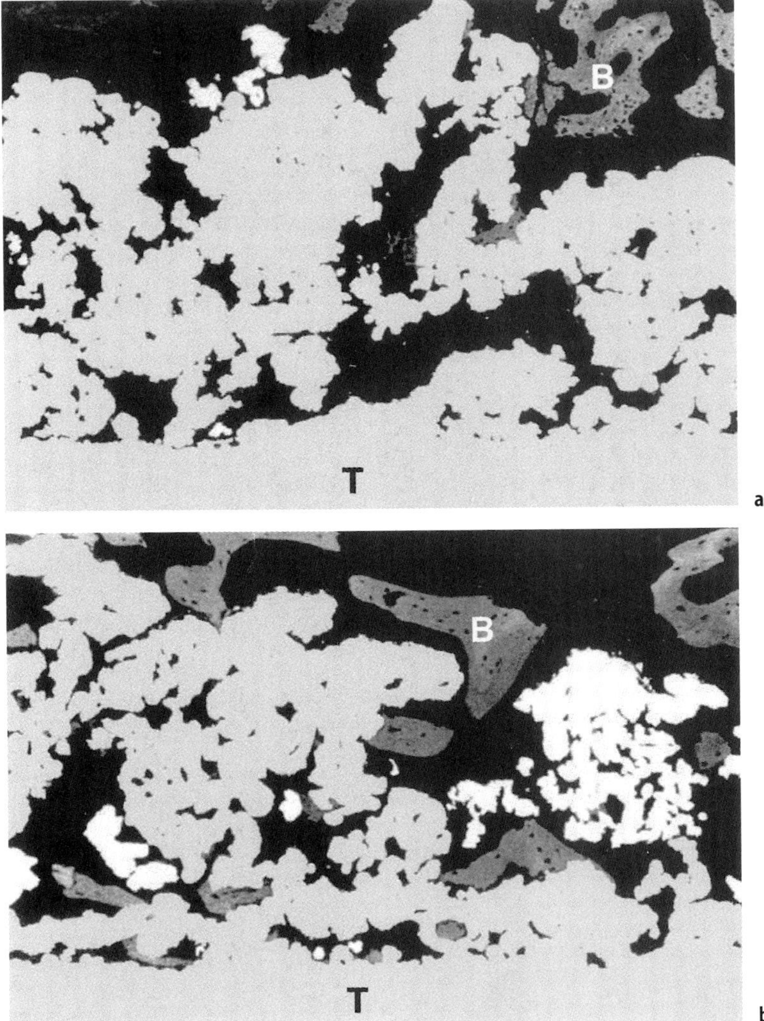

Figure 13.2. SEM images of implant surface in the canine THA study one month after implantation (×80). T: titanium. B: bone. **a** Plasma-sprayed titanium alloy implant (control). **b** AW-GC bottom-coated titanium alloy implant. Note the rapid bone formation into the pores.

artificial bone. AW-GC (apatite–wollastonite-containing glass-ceramic) is one of the bioactive ceramics developed at Kyoto University in 1983 [11,12]. This glass–ceramic has the capability of bonding to bone in vivo and has better mechanical properties than synthesized hydroxyapatite. In an attempt to improve the coating of THA in terms of bioactivity and mechanical strength, AW-GC coating was developed.

The bone-bonding ability of AW-GC-coated titanium plate was evaluated in a rabbit study [13]. In this study the coated plates were implanted in rabbit tibiae under non-weight-bearing conditions and compared with non-coated titanium alloy, AW-GC, and hydroxyapatite plates. The failure load of coated plates was as high as that of AW-GC for all periods and significantly higher at three and four weeks than that of hydroxyapatite. Uncoated Ti alloy showed significantly lower values for all periods.

In the study using coated materials implanted in canine femoral condyles, two kinds of AW-GC coating were investigated [14]. The plasma spray-coated titanium alloy implants were coated by AW-GC and investigated under weight-bearing conditions. The first group was AW-GC full-coated titanium alloy implants and the second group was AW-GC bottom-coated ones (Figures 13.1, 13.2). As a result, thin AW-GC full coating over the plasma spray-coated titanium alloy implants did not increase the interfacial shear strength values significantly at any test point, whereas AW-GC bottom coating significantly increased the values at each test point. In the canine study using total hip prostheses, the AW-GC bottom-coated titanium implant also showed higher bonding strength and rapid bone formation around the implant than simple plasma-sprayed titanium [15].

This glass–ceramic coating was applied for the cementless THA combined with new vanadium-free titanium alloy (KOMLLOY-5, Ti-6Al-2Nb-1Ta) for the stem and the socket back, and zirconia ceramic for the femoral head [16] (Figure 13.3). THA was first clinically used in 1992 in Japan, demonstrating encouraging short to medium-term results.

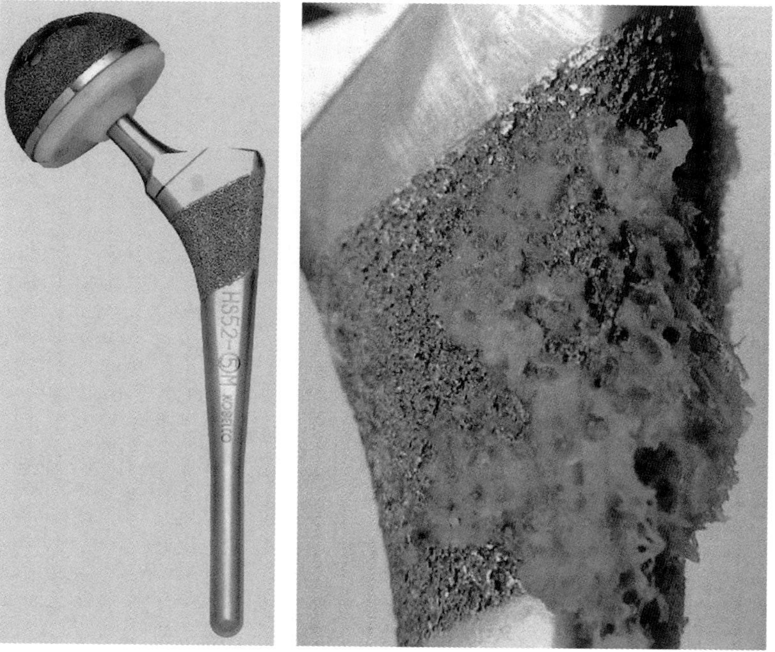

Figure 13.3. Cementless total hip prosthesis with A-W glass–ceramic coating. **a** Kobelco H-5. **b** Coated surface of the retrieved specimen. Note the bone ingrowth on the porous surface.

New Coating: "In Vivo Ceramic Coating"

Plasma-sprayed hydroxyapatite or glass–ceramic coating has been used clinically in THA as mentioned before, and the medium-term results are satisfactory. However, concerns still remain regarding the stability of the coated ceramic surface. In manufacture, it is important to fabricate a thin but stable ceramic layer, however, the control of the composition and structure of the apatite layer on the metal is difficult. Basically, the coated ceramic layer doesn't bond to metal chemically, although it bonds to bone chemically and is gradually absorbed in vivo. Kokubo et al. reported an extremely new type of titanium coating [17,18]. With this

a

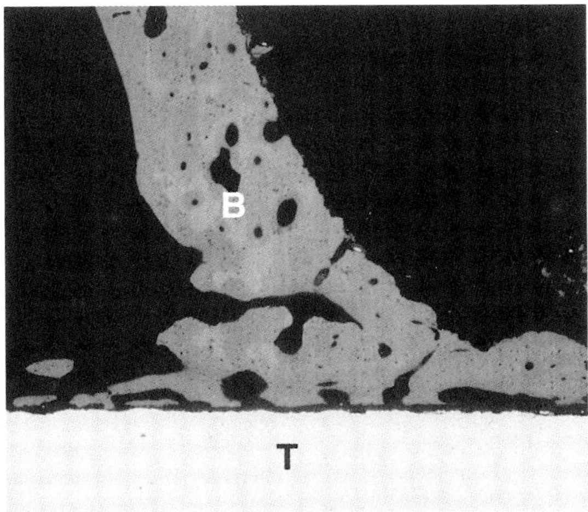

b

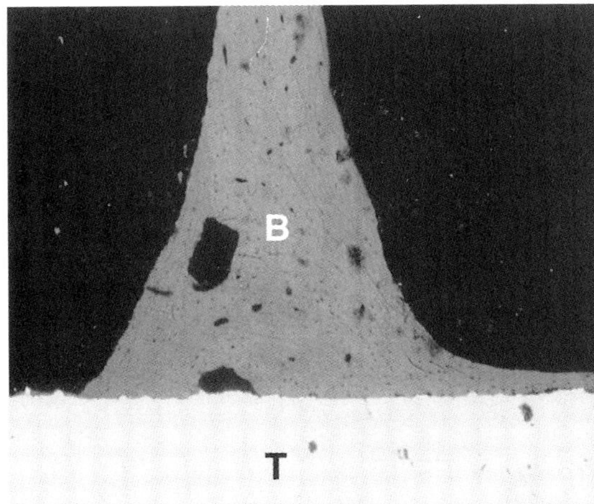

c

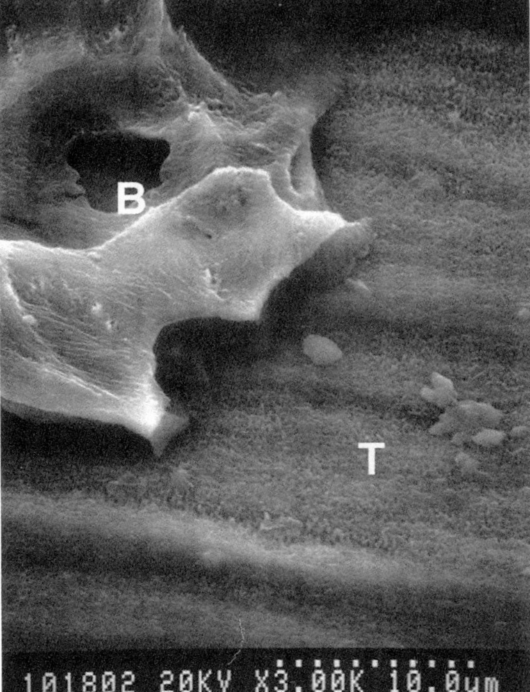

Figure 13.4. SEM images of the bone–implant interface in the rabbit study 16 weeks after implantation (×40). **a** Non-treated titanium alloy plate (control, ×40). **b** Alkali- and heat-treated titanium alloy plate (×40). The bone bonds to the implant directly. **c** Surface image of the treated titanium implant. Bone formation is observed directly on the implant surface. T: titanium. B: bone.

method, titanium was alkali- and heat-treated, providing a "hydroxyapatite nucleus" on the metal surface that will become hydroxyapatite crystals in vivo. In the study in vitro, titanium plates were alkali-treated with NaOH followed by heat treatment below 600 °C. The plates were then soaked in simulated body fluid and examined. It was found that a dense and uniform layer of bone-like apatite was formed on the surface of the alloys. It was thought that the amorphous alkali titanate formed after the treatment transformed into TiO_2 hydrogel after soaking in the fluid, inducing apatite nucleation on its surface.

In animal experiments it was proved that this coating provided high bioactivity to the titanium alloy [19,20]. The treated titanium plates were implanted in rabbit tibiae and the tensile bonding values were measured. The failure load for the treated titanium plates were 7–30 N and non-treated plates were 0 N at eight weeks. The failure loads were measured up to 24 weeks with values of 11–30 N for the treated groups and 0.5–0.7 N for the non-treated groups (Figure 13.4).

Strictly speaking, this coating treatment does not belong with the other "ceramic coating" treatments because the titanium is coated in vivo after implantation. Compared to plasma-sprayed ceramic coatings, however, this treatment is simple and effective, possibly reducing the cost of THA implants. This coating was applied to cementless THA femoral stems, now under clinical trial in Japan.

References

1. Geesink RG, de Groot, Klein CP. Chemical implant fixation using hydroxyapatite coatings. The development of a human total hip prosthesis for chemical fixation to bone using hydroxyapatite coatings on titanium substrates. Clin Orthop 1987;225:147–170.
2. Geesink RG, de Groot, Klein CP. Bonding of bone to apatite-coated implants. J Bone Joint Surg Br 1988;70;17–22.
3. Stephenson PK, Freeman MA, Revell PA, Germain J, Tuke M, Price CJ. The effect of hydroxyapatite coating on ingrowth of bone into cavities in an implant. J Arthroplasty 1991;6:51–8.
4. Oonishi H, Yamamoto M, Ishimaru H, Tsuji E, Kushinami S, Aono M et al.. The effect of hydroxyapatite coating on bone growth into porous titanium alloy implants. J Bone Joint Surg Br 1989;71:213–16.
5. Geesink RG, Hoefnagels NH. Six-year results of hydroxyapatite-coated total hip replacement. J Bone Joint Surg Br 1995;77:534–47.
6. Donnelly WJ, Kobayashi A, Freeman MA, Chin TW, Yeo H, West M et al. Radiological and survival comparison of four methods of fixation of a proximal femoral stem. J Bone Joint Surg Br 1997;79:351–60.
7. Capello WN, D'Antonil JA, Manley MT, Feinberg JR. Hydroxyapatite in total hip arthroplasty. Clinical results and critical issues. Clin Orthop 1998;355:200–11.
8. Bloemaum RD, Beeks D, Dorr LD, Savory CG, DuPont JA, Hoffmann AA. Complications with hydroxyapatite particulate separation in total hip arthroplasty. Clin Orthop 1994;298:19–26.
9. Bauer TW, Geesink RC, Zimmerman R, McMahon JT. Hydroxyapatite-coated femoral stems. Histological analysis of components retrieved at autopsy. J Bone Joint Surg Am 1991;73:1439–52.
10. Buma P, Gardeniers JWM. Tissue reactions around a hydroxyapatite-coated hip prosthesis. J Arthroplasty 995;10:389–95.
11. Kokubo T, Shigematsu M, Nagashima Y, Tashiro M, Nakamura T, Yamamuro T et al. Apatite- and wollastonite-containing glass-ceramic for prosthetic application. Bull Inst Chem Res Kyoto Univ 1982;60: 260–8.
12. Nakamura T, Yamamuro T, Higashi S, Kokubo T, Itoo S. A new glass-ceramic for bone replacement: Evaluation of its bonding to bone tissue. J Biomed Mater Res 1985;19:685–98.
13. Takatsuka K, Yamamuro T, Kitsugi T, Nakamura T, Shibuya T, Goto T. A new bioactive glass-ceramic as a coating material on titanium alloy. J Appl Biomater 1993;4:317–29.
14. Ido K, Matsuda Y, Yamamuro T, Okumura H, Oka M, Takagi H. Cementless total hip replacement. Bio-active glass ceramic coating studied in dogs. Acta Orthop Scand 1993;64:607–12.
15. Yamamuro T, Takagi H. Bone bonding behavior of biomaterials with different surface characteristics under load-bearing conditions. In: Davies JE, editor. The Bone–Material Interface. Toronto: University of Toronto Press, 1991;406–14.
16. Yamamuro T, Nakamura T, Hirokazu I, Matsuda T. A new model of bone-conserving cementless hip prosthesis made of high-tech materials: Kobelco H-5. In: Imura S, Wada M, Omori H, editors. Joint Arthroplasty. Tokyo: Springer-Verlag, 1999;213–24.
17. Kokubo T, Miyaji F, Kim HM. Spontaneous formation of bone-like apatite layer on chemically treated titanium metals. J Am Ceram Soc 1996:79;1127–9.
18. Kim HM, Miyaji F, Kokubo T, Nakamura T. Preparation of bioactive Ti and its alloys via simple chemical surface treatment. J Biomed Mater Res 1996;32:409–17.
19. Nishiguchi S, Kato H, Fujita H, Kim HM, Miyaji F, Kokubo T et al. Enhancement of bone-bonding strength of titanium alloy implants by alkali and heat treantments. J Biomed Mater Res 1999;48:689–96.
20. Fujibayashi S, Nakamura T, Nishiguchi S, Tamura J, Uchida M, Kim HM et al. Bioactive titanium: Effect of sodium removal on the bone-bonding ability of bioactive titanium prepared by alkali and heat removal. Submitted.

14 Ceramics in Orthopedics

B. Masson, R. Rack, G. Willmann, and H. G. Pfaff

Introduction

Historic

First Implantation

In 1970, Dr Boutin developed, in co-operation with the French company Ceraver, a hip prosthesis with cup and ball made of aluminum oxide ceramic (Al_2O_3) that he implanted successfully at Marzet Clinic, in Pau, France [1,2], In 1977 in Paris, Professor Sedel started with a cemented plain alumina cup and a cemented titanium stem.

Clinical Experience

In the 1970s, the German company Friedrichsfeld also began to look into ceramic/ceramic wear couples. This same decade saw the German Professor Mittelmeier in cooperation with the company Osteo to develop a cementless total hip prosthesis with a ball and cup made of alumina ceramic which was implanted for the first time in 1974 in Homburg/Saar [3]. This high-quality bioceramic continued to be developed by the company that became Feldmüle, which is now Ceramtec [2]. Subsequently the German federal government started to fund a R&D program to boost research on ceramic components for hip endoprosthesis. In 1985, zirconia ceramic (ZrO_2) could be found in orthopedics in France and after 1989 in the USA [4].

Number of Implantations

The European Community approved the combination alumina/polyethylene and alumina/alumina bearing couples. In total more than 3.5 million ball heads and, since the 1980s, more than 350,000 inserts made of alumina have been successfully implanted clinically [4]. If one takes into account the worldwide production of zirconia ball heads, one can estimate the number of implantations at more than 350,000 pieces, mainly in Europe and in the United States [2].

Design Concept

Taper Fixation

In cooperation with Feldmüle, Sulzer developed, for the attachment of the ceramic ball head, a conical connection [1,5]. It was used by Professor Mittelmeier immediately and Dr Boutin followed suit. An ISO standard draft in London in 1991 specified the characteristics of the Euro conical spigot as the 14/16 mm type with a spigot angle of 5°43′30″. Inspection of the conical bore of the retrieved heads has been done and there is no indication of any wear, corrosion, or fretting corrosion [6] (see Figure 14.1).

The Cup

Ceramic materials are very strong when used in compression, thus, the first pioneers started with cups made entirely out of ceramics – so-called monolithic or bloc cups. These were either cemented, attached by stressing the bone via press-fit, or screwed into the acetabulum.

Monolithic

First clinical experience showed that the main advantages of using ceramics – low wear rates – were feasible with these designs. But cases of unacceptable wear rates were also reported, especially for screwed monolithic cups. These results were due to the insufficient surface compatibility of the ceramic and also to the design

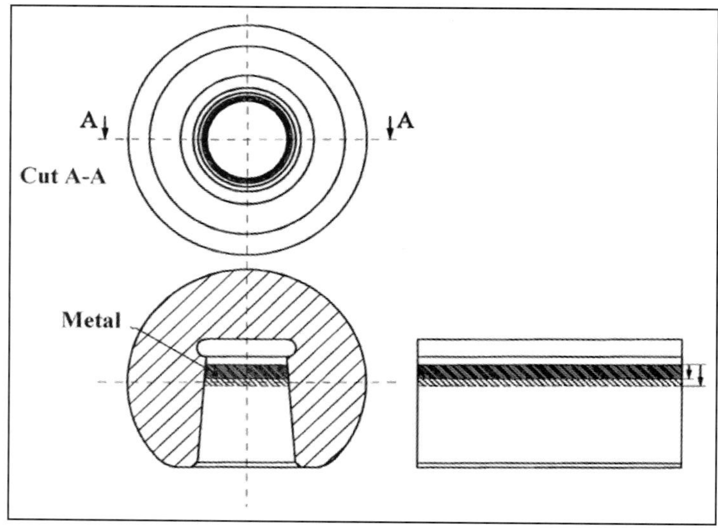

Figure 14.1. Contact zone between metal taper and conical bore in the head. There is always a circular metal print in the bore.

of the external surface of the screwed-in ceramic sockets. Conditions were thus very unfavorable for osteo-integration. The consequence, observed in a number of cases, was migration of the monolithic ceramic socket and frequently tilting of the socket and penetration into the pelvis. This instability also led to very high wear rates [7].

Modular

This non-union between ceramic and bone led to the use of metallic shells, resulting in excellent bony fixation. The aim was to fix the alumina cup inserts into the metal backs of modular prosthetic systems without neglecting surgical requirements and safety aspects of [8]. The standard today is a modular cup consisting of a metal back and a polyethylene (UHMWPE) insert. Polyethylene cups are used in 85% of cases although alumina ceramic cups are used and have yielded positive results (see Figure 14.2).

Fixation

Direct Fixation

The technique used for connecting the ceramic cup insert to the metal back has to be selected to conform with the safety of application as specified by EN 14630. Different kinds of fixations were discussed as possible connecting

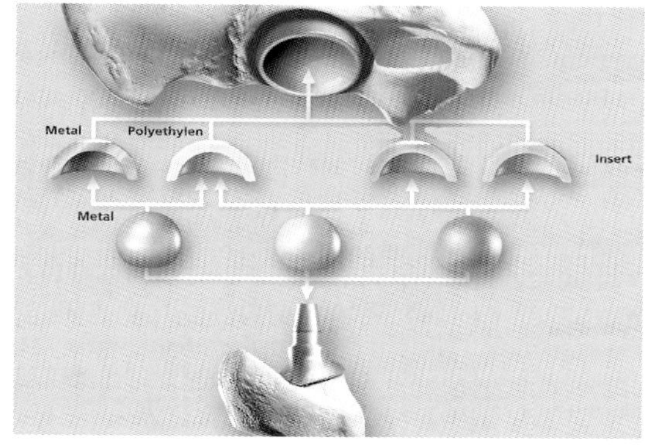

Figure 14.2. Combinations of materials for hip endoprostheses.

techniques. The one that remains is a well-known technique consisting of clamping the insert into the metal back [8]. For this reason, the ceramic cup insert was designed as a truncated cone. The insert is clamped into the conically shaped metal back via the surface areas of the tapers (Figure 14.3).

The surfaces transferring the load applied must ensure that there is no linear application of force in order not to generate any stress concentration which could cause crack formation inside the ceramics due to Hertzian stress. Cup concepts using ceramic inserts with 1:10 (5°42′30″) and 1:3 (18°55″) tapers are used in

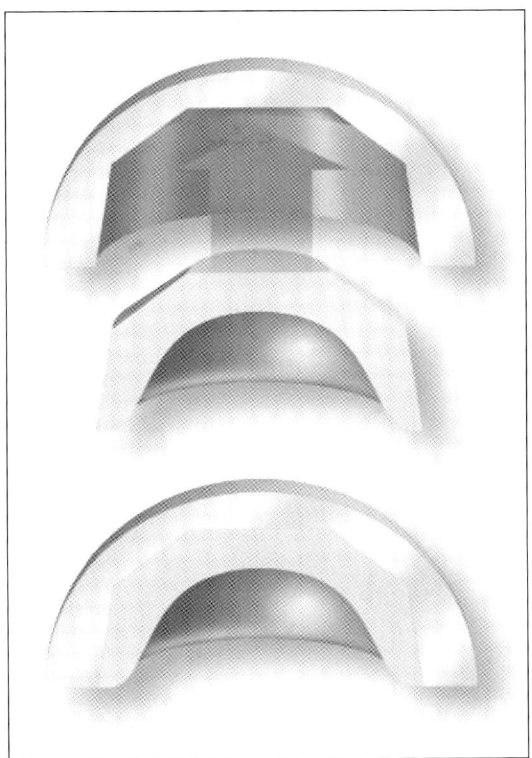

Figure 14.3. Direct fixation for head diameters 26, 28, and 32 mm. Alumina cup inserts can be fixed in the cup shell by means of conical clamping.

clinical application. For both these tapers there is no micro movement and this antitwist fixation is achieved if the tolerances in respect of surface quality, roundness, and straightness of the taper are selected correctly.

Sandwich Fixation

Sandwich fixation was introduced onto the market in 1993. It consists of a conical alumina articular liner. Its external polyethylene coating distributes the dynamic loads. During the operation it is fixed into the acetabular cup with a special hammer (Figure 14.4). Mechanical tests on the ceramic–polyethylene assemblage of conical liners concerning the torsional stability, the dimensional stability of the polyethylene as well as the static and fatigue strength have been performed.

Axe Rotation Design

In order to minimize the probability of dislocation of the ceramic femoral head, the design of this femoral head has changed [8]. The center of motion of the femoral head (M1 before, M2 today) has been altered and moved closer to the cup by several millimetres (Figure 14.5). The edge of the cup has also been rounded off and polished in order to facilitate its return to the bed subsequent to dislocation.

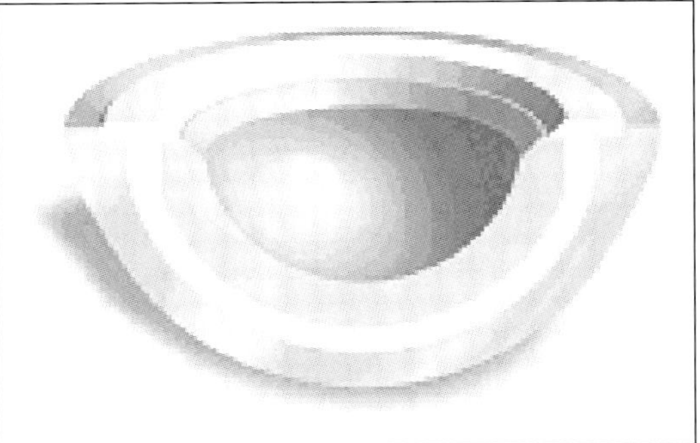

Figure 14.4. Sandwich fixation.

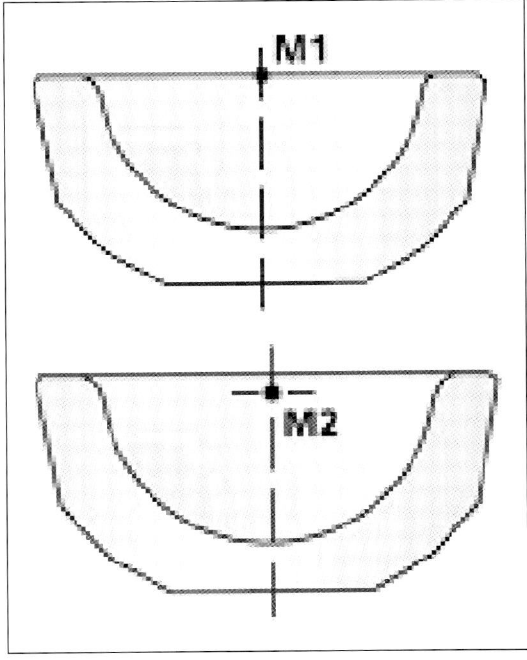

Figure 14.5. Cross-section through alumina ceramic insert, with indication for the positions of the center of rotation of the ceramic balls. M1: The center of rotation identical with center of hole of the socket. M2: Center of rotation of ball shifted inward and edges of insert opening rounded.

Ceramics

Alumina (Aluminium Oxide)

Properties

Today the most widely used bioceramic is aluminum oxide. Common material properties can be found in Table 14.1.

Monophasic

Pure aluminum oxide, Al_2O_3, has one thermodynamically stable phase at room temperature, which is designated as the α phase [9] (Figure 14.6).

Biocompatibility

Ceramic wear particles bear a resemblance to fine grains, mostly about 1 μm in diameter, or slightly larger fragments, between 0.5 and 10 μm in diameter. This description is based on light microscopy of the periprosthetic tissue [10].

In vitro and in vivo studies and research on retrieved implants and pseudomembranes has led to a very rich documentation on the biocompatibility of the bulk Al_2O_3 [11]. Today, alumina is the most bio-compatible material known. The result of this low cellular reaction around alumina advocates that the use of an alumina/alumina bearing couples will decrease

Table 14.1. Material specifications

Material Properties	Units & Standards	Specification Standards ISO 6474, 2nd ed., 1994, Type A
Density (ρ)	g/cm^3 DIN EN 623-3	>3.94
Color	–	–
Alumina content	%	>99.5
Impurities $SiO_2 + CaO + Na_2O$	%	<0.1
Additives (MgO)	%	<0.3
Microstructure mean grain size	μm	<4.5
Elastic modulus	GPa	**
Flexural Strength (4 Point Bending) (σ_{4PB})	Mpa DIN EN 843-1	>250
Fracture Toughness Rate (K_{1c})	Mpam$^{1/2}$	**
Hardness (HV) Vickers	HV0,5 DIN V ENV 843-4	**
Thermal Conductivity (at 25 °C) (λ)	W/mK DIN V ENV 821-2	**
Wetting Angle (α)	°	**

** No values mentioned in the standard.

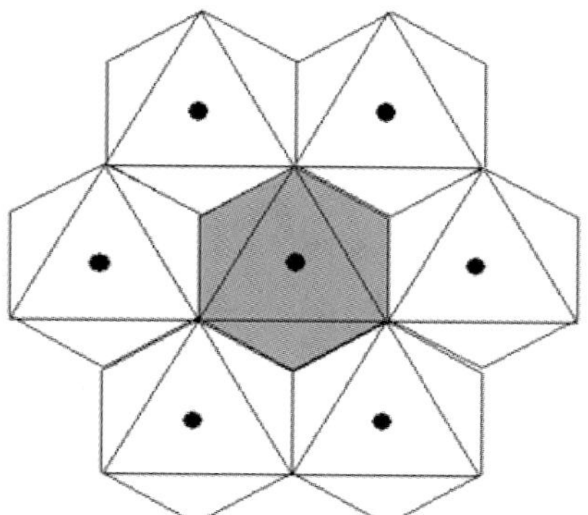

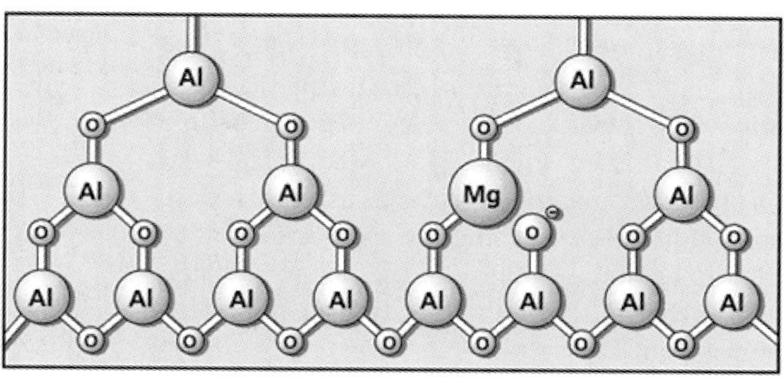

Figure 14.6. Schematic representation of the monophasic alumina alpha structure. Bivalent magnesium ion in the lattice site of a trivalent aluminum ion. The result is an unoccupied valence of an oxygen ion.

and could even suppress biological loosening initiated by wear debris as there is less debris and consequently less reactivity to the debris.

A recent study by the INSERM in Bordeaux (M. F. Harmand et al., 2000, personal communication) shows different in-vitro cell reactions (Macrophage, Interleukine 1β, Pge2, ETNFα) according to the kind of debris. They compare alumina, zirconia, and CoCr debris. There was no cytotoxicity with alumina whatever the size of the particles. In the case of zirconia, macrophage activation was observed with the small and large particles. The cytotoxicity response was according to the quantity and the time of contact.

Scratch Resistance

The hardness of alumina (2,300 HV) is greater than that of diamond. Alumina can only be scratched by diamond. This is a major quality in case of third-body wear (Figure 14.7). If there is a particle (by crescent hardness) of bone, metal, hydroxyapatite (HA), or cement (zirconium oxide addition for opacification) there will be no scratches, and as a result no tribological consequences on the alumina/alumina bearing couple.

Wettability

When the ionic character of the surface increases, the wettability increases. The difference in electronegativity involving the elements that constitute the surface layer form the ionic character Alumina and zirconia are highly ionic, hence their high wettability [12].

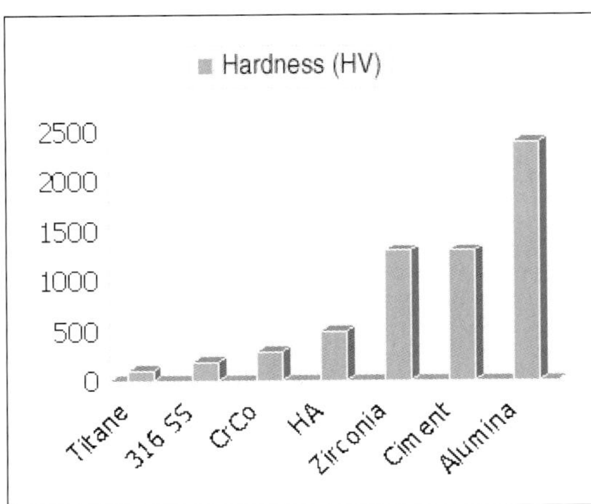

Figure 14.7. Vickers Hardness of different materials.

Thermal Conductivity

Recent studies have shown that the temperature increases on the friction area of the prosthesis when the patient is walking. This temperature depends of the bearing couple and its capacity to eliminated the heat. McKellop showed in a hip simulator that the temperature on top of the ball head with Zr–PE bearing couple was 51.3 °C, with CrCo–PE 40.4 °C, and 35.6 °C with Al–PE [15,16].

The thermal conductivity of alumina (30 W/mK) is high, therefore the transmission of heat is easy. On the other hand, the thermal conductivity of zirconia is low (2.5 W/mK).

Improvements

The raw material has changed over the years. The first generation of alumina was industrially pure at 99.5%. In the 1980s, raw material of better quality and purity was used, the "second generation" of alumina. This is still used today. Contaminants such as silicates and various oxides of alkalis and iron can create glassy phases at the boundaries of the ceramic crystal, leading to degraded properties [13]. A clean room was introduced in order to reach an even higher degree of purity. Today the range of purity ranges between 99.7 and 99.9%. Another improvement is micronization that enables the production of a very homogeneous powder with a highly regular grain size.

Manufacturing Process

In 1995 major changes were introduced in the manufacturing process of what was to become the third-generation alumina.

Before this date the ball head was marked by diamond engraving. The changes in surface geometry were too aggressive for the alumina. Shallow laser marking, as is commonly used today, does not weaken the ceramic components. The fine-grained powder from which oxide-ceramics are produced is compacted into a "green body". This green body is constituted of agglomerated particles whose density is fifty times superior to the final density. The agglom-

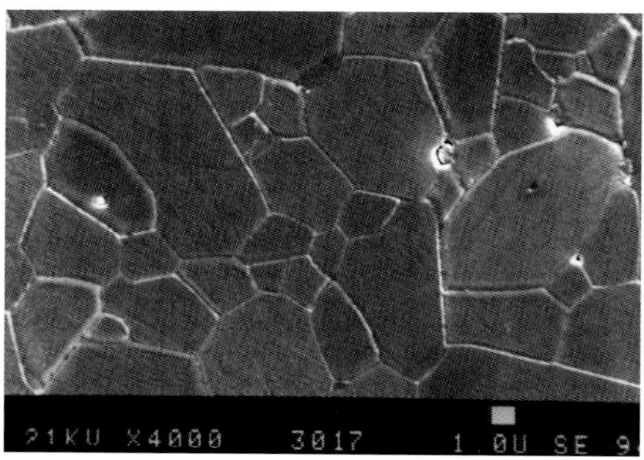

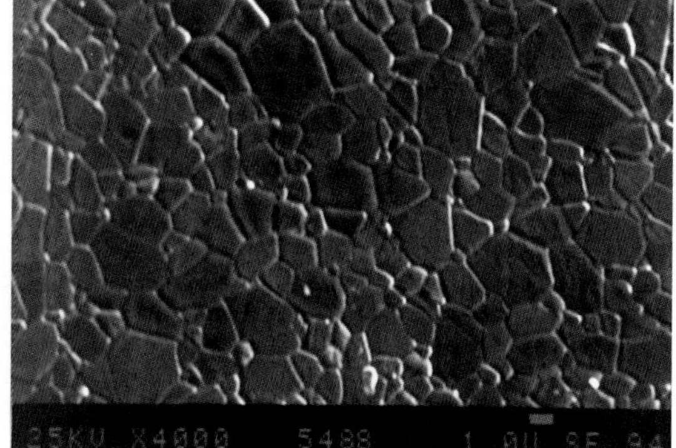

Figure 14.8. **a** SEM picture of alumina before 1995; grain size of 3.2 μ and a bending strength 4p of 500 Mpa. **b** SEM picture of alumina after 1995; grain size 1.8 μm and a bending strength 4p of 580 Mpa.

erated particles are transformed into a dense, polycrystalline body during the firing process (sintering) [14].

A special sintering process called HIP (hot isostatic pressed) is now used to reduce the size of the grain and thus increase the mechanical strength. By precisely controlling the temperature and pressure (1,000 atmospheres) an average grain size below 2 μm can be obtained and an alumina with a high density of 3.98 g/cm^3 is created. The comparison of the two following microradiographs shows the high degree of improvement (Figure 14.8). This treatment also improves the surface finish.

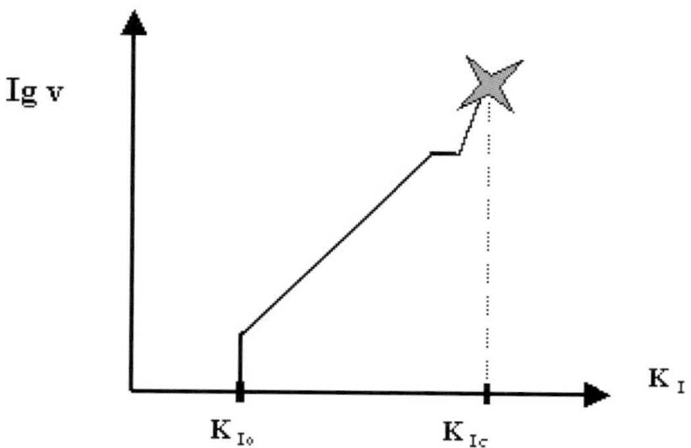

Figure 14.9. Velocity of crack growth versus stress intensity Ig v = Velocity of crack growth. K I = Stress intensity. K Io = Fatigue limit. K Ic = Fracture toughness.

ISO 6474

The ISO 6474 standard for medical grade alumina was introduced for the first time in 1979 and revised in 1994. Table 14.1 summarizes the specifications of the material. The right column shows the properties of the third generation of alumina that is now commercialized [10].

Third Generation

Technical Results

Crack growth, which can subsequently cause fracture, can be the source of an extremely high load on the bearing couple, for example during a trauma. One hundred and seven ceramic ball heads have been retrieved and examined. No loss in fracture load was detected when burst strength tests were undertaken. This can be explained by the relation between the velocity of the crack propagation and the stress intensity (Figure 14.9). The in vivo stress intensity, K_I, of ceramic femoral heads is generally about 0.6 Mpa m$^{1/2}$ when exposed to normal strain. This level of stress intensity ranges below the fatigue limit, K_{Io}, of alumina, which is rated to be approximately 1 Mpa [15].

Clinical Results

Toni et al. [16] have demonstrated how fracture rates have diminished over the years. In the 1970s the high fracture rates were common, due to the low-density alumina manufactured by companies that did not respect the later-introduced legal specifications.

However, the analysis of alumina ceramics and more precisely Biolox® femoral heads shows a fracture rate that has decreased from 0.026% for the first generation to 0.004% since 1994 (Table 14.2) [18].

Zirconia (Zirconium Oxide)

Properties

Material properties can be found in Table 14.3.

Triphasic

The crystal structure of pure zirconia can either be monoclinic, tetragonal, or cubic (Figure 14.10). The different microstructures are controlled by the addition of various cubic oxides such as MgO, CaO, or, more recently, Yttrium (Y_2O_3) [21]. Currently, Yttria-Tetragonal Zirconium Polycrystal (Y-TZP) is manufactured according to the ISO/DIS 13356. The volume of the monoclinic phase is superior by 5% to the tetragonal phase; the mechanical and tribological properties are poor.

Table 14.2. Published revision rates of THRs

Rate of revision	Due to
Approx. 10%	Aseptic loosening
Approx. 2%	Fracture of the prosthetic stem
Approx. 1%	Septic loosening
Up to 10%	Compiled fracture of early alumina heads offered by other manufacturers [17]
0.01%	Compiled fracture of a Biolox head alone; Semlisch's analysis for Sulzer [18]
0.015%	Compiled fracture of a Biolox head alone; Semlisch's analysis for Sulzer [15]
0.026%	Fracture of a Biolox head (first generation); CeramTec analysis [16,19]
0.014%	Fracture of a Biolox head (second generation); CeramTec analysis [16,19]
0.06%	Fracture of standard Biolox heads (first and second generation), Fritsch's investigation is based on the revision in the hospital of Homburg/Saar [20]
0.004%	Fracture of a Biolox forte head (third generation); CeramTec analysis [16,19]

Phase Transformation

The outstanding mechanical properties of zirconia ceramics are due to their phase transformation mechanism [22]. In its normal state, the material is fixed in its metastable tetragonal phase. Should crack formation occur, particles near the crack can transform from tetragonal into monoclinic, increasing their volume by approximately 5%. This can close the advancing crack front and thus "heal" the fracture (Figure 14.11). Unfortunately, if this process is not con-

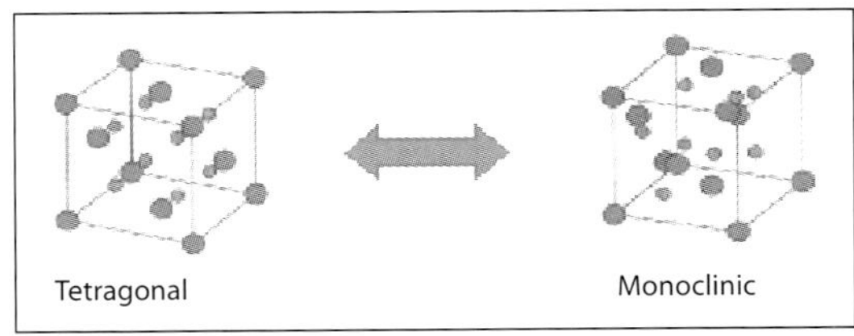

Figure 14.10. Schematic representation of two different zirconia phases.

Table 14.3. Material properties of zirconia

Material Properties	Units & Standards	Zirconia ISO 13356, 1997
Density (ρ)	g/cm^3 DIN EN 623-3	μ 6.00
Color	–	–
Combined zirconia, yttria content	%	>99
Additives (Y_2O_3)	%	4.5–5.4
Monoclinic phase	%	**
Microstructure mean grain size	μm	<0.6
Elastic modulus	GPa	**
Flexural strength (four-point bending) (σ_{4PB})	Mpa DIN EN 843-1	μ 900
Fracture toughness rate (K_{1c})	Mpam$^{1/2}$	**
Hardness (H)	HV0,5 DIN V ENV 843-4	**
Thermal conductivity (at 25 °C) (λ)	W/mK DIN V ENV 821-2	**
Wetting angle (α)	°	**

** No values mentioned in the standard.

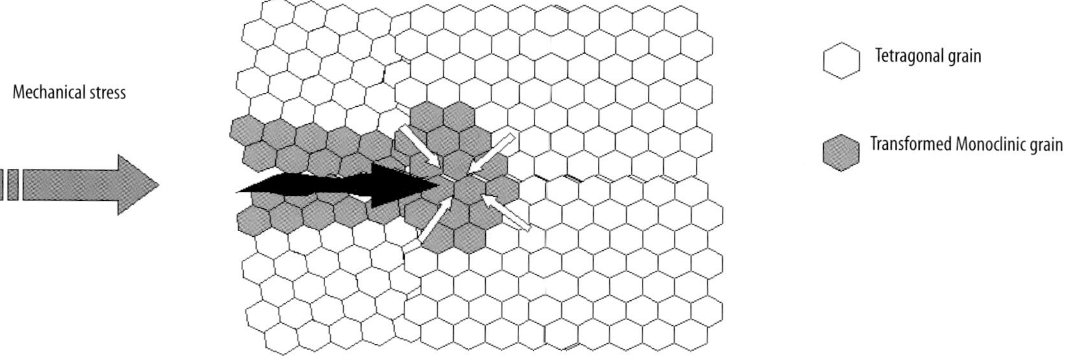

Figure 14.11. Diagram of the phase transformation after compressive stress.

trolled and tightly constrained in the material, so-called low-temperature degradation may occur [23].

Under hydrothermally challenging conditions, as those found in vivo, micro and macro crack formation can start due to a spontaneous phase transformation from the tetragonal to the monoclinic. This phenomenon expands according to growth mechanism laws and is transferred from one grain to the next, until the whole component is penetrated and transformed. The result is an embrittlement by rupture of grain structures and the complete loss of strength in the ceramic.

This degradation can be suppressed by increasing the amount of foreign oxide doping or by reducing the grain size [24] in which case a stable zirconia is produced. Unfortunately, such zirconia doesn't show the above-outlined transformation toughening mechanism. Strength and toughness are good, but fracture healing is very much reduced.

FDA Recommendation

After the MDA (Medical Devices Agency) recommendation [25] about the risk of steam re-sterilization. of zirconia ceramic heads, the FDA (Food and Drug Administration) published the following information in May 21, 1997:

"*This is to inform you that steam sterilization has been associated with surface roughening of zirconia ceramic femoral head components of total hip prostheses. This occurs because exposure to steam and elevated temperatures may lead to a phase transformation in the crystal structure of the zirconia material. As a consequence of this roughening, increased wear on the ultra-high-molecular-weight polyethylene acetabular component may occur, which can cause premature failure and require early revision.*" (Figure 14.12).

Technical Results

Typical zirconia ceramics used in orthopedics follow the ISO 13356 standard. Their mechanical properties can be seen in Table 14.3.

Clinical Results

Very little has been published concerning zirconia ball heads despite their introduction as early as 1985. Today approximately 350,000 pieces have been implanted. The fracture rate, according to reported cases, is 0.01%. This takes into account all fractures of implanted zirconia ball heads between 1985 and 1996.

Future Perspectives

The biggest manufacturer of orthopedic ceramic implants has decided to discontinue to manufacture pure zirconia components, due to the problem of long-term stability of the material. Some other manufacturers of ceramic implants continue to produce zirconia.

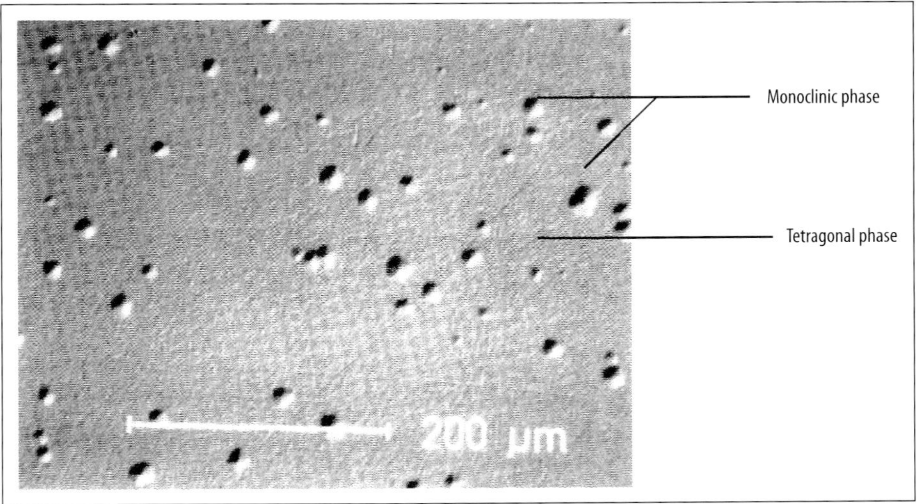

Figure 14.12. Hydrothermally treated zirconia (Y-TZP).

Hybrid Ceramic Compounds

With alumina it is not possible to produce femoral ball heads smaller than 28 mm, certain tapers, or certain shapes such as those used in knee implants. As described in the previous chapters the hydrothermal instability can cause problems in long-term use. In vitro tests of the zirconia/zirconia bearing couple does not show acceptable results for it to be used in orthopedics [26].

Concepts for Improving Ceramics

Ceramics are considered tough when they present a high resistance to crack propagation. In fracture mechanics of ceramics, the relation ship between toughness and strength is given as:

$$\sigma = \frac{K_{Ic}}{\sqrt{a \cdot y}}$$

where ρ = strength, K_{Ic} = fracture toughness, a = flaw size, and y = material constant.

Both the flaw size and the reduction of the grain size must be considered when improving the strength of the ceramic material. The two following concepts have been explored.

Zirconia Toughened Alumina (ZTA). Small particles of metastable zirconia are finely dispersed in an alumina matrix. This gives the material the possibility of transformation toughening, as outlined above, and thus higher toughness and resistance to crack propagation. First mechanical trails were reported and showed promising results. But, like zirconia, ZTA is not always a thermodynamically stable material [27].

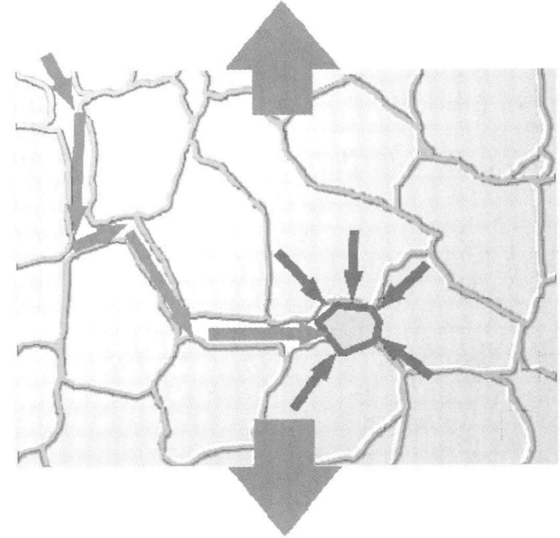

Figure 14.13. The principle of transformation toughening by zirconia particles, which are dispersed in the alumina matrix.

Alumina Matrix Composite(AMC). By the introduction of yttria-coated, nanosized zirconia particles, the existing stability problems of ZTA ceramics can be overcome (see Figure 14.13). Furthermore, these zirconia grains are isolated from each other by a solid and stable alumina matrix phase that constrains these zirconia particles and suppresses their transformation [28].

Another approach to improve strength is the principle of reinforcement by introducing anisotropic whisker-like crystals. The crack energy is dissipated by deviation of the crack around the fiber crystal (Figure 14.14), which is associated with an increase in strength and toughness.

The loss of hardness from alumina by the addition of zirconia can be compensated by another component – chromium oxide. Further composites need to be added to improve toughness.

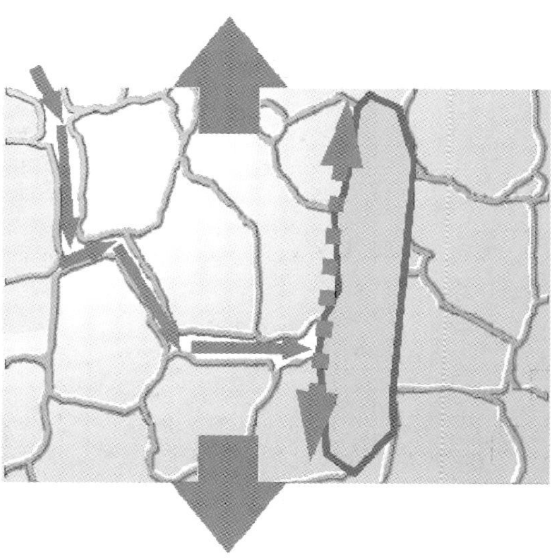

Figure 14.14. The principle of reinforcement by whiskers or platelet-like crystals in an alumina matrix.

Platelet-like crystals are formed in situ during sintering and are therefore homogeneously distributed throughout the ceramic (Figure 14.15); they have a great influence on strength and toughness. Figure 14.16 shows the microstructure of the alumina matrix composite.

New applications such as knee components are technically feasible thanks to this high-strength material (Table 14.4).

Assessment of Wear Couples

Interface Head Insert

The Bearing Couple

The hardness of a material is usually defined as resistance to scratching from other materials. Therefore, a simple rule can be established. The higher the level of hardness of a material, the lower will be the wear of this component. When using two materials as a wear couple in an aqueous environment, a combination of equal hardness materials has proven to be advanta-

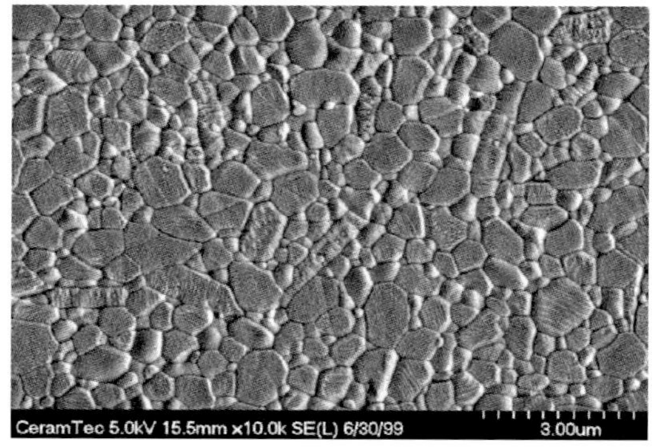

Figure 14.15. Microstructure of the alumina matrix composite.

Table 14.4. Material properties of alumina matrix composite

Material Properties	Units & Standards	AMC Biolox®delta	Test Report
Density (ρ)	g/cm³ DIN EN 623-3	>4.36	*
Color	–	Mauve	
Young's modulus (E)	Gpa DIN EN 843-2	350	PB (TF-A) 004
Flexural strength (four-point bending) (σ_{4PB})	Mpa DIN EN 843-1	1,138	PB (TF-A) 001*
Fracture toughness rate (K_{Ic})	MPam$^{1/2}$ Notched beam	5.7	*
Vickers hardness (H)	HV 1 DIN V ENV 843-4	1,975	PB (TF-A) 005*
Thermal conductivity (at 25 °C) (λ)	W/mK DIN V ENV 821-2	16.7	PB (TF-A) 008
Wetting angle in ringer's (α)	°Sessile drop	2.5	PB (TF-A) 006

Note: * These values are tested regularly with each production batch. Results shown here are from batch CH0201, QF-LA 0130, 25.6.98 if not stated otherwise.

geous. Indeed if this is not the case, the softer material will wear uncommonly quickly [8,29].

The Lubricant

Most wear couples need to be lubricated in order to enhance sliding as the main form of interaction, and to avoid surface contact, leading to abrasion, between the two partners. In mechanical applications an artificial lubricant is usually introduced. Unfortunately, this is not easily accomplished in biomedical applications. The body itself provides a natural lubricating medium to its joint – the synovia. This lubricant will be of major importance in long-term friction applications such as joint replacement. The absence of lubrication or poor lubrication has a considerable effect on the wear rate.

The Load

Load distribution on the contact area of the bearing couple is a main element to be taken into consideration. In a ceramic/ceramic bearing couple the diameter of the ball head has no influence on abrasion and consequently on the wear rate. However, the shape of the two components will have a major action on the tribology.

Different forms of contact have been studied (Figure 14.16). The first possible combination is called "equatorial" or circumferential. This case introduces hoop stresses in the ceramic insert, possibly leading to fracture and chipping of the insert.

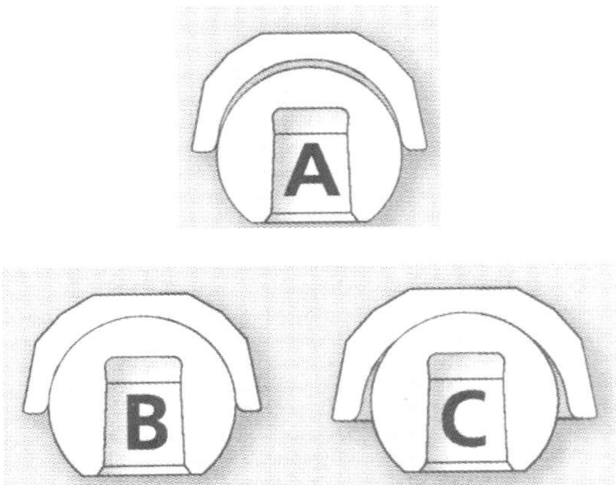

Figure 14.16. The effect of shape on tribology of hip joints. A – equatorial type; B – perfect congruance between head and insert; C – optimal interface between head and insert.

The second possibility is to obtain a perfect congruence between the ball head and the insert, but – aside from technical and geometrical (sphericity) problems – this does not enable correct lubrication and has high frictional resistance, creating abundand wear debris.

The third possibility is the one used today. The diameter of the ball head is slightly inferior to the size of the insert, enabling both good distribution of the load and perfect lubrication. This design demands great care in the manufacturing techniques and also during quality control. The result of the wear couple will therefore depend on the level of competence and the requirements of the company manufacturing the ceramic.

Tribology

Ring-On-Disk

In order to screen the tribologically required materials for hip joint replacement a ring-on-disk test as detailed by ISO 6474 is often performed [30]. This test is standardized – allowing good comparison between results – and stresses the materials higher than their final application as joint replacement components, thus enabling the scientist to reduce testing time.

The flat surface of the ring (inner diameter of 14 mm, outer diameter of 20 mm) and the disk (outer diameter of 30 mm) oscillate (±25°, f = 1 Hz) against each other under a constant load of F = 1,500 N. Most material combinations for hip joint replacement have successfully passed this test. Any new combination should be tested, as the results can sometimes be unexpected. Whereas many ring-on-disk tests with the combination zirconia (Y-TZP)/UHMWPE finished normally, experiments with the combination zirconia (Y-TZP)/zirconia (Y-TZP) and zirconia (Y-TZP)/alumina were both stopped after approx. 50,000 cycles because of massive wear. Figure 14.17 shows the typical wear volumes of a successful test, and the disastrous wear found with the above combinations.

Hip Simulator

Today, an international standard (ISO 14242) exists for hip simulator testing based on the

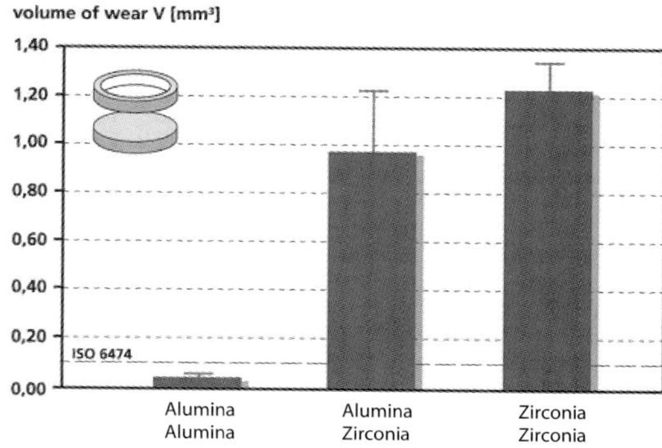

Figure 14.17. Volume of wear of the disk material after ring-on-disk experiment of the combination of ceramic ring and ceramic disc (mean value ± standard deviation, n = 5).

ASTM Guide for Gravimetric Wear Assessment of Prosthetic Hip Designs in simulator Devices (F 1714-97). Therein, the femoral and acetabular components of a test specimen are placed in position in their normal configuration (cup above head) and the apparatus transmits specified time (1 Hz) varying forces (double peak curve, maximum loads 3 kN) between the components together with specified relative angular displacements (flexion/extension +25° to −18°; adduction/abduction +7° to −4°; and inward/outward rotation −10 to +2°). The test takes place in a controlled environment (diluted bovine serum at 37 °C) simulating physiological conditions.

The following methods of measurement are permissible.

The test specimen is assessed for wear by testing for loss of mass in the simulator. A loaded, non-articulating control specimen is intended to allow for fluid sorption (when using UHMWPE cups).

A coordinate measurement machine is used to map the articulating surface of a total hip prosthesis relative to a reference position, direction, and plane prior to the start of the wear test and at suitable intervals during the test. From these data the volumetric change between measurements is determined. Loaded non-articulating controls can be used to allow the effects of plastic flow (when using UHMWPE cups).

The volumetric wear value can be converted from linear wear value by the following formula:

Table 14.5. Commonly found linear and volumetric wear rates of tribological materials from simulator studies

Bearing couple	Linear wear [mm/million cycles]	Volumetric wear [mm³/million cycles]
Metal/UHMWPE	0.2	35 to 200
Zirconia/UHMWPE	0.1 to 0.2	31 to 67
Alumina/UHMWPE	0.1 to 0.2	31 to 79
Alumina/Alumina	<0.002 *	<0.3 *

* denotes typical detection limits, values measured are on this level.

Volumetric wear = $\sigma \times r^2 \times$ linear wear, where r is the nominal radius of the wear couple.

The wear of ceramics has been well documented in the applications in engineering and as biomaterials [10]. The following results have been established after numerous studies on the subject (Table 14.5 and 14.6).

Bearing Couple

Alumina/PE

Time and experience have proven that the wear in the alumina/PE couple is two to five times less than in the metal/PE combination [14]. A consensus has been reached concerning the PE wear in vivo at twice that of alumina. A recent study of retrieval components compare metal-on-polyethylene combinations and ceramic-on-polyethylene combinations after 4–8 years in situ. More than twice as many polyethylene wear particles were produced from prostheses with metallic heads [17].

Alumina/Alumina

The alumina/alumina bearing couple presents the lowest recorded clinical wear. One of many studies done in a hip simulator of the alumina/lumina bearing couple was in Helsinki University of Technology, where it showed a very low wear rate. Visual examination and weighing of a 26, 28, and 32 mm-diameter head after 5 million cycles did not show wear. Scanning electron microscopy showed very slight wear marks, consisting of a removal of grains [31].

Wear rates have been found to be less than 0.005 mm per year after study of retrieved alumina/alumina wear couples [32]. Most authors, such as Nizard, Sedel, and Mittelmeier

Table 14.6. Wear rate data published on new ceramic materials from simulator studies

Wear couple	Linear wear rate per year	Comment	ISO or ASTM Standards	Approved by FDA
Metal/UHMWPE	0.2–0.5 mm	Satisfactory mid-term experience when used for old and not active patients	Yes. Various	Yes Class II, 510(k)
Alumina/UHMWPE	<0.1 mm	Good long-term clinical results	Yes. ISO 6474, ASTM F 603	Yes Class II, 510(k)
Mg-PSZ/UHMWPE		Not used anymore	No standard for Mg-PSZ zirconia	Yes. Class II, 510(k)
Y-TZP/UHMWPE first generation (black)		High revision rate. Not used anymore	No standard	
Y-TZP/UHMWPE second generation	<0.1 mm	Satisfactory mid-term experience	Yes. ISO 13356	Yes Class II, 510(k)
Alumina/Alumina	0.005 mm	Very good long-term experience	Yes. ISO 6474, ASTM F 603	Class III. Six IDEs ongoing in USA. FDA approval Feb. 2003
Y-TZP/Alumina	Controversial	In vitro test, very few clinical cases	No standard for the wear couple	No
Y-TZP/Y-TZP	Controversial	In vitro testing only	No standard for the wear couple	No
ZTA/PE	Same as Al/Al	In vitro testing only	No	No Status: R&D
ZTA/ZTA	Less than Al/Al	In vitro testing only	No	No Status: R&D
AMC/AMC	<0.002 mm *	In vitro testing only	Yes	No IDEs ongoing
AMC/Alumina	<0.002 mm *	In vitro testing only	Yes	No IDEs ongoing

* according to Zirconia content (and thus density).

report satisfactory results, especially in young and active patients after a minimum of ten years. The ceramic/ceramic bearing couple wear rate is 200 times lower than that of the metal/PE couple. The average wear rate of alumina/alumina total hip replacements reported in the Hôpital Saint-Louis in Paris is 0.025 μm per year.

Other Bearing Couples

Because of the positive experience with ceramic as a bearing couple material, new materials and material combinations have been proposed and first information has been reported. The result of the investigation on zirconia/alumina is very controversial; some people report disastrous wear, some not.

Zirconia/Alumina

A recent publication shows three different modes of wear behaviour in full ceramic bearing couples. The optimum case (mode 1) reveals very low wear; the observed phenomenon is limited grain pull-out as well as areas showing mild abrasion wear. If the main influence parameters (time, load, chemistry) change, the wear interface degrades and wear rates increase exponentially. Mode 3 is characterized by a steady-state wear rate at a high level; the wear surfaces are completely destroyed.

From the very widely varying results published on zirconia/alumina couples, it seems likely that these applications run in mode 2, with slight changes in simulator conditions creating very diverging results. The discussion on these wear applications is at the moment very controversial.

A clinical case was reported where a patient had complained about squeaking and noise in his total hip replacement after 37 months. The revision showed the retrieved components to be an alumina cup and a zirconia ball head. The linear wear of the cup was about 12 μm, and 9 μm for the head [33].

Zirconia/Zirconia

Because of possible severe degradation and decreased strength due to temperature and humid environments the Y-TZP material is not widely used in hard–hard applications [22].

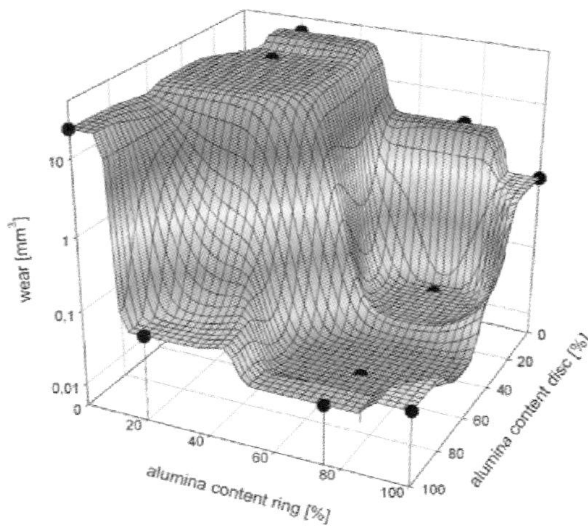

Figure 14.18. Wear volume of different combinations of alumina and zirconia coupling. The difference to 100% represents the zirconia content.

ZTA/ZTA Bearing Couple

A resent study [34] clearly demonstrated the influence of the zirconia content on the tribological behavior of such wear couples, using the ring-on-disc method. A high amount of zirconia at both sides (ring and disc, equivalent to ball head and insert) led to catastrophic failure (up to 45 mm^3) (Figure 14.18). The best results have been achieved for the alumina/alumina coupling with a small amount of zirconia (0.008 mm^3 wear volume).

Combinations

With the release of more and more new materials for bio-tribology, the possibility of assembling different wear couples increases dramatically. It is important to limit these experiments in order to avoid catastrophical failure in the future.

Standardization

No real standardization exists for ceramic components. The shape of the head and the inserts, and the diameter, the tolerance, and the deviation of sphericity are specified but no standard exists. Each manufacturer will determine its own criteria. For this specific reason the combination of two ceramics from two different manufacturers will cause problems with the tribology and important wear rates are to be expected. All combinations that have not been examined and approved will cause legal problems if used.

Legal Risks

According to the most recent jurisdiction, surgeons using prosthetic components from one manufacturer together with components from another manufacturer, which have not been approved explicitly for this purpose by the supplier of such components, shall be held responsible for the quality of the components, and for any technical problem resulting from the combinations [35,36]. Therefore, the leading manufacturers dictate "Never mix and match". Only the following combinations are safe with regard to the above criteria (Table 14.7).

It is thus good practice to consult with manufacturers of all components before starting any tests, trails, or even implantations.

FDA Approval

The strictest criterion today is that of the FDA, which classes its medical products [37]. Under class II come ball heads made of aluminum and zirconium oxide ceramics that articulate against the more usual polyethylene (UHMWPE) surface. Under class III come the wear couples that articulate against a ceramic cup. In the USA, before a product can gain Class III certification, the manufacturer has to submit a premarket approval application (PMA), which revolves around laboratory data and clinical data [38]. The clinical data is then collected in the form of a clinical trial performed by the manufacturer with FDA support and input – this is the Investigational Device Exemption (IDE).

Table 14.7. Approval combinations

Ball head	Cup
Alumina made from one manufacturer	Polyethylene
Zirconia made from one manufacturer	Polyethylene
Alumina made from one manufacturer	Alumina made from the same manufacturer

Bibliography

Willmann G. Ceramic cup inserts for hip endoprostheses. Biomed Technik 1996;41:98–105.

Willmann G. Ceramic cup inserts for hip endoprostheses. Biomed Technik 1997;42:256–63.

Liao Y, McKellop H. Effect of forced cooling on the wear of UHMWPE cups against CoCr or zirconia balls. Sixth World Biomaterials Congress. Hawai, 2000.

Lu Z, McKellop H. Frictionnal heating of bearing materials tested in a hip joint wear simulator. Proc Instn Mech Eng 1998;211:101–8.

Chamier W. MT-W Nr. 97115 der CeramTec, 1997.

Willmann G. Wie sicher sind keramische Kugelköptfe für Hüftendoprothesen? Mat Wiss u Werkstofftechnik 1996; 27:280–6.

Fritsch EW, Gleitz M. Ceramic Femoral Head Fractures in Total Hip Arthroplasty. Clin Orthop Rel Res 1996;328: 129–36.

Willard L, Sauer M, Anthony E. Predicting the clinical wear performance of orthopaedic bearing surfaces. Alternative bearing surfaces in total joint replacement ASTM STP 1346, American Society For Testing and Materials, 2000.

Willmann G. Evaluation of zirconia femoral heads in THR. Third annual symposium on alternative bearing surfaces in total joint replacement. Philadelphia, October 2000.

References

1. Semlitsch M. When should we employ aluminium oxide ceramics as an alternative to metal balls in articulation with UHMW polyethylene cups of total hip prostheses? Sixth Symposium on Biomaterials. Göttingen, Germany, 1994;1–22.
2. Willmann G. Bioceramics in orthopaedics. What we have learned in 25 years? Med Orth Tech 2000;120:10–16.
3. Willmann G, Früh HJ, Pfaff HG. Wear characteristics of sliding pairs of zirconia for hip endoprosthesis. Biomaterials 1996;22:2157–62.
4. Cales B. Fracture ratio of zirconia hip joint heads compared to other ceramic bearing systems. Norton desmarquest fine ceramics, 1999.
5. Willmann G. Examen de 87 têtes fémorales en céramique après utilisation in vivo. In: Enke, editor. Bioceramics on Orthopaedics – New applications. Stuttgart, 1998;13–18.
6. Willmann G. Investigation of 87 retrieved ceramic femoral heads. In: Enke, editor. Bioceramics in Orthopaedics – New Applications. Stuttgart, 1998;13–18.
7. Willmann G. Ceramic cup inserts for Hip endoprostheses. Biomed Technik 1997;42:256–63.
8. Willmann G, Kälberer H, Pfaff HG. Ceramic cup inserts for hip endoprostheses. Biomed Technik 1996;41:98.
9. Jahanmir S. Friction and wear of ceramics – Advanced Ceramics in Tribological Applications 1994;3–12.
10. Lerouge S, Yahia, Sedel L. Alumina ceramic in total joint replacement. Hip Surg 1998;31–40.
11. Thomas P. Assessment of immuno-allergological properties of ceramic and metallic compounds in vitro. Hip Int 2000;10(3):359–62.
12. Davidson JA. Characteristics of metal and ceramic total hip bearing surfaces and their effect on long term ultra-high-molecular-weight polyethylene wear. Clin Ortho Relat Res 1998;294:361–78.
13. Clarke I. Material properties of structural ceramics. Third Annual Symposium on Alternate Bearing Surfaces in Total Joint Replacement. Philadelphia, Oct 2000.
14. Heimke G, Willmann G. Follow-up study based on wear debris reduction with ceramic–metal modular hip replacement. Biomaterials Engineering Devices: Human Applications 2000;2:223–51.
15. Willmann G. Survival rate and reliability of ceramic femoral heads for THA. Mater Sci Eng 1998; 29(10):595–604.
16. Toni A, Sudanese A. Ceramics in Total Arthroplasty. Encyclopedic Handbook of Biomaterials and Bioengineering, Part A, Vol. 2. New York: Marcel Dekker, 1995;1502–44.
17. Bos I. Histological investigation of polyethylene particles in total hip replacement: Ceramic versus metal head. Hip Int 2000;10(3):151–60.
18. Masson B, Willmann G, Von Chamier W. Fiabilité du couple alumine/alumine dans la prothese totale de hanche. Journées Lyonnaise de la Hanche 1999;397–402.
19. Semlitsch M, Weber H, Steger R. 15 Jahre Erfahrung mit Ti-6A1-78Nb-Legierung für Gelenkprothesen. Biomed Technik 1995;40:347–55.
20. Semlitsch M, Dawihl D. Basic Requirements of Alumina Ceramic in Artificial hip Joints Balls in Articulation with Polyethylene Cups. Technical Principles, Design and Safety of Joint Implants. Seattle: Hogrefe & Huber, 1994;99–101.
21. Jahanmir S. Friction and wear of ceramics – Advanced Ceramics in Tribological Applications. 1994;3–12.
22. Cales B. Fracture ratio of zirconia hip joint heads compared to other ceramic-bearing systems. Norton desmarquest fine ceramics, 1999.
23. Piconi C, Maccauro. Zirconia as a ceramic biomaterial. Biomaterials, 20th ed. Elsevier 1999;1–25.
24. Yoshimura M. Phase stability of zirconia. Tokyo institute of technology research lab of engineering materials Ceramic bulletin 1998;67(12):1950–5.
25. UK Medical Devices Agency, Adverse Incidents Centre. Safety Notice MDA SN 9617, "Zirconia Ceramic Heads for Modular Total Hip Femoral Components: Advice to Users on Re-Sterilization".
26. Pfaff HG. A new material concept for bioceramics in orthopedics. Bioceramics in hip joint replacement. Fifth symposium, CeramTec Stuttgart, 2000;136–45.
27. Claussen N. Fracture toughness of Al_2O_3 with an unstabilized ZrO_2 dispersed phase. J American Ceramic Society 59(1–2).
28. Burger W, Richter HG. High strength and toughness alumina matrix composites by transformation toughening and in situ platelet reinforcement (ZPTA) – The new generation of bioceramics. Bioceramics 2000;13:454–548.
29. Masson B. CeraNews Septembre 2000;10, CeramTec.
30. Willmann G. Wear characteristics of sliding pairs of zirconia (Y-TZP) for hip endoprostheses. Biomaterials 1996;17(22):2157–62.
31. Saikko V, Pfaff H-G. Wear of alumina on alumina total replacement hip joints studied with hip simulator. Second Symposium on ceramic wear couples. Stuttgart, 1997;117–122.

32. Bos I, Henssge J, Willmann G. Morphological characterisation of joint capsule around hip prostheses with alumina or alumina combinations. Die Keramikpaarung BIOLOX in der Hüftendoprothetik. Proceedings des 1. CERASIV Symposiums. Stuttgart 1996;24–30.
33. Morlock M. The wear couple Zirconia/Alumina in THR: A case study. Reliability and long-term results of ceramics in orthopaedics. Stuttgart, 1999;102–7.
34. Kaddick C, Pfaff H-G. Wear study in the alumina–zirconia system: reliability and long-term results of ceramics in orthopaedics. Stuttgart, 1999;96–101.
35. Willmann G, Kälberer H. Ceramic cup insert for hip endoprostheses. Biomed Technik 1996;41:98–105.
36. Willmann G. Ceramic sockets for total hip replacement; Never mix and match. Biomed Technik 1998;43:184–6.
37. Willmann G. Experience on zirconia ceramic femoral heads. Sixth World biomaterial congress. Hawaii, 2000.
38. Garino JP. The status and early results of modern ceramic–ceramic total hip replacement in the United States. Bioceramics in hip joint replacement. Fifth symposium. CeramTec: Stuttgart, 2000;88–91.

15 Usual Mechanical Tests: Mechanical Properties of the Fundamental Elemental Materials

P. Grosbras and A. Junqua

Introduction

Materials are chosen according to the uses that are intended:

does it have to be rigid or ductile?

will it be subjected to sudden, intense, or repeated forces?

will it be subjected to high temperatures?

will it be subjected to friction?

etc.

So, the properties that are required are selected, then we go to the catalogs to look for the best material. It is necessary, however, to understand how the properties listed in the catalogs have been obtained and what they mean; it must also be realized that these properties have been established by standard tests under conditions that might not be the same as when the material is being used.

We will therefore study various tests that enable us to know how a material subjected to an external force will "respond". The results of these tests are the mechanical properties.

The force applied, $\vec{F}$, is, of course, compensated for by an opposite force (reaction), otherwise the material does not deform, it would be displaced. The material is said to be stressed (Figure 15.1).

Definitions

Stress

If a force $\vec{F}$ is applied perpendicularly to the surface S_0; the stress is the ratio of the intensity of the force to the initial section S_0. More precisely, this is called nominal stress, and is expressed in pascals (P_a). Very often, stress has a high value and is expressed in MP_a.

If the force is not perpendicular to the surface, it is broken down into two components: normal force, $\vec{F}_n$, and tangential force, $\vec{F}_t$; in this case, two stresses are defined:

$$\sigma = \frac{F_n}{S_0} \quad \text{and} \quad \tau = \frac{F_t}{S_0}.$$

Shear stress MP_a

The following four states of stress are encountered in practice (Figure 15.2):

State of Uniaxial Stress

Simple traction, or simple compression (the test piece is long in relation to the radius).

State of Biaxial Stress

A spherical vessel containing a fluid under pressure is loaded in two directions; the stresses are not necessarily symmetrical.

Hydrostatic Pressure

The stresses are identical on all the faces.

By convention: $\sigma > 0 \rightarrow$ traction
$\sigma < 0 \rightarrow$ compression

(See Figure 15.3)

Simple Shear (Torsion)

Deformation ε

Materials react to stresses by deforming; a stress σ brings about a deformation ε defined by the

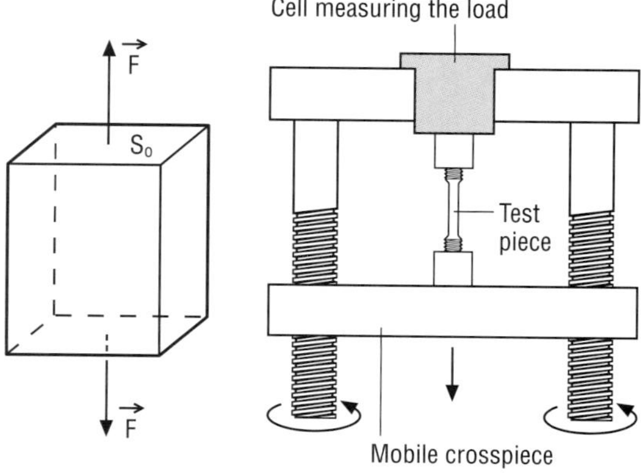

Figure 15.1

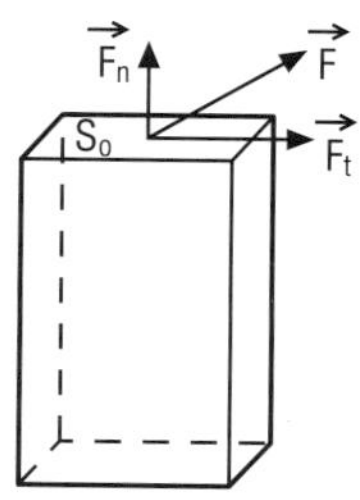

Figure 15.2

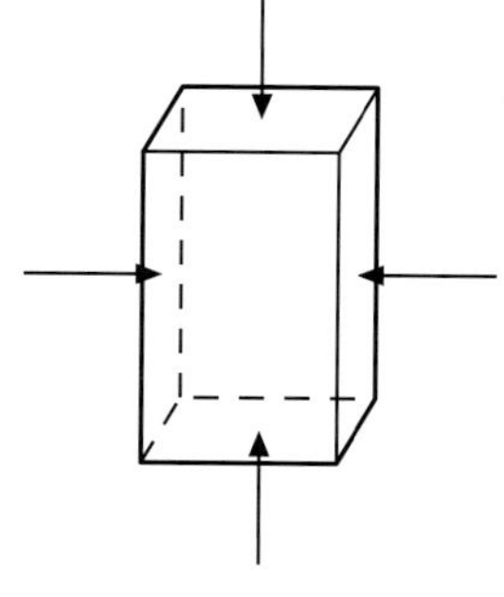

Figure 15.3

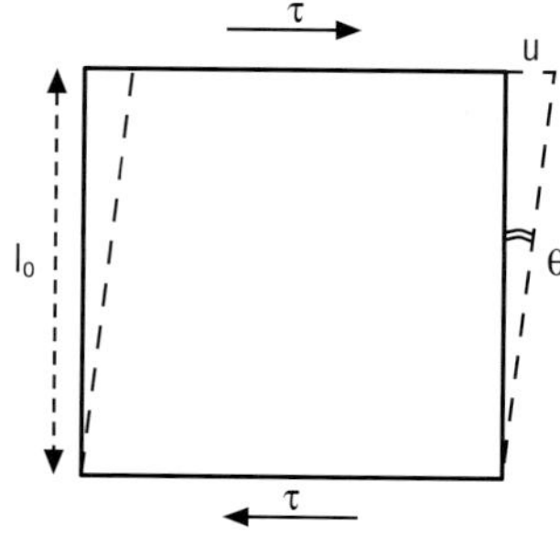

Figure 15.4

ratio of elongation Δl of the material to the initial length l_0.

$$\varepsilon = \frac{\Delta l}{l_0} \quad \Delta l = l - l_0$$

The lengths are measured between two reference points on the test piece. It can be specified that this is *nominal deformation* (or unitary elongation in relation to the initial length); it is a number without dimension expressed as a percentage.

Shearing creates shear deformations (see Figure 15.4). If the shear takes place along a length u (θ: shear angle), the shear (or distortion) is defined by:

$$\gamma = tg\theta = \frac{u}{l_0}$$

In general, the deformations are small: $\gamma\% \approx \theta$ in radians.

Hydrostatic pressure brings about a change in volume called *expansion.*

Resistance

Resistance measures the maximum stress that a material can withstand before breaking. If F_m is the maximum force, the resistance is $\sigma_R = \frac{F_m}{S_0}$. σ_R is therefore the breaking stress.

Ductility

Ductility is the property owing to which a material deforms permanently before breaking.

Toughness

Toughness measures the energy required to cause breakage. It is therefore measured in Joules.

Hardness

Hardness measures a material's resistance to penetration.

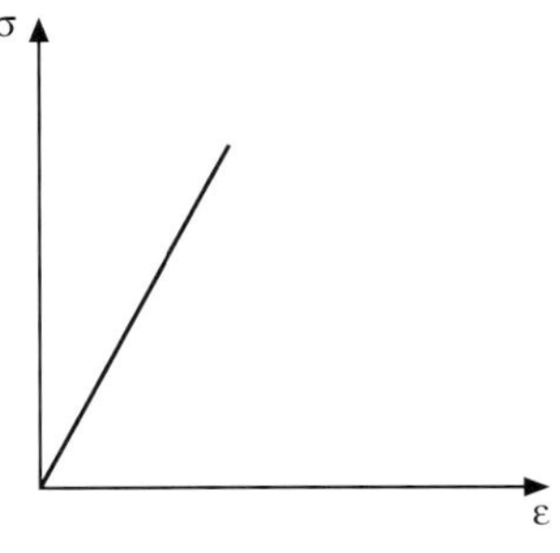

Figure 15.5

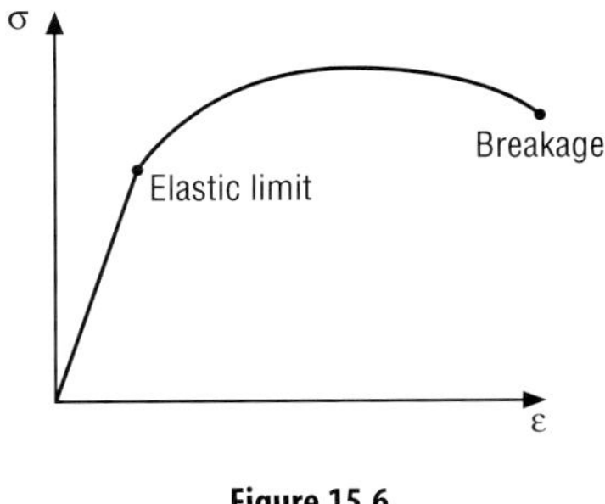

Figure 15.6

Note: For a given application, different values will be specified for these properties; they are often contradictory (for example, resistance and ductility); compromises therefore have to be made.

Deformation Curves

The curve is represented as $\sigma(\varepsilon)$; the shape of this curve and the values of the parameters that govern it can be very different. In general, a distinction is made between a number of ranges:

Elastic Range

Deformation is said to be *elastic* if the test piece regains its initial dimensions when the external force is suppressed. The modulus of elasticity measures the resistance of a material to elastic deformation: materials with a low modulus are flexible and undergo considerable elastic deformation; this effect may be desirable (springs, vaulting poles, etc.), but in many other practical cases the deformation is not at all desirable and materials with a high modulus will be used. For copper, steel, tungsten carbide, and selenite, the range of elastic deformation coincides with the initial linear portion of the deformation curve (see Figure 15.5).

The slope of the linear part of the curve defines the Young's modulus E (or modulus of elasticity). E is expressed in P_a or in MP_a (sometimes, but incorrectly, in kg/mm^2; 1 kg/mm^2 $\approx$ 10 MP_a). $\sigma = E\varepsilon$ (Hooke's law).

In the same way, other moduli are defined: $\tau = G\gamma$ (G = modulus of shear).

P (pressure) $= -\chi \dfrac{\Delta V}{V_0}$, χ = coefficient or modulus of compression.

The moduli of elasticity (E, G, χ) are expressed in MP_a.

Linear elasticity applies only to small deformations ($\varepsilon \leq 0.1\%$) typically. Beyond this, some solids break, others continue to deform, but the deformation is no longer elastic (this is known as plastic deformation); the limit established is called the limit of elasticity or the elastic limit. Finally, some solids (rubber) remain elastic to considerable deformation, but no longer obey the laws of linear elasticity (see Figure 15.6).

Some E values for various materials are:

Diamond	1,000,000	MP_a
Silicon carbide	450,000	
Steel	210,000	
Aluminum	70,000	
Alumina	35,000	
Glass	71,000	
Nylon	2,800	
Rubber	100	
PVC	1	

The usual materials are therefore in a convenient range to choose from for a given use ranging from 1 MP_a to 10^6 MPa, i.e., over six orders of magnitude!

Rigidity is the property a material has for deforming elastically under the action of a stress. Young's modulus is a possible way of measuring the material's rigidity; a material is all the more rigid as the elastic deformation is low for a given level of stress (i.e., E is high). See Figure 15.7.

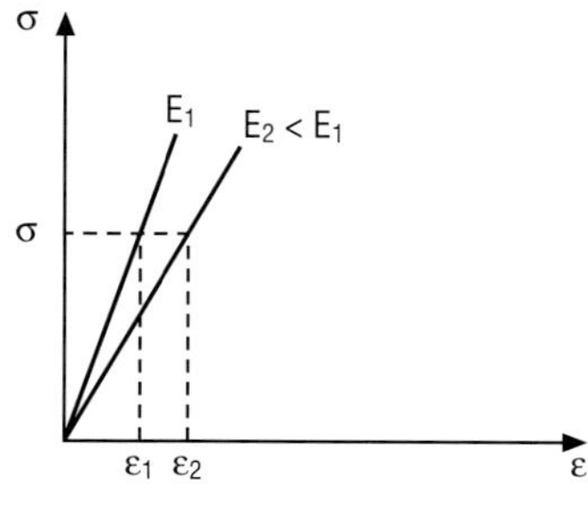

Figure 15.7

Material 1 is more rigid than material 2.

Definition of Poisson's Coefficient

In the case of a uniaxial stress, longitudinal elastic deformation is accompanied by lateral deformation (in the directions perpendicular to the stress), characterized by Poisson's coefficient ν (see Figure 15.8).

This can cause: following the axis z, $\sigma = \sigma_z$ and $\varepsilon_z = \frac{\sigma}{E} = \varepsilon$ and following the axis x, $\varepsilon_x = -\nu\varepsilon$.

For most metals, for example, $\nu \approx 0.33$, and for many materials, $0.25 < \nu < 0.5$.

For example, for a cylindrical rod with an initial radius r_0:

$$\varepsilon_x = \frac{\Delta r}{r_0} \text{ and } \varepsilon = \frac{\Delta l}{l_0} \rightarrow \frac{\Delta r}{r_0} = -\nu.$$

With a small change Δr in the radius, this becomes:

$$\frac{\Delta V}{V_0} = \frac{\pi(r_0+\Delta r)^2(l_0+\Delta l) = \pi r_0^2 l_0}{\pi r_0^2 l_0}.$$

If terms of the order of 2 and greater are ignored, such as Δr^2, $\Delta r\Delta l$ . . . , we have:

$$\rightarrow \frac{\Delta V}{V_0} = \frac{\pi r_0^2 l_0 + 2\pi r_0 l_0 \Delta r + \pi r_0^2 \Delta l - \pi r_0^2 l_0}{\pi r_0^2 l_0}$$

$$\frac{\Delta V}{V_0} = 2\frac{\Delta r}{r_0} + \frac{\Delta l}{l_0} = -2\nu\varepsilon + \varepsilon.$$

Finally, $\frac{\Delta V}{V_0} = \varepsilon(1-2\nu)$.

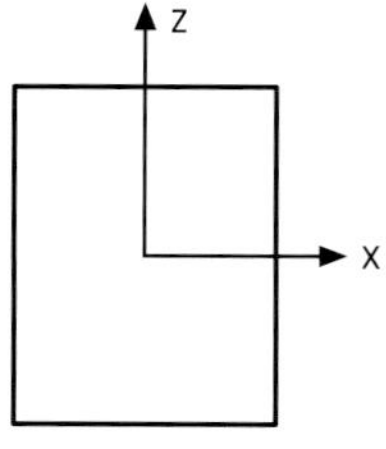

Figure 15.8

There is, therefore, a slight increase in volume under traction; for ν = 0,3 and $\varepsilon = 10^{-4} \rightarrow \frac{\Delta V}{V_0}$ $= 4.10^{-5}$.

$\frac{\Delta V}{V_0} = 0$ *if* $\nu = 0{,}5 \rightarrow$ there is no change in volume.

To summarize, the elastic range is characterized by three properties:

Hooke's law: stress proportional to the deformation; it does not depend on the type of deformation,

reversibility of the deformation,

slight change in volume.

It is therefore shown that: $G = \frac{E}{2(l+\nu)}$ and $\chi = \frac{3(l-2\nu)}{E}$.

Range of Plastic Deformation

The deformation of the material is said to be *plastic* if it does not regain its initial dimensions when the load is removed. Beyond the elastic phase, the application of greater stresses causes *permanent deformation*: the atoms have been displaced irreversibly. In the plastic phase (the part of the curve beyond A) there is no longer proportionality between stress and deformation (see Figure 15.9).

The limit between the two ranges (point A) is called: the *elastic limit*, σ_E, which measures the ability of the material to resist plastic deformation. When the load is removed, for example from B (or when breakage takes place at point D), the state of the material (σ, ε) follows a straight line BB′ parallel to OA (or DD′). OB′ (or

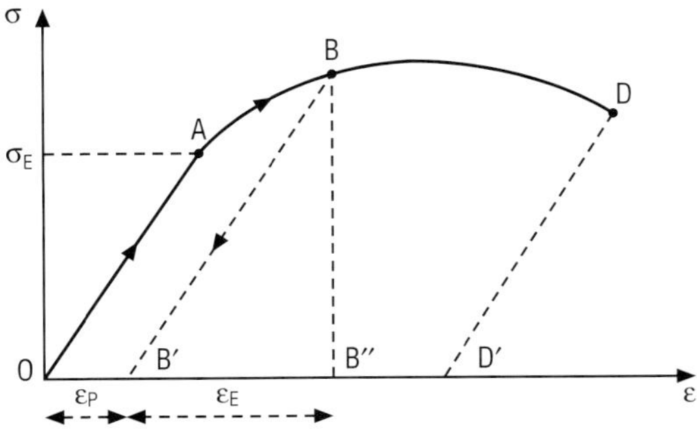

Figure 15.9

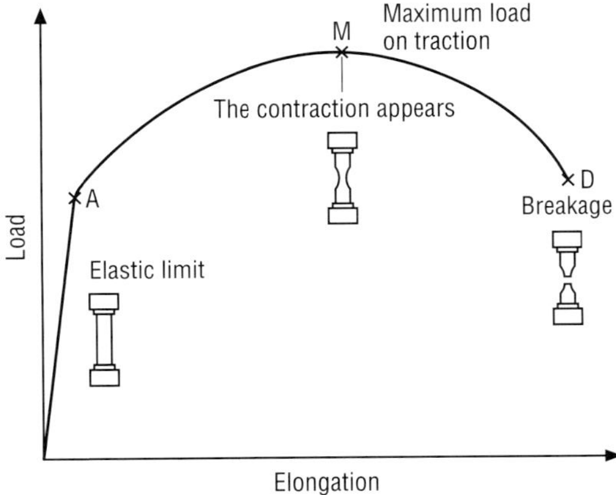

Figure 15.10

OD′) represents permanent elongation; OD′ is the elongation particular to breakage (B′B″ is elastic deformation).

So, if the load is removed completely, i.e., when the stress reaches the value zero, there remains a permanent, so-called "plastic" deformation, ε_p, measured by OB′ and there has been elastic shrinkage ε_E measured by BB′ (see Figure 15.10).

The plastic range consists of two parts:

in the part AM, the stress increases up to a maximum,

from M, a necking appears on the test piece; this is the start of the contraction. The deformation is localized in the contraction zone, which explains why the curve returns towards the ε axis. This phenomenon lasts for either a short or long time and breakage occurs in the contraction zone where the section is weakest.

Different Types of Deformation Curve

They correspond to possible behaviors of the material.

Ductile Material

Breakage takes place after major plastic deformation. There are two properties for measuring ductility:

Elongation at breakage, $A = \dfrac{l_f = l_0}{l_0}$ expressed as a percentage. As deformation is often localized in the contraction zone, the information is only meaningful if the length l_0 which serves as a reference is specified.

Contraction at breakage, $\sum = \dfrac{S_0 - S_f}{S_0}$ is also expressed as a percentage. The reduction in cross section is considerable for very ductile materials. Σ is a property that has the advantage of not depending on a reference length as in the determination of A; it can also be used to determine the true deformation at the break point (true deformation involves S and not S_0). A and Σ vary in the same direction to characterize the different materials; it is difficult to establish the exact relationship between these two properties since the deformation can be extremely localized.

Brittle Material

Breakage takes place without there being a plastic phase (the deformation is purely elastic; σ_E and R_m are merged). A and Σ are very small (nil in theory). Examples include tungsten carbide, selenite, ceramics, gray cast iron, etc (see Figure 15.11).

Important note: Heat treatment affects the mechanical behavior of a material considerably; for example, a ductile metal may become brittle; this is why lead becomes brittle at very low temperatures.

Linear and Non-linear Elasticity; Inelastic Behavior

Linear Elasticity (see Figure 15.12)

The above curve describes the case of a material with linear elastic behavior (Hooke's law).

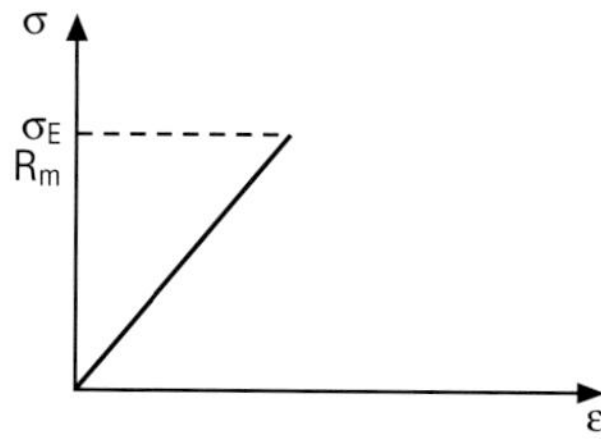

Figure 15.11

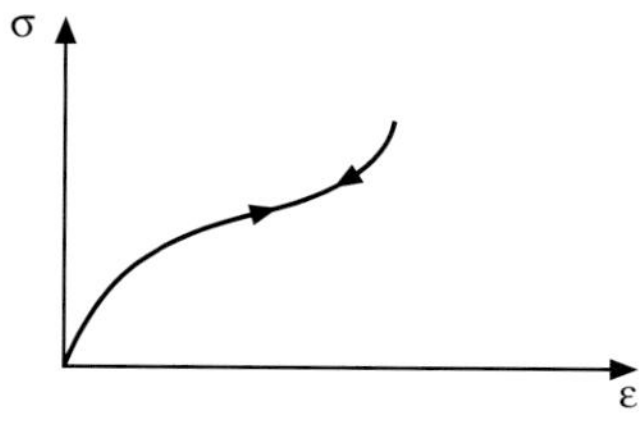

Figure 15.13

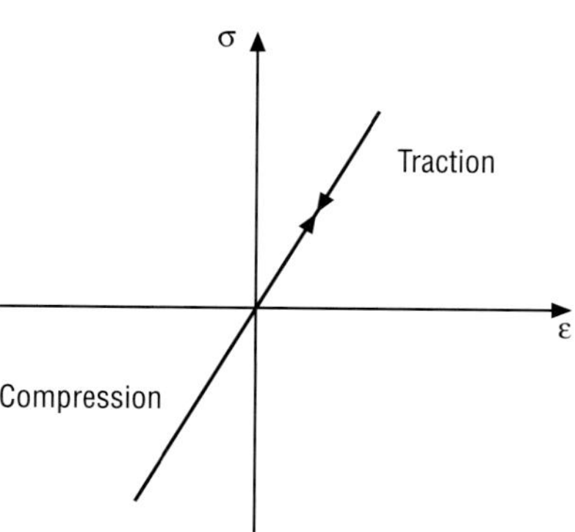

Figure 15.12

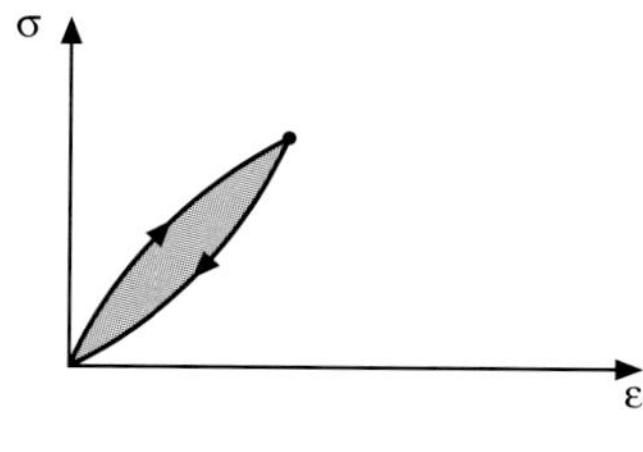

Figure 15.14

The slope E of the deformation curve is the same in traction and in compression. The value of the area under the deformation curve is: $a = \int \sigma d\varepsilon = \int \frac{F}{S_0} \frac{dl}{l_0} = \frac{1}{V_0} \int F\,dl.$

Stored elastic energy by unit of volume (or density of elastic energy). $a = \frac{W}{V_0}$. If the stress is released, this energy is completely regained.

Non-linear Elasticity (see Figure 15.13)

The above deformation curve is that of a non-linear elastic material. This is the case with rubber, for example. The deformation may be considerable and the material remains elastic; when the stress is released, the same curve is described as when Σ was increasing, and all the energy stored during traction is regained when the load is removed (it can be passed to projectiles; for example, arrows, catapults, etc.).

Inelastic Behavior (see Figure 15.14)

Finally, there is a third type of behavior that some materials obey – *inelastic behavior* – the discharge curve does not exactly follow the load curve. Energy is therefore dissipated during a load–discharge cycle (area: dissipated energy by unit of volume). All solids are, in fact, slightly inelastic. Sometimes this is a desirable effect in order to attenuate vibrations or noise (polymers, soft metals such as lead, etc.); but often attenuation is not very desirable (springs, musical instruments, etc.).

Rational Curves

Introduction

For small plastic deformations, the deformation curve in traction and the deformation curve in compression are symmetrical; but they become different with large deformations. The length of the test piece that is being crushed becomes smaller while its cross section increases in order to keep the volume constant; the stress needed to continue the compression increases; there is no contraction and the test piece could be crushed almost indefinitely if cracks did not appear that bring about breakage. The difference is therefore due to geometry. Hence the definition of a new stress to get round this (see Figure 15.15).

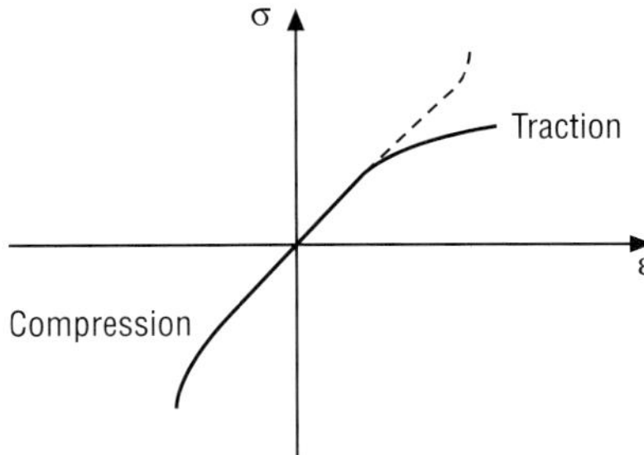

Figure 15.15

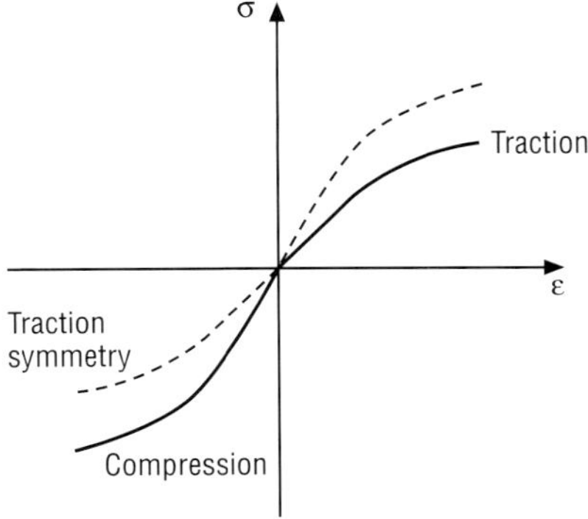

Figure 15.16

Rational Stress (or True Stress)

S is the true cross section (instant). This way of proceeding compensates for the reduction in cross section in traction and its increase under compression. The curves cannot always be superimposed (see Figure 15.16).

$S < S_0$ (traction) $\rightarrow \sigma_r > \sigma$ (see Figure 15.17)

However, in traction an adjustment towards the ε axis can no longer be seen after contraction (σ_r increases until breakage). Knowing σ and ε, σ_r can be calculated:

$$V = cste \rightarrow \quad Sl = S_0 l_0$$

$$S_0 = S\frac{l}{l_0}$$

$$\sigma_r = \frac{F}{S_0}\frac{S_0}{S} = \sigma\frac{l}{l_0}$$

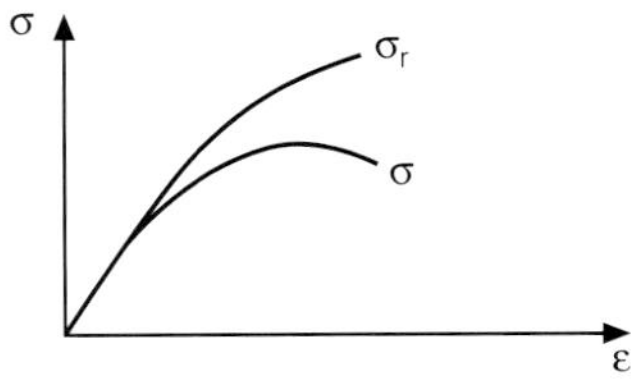

Figure 15.17

$$\text{or} \quad \varepsilon = \frac{l-l_0}{l_0} = \frac{l}{l_0} - 1 \rightarrow \frac{l}{l_0} = 1+\varepsilon$$

where $\sigma_r = \sigma(1+\varepsilon)$

Rational Deformation (or True Deformation)

The curves σ_r (ε) in traction and compression cannot be superimposed because a change in length Δl does not give identical states of deformation in both cases.

For example, if $\Delta l = \frac{l_0}{2}$, on a $1 = 1{,}5 l_0 \underbrace{\left(\times\frac{3}{2}\right)}_{\text{in traction}}$

and $l = 0{,}5 l_0 \underbrace{(:2)}_{\text{in compression}}$.

If the length is doubled (traction) $\varepsilon = \frac{2l_0 - l_0}{l_0} = 1.$

If it is divided by two (compression)

$$\varepsilon = \frac{\frac{l_0}{2} - l_0}{l_0} = -0{,}5.$$

Plastic deformation is much greater in compression than in traction and the two states cannot be compared. States can only be compared by adding elementary deformations end to end because at the limit of infinitesimal deformations the difference is negligible. During an elementary deformation, the deformation has to be calculated from the length of the test piece just before this deformation.

$$d\varepsilon = \frac{dl}{l}$$

For the overall deformation, which becomes the true deformation, the following is obtained:

$$\varepsilon_r = \int_{l_0}^{l} \frac{dl}{l} = L_n \frac{l}{l_0} = L_n \frac{l_0 + \Delta l}{l_0},$$

$$\varepsilon_{r^\cdot} = L_n(1+\varepsilon).$$

ε	0.05	0.10	0.15	0.20	0.25
ε_r	0.04879	0.09531	0.13976	0.18232	0.22314
ε	0.30	0.35	0.40	0.45	0.50
ε_r	0.26236	0.30010	0.33647	0.37156	0.40457

For small deformations ($\varepsilon_r \approx \varepsilon$; $\sigma_r \approx \sigma$); in cases where the length is doubled by traction and where the height is reduced by half by compression, the deformations have different values while the true deformations have opposite values:

	ε	ε
Traction	1	$L_n(1+1) = L_n 2 = 0.693$
Comprssion	-0.5	$L_n(1-0.5) = -L_n 2 = -0.693$

Rational Curve

Curve $\sigma_r = f\,(\varepsilon_r)$; it has the same course as the deformation curve until the contraction appears; beyond this, S is taken; so σ_r increases until breakage. This time, the curves are symmetrical. The true stress enables the forces to be analyzed during deformation; but the nominal stress is more useful in practice because technical specifications are given on the basis of the initial dimensions.

Plastic Deformation Energy

Rolling, forging, and drawing metals are operations which consume energy; the same applies to the injection molding of polymers; the work performed to change its shape irreversibly is *plastic deformation energy* (see Figure 15.18).

Its value by unit of volume is the hatched area under the deformation curve (the line stops at ε_P because the elastic deformation energy is restored on discharge). With the true curve, the same value is obtained:

$$a_r = \int \sigma_r d\varepsilon_r = \int \frac{F}{S}\frac{dl}{l} = \frac{1}{V}\int Fdl = \frac{W}{V}.$$

As $V = V_0$; $a_r = a$

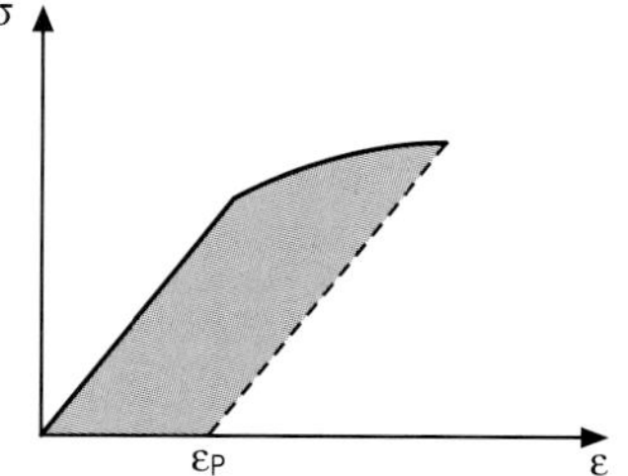

Figure 15.18

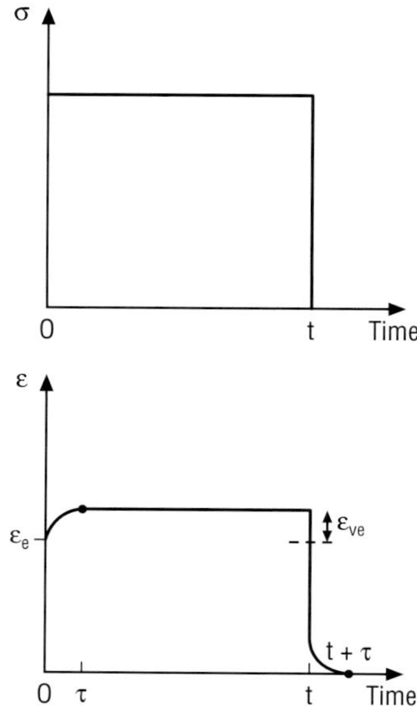

Figure 15.19

It is important to know this energy for shaping materials: it determines the force of the rollers in a steel mill, the pressure of a molding machine, etc.

Viscous Deformation

This is a phenomenon seen in certain materials such as polymers. When subjected to a sudden stress, polymers undergo sudden deformation, followed by a different deformation called *viscous deformation*. There is *viscoelasticity* and *viscoplasticity*.

Viscoelastic Deformation

After suppression of σ, ε becomes nil again. $\varepsilon = \varepsilon_e + \varepsilon_{ve}$. The viscoelastic deformation is only

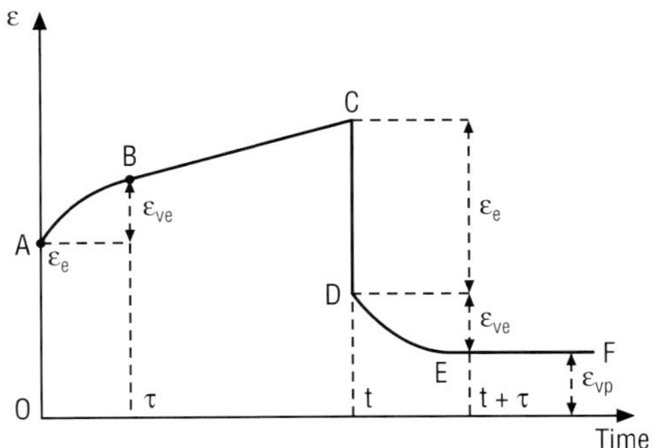

Figure 15.20

fully achieved after a certain time τ – the relaxation time – which manifests itself in the rubber-like state. The modulus of elasticity therefore depends on the period for which the stress is maintained and the speed with which it is applied. This is why the period is specified $E\,(10\,\text{s}) = (\tau)\varepsilon$, measured after the load has been applied for 10 s (see Figure 15.19).

Viscoplastic Deformation

See Figure 15.20.

OA = DC = ε_e = sudden elastic deformation.

AB, DE = viscoelastic deformation (ε_{ve}).

EF = viscoplastic deformation (ε_{vp}); continues after removal of the stress at time $t + \tau$.

Viscosity is one of the properties of the liquid state: it is responsible for the speed of flow, which is slower the higher the coefficient of viscosity η. Flow therefore takes place due to the presence of shear stresses. Polymers which display this phenomenon have a structure typical of the liquid state.

16 Tribology of Endoprostheses

C. Wang and Y. Wang

This chapter will discuss the problems of tribology in endoprosthesis. Tribology, defined as "the science and technology of interacting surfaces in relative motion and the practices related thereto" [1], consists of and stresses three main topics: friction, wear, and lubrication.

Friction in Natural and Artificial Human Joints

Friction

The phenomenon of friction is defined as the tangential resistance as a result of relative motion or motion tendency on the surface. The well-known laws of dry friction, established by Amontons and Coulomb are:

The force of friction, F, is directly proportional to the applied normal load, W.

The force of friction, F, is independent of the apparent area of contact.

The force of friction, F, is independent of the sliding speed, V.

These laws can be written by the formula
$F = \mu W$ (1)

where μ is known as the coefficient of friction. This formula is still widely used in many engineering applications.

Early investigators attributed the force of friction to the effort required to cause interlocking asperities to ride over each other. These early geometrical explanations were inconsistent with the concept of energy loss in the friction process, since the work done in drawing one surface to the tops of the asperities on the opposing surface would be recovered when the surface subsequently moved into the valleys in a non-dissipative system. In addition, when friction surfaces are very smooth, the friction coefficient predicted by this theory will be very small, which is not consistent with the facts. The reason why the friction coefficient between smooth surfaces is large is that the molecular attractive force between surfaces then will be very strong. So, the molecular attractive action between surfaces also contributes to the friction force.

Another two factors that contribute to the force of friction are: adhesion and deformation. The former is associated with the force required to shear asperity junctions formed by cold welding and the latter with the force required to slide the asperities of one surface through or over the asperities on the mating surface. As we know, real surfaces encountered in engineering and bioengineering are rough on a molecular scale. When they are in contact, they are not in real contact everywhere. As shown in Figure 16.1, only a very small fraction of nominal contact area is the real area of contact. The softer the surface material, the greater the real area of contact. Since the real area of contact between friction surfaces is very small, the pressure acting on the real contacting area can be very high. Under the action of this high pressure, contacting materials will adhere together as shown in Figure 16.2a. This phenomenon is called cold welding. When surfaces in cold welding move relatively, junctions formed by cold welding must be sheared. This shearing force comprises part of the friction force.

Adhesion is sometimes referred to as the molecular component of friction, and deformation as the mechanical or ploughing component shown in Figure 16.2b, where F stands for the

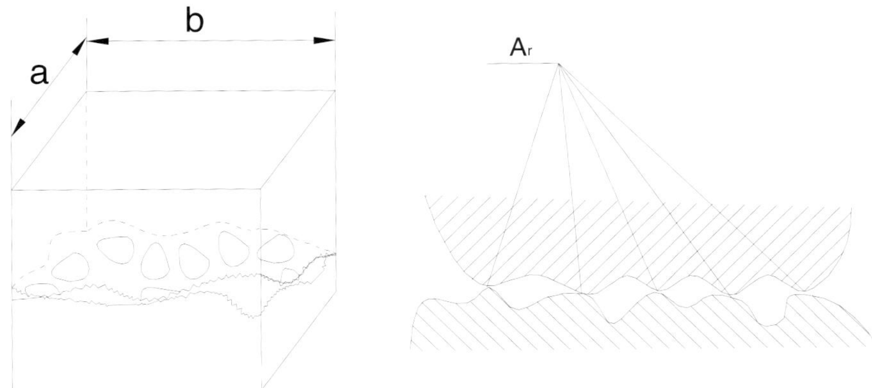

Figure 16.1. The real area of contact between friction surfaces.

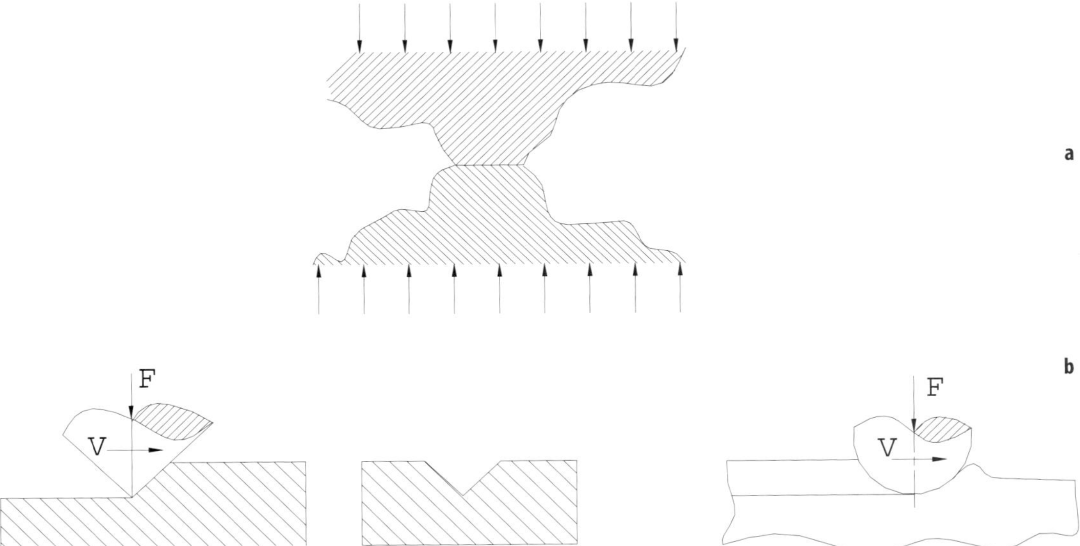

Figure 16.2. Cold welding and ploughing in real contact areas.

normal load, and V stands for the relative velocity. Adhesion normally dominates the friction between metals, particularly if they are of similar microstructure, compositions, and hardness, but the deformation component can be a major factor when a rough, hard material like metal slides over a much softer material like plastic.

The concept of adhesion in relation to friction was first proposed by Desaguliers, but Bowden and Tabor established the whole theory. The adhesion force can be expressed simply as

$$F = A_r \tau_b \tag{2}$$

where A_r is the real area of contact between the solids, and τ_b the shear stress of the softer material. Therefore, if the shear stress of the softer material can be lowered in some way, for example by using plastics as the surface material, the friction force will be reduced. This concept leads to the appearance of low-friction, Charnley-type artificial hip joints.

The calculation of deformation friction force is more complex and should be conducted with a particular geometry model, but this component is relatively small in most cases. Friction between clean metals can be of an extremely high degree, but the presence of an oxide film reduces this considerably by reducing the value of τ_b. Further reduction in friction is afforded by the lubricant of surface-active molecules, such as fatty acids, which further reduce the shear stress of the contacts. In the case of a metal sliding on a plastic, the lubricant often has little effect on friction because of the low contact shear stress, even in dry conditions.

Friction in Natural Human Joints

Joints are friction pairs working in the human body. The friction coefficient of a natural hip joint is very low, only about 0.001 ~ 0.03, due to the important role of articular cartilage. In the middle of a natural hip joint the thickness of articular cartilage is about 4~7 mm, and the thickness of articular cartilage will decrease correspondingly in the boundary of the joint. The matrix of articular cartilage is bone glue sulphate. Around the matrix is filled with a 3D network of collagen fibers and some tissue fluid. Some compositions of the fluid are conjoined with fiber structure, but most compositions are retained in gaps between fibers. Cartilage structure will be compressed when loaded and will expand when unloaded. This will cause squeezing and intake of fluid. The changing rate of the process determines the variation of cartilage thickness with time and has great importance to joint lubrication. The average gap size of cartilage is 60 Å.

The synovial fluid in the joint capsule is a permeable fluid of plasma [2]. It is filled with mucoprotein acid and some small-size, cell-like composition. The synovial fluid is a non-Newtonian fluid. Under pressure, since fluid compositions will penetrate into articular cartilage, the concentrated synovial fluid in the cartilage surface will become a layer of gel composed mainly of mucoprotein acid, which acts as a boundary lubricant film in joint motion. In a normal walking cycle, the lower limb of a person will experience two different situations, i.e., stance phase and swing phase. The loads on hips and knees during walking often vary in such a way that high loads combine with low surface-entraining velocities; low loads, in general, coincide with higher entraining velocities. Natural human joints will undergo different lubrication states in different conditions. Since the lubrication mechanism in natural human joints is very complex, it is not proper to apply general engineering lubrication models in this case. It is generally believed that the good lubrication states of natural human joints are relative to the pumping-in and pumping-out mechanism of synovial fluid in the cartilage.

Friction in Artificial Joints

Joint function might be damaged by certain disease or trauma. One restorable treatment for that is the use of an artificial joint. When an artificial joint is implanted, the surgeon needs to have the primary intention of restoring the architecture of the joint and thereby facilitating the motion and providing the necessary stability. However, the medical doctor ought to be aware that if he can reduce the friction between the two components, then he will reduce the shear stresses in bone–cement or bone–prosthesis interfaces. This would extend the useful life of a prosthesis in the patient. Friction in artificial joints can be divided into two states: film-separated friction and boundary friction. The former is actually a fully lubricated state and will be discussed later. Here we discuss boundary friction only.

The magnitude of the frictional forces in artificial joints depends on several factors. The material combination, the existence of a lubricant, surface roughness, and the loads acting in the joint are all very important. Experiments have indicated that the friction coefficient of metal–metal prostheses is about three times of that of metal–plastic prostheses. The origins of friction resistance in metal–metal prostheses are multiple. Since, in metal–metal prostheses, both of the components are made of the same material and they are of similar microstructure,

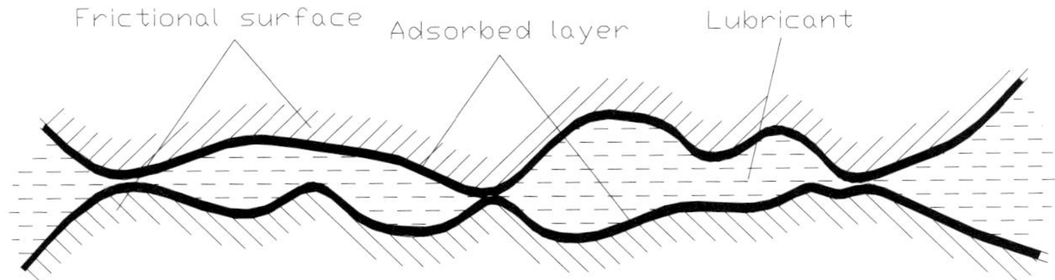

Figure 16.3. Boundary lubrication model.

compositions, and hardness, they have a greater tendency to adhere to each other. In some cases, the strength of the interface bond is so high that when relative motion at the surface occurs, the underlying structure of the material fails rather than the interface adhesion. Once a piece of material has been torn from one component and firmly attached to the other, it causes abrasive damage which will increase the friction force further. With human synovial fluid as a lubricant, friction resistance in metal–metal prostheses will decrease to some degree.

Lubrication in Artificial Joints

Lubrication Mechanism

Lubrication is a process by which the friction and wear between two solid surfaces in relative motion is reduced significantly by interposing a lubricant between them. The role of lubrication is to separate the moving surfaces with a film of solid, liquid, or gaseous material that can be sheared with low resistance without causing any damage to the surfaces. According to the lubrication film thickness and contacting situation in the interface, lubrication models can be classified as boundary lubrication and film lubrication. In boundary lubrication, there is a considerable asperity interaction between the contacting solid surfaces. The friction under boundary lubrication is due partly to shear in the lubricant and partly to shear in the asperity contacts. There are two kinds of film lubrication models in artificial joints: hydrodynamic lubrication and elastohydrodynamic lubrication (EHL).

Boundary lubrication occurs in the later stage of the stance phase. The boundary lubrication model is illustrated in Figure 16.3. The thickness of lubrication film at this time is near the molecular chain length of mucoprotein acid (5,000 ~10,000 Å). Under pressure, the liquid composition of the synovial fluid is squeezed out largely and the mucoprotein acid becomes a boundary lubricant, which separates two joint surfaces. The process will not end until the beginning of the swing phase when two joint surfaces are separated again by the hydrodynamic lubrication mechanism stated below.

Hydrodynamic lubrication is based on the formation of a thick lubricant film of characteristic geometric profile that develops automatically between opposing solid surfaces having relative motion to each other.

Hydrodynamic lubrication can be further divided into two types:

Shearing film lubrication: This is generated by the relative shearing motion between joint surfaces and occurs in the swing phase. The shearing film lubrication model is illustrated in Figure 16.4, where Vi represents the velocity of the inlet, Vo stands for the velocity of the outlet, and V stands for the velocity of the moving surface. The characteristic geometric profile in shearing film lubrication resembles a converging wedge having maximum thickness at its inlet and minimum thickness at its outlet. Compared with the strength of the mating materials the hydrodynamic pressure is low and will not cause appreciable local deformation. The loading capacity of hydrodynamic lubrication depends mainly on the lubricant viscosity and the relative speed of the moving surfaces.

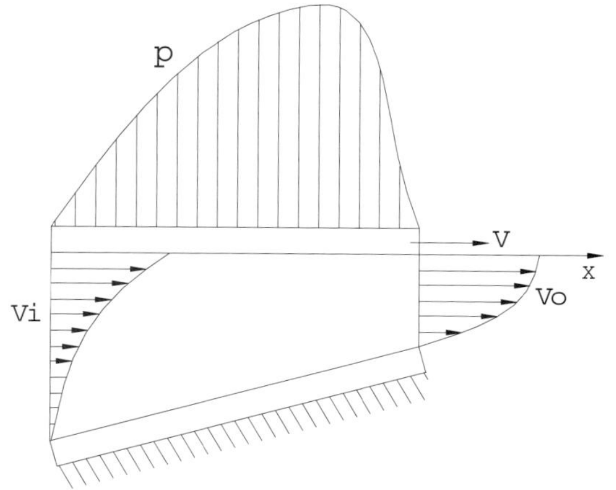

Figure 16.4. Shearing film lubrication model.

Because the joint in the swing phase is in an unloaded state and the relative motion speed between the surfaces is great, the thickness of the hydrodynamic lubrication film is relatively large.

Squeezing film lubrication: This is the main lubrication state of human joints, and occurs over a long period after the beginning of the stance phase. The squeezing film lubrication model is illustrated in Figure 16.5, where Vo stands for the velocity of outlet, and V stands for the vertical velocity of the moving surface. Because joint load at that time is large and the shearing hydrodynamic lubrication effect is very small, the load is borne mainly by the squeezing film.

Elastohydrodynamic lubrication (EHL) applies to hydrodynamic conditions where surface deformation is comparable with the hydrodynamic film thickness and surface deformation affects the hydrodynamic behavior of the interface. The most prominent example of EHL is the Hertzian contact in rolling-element bearings, gears, and cams, in which the contact areas are typically one-thousandth those occurring in journal bearings. The heavy loads cause local elastic deformation of the mating surfaces, which provides a coherent hydrodynamic film and avoids asperity interaction. In joint lubrication, since the elastic modulus of cartilage or polymer is relatively low, local elastic deformation is high even in ball-and-socket contacts.

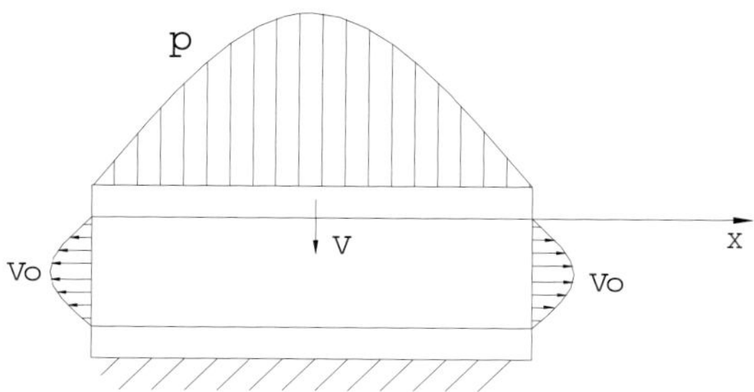

Figure 16.5. Squeezing film lubrication model.

Therefore, EHL is the main lubrication model in joint lubrication.

In the walking cycle, joint surfaces will undergo the above three lubrication states repeatedly. The complete process cycle is from a totally separated state, when the joint cartilage is fully filled with synovial fluid to the boundary lubrication stage. In this case, the squeezing film is the main reason why joint surfaces have low friction coefficients.

EHL Analysis of Artificial Joints

Since the joint load and relative speed between joint surfaces are changing continuously when a human moves, the thickness of the lubrication film in the joint is not constant. The variation curve of film thickness with time in a walking cycle is shown in Figure 16.6 [3]. The results indicate that when a human is walking, the lubrication states of the joint are changing in the walking cycle.

In the above curve, the part concerned most is the squeezing film lubrication stage. This part of the curve was calculated in detail in reference [4], and the influence of different parameters upon film thickness was also analyzed according to the calculation results.

The above results were all obtained on condition that the synovial fluid is a Newtonian fluid. The influence of non-Newtonian properties upon joint lubrication was investigated in reference [5]. The results indicated that non-Newtonian properties would cause film thickness to increase to some degree and that taking the squeezing term into account, under unsteady conditions, artificial hip joints would be in the EHL state largely in a walking cycle.

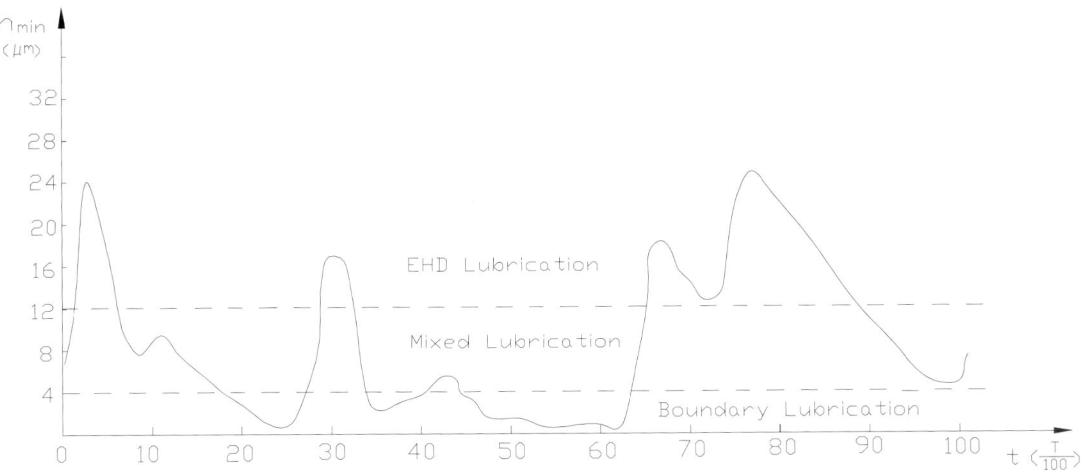

Figure 16.6. The variation curve of film thickness with time in a walking cycle.

The influence of different parameters can be analyzed by the above calculation method:

Diameter of sphere head D: The calculation results indicated that the thickness of the EHL film would increase with greater D.

Radial clearance C: The smaller C, the greater R. Therefore the head and socket should have an accurate fit with small clearance.

Elastic modulus of acetabular cup material E: The smaller the elastic modulus, the greater the thickness of the EHL film. This demonstrates at the same time that a low value of elastic modulus of natural joints plays an important role in increasing film thickness.

Period of walking cycle: According to the dynamic analysis of joint lubrication state, the shorter the time needed for a step, the bigger the film thickness. From this point of view, patients who had replaced artificial joints should walk quickly in short steps rather than the normal walking habit.

Body weight of a patient: The greater the body weight of a patient, the smaller the thickness of the EHL film. Therefore, a large sphere head should be used for this kind of patient.

Viscosity of synovial fluid η: The greater η, the greater the thickness of the EHL film. They are linear in relation. In common cases the joint capsule is not retained after artificial joint replacement. Joint lubrication relies on tissue fluid and the viscosity of tissue fluid is low.

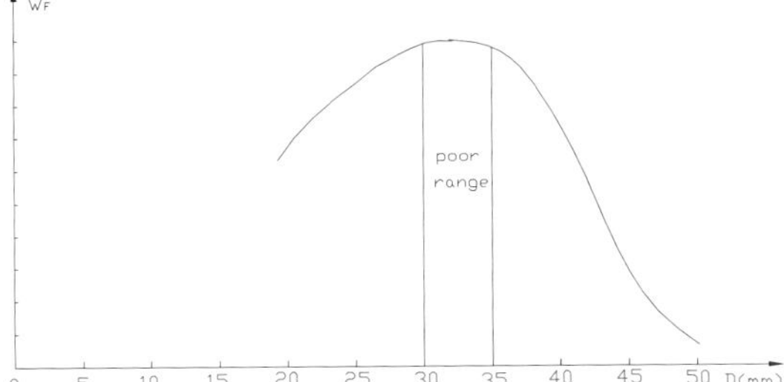

Figure 16.7. The relation curve of friction work and diameter of sphere head.

Although a greater diameter of the sphere head can increase the thickness of the EHL film, it forms a greater frictional torque in the subsequent boundary lubrication stage. A smaller diameter of sphere head will reduce frictional torque in the boundary lubrication stage, but it will also reduce the thickness of the EHL film. So there exists an optimal diameter of the sphere head, which will make an artificial joint produce the minimum frictional work overall. The relation curve of the friction work done in a walking cycle and the diameter of the sphere head is illustrated in Figure 16.7 [6]. The results indicate that there is a range for the value of D (about 30~35 mm) in which frictional work is greater than in other ranges. It was also verified by experiments that the wear volume of Muller artificial joints (D = 32 mm) was greater than Charnley artificial joints (D = 22 mm). Therefore, in design the value of D should be less than 30 mm.

Wear of Artificial Joints

Basic Principle of Wear

The phenomenon of wear is defined as the progressive loss of substance from the operating surface of a body occurring as a result of relative motion at the surface [6]. A common feature of the wear life of most engineering components is that only a relatively small percentage of the volume or weight needs to be removed before correct functioning is impaired. Wear is the main cause of element failure in most cases. However, wear is not always unfavorable and many processes such as walking, polishing, and writing with a pencil rely on wear.

There are five main types of wear.

Adhesive wear: Adhesive wear processes are initiated by the interfacial adhesive junctions which will form if solid materials are in contact on an atomic scale. When a normal load is applied, the local pressure at the asperities becomes extremely high. In some cases, the yield stress is exceeded, and the asperities deform plastically until the real area of contact has increased sufficiently to support the applied load. In the absence of surface films, the surfaces would adhere together, but very small amounts of contaminant may minimize or even prevent adhesion. Continued sliding causes the junctions to be sheared and new junctions to be formed. The chain of events that leads to the generation of wear particles includes the adhesion and fracture of the mating surfaces.

Abrasive wear: Abrasive wear may be described as the damage to a surface by a harder material. There are two general situations in which this type of wear occurs. In the first case, the hard surface is the harder of the two rubbing surfaces, for example, in grinding, cutting, and filing. In the second case, the hard surface is a third body, generally a small particle of wear, caught between the two other surfaces and very hard. In the abrasive wear process, asperities of the harder surface press into the softer surface, with plastic flow of the softer surface. When there is a relative motion between the two surfaces, the hard surface removes the softer material by the combined effects of microploughing, microcutting, and microcracking.

Fatigue wear: In practice, all machines involve the periodic variations in stress. A rotating shaft will be subjected to reversal of bending stress, the race of a rolling contact bearing and the surface of a roller will experience continual application and release of hertzian stress, and the journal surface in a hydrodynamic-lubricated sliding bearing will experience repeating stresses because of shaft rotating. All these repeating stresses in a rolling or sliding contact can give rise to fatigue failure. These effects are mainly based on the action of stresses in or below the surfaces without needing a direct physical contact of the surfaces under consideration. Subsurface-fatigue and surface-fatigue wears are the dominant failure models in rolling bearings.

Corrosive wear: The wear due to adhesion, abrasion, and fatigue can be explained in terms of stress interactions and deformation properties of the mating surfaces, but in corrosive wear, the dynamic interaction between environment and mating material surfaces plays an important role. If the two surfaces react actively with the environment, the rubbing of surfaces together in such an environment results in the continuous formation and removal of reaction products. Since the material of the contacting surfaces is in the reaction products, the material is being removed from the surfaces.

The corrosive wear process has two main steps:

In the first step, the contacting surfaces react with the environment, and reaction products are formed on the surface.

In the second step, attrition of the reaction products occurs as a result of crack formation and abrasion in the contact interactions of the materials.

Clearly, in attempts to study the mechanism of corrosive wear, besides the effects of deformation and adhesion, the chemistry of the reaction product formation must also be considered, taking into account the contribution of frictional energy in these processes.

Another form of wear is fretting wear, which will be discussed in a later section.

Wear of Artificial Joints

The wear areas in an artificial hip joint are: the interface between the artificial femoral head and acetabular cup, the fixing interface of the prosthesis, and the mating surfaces of compound prostheses.

The interface between the artificial femoral head and acetabular cup is the main wear area in an artificial hip joint. It was reported that the wear rate of Charnley artificial joints (D = 22 mm) was 0.1 mm~0.2 mm per year [6]. The wear rate of Muller artificial joints (D = 32 mm) was 0.05 mm, 0.06 mm, and 0.08 mm respectively when artificial acetabular cups were made from polyethylene and the artificial femoral head was made from Co–Cr–Mo alloy, stainless steel, or titanium alloy. Wear limits the life of an artificial hip joint, which is usually 20 years. Wear also limits the youngest age of a patient who eligible for artificial joint replacement, which is usually 50 years. It is an important aim of artificial joint engineering to lower the age at which patients become eligible for joint replacements.

Joint surface wear could result in tissue reaction of the whole body. Investigations after surgery indicated that after titanium femoral head replacements, large areas of a patient's hip might appear black and overdue amounts of titanium element were found in composition analysis of his hair. In recent years, polyethylene particles were found in pelvic lymphatic glands, which indicated that wear particles might result in immune reaction of the whole body.

Diameters of wear particles vary for different materials. For polyethylene, the average diameter is $0.53 \pm 0.3\,\mu$m, and the diameter of 92% of wear particles are less than 1 μm. For metal, the size of wear particles are in the range 0.8~1.0 * 1.5~1.8 μm. Since Co–Cr–Mo alloy has greater hardness and its wear particles are small. For titanium–polyethylene artificial hip joints, both titanium and polyethylene wear particles are not only large in quantity but also large in size. At the same time, bone cement particles are commonly found near prostheses. These wear particles are absorbed by tissue near the prosthesis and a membrane of connective tissue is formed. After histologic study of the membrane, it is found that under stimulation of foreign wear particles, giant cells and macrophages would phagocytize wear particles. Fiber tissue matrix, macrophages, and foreign giant cells form a film of membrane. Wear particles are distributed largely in the matrix and the above cells. Although the mechanism of formation is not yet clear, it is certain that the phenomenon would result in bone solution and loosening of prostheses. There are three main kinds of mechanism of wear particle formation: adhesion action, plough action, and fatigue. Involvement of wear particles might also bring in the fourth kind of wear, i.e., abrasive wear.

Measures for reducing wear in the interface between the artificial femoral head and acetabular cup are as follows:

Reduce roughness of the artificial femoral head to the magnitude of nanometers, which will reduce plough wear greatly.

Use modified polyethylene, such as carbon fiber-reinforced polyethylene.

Use metal–metal and ceramic–ceramic joint mating surfaces. This is a new idea after the phenomenon that wear particles might result in bone solution was found. Muller observed the wear rate of metal–metal joint mating surfaces and found that it was only 1/40 of that of metal–polyethylene joint mating surfaces.

Use permeable materials. Materials of good permeability are beneficial to the formation of boundary lubrication films in the human body, which is conducive to lower wear. Ceramic materials are obviously better than metallic materials in this respect.

In recent years, more and more attention has been paid to the fact that the disinfection method of a polyethylene prosthesis has an influence upon its strength and wear rate. The usually used γ-ray disinfecting method might cause polyethylene to oxygenolyse, lower molecular weight, accelerate aging, and raise wear rate, because the free radical released in γ-ray irradiation would remain in the material for a longer time and gradually react with oxygen. The results of Filscher's study indicated that the wear rates of γ-ray disinfected polyethylene and undisinfected polyethylene were >20 and <10, respectively.

There are two interfaces when fixed by bone cement, i.e., bone cement–prosthesis and bone cement–bone. There is only one interface when fixed by biotic surfaces, i.e., natural bone–prosthesis.

No matter what kind of interface, the joining mechanism of the interface is mechanical joining, such as the inter-embedding of surface asperities and the ingrowth of bone cells into surface pores. Bone cement has no adhesive capacity. Because of the difference in elastic moduli of the different materials, fretting wear takes place at the interface when loaded. A polishing phenomenon could be observed in some areas of the joint shank which was removed in a second surgery. The micro motion between prosthesis and natural bone would impede the ingrowth of bone cells. The porous surface of a joint shank could be obtained by spraying hydroxyapatite or sintering microspheres, titanium fibers, and biotic ceramics. Removal of material in fretting wear, three-body wear in interfacial gaps, the membrane formed by invasive wear particles, and bone solution would finally destroy the fixing interface and make the prosthesis loosen.

In artificial hip joints, mating surfaces refer mainly to the mating surface between the sphere head and the conical mating surface of a joint shank, and the joining surface between the back of a polyethylene acetabular cup and its envelope. Some products may have a metallic cup frame to form a biotic interface between the prosthesis and acetabular cup.

In the mating surface between the sphere head and the conical mating surface of a joint shank, electrochemical wear might also take place besides fretting wear because of the penetration of tissue fluid. Research results indicated that the possibility of wear was less than 6% when both parts were made from Co–Cr–Mo alloy. The possibility of wear was less than 10% when both parts were made from titanium alloy and the possibility of wear was greater than 30% when the two parts were made from different materials. As a rule of joint design, as few mating surfaces as possible should be used and interfaces of different materials should be avoided if possible. The fitting degree of acetabular cups and their envelope has great influence upon the wear rate of the interface. It is usually less than 70% and depends upon the manufacturing quality and system deformation. To guarantee the fitting degree in deformed conditions, the outer surface of the acetabular cup and the inner surface of the envelope should be modified geometrically. Many anti-fretting wear design structures have been adopted in product design.

Fretting Wear

Fretting wear is a typical compound wear, it occurs in situations where there is a confined space or relative micro-motion between components. In an artificial joint there are several interfaces where fretting wear might take place, for example, metal/cement interface, implant/coating interface, coating/substrate interface, coating/bone interface, cement/bone interface, and implant/bone interface [7]. Waterhouse observed some examples of fretting fatigue failures of metallic components in the human body. These failures were the result of the corrosive conditions produced by body fluids. The bone plates are used to fix fractured bones and are screwed into the bone. Failure occurred as a result of fretting between the underside of the screw-head and the countersink of the hole. Also, in total hip replacements, removal of the lesser trochanter enables the surgeon to have better access. When the lesser trochanter is replaced on completion of the operation it is customary to attach the bone by wiring it around the stem of the femoral component. Fretting at this contact has initiated fatigue failure of the latter.

In some of these implant devices, metal-to-metal contacts (for example, screws and bone plates) could suffer fretting owing to the activity of the patient. It is found that the corrosion current increased linearly with the amplitude of the slip at a fixed frequency in potentiostatic experiments. It was also linearly related to frequency at a fixed amplitude.

Mechanisms of fretting wear in metal/metal interfaces have been investigated and explained by a number of investigators. A commonly accepted view is that the first stage of fretting is adhesive contact of the asperities on opposing contact surfaces. These adhesive contacts are important, as they are often thought to be the mechanism by which the majority of the cracks are nucleated. After this stage several things may occur: breakage of the asperities, which cause the production of fretting debris, oxidation/corrosion of "fresh surface" and/or debris. Cracks that nucleate may propagate at various rates and angles to the contacting surfaces and cause premature fatigue failure.

Fretting wear in artificial joints may result in the following forms of damage:

the host body may become infected by the debris,

cracks may form from the conjoint action of fretting fatigue that may lead to component integrity problems,

implants may become loose.

Fretting wear is induced by micro-motion in contacting surfaces. Therefore, any causes of micro-motion may produce fretting wear. The typical causes of micro-motion are as follows:

Tangential shearing force: Micro-motion will arise in a Hertzian contact, when a monotonically increasing or cyclically varying shearing force, less than that necessary to cause complete sliding, is applied.

Elastic mismatch: If contacting components are of a different modulus of elasticity, the application of a normal load alone will cause surface particles within the contact patch to displace laterally by different amounts [8].

Geometry reason: For example, if the contact is conforming, then, even under conditions of elastic similarity, it will be found that surface particles will displace tangentially by different amounts, because a formulation for a disc and for an infinite plane containing a hole must be used for the contacting elements.

It is difficult to overcome fretting wear completely at present because its mechanism has not been established. On the basis of recent studies, however, we can take measures from the following aspects to reduce fretting wear:

Design: The most effective way in design is to reduce interfaces between components; that is, as few as possible components should be used. Reducing the combined normal and tangential stress on the contacting surface is also beneficial to lowering fretting wear. In addition, proper fit of prostheses in bone is also very important. For this purpose, CAD/CAM technology has been used to make custom prostheses.

Materials: The most important factor in reducing fretting wear of artificial joints is to use materials of low elastic modulus. In this way, the elastic modulus of prosthesis materials should be close to that of natural bones. Therefore, titanium alloy is better than steel in this respect. In addition, materials of high hardness, high resilience, high heat conductivity, and high resistance against adhesion are beneficial to lowering fretting wear.

Surface treatment: Effective surface treatments for reducing fretting wear of artificial joints are ion implantation and ion nitriding. In addition, residual stresses in contacting surfaces are also beneficial to lowering fretting wear.

Conclusion

An artificial joint is a friction pair in the human body. Its design is a typical tribology design and would involve some key problems like friction, lubrication, and wear life of joints; joint loosening after surgery and so on. Since an artificial joint operates in the environment of a human body, many of its tribological performances are influenced greatly in the area of life science. Therefore, it is one of the focused research objects of interdiscipline – biotribology. Research in biotic material, tissue engineering, and so on are still in their beginning stages. Developments in the research of prosthesis CAD/CAM, virtual surgery, etc. are very rapid. A new subject – prosthesis engineering – is being formed gradually both in academic systems and methodology systems. It will certainly play an important role in the twenty-first century.

References

1. Dowson D, Wright V. An introduction to the biomechanics of joints and joint replacement. London: Mechanical Engineering Publications, 1981.
2. Dumbleton JH. Tribology of natural and artificial joints (Advances in Tribology Series, Vol. 3). New York: Elsevier Science, 1981.
3. Jifei S, Chengtao W, Yeping W. Discussion on the lubrication mechanism of natural and artificial human joints. Journal of Shanghai Jiao Tong University 1986;20:103–17 (in Chinese).
4. Wang CT, Yang MR. Calculation of elastohydrodynamic lubrication film thickness for hip prostheses during normal walking. Tribology Transactions 1990;33:239–45.
5. Wang Yeping, Shen Jifei. Study on the lubrication mechanism of human joints under non-Newtonian properties. Journal of Shanghai Institute of Railway Technology 1991;12:53–61 (in Chinese).
6. Halling J. Principles of Tribology. London: The Macmillan Press, 1975.
7. Hoeppner DW, Chandrasekaran. Fretting in orthopaedic implants: a review. Wear 1994;173:189–97.
8. Hills DA, Urriolagoitia Sosa G. Origins of partial slip in fretting – a review of known and potential solutions. Journal of Strain Analysis 1999;34:175–81.

17 Shape Memory Alloys and their Medical Application

K. Dai and C. Q. Ning

Introduction

Shape memory alloys (SMA) are a novel kind of metallic materials which have the ability to return to their previously defined shape when subjected to some appropriate thermal procedure. Due to such functional properties, and in particular the shape memory effect (SME) and superelasticity (SE), shape memory alloys have attracted much attention in recent years. It was Chang and Read who first observed the unique memory effect of shape memory alloys in Au-47.5 at % Cd alloy early in 1951 [1]. But it was not until 1963 when Buehler and his co-workers [2] rediscovered the SME in equiatomic Ni–Ti that SMAs actually began to cause a great deal of commercial interest, especially after they were widely put into use in the field of medicine [3–5]. Although quite a number of alloys are known to show shape memory behavior, only those that could generate substantial amounts of strain or could generate significant force upon the changing shape so as to make it recover are of commercial value. In the medical field, the family of Ni–Ti alloys is the most popular one that has been widely put into medical use because of its good biocompatibility, substantial resistance to corrosion and fatigue, and the fact that its elastic modulus is quite close to that of human bone. In some cases, Ni or Ti (only a few per cent) in Ni–Ti alloys can be partially replaced by Cu, Co, Fe, Nb, or Mo to improve the hysteresis (stress and/or temperature hysteresis), corrosion behavior, control of transformation temperatures, fatigue behavior, etc [6].

Martensitic Transformation

The functional properties of shape memory alloys are closely related to a solid–solid phase transformation named Martensitic transformation. For many metallic materials, as their cooling speed reaches a certain extent, they will undergo a phase transformation in their crystal structure while cooled from a high temperature form (Austenite) to a low temperature form (Martensite). This procedure is known as Martensitic transformation, which is a non-diffusion phase transformation. The transformation from Martensite to Austenite upon heating is known as the reverse transformation. The characteristic transformation temperatures are defined as follows [7]:

M_s **Temperature:** The temperature at which a shape memory alloy starts transforming to Martensite upon cooling.

M_p **Temperature:** The temperature at which a shape memory alloy is about 50% transformed to Martensite upon cooling.

M_f **Temperature:** The temperature at which a shape memory alloy finishes transforming to Martensite upon cooling.

A_s **Temperature:** The temperature at which a shape memory alloy starts transforming to Austenite upon heating.

A_p **Temperature:** The temperature at which a shape memory alloy is about 50% transformed to Austenite upon heating.

A_f **Temperature:** The temperature at which a shape memory alloy finishes transforming to Austenite upon heating.

These transformation temperatures can be adjusted by slightly changing the alloy compo-

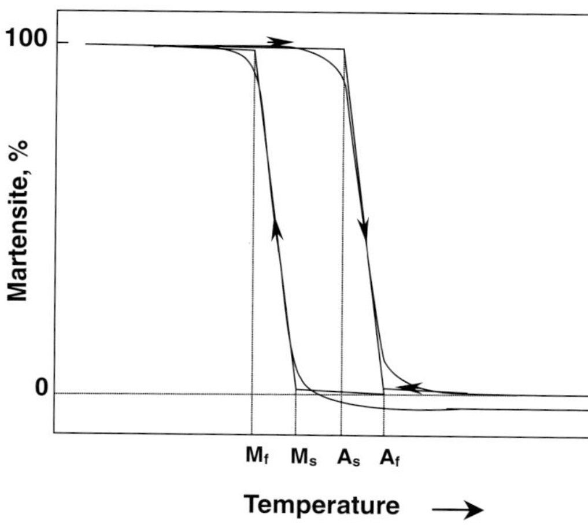

Figure 17.1. Schematic representation of the volume transformed as a function of the temperature.

sitions or by thermo-mechanical treatment. Usually, the illustration of the Martensitic transformation and its reverse transformation is shown as Figure 17.1 [6].

According to its characteristics, Martensite transformation can be classified into two types – thermoelastic Martensite transformation and non-thermoelastic Martensite transformation. What distinguishes shape memory alloys from conventional materials is none other than their ability to form thermoelastic Martensite. During the process of thermoelastic Martensite transformation, while below the transformation temperature, the deformation of the shape memory alloy is caused by simply adjusting the orientation of the crystal structure through the movement of twin boundaries (a twinning mechanism), instead of slipping and dislocation movement. In other words, the shape change resulting from Martensitic transformation can be accommodated by a crystal lattice distortion, and the boundary between Martensite and the parent phase can be driven by slight changes of the temperature or stress. That is to say, thermoelastic Martensite is completely crystallographically reversible. Whereas on the contrary, the growth rate of the non-thermoelastic Martensite is so quick because of the larger driving force from Martensite transformation that the boundary between Martensite and the parent phase is destroyed during the transformation, which results in an irreversible parent-Martensite boundary.

The Martensite transformation occurs not at a single temperature but within a range of temperatures (as shown in Figure 17.1), which varies according to different alloy composition and microstructure constitution, the latter being determined mainly by the thermomechanical treatments. Since the phase transformation temperatures during heating and cooling do not overlap, a temperature hysteresis appears, which also varies according to different alloy systems. This temperature hysteresis is generally illustrated as the difference between A_f and M_s (i.e., $\Delta T = A_f - M_s$) or the difference between A_p and M_p (i.e., $\Delta T = A_p - M_p$).

Functional Properties

Shape Memory Effect

Usually, under external forces, a common metallic material deforms elastically first, then plastic deformation occurs after its yield point, and finally, even if the force is removed, the permanent deformation will be reserved. But for some other alloys, even when a plastic deformation occurrs, they can still return to their original shapes after being heated up to a certain temperature. Such a shape recovery phenomenon is called the Shape Memory Effect (SME), which is due to the Martensitic transformation in these alloys. When an alloy with a given shape cools from the Austenite form to the Martensite form, it is easily deformed to a new shape (the restriction is that the deformations must not exceed a certain level), but if the same alloy is heated up to its transformation temperature, it will recover its previous shape due to the reversible reverse transformation.

After being deformed, Martensite can recover its parent shape via reverse transformation. This effect is called the one-way memory effect (Figure 17.2 (a)). After given proper training, some alloys can memorize to return to not only the parent shape during heating, but also the deformed Martensite shape during re-cooling. This effect is called the two-way memory effect

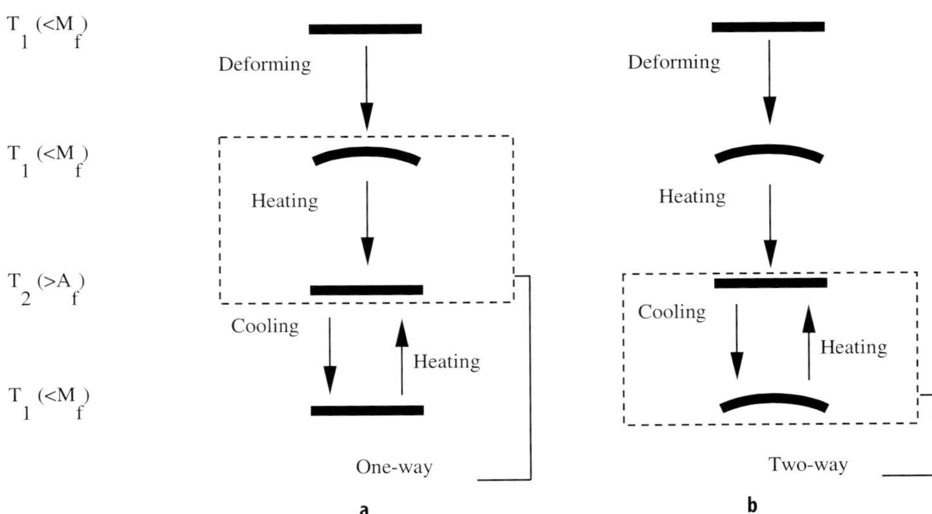

Figure 17.2. Schematic representation of one-way **a** and two-way **b** memory effects.

(Figure 17.2 (b)). The latter can be obtained only after a specific thermomechanical treatment, which is usually called "training". The amount of shape change that can be obtained by the two-way memory effect is always significantly less than that by the one-way memory effect.

Generally, the shape memory effect can be expressed by the shape recovery ratio (i.e., η). If the initial shape of the alloy in Austenite form is l_0 (expressed as length), the shape of deformed Martensite (e.g., tension) is l_1, and the shape after reverse transformation at high temperature is l_2, the η can be expressed as [8]:

$$\eta(\%) = (l_1 - l_2)/(l_1 - l_0) \times 100\%$$

Figure 17.3. Two kinds of superelastic behavior at constant temperature.

Superelasticity

Superelasticity, as the name implies, refers to a phenomenon that the alloy can exhibit strain far beyond its elastic limit upon loading, whereas once the stress is removed, the original strain will be returned completely. According to the characteristics of the stress–strain curves, the superelasticity can be classified into two types: linear and non-linear superelasticity. The latter is caused by a stress that occurs during a loading and unloading process, which leads to Martensitic transformation and its reverse transformation at a temperature range above A_f. The former is probably related to the contribution of microtwins to the deformation. The two kinds of superelastic behavior are shown in Figure 17.3. As for the non-linear superelasticity, when the stress reaches a critical level, the alloy will start to transform into Martensite, accompanied by an increasing strain at constant stress until the alloy is fully transformed into Martensite (see Segment A to B in Figure 17.3). When the stress is removed, the reverse transformation will occur at a lower stress level (see Segment C to D in Figure 17.3). The SME described above depends on temperature changes. In contrast, the superelastic effect of shape memory alloys is a kind of isothermal phenomenon and the temperature changes are not necessary. The critical stresses are dependent on the alloy itself and its temperature. In general, the stress levels increase linearly with increasing temperatures [6]. Reversible strain obtained by the superelastic effect is always up to 8%, which is 10–20 times higher than the normal elastic strain of conventional metallic materials. As shown in Figure 17.3, the stress upon loading and unloading does not overlap and shows a hysteresis as well.

High Damping Capacity

The shape memory alloys have a high damping capacity in the Martensite state or two-phase state. The high damping capacity of SMA is related to the hysteretic movement on interfaces (Martensite variants interfaces, twin planes, parent-Martensite interfaces) whereas a contribution of dislocations is not excluded [5].

Applications

Shape memory alloys have now been widely used in many fields of industry such as fasteners, actuators, satellite antenna, and decorations, etc. This chapter only introduces several examples of shape memory alloy products applied in medical fields.

Application of the Shape Memory Effect

Free Recovery: This refers to the memory effect that after an SMA component is deformed in the Martensitic state, the component can recover its previous shape upon heating without any restrictions. A prime application of this kind of SME is the *blood-clot filter* developed by Simon [9]. It is made from Ni–Ti wire. After being chilled to make it collapse, the filter is inserted into the vein, and the temperature of body heat is high enough to help it return to its previous functional shape, enabling it to anchor itself in a vein and catch passing clots.

Constrained Recovery: If a deformed shape memory alloy or its components are subjected to an external constraint during the heating process, it will induce a recovery force, which is a function of the temperature, and varies with the constraint strain. The larger the constraint strain becomes, the greater the recovery force will be. This recovery force can be used for the purpose of clamping, fixation, or stiffening. In biomedical fields there are many such products: dental root implants, stents, and fixators for bone fractures. The shape memory compression staple will be described as an example here: the staple was designed by Dai and it is the first SMA device used inside the human body [10,11]. It is U-shaped, having two straight legs connected by a transverse wave-like segment with angles of 70° (Figure 17.4 (a)). At low temperature, i.e., below 4 °C, the wavy segment is expanded to increase the length. At the same time, the included angles are expanded from 70° to 90° to elongate the span between the ends of the legs (Figure 17.4 (b)). After the fracture is reduced, the expanded staple is placed across the fracture line by inserting the two legs into the holes prepared in the proximal and distal fracture fragments respectively. When the local temperature is raised by hot compresses with hot saline gauze, the staple(s) will tend to restore its original shape but at the same time being constrained by the walls of the holes so that they will generate a constrained recovery force on both side of the fracture line to hold the fracture fragments in place, exerting sustained compression on the fracture and resisting muscular pull (Figure 17.4 (c) and (d)). The compression force will be significantly decreased if the distance between the two holes is not wide enough.

By the same mechanism, the inner diameter of the shape memory saw-tooth arm fixator, which is introduced in Chapter 22, should be less than the outer diameter of the fracture bone in order to constrain the recovery of the fixator so that a reliable stability of the fracture can be obtained.

Up to now, nickel–titanium shape memory alloy has been used for the manufacture of several kinds of fracture fixators, scoliosis correction devices, prostheses for hip resurfacing, intervertebral prostheses, arterioembolizators,

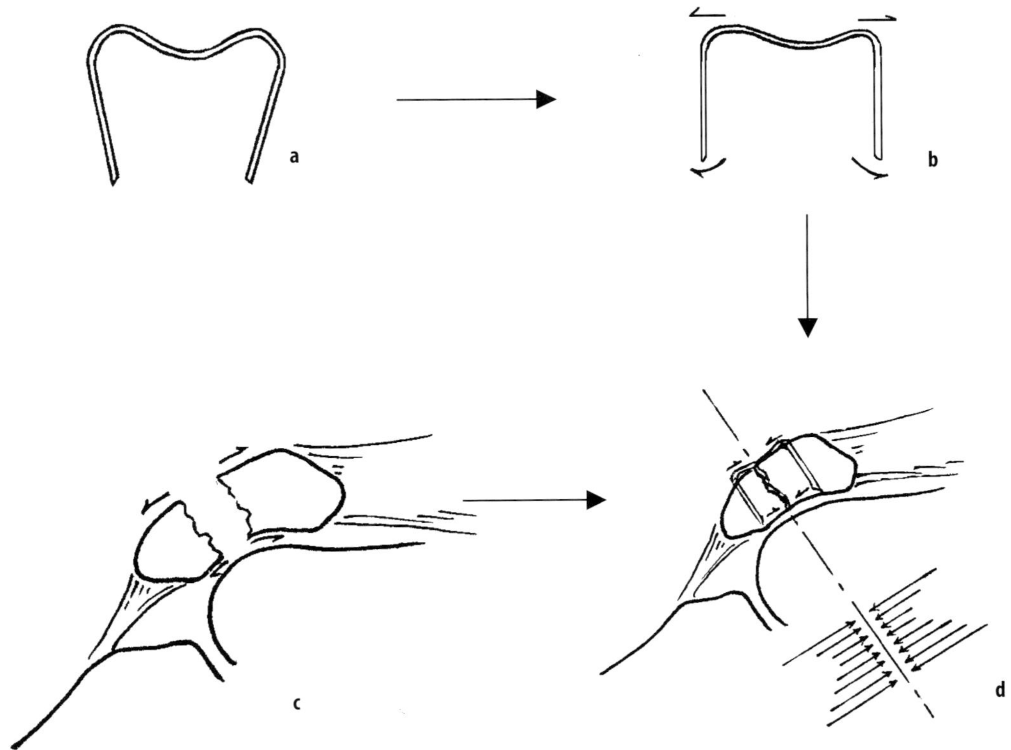

Figure 17.4. **a** Schematic diagram of the shape memory compression staple. **b** The staple distracted at the wavy segment and the included angle are expanded from 70° to 90°. **c** Patella fracture with a separated tendency. **d** After fracture reduction and staple fixation, the staple can not restore to its original shape and a large recovery force is created at both sides of the fracture to fix the fracture against the distraction forces produced by the muscle and joint flexion.

dental root implants, and stents of hollow organs [3,4].

Superelastic Applications

Arch wires made of Ni–Ti for orthodontic correction have been used for many years and are especially helpful for correction of tooth deformation.

Self-expanding Stent

The self-expanding stent is already widely used in the treatment of stenosis of certain hollow structures such as vessels, trachea-bronchus, urethra, biliary tract, esophagus, etc. Within the catheter, the compressed stent made of shape memory alloy is inserted into the duct by interventional techniques and is left in the stenosis area. In this way it will not only constantly expand the duct or vessel with its superelastic property or shape memory effect but it will also avoid the shifting of the self-expanding stent [3,4].

References

1. Chang LC, Read TA. Plastic deformation and diffusionless phase changes in metals – the gold-cadmium beta phase. Trans AIME 1951;189:47–52.
2. Buehler WJ, Gilfrich JW, Wiley RC. Effect of low-temperature phase changes on the mechanical properties of alloys near composition TiNi. J Appl Phys 1963;34:1475–7.

3. Dai KR, Chu YY. Studies and applications of NiTi shape memory alloys in the medical field in China. Bio-Med Mater Engin 1996;6:233–40.
4. Chu YY, Dai KR, Zhu M, Mi X. Medical application of NiTi shape memory alloy in China. Mater Sci Forum 2000;327:55–62.
5. Van Humbeeck J. Non-medical applications of shape memory alloys. Mater Sci Eng 1999;(A)273–5:134–48.
6. Van Humbeeck J, Stalmans R, Besselink PA. Shape memory alloys. In: Helsen JA, Breme HJ, editors. Metals as Biomaterials, Ed. 1. Chichester: Wiley & Sons, 1998;73–100.
7. Otsuka K, Ren X. Recent developments in the research of shape memory alloys. Intermetallics 1999;7:511–28.
8. Zuyao Xu. Shape Memory Materials. Ed. 1. Shanghai: Shanghai Jiaotong University Press, 2000;2–5.
9. Simon M, Athanasoulis CA, Kim D. Simon nitinol inferior vena cava filter: initial clinical experience. Radiology 1989;172:99–103.
10. Dai KR, Hou XK, Sun YH, Tang RG, Qiu SJ, Ni C. Treatment of intra-articular fractures with shape memory compression staples. Injury 1993;24:651–5.
11. Dai KR, Zhang XF, Yu CT. Orthopaedic application of shape memory compression staple. Chin J Surg 1983; 21:343–5.

II

Tissue Biomechanics and Histomorphometry

II – Tissue Biomechanics and Histomorphometry

II A – Histology and Bone Architecture

18 Normal Histologic Architecture of Tissue

M. Péoc'h

Introduction

Bone and articular cartilage are highly specialized tissues of the skeletal system. Bone serves five main functions:

Mechanical support (ribs)

Movements, as the site of muscle attachments (long bones)

Protective, as encasement of organs (skull)

Metabolic, as a reserve pool of various ions especially calcium and phosphorus

Hematopoietic: bones provide host sites for the hematopoietic tissue (the bone marrow).

The human skeleton comprises 206 individual bones divided into two main groups, the flat bones (skull, scapula, vertebrae, pelvic bones) of the axial skeleton and the predominantly tubular bones of the appendicular skeleton. They are composed of multiple tissues including bone, cartilage, fat, connective tissues, hematopoietic bone marrow, nerves, and vessels. Long bones have a wider portion at each end (the epiphysis), a cylindrical tube in the middle (the diaphysis) and a transition zone between them (the metaphysis). In a long bone that is growing, the epiphysis and metaphysis are separated by the epiphyseal cartilage (the growth plate) which becomes entirely ossified after the end of skeletal growth. Bone tissue is classified on the basis of its gross and microscopic structure in two forms: cortical (compact bone), which constitutes the cortex, and trabecular (cancellous or medullary), which forms the central regions.

Different Forms of Bone

Structural stability is provided by the cortical bone, comprising approximately 80% of skeletal bone in weight. Trabecular bone forms a complex intramedullary network that provides a huge surface area for its role in mineral metabolism.

Cortical (Compact) Bone

The bone cortex, except in the region of the articular cartilage, is surrounded by the periosteum, which consists of an outer fibrous layer and an inner cellular layer of osteoprogenitor cells, fibroblast, and osteoblasts. The cellularity of this layer depends on the age and bone remodeling activity in the particular region. The collagen of the outer fibrous layer is continuous with that of the joint capsule, tendons, and muscle fascia. Where the tendons insert into the periosteum, collagen fibers (Sharpey's) pass through the periosteum into the bone lamellae. The cortex is a dense or compact bone composed of longitudinal, circumferential, and concentric lamellae (Figure 18.1, plate section), best demonstrated by their birefringence with plane polarized light (Figure 18.2, plate section). The cortex has Haversian canals, through which blood vessels pass in a longitudinal arrangement, and horizontally distributed Volkmann canals with smaller vascular branches. Haversian canals are surrounded by concentric lamellar bone called an osteon. This is the structural and remodeling unit for cortical bone.

Trabecular Bone

The centers of most adult bone are composed of trabecular bone forming a complex intramedullary network that provides a surface area for its role in mineral metabolisms [1]. The lamellae parallel the long axis of the trabeculae (Figure 18.3, plate section). Most trabeculae are avascular. Separating the bone from the marrow space is the endosteum, a layer composed of osteoclasts, macrophage-like cells, and active and inactive osteoblasts. In some areas, a thin, loose fibrous tissue is present around the cells of the endosteum.

Woven or Immature Bone

Woven bone is generally a fine cancellous or spongy bone, organized in trabeculae and spicules. The collagen fibers are arranged in a meshwork or felt-like pattern (Figure 18.4, plate section). The noncollagenous elements form a greater percentage of the matrix than in mature bone. Mineralization is intense. The osteocytes are randomly and unevenly distributed within the trabeculae, with large spherical to plump spindle-shaped lacunae; their number is higher per unit volume than in lamellar bone. Woven bone is seen in the embryo, growing prepubertal bone, malformed bone, fracture callus, or other sites of periosteal injury, at sites of bone repair or at the reaction to bone tumors (Codman's triangle).

Bone Composition

Bone comprises an organic matrix mainly composed of collagen (90%). The remaining 10% consists of varying lipids and proteins, including a calcium-binding protein called osteoclastine. The deposition of osteoide, composed of type 1 collagen, is the first step in bone formation. Osteoide is the framework upon which spindle-shaped crystals of hydroxyapatite [$Ca_{10}(PO_4)_6(OH)_2$], are deposited [2]. Bone also has a ground substance of glycoproteins and proteoglycans that has a high ion-binding capacity and is thought to play an important role in calcification and the adherence of hydroxyapatite to the collagen. When hydroxyapatite is deposited, the osteoide becomes bone. The bone tissue is categorized according to its collagenous organization into woven and lamellar bone. In woven bone (seen in areas of rapid bone growth, such as primary bone of the embryo, fracture callus, or tumor bone) the collagen is arranged in an irregular feltwork, while in lamellar bone the collagen consists of regularly arranged sheets. Since the mineral and fibers are well organized and closely associated, lamellar bone has greater rigidity and tensile strength and less elasticity than woven bone.

Three types of specialized bone cells exist: osteoblasts, osteocytes, and osteoclasts.

Osteoblasts

The osteoblasts, derived from the osteoprogenitor cells, first become recognizable as cells with round or ovoid nuclei and fairly abundant cytoplasm. They synthesize the collagen of the osseous matrix and are principally responsible for bone mineralization [3]. These cells cover the bone-forming surface and consist of a row of columnar cells with amphophilic to basophilic cytoplasm and eccentrically located nuclei, often with a prominent nucleolus (Figure 18.5, plate section). There is usually a perinuclear halo. Osteoblasts have an extensive granular endoplasmic reticulum, making their cytoplasm basophilic, and contain abundant alkaline phosphatase, which plays a role in the subsequent mineralization of the collagen matrix. The ultrastructure of these cells reveals a large, prominent Golgi apparatus, and numerous mitochondria and lysosomes. The cytoplasmic surface demonstrates multiple processes in contact with adjacent osteoblasts and osteocytes. As the rate of bone formation diminishes, osteoblasts become flattened, have less cytoplasm, and finally are inconspicuous cells covering the resting bone surface, which appear spindle-shaped in histological sections (Figure 18.6, plate section). The osteoblasts respond to mechanical stimuli to mediate the changes in bone growth, size, and shape.

Osteocytes

Osteocytes are osteoblasts that have become surrounded by osseous matrix. They are responsible for the maintenance of the bone. The cells have ovoid nuclei smaller than those of osteoblasts, and only a small amount of cytoplasm. The main body of the osteocyte is located in small soft-tissue areas of the mineralized bone – the lacuna. They have delicate cytoplasmic processes (dendrites) that run through the matrix to abut one another. In woven bone, they are randomly distributed throughout the matrix, are large and spherical or plump spindles (Figure 18.4), while in lamellar bone they are relatively rarer, smaller and spindle-shaped, and evenly distributed along the bone lamellae.

Osteoclasts

Osteoclasts are multinucleated cells largely responsible for bone resorption (Figure 18.7, plate section). These cells are derived from hematopoietic progenitor cells related to the monocyte/macrophage cell lineages. Osteoclasts, resulting from the fusion of mononucleated cells, have two to six nuclei and are juxtaposed to the bone surface, where they lie in small reabsorption bays (Howship lacunae). The osteoclast binds to the bone surface by numerous cytoplasmic extensions which give the cell a villous-like, ruffled border. Secreted proteases permit the removal of the organic matrix [4] and simultaneously dissolve hydroxyapatite crystals.

Bone tissue is a hard, biphasic tissue composed of a mineralized organic matrix and at least three identifiable specialized cell types. The biphasic structure provides bone with ideal hardness, flexibility, and tensile strength without its being excessively brittle [5].

References

1. Teitelbaum SL, Bullough PG. The pathophysiology of bone and joint disease. Am J Pathol 1979;96:283–354.
2. Posner AS. The mineral of bone. Clin Orthop 1985;20: 87–99.
3. Marks SJ, Popoff SN. Bone cell biology: the regulation of development, structure, and function in the skeleton. Am J Anat 1988;83:1–44.
4. Delaisse JM, Boyde A, Maconnachie E et al. The effects of inhibitors of cysteine-proteinases and collagenase on the resorptive activity of isolated osteoclasts. Bone 1987;8: 305–13.
5. Gurley AM, Roth SI. Bone. In: Sternberg SS, editor. Histology for Pathologists. New York: Raven Press, 1992; 63–79.

19 Histology of Bone Callous

C. Bouvier

Factures are breaks in the continuity of bone. Then a reparation occurs which consists of growth of new tissue developing around and between the ends of the bone fragments. This new tissue, which forms a bridge between the fragments, is termed a callus. Like any bony structure, bone callus is remodeled as it grows, so that histologic features vary greatly with time.

In the very first hours following a bone injury, a hematoma forms between the ends of the bone because blood vessels of the soft tissue, of the periosteum, and of the spongy bone have been torn. On microscopic examination, only diffuse hemorrhage is seen within the intertrabecular spaces and in the periosseus tissues (Figure 19.1). Then, this blood coagulates to form a clot which serves as a matrix for inflammatory cells and fibroblasts as well as neoformed capillaries (Figure 19.2). Vessels of the Haversian systems are also torn at the fracture line. This results in the death of the osteocytes for a certain distance on each side of the fracture line. Dead bone undergoes lysis and is formed of pycnotic osteocytes. Osteoclastic resorption of devitalized bone fragments then occurs.

The inflammatory cells which colonize the hematoma in the early stage of repair secrete various cytokines; growth factors such as PDGF, FGF, TGFβ. These substances activate bone progenitor cells, especially those present in the inner layer of the periosteum. After a week,

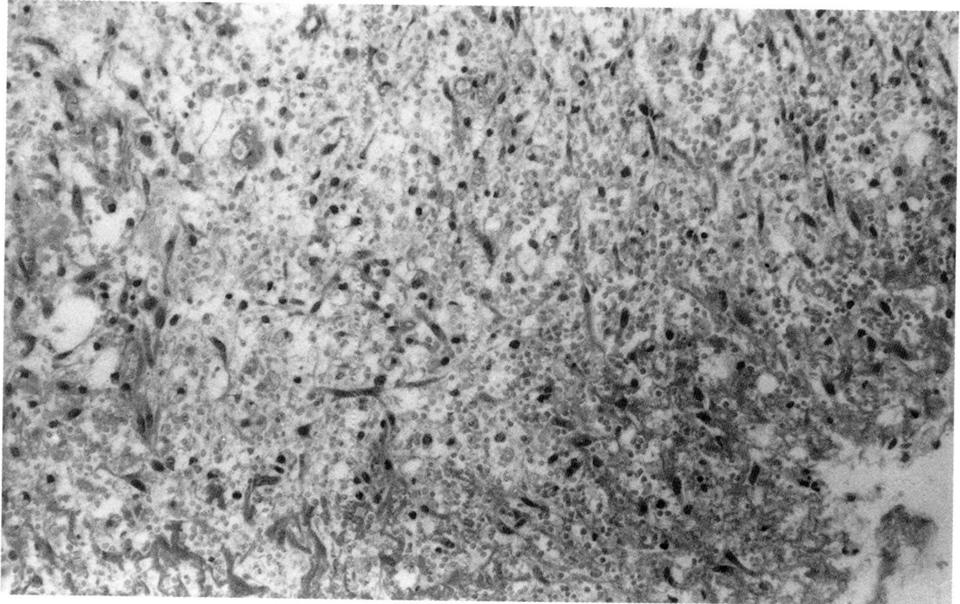

Figure 19.1. Following a bone injury, only diffuse hemorrhage is seen within the intertrabecular spaces and in the periosseus tissues.

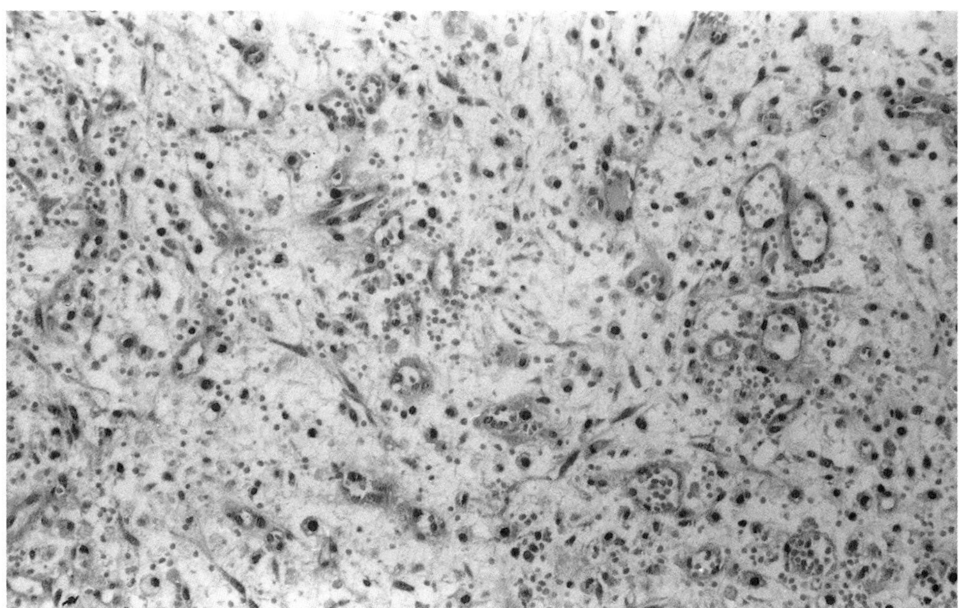

Figure 19.2. Blood coagulates to form a clot which serves as a matrix for inflammatory cells and fibroblasts as well as neoformed capillaries.

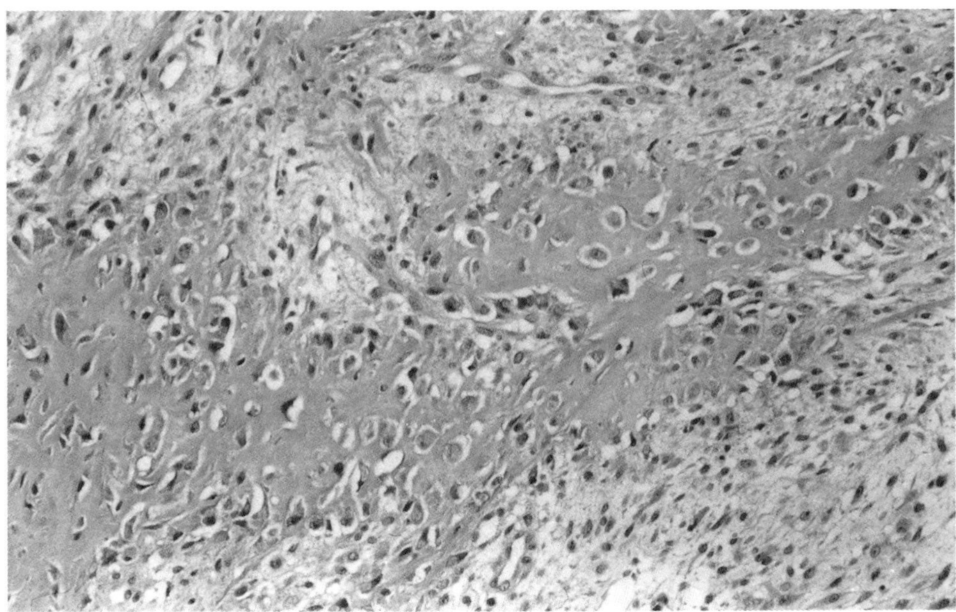

Figure 19.3. Approximately one week after injury, plump osteoblasts differentiate and produce immature osteoid deposits.

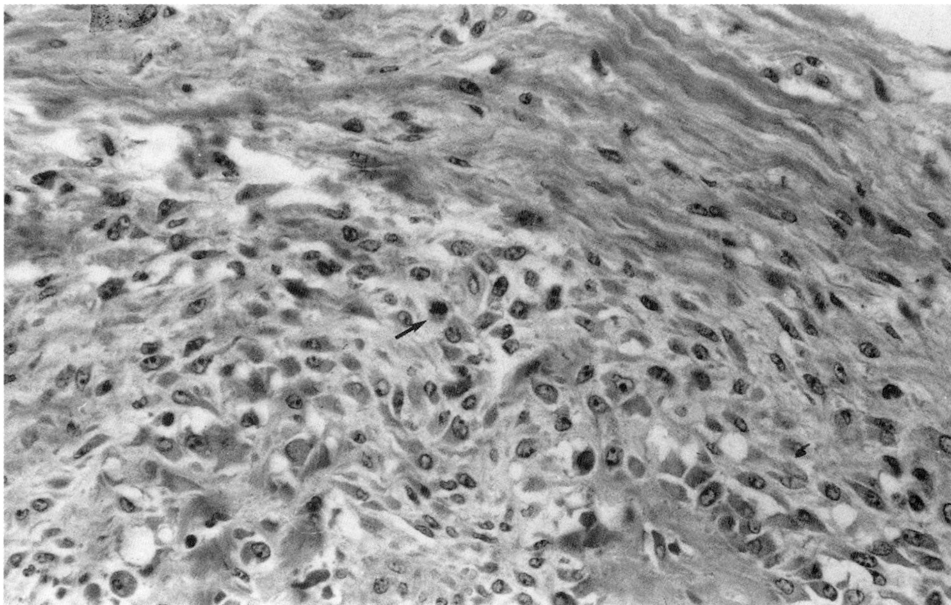

Figure 19.4. Approximately one week after injury, cellularity is usually high and numerous but normal mitotic figures are seen.

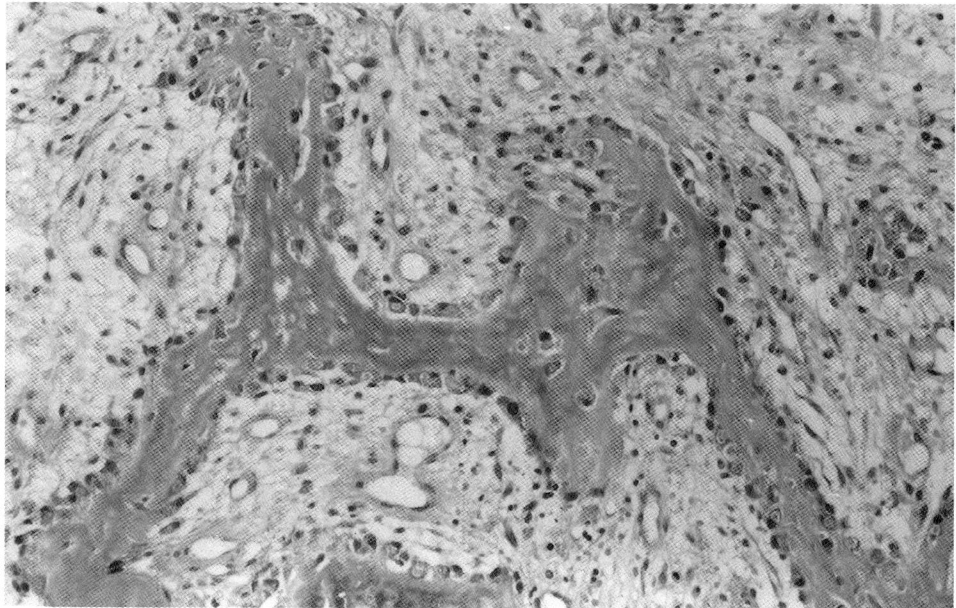

Figure 19.5. After three weeks, a trabecular pattern is more obvious in the neoformed bone with well-defined osteoblastic rimming.

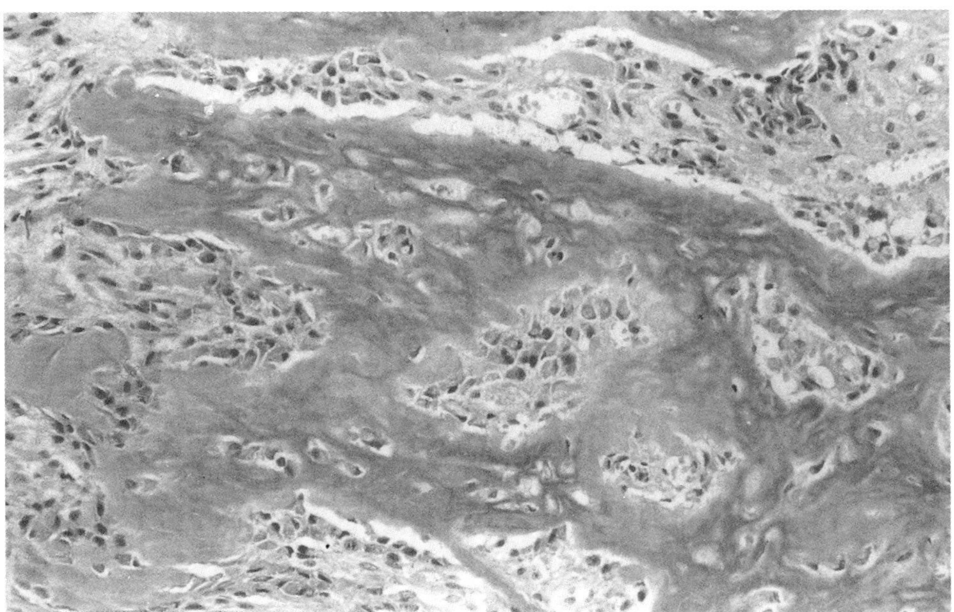

Figure 19.6. Calcium deposits can also be seen at three weeks.

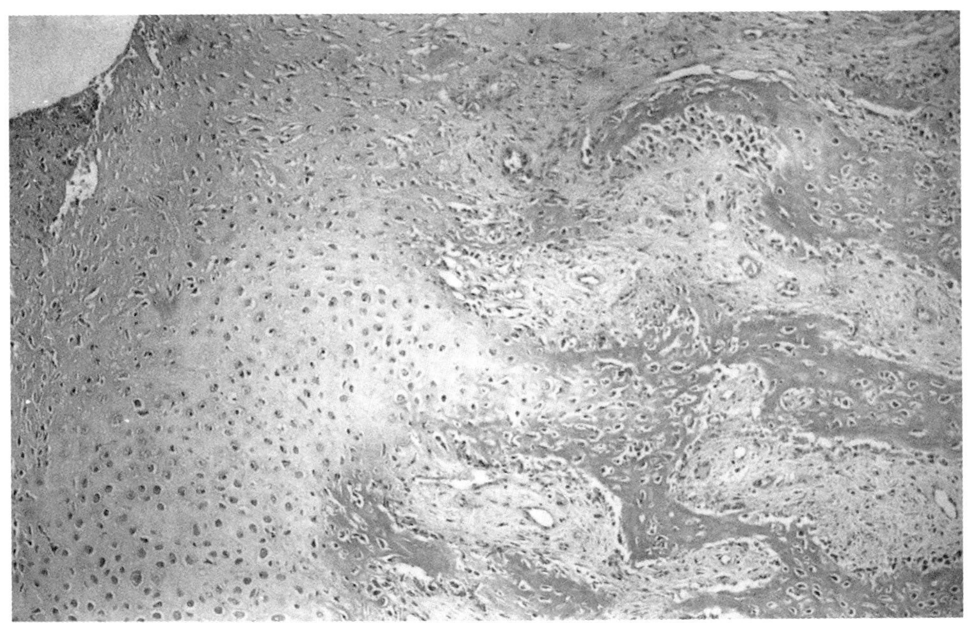

Figure 19.7. When present, cartilaginous areas may look florid and atypical.

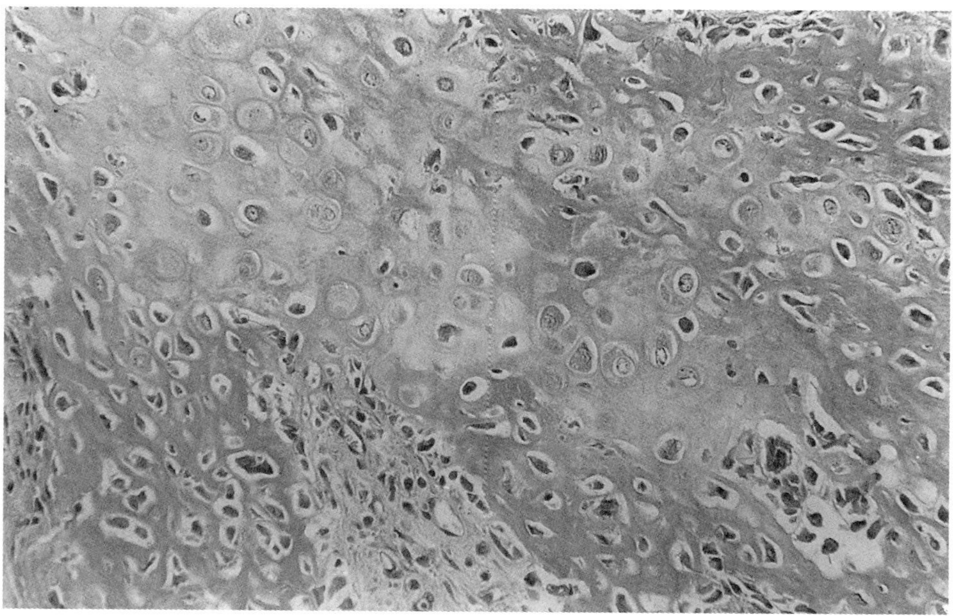

Figure 19.8. Enchondral ossification.

plump osteoblasts differentiate and produce immature osteoid deposits (Figure 19.3). At this stage cellularity is usually high and numerous but normal mitotic figures are seen (Figure 19.4). Also, osteoid and chondroid material may not show the obvious functional arrangement of a reactive process. At this stage, if a biopsy is performed without prior knowledge of the bone healing process, osteosarcoma can be misdiagnosed. Later, maturation of these substances resolves the problem. Patients with pathological conditions such as osteogenesis imperfecta can produce exuberant hypertrophic callus. In such cases, it is important to correlate the histologic features with the roentgenographic appearance.

After three weeks, a trabecular pattern is more obvious in the neoformed bone with a well-defined osteoblastic rimming (Figure 19.5). Calcium deposition also begins (Figure 19.6). Cartilage is an inconstant component of bone callus and has a temporary existence only. The formation and presence of cartilage depend on mechanical factors. When present, cartilaginous areas may look florid and atypical (Figure 19.7). Enchondral ossification then occurs (Figure 19.8).

After six weeks and for several months, bone remodeling will continue according to mechanical forces and will produce a solid lamellar bone.

20 Tissue Reactions to Products of Wear and Corrosion

A. Nehme and J. Puget

Introduction

Prosthetic joint replacement is now one of the most common surgical operations performed in Western countries. With the passage of time, some implants undergo corrosion [1]; or generate particulate debris from abrasion at the articulating surfaces or from micromotion at their interfaces with the bone [2]. The nature and ultimate fate of these products as well as their implication after long-term systemic exposure are among the least understood aspects of arthroplasty of the hip and knee.

Wear debris is not biologically inert. Its accumulation in local synovial tissue is associated with a chronic inflammatory reaction, the nature of which depends on the type and size of the particles [3], the concentration and duration of exposure [4], and the surface characteristics of the implant [5].

Polyethylene particles are recognized as a major factor in the survival of joint prostheses [6]; and metallic particulate species can also play an important role [7,8]. Furthermore, elevated levels of metallic elements from which implants are made have been reported in the distant organs and body fluids of patients with joint replacements [9–11].

The aim of this chapter is to discuss the different local and distant tissue reactions to the numerous products of wear and corrosion in relation to their characteristics and to their association with friction.

Characterization of Wear

Wear can be defined as the loss of material from the surfaces of the prosthesis as a result of motion between those surfaces. Material is lost in the form of particulate debris. Wear must be distinguished from plastic deformation or creep, which changes the shape of the implant without causing loss of material or producing particulate debris. There are three types of wear:

abrasive, in which the harder surface produces grooves in the softer material,

adhesive, in which the softer material is transferred as a thin film onto the harder surface,

fatigue, in which repetitive loading produces subsurface cracks and particles, or sheets of materials subsequently delaminate and are lost from the surface.

With the highly conforming surfaces in total hip arthroplasties, abrasive and adhesive mechanisms appear to be far more important than the fatigue wear mechanism.

The factors that determine wear are; 1) the coefficient of friction of the materials and their surface finish, 2) the hardness of the materials, 3) the applied load, 4) the sliding distance for each cycle depending on the diameter of the head and the motion of the hip, and 5) the number of cycles that occur over time.

Wear is difficult to measure accurately. In vivo wear measurements have varied considerably, and in vitro wear simulators have usually underestimated wear rates. This implies that simple unidirectional wear tests, which do not reproduce physiological hip motion, are of limited value in predicting wear with new designs or alternative bearing surfaces.

Clinical roentgenographic studies usually report linear wear rates whereas in vitro studies often report volumetric wear. Volumetric wear can be assessed in vitro by measuring the volume of particles released or by a change in the weight of the implant. Neither technique is applicable for retrieved implants because debris cannot be recovered easily, and polyethylene can absorb water during use, slightly altering its weight. Therefore, wear is commonly expressed as the linear distance of penetration of the head into the polyethylene and is measured by comparing initial and follow-up roentgenograms.

Very smooth surfaces such as those used in total hip components still have undulating peaks and valleys. When two surfaces first slide against each other under load, many of these asperities are removed, producing a high initial wear rate referred to as the "wearing-in" period. As the surfaces adapt to each other, contact areas increase and wear rates decrease, eventually reaching a steady state. The more complex prostheses become, by modularity and supplementary fixation of the components, the more likely the generation of metal particles from their junctions.

Products of wear and corrosion are characterized by their type, their size, and their amount, the latter being difficult to measure accurately as we have seen. Their accumulation in the local synovial tissue is associated with a chronic inflammatory reaction, the nature of which depends on the size and type of the particles [3], the concentration and duration of exposure [4], and the surface characteristics of the implant [5].

Lee et al. [3] have measured the size of metallic and polyethylene particulate debris around failed cemented arthroplasties. Femoral components were from equal numbers of titanium alloy stems, cobalt–chromium alloys, or stainless steel. The mean size of the metallic particles was 0.8–1.0 micrometers by 1.5–1.8 micrometers. The particle sizes of the three metals were similar. The mean size of polyethylene particles was 2–4 micrometers by 8–13 micrometers. They were larger in tissue retrieved from failed titanium alloy implants than from cobalt–chromium and stainless steel. Their results suggest that factors other than the size of the metal particles – such as the constituents of the alloy and the amount and speed of generation of debris – may be more important in the failure of hip replacements.

Kadoya et al. [12] have also found that polyethylene particles accumulated in the interface tissues are extremely small (mean size 0.82 micrometers) and that they are present in large amounts (range 5.2 × 10 (8) to 9.2 × 10 (10) per gram of tissue: mean = 1.4 × 10 (10). A more important point shown in their quantitative extraction is that a significant difference is observed in the number of polyethylene particles between osteolysis positive and negative cases, but with a similar size of particles in these two groups. The critical number of polyethylene particles is around 1.10 (10) particles per gram of tissue, which could be the prerequisite for the progression of osteolysis.

Furthermore, Willert and Semlitsch [13] have shown that prostheses with polyethylene cups and metal balls are associated with a polyethylene wear rate of 100–300 micrometers per year, Ceramic balls with polyethylene cups are associated with an expected wear rate of 50–150 micrometers per year, and that Co–Cr–Mo–C metal-on-metal and Al_2O_3 ceramic-on-ceramic pairings have the lowest wear rate of 2–20 micrometers per year. But, whereas the total volume of particles produced in metal-on-metal total hip replacements (THRs) is smaller than in metal-on-polyethylene THRs, the number of these particles is larger in metal-on-metal THRs due to their small size (0.1–400 micrometers) [14].

Nature of the Local Inflammatory Reaction

Total hip arthroplasty failure caused by prosthetic loosening in the absence of infection has been well documented in early reports of total hip surgery [15,16,17]. Gross and histologic examinations of failed total hip prostheses have described the presence of fibrous tissue and wear debris at the cement–bone interface [13]. More detailed examination of these tissues found sheets of macrophages in a fibrous tissue stroma with giant cells, polymethylmetacrylate particles, and metal debris in an apparent foreign-body-type reaction [18].

Initial reports attributed this histologic reaction to the presence of polymethylmetacrylate leading to the concept of cement disease [19]. More recently, however, reports of similar tissue at the bone implant interface in uncemented total hip arthroplasties have led to the concept of particle disease to describe the foreign-body reaction noted in association with metal and polyethylene debris with or without polymethylmetacrylate debris. Numerous reports have also supported the association between interface membranes and the occurrence of periprosthetic osteolysis [4,20–23]. These endosteal erosions have been described around cemented and uncemented implants and loosened and well-fixed implants [24]. Cobalt–chrome alloy, titanium–aluminum–vanadium, and ultra-high-molecular-weight polyethylene (UHMWPE) have been associated with the activation of periprosthetic tissues leading to osteolysis and loosening.

Experimental evidence supporting the association between particulate wear debris and osteolysis has been based on two primary observations:

Micron-sized metal and polyethylene debris induces a foreign-body-type reaction when placed in various animal tissues [17,25,26].

These tissues, experimentally induced, and human retrieval specimens, produce various enzymes and cytokines capable of stimulating bone resorption.

Goldring et al. [27,28] showed that interface membranes from failed human THRs were capable of producing PGE2 and collagenolytic factors, both of which have been shown to stimulate bone resorption in vitro. The results of similar studies of periprosthetic membranes in tissues cultures have shown the release of collagenase, gelatinase, IL1, IL6, and TNF [20,29,30], all implicated in the stimulation of bone resorption in vitro and in vivo [27,28,31,32].

Role of PMMA, UHMPE, and Metal Particles in the Inflammatory Reaction

Kadoya et al. [12] have shown that polyethylene particles are primarily responsible for macrophage recruitment and attachment onto the bone surface, which leads to osteolysis. Secondly, the significance of macrophages in the context of bone loss is not only as cells producing inflammatory mediators which could activate osteoclasts, but as cells which could resorb bone directly as well. Thirdly, polymethylmetacrylate (PMMA) particles cannot be regarded as a potent stimulatory factor for bone resorption. They suggest that the importance of fragmented PMMA in the context of osteolysis may have been over-emphasized in previous studies without paying enough attention to their actual size and distribution.

Several studies have demonstrated the significant role of metal particles in the failure of implants [33–35]. An enhanced influence of metal particles on osteoclastic bone resorption is observed [12] and can be interpreted in two ways: metal particles might be produced predominantly in more unstable implants, thus enhanced osteoclastic resorption could be explained as being related to the enhanced remodeling around more unstable implants. Another possibility is that metal particles have different effects on osteoclast recruitment and activation; especially that the production of inflammatory mediators such as IL2-B2 is more apparent in the presence of metal particles. If this is the case, it is conceivable that these mediators change the coupling process into a resorption-dominant state, which might explain extensive bone loss in metallosis.

Nonetheless, in the comparison of metal-on-polyethylene with metal-on-metal hip prostheses, a major focus is the lower amount of volumetric wear and overall tissue histiocytic reaction, particularly multinucleated giant cells in the latter [36]. The same authors [36] compared the intensity of the tissue reaction around metal-on-metal THRs to the reaction around metal-on-polyethylene THRs and found that the number of histiocytes is one grade lower around metal-on-metal THRs and that there are no giant cells reacting to metal particles. However, their result does not assess the relation between the intensity of the inflammatory reaction and the amount of the associated osteolysis.

Similarly, there is considerable interest in ceramic implants because of the increased awareness of its excellent mechanical and sliding characteristics. Nevertheless, some cases of fracture of the ceramic femoral head have been reported [37,38,39], and the revision operation after this complication may be problematic in terms of the choice of the type of the femoral head to be inserted [40].

All these accumulated data on metal-on-metal and ceramic-on-ceramic hip implants seem to be very promising and would support continued efforts in metal-on-metal and ceramic-on-ceramic designs.

Dissemination of Wear Particles

The importance of wear particles generated by wear and corrosion of joint replacement prostheses has been understood primarily as we have seen in the context of the local effects of particle-induced periprosthetic osteolysis and aseptic loosening. The fate of wear particles in the body remains uncertain. Lankamer et al. [41] have reported histologic and electron-microscopic evidence from two cases which shows that metallic debris can be identified in the lymphoreticular tissues of the body distant from the hip some years after joint replacement.

Urban et al. [42] recently reported that the systemic distribution of metallic and polyethylene particles was a common finding both in patients with a previously failed implant and in those with a primary total joint prosthesis. The prevalence of particles in the liver or spleen was greater after reconstructions with mechanical failures.

In the majority of patients the concentration of wear particles in the organs (para-aortic lymph nodes, spleen, and liver) was relatively low and without apparent pathological importance. However, in one rare case, granulomas formed in the liver, spleen, and abdominal lymph nodes, and compromised hepatic function in response to heavy accumulation of wear debris from a hip prosthesis with a mechanical failure.

Carcinogenicity

As early as the 1950s, laboratory investigations into the carcinogenicity of modern dental and orthopedic alloys were undertaken. Such studies were prompted by the observations that workers, particularly in nickel and chromate refining, have increased risks of nasal and lung tumors. For the past 30 years sporadic case reports have documented the development of malignant neoplasms proximate to an orthopedic implant. But the results of epidemiologic studies have not shown yet any excessive number of tumors in patients receiving stainless steel or superalloy implants. Some preliminary observations have suggested some possible premalignant changes in bone marrow adjacent to worn total hip arthroplasty implants [43]. They suggest, however, longer postoperative intervals and more epidemiologic data to see if there is any risk that would be pertinent to a young patient at primary arthroplasty.

Conclusion

Ideally, neither articulation, modularity of devices, nor implant anchorage should be the source of wear products, but the articulation will always produce wear on physical principles. One should therefore take advantage of all possibilities to reduce the production of wear particles in the joint bearing as well as at the anchoring surfaces. The surgeon should also consider revision in patients in whom large amounts of particulate debris may be generated. Serum and urine trace metal analysis may provide early confirmation of failure and aid in the timing of revision operations in patients with a symptomatic or failed device. Administration of drugs to inhibit particle-induced bone resorption should be considered for the present as a second line of defense, since this is directed against an aftermath and does not influence the continuing wear process.

References

1. Meachim G, Williams DF. Changes in non-osseous tissue adjacent to titanium implants. J Biomed Mater Res 1973;7:555–72.
2. Galante JO, Lemons J, Spector M, Wilson PD Jr, Wright TM. The biologic effects of implant materials. J Orthop Res 1991;9:760–75.

3. Lee JM, Salvati EA, Betts F. Size of metallic and polyethylene debris particles in failed cemented total hip replacements. J Bone Joint Surg (Br) 1992;74-B:380–4.
4. Willert HG, Bertram H, Buchhorn GH. Osteolysis in alloarthroplasty of the hip. The role of ultra-high-molecular-weight polyethylene particles. Clin Orthop 1990;258:95–107.
5. Witt JD, Swann M. Metal wear and tissue response in failed titanium alloy total hip replacements. J Bone Joint Surg (Br) 1991;73-B:559–63.
6. Goodman S, Lidgren L. Polyethylene wear in knee arthroplasty. A review. Acta Orthop Scand 1992;63:358–64.
7. Ballard WT, Shanbag AS, Jacobs JJ. Particle debris. In: Callahan AG, Rosenberg, Rubash HE, editors. The Adult Hip. Philadelphia: Lippincot-Raven, 1998;267–77.
8. Jacobs JJ, Skipor AK, Patterson LM, Hallab NJ, Paprosky WG, Black J et al. Metal release in patients who have had a primary hip arthroplasty. A prospective, controlled longitudinal study. J Bone Joint Surg 1998; 80-A:1447–58.
9. Bartolozzi A, Black J. Chromium concentration in serum blood clot and urine from patients following total hip arthroplasty. Biomaterials 1985;6:2–8.
10. Jacobs JJ, Skipor AK, Black J, Urban RM, Galante JO. Release and excretion of metal in patients who have a total hip replacement component made of titanium-base alloy. J Bone Joint Surg 1991;73(A):1475–86.
11. Case CP, Langkamer VG, James C, Palmer MR, Kemp AJ, Heap PE et al. Widespread dissemination of metal debris from implants. J Bone Joint Surg 1994;76(B): 701–12.
12. Kadoya Y, Kobayashi A, Ohashi H. Wear and osteolysis in total joint replacements. Acta Orthop Scand 1998; (Suppl 278):69.
13. Willert HG, Ludwig J, Semlitsch M. Reaction of bone to methacrylate after hip replacement arthroplasty. A long-term gross-light and electron microscopic study. J Bone Joint Surg 1971;56A:1368–82.
14. Savio III JA, Overcamp LM, Black J. Size and shape of biomaterial wear debris. Clin Mater 1993;12:1.
15. Charnley J. The histology of loosening between acrylic cement and bone. J Bone Joint Surg 1975;57(B):245.
16. Charosky CB, Bullough PG, Wilson PD. Total hip replacement failures: A histological evaluation. J Bone Joint Surg 1973;55A:49–58.
17. Evans EM, Freeman MAR, Miller AJ, Vernon-Roberts B. Metal sensitivity as a cause of bone necrosis and loosening of the prosthesis in total joint replacement. J Bone Joint Surg 1971;56B:66–642.
18. Mirra JM, Amstutz HC, Matos M, Gold R. The pathology of the joint tissues and its clinical relevance in prosthesis failure. Clin Orthop 1976;117:221–40.
19. Jones LC, Hungerford DS. Cement disease. Clin Orthop 1987;225:192–206.
20. Chiba J, Iwaki Y, Kim KJ, Rubash HE. The role of cytokines in femoral osteolysis after cementless total hip arthroplasty. Trans Orthop Res Soc 1992;17:350.
21. Chiba J, Rubash HE, Kim KJ, Iwaki Y. The characterization of cytokines in the interface tissue obtained from failed cementless total hip arthroplasty with and without femoral osteolysis. Clin Orthop 1994;300: 304–12.
22. Kim KJ, Chiba J, Rubash HE. In vivo and in vitro analysis of membranes from hip prostheses inserted without cement. J Bone Joint Surg 1994;76A:172–80.
23. Kim KJ, Rubash HE, Wilson SC, D'Antonio JA, McClain EJ. A histologic and biochemical comparison of interface tissues in cementless and cemented hip prostheses. Clin Orthop 1993;287:142–52.
24. Maloney Wl, Iasty M, Harris WH, Galante IO, Callaghan LL. Endosteal erosion in association with stable uncemented femoral components. J Bone Joint Surg 1990; 72A:1025–34.
25. Goodman SB, Chin RC, Chiou SS. A Clinical pathologic-biochemical study of the membranes surrounding loosened and non-loosened total hip arthroplasties. Clin Orthop 1989;244:182–7.
26. Goodman SB, Fornasier VL. Clinical and experimental studies in the biology of aseptic loosening of joint arthroplasties and the role of polymer particles. In: St. John KR, editor. Particulate Debris From Medical Implants: Mechanisms of Formation and Biological Consequences. Philadelphia: American Society of Testing Materials STP 1114, 1992;27–37.
27. Goldring SR, Schiller AL, Roelke M. The synovial-like membrane at the bone-cement interface in loose total hip replacements and its proposed role in bone lysis. J Bone Joint Surg 1983;65A:575–84.
28. Goldring SR. Jasty M, Roelke MS. Formation of the synovial-like membrane at the bone-cement interface: Its role in bone resorption and implant loosening after total hip replacement. Arthritis Rheum 1986;29: 836–42.
29. Dayer JM, Breard J, Chess L, Crane SM. Participation of monocyte-macrophages and lymphocytes in the production of a factor that stimulates collagenase and prostaglandin secretion by rheumatoid synovial cells. J Clin Invest 1979;64:1386–92.
30. Horowitz SM, Doty SB, Lane JM, Burstein AH. Study of the mechanism by which mechanical failure of polymethylmethacrylate leads to bone resorption. J Bone Joint Surg 1993;75A:802–13.
31. Murray DW, Rushton N. Macrophage stimulate bone resorption when they phagocytose particles. J Bone Joint Surg 1990;72B:988–92.
32. Dowd JE, Schwendeman LJ, Macaulay W, Doyle JS, Shanbhag AS, Wilson S et al. Aseptic loosening in uncemented total hip arthroplasty in a canine model. Clin Orthop 1995;319:106–21.
33. Agins HJ, Alock NW, Bensal M. Metallic wear in failed titanium-alloy total hip replacements. J Bone Joint Surg 1988;70A:347–56.
34. Kelly SS, Johnston RC. Debris from cobalt-chrome cable may cause acetabular loosening. Clin Orthop 1992;285: 140–6.
35. Lombardi AV, Mallory TH, Vaughan BK, Drouillard P. Aseptic loosening in total hip arthroplasty secondary to osteolysis induced by wear debris from titanium-alloy modular femoral heads. J Bone Joint Surg 1989;71(A): 1337–42.
36. Doorn PF, Mirra JM, Campbell PA, Amstutz HC. Tissue reaction to metal-on-metal total hip prosthesis. Clin Orthop 1996;329S:187–205.
37. Cooke FW. Ceramics in orthopedic surgery. Clin Orthop 1992;276:135–46.
38. Sedel L. Editorial. Ceramic hips. J Bone Joint Surg 1992; 74-B(3):331–2.
39. Sedel L, Kerboull L, Christel P, Meunier A, Witvoet J. Alumina-on-alumina hip replacement. Results and survivorship in young patients. J Bone Joint surg 1990;72B: 658–63.

40. Allin J, Goutallier D, Voisin MC, Lemouel S. Failure of a stainless steel femoral head of a revision total hip arthroplasty performed after a failure of a ceramic femoral head: a case report. J Bone Joint Surg 1998;80A: 1355–60.
41. Langkamer VG, Case CP, Heap P, Taylor A, Collins C, Pearse M et al. Systemic distribution of wear debris after hip replacement: a cause for concern? J Bone Joint Surg 1992;74(B):831–9.
42. Urban RM, Jacobs JJ, Tomlinson MJ, Gavrilovic J, Black J, Peoc'h M. Dissemination of wear particles to the liver, spleen, and abdominal lymph nodes of patients with hip replacement. J Bone Joint Surg 2000;82A(4):57–77.
43. Case CP, Langkamer VG, Howell RT, Webb J, Standen G, Palmer M et al. Preliminary observations on possible premalignant changes in bone marrow adjacent to worn total hip arthroplasty implants. Clin Orthop 1996;329: 269–79.

21 The Influence of Hyperbaric Oxygen Treatment on Cell Multiplication

B. Clouet D'Orval

Hyperbaric oxygen (HBO) therapy is successfully used in treating anaerobic infection that results in gas gangrene as well as severe aerobic infections such as necrotizing fasciitis and chronic refractory osteomyelitis. In addition, it reduces morbidity and mortality resulting from carbon monoxide intoxication. Many protocols using HBO to promote tissue and bone healing by increasing oxygenation have been investigated and therapeutic indications are better defined for using HBO in selected orthopedic situations.

Physiopathology of Hyperbaric Oxygen Treatment

The use of hyperbaric oxygen treatment rests on the principle of the solubility of gases in liquids. Oxygen is transported towards tissues by blood flow in dependent or dissolved form. In the event of, in particular, hypoxic tissue suffering, tissue hyperoxygenation is possible by increasing the dissolved quantity of oxygen per oxygen inhalation under pressures higher than the atmospheric pressure.

Oxygen Transport to Tissues

In Normobaric Oxygen (20% oxygen)

Alveolar stage: The diffusion of oxygen between alveoli and capillaries of the lungs is carried out according to a low gradient of pressure between alveoli and capillaries.

Blood stage: In blood, oxygen is transported in two forms: the major part is combined with hemoglobin (98.5%), and the other part, proportionally weaker (1.5%), is dissolved in plasma. The quantity (q) of gas that dissolves is directly proportional to the arterial partial oxygen pressure (PaO2), and to the coefficient of solubility (α) in blood according to the law of Henry ($Q = \alpha PaO2$), which is 0.3 ml/100 ml of plasma. This intermediate dissolved form is essential to allow the fixing of oxygen on hemoglobin. In addition, the degree of saturation of hemoglobin depends on PaO2 and varies according to a non-linear curve of dissociation. The maximum oxygen capacity fixed can reach a level of 20.1 ml/100 ml of blood, that is to say, a total quantity of oxygen transported of 20.4 ml/100 ml.

Tissue stage: The distribution of oxygen in tissues relies on the same principle of diffusion according to a gradient of pressure between capillary blood and the cells. It depends on the tissue partial oxygen pressure (increasingly lower than PaO2), of the metabolic request, of the tissue solubility of oxygen, and the local vasoconstriction.

In Hyperbaric Oxygen (100% oxygen)

The increase in the arterial partial oxygen pressure has several consequences.

The increase in the quantity of oxygen dissolved can be enough to meet the tissue needs (6 ml/100 ml) with 3 atmospheres (3 ATA). The affinity to hemoglobin for oxygen is decreased.

A systemic arterial hypertension related to a peripheral vasoconstriction and a bradycardia reflex results, related to a vasoconstriction reflex to tissue hypoxia and not to a general reaction which would involve a fall of the oxygen delivered with fabrics. The quantity of oxygen exchanged is, in fact, increased by pericapillary diffusion.

A persistent increase in the deformability of red cells.

Effects of HBO on Tissue and Cell Growth

HBO increases dissolved oxygen in the blood and results in a high partial pressure of oxygen (PaO_2). An increase of PaO_2 maintains tissue oxygenation in the absence of hemoglobin [1] and also has an antiedematous effect. In fact, one finds in the microcirculatory sector a reduction in the blood flow of about 20% by the phenomenon of vasoconstriction reflex, which decreases the capillary flow of transudation and thus the formation of edemas. In addition, it stimulates fibroblast growth, increases collagen and intracellular adenosine triphosphate synthesis, and promotes more rapid growth of capillaries (angioneogenesis) [2,3] and osteoblastic and osteoclastic activity [4]. The stimulation of osteogenesis by HBO has been reported in animal experiments and clinical cases.

Effects of HBO on Wound Healing

The cicatrization of a wound is a complex phenomenon which brings into play several cellular types in a process of detersion, angioneogenesis, and cellular repair. After a first inflammatory phase with polynuclear surges of neutrophiles and macrophages, a second phase of granulation with conjunctive proliferation of neocapillaries, fibroblasts, and collagen synthesis begins. A last phase of epithelialization finishes the process. It has for a long time been recognized that delays or failures of cicatrization are primarily due to the existence of an infection or a local ischemia. The hypoxic area involves a reduction or even a stop of the cellular proliferation and collagen synthesis; from there a stop of neocapillary formation, just as a significant reduction in the macrophagic activity and bactericidal capacity of the polynuclear cells. In addition, a sensitivity particular to the infections of ischemic tissues deteriorates the local conditions.

An additional oxygen contribution, such as carried by hyperbaric oxygen treatment, involves a series of eutrophic and healing effects [2]:

Increase in the distance of pericapillary diffusion of oxygen.

Angiogenesis stimulation by accelerating growth of neocapillaries.

Fibroblastic proliferation by increasing local oxygen pressure.

Quantitative and qualitative increase (hydroxylation of the proline) of collagen synthesis.

Epithelialization by increasing mitotic index, and mobility of the epithelial cells.

Effects of HBO on Bone Healing

The process of osseous repair is comparable with conjunctive repair. As soon as the fracture hematoma is formed an inflammatory reaction with detersion involves the osteolytic aspects of the distal ends of the fracture. Cartilage and osseous cellular differentiation occurs after a proliferative phase within young conjunctive tissue of granulation. It is performed, in fact, by a metamorphosis of the local fibroblasts which leads to two types of ossification; on one hand, enchondral, central, caused by the development of a cartilage matrix, and on the other hand, periostal, which produce an osseous structure directly. The two processes lead initially to osteoid plates, which will be the object of a secondary replanning leading to complete repair.

Hyperbaric oxygen treatment by the increase of the partial pressure of oxygen promotes the capillary proliferation to reappear, just as it restores normal capacities of synthesis and proliferation of osteoblasts. Osteogenesis and collagen synthesis are decreased by hypoxia, and a moderate increase of oxygen supply (80 mm Hg) stimulates their appearance [5]. The osteoclastic activity begins again, and allows resorption of the infected and necrosed osseous residues [4].

Using a bone harvest chamber, Nilsson et al. [6] reported that HBO treatment caused a significant increase in bone formation and histologically demonstrated that lamellar bone developed in the chamber canal. In a rat bone fracture model, HBO accelerated healing of bone and induced a greater amount of new bone formation. Many factors explain the promotion of bone healing by HBO. The differentiation of osteoblasts from mesenchymal cells may be influenced by HBO [7]. Shaw and Basset [8] have shown that increased oxygen tension causes cellular differentiation to osseous tissue, whereas decreased oxygen tension results in cartilage formation. It has also been shown that osteoclastic and osteoblastic activity is increased by increased oxygen tension. Collagen synthesis is also increased by HBO. The role of the hypoxic area (injury area) in intermittence with HBO therapy is important to promote angiogenesis [7]. The return to the hypoxic state with high lactate levels stimulates secretion of a variety of biochemical messengers, including a chemotactic wound angiogenesis factor, by the macrophages. Neovascularization supplies more oxygen to the central hypoxic area and HBO therapy, with a moderate increase of oxygen tension (80 mm Hg), is especially effective in promoting angiogenesis and osteogenesis. Tuncay et al. [9] investigated the effects of ambient hypoxia and hyperoxia on osteoblast function in vitro. In low ambient oxygen tension, cellular proliferation increases, whereas alkaline phosphatase (AP) activity, collagen synthesis, media PaO_2, and PCO_2 decrease. In contrast, in hyperoxy conditions, cellular proliferation is suppressed with concomitant increases in: AP activity, collagen synthesis, and partial pressures for oxygen and carbon dioxide. In crossover experiments, where cells were initially grown in hypoxic conditions and were switched to hyperoxic conditions, their metabolic activities were abruptly reversed. In addition, various oxygen concentrations influence the metabolism of bone cells [8]. Maximum osteogenesis is observed with an oxygen concentration of 35% with little or no osteogenesis at 5%. Exposure to high oxygen concentrations (95%) is known to induce damage to cells, possibly due to an increased oxygen radical production. In addition, it appears to induce less osteogenesis and cause more osteoclasia and chondroclasia. With regard to collagen synthesis, low oxygen tension suppresses collagen fiber formation, whereas the effect of high oxygen tension is found to vary according to the duration of exposure to HBO. Short-term exposure (6 hours) causes maximum collagen synthesis, while long-term exposure (2 weeks) produces less fibers.

Therapeutic Indications

Effect of HBO on Autogenous Bone Grafts

Sawai et al. [7] reported a histologic study concerning free bone graft osseointegration. The interface between a grafted iliac corticocancellous bone and a bone defect of the mandible was histologically examined in rabbits with or without HBO therapy. At one week after grafting, osteoid formation in the experimental group was much greater than in the control group. Union between the grafted and the host bone was observed in the experimental group at two weeks after grafting, but it was not observed in the control group until four weeks. Although it was difficult to differentiate grafted from host bone in the experimental group at four weeks, it was readily distinguishable in the control group. These results indicate that HBO accelerates the union of autogenous free bone grafts.

Effect of HBO on Bone Healing of Tibial Lengthening

Ueng et al. [10] reported that bone healing of tibial lengthening is enhanced by HBO therapy in a study of bone mineral density (BMD) and torsional strength in rabbits. In a first group exposed to 2.5 atmospheres absolute of hyperbaric oxygenation for two hours daily, it would appear that the BMD was increased significantly (96.9% at six weeks) in the distraction segment compared with the second group who did not go through HBO (79.2% at six weeks). In addition, torsional strength of lengthened tibia of the HBO group was increased significantly

(88.6% of maximal torque at six weeks) compared with the non-HBO group (76% of maximal torque at six weeks). In clinical lengthening, the central zone of the distraction segment is usually the poorest part of callus formation. Yasui et al. [11] found that as distraction proceeded, the poor bone formation in the central zone was due to its hypovascularity via an investigation of microangiography. HBO is known to promote the development of new blood vessels in the hypovascular tissue; therefore, there is a good chance that it will increase the rate of vascular formation in the distraction segment and then promote bone formation. These results suggest that the bone healing of tibial lengthening is enhanced by intermittent hyperbaric oxygen therapy.

Effect of HBO on Avascular Necrosis of the Femoral Head

Levin et al. [12] reported treatment of experimental avascular necrosis of the femoral head with hyperbaric oxygen in rats in a histological evaluation of the femoral head during the early phase of the reparative process. The healing of vascular deprivation-induced necrosis of the femoral head of rats exposed to hyperbaric oxygen was compared with that in untreated rats. Newly formed appositional and intramembranous bone was more abundant and remodeling was more advanced in the femoral heads of the hyperbaric oxygen-treated than untreated rats sacrificed on the 42nd postoperative day; there was also less necrotic debris in the femoral heads of the treated rats. There were no differences in the severity of the degenerative changes of the articular cartilage of the treated and untreated rats. Exposure of rats to hyperbaric oxygen does not preserve tissue viability after all arteries supplying the femoral head are severed. Yet, resulting in an increased oxygen tension of the tissues, it seems to provide the optimal settings for reparative processes. The results suggest that hyperoxygenation-mediated relief of ischemia enhances the fibroblastic, angioblastic, and osteoclastic activities such that healing of the rats' necrotic femoral heads is expedited.

Adjunctive HBO to the Management of Radionecrosis

Irradiation induces hypocellularity, hypovascularity, and hypoxemia in tissue, and radionecrosis can occur after irradiation treatment, particularly osteoradionecrosis. Some clinicians do not use implants (titanium, hydroxyapatite implants) because of the risk of post-radiation osteonecrosis decreasing the success of a prosthetic restoration. In the maxillofacial region, HBO therapy has been useful in both the treatment and prevention of osteoradionecrosis and in reconstruction involving irradiated tissue. Experimental studies have revealed the effects of HBO therapy on the osseointegration and healing around titanium [13] or hydroxyapatite-coated [14] implants placed in irradiated bone. HBO therapy seems to slightly improve the cell population for bone regeneration capacity and then encourages trabecular bone around the implant to remodel at a level close to that of nonirradiated bone. On the other hand, HBO accelerates bone remodeling in non-irradiated bone and improves HA–bone contact in both irradiated and nonirradiated bone.

Conclusions

HBO enhances the fibroblastic, angioblastic, osteoblastic, and osteoclastic activities by increasing local oxygen pressure. It stimulates angiogenesis, collagen synthesis, accelerates healing of bone, and induces a greater amount of new bone formation. Adjunctive hyperbaric oxygen treatment seems to be useful on healing and regeneration of bone defects in reconstructive bone surgery. It is a very helpful tool and a new approach in the management of osteonecrosis, assisting the classical surgical principles.

References

1. Niinikowski J, Hunt TK. Oxygen tension in healing bone. Surg Gynecol Obstet 1972;134:746–50.

2. Tompach PC, Lew D, Soll JL. Cell response to hyperbaric oxygen treatment. Int J Oral Maxillofac Surg 1997;26(2): 82–6.
3. Kindwall EP. Uses of hyperbaric oxygen therapy in the 1990s. Cleve Clin J Med 1992;59(5):517–28.
4. Mainous EG. Osteogenesis enhancement utilizing hyperbaric oxygen therapy. HBO Rev 1982;3:181–5.
5. Boyne PJ. Effect of increased oxygenation on osteogenesis enhancement. In: Davis JC, Hunt TK, editors. Hyperbaric Oxygen Therapy. Bethesda: Undersea Medical Society, 1977;205–16.
6. Nilsson P, Albrektsson T, Granström G. The effect of hyperbaric oxygen treatment on bone regeneration. Int J Oral Maxillofac Implants 1988;3:43.
7. Sawai T, Niimi A, Takahashi H, Ueda. Histologic study of the effect of hyperbaric oxygen therapy on autogenous free bone graft. J Oral Maxillofac Surg 1996;54(8): 975–81.
8. Shaw JK, Basset CA. The effect of varying oxygen concentration of osteogenesis and embryonic cartilage in vitro. J Bone Joint Surg 1967;49(A):70–3.
9. Tuncay OC, Ho D, Barker MK. Oxygen tension regulates osteoblast function. Am J Orthod Dentofacial Ortrhop 1994;105(5):457–63.
10. Ueng SW, Lee SS, Lin SS, Wang CR, Liu SJ, Yang HF et al. Bone healing of tibial lengthening is enhanced by hyperbaric oxygen therapy: a study of bone mineral density and torsoinal strength on rabbits. J Trauma 1998;44(4):676–81.
11. Yasui N, Kojimoto H, Sakaki K. Factors affecting callus distraction in limb lengthening. Clin Orthop 1993;293: 55–60.
12. Levin D, Norman D, Zinman C. Treatment of experimental avascular necrosis of the femoral head with hyperbaric oxygen in rats: histological evaluation. Exp Mol Pathol 1999;67(2):99–108.
13. Ueda M, Kaneda T, Takahashi H. Effect of hyperbaric oxygen therapy on osseointegration of titanium implants in irradiated bone. Int J Oral Maxillofac Implants 1993;8:41–4.
14. Chen X, Matsui Y, Ohno K, Michi K. Histomorphometric evaluation of the effect of Hyperbaric Oxygen Treatment. Int J Oral Maxillofac Implants 1999;14:61–8.

22 Rational Utilization of the Stress Shielding Effect of Implants

K. Dai

Fracture healing is an extremely complicated biological process. The primary factor affecting fracture healing is the blood supply to the fracture site, which is the basic assurance of successful treatment, and as well as sustained blood supply to the fracture site, a favorable local mechanical environment is also an essential requirement for fracture healing. Research in past years has shown that the mechanical environment favorable to fracture healing includes two aspects: stability of the fracture ends (exclusion of harmful movement and shear stress) and stimulation of physiological stress. In the past, clinical treatment of fractures focused on the means to ensure the stability of the fracture ends, while modern treatment for fracture gives more regard to the physiological requirements for fracture healing.

The earliest treatment for fracture mainly consisted of external fixation with plaster immobilization, which is unfavorable for early functional exercise and may lead to joint stiffness and muscular atrophy, giving rise to the so-called fracture disease. The application of rigid internal fixation is a major step forward in the treatment of fracture. Various devices for internal fixation of fracture involve the principle of utilizing the stress shielding effect of the implant to ensure the stability of the fracture ends. Stress shielding is a mechanical concept, i.e., when two or more components with different elastic moduli make up a mechanical system, the phenomena of redistribution of load, stress, and strain will take place: the component with a higher elastic modulus will bear more load and the component with a lower elastic modulus will bear less load, while stress and strain will be correspondingly less. Rigid internal fixation chiefly uses a bone plate of a higher elastic modulus to fix the fracture, imposing certain stress-shielding effects on the bone, which is of a lower elastic modulus, thereby ensuring the stability of the fracture ends.

Prior to the advent of the concept of rigid internal fixation (brought about mainly through the action of compression plates) was put forward, the stress-shielding effect of an ordinary bone plate was insufficient for the maintenance of absolute stability of the fracture ends. With its strong stress-shielding effect, the compression plate amply ensures the stability of fracture ends and close contact between them; primary healing is then obtained. Rigid fixation is also advantageous for early exercise of the involved limb, getting rid of the complications of muscular atrophy and stiff joints.

However, the application of compression plates brings in another problem, i.e., the incidence of osteoporosis of bone plate origin. Means of observation, including quantitative histology, polar microscopy, transmission electron microscopy, scanning electron microscopy, microangiography, mechanical tests, and molecular biology investigations have been employed to study what bearing the stiffness of internal fixation and stress stimulation has on fracture healing and bone remodeling. The results showed that in a situation of identical blood supply, bone plates of different stiffness can lead to osteoporosis to varying degrees (in terms of bone loss, erosion, breakage and disorganization of collagen fibers in bone matrix, and the decrease in the mechanical strength of bone) and that the higher the stiffness of the bone plate and the longer the duration of fixation, the more serious osteoporosis becomes. Only after the removal of the bone plate can osteoporosis be gradually reversed. The incidence of such a complication is related to the changes in the metabolism of bone cells resulting from stress shielding and the proliferation and dilation of micro-vessels following bone plate fixation. The effect of stress shielding can intensify the proliferative reaction of the micro-vessels [1,2,3].

It is clear that the stress-shielding effect is the essential means of post-fracture stabilization of fracture ends, yet the resultant decrease of stress stimulation will considerably influence bone healing, especially bone remodeling in the later stage of healing. For this reason, the progress of clinical treatment of fracture lies in the resolution of the contradiction of fixation and stress stimulation, and the combination of the effective and rational utilization of the stress-shielding effect of internal fixation and the elimination of its unfavorable effects.

The Study of Low-stiffness Bone Plates

In the late 1970s, Woo et al. initiated the concept of nonrigid internal fixation, i.e., the use of low-stiffness bone plates made from a composite material consisting of plastic, carbon fiber, epoxy resin, etc. Animal experiments showed that the low-stiffness bone plates could markedly lessen the osteoporosis consequent upon stress shielding, but the stability provided by the composite bone plate for the fracture site was rather unsatisfactory, resulting in a higher incidence of delayed union and non-union of the fracture [4]. Alongside the further exploration of biodegradable materials, the use of resorbable, biodegradable materials, such as polylactic acid and polyglycolic acid, to make bone plates began, conceiving that with the degradation of the material in the body, the stress-shielding effect of internal fixation on the fracture site would be gradually lowered, so that the fracture site would gradually take up the stimulus of physiological stress, and in addition, it would not be necessary to remove the plate in a second operation. Zimmerman [5] and other scholars used a composite material incorporating carbon-reinforced poly-L-lactic acid (PLA) to make a laminated, degradable bone plate, and used it to fix an experimentally osteotomized dog femur, to find that, owing to the hydrophobicity of the material and infiltration of water, the stiffness of the fixation system rapidly lessened before the internal stability for bone healing was accomplished, leading to hypertrophic non-union. Therefore, PLA can only be applied to certain fractures, where the implant does not have to maintain a high stress. Hanafusa et al. [6] used PLA acid to make composite, low-rigidity, biodegradable bone plate and achieved certain success in osteotomy models of rabbit femur. In eight weeks, bone healing rates reached 64%. And in 25–40 weeks, no difference was found in bone structure under the plate and the mechanical strength of the healing bone, as compared with normal bone, the results being markedly superior to the controls using stainless steel bone plates for fixation. However, 14% of the degradable plate fixations failed. In addition, the histological reaction induced by degradation products is still a problem pending a solution. For instance, polyglycolic acid is apt to cause the formation of aseptic sinus with an incidence as high as 25%, and regional toxicity will appear following hydrolysis of the material in 10–28 days [7]. It is therefore evident, as far as the problems of the initial rigidity of biodegradable materials is concerned, that control of the degradation speed and the histological reaction induced by degradation products have not yet been solved, and this material can hardly meet the requirements of the internal fixation of long-bone fractures.

The Study of Stress-relaxation Bone Plates

In the late 1980s, Korvick et al. [8] and Tomita et al. [9], in animal experiments, placed a silicone rubber cushion between the bone plate and the bone surface to lessen the effect of stress shielding on fracture healing, and put forward the concept of cushioned plate fixation. Jasmine et al. [10] in their studies also used a layer of cushion made of polyethylene under the bone plate to relax the screws, and the cushion reduced the stress-shielding effect of the fixation system. Unfortunately, the abovementioned studies did not investigate the change of stiffness and the method of control, and no further application has been reported. In 1994, Kostopoulos et al. [11] used 315LVM stainless steel to design and make six-hole, two-part sliding plates (SP), and applied them to the fixation of osteotomized radius in sheep with dynamic compression plates (DCP) as controls. Free weight bearing was allowed after the operation and the progress was observed for 2–24 weeks. At different postoperative stages, the dynamic modulus of elasticity and static ultimate bending strength of the callus were measured. The test results showed that the sliding plate, with its unique sliding mechanism, exerts stimulation of intermittent compressive load to the fracture site, thus promoting bone healing with callus of clearly better quality than that of the DCP group. This new design demonstrated the promising prospect of improvement in bone plate fixation devices. Further studies regarding the control of the extent of movement and the choice of duration and volume of weight bearing have yet to be carried out.

In the 1980s, The Ninth People's Hospital affiliated to Shanghai Second Medical University and the University of Iowa, USA cooperated in the design and making of a stress-relaxation bone plate system (Figure 22.1), which incorporates a traditional rigid bone plate with viscoelastic washers capable of creeping in the screw holes. Animal experiments and in vitro observations [12,13,14] found that with the action of constant loading, creep and fatigue failure of the viscoelastic washers gradually took place with the lapse of time, and this gradually lowered the general stiffness of the bone plate system, or its stress shielding rate, leading to apparently less underplate resorption and disorganized bone structure than those of the rigid plate group. The bone plates with or without viscoelastic washers had the same stress-shielding rate in the early stage, while the viscoelastic washers began to be creep and become damaged from the second week of fixation, and in 8–12 weeks, the damage became evident, resulting in a statistically significant lowering of the stress-shielding rate of the bone plate system (Figure 22.2).

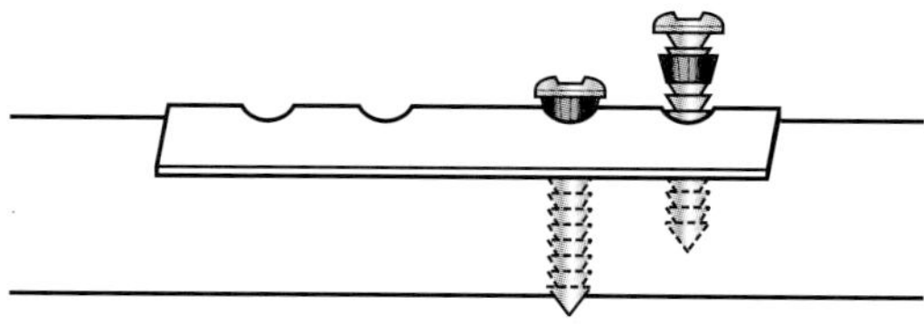

Figure 22.1. Viscoelastic washers placed between screws and plate to form a stress-relaxation plate system.

The results of experiments with New Zealand rabbits showed that in the early stage of bone plate internal fixation of the tibia following osteotomy, the stress-shielding rate was about 70% both in the stress-relaxation bone plate group with cushions (SRP) and the rigid bone plate group without cushions (RP). Then, the rate gradually decreased in the SRP group and was markedly lower than that in the RP group within eight weeks post-fixation. In 36~48 weeks after surgery, the rate in the SRP group remained stable at around 27%, while in the RP group it remained between 65% and 70% (Table 22.1).

Microscopic and transmission electron-microscopic observations showed that within two weeks after the operation, there emerged many functionally active osteoblasts and chondroblasts, expressing high levels of type I and type II collagen mRNA respectively (Figure 22.3). By four weeks after the operation, more and more osteoblasts appeared without chondroblasts; the level of expression of type I collagen mRNA also mounted to a peak. The active osteoclasts were also observed by then. Eight weeks after surgery, there were still many functionally active osteoblasts, osteoclasts, and expression of type I collagen mRNA. In week 24 post-operation, normal osteocytes were found in the matrix of the healing site, while osteoclasts were seldom present. However, in the RP group two and four weeks post-operatively, the number and functional activity of osteoblasts were already behind the SRP group, and in eight weeks the function of osteoclasts was active, accompanied by osteocytic osteolysis, which lasted until 24 weeks after surgery, indicating

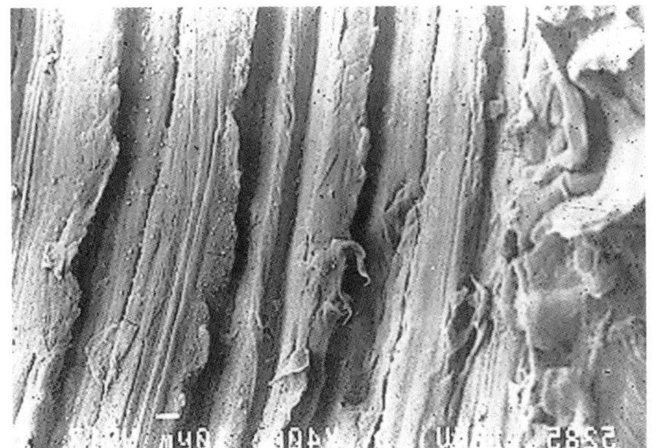

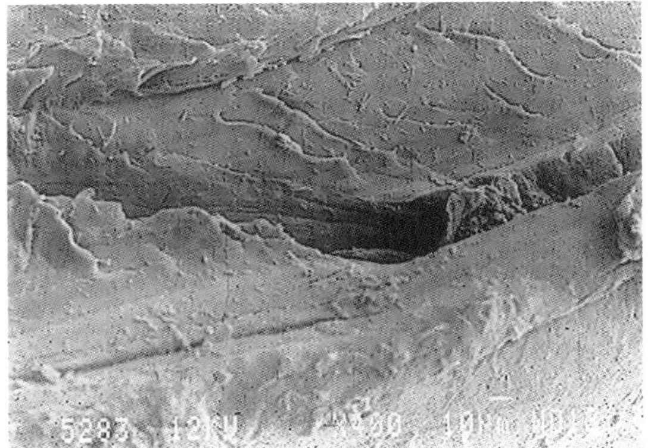

Figure 22.2. Viscoelastic washers seen under scanning electron microscope. **a** Preimplantation, the surface of the washer was smooth and even and regular traces of machining were visible (original magnification 400 ×). **b** Four weeks post-implantation, the washer is deformed and split (original magnification 400 ×). **c** Twelve weeks post-implantation, the washer is largely deformed and the surface is split into layers (original magnification 400 ×).

Table 22.1. Comparison of the stress-shielding rate of two kinds of plates ($x \pm s$, %)

Time (week)	SRP	RP	P Value
0	70.3 ± 2.2	71.4 ± 2.9	>0.05
2	68.7 ± 5.2	70.8 ± 4.3	>0.05
4	64.6 ± 4.8	68.1 ± 4.6	>0.05
8	54.9 ± 3.7	67.3 ± 3.8	<0.05
12	45.9 ± 4.8	68.4 ± 5.0	<0.05
24	31.8 ± 3.2	66.5 ± 4.8	<0.01
36	28.3 ± 2.9	65.7 ± 3.3	<0.01
48	27.2 ± 1.7	65.3 ± 3.0	<0.01

the prevailing resorption process in the remodeling stage. In addition, there was only the expression of type I collagen mRNA during the different periods after the operation, with levels of expression lower than that in the SRP group. Stress relaxation of SRP stimulated the differentiation of mesenchymal cells into osteoblasts and chondroblasts with high-level expression of type I and II collagen mRNA respectively, and benefited callus remodeling.

In the early stage of internal fixation, the changes in callus and in cortex structure beneath the plate were similar between the SRP group and the RP group; the collagen fibers in the callus were oriented in a disorderly pattern, and the cortical bone under the plate became thinner and resorption cavities could be seen there. In week 12 post-operatively, the callus in the SRP group started to transform to lamellar bone and the resorption cavities reduced in size; the osteoblasts lying on the surface of the cavities expressed and synthesized type I collagen mRNA and secreted type I collagen (Figure 22.4), of which the array approached a regular pattern. In week 36 the resorption cavities virtually vanished (Figure 22.5A), the cortical bone became thicker, and the CT values were near normal. A different picture was seen in the RP group, where in week 12 the callus and resorption cavities in the cortex under the plate continued to grow in number and size, and finally turned into cancellization. The expression and synthesis of the collagen gene had always been rare and the cortical bone became thinner, accompanied by lower CT values. Bone resorption was still evident in week 24~36, and the cortical bone was cancellated (Figure 22.5B).

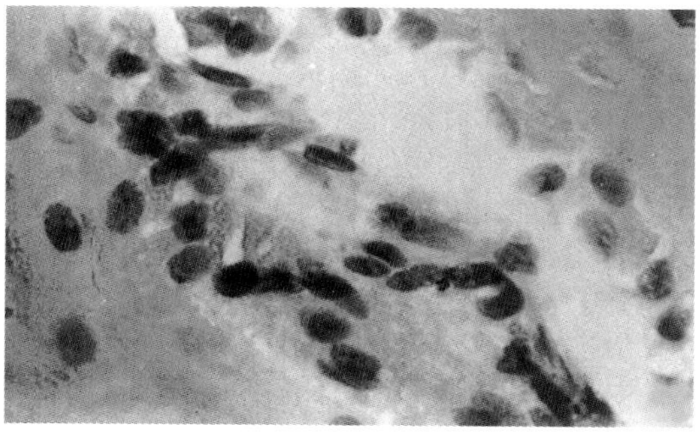

Figure 22.3. Two weeks after fixation of stress-relaxation plate, osteoblast lying on the surface of trabecular expressing collagen type I mRNA. In situ hybridization × 400.

The three-point bending test of tibia specimens with the bone plate removed showed results in favor of the SRP group, of which the bending strength was 16.7% of normal in week two and rose to 93.6% in week 48. However, the improvement in bending strength in the RP group was rather slow: it was markedly lower than that in the SRP group in week eight ($p < 0.05$), and only 58% in week 48 ($p < 0.01$) (Table 22.2). These findings demonstrate that with the presence of the stress-relaxation bone plate system, the gradual lowering of the stress-shielding rate could lead to the recovery of the morphology, structure, and mechanical properties of the bone (once porous) under the plate.

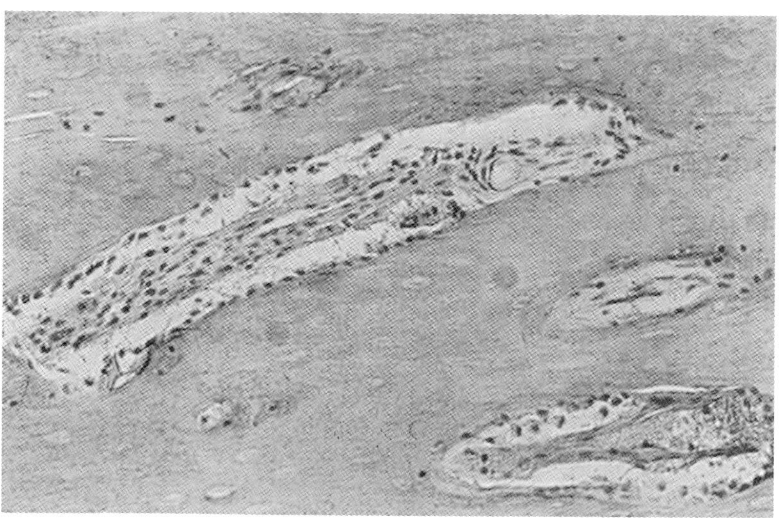

Figure 22.4. Twelve weeks after fixation of stress-relaxation plate, the absorption cavities were surrounded by the positively stained collagen type I matrix. Immunohistochemistry × 100.

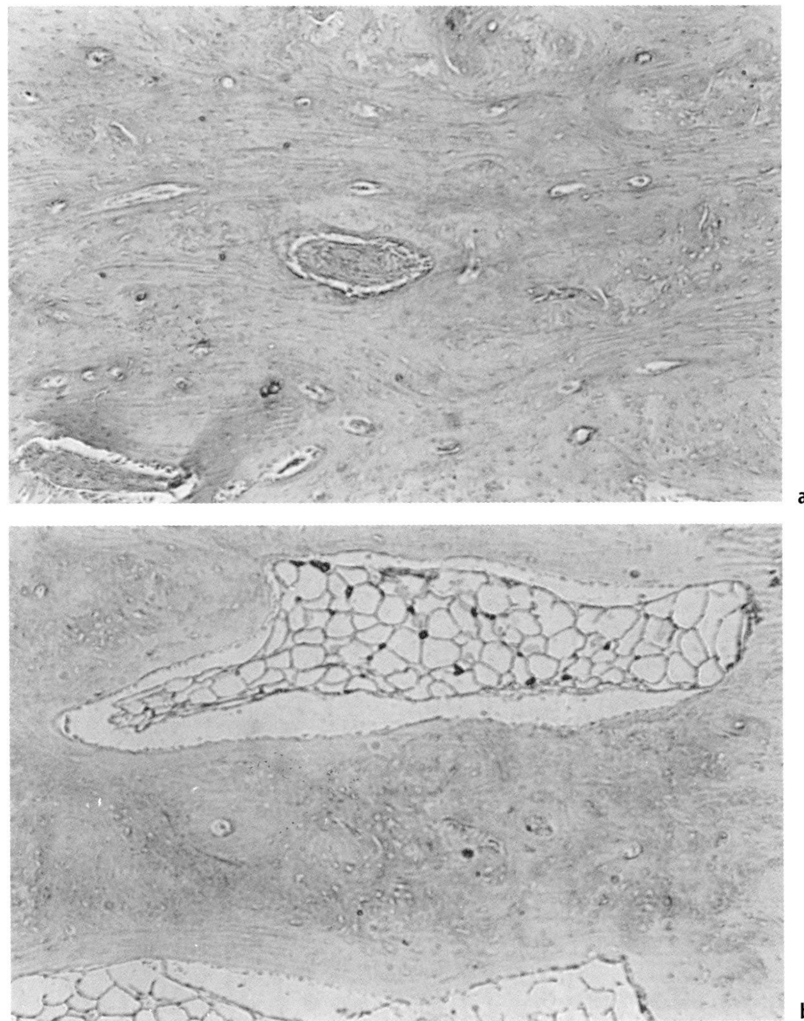

Figure 22.5 **a** 24–36 weeks after fixation of stress-relaxation plate, the absorption cavities were decreased in size and filled and repaired by bone tissue. Immunohistochemistry × 40. **b** 24–36 weeks after fixation of stress-relaxation plate, the absorption cavities were still evident and no repairing process could be found. Immunohistochemistry × 40.

Table 22.2. Comparison of bending strength of bone under two kinds of plates (x ± s, %)

Time (week)	SRP	RP	p Value
2	16.70 ± 1.23	16.38 ± 1.73	>0.05
4	45.24 ± 4.42	41.24 ± 5.71	>0.05
8	58.90 ± 7.50	47.39 ± 4.80	<0.05
12	78.69 ± 6.77	64.46 ± 7.19	<0.05
24	88.38 ± 9.36	60.00 ± 8.25	<0.05
36	89.48 ± 6.71	58.23 ± 5.91	<0.01
48	93.63 ± 5.24	57.95 ± 4.63	<0.01

In the RP group, however, the case continued to aggravate, and the porous change of the callus and the bone under the plate would be irreversible unless the bone plate was removed [3].

Microangiography and vascular volume determinations revealed that both osteotomy and bone plate internal fixation could lead to regional ischemia in cortical bone[17]. In week 2~4, the dilation of blood vessels arising from the medullary side could be seen in the tibia in both groups; these vessels invaded into the ischemic zone, which largely disappeared in week 8~12. However, a difference was observed after 16 weeks. The vascular volume in the SRP group did not change further, while the vascular dilation in the RP group proceeded, even with the formation of blood sinuses, causing the volume to further increase. The result was a statistically significant difference as compared with the SRP group. Obviously, the impairment to blood circulation in the early stage was relevant to trauma and surgery, while in the later stage the difference between the two groups was attributable to the increasing difference in the stress-shielding rate of the different bone plates employed in the two groups.

The stress-shielding rate of the stress-relaxation bone plate was similar to that of the rigid bone plate in the early stage of fixation, acquiring the same stability of the fracture site in 8~12 weeks. The subsequent lowering of the stress-shielding rate was of benefit to the reversal of osteoporosis in the fixed segment and to the remodeling of callus. The speed of attenuation of the bone plate stress-shielding effect could be adjusted by the thickness and number of viscoelastic cushions. By adding a certain number of viscoelastic cushions of various thicknesses to the traditional bone plate systems, the total rigidity of the system could be gradually attenuated, without any alteration in the design of the bone plate itself, and without any additional surgical time and trauma. With negligible addition to the cost of surgery, the demand for firm fixation in the early days of fracture and for ample stress stimulation in the later stage could be met. Therefore, it is believed that the gradual decrease of rigidity brought about by the stress-relaxation bone plate system led to the gradual increase of mechanical strength at the fracture site almost synchronously, and facilitated the building up of an optimal and gradually changing mechanical environment in the fracture area, effectively preventing the onset of osteoporosis of bone plate origin and re-fracture.

Selective Stress Shielding

Selective stress shielding is another approach to solve the problem of osteoporosis of stress-shielding origin. The postulated mechanism is that the selective stress-shielding effect counters bending, torsion, and shear stress adequately, but counters compressive stress fractionally. As the latter can stimulate the remodeling of callus, the mechanical properties of the fracture site can be improved. The intramedullary nail is a striking example. But it further impairs the blood supply in the medullary cavity and its torsional strength has yet to be improved. In this respect, the shape memory alloy saw-tooth embracing fixator we developed can basically satisfy this requirement.

The embracing fixator is made of Ni–Ti shape memory alloy, the strength of which is adequate to meet the requirements for internal fixation of a long bone shaft, and the elastic modulus at 37 °C is only 54.18 Gpa (200 Gpa for 316L stainless steel). Its shape memory effect makes the placement of the fixator during operation quite simple and ensures the maintenance of saw-tooth arms firmly embracing the fracture segment after surgery.

The deformation temperature of the fixator is 4 °C and the recovery temperature 37 °C. For the operation, a fixator of the appropriate size and shape is chosen and immersed in disinfected ice water to reduce the temperature and to expand and open the arms. After reduction of the fracture, the fixator is implanted with the body of the fixator on the tension side, then a wet compress with hot saline gauze is applied to raise the temperature. With the shape memory effect, the saw-tooth arms close, and by the recovery force, exerts a firm grasp on the distal and proximal ends of the fracture.

In vitro mechanical analyses and in vivo animal experiments have been done to evaluate the embracing fixator [19]. In vitro experiments found that the difference in bending strength of the embracing fixator and the bone plate is not significant, but the compressive stress-shielding rate of the embracing fixator is markedly lower than that of the bone plate. While the torsion strength of the fixator is significantly higher than that of the intramedullary nail, no harm is done to the medullary blood vessels. Torsion experiments showed that the yielding torsional moment of the fixator is markedly higher than that of the intramedullary nail group, the ratio of the averages being 6.85 : 1; but lower than that of the bone plate group, the ratio of the averages being 1 : 1.28, not much different. Animal experiments demonstrated that when fixed with an embracing fixator, the fracture shows satisfactory secondary union with the formation of a certain amount of outer callus. In addition, the disorganization of collagen fiber and the degree of bone resorption are lower than in the rigid bone plate group. The shape memory saw-tooth embracing fixator has been put to clinical use to treat fractures of the shaft of the femur, humerus, tibia, and other long bones, and is particularly good for the fixation of periprosthetic fractures of the femur, the result of fixation being definite and reliable, and the placement and removal quite easy [20].

As the keynote of this book is devoted to the discussion of biomechanical and biological material problems in orthopedics and traumatology, this chapter directs our attention to the effect of the mechanical characteristics of internal fixation on the treatment of fracture. The author would like to emphasize that the blood supply in the fracture area is the basic guarantee for the healing of fracture. When the blood supply in the fracture area is good, the mechanical environment of the fracture area will be of great importance to the quality of fracture healing and the mechanical properties of the bones. In the treatment of fracture, blood supply and mechanical conditions should complement each other. Only if both of them are managed can good results be achieved.

References

1. Chen YQ, Dai KR, Qiu SJ, Zhu ZA. Bone remodeling after internal fixation with different stiffness plates: Ultrastructural investigation. Chinese Med J 1994;107:766–70.
2. Wang Y, Dai KR. Effect of internal fixation plates on microcirculation in under-plate cortical bones: microangiography and scanning electron microscopy. Chinese Med J 1994;107:929–33.
3. Zhu ZA, Dai KR, Qiu SJ. Repair of regional osteoporosis after removal of rigid fixing plate: an experimental investigation. Chinese Med J 1994;107:364–7.
4. O'Sullivan ME, Chao EY, Kelly PJ. The effects of fixation on fracture healing. J Bone Joint Surg 1989;71(A):306–10.
5. Zimmerman M. The design and analysis of a laminated partially degradable composite bone plate for fracture fixation. J Biomed Res 1987;21(Suppl):345–61.
6. Hanafusa S, Matsusue Y, Yasunaga T, O Ka M, Shi Kinami Y, I Kada Y. Biodegradable plate fixation of rabbit femoral shaft osteostomies. Clin Orthop 1995; 315:262–71.
7. Strycker ML. Biodegradable internal fixation. J Foot Ankle Surg 1995;34:82–9.
8. Korvick DL, Newbrey JW, Bagby GW, Pettit GD, Lincoln JD. Stress shielding reduced by a silicon plate bone interface. Acta Orthop Scand 1989;60:611–16.
9. Tomita N, Kutsuna T. Experimental studies on the use of a cushioned plate for internal fixation. Int Orthop 1987;11:135–9.
10. Jasmine MS, Dahners LE, Gilbert JA. Reduction of stress shielding beneath a bone plate by use of a polymeric underplate. Clin Orthop 1989;246:293–9.
11. Kostopoulos V, Vellios L, Fortis AP, Panagiotopoulos E, Milis Z, Lambiris E. Comparative study of callus performance achieved by rigid and sliding plate osteosynthesis based upon dynamic mechanical analysis. J Med Eng Technol 1994;18:61–6.
12. Tang TT, Dai KR, Xue WD. Mechanical analysis of reaction of changing-stiffness plate to cyclic vertical compressive loading. J Appl Biomech 1996;11(1):42–6.
13. Dai M, Dai KR, Qiu SJ. The effects of stress-relaxation plate on bone remodeling: an experimental study. Chinese J Surg 1995;33:698–700.
14. Dai KR, Dai Min, Wang KY, Xue WD. The influence of stress-relaxation plate on the geometry configuration and mechanical property of bone: an experimental study. Chinese J Med 1995;75:414–16.
15. Zhang XL, Dai KR, Tang TT. The influence of stress-relaxation plate on collagen gene expression and cellular ultrastructure of fracture healing. Chinese J Orthop 2000;20(6):362–5.
16. Zhang XL, Dai KR, Tang TT. Effects of stress-relaxation plate on the disorganization and reparation of regional bone structure. Acta Univ Med Second Shanghai 2000; 20(6):488–90.
17. Dai M, Dai KR. An experimental study of the effect of stress-relaxation plate fixation on cortical bone microcirculation. Chinese J Orthop 1998;18(8):484–7.
18. Zu XS, Dai KR, Wu XT, Xu XL, Xue WD. Investigation on stability of shape memory sawtooth-arm embracing internal fixator. J Appl Biomech 1995;10(2):40–6.
19. Dai KR, Wu XT, Zu XS. An investigation of the selective stress-shielding effect of shape-memory sawtooth-arm embracing fixator. Mater Sci Forum 2002; 394-395:17–24.
20. Dai KR, Ni C, Wu XT, Qiu SJ, Xu XL, Zhu XS. An experimental study and preliminary clinical report of shape-memory sawtooth-arm embracing internal fixator. Chin J Surg 1994;32:629–32.

II – Tissue Biomechanics and Histomorphometry

II B – Pathology of the Cartilage and Articular Physiology

23 Cartilage Cells

P. Frayssinet, J. L. Jouve, and E. Viehweger

Cartilage is a specialized connective tissue consisting of a particular extracellular matrix which encases the cells or chondrocytes which synthesize it. It has a remarkably low cell density and is unable to heal when subjected to a lesion.

The cartilage cells synthesize the extracellular matrix (ECM) and organize its pattern. They originate from a common mesenchymal stem cell with osteoblasts, adipocytes, fibroblasts, or pericytes (Figure 23.1). Different kinds of cartilage can be evidenced during development. The primary cartilage forms the template of the long bones. The secondary cartilage can be divided into hyaline, elastic, and fibrocartilage.

The Cells of the Embryo Cartilage

The cells of the primary cartilage are derived from the mesoderm in the case of the appendicular and spinal cartilage and from the neural crest for most of the bones of the head except for some which are of mixed origin [1].

The first stage is the migration of cells to the site of future skeletogenesis. The second is the tissue (epithelium-mesenchymal) interaction that results in cell condensation [2].

Condensation Zone

This takes place in the mesenchyme when a previously dispersed population of cells forms aggregates. Patterning within condensations can be complex, forming more than one cartilage anlage or bone element. Condensations also involve elevated levels of extracellular matrix or cell surface molecules such as hyaladherins, versican, tenascin, syndecan, N-CAM (neural-cell adhesion), Hedgehog proteins, heparan, and chondroitin sulfate [2,3].

Most of these non-tissue-specific molecules are stage specific and disappear in the next step of cartilage differentiation. Versican is a hyaluronic acid-binding chondroitin sulfate proteoglycan [4]. Syndecan is a cell surface proteoglycan receptor. Syndecan binds to tenascin which is a large glycoprotein binding to chondroitin sulfate proteoglycans [5,6]. N-CAM could mediate cell adhesion in prechondrogenic condensation [7].

Primary Cartilage

This results from the synthesis of a cartilaginous extracellular matrix by the condensed cells. The cells are first dispersed at random in the extracellular matrix. The anlage is surrounded by a perichondral membrane. The cells of the anlage then become organized in columns, after which a zone of hypertrophic cartilage appears in the center of the diaphysis. Most of the hypertrophic cells die by apoptosis. This zone is then invaded by a vascular connective tissue. Chondroclasts excavate a cavity in the chondral matrix and resorb the matrix between the hypertrophic cells. Mesenchymal cells then invade the cavity and synthesize an osteoid matrix at the surface of the cartilage matrix separating the previous hypertrophic cell columns. Two ossification fronts progressing from the diaphysis center toward the epiphysis are then organized (Figure 23.2). Five zones, resting, proliferating, maturating, hypertrophic, and ossification zones can be then evidenced at the front level. These five zones will be present in the subsequent growth plate once a second center of ossification has formed in the epiphysis (Figure 23.2).

The perichondral membrane is continued at the level of the epiphysis by a membrane which furnishes chondrocytes at its internal layer. The cells of the perichondral membrane directly form a layer of mineralized tissue at the mid diaphysis. This collar bone then progresses toward the epiphysis and enlarges by apposition of several bone layers while the inner part is resorbed. Osteoprotegerin transcripts are detectable in the areas of ossification in the fetal cartilage.

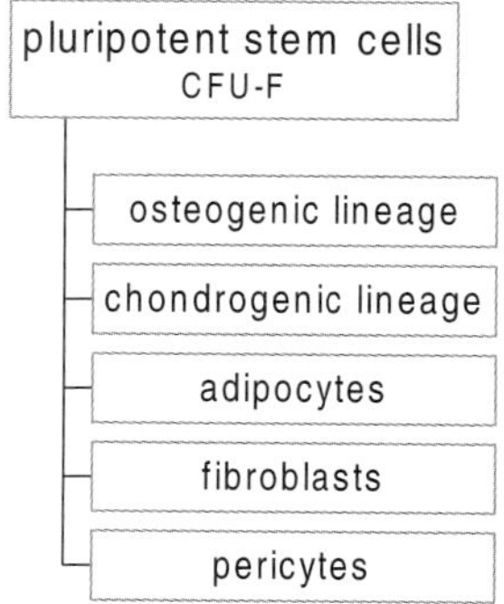

Figure 23.1. Origin of cartilage cells.

Growth Cartilage

The growth plate is a primary cartilage zone persisting when the rest of the primary cartilage has been ossified. This zone is located at the junction between the primary and secondary ossification centers. The function of this cartilage structure is to permit growth in bone length. Chondrocytic activity in the growth plate results in matrix calcification, and this calcified cartilaginous matrix becomes the scaffold for new bone formation by osteoprogenitor cells in the metaphysis. This structure is useful for the study of the chondrocyte differentiation as

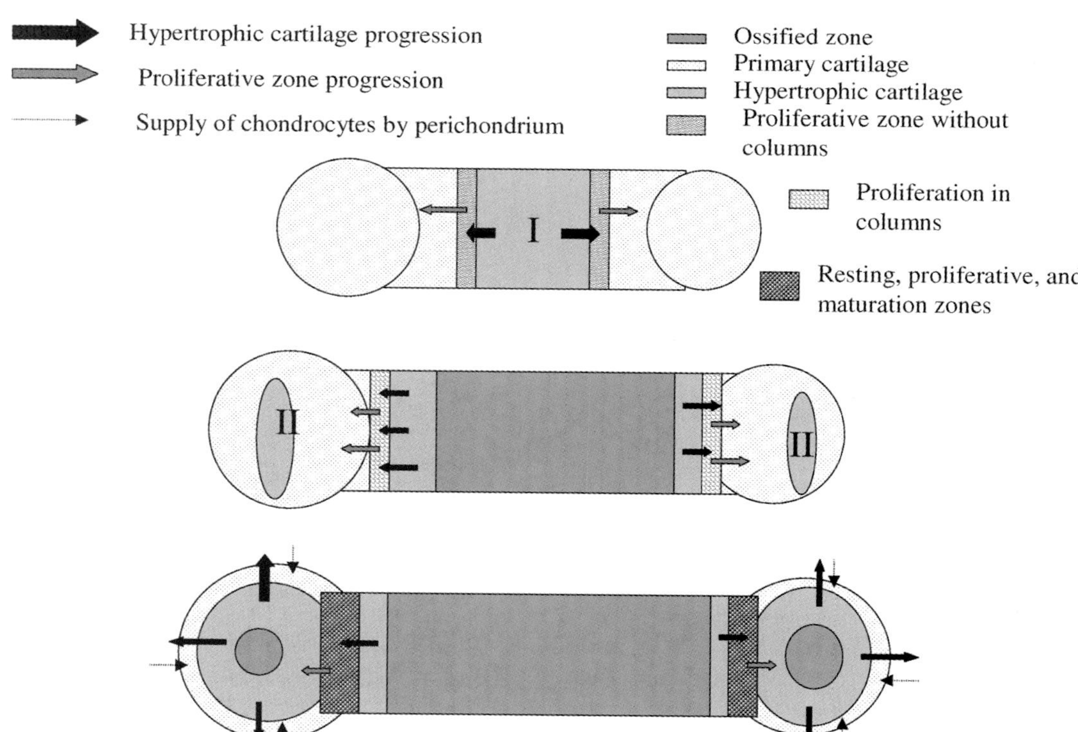

Figure 23.2. Primary cartilage and ossification.

all the stages are present and grouped in specialized zones depending on the differentiation stage. Five different zones can be evidenced between the epiphysis and the metaphysis. The first is the rest zone, in which the chondrocytes are uniformly dispersed. This layer is irregularly shaped and the cells proliferate supplying the following zone with chondrocytes. These cells have very long cell-cycle times compared to cells in the proliferative pool [8,9]. Cells with the characteristics of stem cells are contained in this zone. A subpopulation of cells forming a thin cell layer close to the proliferative zone which have different lectin staining abilities (Con A, RCA-I) are probably made of stem cells [10].

The proliferative cell zone contains cells, from the beginning of clonal expansion, until the cell exits the cell cycle and begins terminal differentiation. The number of cells in the proliferative zone is positively correlated with the rate of growth [11]. However, it is not clear whether all the cells divide or how many times. The number of cells involved in the proliferative zone differs over time and in different growth plates growing at different times.

The cells of the transition zone are spatially distal to the last chondrocytes that incorporate BrdU on pulse labels but spatially proximal to cells demonstrating the morphologic changes consistent with the rapid volume increase of cellular hypertrophy [12].

Buckwalter et al. and Hunziker et al. [13,14] have demonstrated that there is a strong correlation in the hypertrophic zone between cell hypertrophy and growth rate. Moreover, the height of the chondrocytes increases more than the width. Directed shape change accompanying volume increase is a major determinant of overall growth production. The fate of hypertrophic cells is unclear. There are many arguments indicating that most of these cells die by apoptosis. Other authors have suggested that some of them could transdifferentiate into bone-forming cells. It is possible that hypertrophic cells could undergo an asymmetric division, one cell being able to transdifferentiate, the other dying by apoptosis [15]. The hypertrophic cells in the lower part of the hypertrophic zone initiate a mineralization of the cartilage matrix (Figure 23.2).

In the ossifying zone, the mineralized cartilage is resorbed by osteoclasts forming columns of cartilage perpendicular to the growth plate. Bone-forming cells adsorb at the surface of these columns and synthesize a bone extracellular matrix. A vascular invasion occurs in the ossifying zone between the cartilage columns. A variety of angiogenic factors are expressed in the growth plate including members of FGF and TGF-β families, IGF-1, EGF, PDGF-A, Cyr61, and transferrin. Their importance and role is not known.

Hypertrophic chondrocytes synthesize VEGF and the administration of mFlt(1–3)-IgG which sequesters this protein and makes it inactive results in a considerable expansion of the hypertrophic zone and the arrest of length growth in postnatal mice [16]. The hypertrophic chondrocytes in the expanded hypertrophic zone are not apoptotic. VEGF, produced by hypertrophic chondrocytes, recruits endothelial cells which induce and maintain blood vessels. These blood vessels bring in chondroclasts, osteoblasts, and pro-apoptotic signals.

The hypertrophic chondrocytes of the growth plate also synthesize osteoprotegerin ligand (OPGL) which induces osteoclast differentiation and activates osteoclasts [17].

Articular Cartilage

This results from disappearance of the primary cartilage due to the progression of ossification in the secondary ossification center occurring in the physis. It can be compared to the growth plate cartilage. The hypertrophic and maturation zones are identical. There is no rest zone. Instead, a perichondral membrane furnishes cells to the underlying zones which do not show clearly oriented columns. In the adult cartilage, four zones from the articular surface to the subchondral bone can be evidenced [18].

The superficial zone consists of two layers: the first is a sheet of fine cell-free fibrils, the lamina splendens, and a deeper sheet of flattened chondrocytes synthesizing a matrix rich in collagen fibrils parallel to the joint surface, which provides great tensile strength and increases the resistance to shear forces.

Table 23.1. Noncollagenous proteins of the cartilage. The effect of the gene mutation is known in certain cases

Protein	MW	Human mutation phenotype	Presumed role
CDMP-1 (GDF5)		Acromesolic chondrodysplasia Chondrodysplasia, brachydactyly	Close to BMP 5 and 6, stimulates the production of proteoglycans
CD-RAP	12	Unknown	Unknown
Chondroadherin	36	Unknown	Promotes chondrocyte attachment to plastic dishes
Chondrocalcin	35	No specific defects but loss of C-terminal region by a frameshift mutation results in arthro-ophtalmopathy	Negative feedback on the biosynthesis of type II collagen
Chondromodulin I	18–24	Unknown	Potentialization of FGF-2 effect on chondrocytes, inhibition of proliferation of endothelial cells
Chondromodulin II	16	Unknown	Enhancement of chondrocyte and osteoblast proliferation, promotes osteoclast differentiation
CILP	91, 5	Unknown	Unknown
CLECSF1	18	Unknown	Unknown
CMP/matrilin 1	148	Unknown	Immobilization of chondroitin sulfate-containing proteoglycan fragments after their release by aggrecanase
COMP	320	Pseudoachondroplasia	
Pleitrophin	18	Unknown	Enhancement of neural outgrowth in developing neurons, stimulation of angiogenesis in tumors
PRELP	55–58	Unknown	Cartilage-specific gene regulation
PARP	24	Unknown	Unknown
Matrix Gla protein	8.5	Unknown	Maintenance of calcified matrix

The transitional zone, which is several times thicker, contains spheroid chondrocytes. In the middle (radial) zone, the spheroid chondrocytes align in columns. A thin basophilic line constitutes the interface between the uncalcified and the calcified matrix. The chondrocytes in the calcified zone separating the radial zone from the subchondral bone are smaller.

The Extracellular Matrix

Cartilage can be considered as aggregates of the chondroitin sulfate proteoglycans, aggregan, trapped within a mesh of collagen fibrils that are comprised of type II and low amounts of type IX and XI collagens. The structure and role of aggregans in joint tribology have already been largely described and are not the subject of this paper. Aggregans and aggregates of the epiphysis and physis have the same composition and structure as aggregans and aggregates from other cartilage. The size of aggregan chondroitin-6 sulfate chains increases between the reserve and hypertrophic zones of the growth plate as does the proportion of chondroitin-6 sulfate disaccharides [19].

Large, intact aggregates may inhibit cartilage matrix mineralization in the growth plate [20,21]. Their enzymatic degradation can suppress this protective role [22]. There is electron microscopy evidence that aggregate size decreases in the hypertrophic zone [23].

Cartilage also contains nonspecific and specific noncollagenous molecules [24] (Table 23.1). The nonspecific molecules include members of the TGF-β and FGF families. The specific protein and glycoprotein contents of the cartilage vary with time and location. Their role is often unknown and not unique. For example, matrix GLA protein, a γ-carboxy glutamic acid-rich, vitamin K-dependent and apatite-binding protein, is a regulator of hypertrophic cartilage mineralization during development and can block endochondral and intramembranous ossification in the chick limb [25].

The extracellular matrix can be divided into two different sectors. The pericellular matrix

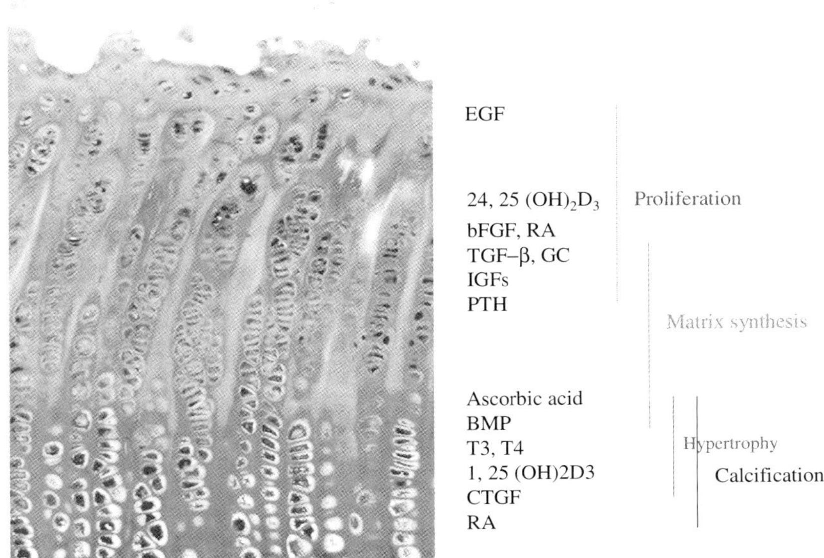

Figure 23.3. Paracrine, autocrine, and systemic factors active on the chondrocytes at different stages of their maturation.

surrounding the chondroblast cavity is rich in proteoglycans, noncollagenous proteins (anchorin CII), and nonfibrillar collagens (type-IV collagen) [26]. The chondrocyte, together with this pericellular region and a surrounding capsule, is termed the chondron. Chondrons are rich in proteoglycan and type II, VI, and IX collagen [27,28].

Studies of isolated chondrocytes with their surrounding matrix showed that application of a stress to the chondrocyte in situ produced large changes in shape and cross-sectional area and modifications in collagen orientation which recovered after removal of the compression [29,30].

The interterritorial region exhibits a more parallel orientation of the collagen fibrils of largest diameter.

Autocrine and Paracrine Factors Active on the Cartilage Cells (Figure 23.3)

The extracellular matrix of the cartilage contains many autocrine and paracrine factors, most of which are growth factors. TGF-β has been found to be a potent mitogen for growth plate chondrocytes [31]. The mitogenic response increases with the chondrocyte maturation [32]. In nonhypertrophic chondrocytes, the dose–response curves are biphasic at high dosage. TGF-β1 and β2 also stimulate proteoglycan synthesis. In vivo, FGF2 inhibits longitudinal bone growth [33]. FGF2 stimulates proliferation in the perichondrium and inhibits it in the proliferative, epiphyseal, and hypertrophic chondrocytes [33]. FGF2 is synergistically mitogenic with TGF-β [34,35]. FGF2 is the most potent stimulator of TGF-β secretion by growth plate chondrocytes. TGF-β also regulates PTHrP expression in chondrocytes from growth plate or epiphysis of chicken embryo [36].

BMPs stimulate growth plate chondrocyte maturation. BMP6 has a major role in this process. It is expressed by sternal chondrocytes prior to type X collagen expression. Isolated growth plate chondrocytes committed to undergo maturation express BMP6 [37]. BMP7 can reverse the suppressive effect of PTHrP on maturation and is localized to human hypertrophic cartilage. BMP7 was shown to regulate chondrocyte maturation in the growth plate. Regulation of BMP7 by retinoic acid may be important in normal growth [38]. BMP2 acts on pre- and post-condensation chondrogenesis. BMP2 promotes chondrogenesis and leads to the accumulation of extracellular matrix [39]. BMP4 is present in condensing mesenchyme of limb buds [40]. BMP5 deficiency prevents age-related decelerations in chondrocytic proliferation and the initiation of hypertrophic differentiation [41]. GDF5, which belongs to a new subgroup of the BMP family, stimulates primary cartilage development. The response is stage dependent [42].

Growth plates are sensitive to vitamin D [43]. 24,25-$(OH)_2D_3$ affects less mature cells, particularly those in the resting zone [44]. 1,25-$(OH)_2D_3$ affects activity in the maturation and upper hypertrophic cartilage [44]. Receptors are present for both vitamin D. Cartilage cells can synthesize their own 25-$(OH)_2D_3$ [45]. Growth plate cells of different species respond to 1,25-$(OH)_2D_3$ by going further in maturation. TNF-α can induce apoptosis of chondrocytes in vitro and inhibits the differentiation of stem cells into chondrocytes [46].

Midkine is a heparin-binding growth/differentiation which is expressed by hypertrophic chondrocytes. It increases the synthesis of sulfated glycosaminoglycans, aggregan, and type II collagen when transfected into ATDC5 chondrogenic cells [47]. Connective tissue growth factor (CTGF) is a member of an emerging family of extracellular proteins known as the CNN family, characterized by significant sequence homology and the conservation of all 38 cysteine residues. Recombinant CTGF protein promotes the proliferation, maturation, and hypertrophy of cultured chondrocytes [48]. Moreover, rCTGF induces the adhesion, proliferation, and migration of vascular endothelial cells. Furthermore, it stimulates the proliferation and differentiation of cultured osteoblastic cells [49].

Systemic Factors

In the 1950s, Ray et al. [50] evidenced the role of thyroid and growth hormones in the regulation of longitudinal bone growth and maturation in the proximal rat tibia. It was suggested that thyroid hormone increased maturation of the secondary center of ossification. It was then demonstrated that, in vitro, the administration of thyroid hormone increased type X collagen secretion, alkaline phosphatase activity, and cellular hypertrophy [51–53], which are markers of growth plate cell maturation.

IGF-I has been shown to be capable of stimulating skeletal growth [54]. IGF-II was less active. IGF-I appears to mediate most of the actions of the growth hormone on the growth plate [55]. IGF-I and II exert a synergistically mitotic effect on the growth plate chondrocytes. IGF-I and FGF2 seem to have an additive effect on this mitotic activity and other factors such as EGF could modulate this action [56,57].

PTH and PTHrP show a mitogenic effect on growth plate chondrocytes, a selective suppression of type X collagen expression, and stimulation of proteoglycan synthesis [58–60]. PTHrP has been shown to be produced by various cell types such as periosteum and perichondrium.

Retinoic acid has been shown to be necessary for chondrocyte maturation [61].

Influence of Physical Factors on the Chondrocytes

Mechanical Stimulation

Insufficient or excessive mechanical loads may induce cartilage degeneration both on articular and growth cartilage. Mechanotransduction may be resolved into extracellular factors such as cell and matrix deformation, hydrostatic pressure, fluid flow, and streaming potential.

Although very different systems were used to perform the experiments on overloading of the growth plate, it seems that:

One early change in chondrocyte metabolism consists of a reduction in the rate of DNA synthesis [62].

This results in a decreased number of chondrocytes in the growth plate [63].

The extent of cellular hypertrophy may be reduced together with the number of hypertrophic cells [64].

The growth potential of the physis will be restored after removal of the presence of the load.

It seems, however, that the effects on cell metabolism depend on the magnitude of the load applied to the growth plate.

At the articular cartilage level, abnormally high physical loads initiate proteoglycan degradation, increase chondrocyte metabolism, and result in joint destruction. The chondrocyte reaction to hydrostatic pressure depends on culture conditions and testing regimens. Cyclic hydrostatic pressure (5 MPa) decreases sulfate incorporation in the chondrocyte monolayers when applied at 0.0167, 0.05, 0.25, and 0.5 Hz. It stimulates sulfate incorporation when applied to cartilage explants at 0.5 Hz. The duration of application also has an influence. Twenty seconds or five minutes of compression (5–15 MPa) stimulates sulfate and proline uptake in adult bovine cartilage. At 30 MPa, continuous but not cyclic hydrostatic pressure alters the chondrocyte Golgi apparatus and decreases proteoglycan synthesis [65]. Dynamic strain at 0.3, 1, or 3 Hz produced a significant reduction in nitrite (NO) production [66] by chondrocytes, NO being able to influence metalloproteinase activity, apoptosis, and cytoskeletal organization in these cells.

Electric and Magnetic Fields

The flow of ions through the extracellular matrix produced a number of electrokinetic phenomena. The endogenous electric current densities produced by mechanical loading under physiologic conditions were approximately 1 Hz and 0.1–1.0 mA/cm^2. Exogenous applied electric fields have been widely used to stimulate bone and cartilage formation in vivo. The response of the skeletal tissue to electric and magnetic stimulations may be influenced by several variables such as amplitude, frequency and exposure duration or intensity, and magnetic induction.

Electric stimulation can lead to the secretion of various growth factors by different cells, the TGF-β family is particularly involved. Electric stimulation can be obtained by three different techniques: direct current (DC), capacitive coupling (CC), and inductive coupling (IC). IC is interesting for biological systems because it does not require any physical contact. A current-carrying coil produces a time-varying magnetic field that induces a secondary electric field in the biological sample according to Faraday's laws. IC produces a temporal acceleration and quantitative increase in chondrogenesis. Subsequent studies have demonstrated that IC fields promote chondrogenic differentiation and cartilage maturation. These studies employed a very wide range of current characteristics: between 1 and 100 Hz and 1 mG and 1 G [67–69].

Using IC stimulation of rabbit chondrocytes at 15.4 Hz and 2 G, Hiraki et al. demonstrated an increase in sulfate incorporation, enhancement in cAMP levels, and ornithine decarboxylase activity [70]. Two studies related the exposure duration of a 60 kHz CC field to cAMP, proliferation and sulfate incorporation in bovine growth plate chondrocytes. Cell proliferation was evidenced with a field strength of 1.5 to 3.0×10^{-2} V/cm. Higher and lower fields did not produce stimulation [71]. Short-term exposure (up to 5 mn) increased cAMP levels fourfold over the control but longer exposure (10–20 mn) did not increase cAMP levels [72].

The effects of EMC fields on cartilage and endochondral bone formation are unclear despite the fact that these effects exist, probably because the characteristics of the applied EMC differed considerably in each experiment.

References

1. Hall BK, Miyake T. Divide, accumulate, differentiate: cell condensation in skeletal development revisited. Int J Dev Biol 1995;39:881–93.
2. Solursh M. Differentiation of cartilage and bone. Curr Opin Cell Biol 1989;1:989–94.
3. Iwamato M, Enomoto-Iwamoto M, Kurisu K. Actions of Hedgehog proteins on skeletal cells. Crit Rev Oral Biol Med 1999;10:477–86.
4. Yamagata M, Shinomura T, Kimata K. Tissue variation of two large chondroitin sulfate proteoglycans (PG-M.versican, PG-H.aggregan) in chick embryos. Anat Embryol 1993;187:433–44.
5. Bernfield M, Hinkes MT, Gallo RL. Developmental expression of the syndecans: possible function and regulation. Development 1993;(suppl):205–12.
6. Dunlop L-LT, Hall BK. Relationships between cellular condensation, preosteoblast formation and epithe-

lial-mesenchymal interactions in initiation of osteogenesis. Int J Dev Biol 1995;39:357–71.

7. Tavella S, Raffo P, Taccheti C, Cancedda R, Castagnola P. N-CAM and N-cadherin expression during in vitro chondrogenesis. Exp Cell Res 1994;215:354–62.
8. Farnum CE, Wilsman NJ. Determination of proliferative characteristics of growth plate chondrocytes by labeling with bromodeoxyuridine. Calcif Tissue Int 1993;52: 110–19.
9. Apte SS. Application of monoclonal antibody to bromodeoxyuridine to detect chondrocyte proliferation in growth plate cartilage in vivo. Med Sci Res 1988;116: 405–6.
10. Farnum CE, Wilzman NJ. In situ localization of lectin-binding glycoconjugates in the matrix of growth plate cartilage. Am J Anat 1986;176:65–82.
11. Farnum CE, Wilzman NJ. Growth plate cellular function. In: Buckwalter JA, Ehrlich MG, Sandell LJ, Trippel SB, editors. Skeletal Growth and Development. Clinical Issues and Basic Science Advances. Rosemont, CA: American Academy of Orthopedic Surgeons, 1998.
12. Hunter JK. Role of proteoglycan in the provisional calcification of cartilage:a review and reinterpretation. Clin Orthop 1991;262:256–80.
13. Buckwalter JA, Mower D, Ungar R, Schaeffer J, Ginsberg B. Morphometric analysis of chondrocyte hypertrophy. J Bone Joint Surg 1986;68A:243–55.
14. Hunziker EB, Schenk RK. Physiological mechanisms adopted by chondrocytes in regulating longitudinal bone growth in rats. J Physiol Lond 1989;414:55–71.
15. Roach HI, Erenpreise J, Aigner T. Osteogenic differentiation of hypertrophic chondrocytes involves asymmetric cell divisions and apoptosis. J Cell Biol 1995;131:483–94.
16. Gerber H-P, Vu TH, Ryan AM, Kowalski J, Werb Z, Ferrara N. VEGF couples hypertrophic cartilage remodeling, ossification and angiogenesis during endochondral bone formation. Nature Medicine 1999;5:623–8.
17. Lacey DL, Timms E, Tan H-L, Kelley MJ, Dunstan CR, Burgess T et al. Osteoprotegerin ligand is a cytokine that regulates osteoclast differentiation and activation. Cell 1998;93:165–76.
18. Buckwalter JA, Mankin HJ. Articular cartilage, part I: tissue design and chondrocyte-matrix interactions. J Bone Joint Surg 1997;79A(4):600–11.
19. Buckwalter JA. Epiphyseal and physeal proteoglycans. In: Buckwalter JA, Ehrlich MG, Sandell LJ, Trippel SB, editors. Skeletal Growth and Development. Clinical Issues and Basic Science Advances. Rosemont, CA: American Academy of Orthopedic Surgeons, 1998; 225–40.
20. Cuervo LA, Pita JC, Howell DS. Inhibition of calcium phosphate mineral growth by proteoglycan aggregate fractions in synthetic lymph. Calcif Tissue Res 1973; 13:1–10.
21. Hulth A. Experimental retardation of endochondral growth by papain. Acta Orthop Scand 1958;28:1–21.
22. Hjertquist SO, Westerborn O. The effect of papain on epiphysial cartilage in rachitic rats:histologic, autoradiographic, and microradiographic studies. Wirchows Archiv 1962;335:143–58.
23. Buckwalter JA. Proteoglycan structure in calcifying cartilage. Clin Orthop 1983;172:207–32.
24. Neame PJ, Tapp H, Azizan A. Noncollagenous, nonproteoglycan macromolecules of cartilage. Cell Mol Life Sci 1999;55:1327–40.
25. Yagami K, Suh J-Y, Enomoto-Iwamoto M, Koyama E, Abrams WR, Shapiro IM et al. Matrix GLA protein is a developmental regulator of chondrocyte mineralization and, when constitutively expressed, blocks endochondral and intramembranous ossification in the limb. J Cell Biol 1999;147:1097–108.
26. Benninghoff A. Form und bau der gelenkknorpel in irhen beziehungen zur funktion. In: Zellforsh, Z, editor. Zweiter Teil:der aufbau des gelenkknorpels in seinen beziehungen zur funktion. Vol 2 1925:783.
27. Poole CA, Ayad S, Schofield JR. Chondrons from articular cartilage: I. Immunolocalisation of type VI collagen in the pericellular capsule of isolated canine tibia chondrons. J Cell Sci 1988;90:635–43.
28. Poole CA, Flint MH, Beaumont BW. Chondrons extracted from canine tibial cartilage: preliminary report on their isolation and structure. J Orthop Res 1988;6:408–19.
29. Broom ND, Myers DB. A study of the structural response of wet hyaline cartilage to various loading situations. Connect Tissue Res 1980;7:227–37.
30. Freeman PM, Natarjan RN, Kimura JH, Andriacchi TP. Chondrocyte cells respond mechanically to compressive loads. J Orthop Res 1994;12:311–20.
31. O'Keefe RJ, Crabb ID, Puzas JE, Rosier RN. Effects of transforming growth factor-beta 1 and fibroblast growth factor on DNA synthesis in growth plate chondrocytes are enhanced by insulin-like growth factor-I. J Orthop Res 1994;12:299–310.
32. O'Keefe RJ, Rosier RN, Puzas JE. Differential expression of biological effects in maturationally distinct subpopulations of growth plate chondrocytes. Connect Tissue Res 1990;24:53–66.
33. Mancilla EE, De Luca F, Uyeda J, Czerwiec FS, Baron J. Effects of fibroblast growth factor-2 on longitudinal bone growth. Endocrinology 1998;139:2900–4.
34. Crab ID, O'Keefe RJ, Puzas JE, Rosier RN. Synergistic effect of transforming factor beta, and fibroblast growth factor on DNA synthesis in chick growth plate chondrocytes. J Bone Miner Res 1990;5:1105–12.
35. Wu LN, Genge BR, Ishikawa Y, Wuthier RE. Modulation of cultured chicken growth plate chondrocytes by transforming growth factor -beta 1 and basic fibroblast growth factor. J Cell Biochem 1992;49:181–98.
36. Pateder D, O'Keele RJ, Schwartz EM, D'Souza M, Reynolds PR, Puzas JE et al. TGFβ regulates PTHrP expression in chondrocytes. Transactions of the 46th Annual meeting of the Orthopedic Research Society. March 12–15, 2000. Orlando, p 37.
37. O'Keefe RJ, Grimsrud C, D'Souza M, Hicks DG, Puzas JE, Reynolds et al. Bone morphogenetic proteins commit growth plate chondrocytes to maturation. In: Buckwalter JA, Ehrlich MG, Sandell LJ, Trippel SB, editors. Skeletal Growth and Development. Clinical Issues and Basic Science Advances. Rosemont, CA: American Academy of Orthopedic Surgeons, 1998; 319–32.
38. Grimsrud CD, Rosier RN, Puzas JE, Reynolds PR, Reynolds SD, Hicks DG et al. Bone morphogenetic protein 7 in growth plate chondrocytes: regulation by retinoic acid is dependent on the stage of chondrocyte maturation. J Orthop Res 1998;16(2):247–55.
39. Roark EF, Greer K. Transforming growth factor-β and bone morphogenetic protein-2 act by distinct mechanisms to promote chick limb cartilage differentiation in vitro. Dev Dynamics 1994;200:103–16.

40. Jones CM, Lyons KM, Hogan BLM. Involvement of bone morphogenetic protein-4 (BMP-4) and Vgr-1 in morphogenesis and neurogenesis in the mouse. Development 1991;111:531–42.
41. Bailon-Plaza A, Lee AO, Veson EC, Farnum CE, van der Meulen MCH. BMP-5 deficiency alters chondrocytic activity in the mouse proximal tibial growth plate. Bone 1999;24:211–16.
42. Storm EE, Kingsley DM. GDF5 coordinates bone and joint formation during digit development. Developmental Biology 1999;209:11–27.
43. Boyan BD, Dean DD, Sylvia VL, Schwartz Z. Genomic and nongenomic regulation of cartilage by 1,25- and 24, 25- dihydroxy vitamin D_3. In: Buckwalter JA, Ehrlich MG, Sandell LJ, Trippel SB, editors. Skeletal Growth and Development. Clinical Issues and Basic Science Advances. Rosemont, CA: American Academy of Orthopedic Surgeons, 1998;33–58.
44. Takigawa M, Enomoto M, Shirai E, Nishii Y, Suzuki F. Differential effects of 1 alpha, 25-dihydroxycholecalciferol and 24R, 25 dihydroxycholecalciferol on the proliferation and the differentiated phenotype of rabbit costal chondrocytes in culture. Endocrinology 1988; 122:831–9.
45. Schwartz Z, Brooks B, Swain L, Del Toro F, Norman A, Boyan B. Production of 1, 25-dihydroxyvitamin D_3 and 24, 25-dihydroxy vitamin D_3 by growth zone and resting zone chondrocytes is dependent on cell maturation and is regulated by hormones and growth factors. Endocrinology 1992;130:2495–504.
46. Aizawa T, Kon T, Einhorm TA, Gerstenfeld L. Induction of apoptosis in chondrocytes in vitro by tumor necrosis factor-α. Transactions of the 46th Annual meeting of the Orthopedic Research Society. March 12–15, 2000. Orlando, p 39.
47. Ohta S, Murumatsu H, Senda T, Zou K, Iwata H, Muramatsu T. Midkine is expressed during repair of bone fracture and promotes chondrogenesis. J Bone Miner Res 1999;14:1132–44.
48. Takigawa M. Physiological roles of connective tissue growth factor (CTGF. Hcs24): promotion of endochondral ossification, angiogenesis and tissue remodeling. Tissue Engineering for Therapeutic Use 4. Proceedings of the Fourth International Symposium on Tissue Engineering for Therapeutic Use, Kyoto, Japan. 23–24th September 1999. Y. Ikada and Y. Shimizu. Anonymous. Amsterdam: Elsevier, 2000;1-13.
49. Nishida T, Nakanishi T, Asano M, Shimo T, Takigawa M. Effect of CTGF. Hsc24, a hypertrophic chondrocyte-specific gene product, on the proliferation and differentiation of osteoblastic cells in vitro. J Cell Physiol 2000;184:197–206.
50. Ray RD, Asling CW, Walker DG, Simpson ME, Li CH, Evans HM. Growth and differentiation of the skeleton in thyroidectomized–hypophysectomized rats treated with thyroxin, growth hormone and the combination. J Bone Joint Surg 1954;36A:94–103.
51. Ballock RT, Reddi AH. Thyroxine is the serum factor that regulates morphogenesis of columnar cartilage from isolated chondrocytes in chemically defined medium. J Cell Biol 1994;126:1311–18.
52. Burch WM, Lebovitch HE. Triiodothyronine stimulates maturation of porcine growth-plate cartilage in vitro. J Clin Invest 1982;70:496–504.
53. Ohlsson C, Nilsson A, Isaksson O, Bentham J, Lindahl A. Effects of tri-iodothyronine and insulin-like growth factor-I (IGF-I) on alkaline phosphatase activity, [3H] thymidine incorporation and IGF-I receptor mRNA in cultured rat epiphyseal chondrocytes. J Endocrinol 1992;135:115–23.
54. Schoenle E, Zapf J, Hauri C, Steiner T, Froesch ER. Comparison of in vivo effects of insulin-like growth factors I and II and of growth hormone in hypophysectomised rats. Acta Endocrinol (Copenh) 1985;108:167–74.
55. Salmon WD Jr, Daughaday WH. A hormonally controlled serum factor which stimulates sulfate incorporation by cartilage in vitro. J Lab Clin Med 1957;49: 825–36.
56. Hiraki Y, Inoue H, Kano Y, Fukuya M, Suzuki F. Combined effects of somatomedin-like growth factors with fibroblast growth factor or epidermal growth factor in DNA synthesis in rabbit chondrocytes. Mol Cell Biochem 1987;76:185–93.
57. Bonassar LJ, Trippel SB. Interaction of epidermal growth factor and insulin-like growth factor-I in the regulation of growth plate chondrocytes. Exp Cell Res 1997;234:1–6.
58. Lovey LS, Gelb D, Hurwitz SR, Puzas JEE, Rosier RN. Effects of parathyroid hormone-related peptide on chick growth plate chondrocytes. J Orthop Res 1993; 11:884–91.
59. Crabb ID, O'Keefe RJ, Puzas JE, Rosier RN. Differential effects of parathyroid hormone on chick growth plate and articular chondrocytes. Calcif Tissue Int 1992; 50:61–6.
60. Suda N, Shibata S, Yamazaki K, Kuroda T, Senior PV, Beck F et al. Parathyroid hormone-related protein regulates proliferation of condylar hypertrophic chondrocytes. J Bone Miner Res 1999;14:1838–47.
61. Koyama E, Golden EB, Kirsh T, Adams SL, Chandraratna RAS, Michaille J-J et al. Retinoid signaling is required for chondrocyte maturation and endochondral bone formation during limb skeletogenesis. Dev Biol 1999;208:375–91.
62. Ehrlich MG, Mankin HJ, Treadwell BV. Biochemical and physiological events during closure of the stapled distal femoral epiphyseal plate in rats. J Bone Joint Surg 1972;54A:309–22.
63. Farnum CE, Wilsman NJ, Nixon A, Belanger L. Chondrocytic response to pressure applied across the growth plate: uniaxial stapling of the rat proximal tibia. Trans Orthop Res Soc 1993;18:701.
64. Parkkinen JJ, Ikonen J, Lammi MJ, Laakkonen J, Tammi M, Helminen HJ. Effects of cyclic hydrostatic pressure on proteoglycan synthesis in cultured chondrocytes and articular cartilage explants. Arch Biochem Biophys 1993;300:458–65.
65. Parkkinen JJ, Lammi MMJ, Pelttari A, Helminen HJ, Tammi M, Virtanen I. Altered Golgi apparatus in hydrostatically loaded articular cartilage chondrocytes. Ann Rheum Dis 1993;52:192–8.
66. Lee DA, Frean SP, Lees P, Bader DL. Dynamic mechanical compression influences nitric oxide production by articular chondrocytes seeded in agarose. Biochem Biophys Res Comm 1998;251:580–5.
67. Aaron RK, Ciombor DM. Acceleration of experimental endochondral ossification by biophysical stimulation of the progenitor cell pool. J Orthop Res 1996;14: 582–9.
68. Archer CW, Ratcliffe NA. The effects of pulsed magnetic fields on bone and cartilage in vitro. Trans BRAGS 1981;1:1.

69. Delgado JM, Leal J, Monteagudo JL, Gracia MG. Embryological changes induced by weak, extremely low-frequency electromagnetic fields. J Anat 1982;134:533–51.
70. Hiraki Y, Endo N, Takigawa M, Asada A, Takahashi H, Suzuki F. Enhanced responsiveness to parathyroid hormone and induction of functional differentiation of cultured rabbit costal chondrocytes by a pulsed electromagnetic field. Biochem Biophys Acta 1987; 931:94–100.
71. Armstrong PF, Brighton CT, Star AM. Capacitively coupled electrical stimulation of bovine growth plate chondrocytes grown in pellet form. J Orthop Res 1988;6:265–71.
72. Brighton CT, Townsend PF. Increased cAMP production after short-term capacitively coupled stimulation in bovine growth plate chondrocytes. J Orthop Res 1988;6:552–8.

24 Articular Cartilage: Biomechanics, Injury, and Surgical Treatment of Defects

J. A. L. Hart and R. K. Miller

Articular cartilage (AC) has a bland macroscopic appearance that belies its complex structure. Loss of this tissue by early wear or traumatic injury can have severe health consequences for the patient and economic consequences for the community. In certain circumstances this tissue may be replaced. Before considering replacement of this tissue it is essential to have an understanding of normal cartilage structure and biomechanics and to evaluate the biomechanics of the joint into which the tissue is to be placed.

AC plays a vital role in the musculo-skeletal system, performing several different mechanical functions. It provides an almost frictionless surface in synovial joints [1]. It deforms under load, thereby increasing the contact area and reducing the contact stress that is applied to the cartilage itself and to the underlying bone. This deformation is associated with energy transfer and dissipates some of the energy associated with load transfer [2].

Cartilaginous tissue behaves as a fiber-reinforced composite material containing two distinct phases: a solid phase made up of two structural molecules (collagen and proteoglycan) and cells and a fluid phase containing water and solutes (ions and nutrients). The highly complex biomechanical behavior is determined by the elaborate construction of the extracullar matrix. With these solid and fluid elements the matrix behaves in a biphasic way and this concept is central to an understanding of cartilage biomechanics.

Structure and Function of Articular Cartilage

Composition of Cartilage

Normal AC consists of chondrocytes which account for approximately 10% of the wet weight of AC. These highly specialized metabolically active cells are responsible for the development and maintenance of the extracellular matrix (ECM). They have a limited ability to replicate and are unable to migrate to adjacent areas of cartilage as they are effectively trapped in the matrix they produce. The incapacity of the cells to migrate to damaged areas has important implications for the repair process of cartilage and this will be discussed later.

Collagen is the most abundant macromolecule in the ECM and accounts for 10–30% of its wet weight. The collagen, mostly type II, is dispersed throughout the ground substance in the ECM and is formed by three polypeptide chains that have strong covalent cross-links. Collagen fibers are thin and slender and therefore alone offer little resistance to compression forces. They are, however, very strong in tension and are responsible for providing the tensile properties of articular cartilage.

The orientation of the collagen fibers differs at different depths of AC. In the superficial zone the collagen fibers are densely packed and run parallel to the articular surface. In the intermediate zone the fibers are randomly orientated and in the deep layer the fibers are orientated perpendicular to the surface layer and anchor the matrix to the tidemark region and subchondral bone plate (SBP).

Proteoglycans make up the second largest group of macromolecules in the ECM and account for 5–10% of the wet weight. They consist of extended polysaccharide units called glycosaminoglycans which are attached by covalent bonds to a protein core. A globular region at the end of the protein core binds to hyaluronan, facilitated by a specialized link protein. Via this mechanism, large numbers of proteoglycans, known as aggrecans, aggregate to form large macromolecules with molecular weights of the order of 20,000. These large molecules become mechanically entrapped within the meshwork of collagen fibrils forming a strong porous-permeable composite. This represents the solid phase of cartilage structure.

Proteoglycans contain repeating sulfate and carboxylate groups along their chains, which become negatively charged when placed in an aqueous solution. This creates a high fixed charge density within the AC. The dense concentration of negatively charged proteoglycans exerts a large swelling pressure which is resisted by the tensile stress on the surrounding collagen network. The balance of expanding total swelling pressure exerted by the proteoglycans and the constraining tensile force within the collagen network determines the degree of hydration in cartilage.

Disruption of this balance by damage to either component causes an increase in tissue hydration and significantly alters the ability of the cartilage to bear load [3,4].

Biomechanical Behavior

AC consists of a fluid component within a solid porous-permeable matrix. This biphasic nature profoundly influences its mechanical behavior when loaded. When loaded, a pressure gradient acts on articular cartilage causing the fluid (water and ions) to flow through the solid, permeable matrix, generating a large frictional drag on the matrix [1,5–7]. The resulting transfer of load between the solid and liquid phases and pressurization of the interstitial fluid give rise to the predominant mechanical properties of AC.

AC is permeable and fluid may flow through it [8,9]. Permeability is a measure of the ability of fluid to flow through a porous-permeable material such as an ECM and is inversely proportional to the frictional drag exerted by the fluid [10]. The matrix of AC generates very large drag forces at very small flow speeds. The resisting drag forces associated with interstitial fluid flow appears to be the major mechanism for the load response of AC.

The permeability of AC varies in a non-linear way with the applied load (and thus the generated strain). Thus, as the cartilage is compressed under load, fluid extrudes from it and this reduces the pore size and permeability of the ECM. As the permeability of the ECM decreases the drag forces on fluid movement increase [10–13].

The overall result of this is to provide AC with a self-protective mechanical feedback mechanism which stiffens the cartilage by limiting rapid fluid flow in response to high and increasing load.

The fluid flow through the solid matrix of cartilage is responsible for its viscoelastic behavior, which has important implications for cartilage function. Viscoelasticity is defined as the time-dependent response of a material that has been subjected to a constant load or deformation. This behavior is exhibited as creep and stress relaxation. Creep occurs when a material undergoes constant loading. Typically, the material responds by initially deforming rapidly and then deforming more slowly over time. This time-dependent deformation continues until equilibrium is reached. When a viscoelastic material undergoes constant deformation it typically responds with high initial stress which progressively diminishes with time. This time-dependent stress response is known as stress relaxation.

AC has been shown to exhibit both creep and stress-relaxation behaviors which are due to fluid-flow-dependent and fluid-flow-independent mechanisms [1,2].

As AC is compressed it undergoes volumetric change, which causes a pressure gradient in the tissue and results in the flow of interstitial fluid. This flow through the porous ECM causes the generation of a significant frictional resistance within the tissue.

The biphasic nature of AC may be quantified by describing the three major forces which act to balance the externally applied load at the surface. The externally applied load at the surface will be balanced by 1) the stress developed within the solid phase, 2) the pressure developed within the fluid phase, and 3) the frictional drag due to fluid flow through the solid phase. The interaction among these three forces gives rise to the viscoelastic effects exhibited in AC. The significance of the creep response is that when the joint is loaded for a long period of time, the contact area will greatly increase with time, spreading the compressive load over an ever-increasing area. During this phase the cartilage interstitial fluid is extruded into the joint space, creating a natural circulation required for cell nourishment and providing a source of lubricant for fluid film lubrication.

An applied load is balanced, initially by the compressive stress of the ECM and the frictional drag of the extruding fluid; significant fluid pressurization transmits the load to the subchondral bone (SCB). Fluid extrudes from the cartilage and deformation increases, leading to increased stress developing in the solid matrix. When the stress within the matrix eventually balances the applied load both fluid flow and deformation cease and the tissue is in a state of mechanical equilibrium.

The compressive forces applied to AC generate significant tensile forces within the solid phase of that tissue. These tensile stresses are known as hoop stresses as they resemble the force applied by the hoops of a barrel to keep the barrel contained [14]. The network of collagen fibers in the ECM are the primary determinants of the tensile behavior of these tissues and therefore the orientation of the collagen network influences their tensile behavior. Typically, collagenous fibrous tissues show nonlinear tensile loading deformation. Initially, a small load causes a large deformation, probably as the collagen fibers uncrimp and assume a more uniform orientation. As the collagen fibers eventually become taut and assume a uniform orientation, they continue to absorb the tensile load, but in a linearly elastic way [15,16]. The orientation of the collagen fibers varies at different depths of AC and therefore the tensile properties will also vary at different depths.

Shear stress occurs when forces are applied parallel to the surfaces of a material. Although the predominant load on AC is compressive, significant shear stresses are developed within the tissue, particularly in the deep zone near the tidemark. These stresses can be particularly damaging at the junction of zones of different cartilage stiffness. For example, at the junction of the soft AC and the harder SCB these shear forces can be very high, leading to the interruption of this interface [17–19]. The overall stiffness of AC in shear is directly proportional to the amount of collagen in the tissue [20].

When AC is damaged there is an increase in its water content from the average normal of 60–85% to greater than 90% [21–23]. The increase in water content in turn has a great effect on the permeability, fluid flow, and thus the compressive stiffness of AC [7,24,25]. These changes reduce the ability of articular cartilage to bear load and to dissipate the energy produced by loading the joint. The increase in hydration when accompanied by damage to the collagen network and loss of the tensile stiffness of AC in the superficial zone is the initial step in an irreversible progression of cartilage degeneration. The swelling behavior of AC is caused by the interaction between its fixed charge density, caused by the fixed negative charges on the keratin sulfate and chondroiton sulfate molecules and the ionic constituents of the synovial fluid [11,26,27]. In order for the cartilage to remain electro-neutral, a large number of counterions must be present within the ECM.

The resulting ionic concentration in the articular cartilage is higher than in the surrounding synovial fluid. This excess of ion particles within the matrix creates a pressure referred to as the Donnan osmotic pressure, and fluid will flow into the tissue to maintain osmotic equilibrium [6,10,27].

Swelling of AC also arises from the repulsion between closely spaced, negatively charged sulfate and carboxyl groups fixed along the chondroitin sulfate and keratin sulfate groups. This effect, the chemical-expansion stress, also depends on the internal concentration of ions in solution floating around the proteoglycans, because these ions shield the charged groups from interacting with each other. The total swelling pressure in cartilage is therefore equal to the sum of the chemical expansion stress and the Donnan osmotic pressure [11,12,28].

Cartilage Wear and Injury

Diffuse Wear

Throughout a lifespan of many decades most diarthrodial joints are subjected to millions of load cycles and, despite this, show little or no evidence of wear [29,30]. Simple cyclical wear does not account for many instances of AC damage and other factors are likely to be operating. Cartilage possesses an effective mechanism to protect itself from cyclical loading and it is likely that the biphasic nature of cartilage is the centerpiece of this protective mechanism.

Obesity plays a role in the development of osteoarthritis as epidemiological studies indicate [31]. This may act simply by abnormally increasing the loads to which the joint is subjected. Alternatively, there may be a common biological mediator leading to obesity and joint degeneration although none has yet been identified. Immobility of joints can lead to osteoarthritis. Experimental immobilization of dog limbs has been shown to lead to compositional change, mechanical change, and thinning of articular cartilage [32].

Sport as a risk factor is controversial. The repetitive loading associated with activities such as running have not been associated with an increased incidence of osteoarthritis in the knee. The degenerative changes associated with particular sports such as soccer and football more likely relate to specific damage to the articular surface with or without associated ligament damage. These focal articular cartilage defects are dealt with in detail in the next section.

A further mechanism of accelerated wear is when damage occurs at or near the junction of cartilage and calcified cartilage–subchondral bone. Cracks in this area lead to rapid fluid exudate when the cartilage is loaded and therefore greatly increased stress in the solid matrix.

Focal Articular Lesions

Mankin [29] classified AC lesions as impact injuries and lacerations. Impact injuries involved the SCB and were dose dependant. Lacerations were superficial or deep. Deep lacerations involved the SCB and healed by a mesenchymal cell-generated, fibrous repair which ultimately broke down, whereas superficial lesions remained dormant, neither progressing nor undergoing repair. This latter concept was based on the experimental work of Meachim [33] in rabbits, where articular cartilage was incised. The clinical situation in humans is not analogous, however. Although lacerations do occur, they are usually iatrogenic. Traumatic lesions are usually abrasions not lacerations and fray at the margins, releasing AC fragments into the joint. Grande et al. [34] have demonstrated that in rabbits, superficial AC lesions do progress by shredding, inducing a synovitis eventually leading to the development of osteoarthritis by the production of lysomal enzymes by the inflamed synovium. AC fragments can induce AC breakdown directly [35]. Clinically, 50% of adolescents with unipolar, single-compartment AC lesions developed joint-space narrowing at 14 year follow-up [36] Osteochondritis dissecans of the medial femoral condyle results in moderate to severe arthritis in 32% [37]. Anterior cruciate ligament (ACL) deficient knees develop degenerative changes [38,39].

Focal lesions of articular cartilage in the knee joint arise from a number of different mechanisms (Table 24.1). Many of these lesions occur in young individuals and are quite distinct from isolated lesions seen in conjunction with osteoarthritis [40].

Table 24.1. Etiology of focal articular cartilage lesions

Direct Trauma	Secondary	Idiopathic
Shear	Patellofemoral disease	
	Acute	
	Chronic	
Impact	Ligamentous injuries	
	ACL, PCL, Collaterals	
	Acute	
	Chronic	
Intra-articular fractures	Meniscal tears	
	Lunge lesions	
	Osteochondritis dissecans	

The increased recognition of early lesions by arthroscopy and improved magnetic resonance imaging, together with the increased development and intensity of sports in the last decade, have led to a resurgence of interest in joint resurfacing techniques. Curl et al. [41], in a retrospective survey of 31,516 arthroscopies, found that 5% of patients under the age of 40 had unipolar lesions with exposed SC bone on the medial femoral condyle (MFC). Approximately 10% of acute hemarthroses are associated with full-thickness chondral defects [42].

Classification of AC Defects

Walker [43] reviewed the histological and macroscopic systems which had been developed for grading AC defects. Most of the histological systems are based on animal studies [44–47] and the macroscopic systems on human postmortem studies [48,49]. These studies are at best descriptive, semiquantatitive, and reliability is rarely given [43].

Mankin et al. [50] devised a histological system based on structure, cells, Safranin-O staining and tidemark integrity. They also introduced a scoring system with 0 as normal cartilage and 12–15 as severely disrupted. This is the best and most widely used of the histological grading systems, but its value is mainly as a research tool.

The most widely used classification system for the classification of AC defects in clinical practice is that described by Outerbridge [51]. He graded AC defects into four groups according to macroscopic appearance:

Grade 1	Blister
Grade 2	Fissures and clefts <1 cm^2
Grade 3	Fissures, clefts, and flaps >1 cm^2 extending to subchondral bone plate (SBP)
Grade 4	SBP exposed

This classification is very useful for treatment of AC defects, because it differentiates those lesions which are likely to require surgical treatment and which treatment is most appropriate, where size is a major consideration.

The International Cartilage Repair Society, founded in October, 1997, formed a working group at the First International Symposium in Freibourg, Switzerland to develop a consensus on a classification system for AC injuries and repairs.

A Cartilage Standard Evaluation Form/Knee [52,53] classes AC defects into four grades; Grade 1, (normal) to Grade 4, (severely abnormal). The following information is recorded:

Size, depth, location and condition of defect and opposing surface.

Pre and post debridement measurements.

Patient history including a patient self-assessment sheet.

Objective assessment by surgeon.

X-ray and MRI findings.

Harvest site pathology.

Evaluation of tissue repair based on macroscopic arthroscopic appearance:

- Degree of filling.
- Marginal integration.
- Macroscopic appearance tissue analysis.

Hopefully, this form will be used by those carrying out AC resurfacing and will allow comparison between different treatments.

Other Methods of Evaluation of AC

Although clinical and arthroscopic evaluations remain the mainstay of assessing AC repair, biomechanical testing and MRI are alternative techniques, which are less invasive. Indentometers can measure physical characteristics of AC and repairs, but are expensive and not totally reliable.

Ultrasound probes may be able to give more accurate information [54]. Development of new MRI techniques have greatly facilitated the ability to image AC and will aid in both the diagnosis of AC defects and evaluation of repairs. According to Bobic, it may soon be possible to distinguish between hyaline and fibrocartilage [54].

Treatment of Focal Articular Cartilage Defects

The lack of a blood supply, the inability of cells within a rigid matrix to move to the site of

Table 24.2. Current clinical repair techniques for focal AC defects

Bone Stimulation	Autologous Tissue Grafting	Allografting	Scaffolds
Drilling	Perichondrium	Osteochondral blocks	Synthetic
Open Abrasion	Periosteum	Osteochondral shells	Biological
Arthroplasty	Meniscus		
Arthroscopic Abrasion Arthroplasty	Autologous Chondrocyte Implantation (ACI)		Carbon Fiber
Microfracture	Osteochondral Autografting (OCG)		Seeded
	Bone Paste		

injury, and the limited potential of chondrocytes to proliferate greatly limits the process of repair [35,52]. William Hunter in 1743 [55] recognized that damaged AC had a poor capacity to heal.

Two fundamental steps are required in the treatment of focal AC defects: stabilization and repair. Stabilization, by removing loose fragments from the margins of defects using mechanical instruments or shavers, reduces the debris load in the joint and improves symptoms by reducing synovitis [56]. However, as Jackson [57] pointed out, the effects are short-lived and symptoms recur within two years.

Repair of the defect can be achieved utilizing one of two basic concepts; bone marrow stimulation or transplantation of autogenous or allogenic tissues. With bone marrow stimulation techniques the subchondral bone plate is breached, allowing undifferentiated mesenchymal cells to enter the defect and participate in the healing process. With transplantation of autogenous or allogenic tissue, well-differentiated chondrocytes are grafted into the area to participate in the healing process. However, the majority of these techniques except autologous chondrocyte implantation (ACI) breach the subchondral bone plate, allowing mesenchymal cells to also enter the area.

The currently clinically available techniques are listed in Table 24.2. Experimental techniques, many of which have not as yet been used in humans, are discussed later.

All repair techniques aim to produce a stable tissue where large aggrecan molecules can be synthesized long-term and are attached to a classic Type II collagen matrix and be integrated to the pre-existing matrix. The complexity of this tissue has been previously discussed and no technique has thus far been completely successful in achieving this goal [58].

Bone Stimulation Techniques

Drilling the SBP was first described by Pridie [59] in 1959. This allowed a fibrin clot to form in the defect which was later invaded by tissue containing mesenchymal cells, derived from the bone marrow. Penetration of the SBP was required [60]. Insall [61], reporting on 62 procedures performed by Pridie, found good results subjectively in 77% and objectively in 64%. In adult rabbits the defect was filled with tissue resembling fibrocartilage and the repair was never complete [62]. Spongioplasty [63], where the subchondral bone was abraded, and abrasion arthroplasty [64,65] are modifications of the original drilling technique and gave encouraging results in the short term. However, Altman et al. [66] concluded that in animals, the repair tissue was not ideal and unlikely to last, particularly with large defects and in older adults [65,67]. Gillogly et al. [68] stated that numerous reports with these techniques indicated that results deteriorated over a short period postoperatively. The results were better in patients less than 40 years of age [69].

Steadman [70] introduced the procedure of microfracture, which is another variant of the Pridie procedure. All abnormal AC is removed, including the calcified zone. The SBP is preserved but penetrated with an awl. He reported his experiences with 235 consecutive patients at the ICRS meeting in Boston, 1998 [54]. Follow-up was from two to 12 years. At seven years, 75% of patients had less pain and 20% had similar pain prior to surgery. No further surgery was required in 92%. Gill [54] presented a prospective series of 103 patients treated by microfacture at the same meeting. All patients were athletes and aged under 35; one third of the repairs were performed within three months of injury. Average follow-up was six years when the "defect score" was 75 compared to 45 preoperatively.

In 40 second-look arthroscopies, 50% had "normal" cartilaginous material, the remainder having variable amounts of regeneration. Results were better with small, acute, isolated lesions. Steadman claims that the results are better than other penetration procedures because the awl does not necrose bone and the microfracture augments the healing response. The strict postoperative program, emphasizing continuous passive motion (CPM) [71] and restricted weight bearing, allows tissue regeneration. Salter demonstrated that CPM augmented healing of small defects in rabbits [72].

This technique, if it can be shown to be superior to other penetration techniques, is an attractive option because it can be performed arthroscopically, is readily available, and is relatively inexpensive. It would, therefore, be particularly indicated in small, primary lesions in the young, seen in association with acute ligamentous injuries.

Tissue Grafting

A variety of tissues with chondrogenic potential have been used to repair AC defects and are listed in Table 24.2.

Periosteum has been used to repair AC defects in animals [73,74,75] and in a small number of human studies [3,76,77] where the results have been disappointing in the long term with 75% fair or poor results at eight-year follow-up for the MFC [78] and 64% with increased pain at nine years after patellar grafting [76,79]. O'Driscoll and Salter [74,75] demonstrated that healing of 3 mm defects on the rabbit trochlea was improved with the use of CPM. This has been supported clinically with periosteal grafting to the patella [80].

O'Driscoll reported his experience with a small number of human autologous periosteal grafts at the Annual AAOS meeting in 1997 [81,82]. He emphasized the importance of preserving the cambium layer by careful dissection. The subchondral bone plate was penetrated and the graft was sutured to the base of the defect with the cambium layer facing upwards into the joint. Only nine of 16 patients had satisfactory clinical results. However, Jaroma and Ritsila (1987) found experimentally that the orientation of the cambium layer made no difference [83].

Perichondrium as a source of chondrogenesic tissue is logical and reduces donor site morbidity. Numerous experimental studies have been performed, mainly in rabbits, and the ability to fill AC defects with tissue resembling hyaline cartilage and containing Type II collagen has been frequently demonstrated [4,84,85,87,88,96]. Coutts et al. [88] achieved effective chondrogenesis in 60% of defects in rabbit knees, but experienced problems with adherence of the graft to the base of the defect. They have not reported on use of their technique clinically. Homminga [89] implanted perichondrial grafts in 25 human knees with 30 defects. Arthroscopically, 90% of the lesions were filled at three to nine months but ossification occurred in 2/3 of the grafts by two years. Minas and Nehrer [90] also had failures due to ossification and delamination in 70%. Bouwmeester et al. reported only 43% effective results when perichondrium was implanted and fixed with fibrin [91].

The undoubted chondrogenic potential of periosteum and perichondrium, the low donor-site morbidity, the ability to cover large defects, the cost effectiveness, and the opportunity to add growth stimulants such as TGFβ1 [81] make the use of these tissues attractive. Periosteum seems to be more effective than perichondrium [81,92]. Results in humans have been variable, however. This may be due to graft damage during tissue handling [81,93,94], poor fixation of the graft [79,88], or ossification of the graft in the long term [89,90]. Sumen et al. have used meniscal pads experimentally [95] and clinically [96] and demonstrated chondrogenesis.

Brittberg et al. [97] pointed out that many of the experimental studies using periosteum and perichondrium have been carried out in small, immature animals. Breaching of the SBP in most studies makes interpretation of the results difficult and may be responsible for the ossification seen clinically. At present, despite the potential advantages, periosteal and perichondrial grafts do not seem to be an appropriate clinical solution.

Other techniques using meniscal fragments [93] and bone paste [94] as sources of cartilage cells have been described.

Autologous Chondrocyte Implantation (ACI)

AC chondrocytes were first isolated in tissue culture in 1965 [98]. A series of studies between 1971 and 1982, mainly in the United Kingdom, demonstrated that chondrocytes could be successfully cultured and transplanted into AC defects [99–102]. Aston and Bentley [103] implanted cultured allograft chondrocytes into AC defects in rabbit knees.

At one year, 64% were successful, the failures due to inadequate fixation rather than rejection. The problem of fixation was overcome by a group in Gothenberg who developed a technique of culturing autologous chondrocytes in rabbits and implanting them beneath a periosteal flap, sewn to the margins of the defect [104]. They found that one year after implantation into rabbit patellae, 82% of the defect was covered by cartilage-like tissue, compared to 19% coverage in control defects. They claimed that the matrix was largely type II collagen on the basis of polarized light microscopy [34]. Brienan et al. [105] were unable to confirm regeneration of articular cartilage beneath periosteal flaps in dogs without ACI. The Gothenberg group applied this technique to humans and published their results in 1994 [106]. The procedure is performed in three stages.

Cartilage tissue is removed from the knee at arthroscopy. Cells are isolated by mincing the AC and digesting the matrix with collegenase. The cells are then cultured in flasks containing autologous serum as a monolayer over a period of 14 to 21 days, resulting in a tenfold increase in cells. At a second open operation the defect is excised removing *all* damaged tissue, leaving intact vertical margins. A periosteal flap taken from the tibia or the femur is then sewn over the defect with the cambium layer facing into the defect and the margins sealed with fibrin sealant. After testing the integrity of the seal, the chondrocyte suspension is injected beneath the flap and the final sutures and sealant placed. Care is taken not to violate the SBP to prevent bleeding and migration of mesenchymal cells from the bone marrow. Twenty-three patients were treated with a mean age of 27 years with follow up from 1.5 to 5.5 years, with a mean of 3.25 years. Sixteen defects were on the MFC (three osteochondritis dissecans [OCD]) and seven on the patella. The results were assessed clinically and arthroscopically. At two years, 14 of the 16 patients with MFC lesions had good or excellent results clinically and at arthroscopy, the defects were filled with whitish tissue level with the articular surface, which was firm on probing. Biopsy of 15 of the 16 femoral grafts showed an intact surface in 11 with a hyaline appearance with positive immunological testing for Type II collagen. The results for the patella were not so good, with two good or excellent, three fair, and two poor, clinically. The arthroscopic and biopsy findings were correspondingly worse than the MFC lesions. In a subsequent presentation to the A.A.O.S. in Atlanta in 1996 [107], Peterson stated that he had treated 300 defects in 251 patients. At two to seven year clinical follow-up of 47 procedures in 44 patients, isolated MFC lesions (19) were improved in 84%, MFC lesions with ACL lesions (9) in 71%, and patellar lesions(16) in 63%. The comment was made that correction of patellofemoral joint biomechanics had improved the results for the patella. In a further presentation to the 2nd International ICRS Semi Annual Meeting [30] in 1998, Peterson reported on a larger series. Good or excellent clinical results were seen in the following sites in the knee: Isolated MFC 90%; MFC + ACL 74%; OCD 84%; Patella 69%; Trochlea 58%; Salvage/Multiple 75%.

He also presented a long-term follow up assessment of a subset of 38 patients, 31 of whom had good or excellent results at two years. At subsequent reviews, 5–10 years after surgery, 30 of the 31 still had good or excellent results, suggesting that the repairs were effective in the long term. Peterson et al. [108] recently published their results of the first 101 patients healed by ACI using their techniques. Ninety-four patients were reviewed with 2–9 year follow-up. The results were similar to previous reports. Graft failure occurred in 7%. Graft hypertrophy was seen in 26 of 53 arthroscopies, 7 of which were symptomatic. The symptoms resolved after trimming. Biopsy of 37 repairs showed type-II collagen in a hyaline matrix.

Minas [109] reported a series of 70 patients, aged between 14 and 55 years (mean 35.8) with defects ranging in size from 1.5 to 21 cm² treated by ACI. The knees were classified as simple, involving the femoral condyles, complex, involving the PFJ or tibia or where an associated biomechanical procedure was required and salvage where there was early evidence of OA or a bipolar lesion was treated. An average of two defects per knee were treated. Significant improvement was seen in WOMAC scores at 12 months and knee society scores at two years for the whole group. The simple and complex groups recovered quicker and had better outcomes than the salvage group.

There were five treatment failures in the whole group; three of these had falls in the post-operative period, resulting in delamination and the other two developed adhesions. 26 complications required surgical intervention in the first eight months for periosteal adhesions or graft hypertrophy or for joint stiffness. These results are short term but the overall clinical success rate of 93% in a difficult group of patients is encouraging.

The importance of correcting associated biomechanical abnormalities, particularly in more advanced disease, cannot be overemphasized [68,90]. Gillogly [68] treated 53 defects in 41 knees ranging in age from 14 to 52 years with an average of 36.2. Nineteen of 41 patients had concomitant biomechanical procedures; 12 of 13 PFJ grafts had antero-medialization of the tibial tubercle. Follow up was short at one year (25 patients) but 88% rated as good, very good, or excellent using the Knee Society or Modified Cincinnati Scores. Volume 5 ACI International Cartilage Repair Registry Report, February, 1999 [110] recorded information from 583 sites in the USA, Europe, and Israel. Ages ranged from 15 to 55 years in 97% with an average of 35 years where 71% had had prior surgery. Ninety percent of the defects were on the distal femur where the average size was 4.6 cm². Follow up was three years for 40 patients, two years for 220, and one year for 55. Eighty percent of patients with MFC lesions were improved, 90% reported no adverse effects, and 9% required re-operation, 3% for arthrofibrosis and 2% for hypertrophy. Robinson et al. [111] demonstrated that MRI imaging studies in eight patients followed for six months to five years following ACI had defects filled with tissue having similar signal characteristics to cartilage.

The advantages of ACI are that it utilizes autologous cells, produces a hyaline cartilage repair in humans [104,112], and can be used for large defects. Good results have been reported in a number of series [54,68,90,10,107,113]. The disadvantages are that it requires two operations, one of which is open, and the use of a sophisticated sterile laboratory. The donor site is relatively innocuous but is an invasive procedure. The technique requires an intact SBP and surgical access can be difficult. The technique is demanding and needs to be meticulous.

The cost is significant although it varies from site to site. Failures have been due to graft hypertrophy, graft displacement [113], or, particularly in the patello-femoral joint, failure to correct associated biomechanical problems. Lesions on the MFC heal better than lesions on the patella and tibia [52]. Graft displacement can be independent or due to the early use of CPM, if the graft is proud [80]. Suturing may cause damage to the surrounding AC [104].

These results for ACI are very encouraging but as Newman [35] points out, lack of controls, randomization and outcome analysis, and lack of biochemical and biomechanical data are a matter for concern. These concerns were also voiced by Messner and Gillquist [114]. Graft fixation remains an issue. The great advantage is that if the procedure fails the patient is not biologically worse off compared to those techniques that breach the SBP.

Osteochondral Autografting

Wirth and Rudert [115] stated that osteochondral autografting (OCAG) for AC defects in the knee had been reported in the German literature in the 1930s with a 30% failure rate. Matsusue et al. in 1993 [116] described a patient where three 5 mm autogenous osteochondral grafts were used to repair an AC defect on the MFC in association with an ACL reconstruction. Arthroscopy at two years revealed an effective repair. Outerbridge et al. [117] used the lateral patellar facet to resurface the MFC for OCD and reported improved symptoms in 10 patients at an average follow up of 6.5 years. Hangody first reported his technique of osteochondral mosaicplasty in 1994 [118]. In this technique, small cylindrical osteochondral grafts up to 5.5 mm diameter are harvested from the periphery of the trochlear and inserted into the prepared recipient site. The holes in the recipient site are prepared with a drill guide to keep the drill holes 1 mm apart and perpendicular to the articular surface. The grafts are then introduced into the prepared holes by finger pressure and gently tapped home. Bobic [119] has used this method arthroscopically in 12 patients with defects from 10 to 22 mm with 10 excellent results at one year. Hangody presented results of 113 of a total of 473 procedures with a three-year follow up at the ICRS meeting in Boston in 1998 [54]. An average of eight grafts were used for each defect; 27% were performed arthroscopically. Associated procedures were performed in 78%. Average modified HSS scores were 91.1.

OCAG has the advantages of being readily accessible, cost effective, and feasible as an arthroscopic procedure. The problems of rejection and disease transmission are eliminated. The early results seem good up to three years. The donor site is a major concern; the procedure creates two osteochondral defects to treat a chondral defect and is a classical case of robbing Peter to pay Paul. There are real possibilities that aggressive harvesting could lead to SBP collapse or OA. There is also concern that the transplanted cartilage could undergo degeneration as Czitrom [120] has shown in osteochondral allografts; this may be aggravated by the tapping necessary to seat the plugs. Contouring is a problem on the femoral condyles, which limits the method to small defects 20 mm in diameter, and the cobblestone surface may be abrasive [121]. Finally, the gaps between the plugs and the walls of the defect fill with fibrous tissue and it remains to be seen whether this will undergo metaplasia to hyaline cartilage in the long term.

Osteochondral Allografting

Osteochondral allografts may be used as joint replacements [122] or as shell allografts. Lexer [123] first reported the use of osteochondral shell allografts to treat articular cartilage injuries in 1925, with a success rate of 50% in 23 patients with upper limb injuries.

Since then the technique has been limited in North America to a small number of centers. Long-term follow up studies from Toronto [124,125,126] in 100 patients demonstrated successful results in 75% at 5 years, 64% at 10 years, and 63% at 14 years. Results were improved when fixation was adequate and an unloading osteotomy was performed. Survival rates for unipolar grafts were much better than for bipolar grafts. Bugbee and Convery [127] reported a success rate of 86% with unipolar lesions and 53% with bipolar grafting in 92 knees with a two-year follow up. Collapse of the bone graft was a major cause of failure in the Toronto series [128].

Garrett [129] also ascribed failure to collapse and fragmentation of the bone in a series of patients with OCD, where the success rate was 85%.

Despite these encouraging reports, this method has been used relatively infrequently. Fresh allograft has been shown to be superior to frozen material due to diminished chondrocyte viability in the latter [128,130]. The use of fresh allografts makes procurement difficult [131] and creates the possibility of disease transmission [127,132]. Although the risk of disease transmission is 1 : 1,600,000 in properly screened and tested donors, a case of HIV transmission has been recorded [133]. The early problems with fixation have been largely overcome by innovative techniques, specifically designed for each site [134]. Sepsis (5–10%) [124] is also a significant problem.

Rejection is not a major issue with these shell allografts as AC is "immunologically privileged"

Table 24.3. Experimental scaffolds for AC defects

Biodegradable Synthetic	Biodegradable Natural	Non-biodegradable	Seeded Scaffolds
Polyglycolic acid (PGA) Freed, 1994 [135]	*Collagen sponge* Speer, 1979 [136]	*Nylon* Kuhns, 1993 [137]	*Collagen sponge* Nixon, 1993 [164], Sams & Nixon, 1995 [146], Sams et al. 1995 [147], Fujisato, 1996 [149] *Collagen gel* Wakitani, 1989 [138]
Polyurethane Klompmaker, 1992 [165]	*Fibrin clot* Paletta et al. 1992 [140]	*Carbon fiber* Minns & Flynn, 1978 [153], Minns et al. 1982 [139]	*PGA* Ruuskanen, 1991 [167], Vacanti, 1991 [143], Freed, 1993 [144], Freed, 1994 [135], Ratcliffe, 1998 [54] *PLA* Freed, 1993 [171], Chu, 1995 [142]
Polylactic acid (PLA) von Schroeder, 1991 [138]		*Dacron* Messner, 1993 [166]	*Carbon Fiber* Brittberg et al. 1996 [97], Robinson, 1993 [168]
	Meniscus Sumen, 1995 [95]		*Hyaluronic acid* Robinson, 1990 [171]

and the reaction to the bone is considered to be clinically insignificant [133]. Cell viability, although well maintained initially, does reduce to 37% at six years, Czitrom [120]. Finally, cost is a significant factor as it is with all major allografting procedures. At present, osteochondral allografting should be limited to major lesions which cannot be treated with other methods where bone and cartilage loss coexist or as a salvage procedure where other techniques have failed.

Scaffolds

The use of scaffolds to replace AC defects seems an attractive idea, particularly in view of the structured matrix of AC originally described by Benninghoff [8] and the need to retain the cells in the defect [100]. There has been considerable experimental work and a small number of clinical studies using scaffolds. Scaffolds are classified in Table 24.3.

The techniques in Table 24.3 are experimental. Studies using scaffolds alone have been disappointing [95,135–140,165,166].

With seeded scaffolds the results have been more encouraging, with filling of defects with hyaline-type tissue containing variable amounts of type II collagen. Some biocompatibility problems with the carrier have occurred [141]. Chu et al. [142], using PLA, reported that type II collagen only comprised 19% of the matrix. Vacanti's group in Boston [143] developed a three-dimensional mesh of PGA containing cultured chondrocytes, which has been able to generate a hyaline cartilage repair. Freed and his colleagues [135,144] seeded tissue-cultured chondrocytes into scaffolds of PGA and PLA and inserted them as disks into nude mice, where they retained their original shape or as implants into AC defects in adult rabbit knees, where filling of the defects was achieved and where the results were better than with PGA alone. PGA was more effective than PLA. The incorporation of allogenoic cultured chondrocytes into PGA scaffolds has been further developed at The Advanced Tissue Sciences Laboratory in San Diego, USA where the composites are produced in bioreactors which produce better-quality grafts [35,54,145]. Skin composites are already being used for skin grafting clinically [54].

These tissue engineering techniques are exciting and offer tremendous potential for the future. Freed et al. [135] pointed out that seeded scaffolds produce a hyaline cartilage repair, are better than a scaffold alone, theoretically can cover large areas, and have structural integrity which make them easier to handle, allowing an arthroscopic insertion [76]. They can also act as a carrier for growth factors which can accelerate repair [62]. In animal studies to date the constructs have been press-fitted into small defects, leaving the SBP intact. However, in larger animals, problems have been encountered with fixation without breaching the SBP. In large defects in horses allogeneic chondrocytes in collagen scaffolds fared no better than controls [146,147]. Marginal integration has not occurred [54]. Cost is a significant factor with these implants.

Hunziker and Rosenberg [148] reported the use of tissue growth factors such as TGF-β1 to recruit mesenchymal cells from the synovium and produce a repair in AC defects. They also used chondroitinase ABC to digest the host matrix to enhance integration. Growth factors have been used by others to enhance repair [81,149]. Evans [54] has suggested that gene therapy may be used both to promote tissue growth and repair and prevent cartilage degeneration as an adjunct to cartilage repair procedures.

Mesenchymal repairs using fracture callus [150], osteochondral paste [151], or bone morphogenic collagen mesh have been used clinically and may be more appropriate in treating osteochondral defects.

Carbon Fiber Resurfacing

The Matrix Support Prosthesis (MSP) was developed to provide a strong, biocompatible scaffold which was porous and allowed host ingrowth [152]. The MSP was constructed with multi-directional fibers to simulate the Benninghoff arcades and is available as rods or pads. Studies in rabbits showed that the MSP was biocompatible and induced a repair which was superior to untreated defects in controls [139,153]. Clinical studies [152,154–158] on patients treated with carbon fiber resurfacing (CFR) for AC defects revealed good or excellent results in 77 to 91% subjectively and 85 to 90% objectively. Pongor [156] independently analyzed Bentley's series where MSP pads were used to resurface lesions on the MFC and the patella.

No associated biomechanical corrections were carried out. The overall results were 71% good or excellent subjectively and objectively. However, 10/11 results for OCD lesions on the MFC were good or excellent, whereas on the patella, only 6/17 were good or excellent. The poor results for the patella in this study certainly influenced the overall result and may have been due to non-correction of maltracking or the use of pads rather than rods for the larger lesions. Similar poor results were reported for the patella with pads and without biomechanical correction [159]. Peterson [107] found that his results for ACI on the patella were improved when he corrected PFJ biomechanics.

We have recently reviewed 128 knees treated with MSP rods and pads at an average follow up of 22.6 months. The ICRS score (max.12) for the AC repairs were: MFC 10.5; LFC 9.8; Trochlea 9.9; Patella 9.4. Repairs with rods on the MFC were 95.1% Grade 1 or 2 compared to 89.7% for patellar rods. The results were significantly better for the patella when realignment and anteriorization were carried out (96.7% cf. 66%). Pads on the patella fared worse than rods in the mechanically corrected group. Small lesions fared better than large lesions but there was no difference in patients aged above and below 40 years.

CFR is an effective method of resurfacing AC defects. Rods are more effective than pads and correction of biomechanics improves the results. The repair is initially by mixed fibrous and fibro-cartilaginous tissue, but a biopsy in a patient eight years following surgery showed hyaline cartilage.

CFR implants are biocompatible [160] and readily available, sterility is not an issue and only one procedure is required. It is less age-dependent and is suitable for all surfaces. Very large defects can be treated. The disadvantages are the cost of the implant and the fact that in most cases a fibrous repair occurs which may, however, ultimately change to hyaline cartilage. As Coutts has pointed out [54], it may not be necessary to achieve a hyaline articular cartilage repair to achieve a good result. Reactive synovitis with MSP implants does not appear to be a problem [54,154,155]. Carbon fiber has been used as a scaffold loaded with chondrocytes and produced results superior to PGA and controls when implanted into AC defects in rabbits [97].

Rehabilitation

Most authors have recommended a period of non-weight bearing between six and twelve weeks combined with early movement. Some, such as Steadman, insist on CPM. Return to sport varies widely between three and twelve months. Apart from the studies by O'Driscoll and Salter on CPM, there is little evidence to support which program is best. Rehabilitation will often be determined by the associated procedure; each patient requires a specific program.

Loading does appear to be essential to maintain healthy articular cartilage but the physiological limits are not defined [43]. A program combining early movement utilizing CPM and intermittent loading with the unloaded periods predominating the early phase should be the basic tenet [43]. Passive motion should be combined with gentle traction to avoid compression of the treated surfaces. Eccentric loading should be avoided [43]. Isometric exercises, performed at angles that do not engage the AC lesion, are recommended to improve muscle function. Shear loading to the joint surfaces that have been repaired should be avoided [161].

The Future

Tissue engineering with hormonal and genetic manipulation [162] is the way for the future. Major strides have been made in the last decade in the area of joint resurfacing and with the new technologies that are emerging, further progress will proceed rapidly to help to overcome what is likely to become an epidemic in future years. Ideal constructs should be bio-absorbable and be inserted arthroscopically [52]. There are still a number of unsolved problems: Can results from animal experiments be applied to humans [162]? Are free cells preferable to cells in scaffolds? Are bio-absorbable scaffolds preferable? Should the SBP be penetrated and what are the best methods of fixation? We require more information on the significance of defect size, site, and whether some donor sites are preferable. Is donor site morbidity significant and can bipolar lesions be treated? The role of CPM needs to be defined and precise rehabilitation protocols established.

Reproduction of normal articular cartilage that is able to withstand normal mechanical loading is a goal that may never be fully achieved. It is imperative to correct any underlying biomechanical abnormality that would lead the joint to abnormal loading conditions. Patient selection is paramount; these techniques are applicable to focal defects and will rapidly fall into disrepute if used to treat joints where the disease is too advanced [163]. Disease progression is a devastating event.

References

1. Mow VC, Holmes MH, Lai WM. Fluid transport and mechanical properties of articular cartilage: a review. J Biomech 1984;17:377–94.
2. Mow VC, Kuei SC, Lai WM, Armstrong CG. Biphasic creep and stress relaxation of articular cartilage in compression? Theory and experiments. J Biomech Eng 1980;102:73–84.
3. Hoikka VEJ, Jaroma HJ, Ritsila VA. Reconstruction of the Patella Articulation with Periosteal Grafts. Acta Orthop Scand 1990;61:36–9.
4. Homminga GN, van der Linden TJ, Terwindt-Rouwenhorst EA, Drukker J. Repair of articular defects by perichondrial grafts. Experiments in the rabbit. Acta Orthop Scan 1989;60:326–29.
5. Frank EH, Grodzinsky AJ. Cartilage electromechanics – I. Electrokinetic transduction and the effects of electrolyte pH and ionic strength. J Biomech 1987; 20:615–27.
6. Maroudas A. Physicochemical properties of articular cartilage. In: Freeman MAR, editor. Adult Articular Cartilage. Kent, UK: Pitman Medical, 1979;215–90.
7. Mow VC, Ratcliffe A, Poole AR. Cartilage and diarthrodial joints as paradigms for hierarchical materials and structures. Biomaterials 1992;13:67–97.
8. Benninghoff, A. Form und Bau der Gelenkknorpal in ihren Beziehungen zu Funktion. II Teil: Der Aufbau des Gelenkknorpel in seinen Beziehungen zu Funktion, Z Zellforsch 1925;2:783–862.
9. McCutchen CW. The frictional properties of animal joints. Wear 1962;5:1–17.
10. Maroudas A. Bullough P. Permeability of articular cartilage. Nature 1968;219:1260–1
11. Lai WM, Hou JS, Mow VC. A triphasic theory for the swelling and deformation behaviors of articular cartilage. J Biomech Eng 1991;113:245–58.
12. Lai WM, Mow VC. Drag-induced compression of articular cartilage during a permeation experiment. Biorheology 1980;17:111–23.
13. Mansour JM, Mow VC. The permeability of articular cartilage under compressive strain and at high pressures. J Bone Joint Surg 1976;58A:509–16.
14. Popov EP. Mechanics of Materials. Englewood Cliffs, NJ: Prentice Hall, 1976.
15. Schmidt MB, Mow VC, Chun LE, Eyre DR. Effects of proteoglycan extraction on the tensile behavior of articular cartilage. J Orthop Res 1990;8:353–63.
16. Woo SLY, An KN, Arnoczky SP, Wayne JS, Fithian DC, Myers BS. Anatomy, biology and biomechanics of tendon, ligaments and meniscus. In: Simon SR, editor. Orthopaedic Basic Science. Rosemont IL:

American Academy of Orthopedic Surgeons, 1994; 45–87.

17. Armstrong CG, Mow VC, Wirth CR. Biomechanics of impact-induced microdamage to articular surface – A possible genesis for chondromalacia patella. In: Finerman G, editor. AAOS Symposium Sports Medicine: The Knee, St. Louis: C.V. Mosby Company, 1985;54–69.
18. Donohue JM, Buss D, Oegema TR Jr, Thompson RC Jr. The effects of indirect blunt trauma on adult canine articular cartilage. J Bone Joint Surg 1983;65A:948–57.
19. Nagel D, Burton D, Manning J. The dashboard injury. Clin Orthop 1977;126:203–8.
20. Zhu W, Mow VC, Koob TJ, Eyre DR. Viscoelastic shear properties of articular cartilage and the effects of glycosidase treatments. J Orthop Res 1993;11:771–81.
21. Hori RY, Mockros LF. Indentation tests of human articular cartilage. J Biomech 1976;9:259–68.
22. Jaffe FF, Mankin HJ, Weiss C, Zarins A. Water binding in the articular cartilage of rabbits. J Bone Joint Surg 1974;56A:1031–9.
23. Mankin HJ, Thrasher AZ. Water content and binding in normal and osteoarthritic human cartilage. J Bone Joint Surg 1975;57A:76–80.
24. Mow VC, Setton LA, Ratcliffe A, Buckwalter JA, Howell DS. Structure–function relationships for articular cartilage and effects of joint instability and trauma on cartilage function. In: Brandt KD, editor. Cartilage Changes in Osteoarthritis, Indianapolis, IN: Indiana University School of Medicine Press, 1990;22–42.
25. Setton LA, Zhu W, Mow VC. The biphasic poroviscoelastic behavior of articular cartilage: role of the surface zone in governing the compressive behavior. J Biomech 1993;26:581–92.
26. Grodzinsky AJ, Roth V, Myers E, Grossman WD, Mow VC. The significance of electromechanical and osmotic forces in the nonequilibrium swelling behavior of articular cartilage in tension. J Biomech Eng 1981;103:221–31.
27. Maroudas A. Biophysical chemistry of cartilaginous tissues with special reference to solute and fluid transport. Biorheology 1975;12:233–48.
28. Mow VC, Tohyama H, Grelsamer RP. Structure–function of knee articular cartilage. Sports Med Arthroscopy 1994;2:189–202.
29. Mankin HJ. The response of articular cartilage to mechanical injury. J Bone Joint Surg 1982;64A:460–6.
30. Buckwalter JA. Articular cartilage: injuries and potential for healing. J Orthop Sports Phys Ther 1998;28: 192–202.
31. Felson DT, Anderson JJ, Naimark A, Walker AM, Mennan RF. Obesity and knee osteoarthritis: The Framingham Study. Ann Intern Med 1988;109:18–24.
32. Jurvelin J, Kirivanta I, Tammi M, Helminen HJ. Softening of canine articular cartilage after immobilisation of the knee joint. Clin Orthop 1986;207:246–52.
33. Meachim G. The effect of scarification on articular cartilage in the rabbit. J Bone Joint Surg 1963; 45-B1:150–61.
34. Grande DA, Pitman MI, Peterson L, Menche D, Klein M. The repair of experimentally produced defects in rabbit articular cartilage by autologous chondrocyte transplantation. J Orthop Res 1989;7:208–18.
35. Newman AP. Articular cartilage repair. Am J Sports Med 1998;26:309–24.
36. Messner K, Maletius W. The long-term prognosis for severe damage to weight-bearing cartilage in the knee: a 14-year clinical and radiographic follow-up in 28 young athletes. Acta Orthop Scand 1996;67:165–8.
37. Twyman RS, Desia K, Aichroth PM. Osteochondritis dissecans of the knee: A long-term study. J Bone Joint Surg 1991;73B:461–4.
38. Cameron M, Buchgraber A, Passler H, Vogt M, Thonar E, Fu F et al. The natural history of the anterior cruciate ligament-deficient knee. Changes in synovial fluid cytokine and keratan sulfate concentrations. Am J Sports Med 1997;25:751–4.
39. Sherman MF, Warren RF, Marshall JL, Savatsky GJ. A Clinical and Radiographic Analysis of 127 Anterior Cruciate Insufficient Knees. Clin Orthop 1998;227: 229–37.
40. Buckwalter JA. Regenerating Articular Cartilage: Why the sudden interest? Orthopaedics Today 1996;19: 26:1–6.
41. Curl WW, Krome J, Gordon ES, Rushing J, Smith BP, Poehling GG. Cartilage injuries: a review of 31,516 knee arthroscopies. Arthroscopy 1997;13:456–60.
42. Noyes FR, Bassett RW, Grood ES, Butler DL. Arthroscopy in acute traumatic hemarthrosis of the knee. Incidence of anterior cruciate tears and other injuries. J Bone Joint Surg 1980;62A:687–95.
43. Walker JM. Pathomechanics and Classification of Cartilage Lesions, Facilitation of Repair. J Orthop Sports Phy Ther 1998;28:216–31.
44. Lanier R. The Effects of Exercise on the Joints of Inbred Mice. Anat Rec 1946;94:311–21.
45. Jurvelin J, Kuusela T, Heikkila R, Peltari A, Kiviranta I, Tammi M et al. Investigation of Articular Cartilage Surface Morphology with a Semiquantative Scanning Electron Microscopic Method. Acta Anat 1983; 116:302–11.
46. Walker JM. Exercise and Its Influence on Ageing in Rat Knee Joints. J Orthop Sports Phys Ther 1986;8:310–19.
47. Shortkroff S, Barone L, Hsu H-P, Wrenn C, Gagne T, Chi T et al. Healing of Chondral and Osteochondral Defects in a Canine Model: The Role of Cultured Chondrocytes in Regeneration of Articular Cartilage. Biomaterials 1996;17:147–54.
48. Bennett GA, Waine H, Bauer W. Changes in the Knee Joint at Various Ages, with Particular Reference to the Nature and Development of Degenerative Joint Disease. New York: The Commonwealth Fund, 1942.
49. Mitrovic D, Borda-Iriate O, Naveau B, Stankovic A, Uzan H, Quintero M et al. Results of Autopsy Examination of Knee Cartilage of 120 Patients Dying in the Hospital. II. The Femoro-Tibial Joint. Rev Rheum Mal Osteoartic 1989;56:505–10.
50. Mankin HJ, Radin EL. Structure and Function of Joints. In: McCarty DJ, Koopman WJ, editors. Arthritis and Allied Conditions (12 ed.), Vol 1. Philadelphia: Lea and Febiger, 1993;181–97.
51. Outerbridge RE. The aetiology of chondromalacia patellae. J Bone Joint Surg Br 1961;43:752–7.
52. Mandelbaum BR, Browne JE, Freddie F, Lyle M, Mosely JB Jnr, Erggelet C et al. Articular Cartilage Lesions of the Knee. Am J Sports Med 1998;26:853–61.
53. International Cartilage Repair Society Documentation and Classification System. Freibourg, Switzerland, Newsletter 1, March 1998;5–8.
54. Presentations in Focus. "Excerpts from the International Cartilage Repair Society Semi-Annual Meeting, November 16–18, 1998, Boston, Massachusetts". Medical Education Network, New York.

55. Hunter W. Of the structure and disease of articulating cartilages. Phil Trans R Soc (Lond) 1743;42:514–22.
56. Ogilvie-Harris DJ, Fitsialos DP. Arthroscopic Management of the Degenerative Knee. Arthroscopy 1991;7: 151–7.
57. Jackson RW. Arthroscopic Treatment of Degenerative Arthritis. In: McGinty JB, editor. Operative Arthroscopy. New York: Raven Press, 1991;319–23.
58. Gilbert JE. Current Treatment Options for the Restoration of Articular Cartilage. Am J Knee Surg 1998;11: 42–6.
59. Pridie KH. A Method of Resurfacing Osteoarthritic Knee Joints. J Bone Joint Surg 1959;41B:618–19.
60. Vachon A, Bramlage LR, Gabel AA, Wisebrode S. Evaluation of the Repair Process of Cartilage Defects of the Equine Third Carpal Bone With and Without Subchondral Bone Perforation. Am J Vet Res 1986;47: 2637–45.
61. Insall J. Intra-articular surgery for degenerative osteoarthritis of the knee. J Bone Joint Surg 1967;49B: 211–12.
62. Mitchell N, Shephard N. The Resurfacing of Adult Rabbit Articular Cartilage by Multiple Perforations Through the Subchondral Bone. J Bone Joint Surg 1976;58A:230–3.
63. Ficat RP, Ficat C, Gedeon P, Toussaint JB. Spongialization: a new treatment for diseased patellae. Clin Orthop 1979;144:74–83.
64. Johnson LL. Arthroscopic Abrasion Arthroplasty in Operative Arthroscopy. New York: Raven Press, 1991; 341.
65. Kim HK, Moran ME, Salter RB. The potential for regeneration of articular cartilage in defects created by chondral shaving and subchondral abrasion. An experimental investigation in rabbits. J Bone Joint Surg 1991;73A:1301–15.
66. Altman RD, Kates J, Chun LE, Dean DD, Eyre D. Preliminary observations of chondral abrasion in a canine model. Ann Rheum Dis 1992;51:1056–62.
67. Hanie EA, Sullins KE, Powers BE, Nelson PR. Healing of full-thickness cartilage compared with full-thickness cartilage and subchondral bone defects in the equine third carpal bone. J Equine Vet 1992;24:382–6.
68. Gillogly SD, Voight M, Blackburn T. Treatment of articular cartilage defects of the knee with autologous chondrocyte implantation. J Orthop Sports Phys Ther 1998;28:241–51.
69. Friedman MJ, Berasi CC, Fox JM, Del Pazzo W, Snyder SJ, Ferkel RD. Preliminary results with abrasion arthroplasty in the osteoarthritic knee, Clin Orthop 1984; 182:200–5.
70. Blevins FT, Steadman JR, Rodrigo JJ, Silliman JF. Treatment of articular cartilage defects in athletes: an analysis of functional outcome and lesion appearance. Orthopaedics 1998;21:761–7.
71. Rodrigo JJ, Steadman JR, Silliman JF, Fulstone HA. Improvement of full thickness chondral defect healing in the human knee after debridement and microfracture using continuous passive motion. Am J Knee Surg 1994;7:109–16.
72. Salter RB, Simmonds DF, Malcolm BW, Rumble EJ, MacMichael D, Clements ND. The biological effect of continuous passive motion on the healing of full-thickness defects in articular cartilage. An experimental investigation in the rabbit. J Bone Joint Surg 1980;62A:1232–51.
73. Rubak JM, Poussa M, Ritsila V; Effects of joint motion on the repair of articular cartilage with free periosteal grafts. Acta Orthop Scand 1982;53:187–91.
74. O'Driscoll SW, Salter RB. The induction of neochondrogenesis in free intra-articular periosteal autografts under the influence of continuous passive motion. An experimental investigation in the rabbit. J Bone Joint Surg 1984;66A:1248–57.
75. O'Driscoll SW, Keeley FW, Salter RB. The chondrogenic potential of free autogenous periosteal grafts for biological resurfacing of major full-thickness defects in joint surfaces under the influence of continuous passive motion. An experimental investigation in the rabbit. J Bone Joint Surg 1986;68A:1017–35.
76. Hendrikson DA, Nixon AJ, Grande DA, Todhunter RJ, Minor RM, Erb H et al. Chondrocyte–fibrin matrix transplants for resurfacing extensive articular cartilage defects. J Orthop Res 1994;12:485–97.
77. Korkala OL, Kuokkanen HO. Autoarthroplasty of knee cartilage defects by osteoperiosteal grafts. Arch Orthop Trauma Surg 1995;114:253–6.
78. Angerman P, Riegels-Nielsen P. Osteochondritis dissecans of the femoral condyle treated with periosteal transplantation: A preliminary clinical study of 14 cases. Orthop Int 1994;2:425–8.
79. Sandelin J. Long-term results of reconstruction of the patellar articulation with periosteal grafts. Cartilage Repair Symposium, Bermuda, August, 1997.
80. Alfredson H, Lorentzon R. Superior results with continuous passive motion compared to active motion after periosteal transplantation. A retrospective study of human patella cartilage defect treatment, Knee Surg Sports Traum Arthrosc 1999;7:232–8.
81. O'Driscoll S, Meisami B, Fitzsimmons JS, Miura Y. Viability of periosteal tissue obtained post-mortem, Trans Orthop Res Soc 1995;20:108.
82. O'Driscoll S, Periosteal transplantation. Articular cartilage regeneration: chondrocyte transplantation and other technologies. Presented at the Annual Meeting of the American Academy of Orthopaedic Surgeons, San Francisco, CA, February, 1997.
83. Jaroma HJ, Ritsila VA. Reconstruction of patellar cartilage defects with free periosteal grafts. An experimental study. Scand J Plast Reconstr Surg Hand Surg 1987;21:175–81.
84. Skoog T, Ohlsen L, Sohn SA. Perichondrial potential for cartilaginous regeneration. Scand J Plast Reconstr Surg 1972;6:123–5.
85. Ohlsen L. Cartilage formation from free perichondrial grafts: an experimental study in rabbits. Br J Plast Surg 1976;29:262–7.
86. Amiel D, Coutts RD, Harwood FL, Ishizue KK, Kliener JB. The chondrogenesis of rib periochondrial grafts for repair of full-thickness articular cartilage defects in a rabbit model: a one year postoperative assessment. Connect Tissues Res 1988;18:27–39.
87. Homminga GN, Bulstra SK, Kuijer R, van der Linden AJ. Repair of sheep articular cartilage defects with a rabbit costal perichondrial graft. Acta Orthop Scand 1991;62:415–418.
88. Coutts RD, Woo SL, Amiel D, von Schroeder HP, Kwan MK. Rib perichondrial autografts in full-thickness articular cartilage defects in rabbits. Clin Orthop 1992;275:263–73.
89. Homminga GN, Bulstra SK, Bouwmeester PS, van der Linden AJ. Perichondral grafting for cartilage

lesions of the knee. J Bone Joint Surg 1990;72B: 1003–7.
90. Minas T, Nehrer S. Current concepts in the treatment of articular cartilage defects. Orthopaedics 1997;20: 525–38.
91. Bouwmeester SJ, Beckers JM, Kuije R, van der Linden AJ, Bulstra SK. Long-term results of rib perichondrial grafts for repair of cartilage defects in the human knee. Internat Orthop 1997;21:313–17.
92. Vachon A, McIlwraith CW, Trotter GW, Norridin RW, Powers BE. Neochondrogenesis in free intra-articular, periosteal, and periochondrial autografts in horses. Am J Vet Res 1989;50:1787–94.
93. Fitzsimmons JS, O'Driscoll SW. Technical experience is important in harvesting periosteum for chondrogenesis. Trans Orthop Res Soc 1998;23:914.
94. Gallay SH, MiuraY, Commisso CM, Fitzminnons JS, O'Driscoll SW. Relationship of donor site to chondrogenic potential of periosteum in vitro. J Orthop Res 1994;12:515–25.
95. Sumen Y, Ochi M, Ikuta Y. Treatment of articular defects with meniscal allografts in a rabbit knee model. J Arthro Rel Surg 1995;2:185–95.
96. Ochi M, Sumen Y, Jitsuiki J, Ikuta Y. Allogenic deep-frozen meniscal graft for repair of osteochondral defects in the knee joint. Arch Orthop Trauma Surg 1995;114:260–6.
97. Brittberg M, Nilsson A, Lindahl A, Ohlsson C, Peterson L. Rabbit articular cartilage defects treated with autologous cultured chondrocytes. Clin Orthop 1996;326: 270–83.
98. Smith AU. Survival of frozen chondrocytes isolated from cartilage of adult mammals. Nature 1965;205: 782–4.
99. Chesterman PJ, Smith AU. Homotransplantation of articular cartilage chondrocytes. An experimental study in rabbits. J Bone Joint Surg 1968;50B:184–97.
100. Bentley GA, Greer RB III. Homotransplantation of isolated epiphyseal and articular cartilage chondrocytes into joint surfaces of rabbits. Nature 1971;230:385–8.
101. Green WT Jr. Articular cartilage repair: Behavior of rabbit chondrocytes during tissue culture and subsequent allografting. Clin Orthop 1977;124:237–50.
102. Aston JE, Bentley GA. Culture of articular cartilage as a method of storage: Assessment of maintenance of phenotype. J Bone Joint Surg 1982;64B:384.
103. Aston JE, Bentley GA. Repair of articular surfaces by allografts of articular and growth in plate cartilage. J Bone Joint Surg 1986;68B:29–35.
104. Peterson L, Menche D, Grande D. Chondrocyte transplantation – an experimental model in the rabbit. In: Transactions from the 30th Annual Orthopaedic Research Society, Atlanta, February 7–9, 1984:218.
105. Brienen HA, Minas T, Hsu HP, Nehrer S, Sledge CB, Spector M. Effect of cultured chondrocytes on repair of chondral defects in a canine model. J Bone Joint Surg 1997;79A:1439–51.
106. Brittberg M, Lindahl A, Nilsson A, Ohlsson O, Peterson L. Treatment of deep cartilage defects in the knee with autologous chondrocyte transplantation. N Engl J Med 1994;331:889–95.
107. Peterson L. Symposium on articular cartilage repair. Presented at the Annual Meeting of the American Academy of Orthopedic Surgeons, Atlanta, GA February, 1996.
108. Peterson L, Minas T, Brittberg M, Nillsson A, Sjogren-Jannson E, Lindahl A. Two to nine year outcome after autologous chondrocyte transplantation of the knee. Clin Orthop 2000;374:212–34.
109. Minas T. The role of cartilage repair techniques, including chondrocyte transplantation, in focal chondral knee damage. Instr Course Lect 1999;48:629–43.
110. ACI International Cartilage Repair Registry. Periodic Report, 5: February, 1999.
111. Robinson D, Ash H, Aviezerb D, Agar G, Halperin N, Nevo Z. Autologous chondrocyte transplantation for reconstruction of isolated joint defects: The Assaf Harofeh experience. Isr Med Assoc J 2000;4:290–5.
112. Richardson JB, Caterson B, Evans EH, Ashton BA, Roberts S. Repair of human articular cartilage after implantation of autologous chondrocytes. J Bone Joint Surg 1999;81B:1064–8.
113. Minas T. The role of cartilage repair techniques, including chondrocyte transplantation in focal chondral knee damage. Instr Course Lect 1999;48:629–43.
114. Messner K, Gillquist J. Cartilage repair: A critical review. ACTA Orthop Scan 1996;67:523–9.
115. Wirth CJ, Rudert M. Techniques of cartilage growth enhancement: a review of the literature. Arthroscopy 1996;12:300–8.
116. Matsusue Y, Yamamuro T, Hama H. Arthroscopic multiple osteochondral transplantation to the chondral defect in the knee associated with anterior cruciate ligament disruption. Arthroscopy 1993;9:318–21.
117. Outerbridge HK, Outerbridge AR, Outerbridge RE. The use of a lateral patellar autologous graft for the repair of a large osteochondral defect in the knee. J Bone Joint Surg 1995;77A:65–72.
118. Hangody L, Karpati Z. New possibilities in the management of severe circumscribed cartilage damage in the knee. Magyar Traumatologia Ortopedia Kezsebeszet Plasztikai Sebeszet 1994;37:237–43.
119. Bobic V. Arthroscopic osteochondral autograft transplantation in anterior cruciate ligament reconstruction: A preliminary clinical study. Knee Surg Sports Traumatol Arthrosc 1996;3:262–4.
120. Czitrom AA, Keating S, Gross AE. The viability of articular cartilage in fresh osteochondral allografts after clinical transplantation. J Bone Joint Surg 1990;72A: 574–81.
121. Akeson W. Current status of cartilage grafting. West J Med 1998;168:121–2.
122. Tofe MH, Gebhardt M, Tomford W, Mankin HJ. Reconstruction of defects of the proximal femur using allograft arthroplasty. J Bone Joint Surg 1988;70A:507–16.
123. Lexer E. Joint transplantations and arthroplasty. Surg Gynecol Obstet 1925;40:782–809.
124. McDermott AG, Langer F, Pritzker KP, Gross AE. Fresh small-fragment osteochondral allografts. Long-term follow-up study on first 100 cases. Clin Orthop 1985; 197:96–102.
125. Zukor DJ, Oakeshott RD, Gross AE. Osteochondral allograft reconstruction of the knee. Part 2. Experience with successful and failed fresh osteochondral allografts. Am J Knee Surg 1989;2:182–91.
126. Beaver RJ, Mahomed M, Backstein D et al. Fresh osteochondral allografts for post-traumatic defects in the knee: a survivorship analysis. J Bone Joint Surg 1992; 74B:105–10.
127. Bugbee WD, Convery FR. Osteochondral allograft transplantation. Clin Sports Med 1999;18:67–75.

128. Oakeshott RD, Farine I, Pritzker KP, Langer F, Gross AE. A clinical and histologic analysis of failed fresh osteochondral allografts. Clin Orthop 1988;233:283–94.
129. Garrett JC. Osteochondral allografts for reconstruction of articular defects. In: McGinty JB, Caspari RB, Jackson RW, Poehling GG, editors. Operative Arthroscopy, 2nd ed. Philadelphia; Lippincott-Raven, 1996;395–403.
130. Rodrigo JJ, Thompson E, Travis C. Deep freezing vs 4°C preservation of avascular osteocarilaginous shell allografts in rats. Clin Orthop 1987;218:268–75.
131. Shelton WR, Treacy SH, Dukes AD, Bomboy AL. Use of allografts in knee reconstruction: II. Surgical considerations. J Am Acad Orthop Surg 1998;6:169–75.
132. Shelton WR, Treacy SH, Dukes AD, Bomboy AL. Use of allografts in knee reconstruction: I. Basic science aspects and current status. J Am Acad Orthop Surg 1998;6:165–8.
133. Simonds RJ, Holmberg SD, Hurwitz R et al. Transmission of Human Immunodeficiency Virus Type 1 from a seonegative organ and tissue donor. N Engl J Med 1992;326:726–32.
134. Garrett JC. Osteochondral allografts for reconstruction of articular defects of the knee. AAOS Instr Course Lect 1998;47:517–22.
135. Freed LE, Grande DA, Lingbin Z, Emmanuel JC, Marquis JC, Langer R. Joint resurfacing using allograft chondrocytes and synthetic biodegradable polymer scaffolds. J Biomed Mater Res 1994;28:891–9.
136. Speer DP, Chvapil M, Volz RG, Holmes MD. Enhancement of healing in osteochondral defects by collagen sponge implants. Clin Orthop 1979;14:326–35.
137. Kuhns JG, Potter TA, Hormell RS, Ellerton WA. Nylon membrane arthroplasty of the knee in chronic arthritis. J Bone Joint Surg 1953;35A:929–36.
138. Wakitani S, Kimura T, Hirooka A, Ochi T, Yoneda M, Yasui N et al. Repair of rabbit articular surfaces with allograft chondrocytes embedded in collagen gel. J Bone Joint Surg 1989;71B:74–80.
139. Minns RJ, Muckle DS, Donkin JE. The repair of osteochondral defects in osteoarthritic rabbit knees by the use of carbon fibre. Biomaterials 1982;3:81–6.
140. Paletta GA, Arnoczky SP, Warren RG. The repair of osteochondral defects using an exogenous fibrin clot. An experimental study in dogs. Am J Sports Med 1992;20:725–31.
141. Itay S, Abramovici A, Nevo Z. Use of cultured embryonal chick epiphyseal chondrocytes as grafts for defects in chick articular cartilage. Clin Orthop 1987; 220:284–303.
142. Chu C, Coutts RD, Yoshioka M et al. Articular cartilage repair using allogenic perichondrocyte-seeded biodegradable porous polylactic acid (PLA): A tissue re-engineering study. J Biomed Mater Res 1995;29: 1147–54.
143. Vacanti CA, Langer R, Schloo B, Vacanti JP. Synthetic polymers seeded with chondrocytes provide a template for new cartilage formation. Plast Reconstr Surg 1991;88:753–9.
144. Freed LE, Vunjak-Novakovic G, Langer R. Cultivation of cell-polymer cartilage implants in bioreactors. J Cell Biochem 1993;51:257–62.
145. Freed LE, Vunjak-Novakovic G. Tissue culture bioreactors: chondrogenesis as a model system. In: Lanza R, Langer R, Chick W, editors. Principles of Tissue Engineering. Austin, TX: RG Landes Company, 1997; 151–65.
146. Sams AE, Nixon AJ. Chondrocyte-laden collagen scaffolds for resurfacing extensive articular cartilage defects. Osteoarthritis Cartilage 1995;3:47–59.
147. Sams AE, Minor R R, Wootton JA, Mohammed H, Nixon AJ. Local and remote matrix responses to chondrocyte-laden collagen scaffold implantation in extensive articular defects. Osteoarthritis Cartilage 1995; 3:61–70.
148. Hunziker EB, Rosenberg LC. Repair of partial-thickness defects in articular cartilage: cell recruitment from the synovial membrane. J Bone Joint Surg 1996; 78A:721–33.
149. Fujisato T, Sajiki T, Liu Q et al. Effect of basic fibroblast growth factor on cartilage regeneration in chondrocyte-seeded collagen sponge scaffold. Biomaterials 1996;17:155–62.
150. Takahashi S, Oka M, Kotoura Y, Yamamuro T. Autogeneous callo-osseous grafts for the repair of osteochondral defects. J Bone Joint Surg 1995;77B:194–204.
151. Stone K. "Osteochondral slurry" method of cartilage transplantation. Proceedings of the Second ICRS International Symposium on Cartilage Repair. Freibourg, Switzerland, October 1997.
152. Minns RJ, Muckle DS, Betts JA. Biological resurfacing using carbon fibre. Orthop Inter 1993;1:414–24.
153. Minns MJ, Flynn M. Intra-articular implant of filamentous carbon fibre in the experimental animal. J Bioeng 1978;2:279–86.
154. Muckle DS, Minns RJ. Biological response to woven carbon fibre pads in the knee. J Bone and Joint Surg 1989;71B:60–2.
155. Hart JAL, Butorac RB. Articular resurfacing with carbon fibre implants, J Bone Joint Surg 1990;72B: 1103.
156. Pongor P, Betts J, Muckle DS, Bentley G. Woven carbon surface replacement in the knee: independent clinical review. Biomaterials 1992;13:1070–6.
157. Brittberg M, Faxen E, Peterson L. Carbon fiber scaffolds in the treatment of early knee osteoarthritis. A prospective four-year follow-up of 37 patients. Clin Orthop 1994;307:155–64.
158. Minns RJ, O'Brien TK, England P, Betts JA, Muckle DS. Biological resurfacing of the knee: A review of surgical management. The Knee 1995;1:197–200.
159. Meister K, Cobb A, Bentley G. Treatment of painful articular cartilage defects of the patella by carbon-fibre implants. J Bone Joint Surg 1998;80B:965–70.
160. Mendes DG, Angel D, Grishkan A, Boss J. Histological response to carbon fibre. J Bone Joint Surg 1985; 67B:645–9.
161. Irrgang JJ, Pezzullo DD. Rehabilitation following surgical procedures to address articular lesions in the knee. J Orthop Sports Phys Ther 1998;28:232–40.
162. Doherty PJ, Zhang H, Tremblay L, Manolopoulos V, Marshall KW. Resurfacing of articular cartilage explants with genetically-modified human chondrocytes in-vitro. Osteoarthritis Cartilage 1988;6:153–9.
163. Buckwalter JA, Mankin HJ. Articular cartilage: degeneration and osteoarthritis, repair, regeneration and transplantation. Instr Course Lect 1998;47:487–504.
164. Nixon AJ, Sams AE, Lust G, Grande D, Mohammed HO. Temporal matrix synthesis and histologic features of a chondrocyte-laden porous collagen cartilage analogue. Amer J Vet Res 1993;54:349–56.

165. Klompmaker J, Jansen HW, Veth RP, Nielsen HK, de Groot JH, Pennings AJ. Porous polymer implants for repair of full-thickness defects of articular cartilage: an experimental study in rabbit and dog. Biomaterials 1992;13:625–34.
166. Messner K. Hydroxylapatite-supported Dacron plugs for repair of isolated full-thickness osteochondral defects of the rabbit femoral condyle: mechanical and histological evaluations from 6–48 weeks. J Biomed Mater Res 1993;27:1527–32.
167. Ruuskanen MM, Kallioinen MJ, Kaarela OI, Laiho JA, Tormala PO, Waris TJ. The role of polyglycolic acid rods in the regeneration of cartilage from perichondrium in rabbits. Scand J Plast Reconstr Surg Hand Surg 1991;25:15–18.
168. Robinson D, Efrat M, Mendes DG, Halperin N, Nevo Z. Implants composed of carbon fiber mesh and bone marrow-derived, chondrocyte-enriched cultures for joint surface reconstruction. Bull Hosp Joint Dis 1993;53:75–82.
169. Nixon AJ, Hendrikson DA, Lust G, Grande D. Chondrocyte-laden fibrin polymers effectively resurface large articular cartilage defects in horses. Trans Orthop Res Soc 1992;17: 17.
170. Sims CD, Butler PE, Cao YL, Casanova R, Randolph MA, Black A et al. Tissue engineered neocartilage using plasma derived polymer substrates and chondrocytes. Plast Reconstruct Surg 1998;101:1580–5.
171. Robinson D, Halperin N, Nevo Z. Regenerating hyaline cartilage in articular defects of old chickens using implants of embryonal chick chondrocytes embedded in a new natural delivery substance. Calcif Tissue Int 1990;46:246–53.

25 Massive Allografts in Hip Reconstructions

D. G. Poitout, P. Tropiano, B. Ripoll, and G. Marck

Revision arthroplasty of the hip and large-scale excision due to tumors are reasons for large losses of bone [1–3]. The head of the femur can only be replaced by a bone graft [4]. Only autografts are osteo-inducing and, when assimilated, have undeniable mechanical properties [5]. However, the amount that can be removed is limited [6] and other solutions have to be sought when massive reconstruction of the acetabulum [7] and of the proximal femur [8] are needed.

Massive allografts make muscle refixation easier [9] and provide the volume of bone needed for reconstruction of the loss of substance [10,11], which is why we prefer them to massive metal prostheses [12]. However, their biological characteristics, their fragility on exposure to fatigue stresses [13,14], and their immunological properties [15–17] are subjects for discussion. Since 1982, we have been using massive grafts preserved in the Bone Bank in Marseilles.

Etiologies

In all cases it was a question of replacing large losses of bone, either due to tumors [18], infection [19], or after repeated loosening of total hip prostheses.

Tumoral Etiologies

Chondrosarcomas

Fibrosarcomas

Ewing's tumors

Plasmocytomas

Rhabdomyosarcoma

Giant cell tumor

Infectious Etiologies

Echinococcosis

Sepsis

Hip Reconstruction after Revision Surgery

Acetabulum

Proximal femur

Allografts

All the grafts were taken from subjects who were brain dead. The bone grafts were removed under the same aseptic conditions as for any orthopedic operation. The team which removed the graft recorded the nature of the graft, its size, its side, and the presence or otherwise of tendon insertions. Once it has been frozen in a bag containing a cryoprotector and antibiotics, the graft is no longer visible. It is a block of ice which can no longer be identified other than by measuring radiographs.

Surgical Techniques

Reconstructions of the Pelvis

The surgical technique has always been the same. Only the methods used to anchor implants have evolved, resulting in osteosynthesis being more rigid.

Patient Position

The patient is laid in a ¾ dorsal decubitus position with a support on the controlateral, antero-superior iliac spine and another on the upper lumbar spine, opposite the second lumbar vertebra. The controlateral leg is flexed and immobilized by sparadrap. The homolateral leg is left free so that it can be mobilized.

Approach

We have described a wide approach [20] allowing the whole of the hemi-pelvis to be excised. The same approach was selected for more limited excisions in its useful part. The incision was started in the inguinal region taking the vessels into account, then extended downwards and upwards:

downwards, towards the homolateral ischium, along the inguinal crease, reaching the controlateral pubic region, then following the base of the thigh up to the tuberosity of the ischium.

upwards, along the crural arcade up to the antero-superior iliac spine.

It continues along the superior border of the iliac wing up to two finger-breadths inside the sacroiliac joint, to descend vertically to the tip of the sacrum. A vertical incision is made ascending in a Y, branched at the postero-superior angle of the posterior incision, climbing along the transverse apophyses of the lumbar vertebrae, allowing a wide approach to the lumbar region, if needed.

Dissection

The operation starts by dissecting the femoral and iliac vessels which are located on sacs, then the spermatic cord and crural nerve which are protected by the psoas (the sac takes the psoas and the crural nerve together). The peritoneum is usually easily detached from the psoas and iliac muscles, because the tumor pushes back the muscle masses and only at a very late stage does it invade these muscles and adhere to the peritoneum.

The pubis is then freed of its muscle insertions, and the bladder, detached from the pubis, is pressed back with the peritoneum. The pubis has to be dissected right up to the controlateral obturator opening. The hypogastric artery and veins are dissected and severed where they bifurcate. Ligating them does not lead to any problems in the urogenital or rectal regions. This vascular stage and particularly the venous stage is often long and difficult with the tumor projecting over the vessels.

The lumbar and sacral nerve roots are dissected up to their origin in the spine. Nerves coming into contact with the tumor or contained within it have to be severed.

There has to be no doubt about the quantity of malignant tumor excised, as if even the smallest amount of neoplastic tissue were to remain in the periphery of a nerve or a vessel, the tumor would definitely recur at the site, resulting in this large-scale surgery being in vain.

Dissection is generally difficult on the anterior surface of the sacrum. Meticulous hemostasis of the presacral vessels has to be achieved and the lumbar arteries and veins have to be ligated unilaterally, as required, which will make it easier to mobilize the primitive iliac vessels, to push back the nerve roots, and will facilitate access to the anterior and posterior surfaces of the ala of the sacrum.

The muscles attached to the iliac wing are cut depending on the site of the tumor. The iliac muscle and the gluteus minimus are generally left in contact with the tumor and resected with the bone. If possible, the innervation of the gluteus medius and gluteus maximus should be retained. The ischiopubic branch is dissected step by step and, in man, it is necessary to remain in contact with the periosteum, to isolate this from the cavernous bodies which bleed easily when they are injured and are intimately attached to the inferior border of the pubis and to the first centimeters of the ischiopubic branch. This retropubic and subpubic dissection has to be performed very meticulously with a cold bistoury. It is then necessary to sever, step by step, the various muscles which are attached to the ischiopubic branch and the ischium, both at its external surface and at its internal surface (the levator ani muscle and the obturator and adductor muscles).

This dissection is only possible if the approach route follows exactly the inferior internal border of the ischiopubic branch up to the posterior surface of the ischial tuberosity. The obturator nerve generally has to be cut, unless it is possible to sever the iliopubic branch without running the risk of spreading the cancer. If the tumor reaches the coxofemoral joint or even merely comes into contact with the round ligament, it may be necessary to cut the neck of the femur outside the capsule and resect the femoral head, the neck, the capsule, or even the upper extremity of the femur in a block with the tumor.

Osteotomies

They are performed at the end of the dissection. The pubis is cut using the oscillating saw, at the controlateral pubic symphysis in the middle of the healthy spongy tissue. The posterior section is made into the healthy spongy tissue in the ala of the sacrum, using a flat-headed chisel or the oscillating saw, protecting the nerve roots. The iliac bone is then removed and all that remains between the trunk and the lower limb are:

in front, the femoral vessels and the crural nerve;

behind, the branches of the sciatic nerve.

Reconstruction

An acetabular cup is cemented directly on the bone from the bone bank before it is fastened. The sacroiliac joint or ala of the sacrum left in place has been completed, as well as the pubic symphysis. The hemi-pelvis is strengthened by osteosynthesis plates and fixed in the upper part of the pubis by self-tapping screws. The posterior part of the pelvis is fixed in the ala of the sacrum using spongy bone screws. Care will have been taken before implanting the graft to perforate it with many holes in order to allow muscular refixation by trans-bone sutures.

In order to avoid an obturating hernia occurring, the obturator frame is closed with a silastic plate fixed in the perforated orifices of the bone.

The femoral prosthesis is cemented in the desired position in the diaphyseal shaft of the femur after the canal has been prepared. Once the graft has been fixed, the gluteal muscles are inserted by numerous trans-bone sutures. The ischiatic and adductor muscles will be fixed on the ischium and the ischiopubic branch, as will the levator muscles of the anus. The muscles of the anterior and lateral abdominal wall are reinserted on the iliopubic branch, avoiding a hiatus being left behind which could lead to a hernia.

Surgical Sequelae and Rehabilitation

For two days the patient will remain anesthetized in the intensive care unit in order to reduce pain, facilitate restoration of hydroelectrolyte balance, and to gradually wean the patient off assisted ventilation.

Walking and partial weight-bearing on a standing table are permitted from the eighth day after the operation. Total weight-bearing is possible by 15–21 days postoperatively.

For Isolated Reconstructions of the Acetabulum

The approach used in revision surgery of the hip is a lateral external transgluteal route, creating a digastric muscle from the gluteus medius and with the vastus externus left connected to each other by their trochanteric attachments. The femur is dislocated and freed from its acetabular attachments.

The second stage involves the freeing of the acetabulum: ablation of the implant, of all the cement, and of any fibrosis. Drilling is performed initially to remove all the necrotic cortical bone. The underlying spongy bone is exposed. It is onto this live bone that the femoral head allografts are screwed. The femoral heads are modeled in such a way as to reconstruct the acetabulum perfectly and to make it resemble normal anatomy as much as possible. We do not implant large cups as we prefer to reconstruct the bone of the acetabulum rather than to insert too large an implant.

Once the allografts are screwed solidly into the host bone, further drilling is performed.

Anchorage points are drilled into the roof of the acetabulum, the pubis, and the ischium. These points cross the allografts and go right into the patient's bone in such a way that the cement fixing the cup rests directly on the allografts as well as on the host bone. The cement is applied when it is in its pasty phase so as not to creep between the allograft and the spongy bone of the host's acetabulum.

For Sheathed Prostheses

The patient is laid in a lateral decubitus position. The approach in these reconstructions has also always been through the digastric muscle. When tumors are involved, excision of the cancerous growths has often required considerable resection of the muscle.

Where there have been large losses of bone, after prosthetic loosening, depending on the amount of bone remaining, the technique has been different:

resection in principle of the proximal extremity of the femur to reconstruct the Merckel spur and the large trochanter, also removing all the cement debris.

impaction of the allograft in the sheath of remaining bone.

We have not used the spongy allograft as a sheath, preferring to support the shaft of the prosthesis on a massive cortical allograft.

The Prosthesis

Designed to mold itself onto the anterior convexity of the femoral diaphysis, the shaft of the prosthesis is adapted to the anatomical curvature of the femur. This curvature gives it its good diaphyseal stability in rotation. We used to only cement the distal femoral part, but now the allograft and the host bone are cemented in order to obtain better stability.

The allograft–host bone union is cut like the step of a staircas,e perfectly adjusted to allow a greater contact surface area and to counteract rotation stresses. In all cases, whether on account of tumors or not, spongy autografts taken at the expense of the iliac crest are arranged all around the allograft–host bone union.

Weight-bearing is permitted after the same intervals for those of an ordinary prosthesis. The mechanical strength of the allograft is at its maximum the day it is implanted. If this strength were to decrease, this would be during peripheral rehabilitation, 12 to 18 months later.

In Cases of Sepsis

Surgery is always performed in two stages:

in the first stage there would be cleaning and excision of the cancerous parts and of all the debris [32]. Many biopsies are performed to search for the causative organism or organisms. Depending on the amount of bone lost, a cement support mixed with antibiotics suitable for the pathogen, if it is known, is left in situ.

The role of this support is to:

keep the site of the resected femur free, which will make subsequent replacement with an allograft easier;

to deliver antibiotics at a bactericidal dose in situ.

a second irrigation–drainage stage, using suitable antibiotics in perfusion. The limb is placed in traction;

a third reconstruction stage after the biological markers of inflammation (sedimentation rate, C-reactive protein) have been normal for at least two months.

Results

For sheathed prostheses used during revision surgery on the hip, the results were judged from the functional point of view, according to the Harris score. The results were:

good and very good in 70% of cases,

average in 5% of cases,

poor in 25% of cases.

For tumors, the necessary muscle resections lowered the functional score. The results were analyzed from the oncological point of view, and compared with other techniques, metal reconstruction prostheses, massive techniques, and amputation. The functional results during patient survival were good and very good in 90% of cases.

From the carcinological point of view, the chondrosarcomas which were the main reasons for performing our pelvic grafts were always resected on a large scale. We are pleased to report that there were only very few relapses. On the other hand, the other sarcomas all resulted in the death of the patient after a longer or shorter interval.

Discussion

Oncological Indications

The results were good for the massive allografts on all or part of the pelvis when the reason for the excision was a chondrosarcoma. For the other sarcomas: fibrosarcoma, rhabdomyosarcoma, or Ewing's tumor, the patients survived for no longer than one year in spite of pre-operative, per-operative, and postoperative chemotherapy, which confirms the severity of the siting of these tumors in the pelvis which are often diagnosed at too late a stage.

Infectious Indications

The hemipelvic graft in one case of echinococcosis developed into the hydatidosis recurring. This indication was adopted because of the condition extending to the whole of the hemi-pelvis and to the chamber of the adductor muscles but this extensive hydatidosis was not operated on sufficiently radically and there were relapses from the recesses in the muscles of the thigh. Septic loosening of prostheses appear to us to be a good indication for surgery in two stages. We no longer have any septic relapses with allografts apart from with metal reconstruction prostheses [21,22].

Sheathed Prostheses

Since we have been using sheathed reconstruction prostheses, the functional results have improved because of the ease of fixing the muscles on the allograft [23]. The cortical sheaths are rehabilitated on the periphery and allow the muscle masses to be fixed.

In All

A fracture of the allograft may heal completely. The allograft–host bone union has to be fastened by stable osteosynthesis and surrounded by spongy autografts implanted during the operation [24]. Osteosynthesis of the hemipelvic grafts has evolved: initially we only secured the graft to the sacrum with three screws. Now we use a screwed plate attached to an autograft. For the same reasons of stability, the shafts of the prosthesis are cemented into the host femur as well as into the allograft. We only found one case of pseudarthrosis at the host–graft union.

Healing may take a very long time, up to four years [25]. From the point of view of biomechanics, the sheathing of the prosthesis reduces the stresses on flexion on the shaft, which explains why none of the shafts of the prostheses broke. The number of septic complications is comparable in our series to that of the series using massive metal prostheses [3,13,26,27]. On the other hand, a number of serous effusions (always aseptic) led us to review the immunology of the bone allografts. To date, allografts are performed without taking the HLA or even the ABO Rhesus groups into account [24]. Perhaps more importance should be attached to these compatibility phenomena.

Enneking [28] demonstrated the benefits of reducing the inflammatory reaction on contact with a massive allograft of the femur of a dog. The host–graft union healed more quickly when this non-specific reaction was suppressed. In rats of different strains, Muscolo [29] found immune responses of the humoral type after bone allografts. Friedlander [30] showed, by using microcytotoxicity reactions in the rabbit, that cryopreserved cortical bone caused fewer reactions than cortico-spongy bone and fresh bone, but that this reaction existed nevertheless.

Bibliography

Enneking WF, Eady JL, Burchardt H. Autogenous cortical bone grafts in the reconstruction of segmental skeletal defects. J Bone Joint Surg (Am) 1980;62A:1039–58.

Friedlaender GE. Current concepts review. Bone grafts. The basic science rationale for clinical applications. J Bone Joint Surg (Am) 1987;69A:786–90.

Friedlaender GE, Strong DM, Sell KW. Studies on the antigenicity of bone. II. Donor-specific anti-HLA antibodies in human recipients of freeze-dried allografts. J Bone Joint Surg (Am) 1984;66A:107–12.

Gérard Y. Banque d'os: introduction. Rev Chir Orthop 1988;74:110–1.

Hernigou P. Conservation des allogreffes osseuses. Rev Chir Orthop 1988;74:114–6.

Hernigou P, Delépine G, Goutallier D. Allogreffes massives cryopréservées et stérilisées par irradiation. Rev Chir Orthop 1986;72:267–76.

Jofe MH, Gebhardt MC, Tomford WW, Mankin HJ. Reconstruction for defects of the proximal part of the femur using allograft arthroplasty. J Bone Joint Surg (Am) 1990;70A:507–16.

Ottolenghi CE. Massive osteo and osteo-articular bone grafts: techniqe and results of 62 cases. Clin Orthop 1972;87:156–64.

Petty W, Goldsmith S. Resection arthroplasty following infected total hip arthroplasty. J Bone Joint Surg (Am) 1980;62A:889–96.

Poitout D, Novakovitch G. Utilisation des allogreffes en oncologie et en traumatologie. Int Orthop 1987;11: 169–78.

Poitout D. Biologie des allogreffes osseuses. Rev Chir Orthop 1988;74:112–4.

References

1. Allan DG, Lavoie GJ, Mc Donald S, Oakeshott R, Gross AE. Proximal femoral allografts in revision hip arthroplasty. J Bone Joint Surg (Br) 1991;73B:235–40.
2. Borjaz FJ, Mnaymneh W. Bone allografts in the salvage of difficult hip arthroplasties. Clin Orthop 1985;197: 125–31.
3. Burrows HJ, Wilson JN, Scales JT. Excision of tumours of the humerus and femur, with restoration by internal prosthesis. J Bone Joint Surg (Br) 1975;57B (2):148–59.
4. Delloye C, De Nayer P, Allington N, Munting E, Coutelier L, Vincent A. Massive bone allografts in large skeletal defects after tumor surgery: A clinical and microradiographic evaluation. Arch Orthop Traum Surg 1988;107:31–41.
5. Loty B. Irradiation des allogreffes osseuses. Rev Chir Orthop 1988;74:116–7.
6. Enneking WF, Morris JL. Human autologous cortical bone transplants. Clin Orthop 1972;87:28–35.
7. Emerson RH Jr, Head WC, Berklacich FM, Malinin TI. Non-cemented acetabular revision arthroplasty using allograft bone. Clin Orthop 1989;249:30–43.
8. Gross AE, Lavoie MV, McDermott P, Marks P. The use of allograft bone in revision of total hip arthroplasty. Clin Orthop 1985;197:115–22.
9. Loty B. Allogreffes osseuses massives. Rev Chir Orthop 1988;74:127–31.
10. Mankin HJ, Doppelt S, Tomford W. Clinical experience with allograft implantation: the first ten years. Clin Orthop 1983;174:69–86.
11. Poitout D, Lempidakis M, Loncle X. Allogreffes ostéo-cartilagineuses massives dans le traitement des nécroses ou des pertes de substance articulaires. Chirurgie 1991;117:193–8.
12. Poitout D. Indications classiques des allogreffes osseuses. Rev Chir Orthop 1988;74:118–9.
13. Berry DJ, Chandler HP, Reilly DT. The use of bone allografts in two-stage reconstruction after failure of hip replacements due to infection. J Bone Joint Surg (Am) 1991;73A:1460–8.
14. Enneking WF, Mindell ER. Observations on massive retrieved human allografts. J Bone Joint Surg (Am) 1991;73A:1123–42.
15. Bonfiglio M, Jeter WS. Immunological responses to bone. Clin Orthop 1972;87:19–27.
16. Friedlaender GE. Immune responses to osteochondral allografts, Current knowledge and future directions. Clin Orthop 1983;174:58–68.
17. Stevenson S, Horowitz M. Current concepts review. The response to bone allografts. J Bone Joint Surg (Am) 1989;71A:1297–307.
18. Enneking WF, Spanier SS, Goodman MA. Current concepts review. The surgical staging of musculoskeletal sarcoma. J Bone Joint Surg (Am) 1980;62A:1027–30.
19. Merle d'Aubigné R, Méary R, Thomine JM. La résection dans le traitement des tumeurs des os. Rev Chir Orthop 1966;52:305–24.
20. Poitout D, Gaujoux G, Lempidakis M. Reconstructions iliaques totales ou partielles à l'aide d'allogreffes de banque. Int Orthop 1990;14:111–9.
21. Wilson PD Jr, Aglietti P, Salvati EA. Subacute sepsis of the hip treated by antibiotics and cemented prosthesis. J Bone Joint Surg (Am) 1974;56A:879–98.
22. Wroblewski BM. One-stage revision of infected cemented total hip arthroplasty. Clin Orthop 1986;211: 103–7.
23. Huten D. Utilisation des allogreffes osseuses dans les reconstructions fémorales au cours des reprises de prothèse totale de hanche. Rev Chir Orthop 1988;74:122–4.
24. Hernigou P. Evolution des allogreffes massives. Rev Chir Orthop 1988;74:131–4.
25. Hedde C. La réparation des lésions cotyloïdiennes par allogreffe de tête fémorale de banque. Rev Chir Orthop 1988;74:120–2.
26. Pellicci PM, Wilson PD Jr, Sledge CB, Salvati EA, Ranawat CS, Poss R et al. Long-term results of revision total hip replacement. A follow-up report. J Bone Joint Surg (Am) 1985;67A:513–6.
27. Sim FH, Chao EYS. Hip salvage by proximal femoral replacement. J Bone Joint Surg (Am) 1981;63A:1228–39.
28. Enneking WF. Histological investigation of bone transplants in immunologically prepared animals. J Bone Joint Surg (Am) 1957;39A:597–615.
29. Musculo DL, Caletti E, Schajowicz F, Araujo ES, Mkino A. Tissue-typing in human massive allografts of frozen bone. J Bone Joint Surg (Am) 1987;69A:583–95.
30. Friedlaender GE, Strong DM, Sell KW. Studies on the antigenicity of bone. I. Freeze-dried and deep-frozen allografts in rabbits. J Bone Joint Surg (Am) 1976;58A: 854–8.

26 Deep-frozen Osteochondral Allografts

D. G. Poitout, P. Tropiano, B. Ripoll, and G. Marck

Reconstructive megaprostheses are commonly used but they might be responsible for mechanical failures in young patient cases. On the other hand, allografts allowing vascular recolonization and a rapid muscle tightening give better functional results. We are only just beginning to receive results concerning graft integration and cartilage evolution (13 years on). One of the main problems concerning articular grafts is the biomechanical behavior of ligamentary allografts, or the refixation of the receiver's capsule onto the graft.

Materials and Methods

Cartilaginous Allograft

Biology and Immunology

The nutritive substances necessary to the survival of chondrocytes can only be supplied from the synovial liquid or from the blood vessels of the sub-chondral bone. This underlines the importance of the permeability of the matrix depending on the degree of hydrophilicity of proteoglycans. Cartilaginous cells, like collagen, have a very long turnover (over 300 years) so the cartilage allografts do not have to be replaced or rebuilt after grafting. Immunologically speaking, cartilage is not responsible for the immunological response because of its matrix, which isolates chondrocytes and immunologically active cells from the receiver's antigenic activity.

The grafting of a normal cartilage does not induce an immunological reaction, although each of its composite parts induce an immunological response.

Preservative Process

These are fundamental to maintaining the nutrition of articular graft. As chondrocytes cells must be kept alive we have to avoid methods which lead to their destruction. Freeze drying and irradiation can not be used for this reasons, which is why, beginning in 1981, we have used cryopreservation in liquid nitrogen at −196 °C for bone and cartilage allografts. This method allows the indefinite conservation of massive osteo-cartilaginous pieces if a cryopreservative such as dimethyl sulfoxide (DMSO) is used.

Clinical Aspect of Cartilage A

Bone and Cartilage

Since 1979 we have used fresh allogenic bone grafts and, since 1981, deep-frozen, allogenic grafts to rebuild the skeleton. Massive osteochondral allografts have been commonly used since 1985. Between 1978 and 2000, there have been 185 massive osteochondral allografts.

If we want to transplant an osteocartilaginous graft we can use a massive allograft with the cartilage surface, so that revascularization can take place slowly. Articular allografts can only succeed if the biomechanical behavior of the joint is borne in mind. Histologically speaking, chondrocyte cells remain alive in most cases. Biopsies show that cartilage architecture is frequently impaired, especially regarding the superficial levels. Also, the lack of pain, the normal radiological aspect, and the lack of necrosis seen in most of our cases is appreciable.

We will discuss five possible methods of osteochondral grafting.

Cartilage alone without ligamentary fixation, which gives generally excellent results.

Osteochondral allografts without ligament or capsule. It appears that refixation of the receiver's ligaments on the allografts occurs rapidly and gives, within a few weeks, an excellent stability to the joint.

Osteochondral allografts with a ligament or capsule. Tendons and ligamentary allografts induce in 10% to 20% of cases an immunological response but the biomechanical behavior of the joint after grafting seems not to be as good as refixation of the receiver ligaments on the grafts after 1–2 years, because of the revascularization of the transplanted ligament which is very slow and induces instability in the joint. The allograft ligaments need to be protected by an artificial ligament during this revascularization period.

Total joint reconstruction. In bone tumors when muscles and ligaments have to be removed, as in some traumatological cases, the use of massive whole joint allografts can be indicated (knee, elbow) but in most cases instability appears within a few months.

Massive prosthesis surrounded by allograft. These lead to poor vascular surrounding of the graft and problems linked to articular instability. We propose the use of massive metallic articular prostheses surrounded by one allograft. Prostheses give immediate stability and allow the patient to walk within a few days, while the graft will permit the remaining muscles to reattach themselves.

This latter is the procedure of choice when a reconstruction of the upper part of the femur is required. Between 1978 and 2000 there were 356 cases.

For distal femur and proximal tibia cases, it is necessary to remove the ligaments, capsule, and large part of the muscles. Between 1978 and 2000 there were 67 cases. In the case of reconstruction after some weeks of the lower part of the femur, we have to use an expending bag placed under the muscles and the skin to avoid tension of the skin and the risk of necrosis.

Results

Complications have arisen related to cases of massive osteocartilaginous allograft (185 cases):

Six tumoral recurrences leading to amputation.

Four cases of sepsis in patients with heavy chemotherapy, also leading to amputation.

Six cases of sepsis where traumatic lack of substance with skin and muscle defects led to do an arthrodesis.

Two fractures of the internal tibial plate obliged us to use a megaprosthesis surrounded by allograft.

One case of sepsis two years after the operation, forcing us to remove the first prosthesis and replace it six months later after having made a muscular and skin reconstruction. The functioning is now fair.

Two total knee prostheses have been used to replace the articular shaft destroyed by articular laxity due to ligamentary lengthening.

Discussion and Conclusion

Histological, immunological, biological, and biomechanical studies have shown that a cartilage allograft can be used if:

an entire and normal cartilage is grafted,

if the structure is not impaired,

if the preservation process has not killed the cells or destroyed the proteinic matrix, the collagen, or the proteoglycans,

and if the biomechanical behavior of the joint is considered.

In other cases, the use of massive articular metal prostheses surrounded by allograft seems to be more appropriate, especially in bone tumor cases.

III

Biomechanics of the Bone Growth

27 Biomechanics of the Spine During Growth

J. Dubousset

Introduction

The biomechanics of the spine starts at conception. Soon after the gametes meet, their DNA combine to combine all the genes necessary to determine the unique aspects, both static and dynamic, of the spine that is so important for erect bipedal posture – the basic characteristic of the human skeleton.

Even if there are general characteristics common to most, every person is unique. It is the multiplicity of variations that so define normality within the population.

The growth of the spine starts long before birth, then continues throughout infancy and childhood until the end of adolescence with the so-called "growth spurt" completing the entire program. Each phase of this time is different from the previous one due to not only the size of the elements involved but also the physical and biological characteristics of the tissues involved.

Finally, according to all the genetics, the young adult will benefit from functional harmonization between organ systems throughout the growth period. For example, the nervous system plays a major role as part of the "spinal organ" and develops contiguously with the spine itself.

The spinal organ is from an anatomical point of view made of bone structures – the vertebrae – linked by intervertebral disks and ligamentous and muscular formations. Its goals are two-fold: first to give static and dynamic support for the trunk, considering that the first vertebra – from a biomechanical point of view – is the head in its proper weight and location (being that the occipital bone develops from the first three cephalad somites), so we can call it cephalic vertebra. In doing so, we must consider that the entire pelvis should be considered the last vertebra (pelvic vertebra). It acts as *intercalary* bone between trunk and lower limbs with a strategic role in both standing and sitting. Secondly, the spine is the supportive and protective structure surrounding the central nervous system structures, including specifically the spinal cord.

Embryology

It is important to remember the various embryologic stages during the first month of life, because as soon as the 15th day after conception, the embryo (1.5 mm in length) is an embryonic disk made up of three distinct structures: ectoderm, mesoderm, and endoderm. From the 15th day, the neural fold starts to establish on the ectoderm. By day 20 it starts to close at the middle, giving rise to the neural tube. Closure of this neural tube occurs more quickly on the cephalad side (26th day) than on the caudal side (28th day). It is during this time that (under the influence of the notochord) the mesoderm will organize. The notochord is the longitudinal axis and, around it, vertebral bodies will organize. The paraxial mesoderm becomes segmented, giving rise to the first somites that develop centrally within the embryo. In reality, because of the important development of the cephalic part of the embryo, this level fits with the occiput. Segmentation then continues caudally. By the end of the fifth week, the human embryo consists of 42 pairs of somites, half on the right, half on the left. These somites will give the future vertebrae and the muscles (myotome).

At the beginning of the fourth week, it is easy to see cells from the medial border of the somites migrating around the notochord to build the outline of the vertebra, the remaining cells spread out laterally to give rise to the muscles, notably the paravertebral muscles. By the 28th day, when the caudal neuropore is closed, spinal segmentation is already perfectly apparent. At this purely mesenchymal stage, all the vertebrae are organized anatomically. At the same time, developing from the mesoderm, the future nephrotome (that goes on to develop into much of the urogenital system) will develop, as well as the heart and great vessels.

At approximately two months of gestation, chondrification and them ossification will begin. These continue until adulthood at around 18–20 years old.

Biomechanical Consequences

It is clear that any problems occurring during this embryologic phase will disturb the future biomechanics of the spine. For example, congenital malformations occur early during development. Any missing levels of bone, cartilage, or soft tissues leading to instability or any malalignment with kyphosis or lateral curvature (with related consequences as failure of segmentation) at an early stage of development will disturb growth and development in later embryonic or postembryonic life.

Vertebral Growth in Space and Time (Figure 27.1)

Vertebral Growth from Enchondral Ossification

The model of the cartilage starts to ossify as soon as the second gestational month. The ossification points exist at the level of vertebral bodies, posterior arches, laminae spinous, and transverse processes. They determine the real growth plate that is already visible at birth, by which time only 30% of spinal ossification is achieved.

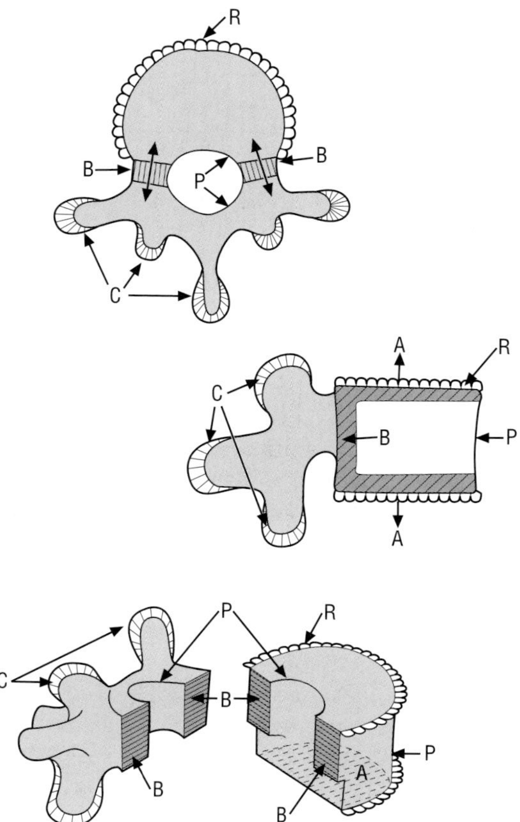

Figure 27.1. Growing cartilage of a vertebra. A. Vertebral plateau. B. Bipolar cartilage = neurocentral cartilage. C. Posterior elements apophyseal cartilage. D. Periosteum. E. Ring apophysis.

Vertebral Growth Plates

Each vertebra has many growth plate levels, but these are coordinated, so that growth is synchronous in all three dimensions.

Each vertebral plateau has its own growth plate (one superior and one inferior) for each vertebra, providing a vertebral growth in length of 1.2 mm/year per vertebra in the lumbar region and 0.9–1 mm in the thoracic region.

The neurocentral cartilage is very important in that it represents the junction between the anterior and posterior body. It is a bipolar cartilage growing in both antero-posterior and horizontal directions. Each is located vertically to and on each side of the neural canal between the two plateaus. The junction between vertebral plateau, cartilage, and neurocentral cartilage is really a unique region when three-dimensional growth is determined. These will close between 8 and 12 years and sometimes older. It is clear that the unilateral closure of these cartilages leads to horizontal deformity, typically a scoliosis.

The growth around the posterior arch has growth that is concurrent with neural tube evolution. Secondary ossification centers appear at the tip of spinous process transverse, facet joints and work under the influence of ligaments and muscle insertions. It is important to note that the *pars interarticularis* is already ossified at birth.

Generally, the fusion of spinous processes on the middle line occurs at age one, first in the thoracic area and then lumbar and sacrum. Reverse ossification is active for the atlas at age two and the axis at age four.

The size of the spinal canal is remodeled throughout the construction–destruction process and has reached adult size at age five. This is an essential point for pathology, allowing, for example, the performance of circumferential fusion on a small child without worry [1–3].

Finally, at age ten a ring apophysis appears at the superior and inferior plateau border that ossifies slowly and fuses with the vertebral body between 18 and 25 years old.

Vertebral Growth in Time (Figure 27.2)

As it very well represented in the excellent book of Alain Dimeglio [4], spinal growth is exponential from 0 to 5 years old. At that age, it should be noted that 70% of sitting height is achieved in girls and 66% in boys. The gain for the upper segment is 27 cm in five years (12 cm for the first year, 5 cm for second year, 4 cm for third year and 3 cm for fourth and fifth years. Between 5 and 10 years old, growth is slower at 2–3 cm/year. At 10 years old, 80% of final height is acquired. The growth spurt then occurs with a mean growth of 4.5 cm/year for the upper segment, which is then variable from one

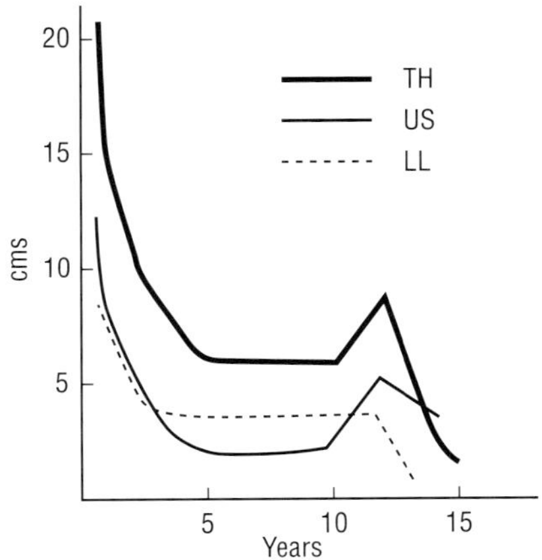

Figure 27.2. Growth in time (speed). TH: Total height. US: Upper segment height (sitting height). LL: Lower limb height.

patient to another. The amplitude of such growth spurts is higher in boys (2 cm/year more) than in girls, but growth spurts are always two years later in boys. At the time of the growth spurt, the individual speed of growth for each vertebra is 1.2 mm for thoracic and 1.6 mm for lumbar between 13 and 16 years.

Another factor about T1–S1 size is important for biomechanics. The thoracic segment represents two-thirds and the lumbar segment one-third, but T1–S1 represents 50% of sitting height, which is why it is important to know the maturation of bone and cartilage for spinal growth. The bone age of the hand is less helpful than Risser sign that remains an important marker: linked with the closure of the triradiate cartilage (corresponding more or less to Risser +), linked and correlated to closure of the cartilage of the elbow, but also correlated to sexual maturation (pubic hair, breasts in girls, testis volume and beard in boys). These signs give an idea of the status of the patient at the time of examination, allowing the real vertebral age to be assessed, because none of these signs is absolute by itself. Large variation may exist from one individual to another and if we multiply the signs a closer measurement can be made.

Finally, it is very important to remember that thoracic cage growth is very much related to spine development. From newborn infants to adults, thoracic cage volume enlarges by a factor of 14, and 50% of the final volume is achieved by age 10. However, it should be noted that the thorax continues to increase in volume for two years after spinal growth has finished.

Biomechanical Consequences

We have many growth tables and charts to help to evaluate the development of the spine and the thorax according to age. This helps to decide, for example, when it is a good opportunity to perform fusion for spinal deformity and it is important to recognize the resultant loss of height that can be expected, especially if it is extended to the thoracic area. If the entire thoracic area in a boy of 10 years old were fused, 6 cm of thoracic height would be removed, and if the same were to occur in the lumbar area, 3.6 cm would be lost.

A typical consequence for biomechanics of spinal growth in space as well in time is the crankshaft phenomenon (Figure 27.3), which is the phenomenon complicating some treatments of the pediatric spine. It is seen when isolated posterior fusion is realized on a scoliotic immature spine. Despite a solid posterior fusion, a progressive bending of the fusion mass is observed because of the continuation of anterior growth. The biomechanical explanation is that the growth plate on the front is deviated laterally but continues to grow and cause further deformity until growth is completed. This is why treatment is rooted in biomechanics. It is a preventive anterior epiphysiodesis performed at the same time as posterior fusion and on the same levels [5–7].

Importance of the Soft Tissue Components of the Spinal Organ

Even if soft tissues (disks, ligaments, and muscles) comprise almost half of the anatomic structures of the spinal organ, research in them is less than in bone and cartilage. Nevertheless, their quality appears very important when considering flexibility or stiffness and even in some case laxity and instability.

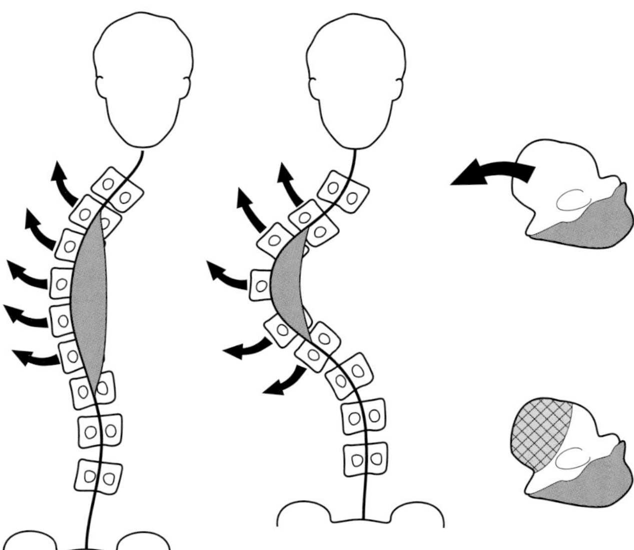

Figure 27.3. Crankshaft phenomenon. A solid posterior fusion on a growing scoliotic spine plays the role of a posterior tether on a distorted element and the continuation of anterior growth (arrow) increases the rotational deformity. Prevention is anterior convex epiphysiodesis (at the opposite side of the posterior fusion).

The Definition of Instability

The definition of instability is still controversial and involves consideration of the bone and cartilage in the joint facets and disk space structures as well as the soft tissues in disk space structures, capsules, ligaments, and muscles. The quality of these tissues, especially connective tissues, collagen, and elastic tissues, depends mainly on the genetic aspects of their basic molecular structure. From early childhood, one can recognize the elastic component of the tissues with hyperextension of the thumb or hyperlaxity of the joints. This may contrast to joint stiffness with a decrease of normal motion. The same variability occurs in muscle displacement and range of motion; everything is individual.

It is clear that with increasing age, maturation of the soft tissues occurs, but this phenomenon is much less studied than bone and cartilage maturation. For example, hypermobility is more frequent at an early age, occasionally allowing hypermotion, which suggests instability, such as the C2–C3 flexion extension test where we can find such over-alignment considered physiological. On the other hand, one can have a large amount of displacement in severe neck injury in infants giving spinal cord injury without any bone fracture.

Of course, the relative higher weight of the head related to the entire body in an infant can play a role in these problems, but it is also certain that with time this hypermobility decreases unless a real pathological connective tissue pattern remains such as in Ehlers–Danlos or Marfan disease.

The Biomechanical Importance of Spine Stability

The biomechanical importance of spine stability and the permanent mixture between bone and cartilage and soft tissue conditions is well demonstrated by the consequence of laminectomy in a growing child. A biomechanical study has been made in the spinal columns of neonates and children by measuring intradiscal pressure of a specimen submitted to constant load in the normal specimen first, and then after removal one by one of the posterior structures. Starting with only removal of the interspinous ligament, the pressure increases; then as one side of the ligamentum flavum then both sides, then one facet joint and then the second facet is removed, the pressure increases further. This clearly demonstrates that the pressure inside the disk space was increasing constantly proportional to increasing posterior destruction of the stabilizing elements with a subsequent increase in kyphosis (Figure 27.4, 27.21) [8–11].

This experiment corroborates with our observation of a direct relationship between increasing instability and kyphosis in pediatric pathology. Finally, the hyper pressure on the disk space transmits to the growth plate and delays growth potential because the hyper pressure completely changes the aspect of the body, becoming cuneiform with a decrease in anterior growth and progressing kyphosis.

Prevention by laminoplasty instead of laminectomy when possible – and consisting of replacing the posterior elements removed en

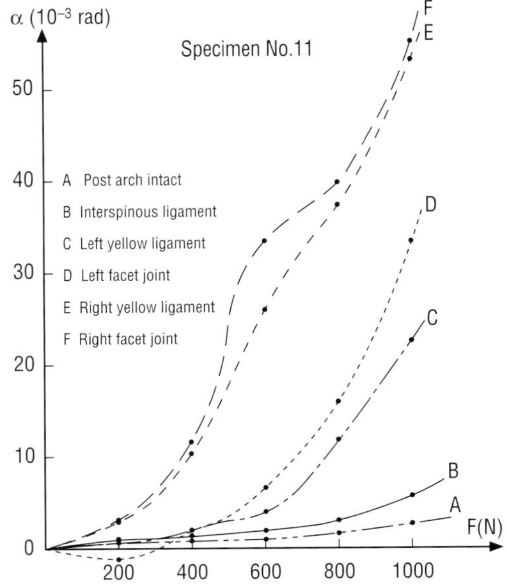

Figure 27.4. Evolution of kyphosis after experimental removal of the successive elements of the posterior arch in a spinal specimen. α = sagittal angle. F = force applied. After D'après H. Robert. J Biophys Med Nucl 1984;8(4):243–9.

bloc after the neurosurgical intraspinal work associated with proper immobilization giving proper healing – prevents kyphosis and secondary growth disorders.

The Importance of Soft Tissue Maturation

The importance of soft tissue maturation is well demonstrated in a study of a group of patients with scoliotic curves where nonsurgical treatment was started after the end of most vertebral growth (Risser +++ or ++++) and continued to one year after completion of bone maturation. Despite complete bone maturation, improvement of the curve was maintained after removal of the brace, so the only explanation can come from the maturation and stiffness seen on the soft tissues of such scoliotic spines [12].

The Importance of Soft Tissue Factor in the Growing Spine

The importance of soft tissue factor in the growing spine with a combination of ligamentous, muscular, and nervous components is also easy to demonstrate by studies of patients simultaneously injured during the treatment of some tumoral disease of the chest, such as neuroblastoma, in the paraspinal area.

When ribs are removed in a child with subsequent loss of intercostal space and then an intercostal neurovascular bundle is removed, the result is invariably a spinal deformity convex toward the side of the injury.

It is not only a question of bone but it is also possible for a soft tissue defect to give slight instability on the ligament linking two or three spinal segments. This leads to a localized difference in bone pressure of the growth plate secondary to paralysis of posterior muscles leading to deformity. Even if radiation therapy is given with the well-known disastrous effects on the growth plate, the deformity is still convex to the side of the soft tissue injury. Conversely, for Wilms tumor in a child treated with surgery and radiation, the deformity is secondary to asymetrical radiation because there is no involvement of paraspinal muscles and the effect of asymetrical radiation on the growth plate results in a deformity concave to the side of the injury. These two examples demonstrate the biomechanics of the bone as well as biomechanics of soft tissue on a growing spine [5,10].

Setting up of Erect Posture and its Consequences

Development of Erect Posture in Humans [13]

At birth, the shape of the sagittal spine is like a gentle C curve. When the child is in a prone position and then starts to lift the heavy head, cervical lordosis develops. The next step is crawling on the upper limbs and knees, during which time cervical lordosis increases and lumbar lordosis initiates. Then when the child stops, he will generally lift up the upper limbs and raise his back on the knee flexed, but with hip extended, leading to further lumbar lordosis progression. Then the child learns to stand up with proper cervical lordosis, thoracic kyphosis, and lumbar lordosis. The erect posture is so acquired and will develop throughout childhood and adolescence into adulthood.

Maturation of the Central Nervous System

It is during the child's learning of how to achieve erect posture that the maturation of the nervous system takes place. At birth, the nervous system is still immature, especially when considering the myelinization of the central nervous area.

The coordination and achievement of balance continues to extend throughout the majority of childhood. The postural balance integrates sensory stimuli coming from the eyes, ears, vestibular and proprioception with proper reflex function from the motor reaction of muscles surrounding the joints themselves getting information from receptors inside the capsules or tendons. The action of the neurohormonal transmitters working on the postural balance is still largely unknown in terms of the relationship between right and left brain and the function of hypothalamus.

There is a research methodology developing around the etiopathogeny of idiopathic scoliosis with an experiment performed on the pineal gland, neurotransmitter function and a bipedal animal where scoliosis exists, which can be reproduced, and quadrupedal animals where scoliosis does not exist and cannot be reproduced experimentally with neurological lesion performed distant from the spine itself (14).

The maturation of this nervous system is probably greatly involved in idiopathic scoliosis and may explain why some severe infantile curves can regress completely with a pure nonsurgical treatment (brace or cast, for example) and some even spontaneously, despite some remnant of the deformity (cuneiform vertebrae in a straight spine). This may also explain, in cases of persistent immaturity of the nervous system, why some other malignant cases cannot be corrected with the same kind of treatment.

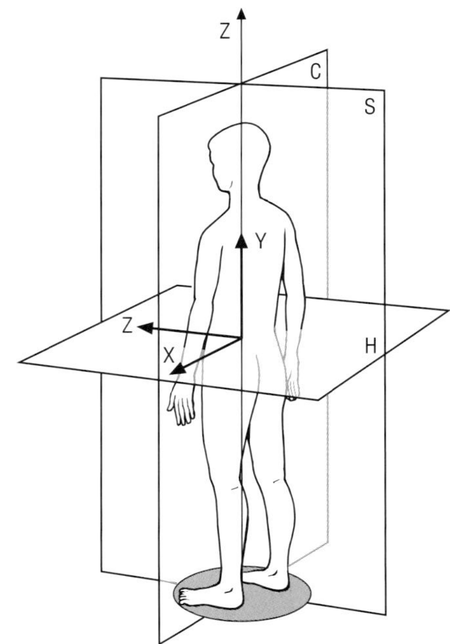

Figure 27.5. Standing posture planes of reference and polygon of sustentation. 1. Polygon of sustentation. 2. Orthogonal gravity line. R: Sagittal plane. H: Horizontal plane. C: Coronal plane.

Biomechanical Consequences of Erect Posture in Children and Adolescents

Note: the concepts recognized and developed here for the growing spine are easily adaptable for the adult spine.

Static and Dynamic Three-dimensional Balance (Figure 27.5)

If we consider the erect posture in a human, in a standing position both feet are pushing on the ground and delineate a surface ellipsoidal called the polygon of sustentation. If you consider the same human at any age seated, you have the same polygon of sustentation now looking like a frame with both thighs and ischial tuberosities.

From the center of this polygon of sustentation either standing or sitting, if you draw an orthogonal line, all of the body including the head is aligned harmoniously within this line with a patient in a good three-dimensional balance. Anatomically on the sagittal plane this line goes from the tragus to the center of this polygon and the spine is harmoniously aligned with cervical lordosis, thoracic kyphosis, lumbar lordosis, sufficient pelvic tilt, hips extended, and knee extended with the ankle joints at approximately 90° to the ground. This is defined as a balanced spine where the spinal curvatures balance themselves in opposite directions to achieve harmony and function with the head projecting inside this polygon of sustentation.

When the patient is seated, the same rules of sustentation apply with the head entirely balanced at the center of the sitting frame and the spinal curves adapting to the proper pelvic tilt necessary to compensate for the absence of lumbar lordosis or kyphotic lumbar deformity. This balance is defined as static when no motion at all is observed, but in humans there is in reality constant motion within this polygon in a three-dimensional manner.

Cephalic and Pelvic Vertebra Concept (Figures 27.6, 27.7)

Bearing in mind the three-dimensional balance, it is easy to consider the entire head with its own

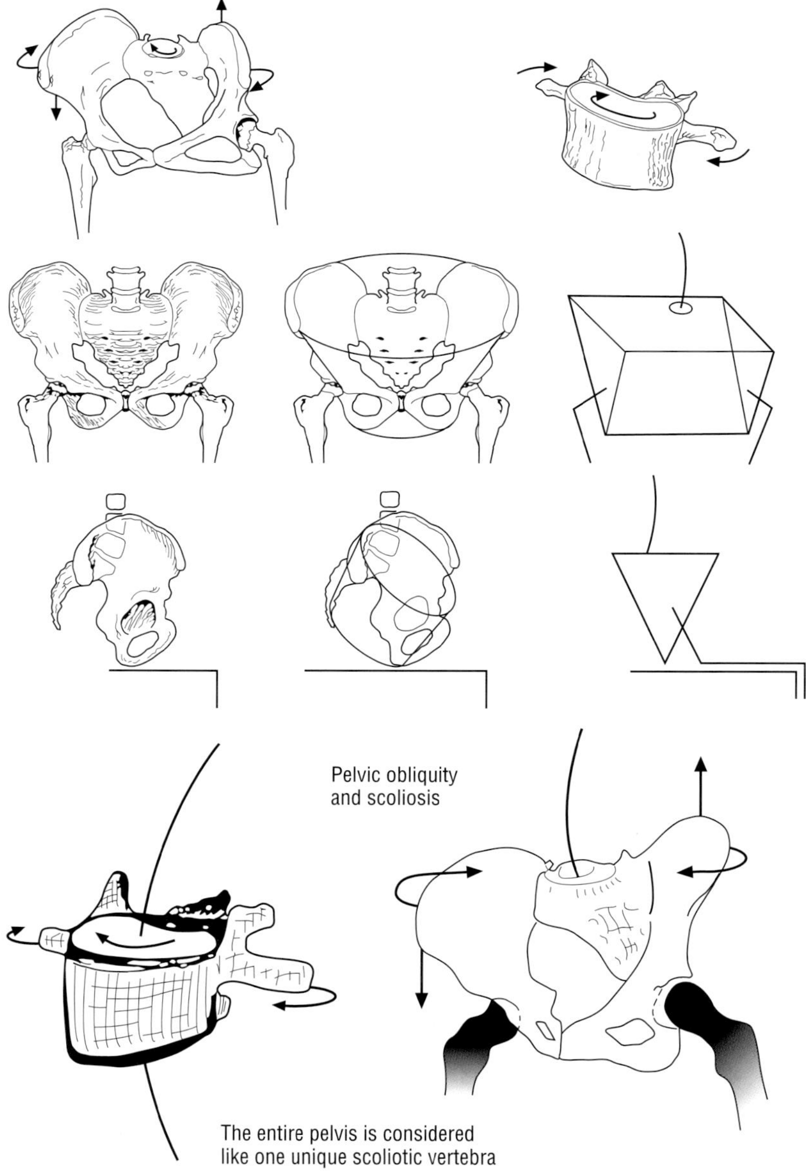

Figure 27.6. Coming from the study of pelvic obliquity and scoliosis, the entire pelvis must be considered as one single vertebra.

mass and weight (4.5–5 kg) as the first vertebra. This weight lying at the top of the spinal construct is almost always moving and, like a gyroscope, maintaining balance [13,15,16].

Simultaneously, the entire pelvis (because of the minimal movement inside the sacro-iliac joint – less than 1.5° motion, unless during pregnancy when it can reach 3.5° of motion) can be considered as one unique vertebra, the last in the spine, and acting like an intercalary bone between trunk and lower limbs.

Doing so when a global thoracic kyphotic deformity occurs, the pelvic vertebra can compensate by an anterior tilt as compensation, lordosis at the lumbar level, etc. This pelvic vertebra has 6° of freedom for each hip joint and

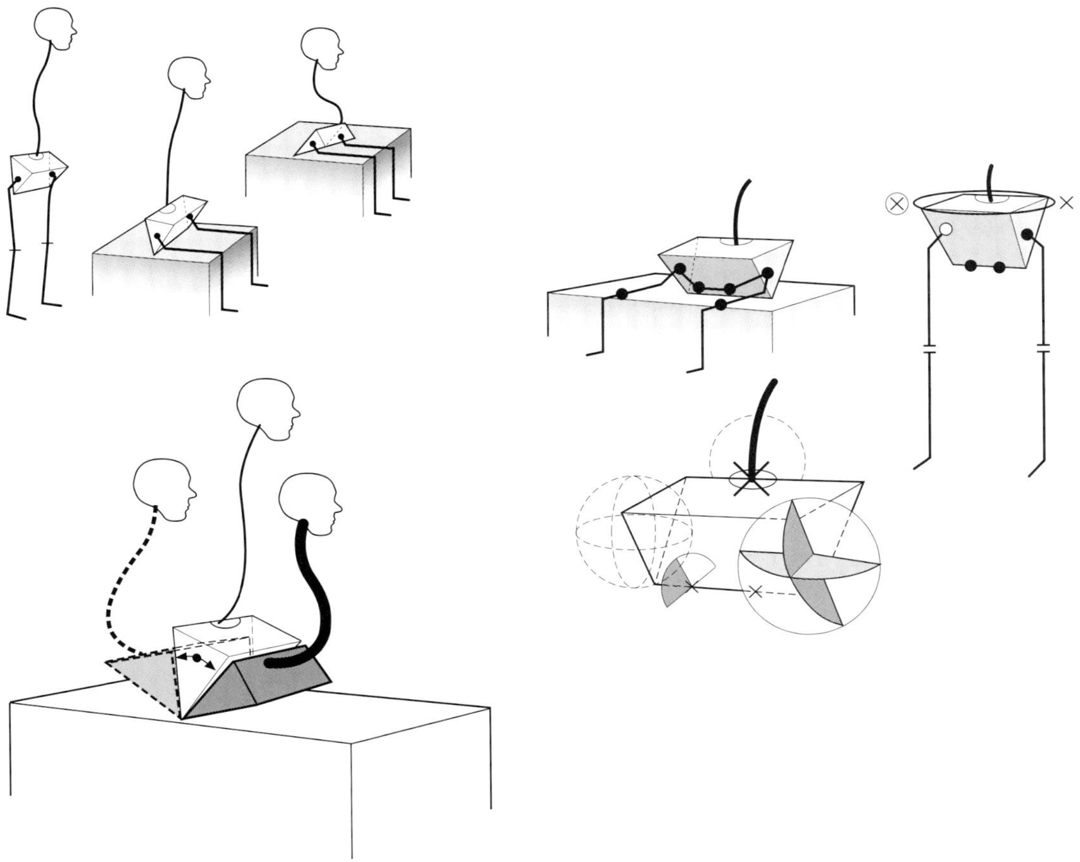

Figure 27.7. The entire head must be considered as the cephalic vertebra and the pelvis as the pelvic vertebra acting as an intercalary bone to achieve balance.

the same for the lumbo-sacral junction. It is interesting to note the various work about this pelvic vertebra, because it has been proven that the proper anatomy of this pelvic vertebra plays a major role in the amount of lordosis necessary to achieve balance, especially in the sagittal plane. The angle between the center of the femoral head and the orthogonal projection to the center of the S1 superior plateau determine the incidence. Variation in this angle is observed generally between various patients of no more than 12° to 15°. It is also remarkably stable during life with very few or almost no changes after age five. This angle determines the amount of lordosis necessary to maintain good sagittal balance. The incidence represents in reality the positioning of the sacroiliac (SI) joint in the space regarding one of the femoral heads. This explains the wide variation seen between males and females and the morphotypes of patients (thin and tall or short and wide).

The Concept of the Conus of Economical Consumption and Economical Function (Figure 27.8)

If we consider the human body in a standing (or sitting) posture, the feet are located within the polygon of sustentation, the body can, under the influence of muscle function, move in a conical fashion without moving the feet. The maximum variable amplitudes are at the pelvic and the head levels; we can determine the maximum dimensions of this "cone of movement" when

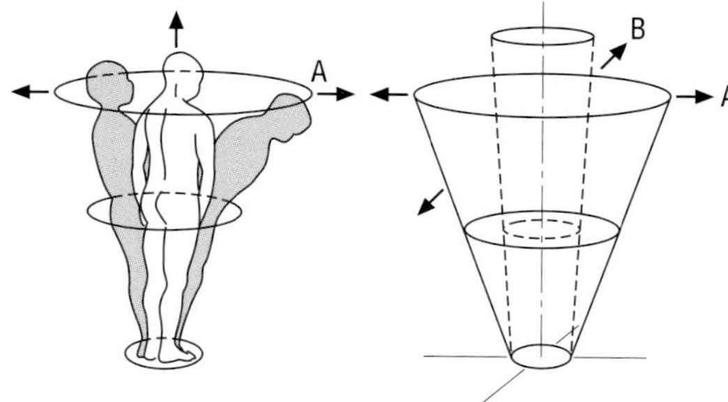

Figure 27.8. The concept of the conus of economical function in a standing position. A. Maximum of function before falling down. B. Economical function where the spine is balanced with almost no muscle action.

muscles are working at their maximum excursion and strength to maintain balance. But also we can determine a smooth cone when muscles are working at their minimum – in some cases doing almost nothing to maintain balance, which is mainly achieved by the passive balanced stretching of discs, ligaments, and soft tissue structures with the minimum of muscle action. This is seen when the body is well-balanced within this smooth cone, and muscle function is economical. When the body is out of balance, we must anticipate permanent costly muscle function [13,15,16].

Biomechanics of Ligaments, Aponevrosis and, Muscles Surrounding the Spine for Static and Dynamic Function During Erect Posture

One must distinguish two groups of soft tissues structures:

Ligaments and aponevrosis (not only within the spine itself, but also concerning the entire trunk of the body) will determine rigidity within the large thoracolumbar fascia. When all these fibers are stretched then the body is rigid and stable. When these fibers are relaxed then potential movement can occur, especially considering that motion and passive motion is only limited by the passive tension of these structures [16,17].

The muscles around the spine consist of two types, the ones very close to the spinal body structures like interspinous, paraspinous, multifidus, and transverse, and the ones relatively distant from the spine, like abdominal muscles, ilio-costalis, longissimus dorsi, longissimus lumborum. Each one of these muscles may have two main functions: some work like a spring pushing or pulling two elements one against the other; and some work like a ball expanding inside a net and tensioning the inextensible fibers of the network.

The first function (spring or pump) is mainly devoted to motion and voluntary maneuvers, the second is mainly devoted to posture and especially economical posture and are mainly reflex (Figure 27.9).

One can understand why the automatic reflex system is of importance when considering the necessity for humans to maintain or recover the horizontal gaze whatever the spinal deformity. There is a complex net of short autonomic reflexes from visual vestibular as well as mechanical extensor flexor muscles and ligaments of receptor origin to achieve erect posture.

It is probably a similar system that has allowed humans to acquire the skill of bipedal walking, because walking is a permanent search for balance after the instability induced by a first step. This is why patients who are not able to stay seated in an erect posture on the side of a table are also not able to walk, because the muscle function of their erect posture giving

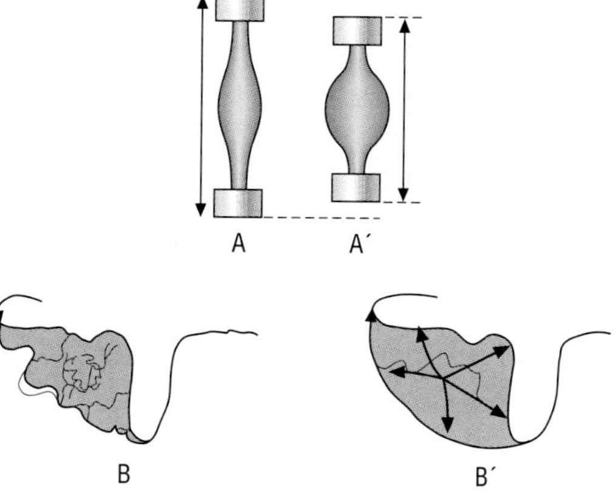

Figure 27.9. Concept of "string" and "pump" function of the muscles surrounding the spine, A, A′: String function moving the bone. B, B′: Pump function expanding and tensioning the net of aponevrosis of the postero-lateral spinal muscles.

stability and motion of their spine is not achieved. For the same reason, some diplegia cerebral palsy patients are unable to stop while walking in the middle of the room without the help of the upper limbs or a stable support.

Application to the Common Pathology of the Growing Spine

Sagittal Deformity

Pure sagittal deformity is easy to understand because there is no deviation either in the horizontal plane or in the coronal plane.

Deformity may be regular in abnormal kyphosis or lordosis, but occurring at every level, resulting in an apical zone that is the most deviated from the gravitational axis and yet is smooth because there is very little abnormality between one vertebra and the next or at the junctional zones – zones of transition from one direction (for example, kyphosis) to another (for example, lordosis). This regular kyphosis with a smooth apical zone generally does not create apical compression at the level of the spinal canal.

The main pathology for this is Scheuermann kyphosis with various apex, thoracic, or thoracolumbar, but very seldom lumbar, where hyperkyphosis results in a subsequent hyperlordosis above and below the kyphosis in order to maintain the sagittal balance. In contrast, thoracic hypokyphosis leads to a decrease in normal cervical and lumbar lordosis and creates a flat back. From a three-dimensional point of view, thoracic hyperkyphosis is generally associated with an increase in the antero-posterior diameter of the chest, the vertical orientation of the ribs, and very few respiratory consequences. In contrast, thoracic hyperlordosis is associated with a decrease of the antero-posterior diameter, horizontal ribs, and more respiratory compromise, leading to intermittent compression of the airway in some cases and atelectasia in major cases. This has allowed description of the spinal penetration index (Figure 27.10), which is the amount of protrusion of a real vertebral intrathoracic hump, which in a normal spine is around 10% of the surface or thoracic volume, and can reach 50% in some severe lordotic cases (Figure 27.10). This allows data to be established on the useful thoracic volume devoted to the lungs and mediastinal structures.

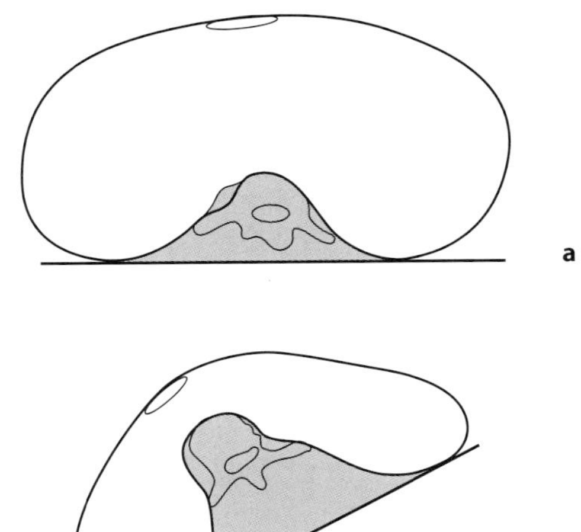

Figure 27.10. Spinal penetration index. The surface or volume occupied by the spinal structures inside the thoracic cage. Normally around 10% (**a**), it can reach 50% in severe cases (**b**).

Sagittal deformity may be angular; that is, the deformity lies in the sagittal plane only, but on a few levels (sometimes only one), and creates a sudden change in the alignment of the apical zone with possible compression of the spinal canal. The angular kyphosis may be especially associated with instability that may be immediate when the posterior alignment of the vertebral bodies are demonstrating a sudden abnormal motion when dynamic imaging is performed. It may be potential when dynamic imaging fails to demonstrate abnormal motion, but where malalignment exists from the beginning. In such cases, either progression of the deformity with growth or sudden mild injury can create a spinal cord traumatic lesion [1–3,18,19].

The biomechanical consequence for surgery of such angular kyphotic spines is that when an anterior empty space exists in front of the spine, it is mandatory to supplement the anterior pillar by anterior fusion to fill the gap and stabilize the

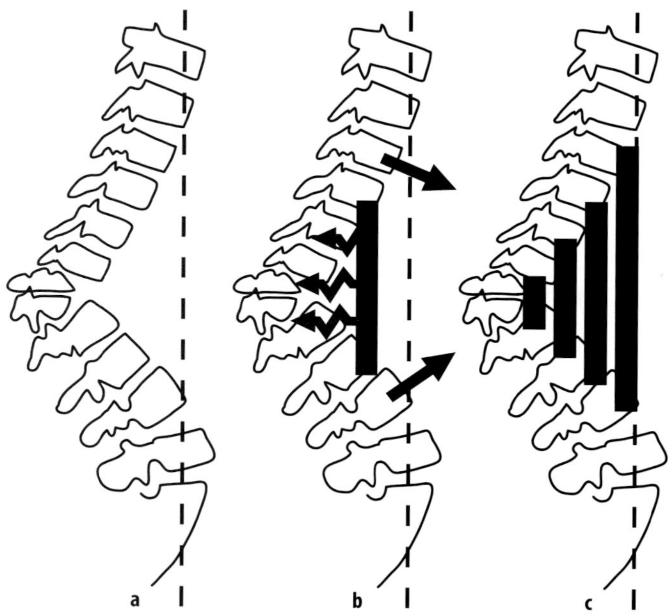

Figure 27.11. **a** Angular rigid kyphosis. Gravity line is represented by the dotted line. **b** Bypass strut graft kyphosis forces (straight arrows) remain straight and a gap opens toward the apex (z arrows). **c** To suppress it, it is necessary to perform a palissade strut graft from the apex of the curve toward the gravity line.

spine. This stability can be achieved if there is a large kyphosis by an anterior strut "palissade" graft inserted from the bottom of the apex going ideally to the gravity line and so suppressing the kyphosing forces (Figure 27.11).

Scoliotic Deformities

The basic structural scoliotic segment [12,16] is represented by a succession of vertebral units that are always located in extension or lordosis from one to the next with axial rotation of each vertebra always in the same direction. This begins with one neutral vertebra (without axial rotation) and continues with a successive increase in axial rotation, achieving maximum axial rotation at the apical vertebra of the segment. This then runs down with decreasing axial rotation in the same direction to the end at the next neutral vertebra. This movement is called torsion (Figure 27.12), the real basis for deformity, that cannot be located in one plane, but has an infinite number of planes. This phenomenon is secondary to a rotatory movement that is thwarted by the orientation of the pelvis, shoulder, and head, and involving subsequently all bone and soft tissue structures of the spine (and perhaps each molecule). Three-dimensional computerized reconstruction demonstrates well the anatomy and also the progression in space of the deformity with time. This also demonstrates that the Cobb angle so widely used as a measure in fact only measures the collapsing of the spine and not the real three-dimensional deformity (Figure 27.12, 27.13, 27.14).

So, a scoliotic spine is composed of either multiple scoliotic structural segments or by just one scoliotic structural segment followed above and below by a compensatory (not structural) segment in order to achieve alignment and balance for either standing or sitting. Each one of these segments is linked to another with a junctional zone representing either only one disk space, the zone of one disk and the vertebra above and the vertebra below, or at maximum a group of two consecutive vertebrae (the middle element has a neutral axial rotation while the adjacent has an opposite side axial rotation from above to below). This enables a balanced spine in three dimensions (Figure 27.14) and is why the maximal axial rotation in a scoliotic spine is located at the apex, but with the minimal intervertebral rotation (that is the

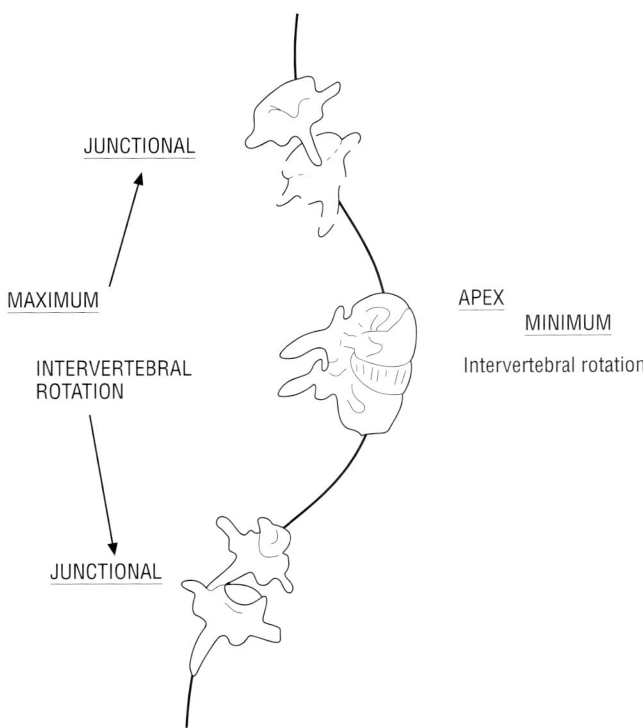

Figure 27.12. The basic structural scoliotic segment where the anterior length is greater than the posterior length at the level of two consecutive vertebral units.

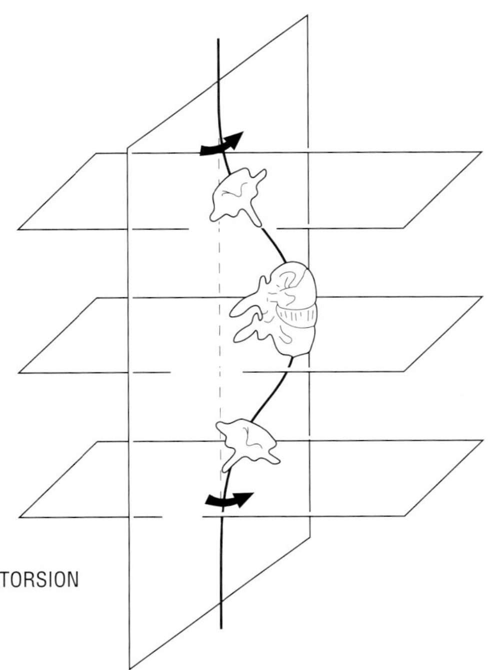

Figure 27.13. Torsion is the 3D characteristic deformity of a structural scoliosis.

axial rotation of two adjacent vertebrae), so the apex is the stiffest part. Conversely, the junctional zone has a minimum of axial rotation, but maximal intervertebral rotation, so is the most mobile (Figure 27.12, 27.16) [16].

This concept was very important in the biomechanical design and use of spinal instrumentation, especially for CD instrumentation of the many multiple hook, screw, and rod systems that are extensively used worldwide. It is why, in reality, for correction with these systems, compression, distraction, rotation, and translation are always used for any correction of a scoliotic curve and that all discussions about this problem of using more rotation than translation or compression are negated, because all mechanisms of correction are obligatory due to the three-dimensional shape of the scoliotic curve. These must be analyzed simultaneously in both antero-posterior and lateral planes, each segment replaced in the entire spine (Figure 27.15, 27.16).

There are two consequences of this three-dimensional analysis derived from the following concepts.

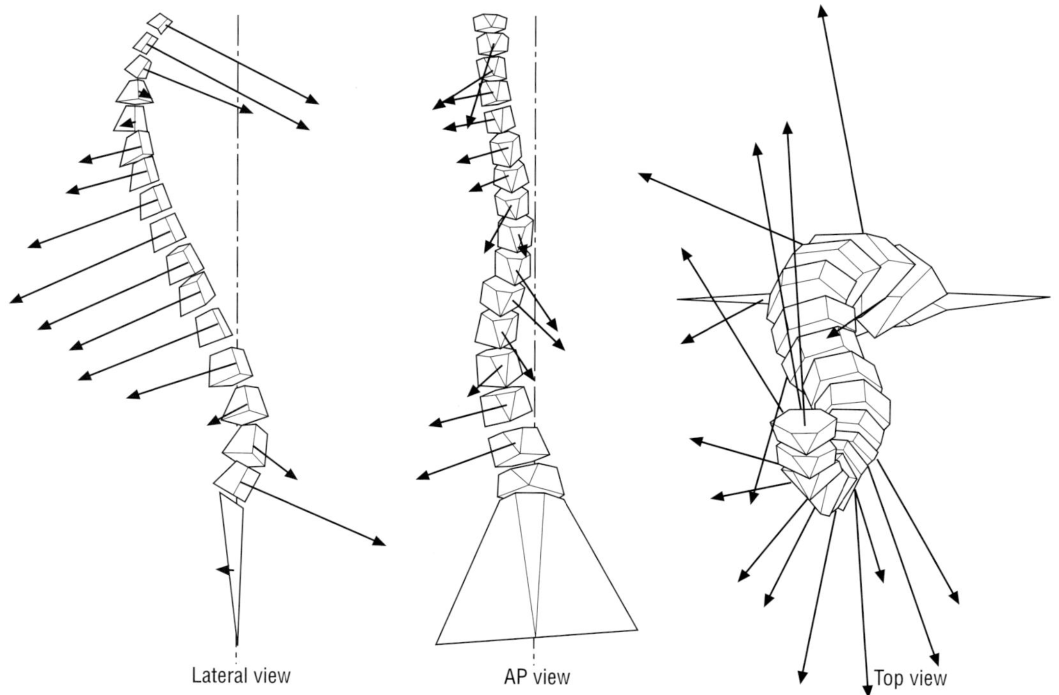

Figure 27.14. Three-dimensional computerized reconstruction of a scoliotic curve. The top view is particularly well suggestive of the real deformity not always recognized from X-rays.

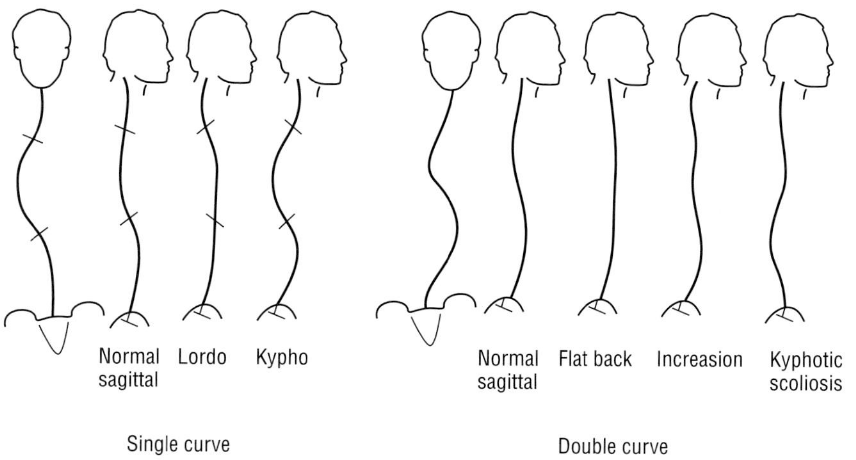

Figure 27.15. A practical approach to the surgical treatment of a scoliotic spine either for a simple curve or a double curve. Global analysis demonstrates well for a similar AP view the various possible sagittal alignments. To perform the proper fitting of the instrument, it is mentally necessary to match the AP and lateral aspects globally as well as for each segment of the spine and determine the apical zone and junctional zone.

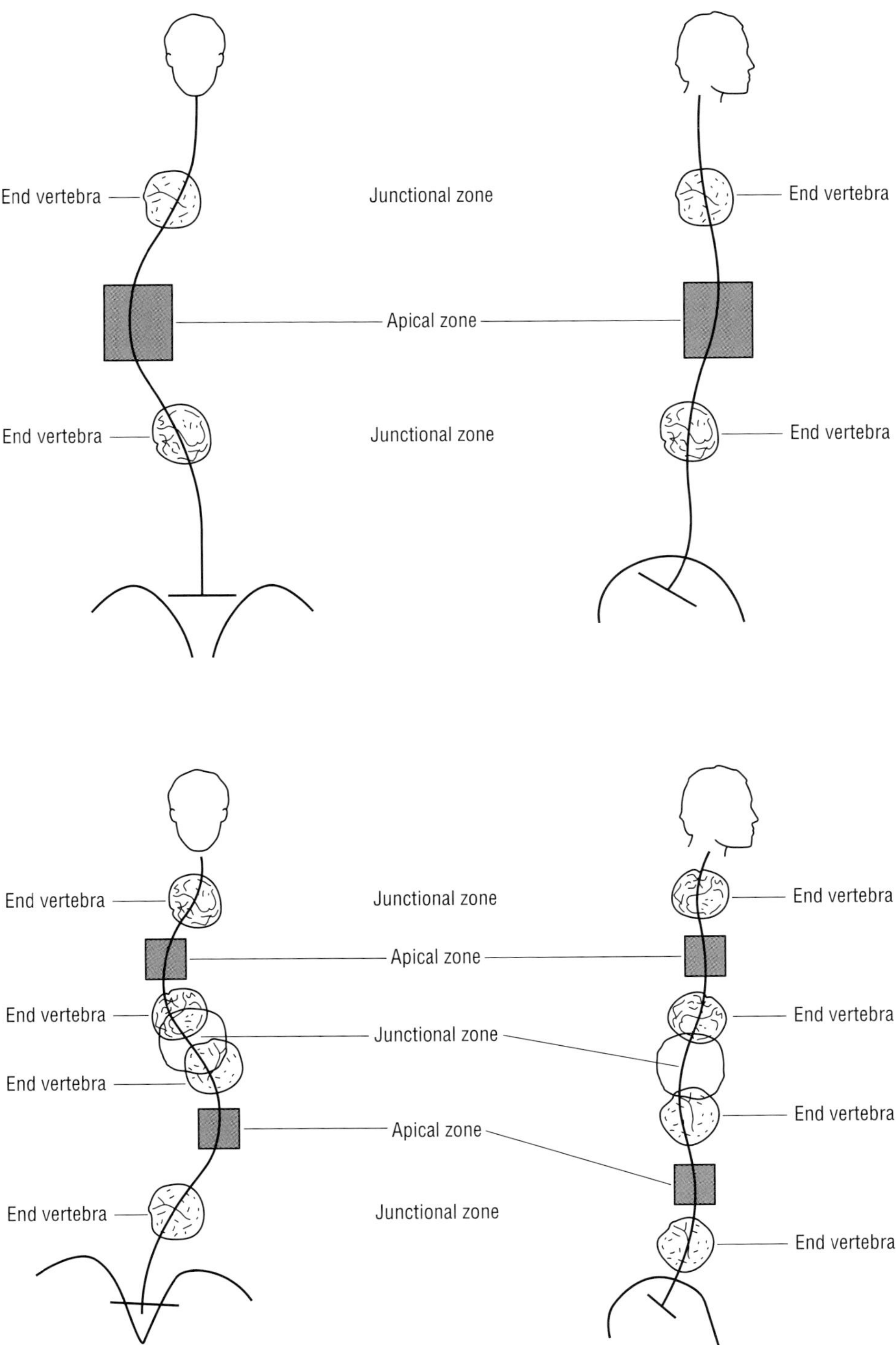

Figure 27.16. Analysis to determine the apical zone and junctional zone for idiopathic curves.

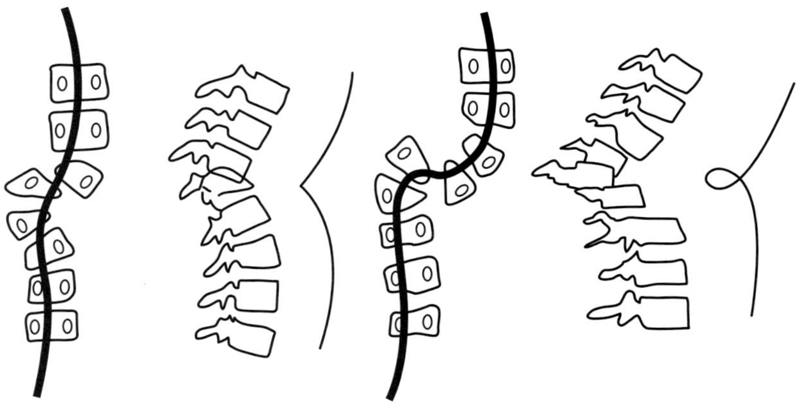

Figure 27.17. A. The rotatory dislocation of the spine concept. The apex of the kyphosis coincides with the junction of the two curves on the coronal plane. B. An increase in the angle of both curves in the coronal plane means an increase in the kyphosis.

The rotatory dislocation of the spine concept (Figure 27.17) [20] that was described in 1972 for some dystrophic, congenital, or idiopathic conditions [21] and explained by the existence of a junctional zone between two scoliotic segments occurring in weak bone tissue. The basic mechanism is that at the level of the most dysplastic area, when it is located at the junction of two lordotic segments, a rotational movement in the opposite direction occurs in front of the spinal canal initiating the deformity. As the deformity increases, an expulsive force generated by the rotational movement leads to increased deformity in the three planes and in some cases may compromise the spinal cord. However, in most cases the canal is still continuous without any dislocation, especially if the dystrophy exists on multiple successive levels (2, 3 or 4). Then, because the canal is in continuity and the rotating deformity is progressively increased, it is obvious that the flexibility of the spine can be used in the opposite direction. Progressive but constant traction can reduce the deformity by a large amount. In addition, when progressive neurological problems appear subsequent to this kyphosis, they can be reversed completely by this traction method, because of the relaxation of the tension of the neural structures. Subsequently, stabilization of the spine must also be achieved by circumferential front and back fusion. The anterior fusion must be done from the concavity (Figure 27.18) of the coronal deformity in order to be in alignment with the gravity line, which cannot be achieved if the approach is from the convex side. This convex approach additionally causes more instability, because removal of the convex ligaments working in tension will disrupt the tension band provided by the ligaments. Here, the concave strut grafts will work in compression and stabilize the spine.

For idiopathic conditions (Figure 27.16), it is exactly the same mechanism occurring on good bone vertebrae. The deformities are only slightly important and allow determination of the junctional zone that is especially critical to avoid the so-called junctional kyphosis factor of poor balance resulting from instrumentation or from spontaneous evolution that occurs mainly in the upper thoracic or thoracolumbar areas [12,16].

It is easy to recognize on antero-posterior and sagittal X-rays because the apex of the kyphosis lies at the junction of two scoliosis curves which is anatomically lordotic.

Hyper-rotatory Kyphosis

Hyper-rotatory kyphosis is another basic deformity of the scoliotic spine that occurs especially in infants. It is in reality a paradoxical kyphosis, corresponding to intervertebral lordosis but with an axial rotation so marked that the lateral scoliotic collapse appears on the sagittal plane and the patient looks kyphotic. It is apparent when

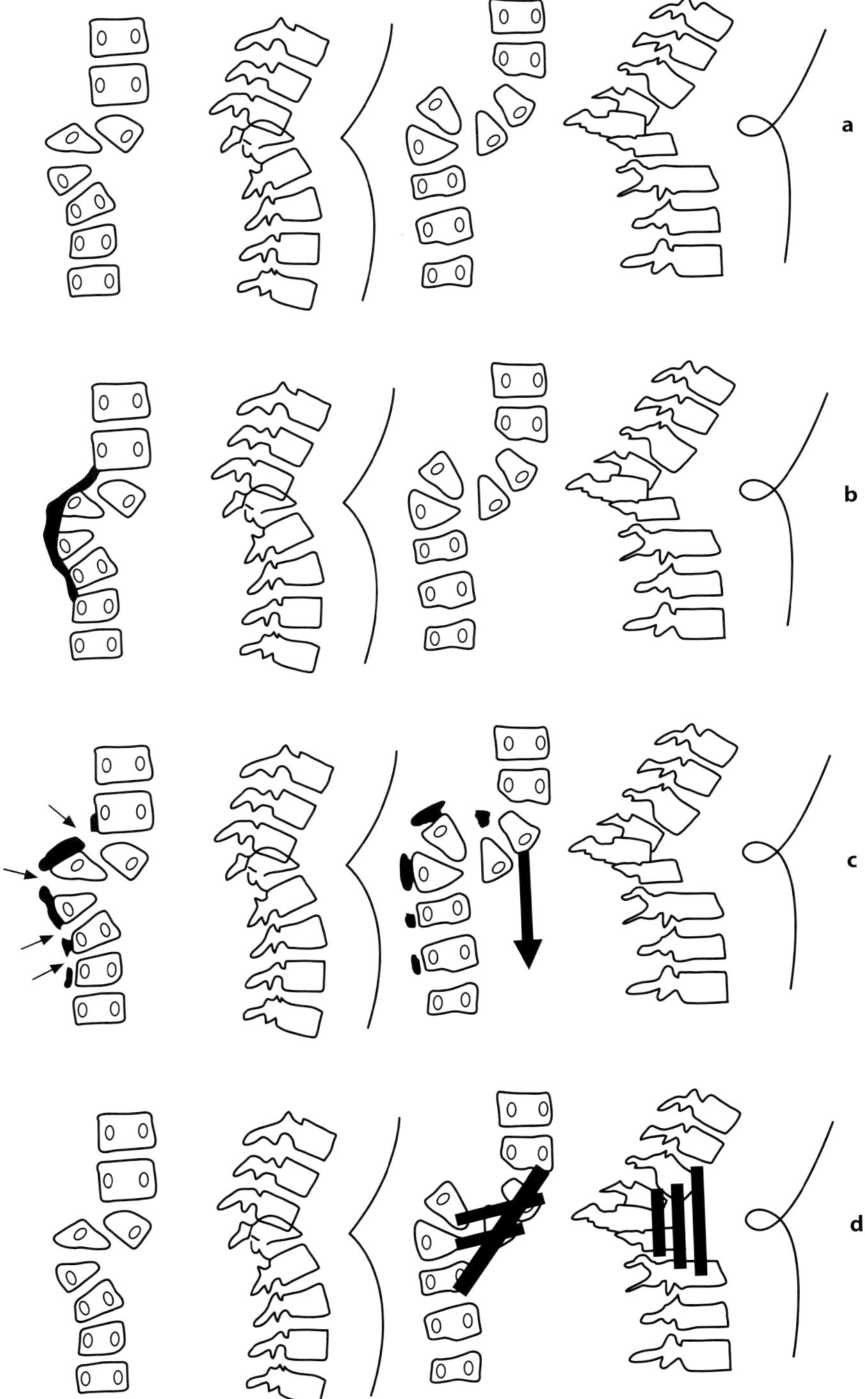

Figure 27.18. **a** Rotatory dislocation pattern. **b** Notice the tension band on the convexity. **c** When the convex approach is performed, the tension band of the ligaments on the anterior part of the spine is disrupted and instability is increased, so curve must worsen (arrow). **d** The concave side approach with strut grafts stabilized. Notice that they are always oblique in the coronal plane.

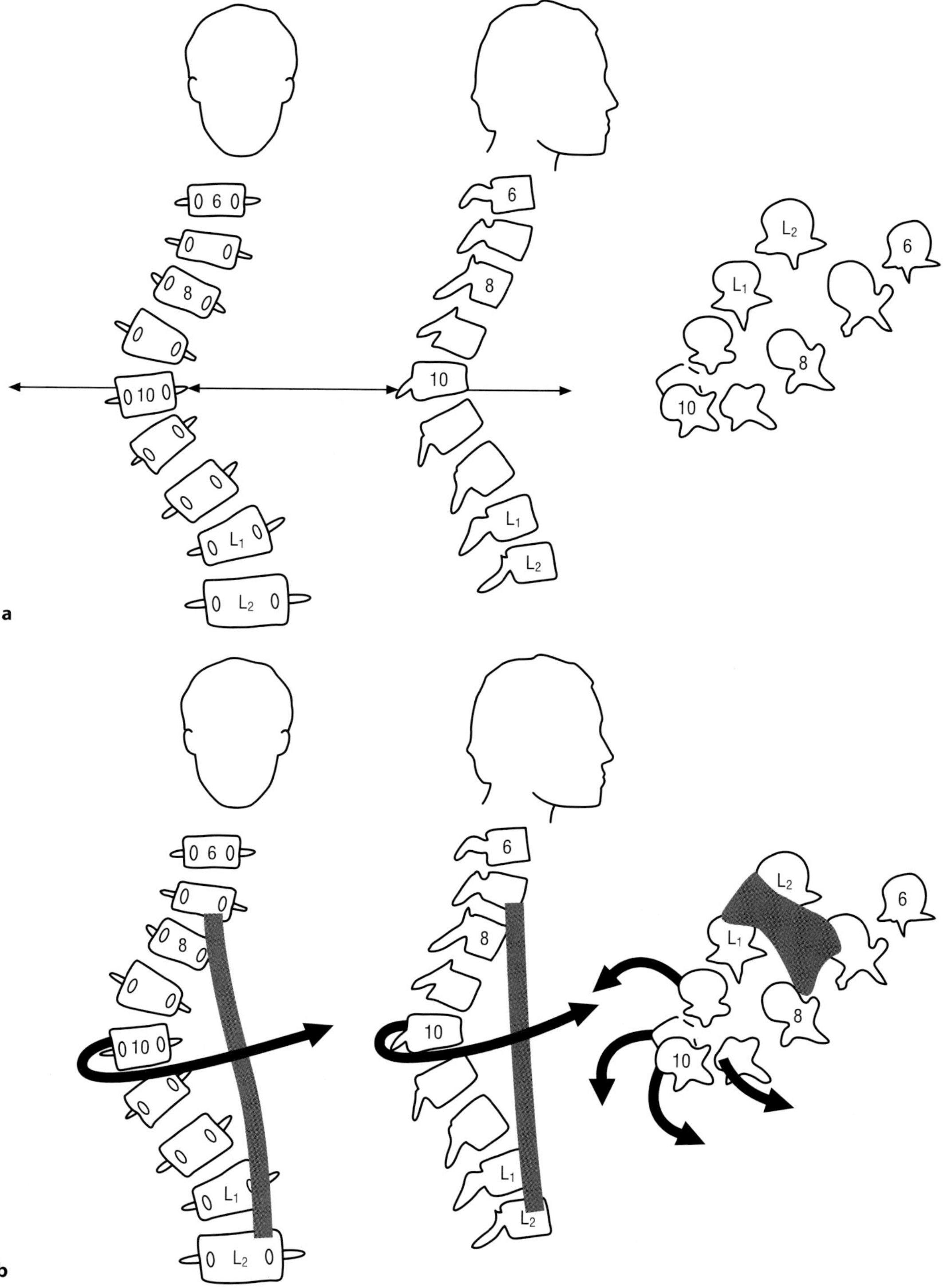

Figure 27.19. Apparent kyphoscoliosis. **a** Hyper-rotatory type. Apex in sagittal and coronal projection coincide. **b** If anterior fusion from the concavity with bypass the anterior growth will continue. **c** If posterior convex, the anterior growth will continue and create progression (crankshaft). Only anterior convex fusion will prevent the crankshaft.

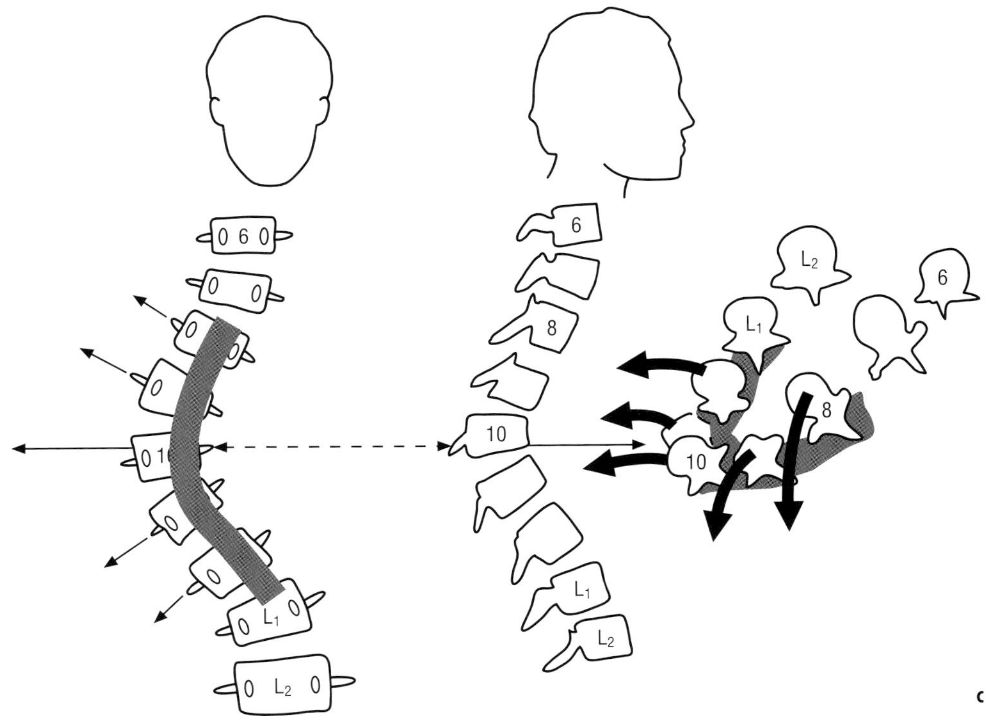

Figure 27.19. *(continued)*

viewed from the top in three-dimensional computerized reconstruction and easy to recognize on antero-posterior and sagittal views because the apex of the kyphosis is at the same level as the apex of the scoliotic curve (Figure 27.19 a, b).

This deformity is common not only in infantile and very progressive juvenile scoliotic idiopathic curves, but also in lumbar kyphosing scoliosis in adult and elderly spines. In these latter cases it is mixed with degenerative change in capsules, disks, and ligaments. For this hyper-rotatory kyphoscoliosis, the concept of three horizontal columns is very important, because if the anterior column disappears or was ejected laterally and backward from the gravity line, nothing would hold the spine in the front and the collapse would appear kyphotic. This type of hyper-rotatory curve can be corrected through an anterior convex approach with detorsion devices following anterior real release and instrumentation. The realignment of the spine is achieved by correction of axial rotation and lateral ejection, automatically correcting the sagittal plane [12,16].

This hyper-rotatory kyphoscoliosis is observed in the crankshaft phenomenon already described above, because introduced to the three-dimensional concept is a fourth dimension, time (Figure 27.19 b, c).

Paralytic Pelvic Obliquity

Paralytic pelvic obliquity is also a great application of spinal biomechanics in the growing spine (Figure 27.20), coming directly from the pelvic vertebra concept. This pelvic vertebra intercalary between spine and lower limbs will be displaced in three dimensions according to the imbalance created by the paralysis. It is why obliquity of the pelvis must be considered in three dimensions and displaced from the physiologic and anatomic positioning in relation not only to the amount and quality of the paralysis but also to the subsequent symmetrical or asymmetrical contracture coming from the paralysis. Obliquity of the pelvis must be considered in the coronal, sagittal, and horizontal planes. Each displacement in one direction results in a change in the other two. We must distinguish the three levels of disorder and deformity able to move the pelvic posture: they can exist above the pelvis, below the pelvis, and inside the pelvis itself [7,16,22,23].

The reasons for disorders *above the pelvis* occur in the evolution of the lumbar spine. Large differences exist between the deformities of the pelvis in the same direction as the lumbar spine whatever the plane of study. It is the regular pelvic obliquity. Conversely, the pelvis can be displaced in the opposite direction (whatever plane) which is termed opposite pelvic obliquity.

Below the pelvis, reasons involve contractures and symmetrical or asymmetrical paralysis giving rise to a displacement of the pelvic vertebra due to these contractures (for example, hip flexor contractures giving hyperlordosis and pelvic anteversion, adductum, or abductum), which leads to displacement in the coronal plane of the pelvis.

Finally, if the paralysis occurred in a young growing child, exactly the same thing as in other joints or bones appears and the deformity occurs *inside the pelvis* and can result in the bone of the pelvis twisting one side around the other. Correction of disorders above the pelvis can be achieved with correction of the spine itself, while correction of disorders below the pelvis is achieved with hip surgery. Internal pelvic correction can only come from sophisticated pelvic osteotomies. This also leads to cases of paralytic scoliosis in which fusion to the sacrum is recommended. This not only has to be done every time the pelvis is displaced in the same direction as the lumbar curve or when this displacement is even more pronounced, but also when permanent asymetrical contractures or paralysis remain at the lumbo-pelvic junction. On the other hand, it is not necessary to fuse if L5 is perfectly aligned with the pelvis and when no asymmetry exists in the three-dimensional muscle balance at the lumbo-pelvic junction. These concepts must be considered if surgical treatment is to be achieved in a logical manner (Figure 27.19b) [15].

Post-laminectomy Disorder

When the resulting kyphosis lies at only one or two successive levels, there is angular kyphosis. When on a long segment with many involved levels (for example, the entire thoracic spine), this results in a long, smooth, regular kyphosis. However, if the location is cervico-thoracic (for example a C5–D5 laminectomy), then the patient starts to develop kyphosis because of the failure of the stabilization, especially just after the junction with the normal spine, as the combination of the weight of the head and the necessity for horizontal gaze leads to the well-known swan neck deformity that progresses throughout development if not treated and protected by a brace and cast (Figure 27.21 a, b) [8–10].

Pathologic Examples of Biomechanics Related to Spinal Balance in Childhood and Adolescence

Fusion to the sacrum [15] can restore balance and favor walking in paralytics; for example, a paralytic post-polio spine fixed to L4 for scoliosis in a patient with complete paralysis of both lower limbs. Despite two lower-limb calipers, the standing posture can not be achieved without the support of two crutches. It appears as a progressive flexion of L5 and pelvic vertebra to the front. Extending the fusion in a lordotic manner to the sacrum allows balance and standing position to be restored with two below-hip calipers and without any additional support from the upper limbs.

Fusion to the Sacrum Without Sufficient Lumbar Lordosis (Flat Back)

This can destroy the possibility of standing in some muscular dystrophy patients. This flat back gives rise to anterior tilting of the trunk and instability. The gravity axis runs in front of the femoral head instead of going behind. Correction can be made by osteotomy of the lumbar fusion giving rise to lordosis or by bilateral anterior opening pelvis osteotomy allowing

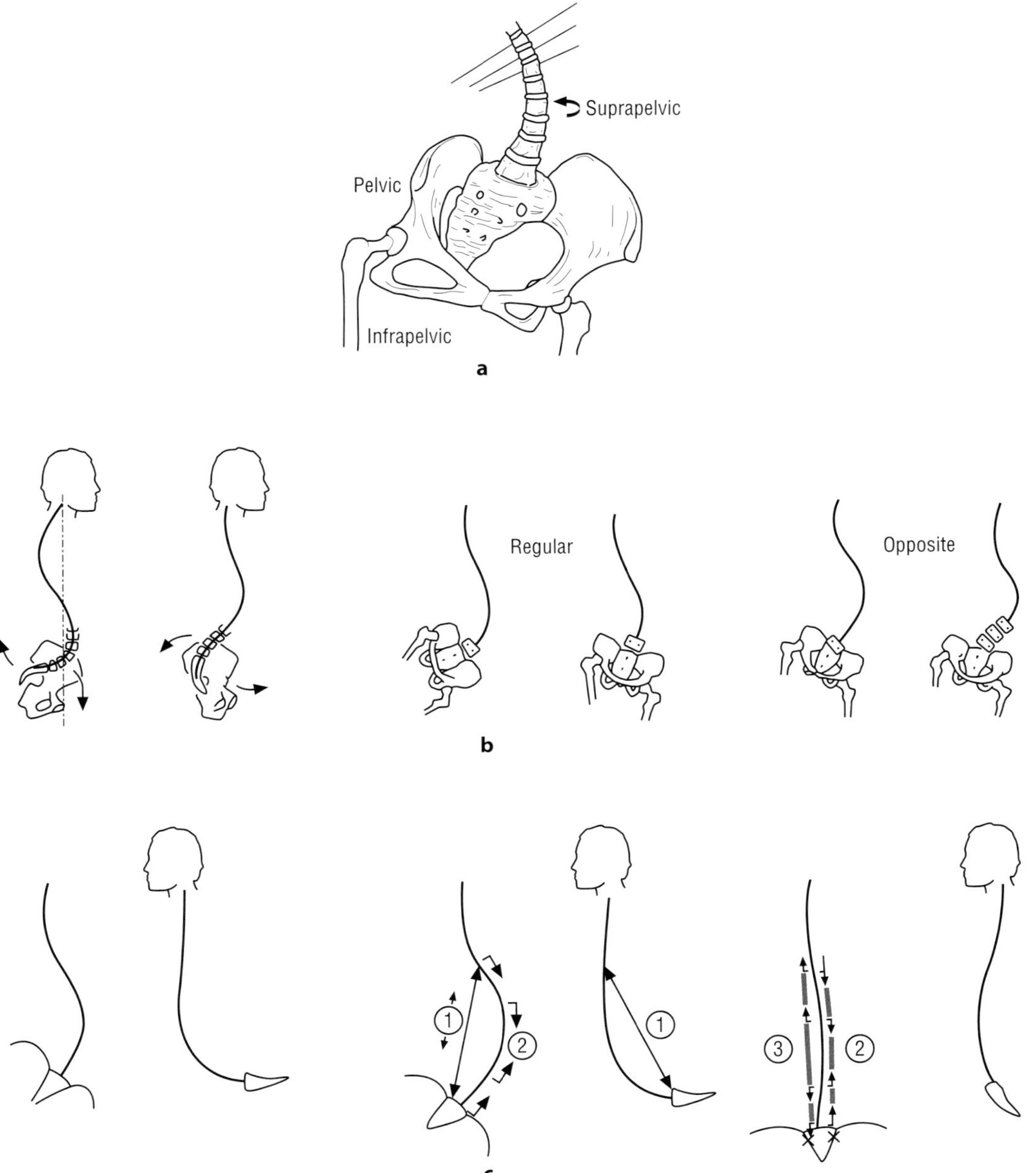

Figure 27.20. **a** In paralytic pelvis obliquity, the reasons for distorsion arise from three levels: above (supra pelvic) = spine deformity, below (infra pelvic) = lower limb contracture, inside (pelvic) = proper distorsion of the pelvic unit. **b** We must think, too, on the sagittal plane with lumbosacral lordosis or kyphosis as in the coronal plane. This allows regular pelvic obliquity where the lumbar spine and pelvis are going in the same direction or in the opposite direction when the pelvis is displaced in the opposite direction of the lumbar spine. **c** The logical way to analyze and treat pelvic obliquity with lordosis.

posterior translation of the gravity line (until lying just behind the center of the femoral heads) and achieving sagittal balance [15,17].

L5–S1 spondylolisthesis is a perfect example of a genetic/traumatic growing spine disorder. Two types of spondylolisthesis in children can be recognized (Figure 27.22 a, b).

One type is secondary to repeated traumatic stress fracture of the *pars interarticularis* arising from overuse and is often seen in gymnasts and other repetitive sporting activities in young children. The lysis may remain without slippage, and when the slippage starts, the sacrum is still horizontal. From this point it remains almost always horizontal. This occurs in reality with a genetic factor probably on the shape of the pelvis and lumbo-sacral area as we can find examples in one or both parents.

The other type of pediatric spondylolisthesis is sometimes called congenital, even if spondylolisthesis generally does not exist at birth. Very quickly a congenital kyphosis occurs at the lumbo-sacral area and creates a vertical sacrum with common local congenital anomalies like spina bifida occulta as well as very short posterior lever arm for soft tissue and ligamentous structure insertions.

The horizontal sacrum spondylolisthesis, where the sagittal angle constructed by the upper part of L5 and posterior part of S1 is over 110°, generally does not progress and very rarely requires surgical treatment. However, congenital spondylolisthesis with a vertical sacrum (where the angle is below 90° and describing a true lumbo-sacral kyphosis) is always progressing and almost always requires surgery to prevent further progression. This surgery will also correct the kyphosis to lordosis and fuse the lumbo-sacral junction into the correct position, restoring an angle of at least 110° and ensuring permanent stability when fusion is achieved [24].

In vivo dynamic evaluation of scoliosis before and after surgery and spinal fusion is another demonstration of spinal biomechanics [25,26]. This is done with a three-dimensional Vicon camera and by studying posture as well as motion in the three directions for various levels of fusion in idiopathic scoliosis. This study has demonstrated that:

Pelvic and head posture in a thoracic scoliotic curve group of children is very different from the control group of children of the same age and without spinal deformities.

After surgery, the three-dimensional position of the pelvis and the head is very different from the preoperative posture, showing that compensation for the change of the spinal contour in a segmental region (thoracic) develops, changing head and pelvic positioning to achieve balance. This is why motion of the hip joints, especially for possible hyperextension, is so important, allowing motion of the pelvic vertebra in the sagittal plan to ensure this compensatory tilt of the intercalary bone in the sagittal balance chain.

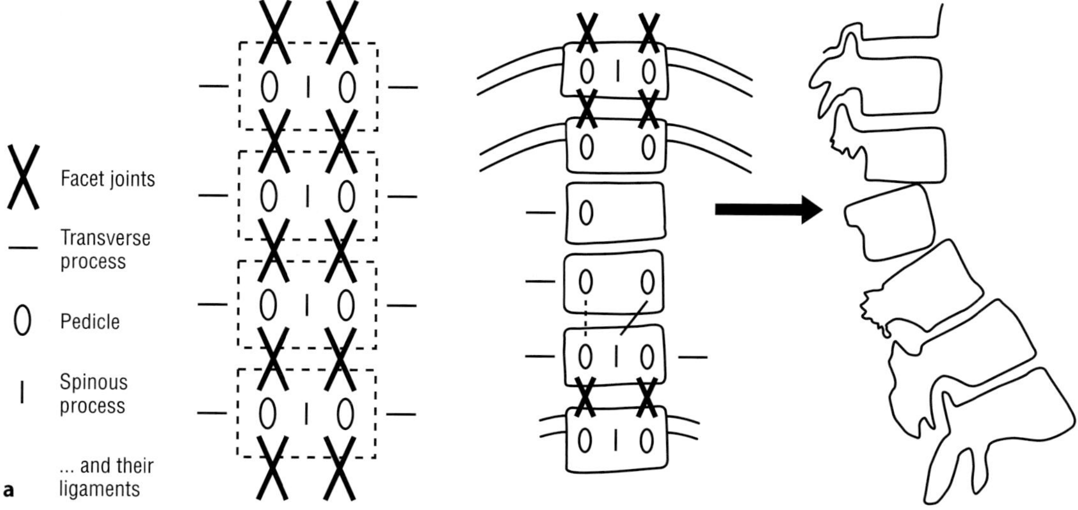

Figure 27.21. Post-laminectomy disorders. **a** Analysis of removal of stabilizing structures and consequences. **b** Two types of post-laminectomy kyphosis, i) acute cervical kyphosis after short laminectomy, ii) swan-neck deformity after long cervico-thoracic laminectomy.

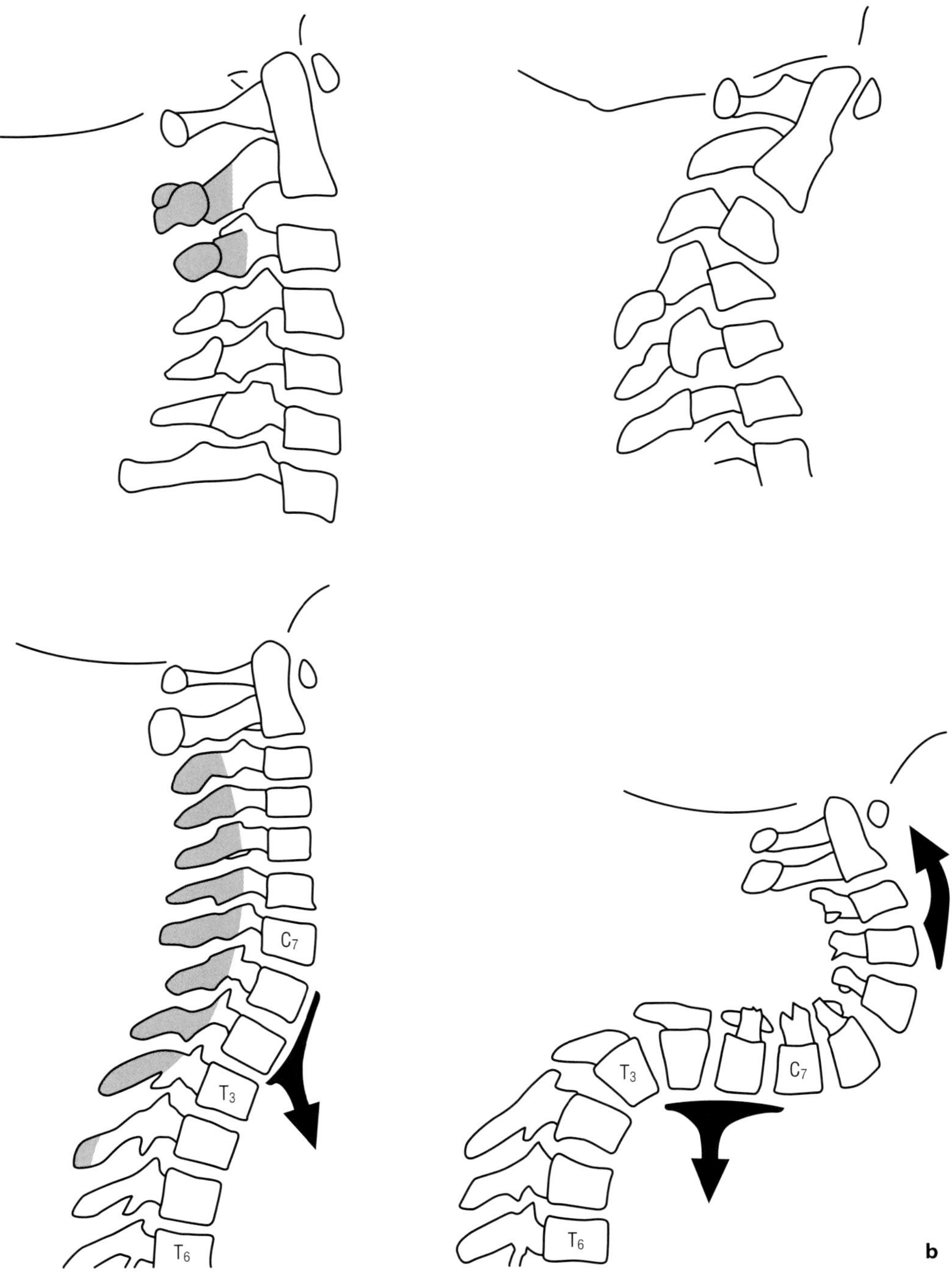

Figure 27.21. *(continued)*

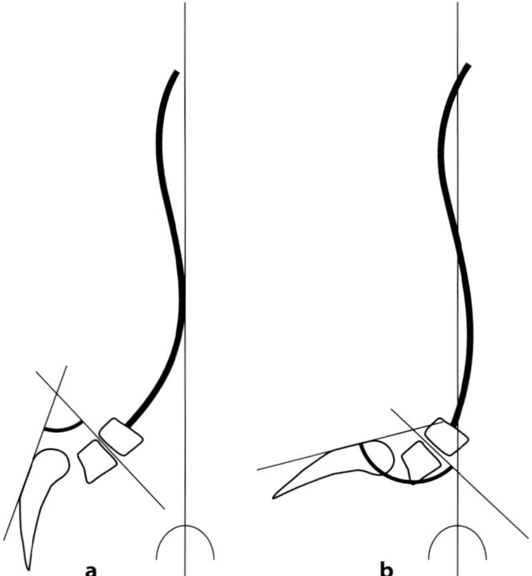

Figure 27.22. Spondylolisthesis in children. The difference between the horizontal (**b**) and vertical sacrum (**a**) is very important for prognosis.

Genetic Factors

Genetic factors are involved in all previous characteristics of the growing spine. This is well known in the study of bone and cartilage structures and very much confirmed when evaluating the quality of the soft tissues (hyperlaxity, borderline chondrodystrophy). Moreover, very seldom do short-stature parents give genes for height to their children. It is likely that even maturation of the nervous system is genetically predicted. However, each person is unique because of the amount of characteristics that have been delivered genetically, bearing in mind that some changes are resultant on the interaction of adjacent genes.

This results in some specific factors that affect the growing spine and differentiates some children from the adult population or even from other children of the same age. This provides them with their own unique growth potential, hypermobility, and adaptation to malleability of the spinal organ.

The changing mechanical properties of the spinal structures with age depends also very much on genetics as does the shape of the sagittal contour that determines thoracic kyphosis and subsequent lumbar lordosis, or the expansion of the shoulders and full span of the upper limbs. Some children grow and develop quickly, others more slowly. It is amazing to find that a Mediterranean girl has menarches at around 11 years of age, but it is four years later in a Scandinavian girl (at around 15 years of age). The variation is individual but genetically predicted. It is the same for the quality of regeneration after bone or articular cartilage injury, and finally also for maturation of neuromuscular control.

Rather than comparing one child to another, giving difficult references for normality, we must understand that normality covers a wide range of variations and that each child must be analyzed individually for every item relating to the definition of spinal biomechanics.

Conclusion

Biomechanics of the spinal organ during the growth period is of importance for the function of the spine during adulthood, because it is the time when the skeletal structures are being completed prior to adulthood and aging. However, it is also the time when maturation of neuromuscular systems occur giving the best possible adaptation to the genetic morphotype of the person to allow erect posture for standing or sitting or to benefit moving performance.

Before growth has ended, aging begins, giving rise to other pathological changes that will influence the spinal organ and subsequent posture and function. The influence of genetics in aging is just as great. We can confidently say that for biomechanics in orthopedics, and especially for the spine, genetic and traumatic factors are the only factors working on developmental growth as well as the degenerative aging spine.

Finally, orthopedics in general but especially seen within the scope of spinal physiopathology is really a three-dimensional science and must be understood theoretically but also practically within these dimensions.

Bibliography

Dubousset J. Cyphose angulaires de l'enfant. Scoliose und Kyphose; sous la direction de K. Zielke. Stuttgart: Hippokrates Verlag, 1976.

Dubousset J. Torticolis in children caused by congenital anomalies of the atlas. JBJS 1986;68A(2):179–88.

Lavaste F, Skalli W, Robin S, Diop A, Dubousset J. Modélisation tridimensionnelle par éléments finis du rachis lombaire. Rachis 1991;3(6):475–84.

References

1. Dubousset J, Gonon GP. Les cyphoses et cyphoscolioses angulaires. Rev Chir Orthop 1983;69 (supp II).
2. Dubousset J. Congenital kyphosis. The Pediatric Spine, Bradforf Hensinger, 1986.
3. Dubousset J. Congenital kyphosis and lordosis. The Pediatric Spine. Weinstein Raven Press, Ch. 10, 1994.
4. Dimeglio A. Le rachis en croissance. Sauramp 1, 1 vol.
5. Dubousset J. Déformations rachidiennes post-radiothérapiques après traitement du néphroblastome chez l'enfant. Rev Chir Orthop 1980;7(66):441–51.
6. Dubousset J, Herring T, Shufflebarger H. The crankshaft phenomenon. Journal of Ped Orthop 1989;5:541–50.
7. Dubousset J, Katti E, Seringe R. Epiphysiodesis of the spine in young children for congenital spinal malformation. J. Pediatr Orthop 1993;1:123–30.
8. Dubousset J, Guillaumat M, Mechin JF. Cyphoscoliose après laminectomie chez l'enfant. Réunion commune GES-SRS Lyon, Bosq, 1963;67.
9. Dubousset J, Guillaumat M, Mechin JF. Séquelles des laminectomies de l'enfant. Les compressions médullaires non traumatiques chez l'enfant. Sous la direction de J. Rougerie. Paris: Massons, 1973.
10. Dubousset J. Instability of the spine secondary to the treatment of intra spinal tumors in children: diagnosis, cure and prevention. Spinal Instability. Springer Verlag, 1991.
11. Dubousset J. Spinal deformities secondary to traumatic lesions involving the spine and spinal cord in children. Thoracolumbar Spine Fracture. Raven Press, Ch. 245, 1993.
12. Dubousset J. La scoliose idiopathique – Perspectives actuelles et futures. In: Kohler R, Berard J, editors. La scoliose idiopathique. Sauramps, 1997.
13. Dubousset J, D'Amico M et al. Three-dimensional analysis of spinal deformities. Ios Press, 1992;232–3.
14. Dubousset J. La scoliose idiopathique est originaire d'un désordre du système nerveux. Minerva Ortopedica e Traumatologiia 1991;11:875–80.
15. Dubousset J. Treatment of paralytic scoliosis with special reference for pelvic obliquity. In: Bridwell, Dewald, editors. Surgery of the Spine. Vol 1 & 2, Ch. 54. Lippincott First Edition 1992, Second Edition 1997.
16. Dubousset J. Three dimensional analysis of the scoliotic deformity. The Pediatric Spine. Weinstein Raven Press, Ch. 22, 1994.
17. Dubousset J. Balance – Considerations in revisions in children. Revision Spine Surgery. Mosby, Ch. 6, 1999.
18. Dubousset J, Duval-Beaupère G, Anquez L. Déformations vertébrales et paralysie – le rachis luxé congénital. Les compressions médullaires non traumatiques de l'enfant; Sous la direction de J. Rougerie. Paris: Masson, 1973.
19. Zeller R, Ghanem I, Dubousset J. The congenital dislocated spine. Spine 1996;21:1235–40.
20. Zeller R, Dubousset J. Progressive rotational dislocation in kyphoscoliotic deformities – Presentation and treatment. Spine 2000;25(9):1092–7.
21. Duval-Beaupere G, Dubousset J. La dislocation rotatoire progressive du rachis. Rev Chir Orthop 1972;4:323.
22. Dubousset J. CD instrumentation, pelvicobliquity. Der Orthopäde 1990;19:5.
23. Dubousset J. Pelvic obliquity correction. Lumbo-sacral and spino-pelvic fixation. Lippincott Raven, Ch. 4, 1996.
24. Dubousset J. Treatment of sondylolysis and spondylolisthesis in children and adolescents. Clin Orthop 1997;337:77–85.
25. Ployon A, Lavaste F, Maurel N, Skalli W, Roland Gosselin A, Dubousset J et al. In vivo experimental research in the pre and post operative behavior of the scoliotic spine. Human Movement Science 1997;16(2,3):299–308.
26. Ployon A, Lavaste F, Maurel N, Skalli W, Dubousset J, Zeller R et al. Protocole pour l'évaluation expérimentale 3D in vivo de la posture et de la cinématique globale du rachis. Rev Chir Orthop 1997;83:719–29.

28 Biomechanics of the Growth Plate

S. Saito, A. Kusaba, and U. K. Lewandrowski

The growth plate exists between the physis and epiphysis in tubular bones and distributes the force and load placed on osseous units. A growth plate has different mechanical properties from bone [1–6]. It is important to study the mechanical properties and biomechanics of the growth plate to understand the mechanism of its injury (i.e., epiphyseal slip of the femoral head, epiphyseal fracture of long bones) and some authors have studied the property by experimental methods [1,7–9].

The growth plate has three-dimensional unevenness, which has been confirmed by scanning electron microscopic studies. This unevenness gives the growth plate a multi-directional endurance against compression, strain, and shear stress. The yield strength of a growth plate depends on the degree of this three-dimensional unevenness [10,11].

Bone, cartilage, and epiphyseal plates have elastic and viscoelastic properties [12]. Methods to study the viscoelastic materials include stress absorbing tests creep tests, histeresis and dynamic stress tests. Recent studies reveal that measurement should be performed under dynamic load to evaluate this property [12–19]. The tangent delta value represents the viscoelasticity as the dynamic ability of shock absorption calculated from the phase lag (delta) between dynamic stress and the response of the specimen to this stress to indicate the viscoelasticity. This tangent delta value represents both energy loss (i.e., total amount of the absorbed stress by a specimen) and phase lag. It also demonstrates the softness of the specimen as the "ability for storing the stress" [14]. When this value is larger, it means that the specimen is softer, e.g., jelly has more tangent delta value than rubber. Kobayashi analyzed the mechanical property of maturative callus by both static and dynamic tests [20]. According to his results, static tests only indicated the mechanical property of the weakest portion in uneven callus, while dynamic tests indicated the total mechanical property of callus quantitatively. Yoshida pointed out that the minimum change of tangent delta was observed under a dynamic stress of 1 to 10 Hz. This result indicates that the epiphyseal plate has viscoelastic properties rather than viscotic properties [21]. Mow produced a dynamic viscoelasticity test by applying shear stress on articular cartilage. According to this experiment the dynamic elasticity of the cartilage was 0.2 to 2.5 MPa (under 0.01 to 20 Hz dynamic load), and its tangent delta value was 0.6 to 2.6. These results infer that epiphyseal plates have shock-absorbing abilities [22].

The yield strength of the growth plate has been variously reported as it depends on the material of the experiment [23]. According to Bright, the strength of the force which made a crack in the growth plate was one-half of the yield strength of the growth plate. On the other hand, Bright and Nakada reported that there was no relation between these two strengths [1,11]. Chung reported that force within the strength of physiological range was enough to cause epiphyseal slipping in overweight children and suggested that purely mechanical factors may play a major role in the etiology of slipped capital femoral epiphysis [24]. Williams made an experimental study and found that the shear strength of the physis varies with anatomic location. According to this experiment, the posterior region of the tibia physis had the greatest strength and stiffness, lowest physeal thickness, and steepest inclination [23].

Kim pointed out that the strength depends on the existence of the cartilage membrane [10]. The membrane seems to reinforce the growth plate in cooperation with its three-dimensional structure. In his report, the ratio of fracture strain between the growth plate without/with cartilage membrane was 61% (tensile strength) and 37% (strain strength). Chung made post-mortem studies of hips in children to determine the shear strength and modes of failure of the capital femoral growth plates [24]. In this experiment, the shear strength of the human growth plate varied with age and was greatly dependent on the surrounding perichondrial complex in infancy and early childhood, but less so in adolescence. When this complex was excised, the strength of the epiphyseal plate was diminished, especially in younger children. The velocity of stress is also an important factor in growth plate injury, as the growth plate has viscoelastic property. Kim reported that Salter–Harris type I injury is caused by the stress of relatively slow velocity and II and III by high velocity [10]. Histological studies performed concurrently with the yield-strength test varies among authors. This seems to be brought about mainly by differences in the method of experimentation. According to Salter and Amailo, epiphyseolysis occurred in the hypertrophying layer [25,26]. Bright reported that it occurred with a stairway form in the hypertrophying layer and palisading layer [1]. According to Nakada and Brasher, epiphyseolysis occurred in different layers depending on the portion of the growth plate [11,27]. From the clinical point of view, epiphyseolysis mainly occurs in the hypertrophying layer or transient portion to the metaphysis in slipped capital femoral epiphysis [23,28,29].

The mechanical property of the growth plate depends on the maturation. It has been reported that the strength of the epiphyseal plate increases along with maturation [11,24]. Yoshida reported that the viscoelastic property of the epiphyseal plate under dynamic stress decreases according to body weight, i.e., maturation [21]. Guse reported that the stiffness and strength of the growth plate increased with maturation [30]. In his report, it was suggested that this increase of strength depended not only on the growth of the cross-section area of the growth plate but also on the increase of strength of the epiphyseal plate component itself. On the other hand, traumatic epiphyseal injury or femoral head slip often occurs during the growth spurt. Guse explained there reciprocal facts by an ultimate strain theory [30]. Mechanical property change caused by aging also seems to be brought about by the change in structural and biochemical composition of the growth plate. The thickness of the growth plate decreases and the three-dimensional structure is emphasized corresponding to maturity [11,24]. The strength and elastic modulus of cartilage is in proportion to the quantity of collagen, and the viscoelasticity is to that of proteogrycan. It has been known that collagen increases, and moisture and proteogrycan decrease, in the cartilage matrix corresponding to maturity [11].

The growth plate is subject to some local and systematic factors; such as insulin-like growth factor, epithelial growth factor, proteoglycans, androgen and estrogen steroids, thyroid hormones, and growth hormone [28,31–34]. More especially, growth hormone affects the mechanical property of the growth plate by acting on the germinal layer. Lack of growth hormone causes shortening of the tubular bone, thinning of the epiphyseal plate, and the increase of ultimate strength or strain. However, lack of growth hormone makes no change in the longitudinal elasticity of the bone [32].

As mentioned above, mechanical stress affects the growth plate and the longitudinal growth of the bone. Excessive mechanical stress, i.e., traumatic injuries to the epiphyseal cartilage, may cause a pathologic interference with osteogenesis that is often irreversible [26,35,36]. On the other hand, some degree of mechanical stress is necessary to activate the growth plate. Greco explained these effects of mechanical stress from the viewpoint of cartilage metabolism; single, high-compressive force induced an irreversible [^{35}S] sulfate uptake [8]. This result indicates that acute compressive trauma may cause a permanent inhibition of growth plate chondrocyte proliferation and matrix synthesis. Multiple, intermittent, low-compressive force, grossly as in physiologic load-bearing, inversely enhances the metabolism of the growth plate by increasing both mitotec activity and matrix synthesis [8]. Klein demonstrated that intermittent compressive force enhanced in vitro Ca^{2+} and HPO_4^{2-} precipitation in the growth plate [37]. These reports infer that appropriate mechanical stress can modulate not only the direction but also the extent of bone growth. Clinically, it is well known that insufficient load stimulation causes inadequate growth in bones [38,39]. The growth plate behavior observed in the proximal femur of a congenital hip dislocation is a good example. The authors have experimented on canines to clarify the abnormal shape of the femur in congenital hip dislocation. According to this experiment, physiological antetorsion of the femur is formed mainly in the distal one-third of the femur, and proximal femur rotates at the growth plate in the direction of slight retroversion rather than anteversion in maturity. The total antetorsion angle is given by the difference between this distal antetorsion and slight proximal retroversion. Adequate load seems to be necessary to get this proximal retroversion so that the antetorsion angle will be in excess in congenital hip dislocation [38,39].

Growth speed also depends on the continuous force given to the growth plate [5,7]. Epiphyseodesis is a clinical application of this phenomenon [40–45]. Continuous compression stress can stop longitudinal growth in bones. It causes a decrease in height of the column cells, followed by hyperplasia of the column cells, which reveals an inhibition of the mineralization with continuity of cell production. This process of mineralization seems to be more sensitive to compression while the development of cartilaginous cells is scarcely affected [7]. Bonnel, using external fixators, measured compression forces through the growth plate of the lower end of the femur in rabbits, to calculate the degree of compression force necessary to slow down or to stop growth. According to this experiment, growth may become normal following release of compression provided that the compression was not excessive, and a histological study showed lesions varying from a simple diminution in the height of cellular columns to the development of bony bridges depending on the intensity of the compression [7]. Distracting force has been described as an accelerating factor. However, experimental studies have brought contradicting results. A distraction on the superior epiphysis of rat tibia led to lengthening of it, followed by a complete destruction of the histologic structures of the growth plate, which was replaced by fibrous tissue. The lengthening of sheep tibial growth plates resulted in epiphyseolysis, rapidly followed by an epiphysiodesis [46].

The growth plate is affected not only by direct force but also by remote force, such as electromagnetic fields or radiation [47–49]. Brighton has made a series of experiments and reported that the application of a proper electrical field stimulated the rabbit growth plated and promoted longitudinal bone growth. He referred the possibility of an application of this stimulation to the human body [48,50]. However, the mechanism of this growth stimulation is still unknown [7]. The growth plate is radiosensitive, and the proliferating zone is the most sensitive [51]. In animals, 1,200–3,000 rads of irradiation results in maximum damage to the proliferating zone, with cessation of growth. Irradiation causes not only growth disturbance but sometimes also causes slipped capital femoral epiphysis [51].

Phenomena which occur from diseases or physical behaviors of the growth plate are relatively well understood, as most of their results are visible. However, most of their mechanisms are hidden by a veil of mystery and there are still many contradictions and conflicts in the literatures as shown in this article. Investigations of new approaches, such as biomechanics, biochemistry, and molecular biology, are expected to reveal the mysteries of the growth plate [52,53].

Acknowledgement

The authors wish to thank Dr. W. C. Kim in the Department of Orthopaedic Surgery, Kyoto Prefectural University of Medicine, Dr. Y. Shirasaki in the National Engineering Laboratory, and Dr. J. Scholz in the Department of Orthopaedic

Surgery and Biomechanics, Berlin Free University, Neukölln Teaching Hospital for their cooperation and valuable advice.

References

1. Bright RW, Virginia R, Burstein AH. Epiphyseal-Plate Cartilage. J Bone Joint Surg 1974;56A:688–703.
2. Cohen B, Chorney GS, Phillips DP, Dick HM, Buckwalter JA, Racliffe A. et al. The microstructural tensile properties and biomechanical composition of the bovine distal femoral growth plate. J Orthop Res 1992;10:263–75.
3. Cohen B, Chorney GS, Phillips DP, Dick HM, Mow VC. Compressive stress-relaxation behavior of bovine growth plate may be described by the nonlinear biphasic theory. J Orthop Res 1994;12:804–13.
4. Frost HM, Jee WSS. Perspectives: A vital biomechanical model of the endochondral ossification mechanism. Anat Rec 1994;240:435–46.
5. Hunziker EB. Mechanism of longtudinal bone growth and its regulation by growth plate chondrocytes. Microscopy Res Tech 1994;28:505–19.
6. Inoue H, Hiasa K, Samma Y, Nakamura O, Sakuda M, Iwamoto M. et al. Stimulation of proteoglycan and DNA syntheses in chondrocytes by centrifugation. J Dent Res 1990;69:1560–3.
7. Bonnel F, Dimeglio A, Baldet P, Rabischong. Biomechanical activity of the growth plate. Anat Clin 1984;6:53–61.
8. Greco F, de Palma L, Specchia N, Mannarini M. Growth-plate cartilage metabolic response to mecanical stress. J Pediatr Orthop 1989;9:520–4.
9. Moen CT, Pelker RR. Biomechanical and histological correlation in growth plate failure. J Pediatr Orthop 1984;4:180–4.
10. Kim WC. Biomechanical properties of growth plate. J Jpn Paed Orthop Ass 1996;6:128–32.
11. Nakada D. Torsional strength of the eipphyseal plate and fracture patterns with aging, three-dimensional analysis with SEM. J Jpn Orthop Assoc 1993;67:1045–54.
12. Black J. Properties of natural materials. In: Black J, editor. Orthopaedic biomaterials in research and practice. New York: Churchill Livingstone, 1988; 105–31.
13. Akai M, Oda H, Shirasaki Y, Tateishi T. Electrical stimulation of ligament healing. Clin Orthop 1988;235:296–301.
14. Kusaba A, Kuroki Y, Kondo S. In vivo mechanical property change of polyethylene cups. Trans CORS 1995;2: 105.
15. Rohl L, Linde F, Odgaard A, Hvid I. Simultaneous measurement of stiffness and energy absorptive properties of articular cartilage and subchondral trabecular bone. Proc Inst Mech Eng 1997;211:257–64.
16. Sasaki N, Enyo A. Viscoelastic properties of bone as a function of water content. J Biomech 1995;28: 809–15.
17. Setton LA, Zhu W, Mow VC. The biphasic poroviscoelastic behavior of articular cartilage: Role of the surface zone in governing the compressive behavior. J Biomech 1993;26:581–92.
18. Takahara A, Yamada K, Kajiyama T, Takayanagi M. Evaluation of fatigue lifetime and elucidation of fatigue mechanism in plasticized vynyl chloride in terms of dynamic viscoelasticity. J Appl Polym Sci 1980;25: 597–614.
19. Takahara A, Yamada K, Kajiyama T, Takayanagi M. Analysis of fatigue behavior of high-density polyethylene based on dynamic viscoelastic measurements during the fatigue process. J Appl Polym Sci 1981;26: 1085–104.
20. Kobayashi A. Study of the biomechanical characteristics of fracture callus. J Jpn Orthop Assoc 1987;61:1273–84.
21. Yoshida M, Kim WC, Arai Y, Inoue N, Watabe K, Takai N. et al. The effect of maturation on dynamic visco-elastic properties of epiphyseal plate in rabbit. J Jpn Clin Biomech 1994;15:147–50.
22. Mow VV, Zhu V, Ratcliffe A. Structure and function of articular cartilage and meniscus. In: Mow VC, Hayes VC, editors. Basic orthopaedic biomechanics. New York: Raven Press, 1991;143–73.
23. Williams JL, Vani JN, Eick JD, Petersen EC, Schmidt TL. Shear strength of the physis varies with anatomic location and is a function of modulus, inclination and thickness. J Orthop Res 1999;17:214–22.
24. Chung SM, Batterman SC, Brighton CT. Shear strength of the human femoral capital epiphyseal plate. J Bone Joint Surg 1976;58A:94–103.
25. Amadio P, Ehrlich MG, Mankin HJ. Matrix synthesis in high density cultures of bovine epiphyseal chondrocytes. Connect Tissue Res 1983;11:11–19.
26. Salter RB, Harris WR. Injuries involving the epiphyseal plate. J Bone Joint Surg 1963;45A:587–622.
27. Brasher HR Jr. Epiphyseal fractures. A microscopic study of the healing process in rat. J Bone Joint Surg 1959;41A:1055–4064.
28. Harris R, Hobson KW. The endocrine basis for slipping of the upper femoral epiphysis. J Bone Joint Surg 1950; 32B:5–11.
29. Hurley JM, Betz RR, Loder RT, Davidson RS, Alburger PD, Steel HH. Slipped capital femoral epiphysis. The prevalence of late contralateral slip. J Bone Joint Surg 1996;78A:226–30.
30. Guse RJ, Connolly JF, Alberts R, Lippiello L. Effect of aging on tensile mechanical properties of the rabbit distal femoral growth plate. J Orthop Res ••;7: 667–73.
31. Carpenter JE, Hipp JA, Gerhart TN, Rudman CG, Hayes WC, Trippel SB. Failure of growth hormone to alter the biomechanics of fracture-healing in a rabbit model. J Bone Joint Surg 1992;74A:359–67.
32. Kanemitsu K, Takai N, Arai Y, Fujii T, Watanebe N, Takenaka N. et al. Effects of growth hormone on the structural properties of the epiphyseal plate. Proceedings of 23rd. Annual Meeting of Clinical Biomechanics, 1996;115.
33. Koskinen E, Katila T. Effect of 19-norandrostenololylaurate on serum testosterone concentration, libido, and closure of distal radial growth plate in colts. Acta Vet Scand 1997;38:59–67.
34. Trippel SB, Wroblewski J, Makower AM, Whelan MC, Schoenfeld D, Doctrow SR. Regulation of growth-plate chondrocytes by insulin-like growth-factor I and basic fibroblast growth factor. J Bone Joint Surg 1993;75A: 177–89.

35. Fell HB. Skeletal development in tissue culture. In: Burne GN, editor. The biochemistry and physiology of bone. New York: Academic Press, 1956.
36. Sissons HA. The growth of bone. In: Burne GN, editor. The biochemistry and physiology of bone. New York: Academic Press, 1971.
37. Klein-Nuland J, Veldhuijzen JP, Burger EH. Increased calcification of growth plate cartilage as a result of compressive force in vitro. Arthritis Rheum 1986;29:1002–9.
38. Saito S, Kuroki Y, Uchida T. An Experimental Study on the Change of the Femoral Antetorsion. J Jpn Orthop Assoc 1978;52:1185.
39. Saito S, Kuroki Y, Uchida T, Mori Y. Experimentelle Untersuchungen ueber Entstehung der Untersuchungen der Antetorsion am Femur. Z Orthop 1980; 118:162.
40. Eastwood DM, Cole WG. A graphic method for timing the correction of leg-length discrepancy. J Bone Joint Surg 1995;77B:743–7.
41. Gabriel KR, Crawford AH, Roy DR, True MS, Sauntry S. Percutaneous epiphyseodesis. J Pediatr Orthop 1994; 14:358–62.
42. Givon U, Ishikawa S, Dabney KW, Hacke HT, Bowen JR. Growth cartilage arrest with staples. Experimental study. In: Fernándes P, de Pablos J, editors. Surgery of the growth plate. Madrid: Ediciones Ergon 1998; 54–63.
43. Herranz JG. Growth cartilage arrest with staples. Experimental study. In: Fernándes P, de Pablos J, editors. Surgery of the growth plate. Madrid: Ediciones Ergon 1998;33–53.
44. Joseph B, Srinivas G, Thomas R. Management of Perthes disease of late onset in southern India. The evaluation of a surgical method. J Bone Joint Surg 1997;78B:625–30.
45. Synder M, Harcke HT, Bowen JR, Caro PA. Evaluation of physeal behavior in response to epiphyseodesis with the use of serial magnetic resonance imaging. J Bone Joint Surg 1994;76A:224–9.
46. Monticelli G, Spinelli R. Distruction epiphysiolysis as a method of Circel lengthening. Clin Orthop 1981;154: 254–77.
47. Armstrong PF, Brighton CT, Star AM. Capacitively coupled electrical stimulation of bovine growth plate chondrocytes grown in pellet form. J Orthop Res 1988; 6:265–71.
48. Brighton CT, Pfeffer GB, Pollack SR. In vivo growth plate stimulation in various capacitively coupled electrical fields. J Orthop Res 1983;1:42–9.
49. Iannacone WM, Pienkowski D, Pollack SR, Brighton CT. Pulsing electromagnetic field stimulation of the in vitro growth plate. J Orthop Res 1988;6:239–47.
50. Brighton CT, Jensen L, Pollack SR, Tolin BS, Clark CC. Proliferative and synthetic response of bovine growth plate chondrocytes to various capacitively coupled electrical fields. J Orthop Res 1989;7:759–65.
51. Loder RT, Hensinger RN, Alburger RD, Aronson DD, Beaty JH, Roy DR. et al. Slipped capital femoral epiphysis associated with radiation therapy. J Pediatr Orthop 1998;18:630–6.
52. Caruso EM, Lewandrowski KU, Ohlendorf C, Tomford WW, Zaleske DJ. Repopulation of laser-perforated chondroepiphyseal matrix with xenogeneic chondrocytes. An experimental model. J Orthop Res 1996;14: 102–7.
53. Schollmeier G, Uhthoff HK, Lewandrowski KU, Fukuhara K. Shaping of the bone bark during growth in width of tubular bones. An invistegation in human fetuses. Clin Orthop Res, in press, 1999.

29 Biomechanics and the Growth Plate

G. Bollini, E. Viewegher, J. M. Guillaume, F. Launay, and J. L. Jouve

The function of growth with which the physis is concerned is subject to many genetic, hormonal, metabolic, vascular, and mechanical factors.

From Delpech [1] in 1823 who wrote: "By bending the bones in the opposite direction to the direction in which they wanted to go, it (the device for raising a club foot) was able to produce constant compression on certain points of the articular surface and retard the development of the compressed parts . . ." to Blount [2] who suggesting putting staples on growth plates to modify their function, the role of biomechanics on the physis has long being intuitively understood. Since then many experiments have better explained the interactions between biomechanics and growth.

We propose to describe what is currently known about biomechanics as an isolated factor acting on the growth plate. This presentation is artificial insofar as the effects which can be seen have been mediated by the other factors, hormonal, metabolic, or others, referred to above. We will then describe the biomechanics concerned specifically with the growth plate and the biomechanical effects on growth before discussing clinical applications.

The Biomechanics of the Growth Plate

Organization of the Growth Plate and Resistance to Stresses

The growth plate consists, successively from the epiphysis to the metaphysis, of the reserve or germinal zone; the proliferative zone, organized into a seriated column; and of the hypertrophic zone, itself subdivided into a maturation zone and a degeneration and calcification zone. All these zones are made up of chondrocytes and an extracellular matrix.

This matrix consists of 80% water and 20% macromolecules (proteoglycans, collagen, and glycoproteins). The glycosaminoglycans forming part of the proteoglycans are very hydrophilic. Pressure makes them force out the molecules of water which they recapture as soon as the pressure eases.

Collagen is secreted by the chondrocytes. It is collagen which gives the growth plate its cohesion and its strength.

This physis is surrounded by two structures, Ranvier's ossification and Lacroix's perichondrial layer. Of these two, it is mainly the second which, by a collagen–fibrous network continuous with the periosteum to the metaphysis and the fibrous portion of Ranvier's ossification segment to the apophysis, ensures the mechanical stability of the physis against shear stresses in particular.

Another anatomical arrangement of the physis assists its resistance to shearing and these are the mamillary processes. The boundary between the growth plate and the metaphysis is not a flat but an *uneven* surface made up of these mamillary processes enabling better anchorage of the growth plate on the metaphysis.

Finally, the ligamentary formations also help to give the physis mechanical strength.

Experimental Data

Moen [3] subjected calf femurs and tibias to compression, flexion, tension, and shear stresses and studied the "histological journey" traveled by fractures separating the growth plate which have been observed.

Separation occurs mainly in the reserve zone for compression, in the proliferative zone for traction, and between the proliferative zone and the hypertrophic zone for flexion. Shearing does not correspond to a specific breakage zone.

Chung [4] also studied the breakage sites of 50 specimens of the femoral neck and head of children subjected to mechanical stresses. The fractures seen were of Salter and Harris type I for specimens from children under five years of age and of type II in older children. Resection of the perichondrial layer increased the number of type I and type II fractures seen. This last idea is confirmed by Gigante [5] who estimated the loss of strength on shearing of growth plate after excision of the perichondrial layer to be 50%.

Amamilo [6] studied the importance of the axial forces and in shearing on the growth plate of 110 rats necessary for obtaining epiphysiolysis. Two groups were compared according to the absence or presence of the periosteum. The forces were significantly less marked when the periosteum had been removed, in particular in the youngest specimens. Bright [7] studied the relationship between the extent of the shear stresses needed to obtain epiphyseal separation and the rate of application of this force. He concluded that the faster the stress is applied, the more marked it has to be to produce separation. This underlines the viscoelastic properties of the physis. He also demonstrated that as soon as 50% of the shearing load needed for separation is reached, microfractures appear histologically in the physis. If the force applied is removed, these microfractures remain, subsequently reducing the mechanical strength of the growth plate.

Rudicel [8] showed that for a proximal femoral cartilage in the rabbit subjected to a shear stress, the line of separation advances with age from the proliferative zone to the hypertrophic zone of the growth plate. Lee [9], and Williams [10] underlined the anisotropic nature of the mechanical resistance of growth plate to identical stresses exerted during simple shearing, parallel to the growth plate according to the direction from which antero-posterior or postero-anterior force is applied.

Peltonen [11] studied the resistance of the physis in the lamb to torsional stresses. The breaking strength increases with age with a reduction in the phase immediately before puberty. In the young animal, the zone of separation passes into the zone of the hypertrophic layer. In the older animal, the line of separation is more sinuous, sometimes carrying away metaphyseal fragments.

Growth Force Developed by the Physis

It is currently observed that the ends of staples implanted to arrest growth move apart over time. Bylander [12] monitored growth using stereophotogrammetry on markers implanted after epiphyseal stapling of a boy aged $9^1/_2$ years. At this age normal growth ranges between 40 and 60 μm per day. After stapling, growth is initially very slow, then increases over the next three months to reach 10 μm per day, as the widening of the staples increases. In the following six months, growth slows again to reach a minimum value of 2 μm per day, which persists for approximately two years. At the same time, the ends of the staples move apart by a third of the distance separating them.

Bylski-Austrow [13] investigated the forces needed to separate the ends of staples seen in 20 patients by reproducing this separation on a mechanical test bench. He concluded that the static force exerted inside the distal femoral or proximal tibial growth plate is 0.5 KN per physis. The stress developed would be equivalent to 1 MPa.

Bonel [14] evaluated the force developed by a distal growth plate in the rabbit, which can be as much as 38 N before complete inhibition by compression, which, brought to the surface, represents a force of 19 g/f/mm^2.

Characteristics of the Growth Plate

The growth plate is a viscoelastic structure, displaying mechanical anisotropy which, by growth, is capable of developing a force approximately equivalent to the weight of the body. The stresses applied to the growth plate can vary in intensity, direction of application, and production cycle, leading to different responses from the physis, which we will now study.

Effect of Weightlessness on the Growth Plate

The adverse effects of weightlessness on chondrogenesis have been underlined by Duke [15]. The proliferation and differentiation of the chondrocytes are adversely altered as is the extracellular matrix. However, the problem of how to reproduce the results remains because of the many parameters inherent in this type of experimentation. This is why the experiments were conducted in the laboratory by Duke, Montufar-Solis, and Wronski [15–18] in order to be able to analyze more accurately the fate of the growth plate in animals placed in a state of weightlessness. Subjected to weightlessness, the growth plate decreases in thickness, in particular in the hypertrophic zone. When returned to a state of normal gravity, growth appears to be stimulated compared with the control animals. This "rebound" capacity depends on how long the period of weightlessness lasts. Present after 7 days in a weightless environment, it is no longer seen after 12.5 days spent in a weightless state.

It should be noted that opposite experiments conducted by Montufar-Solis [16], in which rats were subjected to hypergravity, revealed a reduction in the height of the hypertrophic layer and a reduction in the number of cells both in the hypertrophic layer and in the proliferative layer.

Sibonga [19] studied the effects of weightlessness on the growth of the length of bones in rats. He saw no difference in growth between the control group and the group subjected to gravity-free situations by using a fluorochrome marker, irrespective of the age or sex of the specimens or the duration of the absence of gravity. On the other hand, he saw changes in the extracellular matrix which could have a negative effect on endochondral ossification.

Effect of Compression on the Growth Plate

Delpech [1] in 1989 and Hueter [20] in 1862 postulated that compression reduces cartilage growth. According to Pauwels [21], a growth plate subjected to stress on flexion, provided that it is biologically intact, will develop differential growth, more marked on the side of the compression than on the traction side, the end result being that the stresses to which the cartilage is subjected are again uniformly distributed over its surface.

Bonnel [22] demonstrated that by applying compression which is constantly readjusted to remain constant on a distal femoral growth plate in the rabbit, it is possible to retain a regular quantity of infra-physiological growth. Bonnel [14] also demonstrated that if compression of less than 32 N is applied for one month to distal cartilages in rabbit femurs and if this compression is then removed, not only does growth resume but at a faster rate than normal for 4–5 days, an "overshoot" phenomenon. These findings perhaps explain what Pauwels saw and seem to contradict the laws of Hueter and Delpech.

Alberty [23] saw a reduction in the height of the hypertrophic and proliferative zones of the physes of distal femurs in rabbits subjected to compression as well as a reduction in the number of chondrocytes in the proliferative zone.

Mankin [24] subjected distal radiuses of 5–7-day-old calves to a compression of 245 N (0.012 M Pa) in an organ culture for 24 hours. The compression reduces the synthesis activity and the secretion of prostaglandin. No change was seen in the incorporation of the thymidine, which would seem to indicate that cell multiplication is not adversely affected.

Arriola [25] subjected 18 lambs to compression of the distal femoral and proximal tibial physes. He found a thicker germinal zone on the tibias which had been operated on. The proliferative zone of the tibias subjected to compression was reduced in height whereas the hypertrophic zone was thicker both in the femoral physes and in tibias which had been operated on.

Cohen [26] carried out an in vitro study on osteo-physo bone test pieces removed from the central and peripheral part of the growth plate in distal femurs in calves. The results show that the aggregation module is not dependent on the permeability of the cartilage–bone interface. The coefficient of permeability is very dependent on it and Poisson's coefficient is dependent on the permeability of the interface. In all, the periphery of the growth plate is more compliant and permeable than the central part.

Gray, Greco, and Klein [27–29] showed that, depending on its intensity, the action of compression can act on the proliferation of the chondrocytes and/or on their syntheses. A low-intensity static force reduces the synthesis of the glycoaminoglycans and protein synthesis in a dose-related manner. If the intensity of the force increases, both cell proliferation and matricial synthesis are adversely affected.

Finally, for supra-physiological forces, the hypertrophic layer doubles itself in thickness with a zone of fibrous tissue, the columnar architecture of the proliferative zone disappears, and vascular tissue increases on the epiphyseal side of the physis. This tissue penetrates through the germinal and proliferative zones with ossification creating Trueta's epiphysiodesis [30].

Asymmetrical Compression

Most of the studies on asymmetrical compression were carried out after having modified the angulation of a diaphysis by osteotomy. Using an external fixator, Collard-Meynaud [31] applied asymmetrical and continuous forces on the distal physis of a lamb's radius for 9 weeks. The force exerted remained within the threshold of irreversibility. Angular deformation occurred due to the lesser growth of the most compressed side. Removing the asymmetrical compression device revealed the same as that demonstrated by Pauwels [21], namely, gradual correction of the deformation.

Effect of Tension on Growth Plate

Symmetrical Distraction

Kenwright [32] studied three types of tension by external fixator on the physes of rabbits close to the end of their growth period. Three tension forces were applied at 0.13, 1.26, and 0.53 mm/day. When the rate of growth was 0.53 mm/g, the force applied rose to 25 N then decreased very rapidly, corresponding to epiphysiolysis. For lesser amounts of daily traction, the force applied culminated in 16 N then remained stationary without epiphyseal separation but without a significant gain in length. This study contradicts that conducted by Pereira [33] who performed distraction of 0.5 mm/day by external fixator in 32 six-week-old rabbits with three sub-groups being distracted for 2, 3, and 4 weeks, respectively. After the fixators had been removed, an increase in length of 3, 3.6, and 4 mm, respectively, was obtained. Monitoring after removal of the distractors over 13 weeks shows that the gain in length obtained is maintained. The contradiction between these two studies is perhaps only apparent because the age of the rabbits is very different, close to the end of the growth period in the first, younger in the second. Two other contradictory studies are reminiscent of these two studies on the gain in length, namely two studies on the effects of tension on the cellular activity of the chondrocytes.

Apte [34] performed distraction in immature rabbits with a distractor reaching a maximum force of 20 N. The proliferative cells were labeled with bromodeoxyuridine (BU-dR) and were revealed by anti BU-dR antibodies. After ten days of distraction with a constant tension force, the seriated columns were disorganized and cell proliferation decreased. There was a lengthening due to the separation between the physis and the metaphysis. Cell proliferation was inhibited. The apparent thickening of the growth plate was not due to an increase in cell proliferation but to inhibition of the endochondral ossification, resulting in an accumulation of hypertrophic cells.

Alberty [23] labeled the cells in the same way with BU-dR. Of 15 rabbits undergoing epiphyseal distraction, 13 displayed epiphysiolysis. In this experiment, the tension did not have an effect on the number of proliferative cells. In another publication produced in the same year, to these findings Alberty [35] added the demonstration of BU-dR-labeled hypertrophic cells, the number of which ranged between 5 and 20 on the epiphyseal side of the physis and suggested that hypertrophic cells are capable of multiplying under certain conditions.

Elmer [36] labeled the chondrocytes with tritiated thymidine and the extracellular matrix with radioactive sulfate. In his view, distraction would not have any effect on the extracellular matrix, cell division, or synthesis but would simply produce viscoelastic stretching of the cartilage.

De Bastiani [37] subjected distal growth plates of rabbit femurs to elongation of 0.5 mm and 1 mm per day. At 1 mm per day, separation occurred on the seventh day with ossification and complete disappearance of the cartilage on the 70th day. At a rate of 0.5 mm per day, he did not see any epiphyseal separation on the 28th day, the day when distraction was stopped. On the 70th day, the growth plate retained its normal thickness.

Mankin [24] subjected the cultures of growth plate organs kept for 24 hours at a tension of 245 N (0.012 m M pa). Cell proliferation was evaluated by labeling with tritiated thymidine. The tension increased the synthesizing activity of the cartilage, assessed by labeling with radioactive sulfate and the production of prostaglandin. There was no change in the labeled thymidine activity.

Alberty [38] found that tension had the same microvascular effects as those found by Trueta [30] for compression. Distraction followed by separation was accompanied by hyperplasia of the epiphyseal vessels. The metaphyseal vessels were damaged. Neoangiogenesis appeared in the zone of separation in the third week and in the sixth week, vascular anastomoses appeared through the physis.

Asymmetrical Distraction

Peltonen [11] applied asymmetrical distraction without epiphyseal separation in lambs. New formation of metaphyseal bone was studied by fluorescence microscopy, microradiography, and histomorphometry. The formation of this new bone is due to two phenomena, peripheral growth from the periosteum, and bone metaplasia of the bridges of collagen with a mixture of lamellar and spongy bone. He saw spontaneous recovery after the removal of the distractor indicating that the growth plate had retained its vitality.

Clinical Applications

Lengthening

Diaphyseal Lengthening by Callotasis

A study conducted by Lee [39] studied the effects on growth plates of diaphyseal distraction by callotasis. The overall activity of the cartilage was evaluated by labeling with oxytetracycline and the cellular activity was estimated by labeling with BU-dR. If the length of the diaphysis is elongated by 20%, whatever the rate of elongation applied, there is no change in the activity of the growth plate. On the other hand, if the diaphysis is lengthened by 30% or up to 50%, the overall activity of the growth plate decreases and premature closure of the growth plate occurs.

Lengthening by Epiphysiolysis

When it is performed at rates of 1 or 1.5 mm per day, physeal distraction invariably leads to premature closure of the growth plate. When this distraction is performed at slower rates, 0.5 mm per day, the results are contradictory (De Pablos, De Bastiani, Kenwright) [32,37,40]. For some, physis can retain its vitality at the end of the period of distraction, for others, there is a functional change in the physis. This technique cannot, therefore, be recommended if the aim is to lengthen by epiphysiolysis the bones in a child who still has considerable growth potential and only at the end of the period of lengthening does the growth plate resume its normal function. It is clear that it is not the epiphysiolysis itself, when it occurs, which is the cause of arrested growth, but the distraction which follows. Epiphysiolysis is, in effect, characteristic of all Salter type 1 and 2 fractures which heal without damage to the physis. We even suggested [41] that this epiphysiolysis technique be used to aid in the resection of the epiphysiodesis bridges with the physis starting to grow again after the interposition of surgical cement. Jouve [11] demonstrated that the ideal was for this interposition material to remain in the desepiphysiodesis zone and not climb into the metaphysis to avoid the differentiated reformation of fibrous or osseous epiphysiodesis.

Periosteum

The growth plate is surrounded by Ranvier's ossification zone and the perichondrial layer but its function is unimaginable without the periosteum. With the growth plate, this forms a prestressed structure which we will not discuss in further detail here as a chapter of this book has been devoted to it. In clinical practice, upper metaphyseal fractures of the tibia in children which regularly evolve into valgus are definitely linked to the rupture of the medial hemiperiosteum, partially freeing this pre-stressing.

Legg–Perthes–Calvé Disease

Yoshida [42] has carried out a histological and microscopic angiogeographical study of the heads and necks of rat femurs subjected to forces ranging from 1 to 3 kg. Plastic deformation of the physis under the head in its lateral portion can compress the vessels leading to the epiphysis. This animal model producing osteochondritis has to be validated in man before putting it forward as an explanation for the occurrence of osteochondritis.

Effect of a Tensioned Tendon Graft Aross the Physis

The repair of the anterior cruciate ligament in the child runs the risk of placing a tendon across a growth plate. In a growing dog, Edwards [43] inserted a tendon formed of the fascia lata tensioned to 80 N, crossing the inferior femoral and superior tibial growth plate. A femoral valgus and a tibial varus occurred without the creation of an epiphysiodesis bridge. Houle [44] used the rabbit and reproduced the same experiment by placing a tendon taken from the patellar tendon and crossing four different sizes of tunnel ranging from 1.95 mm to 3.97 mm in diameter, created across the femoral and tibial physis. Eight of the eleven rabbits which were subjected to this protocol displayed epiphysiodesis which was the more marked the larger the diameter of the tunnel. This technique therefore runs two potential risks, one being misalignment due to the presence of asymmetrical forces on healthy cartilage, the other being the risk of epiphysiodesis.

In summary, the growth plate is a prestressed, viscoelastic structure displaying mechanical anisotropy and controlled by the biomechanical stresses to which it is subjected. A better understanding of the physico-chemical mediators which cause the changes in stresses on the physis is a promising area of research which will undoubtedly have numerous clinical applications.

References

1. Delpech JM. De l'orthomorphie par rapport à l'espèce humaine Thèse Paris 1928.
2. Blount WP, Clark GR. Control of bone growth by epiphyseal stapling. J Bone Joint Surg 1949;31:464–78.
3. Moen CT, Pelker RR. Biomechanical and histological correlations in growth plate failure. J Pediatr Orthop 1984;4:180–4.
4. Chung S, Batterman S, Brighton CT. Shear strength of the human femoral capital epiphyseal plate. J Bone Joint Surg 1976;58A:94–103.
5. Gigante A, Specchia N, Nori S, Greco F. Distribution of elastic fiber types in the epiphyseal region. J Orthop Res 1996;14:810–17.
6. Amamilo SC, Bader DL, Houghton GR. The periosteum in growth plate failure. Clin Orthop 1985;194: 293–305.
7. Bright RW, Burnstein AH, Elmore SM. Epiphyseal plate cartilage. J Bone Joint Surg 1974;56A:688–703.
8. Rudicel S, Pelker RR, Lee KE, Ogden JA, Panjabi MM. Shear fractures through the capital femoral physis of the skeletally immature rabbit. J Pediatr Orthop 1985;5: 27–31.
9. Lee KE, Pelker RR, Rudicel SA, Ogden JA, Panjabi M. Histologic pattern of capital femoral growth plate fracture in rabbit. J Pediatr Orthop 1985;5:32–9.
10. Williams JL, Vani JN, Eick JD, Petersen EC, Schmidt TL. Shear strength of the physis varies with anatomic location and is a function of modulus, inclination and thickness. J Orthop Res 1999;17:214–22.
11. Peltonen J, Aalto K, Karaharju E, Aletalo I, Gronblad M. Experimental epiphyseal separation by torsional force. J Pediatric Orthop 1984;4:546–9.
12. Bylander B, Selvik G, Hunsson LI, Aronson S. A roentgen stereophotogrammetric analysis of growth arrest by stapling. J Pediatr Orthop 1981;1:81–90.
13. Bylski-Austrow DI, Wall EJ, Rupert MP, Roy DR, Crawford AH. Growth plate forces in the adolescent human knee: A radiographic and mechanical study of epiphyseal staples. J Pediatr Orthop 2001;21(6):817–23.

14. Bonnel F, Peruchon E, Baldet P, Rabishong P. Comportement mécanique du cartilage de croissance. Rev Chir Orthop 1980;66:417–21.
15. Duke PJ, Montufar-Solis D. Exposure to altered gravity affects all stages of endochondral cartilage differentiation. Adv Space Res 1999;24(6):821–7.
16. Montufar-Solis D, Duke PJ, D'Aunno D. In vivo and in vitro studies of cartilage differentiation in altered gravities. Adv Space Res 1996;17(6–7):193–9.
17. Montufar-Solis D, Duke PJ. Gravitational changes affect tibial growth plate according to hert's curve. Aviation Space and Environmental Medicine 1999;70(31):245–9.
18. Wronski TJ, Morey ER. Recovery of the rat skeleton from adverse effects of simulated weightlessness. Metab Bone Dis Relat Res 1983;4:347–52.
19. Sibonga JD, Zhang M, Evans GL, Westerlind KC, Carolina JM, Morey Holton E et al. Effects of spaceflight and simulated weightlessness on longitudinal bone growth. Bone 2000;27(4):535–40.
20. Hueter C. Anatomische Studien an den Extremitätengelenken Neugeborener und Erwachsener. Virchow Arch 1862;25:572.
21. Pauwels F. Biomécanique de la hanche saine et pathologique. Springer Verlag, 1977.
22. Bonnel F, Peruchon E, Baldet P, Dimeglio A, Rabishong P. Effects of compression on growth plates in rabbit. Acta Orthop Scand 1983;54:730–3.
23. Alberty A, Peltonen J, Ritsila V. Effects of distraction and compression on proliferation of growth plate chondrocytes. A Study in Rabbits. Acta Orthop Scand 1993; 64(4):449–55.
24. Mankin KP, Zaleste DJ. Response of physeal cartilage to low level compression and tension in organ culture. J Pediatr Orthop 1998;18(2):145–8.
25. Arriola F, Forriol F, Cañadell J. Histomorphometric study of growth plate subjected to different mechanical conditions (compression, tension and neutralization): An experimental study in lamb mechanical growth plate behavior. J Pediatr Orthop Part B 2001; 10(4):334–8.
26. Cohen B, Chorney GS, Phillips DP, Dick HM, Mow VC. Compressive stress-relaxation behavior of bovine growth plate may be described by the non-linear biphasic theory. J Orthop Res 1994;12(6):804–13.
27. Gray ML, Pizzanelli AM, Lee RC, Grodzinsky AJ, Swann DA. Kinetics of the chondrocyte biosynthetic response to compressive load and release. Biochim Biophys Acta 1989;991:415–24.
28. Greco F, Palma LD, Specchia N, Mannarini M. Growth plate cartilage metabolic response to mechanical stress. J Pediatr Orthop 1989;9(5):520–4.
29. Klein-Nulend J, Veldhuijzen JP, Van de Stadt RJ, Van Kampen GP, Kuijer R, Burger EH. Influence of intermittent compressive force on proteoglycan content in calcifying growth plate cartilage in vitro. J Biol Chem 1987;262:15490–5.
30. Trueta J. The vascular contribution to osteogenesis. J Bone Joint Surg 1961;43B:800–13.
31. Collard-Meynaud P. Etudes des déformations angulaires expérimentales du radius chez l'agneau Thèse de science. Toulouse, 2001.
32. Kenwright J, Spriggins AJ, Cunningham JL. Response of the growth plate to distraction close to skeletal maturity. is fracture necessary? Clin Orthop 1990;250:61–72.
33. Pereira BP, Cavanagh SP, Pho RWH. Longitudinal growth rate following slow physeal distraction. The proximal tibial growth plate studied in rabbits. Acta Orthop Scand 1997;68(3):262–8.
34. Apte SS, Kenwright J. Physeal distraction and cell proliferation in the growth plate J Bone Joint Surg Br 1994;76(5):837–43.
35. Alberty A, Peltonen J. Proliferation of the hypertrophic chondrocytes of the growth plate after physeal distraction. An experimental study in rabbits. Clin Orthop 1993;297:7–11.
36. Elmer EB, Ehrlich MG, Zaleske DJ, Polsky C, Mankin HJ. Chondrodiastasis in rabbits: a study of the effect of transphyseal bone lengthening on cell division, synthetic function and microcirculation in the growth plate. J Pediatr Orthop 1992;12(2):181–90.
37. De Bastiani G, Aldegheri R, Renzi Brivio L, Trivella G. Limb lengthening by distraction of the epiphyseal plate. A comparison of two techniques in the rabbit. J Bone Joint Surg 1986;68(4):545–9.
38. Alberty A. Effects of physeal distraction on the vascular supply of the growth area: a microangiographical study in rabbits. J Pediatr Orthop 1993;13:373–7.
39. Lee SH, Szo G, Simpson H. Response of the physis to leg lengthening. J Pediatr Orthop Part B 2001;10(4):339–43.
40. De Pablos J, Canadell J. Experimental Physeal Distraction in Immature Sheep. Clin Orthop 1990;250: 73–80.
41. Bollini G, Tallet JM, Jacquemier M, Bouyala JM. New procedure to remove a centrally located bone bar. J Pediatr Orthop 1990;10:662–6.
42. Yoshida G, Hirano T, Shindo H. Deformation and vascular occlusion of the growing rat femoral head induced by mechanical stress. J Orthop Sci 2000;5(5):495–502.43.
43. Edwards TB, Greene CC, Baratta RV, Zieske A, Willis RB. The effect of placing a tensioned graft across open growth plates: a gross and histologic analysis. J Bone Joint Surg 2001;83(5):725–34.
44. Houle JB, Letts M, Yang J. Effects of a tensioned graft in a bone tunnel across the rabbit physis. Clin Orthop Rel Res 2001;391:275–81.

30 Fractures in Children

A. Kusaba and S. Saito

Characteristic of Fractures in Children

Management of fractures in children is completely different from those in adults. The biological reaction to fracture is characteristic in children because of the anatomic, physiological, and biomechanical properties of skeletal structure. Comprehension of the property of bone in children is essential to treat fractures. Some properties act favorably for bone union and some make it difficult to treat the fractures.

The Role of the Periosteum

The detailed biomechanics of the periosteum is described in another chapter of this book. Periosteum in childhood bone is far thicker than that of adults and thus more callus appears in the early term after the injury. This is one of the reasons why union of the fracture is easily obtained in children. High activity and rich blood supply in the thick periosteum affect new bone formation. New bone formation in children occurs so soon after the injury that reduction of the displacement, if necessary, should be achieved in the early stages. The periosteum in childhood bone is stronger than that of adults [1]. It is well known that the connection between metaphysis and epiphysis is very strong. This strong connection is brought about by the strength of the periosteum. When bending stress is given to childhood bone, the periosteum would be ruptured only on the convex side, not on the concave side. The periosteum on the concave side acts as "intact hinge". This "intact hinge" would be helpful to obtain the reduced position and to maintain the reduced position. On the other hand, if the displacement were not reduced at an early stage, the gap between the exfoliated periosteum and the bone would be filled by new bone and severe deformity would remain [2]. This phenomenon is often observed in unreduced supracondylar fractures of the humerus or distal fracture of the radius.

Characteristic of Child Bone

Bone density and bone porosity in children is lower than that in adults [3]. The greater part of the bone cortex in children is occupied by Haversian canals [4]. Because of both the structural properties of bone and the thick periosteum, the whole bone unit in children has more endurance against stress (e.g., compressive stress, tensile stress, bending stress) than that in adults. Currey made an experimental study concerning the strength of the child femur. Compared with adults, children had a lower modulus of elasticity, a lower bending strength, and a lower ash content. However, children's bone deflected more and absorbed more energy before breaking. It also absorbed more energy after fracture had started. The typical greenstick fracture surface of many specimens of children's bone requires more energy for its production than the smooth surface of adult specimens [5]. Even when a fracture occurs in child bone, the fracture line is simple and is rarely accompanied with severe displacement. Most fractures in children are incomplete fractures in which the partial continuity of bone remains (i.e., subperiosteal fracture). When bending stress beyond the physiological tolerance is given to a bone, angular deformity of fracture occurs less often in children than in adults. In such cases, greenstick fracture more likely occurs in children. This fracture is often observed, especially in the distal third of the radius. When axial compressive stress is given to the bone, bamboo or buckle fracture often occurs at the metaphysis, where the maximum porosity exists. Otherwise, a plastic deformation may occur [6]. The plastic deformation often observed in pediatric long bone fracture is due largely to the complex nature of the molecular and histologic aspects of pediatric bone. Pediatric cortical bone has a lower mineral content than adult bone, accounting in part for its different material properties. Although plasticity allows children's long bones to absorb more energy before fracture, a significant deformity may persist after injury [7,8].

Epiphyseal Injury

Fracture of the growth plate is an injury unique to childhood. The prevalence of this injury in all fractures in children is 10–30% [9]. Salter classified this injury into five groups by the force of stress on the epiphyseal plate and the direction of stress distribution at the plate, and the prognosis: type I, fracture through the growth plate; type II, fracture through the growth plate and metaphysis; type III, fracture through the growth plate and epiphysis; type IV, fracture through the growth plate, epiphysis, and metaphysis, and type V, crush or compression injury of the growth plate [2]. Ogden proposed a new classification scheme of physeal and epiphyseal injuries to allow better estimation of prognosis for normal or abnormal growth. This classification is based partially on the Salter–Harris system, but, additionally, details subclassifications that relate to specific injury patterns [10]. Epiphyseal fracture often occurs on the metaphyseal side of the epiphysis and most of the epiphyseal cartilage cells remain in the epiphyseal side. Most such fractures heal without permanent deformity. A small percentage, however, have subsequent complications such as aseptic necrosis of epiphysis, non-union, premature epiphyseal closure (i.e., growth arrest), joint deformity (angulation, rotation) due to partial epiphyseal closure, and unequal length of the leg or arm. The growth plate has three-dimensional unevenness, which has been confirmed by scanning electron microscopic studies. This unevenness gives the growth plate a multidirectional endurance against compression, strain, and shear stress. The yield strength of a growth plate depends on the degree of this three-dimensional unevenness [11,12]. The reported yield strength of the growth plate varies as it depends on the nature of the experiment [13]. According to Bright, the strength of the force which made a crack in the growth plate was one-half of the yield strength of the growth plate. On the other hand, Bright and Nakada reported that no relation existed between these two strengths [12,14]. Kim pointed out that the strength depends on the existence of the membrane of cartilage [11]. The membrane seems to reinforce the growth plate in cooperation with the three-dimensional structure of the growth plate. In his report, the ratio of fracture strain between the growth plate without/with cartilage membrane was 61% (tensile strength) and 37% (strain strength). Amamilo reported that the periosteum contributed significantly to the stiffness of the epiphyseal plate [15]. Hypertrophying and calcifying layers are more fragile than resting and proliferative layers. However, the portion where cracks in the epiphyseal injury exist varies with authors. Bright reported that the crack was observed not only in the hypertrophying layer but also in the resting and the proliferative layer [14]. Nakada reported that crack existed mainly between the proliferative and hypertrophying layer [12]. Amamilo reported that the crack existed in the calcifying layer [15].

Remodeling after Fractures in Children

When malunion occurs after fracture in children, spontaneous correction of deformity is produced by bone remodeling during growth

[16–21]. A good example of remodeling can be observed in birth fracture. Malunion in children is often observed in supracondylar fracture of the humerus, lateral condylar fracture of the humerus, Monteggia fracture, distal radius fracture, proximal femoral fracture, and femoral shaft fracture. The remodeling is brought about by Wolff's law and the law of Heuter–Volkmann. Both of them have been recognized as bone reaction against stress such as weight bearing, muscle tension, and joint movement. Wolff's law is both periosteal absorption on the convex side and periosteal bone formation on the concave side in the diaphysis. Treharne regarded this law as an effect of piezoelectricity of bone [22]. Abraham made a detailed experimental study concerning this law [23], according to which, the bone absorption on the convex side did not occur so much as bone formation on the concave side. Weight bearing did not have so much influence on remodeling; no significant difference was observed between remodeling after fracture of the upper extremity and of the lower extremity. The law of Heuter–Volkmann is asymmetrical epiphyseal growth. Karaharju et al. made an experimental study to find out in which phase and to what extent asymmetrical epiphyseal growth participated in the correction of an experimentally produced deformity. According to their experiment, epiphyseal growth played an important role in the remodeling process. The greatest correction occurred during the first weeks. Correction of the epiphyseal angle took place with acceleration of growth [24]. The remodeling potential of angular deformities in the coronal and sagittal planes in children is well known. According to Wallace and Hoffman, in children less than 13 years of age, malunion of as much as 25° in such planes will remodel enough to give normal alignment of the joint surfaces. They also pointed out that 74% of correction occurred at the physes and only 26% at the fracture site. On the other hand, there is the poor remodeling potential of significant post-traumatic torsional deformity of the femur in children [25]. The authors made a clinical study concerning three-dimensional remodeling against malunion after femoral shaft fracture [26,27]. Valgus deformity occurred more often than varus deformity when the primary union was achieved. This may have been an influence of traction direction. On the other hand, the angle of deformity at the final follow-up was greater in varus deformity than that of valgus deformity. The convex angle of deformity had no significant difference between the anterior and posterior convex deformity at the final follow-up. Rotational malunion was also compensated for to some extent. The antetorsion angle of the femoral neck against the femoral condyle was around 30° at the final follow-up either in external or internal rotational malunion cases. In any kind of deformity, the younger the child the greater the angle of spontaneous correction is observed. Also, the motion and flexibility of joints compensates for the deformity. The Mikulics line shifted from the medial side to near the center of the knee joint in children with valgus convex and shifted from the lateral side to near the center of the knee joint children with varus convex. Such compensation occurred earlier than compensation by remodeling. Considering these compensations, the deformity within the following angles should not cause problems during daily activities. The maximum angle of acceptable deformity after the primary union is 30° in children aged one or under, 20° in children aged one to five, 15° in children six to ten years old, and 10° in children eleven years old or more [26].

Shortening of bone length can be also compensated for in childhood fracture to some extent. Excessive compensation is often observed especially in femoral shaft fractures. Excessive growth of bone length occurs more likely in surgically treated fractures than conservatively treated fractures. The average difference of length discrepancy in conservatively treated femoral shaft fractures was around ten millimeters in the authors' study. Shortening of the femur showed almost no compensation in most children ten years old or more. In most children younger than ten, excessive growth of bone length continued until two years after the injury. Bone growth two or more centimeters greater than the other side occuring during the growth period may be treated by epiphyseodesis (e.g., stapling) or elongation of the leg using external skeletal fixation [28].

References

1. Deppermann F, Dallek M, Meenen N, Lorke D, Jungbluth KH. The biomechanical significance of the periosteum for the epiphyseal groove (Die biomechanische Bedeutung des Periosts fur die Epiphysenfuge) Unfallchirurgie 1989;15:165–73.
2. Salter RB, Harris WR. Injuries involving the epiphyseal plate. J Bone Joint Surg 1963;45A:587–622.
3. Boot AM, de Ridder MA, Pols HA, Krenning EP, de Muinck Keizer-Schrama SM. Bone mineral density in children and adolescents: relation to puberty, calcium intake, and physical activity. J Clin Endocrinol Metab 1997;82:57–62.
4. Rang M. Children's Fractures (2nd ed). Philadelphia: Lippincott, 1983.
5. Currey JD, Butler G. The mechanical properties of bone tissue in children. Bone and Bones. J Bone Joint Surg 1975;57A:810–4.
6. Mabrey JD, Fitch RD. Plastic deformation in pediatric fractures: mechanism and treatment. J Pediatr Orthop 1989;9:310–4.
7. Connolly JF. Torsional fractures and the third dimension of fracture management. South Med J 1980;73: 884–91.
8. Borden S 4th. Roentgen recognition of acute plastic bowing of the forearm in children. Am J Roentgenol Radium Ther Nucl Med 1979;125:524–30.
9. Wilkins KE. The incidence of fractures in children. In: Rockwood CA, Wilkins KE, Beaty JH, editors. Fractures in Children, 4th. ed. Philadelphia: Lippincott-Raven, 1996;3–15.
10. Ogden JA. Injury to the growth mechanisms of the immature skeleton. Skeletal Radiol 1981;6:237–53.
11. Kim WC. Biomechanical properties of growth plate. J Jpn Paed Orthop Ass 1996;6:128–32.
12. Nakada D. Torsional strength of the epiphyseal plate and fracture patterns with aging, three-dimensional analysis with SEM. J Jpn Orthop Assoc 1993;67:1045–54.
13. Williams JL, Vani JN, Eick JD, Petersen EC, Schmidt TL. Shear strength of the physis varies with anatomic location and is a function of modulus, inclination and thickness. J Orthop Res 1999;17:214–22.
14. Bright RW, Virginia R, Burstein AH. Epiphyseal-Plate Cartilage. J Bone Joint Surg 1974;56A:688–703.
15. Amamilo SC, Bader DL, Houghton GR. The periosteum in growth plate failure. Clin Orthop 1985;194:293–305.
16. Saito S, Kuroki Y, Uchida T, Mori Y. Experimentelle Untersuchungen ueber Entstehung der Antetorsion am Femur. Z Orthop 1980;118:612.
17. Husby OS. Spontaneous correction of femoral torsion. Acta Orthop Scand 1987;58:113–16.
18. Oberhammer J. Degree and frequency of rotational deformities after infant femoral fractures and their spontaneous correction. Arch Orthop Traum Surg 1980; 7:249–55.
19. Uchida T, et al. Experimental study of femoral torsion – spontaneous correction of torsional deformity after femoral fracturs. Seikeigeka kisokagaku 1986;13:459–61.
20. Wakita M et al. A study of the projection method of the femur. J Jpn Orthop Assoc 1988;62:1374.
21. Wakita M. Fundamental and clinical studies of the projection method of the femur. Showa Univ J Med Sci 1996;56:140–52.
22. Treharne RW. Review of Wolff's law and its proposed means of operations. Orthop Review 1981;10:35
23. Abraham E. Remodeling potential of long bones following angular osteotomies. J Pediatr Orthop 1989;9: 37–43.
24. Karaharju EO, Ryoppy SA, Makinen RJ. Remodelling by asymmetrical epiphysial growth. An experimental study in dogs. J Bone Joint Surg 1976;58B:122–6.
25. Davids JR, Maguire MF, Mubarak SJ, Wenger DR. Lateral condylar fracture of the humerus following post-traumatic cubitus varus. J Pediatr Orthop 1994;14: 466–70.
26. Saito S, Kuroki Y, Ohgiya H, Marutani R, Obara S, Hayashi J et al. Changes in the alignment of the lower extremities in children – A study of the cases with fractures of the femur and congenital dislocation of the hip. J Jpn Paed Orthop Ass 1993;3:148–56.
27. Saito S, Kurisaki K, Omata T. A study of fractures among children less than 5 years. The Showa University Journal of Medical Science 1997;9:11–15.
28. Glorion C, Pouliquen JC, Langlais J, Ceolin JL, Kassis B. Femoral lengthening by callotasis. A study of a series of 79 cases in children and adolescents (Alongement de femur par callotasis. Etude d'une serie de 79 cas chez l'enfant et l'adolescent.). Rev Chir Orthop Reparatrice Appar Mot 1975;81:147–56.

31 The Pediatric Hip

S. Saito and A. Kusaba

Congenital Dislocation of the Hip Joint

The disease formerly known as Congenital Dislocation of the Hip has more recently been described as Developmental Dislocation (or Dysplasia) of the Hip. It is not clear whether the causes of this disease are congenital, developmental or both. True congenital dislocation is often intractable, as it involves severe dislocation and is accompanied by considerable anomalies. In true congenital dislocation, the acetabulum is remarkably shallow and steep, and antetorsion of the femur is severe. In developmental dislocation, environmental factors have more influence on the occurrence of the dislocation than congenital ones. The kind of diaper is one of the most important factors. The prevalence of dislocation is high in some colder areas. In such areas, diapers or clothes are often fitted down to the knee with the hip in the extended position to keep the children from the cold. In hemilateral dislocation after birth, the lying position has an influence on its occurrence. Joint instability, which is caused by hormones around the time of birth, and contracture due to the position may cause dislocation. In the left wryneck position (i.e., the face rotates to the right and the neck flexes to the left), the trunk often takes a right lateral position. In such a position, the left hip joint takes an adducted position and tends to have adduction contracture, and dislocation may occur in the left hip. The degree of the dislocation varies between children. Hamstring muscles have a strong influence on the occurrence of the dislocation. These muscles push the femoral head proximally. When secondary dysplasia of the acetabulum exists besides the dislocation, spontaneous rectification rarely occurs and the hip is still unstable even after manual reduction. After the occurrence of the dislocation, contracture of the iliopsoas muscles and varus position of the labrum are inhibitors against reduction.

Posture or motion of the fetus in the uterus also has a strong influence on the occurrence. According to Vartan [1] and Tompkins [2], 56–75% of breech delivery fetuses have extended knee posture in the uterus while only 3% of cephalic delivery fetuses have such posture. Wilkinson [3] studied 866 cephalic delivery children and 123 breech delivery children. Fully flexed knee posture was observed in 83% of cephalic delivery children while it was only observed in 35% in breech delivery children. The fully flexed knee posture was observed only in 20% of children with hip dislocation. From these results, he concluded that disturbance of the leg-holding mechanism is an important factor to the dislocation. Vartan [1] reported that disturbance of the leg-holding mechanism causes extended knee posture, which disturbs posture change in the uterus, and causes breech birth. Michelson and Langenskiöld [4] made an experimental study using juvenile rabbits. They fixed the knee joints of the rabbits and succeeded in making artificial dislocations or subluxation. In their experiment, no dislocation or subluxation occurred in the rabbits after hamstring myotomy. They suspected that continuous spasm of the hamstrings was the main cause of the dislocation. The hamstring is a diarthric muscle that affects the hip and the knee joint. Thus, when the hip joint is forced to flex with the knee joint extended, severe axial load is placed on the femur. Michele [5] reported that contracture or shortening of the iliopsoas muscle may cause anthropologic dislocation. He emphasized that the iliopsoas muscle is the most important etiologic factor of congenital dislocation of the hip joint. Mckibbin [6] made a postmortem examination of neonates. According to his report, an extended position of the hip caused a tendency to hip dislocation in the neonates with iliopsoas contracture. After myotomy of the iliopsoas, such a phenomenon declined. Yamamuro [7] made an experimental study and reported that hip dislocation was caused by reciprocal force of the hamstring and the iliopsoas. According to his report, when the knee is fixed in the extended position, the hip is extended due to the excessive tension of the hamstring. The extended hip position causes excessive tension of the iliopsoas and the tension forces the hip to be flexed against the tension of the hamstring. In such a situation, the iliopsoas acts to rotate the femur externally and to cause the dislocation. From these reports, both the hamstring and the iliopsoas seem to have an influence on hip dislocation. In the prenatal period, the hamstring has more influence than the iliopsoas, as most fetuses have flexed posture. After birth, the iliopsoas has more influence than the hamstring, as most neonates have extended posture.

Treatment for Congenital Hip Dislocation

The frog position (flexed, abducted, externally rotated) prevents the hip from dislocation in neonates, even if the hip is unstable and the joint laxity exists. Some methods exist to arrange such a position. The most common method is Pavlik harness. Taking the frog position, the dislocated femoral head is brought near to the acetabulum. The weight of the lower extremity reduces the tension and contracture of the adductor muscles, consequently the femoral head slips over the acetabular rim and is reduced into the acetabulum. This reduction force is given not by the motion of the lower extremity through the stirrup but by the weight of it. The authors made an above-knee Pavlik harness (without the stirrup) and succeeded in the reduction. Generally, the success rate of reduction is around 80% of dislocations in infancy. Pavlik harness is effective for mild dislocation. Traction, manual reposition, or open reduction is necessary for a dislocation that is irreducible in both extended and frog position. Such treatment is also necessary for dislocation accompanied by varus labrum.

The overhead traction method diminishes the spasm of the surrounding muscles and capsular ligaments to enlarge the acetabular inlet, and expands the spasm of obstructive muscles. First, horizontal traction reduces contracture of the iliopsoas and rectus femoris and expands the capsular ligament inferiorly. Next, gradual flexion of the hip with extended knee position reduces the contracture of the hamstring. As the iliopsoas, which presses the acetabular inlet anteriorly, becomes loose, the inlet spreads. Finally, traction in the fully abducted position reduces contracture of the adductor muscles to reduce the femoral head into the acetabulum without pressing the head to the labrum.

Configuration of the Hip Joint and Alignment of Lower Extremity

The femur in congenital hip dislocations has more antetorsion angle and neck shaft angle than anatomical ones. The acetabulum in congenital hip dislocations is steep and shallow. Because of this configuration, the hip is unstable just after the reduction. The steep and shallow acetabulum is improved gradually. When the improvement is insufficient, arthrosis may occur after the residual subluxation. For residual subluxation, pelvic osteotomy and/or varus-derotation osteotomy is necessary. Today, pelvic osteotomy is preferable. Pelvic osteotomy brings about better containment and stability of the hip. However, some pelvic osteotomies that enlarge the ilium superoinferiorly cause hemodynamic changes of the femur and increased growth of the femur. As the result, functional dysplasia may occur in the standing position. Recurrence of valgus sometimes occurs after the varus-derotation osteotomy of the femur.

Atrophy of abductor muscles, residual dysplasia of the acetabulum, or damage of trochanteric apophysis is thought to be the cause of this recurrence of valgus. Excessive recurrent valgus causes recurrence of subluxation and instability of the hip. The degree of recurrent valgus is higher in younger children. Thus, a greater varus angle of the osteotomy is necessary for younger children. Pelvic osteotomy combined with varus-derotation osteotomy is adopted when simple pelvic osteotomy does not bring about sufficient containment. Generally, the combined osteotomy is necessary for the femur with an antetorsion angle of 60° or more. Otani [8] made an experimental study using three-dimensional elasto-optic methods to support such indications. The authors made a study of alignment after the osteotomy. FTA increased just after the surgery [9], however, because of recurrent valgus and the remodeling in the distal femur, FTA decreased in the postoperative course. The functional axis of the lower extremity (Mikulicz Line) returned to go through the midpoint of the knee one year after the surgery.

Femoral Antetorsion and Cooperated Factors of Torsion

The angle of the femoral antetorsion angle is the angle between the mediolateral femoral condyle line and the axis of the femoral neck in a horizontal plane. Many measuring methods of the antetorsion angle have been performed [10–19]. The femoral antetorsion angle in the prenatal period has individual differences and varies between reports [20–22]. During the early prenatal period, the femoral neck has retrotorsion. From the middle stage in the prenatal period, the femoral neck begins to have antetorsion. The position of the legs in the uterus and the tension of muscles attached to the trochanters have a strong influence on this phenomenon. According to Michele, from the sixth to ninth month of intrauterine life, the lower extremities rotate internally and this phenomenon brings about the femoral antetorsion [23]. Some authors point out the relation between congenital hip dislocation and the antetorsion angle in this stage [20,24,25]. In this stage, contracture of the iliopsoas muscle can cause excessive antetorsion or superior displacement of the femoral head. The antetorsion angle is 30° or more at perinatal period. The angle at one year old is around 50° and decreases with growth [25–27]. Such a phenomenon seems to have a relation to walking with an erect posture. According to Lanz, the angle is 15–57° (average, 32°) in neonates, 20–50° (average, 34°) in one- to three-year-old children, 12–38° (average, 25°) in four- to six-year-old children, and –25–37° (average, 12°) in adults [22].

The femur antetorsion angle can increase in patients with some diseases such as congenital dislocation of the hip or Perthes' disease. The increased antetorsion angle may influence the congruency or stability of the hip joint. In residual subluxation after the treatment for congenital dislocation, increased antetorsion angle of the femoral neck is often observed. Better congruency in the abduction-inner rotation position aids in diagnosing the existence of the increased antetorsion angle. As far as the contracture of the iliopsoas muscle exists, the increased antetorsion angle can be improved in the natural course after first beginning to walk [28,29].

Experimental Study of Femoral Antetorsion

The authors made an experimental study concerning femoral antetorsion in congenital dislocation of the hip joint [30,31]. Twenty young mongrel dogs with experimental complete dislocation of the hip joint were studied. Seven or eight Kirschner wires were inserted vertically to the longitudinal axis of the femur. They were inserted taking care not to injure the growth plate, and were inserted so that each wire was just parallel and spaced equally. The most distal wire was inserted in the distal epiphysis. The wire second to the most distal one was inserted in the distal portion of the femoral metaphysis. Thus, the distal epiphyseal line existed between the two distal wires. They were cut and buried in the femur. Increased antetorsion was observed just one week after the surgery. Another Kirschner wire was inserted along the

Table 31.1. Summary of results. Location of wire in femur, time elapsed since dislocation and relative increase or decrease of antetorsion. Dark arrows indicate an increase of antetorsion. Light arrows indicate a decrease of antetorsion. The thicker arrows mark greater changes

Post op. Period (weeks) \ Location	Condylen – Supracondylen	Distal third of femoral shaft	Proximal third of femoral shaft	Subtrochanter	Femoral neck – Femoral head
1–3	→	→	→	→	
4	↑	→	→	→	
6	↑	→	→	→	→
8	⬆	↑	→	⇓	
9	⬆	↑	→	→	
10	⬆	↑	→	⇓	
13			→	⇓	
14	⬆	↑	→	→	⇩
16	⇓	↑	→	⇓	⬆
21	⬆	⬆	→	⇓	⇓

⬆ Increase ↑ of antetorsion; ⇩ Decrease ⇓ of antetorsion; → No change of antetorsion

axis of the femoral neck, and the femurs were examined after retrieval. From the position of each wire, the rotation of the femur was evaluated. The most distal wire rotated internally against the next wire. At this time, all wires in the metaphysis, diaphysis, and the wire in the neck were parallel each other. This phenomenon indicates that the distal epiphyseal line brings about antetorsion. When this distal rotation was excessive, compensative retrotorsion occurred between the proximal two wires (in the portion between the wire in the neck and the next wire) after the occurrence of the antetorsion. (Table 31.1, Figures 31.1–31.6). In conclusion, the femoral torsion is mainly brought about by the distal epiphyseal line and additionally by the proximal portion of the femur. Adequate load seems necessary to get this proximal retroversion so that the antetorsion angle will be excess in congenital hip dislocations [30,31].

Slipped Capital Femoral Epiphysis

Slipped capital femoral epiphysis is divided into three types. The acute type is a fracture in the proximal femoral growth plate. This type is generally caused by significant trauma. In the chronic type, the femoral head slips gradually because of the brittleness of the growth plate. The chronic type is the most common. Slip progresses acutely during the chronic course in acute or chronic types. The most common slip direction is posteroinferior. The direction has a close relation to the angle of the neck shaft. Wilson [32] reported that 85% of posteroinferior slip had 140° of neck-shaft angle, and according to Inhäuser [33], the head slips posterolateralinferiorly when severe coxa valga, such as 160° of neck-shaft angle exists. In such coxa valga, the growth plate in the standing position is almost parallel to the floor. Ordinary anteroposterior radiographs of the hip show that the head slips posteromedialinferiorly because of the externally rotated femur. Posteroinferior slip is clear on the radiograph taken without rotation of the femur. Three-dimensional CT scan studies support this fact. Lateral view radiography is thus essential in measuring the slip angle. Imhäuser's method is quite useful to measure the angle [33]. The anatomical angle in his method is 10° or less. In his criteria, 30° or less is slight slip, 30–70° is moderate, and 70° or more is severe.

According to Harris [34], growth hormone accelerates the mitosis of the proliferating layer

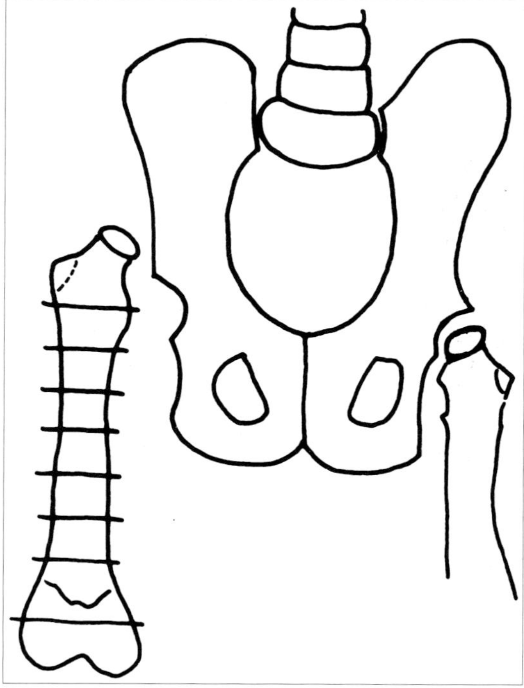

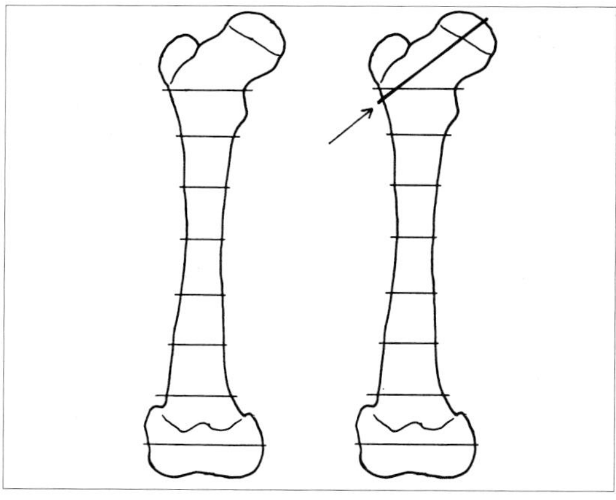

Figure 31.1. The method of experiment. The right femoral head of the dog was dislocated surgically and Kirschner wires were inserted parallel into the femur. Left: Parallel wires were inserted into the femur. Right: One extra wire was inserted through the proximal neck and head still in the same plane as that of the parallel wires along the femur.

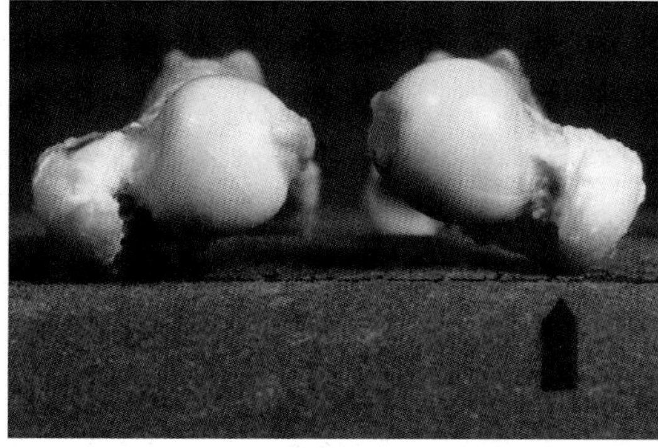

Figure 31.2. One week after dislocation. More antetorsion existed on the right femur. Arrow shows dislocated femur.

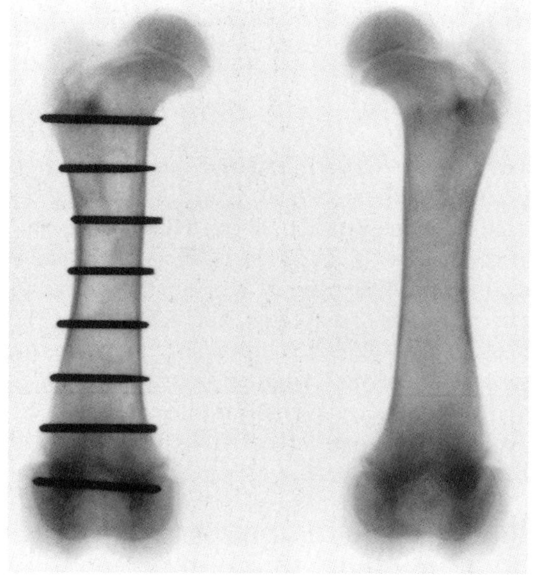

Figure 31.3. Radiograph taken one week after dislocation. All wires were still parallel.

and causes hypertrophy of the hypertrophied cartilage layer. Growth hormone decreases the yield strength of the growth plate while sex hormones increase the strength. His report suggests that insufficient sex hormone or excessive growth hormone may cause the slip. Relative excessive growth hormone can occur when secretion of the growth hormone continues after the growth period.

The growth plate has three-dimensional unevenness, which has been confirmed by scanning electron microscopic studies. This unevenness gives the growth plate a multidirectional endurance against compression, strain, and shear stress. The yield strength of a growth plate depends on the degree of this three-

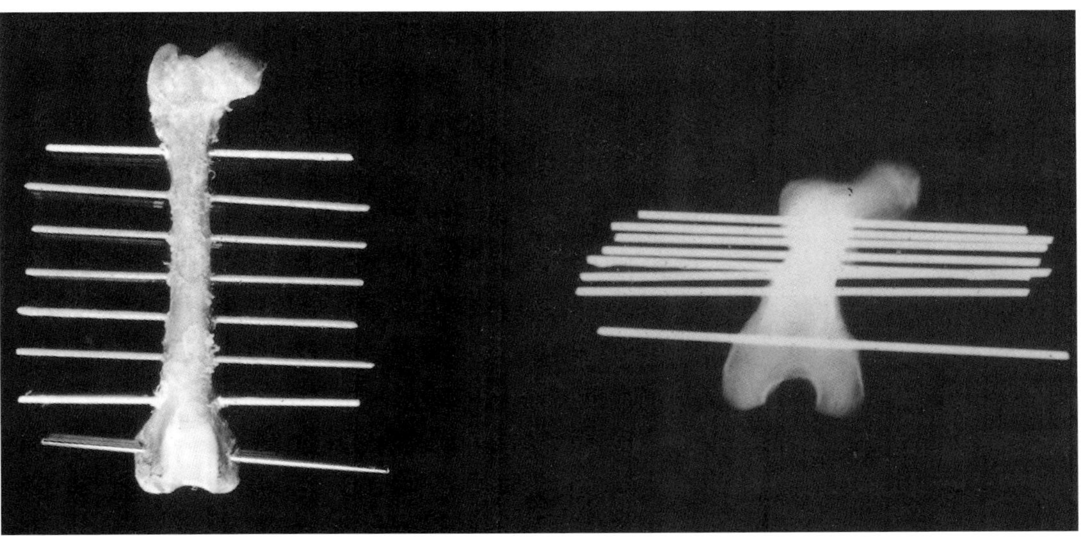

Figure 31.4. Six weeks after dislocation. All wires but the most distal were still parallel.

Figure 31.5. Ten weeks after dislocation. The wire inserted into the uppermost part of the femur rotated slightly inwards (arrow).

Figure 31.6. The wire inserted through the femoral neck and head rotated inward. This indicated decrease of antetorsion.

dimensional unevenness [35,36]. Acute slip is a kind of fracture and thus complete rupture occurs in the growth plate, or Salter–Harris type II injury occurs. In chronic slip, repetition of stress causes gradual disruption and after that slip occurs when the stress exceeds the yielding strength. External rotation contracture occurs when the slip angle is 30° or more. Gait in the externally rotated position due to the pain and weight bearing accelerates the disruption. Morscher [37] analyzed the posterior tilting angle and concluded that the posterior stress force in the internally rotated position is the main cause of slip. In slipped capital femoral epiphysis, the motion range of the hip increases in external rotation and decreases in internal rotation. Flexion range is also limited. Drehmann's phenomenon is obvious in slipped capital femoral epiphysis; as the thigh is flexed, it tends to roll into external rotation and abduction. After progress of the slip, varus deformity of the hip occurs and both Trendelenburg's phenomenon and gait become obvious.

The yield strength of the growth plate has been variously reported as it depends on the nature of the experiment [38]. According to Bright, the strength of the force which makes a crack in the growth plate is one-half of the yield strength of the growth plate. On the other hand, Bright and Nakada reported that there was no relation between these two strengths [36,39]. Chung reported that force within the strength of physiological range was enough to cause the epiphyseal slipping in overweight children and suggested that purely mechanical factors may play a major role in the etiology of slipped capital femoral epiphysis [40]. Williams made an experimental study and found that the shear strength of the physis varies with anatomic location. According to this experiment, the posterior region of the tibia physis has the greatest strength and stiffness, lowest physeal thickness, and steepest inclination [38]. Kim pointed out that the strength depends on the existence of the cartilage membrane [35]. The membrane seems to reinforce the growth plate in cooperation with the three-dimensional structure of the growth plate. In his report, the ratio of fracture strain between the growth plate without and with the cartilage membrane was 61% (tensile strength) and 37% (strain strength). Chung made postmortem studies of hips in children to determine the shear strength and modes of failure of the capital femoral growth plates [40]. In this experiment, the shear strength of the human growth plate varied with age and was greatly dependent on the surrounding perichondrial complex in infancy and early childhood, but less so in adolescence. When this complex was excised, the strength of the epiphyseal plate was diminished, especially in younger children. In acute slip, the membrane is disrupted while it is preserved in chronic slip.

The velocity of stress is also an important factor in growth plate injury, as the growth plate has viscoelastic properties. Kim reported that Salter–Harris type I injury is caused by the stress of relatively slow velocity, and type II and III by high velocity [35]. Histological studies performed concurrently with the yield-strength test varies among authors. This may be brought about mainly by differences in experimental methods. According to Salter and Amailo, epiphyseolysis occurred in the hypertrophying layer [41,42]. Bright reported that it occurred in a stairway form in the hypertrophying layer and palisading layer [43]. According to Nakada and Brasher, epiphyseolysis occurred in different layers depending on the portion of the growth plate [36,43]. From a clinical point of view, epiphyseolysis mainly occurs in the hypertrophying layer or transient portion of the metaphysis in slipped capital femoral epiphysis [34,38,44].

References

1. Vartan CK. The behaviour of the foetus in utero with special reference to the incidence of breech presentation at term. J Obstet Gynec Brit Emp 1945;52:417–34.
2. Tompkins P. An inquiry into the cause of breech presentation. Am J Obstet Gynec 1946;51:595–606.
3. Wilkinson JA. A postnatal survey for congenital dislocation of the hip. J Bone Joint Surg 1972;54B:40–9.
4. Michelson JE, Langenskiöld A. Dislocation or subluxation of the hip, regular sequels of immobilization of the knee in extension in young rabbits. J Bone Joint Surg 1972;54A:1177–86.
5. Michele AA. Iliopsoas. Springfield IL: Charles C Thomas, 1962.
6. McKibbin B. Anatomical factors in the stability of the hip joint in the newborn. J Bone Joint Surg 1970; 52B:148–59.
7. Yamamuro T, Hama H, Takeda T, Shikata J, Sanada H. Sexual hormone in the experimental hip dislocation. Cent Jpn J Orthop Traumat 1976;19:770–1.
8. Otani T. Studies on photoelastic stress analysis of congenital dislocation of the hip. J Jpn Orthop Assoc 1964;37:1001–26.
9. Saito S, Kuroki Y, Ohgiya H. Obara S, Hyashi J, Yamazaki K. Changes in the alignment of the lower extremities in children. A study of the cases with fractures of the femur and congenital dislocation of the hip. J Jpn Paed Orthop Assoc 1993;3:148–56.
10. Schneider B, Laubenberger J, Jemlich S, Groene K, Weber HM, Langer M. Measurement of femoral antetorsion and tibial torsion by magnetic resonance imaging. Br J Radiol 1997;70:575–9.
11. Drehmann F, Becker W. A simple clinical investigation method for the approximative rapid determination of the antetorsional angle of the neck of femur. Z Orthop 1980;118:236–40.
12. Grote R, Elgeti H, Saure D. Determination of the antetorsional angle at the femur with axial computer tomography. Rontgenblatter 1980;33:31–42.
13. Tomczak RJ, Guenther KP, Rieber A, Mergo P, Ros PR, Brambs HJ. MR imaging measurement of the femoral antetorsional angle as a new technique: comparison with CT in children and adults. AJR Am J Roentgenol 1997;168:791–4.
14. Haspl M, Bilic R. Assessment of femoral neck-shaft and antetorsion angles. Int Orthop 1996;20:363–6.
15. Bruckl R, Grunert S, Rosemeyer B. Roentgenologic determination of the actual femoral neck-shaft and antetorsion angle. 2: Alternatives to the Rippstein and Muller procedure. Radiologe 1986;26:305–9.
16. Wissing H, Spira G. Determination of rotational defects of the femur by computer tomographic determination of the antetorsion angle of the femoral neck. Unfallchirurgie 1986;12:1–11.
17. Gunther KP, Kessler S, Tomczak R, Pfeifer P, Puhl W. Femoral anteversion: significance of clinical methods and imaging techniques in the diagnosis in children and adolescents. Z Orthop Ihre Grenzgeb 1996;134:295–301.
18. Gormand E, Barral F, Roussille M, Bochu M, Fournet-Fayard J, Kholer R. Comparison between ultrasonic and x-ray computed tomographic measurements of femoral antetorsion in children. J Radiol 1985;66:789–92.
19. Clarac JP, Pries P, Laine M, Richer JP, Freychet H, Goubault F et al. Measurement of antetorsion of the

femoral neck by ultrasonics. Comparison with x-ray computed tomography. Rev Chir Orthop 1985;71: 365–8.
20. Badgley CE. Correction of clinical and anatomical facts leading to a conception of the etiology of congenital hip dysplasias. J Bone Joint Surg 1943;25:503.
21. Badgley CE. Etiology of congenital dislocation of the hip. J Bone Joint Surg 1949;31A:341.
22. Lanz T, Mayet A. Die Gelenkkorper des menschlichen Huftgelenks in der progredienten Phase ihrer umwegigen Ausformung. Z Anat 1953;117:317.
23. Michele AA. Iliopsoas. Thomas CC, editor. Springfield, IL, 1962.
24. Le Damany P. Die angeborene Hüftgelenkverrenkung. Ihre Ursachen: ihre Mechanismus: ihre arthropologische Bedeutung. Z. Orthop Chir 1908;21:129.
25. Somerville EW. Development of congenital dislocation of the hip. J Bone Joint Surg 1953;40A:803.
26. Chandler F. Anatomical study of congenital dislocation of the hip. J Bone Joint Surg 1929;11:546.
27. Shands AR Jr, Steele MK. Torsion of the femur. A follow-up report on the use of the Dunlop method for its detamination. J Bone Joint Surg. 1958;40A:1147.
28. Sylkin NN. Developmental tendency of the femur head following femoral head necrosis due to conservative treatment of a dislocated hip (2nd report). Z Orthop Ihre Grenzgeb 1995;133:367–73.
29. Lingg G, Nebel G, Thomas W, Hering L. Value of computed tomography in congenital hip dysplasia and hip luxation. Rontgenblatter 1983;36:407–13.
30. Saito S, Kuroki Y, Uchida T, Mori Y. Experimentelle Untersuchungen uber die Entstehung der Antetorsion am Femur. Z Orthop 1980;118:612.
31. Saito S, Kuroki Y, Uchida T. An experimental study on the change of the femoral antetorsion. J Jpn Orthop Assoc 1978;52:1185.
32. Wilson PD, Jacob B, Schecter L. Slipped capital femoral epiphysis. J Bone Joint Surg 1965;47A:1128–45.
33. Imhäuser G. Die jugendliche Hüftkopflösung bei steilem Schenkelhals. Z. Orthop 1959;91:403–13.
34. Harris R, Hobson KW. The endocrine basis for slipping of the upper femoral epiphysis. J Bone Joint Surg 1950;32B:5–11.
35. Kim WC. Biomechanical properties of growth plate. J Jpn Paed Orthop Ass 1996;6:128–32.
36. Nakada D. Torsional strength of the epiphyseal plate and fracture patterns with aging, three-dimensional analysis with SEM. J Jpn Orthop Assoc 1993;67:1045–54.
37. Morscher E. Zur Pathogenese der Epiphyseolysis capitis femoris. Archiv für orthopädische und Unfall-Chirurgie 1961;53:331–43.
38. Yoshida M, Kim WC, Arai Y, Inoue N, Watabe K, Takai N et al. The effect of maturation on dynamic visco-elastic properties of epiphyseal plate in rabbit. J Jpn Clin Biomecha 1994;15:147–50.
39. Bright RW, Virginia R, Burstein AH. Epiphyseal Plate Cartilage. J Bone Joint Surg 1974;56A:688–703.
40. Chung SM, Batterman SC, Brighton CT. Shear strength of the human femoral capital epiphyseal plate. J Bone Joint Surg 1976;58A:94–103.
41. Amadio P, Ehrlich MG, Mankin HJ. Matrix synthesis in high density cultures of bovine epiphyseal chondrocytes. Connect Tissue Res 1983;11:11–19.
42. Salter RB, Harris WR. Injuries involving the epiphyseal plate. J Bone Joint Surg 1963;45A:587–622.
43. Brasher HR Jr. Epiphyseal fractures. A microscopic study of the healing process in rat. J Bone Joint Surg 1959;41A:1055–4064.
44. Hurley JM, Betz RR, Loder RT, Davidson RS, Alburger PD, Steel HH. Slipped capital femoral epiphysis. The prevalence of late contralateral slip. J Bone Joint Surg 1996;78A:226–30.

IV

Applications of Biomechanical Principles to Orthopedics and Traumatology

32 Custom-made Hip Prostheses

J.-N. Argenson, J.-P. Aubaniac, P.-H. Vallotton, T. Clerc, P. J. Rubin, and P. F. Leyvraz

Introduction

The custom-made hip prosthesis itself is now integrated in a whole concept of computer-assisted hip arthroplasty which includes: computer-assisted preoperative planning, computer-assisted designing of custom hip prostheses, and computer-assisted hip surgery. This paper will focus on custom-made hip prostheses, describing the elements which lead to the design and the clinical use of custom stems. It will present a specific philosophy, describing the planning, design, fabrication, and biomechanical evaluation of custom-made hip prostheses.

The Rationale for a Custom Stem and a Custom Neck

Why Custom?

The loosening rate of conventional cemented stems reported in the literature ranges from 2 to 56% [1,2,3,4,5,6,7,8]. For young patients in a large group of 8,406 cases collected from the national Swedish register, the failure rate was 30% at eleven years [9], for patients younger than 55 years old. The high level of activity in such patients significantly increases the stresses applied on the component and may explain the higher failure rate [1,2,6,10,11]. Since Judet in 1978 [12], cementless fixation of femoral stems has been proposed as an alternative to cement fixation. The results of uncemented stems reported from the Swedish register in 1998 were not encouraging [9], for a number of reasons, including for a large part the old types of design in cementless fixation.

Nevertheless, cementless fixation for femoral stem prostheses has its own requirements which include proximal adaptation and avoidance of micromovements in order to obtain an optimal load transmission to the bone [13]. These principles are advocated in order to avoid stress shielding and thigh pain, the two complications often reported with cementless stems [14,15]. The specificity of young patients is not only the level of activity, but the wide range of proximal femoral anatomy [16]. This femoral anatomy can consist of a high canal flare index with a champagne-fluted femur, described by Noble [17], but for congenital or traumatic reasons it can also lead to a narrow, curved, and excessively ante or retroverted upper femur. These anatomical reasons lead some authors to avoid cementless fixation in young patients because of the impossibility of matching the requirements of proximal femoral adaptation in such cases [2,11].

Facing these two conflicting goals of cementless fixation and adaptation to all types of proximal femoral anatomy, the logical answer was to obtain a custom stem for each case.

Why a Custom Stem and a Custom Neck?

For many clinicians, custom-made prostheses correspond to a maximal filling of the femoral cavity associated with a conventional prosthetic neck in the upper part of the prosthesis. The clinical experience of such design showed us that the extramedullary part of the stem, i.e., the prosthesis neck, was at least of equal importance in order to restore correct hip function.

The design of a three-dimensional custom neck allows correction of length, lever arm, and anteversion. The clinical consequence of such design is the restoration of leg length, abductor function, and proper lower limb rotation. The appropriate anteversion of the neck may also contribute to reduce dislocation rate. The mechanical consequence of such neck design is also to optimize load transmission to the bone stem interface and finite element analyses have shown the influence of the extramedullary parameters on the stem stability and stress transfer [18].

If the intramedullary stem design is based upon the reconstruction of the proximal femoral anatomy, the design itself does not match the whole internal femoral anatomy. The priority areas of contact are proximal to obtain stability in rotation, and the distal diameter of the stem is reduced to avoid any cortical impingement distally – apossible source of thigh pain with maximal canal filling stems. It is thus of high importance to preserve all the cancellous bone around the whole stem from proximal to distal by the use of a smooth compactor of identical shape to the final prosthesis.

This whole concept, addressed by Aubaniac and Essinger in 1987, led to the development of software for cancellous bone density evaluation and three-dimensional custom neck design, which is the rationale of the Symbios® custom concept (Symbios Inc., Yverdon, Switzerland). The concept and the first results were published for the first time in 1992 [19], and the use of this technology for solving the problems faced in osteoarthritis following high congenital dislocation of the hip was presented at the American Academy of Orthopedic Surgeons in 1993 [20]. The preoperative radiologic data requires both X-rays and CT-scan and the preoperative planning relocates the new center of rotation and accordingly the new position of the greater trochanter in both craniopodal and mediolateral planes [21].

The Design and Fabrication of Custom-made Hip Stem Prostheses

Preoperative Data

X-ray Data

The radiographic analysis is based on several X-ray views. A full view of the two limbs using scanography is needed to assess the global pelvis and limb anatomical status, and to evaluate the extent of disturbance of the pelvic balance by assessing bilaterally the position of the hip rotation centers (in the vertical axis). A *frontal pelvis view* is used to determine the extent of lever arms between the rotation centers and the corresponding femoral axes. Discrepancies are recorded and will be used later in the preoperative planning to correct the anatomy of the diseased joint such that full restoration of the pelvic balance can be achieved. Eventually, *frontal and lateral X-ray views* of the diseased joint are necessary to complete the X-ray data set (Figure 32.1).

CT Data

Data obtained from a computerized tomography scanner are necessary both for the design of the intramedullary femoral stem and for the planning of the extramedullary part of the joint reconstruction. Except in special cases, the CT data acquisition must follow an established protocol elaborated by Symbios. However, in special cases such as, for instance, very severe congenital dislocations, the radiologist may have to select a modified protocol based on the X-ray status.

The intramedullary femoral anatomy is assessed by CT views taken every 5 mm from the acetabular summit down to the bottom of the lesser trochanter, then every 10 mm until the femoral isthmus. The extramedullary planning requires CT views taken at three different levels: 1) at the base of the femoral neck (assessment of helitorsion axis), 2) at the knee level, across the femoral condyles (assessment of posterior bicondylar axis), 3) at the foot level, by the

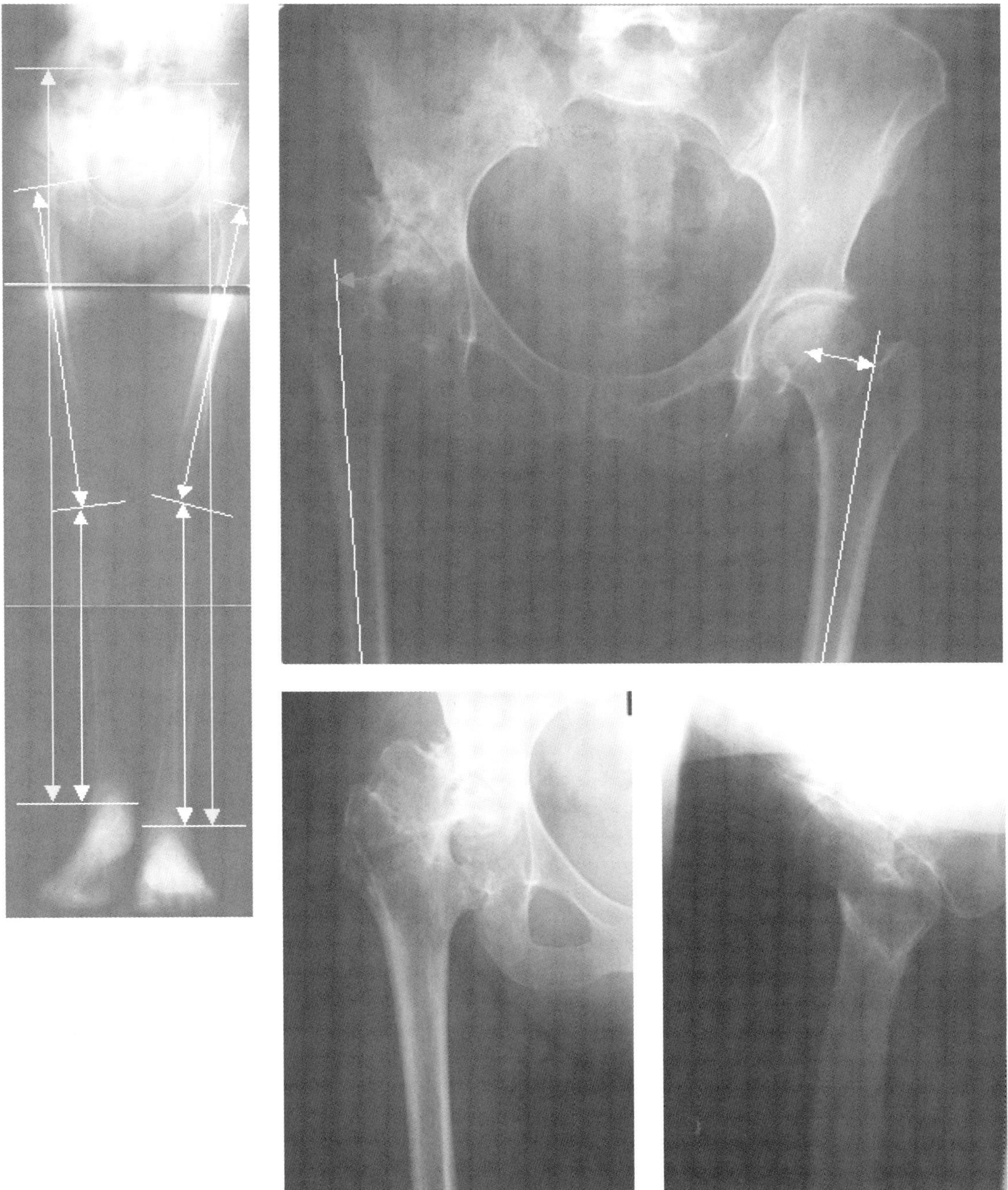

Figure 32.1. Typical set of preoperative X-ray data including scanography, frontal pelvic, and hip frontal and lateral views.

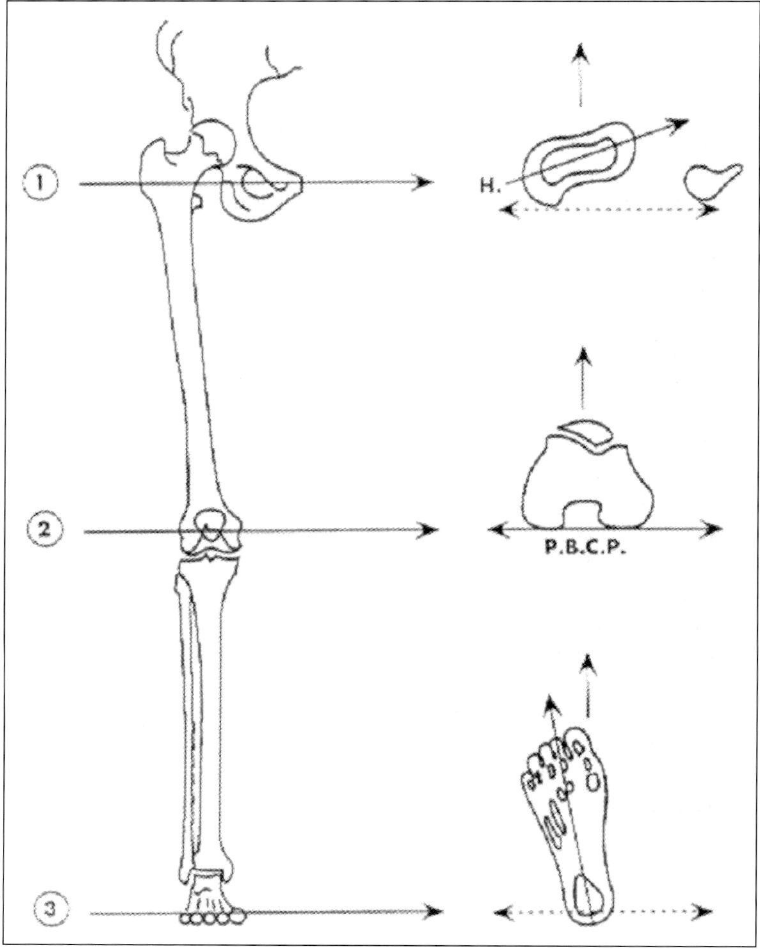

Figure 32.2. CT views required for the extramedullary planning. 1) above the lesser trochanter, 2) knee, 3) foot.

second metatarsus axis (assessment of foot axis) (Figure 32.2).

Preoperative Planning

Acetabular Cup

If the contralateral hip is healthy, planning the rotation center of the replaced joint and the socket size is performed by reproducing the contralateral geometry on the X-ray frontal pelvis view (Figure 32.3). In the presence of a bilateral lesion and in most high dislocation cases, the position of the rotation center and the size of the acetabular socket are decided together with the surgeon. In certain cases, the size is determined using the CT view passing through the center of the true acetabulum (which allows furthermore assessment of bone stock) (Figure 32.4), then by reporting the result on the X-ray pelvic view.

New Position of the Femur

The future position of the femur (as determined, for instance, by the location of the greater trochanter) is determined on the frontal view based on the position of the acetabular socket, on the desired lengthening as determined from the scanogram, and on the neck lever arm (Figure 32.3). This position will deter-

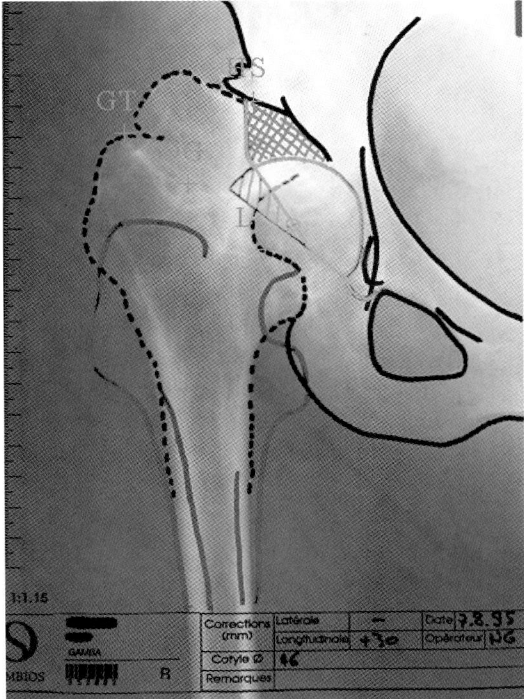

Figure 32.3. Preoperative planning on the X-ray frontal view with (a) anatomical landmark registration (in red), planning of the acetabular socket, and positioning of the greater trochanter.

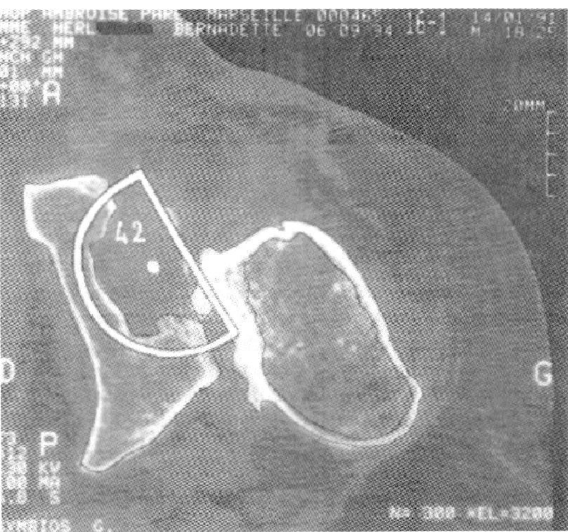

Figure 32.4. Planning of the acetabular socket using CT image.

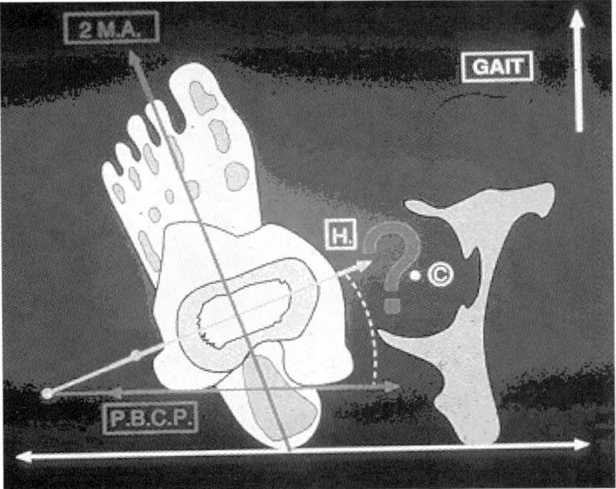

Figure 32.5. Normal gait anatomy.

mine the level of the femoral cut and assess the correct neck lever arm on the frontal view. However, osteotomy of the greater trochanter may be necessary in cases where extensive lengthening is required, associated to a wrong anteroposterior position of the greater trochanter due to excessive anteversion.

Neck Anteversion

The anteversion angle of the prosthesis neck must be set such that normal gait anatomy can be restored. The normal gait anatomy requires three conditions: 1) foot axis showing 10–20° of external rotation, 2) posterior bicondylar axis perpendicular to the gait direction, 3) anteversion of the femoral neck between 15° and 20° with respect to the bicondylar axis (Figure 32.5). It has been shown that in most cases of congenital dysmorphism, the upper femur axis, also called helitorsion axis and defined as the axis passing across the longer diameter at the level of osteotomy, is not aligned with the neck axis [22]. This phenomenon is usually not taken into account in standard prostheses. This results most frequently in such cases in an over- or under-correction of the prosthetic anteversion angle, thus preventing the full restoration of the normal gait. By superimposing the three CT views of the osteotomy level (usually above the lesser trochanter), and of the knee and foot levels, it is possible to calculate the correction angle to add (or subtract) to the helitorsion angle such that a final prosthetic anteversion angle of 15–20° is achieved. An example of such a correction is given in Figure 32.6a, whereas

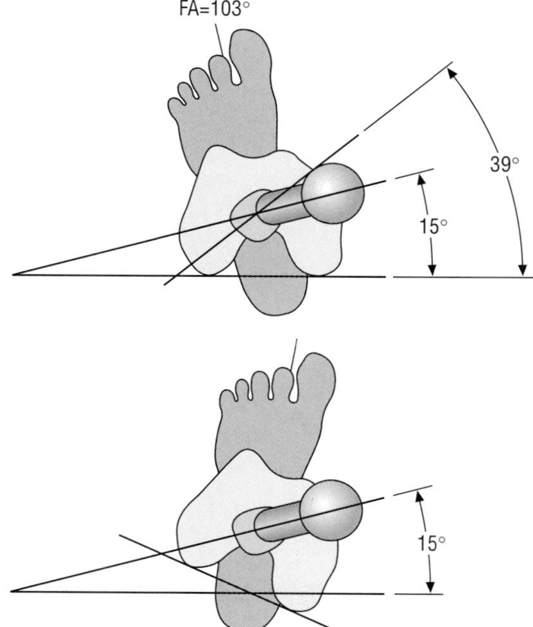

Figure 32.6. **a** Restoration of normal gait anatomy based on the correction of the helitorsion angle, **b** same case without this correction.

Figure 32.6b shows the same case without the correction for helitorsion.

Design of the Intramedullary Section

Contouring

Upon reception, raw CT data is processed by numerical thresholding such that non-bony structures are excluded from the images. Following this "image filtering" step, the design engineer runs an image analysis program to select both the internal and external contours of the bone section on each femoral CT slice (Figure 32.7, plate section). This *contouring* process is normally performed fully automatically, except in the area of the femoral neck and in cases of important artifacts on CT images for which manual intervention is needed.

Matching CT and X-ray Data

Anatomical landmarks on the diseased joint must be first registered. These landmarks will be used later for the definition of the osteotomy. The summits of the greater and lesser trochanters (GT and LT), the digital gap (DG), and the femoral head summit (HS) are localized and indicated on the X-ray frontal view (Figure 32.3).

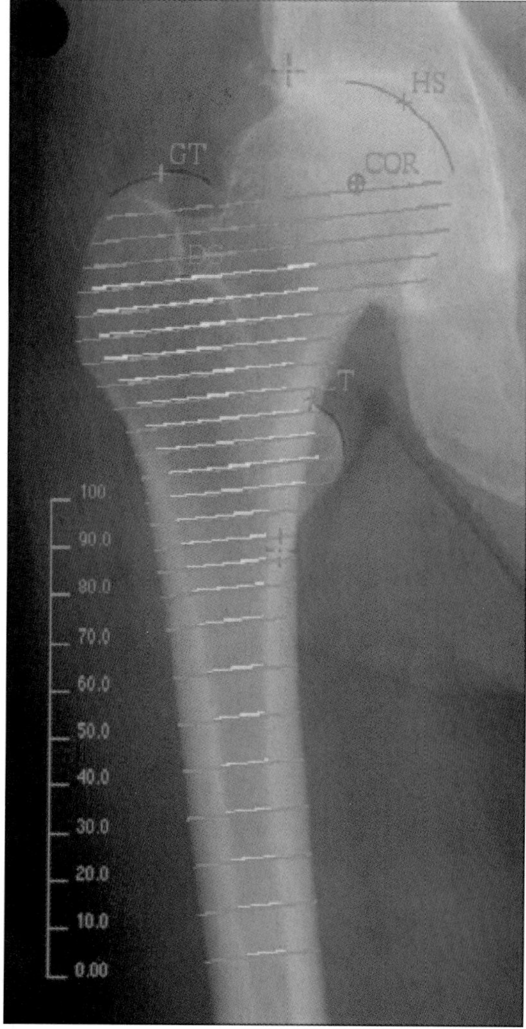

Figure 32.8. Matching of CT and X-ray data in the frontal plane.

The next step in the design process consists of superimposing the CT and X-ray data on the same image file. For this, frontal and lateral radiographic views of the diseased hip are first digitalized using an X-ray-compatible image scanner. The contouring data obtained during the previous step is numerically added to the digitalized X-ray views. A manual fitting of the two types of image is then performed independently on the frontal and lateral view (Figure 32.8).

Definition of Osteotomy Geometry

Once merging of CT and X-ray data is completed, osteotomy directions are calculated and added to the image file. The level of the osteotomy is defined such that neck length, optimized stability in rotation, and optimized bone stock preservation are taken into account.

Generation of the Initial Stem and Extraction

Based on the internal contouring data, the design software uses numerical interpolation procedures to generate a first stem shape limited to the intramedullary zone. However, the very precise reproduction of the femoral internal contour on this first draft makes it most often useless without modifications, as local protrusions and depressions at the bone surface would prevent any movement of the stem within the femur (Figure 32.9a). It is therefore necessary to simulate numerically the extraction of the stem from the femur. This is done by successive iteration steps during which the stem is extracted incrementally by rotations and translations in the three main orthogonal axes. During each iterative step, incremental stem shape modifications are performed by the software in order to allow the extraction while maintaining the contact zones necessary for an optimized mechanical support of the stem in the femur. Optimized support is sought in medial, lateral, and anterior metaphyseal areas. At the end of the simulation, a new, modified version of the stem is obtained that can be implanted into the femur with a very restricted degree of freedom for the insertion path (Figure 32.9b).

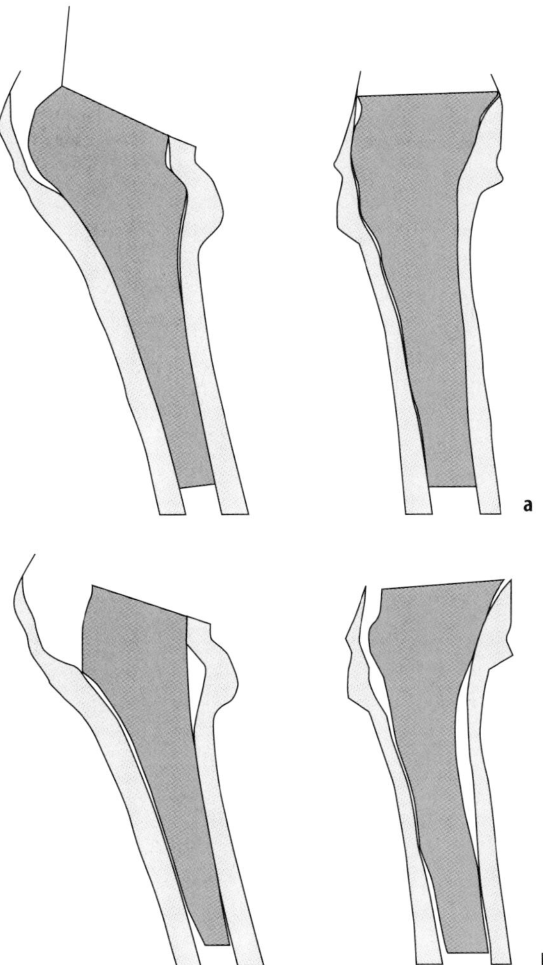

Figure 32.9. a Generation of initial stem with protrusions preventing insertion, **b** modified stem after the extraction process.

Final Corrections

At the end of the extraction process, numerical integration of the new stem shape in the CT data is performed. It enables the design engineer to view each CT section together with the corresponding stem section ("composite" view, Figure 32.10, plate section). By switching to the editor mode of the software, the engineer can also perform a final design "tune up", during which he can still implement slight modifications on each stem section to further optimize bone–prosthesis adjustment.

Stem Validation

The final step in the design of the intramedullary part of the femoral stem consists of simulating a subsidence of the stem in the femoral canal in order to be sure that the stem is at worst in contact with the cortical bone in this shifted position. A numerical three-point bending simulation test is then performed to validate the mechanical resistance of the stem.

Design of the Extramedullary Section

The design of the extramedullary part of the stem is performed as well in the frontal and lateral as in the sagittal plane. The determination of the anteversion angle of the prosthesis neck, taking into account the correction for helitorsion, was explained earlier (see Figure 32.6). With the intramedullary stem integrated in the X-ray frontal view, the design engineer calculates the optimized combination of CCD angle, neck length, and head offset such that the planned rotation center and lever arm are respected (Figure 32.11).

Planning and Prosthesis Validation

The preoperative planning of a custom-made prosthesis is performed by the surgeon and the design engineer together at Symbios. Following the planning, the design of the stem is done entirely by the design engineer. Therefore, the final design must be validated by the surgeon before the fabrication of the prosthesis can be launched. For this, Symbios provides the surgeon with a patient file including the CT composite view (Figure 32.10), the normal gait restoration scheme (Figure 32.6a), and the X-ray frontal (with osteotomy parameters, Figure 32.11) and lateral view with the designed stem.

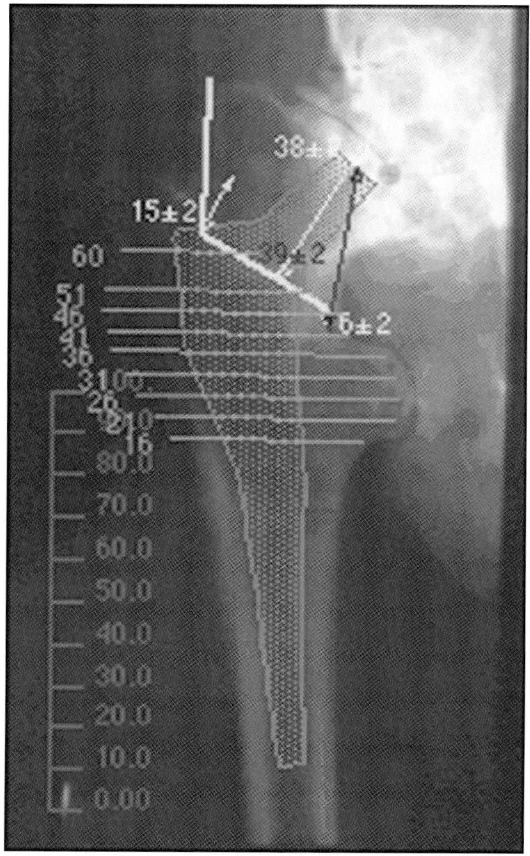

Figure 32.11. Composite X-ray frontal view with integration of intra- and extramedullary stem sections.

Fabrication

Stem Machining

Upon validation of the stem design and preoperative planning by the surgeon, the fabrication of the prosthesis can proceed. For this, the stem CAD data is transferred into a CAM software that pilots a five-axis milling machine. In parallel, a compactor with a smooth surface is machined with the same design as the stem itself. It is used for compaction of the cancellous bone before the stem itself is introduced (Figure 32.12).

Materials and Coatings

Wrought Ti_6Al_4V titanium alloy is used most of the time for the fabrication of the stem. In very few cases, stainless steel stems are produced upon request of the surgeon. The rasp itself is made out of wrought stainless steel. After machining, the prosthesis stem undergoes a surface plasma spray coating procedure which can vary from one stem to the other, depending again on the surgeon's request. In most cases a first layer of ~300 μm of porous titanium followed by a ~80 μm layer of porous hydroyapatite (HA) are coated on the intramedullary section of the stem, from the osteotomy level down to the distal level at which the transition from an elliptic to a circular section takes place.

Sterilization and Packaging

The final steps in the production of the prosthesis are the gamma sterilization and the final

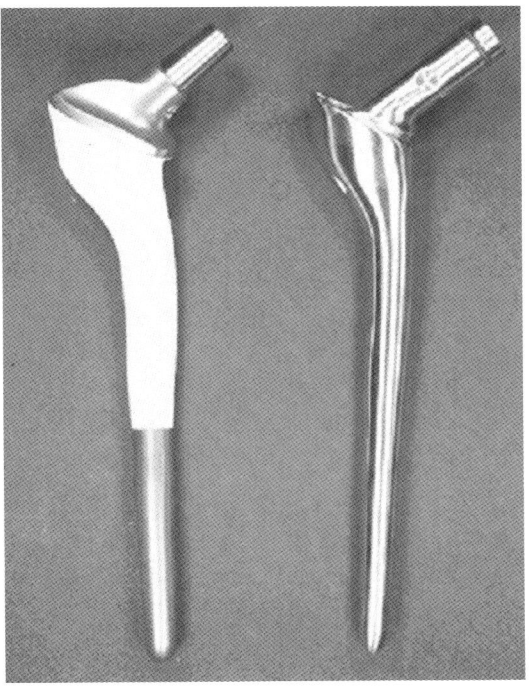

Figure 32.12. Example of porous-coated, custom-made prosthesis together with the corresponding "rasp" for compaction of cancellous bone and implant preparation.

packaging procedure which is performed in clean room conditions.

All in all, the surgeon planning a custom-made hip stem prosthesis can expect a duration of five weeks between the delivery of patient's X-ray and CT data and the delivery of the prosthesis and ancillaries to the surgeon.

The Biomechanical Evaluation of Custom Hip Prostheses by Numerical Modeling

The Biomechanical Model

Due to the development of data processing, numerical methods have become an essential complement to traditional experimental methods of analysis of the movements of deformable solids. They are particularly powerful when:

geometrical shapes of deforming bodies are complex,

deformations are large,

the constitutive materials exhibit non-linear behavior, and

applied loads are dynamic.

These methods are well adapted to the solution of biomechanical problems – in particular those raised by articulating hip prostheses – since the deformable solids (bone structure and implant) have complex geometries, the mechanical behavior of the bone–implant interface is highly non-linear, and the loads which are applied are dynamic.

Numerical modeling thus makes it possible to represent geometrically a bone–prosthesis configuration and to apply the mechanical laws which govern its behavior as a deformable solid subjected to a set of forces. Its principal tool is a data-processing software using the finite element method.

Modeling by numerical methods requires a precise description:

of the prosthesis (3D geometry, mechanical properties of the bone–prosthesis interface),

of the bone structure (3D geometry, distribution of the bone densities, mechanical properties, constitutive laws),

of the loading conditions of the system (articular contact forces, muscular forces).

This description and the use of the finite element method allows the determination of biomechanical variables such as the stress distribution within the bulk of solid bodies, the stress and micromotion distribution at the bone–prosthesis interface, and the temporal evolution of bone structure.

In the design process of a custom hip prosthesis, this approach can be used to validate the stem geometry as well as the neck parameters (CCD angle, anteversion and offset).

The Modeling

The use of a biomechanical model and the numerical methods included makes possible the model of the following structures.

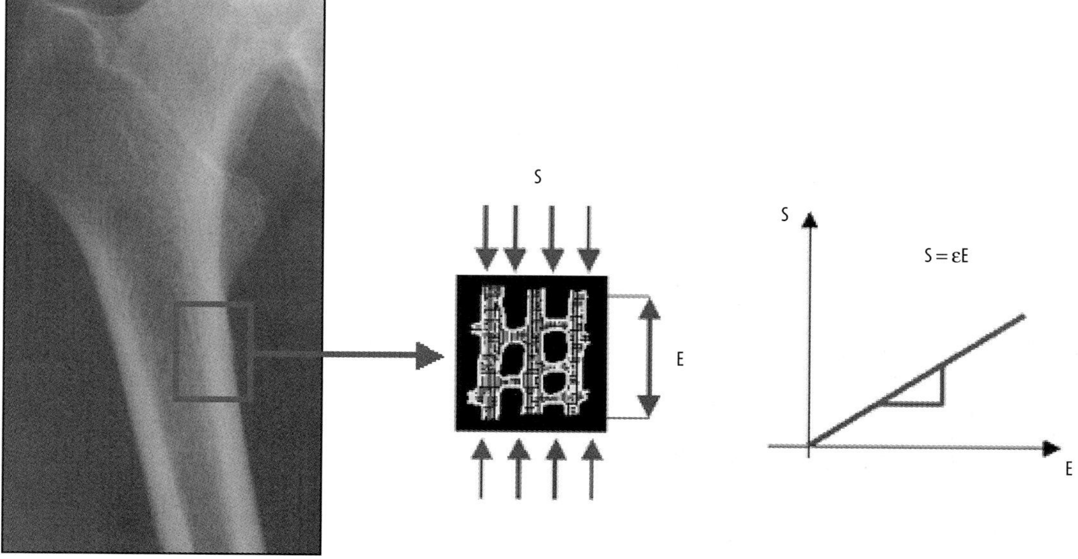

Figure 32.13. Definition of the constitutive law as the relation between stress and strain.

Bone Structure

This structure is complex. The material is primarily elastic but presents also plastic, nonhomogeneous, and anisotropic behaviors. To model it, one takes into account the following parameters:

The *stress* (S) exerted on each finite element; this has the form of a force per unit of area (similar to a pressure) and is expressed in Pascals (Pa).

The *strain* (E), which represents the increase in length of a solid element, normalized to its initial length. For a given element, the relation between stress and strain is characterized by its Young's modulus (Figure 32.13). Bone is an inhomogeneous structure, however, which does not have the same density everywhere. In order to integrate the effect of inhomogeneity in the model, an *apparent density* (ρ) for each node of the grid is determined (Figure 32.14).

The structure of the bone, moreover, is not isotropic: it becomes less deformed when a force is exerted in the longitudinal direction, than for the same force applied in the transverse direction. This difference may reach a factor of 2 (Figure 32.15). It is therefore necessary to introduce a *privileged direction of anisotropy* (v) into the model, which amounts to defining a *tensor of structure* (M).

The constitutive law describing the stress–strain relationships within the bone depends thus on the Young's modulus, the apparent density ρ, the tensor of structure M, and the tensor of deformation (Figure 32.16).

Loading Conditions

The definition of the prosthesis and bone geometries, associated to the constitutive laws, are not sufficient to evaluate the biomechanical behavior of a bone–prosthesis configuration. It is still necessary to introduce the loading conditions induced by the action of muscles, which may vary from one patient to another, in particular according to their weight.

The Simulations

The biomechanical model enables simulation of the following clinical situations.

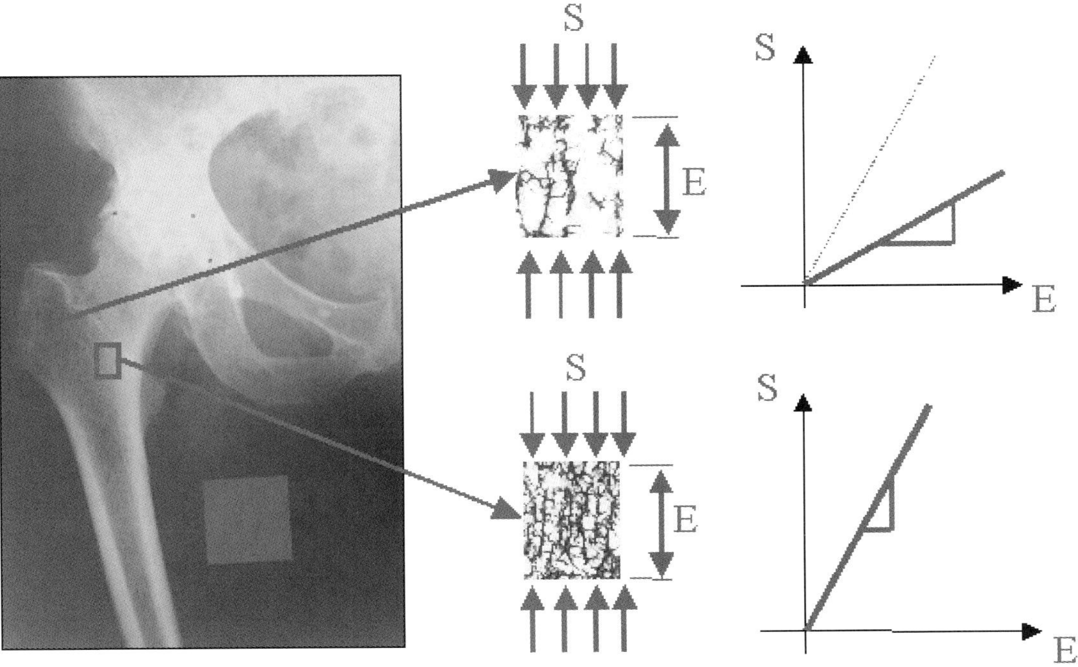

Figure 32.14. Bone inhomogeneity illustrated by the difference of stiffness between two regions of the proximal femur.

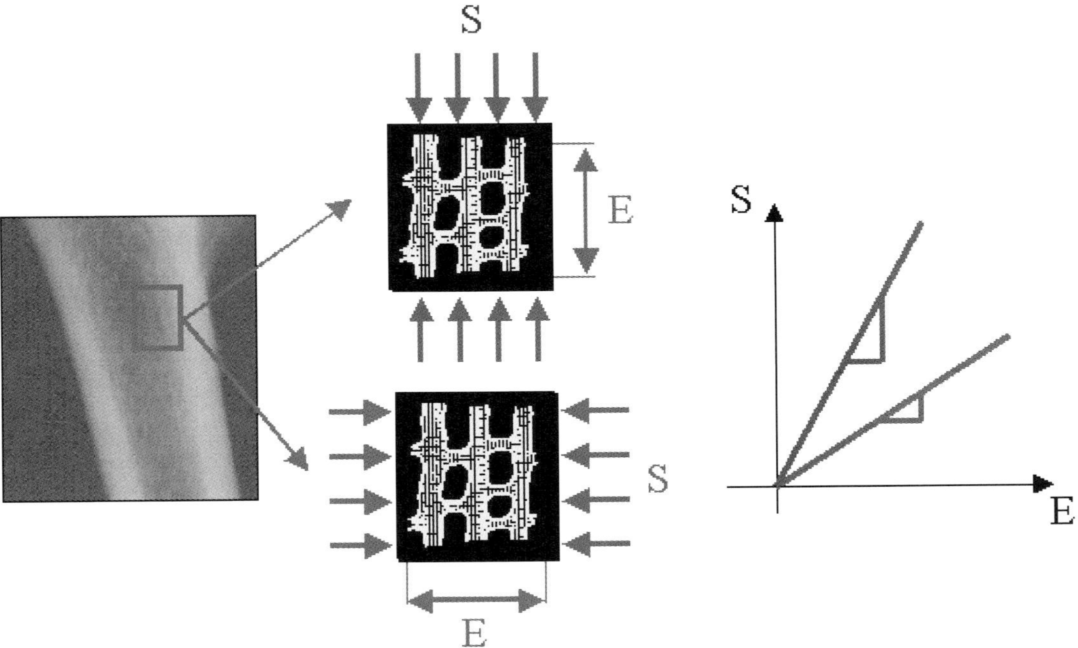

Figure 32.15. Difference in stiffness due to bone anisotropy.

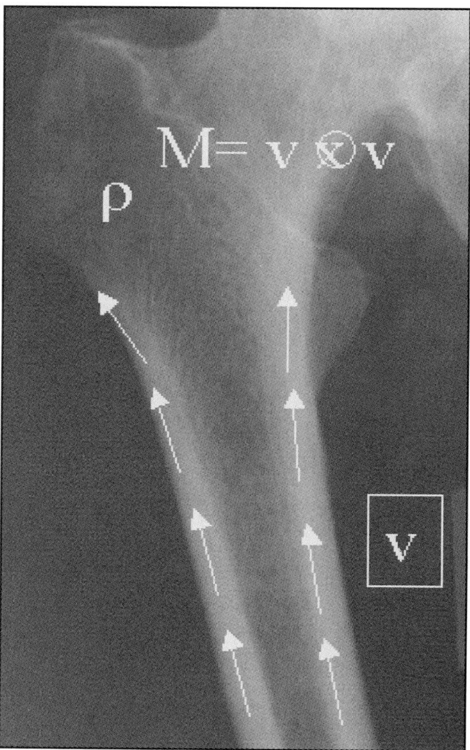

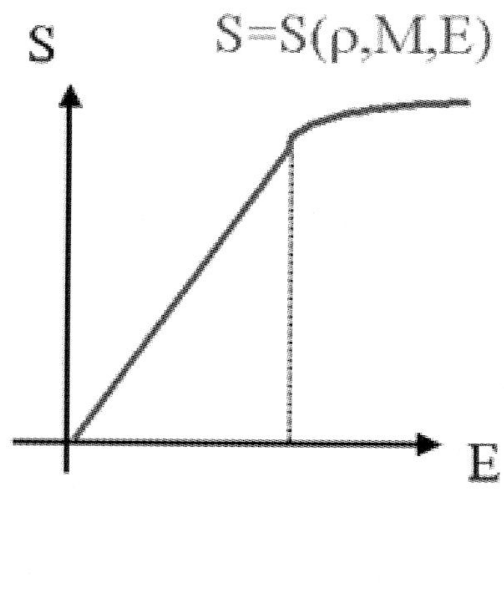

Figure 32.16. Parameters of the constitutive law. v: privileged direction of anisotropy, M: structural tensor, ρ: bone density.

Primary Stability of the Femoral Stem

In the design process of a custom stem, the search for a geometry likely to optimize primary stability is of primary importance: good primary stability ensures a durable secondary fixation. The quality of primary stability depends on the stresses and micromotions which are exerted at the bone–prosthesis interface. We distinguish between:

the micro-slipping (d_t) and the shear stress (p_t), whose directions are parallel with the interface,

the micro-debonding (d_n) and the compressive stress (p_n), whose directions are perpendicular to the interface (Figure 32.17).

Bone Remodeling

The evaluation of primary stability allows for the establishment of certain criteria of implant performance. However, bone is a living tissue which evolves and remodels itself in the course of time: it changes, in particular, under the effect of the stresses to which it is subjected. To supplement the biomechanical evaluation of the bone–implant system it is thus necessary to introduce into the model a function describing bone adaptation (or bone remodeling). The physical activity of a patient, just as the modification of the bone configuration accompanying an arthroplasty, leads to changes in local bone structure. This change induces alterations in the stress distribution (compression and shearing), which results in a *mechanical stimulus* able to modify the *bone density* (Figure 32.18).

The law of bone remodeling is illustrated in Figure 32.20 (plate section). Below a threshold value of the stimulus, there is reduction of bone density, therefore resorption. On the other hand, as soon as the stimulus exceeds a limit value, there is an augmentation of bone density. Inside the interval defined by these two values, there is no modification of density: this interval delimits the zone of balance. The model of bone remodeling thus requires:

measurement of the initial density using a CT scanner, and

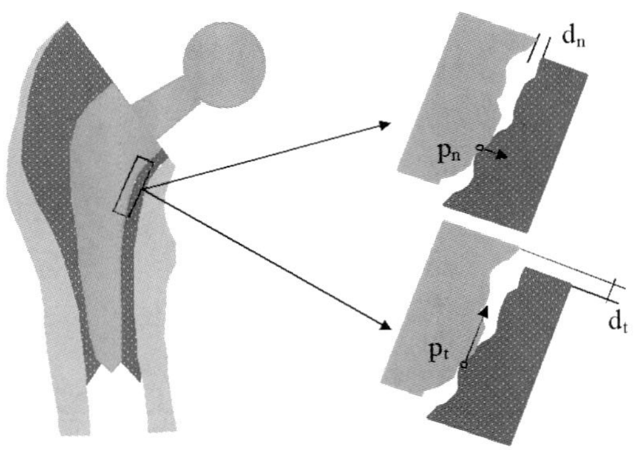

Figure 32.17. Biomechanical variables at the interface. d_t: micro-slipping, p_t: shear stress, d_n: micro-debonding, p_n: compressive stress.

establishment of the relation between bone density and mechanical stimulus before it is possible to calculate the final bone density at equilibrium.

Practical Applications

The biomechanical model offers a wide field of application. An example is presented here, which is the measure of the influence of the extramedullar parameters on the primary stability of a custom hip prosthesis.

Material and Methods

The studied prosthetic stem is a non-cemented, custom stem (Symbios Inc., Yverdon, Switzerland), with an optimal filling of the proximal metaphysis. It is constructed of titanium alloy with a double coating of porous titanium and hydroxyapatite on the proximal 2/3 of the stem.

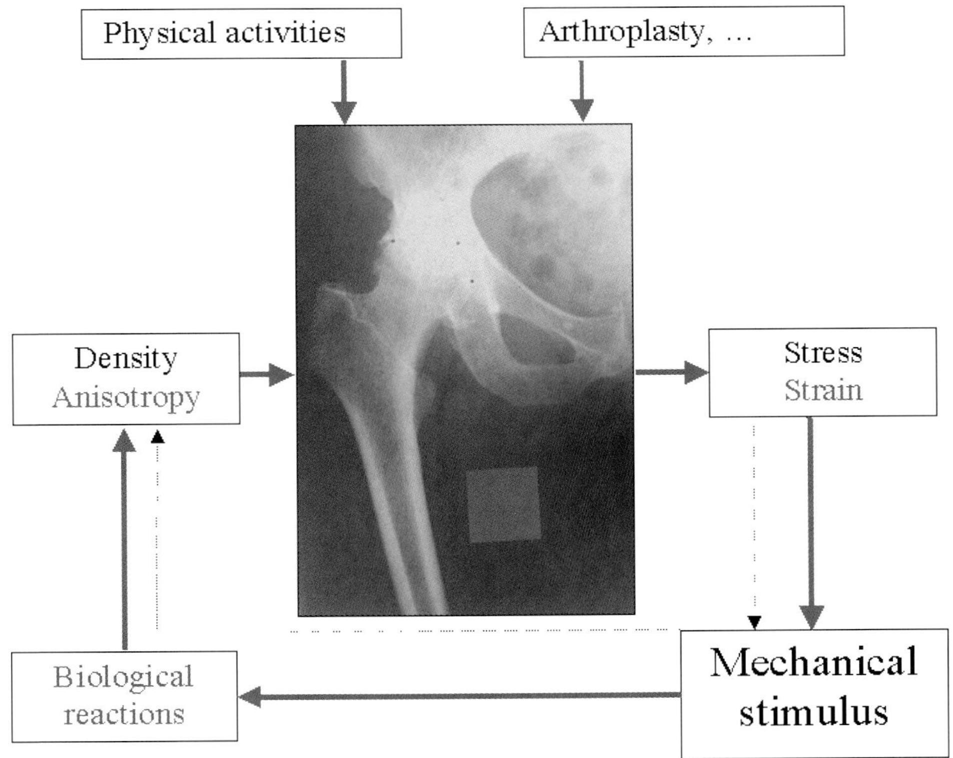

Figure 32.18. Bone remodeling loop.

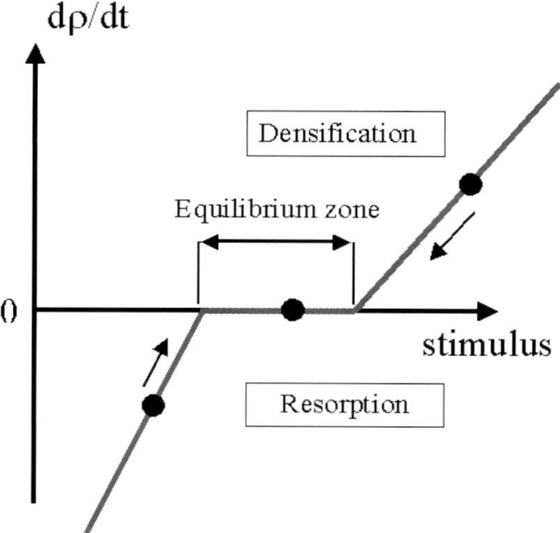

Figure 32.19. Bone remodeling law. Stimulus higher than equilibrium leads to densification, stimulus lower than equilibrium leads to resorption.

Seven configurations of prosthetic neck were investigated:

an anatomical configuration with a normal and ideal center of rotation of the hip,

an "anteverted" configuration with an anteversion of the neck higher by 15° than the anatomical configuration,

a "retroverted" configuration with a retroversion of the neck lower by 15° than the anatomical configuration,

a "lateralized" configuration with a length of neck greater by 10 mm than the anatomical configuration, which leads to a lateralized configuration of 7 mm, and lengthened,

a "medialized" configuration with a length of neck smaller by 10 mm than the anatomical configuration, which leads to a medialized configuration of 7 mm, and shortened,

a configuration in varus with a CCD angle smaller by 15° than the anatomical configuration,

a configuration in valgus with a CCD angle greater by 15° than the anatomical configuration.

The conditions of load correspond to those of monopodal support of a gait cycle, including the principal muscular forces (gluteus maximus, gluteus medius, psoas) applied to the femur. The computed values for each of the seven configurations are the micromotion and stress at the bone–prosthesis interface after implantation, during the immediate postoperative time.

Results

The results show that the distribution of micro-debonding on the bone–prosthesis interface varies by a small amount with the extramedullar parameters (Figure 32.20, plate section). For the seven configurations, interfacial micro-debonding is negligible (blue area) over almost all of the interface. It exceeds 20 μm at the tip of the stem and in the proximal lateral zone. On the other hand, the maximum value of micro-debonding varies according to the various parameters: it is weakest (28 μm) for fixation in an anatomical position and reaches a peak of 35 μm whenever the implant is either anteverted too much, or lengthened and lateralized.

The distribution of micro-slipping does not vary significantly with the extramedullar parameters (Figure 32.21, plate section). Interfacial micro-slipping is higher than 20 μm over all of the interface. Its value exceeds 60 μm at the tip of the stem and in the proximal medial area (calcar region). The area of the interface where micro-slipping exceeds 60 μm is of more significant size for the prosthesis whose neck is lengthened and lateralized. Just as for micro-debonding, the peak of micro-slipping is weakest (68 μm) for an anatomical configuration and is maximum (87 μm) for a configuration either lengthened and lateralized, or anteverted. The stress distribution (compressive and shear) at the interface also varies by a small amount with the extramedullar parameters. The highest compressive and shear stress distribution appears in the distal region and in the proximal medial and lateral regions. However, as for the micromotions, the peaks of stress vary according to the extramedullar parameters; they are minimal for the anatomical configuration and maximum in the case of a lengthened and lateralized neck or anteverted neck.

Conclusion

The extramedullar part of a cementless, custom prosthetic stem influences significantly the biomechanical quality of the fixation of the femoral component. Results therefore indicate that the best initial stability of the stem is obtained when the center of rotation of the hip is replaced in its anatomical position. This requirement is essential when the neck parameters of the custom hip prosthesis are defined.

We thank Dr. T. Quinn for review of the manuscript.

The Clinical Experience

Surgical Consideration

The extramedullary custom neck has been able to solve many of the surgical difficulties faced in excessively anteverted upper femurs, often found in dysmorphic or dysplasic hips. We found by CT-scan measurements extremes values, up to 85° in some etiologies [23]. In those cases the retroversion included in the custom neck offset was able to restore an appropriate anteversion of 15–20° on the knee condylar plane [24]. Some authors have described in such cases the association of a derotational osteotomy to a conventional stem [14], but the restriction in postoperative weight bearing and the incidence of nonunion of the osteotomy may increase the morbidity of the procedure. Another solution would be the use of a modular neck but in cases of large anteversion the possibilities of correction are limited by the risk of fretting.

The intramedullary custom stem aims for a preservation of the dense cancellous bone compacted towards the inner cortical femur by the use of the smooth compactor. This compactor of identical intra and extramedullary shape to the final prosthesis is used as a trial prosthesis during surgery. The preservation of this cancellous bone is of high importance for secondary biological fixation to the hydroxyapatite (HA) covering the final prosthesis. Clinical and radiologic experience led us to move from a proximal HA to a full HA coating [25].

Clinical Indications

When conditions of normal hip anatomy cannot be restored with a standard neck and a standard stem encountered in dysmorphic and dysplasic hips (Figure 32.22).

When full and quick recovery of function is required and when high and long-term solicitations of the implant will occur; this is expected in young age patients (Figure 32.23). The significant change in the results reported in the Swedish register seems to be located around 65 years old, with a high failure rate of conventional implants under that age [9]. The increasing lifetime expectancy will overload this tendency in the future, with at least a twenty-year lifetime expectancy for a 65-year-old patient [26].

Clinical Results

The clinical implantation of this Symbios custom concept (Symbios, Yverdon, Switzerland) started in our department in January, 1990. Between January, 1990 and January, 2000, 1,156 cementless, custom stems have been implanted.

Focusing only on patients of 65 years old or less, and excluding revision of another prosthesis, the series consists of 726 hips. The mean age of the patients was 52 years (range 17–65 years) and the mean weight 72 kg (range 49–147 kg). The etiologies included: osteoarthritis in 273 cases (38%), avascular necrosis in 101 cases (14%), congenital dislocation of the hip in 200 (18%), and dysmorphy in 152 hips (20%). After a 1 to 10 year follow-up, eight patients were dead, 28 lost (3.8%), and 11 excluded for less than one year follow-up, leaving 680 hips to study at an average of 5.6 years of follow-up. The clinical Harris hip score averaged 99 points (range 84–100) for the 387 enthusiastic patients and averaged 95 points (range 83–100) for the 279 satisfied patients. At the time of follow-up, 98% of the patients ranked their result as excellent or good, eight patients (1.2%) found no change, with a mean objective Harris score of 12%, and six patients were disappointed (0.8%) with a mean Harris score of 80 points.

Seven hips were revised for sepsis (1%) and eleven for aseptic failure (1.6%). These revisions for aseptic failure consisted of nine loosening, one fracture, and one persistent pain. Nine of

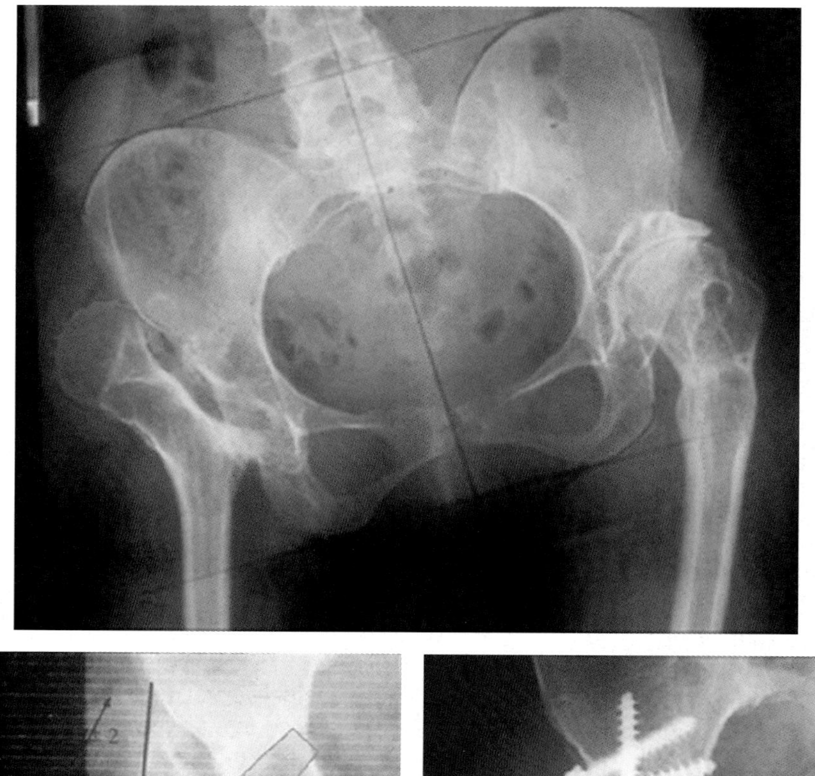

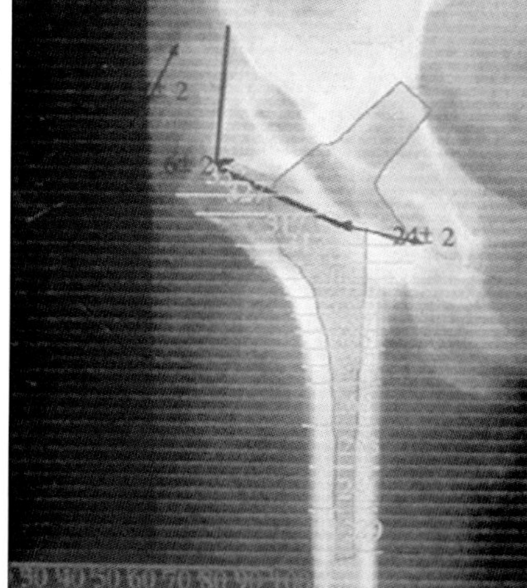

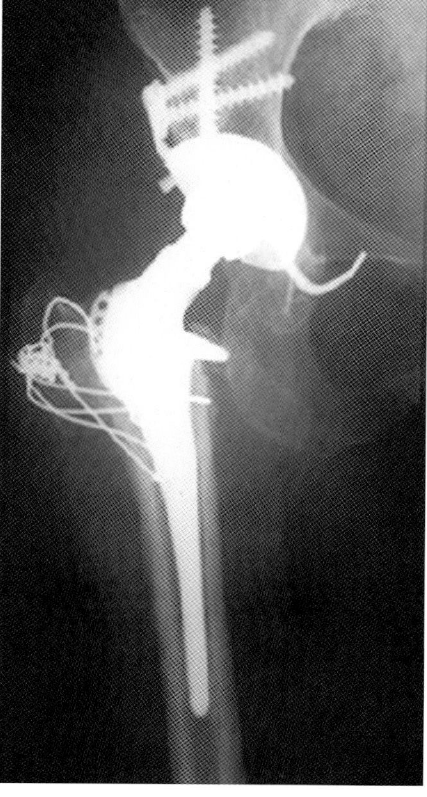

Figure 32.22. **a** Preoperative pelvis A/P view of bilateral CDH, with high dislocation on the right hip. **b** Custom stem planning. **c** Postoperative A/P view of the right hip after five-year follow-up of the prosthesis.

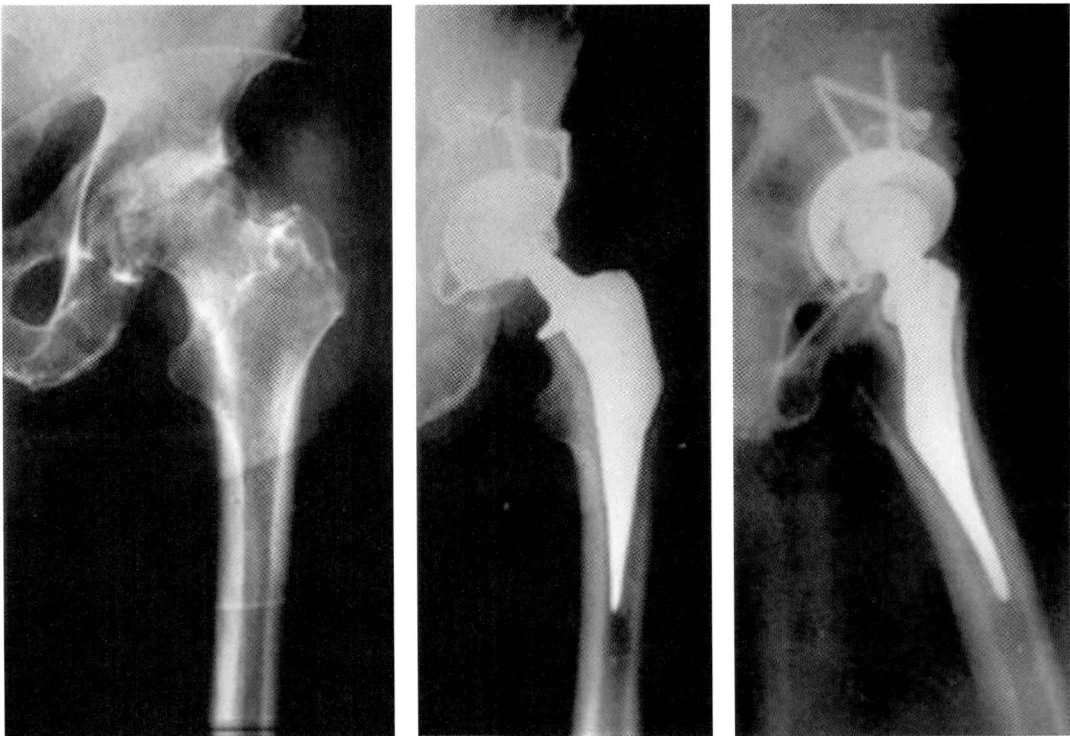

Figure 32.23. Postoperative view of 45-year-old patient with post-traumatic degeneration of the hip joint and the intramedullary adaptation of the stem both on the A/P and M/L view at nine years of follow-up.

these eleven aseptic failures occurred with the proximal HA coating used originally.

Considering stem revision for aseptic failure, as an end point, the Kaplan–Meier survivorship analysis showed a 96.7% survival at 10 years, with a 95% confidence interval (Figure 32.24). The dislocation rate for all etiologies was 1.7% (12 cases) and considering only patients with primary osteoarthritis 0.04% (3 cases).

Discussion

These clinical results at 10 years are encouraging, and are at least similar or better to the results previously reported with conventional cemented implants in young age groups using modern cementing techniques [3,7,11,27,28,29, 30,31,32,33], or with standard cementless prostheses [14,15]. The goals fixed in 1990 seem to have been reached in 2000 with an increased stem longevity for patients under 65 years old, a reduced dislocation rate regarding the 0.6–15% reported in the literature [34], and a return to full social and sport activities.

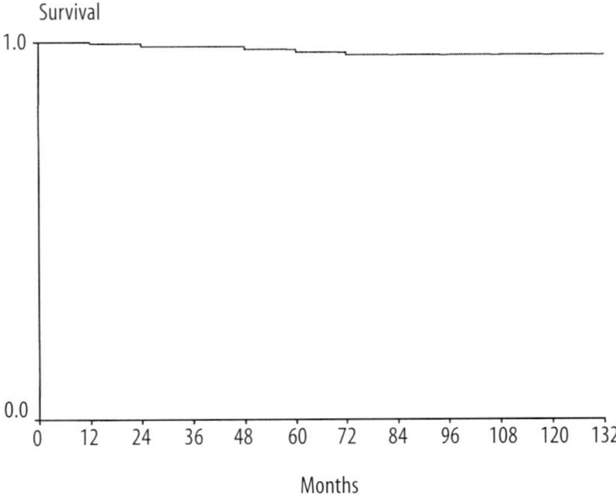

Figure 32.24. The survival curve of the custom stem hip prosthesis, with revision for aseptic failure as an end point.

The remaining problems for the current use of custom stems are: higher price, delay for conception, and surgeon adaptation. The price difference regarding conventional implant moved from a factor to five to a factor to two during the ten last years and this process must continue in the coming years. The five-weeks delay for stem fabrication may be significantly reduced to three weeks in the near future with the regular use of teleradiology. Finally, the orthopedic surgeon, accustomed to having a large number of different-sized prostheses in the operating room, has only to deal with one compactor and one final prosthesis with the custom concept. This requires a learning curve, quickly achieved by the full, computerized, preoperative planning helpful during surgery, and once this adaptation is achieved this custom concept may be able to solve a number of surgical difficulties previously encountered with conventional implants.

Conclusion

Both the biomechanical evaluation and the clinical experience of this intra and extramedullary custom concept are promising, after ten years of clinical use. Another decade will be necessary to evaluate clinically and radiographically the prostheses after 20 years of implantation in young patients. Further research in the biomechanical field, including the expected bone remodeling around the stems by finite element analysis and the evaluation of the patient hip function after total hip arthroplasty using gait analysis or accelerometry during everyday activities, will also be necessary.

Computer-assisted hip arthroplasty is certainly a step forward in the future for restoring function and improving implant longevity for patients with high activity and/or modified anatomy.

References

1. Chandler HP, Reineck FT, Wixson RL, McCarthy JC. Total hip replacement in patients younger than thirty years old: a five years follow-up study. J Bone Joint Surg (Am) 1981;63A:1426–34.
2. Collis DK. Cemented total hip replacements in patients who are less than fifty years old. J Bone Joint Surg (Am) 1984;66A:353–9.
3. Dorr LD, Luckett M, Conaty JP. Total hip arthroplasties in patients younger than 45 years. Clin Orthop 1990;260:215–19.
4. Dorr LD, Takei GK, Conaty JP. Total hip arthroplasty in patients less than forty five years old. J Bone Joint Surg (Am) 1983;65-A:474–9.
5. Halley DK, Wroblewski BM. Long-term results of low-friction arthroplasty in patients 30 years of age or younger. Clin Orthop 1986;211:43–50.
6. Sharp DJ, Porter KM. The Charoley total hip arthroplasty in patients under age 40. Clin Orthop 1985; 201:51–6.
7. Stauffer RN. Ten-year follow-up study of total hip replacement with particular reference to roentgenographic loosening of the components. J Bone Joint Surg (Am) 1982;64A:983–90.
8. White SH. The fate of cemented total hip arthroplasty in young patients. Clin Orthop 1988;231:29–34.
9. Malchau H, Herberts P. Prognosis of total hip replacement in Sweden. Proceedings of the 65th annual meeting of the American Academy of Orthopedic Surgeons, 1988.
10. Boeree NR, Baniister. Cemented total hip arthroplasty in patients younger than 50 years of age. Clin Orthop 1993;287:153–9.
11. Collis DK. Long-term (twelve to eighteen-year) follow-up of cemented total hip replacements in patients who were less than fifty years old. A follow-up note. J Bone Joint Surg (Am) 1991;73-A:593–7.
12. Judet R, Siguier M, Brumpt B, Judet T. A non-cemented total hip prosthesis. Clin Orthop 1978;137:76–84.
13. Robertson DD, Walker PS, Hirano SK. Improving the fit of press-fit stems. Clin Orthop 1988;228:134–40.
14. Mont MA, Maar DC, Krackow KA, Jacobs MA, Jones LC, Hungerford DS. Total hip replacement without cement for non-inflammatory osteoarthritis in patients who are less than forty-five-years old. J Bone Joint Surg (Am) 1993;75A:740–51.
15. Glassman AH. Porous coated total hip replacement in young patients. Read at the annual meeting of the American Academy of Orthopedic Surgeons, New Orleans, Louisiana, Feb 8, 1990.
16. Rubin PJ, Leyvraz PF, Aubaniac JM, Argenson JN, Esteve P, Deroguin B. The morphology of the proximal femur: a three dimensional radiographic analysis. J Bone Joint Surg (Br) 1992;74B:28–32.
17. Noble PC, Alexander JW, Lindahl LJ. The anatomic basis of femoral component design. Clin Orthop 1988; 235:148–65.
18. Ramaniraka N, Rakotomanana L, Rubin PJ, Leyvraz PF. Influence of the extramedullary parameters on the stem stability and the stress transfer. Proceedings of the 11th annual symposium of the International Society for Technology in Arthroplasty, 1998.
19. Argenson JN, Pizzetta M, Essinger JR, Aubaniac JM. Symbios custom hip prosthesis: Concept, realization and early results. J Bone Joint Surg (Br) 1992;74B (Suppl 2):167.
20. Argenson JN, Simonet JY, Aubaniac JM. The indications for cementless custom prostheses in congenital hip dislocation. J Bone Joint Surg (Br) 1993;75B (Suppl 1):113.
21. Argenson JN, Aubaniac JM. Preoperative planning of total hip reconstruction for congenital dislocation of the

hip using custom cementless implants. Journal of the Southern Orthopaedic Association 1994;3:11–18.
22. Husmann D, Rubin PJ, Leyvraz PF, DeRoguin B, Argenson JN. Three-dimensional morphology of the proximal femur. J Arthroplasty 1997;12(4):444–50.
23. Argenson JN, Hostalrich FX, Essinger JR, Aubaniac JM. Preoperative planning in designing custom-made hip prostheses. J Bone Joint Surg (Br) 1992;74B(Suppl 2): 180.
24. Aubaniac JM, Argenson JN, Pizzetta M. Addressing the anteversion problem in severe CDH and primary or secondary dysmorphic, with Egoform and Symbios custom-made prosthesis. Third Annual International Symposium of Custom-made Prostheses, 3–5 October, 1990.
25. Argenson JN, Ettore PP, Aubaniac JM. Revêtement des tiges fémorales non cimentées. Etude comparative clinique et radiographique. Revue de Chirurgie Orthopédique 1997;83(Suppl 2):44–5.
26. Kerjosse R, Tamby I. La situation démographique en 1999. Mouvement de la population. Démographie société in INSEE. Résultats 1999.
27. Amstutz HC, Markolf KL, McNeice GM, Gruen TA. Loosening of total hip components: cause and prevention. The Hip. Proceedings of the Fourth Open Scientific meebog of the Hip Society. St Louis: Mosby, 1976; 102–16.
28. Ballard WT, Callaghan JJ, Sullivan PM, Johnston RC. The results of improved cementing techniques for total hip arthroplasties in patients less than fifty years old. J Bone Joint Surg (Am) 1994;76A:956–64.
29. Harris WH, McCarthy JC, O'Neill DA. Femoral component loosening using contemporary techniques of femoral cement fixation. J Bone Joint Surg (Am) 1982;64A.
30. Joshi AB, Porter ML, Trail IA, Hunt LP, Murphy JC, Hardinge K. Long-term results of Charnley low-fraction arthroplasty in young patients. J Bone Joint Surg (Br) 1993;75B:616–23.
31. Mulroy RD, Harris WH. The effect of improved cementing techniques on component loosening in total hip replacement. An 11 year radiographic review. J Bone Joint Surg (Br) 1990;72B:757–60.
32. Indong OH, Carlson CE, Tomford WW, Harris WH. Improved fixation of the femoral component after total hip replacement using a methacrylate intramedullary plug. J Bone Joint Surg (Am) 1978;60A:608–13.
33. Solomon MI, Dall DM, Learmonth ID, Davenport MD. Survivorship of cemented total hip arthroplasty in patients 50 years of age or younger. J Arthroplasty 1992;7(Suppl).
34. Huten D. Luxation et subluxation des prothèses totales de hanche. In Cahiers d'Enseignement de la SOFCOT:19–46. Expansion Scientifique Française, Paris, 1996.

33 The Mega-hip Prosthesis Surrounded by Allografts

B. Ripoll and D. G. Poitout

Reconstructive metal mega-hip prostheses are commonly used, but they might be responsible for mechanical failure or instability due to muscle non-fixation. One of the actual problems concerning articular allografting is the biomechanical behavior of the ligaments and the revascularization of the cartilage. We do not yet know the latest results concerning osteochondral allograft integration and cartilage evolution, especially at the hip side. Autologous grafts, unlike allogenic grafts, have an important osteogenic potential. But, inasmuch as the procurement volume is limited, they do not permit massive bone or joint reconstruction when there has been partial or total resection as a result of either a bone tumor or a post-traumatic lack of substance. For these reasons, we have elected since 1981 to use deep-frozen allogenic grafts to rebuild the skeleton. We have used fresh allogenic bone grafts since 1979 and, since 1981, deep-frozen, allogenic grafts to rebuild the skeleton:

1979–2002: 5,829 cases (spongious and massive allografts)

1981–2002: 435 massive bone allografts

1983–2002: 423 hip reconstructions

1978–2000: 356 reconstructions of the upper part of the femur

1978–2000: 185 massive osteochondral allografts

Deep-freezing alone allows the preservation of voluminous bone pieces in satisfactory conditions of preservation. This type of preservation keeps the bone architecture in an optimal biological and biomechanical state. Allografts allowing vascular recolonization and rapid muscle tightening seem to give good functional results, which is why we think that, in some indications, it is better to use a mega prosthesis surrounded by allograft.

Allograft Biology

Immunology

Allografts are well incorporated by the skeleton. If osteoid or blood cells (mostly leukocytes) as well as blood vessels and nerves have an inner antigenic potency, leading to immunoreaction, the proteinic matrix and the minerals fixed on it become either non-antigenic or less antigenic. Clinically speaking, these reactions are almost nonexistent with the use of massive allografts.

Biological Integration

Spongious Allografts

Spongious allografts are evaluated in two successive phases. For about three weeks after the grafting, an osteogenic phase induced by the grafted cells may be observed. This is followed by a halt of several months (6–8) and a return of osteogenesis under the dependence, after this time, of the guest cells. Spongious allografts are different from cortical grafts, principally through the fact that the mechanism of their revascularization is effectuated by "creeping substitution" and that the integration of this graft will be complete. The time will be much shorter (about three weeks) for the vessels to penetrate into a spongious graft than to recolonize a cortical graft. The complete revascularization requires about two months.

Cortical Allograft

The rehabilitation of cortical allografts begins with a first phase of active resorption, which is quite normal during the first two weeks, intense for the following four weeks, and it is only about the ninth week that the osteoblasts will appear, inducing the beginning of regeneration. During this period the biomechanical behavior of the graft decreases (50% less at the 108th month) so the stem has to be strong enough and well fixed into the receiver bone to support the body weight. When we compare in the cat the evolution of massive autologous and osteochondral allografts, we see that these the processes are quite similar but the duration of the revascularization process is different. The two types of grafting are revascularized and the dead bone replaced by new bone through creeping apposition. The intensity of the reconstruction is more important between the sixth and the ninth months in the case of allografts, and between the third and the sixth months for autografts.

Computerized, Custon-made Mega Hip Prostheses

Shape

To rebuild the upper part of the femur destroyed by a tumor of after numerous operations we have had to study first the inner shape of the femur. The prosthesis has to have the same anatomical shape as the medullar bone. The form which has been given by CAO is an italic S with a long inferior curve. It is important to rebuild exactly the inside anatomical aspect of the bone for the prosthesis to be directly in contact with the cortical bone. If using a straight stem, good contact everywhere can not be made, and stress shielding will be present in some areas of the bone. Also, massive allografts can not be used because the stem can not be introduced into in it.

Fixation

The mega prosthesis has to be introduced in a massive diaphyseal allograft without risks of breakage. It can be used with or without cement, but we think that the best utilization is with cement in the upper part next to the allograft and also with cement in the lower part of the stem which is fixed into the receiver bone. The muscles surrounding the allograft will fix themselves directly to it. Some special fixation can also be used for the trochanteric muscles.

Clinical Applications

In reconstructive surgery, for the first time we can increase the bone by using a prosthesis surrounded by allograft. In bone tumors when muscles and ligaments have to be removed, as in some traumatological cases, the use of massive osteocartilaginous allografts is not indicated because of the poor vascular surrounding of the graft and problems linked to articular instability. We consider the best solution in these cases is to use massive metallic articular prostheses surrounded by one allograft.

Prosthesis brings immediate stability and allows the patient to walk after a few days, while the graft will permit the remaining muscles to fix themselves to it. This solution is the one to choose where is reconstruction of the upper part of the femur.

Conclusion

The use of a mega hip prosthesis surrounded by allograft has a many advantages. It allows muscular or ligamentary fixation and better biomechanical behavior, decreasing the risks of articular instability. There is no risk of articular necrosis andearly loading can be permitted. In particular, the volume of skeletal bone is increased instead of decreasing – a very important point.

34 Biomechanics of Osteosynthesis by Screwed Plates

E. Gautier and R. P. Jacob

Introduction

Internal fixation of fractures using plates was developed more than a hundred years ago following the widespread use of radiographs. Hansmann reported the technique of plating lower limb shaft fractures as early as 1886 and presented a plate design allowing subcutaneous insertion of the plate and percutaneous insertion of the plate screws [1]. Lambotte, Lane, and Sherman experimented in the first decades of the twentieth century with new implant materials and improved the plate and screw designs to decrease the risk of corrosion and mechanical failure of the implants [2–4]. Danis, Bagby, and Müller introduced the concept of compression plating to improve the stability of fixation and to protect the implant from mechanical overload [5–7].

Since then, conventional plating techniques and plate designs have evolved constantly. This evolution is based on an improved understanding of the biology of fracture healing, of the biomechanics of fracture fixation, and on experience analyzing previous failures – such as fatigue failure of the implants, deep infection, and delayed union or non-union. It has involved the implants, the technique of its application, and the surgical technique with bone and soft tissue care or reconstruction [8–32].

Biological Aspects of Plate Fixation

Blood Supply of Cortical Bone

The three primary components of the afferent vascular system to bone tissue are the principal nutrient artery, the metaphyseal arteries, and the periosteal arterioles. The nutrient and metaphyseal arteries together compose the medullary arterial system, which is the major afferent supply nourishing about the inner two thirds of the bone cortex. The periosteal vessels enter the cortex mainly at sites of fascial and muscle attachment and appear to supply the outer third of the bone diaphysis [33–38]. Cortical circulation usually flows in a centrifugal direction. In the diaphysis, the inner cortical layers are drained through venous channels, the periosteal layers directly by periosteal capillaries. In case of damage to the medullary system following trauma or operation a compensatory flow reversal occurs to some extent [33,38–42].

Vascular Disturbance Due to Trauma, Surgery, and Implant

As a result of bone fragmentation and displacement of fragments, periosteal, intracortical, and endosteal vessels are ruptured [13,35–37,41,43]. At each fracture line all intracortical vessels are disrupted due to the direct damaging of its surrounding osteons. Major displacement of the fracture fragments may disrupt larger vessels like the nutrient artery, the central artery, or its intramedullary branches. This disruption of the medullary blood supply in turn leads to avascularity and devitalization of a large amount of the bone cortex. The stripping of the periosteum with its vascular network during injury is of particular importance, because disruption of the periosteum may be severe or total between fragments, leaving smaller fragments completely devascularized.

The surgical approach to the fracture leads to an additional considerable vascular damage to the bone tissue by the soft tissue retraction. Additional damage is added by subperiosteal exposure that leads to more damage compared to careful epiperiosteal exposure [44]. Fragment manipulation by reduction clamps and the plates itself result in further damage to the blood supply of bone. Complete visualization of the fracture area is needed neither for fracture reduction nor for positioning of the plate and insertion of the screws [15,45,46]. In conventional plating techniques, some amount of contact between plate and bone is needed for stability reasons to allow load transmission by friction at the interface. The axial screw force generated by tightening the screws and the compressive strength of cortical bone gives the minimum area required for load transfer. Shaping the plate to the bone surface as exactly as possible is mandatory so as not to be faced with the problem of secondary fracture dislocation when the fragment is pulled towards the plate by tightening the screws. The biological disadvantage of the conventional contact plating concept is the appearance of a relatively large zone of blood supply disturbance directly underneath the implant (Figures 34.1a, b, plate section). This deficiency of perfusion is caused by direct compression of the periosteal vascular network under the plate and leads to necrosis of cortex adjacent to the plate [47–55]. Dead bone can only be revitalized by removal and replacement (creeping substitution), a biological process which takes a long time to be completed. During the recovery of the blood supply, a temporary porosis of the bone is observed as a result of the tremendous intracortical remodeling. The remodeling activity starts at the boundary between vital and initially devascularized bone and is usually directed towards the implant (Figures 34.2a, b, plate section). It is accepted that necrotic tissue disposes to and sustains infection [56]. The recovery of the original bone structure and vitality generally takes more than one year. In the past, many authors have tried to explain this temporary bone porosity as a functional adaptation of the bone structure to the unloading effect of the plate according to Wolff's law [57–69]. The newer generations of plates (limited contact, no contact implants) decrease the amount of devascularization of cortical bone due to a reduction or the complete absence of implant–bone contact.

Fracture Healing and Stability of Fixation

Fracture healing is the recovery of the biological and mechanical integrity of the osseous tissue, i.e., return of the prefracture tissue vitality and structure as well as the prefracture stiffness and strength of the injured bone segment [70]. The amount of stability achieved by implants is the mechanical input for the biological response of bone healing. Beside the injury itself, the healing process additionally is modulated by the additional surgical damage to the bone and surrounding soft tissue envelope during the process of reduction and fixation [25,71–73]. In plate osteosynthesis the importance of the amount of mechanical stability to achieve direct bone healing was overestimated for a long time. Forcing precise reduction to improve the postoperative radiological appearance was likely to be linked to additional and sometimes extensive surgical trauma with stripping and denuding of bone fragments. Because dead bone is unable to heal, some of the possible complications such as deep infection, nonunion, delayed union, and refracture have to be attributed to the iatrogenic surgical tissue damage during the operative procedure. Radiographically and histologically, different healing patterns can be differentiated depending on the local mechanical environment [3,6,74].

Absolute stability is present when the fracture is stabilized by a stiff implant which maintains the fracture reduction with no or minimal displacements occurring under functional loading. As a biological consequence, primary bone healing without radiographically visible callus formation occurs. It can be assumed that fragment end necrosis induces internal remodeling of the bone, which repairs the fracture by the effect of crossing osteons.

Flexible fixation allows the fracture fragments to displace in relation to each other when load is applied. The external load results only in reversible deformation of the splint. After unloading, the fracture fragments move back into their former relative position. When the load results in an irreversible deformation of the splint, the fragments remain permanently displaced. Such a situation with plastic deformation of the implant is called unstable fixation. All fracture fixation devices possess different degrees of implant stiffness and lead to fixations of gradually differing flexibility depending on how they are applied and loaded. It appears likely that some flexibility of fixation is the most important mechanism triggering and inducing callus [75].

In bone healing, the strain conditions of the involved tissues have to be taken into account when judging under which condition bridging by bone formation will occur or a non-union develop. The comminution reduces the strain magnitude of the interfragmentary tissues in each gap for a given amount of overall displacement, thus allowing its safe differentiation and ossification. On the other hand, a small gap with some instability still present increases the strain of the interfragmentary repair tissue with inherent risk of non-union [76].

Mechanical Aspects of Plate Fixation

Basic Mechanical Principles of Internal Fixation

There are two basic mechanical principles how a fractured bone can be stabilized: interfragmentary compression and splinting. Interfragmentary compression functions by the elastic preload of both the bone and the implant. Thereby, the plate is loaded in tension and the bone in compression, creating high amounts of friction between the bone fragments. Interfragmentary compression can either be static, i.e., induced as a result of a pretensioned implant, or dynamic, i.e., generated by means of the functional load allowing coaptation of the fragments along a non-locked internal or external splint. Interfragmentary compression can be accomplished only in the case of at least a partial bony contact between the main fragments. Additionally, interfragmentary compression is very sensitive to minimum amounts of motion-induced bone resorption, diminishing the preload of both the implant and the bone, with consecutive loss of stability [77,78].

Splinting consists of the connection of an implant to a broken bone. The stability of this composite system depends on the stiffness of the splint itself, the quality of coupling between the splint and the bone, and the presence or absence of fracture comminution and bone defects. Depending on the localization of the implant, splinting can either be external or internal. Internal splints can be positioned inside or outside the medullary cavity. Depending on the mechanical use of the implant, a splint can be either gliding (non-locked or dynamically locked) or non-gliding (statically locked). The plate design itself does not define its later mechanical function; it can be used for splinting and/or for compression osteosynthesis.

Plate Designs and New Plate Developments

The Dynamic Compression Plate (DCP) was introduced in 1969 [8]. The idea of this plate was to enhance stability of fixation by interfragmentary compression and load transfer by friction (Figures 34.3a–c). To improve periosteal vascularity underneath the plate, in a first step the plates were undercut as far as safe application of the plate screws would allow without exceeding the compressive strength of the underlying cortical bone. The Limited Contact Dynamic Compression Plate (LC-DCP), which is currently in clinical use, was the first implant modified to preserve the bone circulation [12,79,80]. Nevertheless, some contact between bone and implant is still needed to allow load

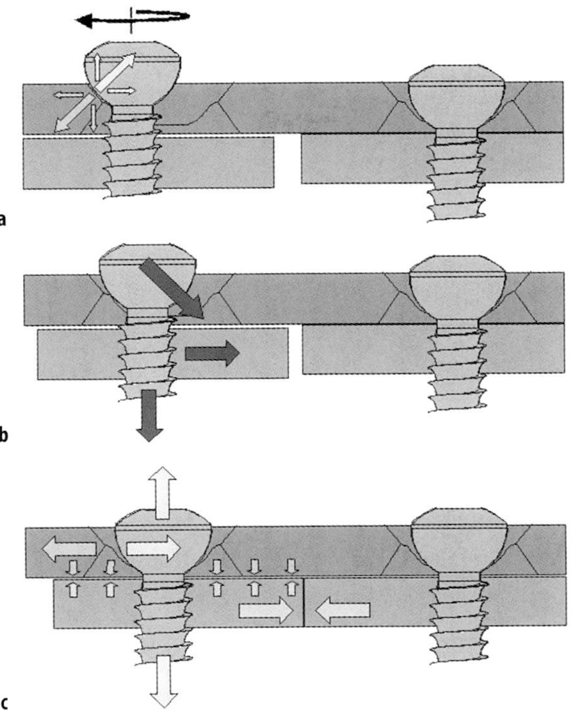

Figure 34.3. Mechanism of dynamic compression. When a plate screw is inserted eccentrically (**a**) the screw head slips down along the oblique part of the DCP hole (**b**). By tightening the screw compression in the fracture plane and friction at the bone–implant interface is created (**c**).

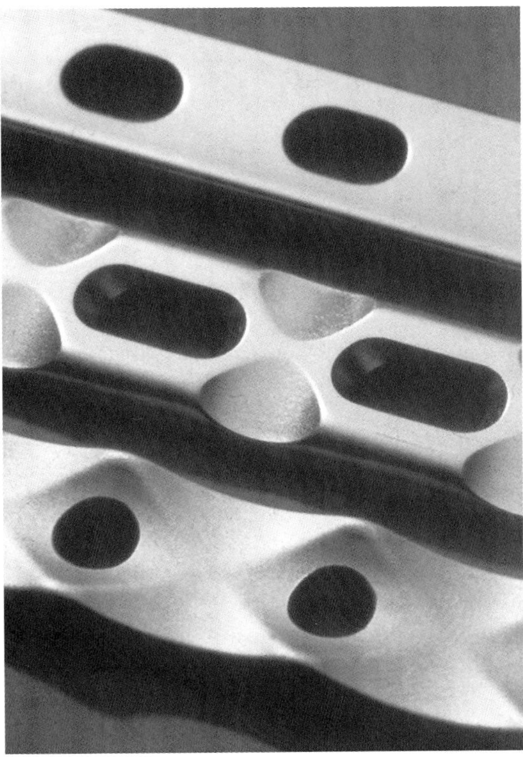

Figure 34.4. Undersurface of different plates. The undersurface of a conventional dynamic compression plate (DCP), the limited contact dynamic compression plate (LC-DCP), and the point contact fixator (PC-Fix) is shown.

transfer by a friction force at the interface created by tightening the screws.

The next step consisted of minimizing the plate–bone contact to isolated points only. With the Point Contact Fixator (PC-Fix) the plate is not compressed towards the bone. The isolated contact points between the plate holes serve only to hold the plate at a distinct distance from the bone surface until the screw heads engage in a conical non-threaded plate hole [81]. With the PC-Fix the load transmission is based on the partial interlocking of the screw heads in the plate holes. Thus, the amount of contact of the implant remains very small and without any adverse biological or mechanical effects to the underlying bone (Figure 34.4).

Nowadays, conventional plating is being increasingly replaced by using internal fixators. Internal fixators are "plates" (splints) with completely locked screw heads. These implants are not pressed onto the bone and do not need any contact between implant and bone. Further advantages are the possible reduction of the screw length to monocortical dimensions with the advantage of not needing screw length measurements and the possibility of using self-drilling and self-tapping screws.

The Less Invasive Stabilization System (LISS) was the first internal fixator of the AO-ASIF conceived for the meta- and epiphyseal regions of the distal femur and proximal tibia (Figure 34.5). Its shape conforms to the anatomical contours of the specific area of the bone and is designed for application via a minimally invasive submuscular approach. The first step is the anatomic reconstruction of the articular component followed by the restoration of the correct bone axis in all planes using the femoral distractor [82,83].

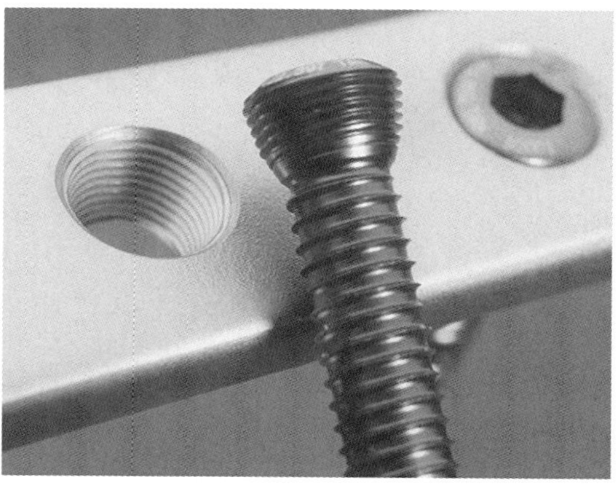

Figure 34.5. Less invasive stabilization system (LISS). The screw head and the plate hole have a conical thread. This construction gives a tight and fixed angle connection between plate and screw.

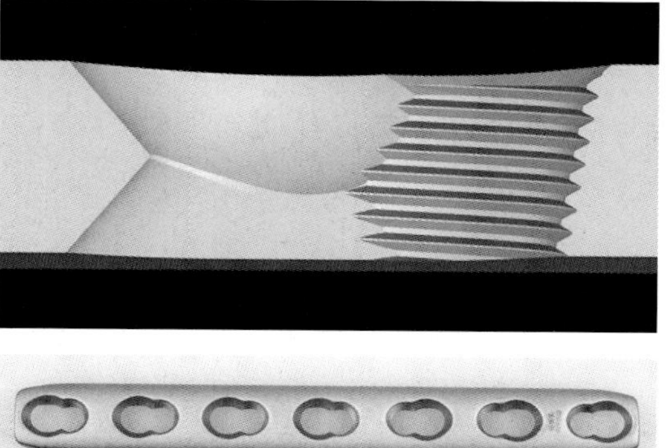

Figure 34.6. Locking compression plate (LCP). The locking compression plate is a further modification of the dynamic compression hole. One half consists of a regular dynamic compression hole; the other half is conical and threaded (**a, b**). Standard cortical or cancellous bone screws as well as locked head bone screws can be inserted according to the surgeon's preference and the mechanical demands of the fixation. The LCP is an asymmetrical plate with a defined middle section (**c**).

The newest development is the Locking Compression Plate System (LCP) which offers the advantage that the surgeon can choose during the operation whether to use it with conventional (non-locked) screws, with locked screws, or with a combination of both [84]. The specific design of the plate hole shows two functional elements: The first half of the hole comprises a dynamic compression unit that is intended for a standard cortical or cancellous bone screw. As in the standard DC-plating technique, the eccentric screw insertion allows an axial compression at the fracture site to be achieved. Additionally, the screw can be angled with respect to the longitudinal and transverse plate axis. The second half of the hole is threaded and conically shaped, permitting the locking of the special locking head screws (Figure 34.6a–c). The conical shape of the threaded screw head stops the tightening well before the thread within the bone sustains critical loads. In addition, this eliminates plastic deformation of the screw while tightening and equally excludes the reported high incidence of screw failures during insertion of conventional screws. No contact between bone and implant is needed for load transmission. The angular stability of the screws results in a lower incidence of screw loosening and secondary displacement of the fracture fragments. In addition, the protection of the periosteal blood supply is superior and in subcutaneous and submuscular plating techniques there is no need for accurate contouring of the plate.

Load Transfer in Conventional Plating

In a conventional plating technique, insertion and tightening of the plate screw generates an axial screw force (F_a), which compresses the plate onto the bone surface. This compression leads to a friction force (F_f) at the bone–implant interface. The friction force is proportional to the amount of compression with the specific bone-implant coefficient of friction (ρ) as a constant of proportionality. Under functional

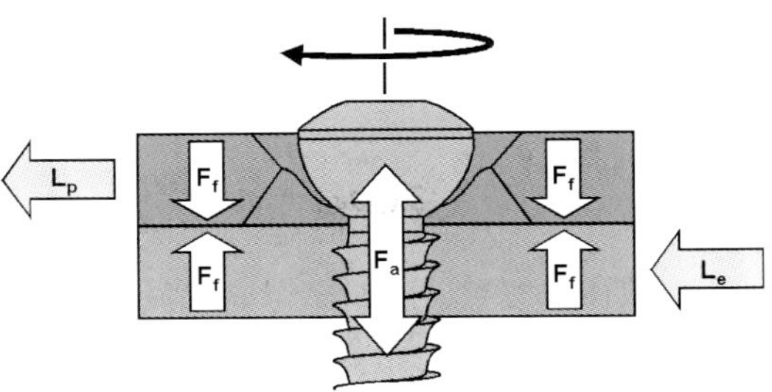

Figure 34.7. Load transfer in conventional plating. Tightening of a conventional plate screw results in tensile load of the screw and in compression at the bone–implant interface. This in turn creates friction at the interface able to withstand external loads tending to displace the plate on the bone surface.

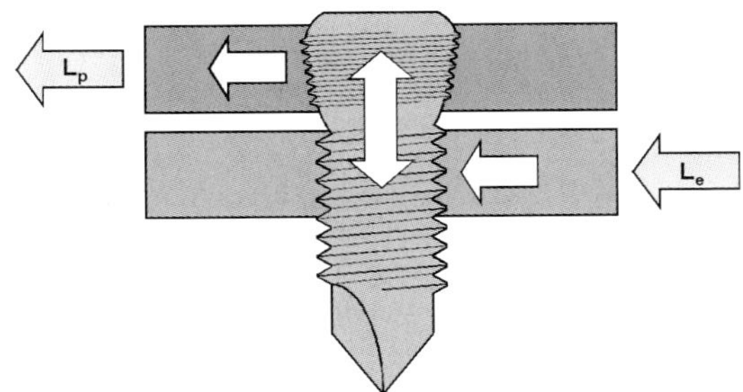

Figure 34.9. Load transfer in locked screw head plating. When the locking screw is tightened, the thread of the screw head engages in the conical threaded hole of the plate providing a stable construct between the plate and the screw. The external load is not transferred by friction force at the interface but by interlocking. This in turn results in a bending moment at the screw–head–screw–shaft junction.

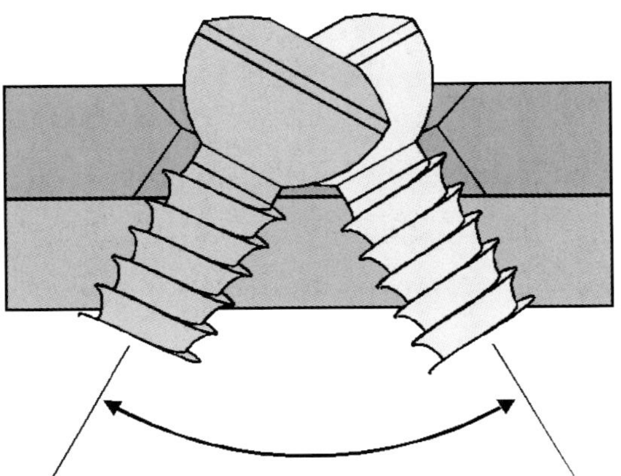

Figure 34.8. Adjustment of plate screws in conventional plating. The advantage of conventional plate screws is the potential to be angled with respect to the longitudinal and transverse axis of the plate.

loading, a plate osteosynthesis is loaded mainly in bending and in axial load. These loading patterns tend to displace the plate with regard to the bone. As soon as the external load (L_e) exceeds the amount of friction force installed at the plate–bone interface, the plate will slip on the bone (Figure 34.7). Under stable conditions, the screw is mainly loaded in tension. Once the plate starts slipping on the bone surface an additional bending moment at the screw head–screw shaft junction is present [78]. The advantage of the conventional plating is the possibility to angle the screws with respect to the plate surface (Figure 34.8). The disadvantage is that with time the axial screw force is diminished due to bone remodeling around the screw threads, leading to a corresponding decrease of the friction force at the bone–implant interface. In addition, in osteoporotic bone the maximum screw force that can be obtained by screw tightening can be very low from the beginning, resulting in a low friction force and later instability.

Load Transfer in Locked Screw-head Plating

In contrast to the conventional plating technique, in "locked screw-head" plating no contact between implant and bone is needed for load transfer from one main fragment to the other. Therefore, no friction force is generated by axial screw force at the bone–implant interface to withstand the displacement forces. Mechanically, this so-called internal fixator is defined as a construct in which the screws are the principal load-transferring elements from the main bone fragments to the implant. In such a construct the screws are firmly locked to the internal fixator to allow for moment and force transfer (Figure 34.9). Thus, the longitudinal displacement forces acting on the construct are directly transferred from the bone to the implant by bending and shear across the screw neck [45,81]. The advantage of the locked screw head is the change of the mechanical loading condition of the screw, which now is mainly in bending and less in axial pull-out. The disadvantage is the fixed angle screw direction inside the plate and the complete loss of the surgical feeling of screw tightening. Even in weak bone,

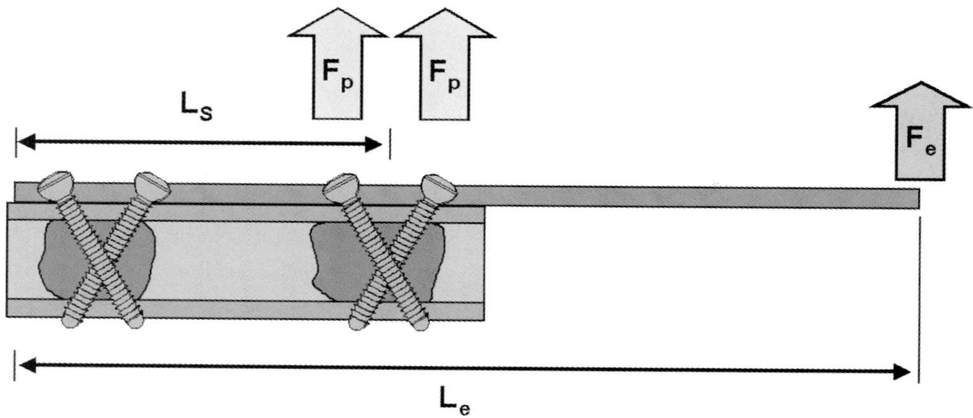

Figure 34.10. Screw fixation in osteoporotic bone. In osteoporotic bone the holding power of conventional screws is weak, leading to only a small amount of friction at the interface with later risk of secondary displacement and screw pullout. To enhance the holding power of the screws, two paired screws should be angled and the intramedullary cavity in between filled with bone cement. F_e external load; F_p pullout force; L_s lever arm screw; L_e lever arm external force.

the threaded screw head engages firmly inside the plate giving false information about the holding of the screw inside the bone.

Plate Fixation in Osteoporotic Bone

In osteoporotic bone the commonly recommended fixation concepts fail because bone quality is poor. The holding power of each screw in conventional plate osteosynthesis is decreased leading to the problem of early pullout of screws and secondary fracture displacement. A possible method of enhancing the quality of fixation is the use of bone cement. Bone cement can be used around blade plates to diminish the danger for cut-out and around the intramedullary part of the screws to enhance the holding power. In such a case, two screws should be angled toward each other and the intramedullary space in between should be filled with bone cement (Figure 34.10). After polymerization the screws are tightened. The pair-wise obliquely inserted screws give a very stable fixation of the plate.

In osteoporotic bone fracture fixation, using the internal fixator concept seems to be advantageous because stability of fixation does not rely on an axial screw force creating a friction force at the implant–bone interface, but on locked screws loaded mainly in bending during functional loading.

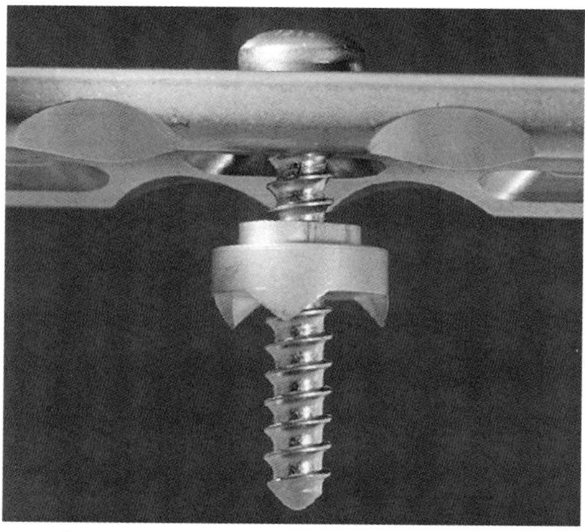

Figure 34.11. Schuhli nut. The schuhli nut is a sort of a threaded washer, which is positioned underneath a conventional DC plate. By tightening of the screw the nut is pulled towards the plate creating an angular stable system with enhanced holding.

Another possibility to increase the holding of screws in osteoporotic bone is the use of "schuhli nuts". Schuhli nuts are placed underneath the plate and have an identical thread like conventional cortical screws. Tightening the screw locks it within a plate hole creating an angular stability of the screw–plate–schuhli construct with enhanced holding power (Figure 34.11). The advantage of the schuhli nuts is avoiding the potential adverse effects of bone cement on fracture healing with extravasation and thermal necrosis [85–87]. But, the fiddling factor of the schuhli nuts is more important than with the use of a locking screw head system (LCP, LISS).

Mechanical Characteristics of Plates

Most plates used for osteosynthesis are still metallic implants (stainless steel or titanium). The availability of the material and, more importantly, the excellent mechanical and biological properties of metals are the main reasons for its widespread use in internal fixation. Metals offer on the mechanical side high stiffness and strength to withstand deformation or fatigue failure, sufficient ductility to allow shaping of the implant, and corrosion resistance, and on the biological side good tissue compatibility without localized toxic reactions [88–91].

According to Hooke's law the relationship between stress and strain is linear for all materials up to the proportional limit ("ut tensio sic vis", Robert Hooke, 1676). The slope of the curve is a measure of the stiffness of the material; it is high in rigid implants and low in flexible implants. In the following short part of the stress–strain diagram, deformation of the material is still elastic, but disproportional until the elastic limit is reached. At the yield point, plastic deformation (creeping) of the material occurs without a substantial change of the stress magnitude. Elasticity means that the material regains completely its original dimensions upon removal of the applied forces. With further loading a plastic and irreversible deformation of the material occurs. Materials capable of withstanding large plastic deformation are referred to as ductile materials; the opposite applies for brittle materials. In osteosynthesis, ductile materials are needed to allow shaping and contouring of the implants. The highest point of the diagram is the ultimate strength of the material. The endpoint of the diagram is given by the failure of the material where the elongation at rupture can be defined. The term strength defines the limit of stress that a material can withstand without rupture; it determines the level of load up to which the implant remains intact (Figure 34.12).

Regularly, all implants are repetitively loaded well below the ultimate strength of the material. Thus, much more important than the ultimate strength of the material is the fatigue behavior of the implants. The relationship between stress magnitude and number of loading cycles is described by Wöhler's curve (Figure 34.13).

Beside the pure mechanical characteristics of implants, its surface structure is important with respect to tissue adherence, which may allow formation of a fluid-filled dead space surrounding the implant promoting growth of bacteria [90,92–94].

Mechanics of Osteosynthesis Using Plates or Internal Fixators

Effect of Plate Length on Screw Loading

In plate osteosynthesis the length of the plate and the position of the plate screws play an important role with respect to plate and screw loading conditions. The longer the splint is, the less pull-out force is created on the screws due to improvement of the working leverage of the screws (Figures 34.14a–d). Thus, when using a plate as a splint the use of a very long plate is important from a mechanical point of view [26,95,96]. With the newer subcutaneous and submuscular plating techniques, the surgical dissection to insert long plates is not increased, thus the mechanical advantage of the use of longer plates has no biological disadvantage [11,15,18,21,26,30,32,97]. Using an internal fixator with locked screw heads, the screw loading is mainly in bending and not in pullout. With an external bending moment all the screws are loaded at the same time and, thus, failure is less frequent (Figures 34.15a, b). Nevertheless, the working leverage of an internal fixator should also be kept long and spacious.

Effect of Screw Position on Plate Loading

Bridging a longer bone segment in the middle of the plate over the fracture area reduces the implant strain due to bending. Bending a plate over a short segment enhances the local strain inside the implant. Bending over a longer segment reduces the local strain that results in

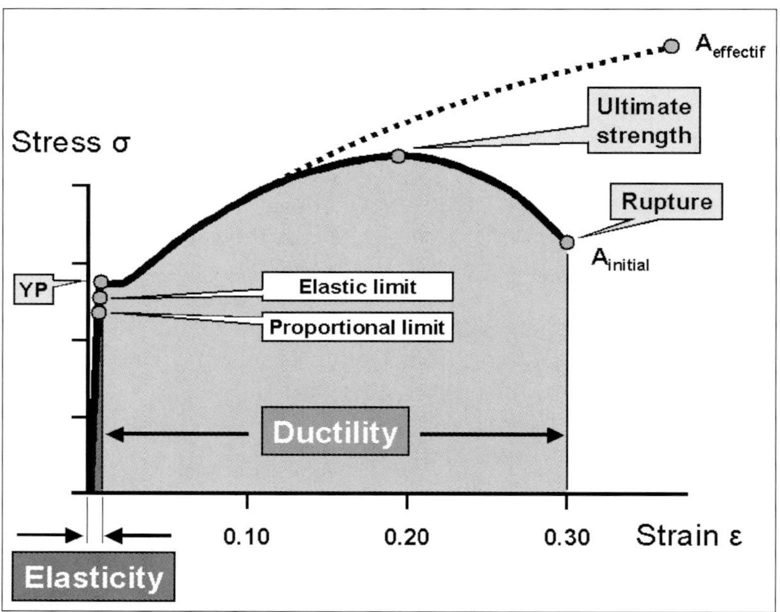

Figure 34.12. Stress-strain diagram. The diagram describes the behavior of a material under tensile load. Stress and strain are proportional up to the proportional limit. The elastic modulus of the material is the constant of proportionality. Up to the elastic limit, a short part of further elastic, but disproportional, deformation of the material is observed. With further loading, plastic and irreversible deformation occurs. The yield point is characterized by creeping of the material, i.e., elongation without an increase of the stress. The ultimate strength is defined as the highest stress tolerated by the material without failure (calculated for the initial cross-section, $A_{initial}$). Failure occurs after further plastic deformation; at that point the elongation at rupture can be determined.

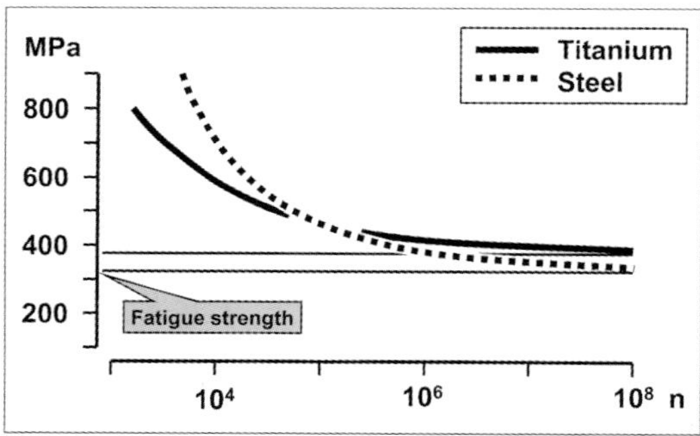

Figure 34.13. Fatigue strength of materials. The fatigue behavior of an implant is much more important in internal fixation than the ultimate strength. During functional rehabilitation until the bone has healed, cyclic loading of the implant occurs with the risk of fatigue failure of the implant. The higher the stress the less loading cycles are tolerated by a given implant. The fatigue strength is defined as the asymptotic line of the so-called Wöhler curves (stress-loading cycle diagram). For high stresses a steel plate withstands more loading cycles than a titanium plate. For low stresses the resistance of titanium against fatigue failure is higher.

a protection effect against fatigue failure of the implant (Figures 34.16a–d). Thus, in compression plating with the load-sharing condition of plate and bone – the inner plate screws can be inserted as close as possible to the fracture; the peripheral screws are inserted at each plate end. With locked screw-head plating and the mechanical condition of splinting, a longer distance between the two screws adjacent to the fracture is needed to obtain a longer distance and a lower elastic plate deformation [28,96,98,99]. When the stress of the plate due to external loads is small, more loading cycles are tolerated without fatigue failure of the implant. The fatigue strength of the implant is more important than the pure ultimate strength and the stress at rupture of the implant [89].

Effect of Plate Position on Rigidity of Fixation

The concept of load sharing is important for the loading condition and endurance of a plate. When bony contact between the main frag-

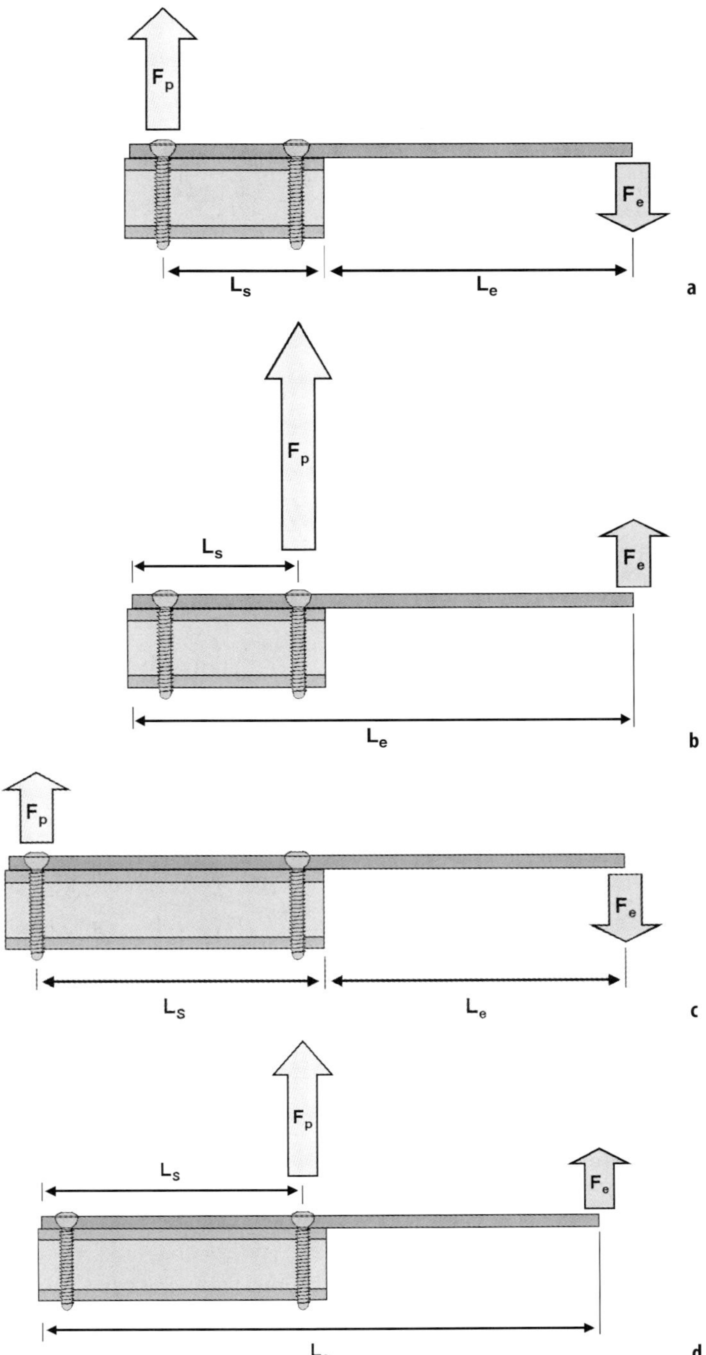

Figure 34.14. Effect of plate length on screw loading. The bending moment acting on an osteosynthesis plate tends to pull out the plate screws. With a short plate the screw loading is relatively high due to a short working lever arm of the screws (**a**, **b**) for both directions of bending moment (force acting towards the plate or away from the plate). The use of a long plate increases the working lever arm of each screw. Thus, under a given bending moment, the pullout force of the screws is decreased (**c**, **d**). L_s = Lever arm screw; L_e = lever arm external load.

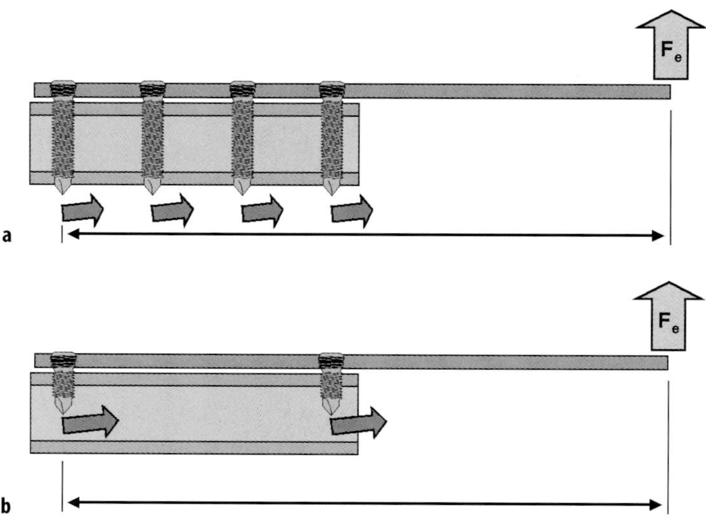

Figure 34.15. Screw loading in locked head screws. The bending moment acting on an internal fixator with locked plate screws leads to simultaneous bending of all the screws and not to an isolated pullout force on each individual screw (a). Thus, resistance against screw failure is improved. This holds true also with the use of monocortical self-drilling screws (b).

ments is achieved after fracture reduction, even broken bone is able to withstand compressive loads, leading to partial unloading of the plate. In the composite mechanical system of plate osteosynthesis the plate is firmly connected to bone either by a friction force at the interface or by locked plate screws. Composite structures, such as a composite bone–metal beam, show a different mechanical behavior compared to individual beams. The rigidity of such a system is much greater than the sum of the rigidity of the individual beams. The composite structure bends in a new common neutral axis lying in between the neutral axes of the two individual beams (Figure 34.17a, b). In case of different modules of elasticity and different cross-sections of the beams, the new neutral axis is found where the products formed by the axial stiffness and the distance to the new neutral axis of each beam is equalized [88,100,101]:

$$(EA)_{plate}(h - z) = (EA)_{bone}\, z$$

E Elastic modulus;
A Area of the cross-section;
h Distance between the two area centers of gravity;
z Shift of the neutral axis of the bone;
h–z Shift of the neutral axis of the plate.

The composite beam theory determines the mechanical behavior of two beams under pure bending or eccentric axial load. The axial rigidity of a beam is the product of its modulus of elasticity and the cross-section (EA). The bending rigidity of a beam is the product of its modulus of elasticity and the area moment of inertia with respect to the specific bending direction (EI). The stiffness of a composite beam can be calculated using the following formula:

$$EI_{composite} = (EI)_{bone} + (EI)_{plate} + (EA)_{bone} z^2 + (EA)_{plate}(h - z)^2$$

According to the Steiner's principle, the increase in the bending stiffness of the composite beam with respect to the bending stiffness of the individual unconnected beams is mainly caused by the shift of the neutral axis with dramatic enhancement of the area moment of inertia of both the bone and the implant [88,91].

A plate has a different bending rigidity according to the individual bending direction (Figure 34.18). Thus, plate position and bending direction influence the overall stiffness of a plate osteosynthesis. Figure 34.19 shows the bending rigidity of an intact tubular bone

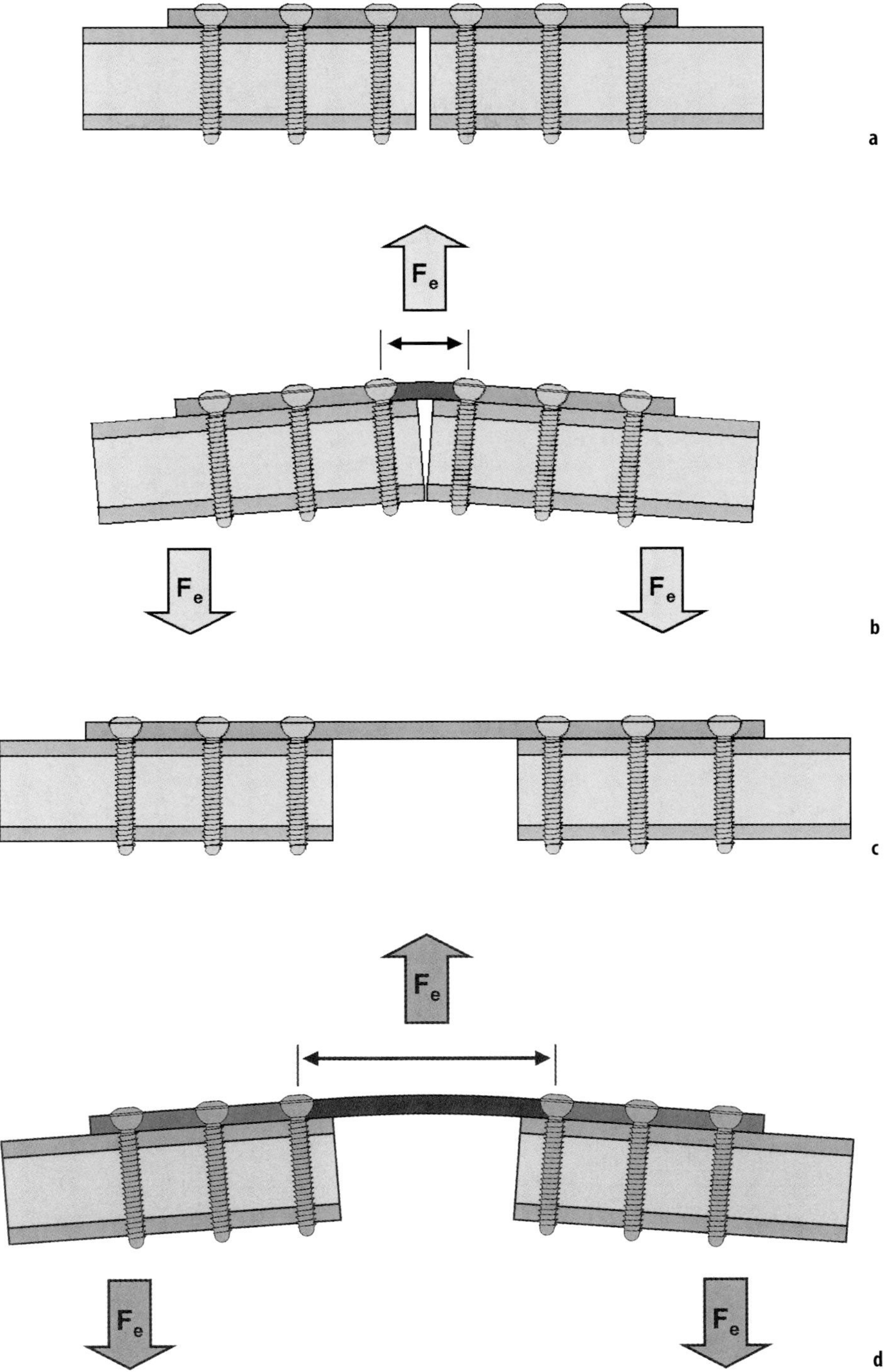

Figure 34.16. Effect of screw position on plate loading. A small gap after internal fixation using a plate leads to a stress concentration of the implant at the gap site when the plate screws are inserted close to the gap (**a**, **b**). When a larger gap is present after osteosynthesis the external load results in a stress distribution over a longer segment of the plate with concomitant decrease of the plate loading (**c**, **d**).

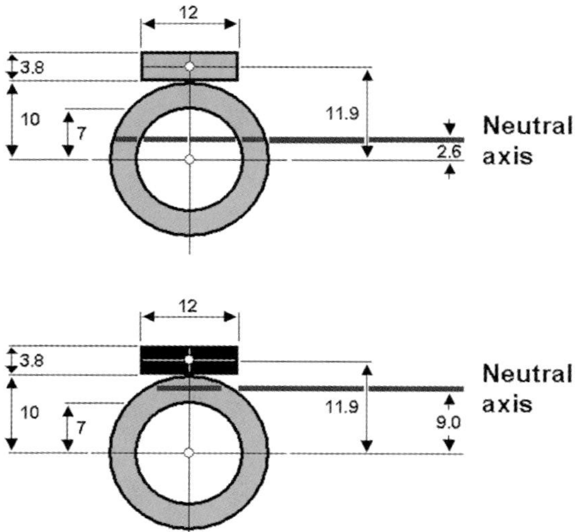

Figure 34.17. Shift of the neutral axis of a composite beam. The tight connection of two beams (plate and bone) results in a shift of the neutral axis of the composite structure. Using a plate with the elastic modulus of bone leads to only a small shift (**a**). A steel plate with high elastic modulus increases the shift of the new common neutral axis. This shift enhances the area moment of inertia – and with that also the stiffness – of both the plate and the bone (**b**).

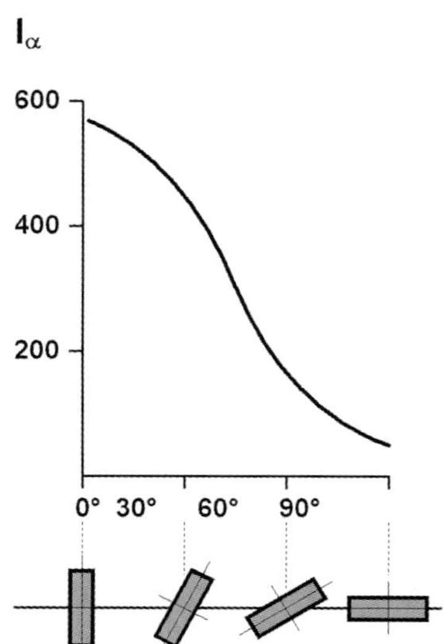

Figure 34.18. Bending stiffness of plate under different bending directions. The bending stiffness of a plate depends on the direction of the bending moment. Bending towards the highest dimension or bending towards the smallest dimension makes a difference of about a factor of eight in the bending stiffness of the plate.

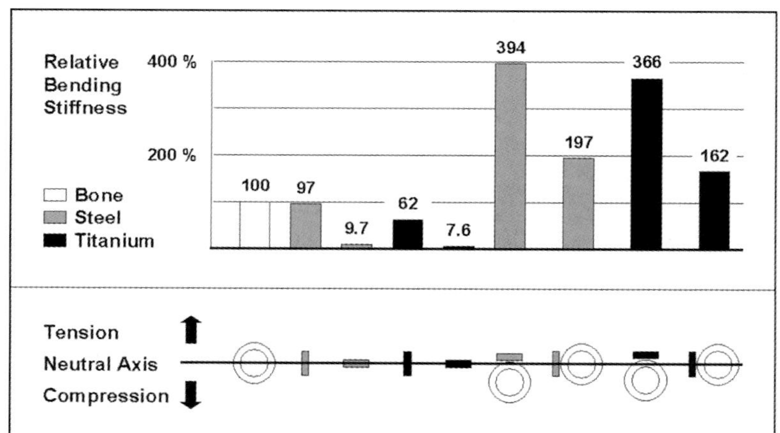

Figure 34.19. Bending stiffness of plate osteosynthesis. The bending stiffness of a plate osteosynthesis depends mainly on the position of the plate with regard to the bending direction. When the bone can withstand compressive loads the highest stiffness of the composite beam is reached when the implant is positioned on the tension side. The elastic modulus of the plate is unimportant when comparing steel and titanium plates (elastic modulus of steel 190 GPa, of titanium 110 GPa).

(taken as 100%), steel, and titanium plates and osteosynthesis using steel or titanium plates under different bending directions [101]. It is remarkable that the stiffness of the plate itself is relatively small compared to the bending stiffness of the bone alone. Using either steel or titanium plate, the differences in the modulus of elasticity of the plates are unimportant with regard to the changes in the overall rigidity due to different bending directions. It is evident that a rigid composite system leads to unloading of the plated bone segment with the unloading being highest in the cortex directly underneath the implant [102].

The shift of the common neutral axis in bending can be even more important in wave plate osteosynthesis where the bending stiffness is higher than in conventional plating techniques when the cortex opposite the wave plate is in tight contact and loaded in compression. Biologically, the wave plate osteosynthesis is advantageous in case of disturbed vascularity of the bone cortex underneath the plate to allow revascularization of the bone despite a bone plate in situ [19,103–105].

Effect of Plate Position and Plate Contouring on Plate Loading

The position of a plate influences the amount of plate loading. This is essential when plating is performed in a diaphysis, which is eccentrically loaded such as in the femur. Each eccentric axial load on a beam results in a bending moment of the beam with tensile stress on one side and compressive stress on the other. For anatomic and mechanical reasons, in the femur the plate is positioned regularly on the lateral side. In case of contact between the main fracture fragments the bone is loaded in compression and the plate in tension. In such a load-sharing situation the load on the plate is low and mainly in tension, which is well tolerated by the implant (Figure 34.20). In case of a comminution without bony contact between the main fragments, the laterally positioned plate is the only load carrier and undergoes high tensile and compressive stress due to pure bending (load-bearing situation). Only the rapid formation of callus with integration of the bony fragments opposite the plate (biobuttress) protects the implant from mechanical overload and fatigue failure. Theoretically, in such a case the loading of the plate can be decreased when positioning the plate on the medial aspect of the femoral diaphysis. In this position the plate is close to the weight-bearing axis of the leg and, therefore, the load on the implant is mainly axial (compressive) and less bending (Figure 34.21). But, clinically, the medial approach to the femur and a medial plating technique is used only when the fracture is associated with a vascular injury needing repair. In osteosynthesis, using a wave plate load sharing between implant and bone is even more important than with a regular straight plate. When there is no osseous contact between the main fragments only the lever arm of the externally applied axial load is increased, creating an even higher bending moment of the plate than with the use of a straight plate. In case of at least partial osseous contact, in addition the lever arm of the wave plate is increased, which leads to a reduction of the implant loading (Figure 34.22).

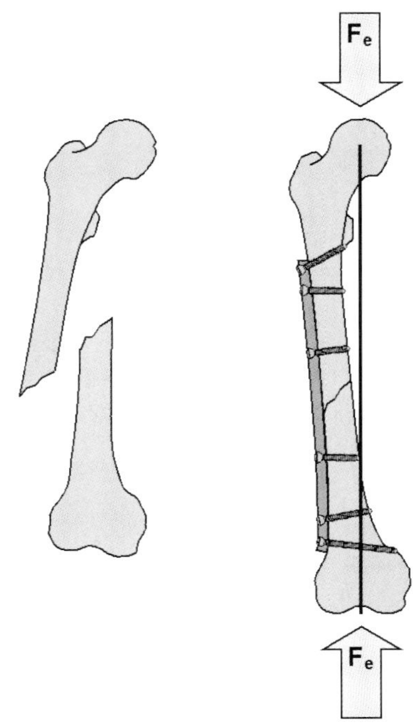

Figure 34.20. Load-sharing situation after femoral plating. In plate osteosynthesis the plate loading due to an eccentric axial load is comparably small when the main fracture fragments are in contact and the bone able to withstand compressive loading (tension band plating).

Clinical Aspects of Plate Fixation

Plate Functions

In clinical practice, plates can be used with different mechanical functions. But, often, a combination of different mechanical functions is

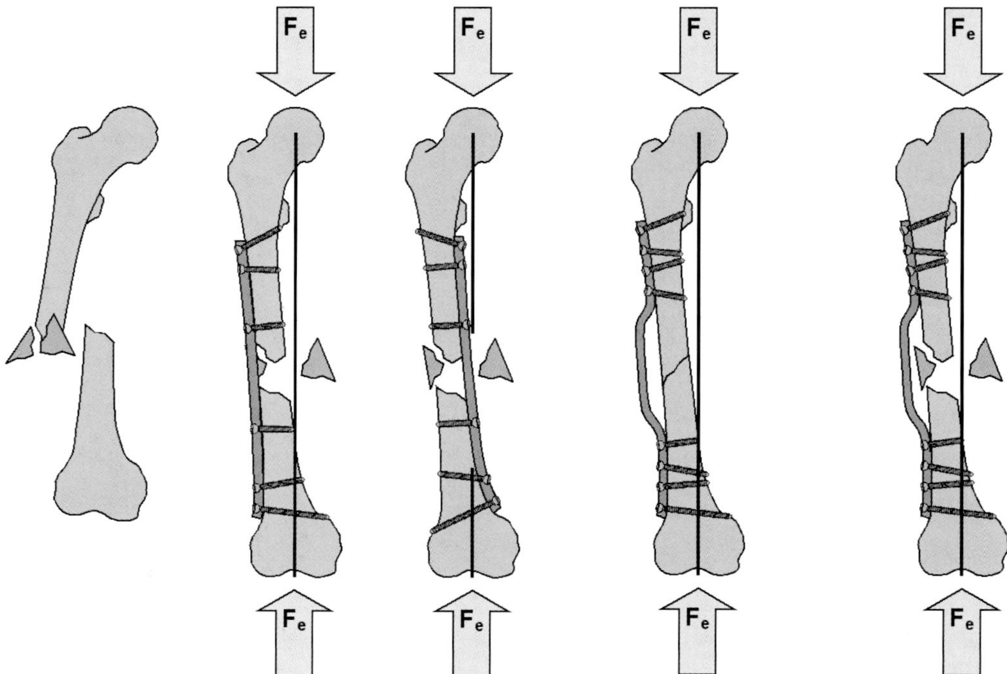

Figure 34.21. Load-bearing situation after femoral plating. In a comminuted fracture pattern the plate is the only load carrier of the osteosynthesis. Due to the distance between the plate and the weight-bearing axis a high bending moment is present, resulting in high bending stress of the plate. In such a situation the plate loading can be substantially decreased when the plate is positioned as closely as possible to the weight-bearing axis. This decreases the eccentricity of the axial load and the plate is mainly loaded in compression.

Figure 34.22. Stiffness of a wave plate osteosynthesis. With at least unicortical contact between the main fracture fragments, a wave plate osteosynthesis is a very stiff composite construct. In a comminuted fracture the wave plate is critical from a mechanical point of view because the distance between the plate cross-section under load and the external axial load is increased. This enhances the bending moment and the plate loading.

possible in plate osteosynthesis. Plate functions include:

Plate as an internal splint. This is the main function of each plate. It is each time present when a plate is used for internal fixation of a fracture. The efficacy of the splint function depends on the mechanical properties of the splint and the quality of anchorage of the splint to the bone.

Bridge plate. The plate spans a comminuted fracture area and provides elastic fixation with the mechanical function of a pure splint.

Tension band plate. The plate is applied on the tension side of the bone (e.g., lateral side of the femoral shaft). The bone is loaded in static and/or dynamic compression. Thus, load sharing between the plate and the bone is present.

Compression plate. Compression can be exerted by the use of eccentric screw holes, the tension device, or by overbending of the plate.

Neutralization or protection plate. In case of a lag screw fixation of a simple fracture, the additional neutralization of the fracture area by means of a protection plate is recommended. The technique of lag screw fixation and compression osteosynthesis is being more and more abandoned due to its inherent risk for surgically induced vascular damage to bone and soft tissues.

Buttress plate. The plate supports a piece of bone mainly in the meta- or epiphyseal area.

Antiglide plate. The plate is positioned in a such a way as to push a fragment into its anatomical position and to inhibit secondary displacement by direct interference with a meta- or diaphyseal bone fragment.

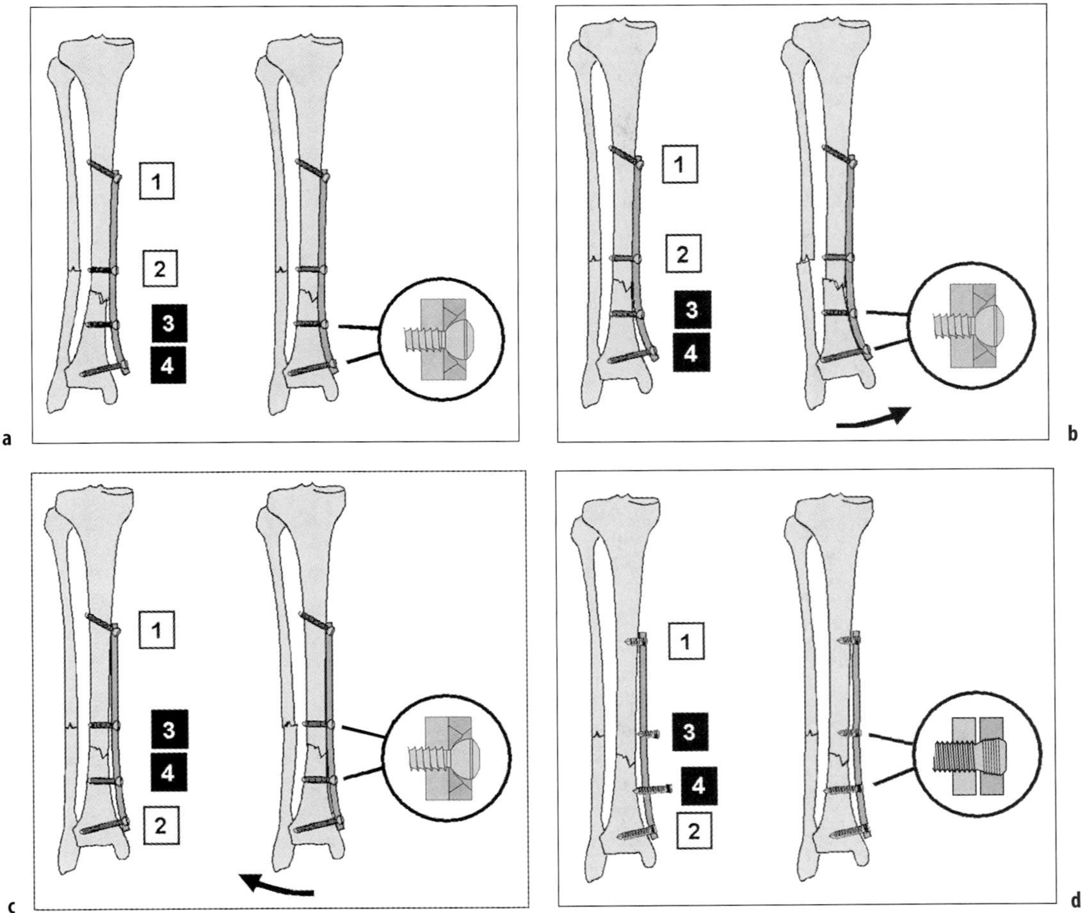

Figure 34.23. Reduction in plate osteosynthesis. In conventional plating the plate should be contoured exactly to the shape of the surface of the bone. After having fixed the plate to the one main fragment, the other fragment can be pulled towards the plate by inserting screws leading to proper reduction and fixation (**a**). When the plate is not properly contoured, reduction on the plate results in an axial misalignment (**b**, **c**). Using an internal fixator with locked screw heads poses no risk for loss of reduction during screw tightening because the screws engage inside the threaded plate hole before pulling the bone fragment towards the plate (**d**).

Reduction Technique

The technique of direct reduction of a fracture by visualization of the fragments is an invasive and devitalizing way to achieve "anatomical" reduction of a fracture. In diaphyseal fractures, today's reduction technique is indirect with a minimal exposure of the fracture and the fragments [46].

Any relatively straight portion of any bone may be reduced by the application of a straight plate. The plate application precedes reduction; it acts as a splint to restore alignment. Distraction of the fracture increases the tension in the soft tissues. This tends to recentralize the fragments, causing them to approximate their previous location in the fractured bone [23]. The disadvantage of this technique is the potential for incorrect reduction when the plate used for reduction onto the implant was initially not shaped accurately to the bone surface (Figure 34.23a–d). Another mechanism is the reduction through interference along the external surfaces of bone and plate. This principle can be demonstrated during reduction and fixation of a malleolar fracture using the antiglide function of the plate [106]. In the case of an oblique fracture, the application of the plate to the proximal fragment results in interference with the distal fragment. As the plate pushes against the displaced distal fragment, it forces the reduction along the oblique fracture surfaces. When using the locked screw-head system no reduction of bone fragments onto the implant is possible. The fracture needs to be reduced properly before the internal fixator is inserted to maintain the achieved reduction. In such a procedure, it is advantageous that fracture fragments do not tend to redisplace during screw tightening because the screws engage inside the threaded plate hole without pulling the fragment towards the plate.

Fixation Technique

Today, the principles of fixation techniques are minimal, but optimal. That means longer splints acting with optimal lever arms for the screws are preferable [14,16,26]. The main fragments are held in place using balanced fixation on both sides of the fracture. In the case of a simple fracture, the use of interfragmentary compression is still safe from a mechanical point of view. A small fracture gap in a simple fracture configuration combined with elastic fixation leads to high interfragmentary tissue strain, which sometimes can inhibit fracture consolidation. In such short fractures, repetitive bending stresses will be concentrated and centered on a short segment of the plate, which thereby can break more easily due to fatigue. The risk of mechanical implant failure can be considerably reduced if longer plates are used despite a short fracture zone, so that stresses are distributed over a longer section of the plate.

In multifragmentary fractures a bridge plate technique is advantageous. The implant spans over a longer bone segment decreasing the strain of the granulation tissue which will be formed between the fracture fragments, allowing its safe differentiation into bone tissue. Additionally, the plate itself undergoes low deformation during functional aftercare as bending stresses are distributed over a long segment of the plate with reduced risk of plate failure.

Conclusions and Outlook

Within the last two decades new thinking on operative fracture treatment using plates has been established. Using plates for internal fixation, the advantages of operative and conservative treatment have to be combined: proper alignment of the injured bone segment and sufficient stability of fixation allowing functional aftercare and an undisturbed natural course of bone healing. Thus, in shaft fractures, the exact reduction of each bone fragment is no longer a goal in itself. Rather, the overall restoration of length, axial alignment, and rotation are the goals.

Plate osteosynthesis keeps its important and well-established place in the treatment of certain fractures. Classical indications for an osteosynthesis using plates or internal fixators are articular fractures, metaphyseal fractures, and some diaphyseal fractures, such as forearm fractures, diaphyseal fractures with associated articular fractures, diaphyseal fractures in polyfractured or polytraumatized patients, narrow medullary canal not suitable for intramedullary rodding, and some diaphyseal fractures in children.

The understanding of bone biology caused not only a change in surgical tactics when performing internal fixation, but it stimulated research to modify and improve the existent implants, aiming to reduce vascular damage to the bone tissue caused by the implant. Internal fixators with locked plate screws mechanically used as pure splints and inserted in a minimally invasive subcutaneous or submuscular way replace more and more the older conventional plating concepts. The new technique minimizes the surgical devascularization and the implant-inherent vascular insult to the bone tissue. Preserving the viability of all fracture fragments by protecting the soft tissue envelope is more important than the primary stability of an osteosynthesis. Healing by means of callus formation is potentially faster than the one based on inherently slow cortical remodeling. The good healing capacity of viable fragments results in stable conditions after a short time period, protecting the implant from fatigue failure. The surgeon's task is to make the sound synthesis between mechanical demands of the fracture and biological competence of the involved tissues.

Common Mechanical Terms and Formula

Brittle material:
Material with no or low capacity of plastic deformation.

Ductile material:
Material with large capacity of plastic deformation.

Elasticity:
Reversible deformation of a material. The material regains its original dimensions and shape after unloading.

Elastic modulus E:
Constant of proportionality between stress and strain. Slope of the initial curve of a stress-strain diagram. Represent the stiffness of a material to an imposed load. Also called modulus of elasticity, Young's modulus:
$E = \delta\sigma/\delta\varepsilon = \tan\varphi$

Friction:
Friction force at an interface. Product of a force perpendicular to a surface and a proportional constant (friction coefficient). $F_f = \rho\ F_a$

Plasticity:
Irreversible deformation of a material. Materials capable of withstanding large strains are referred to as ductile materials; the converse applies to brittle materials.

Stability:
Degree of relative movement between fragments. Absolute stability means no motion between fragments under given load. Relative stability means that the fragments displace under load, but go back to the initial position with unloading.

Stiffness:
The resistance of a material to deformation under load. The higher the stiffness of a material the smaller its deformation under a given load. The product of the cross-sectional area and the elastic modulus expresses axial stiffness: $R_{ax} = A\ E$.
Bending stiffness is defined as the product of the axial area moment of inertia (with respect to the individual bending axis) and the elastic modulus: $R_{be} = I_{ax}\ E$.

Strain ε:
Deformation of a material under a given load. It is expressed as elongation per unit of original length and is a dimensionless quantity $\varepsilon = \Delta l/l_0$.

Strength:
Ability of a material to withstand load without structural failure. It can be reported as ultimate tensile strength, as bending strength or torsional strength. It is expressed in units of force per unit of area (stress) or elongation at rupture (strain).

Stress σ:
Force per unit cross-sectional area. Stress is directly proportional to strain with the elastic modulus (E) as constant of proportionality. Unit of stress is Newton/m^2 (Pa). Normal stress means that the force is acting perpendicular to the surface, shear stress that the force acts parallel to a surface $\sigma = F/A$, $\sigma = E\ \varepsilon$.

References

1. Hansmann M. Eine neue Methode der Fixation der Fragmente bei complicirten Fracturen. Verhandlungen der Deutschen Gesellschaft für Chirurgie 1886;15:134–7.
2. Lambotte A. L'intervention opératoire dans les fractures récentes et anciennes. Paris: Maloine, 1907.
3. Lane WA. The operative treatment of fractures. London: Medical Publishing, 1913.
4. Sherman OWN. Vanadium steel bone plates and screws. Surg Gynecol Obstet 1912;14:629–34.
5. Bagby GW. Compression bone-plating. Historical considerations. J Bone Joint Surg 1977;59A:625–31.
6. Danis R. Théorie et pratique de l'ostéosynthèse. Paris: Masson, 1949.
7. Müller ME, Allgöwer M, Willenegger H. Technik der operativen Frakturenbehandlung. Berlin: Springer 1963.
8. Allgöwer M, Ehrsam R, Ganz R, Matter P, Perren SM. Klinische Erfahrungen mit der neuen Kompression-

splatte "DCP". Acta Orthop Scand Suppl 1969;125: 1–20.
9. Baumgaertel F, Buhl M, Rahn BA. Fracture healing in biological plate osteosynthesis. Injury 1998;29(Suppl): 3–6.
10. Chrisovitsinos JP, Xenakis T, Papakostides KG, Skaltsoyannis N, Grestas A, Soucacos PN. Bridge plating osteosynthesis of 20 comminuted fractures of the femur. Acta Orthop Scand Suppl 1997;275:72–6.
11. Farouk O, Krettek C, Miclau T, Schandelmaier P, Tscherne H. Effects of percutaneous and conventional plating techniques on the blood supply to the femur. Arch Orthop Trauma Surg 1998;117:438–41.
12. Gautier E, Perren SM. Die "Limited Contact Dynamic Compression Plate" (LC-DCP) - Biomechanische Forschung als Grundlage des neuen Plattendesigns. Orthopäde 1992;21:11–23.
13. Gautier E, Perren SM, Ganz R. Principles of internal fixation. Curr Orthop 1992;6:220–32.
14. Gautier E, Ganz R. Die biologische Plattenosteosynthese. Zentralbl Chir 1994;119:564–72.
15. Gautier E, Marti CB, Schuster AJ, Wachtl SW, Jakob RP. Die eingeschobene Femur- und Tibiaplatte. OP-Journal 2000;16:260–7.
16. Gerber C, Mast JW, Ganz R. Biological internal fixation of fractures. Arch Orthop Trauma Surg 1990;109: 295–303.
17. Heitemeyer U, Hierholzer G, Terhorst J. Der Stellenwert der überbrückenden Plattenosteosynthese bei Mehrfragmentbruchschädigung des Femur im klinischen Vergleich. Unfallchirurg 1986;89:533–8.
18. Helfet DL, Shonnard PY, Levine D, Borelli J. Minimally invasive plate osteosynthesis of distal fractures of the tibia. Injury 1997;28(Suppl 1):42–8.
19. Karnezis IA. Biomechanical considerations in "biological" femoral osteosynthesis: an experimental study of the "bridging" and "wave" plating techniques. Arch Orthop Trauma Surg 2000;120:272–5.
20. Kinast C, Bolhofner BR, Mast JW, Ganz R. Subtrochanteric fractures of the femur. Results of treatment with the 95 degree condylar plate. Clin Orthop 1989;238:122–30.
21. Krettek C, Schandelmaier P, Miclau T, Tscherne H. Minimally invasive percutaneous plate osteosynthesis (MIPPO) using the DCS in proximal and distal femoral fractures. Injury1997;28(Suppl 1):20–30.
22. Krettek C, Schandelmaier P, Miclau T, Bertram R, Holmes W, Tscherne H. Transarticular joint reconstruction and indirect plate osteosynthesis for complex distal supracondylar femoral fractures. Injury 1997;28(Suppl 1):S31–41.
23. Mast J, Jakob R, Ganz R. Planning and reduction technique in fracture surgery. Berlin: Springer, 1989.
24. Miclau T, Martin RE. The evolution of modern plate osteosynthesis. Injury 1997;28(Suppl 1):3–6.
25. Perren SM. The concept of biological plating using the limited contact dynamic compression plate (LC-DCP). Injury 1991;22(Suppl 1):1–41.
26. Rozbruch RS, Müller U, Gautier E, Ganz R. The evolution of femoral shaft plating technique. Clin Orthop 1998;354:195–208.
27. Rüedi TP, Murphy WM. AO principles of fracture management. New York: Thieme Stuttgart, 2000.
28. Schmidtmann U, Knopp W, Wolff C, Stürmer KM. Results of elastic plate osteosynthesis of simple femoral shaft fractures in polytraumatized patients. An alternative procedure. Unfallchirurg 1997;100:949–56.
29. Siebenrock KA, Müller U, Ganz R. Indirect reduction with a condylar blade plate for osteosynthesis of subtrochanteric femoral fractures. Injury 1998;29(Suppl 3):7–15.
30. van Riet YE, van der Werken C, Marti RK. Subfascial plate fixation of comminuted diaphyseal femoral fractures: a report of three cases utilizing biological osteosynthesis. J Orthop Trauma 1997;11:57–60.
31. Wenda K, Runkel M, Degreif J, Rudig L. Minimally invasive plate fixation in femoral shaft fractures. Injury 1997:28(Suppl 1):13–9.
32. Weller S, Höntzsch D, Frigg R. Die epiperostale, perkutane Plattenosteosynthese. Eine minimal-invasive Technik unter dem Aspekt der "biologischen Osteosynthese". Unfallchirurg 1998;101:115–21.
33. Brookes M. The blood supply of bone. An approach to bone biology. London: Butterworth, 1971.
34. Ficat P, Arlet J. Ischémie et nécrose osseuses. L'exploration fonctionelle de la circulation intra-osseuse et ses applications. Paris: Masson, 1977.
35. Göthman L. Vascular reactions in experimental fractures. Acta Orthop Scand Suppl 1961;284:1–34.
36. Macnab I, de Haas WG. The role of periosteal blood supply in the healing of fractures of the tibia. Clin Orthop 1974;105:27–33.
37. Rhinelander FW. The normal microcirculation of diaphyseal cortex and its response to fracture. J Bone Joint Surg 1968;50A:784–800.
38. Trueta J. Blood supply and the rate of healing of tibial fractures. Clin Orthop 1974;105:11–26.
39. Kelly PJ. Anatomy, physiology and pathology of the blood supply of bones. J Bone Joint Surg 1968;50A: 766–83.
40. Nelson GE, Kelly PE, Peterson LFA, Janes JM. Blood supply of the human tibia. J Bone Joint Surg 1960; 42A:625–36.
41. Rhinelander FW. Tibial blood supply in relation to fracture healing. Clin Orthop 1974;105:34–81.
42. Rhinelander FW, Wilson JW. Blood supply to developing, mature, and healing bone. In: Sumner-Smith G (ed) Bone in clinical orthopaedics. A study in comparative osteology. Philadelphia: Saunders, 1982;81–158.
43. Moor R, Tepic S, Perren SM. Hochgeschwindigkeits-Film-Analyse des Knochenbruchs. Z Unfallchir Versicherungsmed 1989;82:128–32.
44. Alexander AH, Cabaud HE, Johnston JO, Lichtman DM. Compression plate position. Extraperiosteal or subperiosteal? Clin Orthop 1983;175:280–5.
45. Fernandez Dell'Orca A, Regazzoni P. Internal fixation: A new technology. In: Rüedi T, Murphy WM (ed) AO principles of fracture management. Berlin: Thieme, 2000;249–53.
46. Gautier E, Jakob RP. Surgical reduction and its influence on bone repair: approaches and reduction techniques. In: Rüedi T, Murphy WM (ed) AO principles of fracture management. Berlin: Thieme, 2000;139–55.
47. Gautier E, Cordey J, Mathys R, Rahn BA, Perren SM. Porosity and remodelling of plated bone after internal fixation: result of stress shielding or vascular damage? In: Ducheyne P, van der Perre G, Aubert AE (ed) Biomaterials and Biomechanics 1983. Amsterdam: Elsevier Science, 1984;195–200.

48. Gautier E, Perren SM. Die Reaktion der Kortikalis nach Verplattung - eine Folge der Belastungsveränderung des Knochens oder ein Vaskularitätsproblem? In: Wolter D, Zimmer W (ed) Die Plattenosteosynthese und ihre Konkurrenzverfahren. Berlin: Springer, 1991;21-37.
49. Gautier E, Rahn BA, Perren SM. Vascular remodelling. Injury 1995;26(Suppl 2):11-19.
50. Gunst MA, Suter C, Rahn BA. Die Knochendurchblutung nach Plattenosteosynthese. Helv Chir Acta 1979; 46:171-5.
51. Jacobs RR, Rahn BA, Perren SM. Effect of plates on cortical bone perfusion. J Trauma 1981;21:91-5.
52. Jörger KA. Akute intrakortikale Durchblutungsstörung unter Osteosyntheseplatten mit unterschiedlichen Auflageflächen. MD thesis, University of Bern, 1987.
53. Lippuner K, Vogel R, Tepic S, Rahn BA, Cordey J, Perren SM. Effect of animal species and age on plate-induced vascular damage in cortical bone. Arch Orthop Trauma Surg 1992;111:78-84.
54. Lüthi UK. Auflageflächen von Osteosyntheseplatten und intrakortikale Durchblutungsstörungen. MD thesis, University of Basel, 1980.
55. Perren SM, Cordey J, Rahn BA, Gautier E, Schneider E. Early temporary porosis of bone induced by internal fixation implants. A reaction to necrosis, not to stress protection? Clin Orthop 1988;232:139-51.
56. Arens S, Kraft C, Schlegel U, Printzen G, Perren SM, Hansis M. Susceptibility to local infection in biological internal fixation. Experimental study of open vs. minimally invasive plate osteosynthesis in rabbits. Arch Orthop Trauma Surg 1999;119:82-5.
57. Akeson WH, Woo SL-Y, Rutherford L, Coutts RD, Gonsalves M, Amiel D. The effects of rigidity of internal fixation plates on long bone remodeling. Acta Orthop Scand 1976;47:241-9.
58. Claes L. The mechanical and morphological properties of bone beneath internal fixation plates of differing rigidity. J Orthop Res 1989;7:170-7.
59. Cochran GVB. Effects of internal fixation plates on mechanical deformation of bone. Surg Forum Orthop Surg 1969;20:469-71.
60. Cordey J, Schwyzer HK, Brun S, Matter P, Perren SM. Bone loss following plate fixation of fractures? Helv Chir Acta 1985;52:181-4.
61. Gördes W, Kossyk W, Holländer H. Histologische und histomorphometrische Veränderungen bei Plattenosteosynthesen nach Osteotomien an der Tibia des Kaninchens. Arch Orthop Unfallchir 1975;82:123-33.
62. Matter P, Brennwald J, Perren SM. Biologische Reaktion des Knochens auf Osteosyntheseplatten. Helv Chir Acta 1974;12(Suppl):1-44.
63. Moyen BJ-L, Lahey PJ, Weinberg EH, Rumelhart C, Harris WH. Effects of application of metal plates to bone. Acta Orthop Belg 1980;46:806-15.
64. Strömberg L, Dalen N. Atrophy of cortical bone caused by rigid internal fixation plates. Acta Orthop Scand 1978;49:448-56.
65. Terjesen T, Benum P. The stress-protection effect of metal plates on the intact rabbit tibia. Acta Orthop Scand 1983;54:810-8.
66. Uhthoff HK, Dubuc FL. Bone structure changes in the dog under rigid internal fixation. Clin Orthop 1971;81:165-70.
67. Uhthoff HK, Bardos DI, Liskova-Kiar M. The advantages of titanium alloy over stainless steel plates for the internal fixation of fractures. J Bone Joint Surg 1981;63-B:427-34.
68. Uhthoff HK, Finnegan M. The effects of metal plates on post-traumatic remodelling and bone mass. J Bone Joint Surg 1983;65B:66-71.
69. Wolff J. Das Gesetz der Transformation der inneren Architektur der Knochen bei pathologischen Veränderungen der äusseren Knochenform. Berliner Akademie der Wissenschaften, Reichsdruckerei, Berlin, 1884.
70. Chidgey L, Chakkalakal D, Blotcky A, Connolly JF. Vascular reorganization and return of rigidity in fracture healing. J Orthop Res 1986;4:173-9.
71. Baumgaertel F, Buhl M, Rahn BA. Fracture healing in biological plate osteosynthesis. Injury 1998;29(Suppl): 3-6.
72. Claes L, Heitemeyer U, Krischak G, Braun H, Hierholzer G. Fixation technique influences osteogenesis of comminuted fractures. Clin Orthop 1999;365: 221-9.
73. Tepic S, Remiger AR, Morikawa K, Predieri M, Perren SM. Strength recovery in fractured sheep tibia treated with a plate or an internal fixator: an experimental study with a two-year follow-up. J Orthop Trauma 1997;11:14-23.
74. Schenk RK, Willenegger H. Zum histologischen Bild der sogenannten Primärheilung der Knochenkompakta nach experimentellen Osteotomien am Hund. Experientia 1963;19:593.
75. Terjesen T, Apalset K. The influence of different degrees of stiffness of fixation plates on experimental bone healing. J Orthop Res 1988;6:293-9.
76. Perren S M, Boitzy A. Cellular differentiation and bone biomechanics during the consolidation of a fracture. Anatomia Clinica 1978;1:13-28.
77. Müller ME, Allgöwer M, Schneider R, Willenegger H. Manual of internal fixation. Berlin: Springer, 1990.
78. Perren SM, Claes L. Biology and biomechanics in fracture management. In: Rüedi T, Murphy WM(ed) AO principles of fracture management. Stuttgart: Thieme, 2000;7-30.
79. Perren SM, Klaue K, Pohler O, Predieri M, Steinemann S, Gautier E. The limited contact dynamic compression plate (LC-DCP). Arch Orthop Trauma Surg 1990;109: 304-310.
80. Vattolo M. Der Einfluss von Rillen in Osteosyntheseplatten auf den Umbau der Kortikalis. MD thesis, University of Bern, 1987.
81. Tepic S, Perren SM. The biomechanics of the PC-Fix internal fixator. Injury 1995;26(Suppl):5-10.
82. Babst R, Hehli M, Regazzoni P. LISS-Traktor, Kombination des "less invasive stabilization systems" (LISS) mit dem AO-Distraktor für distale Femur-und proximale Tibiafrakturen. Unfallchirurg 2001;104:503-5.
83. Krettek C, Gerich T, Miclau T. A minimally invasive medial approach for proximal tibial fractures, Injury 2001;32(Suppl):4-13.
84. Frigg R, Frenk A, Haas NP, Regazzoni P. LCP: The Locking Compression Plate System. AO Dialogue 2001;14:8-9.
85. Jazrawi LM, Bai B, Simon JA, Kummer FJ, Birdzell LT, Koval KJ. A biomechanical comparison of Schuhli nuts or cement-augmented screws for plating of humeral fractures. Clin Orthop 2000;377:235-40.

86. Kolodziej P, Lee FS, Patel A, Kassab SS, Shen KL, Yang KH et al. Biomechanical evaluation of the schuhli nut. Clin Orthop 1998;347:79–85.
87. Simon JA, Dennis MG, Kummer FJ, Koval KJ. Schuhli augmentation of plate and screw fixation for humeral shaft fractures: a laboratory study. J Orthop Trauma 1999;13:196–9.
88. Müller H. Festigkeits-und Elastizitätslehre. Carl Hanser, München, 1970.
89. Perren SM, Mathys R, Pohler O. Implants and materials in fracture fixation. In: Rüedi T, Murphy WM (ed) AO principles of fracture management. Stuttgart: Thieme, 2000;33–42.
90. Perren SM, Pohler OEM, Schneider E. Titanium as implant material for osteosynthesis applications. In: Brunette DM, Tengvall P, Textor M, Thomson P (ed) Titanium in medicine. Berlin: Springer, 2001;771–825.
91. Popov EP. Mechanics of Materials. Englewood, NJ: Prentice-Hall, 1976.
92. Arens S, Schlegel U, Printzen G, Ziegler WJ, Perren SM, Hansis M. Influence of the materials for fixation implants on local infection. An experimental study of steel versus titanium DC-Plates in rabbits. J Bone Joint Surg 1996;78B:647–51.
93. Johansson A, Lindgren JU, Nord CE, Svensson O. Local plate infections in a rabbit model. Injury 1999;30: 587–90.
94. Johansson A, Lindgren JU, Nord CE, Svensson O. Material and design in haematogenous implant-associated infections in a rabbit model. Injury 1999;30:651–7.
95. Burstein AH, Wright TM. Fundamentals of orthopaedic biomechanics. Baltimore: Williams & Wilkins, 1994.
96. Stürmer KM. Elastic plate osteosynthesis, biomechanics, indications and technique in comparison with rigid osteosynthesis. Unfallchirurg 1996;99:816–29.
97. Farouk O, Krettek C, Miclau T, Schandelmaier P, Guy P, Tscherne H. Minimally invasive plate osteosynthesis: does percutaneous plating disrupt femoral blood supply less than the traditional technique? J Orthop Trauma 1999;13:401–6.
98. Ellis T, Bourgeault CA, Kyle RF. Screw position affects dynamic compression plate strain in an in vitro fracture model. J Orthop Trauma 2001;15:333–7.
99. Field JR, Tornkvist H, Hearn TC, Sumner-Smith G, Woodside TD. The influence of screw omission on construction stiffness and bone surface strain in the application of bone plates to cadaveric bone. Injury 1999; 30:591–8.
100. Cordey J, Gautier E. Strain gauges used in the mechanical testing of bones. Part III: Strain analysis, graphic determination of the neutral axis. Injury 1999;30 (Suppl):21–25.
101. Gautier E, Perren SM, Cordey J. Influence of the plate position relative to the bending direction onto the rigidity of a plate osteosynthesis – a theoretical analysis. Injury 2000;31(Suppl 3):14–20.
102. Gautier E, Perren SM, Cordey J. Strain distribution in plated and unplated sheep tibia. An in vivo experiment. Injury 2000;31(Suppl):37–44.
103. Blatter G, Weber BG. Wave plate osteosynthesis as a salvage procedure. Arch Orthop Trauma Surg 1990;109:330–3.
104. Ring D, Jupiter JB, Sanders RA, Quintero J, Santoro VM, Ganz R et al. Complex nonunion of fractures of the femoral shaft treated by wave-plate osteosynthesis. J Bone Joint Surg1997;79B:289–94.
105. Ring D, Jupiter JB, Quintero J, Sanders RA, Marti RK. Atrophic ununited diaphyseal fractures of the humerus with a bony defect: treatment by wave-plate osteosynthesis. J Bone Joint Surg 2000;82B:867–71.
106. Brunner CF, Weber BG. Special techniques in internal fixation. Berlin: Springer, 1982.

V

Applications of Biomechanics Principles to Oncology

35 Malignant Bone Tumors: From Ewing's Sarcoma to Osteosarcoma

D. G. Poitout and J. Favre

Introduction

Primitive malignant bone tumors are rare as they represent less than *1% of all cancers.* Osteosarcoma and Ewing's sarcoma occur the most frequently. They affect, in particular, older children, adolescents, and young adults. For many years these tumors could be controlled locally (often involving an amputation) by radical surgery accompanied or not by radiotherapy, depending on the histological type. Unfortunately, most of the patients died within two years from secondary lesions in the lung.

Only recently has it been possible to improve the survival rate. Indeed, the prognosis of these tumors, until then complicated by the occurrence of pulmonary metastases in almost 90% of cases, has been revolutionized by the advent of "heavy" chemotherapies, capable of eradicating infraclinical metastases. At the same time, the progress made in surgical techniques has enabled the number of limbs saved to be increased and therefore an improvement in the future of these young patients in terms of function. First of all we should stress the importance of close multidisciplinary collaboration, from diagnosis to post-therapeutic monitoring through all the stages of treatment, a collaboration upon which the prognosis of the survival of the patient for whom we are responsible depends.

Epidemiology

Frequency

There are approximately one hundred new cases of osteosarcoma per year in France. The incidence of Ewing's sarcoma is estimated at two to three new cases per year and per million children under 15 years of age in the USA and in the United Kingdom. There is a slight preponderance of males suffering from these two tumors with a gender ratio of approximately 1 : 5. Osteosarcoma can occur at any age but is found mainly in young people with an average age of 17 years.

Ewing's sarcoma most often occurs during the second decade of life with an average age of 11. It is exceptional before five years of age and after 30. Finally, it should be noted that Ewing's sarcoma is very rare in people of black African descent.

Risk Factors

There is often a history of trauma in the weeks preceding the discovery of the tumor, but whether this factor is responsible is still being discussed and it could be no more than a factor which makes clinical discovery more likely. Both types of tumor seem to occur more frequently in tall patients. This, no doubt, is related to the hormone changes of rapid growth.

Osteosarcoma

In the child, osteosarcoma may be accompanied by familial retinoblastoma in 4% of cases. Then the same cytogenetic anomalies are found for bone tumors as for the retinal tumors. One percent of Paget's disease of the bone can degenerate into osteosarcoma. A radiation-induced etiology is also possible for this disease (osteosarcomas have been described in particular after anti-inflammatory radiotherapy for aneurysmal bone cysts). In these two latter cases, adults beyond the quarantine period are most often involved.

Tumors of the ethmoid in people exposed to wood dust are classical, with the most frequent histology being adenocarcinomas, of course, but osteosarcomas can also be present.

Ewing's Sarcoma

Few specific risk factors have been described for Ewing's sarcomas. They may occur on pre-existing benign bone lesions.

Pathological Anatomy

Varieties

Osteosarcomas

Osteosarcomas start in the center of the medulla, most frequently in the metaphysis of the long bones. The lower extremity of the femur, and the upper extremity of the tibia or of the humerus are the most frequent sites (80% of cases). At a microscopic level, these tumors are characterized by osteoid production. Osteoblastic and chondroblastic varieties (characterized by the presence of chondroid tissue) have been described. Due to cell differentiation and the quality of the stroma, distinctions can be drawn between the fibrous osteosarcomas (with an abundance of collagen), telangectasic osteosarcomas (considerable vascularization), and anaplastic osteosarcomas (very non-differentiated). These last two varieties have the worst prognosis. Parosteal sarcoma, a highly ossifying variety of osteosarcoma, has a better prognosis.

Anatomical pathological examination can also pinpoint the fusiform or small cell nature of the tumor (the latter have a poorer prognosis). Some authors also suggest a cytological scale of 1 to 4 in order of increasing malignancy. Indeed, the introduction of intensive chemotherapy has removed the role of these prognostic factors. The histological variety of the tumor no longer affects the course of the disease.

Ewing's Sarcoma

Unlike osteosarcoma, in more than half of cases, Ewing's sarcoma develops in the axial skeleton starting from the pelvic girdle. This fact was first described by J. Ewing in 1921 as a bone tumor with small round cells, distinct from osteosarcoma. The origin of the tumor has long been discussed. An endothelial, neural, mesenchymatous starting point has been suggested. It would currently appear, owing to the immunohistochemical and cytogenetic data, that it is one of the tumors deriving from the neurectodermis, of which it would be the least differentiated form, and of which the neuroepithelioma would be the differentiated form.

From an anatomical pathological point of view, it is a monomorphous proliferation of small round cells with fine chromatin, positive PAS, with a fine network of intercellular reticulin. Immunohistochemistry and, in particular, cytogenetics make it possible to differentiate Ewing's sarcoma from other small round cell tumors in the child and in the adolescent. In more than 80% of cases of Ewing's sarcoma, there is specific translocation (11, 22 q24; q12). In the absence of specific translocation, chromosome 22 derives from translocation in 9% of cases.

Spread

Osteosarcomas

It is customary to say that "osteosarcoma is a lung disease starting in the bone". This brief description straight away emphasizes the early and extreme frequency of the pulmonary micrometastases which do, of course, determine the prognosis in terms of survival. Locally, the tumor develops centrifugally, invading and destroying the normal bone up to the cortex and to the adjacent soft tissue. There is often a pseudo-capsule which really consists of inflammatory tissue at the interface between the tumor and normal tissue.

The very frequent presence of satellite tumoral nodules situated on the same bone as the main tumor but not continuous with it has also been described. These are, in fact, "local metastases" which have developed from tumoral emboli via the medullary sinuses. These have to be taken into account when determining the extent of the surgical excision. Ganglionic involvement is unusual for this histological type since it only appears in less than 10% of cases in autopsy series. On the other hand, hematogenic metastases are extremely frequent and develop at an early stage. They occur predominantly in the lungs, and it is estimated that approximately three-quarters of patients have microscopic pulmonary metastases at the time of diagnosis.

The second site for metastases in order of frequency is the bone but it is rare that bone metastases are present if earlier pulmonary lesions are absent. Other types of metastases only occur very exceptionally.

Ewing's Sarcoma

The local development of Ewing's sarcoma is slightly different. There is often complete invasion of the medullary cavity of the bone concerned as well as involvement of the cortex and of the soft tissue. Ganglionic metastases seem to be a little more frequent than for osteosarcomas, but here, too, it is the hematogenous metastases which tend to occur with a frequency of 15–40% at the time of diagnosis. Pulmonary secondaries still arrive there first, followed by bone metastases and invasion of the medulla.

Diagnosis

Circumstances in which it is Discovered

Pain is often the first symptom; it may be intense and accompanied by serious functional infirmity or, on the other hand, not intense and only felt if pressure is applied. Sometimes there is a nocturnal recrudescence. This pain, caused by intraosseous hypertension, is considerable in the forms which start centrally. It may also be accompanied by limping in lesions of the leg.

Its site varies depending on the type of tumor:

In the case of osteosarcoma, it is most frequently found in the metaphysis of the long bones, "close to the knee and far from the elbow". Lower extremity of the femur (50%), upper extremity of the tibia (15%), upper extremity of the humerus (15%).

Ewing's sarcoma tends to develop in the axial skeleton and in shoulder and pelvic girdles. Involvement of the flat bones, although exceptional for osteosarcomas, is very common in Ewing's sarcoma.

The tumor can also be palpated. A careful examination shows that it is one with the bone, with the neighboring joints rarely being affected. It varies in consistency, being hard, firm, or sometimes softer. The tumor may be pulsating. Classically there may be crepitus on pressure but this sign is, in fact, rare. The tumor increases rapidly in size, deforming the bone concerned; the skin has a smooth and stretched appearance, with dilated veins.

The disease may also be manifested by a pseudo-infectious syndrome. Indeed, hemorrhaging and intra-tumoral necrosis may make the tumor fluctuant, with an increase in temperature locally and cutaneous erythema, mimicking an abscess.

The diagnosis may also be made in the presence of a spontaneous fracture or secondary to minimal trauma.

When the tumor is large, the mass can lead to vascular, nervous, and medullary compression of a hollow organ; this rarely reveals the disease except for pelvic or vertebral lesions.

Finally, the diagnosis may be made when prevalent, mainly pulmonary, metastases appear or even when the patient's general condition deteriorates and may be accompanied by fever.

The Components Involved in the Diagnosis

History-taking and Clinical Examination

The person taking the history will look for any risk factors, and the existence of functional or general signs and their development over time. The examination will establish the extent of the lesion, its clinical appearance and the presence of complications. Ganglionic metastases are rare and secondary visceral lesions are usually asymptomatic.

Biology

There are no specific markers of these tumors. At most there may be an inflammatory syndrome without any specific features. Ewing's syndrome may be accompanied by an increase in the LDHs.

Radiological Examinations

X-rays without preparation classically show different images according to the histological type but only an anatomical–pathological study will provide diagnostic proof. The osteosarcomas may be non-ossifying. The X-ray picture then displays non-homogeneous lysis and is poorly defined. On the surface, there is sub-periosteal osteophytosis forming spicules and it is responsible for a "sun ray" appearance. Ossifying osteosarcomas give the same radiological signs, accompanied by intra-tumoral opacities which are sometimes very dense. These ossification images may also exist in extraosseous tumoral zones when the soft tissues are invaded.

The radiological signs of Ewing's sarcoma, such as simple gumming of the bone framework, are sometimes very difficult to see. The typical appearance is that of osteolysis bordered by a considerable periosteal reaction, resulting in the so-called "onion-skin appearance".

Specific incidences may be useful for sites such as the sacrum and the vertebrae.

Whatever type of tumor is involved, bone scintigraphy using Technetium 99 will reveal intense hyperfixation.

Computerized tomography has to be carried out, adhering to certain quality criteria and dual window examination to allow a detailed examination of the bone and soft tissue, before and after injection of the contrast medium. The lesion is irregular, poorly defined, invading the adjacent soft tissue with *often* heterogeneous uptake of the contrast medium.

MRI, used in addition to the scanner and allowing a tri-dimensional study of the tumor, seems to be more successful in evaluating whether it has spread to the soft tissue and to the bone medulla, and establishes its relationship with the vasculo-nervous bundles.

Angiography reveals anarchical vascularization of the tumoral mass. It is particularly useful for establishing the lesion's relationship to the large vessels and for guiding the surgeon in his maneuvers.

Pathological Anatomy

Only an anatomical–pathological examination will make it possible to make a formal diagnosis. Therefore, a biopsy has to be performed as a matter of surgical urgency but it does involve some risk of hematogenic dissemination and local release of growth factors. It involves a real surgical operation and therefore has to be performed by a qualified surgeon who is a member of the multidisciplinary team which will be taking care of the patient. This procedure has to performed under general anesthetic, with a tourniquet at the base of the limb and before any treatment. The technique has to be performed correctly, with surgical opening of the tumor to enable removal accompanied by rigorous hemostasis and careful suturing. Needle or trocar biopsies are only indicated for sites which are difficult to access such as the vertebral bodies. The surgical approach and course of the drains will ultimately have to be removed with the tumor in order to limit the risk of a local relapse.

Differential Diagnosis

There is more than one infection-related problem which may be particularly reminiscent of Ewing's sarcomas, all the other benign bone tumors will also have to be eliminated (osteoma, chondroma, fibroma, angioma, giant-cell tumor, solitary or aneurysmal cyst. But it is, in particular, the other malignant bone tumors which can sometimes pose a problem in differential diagnosis: chondrosarcoma (the prognosis of which is poor because at present there is no treatment apart from radical surgical excision), Parker and Jackson's lymphoma (which really belong to the group of non-Hodgkin's malignant lymphoma and therefore must be treated as such), fibrosarcoma and angiosarcoma (forming part of the group of sarcoma of the soft tissue), and adamantinoma of the long bones (a variety of tumor which does not metastasize much but has a marked tendency to recur locally).

Although the various X-ray examinations may point in the right direction, only a biopsy together with a cytogenetic study allows a reliable diagnosis to be made and therefore to guide the treatment.

Pre-therapeutic Inventory

Inventory

Loco-regional

Clinical examination only allows a rough evaluation of the tumor. Standard X-rays underestimate the extent of the lesion, not only in the bone but also its spread to the soft tissue. Bone scintigraphy is sometimes more precise; it also reveals any satellite tumoral nodules. However, computerized tomography and magnetic resonance imaging in particular clarify the extent of the tumor. Both inside and outside the bone these two examinations are useful because they often provide additional information. The aim of other investigations, and in particular angiography, is to provide us with information on the vascularization of the lesion and on how it relates to the main vessels, therefore enabling the surgeon to select the most suitable operating technique.

Metastatic Spread

There is no point in systematically searching for ganglionic metastases because they rarely occur. On the other hand, secondary pulmonary and bone lesions must be searched for. The standard chest X-ray is insufficient, and a CT scan therefore has to be performed in order to diagnose small lesions. Bone scintigraphy with Technetium allows any bone metastases to be found. For Ewing's sarcomas, myelograms are carried out systematically to look for any medullary invasion. Other visceral lesions only occur in exceptional cases and therefore do not require systematic paraclinical examination.

Classification

The TNM classification is not suitable for primitive malignant bone tumors. At present there is no satisfactory prognostic classification which can be used routinely. However, it is possible to define a number of prognostic factors:

the size of the tumor linked to the prognosis, with a worse prognosis for tumors 10 cm in diameter or larger.

Soft-tissue involvement would also have a worse prognosis but this is so frequent that such a criterion would be of little value.

Osteosarcomas secondary to Paget's disease or radiation-induced have a less favorable course than others.

For Ewing's sarcomas the LDH levels would have a prognostic value.

However, the main factor determining survival is, of course, the response to the treatment.

Treatment

Method

Surgery

Radical Surgery

This is, in fact, amputation of a limb. This has to be performed leaving a safety margin of a few centimeters. Sometimes, in cases where the proximal part of the limb is affected, the only possible procedure is the complete disarticulation of the limb concerned, which of course makes the functional prognosis even worse. Nowadays, with the progress made in surgical techniques and the contribution made by induction chemotherapy, the indications for these methods are becoming fewer and fewer.

Conservative Surgery

This technique has been used successfully in 40–80% of cases depending on the series. Its success depends on close interdisciplinary coordination and strict pre-therapeutic evaluation. This surgery takes place in three stages:

resection of the tumor has to adhere strictly to the rules of oncological surgery in order to limit the risks of a relapse: excision of the whole of the lesion takes place in a single piece with a safety margin of 6–7 cm above the scanographical limit of the involvement in the bone. This resection also includes the muscles adjacent to the extraosseous spread. The part operated on has to include the earlier biopsy sites as well as all the potentially contaminated tissues. The joint and the adjacent articular capsule are also resected.

It is then necessary to reconstruct the bone defect, the average length of which often reaches 15–20 cm. Three main methods can be used:

the implantation of a prosthesis,

bone "graft" from a bone bank,

arthrodesis.

The choice of technique is decided on a case-by-case basis by the surgeon. The operation ends by transposing the muscles and soft tissue. Indeed, adequate muscle and skin cover of the operating area significantly reduces postoperative morbidity.

Radiotherapy

The indications for radiotherapy vary considerably according to the histological type. Osteosarcomas are tumors which respond very little to radiotherapy. The doses needed to obtain an acceptable level of sterilization are greater than or equal to 80 Gy and are the source of major complications. This is why radiotherapy is only used in totally exceptional or palliative cases.

On the other hand, 80–90% of cases of Ewing's sarcoma can be cured by radiotherapy if a dose of 60–65 Gy can be delivered to the whole of the primary tumor. Lower doses (40–45 Gy) are adequate for sterilizing micrometastases or very small postoperative residual lesions.

A rigorous technique with anticipatory dosimetry is necessary in order to limit the risk of complications.

Chemotherapy

There is no doubt that, by enabling the disease in general to be treated, chemotherapy has improved the prognosis of these tumors. Indeed, before this treatment was available, the vast majority of these patients (80–90%) died from metastases in the two years following satisfactory local treatment.

The methods are to be adapted according to the following histological types.

Osteosarcoma

The products used in the treatment of osteosarcomas are the following:

high-dose methotrexate (MTX HD): combines hyperhydration with alkalination followed by injections of folinic acid. At standard doses, methotrexate is not at all effective. Very high doses (of the order of 8–10 g/m^2) are necessary. These doses are potentially lethal or likely to lead to major and, in particular, renal complications. Ideally, "drugs have to be prescribed at plasma doses making it possible to adapt their dosages". In all cases, very strict rules of hydration and alkalination have to be adhered to.

"BCD" combination (bleomycin) cyclophosphamide, actinomycin D) – adriamycin and other anthracyclines.

Iphosphamide, in association with uromitexan and sufficient hydration to limit the risk of hemorrhaging.

CDDP.

Ewing's Sarcoma

Drugs which are effective in the treatment of Ewing's sarcomas are:

anthracyclines,

VAC-type combinations (vincristine, actinomycin O, endoxan),

Bleomycin,

Methotrexate,

IVA combinations (Iphosphamide, vincristine, actinomycin), IVAD combinations (iphosphamide, vincristine, adriamycin). This type of protocol is nevertheless responsible for considerable toxicity which limits the indications for it.

Indications

Osteosarcomas

The first phase of the treatment consists of preoperative chemotherapy, starting as soon as the result of the biopsy is obtained, and consists of courses of treatment with high-dose methotrexate. The patient is then operated on. Whenever possible, conservative surgery is performed. Nevertheless, there are certain contra-indications to this maneuver: major neuromuscular invasion, pathological fracture, poor diagnostic biopsy technique, local infection.

The subsequent chemotherapy methods are decided according to the response to the treatment, evaluated on clinical, radiographic, and particularly anatomopathological criteria. When the response is good, the same type of treatment is continued. If not, the prescription is changed to other potentially effective drugs. Chemotherapy lasts approximately nine months in all. Any delay in the progress of the treatment has to be avoided as much as possible, in particular by prescribing hematopoietic growth factors.

Special Case

Pulmonary metastases, whether they appear first or particularly late, can justify an aggressive approach if the primary tumor is under control: heavy chemotherapy followed by metastasectomy and resuming chemotherapy according to the response to the treatment. This is often necessary if survival is to be prolonged.

Tumors which cannot be eradicated; the treatment combines chemotherapy with irradiation, the dose of which takes account of the adjacent critical organs (50–75 Gy according to the case).

Tumors in elderly subjects are often osteosarcomas for which a risk factor is found (Paget's disease, irradiation, professional exposure, etc.). The treatment, in particular chemotherapy, has to be adjusted according to age, the patient's general condition and predisposition.

Ewing's Sarcoma

In terms of the treatment plan, the treatment resembles that for osteosarcoma: the local treatment is incorporated into the overall strategy, which starts with induction chemotherapy, generally including an anthracycline and a VAC-type combination administered with the dual aim of reducing the tumoral mass and eradicating the micrometastases.

Surgery and radiotherapy are generally accompanied by local treatment. Indeed, surgery alone is often not enough to guarantee satisfactory local control. On the other hand, irradiation alone requires doses of the order of 60–65 Gy, which is sometimes a source of major complications, particularly in children on account of the growth problems that it causes. Furthermore, the level of local recurrence after irradiation remains high (approximately 20%) in spite of chemotherapy. These considerations lead to proposing a combination of radiotherapy and surgery in a number of cases.

Local treatment is followed by a resumption of chemotherapy. Other chemotherapy protocols based on taking cyclophosphamide orally together with other active drugs make it possible to obtain equivalent survival at the price of often less toxicity. The place of intensive chemotherapy with bone marrow autografts is not clearly defined. Metastatic Ewing's sarcomas are caused by systemic chemotherapy. There are fewer indications for metastasectomy than there are for the osteosarcomas. In cases of complete remission, the subsequent therapeutic strategy is the same as in non-metastatic patients.

Course and Monitoring

Osteosarcoma

Nowadays, owing to chemotherapy, the survival rate from osteosarcomas is approximately 50% at five years. Most of the relapses occur within two years but just as survival is being prolonged, there are also cases of late relapses sometimes after eight to ten years.

Metastatic tumors are *mostly fatal* even if prolonged remissions can sometimes be obtained. Monitoring therefore has to be particularly close in the first years but also has to be prolonged. Apart from a careful clinical examination, it also has to include X-rays and a CT scan or MRI examination of the affected region as well as bone scintigraphy and a chest scan to detect secondary lesions.

Ewing's Sarcoma

In spite of all these improvements in therapeutic techniques, the overall results have hardly progressed these last few years. Survival without a relapse is estimated at 60% at three years and overall survival at 70% at three years. The course can be complicated by the occurrence of an osteosarcoma in the area irradiated. In metastatic patients survival rates of more than 50% have been reported after remission and rigorous loco-regional treatment. The practical methods of chemical and paraclinical monitoring are comparable to those for osteosarcoma.

Conclusion

Primitive bone tumors are rare conditions but mainly affect young people. Up until the 1970s, their prognosis was complicated by the appearance of metastases in the vast majority of cases, but they have largely benefited from advances in chemotherapy. However, only close multidisciplinary collaboration makes it possible to define a suitable therapeutic strategy for each patient, to optimize the diagnosis, the treatment, and the subsequent monitoring and hence to increase the chances of a cure.

Therapeutic Orientation in Osteogenic Sarcomas

For a long time the treatment of osteogenic sarcoma has been concerned solely with the local bone lesion which has either been immediately surgically excised, or has been irradiated following an amputation which has been delayed in principle or due to necessity. A successful outcome at five years of 20% is equivalent with these two methods of treatment and, from the functional point of view, although local irradiation allows unnecessary mutilations to be avoided in subjects who are going to die due to the development of pulmonary metastases, it should be recognized that its insufficiencies and sequelae lead, except in rare exceptional cases, to patients who have been cured nevertheless having to undergo an amputation at a later stage.

A better understanding of the factors which affect the prognosis of osteogenic sarcomas has to guide the treatment. From this point of view, the problem of pulmonary metastases dominates that of the bone lesion, and osteogenic sarcoma could be considered to be a lung disorder, in the history of which there is a bone lesion, since death is due to the development of the metastases even though the bone lesion has been removed or could be if its existence had not become contingent from the prognostic point of view.

Lung metastases are present in 98% of patients followed up until their death. The date they appear is known exactly due to the observations made in important series:

85% appear in the first two years,

after eighteen months only 30% of the patients are clear.

But it is necessary to distinguish the development of the metastases linked to an anatomical development which makes it possible to diagnose them and their much earlier presence since the calculation, according to the time of duplication, leads us to think that they already exist in 60–80% of cases at the time when the bone lesion is diagnosed. This idea of an imperceptible disease, upon which the prognosis ultimately depends, is the main element in deciding which of the current treatments to adopt. It allows us to target the treatment in a logical and effective way since it is here that the right or wrong decision is taken. All the more so as many arguments lead us to believe that these lung metastases always occur in osteogenic sarcoma and that even the cures seen with traditional treatments limited to local action on the bone lesion, are only due to the "failure" of these metastases because of the spontaneous role of the natural immune defenses.

The role played by immune reactions, which Tavernier suspected without naming them, is based today on the knowledge acquired in the area of virology, cyto-carcinogenesis, and immunology. We know that osteogenic sarcomas, and certainly other conjunctive sarcomas, are viral in origin (but also that other associated conditions are necessary for oncogenesis). We know that it is the integration or transcription of the viral genome in the tumor cells which is the foreign antigenic element responsible for immune responses involving rejection, due to the involvement of specific tumor or cell-mediated antibodies. We also know that the cytolytic potency of these immune reactions is weak, without possible action on a large bone tumor or large metastasis but that it is perhaps decisive at the stage when lung metastases cannot be seen.

After a long, disappointing time, chemotherapy for osteogenic sarcomas seems to have entered a phase of spectacular progress owing to the products used (high-dose methotrexate, adriamycin).

36 Long Bone Metastasis

B. T. Allende and B. L. Allende

Introduction

Malignant metastatic tumors are the most common bone neoplasms. The skeleton is often affected by metastasic cancer, and the discovery of a long-bone metastasis may be the first symptom of the primary disease. Nearly every malignant neoplasm has been described as having the capability to metastasize to bone; tumors of breast, prostate, thyroid, lung, and kidney are the most common bone-seeking malignant lesions, and between 50 and 85% of affected patients will have bone metastasis at one point of their disease [1,2,3]. Breast and prostate cancer alone account for more than 80% of metastasic bone disease. The capacity of the neoplastic cells to invade bone is related to the histology and to the aggressiveness of the primary tumor. The axial skeleton is where the red marrow is situated in adults, and this pattern of distribution suggests that physical properties of the circulation within the bone marrow could assist in the development of bone metastasis. The outcome of metastasis depends on multiple interactions of metastatic cells with homeostatic mechanisms which the tumor cells usurp. The dissemination of cancer cells to vital organs results in eventual multisystem failure and death.

Pathologic fractures are a relatively late complication of bone involvement. Prostate metastases are usually sclerotic lesions, and as such pathologic fractures are less common than litic breast metastases. Pathologic fractures are devastating complications of metastatic disease; these lesions are associated with considerable morbidity, which includes: unbearable pain, impaired mobility, pathologic fractures, and bone marrow infiltration.

With the advent of improved medical therapies for many types of cancer, life expectancy has dramatically improved in the last two decades, making the orthopedic surgeon more often involved with patient care. This increased patient survival has also driven us to a more aggressive management of patients with bone metastasis. It is estimated that 40% of patients with pathologic fractures survive for at least six months after their fracture, and 30% survive for more than one year.

Patient median survival after the first recurrence of carcinoma in patients without extraosseous sites at the time of diagnosis is approximately 20 months, and it varies according to the primary cancer type: breast 25, prostate 40, thyroid 48, kidney 12, and lung 4 months. Bone metastasis refractory to radiation and chemotherapy has a shorter life expectancy.

Metastasic destruction of bone reduces its load-bearing capabilities, resulting initially in trabecular disruption and microfractures, and subsequently in total loss of bone integrity and pathologic fracture. Clinical trial with bisphosphonates [4] have proved to inhibit tumor-induced bone resorption, correct hypercalcemia, reduce pain, and diminish the development of new osteolitic lesions and fractures; all leading to potential improvements in quality of life. Even though clinical trials with bisphosphonates in patients with multiple myeloma and breast cancer seem promising, no long-term efficacy of these drugs has yet been proved.

The probability of developing a pathological fracture increases with the duration of the disease, and as such they tend to occur more often in tumors with relatively good prognosis. Pathologic fractures have been reported to occur in 9–29% of patients who have bone metastasis. The indications for prophylactic fixation have not been standardized, and will vary according to the size of the lesion, the anatomic location, the bone involved (weight-bearing vs. non-weight-bearing long bones), roentgenographic appearance, and the primary tumor (lung and kidney have a higher incidence of fracture than prostate and breast). There are no biomechanical studies that relate the size and shape of bone defects to the reduction in structural strength in pathologic bone. The most common indications for fixation on impending pathological fractures are increasing pain and cortical destruction (pain is thought to be due to microfractures or to stretching of the periosteum by the increasing tumor size). Sim [2] recommends prophylactic fixation in well-defined lytic lesions that are more than 3 cm in diameter and in which 50% or more of the cortex is destroyed. He also considers pain refractory to radiotherapy an indication for fixation. Mirels [5] developed a scoring system for assessing the risks of pathologic fractures in long bones which takes into account: lesion type, size, site, and associated pain. Even though there is no accurate way to predict the risk of fracture; we have found this scoring system useful. With better and earlier diagnosis and with advances and less aggressive surgical techniques and procedures of internal fixation, we are recommending prophylactic fixation in more cases, with smaller lytic lesions. It is technically easier, and with less morbidity, to stabilize a pathologic bone before it fractures.

The goal of treatment is to prevent bone fracture, and to provide immediate and definitive pain-free usage of the limb so as to preserve function and to improve quality of life. We have to aim at bringing back the patient to the prefracture state; early mobilization and ambulation will ease nursing care, and will shorten hospital stay (realizing life expectancy is short). Relief of pain is also an important goal. Achieving these goals will cause a positive psychological and emotional effect on the patient and their families. This stimulates a more aggressive approach to the management of these patients.

The management of long-bone metastasis varies according to the patient's general conditions, the location of the lesion, the extent of bone compromise, and the primary bone tumor. A complete medical examination in order to evaluate the patient's general condition is very important before deciding the definitive management; surgery in general is safe in this group of seriously ill patients. The location of the metastasis, the extent of bone destruction, and the histologic appearance are also important considerations, and will help decide the type of fixation to be used. Since the estimated length of survival can not be accurately established in this lesion, we do not believe it should be taken into account in deciding the treatment; most of these patients will benefit from surgical treatment.

Biomechanical Considerations

Recognized biomechanical principles used for fracture reduction and stabilization in normal bone do not apply to pathological fractures. In metastatic fractures due to the slow healing capacity of bone, the implants will have to assume a load-bearing role for an extended period of time; this will lead to implant failure if conventional implant fixation techniques are used, consequently different biomechanical guidelines should be applied in the treatment of pathological fractures.

Invasion of metastatic cells in long bones will alter the mechanical properties of bone tissue, reducing the ability of the involved bone to carry ordinary functional loads, which will end in failure and fracture. Bone weakness caused by metastatic lesions is generally more extensive than is evident in X-rays, and will decrease screw or intramedullary nail fixation purchase and strength. Fifty percent of the mineral phase must be resorbed before change is evident in X-rays.

The surgical management should provide enough stability to allow immediate, full weight bearing. We have to keep in mind that the treatment of these tumors is palliative and not curative (oncologic). Surgical techniques for stabilizing pathologic or impending fractures must be individualized for the area of involvement, the quality of bone, and the potential for involvement of adjacent soft-tissue structures. Even though the biomechanical properties are different, implant devices used for pathological fracture fixation are similar to those used in regular fracture management, and include open reduction and internal fixation with plates and screws or endomedullary nails, arthroplasty, allograft plus internal fixation, tumor resection plus segmental bone spacers, and amputations. Fixation devices must be chosen with the understanding that bone union will be delayed or never achieved; in addition, most of these patients are not strong enough to restrain themselves from weight bearing. Due to these two factors the implant should be sufficiently strong to give the patient a definitive, functional, painless limb. The implant chosen will be the one that provides the best fracture stability under the individual circumstance. Ninety-six percent of patients will achieve a good or excellent pain control after adequate fixation of long-bone metastasis [6].

Since allograft replacements have yielded poor results, in cases of tumor patients where immediate, pain-free weight bearing is the goal of treatment, this treatment modality has been abandoned for long-bone metastasis.

Numerous authors [2,7] have proved the advantages of using methylmethacrylate to augment the fixation. The technique recommended for diaphysis plating augmented with cement, consists of:

Tumor curettage.

Cement insertion.

Wait for the cement to harden.

Drill and tap through the cement as in normal bone.

We do not recommend waiting until cement hardening in cases of insertion of a compression screw (DHS or DCS) nor in cases where an endomedullary nail has to be introduced. Care should be taken to avoid interpositioning cement between fragments in order to allow fracture healing (Figure 36.1).

The use of methylmethacrylate to reconstitute large bone defects permits most patients to bear weight immediately. Methylmethacrylate maintains excellent rigidity, especially when compressive load is applied, and is not adversely affected by radiation or by antibiotics. Since the infection rate in these patients is high due to postoperative radiation or to the non-optimal medical general condition, some authors advocate the use of cement impregnated with antibiotics. Even though the effectiveness of the insertion of chemotherapeutic agents into methylmethacrylate is still to be defined, the addition of an antiblastic drug to the acrylic cement may provide sterilization of the residual tumor cells by the local slow and prolonged release of the drug. This technique, along with current anti-mitotic therapy, may provide better local control of the neoplasic lesion.

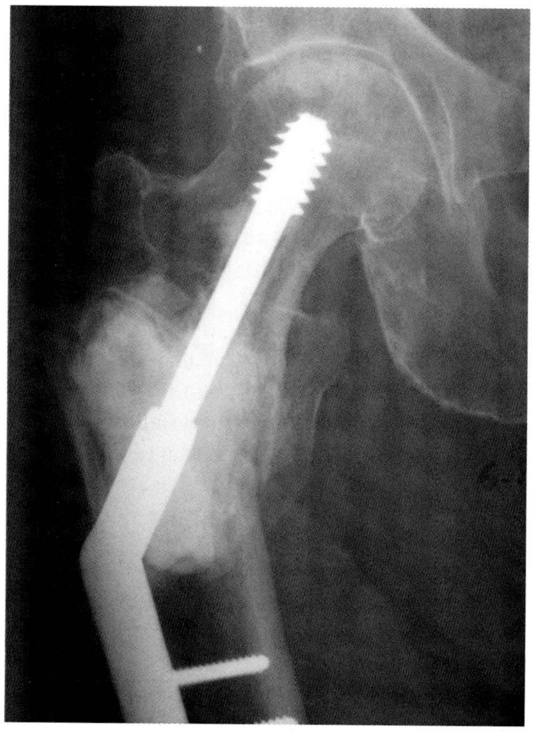

Figure 36.1. Internal fixation augmented with methylmetacrilate, one year postoperative; breast cancer.

Wedin et al. [8], in a retrospective study of 228 metastatic lesions of long bones treated by resection and reconstruction, intralesional curettage, or stabilization only, found a local failure rate of 11%, and the median time to failure was eight months. Local failure was more common in lesion of the femur diaphysis and distal femur (20%), and in kidney cancer (24%). Osteosynthesis had a failure rate of 14% while endoprosthesis only 2% (in proximal femur lesions). The most important risk factor for failure was long survival rate.

We favor postoperative rather than preoperative radiation therapy whenever it is likely that internal fixation will be needed. Postoperative radiotherapy will decrease the risk of local tumor progression but adversely influences union in the absence of rigid internal fixation [9].

Even though surgical stabilization of these lesions is usually indicated, it is not without risks. Thromboembolic complications as well as wound infection risks are higher in these patients. Local as well as systemic dissemination of tumor cells after internal fixation has been proved by Peltier et al. [10]. The importance of the spread of tumor cells along the tract of an intramedullary rod has not outweighed the advantages of this operation. Patients who have a highly vascular lesion, such as metastatic renal carcinoma (hypernephroma), may present a unique problem of uncontrollable bleeding, and should be treated with arterial embolization before surgery to decrease intraoperative complications.

Nonunion occurs in a high proportion of patients who undergo conventional doses of radiation, but most pathological fractures fixed with adequate internal fixation will heal despite adjuvant radiotherapy.

Amputation has few indications in bone metastasis; tumor fungation and infection (common in neglected primary bone tumors) as well as arterial occlusion are rare findings. Amputation is indicated in patients in whom surgical and radiotherapy treatments have failed to alleviate the pain in a functionless extremity, as well as in tumors with significant soft tissue or neurovascular involvement.

Femur

The femur is the most frequently affected long bone by metastasis. Approximately 50% of these lesions affect the proximal femur [11]. The aim of treatment is to obtain a painless limb, and to return the patient to his premorbid state. The decision whether to operate or not depends on the tumor type, location, size, and the patient's general condition. Non-operative treatment for pathological or impending pathological femur fractures will be recommended only in patients who are not expected to survive a surgical insult, and treatment will consist of radiation, pain medication, and immobilization.

Surgical treatment is indicated in most pathological or impending femur fractures. It should provide stable definitive fixation, and should allow immediate weight bearing. Unlike primary lesions, in metastatic lesions a complete tumor resection is not necessary.

Femoral Head and Neck

Due to the high stresses on the proximal femur as well as the low healing potential, replacement arthroplasty is usually recommended [7,11,12,13,14]. Careful examination of the acetabulum should be done in these patients; the choice of hemiarthroplasty versus THA will be given by the involvement of the acetabulum. Commonly, a long-stem femoral component will be used to prophylactically reinforce the remaining proximal femur. Arthroplasties are always cemented. Even though prosthetic complications after tumor resections include a higher incidence of infections and dislocations, reconstruction based on endoprosthetic replacements as opposed to osteosynthetic devices is safer, and is associated with pronounced better results than osteosynthetic devices. Endoprostheses are not dependent in fracture healing, which is often poor in patients with cancer because of systemic and local factors.

Internal fixation in this location has resulted in high failure rates (14–40%), and so is usually avoided; common reasons for failure include: poor initial fixation, improper implant selection, and progression of the disease within the operative field [8,15] (Figure 36.2).

Intertrochanteric Fractures

The treatment varies according to the involvement of the medial cortical bone (lytic lesions that occur in the region of the lesser trochanter are particularly prone to fracture); if this is nil

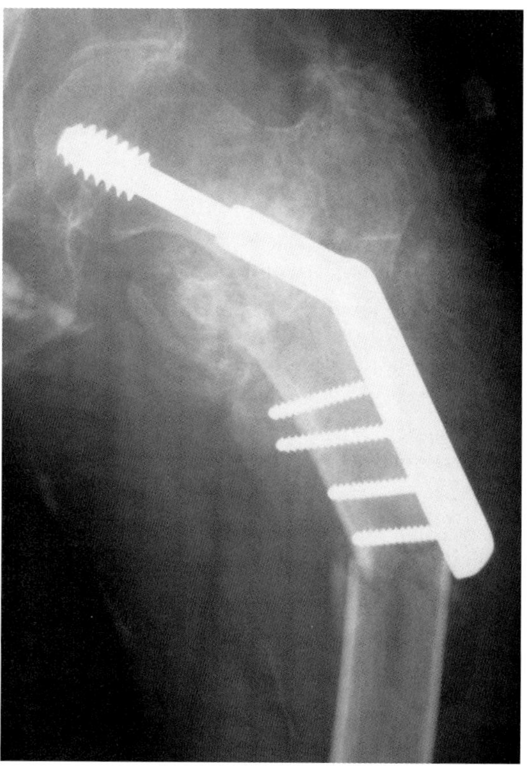

Figure 36.2. Peri-implant fracture in pathologic bone.

or minimal we prefer ORIF with a compression screw or 95° blade plate, always augmented with cement. In order to prevent the screw cutting out of the femoral head, a lateral window should be opened in the lateral cortex, metastatic bone removed from the neck of the femur, liquid cement introduced in the defect, and the compression screw placed before the cement hardens. The same technique is done when the screws are placed in the proximal femur diaphysis. From a biomechanical point of view, cephalomedullary (load-sharing) devices have proved to have some advantages over DHS since a more medial placement closer to the compression side of the femur and away from the lateral tension side is achieved; also, these implants will prophylactically stabilize the rest of the bone, making it more durable.

If the calcar is involved significantly by bone metastasis, or in cases in which internal fixation has failed, or will not achieve a stable fixation due to tumor extension, a cemented, long-stem calcar replacement prosthesis is recommended.

Subtrochanteric Fractures

Cephalomedullary devices (long gamma nails) are the treatment of choice for most of these lesions. Cement augmentation is usually not necessary since the biomechanical properties of these implants will give stable internal fixation that will allow immediate weight bearing. These implants will also address the whole bone in case intracompartment spread of the tumor occurs. The largest nail possible should be used to fill the femoral canal. When the proximal femur has been weakened in such a way by metastatic disease that a reconstruction nail is unable to provide stable fixation, or when previous fixation has failed, a cemented megaprosthesis is indicated.

Diaphyseal Fractures

The use of 4.5 mm dynamic compression plates with cement augmentation for these injuries has practically been abandoned because of the good results obtained with conventional closed, and locked intramedullary nailing (Figure 36.3). The surgical technique is less invasive and, as previously stated, these implants will protect the long bone in case bone disease progresses. Cephalomedullary devices [19] are recommended when tumor progression is suspected (Figure 36.4). Osteosynthetic implants are load sharing and ultimately will fail if the bone does not heal.

Supracondylar Fractures

This is a less-frequent site of femur metastasis, and the most technically demanding to treat. These are the fractures with the highest reported failure rate postoperatively [8,15]. If there is sufficient bone stock, conventional dynamic compression screw/plate devices (DCS) augmented with cement are usually recommended. Distal femoral replacement arthroplasty is also a viable option in lesions with massive destruction of femoral condyles.

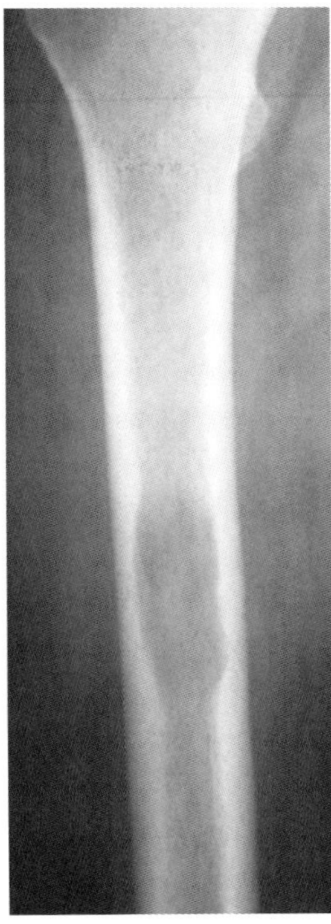

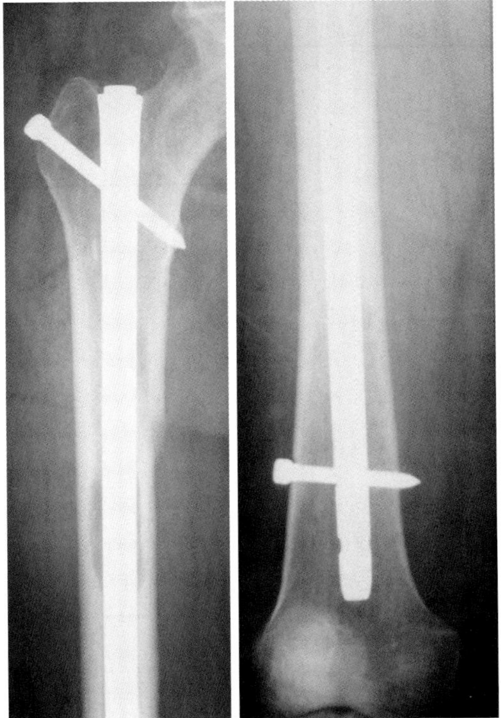

Figure 36.3. **a** Impending pathologic femur fracture. **b** Preventive reamed, thick, locked intramedullary nail.

Humerus

The humerus is the second most common long bone affected by tumor metastasis. Complete fracture is more common in the humerus than in the femur or tibia; since the humerus is a non-weight-bearing bone, the patient does not experience load-related pain suggestive of fatigue before the bone actually fractures.

The treatment of these lesions should aim to restore the patient's function and to relieve pain. Conservative treatment of pathological or impending pathological fractures of the humerus has a more active role than it does in weight-bearing bones. Even though each patient has to be individually planned, radiation therapy plus bracing will relieve pain, and might serve as definitive treatment in some terminally ill patients [2]. Overall analysis has shown few satisfactory results in patients treated nonoperatively for humeral fractures [17].

When bone destruction is so extensive that stability can not be achieved by other methods, surgical treatment is indicated.

Proximal Humerus

Metastasis can be treated conservatively; radiotherapy usually achieves a good degree of pain control and an acceptable degree of function. In cases where conservative treatment has failed, shoulder resection arthroplasty is recommended. Although the range of motion is usually diminished, endoprosthetic replacement of the proximal humerus provides stability and a predictable and reliable method of reconstruction of the upper limb.

Diaphysis

If the patient's general condition allows, most diaphysis pathological fractures should be stabilized, particularly in patients who need their upper extremities for ambulation (crutches, cane). The implant of choice is an antegrade or retrograde locked humeral nail (depending on the fracture site and surgeon's preference). Even though augmentation with cement is usually not necessary, the indication of an open tech-

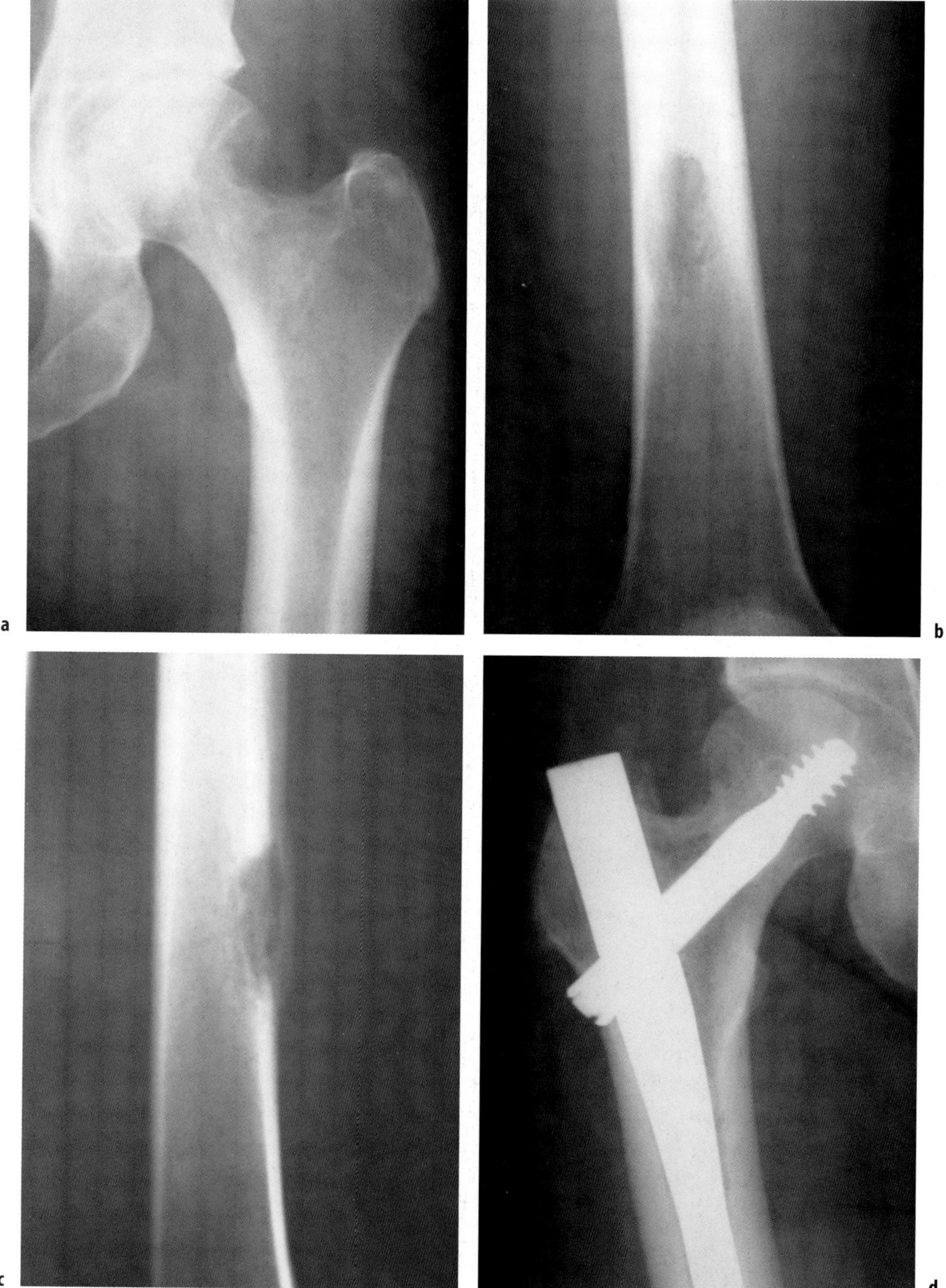

Figure 36.4. **a** Impending pathologic femoral neck fracture. **b** mid femur diaphysis. **c** distal femur diaphysis. **d** Preventive cephalomedullary device; 18 months postoperative.

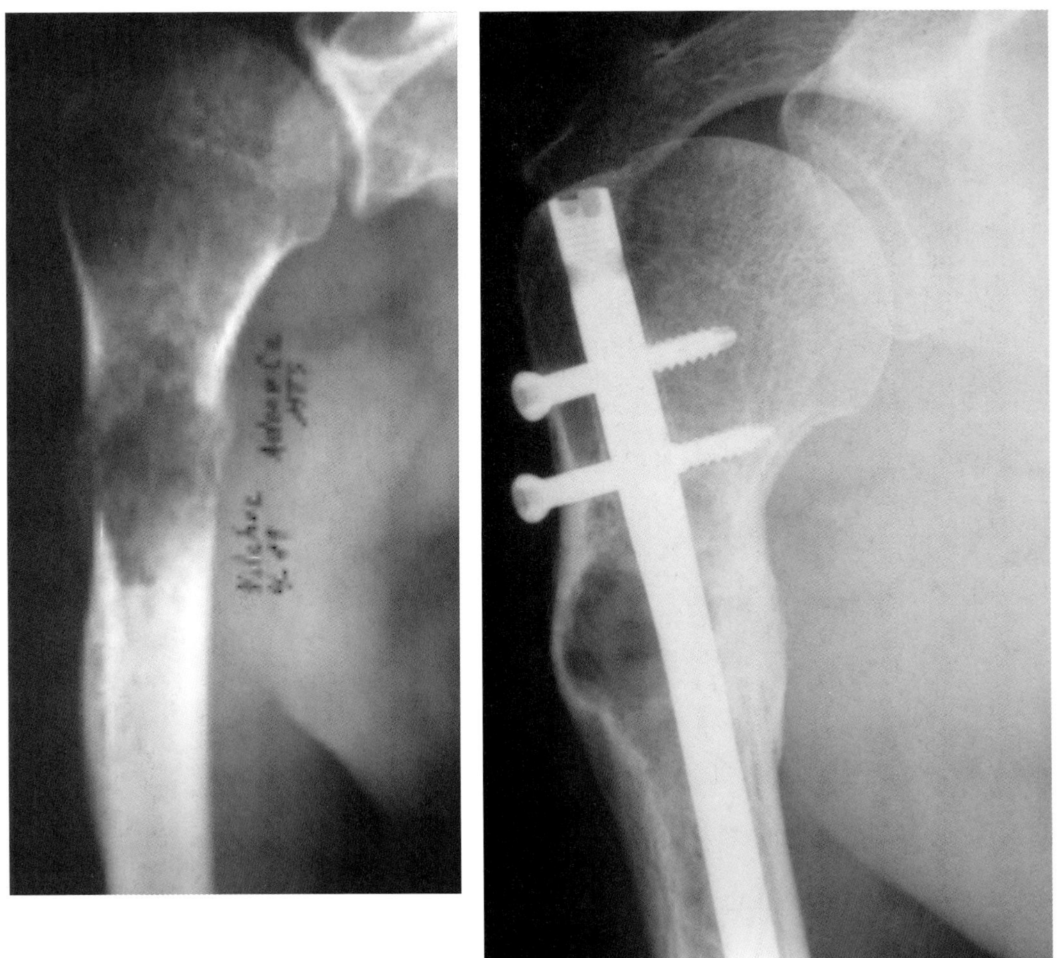

Figure 36.5. **a** Pathologic humerus fracture secondary to thyroid neoplasm. **b** One year postoperative; closed endomedulary nail.

nique with tumor curettage and cement augmentation is a viable option in some lesions. The nail not only stabilizes the fracture, but also acts as prophylactic stabilizer in case of tumor spreading in the humerus (Figure 36.5).

Even though open reduction and internal fixation with plates and screws augmented with cement is advocated by some centers, the bone quality is usually compromised and the screw purchase is poor; this technique has been replaced by the use of closed endomedullary nails [6,7,15].

Tumor resection plus humerus shortening and internal fixation is a viable option when a large segment of bone has been destroyed by the lesion.

Distal Humerus

Metastases are rare and difficult to treat; unfortunately these lesions do not respond to radiotherapy as well as the proximal humerus does. Three options are available for surgical stabilization, and include: a) Tumor removal plus fixation with plates and screws augmented with cement (Figure 36.6), b) tumor removal plus fixation with elastic endomedullary nails augmented with cement, or c) elbow arthroplasty.

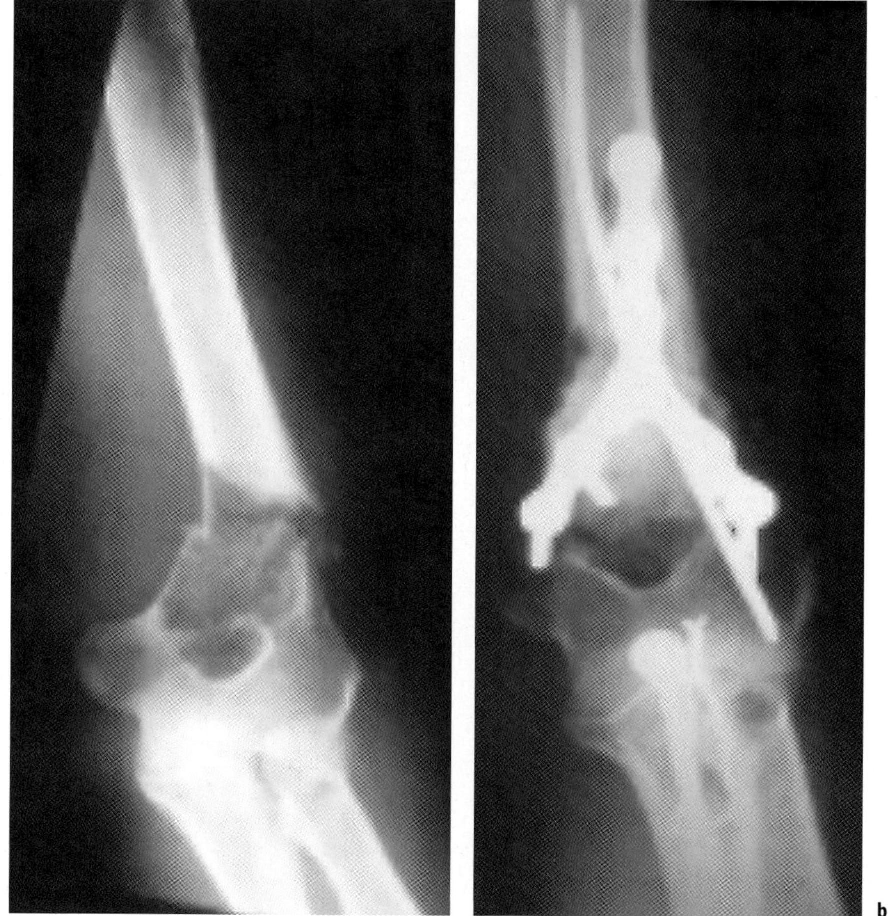

Figure 36.6. **a** Supracondilar extra-articular pathologic elbow fracture. **b** Combined surgical technique.

Even though from the biomechanical point of view the dual plating technique with two 3.5 mm reconstruction plates at 90° to each other is the most stable construct in nonpathological fractures, the poor quality of bone as well as the protection of the rest of the bone in case of tumor progression has made the use of elastic endomedullary nails and cement through a posterior approach the treatment of choice. In cases with important joint infiltration that have failed radiation therapy, a cemented elbow arthroplasty is the last option before amputation. Morrey et al. [16] reported good results in 13 prosthetic replacements for elbow tumors, six of which were metastatic lesions in the distal humerus.

Tibia

The tibia is not frequently involved in metastatic disease, and if it does, it is during the very end stage of the disease. Lung carcinoma is the most frequent primary tumor. Unlike the femur, the tibia can be adequately treated conservatively by radiotherapy plus bracing (PTB).

Surgical Treatment

Proximal Tibia

This is the most common site of litic lesions in the tibia. Tumor curettage plus internal fixation with plates and screws augmented with cement

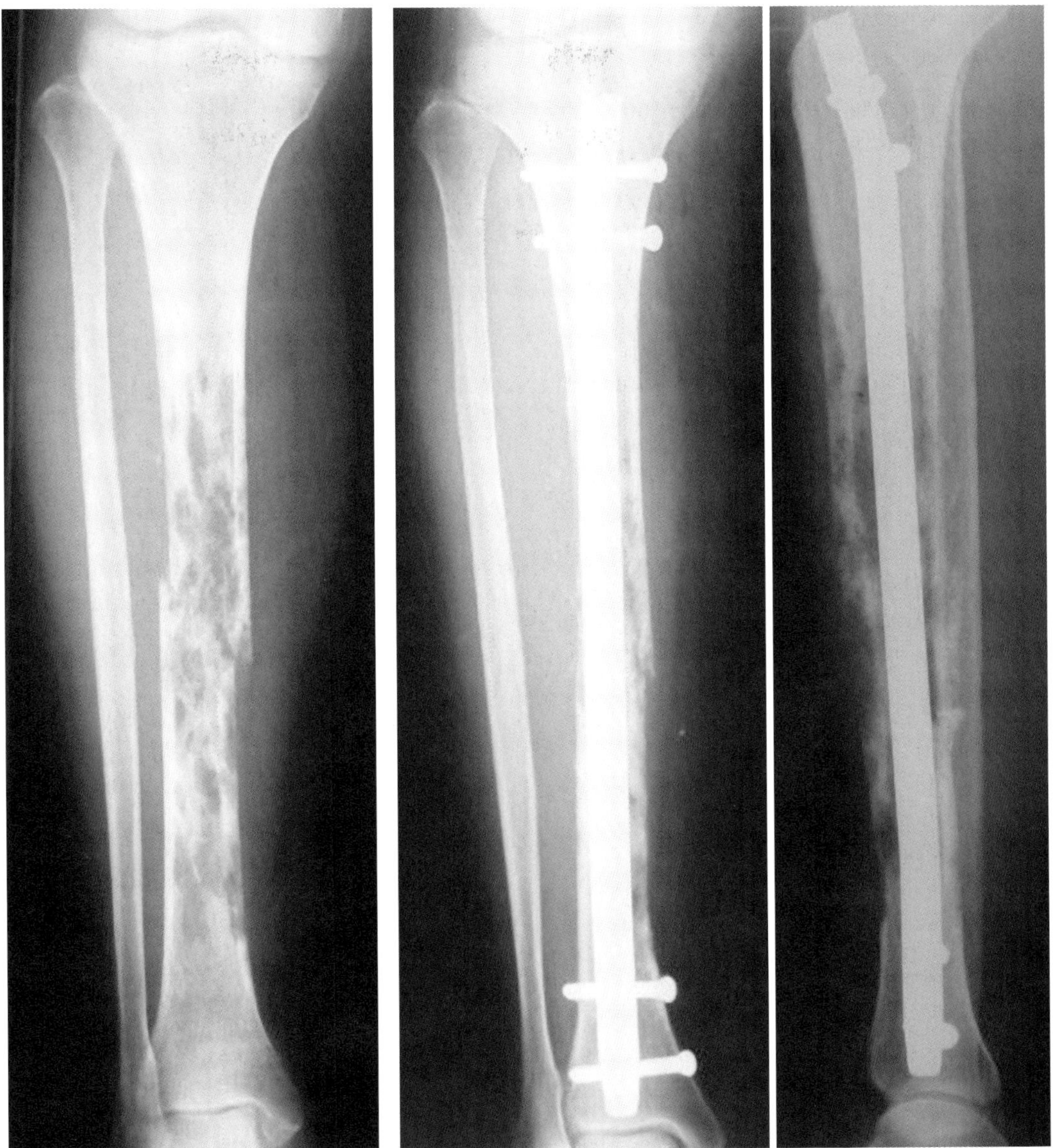

Figure 36.7. a,b

is an excellent option when there is no joint involvement. Tumor resection followed by arthroplasty is recommended in patients with tumor joint involvement as well as in tumor recurrence after previous surgical treatment. Preservation of the extensor mechanism is the most common problem encountered with this technique.

Diaphysis

These lesions have more effect on the mechanical strength of the bone. The use of tumor curettage plus cement, plus fixation with plates and screws is a viable option in isolated lesions with good surrounding bone, in more diffuse lesions the indication of an endomedullary nail is preferred (Figure 36.7). Unlike mid-shaft femur, we believe that the stabilization should be augmented with cement either injected through the reamed medullary canal, or by curettage and cement packing through an open technique.

Distal Tibia

Very few reports of distal tibia metastasis have been published. The principles of distal tibia metaphisis treatment are the same as for the proximal tibia, and includes curettage plus cement and internal fixation.

Amputation is an indication that must be remembered in metastatic cases with large joint involvement of either the knee or ankle.

Radius and Ulna

Metastases of the forearm bones are very rare; only 0.4% of all metastatic bone lesions occur in the radius, and 0.2% occur in the ulna [18]. The most common primary origin is lung neoplasm; in general, intramedullary nailing is the recommended treatment for most diaphyseal lesions. Tension band techniques may be helpful in repairing proximal ulnar lesions; if sufficient bone is present, plating and augmentation with methylmethacrylate can be used. Partial resection with shortening plus plate fixation could be an alternative if it does not produce an important functional deficit.

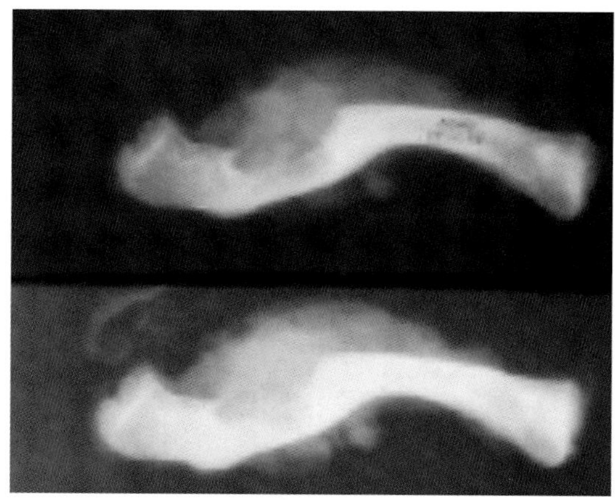

Figure 36.8. Clavicule resection; renal cell carcinoma.

Something similar occurs with the clavicle and fibula, and distal ulna; in many cases resection of the metastasis is recommended. Resection surgery is used in patients considered to have a good prognosis for long survival; mostly patients with kidney cancer with limited metastatic disease (Figure 36.8).

Summary

Although the clinical presentation may vary according to the patient, the tumor, and location of the metastasis, the following surgical treatment guidelines are recommended: 1) cemented prosthesis in proximal femur and humeral head fractures; 2) locked intramedullary nails for femur, humerus, tibia, and forearm diaphysis; 3) in other locations (proximal and distal metaphysis of long bones), combined osteosynthetic devices with cement; 4) wide resection in patients with long life expectancy.

Successful operative treatment is a feasible option in most pathological or impending fractures, and depends on achieving a rigid and durable fixation with either internal fixation or prosthetic devices. Postoperative adjuvant radiotherapy is indicated in most patients to minimize disease progression and possible implant failure.

References

1. O'Connor MI. Symposium: Metastatic bone disease. In: Program and abstracts of the 67th annual meeting of the American Academy of Orthopaedic Surgeons; March 15–19, 2000; Orlando, USA.
2. Sim F. Diagnosis and Management of Metastatic Bone Disease: A Multidisciplinary Approach. New York: Raven Press, 1988.
3. Coleman R. Skeletal complications of malignancy. Cancer 1997;80(suppl 8):1588–94.
4. Diel I, Solomayer E, Costa S. Reduction in new metastases in breast cancer with adjuvant clodronate treatment. N Engl J Med 1998;339:357–63.
5. Mirelis H. Metastasic disease in long bones: A proposed scoring system for diagnosing impending pathologic fractures. Clin Orthop 1989;249:256–64.
6. Harrington K. Orthopedic surgical management of skeletal complications of malignancy. Cancer 1997;80: 1614–27.
7. Harrington K, Sim F, Enis J, Johnson J, Dick H, Gristina A. Methylmethacrylate as an adjunct in internal fixation of pathological fractures: Experience with 375 cases. J Bone Joint Surg Am 1976;58:1047–55.
8. Wedin R, Bauer H, Wersäl P. Failures after operation for skeletal metastatic lesions of long bones. Clin Orthop 1999;358:128–39.
9. Janjan N. Radiation for bone metastasis. Cancer 1997;80:1628–44.
10. Peltier L. Theoretical hazards in the treatment of pathological fractures by the kuntscher intramedullary nail. Surgery 1951;29:466–72.
11. Swanson K, Pritchard D, Sim F. Surgical treatment of metastasic disease of the femur. J Am Acad Orthop Surg 2000;8(1):56–65.
12. Lane J, Sculco T, Zolan S. Treatment of pathological fractures of the hip by endoprosthetic replacement. J Bone Joint Surg Am 1980;62:954–9.
13. Damron T, Sim F. Surgical treatment for metastatic disease of the pelvis and proximal end of the femur. ICL 2000;49:461–70.
14. Aaron A. Treatment of metastatic adenocarcinoma of the pelvis and the extremities -current concepts review. J Bone Joint Surg Am 1977;79:917–32.
15. Yazawa Y, Frassica F, Chao E, Pritchard D, Sim F, Shieves T. Metastasic bone disease: A study of the surgical treatment of 166 pathologic humeral and femoral fractures. Clin Orthop 1990;251:213–19.
16. Sperling J, Pritchard D, Morrey B. Total elbow arthroplasty following resection of tumors at the elbow. Clin Orthop 2000 (in press).
17. Lancaster J, Koman L, Gristina A. Pathologic fractures of the humerus. South Med J 1988;81(1):52–5.
18. Sim F, Pritchard D. Metastatic disease in the upper extremity. Clin Orthop 1982;169:83–94.
19. Allende BL. Lesiones metostésicas de huesos largos. Rev Asos Argent Ortop Traumetol 2003;67(3):161–5.

37 Rebuilding and Prostheses in the Event of Periacetabular Tumors

R. Kotz and E. Schwameis

Surgical treatment of primary malignant periacetabular tumors is one of the most difficult orthopedic surgeries. Tumor resection always leads to significant instability of the affected extremity. Attempting wide resection margins, the femoral head may not be saved in case of small malignant periacetabular tumors. In addition, large pelvic defects lead to considerable shortening and instability of the leg, demanding functional amelioration. Limiting factors for limb salvage surgery are the vicinity of the ventral neurovascular bundle and the dorsal sciatic nerve. This often leads to neural palsies of the leg and to at least venous and lymphatic edema due to resection and replacement of vessels, both resulting in an impaired function of the hip joint. Considering wide resection margins associated with high complication rates and poor functional results, some surgeons prefer hindquarter amputation in the case of large periacetabular tumors. However, since the introduction of sensitive imaging methods and new treatment modalities in chemotherapy leading to decreased local recurrence rates and to an increased survival, limb salvage surgery has almost replaced amputation. The increasing number of pelvic resections gives the orthopedic surgeon complicated pelvic reconstructions [1–10].

From a biomechanical point of view, two main types of pelvic resection are of importance for patients with periacetabular tumors. Firstly type II resection according to Enneking, when the ilium can be preserved, and secondly, the more serious and more difficult type I and type II resection when the acetabulum and the ilium are resected. Whereas a support at the ilium is possible in the first case, partially with a saddle prosthesis or with special reconstruction of the acetabulum fixed at the ilium and sacrum, replacement of the pelvis by a pelvic prosthesis is necessary in the second case, especially when there is an additional type III resection. Many attempts have been made to reconstruct the continuity of the pelvis, however, the ideal pelvic prosthesis does not seem to have been developed yet. All prostheses developed up until now do not provide an anatomic pelvic reconstruction. Additionally, extensive soft tissue resection leads to insufficient coverage of these anatomic pelvic prostheses. For this reason different reconstructions, including a neoacetabulum serving as a support for a hip prosthesis, have been developed. Due to the extensive soft tissue resection and problems with soft tissue reconstruction, the possibility of a significant limp has to be taken into account and many patients have to use crutches for the rest of their lives.

Oncological Procedures

In order to guarantee surgical as well as oncologic standards, meaning oncologically adequate resection margins and the best possible reconstruction of anatomic and functional conditions, it is recommended to limit treatment of patients with malignant periacetabular tumors to tumor centers experienced in complex pelvic resections and reconstructions. A preoperatively established histologic diagnosis is mandatory prior to definitive planning of surgery and further therapy.

Diagnosis starts with determination of the clinical status of the patient including neurologic investigations. Imaging diagnostics are done using conventional radiography, magnetic resonance tomography, scintigraphy, and occasionally computed tomography.

Incisional or closed biopsy is performed aiming at resection of representative material enabling establishment of diagnosis. It is important that the biopsy is performed in the incision line of the planned tumor resection, and fine needle biopsy may require roentgenograms, computed tomography, or magnetic resonance tomography guidance. If fine needle biopsy does not allow definitive diagnosis, incisional biopsy is mandatory. Frozen sections should confirm diagnosis of malignancy before wound closure. All investigations including biopsy should be performed at a tumor center with highly qualified specialists familiar with the evaluation of bone and soft tissue sarcomas.

In the case of diagnosis of a malignant tumor, additional preoperative diagnostic measures are necessary. For depiction of vessels, angiography and phlebography are recommended, which, however, may be replaced by magnetic resonance angiography. A preoperative intravenous pyelography as well as eventual cystoscopy is recommended in case of suspected affection of the adjacent urinary tract, and splinting of the ureter may be performed.

Furthermore, preoperative gynecologic investigation is recommended. In case of a suspected affection of the rectal wall, rectoscopy, sonography, or magnetic resonance imaging of the rectum is performed.

Prior to setting up a therapeutic regimen, tumor staging including computed tomography of the lung and the abdomen, whole-body scintigraphy, and, optionally, positron emission tomography, are performed.

Planning and Determination of Resection Type

For planning of the resection, production of a pelvic model and, if indicated, production of a pelvic prosthesis, a three-dimensional computed tomography of the pelvis is performed. A pelvic model seems to be useful for preoperative planning of resection also in cases of tumors not intended for reconstruction with a custom-made prosthesis.

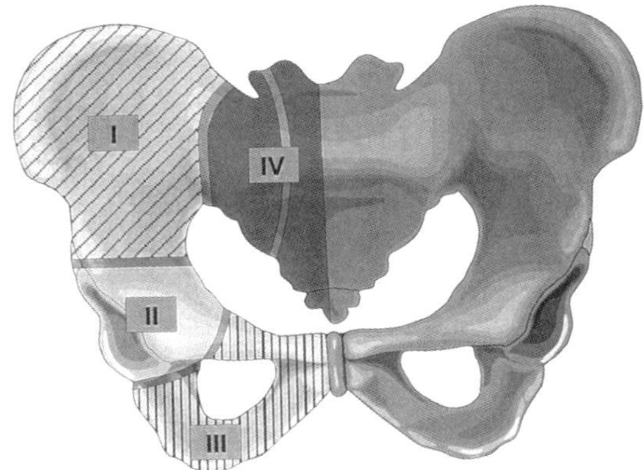

Figure 37.1. Resection types according to Enneking and Dunham.

The type of resection is determined preoperatively according to Enneking and Dunham (Figure 37.1), a procedure enabling judgment of resectability of the tumor [11]. Tumor staging according to Enneking is based on histologic findings, extent of the tumor, and presence of metastases. Prior to surgery, planning of the necessary cooperation with colleagues from other specialties is indicated. Particularly in the case of large tumors, a nephrostomy, vessel and/or nerve reconstruction, partial resection of the urinary bladder, or enterectomy might be necessary and surgery must be planned carefully in cooperation with all involved specialists.

One day of abstinence of nutrition has to be observed prior to surgery. For adequate preparation of the ileum and minimization of germs, retrograde rinsing of the ileum is performed twice on the day before surgery. An orthograde rinsing, a so-called washout, may also be done. Antibiotic therapy for prophylaxis of infection may be performed for two weeks or longer, depending on the clinical status of the patient and the parameters of inflammation.

Resection of larger pelvic tumors and reconstruction of the defect requires the facilities of a large hospital and an experienced team of anesthesiologists. Standard monitoring includes measurement of the arterial blood pressure, pulse oximeter, temperature tube, central venous pressure, transesophageal echocardiography or a flow-directed catheter as well as frequent checks of fibrinolytic parameters. The capacity of the blood bank has to be taken into account and a sufficient number of matching blood units has to be held at disposal. A Rapid Infusion System should the necessity for transfusion of a large number of blood units be required should be available within a short time. Also, a cava filter may be inserted preoperatively; this is especially recommended in patients with a large intrapelvic soft tissue tumor portion, in patients with earlier radiation therapy, and especially in those endangered by thrombosis.

Resectional Procedures

The patient is placed in a lateral position on the contralateral side to allow anterior as well as posterior approached by forward and backward movement of the patient during the procedure. The typical *type I resection* of the ilium is performed if the proximal resection line is running along the iliosacral joint, whereas the distal one goes from the incisura ischiadica towards immediately below the spina iliaca anterior inferior. It is distinguished between resections maintaining the continuity of the pelvis, in most cases not requiring reconstruction (Ia), resections interrupting the continuity of the pelvis requiring reconstruction for stabilization of the pelvic ring (Ib) and those cases in which also partial removal of the sacrum is performed additionally (Ic = type I and IV resection). The surgical approach for type I resections is a curved skin incision along the crista iliaca.

The *type II resection* is a resection of the acetabulum between the spina iliaca anterior inferior and foramen obturatum distally. This procedure may only be performed in case of very small acetabular tumors or in case of tumors of the proximal femur affecting the hip joint and requiring joint resection. The approach is performed via a curved skin incision over the crista iliaca towards dorsally and ventrally the neurovascular bundle towards distal.

If a *type III resection* is performed additionally, the medial resection line goes through the symphysis. The anterior approach is performed via a transverse incision or from the spina iliaca superior via the vessels towards distally, when necessary, combined with a dorsal approach over the tuber ischiadicum or via a medial inguineal approach.

Partial resection of the os sacrum corresponds to *type IV resection* and is applied in case of co-affection of the sacrum within a type I–IV resection.

Type I–IV resection with subsequent reconstruction corresponds to internal hemipelvectomy. The ideal approach is a curved skin incision above the crista iliaca towards dorsally and further distally straight above the neurovascular bundle. In case of amputation, type I–IV resection corresponds to a hindquarter amputation.

Reconstructive Procedures

Reconstruction possibilities after periacetabular resection include endoprosthetic replacement by saddle prosthesis, modular or custom-made endoprostheses, allograft or autoclaved autograft reconstruction, as well as iliofemoral or ischiofemoral arthrodesis. If no reconstruction is possible a flail hip might be taken into consideration [12–16].

Allografts and autoclaved autografts [17–22] where a hip prosthesis is implanted match well the defect; however, they have high rates of infection and graft fractures and in many cases the implant fails and has to be removed afterwards.

Iliofemoral arthrodesis, only possible when the ilium or ischium and an acceptable bone stock for screw fixation can be maintained, has the disadvantage of a stiff hip joint and a shortened limb. Problems may arise from a failed fusion at the small areas of osseous contact and 50% of pseudarthrosis rates have been reported, which are often painless and stable without bony fusion. *Ischiofemoral arthrodesis*, arthrodesis between the proximal femur and the remaining ischium medially, may result in less shortening of the limb and has the advantage of considerable motion in the symphysis, but symphyseal pain is reported.

When a *saddle prosthesis* [23,24] is used, fixation only can be ensured when a large part of the ilium can be preserved. Common complica-

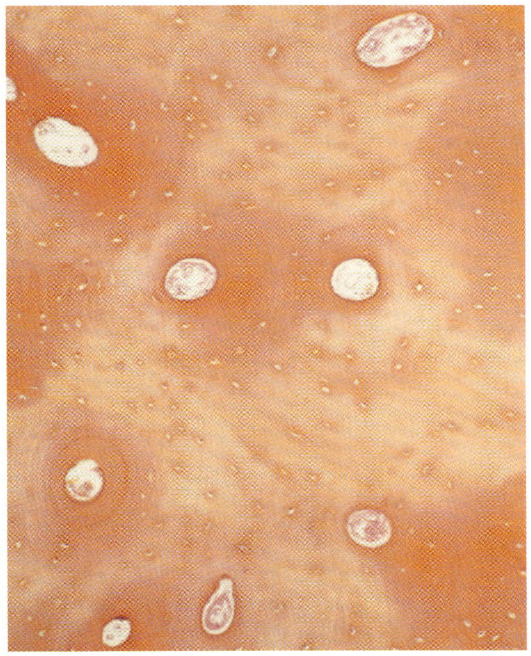

Figure 18.1. Cross section of demineralized cortical bone showing numerous Haversian systems (Hematoxylin erythrosin saffron).

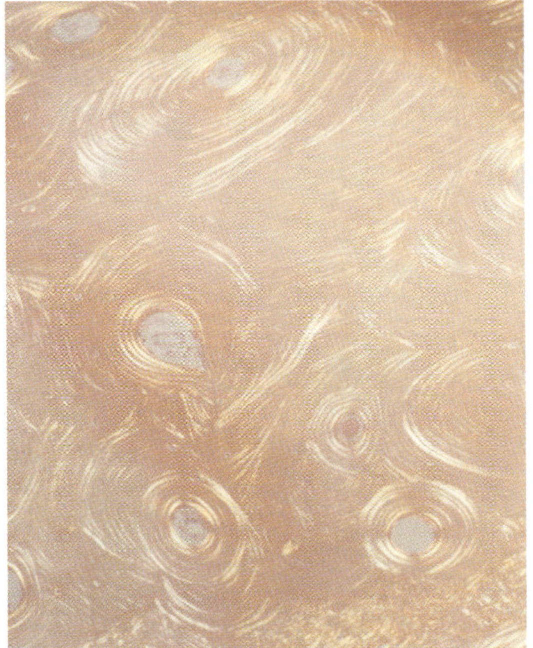

Figure 18.2. Cross section of demineralized cortical bone showing numerous Haversian systems with concentric lamellae (Hematoxylin erythrosin saffron, polarized light).

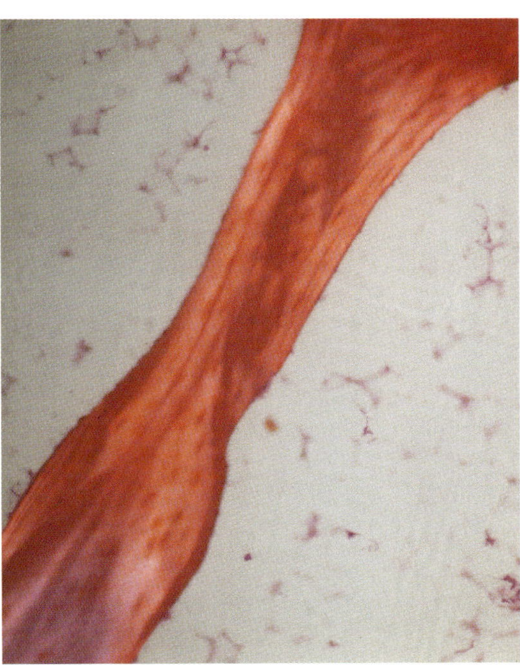

Figure 18.3. Longitudinal section of demineralized trabecular bone showing lamallae parallel to the long axis of the trabeculae (Hematoxylin erythrosin saffron, polarized light).

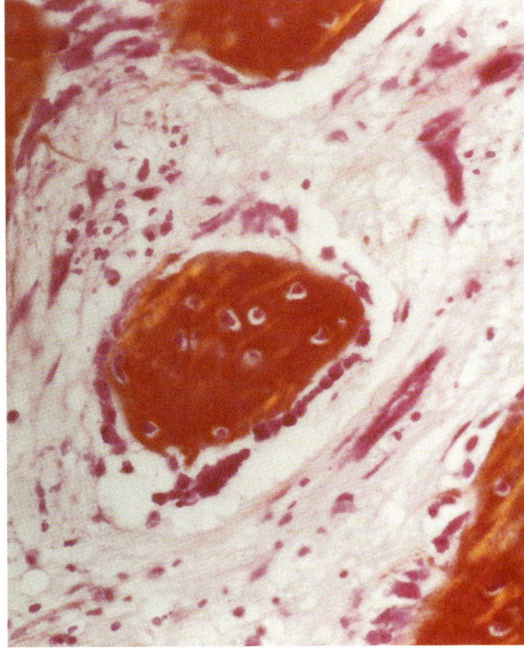

Figure 18.4. Woven bone showing randomly organized trabeculae. The osteocytes are large and irregularly spaced (Hematoxylin erythrosin saffron).

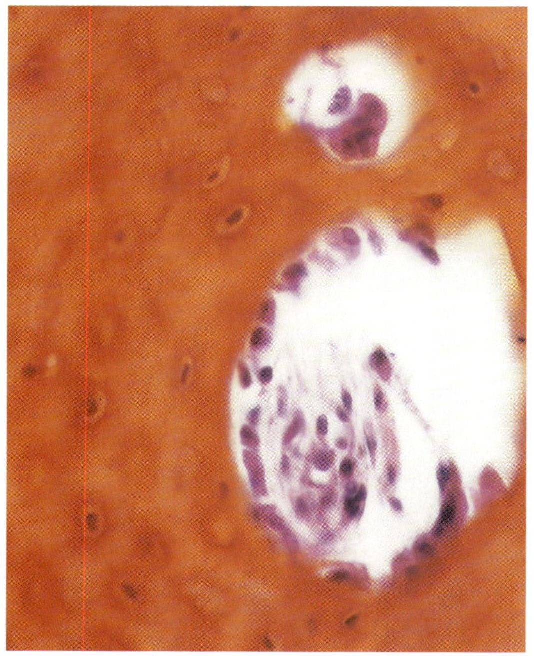

Figure 18.5. Osteoblasts and osteoclasts indicating active resorption as well as formation (Hematoxylin erythrosin saffron).

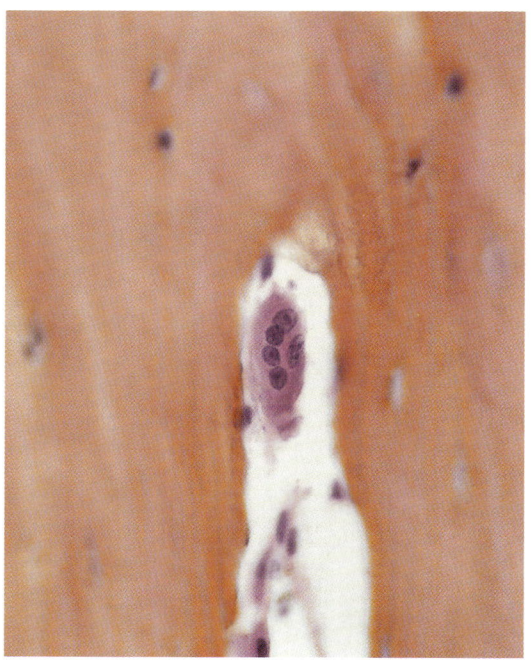

Figure 18.7. Osteoclast, multinucleated cell, responsible for bone reabsorption (Hematoxylin erythrosin saffron).

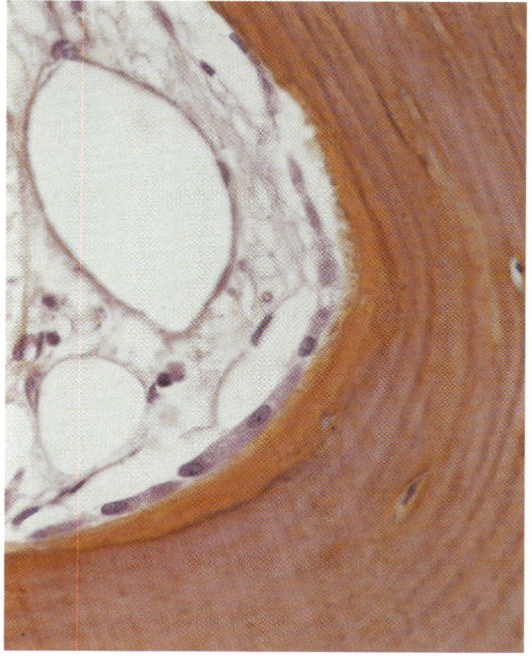

Figure 18.6. Osteoblasts form a columnar layer on the surface of the bone (Hematoxylin erythrosin saffron).

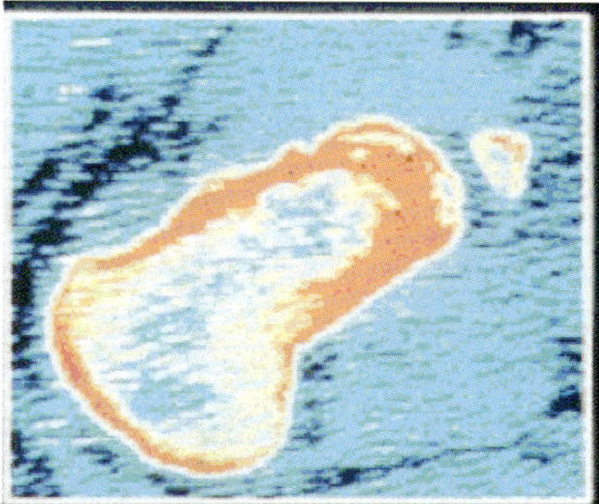

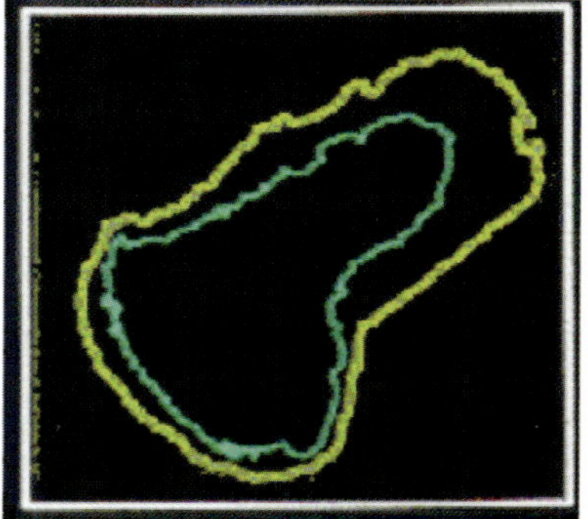

Figure 32.7. Extraction of internal and external femoral contours from CT data.

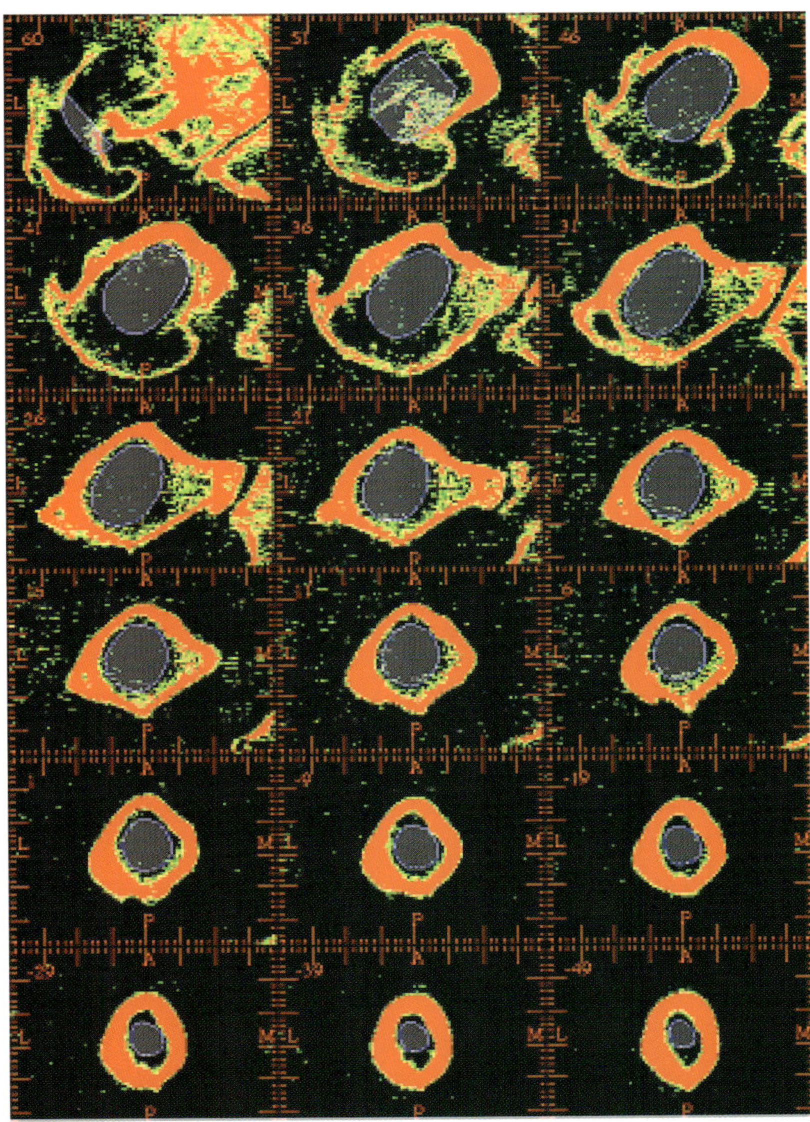

Figure 32.10. Composite CT views with integration of stem sections.

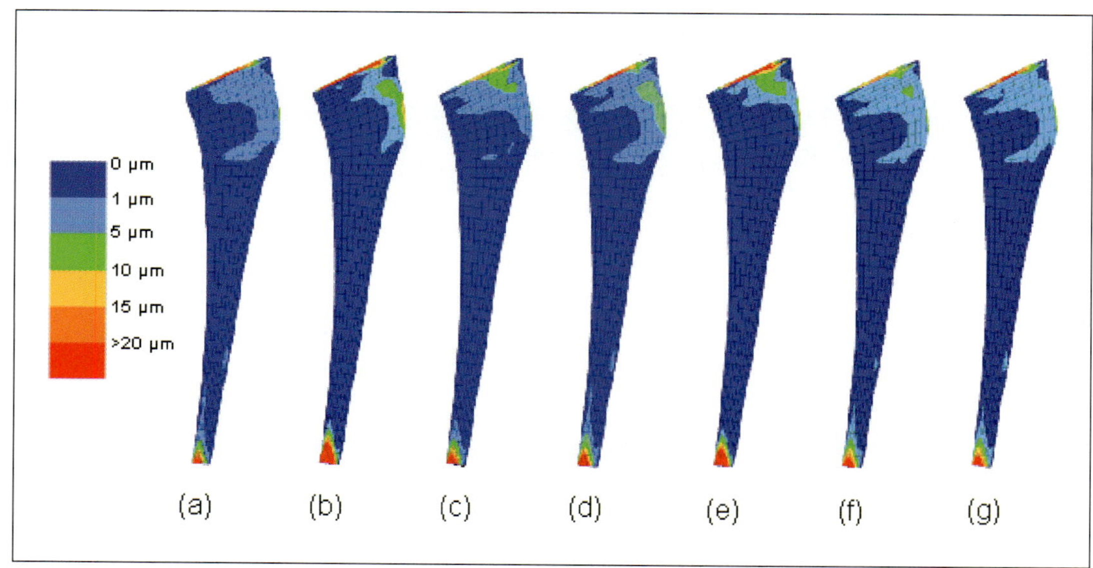

Figure 32.20. Micro-debonding distribution at bone–prosthesis interface in seven neck configurations (a: anatomical, b: anteverted, c: retroverted, d: lengthened and lateralized, e: shortened and medialized, f: in valgus, g: in varus).

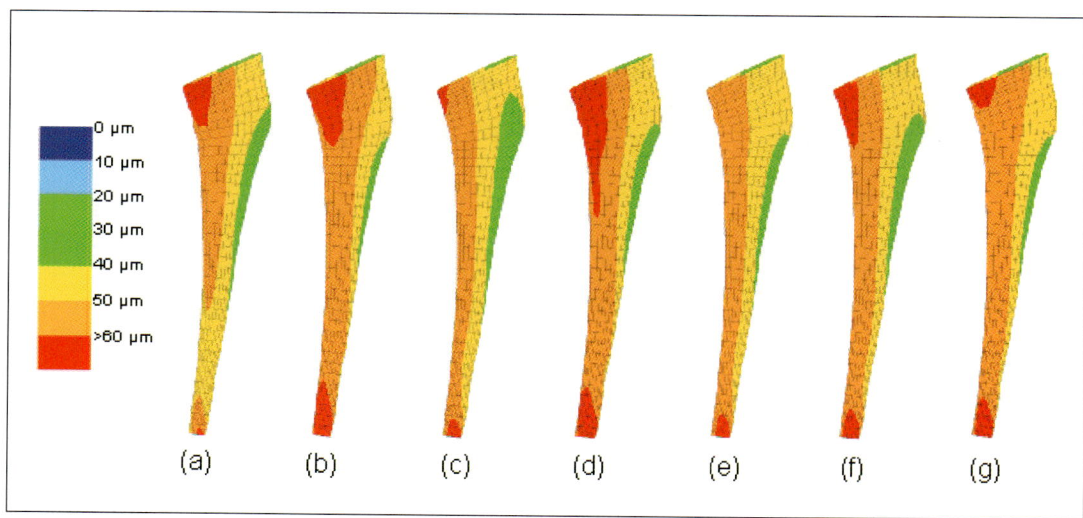

Figure 32.21. Micro-slipping distribution at bone–prosthesis interface in seven neck configurations (a: anatomical, b: anteverted, c: retroverted, d: lengthened and lateralized, e: shortened and medialized, f: in valgus, g: in varus).

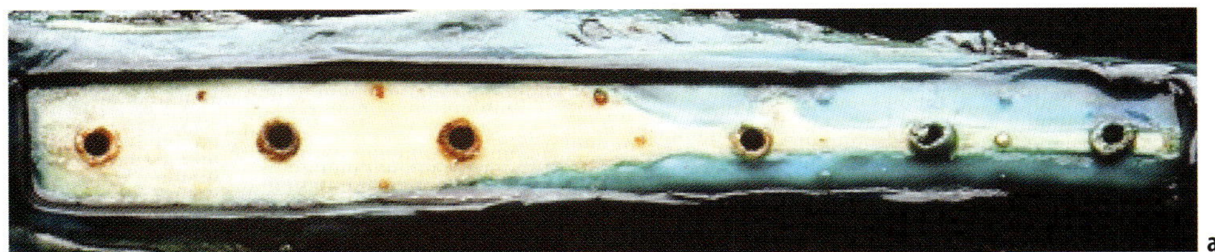

Figure 34.1. Vascular disturbance underneath a conventional plate. The compression of the periosteal vessels results in an avascular area underneath a conventional plate (a). The bone section at four weeks reveals the importance of the disturbance of the blood supply under the plate, which is mainly due to impairment of the venous efflux (b).

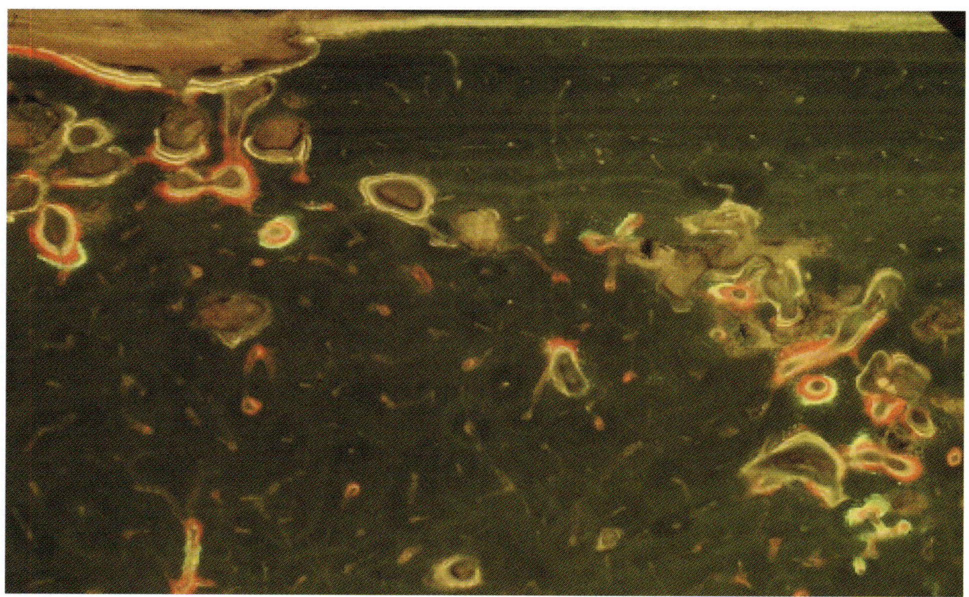

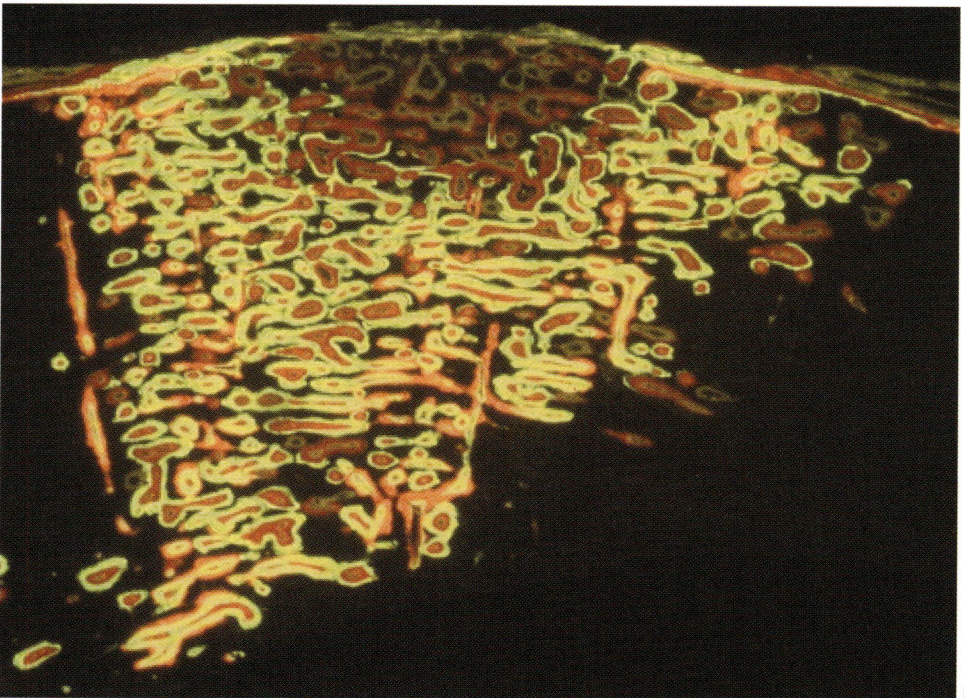

Figure 34.2. Bone remodeling after plating. At ten weeks after plating, fluorescent labeling of cortical bone shows that the remodeling activity starts at the boundary between the zone of avascular bone and viable bone (a). The remodeling front reaches the periosteum at about 20 weeks, resulting in complete revascularization of the formerly avital bone cortex. This creeping substitution of bone results in a temporary porosity (b).

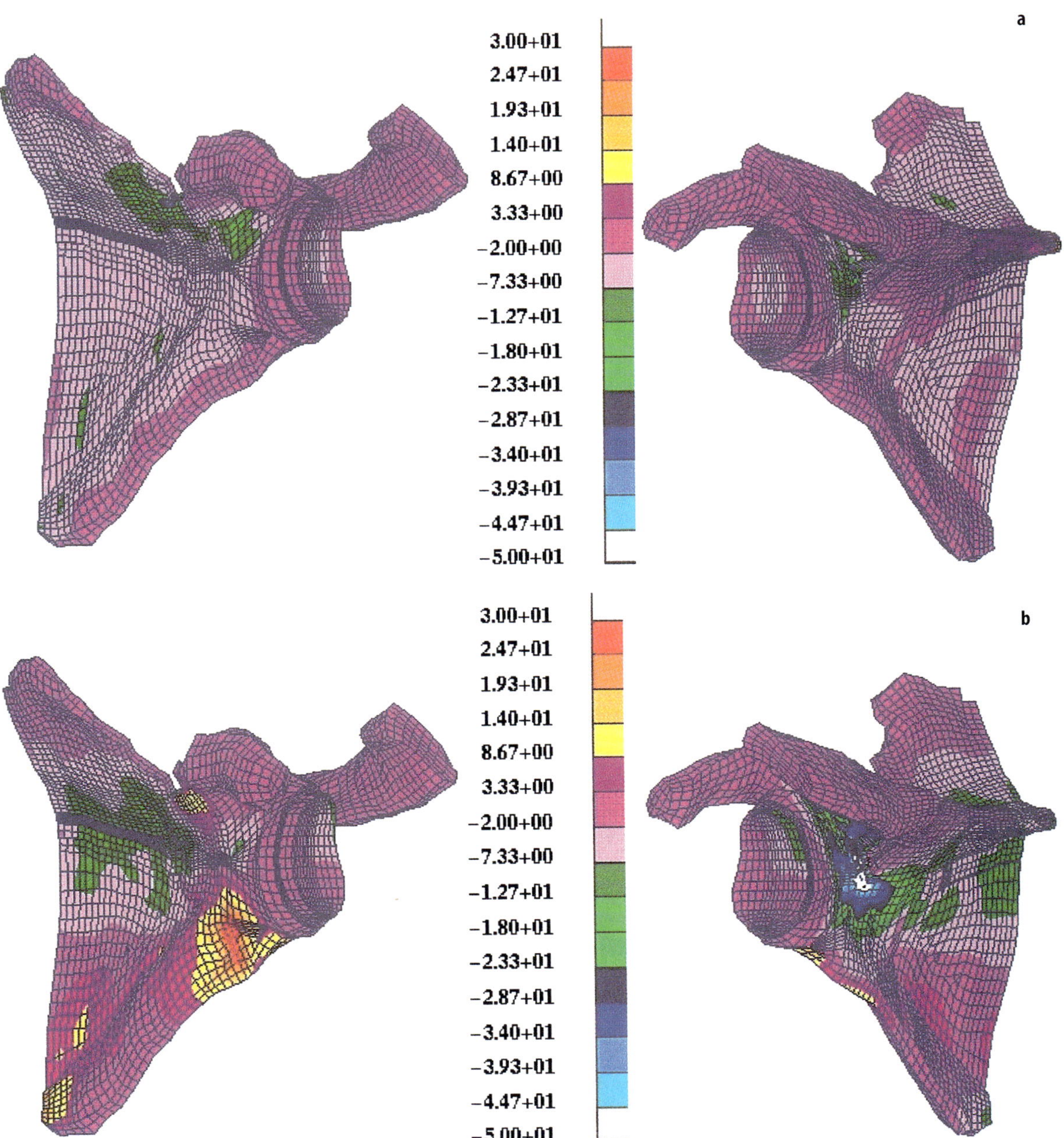

Figure 38.14. Finite element analysis of the bone stresses on the surface of a scapula implanted with a prosthetic glenoid (principle stress value in MPa, compression is negative). (a) A centered articular force with a predominance of compressive stresses. (b) Eccentric articular force producing tensile stresses on the anterior side and compressive stresses at the posterior side brought about by the bending of the scapula.

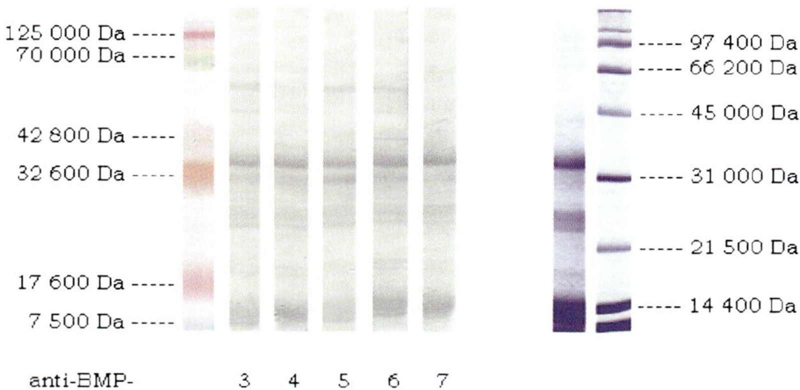

Figure 53.1. Western blot and SDS-Page analysis of native BMPs extracted from bone.

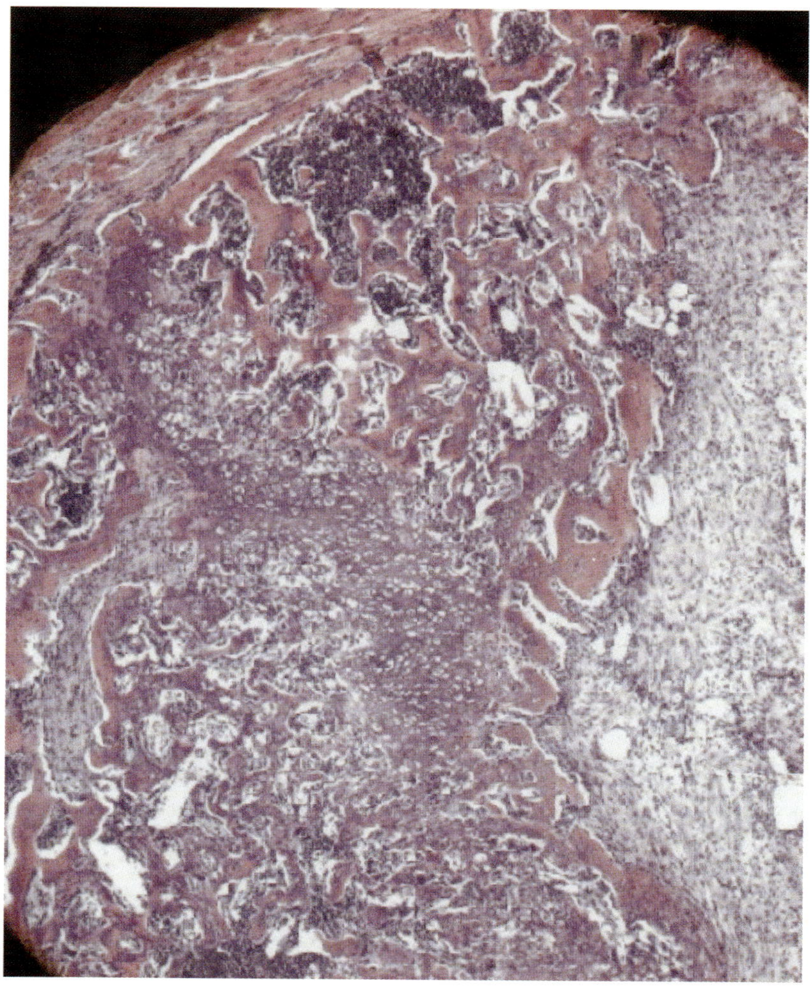

Figure 53.4. A histologic picture of an ectopic bone–ossicle after implantation of native BMP in the muscle pouch of a mouse.

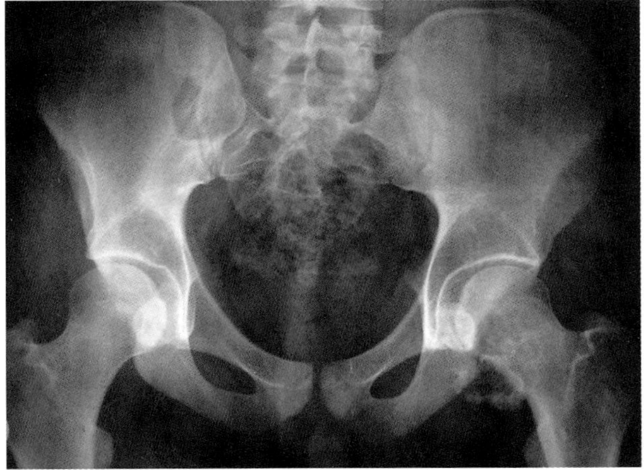

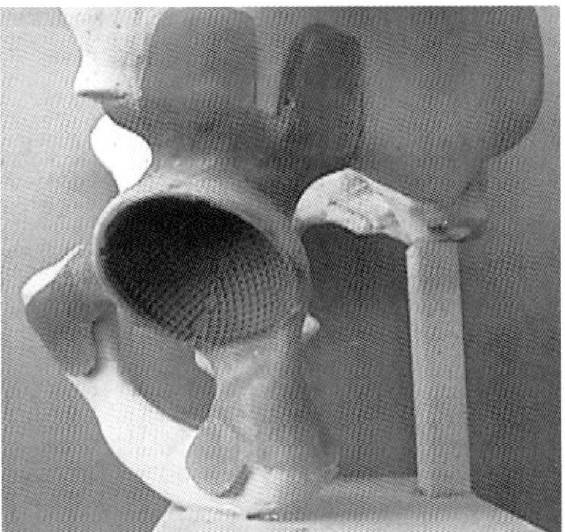

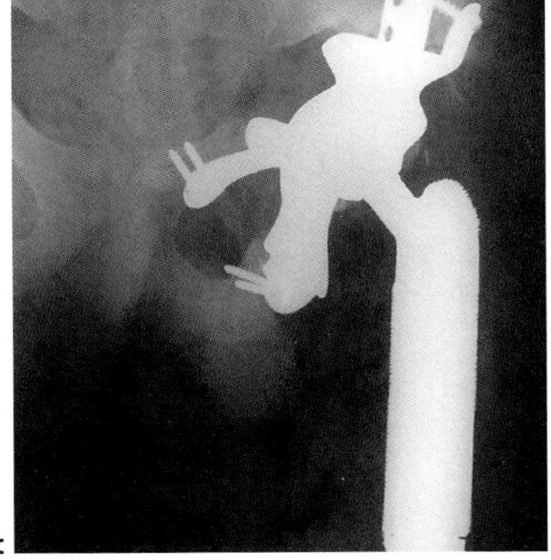

Figure 37.2 **a** Twenty-two-year-old male, chondrosarcoma of the left proximal femur involving the hip joint. Anterior–posterior roentgenogram. **b** Same patient, pelvic model and production of a custom-made endoprosthesis. **c** Same patient, pelvic reconstruction after type II resection and resection of the proximal femur by custom-made endoprosthesis and a KMFTR prosthesis of the proximal femur.

tions are fracture of the remaining ilium and dislocation of the prosthesis.

Implantation of an *endoprosthesis* (Figures 37.2a–c, 37.3a–d, 37.4a–b, 37.5a–c) is a possibility to achieve a functional hip joint without leg-length discrepancy. Several modular and custom-made systems are available [25–29].

According to the complexity of the surgery, size of the defect, preoperative cytostatic chemotherapy and immunosuppression, radiation therapy, duration of surgery, high blood loss, and long preoperative hospitalization, complication rates are high. Commonly, wound-healing disturbances leading to infection have to be treated immediately by necrosectomy, secondary wound closure, and a sufficient soft tissue coverage. Early revision of wound-healing disturbances is useful in order to minimize the risk of infection and delays in chemotherapy. Coverage by myocutaneous flaps is indicated in cases of large defects, e.g., application of a rectus abdominis flap or a free microvascular flap.

In cases of uncontrollable complications the alternatives of hindquarter amputation and flail hip have to be taken into consideration – thereby avoiding numerous attempts at limb salvage. Late complications such as loosening of the prosthesis may be treated by bone grafting (Figure 37.6). When possible, primary bone grafting is recommended during primary implantation of the prosthesis.

Postoperative monitoring at the intensive care unit is recommended. Adequate prophylaxis of thrombosis is a major concern of postoperative treatment as hematomas or a thrombosis may seriously endanger the success

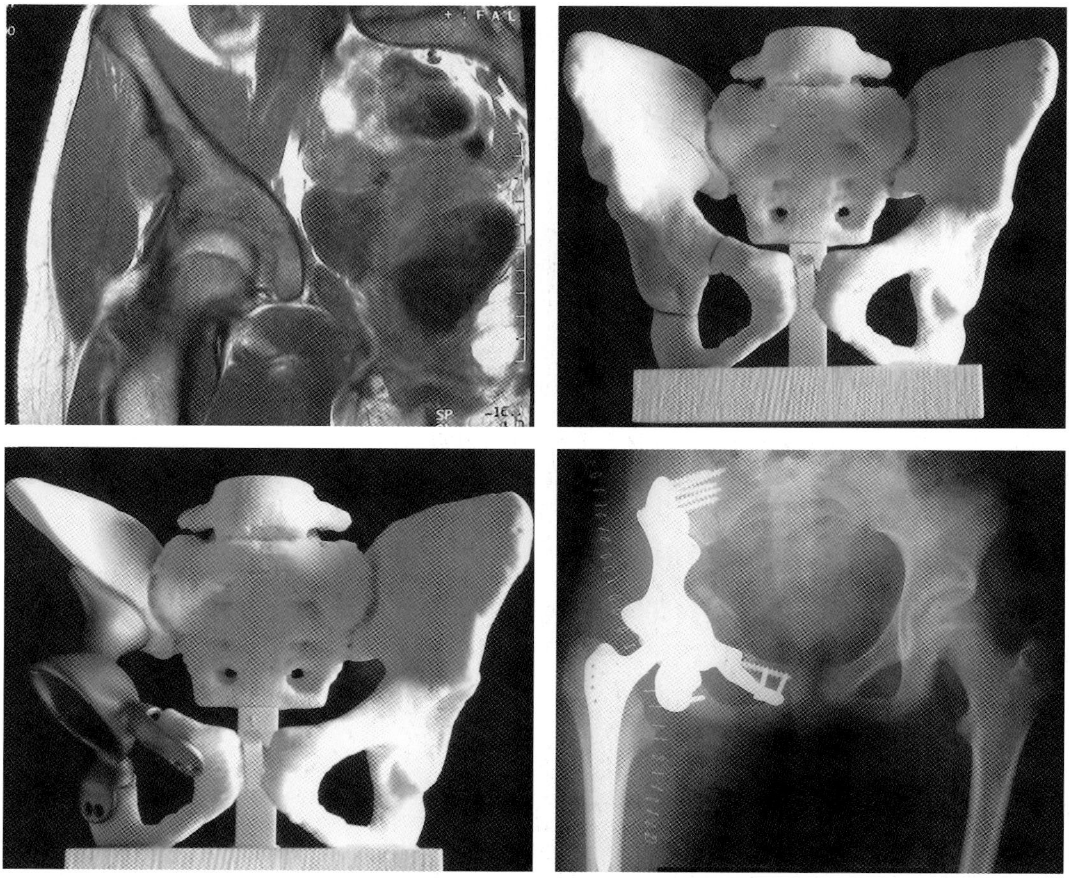

Figure 37.3. **a** Twelve-year-old girl suffering from an osteosarcoma of the right supra-acetabular region – magnetic resonance tomography. **b** Same patient, pelvic model with planning of the type II and partial type I resection. **c** Same patient, pelvic model with custom-made endoprosthesis. **d** Same patient after reconstruction of the defect with a custom-made endoprosthesis.

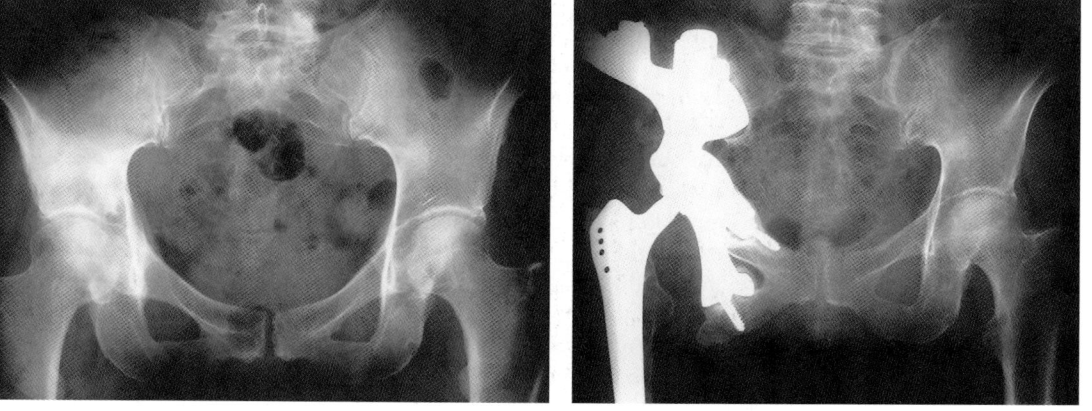

Figure 37.4. **a** Sixty-two-year-old female, chondrosarcoma of the right supra-acetabular region, anterior–posterior roentgenogram. **b** Same patient after type II and III resection and endoprosthetic replacement.

a
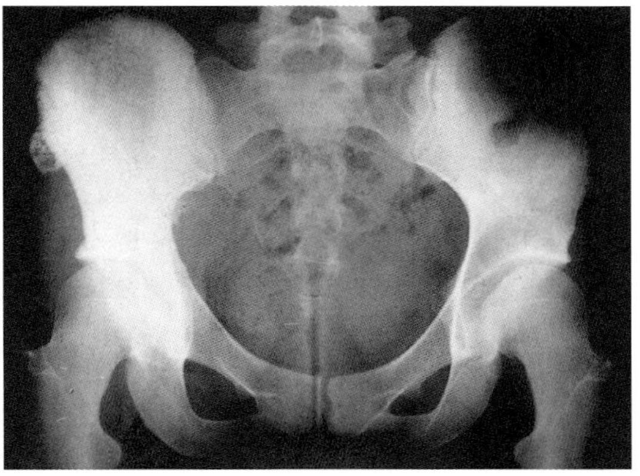

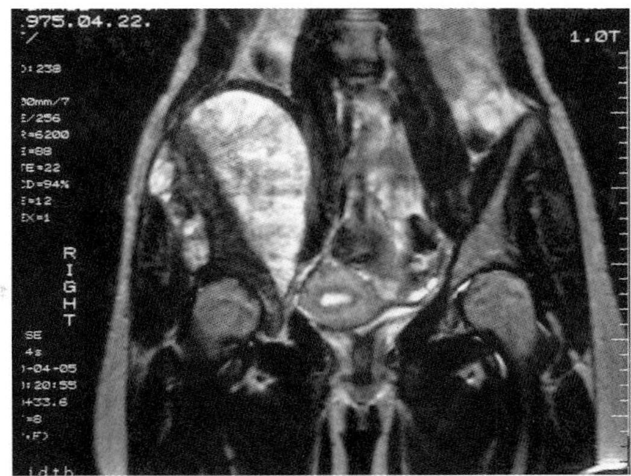

b

c
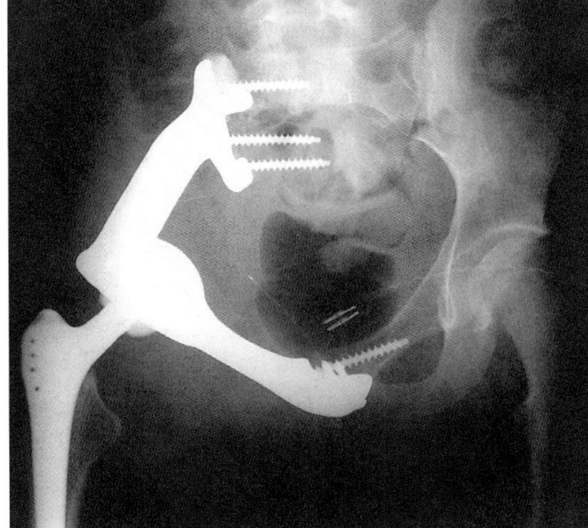

Figure 37.5. **a** Twenty-two-year-old female, chondrosarcoma of the right ilium, anterior–posterior roentgenogram of the pelvis. **b** Same patient, pelvic magnetic resonance tomography. **c** Same patient after type I–IV resection and reconstruction by custom-made endoprosthesis.

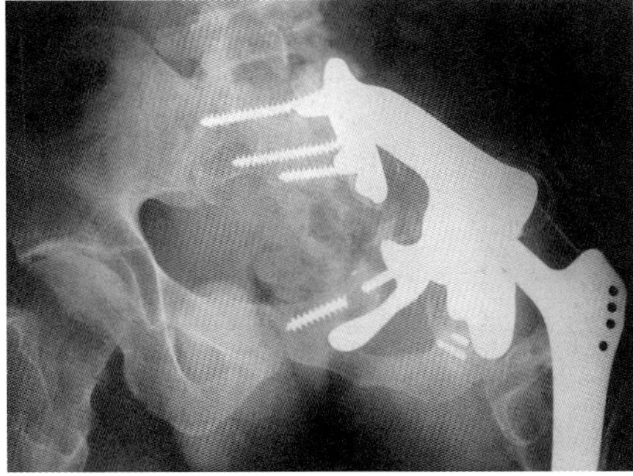

Figure 37.6. Twenty-six-year-old male, reconstruction of the left pelvis after type I, II, and IV resection for a chondrosarcoma by a custom-made endoprosthesis, breakage of screws and loosening of the implant.

of surgery. In most cases immediate postoperative prophylaxis of thrombosis consists of continuous application of heparin. As standard values, a partial thrombin time of 40 seconds for patients without replacement or suture of vessels and of 60 seconds for patients who underwent such procedures seem useful.

Drains may be kept for four days or longer; and 10 to 12 days postoperatively, sonography or computed tomography are indicated for diagnosis of a postoperative seroma or hematoma. Excision of hematoma, lavage, and re-drainage may be performed approximately four days postoperatively.

For each patient an individually designed concept of after-treatment is necessary depending on the stability of the pelvic ring and on the danger of hip dislocation. Pelvic endoprostheses require an early stable fixation of the hip joint by means of a spica cast or an orthosis. Immediately after surgery the patient is positioned without plaster cast or in an open spica cast for the pelvis and the leg. Two weeks postoperatively the patient is provided with a closed spica cast. Afterwards, an orthosis for pelvis, leg, and foot allowing motion of the knee joint and restricted at the hip joint at about 30–40° is recommended, which should be worn for 3–4 months. Stepwise removal of the part for the foot and afterwards opening of the hip arrest may be performed. The orthosis with the short support for the leg and the mobile hip joint is worn for at least one year or for the rest of the patient's life. Mobilization starts two weeks after surgery or later without weight bearing. After six weeks, stepwise weight bearing is allowed. Two arm crutches or a stick are useful walking supports, in most cases for the rest of the patient's life. Physiotherapy should be applied at home as the program in a rehabilitation center is often too strenuous for these patients.

If no reconstruction is possible or in case of complications a flail hip situation may be accepted. Problems are leg length discrepancy and instability of the hip as well as poor functional results.

Hindquarter amputation [30–32] is preferred in cases in which oncologically adequate resection of the tumor and therefore limb salvage is impossible because of involvement of nerves and vessels. The method should be advised to patients with a local recurrence and if the tumor involves the sacral nerve root foramina or the sciatic nerve. Resection of the sciatic nerve leads to a leg without function and patients should be treated by hindquarter amputation. If the femoral neurovascular structures are involved, resection is still possible. The nerve can be resected without major impairment of function and vessels can be resected and replaced.

Malignant tumors located in the periacetabular region are difficult to treat and therefore thorough preparations regarding imaging procedures, oncologic measures, and surgery are mandatory. Surgery of a pelvic tumor leads to instability of the leg, tolerable only in case of small tumors when in exceptional cases no reconstruction of the defect is performed. Large resections in patients with limb salvage require adequate measures for restoration of stability such as saddle prostheses, custom-made and modular tumor prostheses, and allografts. Such patients require special anesthesiologic management intra- and postoperatively as well as long-term antibiotic prophylaxis and immediate surgical management of complications. Patients with large tumors should be considered for hindquarter amputation without hesitation; also, in case of serious complications, hindquarter amputation may become necessary. Due to the complexity of the treatment of periacetabular tumors, such patients should be treated exclusively at specialized tumor centers.

References

1. Frassica FJ, Frassica DA, Pritchard DJ, et al. Ewing sarcoma of the pelvis. Clinicopathological features and treatment. J Bone Joint Surg 1993;75A:1457–65.
2. Ham SJ, Schraffordt Koops H, Veth RP, von Horn JR, Eisma WH, Hoekstra HJ. External and internal hemipelvectomy for sarcomas of the pelvic girdle: consequences of limb-salvage treatment. Eur J Surg Oncol 1997;23:540–6.
3. Hamdi M, Gebhardt M, Recloux P. Internal hemipelvectomy. Eur J Surg Oncol 1996;22:158–61.
4. Kawai A, Huvos AG, Meyers PA, Healey JH. Osteosarcoma of the pelvis. Oncologic results of 40 patients. Clin Orthop 1998;348:196–207.
5. Mercuri M, Capanna R, Manfrini M, et al. The management of malignant bone tumors in children and adolescents. Clin Orthop 1991;264:156–68.
6. Morton DL, Eilber FR, Townsend Jr CM, et al. Limb salvage from a multidisciplinary treatment approach for skeletal and soft tissue sarcomas of the extremity. Ann Surg 1976;184:268–78.
7. Ozaki T, Hillmann A, Winkelmann W. Treatment outcome of pelvic sarcomas in young children: Orthopaedic and oncologic analysis. J Pediatr Orthop 1998;18:350–5.
8. Shin KH, Rougraff BT, Simon MA. Oncological outcomes of primary bone sarcomas of the pelvis. Clin Orthop 1994;304:207–17.
9. Steel HH. Partial or complete resection of the hemipelvis. An alternative to hindquarter amputation for periacetabular chondrosarcoma of the pelvis. J Bone Joint Surg 1978;60A:719–30.
10. Stephenson RB, Kaufer H, Hankin F. Partial pelvic resection as an alternative to hindquarter amputation for skeletal neoplasms. Clin Orthop 1989;242:201–11.
11. Enneking WF, Dunham WK. Resection and reconstruction for primary neoplasms involving the innominate bone. J Bone Joint Surg 1978;60A:731–46.
12. Aboulafia AL, Malawer MM. Surgical management of pelvic and extremity osteosarcoma. Cancer Supp 1993; 71:3358–66.

13. Campanacci M, Capanna R. Pelvic resections: The Rizzoli Institute experience. Orthop Clin North Am 1991;22:65–86.
14. Johnson JTH. Reconstruction of the pelvic ring following tumor resection. J Bone Joint Surg 1978;60A:747–51.
15. O'Connor MI, Sim FH. Salvage of the limb in the treatment of malignant pelvic tumors. J Bone Joint Surg 1989;71A:481–94.
16. O'Connor MI. Malignant pelvic tumors: Limb-sparing resection and reconstruction. Semin Surg Oncol 1997; 13:49–54.
17. Harrington KD. The use of hemipelvic allografts or autoclaved grafts for reconstruction after wide resections of malignant tumors of the pelvis. J Bone Joint Surg 1992;74A:331–41.
18. Langlais F, Vielpeau C. Allografts of the hemipelvis after tumour resection. J Bone Joint Surg 1989;71B:58–62.
19. Mankin HJ, Doppelt SH, Sullivan TR, Tomford WW. Osteoarticular and intercalary allograft transplantation in the management of malignant tumors of bone. Cancer 1982;50:613–30.
20. Mankin HJ, Doppelt S, Tomford W. Clinical experience with allograft implantation: The first ten years. Clin Orthop 1983;174:69–86.
21. Mnaymneh W, Malinin T, Mnaymneh LG, Robinson D. Pelvic allograft. A case report with a follow-up evaluation of 5.5 years. Clin Orthop 1990;250:128–32.
22. Ozaki T, Hillmann A, Bettin D, Wuisman P, Winkelmann W. High complication rates with pelvic allografts. Experience of 22 sarcoma resections. Acta Orthop Scand 1996;67:33–8.
23. Nieder E, Elson RA, Engelbrecht E, et al. The saddle prosthesis for salvage of the destroyed acetabulum. J Bone Joint Surg 1990;72B:1014–22.
24. Van der Lei B, Hoekstra HJ, Veth RPH, et al. The use of the saddle prosthesis for reconstruction of the hip joint after tumor resection of the pelvis. J Surg Oncol 1992;50:216–19.
25. Abudu A, Grimer RJ, Cannon SR, Carter SR, Sneath RS. Reconstruction of the hemipelvis after the excision of malignant tumours. Complications and functional outcome of prostheses. J Bone Joint Surg 1997;79B:773–9.
26. Bruns J, Luessenhop SL, Dahmen G Sr. Internal hemipelvectomy and endoprosthetic pelvic replacement: long-term follow-up results. Arch Orthop Trauma Surg 1997;116:27–31.
27. Gradinger R, Rechl H, Hipp E. Pelvic osteosarcoma, resection, reconstruction, local control and survival statistics. Clin Orthop 1991;270:149–58.
28. Gradinger R, Rechl H, Ascherl R, Plotz W, Hipp E. Partial endoprosthetic reconstruction of the pelvis in malignant tumors. Orthopäde 1993;22:167–73.
29. Windhager R, Karner J, Kutschera HP, et al. Limb salvage in periacetabular sarcomas. Clin Orthop 1996;331: 265–76.
30. Kotz R, Ritschl P, Pongracz N, Zimmermann R. Hemipelvektomie: Indikation, Operationstechnik und Ergebnisse. Acta Chir Austr 1986;18:108–9.
31. Kotz R, Ritschl P, Kropej D, Schiller Ch, Wurnig Ch, Salzer-Kuntschik M. Die Grenzen der Extremitätenerhaltung – Amputation versus Resektion. Z Orthop 1992;130:299–305.
32. Wu KK, Guise ER, Frost HM, Mitchell CL. The surgical technique for hindquarter amputation. A report of 19 cases. Acta Orthop Scand 1977;48:479–86.

VI

Articular Biomechanics

VI – Articular Biomechanics

VI A – Upper Limb

38 The Biomechanics of the Glenohumeral Articulation and Implications for Prosthetic Design

M. Mansat and J. Egan

Introduction

To undertake normal activities of daily living the shoulder joint must allow a quite considerable range of motion [22]. The maximum elevation of the shoulder in the scapular plane generally lies between 170 and 180 degrees (Freedman et al.) [5,3]. To achieve this motion requires contributions from four shoulder components, namely the glenohumeral, acromioclavicular, sternoclavicular, and scapulothoracic articulations.

Muscle forces act across the glenohumeral joint to balance externally applied loads and the large motion at the shoulder involves changes of muscle moment arms. In an analysis of these forces across the shoulder it is convenient to divide the muscles into three groups: thoracoscapular, scapulohumeral, and thoracohumeral muscles. However, the motion is complex, as is the relationship between the muscle actions and the joint movement. Three-dimensional analyses that approximate the real complexity of the shoulder [12,1,24] cannot be solved uniquely due to the large number of active muscle elements. Therefore, some numerical optimization method is required to provide estimates of the applied forces. Electromyography has been used for assessing muscle function, but this technique is limited to a small number of superficial muscles and the signal is not directly related to the muscle force magnitudes.

This paper will follow the two-dimensional analysis used by Poppen and Walker in 1976 and 1978. This considers elevation of the shoulder as a combination of the movement of the scapula across the thorax together with the articulation of the glenohumeral joint. In the normal shoulder the ratio of scapulothoracic to glenohumeral motion is 1:2, though this may reverse to 2:1 in the diseased shoulder, as glenohumeral motion becomes restricted [6].

This two-dimensional analysis is able to establish some quite fundamental biomechanical principles that can be applied to understand the stability of the shoulder in the plane of the scapula. It is justified since it has been shown that the predominant motion of the scapula during the humeral elevation is upward rotation [10]. Biomechanical stability implies that the glenohumeral articulating surfaces should remain in contact throughout the ranges of motion that the joint might naturally encounter.

The analysis presented herein also considers what this requirement for biomechanical stability implies for the design of shoulder prostheses. The first step is to set out some quite fundamental biomechanical principles that can be applied to understand the stability of the shoulder.

Basic Biomechanical Principles

To articulate a shoulder joint, muscle forces must act some distance away from the center of rotation to create a turning moment. As well as rotating the joint, these muscle forces have a tendency to cause the joint surfaces to translate if they are not balanced by an equal opposing force. In particular, the action of the deltoid muscle at the shoulder gives this combination of the abduction rotation of the glenohumeral joint with a tendency for the upward translation of the humeral head across the glenoid surface as shown in Figure 38.1. Excessive translation

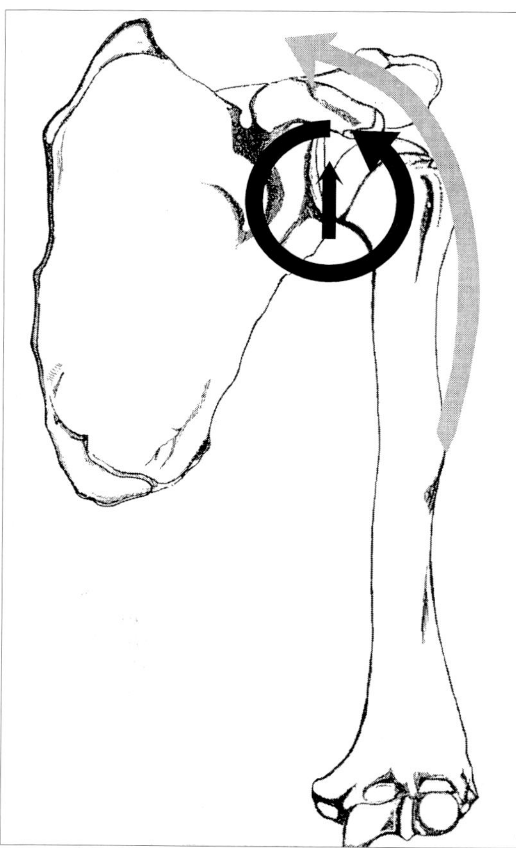

Figure 38.1. The force from deltoid contraction (gray) acts some distance from the center of glenohumeral rotation and therefore acts both to rotate the joint and to translate this superiorly.

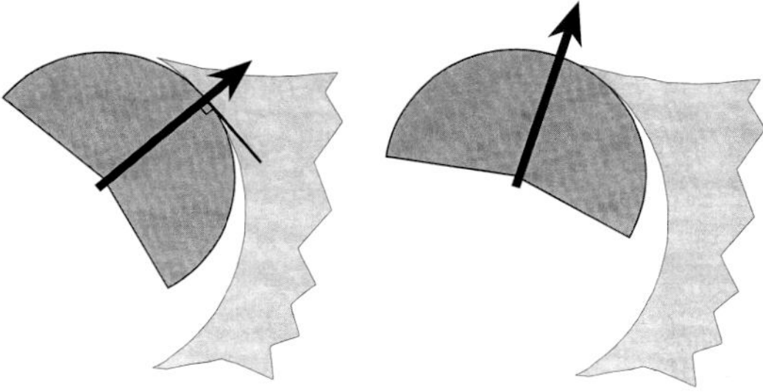

Figure 38.2. Under conditions of zero friction stability of the glenohumeral joint depends on whether the humeral head can slide over the glenoid surface to find a contact that is perpendicular to the applied force. The condition on the left is stable; that on the right is not.

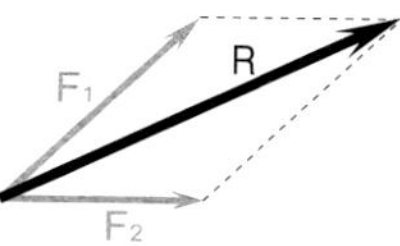

Figure 38.3. The length of the two connecting sides of the parallelogram F1 and F2 represents the magnitude of two forces. The resultant force R is represented in both magnitude and direction by the diagonal across the parallelogram of forces.

must be resisted for the joint to remain stable.

When a force is applied to the proximal humerus, the direction of this force determines whether the shoulder will remain stable or whether subluxation will occur. If we assume in the first instance that the effects of friction are negligible, then the glenohumeral joint force must be exactly perpendicular to the articular surfaces at their point of contact. That is, the position of the humeral head will move across the glenoid surface until the point of glenohumeral contact directly opposes the applied force. If no such point of contact exists then the joint will not be stable as indicated in Figure 38.2. This is the fundamental geometrical contact requirement for joint stability. Indeed, many of the experimental observations on shoulder behavior are a direct consequence of this fundamental principle.

Under the action of the deltoid muscle alone, the shoulder joint would tend to be unstable. The large range of motion requires that joint ligaments remain slack and only apply their passive constraints at extremes of motion. Active stabilization is needed and this is provided by the additional contraction of the rotator cuff muscles. To understand this combined effect of deltoid and rotator cuff contraction we must consider how their forces can be combined into a single "resultant" force. To combine the effects of two or more forces that act in different directions we can use the parallelogram of forces as shown in Figure 38.3. Note that the muscles that give rise to these force do not have to be attached physically to the same point, the parallelogram represents only their directions and not their physical location.

Also, the parallelogram of forces can be used to uncouple a single force R into its horizontal and vertical components, R_x and R_y respectively, as shown in Figure 38.4.

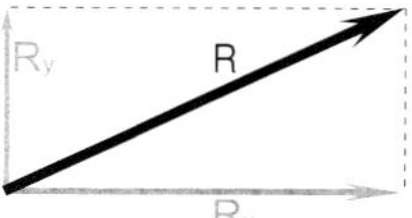

Figure 38.4. A single force R can be separated into its horizontal and vertical components, R_x and R_y using the parallelogram of forces concept. The two forces R_x and R_y will thus produce exactly the same effect as R.

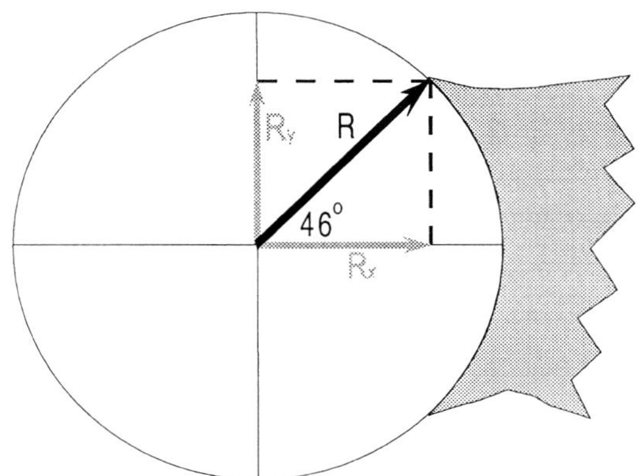

Figure 38.5. This scale drawing shows the mean anatomical supero-inferior dimension of the glenoid and the maximum angle of the joint resultant force R for joint stability. To maintain R within this maximum limit the ratio of $R_x : R_y$ must be less than 1 : 1.04. The angle 46° determines the "constraint", the higher this angle is the more constrained is the joint.

When performing any of the activities of daily living, numerous forces act upon the humerus. These forces are derived from muscle actions, ligament constraints, and any supported weights. If at such a time the arm is held in a steady position, then the humerus itself is said to be in a state of mechanical "equilibrium" in which all applied moments (forces multiplied by the distance to the center of rotation) balance, so the net turning effect is zero. Also, all the applied forces similarly balance, so that all the horizontal and vertical components sum to zero, and consequently the net resultant force on the humerus is zero.

Armed with these three basic concepts of (i) the geometric contact requirements for joint stability, (ii) the parallelogram of forces, and (iii) the balance of turning moments and forces for mechanical equilibrium, we now have the tools necessary for a mechanical analysis of the glenohumeral joint.

Glenohumeral Force and Joint Stability

Before considering the forces that act across the shoulder, let us first examine the stability of the anatomical glenohumeral joint. Any glenohumeral joint force R can be uncoupled using the parallelogram of forces into a force R_x directed centrally into the glenoid and a second shearing force R_y component that is perpendicular to this as shown in Figure 38.4. R_x tends to stabilize the joint whilst the component R_y tends to cause the joint to dislocate. The dominate effect depends on the curvature of the glenoid.

Iannotti et al. [9] have conducted a detailed study into the geometry of the glenohumeral joint and have found the geometry of the anatomical humeral and glenoid surfaces to be approximately spherical. The average radius of curvature of the normal glenoid surface is 27 mm with an average supero-inferior dimension of 39 mm. Therefore, in the scapular plane, this average glenoid surface will subtend an arc of ±46° about the center of the glenoid as shown in Figure 38.5. In the absence of other factors, the requirement for joint stability is that the direction of the joint force R must be within this arc. That is:

$$R_y / R_x < \text{Tan } 46°$$

$$R_y / R_x < 1.04$$

Therefore, whilst this R_y/R_x ratio value will normally vary as the shoulder moves, it must always be less than unity for the shoulder joint to remain stable in the supero-inferior direction. A consequence is that a high R_x force component will increase glenohumeral stability. Subluxation forces will therefore be directly proportional to the applied joint load, as has been observed experimentally by both Fukuda et al. [7] and Severt et al. [8]. Lippit et al. [15] referred to this effect as "concavity compression", noting that increased stability correlated with the depth of the glenoid concavity and indicated that scapular movement may also help centralize the glenohumeral articulation.

Other factors will also play a role in augmenting the stability of the glenohumeral joint. The periphery of the joint is surrounded by a fibrous tissue labrum that further resists joint subluxation by effectively extending the glenoid surface through a greater arc than is shown in Figure 38.5. The labrum consequently increases resistance to subluxation by between 18% and 52% around the glenoid [15]. However, on the other hand, the deformation of the humeral and glenoid articular cartilage will modify surface curvature under loading so that subluxation may begin inside the glenoid periphery. The above theoretical maximum R_y/R_x ratio for the normal glenoid therefore probably overestimates the limit to glenoid stability.

The humeral head radius of curvature is found typically to be 2 mm smaller than the radius of the glenoid [9,15]. Although the deformation of the articular cartilage that covers the opposing surfaces will accommodate some of this mismatch, the curvature difference will give the humeral head the capacity to slide across the face of the glenoid somewhat to find a position of mechanical equilibrium as shown in Figure 38.2.

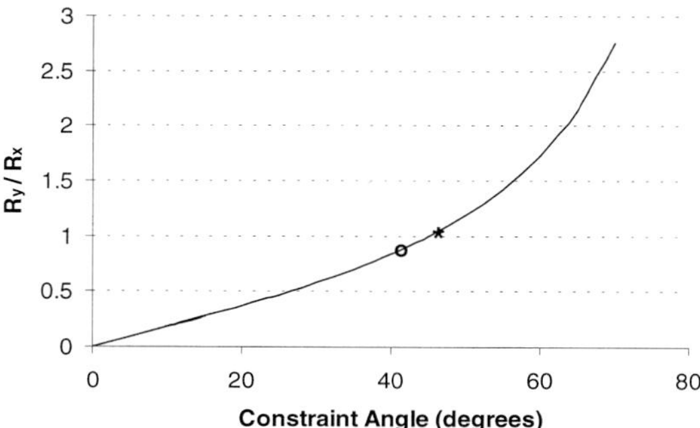

Figure 38.6. Variation of the maximum R_y/R_x value with the geometric constraint of the glenoid. The mark (*) shows the constraint of the average anatomical glenoid and (o) shows the comparable position of the Neer prosthetic glenoid.

Constraint and Conformity of the Glenoid Surface

The angle subtended by the glenoid, as shown in Figure 38.5, determines the *constraint* of the glenohumeral joint. The more the glenoid surface wraps around the humeral head, the greater is the joint stability, but this increased constraint reduces the available range of motion. Therefore, a compromise is needed. In the shoulder, the need for a great range of motion requires a significantly reduced constraint than, for example, in the hip where a greater stability and reduced motion is more suitable.

A second feature that defines the glenohumeral articulation is the *conformity* that is given by the mismatch between the radii of curvature of the opposing contact surfaces. As noted above, the glenoid radius of curvature may typically be 2 mm larger than that of the opposing humeral surface, enabling some degree of translation of the humeral head across the glenoid surface [11]. Therefore, one would expect any increase in conformity to reduce range of motion to some degree, as this sliding motion is more limited, although Harryman et al. [8] found this effect to be small. Reduced conformity does lead to reduced contact areas and thus increased contact stresses, so that again some form of compromise is required.

From the analysis given above, the maximum R_y/R_x ratio defining the limit for the stable shoulder will increase as the constraint increases, varying with the tangent of the glenoid angle as shown in Figure 38.6. Conformity would not be expected to affect the stability in the same way.

Forces in the Abducted Shoulder

Analysis of the kinematics of the shoulder is complex in three dimensions. The joint itself is a mechanically indeterminate system so that, for instance, the same amount of glenohumeral movement can be achieved though different muscle actions [24]. However, an insight into the basic mechanics and the stabilizing action of the rotator cuff can be gained from a simple two-dimensional mechanical analysis. The first such analysis was conducted by Poppen et al. [16] and

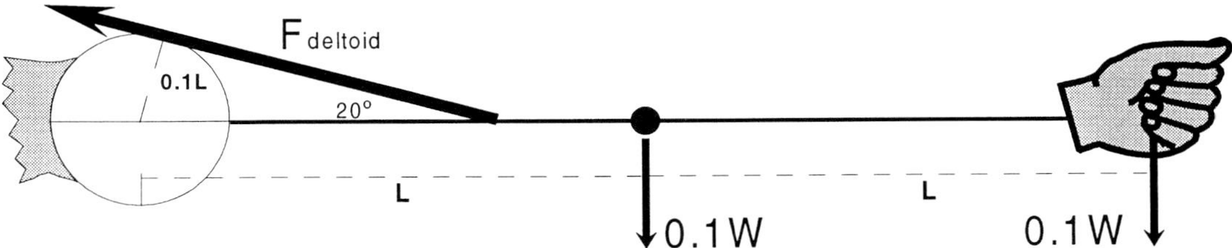

Figure 38.7. Loading example of the horizontal arm. The hand is carrying a weight of 10% body weight (0.1 W) and the weight of the arm itself (0.1 W) acts through the center of the limb. The two loads provide a clockwise turning moment that must be balanced by the anticlockwise moment of the deltoid force for the arm to remain abducted in equilibrium.

the analysis presented here is a derivation of this approach.

Here we will consider the situation shown in Figure 38.1, in which the deltoid muscle alone is acting to abduct the humerus in the scapular plane [1]. The major effect of the rotator cuff musculature in this two-dimensional analysis is to pull the humeral head into the face of the glenoid, though supraspinatus is also able to abduct the shoulder [1]. The role of the rotator cuff musculature in other planes, such as for humeral internal and external rotation, cannot be considered in this two-dimensional analysis.

A first example considers a horizontally outstretched arm of 10% body weight (0.1 W) supporting a further 0.1 W weight in the hand. In this arrangement, the force exerted by the deltoid muscle in order to maintain equilibrium with a horizontal abduction is shown in Figure 38.7. For equilibrium, the clockwise turning moment of the loaded arm must be matched by the anti-clockwise turning moment generated by the deltoid. That is:

$$0.1\text{W} \times \text{L} + 0.1\text{W} \times 2\text{L} = 0.1 \times \text{L} \times \text{F}_{\text{deltoid}}$$

$$\text{F}_{\text{deltoid}} = 3\text{W}$$

So that despite a fairly low physical loading (0.2 W), a deltoid force of three times body weight is needed to maintain equilibrium. This is simply because the deltoid acts through a small moment arm of only 5% of the limb length (2 L), and thus a much higher force is needed to create an equivalent turning moment.

The parallelogram of forces can be used to combine the forces that are now acting on the humerus into a single resultant force R as shown in Figure 38.8. To maintain equilibrium this

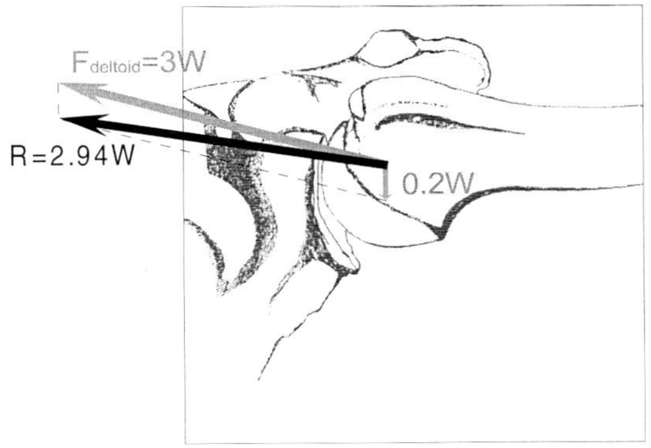

Figure 38.8. The forces applied to the humerus in Figure 38.7 can be collected into a single resultant force that has a magnitude of 2.94 times body weight and acts at 16.3° to the horizontal. Under this loading the shoulder is stable.

joint force is counteracted by an equal opposing reaction force from the glenoid onto the humerus. The single joint force R can be uncoupled using the same technique to give its horizontal and vertical components that are R_x and R_y. In this way R_x is simply the combined horizontal effect of the two original forces and R_y is likewise their combined vertical effect. Therefore, for the mechanical equilibrium shown in Figure 38.8:

$$R_x = 3\text{W}\cos 20^\circ = 2.82\text{W}$$

$$R_y = 3\text{W}\sin 20^\circ - 0.2\text{W} = 0.83\text{W}$$

The parallelogram of forces can be used to calculate the single joint force R where:

$$R^2 = R_x^2 + R_y^2$$

In this case the value of R is 2.9 W and the R_y/R_x ratio value of this resultant force is 0.29.

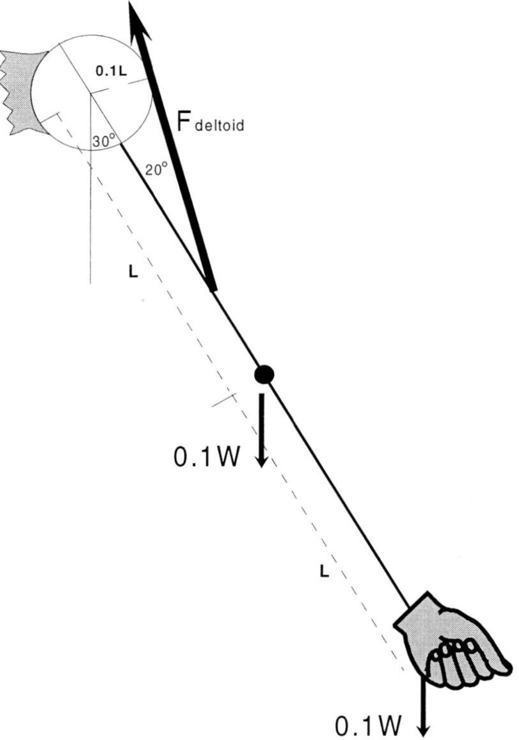

Figure 38.9. Loading example with the arm abducted to 30 degrees. The perpendicular distance of the 10% body weight in the hand and the weight of the arm from the center of rotation is smaller than in Figure 38.7 being 2 L sin 30 and L sin 30 respectively. Therefore, the clockwise turning moments is reduced compared to that in Figure 38.7 and the opposing deltoid force is smaller.

Remember that the shoulder will be unstable if this ratio exceeds a theoretical value of unity. Thus, one may conclude that equilibrium is likely be maintained by the arrangement of forces shown in Figure 38.8.

Now, in a second example, consider the same loading situation with the arm hoisted to only 30° of abduction as shown in Figure 38.9. In this case the balance of rotational moments gives:

$$0.1\text{W} \times \text{L} \times \sin 30^\circ + 0.1\text{W} \times 2\text{L} \times \sin 30^\circ$$
$$= 0.1 \times \text{L} \times \text{F}_{\text{deltoid}}$$
$$\text{F}_{\text{deltoid}} = 1.5\text{W}$$

So now a lesser deltoid force of 1.5 times body weight is required to maintain the equilibrium. However, the direction of this deltoid force is much more vertical, so that the parallelogram of force creates a more vertical resultant glenohumeral force as shown in Figure 38.10. The horizontal and vertical components of this resultant force are now:

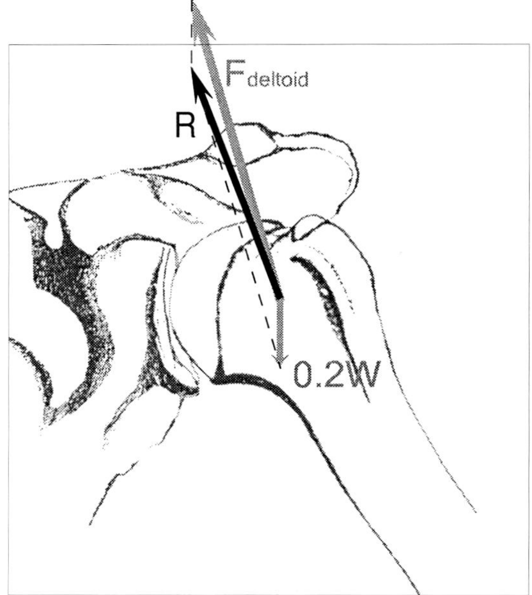

Figure 38.10. The forces applied to the humerus in Figure 38.9 can be collected into a single resultant force that has a magnitude of 1.3 times body weight and acts at 78.5° to the horizontal. Under this loading the shoulder is unstable.

$$R_x = 1.5\text{W}\cos 80^\circ = 0.26\text{W}$$
$$R_y = 1.5\text{W}\sin 80^\circ - 0.2\text{W} = 1.28\text{W}$$

Here the resultant joint force R magnitude is 1.3 W but the R_y/R_x ratio takes a value of 4.92. Under this resultant force, the shoulder would be expected to be unstable in this case and would certainly dislocate superiorly.

Now the action of the rotator cuff musculature can be superimposed onto the resultant joint force calculated using Figure 38.10. Here we will assume that this rotator cuff force acts perfectly horizontally. In combination with the joint force, a new parallelogram can be formed to calculate a new final resultant single force. In Figure 38.11 the R_y/R_x ratio value is brought down to 0.36 (tan 20°), at which point the shoulder may be expected to be stable. Effectively, the rotator cuff is reducing this ratio by increasing the denominator R_x value. However, this stabi-

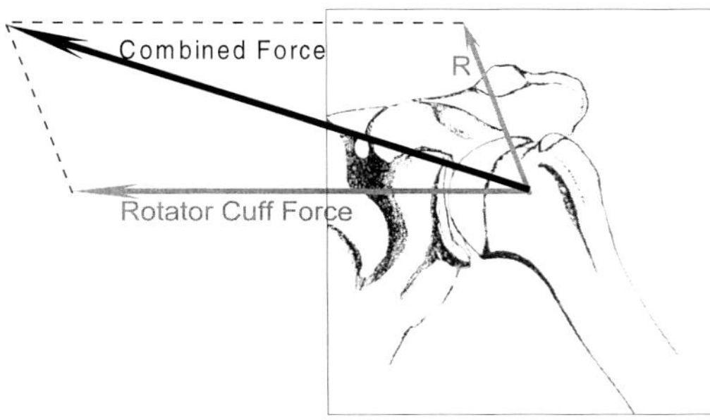

Figure 38.11. The rotator cuff muscles effectively increase the horizontal force component R_x so that the R_y / R_x ratio is reduced to a point where the shoulder can be stable. However, to achieve this with the forces shown requires a rotator cuff force of 3.24 times body weight.

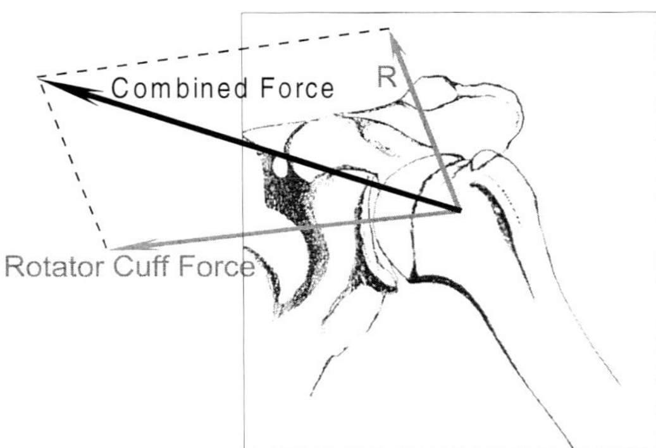

Figure 38.12. Displacing the rotator cuff force downwards somewhat has the effect of reducing the muscle forces and the combined joint reaction force appreciably compared to those in Figure 38.11.

lization requires a large cuff force of 3.24 times body weight and the resultant combined joint reaction force in now 3.72 W. In this case, rotator cuff activity from supraspinatus will be most effective to protect the joint from superior subluxation, although an intact "transverse force couple" (subscapularis, infraspinatus, teres minor) may itself ensure normal glenohumeral motion [20].

In a modification of the above scenario, Figure 38.12 shows the effect when the rotator cuff force is displaced downwards by 10°. With this fairly minor adjustment, the rotator cuff force needed to maintain stability is reduced by 31%.

Because the shoulder system is mechanically indeterminate, it is not possible to be sure that the above picture presents an accurate view of reality, as the same shoulder positions can be achieved through the actions of different muscles. Indeed, supraspinatus may have a role in the initiation of abduction [20]. However, the analysis does demonstrate that high loads can be created at the glenohumeral joint as the rotator cuff contracts to increase the normal force (R_x) to ensure joint stability. The effect of the rotator cuff is required more in early abduction when the direction of the deltoid force is more tangential to the glenoid fossa.

Anterior–Posterior Considerations

The whole of above analysis concerns the abduction of the glenohumeral joint in the scapular plane and movement is entirely supero-inferior. However, the same considerations apply in the antero-posterior (a-p) direction with activities requiring internal and external rotation. In this plane the glenoid fossa is a more complex pear shape with an average superior dimension of 23 mm and a larger inferior dimension of 29 mm [9]. Note that these sizes are smaller than the equivalent supero-inferior dimension, so that the glenoid is less constrained in the antero-posterior direction. The 1:0.8 ratio in the superior and inferior a-p dimension observed by Iannotti exactly matches the 1:0.8 ratio in superior and inferior a-p subluxation forces measured by Lippit et al. [15], once again showing how the glenoid concave geometry is a major determinant of shoulder stability.

In fact, using the theoretical techniques described above, inferior a-p subluxation should require a maximum R_z/R_x ratio value of 0.64. This will again probably be less in practise due to a flattening of the curvature of the articulating surfaces under load. Rotator cuff activity from subscapularis and infraspinatus will be effective in preventing anterior and posterior instability respectively.

The Effect of Joint Friction

The above analysis has been conducted under the assumption that the effect of friction at the glenohumeral articulation is negligible. The coefficient of friction of the cartilage surface of the anatomical joint is about 0.003, which contributes a friction force of only 3 thousandths of the resultant force calculated above. This may accurately be described as negligible. In the case of the prosthetic joint, the coefficient of friction for metal on polyethylene reaches approximately 0.08 so that again the friction force will comprise less than 10% of the joint reaction force. This friction will act in a direction to resist dislocation so that the joint will consequently be more stable by this same amount. Experimental measurement of frictional torque by Severt et al. [18] presents low values of 0.1–0.3 Nm, with the lowest values in this range associated with prostheses with the lowest conformity.

Glenohumeral conformity will have a greater frictional effect with rotation in the plane of the glenoid, as may occur in humeral elevation perpendicular to the scapular plane. Whilst the friction forces will be the same, the frictional torque will then depend on the radius of the joint contact area, which can be much smaller in the less conforming geometry.

The most significant effect of joint friction is likely to occur in hemi-arthroplasty where friction of the metal-on-bone articulation has been estimated to reach 0.3. Therefore, an additional frictional force 30% of the joint reaction force will enter the analysis and make the joint more stable against dislocation. However, the cost of this higher joint friction may be to increase the wear damage of the glenoid cavity which will change its geometry, possibly reducing the constraint and eventually leading to reduced joint stability. This behavior [14] appears in the posterior glenoid wear that may accompany hemiarthroplasty [14]. This loss of glenoid constraint may be restored with the use of a prosthetic glenoid.

Implications for Shoulder Prosthesis Design

The typical share of scapulothoracic to glenohumeral motion is 1 : 1.3 after shoulder arthroplasty rather than the normal 1 : 2 [6], and also the range of motion is often less that the normal shoulder. Therefore, one may expect that normal shoulder kinematics are not usually restored by this surgery. The anatomical position of the glenohumeral articulation is clearly important for rotator cuff function [13]. The balance of moments in Figures 38.7 and 38.9 demonstrates the importance of the deltoid moment arm which must be restored with sufficient lateral offset in a humeral prosthesis, otherwise even larger deltoid forces and joint reaction forces must be generated around the joint to achieve equilibrium.

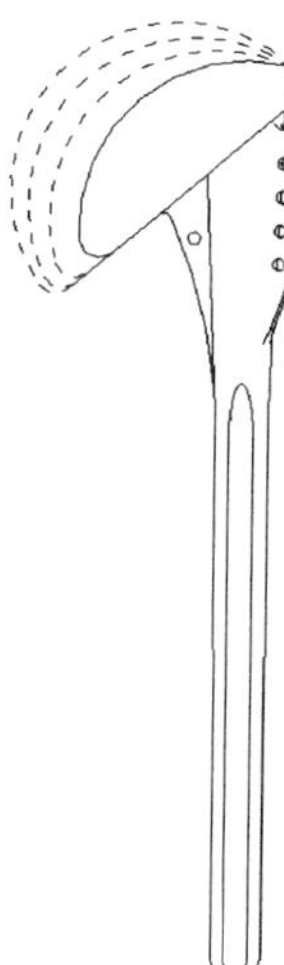

Figure 38.13. In the Neer 3 humeral stem design, the articulation of the larger head sizes is medialized to increase the deltoid moment arm.

In their cadaver study, Severt et al. [18] revealed that the lateral humeral offset that determines the deltoid moment arm increases with humeral head thickness. This anatomical feature is reflected in the design of the Neer 3 humeral component (Figure 38.13) in which the larger humeral head positions are adjusted medially to increase their effective deltoid moment arm.

As discussed above, the constraint and the conformity of the glenohumeral articulation are two principle features that determine shoulder motion and stability. On Figure 38.6 the supero-inferior constraint produced by the standard Neer II prosthetic glenoid component is indicated alongside that of the average anatomical glenoid. From the theoretical calculation as conducted above, the expected maximum R_y/R_x ration at subluxation for the standard Neer II glenoid prosthesis would be 0.89. This is 14% lower than the normal glenoid, although the smaller supero-inferior dimension of the Neer glenoid prosthesis (33.3 mm) is somewhat compensated by a reduced radius of curvature of 25 mm.

In fact, Severt et al. [18] experimentally measured a R_y/R_x ratio of 0.6 at subluxation with the Neer glenoid and this same figure varied from 0.2 to 0.8 with different types of glenoid prostheses depending on the prosthetic design and the direction of loading. Fukuda [7] similarly measured this ratio for the Neer II glenoid at 0.87 decreasing to 0.53 as the central loading (R_x) approaches body weight. These figures are close to the theoretical value shown in Figure 38.6. A difference between these experimental estimates and the theoretical calculation of the maximum R_y/R_x ratio is due to the effects of the deformation of the polyethylene prosthesis under load that reduces its effective curvature and hence its constraint. Indeed, this indicates the importance of achieving the full seating of the all-polyethylene glenoid prosthesis against a prepared glenoid cavity to protect against instability. A ratio value of 0.43 occurs when an all-polyethylene prosthesis is not fully supported on its bone-contacting surface.

Karduna et al. [11] have demonstrated that variation of conformity in a total shoulder prosthesis has a negligible effect on stability as indicated by the maximum R_y/R_x ratio. Therefore, a reduction in conformity with some radical mismatch between the humeral and glenoid articulating surfaces may provide for a more physiological motion with some glenohumeral translation without a penalty of a loss of stability. Severt et al. [18] found that reducing conformity increases the translation to reach the maximum subluxation force. This is to be expected from the analysis presented above (Figure 38.2). In fact, in the absence of polyethylene deformation, a perfectly conforming articulation would not translate at all prior to dislocation.

An increased conformity will increase the frictional torque in elevation rotation in the plane of the glenoid. This will inevitably increase the demands on the fixation of a glenoid prosthesis. On the other hand, a reduced conformity will increase the glenohumeral contact stresses that may take these values beyond the yield stress for the polyethylene and result in an increased wearing of this material [21].

All factors that have the virtue of increasing the stability of the shoulder by increasing the subluxation forces inevitably have the consequence that the higher forces will also place greater demands on the implant fixation. Therefore, it is important to consider how the changes of glenohumeral articulation may affect the stresses transferred from the implant to the bone, which may determine the durability of the fixation of the glenoid prosthesis.

Couteau et al. [2] have used a three-dimensional finite element model to examine the changes in scapular stresses with off-center loading. Figure 38.14 (plate section) shows that a principle effect of the off-center loading is a modification of the bending of the scapula in the vicinity of the posterior notch. One may expect that whilst this bending may normally serve a useful physiological shock-absorbing function, an exaggerated bending may challenge implant fixation. One may conjecture that central glenoid loading is likely to be most conservative for the implant fixation, with a reduction of a so-called "rocking-horse" effect [4]. Following on from the analysis given above, central glenoid loading is dependent on a proper functioning of the rotator cuff musculature. Indeed, rotator cuff dysfunction has been associated with glenoid implant loosening [4].

A second and surprising effect uncovered in the finite-element study of Couteau et al. [2] is the increase of implant bone interface stresses as the conformity of the glenohumeral articulation is reduced.

The scapular bone stock available for glenoid fixation and the condition of the rotator cuff musculature are two factors that must influence prosthesis selection and operative procedure. If the rotator cuff cannot be expected to aid joint stability, then a more conforming articulation that is inherently more stable may be preferred, even though this may place increased demands on glenoid fixation.

Summary

Although the shoulder is a non-weight-bearing joint, high loads can nevertheless be generated across the glenohumeral articulation.

The stability of the glenohumeral articulation, both anatomical and prosthetic, is dependent on the humeral and glenoid articulating geometry, the adequate bony support for a glenoid prosthesis, and also on the function of the rotator cuff musculature.

The mechanical demands placed on glenoid prosthesis fixation are also appreciably affected by the characteristics of the glenohumeral articulation, but the effects here are complex and require a full three-dimensional mechanical analysis to comprehend.

Acknowledgements

Dr Beatrice Couteau of INSERM U305, Toulouse, for biomechanical discussions and permission to use the graphic.

References

1. Bassett RW, Browne AO, Morrey BF, An KN. Glenohumeral muscle force and moment mechanics in a position of shoulder instability. J Biomech 1990;23: 405–15.
2. Couteau B, Hobatho MC, Darmana R, Mansat P. Finite element model of the scapula with an anatomically shaped glenoid implant – Analysis of the joint contact. Proc. XVIIth ISB congress (Calgary), 1999;181.
3. Doody SG, Freedman L, Waterland JC. Shoulder movements during abduction in the scapular plane. Arch Phys Med Rehab 1970;51:595–604.
4. Franklin JL, Barrett WP, Jackins SE, Matsen FA III. Glenoid loosening in total shoulder arthroplasty. Association with rotator cuff deficiency. J Arthroplasty 1998;3:39–46.
5. Freedman L, Munro RR. Abduction of the arm in the scapular plane: scapular and glenohumeral movements. J Bone Joint Surg [Am] 1966;48:1503–10.
6. Friedman RJ. Prospective analysis of total shoulder arthroplasty biomechanics. Am J Orthop 1997; 26:265–70.
7. Fukuda K, Chen CM, Cofield RH, Chao EY. Biomechanical analysis of stability and fixation strength of total shoulder prostheses. Orthopedics 1988;11:141–9.
8. Harryman DT, Sidles JA, Harris SL, Lippitt SB, Matsen FA III. The effect of articular conformity and the size of the humeral head component on laxity and motion after glenohumeral arthroplasty. A study in cadavera. J Bone Joint Surg 1995;77A:555–63.
9. Iannotti JP, Gabriel JP, Schneck SL, Evans BG, Misra S. The normal glenohumeral relationships. An anatomical study of one hundred and forty shoulders. J Bone Joint Surg 1992;74A:491–500.
10. Karduna AR, Williams GR, Williams JL, Iannotti JP. Glenohumeral joint translations before and after total shoulder arthroplasty. A study in cadavera. J Bone Joint Surg 1997;79A:1166–74.
11. Karduna AR, Williams GR, Williams JL, Iannotti JP. Joint stability after total shoulder arthroplasty in a cadaver model. J Shoulder Elbow Surg 1997;6:6–511.
12. Karlsson D, Peterson B. Towards a model for force predictions in the human shoulder. J Biomech 1992;25/2:189–99.
13. de Leest O, Rozing PM, Rozendaal LA, van der Helm FC. Influence of glenohumeral prosthesis geometry and placement on shoulder muscle forces. Clin Orthop 1996;330:222–33.
14. Levine WN, Djurasovic M, Glasson JM, Pollock RG, Flatow EL, Bigliani LU. Hemiarthroplasty for glenohumeral osteoarthritis: results correlated to degree of glenoid wear. J Shoulder Elbow Surg 1997;6:449–54.
15. Lippitt S, Matsen F. Mechanisms of glenohumeral joint stability. Clin Orthop 1993;291:20–8.
16. Poppen NK, Walker PS. Normal and abnormal motion of the shoulder. J Bone Joint Surg Am 1976;58A:195–201.
17. Poppen NK, Walker PS. Forces at the glenohumeral joint in abduction. Clin Orthop 1978;135:165–70.
18. Severt R, Thomas BJ, Tsenter MJ, Amstutz HC, Kabo JM. The influence of conformity and constraint on translational forces and frictional torque in total shoulder arthroplasty. Clin Orthop 1993;292:151–8.
19. Soslowsky LJ, Carpenter JE, Bucchieri JS, Flatow EL. Biomechanics of the rotator cuff. Orthop Clin North Am 1997;28:17–30.
20. Thompson WO, Debski RE, Boardman ND III, Taskiran E, Warner JJ, Fu FH, et al. A biomechanical analysis of rotator cuff deficiency in a cadaveric model. Am J Sports Med 1996;24:286–92.
21. Tomaszewski PR, Ondria JM. Incongruency and its relation to stress in prosthetic shoulder components. Orthopaedic Research Society, 1992.
22. Triffitt PD. The relationship between motion of the shoulder and the stated ability to perform activities of daily living. J Bone Joint Surg 1998;80A:41–6.
23. van der Helm FC. A finite element musculoskeletal model of the shoulder mechanism. J Biomech 1994; 27:551–69.
24. van der Helm FC. Analysis of the kinematic and dynamic behavior of the shoulder mechanism. J Biomech 1994;27:527–50.

39 Biomechanics of the Wrist Joint

Y. Mochizuki, Y. Ikuta, A. Ikeda, and I. Yoshii

Introduction

Man engages in various actions with use of his hands. For the effective and smooth execution of these actions, wrist joints have a unique bearing mechanism not seen in other joints. However, as the bearing mechanism of the wrist joints is complicated, repair of this mechanism once damaged is difficult. Thus, establishment of diagnosis and therapy of diseases of the wrist joints has been delayed for a long period. For the elucidation of the pathology of carpal instability and the etiology of dorsal intercalated segmental instability pattern of scaphoid fracture, which have attracted much attention in recent years, it is extremely important to ascertain the kinematics of normal wrist joints.

The wrist joint is a mixed joint composed of radiocarpal joint, midcarpal joint, and pisotriquetral joint for the radius and eight carpal bones. Furthermore, the carpal bones are divided into proximal carpal row and distal carpal row with scaphoid, lunate, triquetrum, and pisiform included in the former and trapezium, trapezoid, capitate, and hamate included in the latter. Active movement of the two rests with the distal carpal row that is connected by joints with metacarpal bones to which are inserted wrist extensors and wrist flexors forming the power source of wrist joint movement, and the carpal bones of the proximal carpal row excluding pisiform have only passive movement. Thus, the proximal carpal row serves as a bearing for wrist joints and only when this bearing mechanism is normal to various movements of the wrist joints become possible.

A large number of studies have been made on the kinematics of the wrist joints. [1,2,3,4,5,6,7] Based on radiograms obtained by changing body position the authors have reported on the involvement rate on wrist joint movement of radiocarpal joints and midcarpal joints. Arkless and Youm [8,9] with the use of cineradiography have studied the mutual kinematics of various carpal bones. In these previous analytical methods employing radiography and cineradiography, the angle among the various carpal bones on antero-posterior view and lateral view was measured and studied. According to these reports, the changes in the angles formed by radius and lunate were used as indices of movement of radiocarpal joint and the changes in the angles formed by capitate and lunate were used as indices of movement of midcarpal joints. The report was made on the involvement rate in wrist joint movement of radiocarpal joints and midcarpal joints. In all these reports, actual measurement was made on lateral views with measurement being feasible between radius and lunate, between radius and scaphoid, between capitate and lunate, between capitate and scaphoid, and between scaphoid and lunate. Measurement of kinematics around the triquetrum is extremely difficult. It was possible by previous methods of analysis to ascertain the maximum range of movement of the scaphoid at dorsal–volar flexion and at radial–ulnar flexion. On the basis of these findings, most of the reports have supported the columnar theory of Taleisnik [10]. However, by the heretofore employed method of analysis, movement on the ulnar side, particularly the movement around the triquetrum, had been extremely difficult and even by two-directional radiography only the outline of the movement of the carpal bones could be obtained by antero-posterior view and in most cases angle measurement could only be done based on the lateral view, thus limiting analysis of movement from one direction. Furthermore, carpal bones are small with complicated morphology and as their movement is primarily rotational, by angle measurement using radiograms characteristic findings are scanty and measurement has been limited to carpal bones. As carpal bones have multiaxial rotational movement without a fixed axis, the measured values do not necessarily accurately reflect the movement of carpal bones. Furthermore, a weakness has been pointed out that even by a slight twist during radiography the measured values show a great variation. In order to make a more detailed study of the kinematics of the wrist joints, the need has been emphasized that a measurement method to accommodate the three-dimensional movement of the carpal bones showing a multiaxial rotational movement should be employed. We therefore directed our attention to a strain gauge which is used in measuring the strength of machines and structures, and decided to employ strain gauges in analysis of wrist joint movement. A measurement apparatus was fabricated with the central part 5 mm in diameter made arch shaped so that even during multiaxial rotational movement of the carpal bone, strains corresponding to movement could be measured. Therefore, with the use of this measurement apparatus, we analyzed the kinematics of various carpal bones, particularly scaphoid, lunate, capitate, and triquetrum in volar-dorsal flexion of wrist joints, especially the movement of the triquetrum whose analysis had been extremely difficult.

Materials Methods

Cadaver Specimens

As materials, 20 cadavers preserved by arterial embalming without any trauma to the wrist joints, and 31 wrist joints were employed. The cadavers were composed of 10 males and 10 females and the 31 wrist joints were composed of 15 right wrist joints and 16 left wrist joints. Age at time of death ranged from 36 to 76 years with a mean of 65.7 years.

In arterial embalming, coagulated blood within the blood vessels of the cadaver was removed by a special infusion pump, followed by infusion of fixative in the blood vessels. Following embalming, the cadaver was preserved in a cold room at 5 °C. In comparison with cadavers embalmed only by formalin, in cadaver specimens treated by this embalming procedure the elasticity of soft tissue is satisfactorily maintained and the range of motion of the joint is almost equivalent to that of the wrist joint of a living human.

Measuring Apparatus

The measuring apparatus consisted of a board made of $20 \times 6 \times 1$ mm-sized polyethylene that was shaped into an arch with a diameter of 5 mm at the center. The top and sides were 4 mm in diameter, and a double axial strain gauge (Kyowa Dengyo) was attached. The objective of attaching a strain gauge at two sites was to investigate the possibility that measured strains would differ depending on where the gauge was attached.

By fixing both ends of this measuring apparatus with screws and adhesive, the measuring apparatus moved intimately with the carpal bones. In order to measure the strain corresponding to movement even when the carpal bones rotated multiaxially, the center part was processed into an arch shape with a diameter of 5 mm. This allowed a measurement that reflected three-dimensional movement of the carpal bones, and also enabled the analysis of movement on the ulnar side of the wrist, especially the triquetrum periphery, which had conventionally been very difficult to analyze.

Measuring Sites and Procedure

In order to elucidate the kinematics among the carpal bones during wrist movement, we selected the four carpal bones; i.e., the capitate, scaphoid, lunate, and triquetrum.

To avoid the breakdown of the physiological condition, we exposed the dorsal side of the wrist joint of the cadaver specimens. We preserved the ligaments and capsule on the dorsal side. Then we took bidirectional X-ray

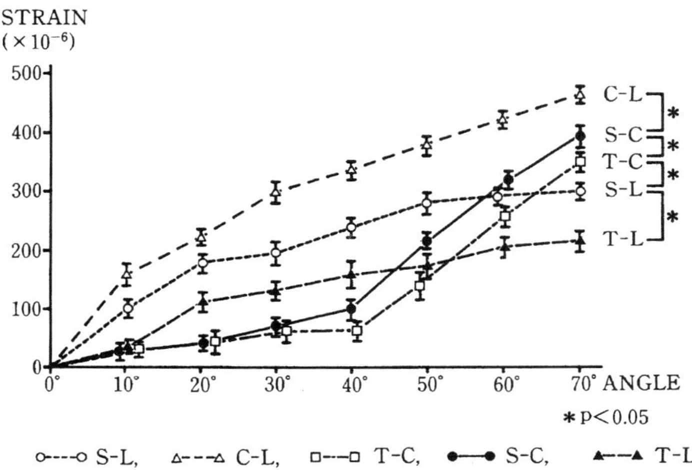

Figure 39.1. The relative motions between two selected carpal bones during dorsal flexion. (C: capitate, L: lunate, S: scaphoid, T: triquetrum).

images, and decided on the sites for inserting the screws.

The regression analysis was used for the statistical analysis.

Results

The measurement results and regression analysis results are detailed for each direction of movement.

Dorsal Flexion

The change of strain during dorsal flexion differed according to the measured site. The strain between the capitate and lunate reached a maximum during dorsal flexion at 70 degrees of dorsal flexion (Figure 39.1). The strain between the scaphoid and lunate was the second highest value up to 50 degrees of dorsal flexion. In particular, the increase in the strain was dramatic up until a dorsal flexion of 20 degrees. The change in strain between the triquetrum and lunate approximated that between the capitate and lunate. The strain between the scaphoid and capitate rapidly increased from 40 degrees of dorsal flexion.

The strain between the triquetrum and capitate also dramatically increased from 40 degrees of dorsal flexion, similar to that between the scaphoid and capitate. When the change in strain at each of the above measurement sites were examined by regression analysis, a significant difference was seen between each of the measured sites.

Volar Flexion

Unlike the dorsal flexion, during the volar flexion there were no measured sites at which the strain markedly increased during the movement (Figure 39.2). The strain increased during the volar flexion in the following order: Between the triquetrum and capitate, between the capitate and lunate, between the scaphoid and capitate, between the triquetrum and lunate, and between the scaphoid and lunate. When the change in strain at each of the above measurement sites were examined by regression analysis, a significant difference was seen between each of the measured sites.

Discussion

In 1976, Taleisnik slightly altered Navarro's concept [11]: The lunate and the entire distal carpal row was included in the central column and this was considered as the flexion–

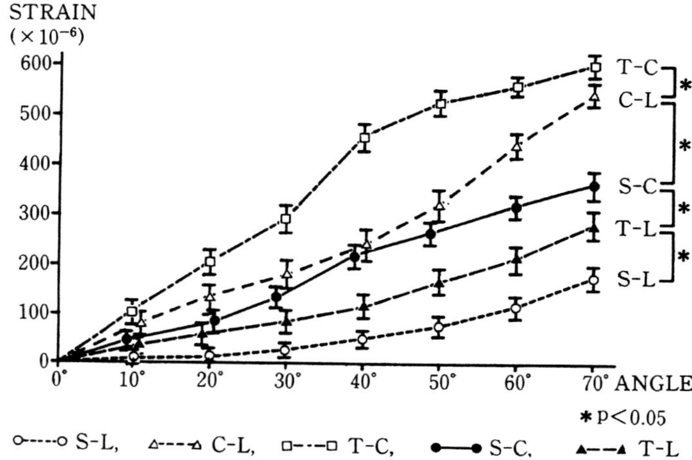

Figure 39.2. The relative motions between two selected carpal bones during volar flexion. (C: capitate, L: lunate, S: scaphoid, T: triquetrum).

extension column, and the lateral column was determined to be only the scaphoid. The distal carpal row and proximal carpal row moved together so this was called the mobile column. The medial column was comprised of only the triquetrum, and this mainly moved during pronation and supination so it was considered to be the rotation column.

However, in 1981 Lichtman [12] proposed the ring theory whereby the carpal bones were viewed as a ring made up of seven carpal bones. The normal movement control is performed by a radial link between the trapezium and scaphoid, and an unlike link between the hamate and triquetrum that are connected relatively weakly. A volar force was applied from the radial link and a dorsal force was applied from the ulnar link. The theory stated that the volar and dorsal forces are offset by the proximal carpal row where the movement is seen to operate mutually among three carpal bones to create stabilization.

These theories are conceptually taken as kinematics of the wrist joint, and in reality are not based on the quantitative data. Does the wrist joint really move as suggested by these theories, or are different kinematics shown? Regarding the kinematics of the wrist joint, we herein discuss the validity of past theories together with an examination of volar and dorsal flexion movement, based on our study results.

Dorsal Flexion

The kinematics differs among each of the carpal bones during dorsal flexion, and the greatest movement in the midcarpal joint was between the capitate and lunate. However, during movement between the scaphoid and capitate, and the triquetrum and capitate, strains were small up to 40 degrees of dorsal flexion, where movement was closely related, but from 40 degrees of dorsal flexion, movement rapidly increased. At 70 degrees of dorsal flexion the strain was shown to be about four times that at 40 degrees. Additionally, at 30 degrees or more of dorsal flexion, there was little increase in movement among the proximal carpal row.

These results suggest the following about dorsal flexion kinematics. The dorsal flexion force added to the distal carpal row is transmitted to the proximal carpal row via the midcarpal joint. The scaphoid and triquetrum that have a ligamentous connection with the distal carpal row dorsiflex while being intimately linked with the distal carpal row. However, a dorsal flexion force is transmitted to the lunate, which does not have a ligamentous connection with the distal carpal row, from the dorsiflexing scaphoid and triquetrum via the intercarpal ligament, and the lunate dorsiflexes due to the capitate pushing on the dorsal side pole of the lunate. Therefore, the greatest movement within the midcarpal joint is seen between the capitate and lunate. However, at a dorsal flexion of 30 degrees or more, the rate of increase in movement between the capitate and lunate is small. This is thought to be because of the radial distal joint surface slopes on the volar side, dorsal flexion of the lunate itself is restricted at 30 degrees. However, despite the limited dorsal flexion among the proximal carpal row, the distal carpal row continues to further dorsiflex; movement between the scaphoid and capitate and between the triquetrum and capitate, which had moved closely linked till that point, rapidly increase.

A comparison of carpal bone movement between the radial and ulnar sides revealed that the change in strain at the midcarpal joint between the scaphoid and capitate was significantly larger than the change in strain between the triquetrum and capitate. Within the proximal carpal row, the strain between the scaphoid and lunate was shown to be 1.5 times the strain between the triquetrum and lunate at 70 degrees of dorsal flexion, so a significant difference was noted between the two. In other words, it is thought that during dorsal flexion, even at the midcarpal joint and within the proximal carpal row, the ulnar side of the wrist moves more closely linked than the radial side of the wrist, and dorsal flexion force is transmitted to the lunate via the triquetrum more than the scaphoid.

Volar Flexion

The kinematics among each of the carpal bones during volar flexion differs as during dorsal flexion. The movement of the midcarpal joint is also always greater than movement among the proximal carpal row, in the following order: between the triquetrum and capitate, between the capitate and lunate, between the scaphoid and capitate, between the triquetrum and lunate, and between the scaphoid and lunate.

This suggests the following about volar flexion kinematics.

The volar flexion force applied to the distal carpal row is transmitted to the proximal carpal row via the midcarpal joint. The transmission of volar flexion force at the midcarpal joint is thought to be due to the shape of the surface of the joint at each carpal bone and friction. During the volar flexion, both at the midcarpal joint and within the proximal carpal row, the radial side of the wrist moves more closely linked than the ulnar side of the wrist.

Consequently, the scaphoid undergoes volar flexion due to the shape of the surface of the joint between the trapezium and scaphoid, which is the radial part of the midcarpal joint, and friction, and this volar flexion force is transferred to the lunate via the intercarpal ligament. However, the lunate volar flexes due to the capitate pushing on the volar side pole of the lunate in a similar way to dorsal flexion. The movement is greater between the capitate and lunate than the scaphoid and capitate due to the existence of these two types of transmission pathways. Furthermore, the results also suggested that together with this volar flexion force that is transmitted from the lunate to the triquetrum via the intercarpal ligament, a volar flexion force also emanates from the distal carpal row, and maximum movement is shown at the midcarpal joint between the triquetrum and capitate.

In previous reports, the kinematics of the ulnar side of the wrist joint have not been able to be analyzed, so it was thought that movement on the same plane would show the same kinematics in the opposite direction. However, the results of our study in which we analyzed the kinematics of the ulnar side of the wrist joint suggest that although dorsal flexion and volar flexion are movements on the same plane, the kinematics of each differs.

Conclusion

In summarizing the foregoing results of analysis, we reached the following conclusion regarding kinematics of wrist joints.

According to the columnar theory, lunate is included in the central column and lunate moves *en block* with distal carpal row. However, in our study, in the movement of wrist joints between capitate and lunate, movement of similar or larger magnitude was observed at other measurement sites. Therefore, we concluded that the columnar theory of considering wrist joint movement ignoring the midcarpal joint is not appropriate.

On the other hand, according the ring theory, unlike the columnar theory, movement of the midcarpal joint is considered important and the distal and proximal carpal rows move connected with radial link and ulnar link, and dorsal and volar flexion show similar kinematics with movement on the same plane. However, in the results of our study through movement on the same plane, the kinematics change according to the movement direction of the wrist joints and thus a conclusion was reached that the ring theory is also inappropriate.

According to our analysis, it is extremely difficult to analyze all the movements of the wrist joints under a single theory and we therefore reached a conclusion that the kinematics of wrist joints differ according to movement direction.

The wrist joint movement of the inclination direction showed a smoother movement than that of the other directions. The wrist joints passively move on the radius distal articular surface by external force transmitted to the midcarpal joint and serves as a type of bearing mechanism. It is therefore considered that the morphology of the radius distal articular surface which serves as base for the movement is a factor influencing wrist joint movement. In fact, in routine medical practice, when the inclination angle of the radius distal articular surface is changed as residual deformation following fracture of the wrist joint, we have experienced many cases whose range of motion of the wrist joint became restricted and impaired daily living. Thus, the morphology of the radius distal articular surface is considered to have a grave effect on wrist joint movement.

References

1. Destot E. Injury of the wrist. Springfield IL: Charles C Thomas, 1926.
2. Fisk GR. Carpal instability and the fractured scaphoid. Ann R Coll Surg Engl 1970;46:63–76.
3. Kapandji A. The physiology of the joints. Baltimore: Williams and Willkins, 1970.
4. Kuhimann JN. Experimenttele Untersuchungen zur Stabilitat und Instabilitat des Karpus. Frakturen, Luxationen und Dissociationen der Karpalknochen. Stuttgart, Hippokrates, 1982.
5. Lange A, Huiskes R. Kinematic behavior of the human wrist joint: A roentgen–stereophoto–grammetric analysis. J Orthop Res 1985;3:56–64.
6. Ruby LK. Relateive motion of selected carpal bones: a kinematic analysis of normal wrist. J Hand Surg 1977;13-A:1–10.
7. Sarrafian SK. Study of wrist motion in flexion and extension. Clin Orthop 1977;26:153–9.
8. Arkless R. Cineradiography in normal and abnormal wrists. Am J Roentogenol 1966;96:837–44.
9. Youm Y. Kinematics of the wrist. J Bone Joint Surg 1978;60-A:426–31.
10. Taleisnik J. Midcarpal instability caused by malunited fractures of the distal radius. J Hand Surg 1984;9-A: 350–7.
11. Navvaro A. Luxaciones del carpo. Anales de la Fracultad de Medicina 1921;6:113–41.
12. Lichtman DM. Ulnar midcarpal instability: clinic and laboratory analysis. J Hand Surg 1981;6:5115–23.

VI – Articular Biomechanics

VI B – Spine

40 Spinal Instability and Imbalance: Definition, Clinical Manifestations, and Biomechanics

J.-Y. Lazennec and G. Saillant

Spinal biomechanics have been the focus of many studies aimed at elucidating the roles of anatomic structures, at analyzing the mechanisms underlying injuries, at interpreting clinical syndromes, and at defining and optimizing therapeutic indications. "Spinal instability" is a somewhat hazy concept used to explain protean clinical syndromes, on the basis of imaging findings that are difficult to interpret and provide only indirect and presumptive information on structural lesions. Most studies of spinal biomechanics investigated the cervical, thoracic, and lumbar segments separately. Recent insights into global spinal balance and its interrelations with the pelvis and lower limbs have prompted substantial changes in rehabilitation techniques and in surgical strategies and procedures.

Spinal Instability

The Mechanical Underpinnings of Spinal Instability

The concept of spinal instability was first used in traumatology, primarily to explain mechanical abnormalities posited from direct evaluations of lesions on plain radiographs and CT scans and from indirect evaluations of adjacent soft tissues. "Spinal instability" was often defined in somewhat unclear terms as any abnormality carrying a risk of secondary displacement in the "short or medium term" [1]. The concept of spinal instability was then extended to a variety of nontraumatic conditions, such as tumors, infections, and, above all, degenerative disease. Thus, spinal instability came to be viewed no longer as an abnormality in itself but rather as a manifestation seen in many conditions so that several subdivisions were created, including trauma-related instability, degenerative instability, and iatrogenic instability.

Half a century ago, Knutson [2] reported that patients with low back pain had abnormal intervertebral mobility, both during flexion and during extension of the spine. This "segmental instability", first described in imprecise terms and on the basis of anecdotal reports, has given rise to controversial interpretations and to considerable speculation [3,4,5]. Marked divergences in opinion continue to exist, regarding both experimental and clinical data.

From the viewpoint of mechanics, a system is unstable if it cannot recover its initial state of equilibrium after being subjected to a disrupting force. The system may either enter a different state of equilibrium (not necessarily associated with a risk of subsequent instability) or remain unstable.

White and Panjabi [6] conducted experimental studies of the functional spinal unit, defined as two adjacent vertebras, the intervening disk, the interlocking facet joints, the anterior and posterior longitudinal ligaments, the ligamentum flavum, and the interspinous ligament. They found that coherence of the coupled elementary translational and rotatory movements was essential to the equilibrium of the system during all movements among the six degrees of freedom of the functional spinal unit. For each elemental movement (flexion–extension, for instance), the neutral zone is the region in which intervertebral motion occurs when the functional spinal unit is subjected to minimal moments (in practice, the moments that occur under normal conditions of spinal function). At each level, the neutral zone repre-

Table 40.1.

Study	L1–L2	L2–L3	L3–L4	L4–L5	L5–S1
Bakke	8.6	11	12	13.9	18.6
Plasmondon	8.1	11.7	13.7	14.3	
Tanz	5.6	76	8.6	12.2	8.2
Allbrook	6	8	13	19	18
Clayson	12.6	15.8	15.9	17.7	18.7
Froning	9	11	13	16	17
Pearcy	13	14	13	16	14
Hayes	7	9	10	13	14
Dvorak	11.9	14.5	15.3	18.2	17
Begg	10	12	14	15	
Cosentino	12	15	15	11	19
Dodd	13	14	13	15.2	
Putto		11.7	12.6	12.3	8.9
Panjabi	8.2	11.8	12.5	16.7	
Luk		10.5	12.5	14.5	16
Frobin	11.8	13.9	14.2	16.4	13.2
White-Panjabi	12	14	15	17	20
Maxi	13	15.8	15.9	19	19
Mini	5.6	7.6	8.6	11	8.2

sents the "safe range of motion" of the functional spinal unit during exposure to physiological loads.

There is a glaring lack of consensus about the criteria and methods appropriate for diagnosing instability [7–9]. The concept of segmental instability is frequently linked to kinematic abnormalities in intervertebral motion [10–14]. Considerable emphasis has been put on atypical centrode position responsible for excessive shearing forces applied to the disk [15] and to abnormally small [16] or high [17–21] degrees of intervertebral rotation.

White and Panjabi [6] suggested a simpler definition of instability as loss of the ability of the spine to maintain normal relations among vertebras during physiological loading.

In the sagittal plane, the relative displacements of the functional units vary widely across individuals and with the general orientation of the spine. Table 40.1 illustrates the substantial divergences in data from the literature. The neutral zone increases when alterations occur in the structures that hold the components of the functional spinal unit together (binding structures) or that limit the movements of these components relative to one another (restraining structures); examples include disk degeneration or rupture, joint distraction, and joint resection. Another cause of neutral zone increase is disturbance of the conditions in which these binding and restraining structures work (abnormalities in sagittal spinal balance). In vivo and under some experimental conditions, mobility is increased as compared to typical values, yet there is no instability: coupled movements remain coherent and the system remains in a state of stable equilibrium. This increased mobility is associated with a broadening of the safe range of motion. Thus "hypermobility" is not synonymous with "instability" [8].

The two vertebras in a functional spinal unit are connected by three "joints": the two facet joints and the intervertebral area linking the two vertebral bodies. These joints are stabilized by components whose effects can be considered passive; some are static (shape and size of the vertebras and orientation of the joint surfaces that act as guides), and others dynamic (viscoelastic structures such as the ligaments, capsules, and anulus). Active stabilization is provided by the muscles, including the main movement-producing muscles (psoas, quadratus lumborum, erector spinae, and abdominal wall muscles) and the postural muscles (multifidi, interspinales, intertransversarii, and rotators).

The instantaneous center of rotation of the functional spinal unit described by Pope and Panjabi [1] is located at the junction of the middle third and posterior third of the nucleus [15]. Various combinations of lesions differ in their effects on the position of this center. Lesions to anterior stabilizing structures, most notably disk degeneration (with intact posterior structures), causes the instantaneous center of rotation to shift backward [17]. During rotation and lateral bending of the trunk, the mechanical function of the posterior structures remains nearly normal, although the relations between adjacent vertebral bodies are abnormal. Conversely, when the posterior structures are more severely altered than the anterior structures, the instantaneous center of rotation tends to shift to the anterior part of the intervertebral space. Rotation or lateral bending of the trunk is associated with frank mechanical dysfunction of the facet joints, which can be shown on dynamic CT scans as paradoxical gaping of the facet joint on the side of bending or rotation [22–26].

Loss of stiffness has been suggested as a better descriptor of instability than increased and/or incoherent mobility [27,28]. With progression of degenerative disk disease, loss of disk height and growth of osteophytes modify the geometry of the intervertebral space and gradually reduce mobility. Alterations in the mechanical properties of tissues soften the anulus and cause laxity, thus reducing the stiffness of the system. Neither Mimura [29] nor Ebara [30] (who performed intraoperative measurements) were able to find evidence of restabilization. This concept casts instability in a new light and suggests that evaluating loss of stiffness may be more relevant than looking for hypermobility or determining centrode position or length [31].

This biomechanical description of the stability of functional spinal units, which is based on structural characteristics and on experimental findings, has been confronted with physical findings in patients, with the goal of defining a "clinical instability syndrome". However, efforts to characterize and interpret functional syndromes in the light of structural lesions or kinematic disturbances have cast considerable ambiguity on the term "instability" [8,32].

We believe destabilization and instability are two different situations [33]. Destabilization is characterized by loss of one or more of the components that ensure spinal stability by keeping the relations between vertebras within the neutral zone, i.e., within the safe range of motion at each spinal segment. Loss of these stabilizing components, which can be categorized as static or dynamic and as active or passive, can radically change the mechanical situation of one or more spinal segments. Instability is only one of the possible mechanical consequences of destabilization. The functional lumbar unit is considered to be unstable when it shows an increase in abnormal movements. The movements can be qualitatively abnormal (abnormal coupling of forces and moments, abnormal acceleration) or quantitatively abnormal (increased mobility, for instance). The instability can be symptomatic or asymptomatic: pain is believed to indicate tissue damage occurring because a mechanical threshold has been reached or transgressed, causing repeated mechanical stress outside the neutral zone.

General Organization of Stabilizing Components within Functional Spinal Units

The vertebras, which are the basic components of the functional spinal units, are connected by a deformable elastic zone, the mobile spinal segment described by Junghans [34] and composed, from front to back, of the anterior longitudinal ligament, the disk, the posterior longitudinal ligament, the facet joint capsules, the ligamentum flavum, and the interspinous and supraspinous ligaments.

Many theories have been put forward about the anatomic organization of stabilizing structures, Roy-Camille [35] suggested that the stabilizing structures could be grouped in three vertical segments and two horizontal segments. The anterior vertical segment includes the anterior longitudinal ligament and the anterior part of the vertebral body; the middle vertical segment the posterior vertebral wall, the posterior longitudinal ligament, the pedicles, the facet joints with their capsules, and the ligamentum flavum; and the posterior vertical segment the spinous processes and laminae. Lesions to the middle vertical segment carry a particularly high risk of destabilization. The two horizontal segments are the vertebra and the mobile spinal segment of Junghans [34] (disk and ligaments).

Denis [10] and Louis [36] each developed a three-column concept. Denis distinguished an anterior column (anterior part of the disks and vertebral bodies), a middle column (posterior part of the disks and vertebral bodies, posterior longitudinal ligament), and a posterior column (posterior ligamentous complex and facet joints). Louis, in contrast, differentiated an anterior column composed of the disks and vertebral bodies and two posterior columns, each extending along the facet joints on one side. The column concept is useful for analyzing segmental lesions but fails to take into account the major influence of the spinal curvatures.

Special Features of Spinal Structures

Many in vivo and biomechanical studies have investigated the effects of the intervertebral disk and facet joints on stability of the lumbar spine motion segments [2,19,24,37–42]. Although degenerative processes in the disk and facet joints may affect motion segment function, little is known about the natural history and tendency to progression of degenerative lesions.

Intervertebral Disk

The intervertebral disk is composed of two synergistic components. The nucleus distributes pressures toward the vertebral endplates and transmits them to the collagen laminae in the anulus; the elastic properties and spatial organization of these lamellae confer shock-absorbing and movement-limiting properties to the anulus. Thus, the anulus strongly links the two adjacent vertebras. The anulus also preloads the disk, serving as a shell that maintains the pressure within the nucleus [43] . At rest, there is a balance between the strong tension from the nucleus that tends to separate the vertebral bodies and the elastic resistance of the anulus that opposes this tension. During flexion–extension and lateral bending of the spine, the vertebral body slides over the nucleus in a movement combining rotation and translation toward the concavity of the spine. The nucleus tends to shift toward the convexity. During axial rotation of the spine, the vertebral body rotates on the nucleus, decreasing intervertebral disk height and increasing intradiscal pressure.

Changes in the mechanical behavior of the disk seen with advancing age mirror changes in physicochemical structure, most notably in permeability to water and concentration in proteoglycans, which govern the water content of the disk [44]. Experimental studies have shown that loading is associated with a decrease in water content and an increase in proteoglycan content within disks.

Studies [45] have shown a strong osmotic pump effect generated by intermittent disk compression during physical activity. This pump effect promotes good nutrition and preserves the mechanical function of the disk. Physical inactivity halts the pump effect and the fluid flows it produces.

Facet Joints and Capsules – Ligamentum Flavum

The posterior facet joints, their capsules, and the adjacent attachment of the ligamentum flavum have been extensively investigated in clinical and experimental studies because they are disrupted during some surgical procedures (such as nerve root decompression by the posterior approach) [46–49].

Nachemson [50] reported that the facet joints carry approximately 18% of the total compressive load applied to a lumbar spine segment, whereas other studies [51] found percentages of 33% to 0% depending on posture. The facets and laminae may carry as much as 70% of the compressive load at narrowed intervertebral disks [46]. Yang and King [52] found the facet loads to be 3% to 25% of the total load, with a maximum of 47%, depending on the severity of disk degeneration and facet joint osteoarthritis. Studies done after bilateral or unilateral facetectomy have shown that absence of the facet joints plays a key role in destabilization (particularly in patients with anterior disk lesions) [2,40,47].

As strongly emphasized by Posner [53] , the medial part of the surface of each lumbar facet joint is located in the coronal plane and the lateral part in the sagittal plane. The lateral part opposes axial twisting motions. The coronal part is the physical abutment that limits sliding forces, conveying them to the adjacent levels. Excision of the medial part of the joint surface (standard hemifacetectomy) reduces the restraining effect on sagittal translation, thereby overloading the remaining structures, most notably the disk. Excision of the posterior structures (laminae, ligamentum flavum, capsules, facet joints, and posterior longitudinal ligament) induces a new anatomic situation associated with accelerated disk degeneration [54].

The components of the functional spinal unit work together with an extraordinary degree of coherence. Many studies have investigated the prevalence of disk or facet joint lesions. Cadaver studies showed that some individuals had disk lesions with intact facet joints or vice versa. Few data are available on the relation between facet joint osteoarthritis and lumbar stability [55,56].

Ligaments [57,58]

The interspinous and supraspinous ligaments powerfully restrain flexion. They are virtually nonexistent at L5–S1. The ligamentum flavum is more flexible and consequently opposes less resistance.

The intertransverse ligaments limit lateral bending and axial rotation.

The anterior longitudinal ligament is a strong band and the only ligamentous restraint to extension of the lumbar spine. It is more fragile at the cervical spine.

The posterior longitudinal ligament theoretically limits flexion. It is strong at the neck but thin at the lower back.

Posterior and Anterior Muscles

The posterior and anterior paraspinal muscles play a crucial role that illustrates the interrelations linking the spine to the pelvis and lower limbs [34]. Improved understanding of the physiology of muscle chains has led rehabilitation therapists to direct their programs not only at strengthening the paraspinal muscles but, above all, at preventing contractures of the pelvispinal and lower limb muscle chains.

Causes of Destabilization

Any lesion of the spine can induce destabilization, which is confined to the segment involved but can modify the balance of the entire spine. Causes of destabilization include injuries (to bone, ligaments, or both), tumors, infections, iatrogenic damage, developmental abnormalities (scoliosis, spondylolisthesis), and degenerative and inflammatory disorders involving the bones and ligaments or the muscles (muscle diseases, neurological disorders). Each of these causes has its own distinctive characteristics. Nevertheless, at a given site, all destabilizing lesions, irrespective of their cause, induce similar effects: thus, infections, injuries, and iatrogenic lesions in a facet joint or disk, for instance, cause similar disturbances in spinal balance, although some of these lesions progress more rapidly than others. There are no intermediate situations: each stabilizing structure is either effective or ineffective.

Many biomechanical studies have investigated relations between degenerative disk disease and segmental instability. In most clinical studies, no significant relation was found between radiographic instability and disk degeneration [59]. Soini et al. [60] found no association between abnormal angular movement on flexion–extension radiographs and diskogram findings suggestive of disk degeneration. Murata et al. [61] reported decreased angular movement in patients with severe disk degeneration. A study by Holmes et al. [23] showed decreased stiffness in disks with moderate degeneration but not in disks with severe degeneration. Panjabi et al. [19] found that the neutral zones increased significantly with disk degeneration, especially in axial rotation and anterior and posterior shear motions. Degenerated disks showed faster creep and less viscoelastic behavior.

Schmidt [62] found that the presence of a high-intensity zone on MRI scans was associated with reduced stiffness of motion segments. Other investigators found no relation between disk degeneration and lumbar segmental instability (Nachemson et al. [63]), Rolander [64] noted that beyond 35 years of age no significant changes occurred in the tensile properties of the lumbar anulus fibrosus. However, these findings from in vitro biomechanical studies fail to take into account the effect of muscles. Furthermore, the precision and accuracy of the measurement methods vary from study to study. Clinically, motion can be evaluated only in the sagittal plane, a major shortcoming because instability during axial rotation or torsion may be marked [63].

Several stabilizing structures can be altered simultaneously, producing complex situations. A lesion predominating in one structure induces dysfunction of the other structures and modifies the spinal curvatures. In degenerative spondylolisthesis, for instance, the combination of vertebral slippage and facet joint lesions alters muscle activity and induces specific adaptation of the pelvic unit that can result in a distinctive posture characterized by marked posterior pelvic tilt. Fujiwara [56] reported interesting relations between facet joint osteoarthritis and segmental instability: facet joint osteoarthritis showed significant negative associations with abnormal tilting and with anteroposterior translational instability, but a significant positive association with anterior translational instability. These abnormalities were common in patients with degenerative spondylolisthesis, particularly those where facet joints were oriented in a more sagittal plane. Fujiwara [56] suggested that the disk may lose its anterior translational stiffness with increasing degeneration, whereas facet joint osteoarthritis may limit abnormal tilting and anteroposterior translation. Sato et al. [65] reported that radiographic instability was detectable before the occurrence of slippage in patients with degenerative spondylolisthesis. A reasonable hypothesis is that in some cases facet joint osteoarthritis may counteract the instability associated with pre-existent disk degeneration, whereas in others it may fail to stabilize the motion segment, thus allowing degenerative spondylolisthesis to occur.

Evolution of Destabilization

Destabilization compromises all spinal functions: strength, limitation of extreme mobilities, and protection of nervous structures. There is a risk of neurological compromise, either in the short term (trauma-related lesions) or in the long term via development of spinal stenosis (caused by changes in spinal curvatures, marked kyphosis, and/or facet joint hypertrophy). Sudden displacement immediately followed by settling can occur in some forms of destabilization, for instance after a vertebral body fracture with damage to the disk and ligaments or a locked facet joint. These severe lesions (classified type A3, type B, or type C in Magerl's scheme) [66] can lead to the erroneous interpretation that no further displacement can occur. Yet, additional and worrisome deformities can develop gradually as a result of soft tissue damage and, in some cases, of poor healing of anterior bony lesions (Figures 40.1, 40.2). Another example is orthopedic or surgical reduction by the posterior route, leaving a disk lesion that can worsen on its own if the anterior defect it creates is not filled. Improved knowledge of these causes of delayed and progressive destabilization occurring even after posterior fixation (Figures 40.3, 40.4) has changed some of the indications of orthopedic treatment of spinal fractures (particularly at the lumbar spine) and has prompted the development of new surgical strategies involving a complementary anterior approach for mini-invasive grafting to fill the defects created by disk lesions. Other lesions cause subtle destabilization with a risk of gradual displacement and insidious progression: examples include overlooked serious sprains, iatrogenic damage, and spondylolisthesis. The displacement occurs little by little. In some cases, a phase of stable "physiological" balance is followed by a new "nonphysiological" balance (Figures 40.5, 40.6) then, finally, by true instability (with hypermobility in some cases) (Figures 40.7, 40.8) or "delayed and permanent settling" (Figure 40.9).

Qualitative and Quantitative Evaluation of Destabilization

Some authors have emphasized the symptoms and signs of instability [7,67], whereas others have focused on radiographic findings [10,12,68–71].

Clinical Considerations

Few sound data are available on the clinical presentation of segmental spinal dysfunction with destabilization. No clear correlations have been identified between specific symptoms and specific types of instability or structural damage. The term "clinical spinal instability syndrome" has been used chiefly in an attempt to improve the homogeneity of functional descriptions of lumbar spine mechanical lesions, which are highly heterogeneous in their type, their combinations with other lesions, and their consequences.

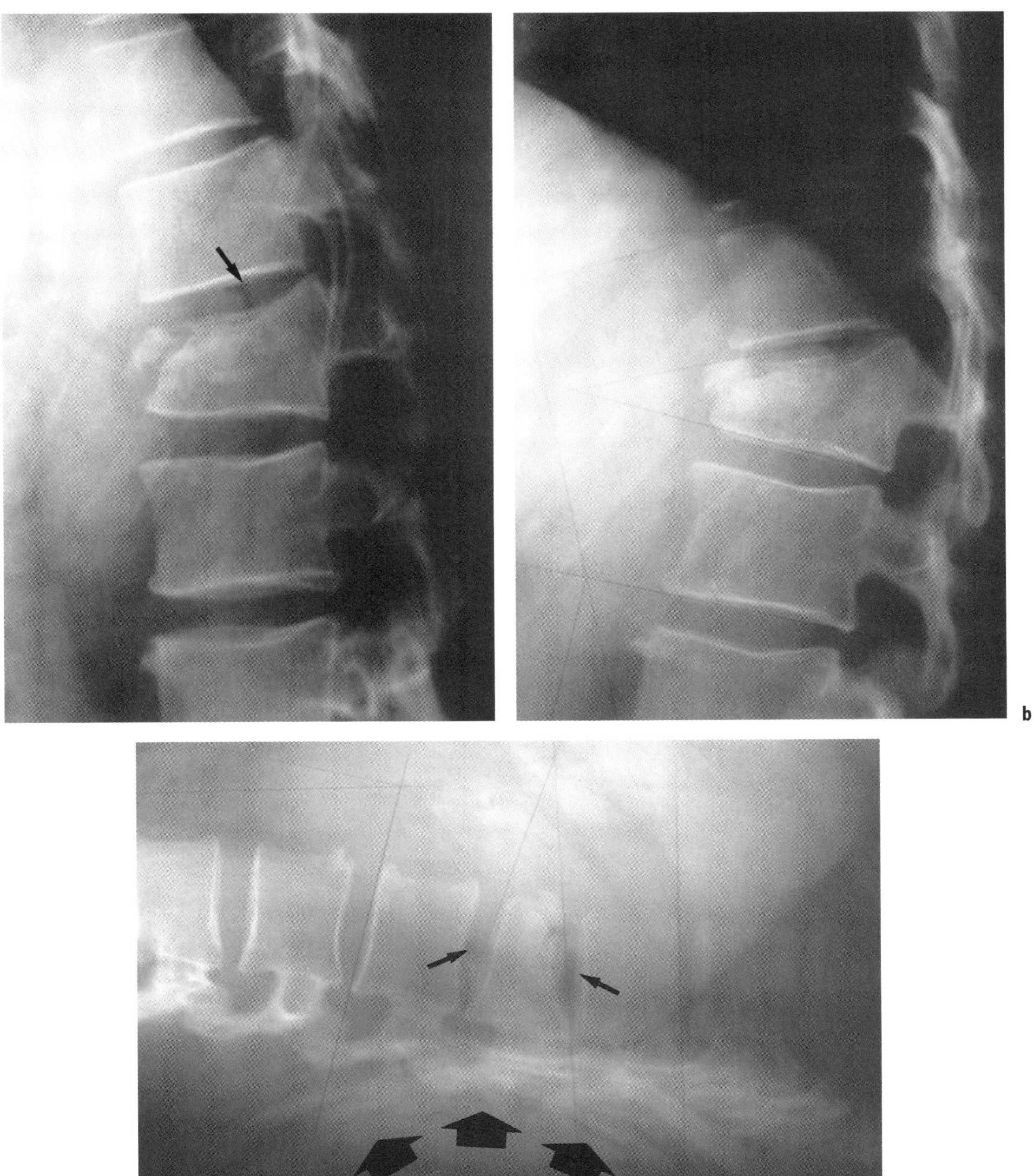

Figure 40.1. Nonoperative treatment after L1 fracture. **a** Immediately after the accident, note the important destruction of the anterior cortex and impaction of the superior endplate with vacuum phenomenon at T12 L1 disk. **b** Destabilization has induced instability with progressive kyphosis both at the bony level with progressive impaction of the anterior cortex and at the diskal level (loss of height and discal kyphosis). **c** Lateral dynamic X-ray with posterior holster: vacuum phenomenon not only at the superior T12 L1 disk but also at L1 L2 level showing a potential lesion of the inferior disk with possible deterioration later on.

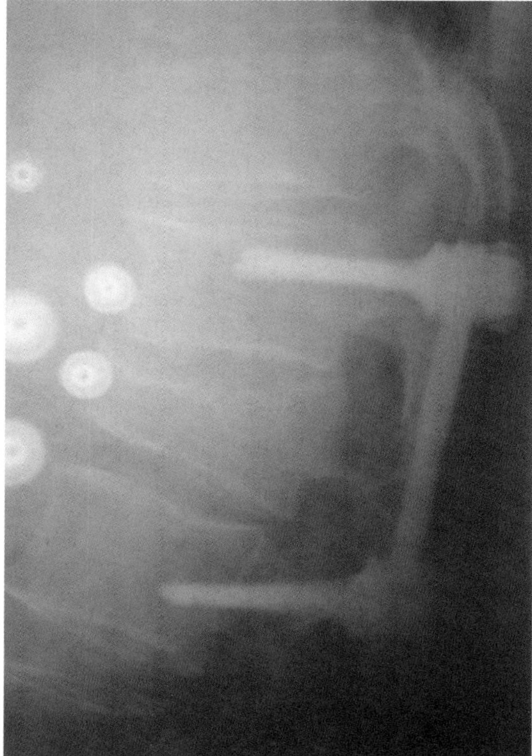
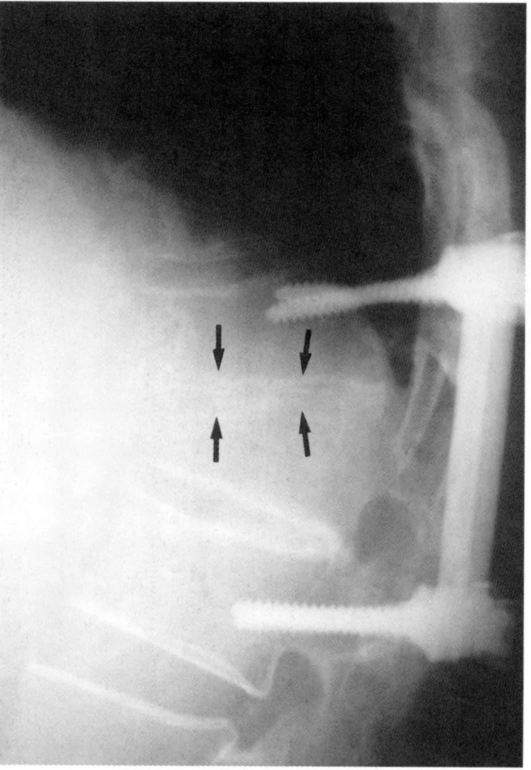

Figure 40.2. **a** L1 fracture treated via a posterior approach with pedicular screwing and graft, despite important anterior lesion and initial destabilisation. **b** Evolution two years later: progressive kyphosis due to diskal height loss at T12 L1 and L1 L2 diskal narrowing. Evolution to a new stabilization with global thoraco-lumbar malalignment. This status cannot be considered unstable.

"Acquired intervertebral derangement" [58] is a somewhat hazy concept because its symptoms are nonspecific and multifactorial. Furthermore, the pain is not correlated with the results of quantitative evaluations of motion. Although the clinical evaluation can be difficult, some signs are highly suggestive, particularly a sudden and unexplained exacerbation of the pain and alleviation of the symptoms by mechanical treatments such as a lumbar, brace or, for the lumbosacral junction, thigh-cuff bracing.

Kirkaldy-Willis and Farfan [3] suggested, in parallel with biomechanical and experimental concepts, that spinal degenerative disease may go through three stages defined on the basis of clinical and structural factors. Stage 1 is characterized by transient dysfunction, stage 2 by instability, and stage 3 by possible restabilization.

Stage 1 disorders [26] or dysfunction is the earliest phase. It can cause pain at one or more spinal levels or referred pain. Although the involved lumbar level does not function normally, pathologic changes and clinical consequences are minimal.

Stage 2, or major dysfunction, has been described mainly at the lumbar spine. There is general agreement that mechanical low back pain with pseudoradicular pain or referred pain [71] is typical, whereas true radicular pain is rare. Some patients report "axial" low back pain that does not extend to the lower limbs and is often felt to denote a "disk lesion". Others present with better-defined pain more marked on one side and accompanied with proximal pseudosciatica often ascribed to facet joint dysfunction. These two patterns of pain can coexist. "Dynamic catching of a nerve root" is the pre-

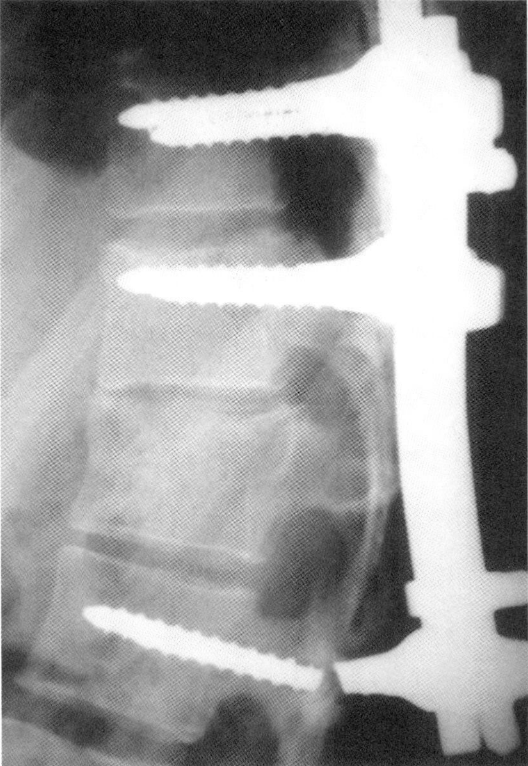
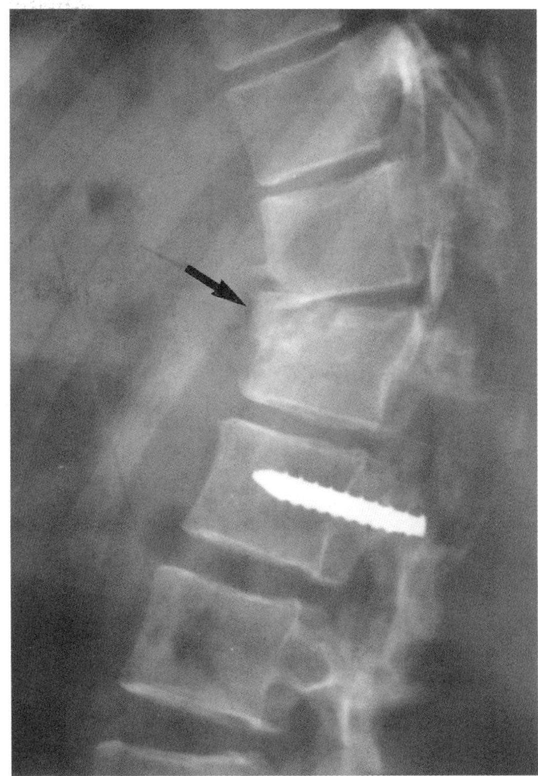

Figure 40.3. Asymetric fixation for L1 fracture. This patient was treated via a posterior approach for apparently mainly bone lesions with correct restitution of the vertebral height. Posterior graft was not sufficient and rupture of the implant occured with progressive loss of height at T12 L1 disk. Unfortunatly, the implants were removed without complementary fixation inducing a dramatic destabilization with evident instability at T12 L1 mobile spinal segment.

senting symptom in some cases. Many patients have misleading symptoms with pain from extraspinal structures (the sacroiliac joint or the muscle and fascia of the quadratus lumborum or piriformis). There have been reports of agonizing but very short-lived shooting pain made worse by sneezing (but not by coughing or defecating) and, above all, by twisting motions of the trunk, which sometimes produce a cracking sensation in the posterior spine [50]. Position has a major influence: prolonged standing, lying down, or flexion are particularly likely to worsen the pain. The changes in spinal curvature induced by sitting on a high or low seat also cause pain. Intra- or periarticular lidocaine injections may have a shorter-lived effect than in patients with common degenerative disease. *True instability* is often characterized by a decrease in disk height. The anulus fibrosus bulges all around the disk, the ligaments and joint capsules are loose, and the joint cartilage is altered.

Stage 3, characterized by the most severe degree of dysfunction, is the last stage. *Restabilization* is a result of fibrosis and formation of osteophytic abutments. In degenerative disease of the lumbar spine, stage 3 is often characterized by restabilization or by development of degenerative pseudospondylolisthesis or rotatory dislocation. The symptoms change, and pain can develop as a result of osteoarthritic changes in neoarticulations between the spinous processes. Restabilization can produce a new stable equilibrium, for instance when vertebral slippage in degenerative spondylolisthesis is stopped by osteoarthritic changes. Another

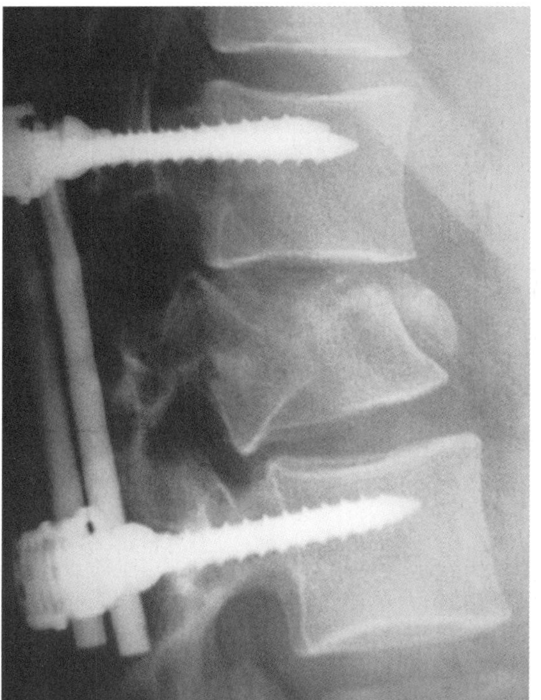
a

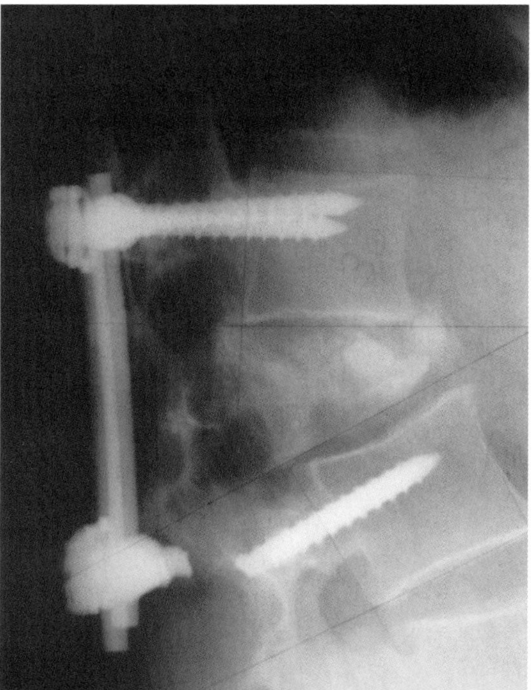
b

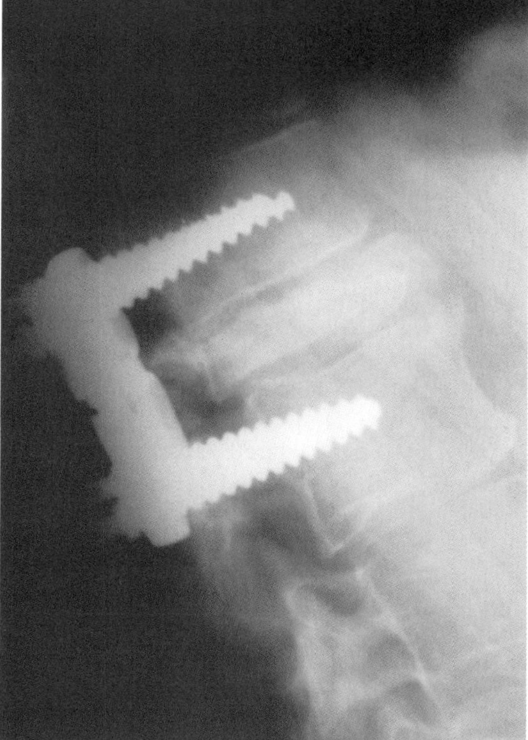
c

Figure 40.4. **a** Comminutive fracture of L2 treated via a posterior approach with pedicular fixation. No graft either posterior or anterior despite severe lesions of the vertebral body and disko-ligamentous structures. **b** Evolution four months later. Complete collapse of the anterior wall with major kyphosis and rupture of the posterior implants: evident destabilization with instability. **c** Treatment via a posterior approach (wedge osteotomy): restitution of better sagittal equilibrium with fusion (compression and grafting).

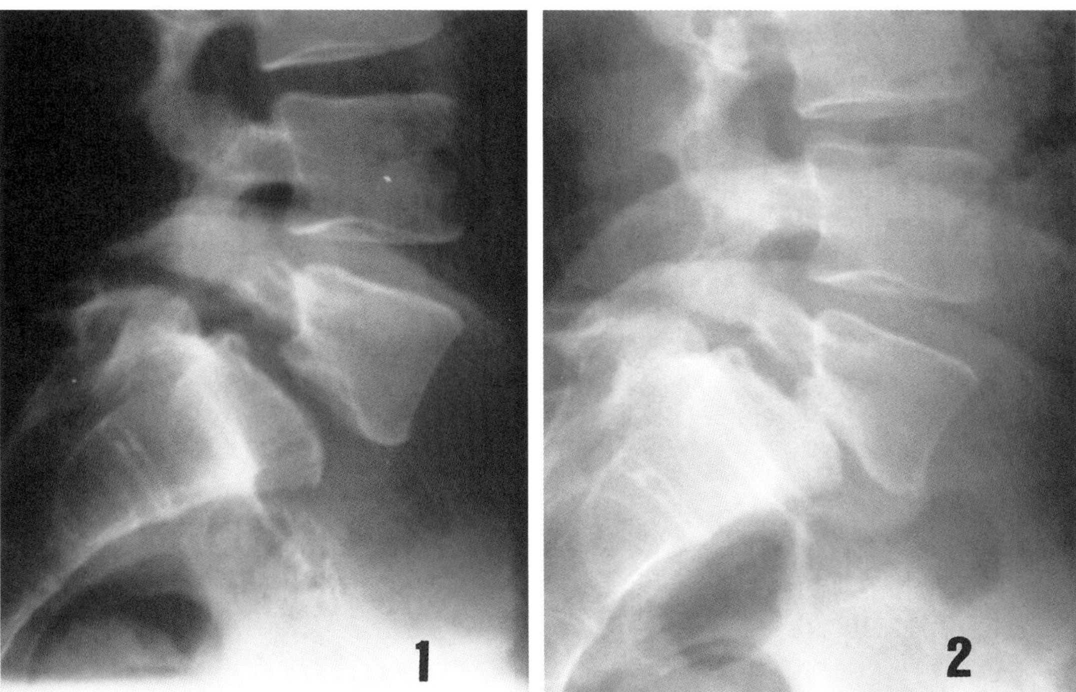

Figure 40.5. L5 S1 spondylolisthesis (L5 and S1 dysplasia). Note progressive displacement with final re-stabilization. In this case, no further significant instability.

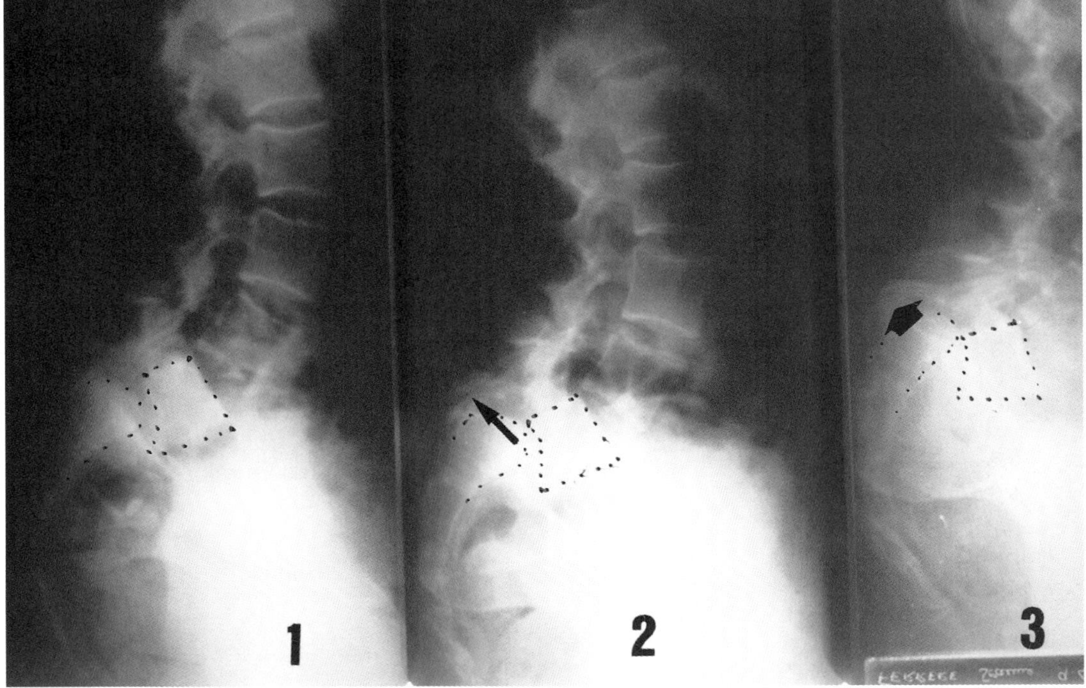

Figure 40.6. L5 S1 spondylolisthesis. Note progressive destabilisation with L5 slipping and sacral retroversion. Between stage 2 and stage 3, six years evolution with lack of restabilization. At ultimate follow-up a new equilibrium is obtained (stable situation with poor clinical status).

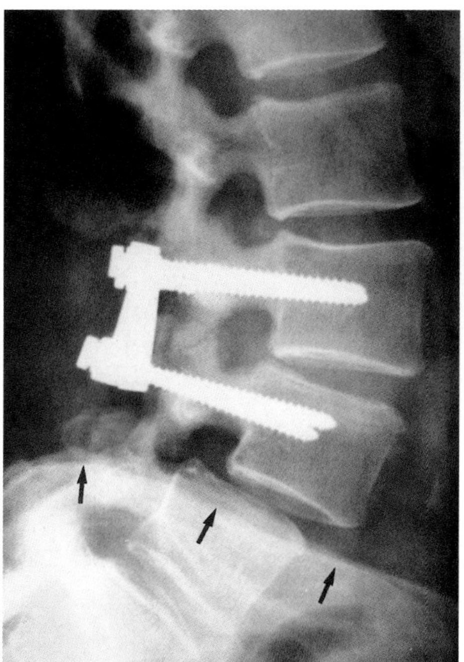

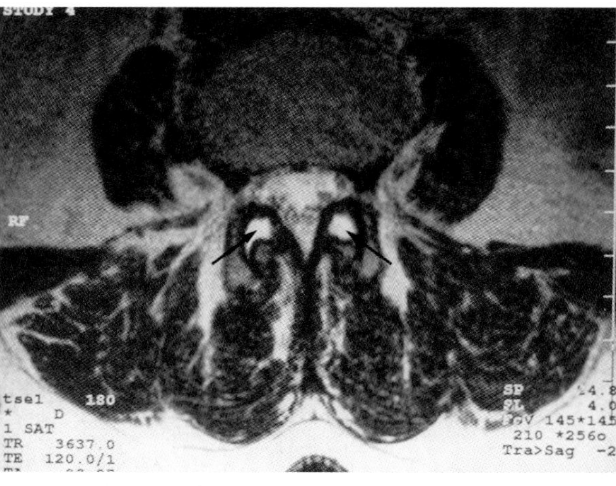

Figure 40.7. Late evolution after L3 L4 posterior floating fusion. Note progressive anterior slippage at L4 L5 level. The destabilized level is just above the iliac crest projection (degenerative articular joints with subluxation (**a**) and lesion of the mobile segment).

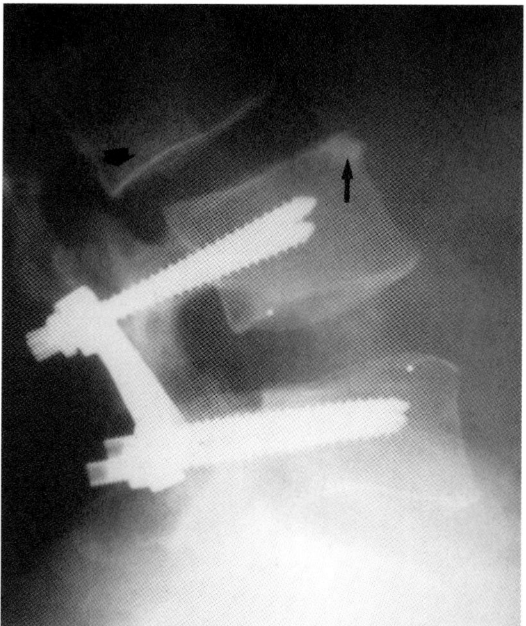

Figure 40.8. Evolution one year after L4 L5 floating fusion with excessive distraction (anterior lumbar interbody fusion). Note L3 L4 retrolisthesis and the reaction of the superior L4 endplate (evolution to MacNab osteophyte). Possible restabilization may be expected.

possibility is a succession of unstable equilibriums, for instance when destruction due to a tumor or infection occurs more rapidly than the changes that tend to stop further displacement.

Destabilization cannot be readily quantified based on clinical grounds because pain is a subjective symptom that varies across patients. Several scoring systems exist for monitoring the course in individual patient's subjective symptoms, physical findings, self-sufficiency, and activities of daily living.

Imaging Studies

The clinical instability syndrome is viewed as a situation in which, in the absence of further lesions, a physiological load induces abnormally marked deformation of the intervertebral space. Movements and loading tests induce changes in vertebral body alignment that are considered characteristic of instability. Most studies of abnormal movements ascribed to instability focused on flexion and extension in the sagittal plane and found two types of abnormalities: atypical changes in the angles reflecting endplate orientation and abnormal parallel dis-

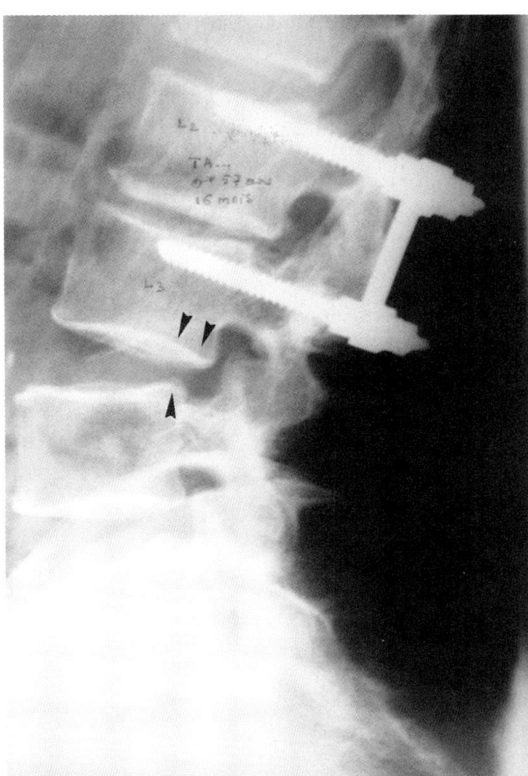
a

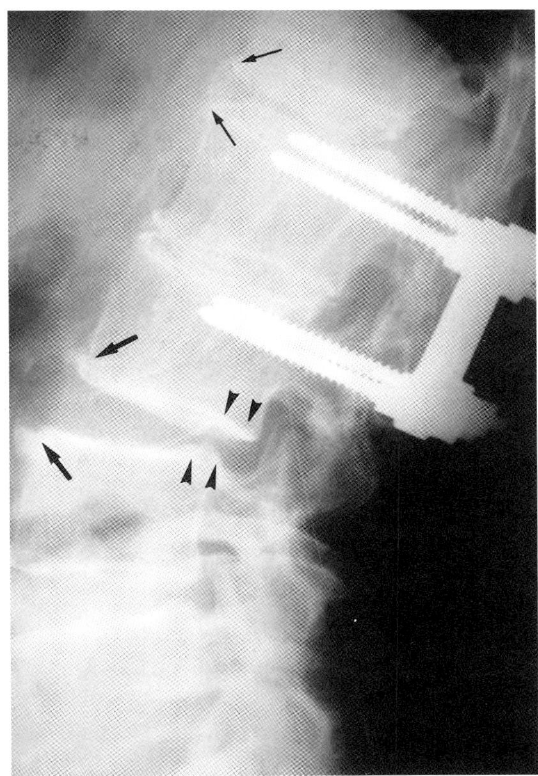
b

Figure 40.9. **a** L2 L3 posterior fusion. a. 16-month follow-up: L3 L4 discal opening and retrolisthesis could be observed. **b** six-year follow-up: note the collapse of the disk at the fused level. No real instability has occured but evolution to some restabilization (MacNab osteophytes) (arrows) associated with progressive anterior bridging of L1 L2 level.

placements of the vertebras, from backward slippage during extension from forward slippage during flexion.

Many methods are available and can be used to obtain a converging set of arguments. Some of these methods are static, others dynamic.

Plain anteroposterior and lateral radiographs remain essential. Loss of disk height cannot be readily standardized because substantial variations in normal disk height exists across individuals. Calculating the disk height as a percentage relative to the adjacent disk has not proved useful. Intradiscal vacuum phenomenon [72], which denotes the presence of gas within the disk, is usually interpreted as a sign of degeneration associated with a risk of destabilization. Similar images can be seen within articular facet joints, the spinal canal, or even in some cases within the vertebral endplates.

Spondylophytes, or traction spurs, normally increase with advancing age. First described by Macnab [32], the traction spur typically arises 2–3 mm from the vertebral endplate and juts out horizontally, showing no tendency to curve toward its neighbor on the adjacent endplate. Traction spurs are often accompanied by facet joint distraction responsible for degenerative pseudospondylolisthesis with loss of disk height. Plain radiographs can be used to evaluate the position of the abnormal vertebral segment relative to the overall spinal curvatures.

Mechanical spinal function can be seriously affected by excessive or inadequate sagittal curvatures or by abnormalities in the position of the pelvis. Comparing the sitting and the standing positions can be extremely useful for detecting atypical angulation or translation seen on lateral radiographs.

Dynamic films are now widely used [62,73–75]. However, extensive experimental and clinical studies have failed to produce standards of reference for evaluating dynamic radiographs. Although the definition of radiographic instability has generated controversy, several types of segmental instability have been described, including abnormal tilting during flexion, rotatory instability (exceeding the range of motion), and translational instability.

Dynamic radiographs are used mainly to evaluate lumbar intervertebral mobilities in the sagittal plane [76]. The angle of rotation can be determined for each vertebral segment as the spine moves from exaggerated lordosis to exaggerated kyphosis. Abnormal tilting during flexion was first described by Kirkaldy-Willis and Farfan [3] as a form of segmental instability. However, these authors did not provide a defining criterion. In a study of the normal range of motion on flexion–extension radiographs in asymptomatic subjects, Hayes et al. [22] found that the mean range of motion was 10° at L3–L4, 13° at L4–L5, and 14° at L5–S1. In a similar study, Boden and Wiesel [77] reported that the normal range of motion was 7.7° at L3–L4 and 9.43° at L4–L5 and L5–S1. Soini et al. [60] defined abnormal motion on flexion–extension radiographs as less than 20° at L5–S1 and more than 15° at higher lumbar levels. Murata et al. [61] defined segmental instability as angular motion greater than 15°. It has been suggested that flexion–extension radiographs should be taken in the recumbent position. Wood et al. [78] noted that taking the films in this position was associated with less pain and with stronger flexion–extension efforts on the part of the patients. Fujiwara [56] defined abnormal tilting during flexion as more than 3° of posterior opening. Given the considerable reading errors that occur because of the many obstacles to reproducibility (pain, difficulties met by patients in performing repeated flexion and extension), this evaluation is of limited usefulness, as pointed out by Dvorak and Panjabi [20,27]. With the classical goniometer technique, measurement error varies from 8° to 10° and can lead to overdiagnosis or underdiagnosis of abnormal rotation (Figure 40.10). The complex method described for determining angulation and translation on lateral radiographs using the Lushka uncinate processes as landmarks does not reduce measurement error [13]. In a study of radiographs obtained in two planes perpendicular to each other, Pearcy et al. [79] found that flexion in patients with low back pain was restricted and accompanied by other movements. There is a striking discrepancy between their results and those reported by Penning [74], who was unable to find evidence on anteroposterior and lateral radiographs of the abnormalities in instantaneous centers of rotation that have been described in the biomechanical laboratory. A study of healthy subjects conducted by Hayes [22] produced disturbing results: angular data varied widely across individuals and across spinal levels.

Although the term "rotatory instability" has been used in the literature [23,80], there is no consensus on the criteria for defining this type of instability. Only translational and sagittal rotation criteria have been recognized as important components of segmental instability [81]. Hayes et al. [22] found that 4 mm of translation occurred in 20% of asymptomatic subjects. However, Boden and Wiessel [77] recommended 3 mm of dynamic slip as the cut-off for translational instability because only 5% of their asymptomatic subjects had values greater than 3 mm. Many other investigators have used dynamic slip greater than 3 mm as the criterion for translational instability [39]. A study conducted in normal subjects with a negative history for low back [82] (pre-employment medical examination) found up to 8 mm of translation at L4–L5 (mean, 3 mm) with 2–20° of mobility (mean, 13°).

Stokes and Frymoyer [75] showed that measurements of spinal angulation, flexion–extension, and translation failed to provide compelling evidence of instability at levels suspected of instability. Morgan and King [83] reported that sagittal translation varied from 3 to 17 mm in a series of patients of whom only 28.6% had low back pain. However, those patients with clear evidence of longstanding degenerative disease had abnormal flexion–extension and translation at the diseased levels, as compared with the other levels.

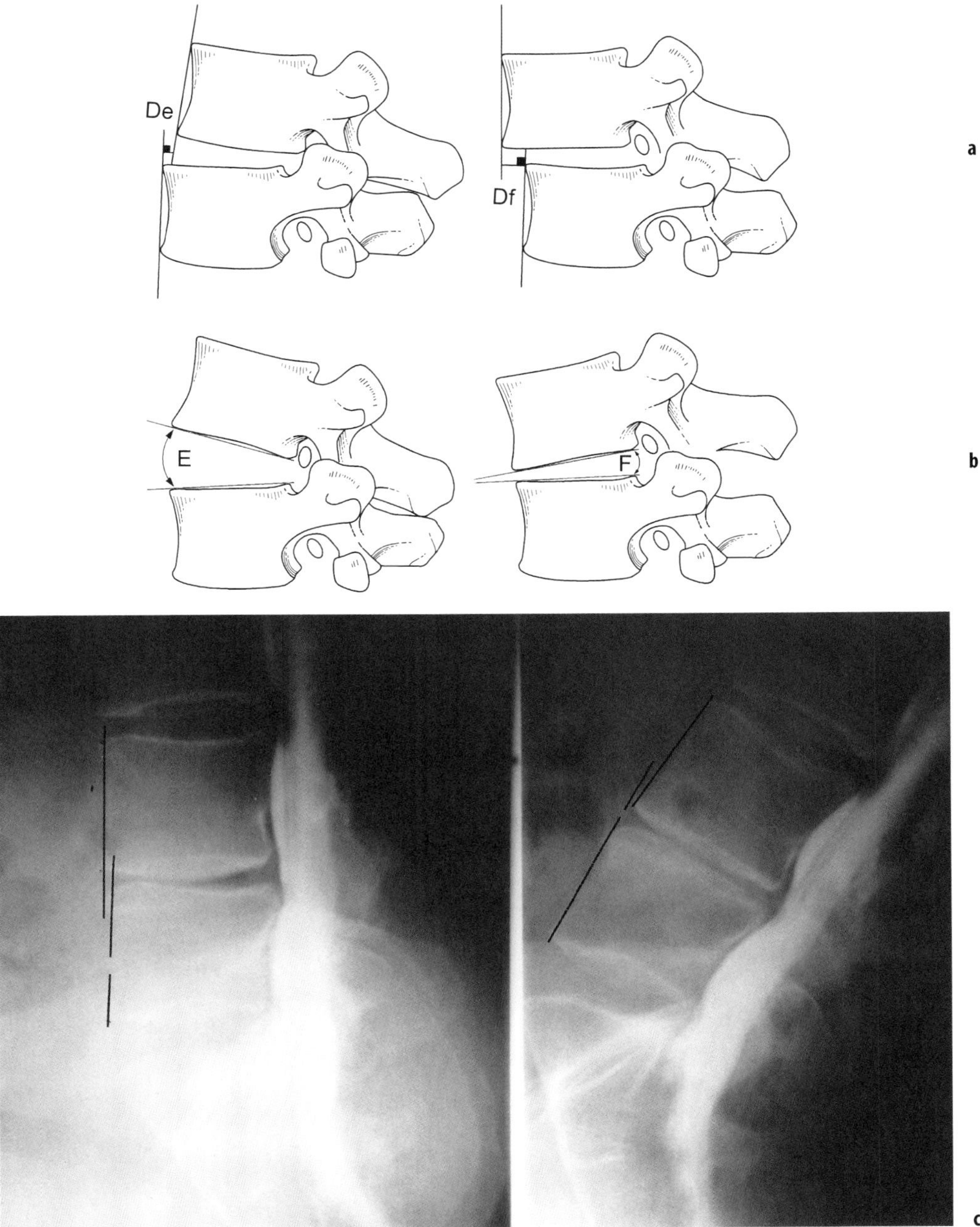

Figure 40.10. Classical dynamic X-rays. **a** Evaluation of antero-posterior translation. **b** Evaluation of diskal angulation (extension–flexion). **c** Dynamic X-rays associated with myelography. Diskal angulation may be measured either by using the vertebral endplate landmarks or by evaluating the relations between the anterior vertebral walls.

The amplitude of the translation that accompanies flexion varies with many factors (including the conditions in which the radiographs are taken and the centering of the incident beam). Measurements of translation, in contrast to those of angulation, are influenced by radiograph magnification. Nevertheless the apparent anterior translation during flexion may be an artifact produced by the axial rotation that accompanies flexion–extension on lateral projections. At lumbar level trigonometric calculations show that 3° of rotation of a vertebra relative to the adjacent vertebra results in about 1.5 mm of apparent relative displacement of the posterior edges, because the distance between the two edges of the vertebra on a lateral projection is about 30 mm and the 3° tangent is around 0.05.

The position of the axis of rotation determines the relative amount of rotation (evaluated according to the angles formed by the endplates) and of translation at the intervertebral disk. If the center of rotation remains within the disk, translation is minimal. On the other hand, if the center of rotation is outside the disk, the geometrical consequences of flexion–extension are different.

Instability, characterized in theory by abnormal or erratic mobilities, may be detectable only during some phases of the passage from flexion to extension. In this situation, a videofluoroscopy dynamic study is needed. However, this method is difficult to validate and is not suitable for routine use as it involves greater radiation exposure than plain radiographs and requires more expensive equipment [84]. These many measurement problems, together with the wide interindividual variations, probably explain the conclusion drawn by Stokes and Frymoyer [75]: dynamic radiographs can provide orientation when confronted to other data.

Computed tomography (CT) is used to evaluate anatomic structures and to look for dysfunction of the posterior spinal components. Twist tests can be performed to determine the orientation of the facet joint surfaces during torsion. They can show gaping or asymmetric narrowing of the facet joint space, or can reveal or create a vacuum phenomenon. Unfortunately, the values vary markedly across individuals, reproducibility of the measurements is poor, and the windows used cause substantial changes in the images. Husson [72] drew attention to the subjective impression of joint distraction and to the asymmetry in facet joint surface orientation at a given vertebral level.

Magnetic resonance imaging (MRI) detects alterations in the mobile spinal segment as incipient nucleus dehydration seen as a decrease in the high signal on T2-weighted images. The disks become gray, then black, and the demarcation between the anulus and nucleus blurs, then disappears. A tiny area of high signal at the center of the posterior part of the anulus has been reported to correlate significantly with annular disruption and, in some studies, with low back pain [85]. These findings have not been replicated consistently, however. Furthermore, the high-intensity zone is a common manifestation of degenerative disease in older patients.

Modic [86] described three types of endplate signal changes, each corresponding to a different histological pattern: type 1, defined as low signal on T1 images with high signal on T2 images, was correlated with inflammatory remodeling and increased vascularization of the endplates (Figure 40.11). Type 2, or high signal on T1 images and isointense or high signal on T2 images, was correlated with fatty degeneration of the vertebral body cancellous bone near the disk, possibly occurring as progression of type 1 (Figure 40.12). Type 3 (low signal on both T1 and T2 images) was correlated with radiological sclerosis.

Modic did not establish correlations between these changes in bone marrow signal and disk degeneration, clinical manifestations, or anatomic derangements. However, changes in endplate signal [87] have been interpreted as evidence of excessive mechanical stress denoting abnormal intervertebral mobility [70]. Further studies are needed to evaluate the correlations between these signal abnormalities and destabilization.

Consequences of Destabilization: Pain

The source of pain throughout the natural history of low back pain is often unclear. Whether mechanical factors are the primary

a b

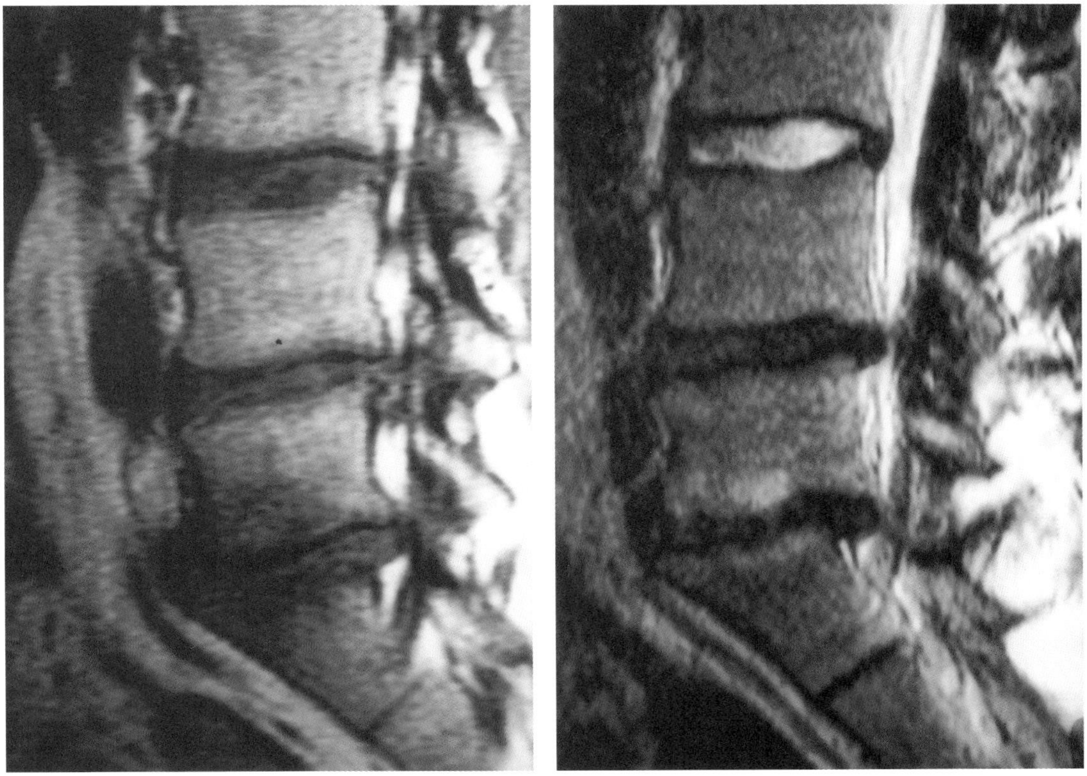

Figure 40.11. Modic 1 MRI imaging. **a** T1. **b** T2.

a b

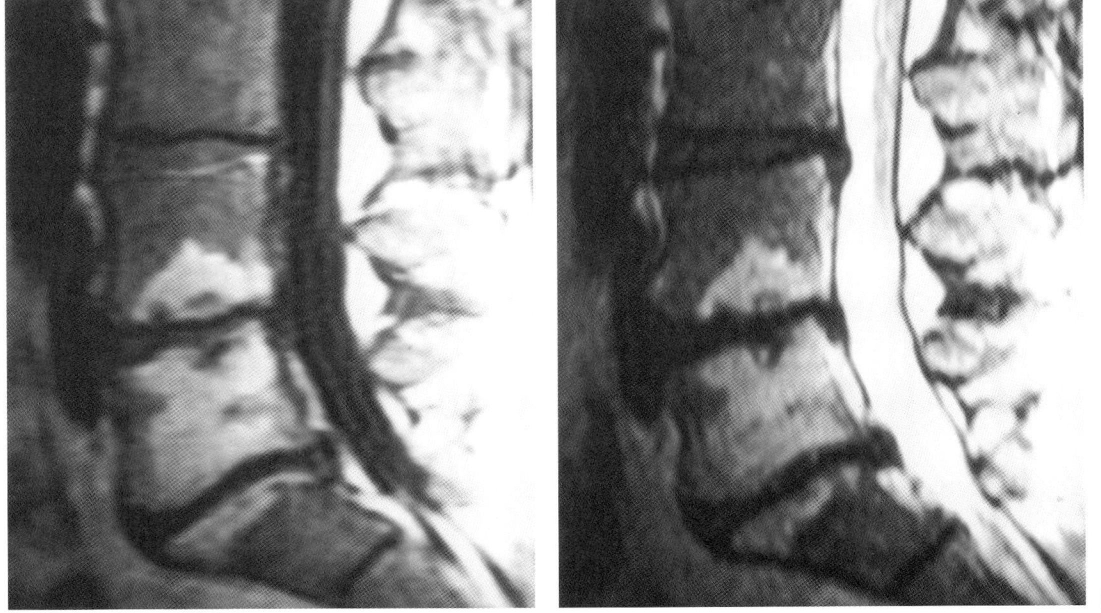

Figure 40.12. Modic 2 MRI imaging. **a** T1. **b** T2.

cause of the pain remains unsettled. Whereas at other joints degenerative lesions caused by overloading are clearly correlated with pain, alterations in the segmental connecting structures of the spine cannot always be considered to be the main source of low back pain.

The nerve supply to the spine is undoubtedly involved [88]. Abnormal stimulation of the nerve endings can cause referred pain. The mobile segment receives an abundant metameric nerve supply via the ventral and dorsal branches of the spinal nerves and the sympathetic system. The pathophysiological importance of the sinuvertebral nerve (arising from the anterior branch of the spinal nerve) and its dural endings is undoubtedly underestimated.

Only the outermost part of the normal intervertebral disk is innervated [77]. The inner disk, including the nucleosus pulposus, is aneural. Some nociceptors may remain silent, explaining why diskography or procedures under local anesthesia cause pain in only part of the abnormal disk. The posterior longitudinal ligament receives a rich nerve supply, as do the capsule and facet joint, which are innervated by three twigs from the posterior branch of the spinal nerve. The interspinous ligament and ligamentum flavum are less abundantly innervated. The surrounding muscles contain numerous nociceptors, of which some are sensitive to changes in temperature and blood flow.

Thus, pain can arise from many structures in the low back, including the vertebral body, muscles, dura mater, ligaments, sacroiliac joints, facet joints, and intervertebral disks. Although the disk has been universally considered to be a source of pain, it is unclear how its relatively meager nerve supply could cause the severe pain experienced by many patients. Many chemicals present in the disk may be capable of stimulating the nociceptors. Furthermore, there is evidence that neoinnervation can occur during degeneration [89,90]. Similarly, neovascularization is a classical finding in anulus tears [30,61].

Absence of an epineurium is a distinctive feature of the nerve roots and is associated with decreased resistance to mechanical stress. Furthermore, mechanical dysfunction can alter the flow of cerebrospinal fluid, which supplies nutrients to the nerve roots. Endoscopic studies have demonstrated the importance of functional changes in arterial and venous networks.

Weinstein [51] and others have reported evidence that the posterior ganglion and a number of physicochemical factors play a role in the genesis of pain. Interestingly, pain referred to the lower limbs can have many sources independent from nerve root or sacroiliac joint disease. Mooney [4] showed that injections into the facet joints could cause pain in the thigh or in the leg down to the ankle, according to the intensity of the nociceptive stimulus. Diskography used as a pain reproduction test has yielded surprising results: in one study, 3% of 225 disks generated pain identical to the spontaneous pain, although their imaging features were normal.

Physical nociceptor stimulation caused by deformities related to biomechanical alterations in the spinal segment directly activates free nerve endings. This explains the importance of mechanical factors in low back pain and the favorable effects of fusion in patients with segmental instability. Displacements, even those too small to be detected by available imaging techniques, can cause excessive mechanical stimulation of the nociceptors. Another possible source of nociceptor activation is variability in intradiskal pressure, although some clearly abnormal disks produce no pain during diskography [43]. Nevertheless, mechanical nociceptor stimulation cannot fully explain the development and perpetuation of low back pain, which require the involvement of suprasegmental factors [68].

Conclusions

In patients with suspected destabilization, all clinical and imaging study data must be examined carefully with the goal of finding several arguments that point in the same direction. A few clinical manifestations, confirmed in some cases by a local injection of anesthetic, suggest a mechanical disorder affecting a single segment. However, the structural data provided by imaging studies often fail to provide clear orientation. In some cases, the analysis of present or potential destabilization is in favor of a given conservative or surgical strategy. In other patients, particularly those with degenerative disease, the complexity of the symptoms and the interrelations with the adjacent levels require a more global approach to the concept of destabilization. An important distinction between an isolated lesion of a bony piece or mobile segment (caused perhaps by an initial or delayed iatrogenic complication) and more extensive problems affecting spinal muscles, spinal curvature organization, and relations between the pelvis and spine. In this situation, an evaluation of the overall balance of the trunk is imperative. Better understanding of standing and sitting positions in the sagittal plane is one of the great challenges, for adjusting lumbar and lumbo-sacral fusion as well as per and preoperative planning for lower-limb arthroplasty.

According to the literature, dynamic tests may be of limited clinical validity because their interpretation rests only on measuring translations in angle variations to look for hyp ermobility. Segmental intervertebral instability has been defined as an abnormal and more or less well-orchestrated response to physiological stress [6]. This response depends on the mechanical properties of the spinal segment involved. It produces the clinical symptoms. However, a reduction in intervertebral mobility has been reported in low-back-pain patients as compared to symptom-free subjects. Nevertheless, this finding is consistent with our observations about "stiff" abnormal disks. Harmony of motion is more important than hypermobility. Harmony of motion can be assessed based on the extent of scatter of mobility axes. Even a disk with limited mobility is considered abnormal if there is substantial scatter in its mobility axes. Pain is caused by local nociceptive signals and by muscle contraction that fails to bend an immobile segment or that seeks to rebalance adjacent segments. "Disk hyperextension" occurs in the segments neighboring "stiff" disks: this local hyperlordosis strives to compensate for deficient posterior curving in the sagittal plane. Correcting the lordosis will improve the symptoms provided the overall balance of the unit formed by the spine and pelvis is maintained. Hyperextension of the spine causes pain by increasing pressures in posterior structures; another source of pain is haphazard wandering of the mobility axis, the muscles being unable to achieve stabilization.

Both the development of postural films (particularly taken in the standing and sitting positions) and improvements in computed tomography (CT) and magnetic resonance imaging (MRI) have increased the precision of disk and joint analysis, thus opening up exciting new possibilities.

Further gains in the objective evaluation of spinal mechanics will probably occur as a result of recent advances in image processing techniques (most notably involving digitization and definition of strict criteria for making documents comparable).

Sagittal Spinal Balance

Spinal curvature abnormalities in the sagittal plane are widely believed to contribute to spinal pain syndromes. Spinal balance in the sagittal plane, which is a consequence of acquisition of the bipedal stance, varies widely across individuals, according to age, the morphotype, and presence of disease. Radiological parameters for evaluating spinal sagittal balance in the standing position have been developed by During [91], and Legaye and Duval-Beaupère [92], who have emphasized the major influence of the position of the pelvis and lower limbs. A more comprehensive evaluation of posture can be achieved by studying not only standing but also sitting and lower limb position [93].

Background

In normal individuals free of musculoskeletal disease, sagittal spinal balance in the standing position is a compromise between the position of the spine and the position of the pelvis and lower limbs. The sitting position has been less studied and raises important problems for standardization in radiological analysis. Sagittal spinal balance is influenced by the morphology of the spine and pelvis and by a number of functional parameters denoting adaptation to the erect position. In normal individuals, a fundamental requirement for satisfactory sagittal spinal balance in the standing position is that the overall axis of gravity run through a zone of tolerance centered on the projection of the femoral heads. The distance between the overall axis of gravity and the line connecting the two femoral heads is the length of the lever on which gravity exerts its force on the axis of rotation through the femoral heads. Roussouly [15] reported that the overall axis of gravity is anterior to the femoral heads, at a mean distance of 1 cm. There is general agreement that the center of gravity of the upper body is on the vertical line that runs through the center of gravity of the entire body. Nevertheless, this condition may be affected by the reference position of lower limbs used for the evaluation.

A plumb line has been suggested as a useful tool for evaluating sagittal spinal balance on lateral radiographs. When held against the prominent C7 spinous process, the line should fall anterior to the thoracic curvature, posterior to the L3 vertebral body and near the femoral heads. Full-length lateral views of the spine, pelvis, and hips obtained in the standing position can be used for this study. However, Van Royen [94] has shown that the position of the plumb line varies considerably with the position of the hips, knees, and ankles. Consequently, this technique is not reliable for evaluating sagittal spinal balance unless a standardized position is used. However, standardized positions are often very different from spontaneous postures.

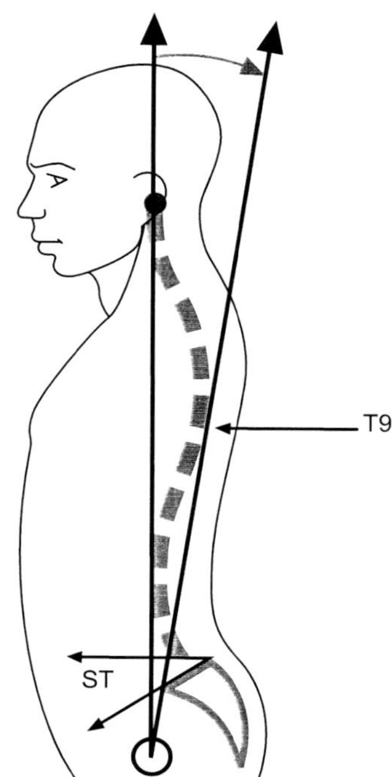

Figure 40.13. Global view of the sagittal equilibrium of the trunk. Definition of T9 sagittal angle and relation with sacral tilt (ST) and position of the pelvis.

Parameters for Evaluating Sagittal Spinal Balance

The Three Classical Angles

The classical angles used to evaluate overall sagittal spinal balance are the angles of *lumbar lordosis*, *thoracic kyphosis*, and *cervical lordosis*. Each of these angles is formed by the horizontal plane and by the vertebral endplate with the greatest degree of tilt relative to this plane.

The T9 Sagittal Angle (Figure 40.13)

This has the center of the line connecting the two hips as its apex, the vertical line through this apex as one of its sides, and the line from the apex to the center of the T9 vertebral body as its other side. T9 is used because it is near the center of gravity of the body segment whose weight is borne by the femoral heads; this center of gravity is usually located anterior to T9, its distance from the vertebral body being dependent on the degree of thoracic kyphosis [95].

Other Parameters Used to Evaluate Relations Between the Spine and Pelvis (Figure 40.14)

The incidence angle (i) is a morphological parameter. It is formed by the line perpendicular to the center of the sacral plateau and by the line running from the center of the sacral plateau to the center of the line that connects the femoral heads. The incidence angle varies widely across individuals, from +20° to +110° in a series by Roussouly [15], although most individuals have values around 54°.

Sacral tilt and pelvic tilt are functional angles that vary with the position of the subject. *Sacral tilt* (ST) is the angle formed by the horizontal

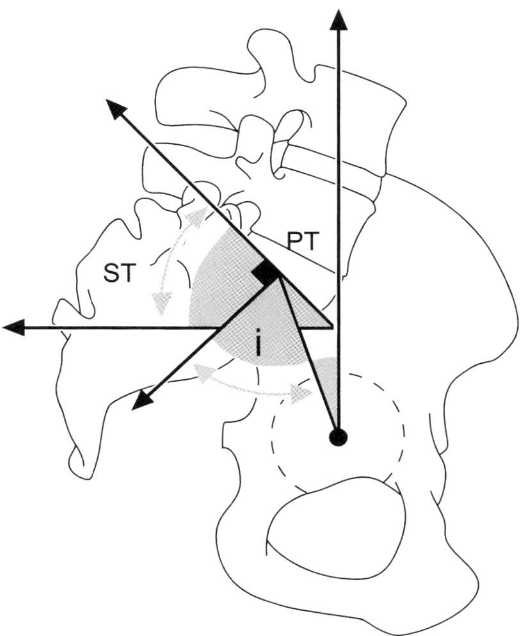

Figure 40.14. Definition of sacral tilt (ST), pelvic tilt (PT), and incidence angle (I).

plane and the tangent to the upper surface of S1. Sacral tilt decreases when the upper surface of S1 moves toward a more horizontal position, i.e., when the entire sacrum becomes more vertical and the pelvis rotates backward (as observed in a sitting position). In contrast, sacral tilt increases when the upper surface of S1 becomes more vertical during movement of the entire sacrum toward a more horizontal plane with anterior rotation of the pelvis (as observed in a standing position) (Figure 40.15). Mean sacral tilt is 42° in the standing position and 29° in the sitting position.

Pelvic tilt (PT) is the angle formed by the vertical plane and the line running from the center of the upper S1 surface to the center of the line connecting the two femoral heads. Pelvic tilt is usually around 12°. By applying the rule of complementary angles, it can be shown that the incidence angle is the sum of sacral tilt and pelvic tilt I = ST + PT (Figure 40.16). Because the incidence angle is fixed in a given individual, this means that changes in sacral tilt are closely linked to changes in pelvic tilt. In a study of normal subjects, Legaye [92] found a chain of

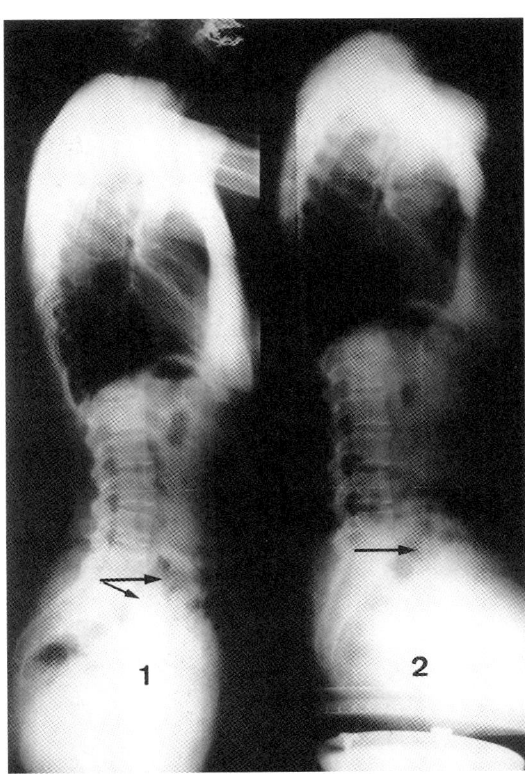

Figure 40.15. Modifications for sacral tilt (ST) during sitting and standing. In standing position (1), ST increases in association with anterior pelvic tilt. In sitting position (2), ST dicreases in association with pelvic retroversion.

significant correlations linking lordosis, kyphosis, T9 sagittal angle, incidence angle, sacral tilt, and pelvic tilt.

S1 overhang. It is defined as the distance in the sagittal plane between the center of the line connecting the two femoral heads and the point on the horizontal line running through this center where the center of the upper surface of S1 projects.

The pelvifemoral angle. Mangione et Senegas [96] added to the above-mentioned parameters the pelvifemoral angle, which measures extension of the hip. This angle is formed by the femoral axis, or a line parallel to the femoral axis, and by a line running from the center of the upper surface of S1 to the center of the line connecting the two hips. The femoral axis is sometimes difficult to determine on full-length lateral radiographs; it is obtained by connecting

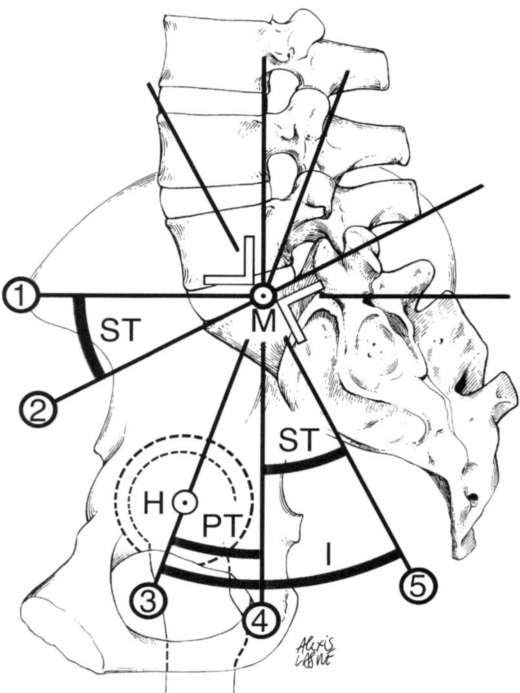

Figure 40.16. Illustration of the complementary angle geometrical construction showing that I = ST + PT.

two points located at the middle of the diaphysis, one at the level of the lesser trochanter and the other 10 cm lower down. A larger pelvifemoral angle indicates greater extension of the hips.

Requirements for Sagittal Balance

The sagittal posture of an erect subject can be modeled as a precarious construction in which several segments are articulated with one another: the trunk is articulated with the pelvis, which in turn is articulated with the lower limbs, where the hips and knees play a key role. Because it is flexible, the spine can adjust to changes in the position and shape of the pelvis. The system is balanced and economical when the level of muscle activity needed to maintain the posture is as low as possible.

Analysis of the sagittal balance parameters described above provides information on spinal morphology in the sagittal plane. Tilting of the sacrum to a more horizontal position and extension of the hips are the two main changes observed during assumption of the standing position. Both changes are closely correlated with the spinal curvatures. Legaye, Duval-Beaupère [92], and During [91], have shown that sacral tilt and lumbar lordosis are related: in the standing position, a greater degree of sacral tilt is associated with more marked lumbar lordosis.

Modifications in spinal sagittal balance for the standing position induces corrective responses (Figures 40.17, 40.18). First, the pelvis can rotate further around the line connecting the hips: the pelvis can swivel around the femoral heads from a position of maximum anterior rotation (where the pelvic tilt angle is negative or zero) to a position of maximum posterior rotation (where the pelvic tilt angle is positive). Thus, pelvic rotation denotes adaptation of the pelvis to the position of the spine. Roussouly [15] reported that pelvic tilt could vary from 0° to 5° (maximum anterior rotation of the pelvis) to 50° (maximum posterior rotation of the pelvis). Beyond this range, no further adaptation can be achieved by rotation around the hips. However, these values are rarely reached because of the stiffness characteristic of the hips and because bony obstacles limit the depth of engagement of the femoral head in the acetabulum.

In patients with an abnormal forward stoop, the pelvis rotates posteriorly around the hips to keep the center of gravity within the narrow zone of tolerance centered by the femoral heads. This results in constant activation of the extensor muscles, sometimes causing pain. The posterior rotation of the pelvis moves the upper surface of S1 backward, decreasing the angle of sacral tilt and thereby inducing additional posterior retropulsion of the trunk.

Analysis of sagittal balance on full-length radiographs provides quantitative information on the limits of motion determined by the anatomic characteristics of each individual. The incidence angle is a fundamental parameter that characterizes the pelvis in each individual. Because the incidence angle is the sum of sacral tilt and pelvic tilt, a smaller incidence angle is

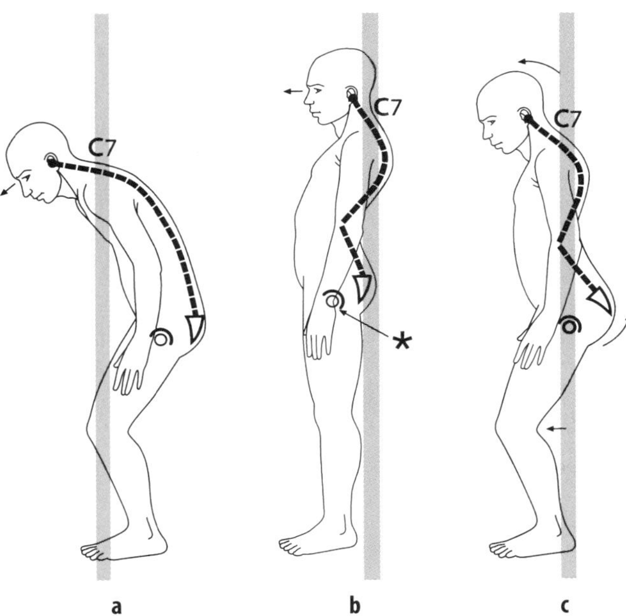

Figure 40.17. **a** Modifications in sagittal balance in case of global kyphosis (as in ankylosing spondylarthritis). **b** Correction with monosegmental wedge osteotomy: side effects on pelvis and lower limbs. Excessive posterior retropulsion of the trunk may induce secondary disequilibrium due to inability of the hips for hyper-extension. **c** Secondary hips and knees flexum may be mandatory for a new equilibration.

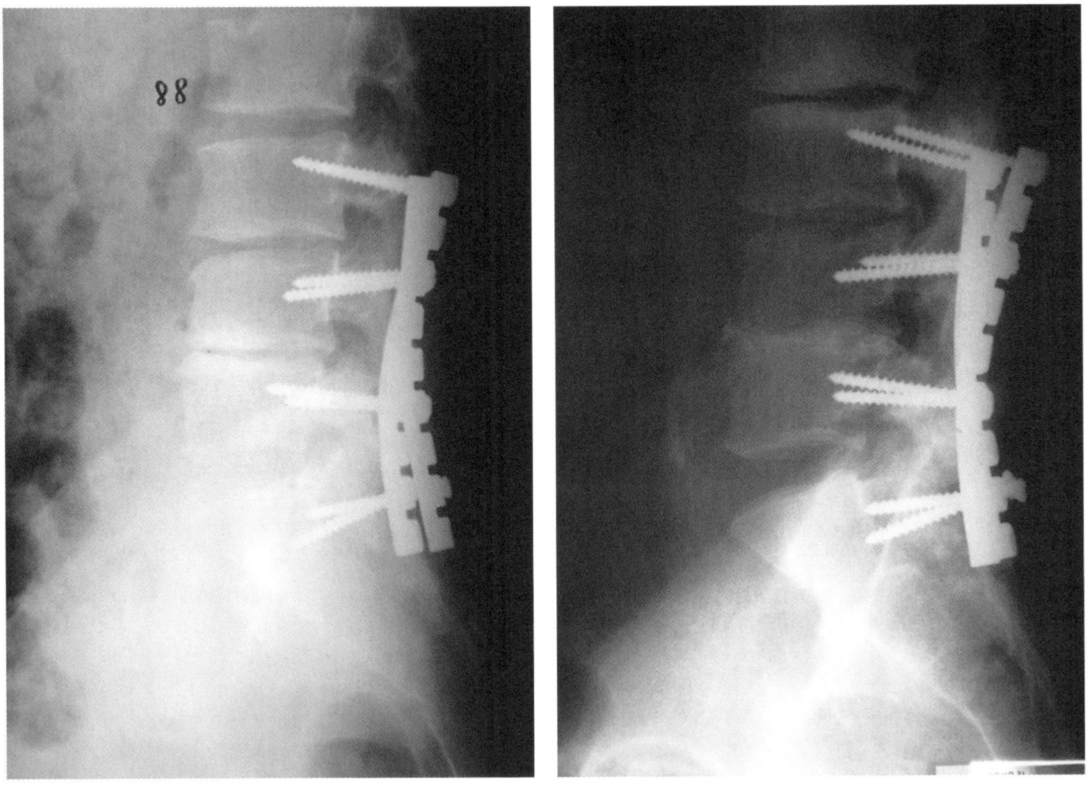

Figure 40.18. **a** Lumbo-sacral fusion with fixation including excessive sacral verticalization (low ST angle), loss of lumbar lordosis (partially due to L3 L4, L4 L5 disk lesion). **b** 12 years later, destabilization at the level above (L2 L3) with retrolisthesis and vacuum sign. Note progressive collapse and spontaneous fusion at L4 L5 disk level.

associated with a narrower range of absolute pelvic tilt and sacral tilt: thus the limits of rotation around the hips are rapidly reached, leaving little room for adjustment to even a minor spinal imbalance. Furthermore, because the range of sacral tilt is small, there is a tendency toward flattening of the lower back.

Conversely, in a subject with a wide pelvis and a large incidence angle, pelvic tilt can increase considerably at the expense of sacral tilt while remaining within the range that can be tolerated by the hips.

When hip extension is limited, the pelvis stays rotated anteriorly and, consequently, the angle of sacral tilt is larger. The degree of lumbar lordosis increases to keep the center of gravity above the femoral heads. This is the morphotype characterized by marked curvatures.

In contrast, when hyperextension of the hips is possible, the pelvis can adapt by rotating backward to a considerable degree, thus decreasing the angle of sacral tilt and flattening the spinal curvatures. Mangione [96] pointed out that muscle contractures are more marked in subjects with this morphotype.

Close correlations exist between pelvic tilt and sacral tilt and between sacral tilt and lumbar lordosis. But, in normal individuals, changes in pelvic tilt make only a small contribution to sagittal spinal balance. If the incidence angle is large, pelvic tilt shows little change, whereas variations in sacral tilt are more marked in proportion to the incidence angle.

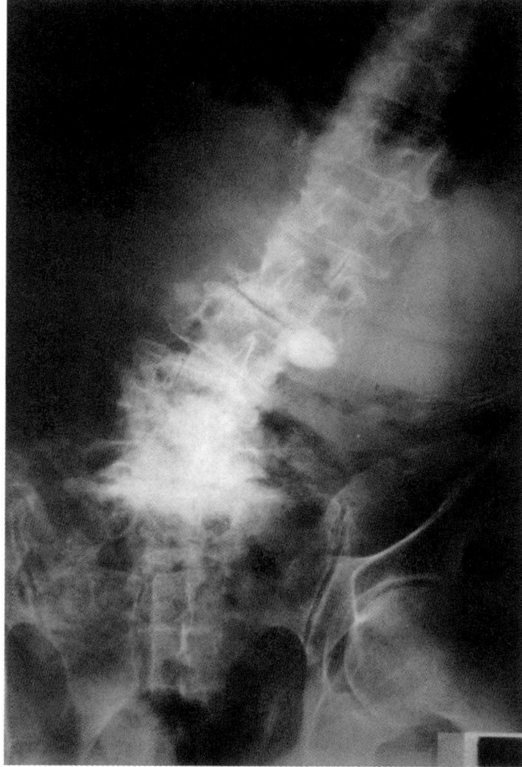

Figure 40.19. Typical lumbar destabilization in aging spine. Frequently, destabilization and instability are due to several problems: neurological dysfunction (Parkinson's disease or extrapyramidal syndrome); muscular degeneration, diskal and articular degenerative lesions. In such cases, the results is a "chewing-gum" spine (as often seen in camptocormia).

Analysis of a Few Specific Situations

Sagittal Balance Disturbances in Elderly Subjects

The decrease in lumbar lordosis seen with advancing age moves the center of gravity of the upper body forward, thus causing marked imbalance in the frontal or in the sagittal plane (Figure 40.19). Muscle abnormalities in degenerative lumbar kyphosis may include weakness of the abdominal muscles and degenerative myopathy within the sacrolumbar muscles. A geometric factor combining disk space narrowing, vertebral body wedging, and fibrosis filling the facet joints has been implicated. Corrective changes include hyperextension of the hips, which causes the pelvis to rotate backward, moving the upper surface of S1 to a more horizontal position. However, the range of hip extension is often limited by osteoarthritis in elderly subjects. To maintain balance, the subject flexes the knees while keeping the hips maximally extended; the balance of the spine is then achieved relative to the line connecting the knees. External rotation of the hips can contribute to improve sagittal balance by moving the lesser trochanter forward, thus further decreasing the tension on the psoas muscle and allowing a greater degree of posterior pelvic rotation.

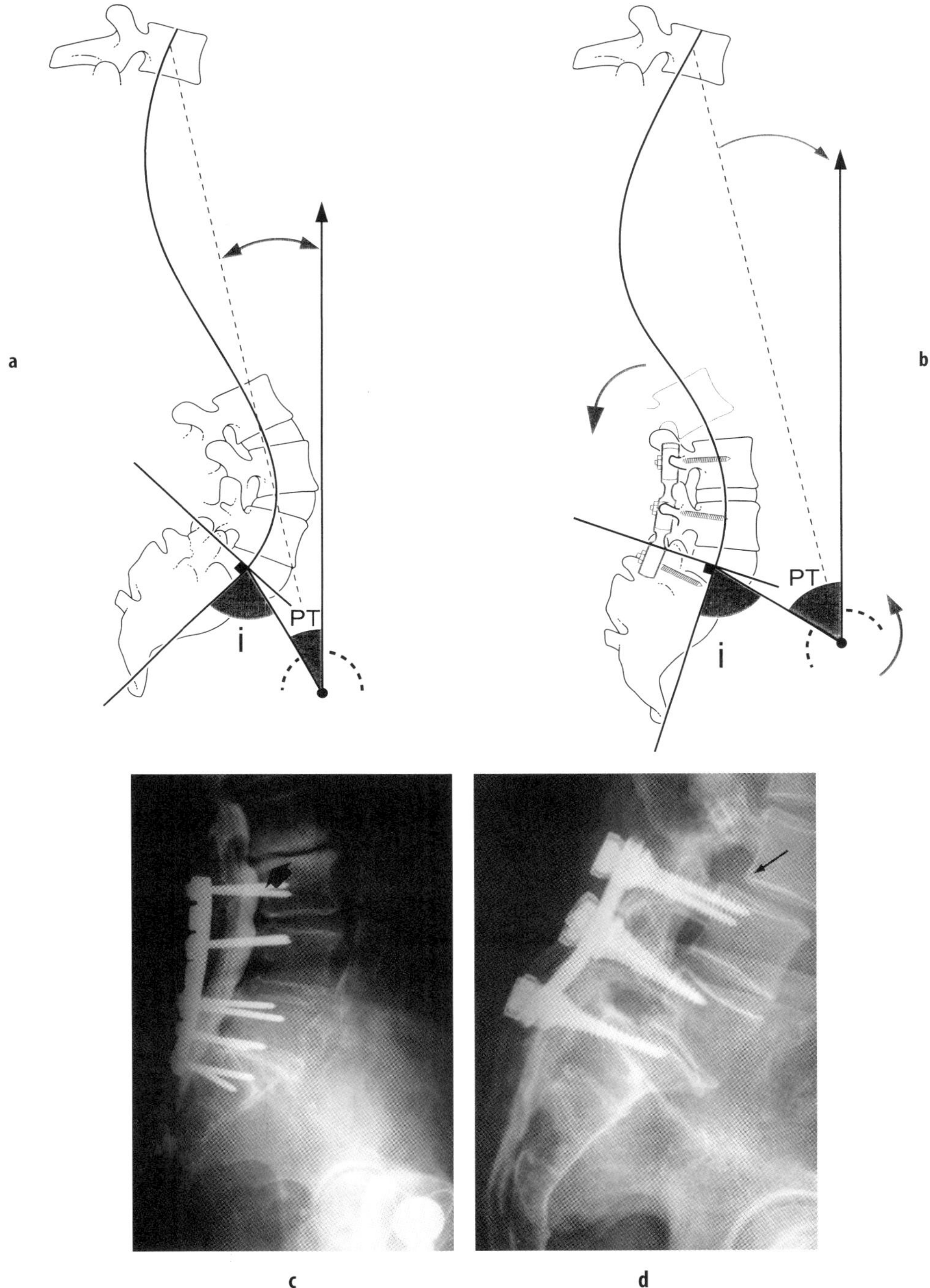

Figure 40.20. Side effects of poor adjustement in lumbo-sacral fusion: fixation with a too vertical sacrum (pelvic retroversion) induces an increase of pelvic tilt (PT). The consequences are different according to the sagittal posture of the upper trunk preoperatively (**a**), postoperatively (**b**). In some cases, retrolisthesis at the upper level may occur (**c**). In other cases, adaptation is obtained via an antelisthesis of the upper block (**d**).

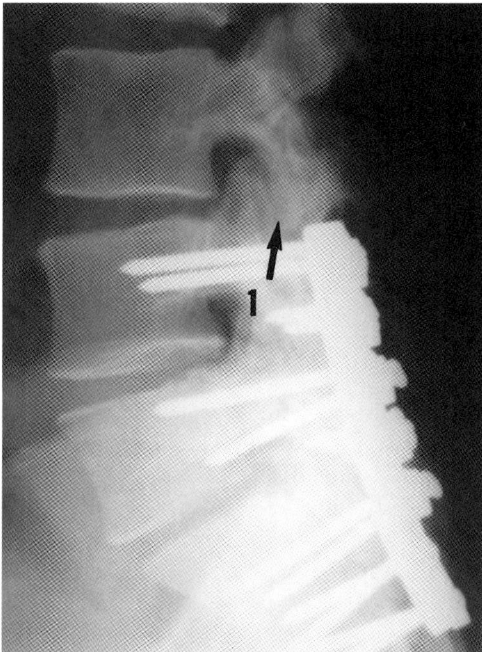

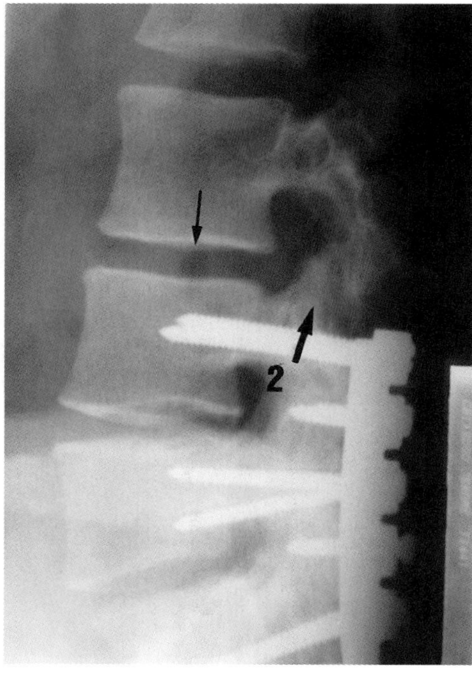

Figure 40.21. Late side effects of L4–S1 lumbosacral fusion (20 years follow-up). Disk L3 L4: reduced height of the disk in standing position (1). This situation induces a collapse of the L3 L4 foramen with secondary radicular pain. In sitting position (2), posterior decoaptation of the articular facets and better posterior opening of the disk with vacuum sign; posterior foramen size is larger. For this patient, the sitting position (with relative retroversion of the pelvis and reduction of lumbar lordosis) provides pain relief.

Spondylolysis and Spondylolisthesis

Lysis of the pars interarticularis is considered to be a consequence of a stress fracture caused primarily by activities in the upright position, particularly with the spine hyperextended. The high prevalence of pars interarticularis lysis [53] in athletes supports this theory. It has been suggested that the lower articular process of the suprajacent vertebra may chisel into the pars interarticularis, causing the fracture. A large incidence angle and marked lumbar lordosis have been associated with an increased risk of pars interarticularis lysis. This is not surprising, as a large incidence angle is associated with greater sacral tilt and therefore with stronger shear forces. In this situation, posterior rotation of the pelvis may be simply a secondary corrective event. Support for this theory can be found in the fact that no cases of pars interarticularis lysis have been reported in patients who had never walked.

Sagittal Spinal Balance and Lumbar Fusion

Lumbosacral fusion combines the lower lumbar spine and the pelvis into a single unit that connects to the rest of the spine at L5 or L4, whereas the normal junction between the spine and pelvis is at S1. The fusion modifies the pelvic tilt angle, and the sacral angle, thus changing the functional conditions imposed on the lumbosacral junction. Furthermore, substantial changes occur at the hips and sacroiliac joints. A narrow pelvis with a small incidence angle raises special challenges for lumbosacral fusion. There is no single optimal degree of lordosis: the value must be determined according to the incidence angle. If the incidence angle is small, a small degree of lumbar lordosis is appropriate, as decreased lumbar lordosis is the rule when the incidence angle is small. Conversely, if the incidence angle is large, the fusion should be adjusted to produce marked lordosis.

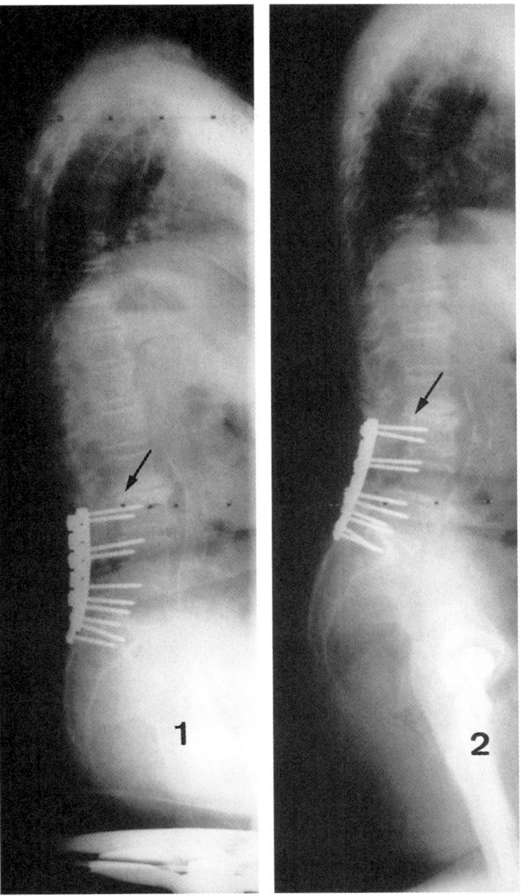

Figure 40.22. Lumbosacral fusion adjusted with a vertical sacrum 15 years follow-up. Note degenerative lesions and instability at the level above with dynamic retrolisthesis and the poor adaptation of the hip with secondary flexum (sitting 1, standing 2).

Faulty adjustment of the fusion with inadequate lordosis and an excessively vertical position of the sacrum results in corrective changes including an artificial increase in lumbar lordosis. This increases stresses imposed on the mobile segment at the junction between the fused and unfused spine and on the sacroiliac joints. Rapidly progressive lesions can develop above the junction (disk disease, ante or retrolisthesis (Figure 40.20, 40.21)). Thus, lumbar fusion should not seek only to afford neurological protection or to correct segmental mechanics. A strong fusion is not the only goal: adjustment in the optimal position is essential [97].

Studies of these factors are important to raise awareness of the difficulties met in adjusting the sacrum. Optimal adjustment can be achieved by carefully positioning the patient and by selecting appropriate fixation material. Lumbosacral fusion can be likened to creation of a new pelvic vertebra whose upper endplate is the upper boundary of the fused unit. This changes the lumbopelvic parameters, including the incidence angle, the angle of pelvic tilt, and above all the angle of tilt of the upper endplate of the fused unit. When the sacrum is initially in a vertical position (a situation often associated with loss of lumbar lordosis particularly in elderly subjects), inadequate fusion of the sacrum in excessive backward rotation replicates the sitting position, causing deleterious effects on the sacroiliac joints and hips (Figures 40.22, 40.23).

Conclusions

Analysis of sagittal balance is essential to understand several pathologies involving either spine or lower limbs. New parameters must be evaluated for planning lumbo sacral fusions or hip and knee prosthesis implantation. In-depth knowledge of the mechanics of the lumbo pelvic complex and a global view of sagittal posture open new fields for more accurate spinal procedures (either conservative or corrective) or for optimization of joint replacement.

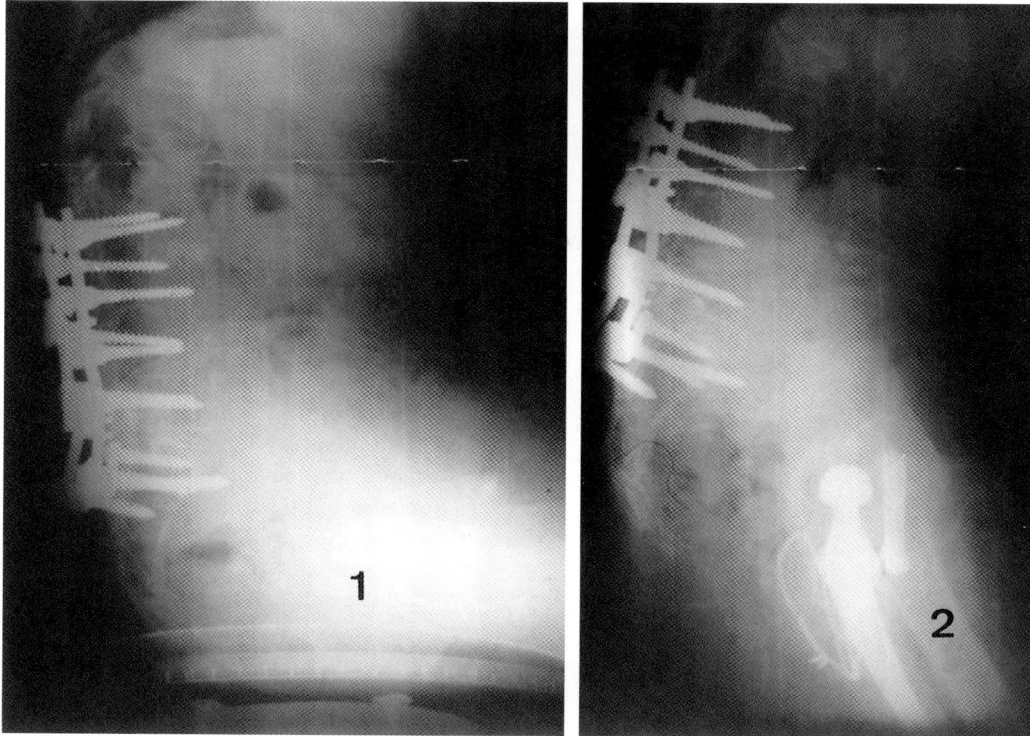

Figure 40.23. Poor adjustment of lumbosacral fusion with global lumbosacral kyphosis (vertical sacrum, pelvic retroversion). Sitting (1), standing (2). This situation induces adaptative hip flexum and secondary perturbation in acetabular cup anteversion (iterative anterior hip prosthesis luxations in this case).

References

1. Pope M, Panjabi M. Biomechanical definitions of spinal instability. Spine 1985;10:255–6.
2. Knutsson F. The instability associated with disk degeneration in the lumbar spine. Acta Radiol 1944;25: 593–609.
3. Kirkaldy-Willis WH, Fargan HF. Instability of the lumbar spine. Clin Orthop 1982;165:110–23.
4. Mooney V. Where is the pain coming from? Spine 1987;12(8):754–9.
5. Putto E, Tallroth K. Extension-flexion radiographs for motion studies of the lumbar spine. A comparison of two methods. Spine 1990;15(2):107–10.
6. White AA, Panjabi MM. Clinical biomechanics of the spine, 2nd ed. Philadelphia: JB Lippincott, 1990.
7. Dvorak J, Panjabi M, Novotny J. Clinical validation of functional flexion-extension roentgenograms of the lumbar. Spine 1991;16:943–50.
8. Mc Fadden KD, Taylor JR. Axial rotation in the lumbar spine and gaping of the zygapophyseal joints. Spine 1990;15:295–9.
9. Nachemson A. Lumbar discography, where are we today. Spine 1989;14:555–7.
10. Denis F. The column spine and signifiance in the classification of acute thoracolumbar injuries. Spine 1983;8:817–31.
11. Dickey JP, Bednar DA, Dumas GA. New insight into the mechanics of the lumbar interspinous ligament. Spine 1996;21:23.
12. Duparc J, Shreiber A, Troisier O. Instabilités vertébrales lombaires. Paris: Exp. Scientifique Française-GIEDA, Paris, 1995.
13. Dupuis PR, Yong Hing K, Kirkaldt-Willis W. Radiologic diagnostis of the lumbar spine: flexion-extension and lateral bending. Spine 1991;16:943–50.
14. Dvorak J, Panjabi M, Chang D. Functional radiographic diagnosis of degenerative lumbar spinal instability. Spine 1985;10:262–76.
15. Roussouly P, Berthonnaud E, Dimnet J. Critères sagittaux de réglage d'une arthrodèse postérieure du rachis. Rachis 1998;10(6):224–6.
16. Palmgren T, Gronblad M, Virri J, Seitsalo S, Ruuskanen M, Karaharju E. Immunohistochemical demonstration of sensory and automatic nerve terminals in herniated lumbar disc tissue. Spine 1996;21:1301–6.
17. Friberg O. Lumbar instability: a dynamic approach by traction-compression radiography. Spine 1987;12(2): 119–29.
18. Nakamura SI et al. The afferent pathways of discogenic low back pain. J Bone Joint Surg 1996;78:602–12.
19. Panjabi M, Goel V, Summers D. Effects of disc degeneration on the instability of a motion segment. Orthop Trans 1983;7:329.

20. Panjabi MM, Chang D, Dvorak J. An analysis of errors in kinematic parameters associated with in vivo functional radiographs. Spine 1992;7(2):200–5.
21. Paris SV. Physical sign of instability. Spine 1985;10: 277–9.
22. Hayes MA, Howard TC, Gruel CR, Kopta JA. Roentgenographic evaluation of lumbar spine flexion-extension in asymptomatic individuals. Spine 1989; 14:327–31.
23. Holmes DC, Browns MD, Eckstein EC, Larta L. Instability assessment of the lumbar functional spinal unit. Orthop Trans 1988;12:619–20.
24. Hoppe E, Tsou P. Postdecompression lumbar instability. Clin Orthop 1988;227:143–51.
25. Hukins DWL, Kirby MC, Sikoryn TA, Aspden RM, Cox AJ. Comparison of structure, mechanical properties and function of lumbar spinal ligaments. Spine 1990;15:8.
26. Husson JL. Instabilité vertébrale à l'étage lombaire. Cahiers d'enseignement de la SOFCOT. Conférences d'enseignement 1995;63–78.
27. Dvorak J, Panjabi M, Chang D, Theiler R, Grob D. Functional radiographic diagnosis of the lumbar spine flexion-extension and lateral bending. Spine 1991; 16:562–71.
28. Mulholland RC. Lumbar segmental instability: clinical definition. In: Szpalski M, Gunzburg R, Pope MH, editors. Lumbar segmental instability. Philadelphia: Lippincott Williams & Wilkins, 1999;55–61.
29. Mimura M, Panjabi M, Oxland TR. Disc degeneration affects the multidirectional flexibility of the lumbar. Spine 1994;19(12):1371–80.
30. Ebara S, Harada T, Hosono N et al. Intraoperative measurement of lumbar spinal instability. Spine 1992; 17(3S):44–50.
31. Saillant G, Lemoine J, Lazennec JY, Benazet JP. La destabilisation du rachis. Rachis 1994;6(65):307–18.
32. McNab I. The traction spur. An indicator of segmental instability. J Bone Joint Surg 1971;53A:663.
33. Poulin P, Daumen-Legre V, Serratrice G. La camptocormie du sujet âgé: myopathie ou dystonie musculaire? Rev Rhum Mal Osteoartic 1993;60:159–61.
34. Junghans H. Spondylolisthesis onhe Spalt in Zwischengelenk-Stuck. Arch Orthop Unfall–Chir 1930;29: 118–27.
35. Roy-Camille R. L'instabilité rachidienne. Rachis 1994; 6(2):107–12.
36. Louis R. Spinal stability as defined by the three-column spine concept. Anat Clin 1985;7:33–42.
37. Benazet JP, Rakover JP Saillant S, Roy-Camille R. Facteurs de déstabilisation chirurgicale du rachis lombaire. Edt Masson 1993:182–90.
38. Butler D, Trafimow JH, Anderson GNJ, McNeill TW, Huckman MS. Discs degenerate before facets. Spine 1990;15:2.
39. Freemont AJ, Peacock TE, Goupille P, Hoyland JA, O'Brien J, Jayson MIV. Nerve ingrowth into diseased intervertebral disc in chronic back pain. Lancet 1997;350:178–81.
40. Ida Y, Kataoka O, Sho T et al. Post-operative lumbar spinal instability occuring or progressing secondary to laminectomy. Spine 1990;15:1186–9.
41. Lindahl O. Determination of the sagittal mobility of the lumbar spine. Acta Orthop Scand 1966;37:241–54.
42. Seenegas J. Stabilisation lombaire souple. Instabilité vertébrale lombaire, GIEDA InterRachis, Paris, Expansion Scientifique Française 1995;209–18.
43. Muggleton JM, Kondracki M, Allen R. Spinal fusion for lumbar instability: does it have a scientific basis? J Spinal Disord 2000;13–3:200–4.
44. Bernhardt M, Bridwell KH. Segmental analysis of the sagittal plane alignment of the normal thoracic and lumbar spines and thoracolumbar junction. Spine 1989;14(7):717–21.
45. Boisaubert B, Montigny JP, Duval-Beaupere G, Hecquet J, Marty. Incidence, sacrum et spondylolisthésis. Rachis 1997;9(4):187–92.
46. Adams MA. The mechanical function of the lumbar apophyseal joints. Spine 1983;8:327–30.
47. Lazennec JY, Ramare S, Arafati N, Laudet CG, Gorin M, Roger B et al. Sagittal alignment in lumbosacral fusion: relations between radiological parameters and pain. Eur Spine J 2000;9:47–55.
48. Luk KD, Chow DH, Evans JH, Leong JCY. Lumbar spinal mobility after short anterior interbody fusion. Spine 1995;20(7):813–8.
49. Maigne R. Dérangements intervertébraux mineurs. Diagnostic et traitement des douleurs communes d'origine rachidienne. Paris: Expansion Scientifique Française, 1989.
50. Nachemson A. Lumbar interdiscal pressure. Acta Orthop Scand 1960;43:1–104.
51. Weinstein J, Claverie W, Gibson S. The pain of discography. Spine 1998;13:1344–8.
52. Yang KH, King AL. Volvo award in biomechanics: mechanism of facet load transmission as a hypothesis of low back pain. Spine 1984;9:557–65.
53. Posner I, White AA, Edwards WT et al. A biomechanical analysis of the clinical stability of the lumbar and lumbosacral spine. Spine 1982;7:374–89.
54. Frymoyer JW, Selby DK. Segmental instability: rationale for treatment. Spine 1985;10:280–6.
55. Frobin W, Brinckmann P, Leivseth G, Biggerman M, Reikeras O. Precision measurement of segmental motion from flexion-extension radiographs of the lumbar spine. Clin Biomech 1996;11(8):457–65.
56. Fujiwara A, Tamai K, An HS, Kurihashi A, Lim T-H, Yoshida H, Saotome K. The relationship between disc degeneration, facet joint osteoarthritis and stability of the degenerative lumbar spine. J Spinal Disord 2000; 13(5):444–50.
57. Cosentino R, Suarez A, Baccani S. Etude radiologique de la mobilité du rachis dorso-lombaire. Rev Chir Orthop 1982;68:91–5.
58. Hanley EN, Matteri RE, Frymoyer JW. Accurate roentgenographic determination of lumbar flexion-extension. Clin Orthop 1976;115:145–8.
59. Bram J, Zaneri M, Min K, Hodler J. Abnormalities of the intervertebral disks and adjacent bone marrow as predictors of segmental instabiliy of the lumbar spine. Acta Radiologica 1998;39:18–23.
60. Sohei E, Takeo H, Noburu H, Masahiro I, Masao T, Yoshibaru M et al. Intraoperative measurement of lumbar spinal instability. Spine 1992;17(suppl):S40–50.
61. Murata M, Morio Y, Kuranobu K. Lumbar disc degeneration and segmental instability. A comparison of magnetic resonance images and plain radiographs of patients with low back pain. Arch Orthop Trauma Surg 1994;113:297–301.
62. Schmidt TA, An HS, Lim TH, Nowicki BH, Haughton VM. The stiffness of lumbar motion segments with a high-intensity zone in the annulus fibrosus. Spine 1998;23:2167–73.

63. Nachemson A. Lumbar spine instability: a critical update and symposium summary. Spine 1985;10: 290–1.
64. Rolander SD. Motion of the lumbar spine with special reference to stabilizing effect of posterior fusion. Acta Orthop Scand 1966;9O:1–144.
65. Sato H, Kikuchi S. The natural history of radiographic instability of the lumbar spine. Spine 1993;18:2075–8.
66. Magerl F, Aebi M, Gertzbein SD, Harms J, Nazarian S. A comprehensive classification of thoracic and lumbar injuries. Eur Spine J 1994;3(4):184–201.
67. Antonietti P, Saillant G, Gagna G, Lemoine J. Dystabilités dégénératives du rachis lombaire. "Instabilités vertébrales lombaires". Paris: Exp. Scientifique Française, 1995;55–62.
68. Benoist M, Boulu P, Deburge A. Physiopathologie et aspects psycho-sociaux de la lombalgie. Cahiers de la SOFCOT 1997;14–18.
69. Bernick S, Walker JP, Pawle WJ. Age changes to the annulus fibrosus in human intervertebral discs. Spine 1991;16:520–4.
70. Bogduk N, Tynan W, Wilson AS. The nerve supply to the human lumbar intervertebral discs. J Anat 1981; 132:39–56.
71. Lorentz M, Patwardhan A, Vanderby R. Load-bearing characteristics of lumbar facets in normal and surgically altered spinal segments. Spine 1983;8:122–30.
72. Husson JL, Poncer R, De Korvin B, Meadeb J. Apport du scanner en twist-test dans la mesure de l'instabilité du rachis lombaire. Rev Chir Orthop 1993;79: 117.
73. Lavaste F, Robin S. Le rachis instable: aspects biomécaniques. Rachis 1995;7(3):173–5.
74. Penning L, Wilmink JT, Van Woerden HH. Inability to prove instability: a critical appraisal of clinical radiological flexion-extension studies in lumbar disc degeneration. Diagn Imag Clin Med 1984;53:186–92.
75. Stokes IA, Frymoyer JW. Segmental motion and instability. Spine 1987;12(7):688–91.
76. Coppes MH, Marani E, Thomeer RT, Groen GJ. Innervation of "painful" lumbar discs. Spine 1997;22:2342–50.
77. Boden SD, Wiesel SW. Lumbosacral segmental motion in normal individuals. Have we been measuring instability properly? Spine 1990;15:571–6.
78. Wood KB, Popp CA, Transfeldt EE, Geissele AE. Radiographic evaluation of instability in spondylolisthesis. Spine 1994;19:1697–03.
79. Pearcy M, Portek I, Shepherd J. Three-dimensional x-ray analysis of normal movements in the lumbar spine. Spine 1984;9(3):582–7.
80. Fernand R, Fox D. Evaluation of lumbar lordosis. A prospective and retrospective study. Spine 1985; 10(9):799–803.
81. Gerzbein SD, Seligman J, Holtby R. Centrode pattern and segmental instability in degenerative disc disease. Spine 1985;10:257–61.
82. Marty C, Legaye J, Duval-Beaupere G. Equilibre rachidien sagittal normal. Ses relations avec les paramètres pelviens. Ses dysfonctionnements: cause de lombalgie. Résonnances Européennes du Rachis, Sept. 1997; 15:21–6.
83. Morgan FP, King T. Primary instability of lumbar vertebrae as a common cause of low back pain. J Bone Joint Surg1957;39B:6–22.
84. Aprill C, Bogduk N. High intensity zone: a diagnostic sign of painful lumbar disk on magnetic resonance imaging. Br J Radiol 1992;65:361–9.
85. Park WM. The place of radiology in the investigation of low back pain. Clin Rheum Disord 1980;6:93–132.
86. Modic MT, Steinberg PM, Rodd JS, Masaryk T, Carter JR. Degenerative disk disease: assessment of changes in vertebral body marrow with MR imaging. Radiology 1988;166:193–9.
87. Seligman JV, Gertzbein SD, Tile M, Kapasouri A. Computer analysis of spinal segment motion in degenerative disc disease with and without axial loading. Spine 1984;9:566–73.
88. Weiler PJ, King GJ, Gerbein SD. Analysis of sagittal plane instability of the lumbar spine in vivo. Spine 1990;12:1300–6.
89. Burton AK, Barrie MC, Gibbons L, Videman T, Tillotson KM. Lumbar disc degeneration and sagittal flexibility. J Spinal Disord 1996;9:418–24.
90. Casey L. Lumbar spinal instability after extensive posterior spinal decompression. Spine 1983;8:429–537.
91. During J, Goudfrooij H, Keessen W, Beeker TW, Crowe A. Toward standards for posture. Postural characteristics of the lower back system in normal and pathologic conditions. Spine 1985;83–7.
92. Legaye J, Hecquet J, Marty C, Duval-Beaupere G. Equilibre sagittal du rachis. Relations entre bassin et courbures rachidiennes sagittales en position debout. Rachis 1993;5(%):215–6.
93. Lazennec JY, Desjardins A, Laudet CG, Lazennec A, Roger B, Saillant G et al. Déstabilisation du rachis lombaire en L4 L5 et L5 S1. Etude expérimentale des conséquences biomécaniques de la discectomie chirurgicale, de l'arthrectomie et de la laminectomie. Edt Masson 1993;192–9.
94. Van Royen BJ, De Kleuver M, Slot GH. Polysegmental lumbar posterior wedge osteotomies for correction of kyphosis in ankylosing spondylitis. Eur Spine J 1998;7: 104–10.
95. Duval-Beaupere G, Schmidt C, Cosson P. A barycentremetric study of the sagittal shape of spine and pelvis. Ann Biomed Eng 1992;20:451–62.
96. Mangione P, Senegas J. Sagittal balance of the spine. Rev Chir Orthop Reparatrice Appar Mot 83:22–32.
97. Kuniyoshi A, Panjabi M et al. Biomechanical evaluation of lumbar spinal stability after graded facetectomies. Spine 1990;15(11):1142–7.

41 Biomechanical Elements of the Scoliotic Spine

P. Tropiano

The management of spinal deformity is evolving at a very rapid pace. Many of the teachings and tenets accepted as dogma in the recent past have gone the way of the medicinal leech, as our understanding of the natural history and biomechanics of scoliosis and kyphosis has expanded.

Abnormal spine morphology was first described in "De Articulaciones" of the Corpus Hippocraticum. Galen (AD 131–201) coined the terms "scoliosis, kyphosis, and lordosis." During the Middle Ages, deformity evoked scorn and derision, disfigurement being considered the work of the devil. Early treatment of scoliosis focused on reversing the environmental factors believed to be causative, such as nutrition and dampness. Later, exercise and a bewildering array of external supports and distraction devices were prescribed. Ignorance spawned a plethora of bizarre management regimens. For example, since scoliosis is unknown in quadrupeds, one clinic in New York in the 1870s admitted children for up to two years simply to assure their walking on hands and knees – there apparently being not much in the way of utilization review back then!

The first spinal fusion for scoliosis was performed in New York by Russell Hibbs in 1914. The results of fusion surgery were generally poor until the advent of internal fixation with the Harrington rod, in the early 1960s. In the past decade, an ever-increasing array of instrumentation systems has been developed, to better realize the goals of improved safety, curve correction, fusion rate, and postoperative patient mobilization.

Over the last 75 years, the cause of idiopathic scoliosis has been sought in investigation of every area that could relate to the deformity. Research has focused on genetics, growth and the endocrine system, postural equilibrium and the central nervous system, all structural elements of the spine, biomechanics, and collagen metabolism. Genetics plays an important role in the development of idiopathic scoliosis. It is eight times more common in girls, and 10 times more likely to be found in close family members. Its transmission can't be defined in classic Mendelian terms, nor have the gene(s) or biochemical sequences been identified. Further, there is evidence that an abnormality exists in the sensory afferents or in the brain stem, disturbing the reflexes that control spinal posture. Whether the pathologic changes are anatomic or biochemical is unknown, and the etiology of scoliosis, while still apocryphal, is probably multifactorial.

Biomechanics of scoliosis attempts to answer three questions in increasing order of difficulty.

What is the position of the scoliotic rachis in articular facet mechanics?
How does scoliotic curvature progress?

What are the mechanisms that enable the appearance or constitution of a scoliosis?

By limiting our discussion to so-called "idiopathic" scoliosis, in other words eliminating scolioses due to known causes (malformation (congenital), neuromuscular, and dystrophic), we will see that the answers are already quite complex.

Mechanical Situation of Scoliotic Rachis

Schematically, there is a difference between lateral scoliotic deviations and anteroposterior deviations, kyphosis, or lordosis. Mechanically, a scoliosis is not a simple lateral deviation. In reality it is a complex deformation that develops in three dimensions in space. The front view of a scoliosis on an X-ray or the side view are only reference points for subsequent examinations. Each of them is only an "oblique view" of the scoliotic deformation. Schematically, a triple deformation is observed: for thoracic curvatures a lateral inflexion in the frontal plane, a vertebral rotation in the horizontal plane, and a lordosis in the sagittal plane (disappearance of the physiological kyphosis). On the other hand, a reduction of the physiological lordosis is observed for lumbar curvatures. It was suggested that the scoliosis was a rotary lordosis. All vertebrae are an extension of each other in the scoliotic curvature [1,2]. When the vertebral rotation exceeds 90°, the rachis collapses in a pseudo-kyphosis but, paradoxically, in this case the vertebra will remain an extension of each other.

Three-dimensional reconstitution of the scoliotic curvature using the GRAF computerized graphics model have enabled us to better understand scoliotic deformation, in particular the "top view" shows the advance of the scoliotic curvature in space. Scoliosis is a torsion of the spinal column in space [3]. This deformation is three-dimensional. The lateral inflexion does not exist in an isolated manner (as is observed on the normal rachis). There is always a combination of a lateral translation of vertebral bodies that is the essential element, and a genuine rotation between some of them. Looking at a scoliosis on a front X-ray, the rotation of a vertebra is defined and measured depending on the position of the pedicles and the spinous process. This is a convenient procedure for creating a reference system; in fact, we do not measure the actual inter-segment rotation of one vertebra with respect to another; we measure a "vertebral torsion" combining lateral displacement and rotation. The unbalance produced by scoliotic torsion modifies the static and dynamic qualities of the spinal column and causes a collapse under load that shortens the trunk of the scoliotic patient when standing up. Finally, there are junction nodes at the ends of the scoliotic curvature that are intended to compensate for the scoliotic curvature and allow subjects to obtain balance of the spinal column. The appearance of these junction nodes is therapeutically very important. These biomechanical scoliotic elements explain interpretation errors in the action of the conservation treatment (kinesitherapy and apparatus), and also in the surgical treatment of a scoliosis. A bipolar instrumentation such as the Harrington apparatus corrected the collapse under load and had little action on the lateral and rotational translation. Transverse detraction (DTT) associated with the harrington apparatus was apparently illogical since it combined bipolar detraction with a transverse traction force bearing on the transverse processes on the convex side (which therefore tended to aggravate the rotation of the apex vertebra). But the DTT considerably improved the angular correction of scolioses. Why? The answer is simple: the DTT combined bipolar detraction with a lateral translation correction of vertebral bodies, without action on the rotation. Do the new Luque instrumentation, the Dove frame, and especially CD and its derivatives have a real vertebral derotation action and when this derotation takes place, is it always beneficial for compensation curvatures? In recent publications there is controversy about the derotation action of posterior instrumentation using screws, hooks, and rods. Some authors estimate this derotation at 40%, others at 20%. In the rest of our discussion we will attempt to explain these divergences.

Costal thoracic deformations that are so important clinically (gibbosity on the convex side of the curvature) with verticalization of the ribs on the convex side and horizontalization of the ribs on the concave ride, are only the consequence of the specific vertebral displacements described above.

How a Scoliosis Progresses

Mechanical Factors

Up to about 25–30° of curvature – structural, vertebral, ligamentous, and muscular – deformations remain moderate and curvatures may stop their progress and even regress, although rarely spontaneously and most frequently under the effect of treatment. The limit of 30° is a genuine mechanical threshold and beyond this angle, the deformation itself contains the factors for further progress. The muscular lever arms become longer on the concave side and shorter on the convex side in the three planes in space (large abdomen, transverse, and spinous muscles). Weight and costal pressures play an aggravating role.

The Growth Factor is Obviously an Essential Element

The scoliosis is progressive until bone maturity and then typically it stabilizes. In reality, work by Ponsetti [4] has demonstrated that when the curvature exceeds 30–40° at the end of growth, aggravation is possible in adults of the order of 1° per year. The growth of a scoliotic vertebra occurs under abnormal mechanical conditions and Delpech's law is applicable as elsewhere. Vertebral epiphysial growth cartilages on the concave side are compressed, causing the vertebral body to become cuneiform and deformation of the entire vertebra in the horizontal plane. According to Stagnara's suggestive image, the vertebra may be compared with a tricycle performing a tight turn, the vertebral body moving to the convex side, the posterior arch to the concave side, and the spinous process in the same direction as the vertebral body. The deformation of disks is equally important: the nucleus pulposus is pushed out on the convex side; MacEwen [5] makes this an essential element in the irreducibility of scoliotic deformation. Vertebral growth and evolution of the scoliotic curvature are related and non-parallel as shown by the curvature defined by Duval-Beaupere [6]. Scoliotic curvature is aggravated in the final years of bone maturity, whereas spinal growth is practically stopped.

The controversy concerning the subject of the derotation action of posterior instrumentation is directly related to these growth factors. In published statistics, all idiopathic curvatures of operated adolescents are normally mixed together (and may even be combined with those for young adults). It appears to us that the examiners have missed an essential concept, namely whether or not evolution of the scoliosis was recent with respect to the operation that is different from the normally described concept of flexible curves and rigid curves. A satisfactory "derotation" can be made on a curvature that developed quickly between 11 and 13 years old by posterior segmentary instrumentation (in other words, correction of the collapse of lateral translation under load and inter-segmentary derotation of the vertebrae).

On the other hand, the same inter-segmentary derotation will not be obtained on a flexible curvature with the same degree, but with an older development, operated on at the same age. Structural deformations, difficult to see on X-rays, are more important in the latter case. It is obvious that the derotation action will be even lower for rigid curvatures. We must also emphasize another concept. Surgical correction by the posterior way, combining metal instrumentation and an arthrodesis, regardless of its quality, will only be really efficient if the subject is close to bone maturity. If arthrodesis correction by the posterior way is done at Risser 0, the posterior arthrodesis will not prevent growth of vertebral bodies at the front and the scoliotic deformation will be aggravated in accordance with the crankshaft phenomenon described by Jean Dubousset, with the grafted zone tilting sideways. Arthrodesis performed long before bone maturity is reached must include both posterior and anterior arthrodesis. Temporary instrumentation processes without arthrodesis have been described to prevent this phenomenon for young children (MOE subcutaneous pin) when the scoliosis cannot be maintained by the orthopedic apparatus. Note that these temporary instrumentations must always be accompanied by a support corset until the final arthrodesis correction operation. Despite these precautions, there are many failures.

What are the Mechanisms that Enable the Appearance and Constitution of an Idiopathic Scoliosis?

Biomechanically, which is the only aspect in which we are interested here, Bisgard and Musselman's works [7] have shown that anterior vertebral structures do not appear to be responsible. However, according to Roaf [8], posterior structures would be inhibited.

This phenomenon is confirmed by Deane and Duthie's work [9]. According to Lindahl and Raedder [10], transverse structures also play a role. Is rotation primitive to lateral inflection, is it secondary or simultaneous?

Despite the large amount of experimental work, including that done by Kay [11], it is still not yet possible to answer these questions, particularly because the same problem arises for the component in lordosis. Can the primum movens, the driving force for the scoliotic deformation, be approached? Experimental and clinical work on the peripheral or central nervous system [12,13] confirms the role of the vestibule and balancing systems. Experimental work carried out by Jean Dubousset in 1987 showed that all lesions located in the pineal gland of the chicken could reproduce all vertebral deviations that can be seen in the child during growth and particularly the right or left laterality of curvatures.

Other works [2,4] draw attention to the existence of metabolic and chemical factors. In any case these various anomalies are related to genetic factors [10,14], as shown by the frequency of family forms that may reach 43%, factors also related to sex, as proved by the fact that 80% of idiopathic scoliosis subjects are girls. Idiopathic scoliosis also appears more and more as a hereditary complaint related to a maturation problem of proprioceptive sensitivity.

Until we can answer this third question with sufficient certainty, the treatment of idiopathic scoliosis will remain symptomatic. The purpose of this treatment is to obtain a curvature of less that 30° if possible at the end of bone maturity [15]. Supports are the only therapeutic means adapted to the severity of curvatures at the present time: muscular re-education support, electric simulation support or nocturnal vertebral traction support, vertebral support in the various orthopedic corsets and surgical methods (Harrington rods, Dwyer Gable and their more recent variants by Luque, Zielke, CD and its derivatives). Therefore, the biomechanical chapter of scoliosis is far from being closed.

Adolescent scoliosis is a fascinating topic which is in a constant state of evolution and updating of opinions and experiences. Rarely in medicine does a surgeon have the ability to make such a dramatic visible alteration to the structure of the human body as in scoliosis corrective surgery. Understanding the topic can seem a daunting challenge. With a thorough understanding of the basic definitions and principles a gradual enlightenment should hopefully follow.

Bibliography

Beals RK. Nosologic and genetic aspects of scoliosis. Clin Orthop 1973;93:23–32.

Belytschko T, Andriacchi T, Schultz A, Galante J. Analog studies of forces in the human spine: computational techniques. J Biomech 1973;6:361–71.

Gonon GP, Demauroy JC, Stagnara P. Journées de la scoliose. 8–10 Fev. 1979. Imp Basc Frères. Lyon, 1979.

Graf H, Hecquet J, Dubousset J. Approche tridimensionnelle des déformations rachidiennes. Rev Chir Orthop 1983;69(5):407–16.

Naasch CL, Moe JH. A study of vertebral rotation. J Bone Joint Surg 1969;51–1:223–9.

Rabiller-Dano M. Approches biomécaniques de la scoliose Idiopathique. Thèse Médecine Tour, 1976.

Stagnara P. Les déformations du rachis, Vol 1. Paris: Masson.

Vercauteren M. Approche étiopathogénique de la scoliose idiopathique. Acta Orthop Bel 1972;38:412–28.

References

1. Dickson R, Lawton JO et al. The pathogenis of idiopathic scoliosis. J Bone Joint Surg 1984;66b:8–15.
2. Nordwall A, Waldenstrom J. Métochromasia of fibroblasts from patients with idiopathic scoliosis. Spine 1976;1:97–8.
3. Perdriolle R. La scoliose – Son étude tridimensionnelle. Maloine SA, editor. Paris, 1979.
4. Ponsetti IV, Pedrini V, Dohrman S. Biomechanical analysis of intervertebral discs in idiopathic scoliosis. J Bone Joint Surg 1973;54A:1972.
5. MacEwen GD, Cowel HR. Familial incidence of idiopathlc scoliosis and its implications in patient treatment J Bone Joint Surg 1970;S2A:405.
6. Duval-Beaupere G, Dubousset J, Quenau P. Pour une théorie unique de l'évolution des scolioses. Nouv Presse Med 1910;78:1141–5.
7. Bisgard J, Musselman H. Scoliosis: its experimental production and growth correction. Growth and fusion of vertebral bodies. Surg Gyn Obs 1940;70:1029–36.
8. Roaf R. The basic anatomy of scoliosis. J Bone Joint Surg 1965;48–8:786–92.
9. Deane G, Duthie RB. A new projectional look at articulated scoliotic spines. Acta Orthop Scan 1973;44: 351–65.

10. Lindahk O, Raedder E. Mechanical analysis of forces involved in idiopathic scoliosis. Acta Orthop Scand 1962;32:27–38.
11. Kay SP. A new conception. and approach to the problem of scoliosis. Clin Orthop 1971;81:125–32.
12. Yamada K, Yamamoto H, Tezuka E. Development of scoliosis under neurological basis, particularly in relation to brainstem abnormalities. J Bone Joint Surg 1974;56A:1764.
13. Nachemson A, Sahlstrand T. Etiologic factors in adolescent idiopathic scoliosis. Spine 1977;2(3):176–84.
14. Cotrel. MK. Le facteur génétique dans la scoliose idiopathique. Paris: Thèse, 1974.
15. Bergoin et al. Critères de dècision et résultats du traitement orthopédique des scolioses idiopathiques de moins de 30°. Rev Chir Orthop 1984;70(Suppl II): 120–3.

42 Pedicle Screw Fixation in Thoracic or Thoracolumbar Burst Fractures

S.-I. Suk and W.-J. Kim

Introduction

When the decision is made to perform a posterior stabilization procedure for a thoracic or thoracolumbar burst fracture, pedicle screw fixation is a valuable assistance. Pedicle screw fixation of the spine, first described by Boucher [1] in 1959 and popularized by Roy-Camille [2,3] in the 1960s, has evolved through an era of pedicle screw and plate to today's modern pedicle screw and rod system. As it offers rigid fixation unparalleled by other fixation methods enabling a reliable fixation even in a vertebra with posterior element defects or severe osteoporosis, it has now become one of the most widely employed fixation devices in the field of spinal surgery, including spine fractures [4–6].

In surgeries for thoracic or thoracolumbar burst fractures, pedicle screw fixation has proven itself to be extremely effective and useful, offering the advantages of rigid fixation, improved segmental control, enhanced deformity correction, and reduction of the fusion extent in both fresh fractures and old fractures with deformity [7,8].

The purpose of this chapter is to introduce the biomechanical, anatomical basis and techniques of treating thoracic or thoracolumbar burst fractures with pedicle screw fixation.

Anatomy and Biomechanics

For a safe and reliable pedicle screw fixation of a thoracic and thoracolumbar burst fracture, a thorough understanding of the pedicle anatomy throughout the thoracic and the lumbar spine and of the biomechanical basis of the pedicle screw fixation is an absolute prerequisite.

Pedicle Anatomy

The pedicles in the thoracic and lumbar vertebrae are two short, thick processes that project dorsally, one on either side from the cranial part of the vertebral body at the junction of its dorsal and lateral surfaces. Medially, they border the spinal dura. Laterally, they are in proximity to the exiting nerve roots and the segmental vessels. Superior and inferior, they form the intervertebral foramina with the pedicles of the adjacent vertebrae, through which the spinal nerve roots exit. The roots traverse the foramina just inferior to the pedicles. This anatomic relationship puts the structures in proximity to dangers of injury in malpositioning of the screws.

The pedicle is an oval-shaped cylinder of cortical bone filled with some cancellous bone in the center. The medial wall of the pedicle is thicker than the lateral wall and for this anatomical reason, the transpedicular screws are more apt to be out of the pedicle laterally than medially.

The pedicle shape, dimensions, and orientations vary from region to region [9–11]. In adults, the vertical diameter of the pedicle increase steadily from 7 mm to 15 mm going down from T1 to L5. The horizontal diameter decreases gradually from 7 mm in T1 to 5 mm in T5 and then gradually increases to 16 mm in L5. The transverse angle of the pedicles, which is the angle formed by the axis of the pedicle and the vertical line, gradually decreases from a mean of 30° in T1 to −5° in T12 and then increases to 30° in L5. The horizontal angle, which is the angle formed by the axis of the pedicle and the horizontal line paralleling the lower vertebral end plate, shows the greatest negative value in T9 and 10 (Figure 42.1. A–D). These facts are important guidelines for an appropriate choice of the screw sizes and directions of screw insertion in the intended instrumentation levels, especially in deformed spines where radiographic guide is made more obscure by the rotational or angular deformity of the spinal column.

In children, the pedicles are smaller, but their relative dimensions and orientations are similar to those found in adults. As the spinal canal reaches 90% of adult size at the time of birth and reaches the adult size by the age of 2 [12], pedicle screw fixation may be carried out without dangers of causing iatrogenic spinal stenosis after this age. Though the pedicles are very small in pediatric patients, their bone is very plastic and the pedicles usually receive a screw larger than the outer diameter of the specific pedicle by plastic deformation of the pedicular cortex when the screw is inserted slowly through the center of the pedicle provided that the pedicular cortex is not violated [13,14].

Biomechanics

Among the many advantages, the greatest offered by the pedicle screw is rigid fixation, far superior to non-pedicle fixations [4–6]. However, obtaining such a reliable, rigid fixation is possible only through a sound technique, with a thorough understanding of the variables which may affect the strength of the fixation.

The influence of various screw-related and insertion-technique related parameters on the rigidity of the fixation offered by a pedicle screw has been extensively studied by nondestructive biomechanical studies and studies measuring the pullout strength of pedicle screws.

Pedicle Screw Diameter

Generally, the pullout strength of a pedicle screw increases with increasing major diameter of the screw as long as the integrity of the pedicle is not violated [15,16]. The pedicles usually accept screws with diameters less than 86% of the isthmic outer diameter of the pedicle without significant structural alterations [10]. A screw of a larger diameter may cause a linear fracture of the pedicles, but the pullout strength is not significantly affected. The pedicles are capable of receiving a screw up to 116% of the isthmic diameter without significant change in the pullout strength. Though some speculate that a screw with a diameter of 3.5–4.5 is sufficient to resist any pullout force generated in the human body, we believe it is much more reliable to use a screw with a diameter of approximately 80% of the pedicle diameter; this offers maximum pullout strength.

Screw Length

Theoretically, the pullout strength of a pedicle screw increases with increasing depth of insertion as the surface area of contact increases. Some even advocate penetration of the anterior vertebral cortex to obtain a bicortical fixation to increase the pullout strength [17]. However, there have been contradictory studies stating that the pullout strength is not significantly affected by the depth of insertion when the screw passes deeper than the posterior one half of the vertebral body [16,18,19]. The authors agree with the latter view and advocate using screws that penetrate 50–70% of the anteroposterior vertebral diameter. This apparent paradox is attributed to the presence of a dense sheet of cortical bone, the so-called neurocentral junction [19], situated in the posterior one third of the vertebral body at the site of former neurocentral synchondrosis, that makes biomechanical bicortical fixation possible (Figure 42.2). The neurocentral junction marks the site of union between the centrum and the two neural arches which develop from separate primary ossification centers and unite at the age of 3–6 years.

Screw Direction

Except in the sacrum where the pedicles are large enough to allow a significant alteration in the directions of the pedicle screws to engage the medial anterior cortex, the upper sacral end plate or the ala, the optimal direction of a pedicle screw seems to be along the axis of the

a

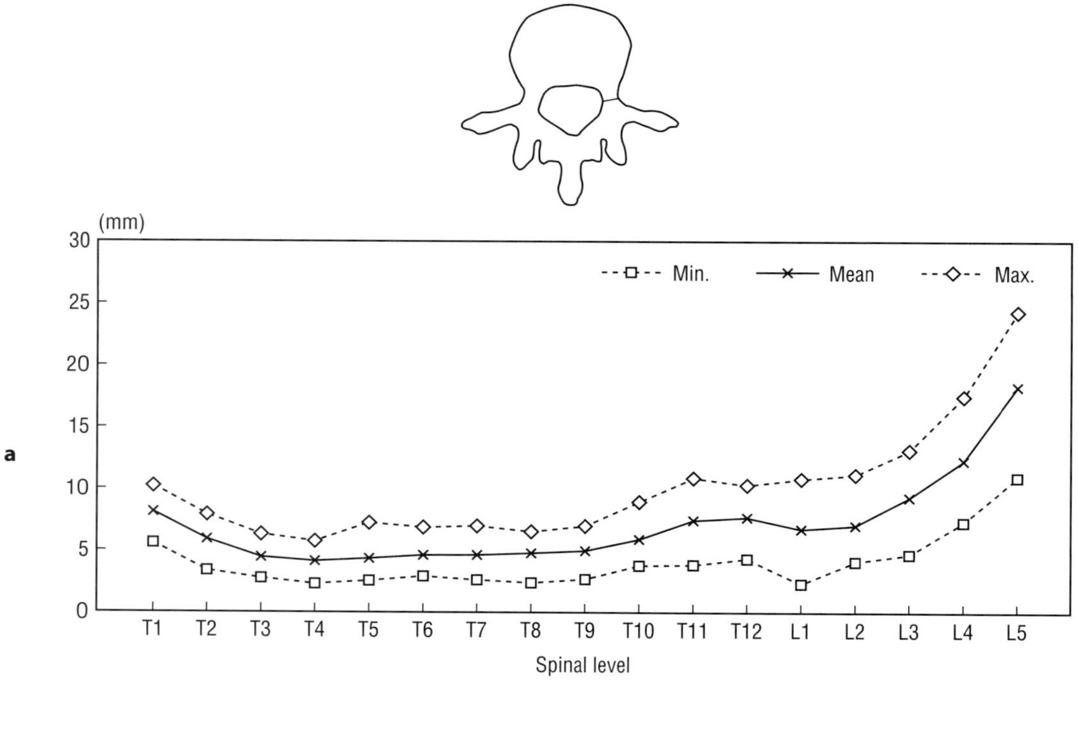

b

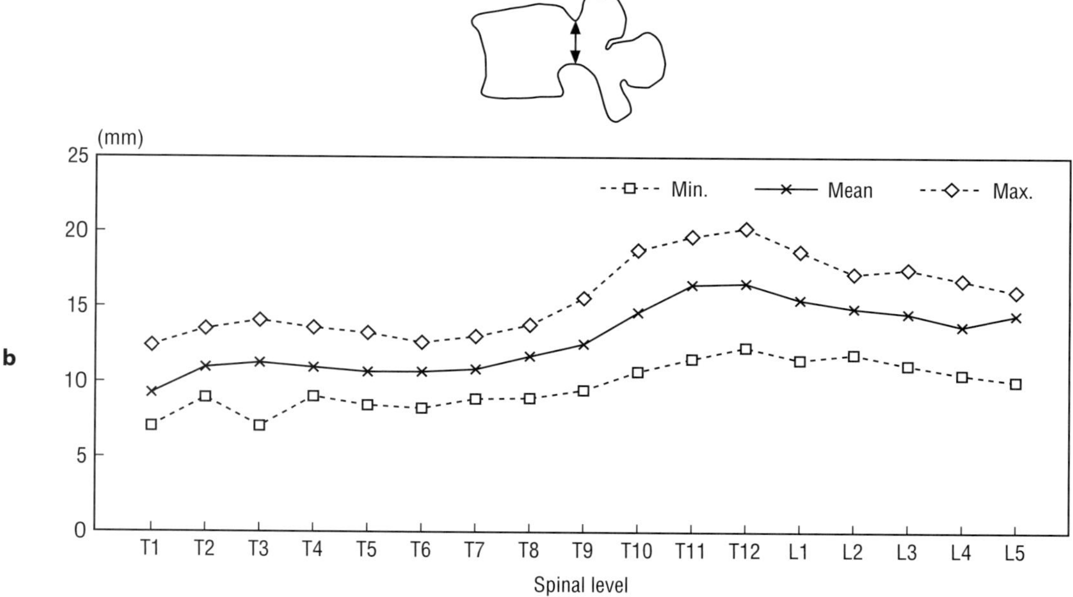

Figure 42.1. **a** Transverse pedicle diameters. **b** Superoinferior pedicle diameters. **c** Anteroposterior pedicle angles. **d** Horizontal pedicle angles.

Anteroposterior pedicle angle (α) and depth to the anterior cortex through line parallel to midline axis (◂- - - -▸) and pedicle axis (◂— —▸) a : midpoint of transverse pedicle diameter.

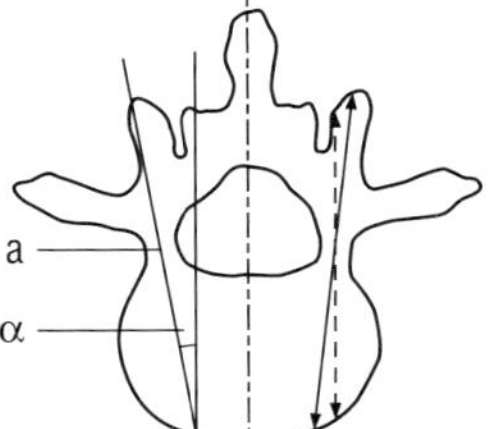

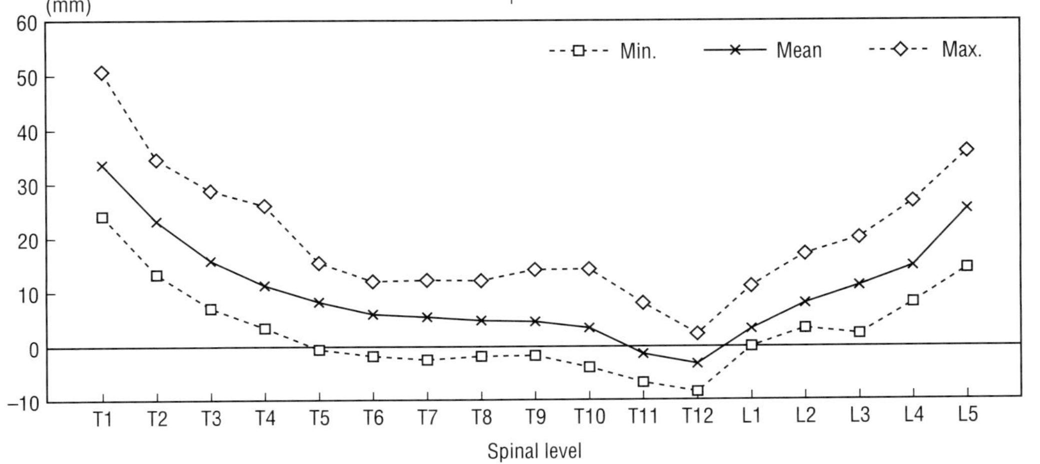

c

Horizontal pedicle angle (β) b : midpoint of superoinferior pedicle diameter.

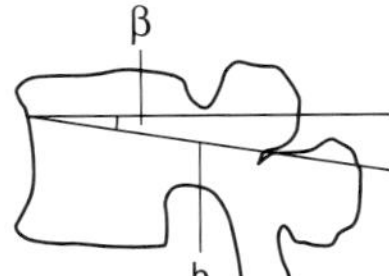

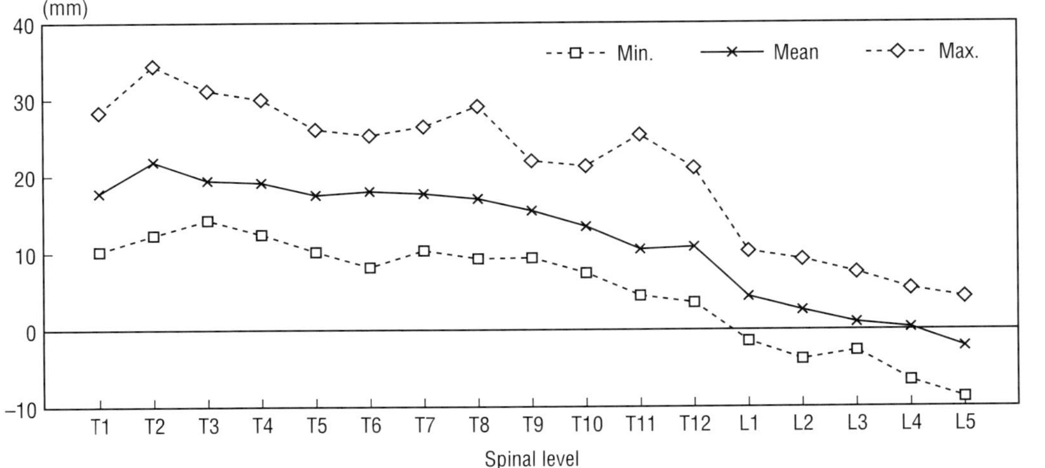

d

Figure 42.1 *(continued)*

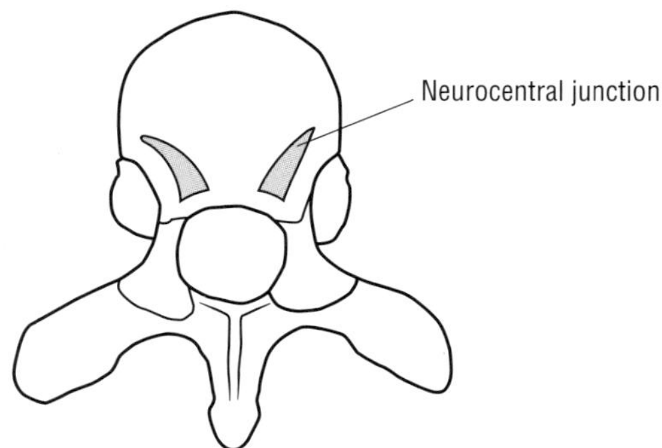

Figure 42.2. Neurocentral junction.

pedicle. Though this does not significantly affect the screw pullout strength in most situations, it reduces the chance of pedicular cortical perforations.

Screw Hole Preparations

Pedicle screws offer greatest pullout strength when the diameter of the screw holes are as large as the minor diameter of the screws inserted, approximately 60% of the pedicular isthmic outer diameter [14]. When the diameter of the holes are smaller, screw insertion is not only difficult but may also cause pedicle fractures. Making a hole of a larger diameter in the pedicle may cause breakage of the pedicle cortex during the hole preparation procedure and reduces the pullout strength [14,20]. As to the method of making the screw holes, there was no significant difference in the pullout strength between holes prepared by drilling and probing [20,21]. However, probing seems to reduce the chance of damage to the pedicular cortex.

Number of Screws

The rigidity of fixation increases with increasing number of screws, thus being most rigid in segmental instrumentation where screws are inserted in every segment fused [5,22]. In a nondestructive study using porcine vertebral columns, the segmental screw fixation construct was significantly stiffer than the nonsegmental screw construct in all the parameters of flexion, extension, lateral bending, and torsion [23]. Though there had been enthusiasm to reduce the instrumentation to one side (unilateral instrumentation technique), the authors advocate bilateral instrumentation to enhance resistance to torsional forces.

Transverse Links

The addition of transverse links to bilateral segmental constructs significantly increases the stability in axial rotation [24–26]. A transverse link is more effective if placed in the proximal part of the construct than if placed distally. Two transverse links show increased torsional stiffness than one transverse link especially in longer constructs.

Methods of Increasing the Stiffness of the Pedicle Screw Construct

When posterior instrumentation with pedicle screws is performed on an unstable spine with an anterior column defect, the lack of normal anterior support significantly reduces the flexion/extension and torsional stiffness of the pedicle screw construct even when a very rigid implant is used [27,28]. In this situation the stability of the pedicle screw construct may be enhanced by restoration of the anterior defect, by expanding the level of instrumentation, or by the use of more rigid external immobilization postoperatively.

Treatment Considerations and Indications

The ultimate goals of treatment for a spinal fracture are restoration of the neurologic and the mechanical stability of the injured spine with minimal sacrifice of motion segments. This same principle applies to the pedicle screw fixation. When restoration of the spinal stability with pedicle screw fixation is contemplated for a thoracic or thoracolumbar burst fracture, both the neurologic and the mechanical aspects need to be considered before finally deciding on the combination of available surgical techniques to successfully accomplish the goals.

As regards biomechanical aspects, the degree of vertebral body comminution and the magnitude of local kyphotic deformity are the major determinants. Though modern biomechanics understand the intact spine as a tricolumnar structure composed of anterior, middle, and the posterior column, the biomechanics concerning the surgical reconstruction of a burst fracture considers the spine basically as a bicolumnar structure, composed of an anterior weight-bearing column and the posterior tension column, whose stability is essentially dependent on the competence of the anterior column. This means that when there is residual, unattended incompetence of the anterior column, the spine will continue to be unstable and is ultimately doomed to fail under repeated loading. This fact is to be always kept in mind in all posterior instrumentation and fusions for thoracic and thoracolumbar burst fractures, including pedicle screw fixation, which inherently restores or reinforces the posterior column but leaves the essential anterior column untouched, relying for its restoration on the bony healing of the fractured vertebral body. The essence of the advantage offered by rigid fixation with pedicle screw fixation lies in the fact that the fractured vertebral body is more effectively protected from the detrimental forces until the fracture union occurs and becomes stable under physiologic loads. By the same token, when the anterior column incompetence is so severe that it is not expected to become a competent weight-bearing structure by itself, a deliberate restoration of the anterior column is warranted. When a significant local kyphosis or comminution of the vertebral body is present, a mere posterior reduction and stabilization with pedicle screw fixation will create an anterior unsupported gap. In the presence of such a gap, the posterior constructs are destined to fail, however stiff they may be, not being able to withstand the flexion moment concentrated at the fracture site devoid of the anterior load-sharing structural support [27,28]. In these situations, elimination of the anterior defect either by restoration of the anterior column or by shortening of the posterior column is warranted. Anterior column reconstruction may be done by an interbody fusion via an anterior or a posterior route, or transpedicular bone grafts.

Posterior Pedicle Screw Fixation

Pedicle screw fixation is applicable and is indicated in all thoracic and thoracolumbar burst fractures in which a posterior fixation and fusion is under consideration. Since it offers a rigid fixation with enhanced segmental control, it is more advantageous than hooks and other nonpedicle fixation devices in the aspect that it offers a better restoration of the spinal sagittal contour and makes a shorter fusion feasible, saving more motion segments [29]. Moreover, as pedicle screw fixation offers a more reliable fixation in the osteoporotic spine and in vertebra with previous laminectomy, they are particularly advantageous in these situations.

Though there are questions about the safety of pedicle screws in the thoracic and thoracolumbar region and though pedicle screws are still under surveillance in the United States (FDA class III), when correctly placed in the pedicles, the screws stay out of the spinal canal and remain insulated from the neural elements by the surrounding pedicular bone, precluding neurologic derangement by the device. The authors have been using pedicle screws in the thoracic spine since 1988 in more than a thousand patients with various conditions including the most severe deformities and revision surgeries, inserting probably more than 10,000 screws. Yet, there was not a single major neurologic or visceral complication attributable to the pedicle screw fixation per se and the authors believe the procedure to be perfectly safe when performed with a sound technique [5,30,31,32].

The indications of a simple posterior stabilization with pedicle screw fixations in thoracic and thoracolumbar burst fractures are:

Cord level fractures with complete paraplegia: Since saving the motion segments is not a crucial problem in this situation, a lengthy fusion with pedicle screws which increases the stiffness of the construct is a suitable choice as it allows an early mobilization without an external support.

Unstable burst fractures without neurologic compromise when they satisfy all of the following.

Spinal canal encroachment <60% on axial CT or MRI.

Local kyphosis <50 degrees.

Without significant comminution of the vertebral body.

Fracture less than 72 hours old.

As these fractures will heal eventually to offer an anterior support, a mere posterior ligamentotaxis and protection from the detrimental forces will suffice (Figure 42.3. a–g).

Fractures with neurologic compromise when they satisfy all of the following.

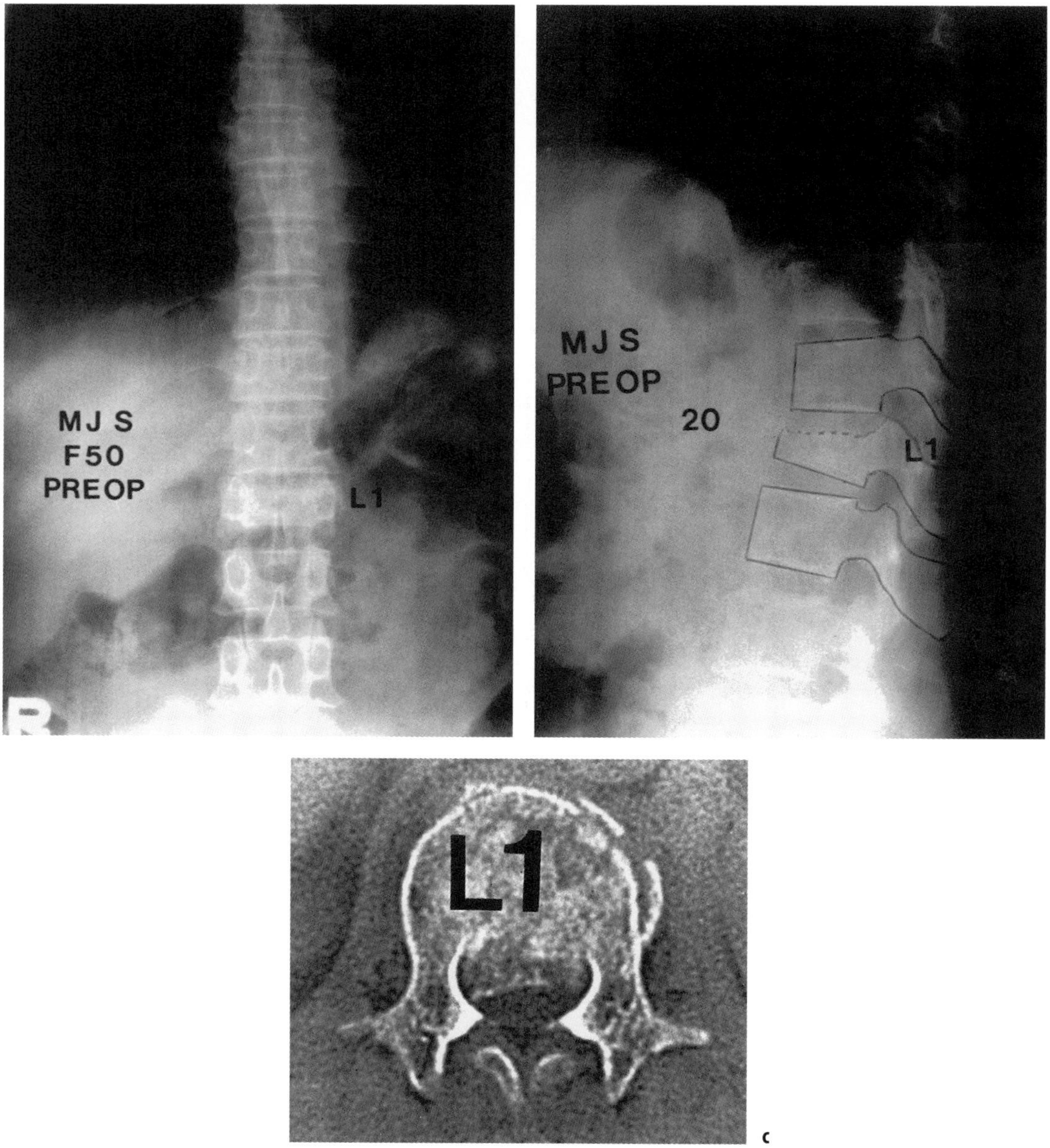

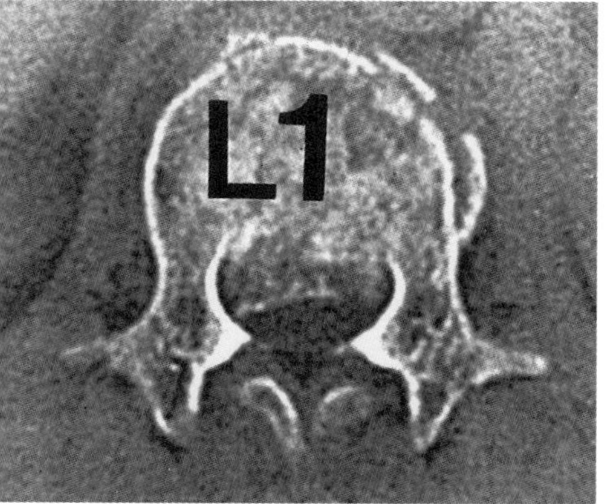

Figure 42.3. **a,b** A 50-year-old female with L1 unstable burst fracture. There was a 20° kyphotic deformity at the thoracolumbar junction. **c** Preoperative CT show 40% canal encroachment. **d,e** She was treated by posterior fusion with pedicle screw fixation from T11 to L2. Postoperatively kyphosis was corrected to 10 °. **f** Postoperative CT show canal encroachment reduced to 10%. **g** Ligamentotaxis. (*continued next page*)

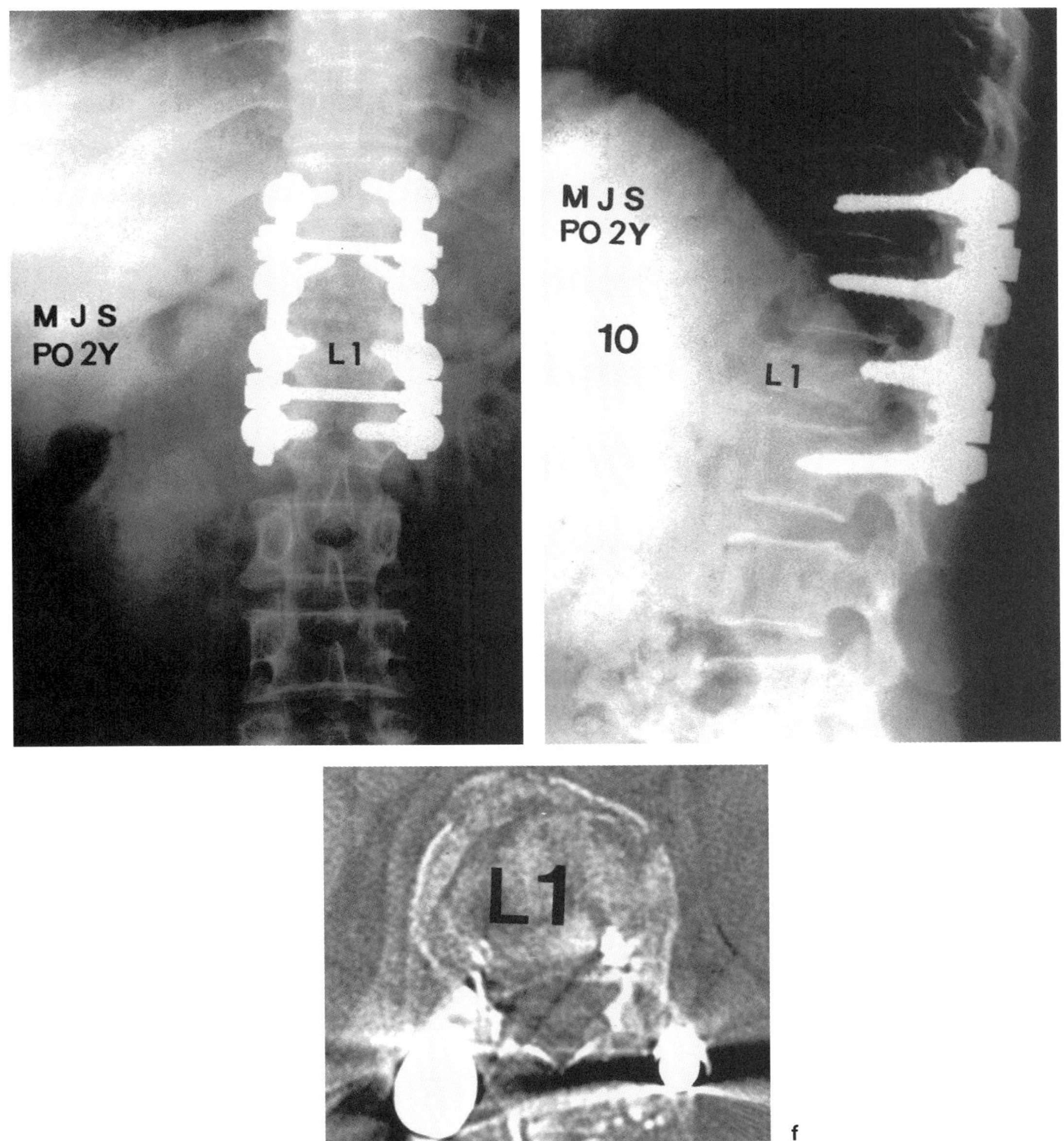

g

Figure 42.3 *(continued)*

Spinal canal encroachment <40% on axial CT or MRI.

Local kyphosis <50 degrees.

Without significant comminution of the vertebral body.

Fracture less than 72 hours old.

In these fractures, a posterior ligamentotaxis may be tried. When there is no neurologic improvement, an additional direct decompression of the neural element is necessary (Figure 42.4. a–h).

Established post-burst fracture kyphosis of less than 50° without significant neurologic compromise: Though the correction of kyphotic deformity is negligible, most patients do well without further progression of the deformity.

Senile burst fractures with pain and progressive deformity without significant neurologic compromise: These osteoporotic burst fractures are best stabilized with a pedicle screw fixation as the pedicles are affected less by the osteoporosis than the laminae. As most of these fractures are without significant canal compromise, an in situ stabilization is usually sufficient.

Posterior Pedicle Screw Fixation with an Anterior Column Reconstruction

In pedicle screw fixation, an additional anterior column reconstruction is indicated when there is significant anterior column incompetence which is not expected to heal sufficiently to function as a stable, weight-bearing structure as in fractures with significant kyphosis, severe vertebral body comminution, and following a corpectomy procedure for direct decompression of the neural elements. Though some favor an anterior stabilization and fusion in these situations, we believe a combined anterior and posterior column reconstruction with posterior pedicle screw fixation has several advantages over the anterior instrumentation/fusion only 33. They are: 1) Better restoration of spinal alignment. When the kyphosis is severe, it is easier to restore a physiologic sagittal profile with a posterior pedicle screw fixation than with an anterior instrument. 2) More reliable rigid fixation. In osteoporotic patients, the pedicle screw fixation offers more reliable fixation than the anterior instruments holding the osteoporotic vertebral bodies. 3) Increased versatility. Pedicle screw fixation offers a rigid fixation in locations where anterior instrumentation is difficult or needs extensive dissection (e.g., cervicothoracic junction). Though an anterior column reconstruction is more commonly performed through an anterior approach, a reliable anterior column reconstruction is also feasible from the posterior, by transpedicular bone grafting [34], by interbody fusion using the costotransversectomy approach [35], or using a modified egg-shell procedure [36]. In the former, cancellous chip grafts are added into the fractured vertebral body and the damaged disc space through the pedicles of the fractured vertebra using a narrow funnel. In the latter, the anterior column may be supported by a structural graft or by autogenous cancellous graft tightly packed into the void in the anterior column. When local kyphosis is severe, a type of posterior closing wedge osteotomy may be performed through the posterior approach to obliterate the anterior unsupported gap. The advantage of the anterior column reconstruction from the posterior in conjunction with pedicle screw fixation is the feasibility of a global fusion through a single approach, saving the operative time and reducing the morbidity of an anterior thoracotomy or thoracolumbotomy [33]. It also permits a direct exploration and repair of the associated dural tear or nerve root entrapment. Its disadvantage is destruction of the relatively intact posterior column, reducing the posterior fusion base. However, this drawback may be overcome by one of the following methods; 1) Saving the lamina and pedicle on one side. 2) By adding a bridging bone graft. 3) By shortening the entire posterior column to achieve a bony contact between the two laminae above and below the laminectomy.

The indications for posterior stabilization with anterior column reconstruction are;

Presence of a neurologic deficit necessitating a formal decompression of the neural tissue:

Spinal canal encroachment more than 60%.

Failure of the indirect decompression by ligamentotaxis

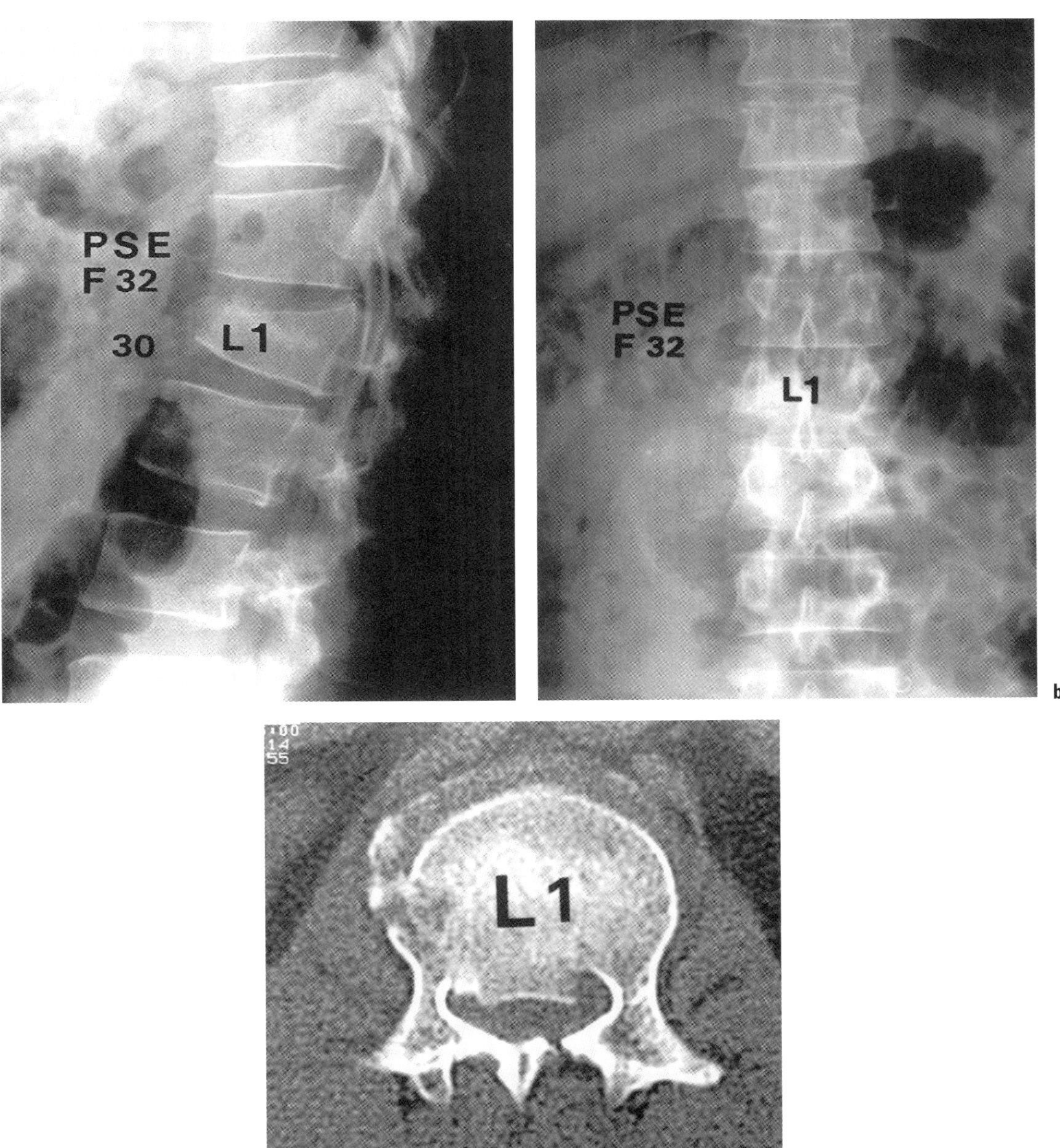

Figure 42.4. a,b A 32-year-old female with L1 unstable burst fracture. Preoperative neurology was intact except for bowel and bladder control. There was a 30° kyphotic deformity at the thoracolumbar junction. **c** Preoperative CT shows 60% canal encroachment. **d,e** She was treated by the posterior fusion with pedicle screw fixation from T11 to L2. But, she had no neurologic improvement. One week later after the posterior surgery, anterior decompression was carried out. Following the anterior surgery, her bladder control improved. **f** CT image after anterior surgery shows no residual canal encroachment and anterior strut graft. **g,h** Postoperative 1-year-follow-up radiographs show satisfactory maintenance of reduction. (*continued next two pages*)

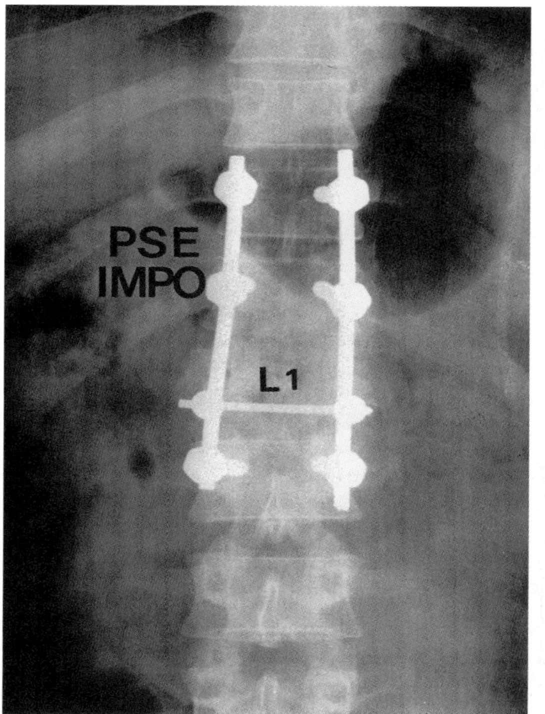

d

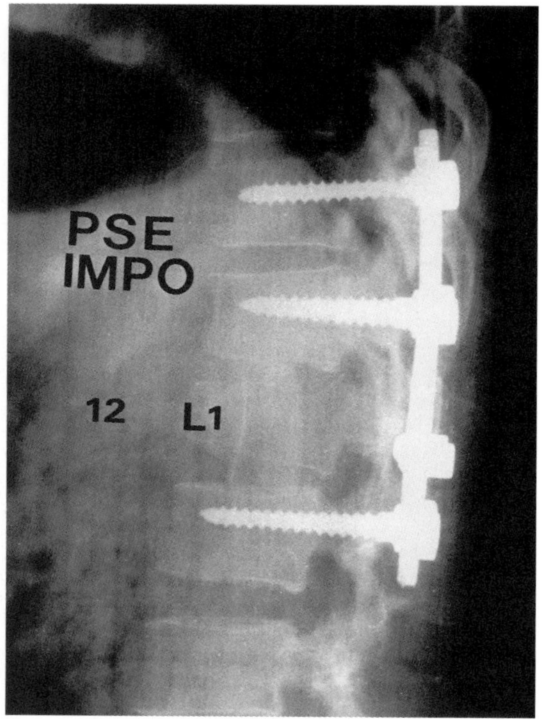

e

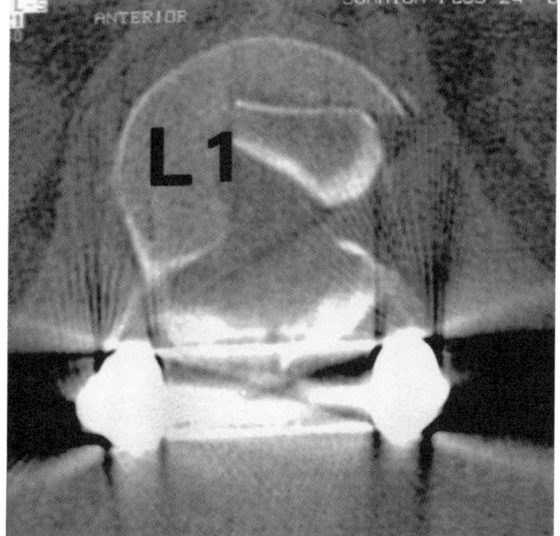

f

Figure 42.4 *(continued)*

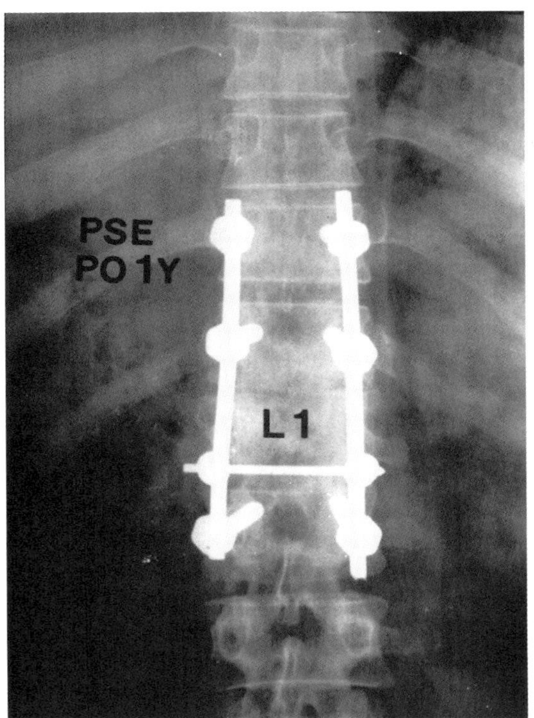

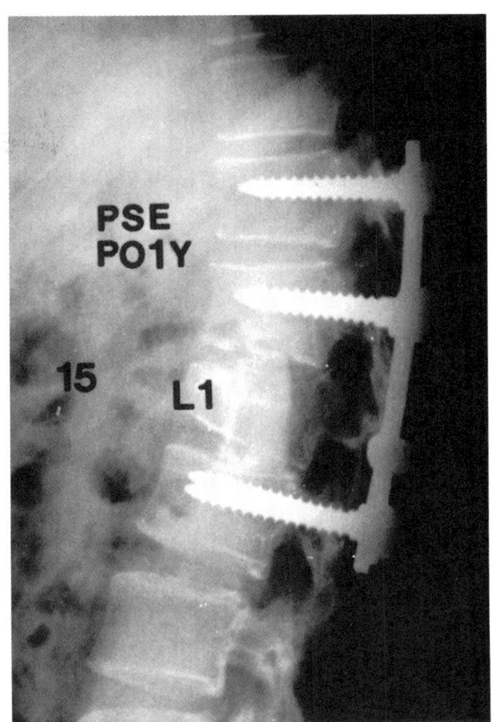

Figure 42.4 *(continued)*

Fractures more than 72 hours old.

Local kyphosis greater than 50°.

Significant comminution of the vertebral body.

Established post-burst fracture kyphosis greater than 50°. (Posterior or anterior fusion alone in this situation is more prone to failure and should be treated by a global fusion.) (Figure 42.5. a–f).

Surgical Techniques

Presurgical Considerations

Anesthesia

For a posterior stabilization with pedicle screw fixation, a general anesthesia with a full monitoring of vital signs including the arterial and the central venous pressure is preferred. As an anterior column reconstruction is often accompanied by a substantial amount of bleeding from the epidural vein and the cancellous bone of the fractured vertebral body, securing a large-bore central venous channel is highly recommended. Hypotensive anesthesia is especially helpful in cases where anterior column reconstruction is considered. When intraoperative-evoked potential monitoring is contemplated, intravenous anesthesia with fentanyl and propopol is preferable to inhalation anesthesia as they affect monitoring less severely than the latter. However, intravenous anesthetics require a substantially longer time for recovery than inhalation anesthetics and may even require several hours of ventilator care in the recovery room.

Intraoperative Monitoring

As we have never had a major neurologic complication related to the pedicle screw placement and are perfectly confident of the safety, we do not use intraoperative neurophysiologic monitoring except in cases in which the spinal cord is directly exposed. However,

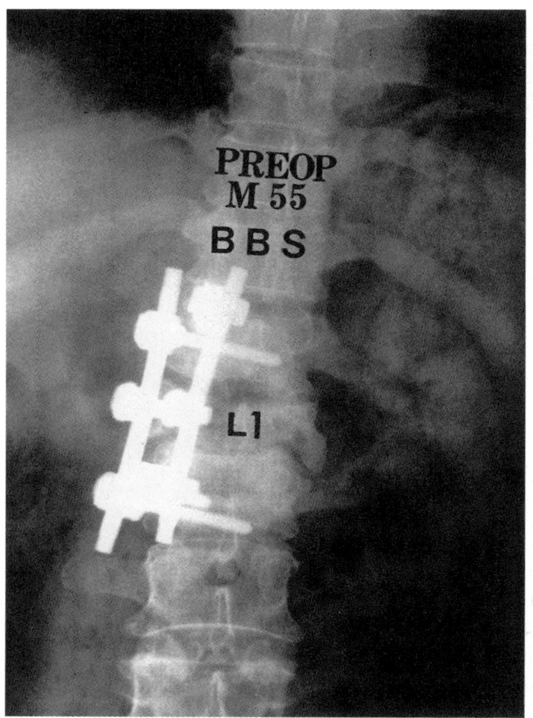

a

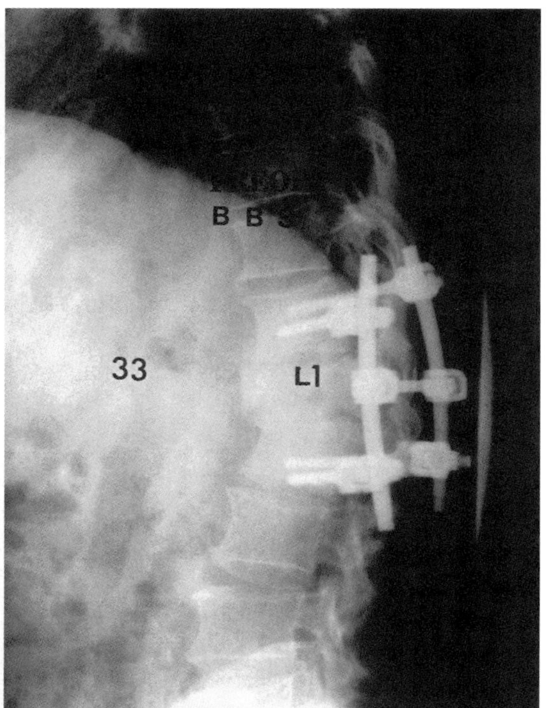

b

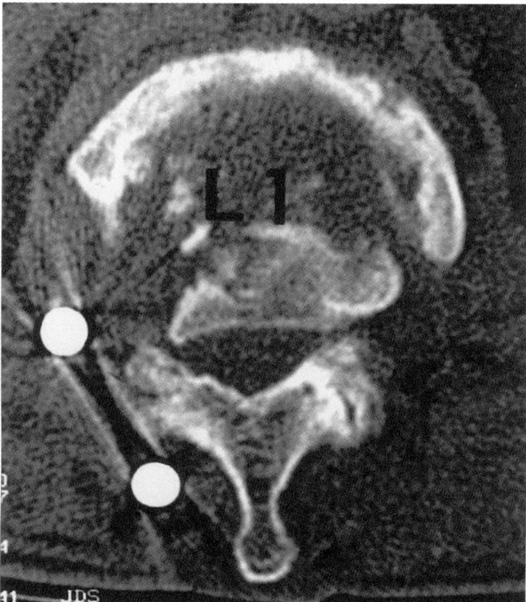

c

Figure 42.5. a,b A 68-year-old male with a post-traumatic kyphosis and cauda equina syndrome. One and a half years prior to the visit, he sustained a burst fracture which was treated by posterior decompression and fusion. **c** Preoperative CT shows retropulsion of fracture fragment with neural compression. **d,e** He was treated by the posterior closing wedge osteotomy at L1 and fused from T11 to L3. Postoperative one-year-follow-up radiographs show satisfactory maintenance of sagittal balance. **f** Posterior closing wedge osteotomy. (*continued next page*)

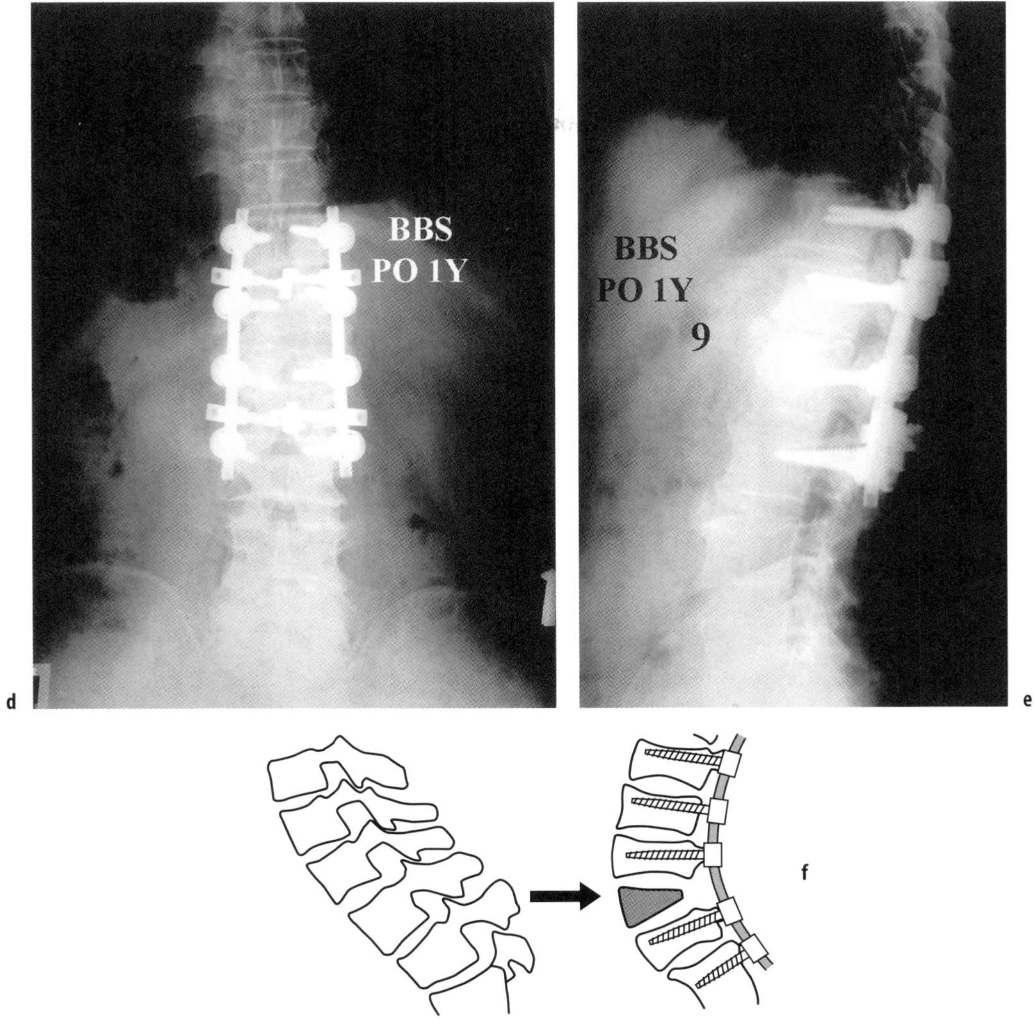

Figure 42.5 *(continued)*

intraoperative monitoring of the neurologic function is a valuable aid to ensure the safety of the procedure and for keeping a record for possible medicolegal problems [37,38]. Although the wake-up test is a reliable monitoring method, we prefer motor-evoked potential or somatosensory-evoked potential, which can be more conveniently performed without extending the operative time [39].

Positioning

The ideal position for a pedicle screw fixation is the standard prone position with the abdomen hung free by means of a pad, four posters, or a special surgical frame to reduce venous bleeding. The operating table should be X-ray penetrable to allow the intraoperative postero-anterior roentgenogram. When posterior column shortening is contemplated, care should to taken to place the osteotomy site over the hinge of the operating table so that closure of the posterior gap created by the osteotomy may be aided by extension of the operating table. For this procedure, we prefer to use roll pads which are flexible enough to allow extension of the spinal column with the table extension.

Fusion/Instrumentation Extent

The instrumentation and fusion should be long enough to offer adequate rigidity of the internal fixation but should be minimized to that absolutely necessary to save as many motion segments as possible especially when the fusion is extended into the lumbar spine. Pedicle screws, which enable rigid fixation, are especially helpful for this purpose. We prefer to fuse all the levels instrumented as the "rod-long, fuse-short" technique of long instrumentation and short fusion not only inflicts more extensive soft tissue damage and increases the risk of damaging healthy joints but also causes degeneration of the articular cartilage of the unfused, immobilized facet joints within the instrumented extent. In thoracolumbar burst fractures, we prefer to fuse two levels proximal and one level distal to the fractured vertebra, securing four and two fixation points proximal and distal to the fracture both for a simple posterior stabilization and posterior stabilization with anterior column reconstruction. The reason for extending two levels higher into the thoracic spine is to increase the stiffness of the construct with minimum sacrifice of the valuable lumbar motion segments.

Choice of Implants

For pedicle screw instrumentation of the thoracic spine, pedicle screws of smaller diameter than the usual lumbar screws may be necessary. For average-sized adults, 6 mm screws are usually sufficient as the thoracic pedicles are large enough. For children and smaller-sized adults whose pedicles are smaller, 4.0 mm screws are advisable in the upper thoracic spine (T 4, 5, 6) where the pedicles are smallest. Concerning the shape of the screws, cylindrical screws are as effective as conical screws in patients with normal bone quality [40]. However, as the fixation strength of conical screws is affected by the mass of the cancellous bone, cylindrical screws may be better for internal fixation of the osteoporotic spine. In patients with a significant local kyphotic deformity or when posterior column shortening is contemplated, use of long-arm or reduction screws may be helpful in gradually correcting the deformity [41,42] (Figure 42.6. a–b).

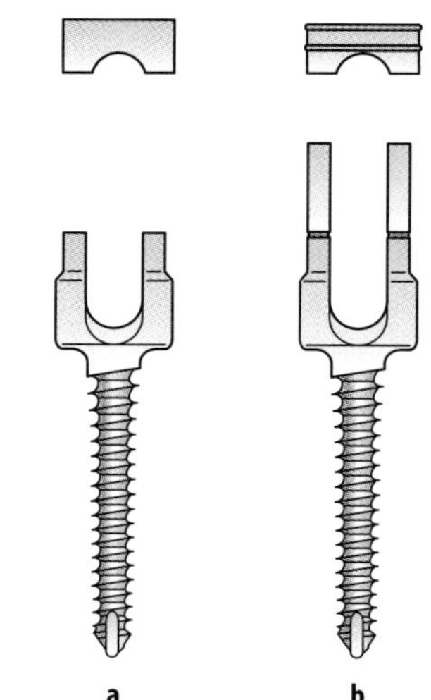

Figure 42.6. **a** Ordinary screw. **b** Long arm screw.

Between the pedicle screw-plate and pedicle screw-rod systems, we prefer the rod system as they are easier to contour to the normal sagittal contour and allow more room for insertion of the bone grafts [43]. Among the many designs of pedicle screw-rod systems, we prefer the top open design similar to the Cotrel–Doubousset, Synergy, and the Diapason system as it is easier to connect the rod to the top open screws than to a side-attaching system. However, we believe the training and the experience of the surgeons are the most important factors in the choice of implants and strongly recommend using the instruments most familiar to the operating staff.

Though there is little difference in the clinical performance as to the metallurgy of the implants, we recommend the use of implants made of titanium in case a postoperative imaging of the spinal canal is needed.

Posterior Pedicle Screw Fixation

Incision and Exposure

As the pedicle screw instrumentation needs a wide exposure to the tip of the transverse processes, the incision should be large enough to avoid excessive retraction on the paraspinal muscles to prevent myonecrosis. For a thoracic fracture, the usual incision spans from the upper end of the spinous process two levels above the uppermost pedicle instrumented in the thoracic spine to the lower end of the lamina of the lowest instrumented vertebra. It is advisable to take an intraoperative roentgenogram to confirm the spinal levels in the course of the exposure to prevent errors of operating on wrong levels.

The spine is exposed in a standard fashion with an electric knife, staying strictly subperiosteal to reduce bleeding. The vertebrae instrumented are exposed to the tip of the transverse processes bilaterally. Care should be taken not to disturb the facets adjacent to the uppermost pedicles instrumented during the exposure as damage to the facets may result in postoperative pain and late instability. On completion of the exposure, the facets within the intended fusion extent are destroyed by inferior facetectomy with thorough removal of the articular cartilage to promote intra-articular arthrodesis. After the facet preparation, pedicle fixation is begun by preparation of the pedicle entry sites. In the thoracic spine, the pedicle entry point is at the junction of the superior margin of the transverse process and the lamina (Figure 42.7. a–b). In the lumbar spine, the point is at the junction of a line drawn through the middle of the transverse process and the lateral margin of the facet joint (Figure 42.8. a–b). The presumed entry sites are decorticated with a rongeur to expose the cancellous bone overlying the pedicles to facilitate insertion of the guide pins for the pedicle screws. In the lumbar spine, in addition to the entry sites, transverse processes and the lateral aspect of the facet joints are decorticated at the same time. They are to be decorticated before the insertion of the pedicle screws as it is difficult to perform a reliable decortication in these bone bases with the screws in place. Decortication of the lumbar transverse processes should be started from the tip towards the base as decortication weakens this very weak bony process. Taking the reverse direction often results in fracture of the transverse process, reducing the bone base available for lateral intertransverse fusion. We prefer to insert pedicle screws even in the fractured verte-

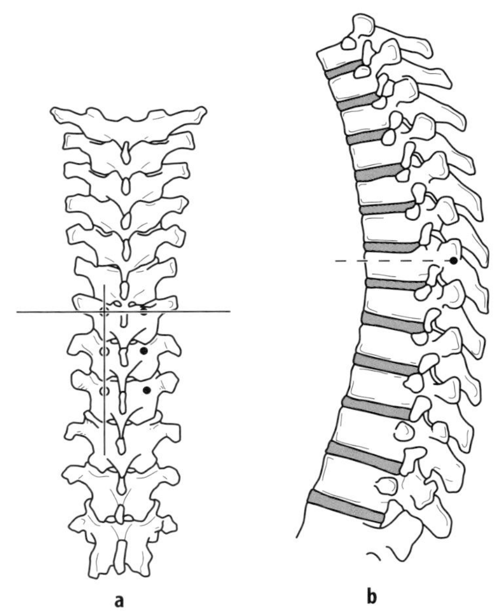

Figure 42.7. Pedicle entrance point in thoracic spine **a** Anteroposterior view. **b** Lateral view (Redrawn from ref. 28).

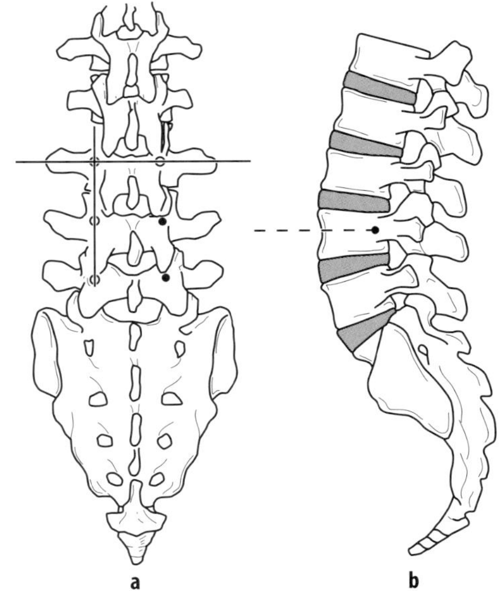

Figure 42.8. Pedicle entrance point in lumbar spine **a** Anteroposterior view. **b** Lateral view (Redrawn from ref. 28).

bra for a simple posterior fusion as such segmental instrumentation increases the stiffness of the pedicle screw construct. However, as there is always a possibility of residual neurologic compromise that might need a direct decompression of the neural elements, we put a short screw that passes just a few screw turns into the pedicle on one side of the fractured vertebra so that the pedicle screws do not become an obstacle for an anterior decompression and reconstruction. For fractures subjected to an anterior column reconstruction from the posterior, we do not put pedicle screws in the fractured vertebra as corpectomy and reconstruction becomes very difficult with the screws in the pedicles.

Pedicle Screw Insertion

Following the preparation of the entry sites the pedicle screws are inserted segmentally into the vertebrae within the fusion extent in the following manner.

Step 1: Insertion of the guide pins. Guide pins, about 15 cm long, made from K-wires are inserted shallowly into the exposed cancellous bone at the presumed pedicle entry point by means of a mallet. To facilitate interpretation of the relative position of the guide pin tips on the intraoperative roentgenograms, the guide pins are directed along the axis of the pedicle in the frontal and the sagittal planes. Following insertion of guide pins on one side, guide pins of a different diameter are inserted on the opposite side pedicles to avoid confusion in reading the intraoperative roentgenogram.

Step 2: Confirming the entry point and screw direction. With the guide pins placed in the presumed entry points, intraoperative PA and lateral roentgenograms are taken to determine the relationship between the presumed entry point and the ideal entry point identifiable on the X-ray, and to determine the direction of the screws. Considering the transverse angle of the pedicles in the thoracic and thoracolumbar spine, the ideal pedicle entry point (IPEP) in a neutrally rotated vertebra is at the junction of the line parallel to the lower or upper end plate bisecting the pedicle and the lateral margin of the pedicle ring shadow on a PA film (Figure 42.9. a–b). In rotated vertebrae, as found in scoliosis, the pedicle on the side of the rotation assumes a more medial position than neutrally rotated vertebra relative to the lateral border of the vertebra, whilst the opposite side pedicle becomes relatively more laterally situated. Naturally, the IPEP on the rotated side (convex side in scoliosis) moves more medially while the IPEP of the opposite side (concave side in scoliosis) moves more laterally with increasing amount of rotation.

On the lateral view, the IPEP is situated at the junction of the line passing through the axis of the pedicle and the posterior border of the facet joints. Screw direction is determined on the lateral X-ray. Ideally, it should be along the axis of the pedicle, sloping about 10° cranially in the thoracic and the upper lumbar spine. However, in clinical practice, insertion of the screws parallel to the superior end plate of the vertebra is preferable to prevent possible penetration of the pedicle screws into the disc space superior to the instrumented vertebra.

Step 3: Pedicle entry. After determining the ideal pedicle entry point and the direction, the entry hole is made with an awl, and the hole is deepened with a small diameter drill or a small curette, taking into considerations the normal transverse angle of the pedicles for the particular level. After passing the pedicle, the drill or curette meets some resistance as it traverses the dense sheet of bone in the neurocentral junction. Over the past several years, various methods of confirming a safe passage through the pedicles have been introduced. Probing, arthroscopic examination of the hole by inserting a narrow scope into the hole [44], measuring the electrical resistance by means of an electrode placed in the hole [45], and measuring the electrical resistance of the inserted screws [46] are among many methods. We use probing to check the entry as a minor pedicular cortical breakage that might occur during pedicle entry does not significantly affect the safety or the strength of fixation of a pedicle screw. A safe entry is confirmed when the probe hole is globally surrounded by bone and meets resistance in all directions, especially when there is a feeling of the spongy cancellous bone giving way to the pressure exerted by the probe at the far end of the hole.

Step 4: Hole preparation. Deep drilling is performed following the probe path using a drill bit

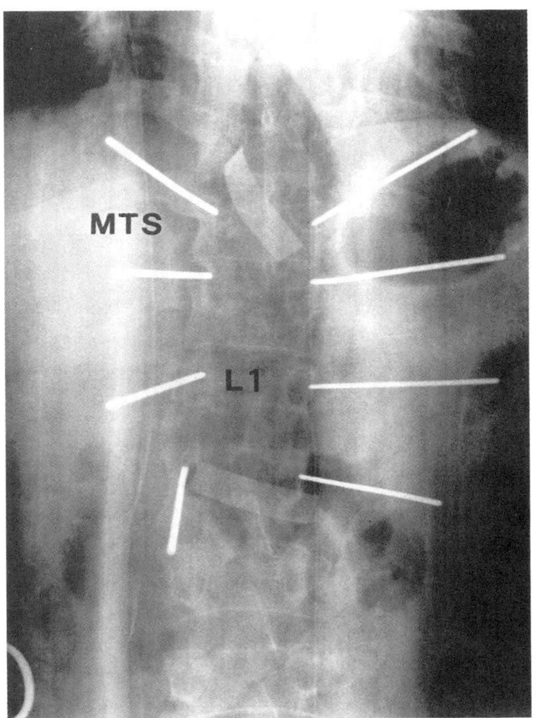

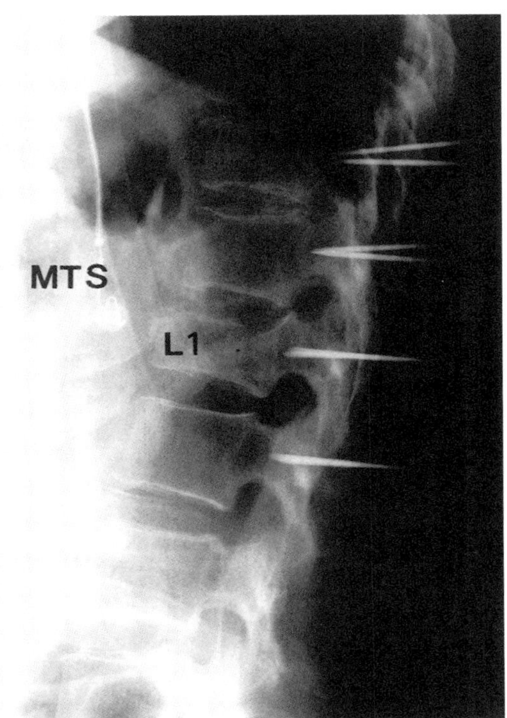

Figure 42.9. Intraoperative roentgenogram. **a** PA roentgenogram. **b** Lateral roentgenogram.

with a diameter the same as the minor diameter of the screw used. When there is a large discrepancy between the sizes of the pilot drill/curette and the final drill, the hole is enlarged by using larger diameter drills sequentially. We do not usually tap the hole, but for some implants, tapping of the hole prior to insertion of the screws may be necessary.

Step 5: Screw insertion. The screw is inserted after reconfirming the bony containment of the drilled screw hole with a blunt probe. When starting to insert the screw, it is absolutely necessary to turn the screw with a very gentle force so that the screw follows the predrilled path. Undue force at the beginning may misdirect the screws in the wrong direction especially in the osteoporotic spine where cortical bone is very weak and offers little resistance. When inserting the screws, the general alignment and depth of the screws should be taken into consideration as misaligned screws make rod attachment extremely difficult. When a screw seems strangely out of alignment compared to other screws, most probably the screw is not in the pedicle properly. When the screws are not in the pedicles, they are not only useless as reliable anchors but also very dangerous, being in proximity with the neural elements medially and inferiorly and the great vessels anteriorly. This requires re-placing the mispositioned screw exactly in the pedicles. If this is not possible, it is much safer to remove the screw than accept a potentially hazardous screw without any function. The depth of the screws are also very important as it is extremely difficult to attach a rod to the screws when there is a large discrepancy. As turning the screw backwards increases the risk of screw loosening, the screws should be inserted a little shallowly at first and then adjusted after completion of screw insertion on the particular side.

Rod Insertion

Following the segmental screw insertion, the rods are attached to the screws. The rod should be long enough to allow reliable locking with the screws at both ends of the construct but not long enough to touch the unfused adjacent facet joints or hinder the motion of the unfused segments. The rod is contoured to the normal sagittal contour of the segments instrumented. Though the thoracolumbar junction is relatively flat, we prefer to add a slightly exaggerated sagittal curve to prevent potential junctional kyphosis.

Ligamentotaxis

After placing the rods on both sides, ligamentotaxis [47] is routinely performed to reduce the fracture fragments. In patients without a significant facet joint instability, the facet joints act as a fulcrum and the compression force over the fracture site via the pedicle screws attached to the normal sagittal contoured rod generates distractive force on the anterior column, effecting some reduction of the fractured fragments by tension in the longitudinal ligaments and the outer annulus. When there is significant facet joint destruction, the facet joints are unable to act as a fulcrum to the posterior compression force and the whole fractured vertebra may collapse under a compressive force, with risk of worsening the canal encroachment. In this situation, the pedicle screws are locked in situ over the fractured segment with some compression over the relatively intact segments. Though some advocate distraction over the fracture site for reduction of the fracture fragments in this situation, we believe distraction in the presence of a significant facet instability will lengthen the spinal column, producing an unsupported gap which increases the risk of implant failure: we are strongly against this idea.

With all the screws tightened, an intraoperative lateral roentgenogram is taken to confirm the restoration of the normal sagittal contour. When the result is acceptable, the rods are connected by cross links. On completion of the instrumentation, posterior fusion is carried out after a wide decortication of the laminae and addition of the local bone gained during the operative procedure. When the amount of the local bone is not sufficient, allograft is added to expand the graft volume. The wound is then closed tightly in the usual manner after careful hemostasis. Suction drains are not used as they only increase postoperative bleeding.

Posterior Pedicle Screw Fixation with Anterior Column Reconstruction

Posterior Pedicle Screw Fixation with Anterior Decompression/Stabilization

For cases for anterior decompression and fusion through an additional anterior approach, the posterior procedure is identical to the simple posterior stabilization. On completion of the posterior procedure, the anterior procedure is carried out. It may be done in the same anesthesia or later as a staged surgery. In the usual patient, we prefer to do the anterior surgery in the same anesthesia as it reduces the morbidity of repeated anesthesia and the duration of the hospital stay.

Posterior Pedicle Screw Fixation with Transpedicular Bone Graft

When transpediular bone graft is chosen for the anterior reconstruction, a pedicle in the fractured vertebral body is used to place the autogenous cancellous bone graft into the fractured vertebral body and the damaged disk space. After placing the bone grafts, a pedicle screw is inserted into the hole used for the bone graft and posterior fixation procedure is carried out in the usual manner.

Posterior Pedicle Screw Fixation with Corpectomy/Anterior Column Reconstruction via the Posterior Approach (Posterior Vertebral Column Resection)

The procedure [42,48] is identical to the posterior pedicle screw fixation until the pedicle screw insertion stage except that pedicle screws are not inserted into the fractured vertebral body. Then the procedure is carried out as described below.

Step 1: Laminectomy. A total laminectomy of the fractured vertebra is performed using a small-bore Kerrison punch. The laminectomy should be very careful especially in patients with a lamina fracture as there might be entrapment of the neural elements between the fracture fragments. When dural laceration is present it is repaired primarily or with a dural graft. The ligamentum flavum beneath the adjacent superior and inferior laminae is also removed during the procedure to prevent compression of the neural elements which might occur with shortening of the vertebral column.

Step 2: Foraminotomy. The bony roof of the neural foramina superior and inferior to the pedicles of the fractured vertebra is removed totally with a Kerrison punch to fully expose the exiting nerve roots. In the thoracic spine, opening up the superior foramina should be accompanied by complete removal of the superior articular facet that lies deep to the lamina of the superior adjacent vertebra as leaving a loose bony fragment in the spinal canal may increase the risk of neurologic compromise with shortening of the vertebral column.

Step 3: Transverse process/rib resection. To isolate the pedicles, an osteotomy is performed at the bases of the transverse process of the fractured vertebra. With a blunt elevator in the osteotomy site and dissecting downward, the lateral aspect of the pedicle and the vertebral body is readily exposed in the lumbar spine. In the thoracic spine, the osteotomy leads to the costotransverse joint. To expose the lateral aspect of the vertebral body, the rib head is removed totally by an osteotomy at the costotransverse junction and disarticulation of the costovertebral joint. In removing the rib, care should be taken not to injure the pleura. When the pleura is inadvertently opened, it is to be closed.

Step 4: Pedicle resection. With the pedicles isolated, the pedicles are resected with a narrow osteotome making a circumferential osteotomy around the base of the pedicles taking care not to injure the exiting nerve roots above and below. The resected pedicles are taken gently out, releasing the soft tissue attachments with a Penfield freer.

Step 5: Corpectomy. Corpectomy is begun by removing the cancellous bone of the fractured vertebral body through the base of the resected pedicles with an angled curette. When enough cancellous bone is removed through the pedicles from both sides, the channel created will meet at the center forming a tunnel with the posterior wall of the vertebral body as the roof. The tunnel is enlarged by removing more cancellous bone anterior and laterally, leaving only the cortical portion of the vertebral body like an eggshell. At this point, a short rod is connected to the screws spanning the fracture site to endow temporary stability to the vertebral column. This temporary stabilization is extremely important as the spine becomes very unstable after the destruction of the cortical shell and inadvertent translation of the vertebral column may occur. Following this temporary stabilization, the remaining posterior wall is imploded into the void created in the vertebral body by means of a downward curette and removed by means of a pituitary forcep. After removing the posterior wall, the damaged endplate and the intervertebral disc is resected with an osteotome and pituitary forceps until the endplate of the adjacent vertebral is exposed. It is crucial to excise the damaged disk as the fractured fragments usually have some attachment to the annulus fibrosus and cannot be completely removed without a formal discectomy. When superior and inferior endplates are both damaged, both endplates and the disks are resected to expose a reliable bone base for fusion. In patients with significant kyphosis in whom posterior column shortening osteotomy is planned, the remaining lateral walls of the subject vertebra are resected by means of a narrow osteotome.

Step 6: Anterior column reconstruction. The anterior column reconstruction may be done by tightly packing the autogenous iliac cancellous chip grafts or by insertion of a structural graft such as an autogenous iliac strut graft or a titanium mesh cylinder. When the anterior gap is small as in posterior column shortening osteotomy, we just pack the gap tightly with cancellous chips. When the gap between the superior and inferior bone base is substantial, we prefer a titanium mesh filled with morcellized autogenous bone from the vertebral body and the iliac crest as taking a strut from the posterior iliac crest is more difficult than from the anterior. After measuring the size of the bone defect, an appropriate-sized titanium mesh is inserted into the gap. When the nerve roots are in the way, in the thoracic spine we choose to sacrifice a nerve root rather than retracting it as a thoracic root is not crucial and retraction of a nerve root may exert tension on the spinal cord. However, in the lumbar spine, as all roots are very important, they are retracted gently to provide room for the mesh. The mesh is impacted deeply into the gap, ideally to rest in the anterior and middle third of the vertebral body. It is important to place the structural support anterior to the vertebral axis of rotation to effectively resist the flexion moment. On completion of the procedure, rods contoured to the desired sagittal contour of the instrumented segments are inserted into the opposite side screws. Then the screws are locked with compression over the mesh. With the rod locked in position, the temporary stabilizing rod is changed to a contoured longer rod and is inserted onto the screws. These are also locked in position with compression over the mesh cylinder.

When anterior reconstruction is complete, an intraoperative PA and lateral roentgenogram is taken to confirm the position of the mesh cylinder. With an acceptable spinal realignment and mesh position, cancellous chips are added beside the mesh cylinder and are tightly packed in position.

Following decortication of the posterior elements, a posterior fusion is performed using a broad sheet of cortical bone from the iliac crest to bridge the posterior defect. Then the wound is closed in the usual manner over double suction drains.

Aftertreatment

For both the simple posterior stabilization and the posterior stabilization with an anterior column reconstruction, the patient is allowed to sit with bed elevation on the first postoperative day. Ambulation is started on the second or third postoperative day with a custom-made TLSO, which is to be kept for three months. The patients are allowed to perform daily activities with the brace and are sent back to work by the first postoperative month.

Complications

Pedicle screw fixation is a safe procedure with very few complications when performed properly.[32] However, potential complications such as the following may occur.

Neurologic complications; may be caused by misplaced pedicle screws. However, complications related to pedicle screws may be prevented by strictly adhering to the proper technique previously described. Of more than 10,000 pedicle screw placed in the thoracic spine, we did not have a single major complication attributable to the pedicle screw placement.

Bleeding; vascular injury due to misplaced screws are extremely rare and are preventable by strictly adhering to the proper screw insertion technique.

References

1. Boucher HH. A method of spinal fusion. J Bone Joint Surg 1959;41B:248.
2. Roy-Camille R, Saillant G, Mazel C. Internal fixation of the lumbar spine with pedicle screw plating. Clin Orthop 1986;203:7.
3. Roy-Camille R, Saillant G, Mazel C. Plating of thoracic, thoracolumbar, and lumbar injuries with pedicle screw plates. Orthop Clin North Am 1986;17(1):147.
4. Boos N, Webb JK. Pedicle screw fixation in spinal disorders: a European view. Eur Spine J 1997;6:2.
5. Suk, SI, Lee CK, Kim WJ, Chung YJ, Park YB. Segmental pedicle screw fixation in the treatment of thoracic idiopathic scoliosis. Spine 1995;20:1399.
6. Vaccaro AR, Garfin SR. Internal fixation (pedicle screw fixation) for fusions of the lumbar spine. Spine 1995; 20:157S.
7. Cheffer MM, Currier BL. Thoracolumbar burst fractures. In: Levine AM, Eismont FJ, Garfin SR, Zigler JE, editors. Spine Trauma. Philadelphia: WB Saunders, 1998.
8. Weidenbaum M, Farcy JPC. Surgical management of thoracic and lumbar burst fracture. In: Bridwell KH, DeWald RL, editors. The Textbook of Spinal Surgery, 2nd ed. Philadelphia: Lippincott-Raven, 1997.
9. Krag MH, Weaver DL, Beynnon BD, Haugh LD. Morphometry of the thoracic and lumbar spine related to transpedicular screw placement for surgical spinal fixation. Spine 1988;13:27.
10. Suk SI, Lee JH. A study of the diameter and change of the vertebral pedicle after screw insertion. Presented at 3rd international meeting SIROT. Boston, MA, 1994.
11. Zindrick MR, Wiltse LL, Doornik A, Widell EH, Knight GW, Patwardhan AG et al. Analysis of the morphometric characteristics of the thoracic and lumbar pedicles. Spine 1987;12:160.

12. Dickson RA, Deacon P. Spinal growth. J Bone Joint Surg Br 1987;69:690.
13. Misenhimer GR, Peek RD, Wiltse LL, Rothman SL, Widell EH Jr. Anatomic analysis of pedicle cortical and cancellous diameter as related to screw size. Spine 1989;14:367.
14. Suk SI, Cha SI, Lee CK. Kim WJ. A study on the pullout strength of pedicle screws in relation to the size of the drill holes and inserted screws. Presented at the 30th Annual Meeting of the Scoliosis Rearch Society, Asheville, NC, 1995.
15. Skinner R, Maybee J, Transfeldt E, Venter R, Chalmers W. Experimental pullout testing and comparison of variables in transpedicular screw fixation. A biomechanical study. Spine 1990;15:195.
16. Zindrick MR, Wiltse LL, Widell EH, Thomas JC, Holland WR, Field BT et al. A biomechanical study of intrapeduncular screw fixation in the lumbosacral spine. Clin Orthop 1986;203:99.
17. Magerl FP. Stabilization of the lower thoracic and lumbar spine with external skeletal fixation. Clin Orthop 1984;189:125.
18. Hirano T, Hasegawa K, Takahashi HE, Uchiyama S, Hara T, Washio T et al. Structural characteristics of the pedicle and its role in screw stability. Spine 1997;22: 2504.
19. Maat GJ, Matricali B, van Persijn van Meerten EL. Postnatal development and structure of the neurocentral junction. Its relevance for spinal surgery. Spine 1996; 21:661.
20. George DC, Krag MH, Johnson CC, Van Hal ME, Haugh LD, Grobler LJ. Hole preparation techniques for transpedicle screws. Effect on pull-out strength from human cadaveric vertebrae. Spine 1991;16:181.
21. Zdeblick TA, Kunz DN, Cooke ME, McCabe R. Pedicle screw pullout strength. Correlation with insertional torque. Spine 1993;18:1673.
22. Resina J, Alves AF. A technique of correction and internal fixation for scoliosis. J Bone Joint Surg Br 1977; 59:159.
23. Lee CS, Suk SI, Sung KS. Biomechanical study on multiple hooks and screws fixation in the long posterior spinal instrumentation. J Korean Spine Surg 1997;4: 212.
24. Dick JC, Zdeblick TA, Bartel BD, Kunz DN. Mechanical evaluation of cross-link designs in rigid pedicle screw systems. Spine 1997;22:370.
25. Lim TH, Eck JC, An HS, Hong JH, Ahn JY, You JW. Biomechanics of transfixation in pedicle screw instrumentation. Spine 1996;21:2224.
26. Suk SI, Kim MD, Shin JW, Lee SJ, Kim WJ. A numerical investigation of the axial rotational stability of transfixation in pedicle screw instrumentation. Presented at 8th international congress of SIROT. Sydney, Australia, 1999.
27. McCormack T, Karaikovic E, Gaines RW. The load-sharing classification of spine fractures. Spine 1994;19: 1741.
28. McLain RF, Sparling E, Benson DR. Early failure of short-segment pedicle instrumentation for thoracolumbar fractures. A preliminary report. J Bone Joint Surg Am 1993;75:162.
29. Suk SI, Shin BJ, Lee CS, Lee MC: Cotrel–Dubousset pedicular screw in the treatment of unstable dorsolumbar fracture. Comparison with Harrington SSI. J Korean Society Fracture 1989;2:91.
30. Suk SI, Kim JH, Kim WJ, Lee SM, Chung ER. Treatment of congenital scoliosis with pedicle screw fixation. Presented at the 33rd Annual Meeting of the Scoliosis Research Society. New York, 1998.
31. Suk SI, Kim WJ, Kim JH, Lee SM. Restoration of thoracic kyphosis in the hypokyphotic spine: a comparison between multiple-hook and segmental pedicle screw fixation in adolescent idiopathic scoliosis. J Spinal Disord 1999;12:489.
32. Suk SI, Kim WJ, Lee SM, Kim JH, Chung ER. Thoracic pedicle screw fixation in spinal deformities: are they really safe? Spine 2001;26:2049.
33. Suk SI, Kim JH, Lee SM, Chung ER, Lee JH. Anterior-posterior surgery versus posterior closing wedge osteotomy in posttraumatic kyphosis with neurologic compromised osteoporotic fracture. Spine 2003;28:2170.
34. Olerud S, Karlstrom G, Sjostrom L. Transpedicular fixation of thoracolumbar vertebral fractures. Clin Orthop 1988;227:44.
35. Maiman DJ, Larson SJ, Luck E, El-Ghatit A. Lateral extracavitary approach to the spine for thoracic disc herniation: report of 23 cases. Neurosurgery 1984;14: 178.
36. Heinig CF, Boyd BM. One-stage vertebrectomy or eggshell procedure. Orthop Trans 1985:9:130.
37. Levy WJ Jr, York DH. Evoked potentials from the motor tracts in humans. Neurosurgery 1983;12:422.
38. Nash CL Jr, Lorig RA, Schatzinger LA, Brown RH. Spinal cord monitoring during operative treatment of the spine. Clin Orthop 1977;126:100.
39. Kim C, Suk SI, Hong KH, Kim JH, Kim WJ, Lee CH et al. Intraoperative monitoring using somatosensory evoked potential during spinal deformity surgery. J Koren Acad of Rehab Med 1999;23:581.
40. Kwok AW, Finkelstein JA, Woodside T, Hearn TC, Hu RW. Insertional torque and pull-out strengths of conical and cylindrical pedicle screws in cadaveric bone. Spine 1996;21:2429.
41. Suk SI. Application of Diapason reduction screws. Presented in Stryker Spinal Invitational Symposium. Cheju, Korea, 1998.
42. Suk SI, Kim JH, Kim WJ, Lee SM, Chung ER, Nah KH. Posterior vertebral column resection for severe spinal deformities. Spine. 2002;27:2374.
43. Puno RM, Bechtold JE, Byrd JA 3rd, Winter RB, Ogilvie JW, Bradford DS. Biomechanical analysis of transpedicular rod systems. A preliminary report. Spine 1991; 16:973.
44. Stauber MH, Bassett GS. Pedicle screw placement with intraosseous endoscopy. Spine 1994;19:57.
45. Darden BV 2nd, Wood KE, Hatley MK, Owen JH, Kostuik J. Evaluation of pedicle screw insertion monitored by intraoperative evoked electromyography. J Spinal Disord 1996;9:8.
46. Darden BV 2nd, Owen JH, Hatley MK, Kostuik J, Tooke SM. A comparison of impedance and electromyogram measurements in detecting the presence of pedicle wall breakthrough. Spine 1998;23:256.
47. Kuner EH, Kuner A, Schlickewei W, Mullaji AB. Ligamentotaxis with an internal spinal fixator for thoracolumbar fractures. J Bone Joint Surg Br 1994;76: 107–12.
48. Suk SI, Kim WJ. Treatment of fixed lumbosacral kyphosis by all posterior vertebral column resection. Presented at the 5th IMAST meeting. Sorrento, Italy, 1998.

43 Biomechanics of Posterior Spinal Instrumentation

S.-I. Suk and W.-J. Kim

Introduction

The main purposes of spinal instrumentation are restoration of stability in an inherently unstable or surgically destabilized spine, and correction and maintenance of spinal deformities via forces effected by the instrumentation [1–6]. Spinal instrumentation may be, at large, divided into anterior instrumentation and posterior instrumentation by the element of the vertebral body utilized to fix the implant to the vertebral column. Those fixing the anterior column (usually the vertebral body proper) are considered anterior instrumentation while those fixing the structures of the posterior column (lamina, facets, pedicles) are considered posterior instrumentation.

Though there are many factors which may affect the choice of the instrumentation method for a specific spinal problem, the two most important factors seems to be the location of the pathology causing the instability or the deformity and the experience of the treating surgeon.

An anterior instrumentation is indicated when the spinal pathology or the instability is located predominantly in the anterior column. However, anterior instrumentation may be difficult to apply when it is necessary to incorporate a large number of spinal segments or when the instrumentation needs to span across the cervicothoracic, the thoracolumbar, or the lumbosacral junction due to the complex anatomy of the junctional regions.

By the same token, a posterior instrumentation is indicated when the spinal pathology or the instability is located predominantly in the posterior column. Though a pathology or instability arising from the posterior column is rare, posterior instrumentation is more commonly used than anterior instrumentation. This is due to the fact that the posterior approach is more often employed than the anterior approach in the practice of spinal surgery due to its relative ease, extensibility, and lower risk of a major visceral injury, which subsequently makes the instrumentations in the posterior column more frequent [7].

The purpose of this chapter is to introduce the biomechanical basis of posterior instrumentation techniques and to offer a guideline for an appropriate choice of technique or techniques for various spinal conditions subject to stabilization or correction by posterior instrumentation.

Implant Characteristics

Materials

The biomechanical characteristic of a spinal implant is often governed by the material properties of the implant. Modern spinal posterior implants are usually made of metal, either in pure or in alloyed forms. Metal is superior to other materials as spinal implants in the aspects of its mechanical strength, relative ease of shaping, possibility of obtaining uniform material property throughout the whole implant, and for mass production.

Metals commonly used for spinal implants are pure unalloyed titanium, Ti-6Al-4V, 316L stainless steel, 22–13–5 stainless steel, cast Co-Cr-Mo, and Vitallium, which is also a Co-Cr alloy [8]. Though pure titanium and the titanium alloys have a lower modulus of elasticity (less stiffness) than the stainless steel or other

Co-Cr alloys, they have the advantage of producing less distortion of CT or MR images [9]. As they allow improved postoperative radiological examination of the spinal canal, they are presently increasing in use as spinal implants (Figure 43.1).

Design

The usual posterior spinal implant is composed of anchoring members, longitudinal members, a kind of component–component connecting mechanism to connect the anchoring members to the longitudinal members, and the transverse members to cross-link the longitudinal members.

Anchoring Members

The anchoring member is the part of the implant system that grips the bony structure of the spinal column and transmits the force effected by the instrumentation to the spinal column. As this component comes directly in contact with the bone, there forms a bone–component interface. Anchoring members for the posterior spinal instrumentation may be divided into a penetrating type and a gripping type by the form of their bone–component interface.

Penetrating-type anchoring members are those engaging the bone by penetration into the bony structure. They make an important component of the cantilever constructs. By the alteration of the component–component connecting mechanism, they may act as a fixed moment arm, non-fixed moment arm, or an applied moment arm cantilever beam (Figure 43.2). Penetrating anchors are divided into two groups; those with pullout resistance and those without pullout resistance. However, penetrating anchors without pullout resistance, called the "posts", are seldom, if ever, used alone in posterior implant systems [8]. Penetrating anchors with pullout strength comprise screws and smooth posts that change shape to offer pullout resistance after penetration into the bone. In present practice, screws are the most commonly employed penetrating anchors in posterior instrumentation [10].

Screws are presently used in the posterior instrumentation system for fixation in the pedicle [6,11], cervical lateral mass [12], sacral ala [13], and iliac wings [14]. They have gained more popularity in recent years as they offer a rigid vertebral grip which is stable immediately after the insertion without need of a force loading and enable reliable fixation in the presence of posterior element defects which preclude the use of gripping types of anchors. But in some situations, employing a penetrating-type anchor may be difficult due to the complex anatomy of the region and risks of causing a major neural element or vascular damage [11,15].

Screws used for posterior instrumentation may be a cortical or a cancellous type. However, the fact that the parts of the vertebra which engage the screws, including the pedicles, are composed of cancellous bone, the use of cancellous-type screws is more common [16]. A screw is made up of five parts: head, core, thread, tip, and the neck that connects the screw head to the core (Figure 43.3). The screw head is the part of the screw opposite the tip and functions as the receiving port to the inserting device. Its main biomechanical function is to resist the inward translation force generated by the rotation of the screw at the terminal phase of screw tightening. When the screw is designed to tighten against a metallic implant, for example a plate, the implant will offer a substantial resistance to pull-through and the screw head needs to be just so big so as not to pass through the screw hole. On the other hand, if the screw is designed to tighten against the bone, the screw head must be substantially larger to offer an effective resistance to pull-through.

The screw core is the part of the screw from which the threads arise. Biomechanically, it provides the strength of the screw per se and resists bending and torsion moments acting on the screw. Since screws are frequently subject to bending moments in posterior instrumentation, the bending strengths of screws have significant clinical importance. The bending strength of a screw is proportional to the section modulus (Z) that is calculated as $Z = \pi D^3/32$, in which D is the core diameter of the screw. As the section modulus changes by the cube of the change in the core diameter, even a slight alteration in the

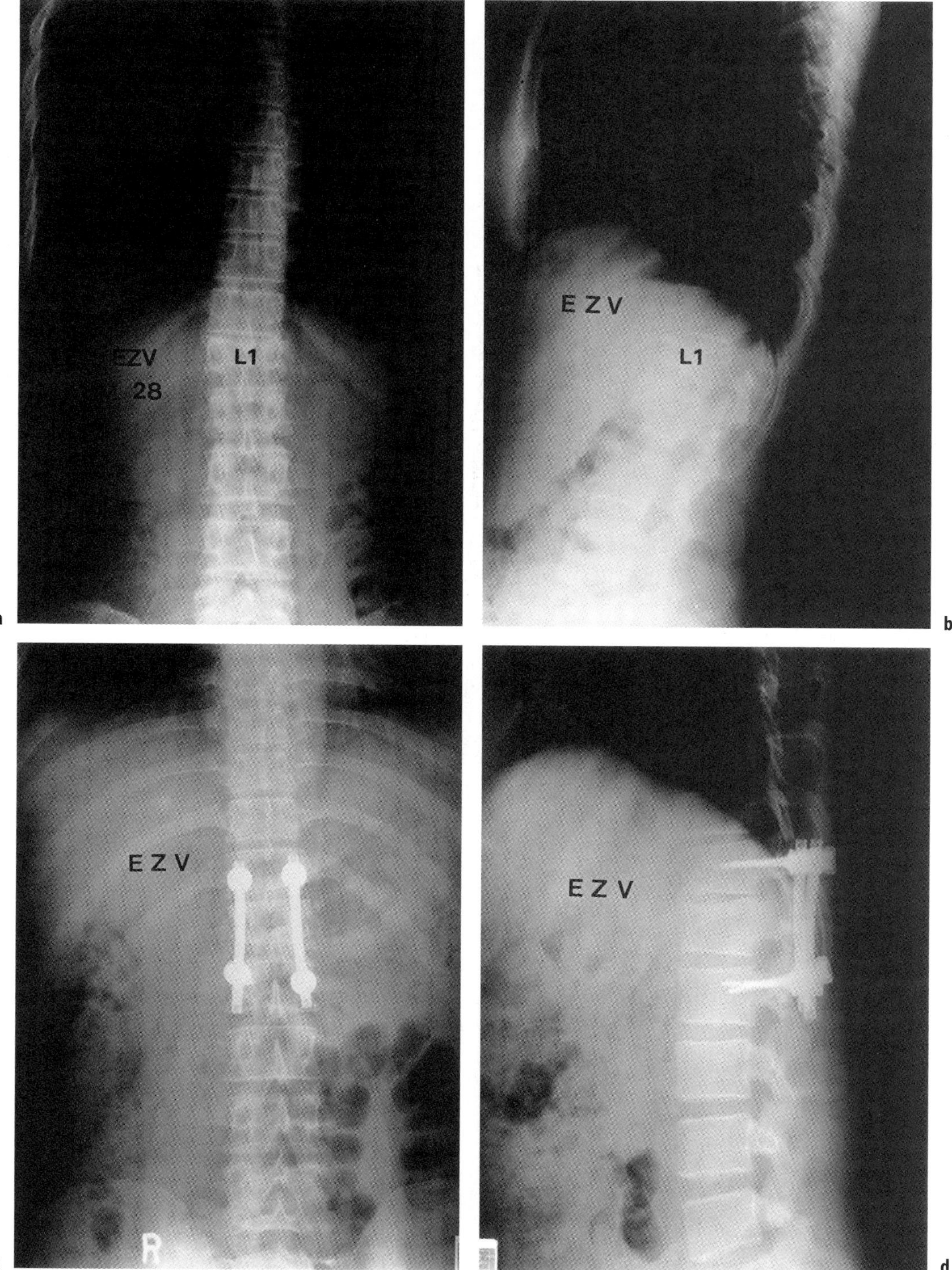

Figure 43.1. **a,b** A 28-year-old male with an unstable bursting fracture of L1. **c,d** He was treated by posterior instrumentation with pedicle screws made of titanium alloy. **e,f** Postoperative sagittal and axial MRI show little metallurgic artifact, allowing view of spinal canal. (*continued next page*)

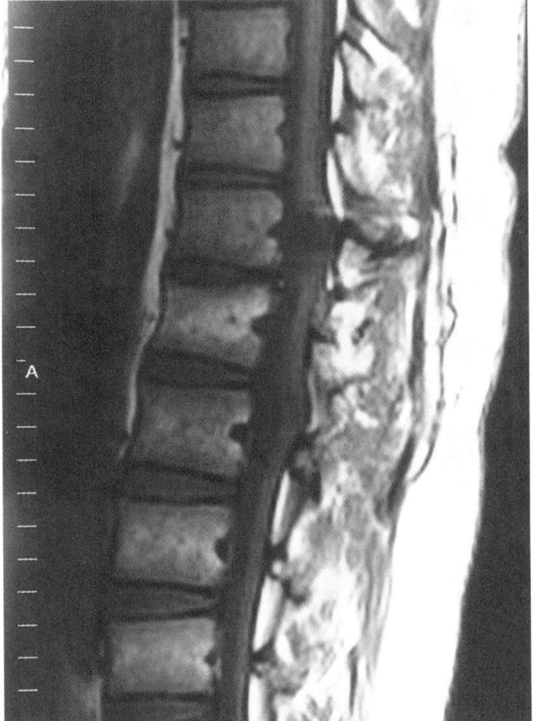

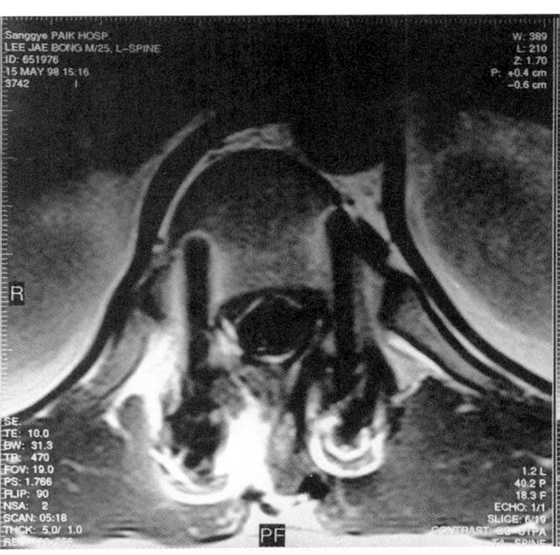

Figure 43.1 (*continued*)

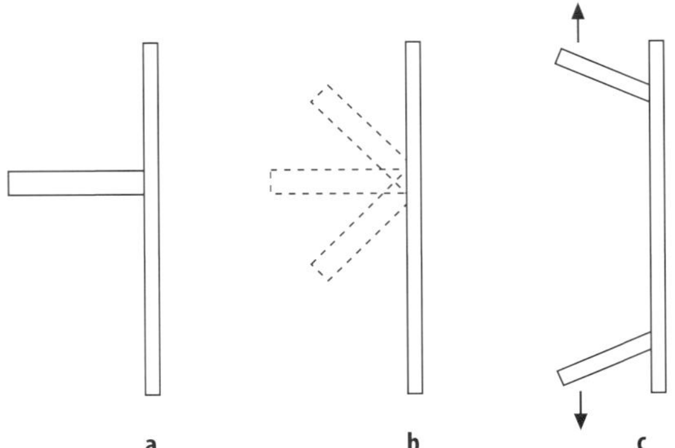

Figure 43.2. Cantilever beam constructs. **a** Fixed moment arm cantilever beam construct. **b** Non-fixed moment arm cantilever beam construct. **c** Applied moment arm cantilever beam construct.

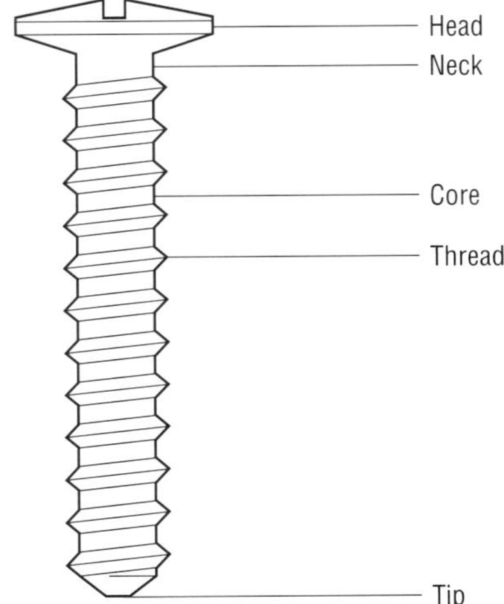

Figure 43.3. A screw and its five parts. The screw is composed of head, core, thread, tip, and neck.

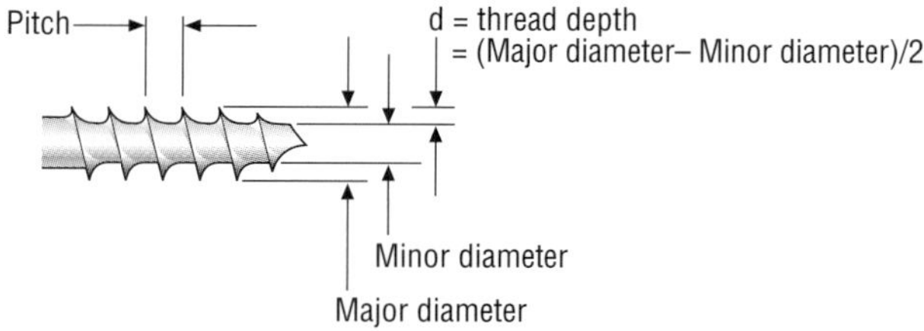

Figure 43.4. Thread depth and pitch.

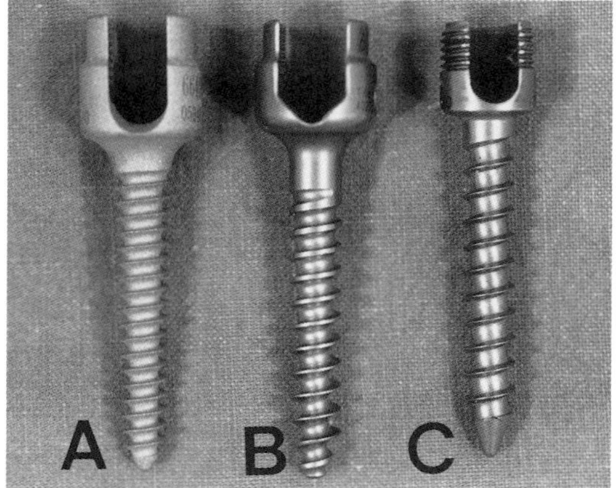

Figure 43.5. Screws of various design. (A) Conical cancellous type screw. (B) Cylindrical screw with tapered conical core. (C) Cylindrical screw.

core diameter greatly affects the bending strength of the screw. This implies the importance of using a screw of the largest permissible diameter when using screws as anchoring members [17].

The screw thread is the part of the screw that provides the pullout resistance against a force directed along the long axis of the screw. As the pullout strength of the screw is proportional to the volume of the bone between the threads, the pullout strength is affected by the major diameter of the screw, thread depth, and pitch, which is the distance between two threads (Figure 43.4). The cancellous-type screws used in posterior instrumentation causes compression of the soft cancellous bone during insertion and increases the density, and hence the amount of bone held within the threads, and are effective in enhancing the pull-out strength of the screws [18]. Screws used in posterior instrumentation have various designs to increase the pull-out strength (Figure 43.5).

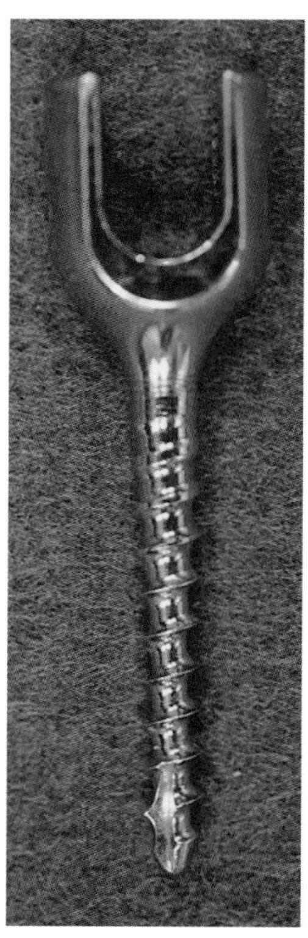

Figure 43.6. A pedicle screw with leading edge flute to act as a self-tapping screw.

The screw tip is the part of the screw that first enters the bone. Though most of the screws used in posterior instrumentation are cancellous-type screws that do not need pre-tapping, some screws have leading-edge flutes like self-tapping screws to facilitate insertion (Figure 43.6).

The neck of the screw is the part connecting the head of the screw to the core. As screws are frequently subject to cantilever bending moments, the neck portion receives most of the bending moment acting on the screw and is the most frequent site of fracture [19]. Some screws are designed to have a reinforced neck to offer more effective resistance to the cantilever bending moments concentrated here.

Gripping-type anchors are the components which "grip" the vertebra without penetrating into the bone. Hooks and wires are gripping-type anchors most commonly employed in posterior instrumentation. The common sites for application of the gripping-type anchor are lamina, pedicle, spinous process, and the transverse process. The pull-out strength of the gripping-type anchors may be substantial as they contact the hard cortical shell of the vertebra. Biomechanically, the pull-out strength of the gripping-type anchor depends on the surface area under the component and the structural integrity of the bony element to which the anchor is attached. Since the bone has to resist the pull-out force by its inherent mechanical strength, even a minor fracture that weakens the part of the posterior element receiving the anchor substantially decreases the pull-out strength. This fact also limits the use of gripping-type anchors in the osteoporotic spine where cortical bone fails to provide enough resistance to the cut through of the components.

Recently, gripping-type anchors have been decreasing in use and are being replaced by penetrating anchors. The main reason for this substitution is the inferiority of holding power when compared to the screws. Additional reasons are prerequisite of an intact posterior element for a reliable fixation, necessity of preloading that inhibits the unconstrained motion of the spinal column under force, and the necessity of intruding the spinal canal that may increase the risk of neurologic injury.

In today's modern posterior spinal instrumentation, implants are designed in such a way that several types of anchoring members may be used in the same instrumentation procedure, allowing the surgeon to choose the anchoring component according to the situation. Gripping-type anchors may be used with penetrating-type anchors in the same instrumentation to share the pull-out strength and hence protect the penetrating-type anchors from excessive pull-out stress.

Longitudinal Members

The longitudinal members are the part of the implant to which the anchoring members are attached. The biomechanical function of the longitudinal member is to resist the principle force applied to the instrument. The longitudinal member of the distraction instrumentation has to resist the bending moment created by the weight of the body above the instrument while the longitudinal member of the compression instrumentation has to resist the tension stress.

Longitudinal members in a posterior implant may be a plate or a rod. Plates are very strong and offer a rigid fixation when combined with a constrained bolt. However, they are gradually becoming less popular as contouring of a plate to conform to the curvature of the spine is difficult and is fraught with technical problems. Additional reasons for this trend are lack of versatility that limits the available anchoring component to screws and the relative bulk of the implant that causes greater extent of soft tissue damage during implantation and takes up room for placement of bone graft.

The rod is presently the most common form of longitudinal member employed in the posterior instrumentation. They are easy to contour and allow attachment of both the gripping and the penetrating-type anchoring members. Similar to screws, their strength is proportional to the section modulus (Z). As $Z = \pi D^3/32$, the strength of the rod is greatly influenced by the diameter. However, as increasing the diameter of the rod also increases the bulk of the implant causing many untoward problems, the manufacturers are trying more and more to produce thinner rods with increased strength which allow the implant to have a low profile [20]. Some are trying to use rods of higher elasticity to prevent mechanical failures,

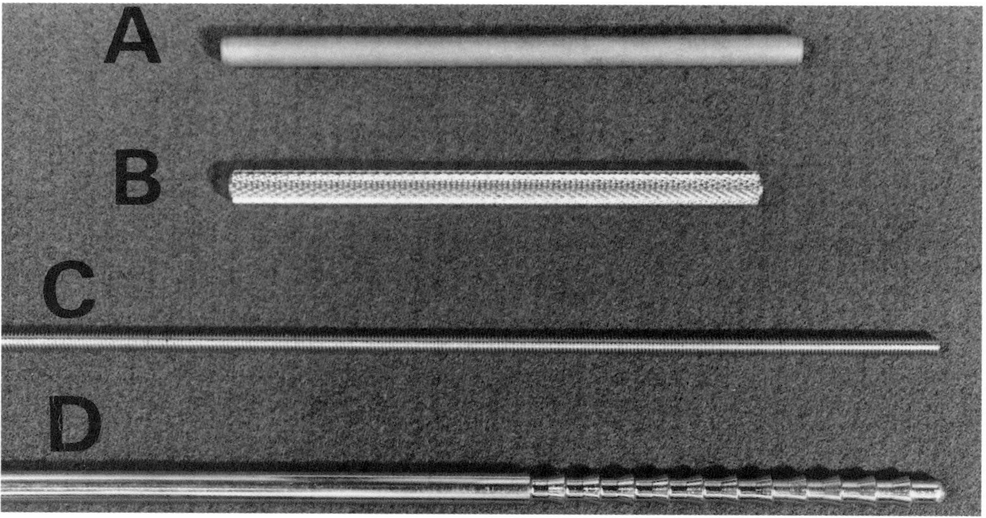

Figure 43.7. Various rods. (A) A smooth rod. (B) A knurled rod to increase surface friction. (C) A threaded rod. (D) A rachetted rod.

but the ultimate result is still to be clarified. Rods may be sliding types with or without a surface finish, a threaded or a rachetted type (Figure 43.7).

The sliding type of rod is the most common type used presently. It allows free sliding of the anchoring members along the length of the rod enabling distraction and compression along the rod. It is suitable for both distraction and compression constructs. While some of these rods are smooth, some have surface alterations to increase the friction between the rod and the component attaching mechanism.

Threaded rods allow powerful, controlled distraction or compression of the anchoring members along the rod. However, as bending of the rod with mechanical benders often results in damage to the threads causing difficulty in tightening of the nuts, they are usually more malleable than the sliding-type rods. This limits their use in posterior instrumentation to compression (tension band) constructs.

Rachetted rods like the one used in the Harrington distraction device are primarily used in distraction constructs. As the rachets act as stress risers, often leading to mechanical failure of the rod, this type of rod is rapidly falling out of favor.

Component–Component Connecting Mechanism

The longitudinal member of the posterior implant is connected to the anchoring members by a component–component connecting mechanism. Except for a wire which is tightened around the longitudinal member to offer a grip, all the connection between the components of posterior implants are one or combinations of the following six fundamental locking mechanisms; (1) three-point shear clamp, (2) lock screw, (3) circumferential grip, (4) constrained screw-plate, (5) semiconstrained screw-plate, (6) semiconstrained anchoring component-rod (Figure 43.8).

The three-point shear clamp is the mechanism of locking employed in the screws in the synergy (Cross Medical, USA) system and the new type DDT of the Cotrel–Dubousset system (Sofamor–Danek, USA). The locking is provided by the force applied at the interface and the friction between the components.

The lock screw mechanism uses the set screw to push the rod to abut the other part of the component and is presently the most common type of locking mechanism used in posterior instrumentation. Examples are the screws

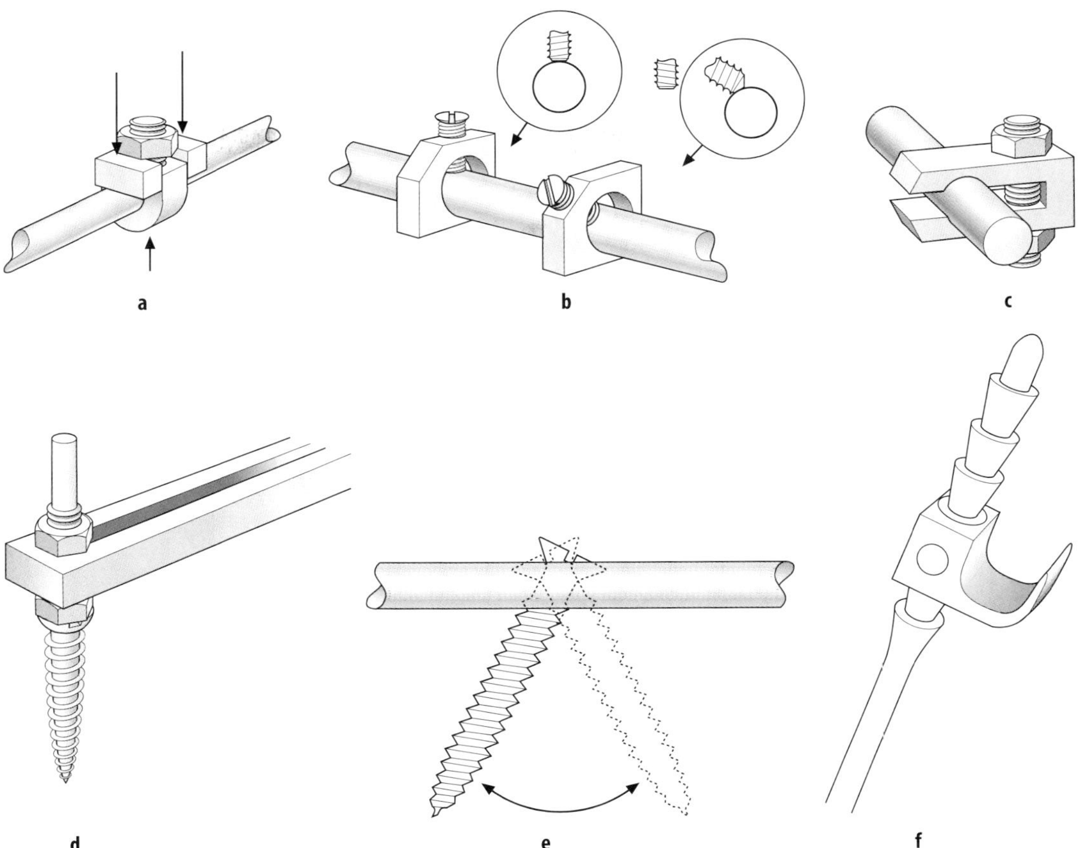

Figure 43.8. Six fundamental locking mechanisms. **a** Three-point shear clamp. **b** Lock screw. **c** Circumferential grip. **d** Constrained screw plate. **e** Semiconstrained screw plate. **f** Semiconstrained anchoring component rod. (Taken from Benzel EC. Component–component interfaces. In: Biomechanics of Spine Stabilization. Principles and Clinical Practice. New York: McGraw-Hill, 1995.)

and hooks of the Cotrel–Dubousset system (Sofamor–Danek, USA) and the Diapason system (Stryker, USA).

Circumferential grip offers connection between the components by friction effected by two halves of the pincer. An example is the old-type DTT in the Cotrel–Dubousset system and the closed clamps in the Colorado system (Sofamor–Danek, USA).

The constrained bolt plate is the type of locking mechanism used in the VSP (Acromed, USA) and connection of the closed clamp with the anchoring components in the Colorado system (Sofamor–Danek, USA). It is very rigid and offers the strongest component–component connection available. However, as it needs a perfect contact between the undersurface of the plate with the upper surface of the screw for optimal function, constrained bolt plate connections directly between the anchoring member and the longitudinal member pose many problems in practice and are also falling out of favor. Newly developed implants employing this component–component interface usually use this mechanism to connect the anchoring member to a clamp that is again connected to the longitudinal member to facilitate the instrumentation procedure.

The semiconstrained screw plate is the type of connection employed in most of the screw plate systems. As the screws are not rigidly fixed to the plates, this connection allows toggling of the screws on the plate and is unable to achieve a true rigid fixation. Although bicortical fixation of the screws increases the rigidity of this kind of connection, obtaining a bicortical fixation from the posterior side of the spine is often difficult and dangerous. In posterior instrumentation, Roy-Camille plates and the cervical lateral mass plates use this component–component interface.

The semiconstrained component-rod is the type of connection that allows toggling of the anchoring member on the rod. A typical example is the Harrington distraction device. As they do not achieve a true rigid fixation and are prone to mechanical failure at the component–rod interface, they are also gradually fading away.

Transverse Members

The transverse member, commonly known as the cross link, is the component that transversely connects two or more longitudinal members of posterior implants to convert the construct into a quadrilateral frame. Its biomechanical function is to enhance the torsion resistance and to resist parallelogram deformation of the construct [21,22]. Transverse members do not increase mechanical resistance to other stresses (e.g., flexion–extension, lateral bending) [21,22].

The optimal number of cross linking is at two sites, as adding more transverse connections does not significantly increase the torsion resistance.

Classification of Posterior Instrumentation

The posterior instrumentation may be classified by the nature of the force imparted by the instrumentation on the spinal column. Though it is sometimes very difficult to define the principle acting force due to the complex three-dimensional nature of the spinal anatomy, the principle forces effected by the posterior instrumentation are distraction, compression, three-point bending, and translation. By the degree of freedom allowed by the instrumentation, they are further divided into rigid and dynamic types. The rigid type is the construct that does not allow motion between the instrumented spinal segments whereas the dynamic types allow some motion in the instrumented segments either by motion at the component–component interface or component–bone interface.

Posterior Distraction Instrumentation

This type of instrumentation exerts a distraction force on the spinal column and is biomechanically characterized by bearing of the axial load created by the weight of the body cranial to the proximal anchoring member by the implant when the patient is in an upright position. It comprises instrumentation applied under active distraction in the operating room and those fixed in so-called "neutral fixation" without any active compression or distraction. Neutral fixation has to be considered a distraction type of instrumentation since the implant has to maintain the length of the instrumented segment per se and has to bear the axial load when the patient assumes an upright position even though there has been no active distraction in its application.

Although many constructs may be used for posterior distraction instrumentation, simple distraction, fixed moment arm cantilever beam fixation, and applied moment arm cantilever beam fixation are the most common forms used for this purpose.

Simple Distraction Construct

This construct applies distraction in a short spinal segment and is typified by the Knodt rod. It is usually a hook-rod system. The characteristic of this type of instrumentation is that the fixation becomes stable only with active distraction by the implant. As distraction is applied posterior to the spinal instantaneous axis of rotation (IAR), the posterior column is lengthened, and a kyphosis is created. This instrumentation system is not employed frequently in present-day posterior instrumentation due to its biomechanical and biological disadvantages of the inability to offer a reliable resistance to the flexion moment and derangement of spinal sagittal contour. However, it may be used in special situations which need correction of local lordotic deformity. When employing this instrumentation, the structural integrity of the anterior column is an absolute prerequisite as elongation of the posterior column in the face of an anterior column incompetence would result in serious exaggeration of kyphosis and ultimate failure of instrumentation.

When simple distraction is applied to a segment longer than four or five spinal segments, the resulting kyphosis from elongation of the posterior column creates abutment of the longitudinal members on the apex of the kyphosis and exerts a three-point bending force.

Fixed Moment Arm Cantilever Beam Fixation Construct

A fixed moment arm cantilever beam construct exerts a distraction force on the spinal column and bears the axial load when applied in an active distraction or in a neutral fixation. However, as active distraction of the spinal column is more effective with the applied moment arm cantilever beam fixation, it is usually used in neutral fixation mode.

As the bending moment is resisted by the fixed moment arm which projects into the vertebral body in front of the IAR, it offers greater resistance to flexion moment even with a shorter-length construct and is much less prone to failure than the simple distraction construct. Constrained screw-plate and most of the pedicle screw-rod systems are typical examples of this type of instrumentation.

As the axial load is borne on the cantilever beam, mechanical failure occurs at the junction of the beam and the longitudinal members [23]. To reduce mechanical failure, adequate reconstruction of the anterior load-bearing ability to share the axial load is advisable.

Contrary to the simple distraction instrumentation that creates a local kyphosis and causes anterior rotatory displacement when applied in a short segment, distraction by the fixed moment arm cantilever results in simple elongation of the segments under distraction along the longitudinal member. This is because of the short effective distance between the point of force application and the component-component interface which acts as the fulcrum and the inherent buttressing effect of the instrumentation reducing resistance to the bending moment.

Applied Moment Arm Cantilever Beam Fixation Construct

The applied moment arm cantilever beam fixation is a variation of the fixed moment arm cantilever beam system and differs only in the sense that an effective torque may be generated posterior to the IAR to offer a flexion or an extension moment to the subject spinal segments. Effective flexion or extension moment is applied by increasing the distance between the point of application of the desired force and the fulcrum located on the longitudinal member and by adopting a component-component locking system which allows connection of the anchoring member and the longitudinal member at various positions. When the components are rigidly locked in the desired position, the construct assumes the biomechanical characteristics of a fixed moment arm cantilever beam construct. The AO internal fixator (AO, Swiss) and the Socon system (Aesculap, Germany) are typical examples of this type of instrumentation.

Posterior Compression Instrumentation

This type of instrumentation exerts a compression force on the posterior spinal column and acts as a tension band resisting the flexion moment generated by application of an axial load. As its primary function is compression, the implant per se is biomechanically unstable in compression and is ineffective in bearing the axial load. So, for the posterior compression system to function adequately, a structure that effectively resists the axial compression force is absolutely necessary [24].

In posterior spinal surgery, the posterior compression system is applied when the anterior weight-bearing column is intact or reconstructed to provide effective load bearing. As application of the compression results in close abutment of weight-bearing structures, the axial load is borne mainly by these structures and little is conveyed to the implant. This allows posterior compression implants to be of smaller bulk than the distraction devices. Posterior compression instrumentation may be applied by a variety of constructs. The most common forms are simple compression fixation, fixed moment arm cantilever beam fixation, and non-fixed moment arm cantilever beam fixation.

Simple Posterior Compression Construct

Typical examples of this type of construct are the Harrington compression system and the Halifax system. As they apply compression force posterior to the IAR, lordosis is created within the instrumented segment. When the structural integrity of the anterior weight-bearing structure is compromised, active application of posterior compression may result in exaggerated lordosis and excessive shortening of the vertebral column until competent anterior structures abut to offer effective resistance to the compression.

As simple compression instrumentation offers little resistance to translational deformation, its present use is practically limited to the cervical spine where the orientation of the facets effectively blocks the anterior translation deformation.

Fixed Moment Arm Cantilever Beam Construct

A fixed moment arm cantilever beam fixation may be used as a posterior compression system by an active application of compression over the instrumented segments. As the compression brings the weight-bearing anterior structures to abut tightly there will be little axial load conveyed to the implants and the implants will function biomechanically just as a tension band.

Using the fixed moment arm cantilever beam fixation for posterior compression instrumentation offers an advantage over the simple posterior compression construct in that it offers a substantial resistance to the translational deformation and has the ability to resist over-compressive axial load, reducing the risk of mechanical failure.

Non-fixed Moment Arm Cantilever Beam Construct

Though this system does not offer a rigid fixation, it may be used to impart stability to the spinal column as a tension band when axial load-bearing structures are competent. Biomechanically, it offers no advantage over the fixed moment arm cantilever beam construct and is fading away. The only advantage of the system is that the bulk of the implant is significantly less than the fixed moment arm type as it does not a need special component–component interface. Its clinical use is primarily in the cervical spine as lateral mass plates.

Posterior Three-point Bending Instrumentation

The three-point bending force is one of the most commonly employed forces in posterior spinal instrumentation. It is the principle force in many situations where translational deformity is reduced. Application of a three-point bending force needs two points onto which forces in same direction are applied, with a fulcrum between the two points. When the construct is stable, the force acting at the fulcrum will be in the opposite direction to the two terminal forces with a magnitude equal to the sum of two terminal forces (Figure 43.9). Three-

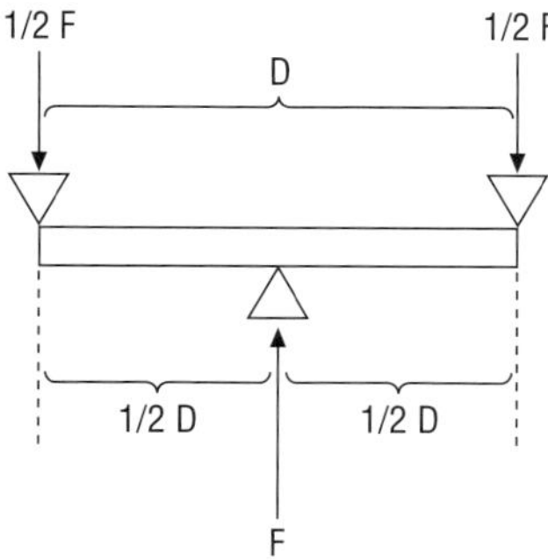

Figure 43.9. If a three-point bending construct is symmetrically placed, the sum of the two terminal forces is of equal magnitude to the force acting on the fulcrum but in opposite directions.

point bending may be applied by various posterior spinal constructs.

Simple Posterior Distraction Construct

The simple posterior distraction construct may be used to effect a three-point bending force on the spinal column. This is attributed to their ability to generate torque posterior to the IAR and create a kyphosis, which will gradually come into contact with the longitudinal member to act as a fulcrum. However, for the simple distraction construct to effect a three-point bending force, a considerable length of instrumentation is necessary and when used in the lumbar spine, due to the length of the spinal segment subject to distraction, a flat back deformity results. Addition of reduction sleeves in the mid-portion of the longitudinal member may hasten the contact of kyphosis to the longitudinal member and also effectively restores lordosis. The use of a simple distraction construct for the attainment of three-point bending force is decreasing due to the necessity of a lengthy instrumentation sacrificing the motion segments.

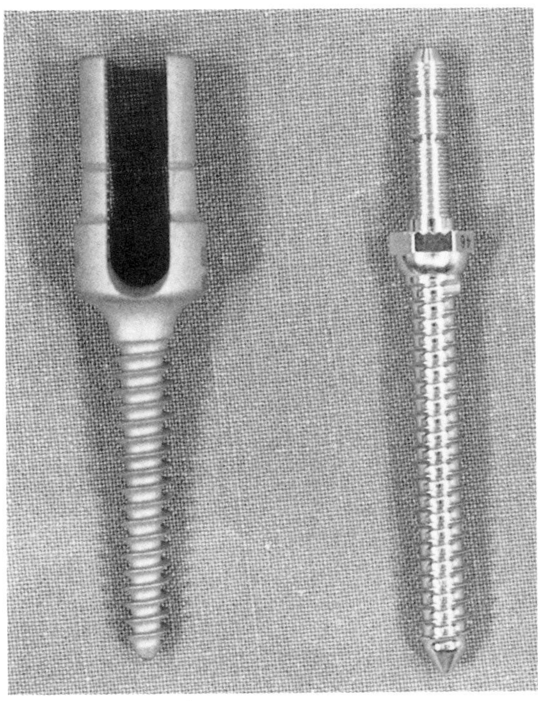

Figure 43.10. Screws with modification to facilitate application of three-point bending force. Extended arms make force application more effective.

Cantilever Beam Constructs

All forms of cantilever beam construct may be applied to generate a three-point bending force on the spinal column. This is attributed to their ability to create a posterior-directed pull-up force on the displaced vertebra by means of screws. They have the advantages over the simple distraction constructs when used in this mode of application as they do not need a lengthy instrumentation and do not cause flattening of the lordosis [25]. Some instrument systems offer specially modified implants to facilitate application of threepoint bending force (Figure 43.10).

Posterior Translation Instrumentation

This type of instrumentation effects translation of the vertebra as the primary force of action and is used mainly for correction of spinal deformities. There are two modes of applying translation to the vertebral column by these instruments. The first is the translation of the vertebra by bringing the anchoring members to the longitudinal members and is called the vertebra-to-rod method [20]. The second is by changing the shape or alignment of the longitudinal member to which the anchoring members have been already attached and may be called the longitudinal member maneuvering method. The rod derotation maneuver and in situ rod bending [26,27] are principle longitudinal member maneuvering methods. Recent investigations suggest possible future employment of shape memory alloys for the latter purpose [28]. There are two construct forms in this category.

Dynamic Translation Construct

This is typified by the Luque segmental sublaminar wiring and is applied in a vertebra-to-rod method. As there is no rigid fixation between the anchoring member and the longitudinal member, some movement will occur between the instrumented spinal segments. Though this is a disadvantage that leads to loss of correction by settling of the spinal column under gravity, it may be exploited in the maintenance of deformity correction in young children as the sliding of the wires on the rods will allow growth of the spinal column.

Rigid Translation Construct

Most of the implants used for spinal deformity fall into this category. As they offer a rigid component–component interface, they, in theory, do not allow intersegmental motion and result in better maintenance of the deformity correction.

The effectiveness of the rigid translation construct is greatly influenced by the position and the fixation strength of the anchoring members and the relative stiffness between the longitudinal member and the spinal deformity. To increase the correction of the deformity, employment of segmental anchors, preferably screws and a stiff rod [6] is advantageous. When the deformity is rigid, increasing the flexibility of the deformity by an adequate releasing procedure is helpful in attaining a better correction. They may be applied in the vertebra-to rod method or the longitudinal member maneuvering method.

Clinical Applications

In practice, the forces applied by the posterior spinal instrumentation is much more complex than those described in the preceding sections due to the three-dimensional nature of the anatomy and coupling actions which inevitably accompany application of forces on a three-dimensional structure. The presence of a multiple number of movable joints acting as hinges and the difference in the mechanical characters between the structures comprising the vertebral column further complicates clear delineation of the forces acting on the instrumented spinal column. This mandates careful preoperative consideration and meticulous planning for every case under consideration for spinal instrumentation.

Fractures and Dislocations

In stabilization of traumatic instabilities of the spine, the degree of structural compromise in the injured vertebral column and the experience of the treating surgeon are the main factors that determine the choice of instrumentation method. When a decision is made to apply a posterior instrumentation for stabilization, the following should be taken into consideration.

Degree of Anterior Column Destruction

When there is severe anterior column destruction, the spinal column is devoid of axial load-bearing ability. Though a posterior rigid distraction instrumentation may be performed, it is not sound biomechanically as the axial load is borne solely by the instrument and often results in mechanical failure of the implant or the bone–implant interface [23]. To resist the bending moment at the fracture site effectively, the instrumentation has to be of considerable length, sacrificing multiple non-injured motion segments. In this situation, adequate reconstruction of the anterior column and application of posterior compression instrumentation with a fixed moment arm cantilever beam construct is preferable (Figure 43.11).

When anterior column destruction is mild or moderate, healing of the fracture effectively restores the axial load-bearing ability of the spinal column. The role of instrumentation here is to prevent further deformation of the spinal column and protect the injured vertebra until healing occurs. In this situation, rigid distraction instrumentation by the fixed moment arm cantilever beam construct or applied moment arm cantilever beam construct is indicated (Figure 43.12).

When the bony structures of the anterior column are relatively intact as in seat-belt-type injury the disks may be brought together to reestablish a competent weight-bearing column using posterior compression

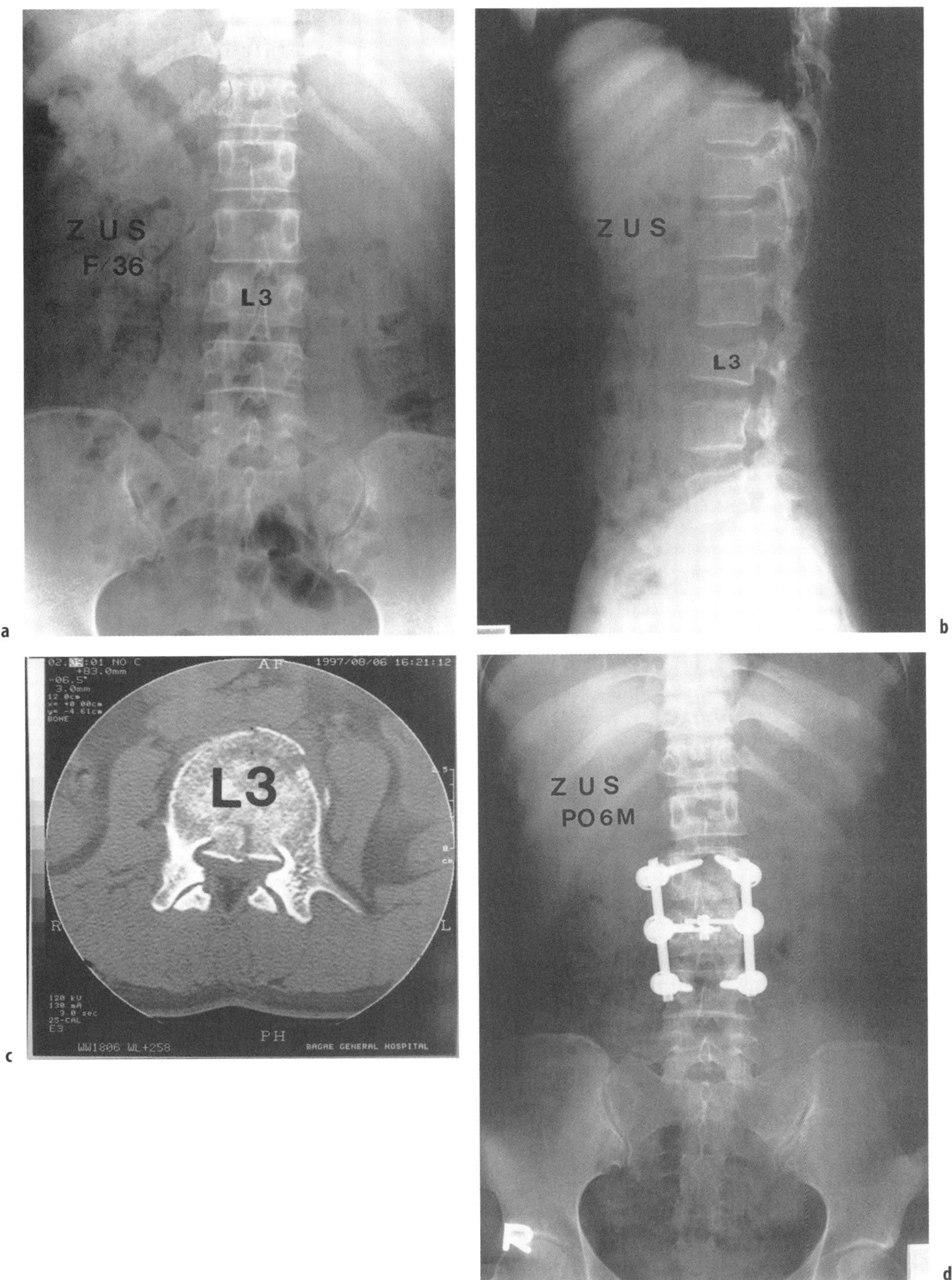

Figure 43.11. **a,b** A 36-year-old female with L3 unstable burst fracture. **c** Preoperative CT shows 50% canal encroachment with severe destruction of vertebral body. **d,e** Treatment consisted of posterior fixed moment arm cantilever fixation. The anterior column was reconstructed with strut bone graft. (*continued next page*)

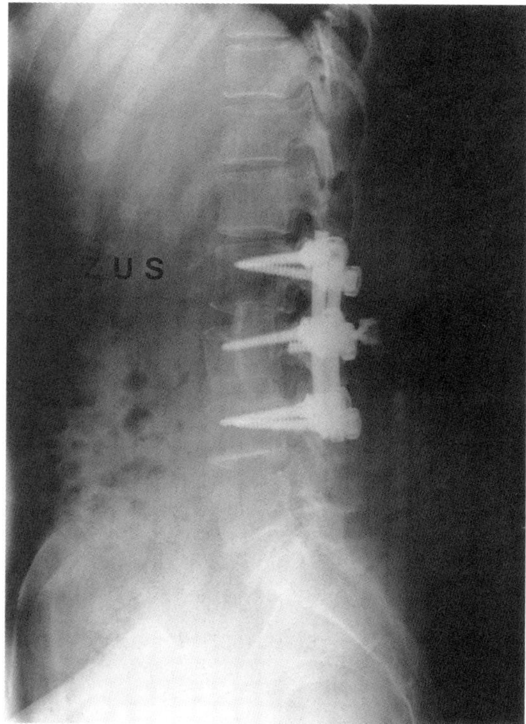

Figure 43.11 (*continued*)

instrumentation [29,30]. In the cervical spine, spinous process wiring or the Halifax clamp may be used. In the thoracic and lumbar spine, posterior compression instrumentation with a type of cantilever beam construct is preferable (Figure 43.13).

Translational Deformity

In the presence of a translational deformity, posterior instrumentation is better than anterior instrumentation as reduction of the facet dislocation is much easier from the posterior side. When translation is significant, three-point bending instrumentation or rigid translation instrumentation is indicated. This is preferably done with a fixed moment arm cantilever or applied moment arm cantilever beam construct. When there is severe destruction of the anterior column, reconstruction of the anterior column after the reduction of translation is indicated (Figure 43.14).

Degenerative Diseases

As most degenerative instabilities have competent anterior weight-bearing columns, posterior compression instrumentation will do the job. For this purpose a fixed moment arm cantilever beam construct is most suitable. When discectomy is performed simultaneously, posterior compression instrumentation with restoration of the anterior column by anterior or posterior interbody fusion is preferable (Figure 43.15).

Deformities

The ideal instrumentation depends on the type, location, and severity of the deformity.

Scoliosis

Historically, correction of scoliosis by posterior instrumentation started in the early 1960s with

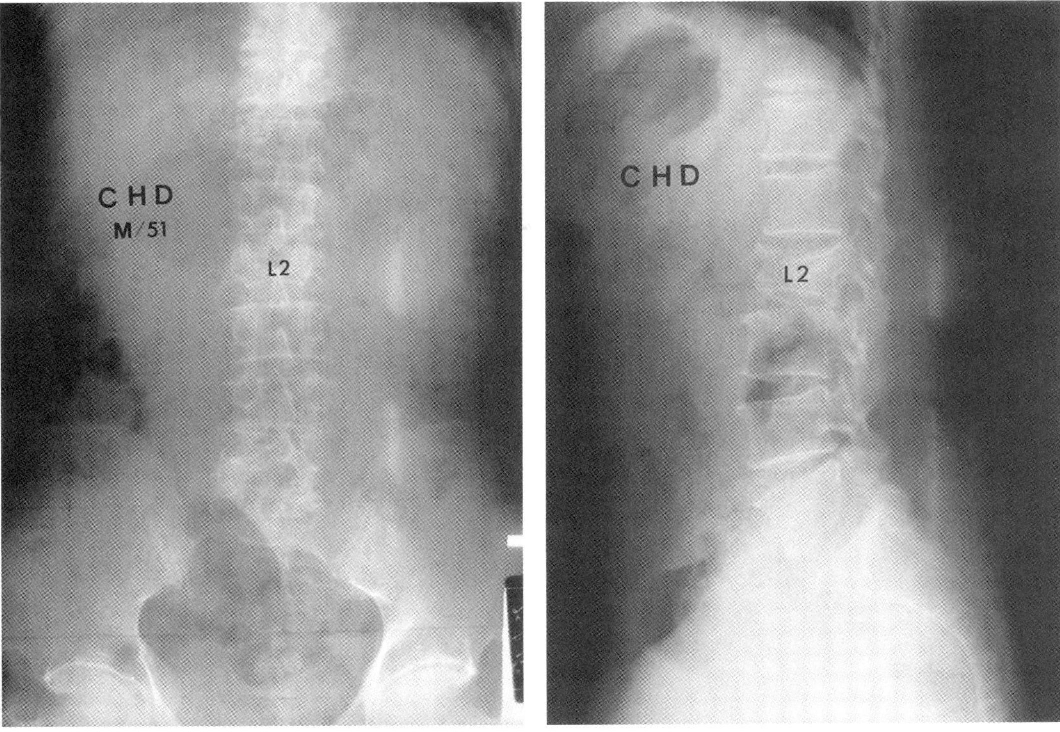

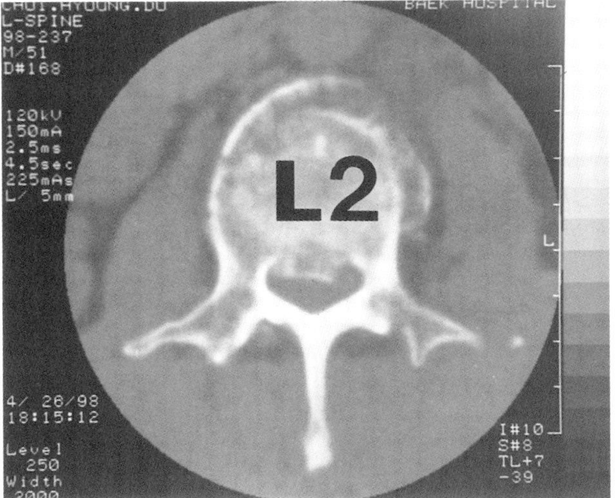

Figure 43.12. **a,b** A 51-year-old male with L2 unstable burst fracture.**c** Preoperative CT shows 40% canal encroachment. **d,e** He was treated by posterior fixed moment arm cantilever fixation. (*continued next page*)

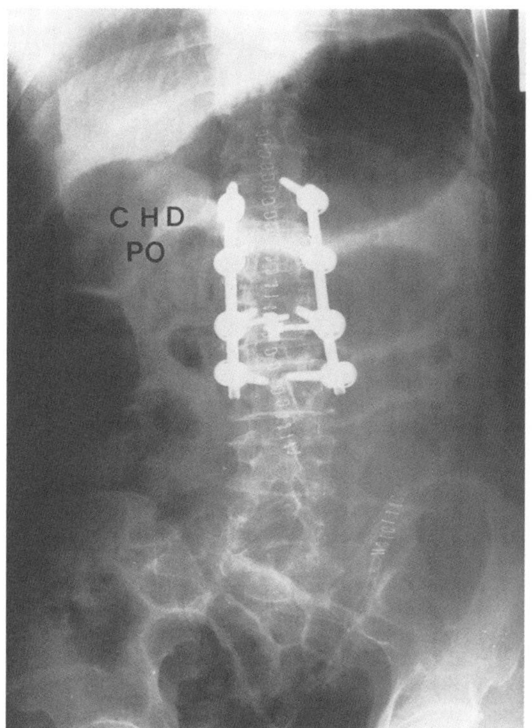

d

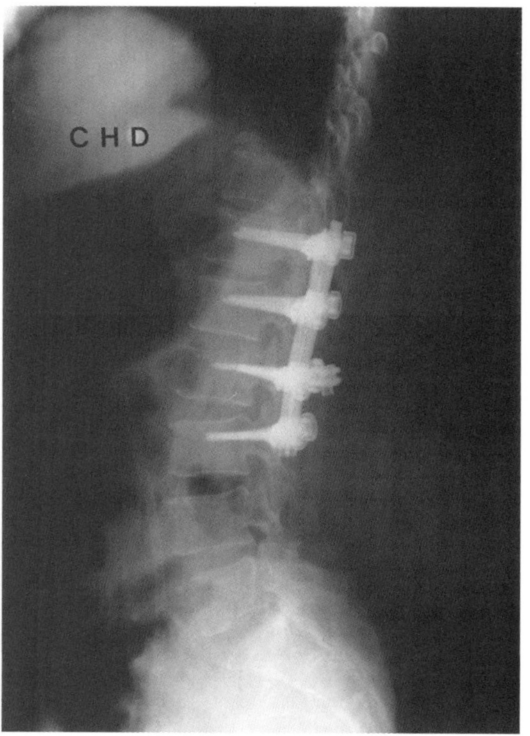

e

Figure 43.12 *(continued)*

the Harrington device, a type of rigid posterior distraction instrumentation [4]. Although a fair amount of correction could be obtained in the coronal plane from elongation of the concave side and associated medial translation of the apex, numerous drawbacks pertinent to the biomechanical characteristics of simple distraction instrumentation were observed including lack of rigidity of fixation necessitating prolonged postoperative external immobilization, high frequency of proximal hook dislodging, mechanical failure at the rachets, and derangement of spinal sagittal contour resulting in loss of normal thoracic kyphosis or flat back syndrome.

Then came the Luque segmental spinal instrumentation in the early 1980s, a type of dynamic translation instrument [5]. This corrected scoliosis employing the vertebra-to-rod method, pulling the vertebra up to the contoured rod by means of wires with a gripping-type anchor holding the lamina. This system solved many of the problems posed by the simple distraction instrumentation. The fixation of the vertebra was more rigid, precluding the necessity of prolonged external immobilization, the derangement of the sagittal contour was much less due to absence of the distraction effect of the instrumentation, and mechanical failures were much less frequent. However, there were also drawbacks pertinent to the biomechanical characteristics of the instrumentation and the anchoring member. As the fixation was dynamic, the vertebral column settling under gravity could not be effectively resisted. The necessity of intruding the canal to pass the sublaminar wires was also a significant problem, increasing the risk of neurologic injury.

Then, in 1983, shortly after popularization of the Luque instrumentation, came the Cotrel–Dubousset system, a rigid translation instrument with multiple anchors that became the prototype for numerous other similar instrumentation systems [3]. It corrected

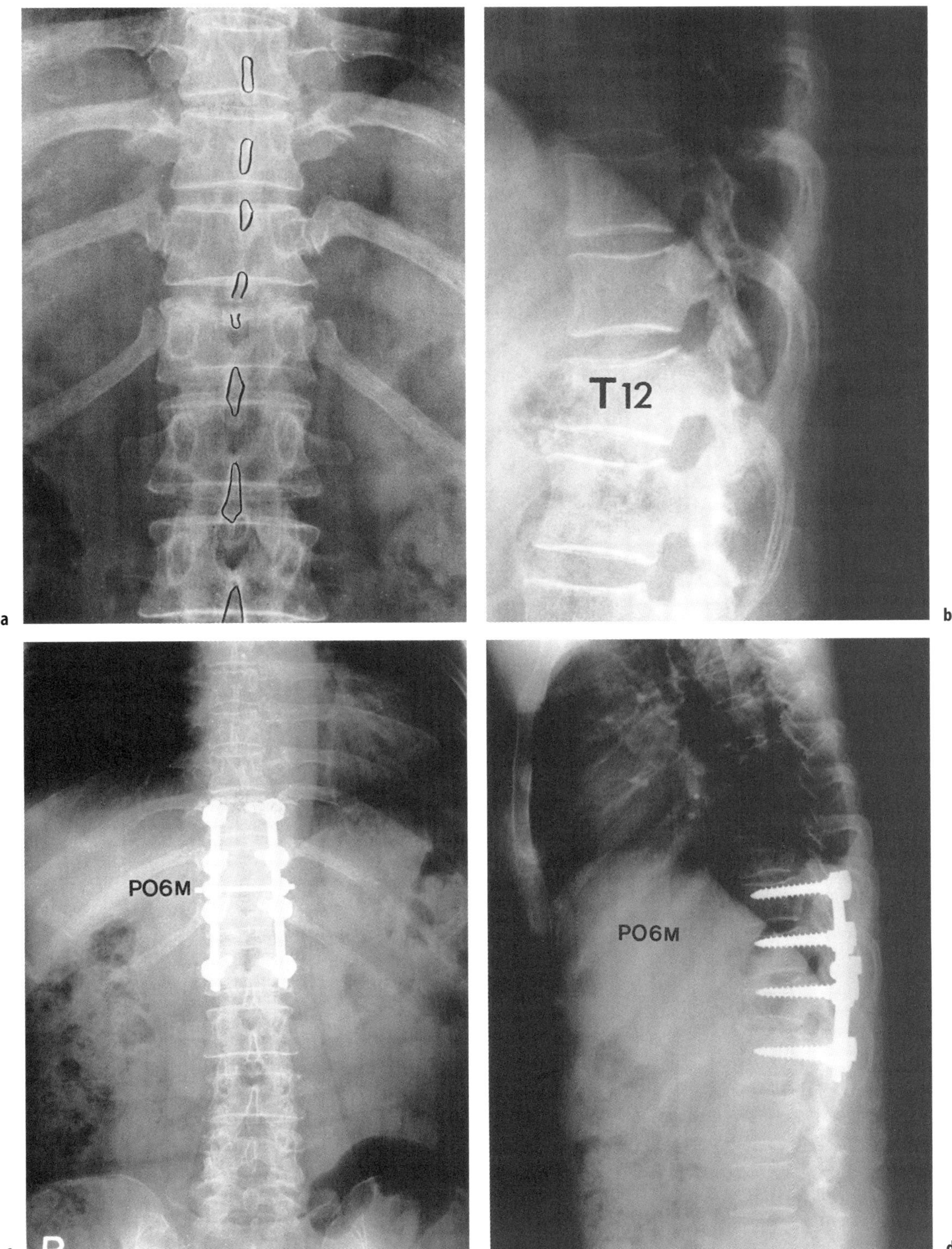

Figure 43.13. a,b A 40-year-old female with T11 chance fracture. **c,d** She was treated by posterior compression instrumentation with pedicle screws.

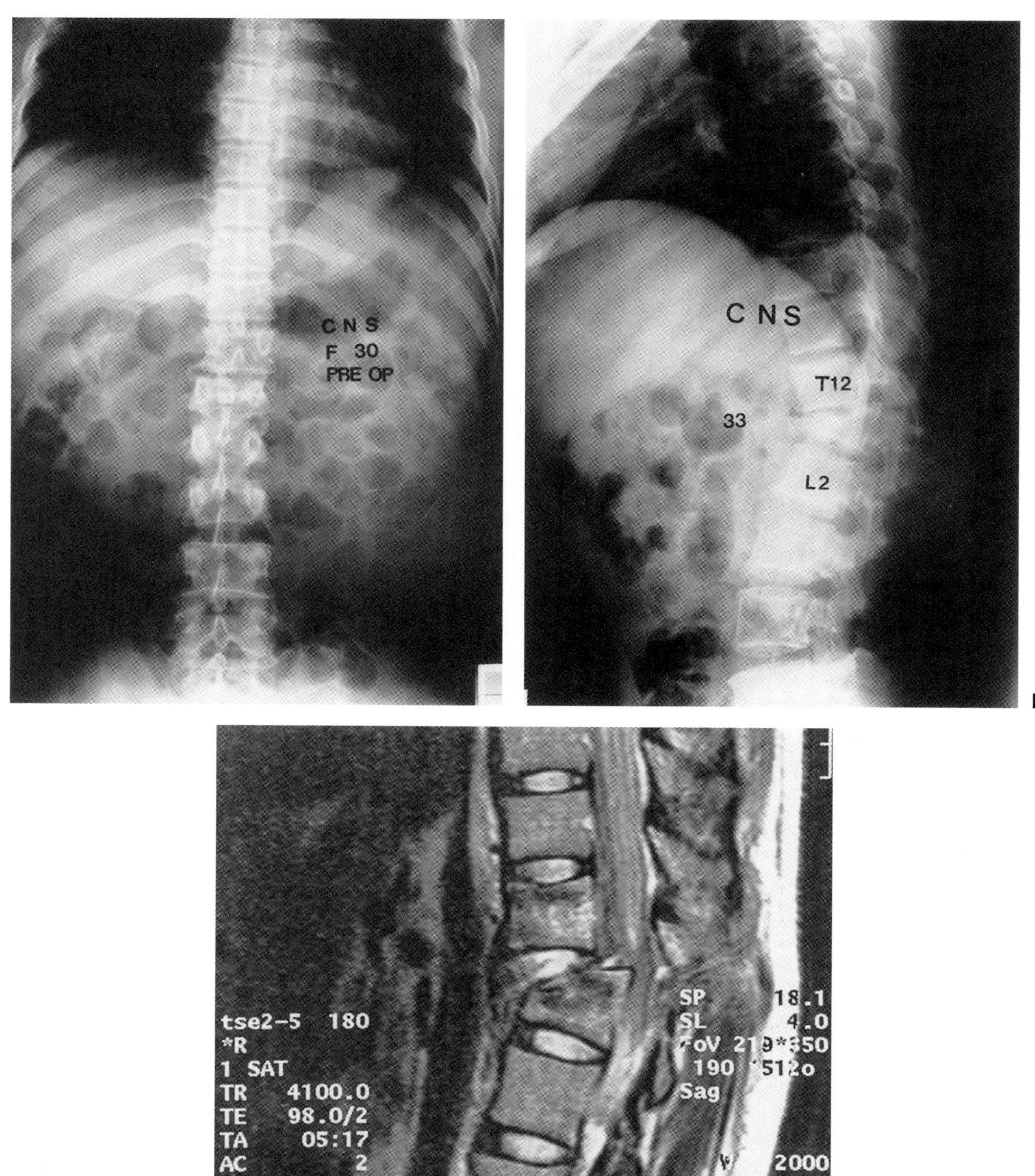

Figure 43.14. **a,b** A 30-year-old female with T12 L1 fracture dislocation. **c** MRI shows gross displacement. **d,e** She was treated by posterior instrumentation in neutral mode and then anterior column reconstruction. (*continued next page*)

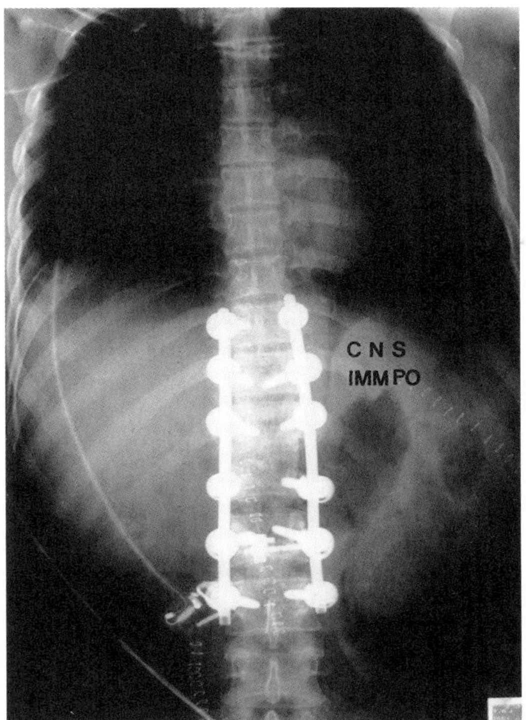

d

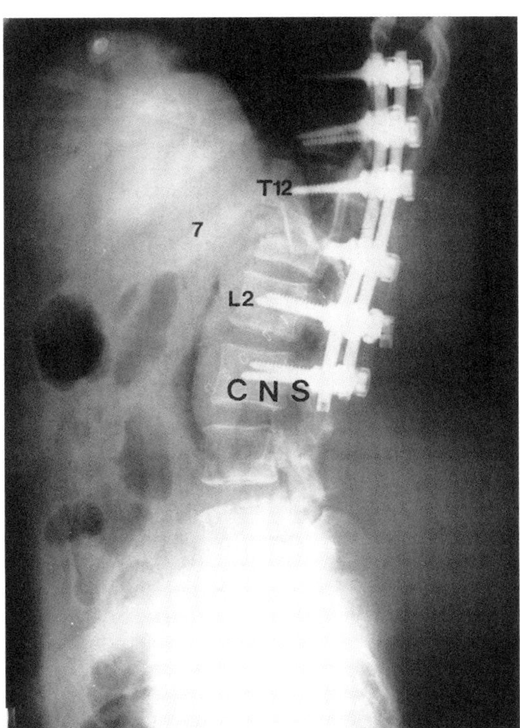

e

Figure 43.14 *(continued)*

scoliosis deformity by translocation of the curve from the coronal plane to the sagittal plane by a rod derotation maneuver, a form of longitudinal member maneuvering method. After the derotation, the system became a combined simple distraction–simple compression instrument by the nature of the anchoring members, generating forces behind the IAR. By their biomechanical characteristics, application of compression produced lordosis. However, application of distraction did not restore as much thoracic kyphosis as desired due to generation of a three-point bending force acting on the apex of developing kyphosis. Inadequate restoration of kyphosis in turn produced very little rotational correction in the horizontal plane. Despite these drawbacks, the system offered a rigid fixation, improved coronal plane correction, and true three-dimensional correction of the scoliosis deformity (Figure 43.16).

In 1992, a major breakthrough in rigid translation instrument was developed by Suk, introducing the use of segmental pedicle screws as anchoring members for the system [6]. The use of penetrating anchors converted the system into a fixed moment arm cantilever beam construct after the derotation maneuver, solving many problems of simple compression–distraction instrumentation.

The increased holding power of the implant using segmental penetrating anchors enhanced the correction of coronal plane deformity and at same time reduced failure at the bone–implant interface. Being converted into a fixed moment arm cantilever beam construct, it did not need any distraction to stabilize the anchoring members. It prevented generation of a ventral three-point bending force, at the same time permitting unconstrained spinal motion, allowing the spinal column to conform more easily to the rod contoured to the normal sagittal contour. In thoracic scoliosis, use of top open screws with a lock screw component–component interface also played a role in

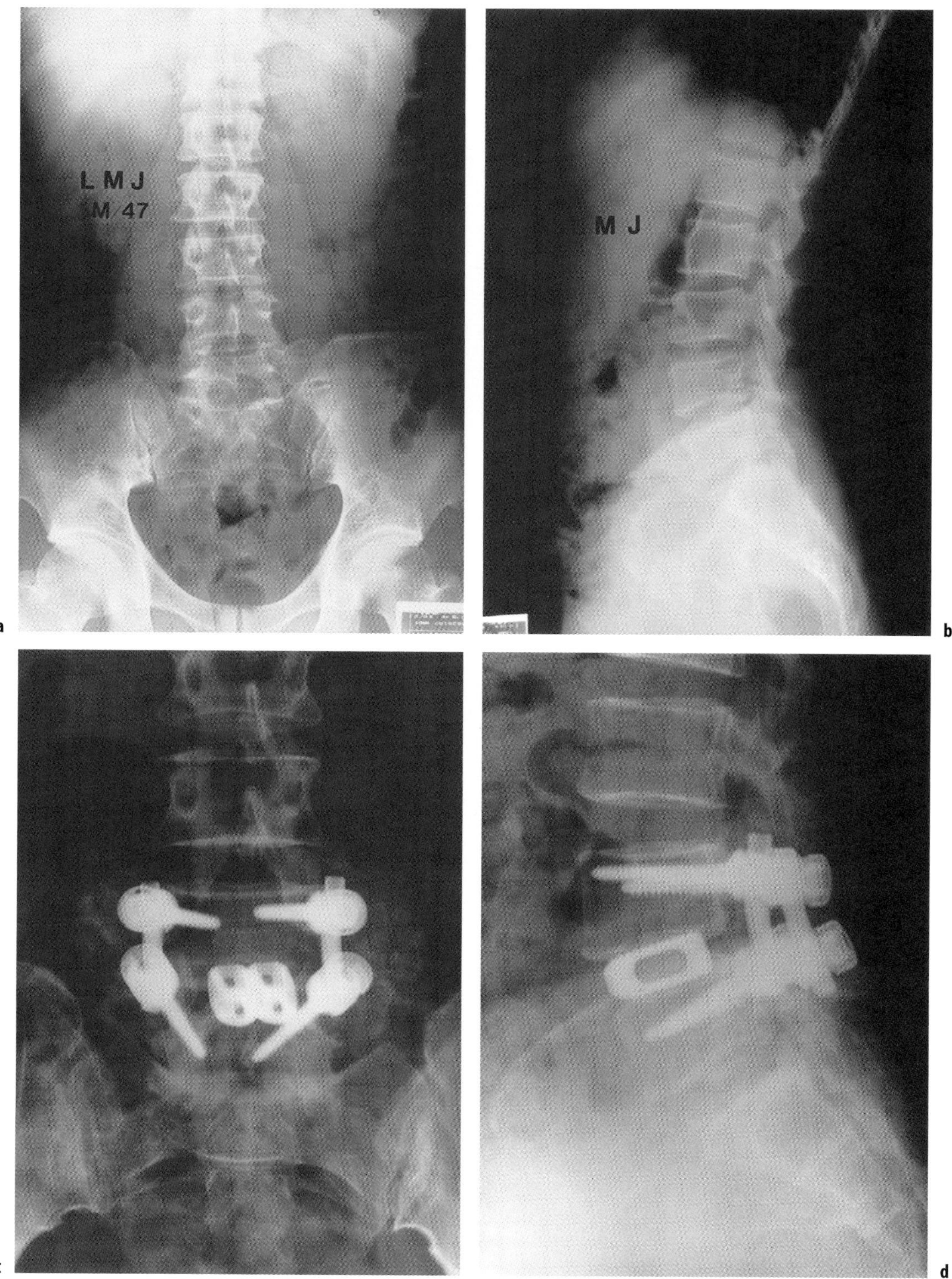

Figure 43.15. a,b A 47-year-old male with L4–5 spinal stenosis. **c,d** He was treated by posterior decompression and posterior compression instrumentation with anterior support.

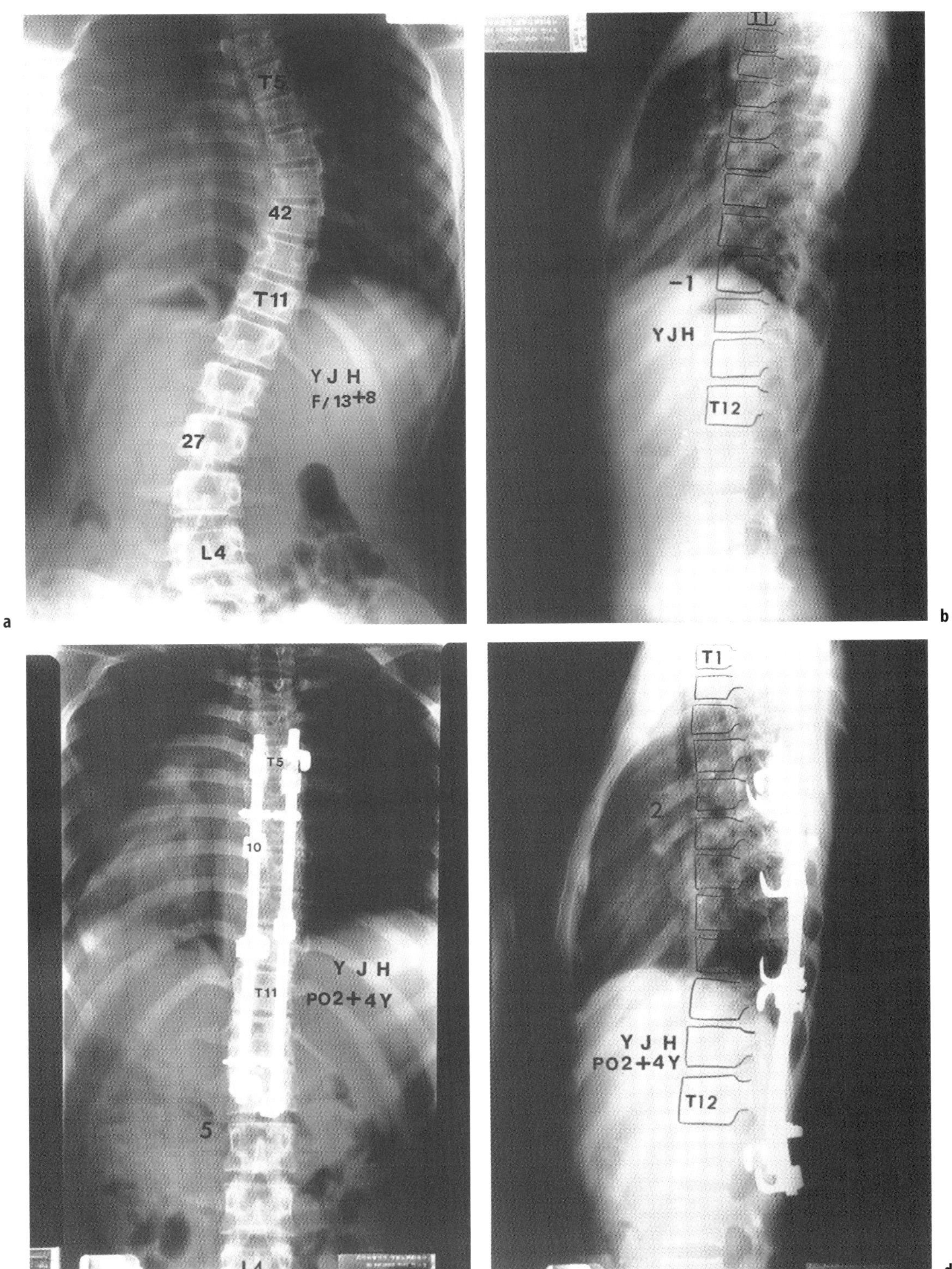

Figure 43.16. **a,b** A 13.7-year-old female with adolescent idiopathic scoliosis. **c,d** She was treated by rigid translation instrumentation using multiple hooks.

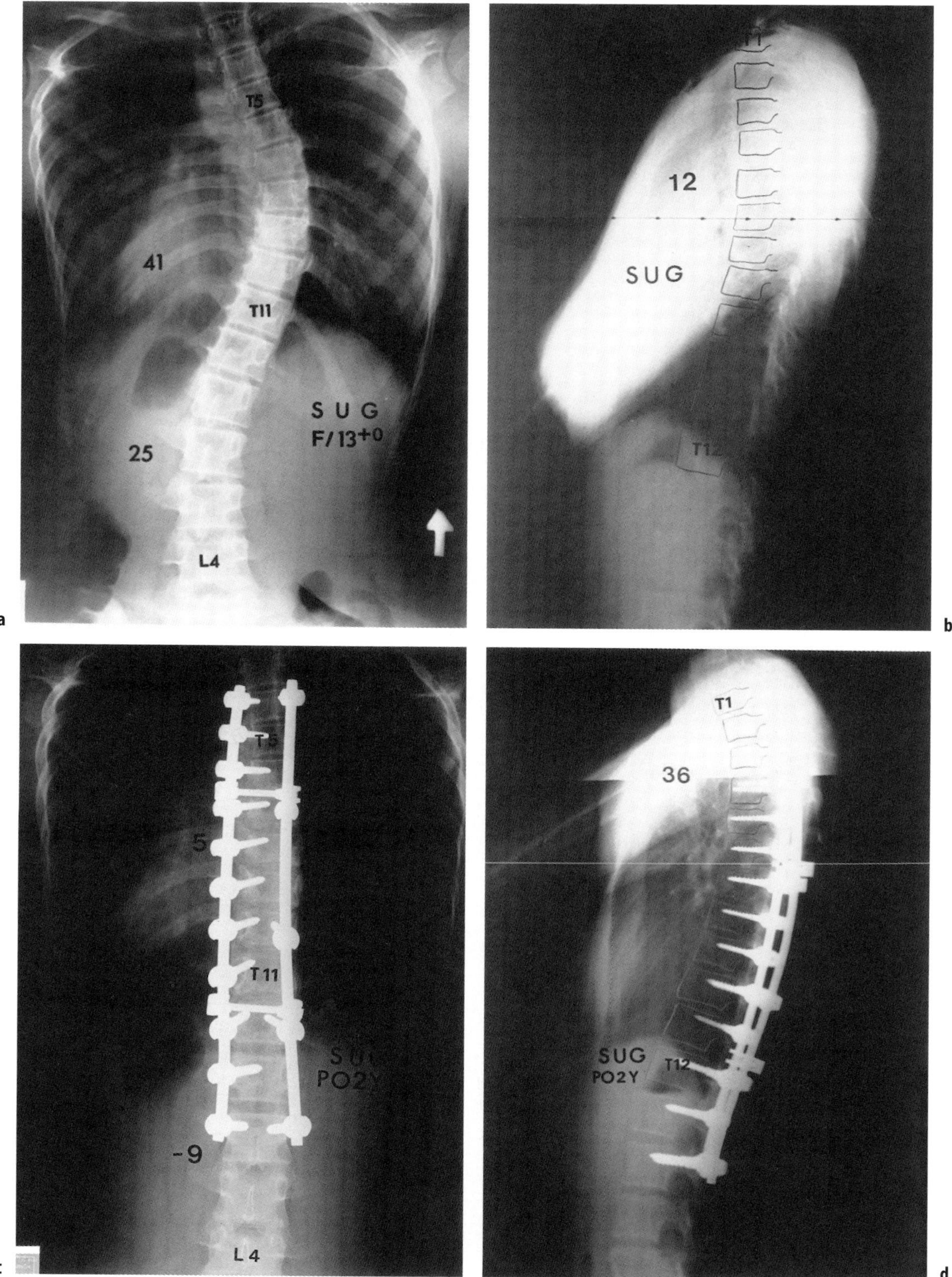

Figure 43.17. a,b A 13-year-old female with adolescent idiopathic scoliosis. **c,d** She was treated by rigid translation instrumentation using segmental pedicle screws. Note the improvement of the sagittal contour.

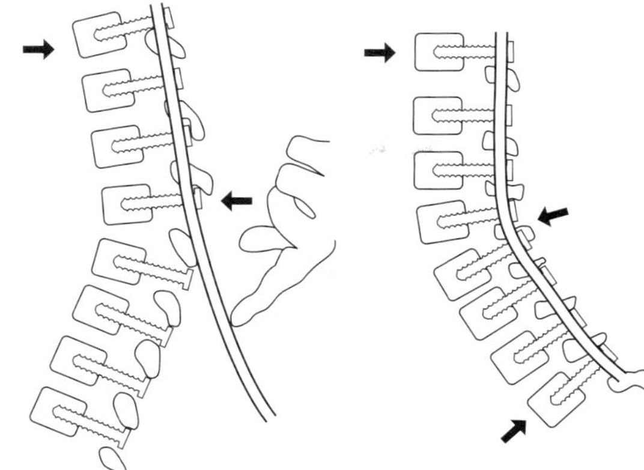

Figure 43.18. Translation instrumentation in kyphosis. Translation instrumentation generates a three-point bending force on the apex of the kyphosis by exerting a posterior displacement force on the vertebra subject to translation.

restoration of thoracic kyphosis, combining the vertebra-to-rod translation to the rod derotation. This restoration of kyphosis in turn improved rotational correction in the horizontal plane, thus truly improving the deformity three dimensionally (Figure 43.17).

In posterior instrumentation for scoliosis, the risk of mechanical failure of the instrument is greatly affected by the flexibility of the deformity and the quality of bone to which the anchoring members are attached regardless of the types of instrumentation used. In rigid deformities, increasing the flexibility of the deformity by an adequate release or an osteotomy prior to posterior correction is advisable to reduce mechanical failures [31,32]. Increasing the number of the anchoring members by segmental application disperses the stress put on each anchoring member and reduces the risk of failure at the bone–component interface especially in the osteoporotic spine.

Kyphosis

The choice of instrumentation method for kyphosis is primarily governed by the flexibility of the deformity. In flexible kyphosis, three-point bending instrumentation offers a satisfactory correction. Although three-point bending forces for correction of kyphosis may be applied by several types of construct, the most commonly used is translation instrumentation. Translation instrumentation generates three-point bending on the apex of the kyphosis by exerting a posterior displacement force on the vertebra subject to translation (Figure 43.18). In flexible deformity, both dynamic and rigid translation instrumentation may yield satisfactory results.

When the kyphosis is rigid, attempts at correction solely by the force of posterior instrumentation are always less than satisfactory due to mechanical failure at the bone–component interface. In this situation, an anterior release or an osteotomy to enhance the flexibility of the deformity is mandatory [31,32]. Following the release/osteotomy, dynamic or rigid translation instrumentation may be applied to yield satisfactory results (Figure 43.19).

When the kyphosis is rigid and so severe as to call for a vertebral column resection, application of three-point bending forces with rigid translation instrumentation with a fixed moment arm cantilever beam construct is most suitable. Employing this type of construct allows compression over the resection site when an anterior structural support was reconstructed. When non-structural cancellous chip

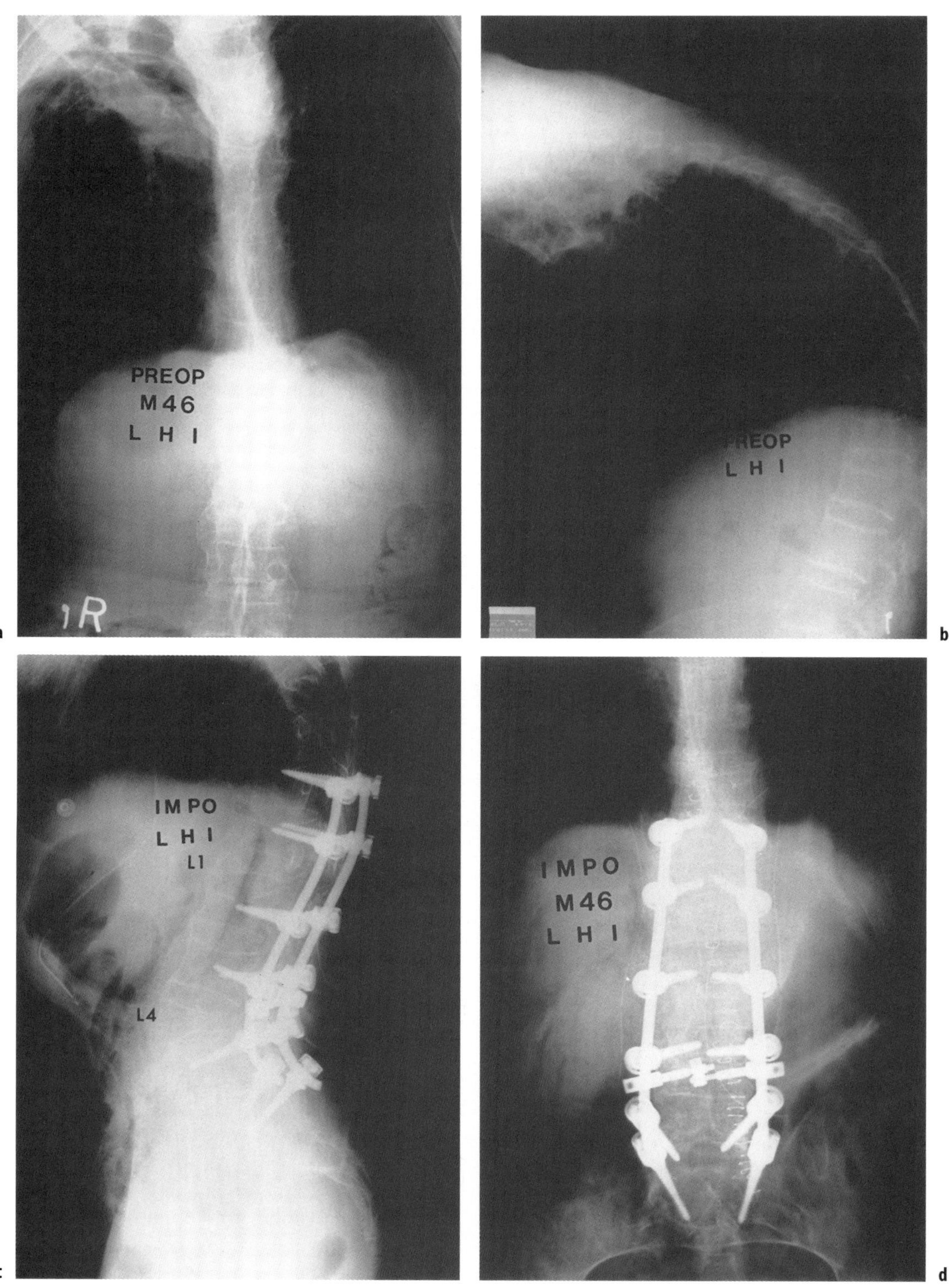

Figure 43.19. **a,b** A 46-year-old male with ankylosing spondylitis. **c,d** He was treated by transpedicular decancellization osteotomy of L1 and L4 with posterior translation instrumentation.

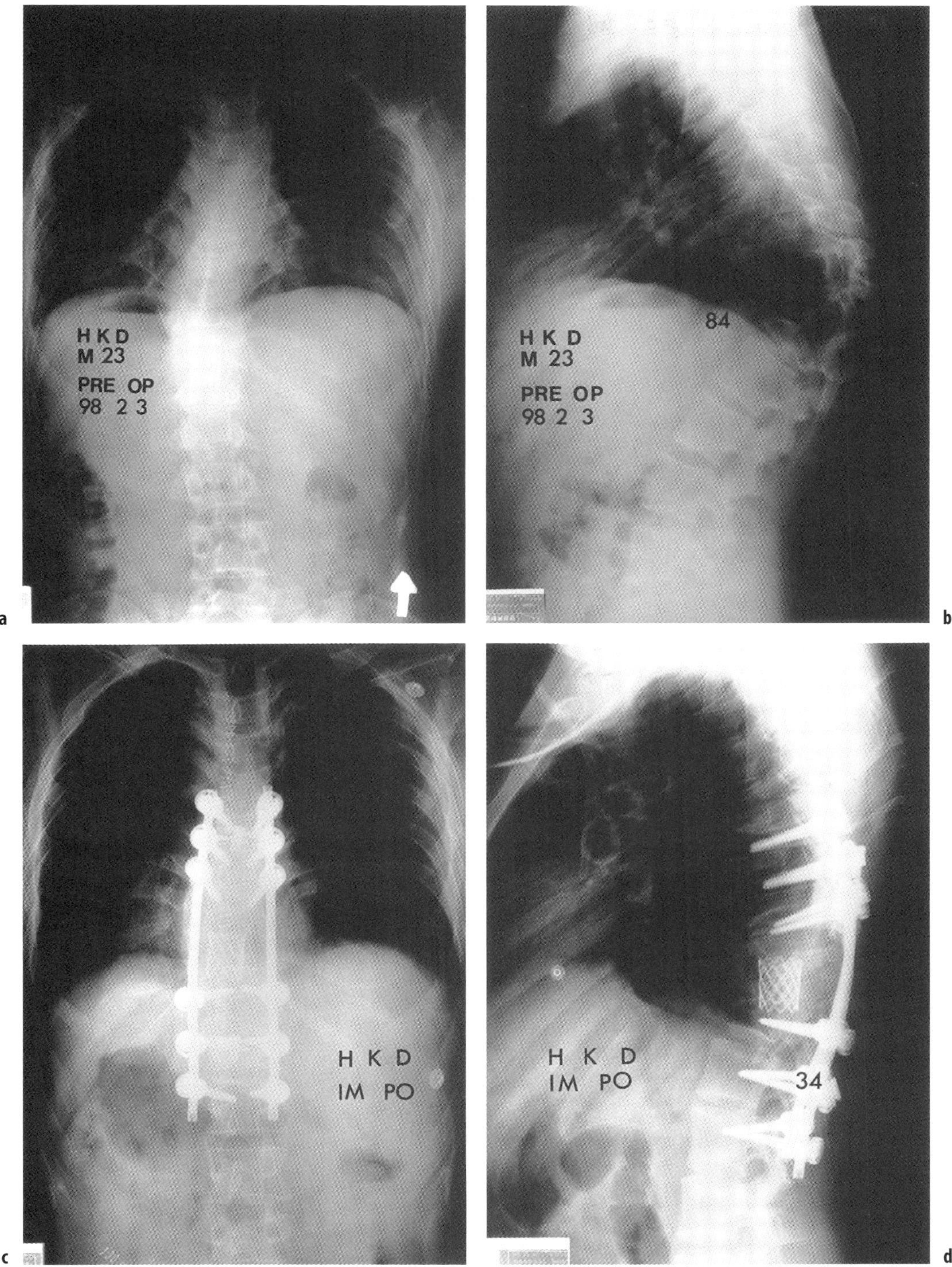

Figure 43.20. a,b A 23-year-old male with congenital kyphosis. **c,d** He was treated by posterior vertebral column resection of T10 and T11 with anterior column reconstruction using titanium mesh and posterior pedicle screw fixation. The posterior instrumentation functions as a translation instrumentation in the first stage of operation, but becomes a fixed moment arm cantilever beam fixation in compressive mode on insertion of anterior strut graft and posterior compression.

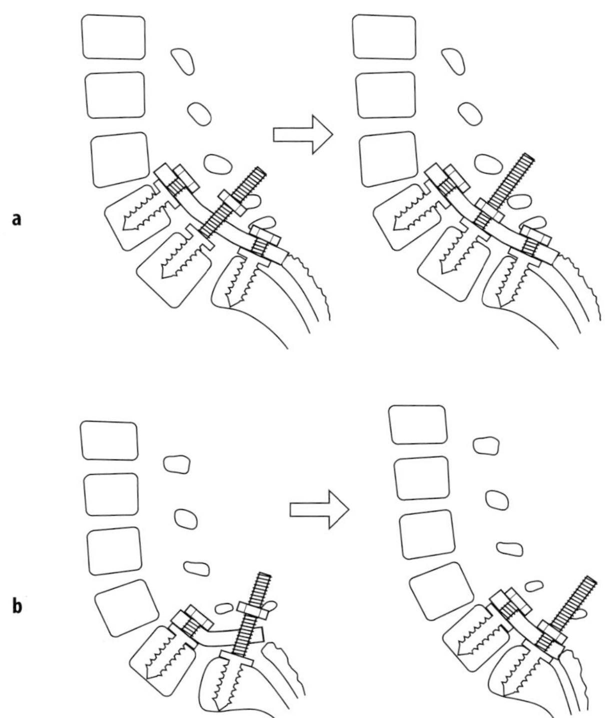

Figure 43.21. **a** In spondylolisthesis, if the displaced segment is intercalary, the construct is a posterior translation construct. **b** In spondylolisthesis, if the displaced segment is at the end of the instrumentation, the construct is a terminal three-point bending construct and the fulcrum is the segment just below.

graft was used to fill the resection gap, the construct, by resisting the axial load, protects the anterior column from collapsing (Figure 43.20).

Spondylolisthesis

Reduction of spondylolisthesis may be achieved either by three-point bending or translation force. Though these two methods are very similar in look, the force acting on the displaced vertebra is quite different. When the displaced segment is at the end of the instrumentation, it becomes a terminal three-point bending construct, the fulcrum being the segment just below. However, when the displaced segment is intercalary, the force for reduction becomes translation (Figure 43.21). Both methods may be effectively carried out by a fixed moment arm cantilever beam construct. Some of the instruments offer screws of special design with elongated connectors for this purpose.

Tumors

The principle of reconstruction of the spinal column by posterior instrumentation in tumor surgery is similar to that for traumatic instabilities. When resection of the tumor results in significant compromise of the anterior column, reconstruction of the anterior column with posterior instrumentation is indicated. When the anterior column following resection is still quite competent of bearing axial loads, posterior compression instrumentation or neutral fixation with distraction instrumentation is sufficient (Figure 43.22).

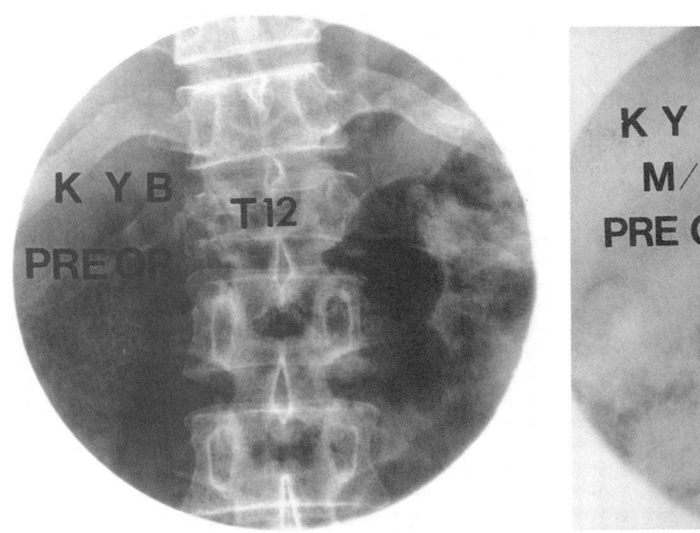

a

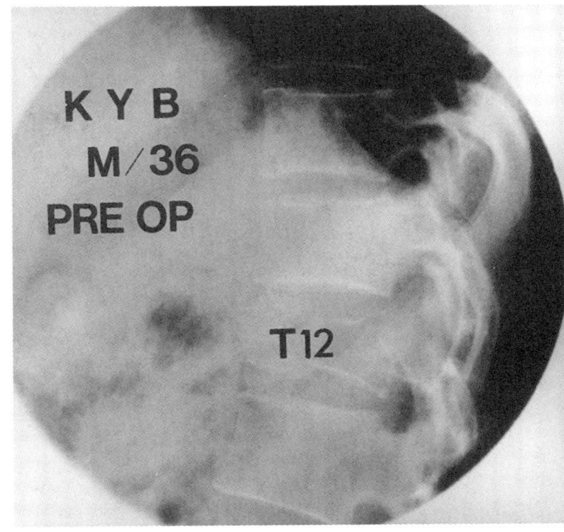

b

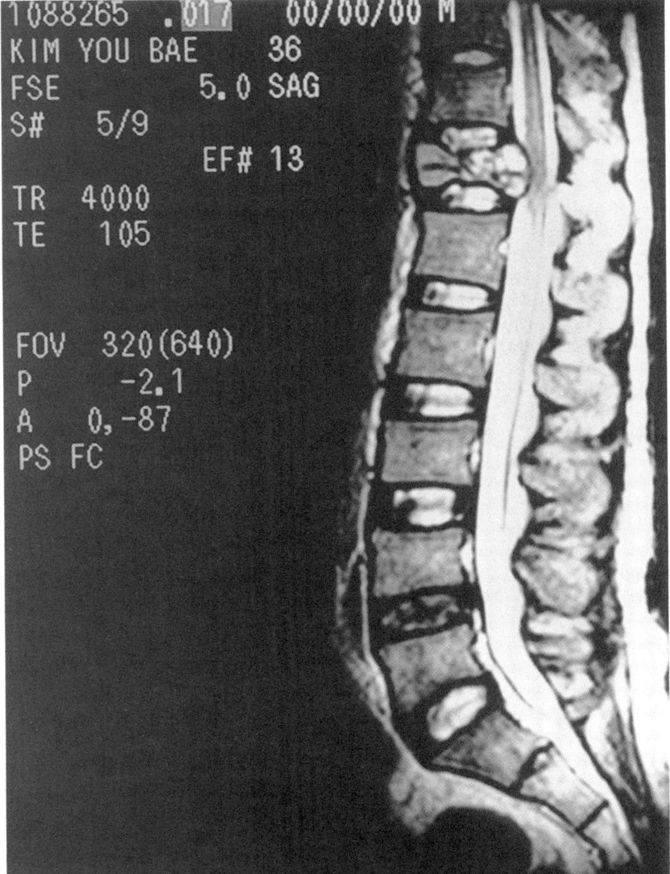

c

Figure 43.22. **a,b** A 36-year-old male with hypernephroma and T12 metastasis. **c,d** Preoperative MRI. The sagittal and axial images show destruction of vertebral body and cord compression by tumor mass. **d,e** He was treated by total resection of the vertebral body with reconstruction of the anterior column using titanium mesh and posterior pedicle screw fixation. The posterior fixed moment arm cantilever instrumentation was applied in compressive mode over the anterior structural graft. (*continued next page*)

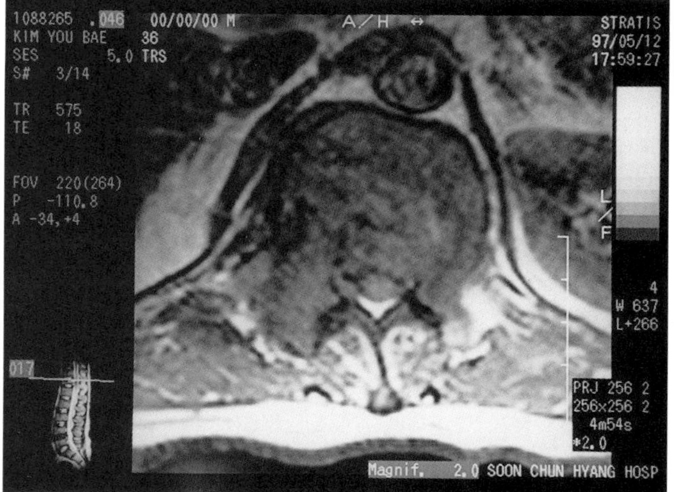

d

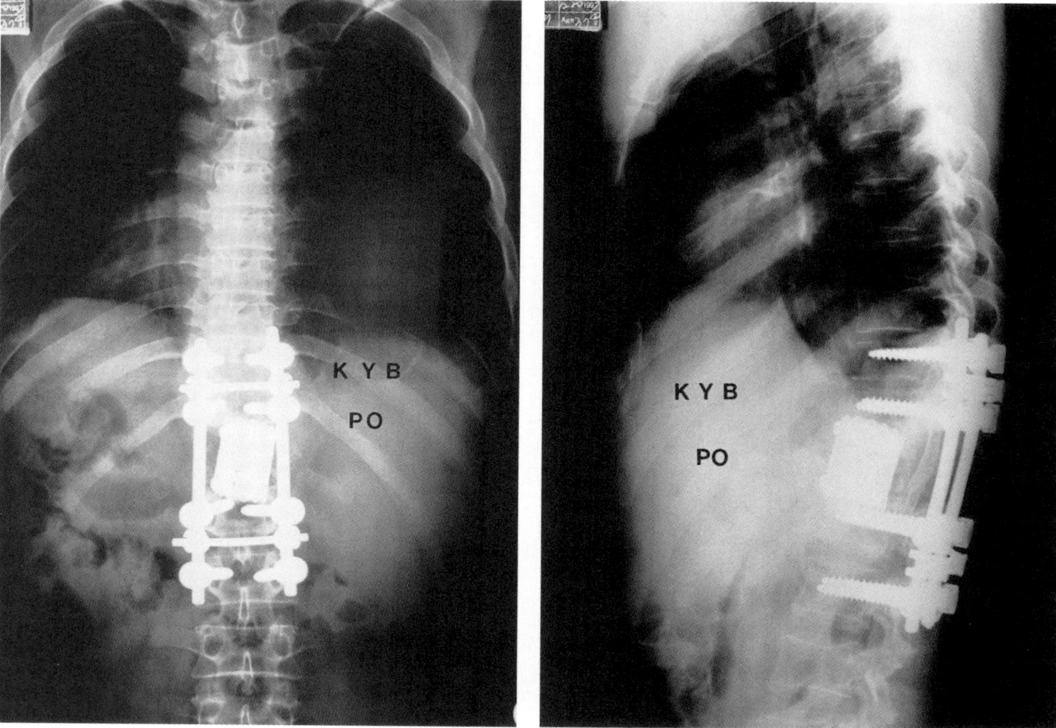

e f

Figure 43.22 *(continued)*

References

1. Abumi K, Panjabi MM, Duranceau J. Biomechanical evaluation of spinal fixation devices. Part III. Stability provided by six spinal fixation devices and interbody bone graft. Spine 1989;14:1249.
2. Ashman RB, Galpin RD, Corin JD, Johnston CE 2nd. Biomechanical analysis of pedicle screw instrumentation systems in a corpectomy model. Spine 1989;14:1398.
3. Cotrel Y, Dubousset J, Guillaumat M. New universal instrumentation in spinal surgery. Clin Orthop 1988; 227:10.
4. Harrington PR. Treatment of scoliosis: correction and internal fixation by spine instrumentation. J Bone Joint Surg 1962;44-A:591.
5. Luque ER. The anatomic basis and development of segmental spinal instrumentation. Spine 1982;7:256.
6. Suk SI, Lee CK, Kim WJ, Chung YJ, Park YB. Segmental pedicle serew fixation in the treatment of thoracic idiopathic scoliosis. Spine 1995;20:1399.
7. Lenke LG. Posterior and Posterolateral Approaches to the Spine. In: Bridwell KH, Dewald RL, editors. The Textbook of Spinal Surgery, 2nd ed. Philadelphia: Lippincott-Raven, 1997:193.
8. Benzel EC. Biomechanics of Spine Stabilization. Principles and Clinical Practice. New York: McGraw-Hill, 1995.
9. Rudisch A, Kremser C, Peer S, Kathrein A, Judmaier W, Daniaux H. Metallic artifacts in magnetic resonance imaging of patients with spinal fusion. A comparison of implant materials and imaging sequences. Spine 1998;23:692.
10. Garfin SR. Spinal fusion: The use of bone screws in the vertebral pedicles. Summation. Spine 1994;19:2300S.
11. Zuk SI, Kim WJ, Lee SM, Kim JH, Chung ER. Thoracic pedicle screw fixation in spinal deformities: are they really safe? Spine 2001;26:2049.
12. Nazarian SM, Louis RP. Posterior internal fixation with screw plates in traumatic lesions of the cervical spine. Spine 1991;16:S64.
13. Licht NJ, Rowe DE, Ross LM. Pitfalls of pedicle screw fixation in the sacrum. A cadaver model. Spine 1992;17: 892.
14. Lonstein JE. The Galveston technique using Luque or Cotrel–Dubousset rods. Orthop Clin North Am 1994;25: 311.
15. Suk SI, Lee CK, Min HJ, Cho KH, Oh JH. Comparison of Cotrel–Dubousset pedicle screws and hooks in the treatment of idiopathic scoliosis. Int Orthop 1994;18: 341.
16. Skinner R, Maybee J, Transfeldt E, Venter R, Chalmers W. Experimental pull-out testing and comparison of variables in transpedicular screw fixation. A biomechanical study. Spine 1990;15:195.
17. Suk SI, Cha SI, Lee Ck, Kim WJ. A study on the pull-out strength of pedicle screws in relation to the size of drill holes and inserted screws. Presented at the 30th Annual Meeting of the Scoliosis Research Society, Asheville, NC, September 13–16, 1995.
18. Kwok AW, Finkelstein JA, Woodside T, Hearn TC, Hu RW. Insertional torque and pull-out strengths of conical and cylindrical pedicle screws in cadaveric bone. Spine 1996;21:2429.
19. Yerby SA, Ehteshami JR, McLain RF. Loading of pedicle screws within the vertebra. J Biomech 1997;30: 951.
20. Schufflebarger HL. Moss Miami Instrumentation. In: Bridwell KH, Dewald RL, editors. The Textbook of Spinal Surgery, 2nd ed. Philadelphia: Lippincott-Raven, 1997:675.
21. Lim TH, Eck JC, An HS, Hong JH, Ahn JY, You JW. Biomechanics of transfixation in pedicle screw instrumentation. Spine 1996;21:2224.
22. Ritterbusch JF, Ashman RB, Roach JW. Biomechanical comparison of spinal instrumentation systems. Orthop Trans 1987;11:87.
23. Cunningham BW, Sefter JC, Shono Y, McAfee PC. Static and cyclical biomechanical analysis of pedicle screw spinal constructs. Spine 1993;18:1677.
24. McCormack T, Karaikovic E, Gaines RW. The load-sharing classification of spine fractures. Spine 1994; 19:1741.
25. Zou D, Yoo JU, Edwards WT, Donovan DM, Chang KW, Bayley JC et al. Mechanics of anatomic reduction of thoracolumbar burst fractures. Comparison of distraction versus distraction plus lordosis in the anatomic reduction of the thoracolumbar burst fracture. Spine 1993; 18:195.
26. Cotrel Y, Dubousset J. A new technique of spine fixation by a posterior approach in the treatment of scoliosis. Rev Chir Othop 1987;70:489.
27. Denis F. Cotrel–Dubousset instrumentation in the treatment of idiopathic scoliosis. Orthop Clin North Am 1988;19:291.
28. Sanders JO, Sanders AE, More R, Ashman RB. A preliminary investigation of shape memory alloys in the surgical correction of scoliosis. Spine 1993;18: 1640.
29. Denis F. The three-column spine and its significance in the classification of acute thoracolumbar spinal injuries. Spine 1983;8:817.
30. Denis F. Spinal instability as defined by the three-column spine concept in acute spinal trauma. Clin Orthop 1984;189:65.
31. Bradford DS, Tribus CB. Current concepts and management of patients with fixed decompensated spinal deformity. Clin Orthop 1994;306:64.
32. Suk SI, Kim JH, Kim WJ, Lee SM, Chung ER, Nah KH. Posterior vertebral column resection for severe spinal deformities. Spine. 2002;27:2374.

44 Biomechanics of Sacral Fixation

J. C. Y. Leong and W. W. Lu

Anatomic Considerations of Sacral Fixation

The wedge-shaped sacrum not only gives support to the vertebral column, it also provides strength and stability to the pelvis. Recent renewed interest in sacral screw fixations together with advancements in mechanical understanding have resulted in increased attention being paid to the surgical anatomy of the sacrum [1–5]. Awareness of the anatomical structure, and the adjacent neurovascular and visceral structures and their configurations will minimize complications and contribute to a successful surgical outcome.

In the adult, the sacrum is composed of five vertebral bodies fused together by four ossified intervertebral disks. The sacrum articulates above with the fifth lumbar (L5) vertebra, below with the coccyx, and laterally from the auricular surfaces with the two iliac bones of the hip to form the sacroiliac joints. The projecting anterior edge of the first sacral vertebra is called the sacral promontory and the two sides are the sacrum alas. The sacral promontory is used as a landmark for making pelvic measurements.

Even though sacral screw fixation techniques vary, the anterior sacral anatomy is particularly important. Bicortical purchase to the sacrum is required to enhance the strength of fixation, but a complex interdigitation of neurovascular, visceral, and urogenital structures lies anterior to the sacrum. These are the common and internal iliac arteries, veins, the lumbosacral and obturator trunks, and the rectosigmoid portion of the large bowel. The common and internal iliac veins are located posterior and lateral to the corresponding arteries (Figure 44.1). It has been shown that the veins lie in the connective tissue immediately in front of the sacral ala [6]. At the level of the first and second sacral vertebrae, the internal iliac veins lie on the anterolateral sacral alar surface; more specifically, the extension of the arcuate line onto the sacrum, medial to the sacroiliac joints. Standard anatomic texts have placed the internal iliac veins directly anterior to the sacroiliac joints [5,7–9]. Other studies [2,6] also reported these vessels medial to that position on the anterolateral surfaces of the sacral ala medial to the sacroiliac joint. The internal iliac artery lies anterolaterally and does not come in contact with the bony sacrum.

Studies have also found that the lumbosacral trunk enters the pelvis deep to the medial margin of the psoas major and descends over the pelvic brim. Within the pelvis itself, the trunk runs on the anterior surface of the sacral ala and rests on the sacral extension of the arcuate line medial to the SI joint. The obturator nerve courses far enough anterior to the sacrum, between the sacroiliac joint and the lumbosacral trunk, that it does not come in contact with the sacrum. Injury to any of these can lead to hemorrhage, neurologic deficit, infection, and chronic pain. Furthermore, sacral osteomyelitis, peritonitis, and sepsis are potential sequelae of bowel perforation with possible life-threatening consequences [4].

For S1 screws, the three structures most commonly at risk are the lumbosacral trunk, the internal iliac vein, and the sacroiliac joint. The internal iliac artery and the obturator nerve are considerably anterior to the sacrum and are not at risk. The descending colon, protected by its mesentery, is not prone to injury at the S1 level. Therefore, two safe zones for S1 screw place-

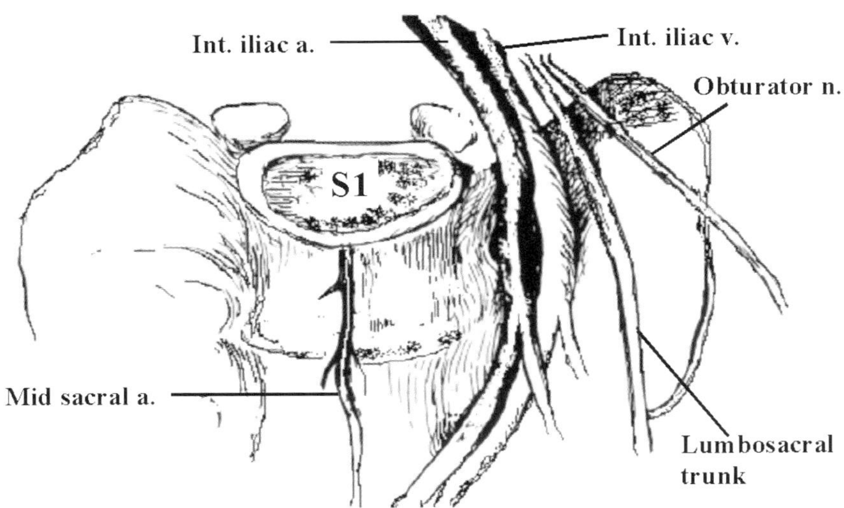

Figure 44.1. Anterior view of sacrum neurovascular anatomy showing the internal iliac veins, internal iliac artery, lumbosacral trunk, etc.

ment have been identified (Figure 44.2a, b): The first is bordered laterally by the SI joint. Its medial border is delineated by the lumbosacral trunk. The second safe zone lies between the sacral promontory medially and the internal iliac vein laterally and is about 22–27 mm wide. There have also been some concerns about the best entry, direction, and depth for sacral screw insertion. By dissecting cadaveric specimens, Esses et al. [10] and Mirkovic and co-authors [6], found that anteromedial bicortical screw fixation had a larger safe zone compared to anterolateral fixation. It has been reported that screws placed anteromedially are stronger than screws inserted anterolaterally [11–13]. Clinically, the anteromedial pathway is more commonly used for sacral screw fixation.

Bone Mineral Considerations of Sacral Fixation

Following the popularity of spinal pedicle screw instrumentation, much attention has been focused on the importance of bone mineral density (BMD) for instrumented spine fusion in clinical practice [14–19]. Clinical reports have shown that sacral screw fixation has a high failure rate. The failure of sacral screw fixation may be due to several factors such as inadequate sacral bone purchase, inappropriate direction or depth of the screw insertion, and BMD variations within the sacrum [12,20–22]. Previous studies [23–25] employed BMD measurements to obtain area density (g/cm^2) using dual energy X-ray absorptiometry (DEXA), or volumetric density (g/cm^3) using quantitative computed tomography (QCT) (Figure 44.3). QCT is advantageous in that it can quantify BMD along the screw insertion pathway, and is therefore a more accurate technique than DEXA for measuring the relationship between BMD and the strength of screw fixation [23,26]. Most investigations into sacral BMD distributions have been carried out on aged specimens, with only a few studies on young cadaveric specimens. Since sacral fixations are commonly applied in patients of young to middle age, BMD measurements of young specimens provide clinically relevant data for the assessment and comparison of sacral fixation. In this text, we will focus on the BMD variations within the S1 body and ala of young patients.

Based on 13 young adult fresh cadaveric sacrum specimens, recent studies [26,27] reported a mean BMD of the S1 vertebral body of 381.9 mg/cm^3, 31.9% higher than that of the sacral ala (mean 296.9 mg/cm^3). Table 44.1 and Figure 44.4 list the BMD at different regions of the S1 body and ala. There are significant differences in BMD, with the regions near the lateral posterior and lateral anterior areas of the vertebral body having higher BMD. This again shows that these areas would provide better purchase for pedicle screw fixation. Furthermore, BMD of the posterior region closest to the spinal canal was lower than that of lateral and central areas. In the ala, the BMD of the internal anterior areas closest to the S1 body were the highest with a BMD of 326.8 mg/cm^3.

The mean BMD at different layers (Figure 44.5) of the S1 body and ala are also listed in Table 44.2. The BMD of the first layer in the S1 body, close to the superior end plate, was significantly higher than the others with a BMD of 516 mg/cm^3. The BMD was lowest in the central layer of the body, and increased again towards the inferior endplate. This suggests that sacral pedicle screws directed close to or cephalad penetrating the superior sacral end plate would achieve better purchase. In the ala, the BMD of the top layer was highest with a BMD of 329 mg/cm^3 and the BMD decreased caudally from the top layer.

a

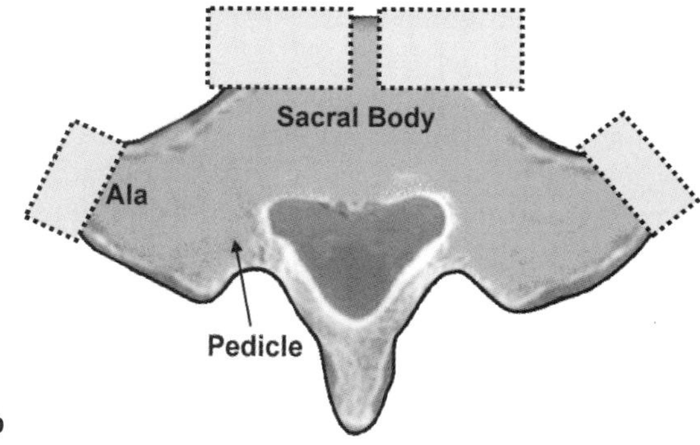

b

Figure 44.2. Two safe zones for each side were identified on the sacrum for screw placements. **a** safe zones shown in white boxes viewed from CT reconstructed 3-D model. **b** top view of the safe zones on the sacrum.

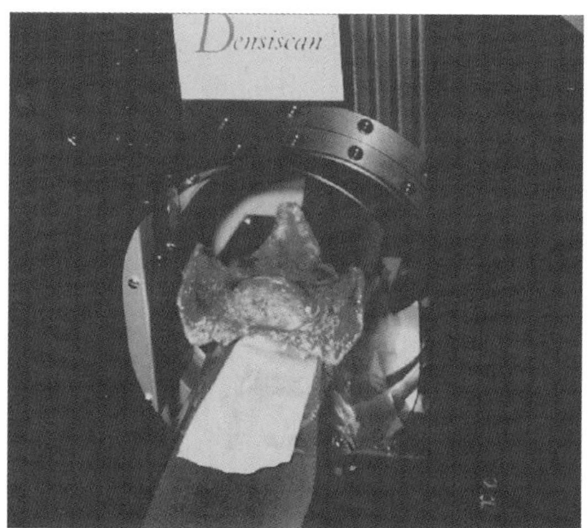

Figure 44.3. The sacral specimen in the QCT device.

Table 44.1. The Mean values (SD) of BMD at different columns of S1 body and ala

S1 Body Columns	Lateral Posterior	Lateral Anterior	Middle Posterior	Middle Middle	Middle Anterior	
The Mean Values of BMD (SD)	381.6 (5)*	368.9 (64.8)*	297.6 (102.6)*	372 (93.9)*	351.1 (68.1)	
Ala Columns	Internal Anterior	Internal Posterior	Middle Anterior	Middle Middle	Middle Posterior	Lateral
The Mean Values of BMD (SD)	326.8 (12)*	226 (135)	205.4 (69.2)	185.1 (91.1)	182 (83.9)	210.1 (77)

* Significant difference found, Unit: mg/cm^3.

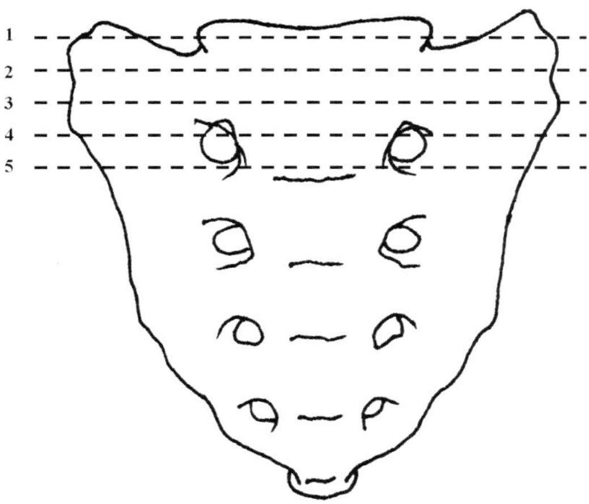

Figure 44.4. Schematic diagram of the front of the sacrum showing the five transverse layers. Lines 1 and 5 represent superior sacral end plate and inferior end plate layers respectively. Line 3 represents the central layer; lines 2 and 4 represent other layers between them.

Based on the BMD distribution within the sacrum [12,26,27], the recommended screw pathway is shown in Figure 44.6. Further to radiographs, CT scans, and specimen observation, it is suggested that the S1 pedicle screw should be directed parallel to the S1 end plate in the sagittal plane, and the ala screw inserted to the tip of the ala along the ala slope. The insertion angle of the S1 pedicle screw in the transverse plane should be between 15 and 25° medially and the angle of the ala screw should be 30–40° laterally for better bone purchase (Figure 44.7 from CT). If the insertion angle is less than 10° for S1 pedicle screws, the screw will not obtain good purchase on the S1 promontory, leading to a greatly reduced purchase of bone and less biomechanical stability. On the other hand, if the insertion angle for the S1 screw is larger than

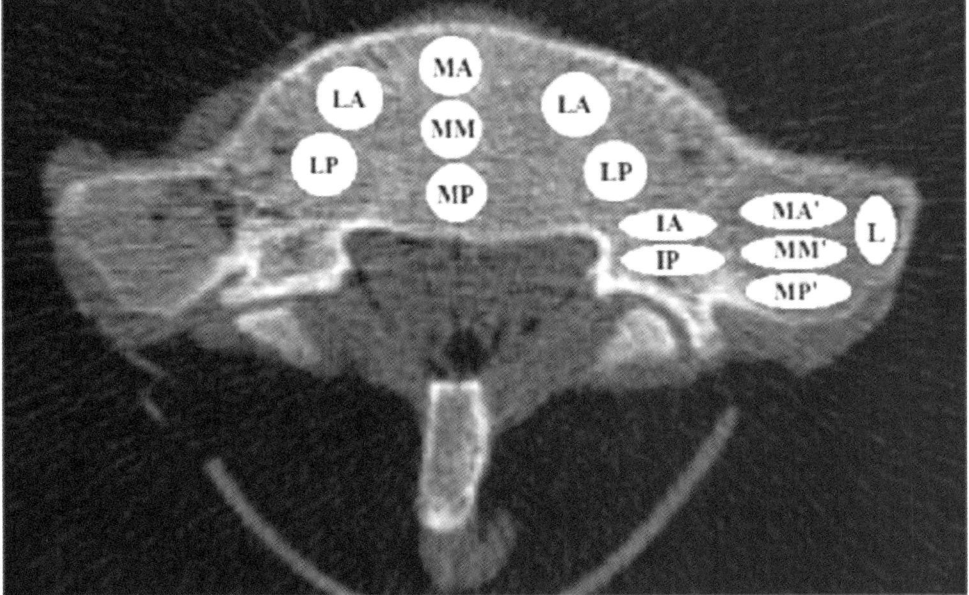

Figure 44.5. Axial view of seven vertical columns in S1 body and six vertical columns in S1 ala. The IA and IP areas are also defined as sacral pedicle.

Table 44.2. The mean values (SD) of bone mineral density at different layers of S1 body and ala

Layers	1	2	3	4	5
S1 Body	516.1 (72)*	376.9 (37)	342.4 (56)	365.6 (56)	397.5 (81)
Ala	N/A	329.1 (196)*	225.6 (128)	177.4 (8)	142.1 (63)

* Significant difference, Unit: mg/cm^3.

25°, the potential for injury to the spinal cord is higher [25]. With respect to ala screw insertion, Edwards and Louis preferred lateral placement into the sacral ala at 35–45° [28,29], while Mirkovic and associates suggested the screw should be angled at 30–40° in order to obtain a better purchase [6]. If the angle of ala screw is greater than 45°, the potential for lumbosacral trunk injury is high [4].

The importance of vertebral BMD for screw fixation has been the subject of several studies to determine screw stability and predict lumbar vertebrae strength [6,10–12,19,30–32]. Some have suggested that preoperative measurement of BMD is necessary for transpedicular screwing in osteoporotic cases [27,33,34]. Other studies [19,35] found that vertebral bodies of human lumbosacral spines had a mean BMD of 92 mg/cm^3 and the mean BMD of the sacrum was 152 mg/cm^3 for the elderly of mean age of 78 years old. Studies also indicated that at a QCT BMD value of less than 90 mg/ml, early loosening of the screw may be expected, while it is less likely with a BMD of more than 120 mg/cm^3. The mean BMD from young adults with mean age of 31 years was approximately 2.5 times greater than that in the above-mentioned group. Since the sacrum mainly consists of cancellous bone and the lumbosacral junction sustains more load than others above the sacrum, failure of sacral screw fixation can occur without significant osteoporosis.

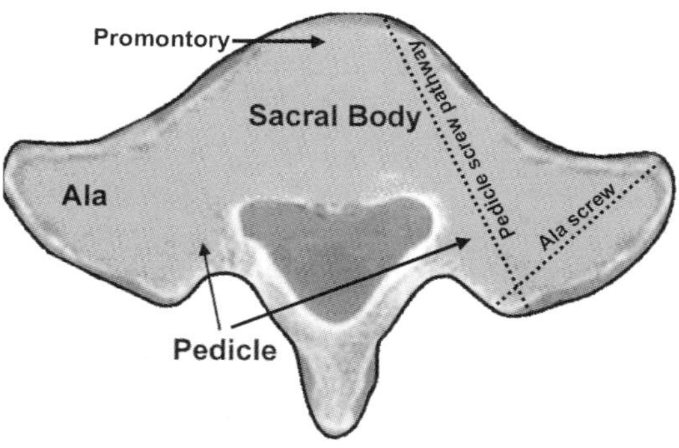

Figure 44.6. Schematic diagram of the sacral body, ala, and pedicle. The pedicle screw pathway and the ala screw pathway are defined based on BMD measurements.

Biomechanical Considerations of Sacral Fixation

Strengths of Varieties of Fixations

The sacral screw can be placed either anteromedially through the S1 pedicle into the promontory or anterolaterally into the sacral

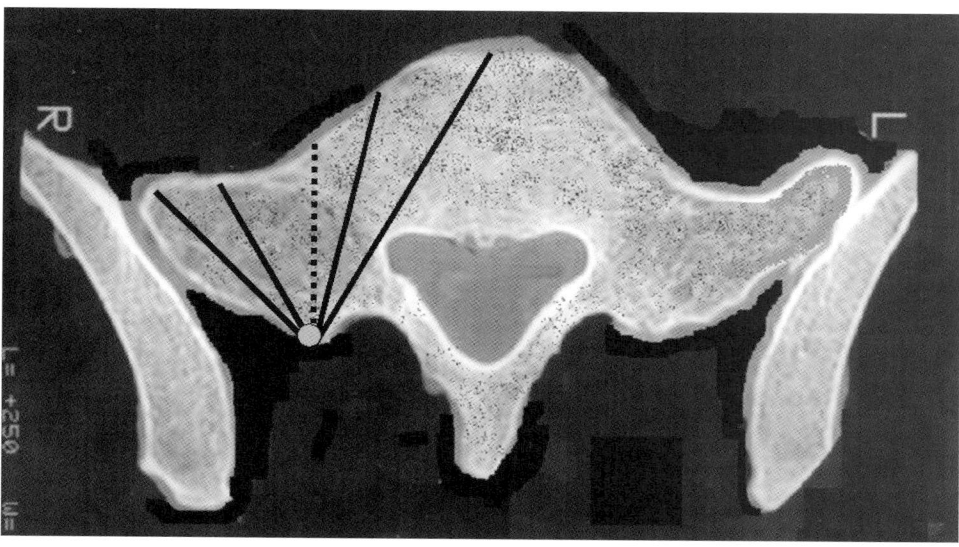

Figure 44.7. CT showing the angles of screw insertion.

ala. There are a few anatomical and biomechanical studies regarding the safety of the screw placement when bicortical insertions are used [2,4,20,36] (see section above). A variety of fixation techniques and instruments have been designed to improve the strength of lumbosacral fixation [7,30,34,37–40]. It is assumed that increasing the number of sacral screws and using a triangulated insertion technique may increase the strength of purchase for lumbosacral fixation. A technique with a sacral Chopin block using S1 pedicle and ala screws with modified CD instrumentation was introduced to clinical practice in the early 1990s [41–43]. This new technique can theoretically increase the strength of fixation due to the anteromedial screw and the added ala screw in divergent triangular orientation. However, a clinical study conducted by Devlin et al. [44] found that the CD system using sacral pedicle and ala screws in the deformity correction of adult patients did not appear to offer any advantages over alternative techniques in achieving arthrodesis.

A recent cadaveric study [12] using a sacral Chopin block with S1 pedicle and ala screws on sacra of different age groups evaluated the stiffness and failure strengths of fixation under different loading conditions. With bicortical screw purchase, the average stiffness of single screw and two divergent triangulated screw fixations (see Figure 44.8) were found to show statistically significant differences under compression, tension, and torsion (see Table 44.3). With one S1 pedicle screw fixation, the average stiffness was 203 N/mm for compression, 147 N/mm for tension, and 2 Nm/deg. for torsion. With two-screw fixation, the average stiffness increased to 255 N/mm (126%), 185 N/mm (126%), and 2.4 Nm/deg (120%) respectively, suggesting that increasing the number of screws and using a triangulated insertion can increase the strength of fixation. Biomechanical tests have also shown that triangulated double-screw instrumented either anteriorly or posteriorly can significantly increase the strength of fixation [38,45,46]. Ogon

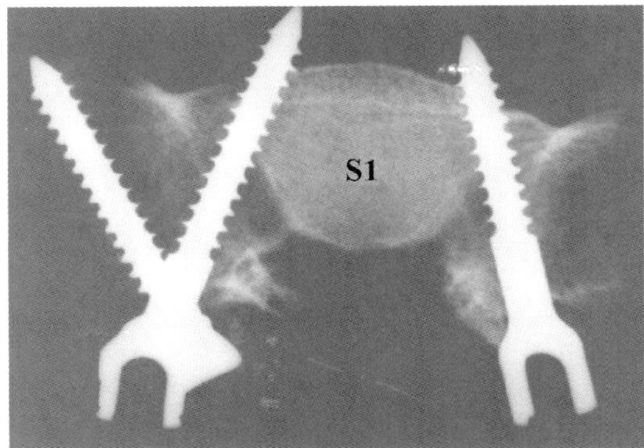

a

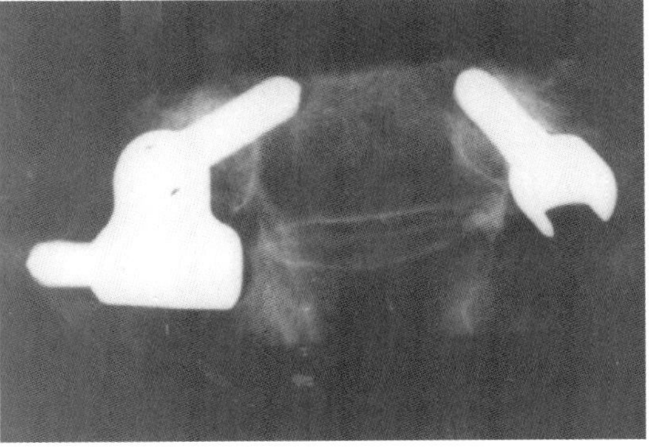

b

Figure 44.8. a,b AP and top view of instrumented specimen. Right side was with one anteromedial S1 pedicle screw and left side with a Chopin Block with anteromedial and anterolateral (alar) screws.

Table 44.3. The average stiffness of younger and aged specimens with one and two screw fixation to the sacrum

	One screw fixation			Two screw fixation		
	Compression (Mean ± SD)	Tension (Mean ± SD)	Torsion (Mean ± SD)	Compression (Mean ± SD)	Tension (Mean ± SD)	Torsion (Mean ± SD)
<30 years	220 ± 34	169 ± 24	2.1 ± 0.8	296 ± 43	209 ± 23	2.7 ± 0.6
>60 years	179 ± 20	112 ± 19	1.9 ± 0.4	195 ± 23	150 ± 27	1.9 ± 0.3
P value	0.003	0.001	0.007	0.001	0.001	0.007
Total	203 ± 35	147 ± 35	2 ± 0.6	255 ± 62	185 ± 38	2.4 ± 0.6

Unit: Compression = N/mm; Tension = N/mm; Torsion = Nm/deg.

and associates [45] found that anterior double-screw fixation can increase fixation strength in vitro by 73%, while Ruland and co-authors [46] found that a doubled pull-out strength can be achieved by using posterior triangulated CD pedicle screws with a transverse plate. By applying the load along the screw axis instead of simulating in vivo loads perpendicular to the screw axis, Zindrick and co-authors [47] conducted a pull-out test to compare the strength of screw fixation in different orientations. They found that the anterolaterally directed screw sustained greater loads to failure. In contrast, other biomechanical studies reported that anteromedial screw fixation was significantly stronger than anterolateral screw insertion [2,20,34,48]. This also correlated with the higher bone density of the S1 centrum, which provides a more rigid bone screw fixation [20].

A negative relation between the specimen age and the stiffness under different loading conditions has also been found [12], and the younger specimens had significantly higher stiffness than the aged ones. The average failure strength under tension load was found to be 1,450 N in the younger specimens (<30 years), but 980 N in the aged specimens (>60 years). Studies have also revealed that the BMD is closely related to the fixation strength of transpedicular screws [33,35,47–49]. Under the same test conditions, it was found that fixation stiffness of the aged specimens were lower than those of younger specimens.

Loosening of Sacral Screw Fixation Under Fatigue Loading

In the early postoperative period, before bony fusion has taken place, the mechanical properties of the fixation will determine the stability of that part of the spine. In some cases, sacral screw fixation may not maintain sufficient stability, eventually resulting in loosening, pull-out, screw breakage, or migration. Cyclic loading is thought to be the main reason leading to change in the biomechanical properties of pedicle screw fixation [50–53] and the major failure mechanism appears to be screw pull-out at the bone–screw interface [44,54,55]. While a few studies have characterized the effects of cyclic loading on pull-out strength of sacral screw fixation, the majority of studies [20,48,56,57] investigating the effects of insertion torque and BMD have used single loading to failure by compression of the screw head perpendicular to the axis of the screw, rather than fatigue loading.

Recent studies [13,58] have simulated cyclic loading of sacral screw fixations in young adult human specimens to investigate the effects of unicortical versus bicortical and medial versus lateral screw fixation on pull-out strength, and also the correlation between pull-out strength, bone density, and insertion torque under fatigue condition. In these studies, human sacrums from young (24–36 years old) cadaver specimens were used. BMD was measured using a peripheral Quantitative Computed Tomography (pQCT) machine with a scan slice thickness of 5 mm. Seven-millimeter Compact CD sacral screws [41,43] were then inserted, randomized into four groups according to the combinations of orientation (medial or lateral) and fixation (unicortical or bicortical). Screws were inserted according to the following depths: unicortically; up to but not engaging the cortex (2 mm short of the anterior cortex); or bicortically, with the threads at least 2 mm anterior to the cortex. The screw holes were not tapped, and the insertion torque was measured throughout screw insertion using a torque driver. The specimens were placed on a servo-hydraulic MTS 858 bionix testing machine and cyclic loading performed with the loading axis directed perpendicularly to the axis of the screw (Figure 44.9a,b). A rigid linkage was used at the screw head, using the CCD rod and linkage system. The cyclic loading ranged from −40 N to −400 N and was conducted at 2 Hz up to 20,000 cycles. Following cyclic loading, the specimen was re-oriented in the testing machine, and a pull-out test was performed along the axis of the screw at a loading rate of 10 N/sec until the screw was completely pulled out of the sacrum. The results from these studies showed that the average pull-out force after cyclic loading was 1271 N for the medial bicortical screw fixation and 778 N for the medial unicortical screw fixation. The average pull-out force after cyclic loading was 570 N for the lateral bicortical screw fixation and 368 N for the lateral unicortical screw fixation. Table 44.4 lists the all the relevant data from this study. The difference between the pull-out force of bicortical and the unicortical fixation was highly significant for the medial screw fixation and significant for the lateral screw fixation, showing that depth of screw penetration significantly affects the pull-out force following fatigue loading. The average pull-out force was significantly greater for the medial bicortical screw fixation than for both bicortical and unicortical lateral (ala) screw fixation [12]. There was also a significant difference of pull-out force between the medial unicortical and the lateral unicortical screw fixation. The pull-out force following fatigue loading is therefore significantly affected by the screw orientation. The maximum mean insertion torque was 1.93 N.m for the medial bicortical screw fixation, which was 35% higher than that for medial unicortical screw fixation (Table 44.4). The insertion torque for lateral bicortical screw fixation was 50% lower than for medial bicortical, but there was no significant difference between the insertion torque for medial unicortical screw fixation and lateral unicortical screw fixation. For medial screw fixation, the pull-out force correlated significantly with insertion torque and BMD. The correlation and linear regression between insertion torque and pull-out force are shown in Figures 44.10 and 44.11 respectively. In this study, the insertion torque also showed a significant linear correlation with BMD. However, no significant linear correlation was seen between pull-out force, insertion torque, or bone density for lateral screw fixation to the ala. Some authors [20,48] also found the strength of medial screws to be stronger than lateral screws under compression, both for bicortical and unicortical fixation, and a similar result was found for the pull-out strength of unicortical screws [2]. In contrast, however, other studies [47,49] have shown that bicortical, laterallyplaced screws sustained higher pull-out loads than medial screws. Based on the anatomical evaluation [4], lateral screw placement, if it were to engage the anterior cortex (bicortical fixation), carries a risk of injury to either the lumbosacral trunk or the internal iliac vein. The potential for trunk injury with laterally bicortical screws is particularly high (55%).

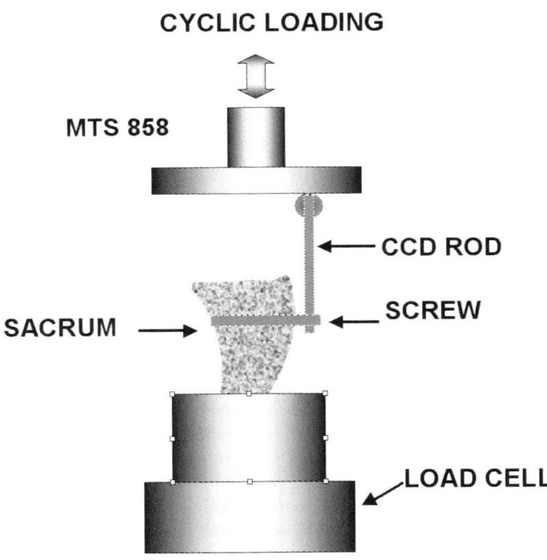

Figure 44.9. Schematic of the specimen loaded in the cyclic loading testing apparatus. The lower part of the sacrum was embedded in a metal cup filled with epoxy. Cyclic axial compressive force was applied to the screw via the CCD rod, allowing translation and rotation of the screw. Force was measured by the load cell, which was situated inferiorly.

Intraspecimen bone densities and insertional torques may account for differences in screw fixation. An in vitro study found the BMD in the

Table 44.4. Average BMD, insertion torque, and pull-out force for the medial and lateral bicortical and unicortical screw insertions. Standard deviations are included in parentheses

Orientation	Depth	No. of specimens	Pull-out force (N)		Insertion torque (N.m)		BMD of body (g/ml)	
Medial	Bicortical	11	1271	(449)	1.93	(0.67)	0.38	(0.08)
	Unicortical	11	778	(443)	1.26	(0.4)	0.4	(0.06)
Lateral (alar)	Bicortical	7	570	(194)	0.96	(0.24)	0.24	(0.05)
	Unicortical	6	368	(171)	0.88	(0.19)	0.26	(0.04)

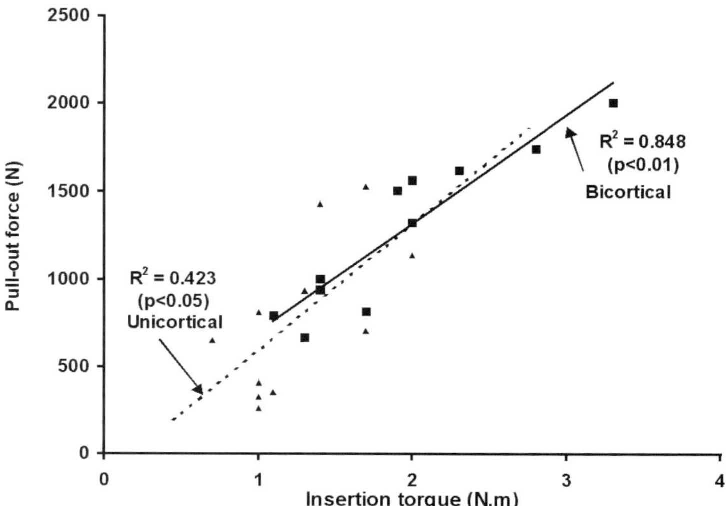

Figure 44.10. A plot of pull-out force versus insertion torque for medial sacral screw insertions. Linear regression lines for both data sets are plotted.

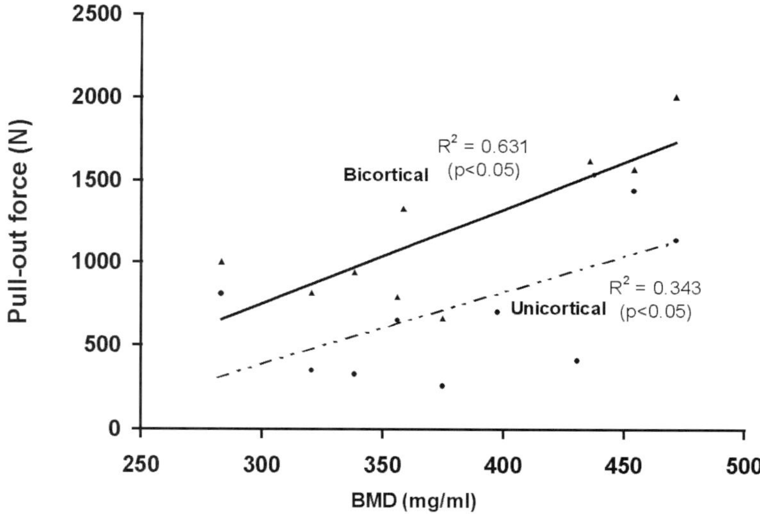

Figure 44.11. A plot of pull-out force versus BMD for medial sacral screw insertions. Linear regression lines for both data sets are also plotted.

S1 centrum region to be 56% denser than in the S1 lateral alar region [48], and this is consistent with other results [13,20]. A recent study [58] further showed the insertion torques for medial bicortical and unicortical screw fixation to be 101% and 43% higher than the insertion torques for lateral bicortical and unicortical screw fixation, respectively. The pull-out forces sustained by medial bicortical and unicortical screw fixation following cyclic loading were significantly higher than those placed laterally, where the BMD and insertion torques are lower.

A previous study of the depth of insertion of transpedicular vertebral screws [34] suggested that even a 5 mm difference in screw length can produce a significant increase in bone screw fixation strength in vertebral bone. However, it has been reported that in an older population, unicortical and bicortical sacral screw fixations sustained similar loads to failure [2,3,48]. The anterior cortex of the sacrum was found to be anatomically thin in the elderly sacrum, and as such has little mechanical effect on screw fixation. For young cadaveric specimens, bicortical screw fixation proved to be stronger than unicortical fixation for both medially and laterally placed screws, despite the overall screw lengths differing by less than 5 mm [12,13]. Engagement of the anterior cortex therefore makes a significant contribution to the pull-out strength of sacral screw fixation following fatigue loading in younger specimens [24,59–61].

Biomechanical findings strengthen the clinical use of preoperative bone density measurement in selecting patients who may be candidates for sacral screw fixation, and suggest intra-operative insertion torque, insertion depth, and orientation as indicators of the fixation strength.

References

1. Banta CJ, King AG, Dabezies EJ, Liljeberg RL. Measurement of the effective pedicle diameter in the human spine. Orthopaedics 1989;12:939–42.
2. Dohring J, Krag MH, Johnson CC. Sacral screw fixation: A morphologic anatomic and mechanical study (abstract). Proc North Am Spine Soc 1990.
3. Farcy JPC, Rawlins BA, Glassman SD. Technique and results of fixation to the sacrum with iliosacral screws. Spine 1992;17 (Suppl):190–5.
4. Mirkovic S, Abitol JJ, Steinman J et al. Anatomic consideration for sacral screw placement. Spine 1991; 16(suppl):289–94.
5. Xu R, Ebraheim NA, Mohamed A, El-Gamal H, Yeasting RA. Anatomic considerations for dorsal sacral plate-screw fixation. J Spinal Disord 1995;8(5):352–6.
6. Mirkovic S, Abitbol JJ, Steinman J, Edwards CC, Schaffler M, Massie J et al. Anatomic consideration for sacral screw placement. Spine 1991;16(Suppl):289–94.
7. Jackson RP. Insertion of intrasacral rods for sacral fixation and spinal correction with in situ rod contouring technique. In: Bridwell KH, Dewald RL, editors. The Textbook of Spinal Surgery, 2nd ed. Philadelphia: Lippincott-Raven, 1997, 2187–209.
8. Morse BJ, Ebraheim NA, Jackson WT. Preoperative CT determination of angles for sacral screw placement. Spine 1994;19:604–7.
9. Xu R, Ebraheim NA, Yeasting RA, Wong FY, Jackson WT. Morphometric evaluation of the first sacral vertebra and the projection of its pedicle on the posterior aspect of the sacrum. Spine 1995;20(8):936–40.
10. Esses SI, Botsford DJ, Huler RJ, Rauschning W. Surgical anatomy of the sacrum. A guide for rational screw fixation. Spine 1991;16(6 Suppl):S283–8.
11. Ashman RB, Bechtold JE, Edwards WT, Johnston CE, McAfee PC, Tencer AF. In vitro spinal arthrodesis implant mechanical testing protocols. J Spinal Disord 1989;12:274–81.
12. Leong JCY, Lu WW, Zheng YG, Zhu QA, Zhong SZ. Comparison of the strengths of lumbosacral fixation achieved with techniques using one and two triangulated sacral screws. Spine 1998;23(21):2289–94.
13. Zhu Qingan, Lu WW, Holmes AD, Zheng YG, Zhong S, Leong CY. The effects of cyclic loading on pull-out strength of sacral screw fixation: an in vitro biomechanical study. Spine 2000;25(9):1065–9.
14. Granhed H, Johnson R, Hansson T. Mineral content and strength of lumbar vertebrae: A cadaver study. Acta Orthop Scand 1989;52:105–9.
15. Hadjipavlou AG, Nicodemus CL, Al-Hamdan FA, Simmons JW, Pope MH. Correlation of bone equivalent mineral density to pull-out strength of triangulated pedicle screw construct. J Spinal Disord 1997;10(1): 12–19.
16. Okuyama K, Sato K, Abe E, Inaba H, Shimada Y, Murai H. Stability of transpedicle screwing for the osteoporotic spine: An in vitro study of the mechanical stability. Spine 1993;18:2240–5.
17. Smit TH, Odgaard A, Schneider E. Structure and function of vertebral trabecular bone Spine 1997; 22:2823–33.
18. Snyder BD, Zaltz I, Hall JE, Emans JB. Predicting the integrity of vertebral bone screw fixation in anterior spinal instrumentation. Spine 1995;20:1568–74.
19. Wittenberg RH, Lee KS, Shea M, White AA 3rd, Hayes WC. Effect of screw diameter, insertion technique, and bone cement augmentation of pedicular screw fixation strength. Clin Orthop 1993;296:278–87.
20. Smith SA, Abitbol JJ, Carlson GD, Anderson DR, Taggart KW, Garfin SR. The effects of depth of penetration, screw orientation, and bone density on sacral screw fixation. Spine 1993;18(8):1006–10.
21. Vesterby A, Mosekilde L, Gundersen H, Melsen F, Mosekilde L, Holm K et al. Biologically meaningful determinants of the in vitro strength of lumbar vertebrae. Bone 1991;12:219–24.
22. Weaver JK, Chalmers J. Cancellous bone: Its strength and changes with aging and an evaluation of some methods for measuring its mineral content. J Bone Joint Surg 1966;48A:289–9.
23. Ericksson S, Isberg BO, Lindgren JU. Prediction of vertebral strength by dual photon absorptiometry and quantitative computed tomography. Calcif Tissue Int 1989; 44:243–50.
24. Imai Y, Sone T, Tomomitsu T, Imai H, Mikawa Y, Watanabe R et al. Precision and accuracy for peripheral quantitative computed tomography evaluated using radial specimens. J Bone Miner Res 1997;12: 263.
25. Rüggsegger P. The use of peripheral QCT in the evaluation of bone remodelling. The Endocrinologist 1994; 4(3):167–76.
26. Zheng YG, Lu WW, Qingan Zhu, Ling Qin, Shizhen Zhong, Leong JCY. Bone mineral density variations of the sacrum in young adults and its significance for sacral fixation. Spine 1999;25(3):353–7.
27. Lu WW, Zheng YG, Holmes AD, Zhu QA, Luk KDK, Leong. Bone JCY mineral density variations along the lumbosacral spine. Clin Orth Rel Res 2000;378: 255–63.

28. Edwards CC. Spinal screw fixation of the lumbar and sacral spine. Early results treating the first 50 cases. Orthop Trans 1987;11:99.
29. Louis R. Fusion of the lumbar and sacral spine by internal fixation with screw plates. Clin Orthop 1986; 203:18–33.
30. Allen BL Jr., Ferguson RL. The Galveston experience with L-rod instrumentation for adolescent idiopathic scoliosis. Clin Orthop 1988; 229:59–69.
31. Glazer PA, Colliou O, Lotz JC, Bradford DS. Biomechanical analysis of lumbosacral fixation. Spine 1996;21: 1211–22.
32. Wittenberg RH, Shea M, Swartz DE, Lee KS, White AA 3rd, Hayes WC. Importance of bone mineral density in instrumented spine fusions. Spine 1991;16(6):647–52.
33. Halvorson TL, Kelley LA, Thomas KA, Whitecloud TS 3rd, Cook SD. Effects of bone mineral density on pedicle screw fixation. Spine 1994;19:2415–20.
34. Krag MH, Beynnon BD, Pope MH, DeCoster TA. The depth of insertion of transpedicular vertebral screws into human vertebrae: Effect upon screw–vertebra interface strength. J Spinal Disord 1989;1:287–94.
35. Wittenberg RH, Shea M, Edwards WT, Swartz DE, White AA, Hayes WC. A biomechanical study of the fatigue characteristics of thoracolumbar fixation implants in a calf spine model. Spine 1992;17(6S):S121–8.
36. Yoganandan N, Larson SJ, Cusick JF, Pintar F, Maiman DJ. Structural strength and kinematics of pedicle screw/plate fixation of the lumbar spine. Presented at the Annual Meeting of the International Society for the Study of the Lumbar Spine. Boston, 1990.
37. Ashman RB, Birch JG, Bone LB. Mechanical testing of spinal instrumentation. Clin Orthop 1988;227: 113–25.
38. Bayley JC, Yuan HA, Fredrickson BR. The Syracuse I-plate. Spine 1991;16:120–4.
39. Krag MH, Beynnon BD, Pope MH, Frymoyer JW, Haugh LD, Weaver DL. An internal fixation for posterior application to short segments of the thoracic, lumbar, or lumbosacral spine: Design and testing. Clin Orthop 1986;203:75–98.
40. Ogilvie JW, Schengel M. Comparison of lumbosacral fixation devices. Clin Orthop 1986;203:120–5.
41. Chopin D. A new device for pelvic fixation for spinal surgery: The sacral block. Presented at 8th International Congress on Cotrel–Dubousset Instrumentation. Minneapolis, 1991.
42. Jackson RP, Hamilton AC. C-D screws with oblique canals for improved sacral fixation: A prospective clinical study of the first fifty patients. 7th Proceeding of the International Congress on Cotrel–Dubousset Instrumentation, 1990;75–86.
43. SOFAMOR. In vitro biomechanical evaluation report of the compact CD fixture in lumbo-sacral. Study done by the Biomechanical Laboratory of Paris-Pr Lavaste Department. 1990 (Unpublished data).
44. Devlin VJ, Oheneba BA, Bradford DS, Ogilvie, JW, Transfeldt EE. Treatment of adult spinal deformity with fusion to the sacrum using CD instrumentation. J Spinal Disord 1991;4(1):1–14.
45. Ogon M, Haid C, Krismer M, Sterzinger W, Bauer R. Comparison between single-screw and triangulated, double-screw fixation in anterior spine surgery. Spine 1996;21:2728–34.
46. Ruland CM, McAfee PC, Warden KE, Cunningham BW. Triangulation of pedicular instrumentation: A biomechanical analysis. Spine 1991;16(Suppl):270–6.
47. Zindrick MR, Wiltse LL, Widell EH et al. A biomechanical study of intrapeduncular screw fixation in the lumbosacral spine. Clin Orth 1986;203:99–112.
48. Carlson GD, Abitbol JJ, Anderson DR et al. Screw fixation in the human sacrum: an in vitro study of the biomechanics of fixation. Spine 1992;17(6S):S197–203.
49. Zdeblick TA, Kunz DN, Cooke ME. Pedicle screw pull-out strength: correlation with insertional torque. Spine 1993;18(12):1673–6.
50. Cunningham BW, Sefter JC, Shono Y, McAfee PC. Static and cyclical biomechanical analysis of pedicle screw spinal constructs. Spine 1993;18(12):1677–88.
51. Lu WW, Luk KDK, Ruan DK, Fei ZQ, Leong JCY. Stability of the whole lumbar spine after multilevel fenestration and discectomy. Spine 1999;24(13):1277–82.
52. Pfeiffer M, Hoffman H, Goel VK, Weinstein JN, Griss P. In vitro testing of a new transpedicular stabilization technique. Europ Spine J;1997,6(4):249–55.
53. Yamagata M, Kitahara I, Minami S et al. Mechanical stability of the pedicle screw fixation systems for the lumbar spine. Spine 1992;17(3 Suppl):S51–4.
54. Perra JH. Techniques of instrumentation in long fusion to the sacrum. Clin Orthop North Am 1994; 25: 287–99.
55. Shea M, Edwards WT, Clothiaux PL, Crowell RR, Nachemson AL, White AA et al. Three-dimensional load displacement properties of posterior lumbar fixation. J Orthop Trauma 1991;5:420–7.
56. McCalden RW, Mcgeough JA, Court-Brown CM. Age-related changes in the compressive strength of cancellous bone. J Bone Joint Surg 1997;79A:421–7.
57. McCord DH, Cunningham BW, Shono Y, Myers JJ, McAfee PC. Biomechanical analysis of lumbosacral fixation. Spine 1992;17:235–43.
58. Lu WW, Zhu QA, Holmes AD, Luk KDK, Zhong S, Leong JCY. Loosening of sacral screw fixation under in vitro fatigue loading. J Orthop Res 2000;18:808–14.
59. Kornblatt MD, Casey MP, Jacobs RR. Internal fixation in lumbosacral spine fusion: A biomechanical and clinical study. Clin Orthop 1986;203:141–50.
60. Lee CK, Langrana NA. Lumbosacral spinal fusion: a biomechanical study. Spine 1984;9:574–81.
61. Myers BS, Belmont PJ, Richardson WJ, Yu JR, Harper KD, Nightingale RW. The role of imaging and in situ biomechanical testing in assessing pedicle screw pull-out strength. Spine 1966;21(17):1962–8.

45 Intervertebral Disk Prosthesis

P. Tropiano and Th. Marnay

Introduction

Sir John Charnley revolutionized modern orthopedics with his development of total hip replacement [1]. Today, hip and knee arthroplasties are two of the most highly rated surgical procedures in terms of patient satisfaction. It is possible that the development of an artificial disk may impact the treatment of degenerative disk disease in a similar fashion. Although the challenges associated with developing a prosthetic disk are great, the potential to improve the lives of many individuals suffering from symptoms of spinal spondylosis is tremendous.

The idea of spinal disk replacement is not new. One of the first attempts to perform disk arthroplasty was undertaken by Nachemson 40 years ago [2]. Fernström attempted to reconstruct intervertebral disks by implanting stainless steel balls in the disk space [3]. In 1966 he published a report on 191 implanted prostheses in 125 patients. Subsidence occurred in 88% of patients over a 4–7-year period of follow-up. These pioneering efforts were followed by more than a decade of research on the degenerative processes of the spine, spinal biomechanics, and biomaterials before serious efforts to produce a prosthetic disk resumed.

Disk Degeneration

The intervertebral disk constitutes a major component of the functional spinal unit. Aging results in deterioration of the biological and mechanical integrity of the intervertebral disks. Disk degeneration may produce pain directly or perturb the functional spinal unit in such a way as to produce a number of painful entities. Whether through direct or indirect pathways, intervertebral disk degeneration is a leading cause of pain and disability in adults [4]. Approximately 80% of Americans experience at least a single episode of significant back pain in their lifetime, and for many individuals, spinal disorders become a lifelong malady. The morbidity associated with disk degeneration and its spectrum of associated spinal disorders is responsible for significant economic and social costs. The indirect economic losses associated with lost wages and decreased productivity are staggering.

Age-related disk changes occur early and are progressive. Almost all individuals experience diminished nuclear water content and increased collagen content by the fourth decade. This desiccation and fibrosis of the disk blurs the nuclear/annular boundary [5]. These senescent changes allow repeated minor rotational trauma to produce circumferential tears between annular layers. These defects, usually in the posterior or posterolateral portions of the annulus, may enlarge and combine to form one or more radial tears through which nuclear material may herniate [6]. Pain and dysfunction due to compression of neural structures by herniated disk fragments are widely recognized phenomena. It should be noted, however, that annular injuries may be responsible for axial pain with or without the presence of a frank disk herniation [7,8].

The spine degenerative process that ends in painful symptoms usually follows a pattern: as the nucleus pulposus loses its ability to retain water and loses disk height, the annulus buckles, permitting torsional delamination of the concentric plies and often radial tears of the annulus. The precursor to nucleus degeneration may be alterations in the permeability of the vertebral end-plate or its microvascularity. Furthermore, many byproducts of anaerobic metabolism normally are potentially toxic to aerobic structures, especially neural tissue. If an annular tear occurs, the accumulating anaerobic intradiskal by-products that normally are pumped out through the pores in the endplate may escape via the tear(s) and reach the polymodal free nerve endings in the outer annular layers, potentially causing severe back pain. These neurotoxins (such as phospholipase, metalloproteinases, or stromalysin) may even reach the spinal nerve dorsal ganglion, causing potentially irreversible dermatomal or myotomal damage with dysesthesias, trigeminal neuralgia (tic douloureux) effect on the nearby ganglion, reflex sympathetic dystrophy, or weakness. The intradiskal pH, normally slightly acid at about 7.2, with progressive degeneration may fall as low as pH 6.8 from the accumulation of lactic acid. With further nucleus disintegration the pH may fall to or below 6.2 at which level regeneration of the gel and repair of the inner annulus fibrosus stop. With severe degeneration, the weight of the body comes to rest on the outer annular ring, effectively removing the force against the disk center, lowering the intradiskal pressure, and allowing fibrous ingrowth and neovascularization to occur [4].

Progression of the degenerative process alters intradiskal pressures, causing a relative shift of axial load-bearing to the peripheral regions of the endplates and facets. This transfer of biomechanical loads appears to be associated with the development of both facet and ligament hypertrophy [9,10]. There is a direct relation between disk degeneration and osteophyte formation [10]. In particular, deterioration of the intervertebral disk leads to increased traction on the attachment of the outermost annular fibers, thereby predisposing to the growth of laterally situated osteophytes [11]. Disk degeneration also results in a significant shift of the instantaneous axis of rotation of the functional spinal unit. The exact long-term consequences of such a perturbation of spinal biomechanics are unknown, but it has been postulated that this change promotes abnormal loading of adjacent segments and an alteration in spinal balance.

Therapeutic Options

The current gold standard therapy for degenerative lumbar disk disease with diskogenic pain that has failed conservative management is surgical arthrodesis. Excellent short-term results for lumbar fusion in the setting of degenerative disk disease have been reported, with clinical success rates of approximately 80% [12–14]. A recent randomized controlled trial demonstrated that at short-term follow-up, lumbar fusion is superior to nonsurgical treatment for degenerative disk disease with chronic diskogenic pain [15]. However, lumbar fusion surgery is not always successful. In an extensive meta-analysis of the literature on lumbar fusion surgery, Turner et al. estimated that 32% of patients had an unsatisfactory outcome. The reported incidence of pseudarthrosis was 14%, and 9% of patients suffered from chronic iliac crest donor site pain. Furthermore, most postoperative protocols for lumbar fusion include bracing for a 3–6 month period, which can be a significant hardship for patients [16].

Furthermore, the results of lumbar fusion in long-term follow-up may be compromised by the development of junctional degeneration at unfused levels adjacent to the fusion. Theoretically, loss of motion at fused segments causes increased stress and motion at adjacent levels, leading to junctional degeneration. In addition, disruption in normal sagittal alignment such as kyphotic collapse of a degenerated disk contributes to the junctional degenerative process [17,18]. Junctional degeneration may take the form of disk degeneration, facet arthrosis, stenosis, or instability. Although the true etiology and incidence is unknown, junctional disk degeneration has been demonstrated adjacent to fused segments in multiple studies [19–21]. The theoretical rationale for the total disk replacement prosthesis as an alternative to arthrodesis is the avoidance of junctional degeneration by preservation of motion and normal sagittal alignment at the instrumented segment. In order to justify its use, a disk replacement prosthesis should demonstrate actual motion in vivo and maintain or improve the sagittal spinal alignment. Without preservation of some motion, a disk prosthesis becomes a costly arthrodesis equivalent. If the theoretical basis of the disk replacement prosthesis is valid, the presence of motion in a disk replacement prosthesis may show some protective effect against the development of junctional degeneration. Finally, a total disk replacement should maintain normal sagittal alignment at long-term follow-up because of the relationship between alterations in sagittal alignment of the lumbar spine and junctional degeneration and low back pain [22,23].

Challenges of Design and Implantation

There are a number of factors which must be considered in the design and implantation of an effective disk prosthesis. The device must maintain the proper intervertebral spacing, allow for motion, and provide stability. Natural disks also act as shock absorbers, and this may be an important quality to incorporate into prosthetic disk design, particularly when considered for multilevel lumbar reconstruction. The artificial disk must not shift significant axial load to the facets. Placement of the artificial disk must be done in such a way as to avoid the destruction of important spinal elements such as the facets and ligaments. The importance of these structures cannot be overemphasized. Facets not only contribute strength and stability to the spine, but they could be a source of pain. This may be especially important to determine prior to disk arthroplasty because it is currently believed that disk replacement will probably be ineffective as a treatment for facet pain. Excessive ligamentous laxity may adversely affect disk prosthesis outcome by predisposing to implant migration or spinal instability.

An artificial disk must exhibit tremendous endurance. The average age of a patient needing a lumbar disk replacement has been estimated to be 35 years. This means that to avoid the need for revision surgery, the prosthesis must last 50 years. It has been estimated that an individual will take two million strides per year and perform 125,000 significant bends; therefore, over the 50-year life expectancy of the artificial disk, there would be over 106 million cycles. This estimate discounts the subtle disk motion which may occur with the six million breaths taken per year [24]. A number of factors in addition to endurance must be considered when choosing the materials with which to construct an intervertebral disk prosthesis. The materials must be biocompatible and display no corrosion. They must not incite any significant inflammatory response. The fatigue strength must be high and the wear debris minimal. Finally, it would be ideal if the implant were imaging "friendly".

All currently proposed intervertebral disk prostheses are contained within the disk space; therefore, allowance must be made for variations in patient size, level, and height. There may be a need for instrumentation to restore collapsed disk space height prior to placement of the prosthesis.

The intervertebral disk prosthesis ideally would replicate normal range of motion in all planes. At the same time it must constrain motion. A disk prosthesis must reproduce physiologic stiffness in all planes of motion plus axial compression. Furthermore, it must accurately transmit physiologic stress. For example, if the global stiffness of a device is physiologic but a significant nonphysiologic mismatch is present at the bone–implant interface, there may be bone resorption, abnormal bone deposition, endplate or implant failure.

The disk prosthesis must have immediate and long-term fixation to bone. Immediate fixation may be accomplished with screws, staples, or "teeth" which are integral to the implant. While these techniques may offer long-term stability, other options include porous or macrotexture surfaces which allow for bone ingrowth. Regardless of how fixation is achieved, there must also be the capability for revision.

Finally, the implant must be designed and constructed such that failure of any individual component will not result in a catastrophic event. Furthermore, neural, vascular, and spinal structures must be protected and spinal stability maintained in the event of an accident or unexpected loading.

Current Prosthetic Devices

Many design types have been performed [25]: 1) low-friction sliding surfaces, 2) spring systems, 3) contained fluid-filled chambers, and 4) disks of rubber or others elastomers, but the majority have not been used clinically. An intervertebral disk prosthesis must be designed to restore disk space height, to restore motion segment flexibility, to generate harmonious motion into the spinal triple joint complex, to be a shock absorber, to prevent disk degeneration at adjacent segments, to reduce or eliminate pain from motion or from nerve compression, and to improve the patient's functional activities. It must be also designed to be biocompatible and durable for more than 40 years. Today, basically two main families of disk may be recognized. The disk nucleus replacement-based devices and the total disk replacement devices.

PDN: Prosthetic Disk Nucleus

Hydrogel disk replacements primarily have hydraulic properties. Hydrogel prostheses are used to replace the nucleus while retaining the annulus fibrosis. One potential advantage is that such a prosthesis may have the capability of percutaneous placement. The PDN implant is a nucleus replacement which consists of a hydrogel core constrained in a woven polyethylene jacket (Raymedica Inc., Bloomington, MN) (Figure 45.1) [4,26].

Ray was the first to indicate the potentiality of a prosthetic nucleus to restore nuclear function and to promote potential healing of the annulus fibrosis. The goal was to re-expand the disk space, tighten the annulus fibrosus, and to restore the functional stability to the spinal segments.

The pellet-shaped hydrogel core is compressed and dehydrated to minimize its size prior to placement. Upon implantation, the hydrogel immediately begins to absorb fluid and expand. The tightly woven ultra-high-molecular-weight polyethylene (UHMWPE) allows fluid to pass through to the hydrogel. This flexible but inelastic jacket permits the hydrogel core to deform and reform in response to changes in compressive forces yet constrains horizontal and vertical expansion upon hydration. Although most hydration takes place in the first 24 hours after implant, it takes approximately 4–5 days for the hydrogel to reach maximum expansion. Placement of two PDN implants within the disk space provides the lift that is necessary to restore and maintain disk space height. This device has been extensively assessed with mechanical and in vitro testing, and the results have been good [4,26]. Clinical trials were first conducted in 1996 and the device was found to be effective in most of the patients that were implanted. Additional trials in 1997–1998 were less successful, with 38% of patients requiring revision surgery because of device migration. Subsequent changes were made to the device shape and to the surgical protocol to facilitate implantation, thereby eliminating the high device-migration rates. Following these modifications, the success rate for the device has improved significantly. The PDN is undergoing clinical evaluation in Europe, South Africa, and the United States. Candidates for PDN would have degenerative disk disease man-

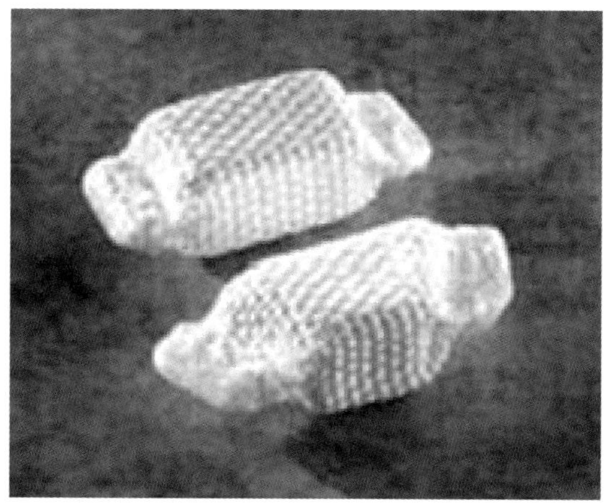

Figure 45.1. The Raymedica prosthetic disk nucleus (PDN) (Raymedica, Inc., Bloomington, MN).

ifested by morphologic changes (internal disk derangements, herniation) and clinical manifestation (back pain with or without leg pain) refractory to conservative therapy. Currently, patients with <5 mm residual disk height are excluded. The physical implantation of the disk is difficult under such circumstances. The PDN imparts stability to the motion segment by applying tension to the annulus. This requires a competent annulus; annular defects may exclude a potential candidate.

Acroflex Disk

Two elastic-type disk prostheses are the Acroflex prosthesis proposed by Steffee and the thermoplastic composite of Lee [25,27]. The first Acroflex disk consisted of a hexene-based polyolefin rubber core vulcanized to two titanium endplates. The endplates had 7 mm posts for immediate fixation and were coated with sintered 250 micron titanium beads on each surface to provide an increased surface area for bony ingrowth and adhesion of the rubber. Osteointegration occurs by means of a rough surface and by small spikes attached to the ventral third of the implant. Transmission of motion only functions properly if there is good osteointegration of the endplates. This prosthesis has largely "constrained" kinematic characteristics. The disk was manufactured in several sizes and underwent extensive fatigue testing prior to implantation. Only six patients were implanted before the clinical trial was stopped due to a report that 2-mercaptobenzothiazole, a chemical used in the vulcanization process of the rubber core, was possibly carcinogenic in rats [28]. The six patients were evaluated after a minimum of three years, at which time the results were graded as follows: two excellent, one good, one fair, and two poor [27]. One of the prostheses in a patient with a poor result developed a tear in the rubber at the junction of vulcanization. The second-generation Acroflex 100 consists of an HP-100 silicone elastomer core bonded to two titanium endplates (DePuy Acromed, Raynham, MA) (Figure 45.2).

Figure 45.2. Acroflex (DePuy Acromed, Raynham, MA).

Mechanical Models: Complete Disk Prostheses

Several articulating pivot or ball-type disk prostheses have been developed for the lumbar spine. Hedman and Kostuik developed a set of cobalt–chromium–molybdenum alloy hinged plates with an interposed spring [29]. These devices have been tested in sheep. At three and six months post-implantation there was no inflammatory reaction noted and none of the prostheses migrated. Two of the three 6-month implants had significant bony ingrowth. It is not clear whether motion was preserved across the operated segments [24].

The Prodisk Implant

The Prodisk intervertebral disk prosthesis (Aesculap AG & Co. KG., Tuttlingen, Germany) is a modular system allowing customization of the device to each patient's unique anatomic and physiologic requirements. It consists of two end plates of titanium alloy and a high-density polyethylene core. The superior end plates are manufactured in four sizes (small, medium, large, extra large), three heights (10, 12, 14 mm) and two geometric configurations: they are either plane-parallel or oblique for the sizes medium, large, and extra large, allowing an even more precise system for attempted surgical reconstruction of the lumbar lordosis curve.

The inferior end plates are manufactured in four sizes and constant obliquity and height. Each end plate is coated with titanium plasma in an attempt to insure improved bony integration and is surmounted by two keels with oblique direction and notches to provide primary stability. The high-density polyethylene core, available in only one height, is composed of an inferior cylindrical part, which fits into the cylindrical case in the inferior titanium plate receptacle, and a spherical part for articulation of the core with the cup in the inferior surface of the upper plate.

In term of kinematics, the Prodisk can be best described as "semiconstrained". This implant was inspired by the work of Fick [30], Lysell [31], and Gonon [32]. It continues the work of René Louis [33] and situates the flexion–extension–rotation center of the mobile spinal unit posteriorly and below with regard to the middle of the inferior end plate of disk space, with a concentric rotation of the posterior facet joints during this movement. Lateral bending on rotation movement occurs in combination with posterior spinal facet rotation. Axial rotation is relatively free, and not constrained, with an axis angled towards the back in the neutral position due to intra-diskal lordosis of the prosthetic element in the neutral position in an effort to be closest to the theoretical axis of rotation [33]. The rotation centers determined in this way are described by White and Panjabi [34].

The Prodisk prosthesis should provide a range of motion of 10° and 10° in flexion–extension and in lateral bending. Good-to-excellent results were reported in the majority of patients receiving this implant [35].

The first-generation Prodisk prosthesis was created in 1989 and implanted in 64 patients. It was an investigational device and its implantation required a long and difficult learning curve. But after a ten-year follow-up of the patients and confirmation of its effectiveness and safety design [35], the need for an implant usable by every surgeon appears essential for its diffusion. The new Prodisk® II (Spine Solution Inc.) implant was designed and launched on the European market in December 1999.

The Prodisk® II (Figure 45.3) was designed to be an evolution of the first generation. It consists of two endplates made of Cr–Co–Mo alloy coated with a titanium Plamapore surface to improve osteointegration. In this prosthesis a monoconvex polyethylene core is used to act as a shock absorber. It is restricted in its base to limit distortion and to avoid any flow. This core is inserted into the caudal endplate of the implant using a multifunctional instrument. Once the polyethylene core is firmly anchored to the caudal endplate, there are two movable parts. Particular importance was placed on minimally invasive insertion technique with easy-to-plan precision insertion and optimal primary anchorage. This is made possible in part by a central anchoring keel having teeth at the top and a trapezoidal shape in profile. The keel is tapped into a groove that has been precisely prepared surgically in the midline of the vertebra using a special chisel. The keel acts secondarily to increase the surface area for anchorage and osteointegration. As a result of the press-fit procedure and the surface teeth, the special profile of the keel increases primary stability. The two endplates of the prosthesis are

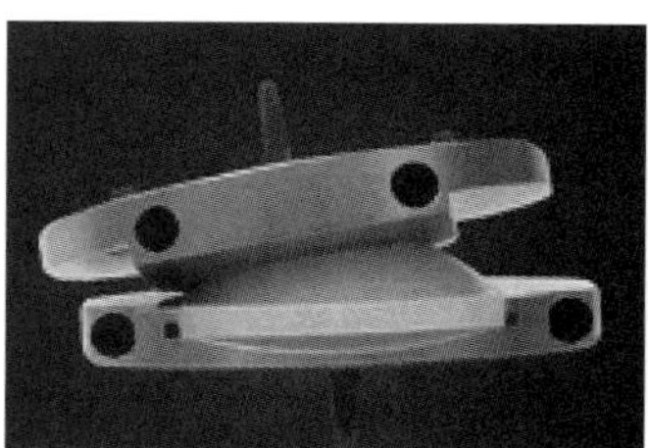
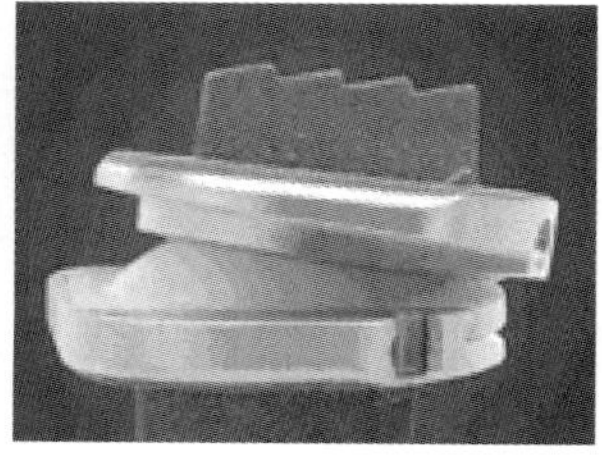
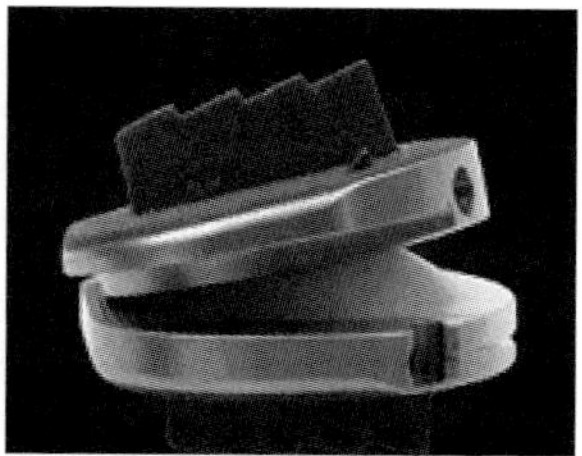

Figure 45.3. Prodisk II (Aesculap AG & Co. KG., Tuttlingen, Germany).

grasped via two anchorage holes on the inside, and the prosthesis is inserted properly into the prepared groove. This allows a high level of precision to be achieved when inserting the prosthesis. The Prodisk® II can be positioned more optimally even if space is limited due to anatomic conditions or if the course of the vasculature is unfavorable.

Prospective randomized studies to compare disk prosthesis to arthrodesis have started in Europe and in the United States comparing the ProDisk® to the "gold standard" of a 360° (front and back) fusion using allograft in the intervertebral space and pedicle screws with autograft posteriorly.

Figure 45.4. Link SB Charité III (Waldemar Link Gmbh and Co., Hamburg, Germany).

Link SB Charité Disk

The Link SB Charité intervertebral prosthesis was developed initially in East Germany in 1987. More than 2,000 patients worldwide have been treated thus far using all three generations of implants [36–38]. The Charité III consists of two endplates of cobalt-chromium alloy and a UHMWPE sliding core that enables near physiological intersegmental movement up to a lordotic inclination of 20°. The end plates are provided in five sizes and two basic configurations of plane as viewed laterally parallel and oblique. The UHMWPE cores are available in three thickness: 7.5, 9.5, and 11.5 mm. The SB Charité prosthesis has "unconstrained" kinematic characteristics (Figure 45.4).

The biomechanical comparison of the range of motion of the normal intervertebral disk vs. The SB Charité III prosthesis in vitro vs. The SB Charité III prosthesis in vivo was measured by Ahrens. The right and left axial rotation, forward flexion, backward extension, and left and right lateral bending were all comparable, but it was noted that the limitation of axial rotation for the artificial disk requires intact facet joints.

Although there is great concern regarding wear debris in hip prostheses in which UHMWPE articulates with metal, this does not appear to occur in the Charité III [39]. This prosthesis has been implanted in over a thousand European patients with relatively good results. In 1994 Griffith et al. reported the results in 93 patients with one-year follow-up [37]. Significant improvements in pain, walking distance, and mobility were noted. Of these patients, 6.5% experienced a device failure, dislocation, or migration. There were three ring deformations, and three patients required reoperation. Lemaire et al. described the results of implantation of the SB Charité III disk in 105 patients with a mean of 51 months of follow-up [38]. There was no displacement of any of the implants, but three settled. The failures were felt to be secondary to facet pain. David described a cohort of 85 patients reviewed after a minimum of at least 5 years post-implantation of the Charité prosthesis [36]. Of these, 97% were available for follow-up, 68% had good or better results, and 14 patients reported the result as poor. Eleven of these patients underwent secondary arthrodesis at the prosthesis level. Despite the concern of many other investigators, it is interesting to note that David treated 20 patients with spondylolisthesis or retrolisthesis with an outcome identical to that of the entire group. Clinical trials using the Charité III prosthesis are ongoing in Europe, the United States, Argentina, China, Korea, and Australia.

Conclusion

Spinal disk replacement is not only possible but is an exciting area of clinical investigation which has the potential of revolutionizing the treatment of spinal degeneration. The development of a prosthetic disk poses tremendous challenges, but the results from initial efforts have been promising. The future for this field, and our patients, is bright.

Bibliography

Bednar JM, Friedenberg ZB, Turner ML. Bipolar femoral endoprosthesis: a study correlating component movement with clinical outcome. J Trauma 1988;28(5):664–8.

Bochner RM, Pellicci PM, Lyden JP. Bipolar hemiarthroplasty for fracture of the femoral neck. Clinical review with special emphasis on prosthetic motion. J Bone Joint Surg Am 1988;70(7):1001–10.

Drinker H, Murray WR. The universal proximal femoral endoprosthesis. A short-term comparison with conventional hemiarthroplasty. J Bone Joint Surg Am 1979; 61(8):1167–74.

Eiskjaer S, Gelineck J, Soballe K. Fractures of the femoral neck treated with cemented bipolar hemiarthroplasty. Orthopedics 1989;12(12):1545–50.

Enker P, Steffee AD. Total disk replacement. In: Bridwell KH, DeWald RL, editors. The Textbook of Spinal Surgery. 2nd ed. Philadelphia: Lippincott-Raven, 1997:2275–88.

Langan P. The Giliberty bipolar prosthesis: a clinical and radiographical review. Clin Orthop 1979;(141):169–75.

Ray CD. The Raymedica prosthetic disk nucleus: an update. In: Keach DL, Jinkins JR, editors. Spinal Restabilisation Procedures. Amsterdam: Elsevier, 2002;273–82.

Vaccaro AR, Silber JS. Post-traumatic spinal deformity. Spine 2001;15;26(24 Suppl):S111–18.

Weinstein PR. Anatomy of the lumbar spine. In: Hardy RW, editor. Lumbar Disk Disease. New York: Raven Press, 1982:5–15.

References

1. Charnley J. Total hip replacement. JAMA 1974;230: 1025–8.
2. Nachemson AL. Challenge of the artificial disk. In: Weinstein JN, editor. Clinical efficacy and outcome in the diagnosis and treatment of low back pain. New York: Raven Press, 1992.
3. Fernstrom U. Arthroplasty with intercorporal endoprothesis in herniated disk and in painful disk. Acta Chir Scand (Suppl) 1966;357:154–9.
4. Rothman RH, Simeone FA, Bernini PM. Lumbar disk disease. In: Rothman RH, Simeone FA, editors. The Spine. 2nd ed. Philadelphia: WB Saunders, 1982: 508–645.
5. Pearce RH, Grimmer BJ, Adams ME. Degeneration and the chemical composition of the human lumbar intervertebral disk. J Orthop Res 1987;5:198–205.
6. Kirkaldy-Willis WH, Wedge JH, Yong-Hing K, Reilly J. Pathology and pathogenesis of spondylosis and stenosis. Spine 1978;3:319–28.
7. Crock HV. Internal disk disruption: a challenge to disk prolapse 50 years on. Spine 1986;11:650–3.
8. Kääpä E, Holm S, Han X, Takala T, Kovanen V, Vanharanta H. Collagens in the injured porcine intervertebral disk. J Orthop Res 1994;12:93–102.
9. Keller TS, Hansson TH, Abram AC, Spengler DM, Panjabi MM. Regional variations in the compressive properties of lumbar vertebral trabeculae. Effects of disk degeneration. Spine 1989;14:1012–19.
10. Vernon-Roberts B, Pirie CJ. Degenerative changes in the intervertebral disks of the lumbar spine and their sequelae. Rheumatol Rehab 1977;16:13–21.
11. Macnab I. The traction spur: an indicator of segmental instability. J Bone Joint Surg 1971;53A:663–70.
12. Kozak JA, O'Brien JP. Simultaneous combined anterior and posterior fusion. An independent analysis of a treatment for the disabled low-back pain patient. Spine 1990;15(4):322–8.
13. Linson MA, Williams H. Anterior and combined anteroposterior fusion for lumbar disk pain. A preliminary study. Spine 1991;16(2):143–5.
14. Moore KR, Pinto MR, Butler LM. Degenerative disk disease treated with combined anterior and posterior arthrodesis and posterior instrumentation. Spine 2002; 27(15):1680–6.
15. Fritzell P, Hagg O, Wessberg P, Nordwall A. 2001 Volvo Award Winner in Clinical Studies: Lumbar fusion versus nonsurgical treatment for chronic low back pain: a multicenter randomized controlled trial from the Swedish Lumbar Spine Study Group. Spine 2001;26(23):2521–32.
16. Schönmayr R, Busch C, Lotz C, Lotz-Metz G. Prosthetic disk nucleus implants: the Wiesbaden feasibility study. Two-years follow-up in ten patients. Riv Neuroradiol 1999;12(Suppl 1):163–70.
17. Katsuura A, Hukuda S, Saruhashi Y, Mori K. Kyphotic malalignment after anterior cervical fusion is one of the factors promoting the degenerative process in adjacent intervertebral levels. Eur Spine J 2001;10(4):320–4.
18. Kumar MN, Baklanov A, Chopin D. Correlation between sagittal plane changes and adjacent segment degeneration following lumbar spine fusion. Eur Spine J 2001; 10(4):314–19.
19. Balderston RA, Albert TJ, McIntosh T, Wong L, Dolinskas C. Magnetic resonance imaging analysis of lumbar disk changes below scoliosis fusions. A prospective study. Spine 1998;23(1):54–8.
20. Danielsson AJ, Cederlund CG, Ekholm S, Nachemson AL. The prevalence of disk aging and back pain after fusion extending into the lower lumbar spine. A matched MR study twenty-five years after surgery for adolescent idiopathic scoliosis. Acta Radiol 2001; 42(2):187–97.
21. Kumar MN, Jacquot F, Hall H. Long-term follow-up of functional outcomes and radiographic changes at adjacent levels following lumbar spine fusion for degenerative disk disease. Eur Spine J 2001;10(4):309–13.
22. La Grone MO. Loss of lumbar lordosis. A complication of spinal fusion for scoliosis. Orthop Clin North Am 1988;19(2):383–93.
23. Turner JA, Ersek M, Herron L, Haselkorn J, Kent D, Ciol MA, et al. Patient outcomes after lumbar spinal fusions. JAMA 1992;268(7):907–11.
24. Kostuik JP. Intervertebral disk replacement. In: Bridwell KH, DeWald RL, editors. The Textbook of Spinal Surgery. 2nd ed. Philadelphia: Lippincott-Raven, 1997: 2257–66.

25. Lee CK, Langrana NA, Parsons JR, Zimmerman MC. Development of a prosthetic intervertebral disk. Spine 1991;16(Suppl 6):S253–5.
26. Ray CD, Schönmayr R, Kavanagh SA, Assell R. Prosthetic disk nucleus implants. Riv Neuroradiol 1999;12(Suppl 1):157–62.
27. Enker P, Steffee A, Mcmillan C, Keppler L, Biscup R, Miller S. Artificial disk replacement. Preliminary report with a 3-year minimum follow-up. Spine 1993;18: 1061–70.
28. Deiter MP. Toxicology and carcinogenesis studies of 2-mercaptobenzothiazole in F344/n rats and B6C3F mice. NIH Pub. No. 88-8, National Toxicology Program, Technical Report Series No. 322. Washington DC: US Department of Health and Human Services, 1988.
29. Hedman TP, Kostuik JP, Fernie GR, Hellier WG. Design of an intervertebral disk prosthesis. Spine 1991;16(Suppl 6):S256–60.
30. Fick R. Handbuch der Anatomie und Mechanik der Gelenke unter Berücksichtigung der bewegenden Muskeln, 1904–1911, Vol 3, Spezielle Gelenk- und Muskelmechanik. Gustav Fischer, Jena, 1911.
31. Lysell E. Motion in cervical spine. Acta Orthop Scand 1969;123:7.
32. Gonon GP, Dimnet J, Carret JP, Courcelles P, Fischer LP, De Mourgues G. Motion picture study of the lumbar spine and lateral inflexion in normal subjects and in patients. Rev Chir Orthop Reparatrice Appar Mot 1978;64(Suppl 2):101–5.
33. Louis R. Chirurgie du Rachis: Anatomie Chirurgicale et Voies d'Abord. Berlin: Springer, 1982.
34. White AA, Panjabi MM. Clinical Biomechanics of the Spine. Philadelphia: Lippincott, 1978;534 pp.
35. Marnay T. The Prodisk: clinical analysis of an intervertebral disk implant. In: Keach DL, Jinkins JR, editors. Spinal Restabilisation Procedures. Amsterdam: Elsevier, 2002;273–82.
36. David TH. Lumbar disk prosthesis: a study of 85 patients reviewed after a minimum follow-up period of five years. Rachis Revue de Pathologie Vertebrale 1999; 11(4–5).
37. Griffith SL, Shelokov AP, Büttner-Janz K, LeMaire J-P, Zeegers WS. A multicenter retrospective study of the clinical results of the LINK® SB Charité intervertebral prosthesis. The initial European experience. Spine 1994;19:1842–9.
38. Lemaire JP, Skalli W, Lavaste F, et al. Intervertebral disk prosthesis. Results and prospects for the year 2000. Clin Orthop 1997;337:64–76.
39. Link HD. LINK SB Charité III intervertebral dynamic disk spacer. Rachis Revue de Pathologie Vertebrale 1999;11.

VI – Articular Biomechanics

VI C – Lower Limb

46 Optimized Treatment of Hip Fractures

K.-G. Thorngren

The Future Problem

Fractures of the proximal part of the femur – hip fractures – are common and costly. The number of hip fractures has increased in all Western countries during recent decades. This has occurred mainly because of an increase in the number of elderly people and also due to an increase in the risk for hip fracture among the oldest persons [1–4,35]. Due to an increased aging population all over the world there will be a geographical shift in the occurrence of hip fractures. The incidence rates of hip fractures are higher in white populations than in others and vary by geographical region. Age-adjusted incidence rates of hip fracture by gender are higher in Scandinavia than in North America and lower in countries of Southern Europe [5,6]. The absolute number of hip fractures in each region is determined not only by ethnic composition, but also by the size of the population and its age distribution. In 1990 one third of all hip fractures in the world occurred in Asia despite lower incidence rates among Asians. Almost half of the fractures occurred in Europe, North America, and Oceania. These populations are smaller but older. It was estimated in the beginning of the 1990s that 323 million people aged 65 years and over were living around the world. This has been estimated to increase to 1,555 million by the year 2050 [7]. The increase will be especially high in Africa, Asia, South America, and the eastern Mediterranean regions. In the USA demographic changes alone will more than double the number of hip fractures from 238,000 1986 to 512,000 in the year 2040 [7]. In another publication [8] the 340,000 hip fractures around the year 2000 will increase to 650,000 in the year 2050. It has been calculated that the now close to two million hip fractures in the world could rise to over 6 million in the year 2050. Of these 71% is calculated to be in Africa, Asia, South America, or the eastern Mediterranean regions [9].

Already today hip fractures are highly resource consuming and strenuous for the organization of medical care. Optimized methods for operation and rehabilitation along with preventive measures are necessary to cope with this increasing problem, otherwise it can become overwhelming.

Fracture Types

Hip fractures consist of different types in the proximal femur. It is very important whether the fracture is located in the femoral neck (cervical fracture, intracapsular) or through the parts of the proximal femur which constitute muscle insertions (trochanteric fractures, extracapsular) because both the treatment and the course of healing are different [28]. Cervical fractures are best classified into undisplaced (Garden I and II) or displaced (Garden III and IV) [11]. Other sub-groupings have proven difficult to reproduce [12]. The trochanteric fractures are for routine use best classified into two-fragmented fractures (stable) or multi-fragment fractures (unstable). The basocervical fractures are a transition form between cervical and trochanteric fractures. They are usually treated as trochanteric fractures, but can in some cases have healing complications similar to the cervical ones. Sub-trochanteric fractures are more comminuted and include the area down to five centimeters below the trochanter minor.

The blood supply to the femoral head is often damaged after cervical fractures, because the vessels either penetrate within the marrow cavity or are positioned sub-periosteally on the femoral neck. Varying degrees of vascular damage caused at the moment of fracture will give varying amounts of healing complications. Extracapsular trochanteric fractures have good vascular supply and few healing complications. Some of them are, however, very shattered with stability problems. Different systems for a more detailed classification of the fractures exist, but these are best suited for specialized research projects, as the reproducibility has been a major problem. In the Swedish national registration of hip fractures (RIKSHÖFT/SAHFE) the following fracture types were registered based on 50,000 cases:

Type I	Undisplaced cervical fractures	17.2%
Type II	Displaced cervical fractures	37.0%
Type III	Baso-cervical fractures	3.1%
Type IV	Trochanteric two-part fractures	23.1%
Type V	Trochanteric multi-fragment fractures	14.6%
Type VI	Sub-trochanteric fractures	5.0%

The diagnosis of a hip fracture is made by ordinary X-ray. From these pictures the fracture type is also classified, and they also give information about circumstances that can influence the choice of operation method, i.e., earlier performed operations. It can also disclose a pelvic fracture, which is a common differential diagnosis for pain from the hip area in elderly patients after a fall. All patients with pain from the hip after a fall, who have a normal, ordinary X-ray, should be further diagnosed with MRI. This can usually disclose undisplaced hip fractures with the risk of displacement and potential functional problems. It is also good for diagnosing undisplaced pelvic fractures, which are not uncommon in the pelvic rami in these age groups. If there is no access to MRI, a CT can also disclose fractures. Scintigraphy performed after 1–2 days can confirm the fracture suspicion if positive with a localized high uptake. If there is a lack of all these facilities, mobilization with weight bearing under supervision is a possibility with repeated X-ray check-ups, but it is a rather costly way of treatment as the patient usually has to be put into a hospital ward. MRI has proven particularly valuable for acute diagnosis of those undisplaced fractures, which are not possible to see on ordinary X-rays. On the STIR sequence an increased signal in the bone marrow is seen and on T1-weighted pictures the fracture edema is seen as a dark line against a light background of trabecular bone marrow.

Early in the pre-operative course increased attention should be given to pain relief for the patient, prevention of pressure sores, and early handling for rapid operation. The treatment should aim at operating as soon as possible, immediate mobilization on the next day with as much weight bearing as can be tolerated as regards pain, but no limitations in weight bearing due to fear of instability in the osteosynthesis. Only in certain very comminuted pertrochanteric or subtrochanteric fractures should non-weight bearing be recommended. In other cases, particularly femoral neck fractures, early weight bearing is a test of the stability of the osteosynthesis and a failure can then be rapidly followed with a re-osteosynthesis or with a hip arthroplasty.

For the fracture types listed above the two major controversial areas are displaced femoral neck fractures and trochanteric multifragment fractures in combination with subtrochanteric fractures. Cervical displaced fractures are a combined biological and biomechanical problem due to the influence of the blood supply on healing whereas the trochanteric/subtrochanteric fractures are predominantly a biomechanical stability problem due to the good vascularization of the bone fragments. There are different philosophies for the treatment of these different fracture types, which will be further discussed below. It is possible to determine the circulation to the femoral head with high accuracy with the use of scintimetry, but this is resource consuming and also tends to delay the operation. MRI, probably with contrast, will possibly in the future become available as a routine tool for the choice of operation method for femoral neck fractures, but these techniques are not yet developed.

Cervical Fractures

The blood supply to the femoral head after a cervical fracture has decisive importance on healing. Healing complications after cervical fracture consist either of re-dislocation (early change of position), pseudarthrosis (non-healing), or segmental collapse (femoral head necrosis after a healed fracture) [10,13]. A segmental collapse is thus rebuilding of the femoral head after vascular damage and needs a healed fracture for the vessels to grow in. At present there is no practical useful method to determine the blood circulation preoperatively. The degree of dislocation of the fracture on an ordinary X-ray picture is not prognostically sufficiently accurate. Preoperative scintimetry is resource consuming, depends on the positioning of the leg, and delays the operation. The goal for the future and a very important area for research is to be able to prognosticate the healing complications preoperatively and, based on that, to choose the primary method of operation. Patients with a good blood supply to the femoral head should then get a primary osteosynthesis and those with a clearly bad circulation, a primary arthroplasty. When waiting for this diagnostic possibility the choice of operating method will be dependent on the grade of dislocation seen on the X-ray combined with the age of the patient, the patient's other medical conditions, and their functional level pre-fracture.

Undisplaced Cervical Fractures

These have little or no displacement of the fracture and usually very little risk for vascular damage to the femoral head and thereby few healing complications. This group of cervical fractures contains the Garden groups I and II [11]. Primary operation with osteosynthesis is advocated all over the world. The methods most often used are either two or more parallel screws or two hook pins. The screws mostly differ by modifications of the configuration of the threading in the top part [10,14]. A few centers have tried not operating on some undisplaced fractures [15], but this leads to increased risks of dislocation and thereby a prognostic deterioration for healing. Non-operative treatment also demands non-weight bearing and increased check-ups, both clinically and with radiography. It is a much safer method to operate on the fracture and allow the patient full immediate weight bearing [16].

Displaced Cervical Fractures

These fractures have been an area of continuous disagreement for the last half century, but gradually agreement is now being reached. There is a geographical difference internationally concerning the treatment principles for displaced cervical fractures. In Scandinavia, particularly in Sweden and Norway, primary osteosynthesis has been performed in all cases of displaced cervical fracture. The basic philosophy has been to perform a small, quick, and for the patient less-burdening operation first and, in case of a healing complication later, as a secondary procedure do a well-planned arthroplasty. This is usually then performed as a total hip arthroplasty where both the femoral and the acetabular parts are exchanged. It is an undisputed fact that the best long-term result after a femoral neck fracture is a healed fracture and preservation of the patient's own femoral head provided no segmental collapse appears. This will give no future problems. An arthroplasty always has the risk of dislocation in the short term and in the long run the risk of loosening, and for hemi-arthroplasties also by deterioration of the acetabular cartilage over the course of several years. When a patient has been fitted with an osteosynthesis and two years have passed since the fracture with no healing complications, there is little risk of further problems from this hip [10,17]. Some patients, however, never regain the full functional level that they had before the fracture. The complications after an arthroplasty increase after 5–10 years and this risk has to be balanced against the expected remaining lifetime of the patient [18]. Therefore, arthroplasty is used mainly in elderly patients with clearly displaced fractures.

In many Western countries the primary choice for a displaced femoral neck fracture is to perform an arthroplasty. The basic principle has been to treat all patients with arthroplasty to avoid healing complications. This treatment philosophy has now been modified and an increasing amount of primary osteosyntheses is performed especially in relatively younger patients and those with less-severe dislocation. Many studies have shown somewhat increased mortality after primary arthroplasty compared to primary osteosynthesis [10,18,19]. At the same time, studies have shown a higher need of re-operation after primary osteosynthesis within the first two years after the fracture compared to after a primary arthroplasty. The complications after a primary arthroplasty develop later and also a re-arthroplasty is a more complex operation and has more inherent complications than a secondary arthroplasty after a failed primary osteosynthesis [14].

The international literature shows that healing problems due to vascular damage of the femoral head by displaced cervical fracture leads to non-union in 10–30% of cases and segmental collapse in a further 10–20% of cases. With an optimized osteosynthesis technique, the healing complications (both non-union and segmental collapse) have been limited to a total of 20–25% for displaced cervical fractures [17,20].

Primary arthroplasty results in dislocation in around 4% of cases following hemi-arthroplasty and in 10% following total hip arthroplasty. Infection consists of 2–5%. Following a hemiarthroplasty, around 20% of cases develop wear and deterioration of the acetabular cartilage in the long term. Loosening is expected in around 10% of cases. Fracture in connection with the arthroplasty amounts to 2–4%. Re-operation with arthroplasty after a primary osteosynthesis has been reported in 20–30% of displaced cervical fractures. A major re-operation within the first years after a primary arthroplasty is expected to be needed in around 10% of cases. These are then rather complex operations [10,14,18].

Unipolar hemiarthroplasty or a total hip replacement give better functional results within the first two years than a primary osteosynthesis. A total hip arthroplasty or a bipolar hemiarthroplasty probably gives better functional results after two years than a unipolar hemiarthroplasty. Cemented stems give better outcomes than uncemented. An uncemented cup is not recommended for osteoporotic patients [10].

Presently, many randomized studies are ongoing both in Sweden and abroad to improve the criteria for the choice between a primary osteosynthesis and a primary arthroplasty [21–24]. Most of these studies have shown a relatively high number of complications for osteosynthesis when compared with previously published consecutive series during the last decades [17,20]). A differentiated treatment protocol results in fewer re-operations [13,25–27].

Based on the preliminary results of these randomized studies the treatment policy in Sweden has changed during the last years so the most displaced fractures in elderly patients now in an increasing number receive a primary arthroplasty. A primary arthroplasty is advocated if the cervical fracture is clearly displaced with lack of continuity both on the frontal and the side view, particularly in patients with high degree of osteoporosis. Also, the patient should have been walking prior to the fracture. The age should be above 70–75 years, where biological age is more important than chronological. Irrespective of the patient's age, primary arthroplasty is recommended in cases with disease to the hip joint such as rheumatoid arthritis or a pathological fracture secondary to malignancy or other destruction of the hip joint – for example, after infection. Also, a lately diagnosed fracture is indicated for arthroplasty, particularly if the scintimetry has shown a low uptake. Arthrosis in the fractured hip joint is also an indication for primary arthroplasty. Primary arthroplasty is, however, not recommended for patients with severe dementia, bedridden patients, or patients with bad muscular function due to the increased risk of dislocation.

The tendency internationally now aims at a differentiated treatment protocol according to the principles given above. In waiting for better diagnostic possibilities of circulation to the femoral head, the principles indicated will probably result in half of the displaced cervical fractures being operated with primary osteosynthesis and the other half with a primary arthroplasty, then preferably with cemented stem and maybe bipolar.

Timing of Operation

Hip fracture patients should be operated on as soon as is practically possible. Directly life-threatening conditions must, of course, have priority over hip fractures, but these elderly patients will have a prolonged rehabilitation and functionally less optimal results if the time between arrival in hospital and the operation is unnecessarily delayed. This in turn leads to more complications and inactivity in these elderly persons. It can also generate increased nursing needs with adverse economic consequences. The goal is to operate on the patient on the day of arrival or, at the latest, within 24 hours. If the patient is operated on with osteosynthesis within six hours of fracture, it has been shown that the risk for blood circulation disturbance to the femoral head and resulting healing complications diminish [28].

Apart from the problems associated with pain and immobilization, a delay of the operation is associated with increased morbidity and mortality. A delay of more than 24 hours between arrival in hospital and osteosynthesis of the fracture has been shown to be associated with increased mortality. Lower mortality has been shown when the operation was performed within 12 hours. If a delay is unavoidable the time should be used to improve the general condition of the patient, particularly the fluid balance [14,18].

Practical Considerations at Operation

Osteosynthesis for cervical fractures is performed with the use of a fracture table allowing traction under the image intensifier. A biplanar image intensifier is preferably used. It is wise to supervise the transferring of the patient to the fracture table, as the injured leg has to be treated with great care to prevent fracture displacement occurring or damage to the retinacular vessels. Manual traction on the leg for straightening during transfer is advisable. Also, to reduce the risk of pressure sores, padding should be applied to any area of pressure such as around the feet, sacrum, and groin. The uninjured leg should be flexed and abducted as much as possible. Positioning of the image intensifier is easier if the hip and knee are flexed to 90 degrees on the uninjured side (Figure 46.1). A displaced cervical fracture is reduced by longitudinal traction followed by inward rotation. A biplanar image intensifier has the advantage that after positioning of the equipment no further movements of the stand or tube are necessary, which avoids jeopardizing the draping and thereby the sterility. Shifting between the views is done on the monitor with a foot pedal, which considerably saves operation time. Also, the easy, rapid shifting between the positions increases the precision in the positioning of the osteosynthesis material. The importance for the circulation to the femoral head of a low traumatic operating technique has been proven [29]). The channel should be predrilled and hammering in of the osteosynthesis material avoided. Also, impaction of the fracture by hammering increases the damage of the circulation to the femoral head. The best way to achieve compression in the fracture is by the patient's own muscle forces at weight bearing. For undisplaced fractures early surgery will allow aspiration of any hematoma within the joint capsule. This may reduce the risk of avascular necrosis caused by ischemia from a tamponade effect on the intracapsular vessels. Cervical fractures are operated with parallel pins or screws to allow axial compression along the axis of the femoral neck perpendicular to the fracture line when the patient is weight bearing. This is a physiological way of compressing the fracture. To prevent slipping out of the osteosynthesis material they are either threaded as screws or have a hook that can be pressed out through a central canal. To facilitate parallel positioning most devices are

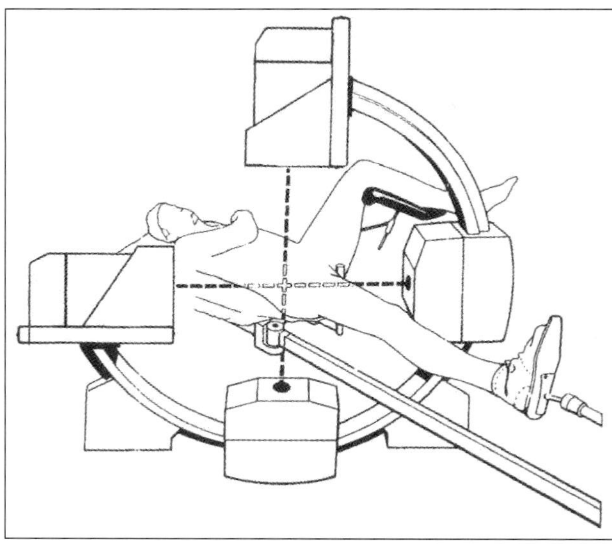

Figure 46.1. Positioning of biplanar image intensifier. Patient supine on traction table.

cannulated and have instruments to enable parallel placement. The most commonly used methods of fixation are two or three parallel cancellous screws, two parallel hook pins, or a dynamic hip screw. The blood circulation to the femoral head via the capsule vessels along the femoral neck is vulnerable. Sudden, forceful movements of the hip during reduction or excessive traction causing fracture diastasis may damage the femoral head circulation. The fracture is usually reduced by applying traction to the outstretched leg, followed by internal rotation. These maneuvers should be checked throughout the procedure in both the lateral and the anterior–posterior radiograph using the image intensifier which should have a large field of view and a good resolution facility. The reduction maneuver is begun by using the fracture table to apply gentle traction to the leg progressively while checking the AP radiograph. Traction is applied until the medial parts of the femoral neck, the calcar region, are approximated with anatomical contact between the bone ends. Next, the lateral view is obtained and the foot is rotated inwards until the dorsal angulation of the femoral neck fracture has been counteracted. This part of the maneuver can be likened to closing an open book. The aim is to restore the alignment of the femoral neck such that a straight line can be drawn to bisect the femoral head, trochanteric region, and shaft. It is essential that there is no residual fracture angulation, as this will increase the risk of re-displacement of the fracture. Quite frequently there is need to apply more than 90 degrees of inward rotation to achieve reduction. Small corrections with ab-adduction and sometimes elevation of the leg may also be needed to obtain an anatomical reduction. Following the reduction maneuver it is advisable to slacken the traction somewhat. This allows some impaction to occur at the fracture site and reduces the risk of the femoral head rotating during drilling.

Open reduction is very seldom indicated. Only in very young patients can it be tried and then combined with insertion of a pedicle graft consisting of a piece of bone with a muscle bridge which is implanted into the fracture site. In all middle-aged and older patients the alternative is rather a total hip arthroplasty if there is inability to obtain an adequate closed reduction. This is also advisable if the fracture is more than one week old or if there is early re-displacement following internal fixation.

Positioning of Two Hook Pins or Screws

Commonly used screws are the Garden screws, Asnis screws, Uppsala screws, and AO screws. In Sweden and Norway the Hansson hook pins are widely used. The screws and pins usually are about 7 mm in diameter and inserted parallel to each other. The aim is to create a three-point fixation, where the first point is the entry hole in the firm lateral cortical bone, the second point is the pin lying on the calcar inferiorly or posteriorly within the medullary cavity of the neck, and the third at the subcondral bone plate (Figure 46.2). The lateral skin incision should be extended distally from a point about 2 cm distal to the greater trochanter for a length of about 5 cm. The exact positioning of the incision is best located using a guide wire or other radio-opaque object on the skin surface and screening with an image intensifier in the AP view. After skin incision a guide wire is introduced. The inferior pin should be inserted through a drill hole at the level of the middle/lower part of the lesser trochanter. If the drill hole is situated distal to this point there is increased risk of fracture of the femur through the distal hole. The distal pin should rest along the calcar femorale and go up into the femoral head until 2–3 mm from the joint line (Figure 46.3 and 46.4). While introducing the Kirschner wire the position is repeatedly checked with an image intensifier in the AP and lateral planes. On the lateral view the guide wire should appear within the center of the femoral head and neck. The second, proximal guide wire is then placed in a position parallel to the first one. It should be spread as far apart as possible from the lower one in the femoral neck. When three screws are used, for example of the Asnis type [30], a triangular pattern is recommended for undisplaced or impacted intracapsular fractures. For displaced fractures, four screws in a diamond pattern is suggested. This is said to give better rotatory stability. Impaction along the femoral neck

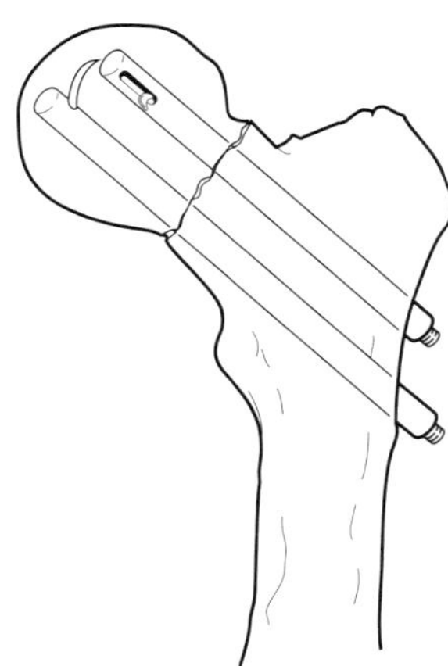

Figure 46.2. Positioning of osteosynthesis material for hook pin osteosynthesis after anatomical position of femoral neck fracture. Anterior/posterior view drawn as seen in image intensifier.

Figure 46.3. Hook pin osteosynthesis. Lateral view as seen in image intensifier.

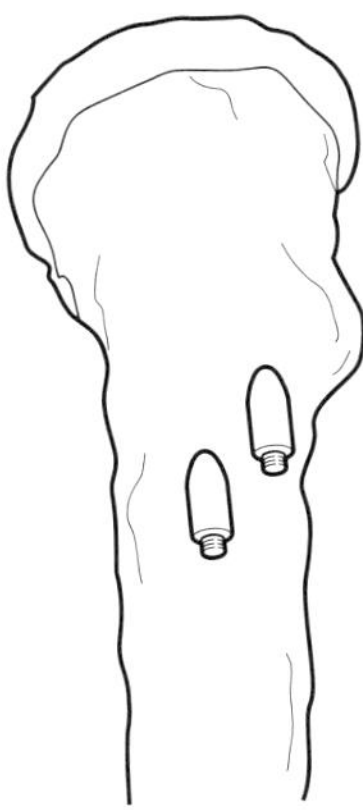

Figure 46.4. Hook pin osteosynthesis. Lateral view as seen at the operation.

combined with a minor rotation gives somewhat angulation of the osteosynthesis material, but still allows further impaction when two pins are used (Figure 46.5 and 46.6).

Considerations at Arthroplasty

The hip fracture patient is usually a woman with osteoporosis and short stature. Extra care should be taken not to cause perforation of the acetabulum by reaming for a total hip arthroplasty or by causing a femoral shaft fracture. Smaller sizes of arthroplasty are usually needed. Postoperative direct weight bearing should be allowed and postoperative restriction should be kept to a minimum. Capsulectomy should be avoided to prevent postoperative dislocation and a posterolateral exposure is usually favored due to limited tissue dissection needed, which gives shorter operation times and less blood loss. The abductors are not damaged by this approach and there is a lower risk of femoral penetration. This exposure has been said to have a somewhat higher risk of dislocation and the sciatic nerve must be carefully watched to prevent damage. The anterolateral approach is possible, with a lower risk of dislocation, but has the disadvantage of a greater amount of tissue dissection and a more restricted access for positioning of straight, long-stem arthroplasties. If a hemiarthroplasty is to be used special care is necessary to avoid damage to the acetabular cartilage. Forceful movements and hammering should be avoided and the femur is preferably prepared only by hand-held reamers. Cementation of the femoral shaft gives better results than the uncemented classical Austin–Moore prosthesis. There is concern about less tolerance of these elderly patients to the cementation procedure so pulse and blood pressure should preferably be monitored and excessive pressure should be avoided even if modern cementing techniques are recommended. There is a risk of cardiac arrhythmia and low blood pressure during the insertion of the cement. To prevent this, a venting catheter to allow air to escape from the femur during cementation has been tried, together with intravenous cortisone. There are no regular studies to approve this.

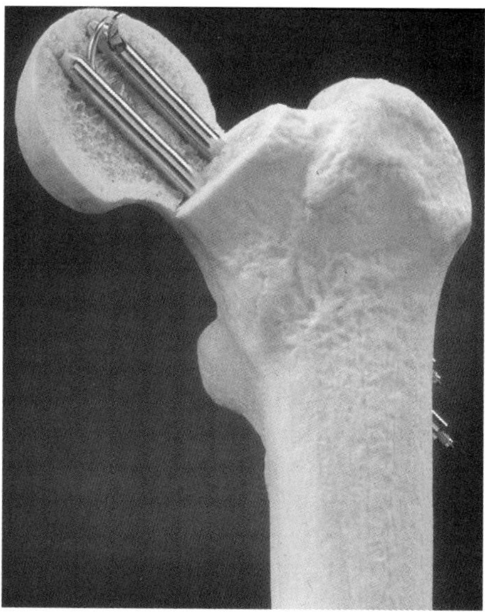

Figure 46.5. Hook pin osteosynthesis.

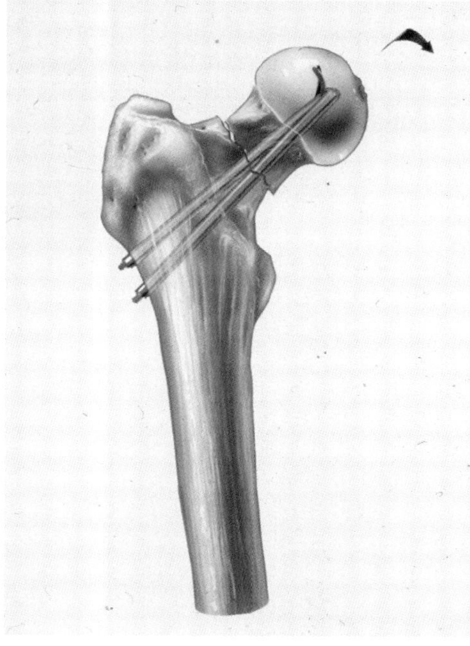

Figure 46.6. Drawing of hook pin osteosynthesis after weight bearing. An axial somewhat rotational compression often occurs resulting in physiological impaction of the fracture and some angulation of the pins still allowing further axial compression.

Depending on the patient's biological age and activity level before the fracture, different types of arthroplasty are chosen. Usually an ordinary, one-block hemiarthroplasty is chosen for the oldest and most disabled patients, whereas a bipolar hemiarthroplasty is used for somewhat younger and fitter patients, and a total hip arthroplasty is given to the youngest and healthiest patients [31].

Trochanteric Fractures

In trochanteric fractures the circulation to both bone ends is undamaged and healing complications are much less usual than for cervical fractures. Instead, osteoporosis increases the risk of fragmentation in trochanteric fractures. A minor percentage of trochanteric fractures can be so comminuted that early direct weight bearing is hindered. The most widely used operation method is a sliding screw plate (Figure 46.7). This is a method that is fairly easy to teach on a large scale and has few complications. Modern metal techniques withstand metal fatigue, apart from cases with longstanding pseudarthrosis where a metal plate fracture can occur, usually after 6–12 months. During the last decade, short, intramedullary devices have been introduced as alternatives to the screw plate. The postulated advantages are shorter operation time, less bleeding due to other exposure, and a biomechanically shorter lever arm for weight bearing on the osteosynthesis material. Randomized studies have not shown any superior results of these intramedullary devices compared to the ordinary extramedullary screw plate. In some cases a significantly increased risk of femoral shaft fracture has been shown. Inadequate reaming of the femur is normally the cause, in conjunction with excessive force when inserting the nail. An alternative reason may be that the lateral cortical bone around the lag screw is not load protected by a barrel, as with the screw plate. A fissure in this area may more readily extend.

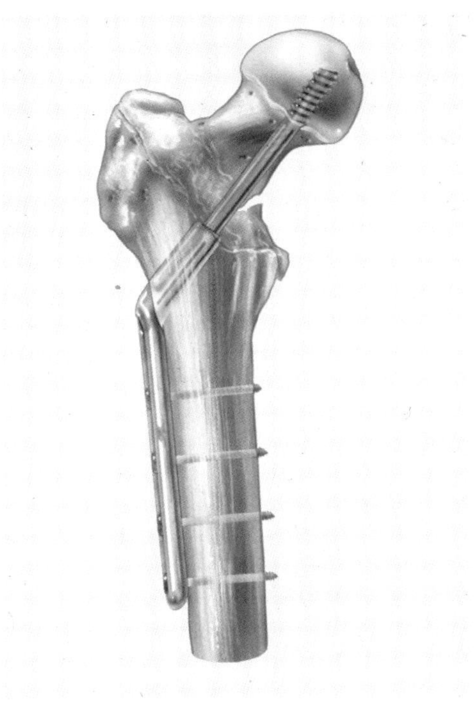

Figure 46.7. Trochanteric fracture operated with screw plate.

The main indications for intramedullary fixation are low trochanteric fractures, hip fractures with associated femoral shaft fracture, and pathological extracapsular fractures.

There is also a modification of the side plate that allows sliding along the femoral shaft combined with that along the femoral neck to impact the fracture more anatomically [32]. The results on consecutive series seem promising with a lowered cut out of the screw in the femoral head.

A basic biomechanical principle for good healing of trochanteric and subtrochanteric fractures is contact between the major weight-bearing bone fragments. Rigid fixation systems counteract this and lead to pseudarthrosis and, in the long run, breakage of the plate due to metal fatigue. At repositioning during the operation and the following mobilization and weight bearing, good contact is aimed at in the major bone fragments, sometimes at the price of a certain shortening of the leg. The main goal is a rapid healing of the fracture. In some cases increasing pain and too much collapse of the fracture make non-weight bearing necessary. This is also the case if the screw through the femoral neck and head threatens to cut through the subcondral bone into the hip joint (Figure 46.8). If the patient cannot support weight bearing with aids such as a walking table, rollator, quatra peds, sticks, or crutches, some weeks of sitting in a chair might be necessary. In the long run all trochanteric fractures heal, usually within 3–5 months. The development of femoral head necrosis is very rare, but there is some percentage of pseudarthrosis development which is higher if a more rigid fixation system has been used.

Dynamic extramedullary osteosynthesis (screw plate) is much better than rigid nail plates [33]. The Ender method, which was previously widespread in Europe, has in several randomized studies shown inferior results compared to the screw plate [34]. The intramedullary type of osteosynthesis with a screw up through the femoral neck and a short intramedullary rod often with transverse screws through the femoral shaft (the first type was called Gamma nail) has in several randomized comparative studies shown the same risk of cut out through the femoral head as the conventional screw plate whereas the intramedullary device has resulted in more re-operations, usually due to fracture at the distal end of the intramedullary nail. The technique is somewhat more demanding to perform [10]. It is, however, used as the only routine method in some centers in Europe. In the literature there are reports of a frequency of cut out of the femoral screw through the femoral head into the acetabulum with the use of a conventional screw plate in up to 10% of cases. This has been diminished with the new axial screw plate (Medoff plate). Reversed, oblique pertrochanteric fractures are especially suited for this type of osteosynthesis.

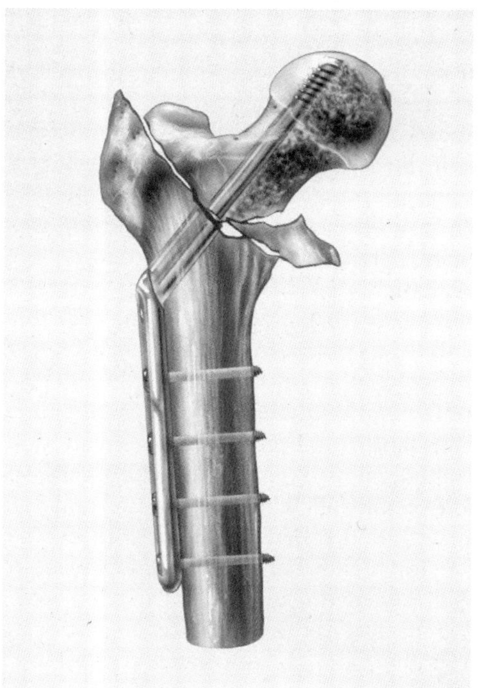

Figure 46.8. Trochanteric fracture with screw plate after compression by weight bearing resulting in cutting through of the femoral head.

Subtrochanteric Fractures

Subtrochanteric fractures have a considerably higher frequency of healing complications compared to trochanteric fractures. This is due to the high mechanical forces acting in this area and the fact that fractures often are very comminuted, giving inferior stability to the osteosynthesis system. One problem with the conventional screw plate for subtrochanteric fractures is that a distal fracture line transfers the dynamic screw plate to become a more static implant as the fracture is situated below the area for the gliding screw. This leads to complications associated with the static fixations such as delayed healing, pseudarthrosis, breakage of the plate, and cutting through of the femoral head by the screw [10,14]. This has led to an increased use of long intramedullary nails with transverse fixation screws in the distal part and a screw through the femoral head and neck in the proximal part. With this device, very long and comminuted femoral shaft fractures can be handled.

Weight Bearing and Rehabilitation

The goal after a hip fracture is to rehabilitate the patient to the same functional level as before the fracture [35]. A stable osteosynthesis system or a well-fixed arthroplasty is a prerequisite for this. The operation should allow direct postoperative weight bearing to start immediately the day after the operation (Figure 46.9). This is possible for the majority of patients operated with osteosynthesis for femoral neck fractures as well as for those receiving an arthroplasty. Weight bearing by walking gives a physiological impaction of the fracture and stimulates the bone healing process. The rule is immediate postoperative weight bearing. Furthermore, elderly patients usually find it difficult to have restricted weight bearing and cannot handle crutches as well as young patients can.

As mentioned above, for trochanteric fractures the majority can have full weight bearing whereas a small part of such fractures need more care due to very comminuted fractures. During recent decades successful rehabilitation programs have spread, consisting of direct mobilization in the hospital and continued walking rehabilitation in the patient's own home [26,36–45].

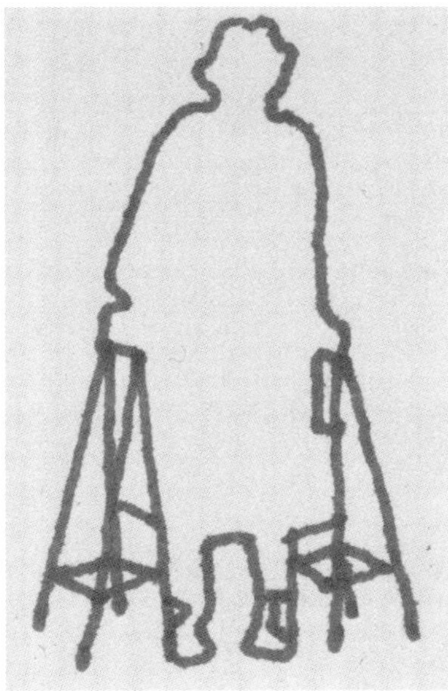

Figure 46.9. Weight bearing with quatrapeds. Direct postoperative mobilization and continued rehabilitation in the patient's own home.

Hip Fracture Audit

Due to the increasing burden on the health care system of osteoporotic fractures in the elderly, particularly hip fractures, it is very important to know the results of everyday treatment on a national basis of these fractures. In Sweden, a national registration of hip fracture treatment was introduced in 1988 [46]. This has spread internationally and in 1995 the Standardized Audit of Hip Fractures in Europe (SAHFE) was started based on the Swedish RIKSHÖFT experience [47].

The pattern of living before fracture and postoperatively up to 4 months after femoral neck (Figure 46.10) or trochanteric fractures (Figure 46.11) shows that of those patients coming from their own home, the majority had returned there after 2–3 weeks of treatment at the orthopedic department. Actually, the mean hospitalization time is now just under 10 days.

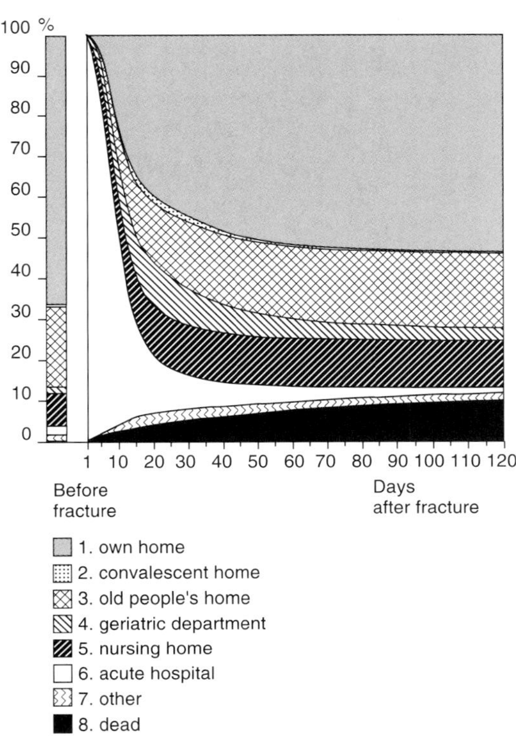

Figure 46.10. Living pattern before and at different time periods after a femoral neck fracture in Sweden based on 28,000 patients.

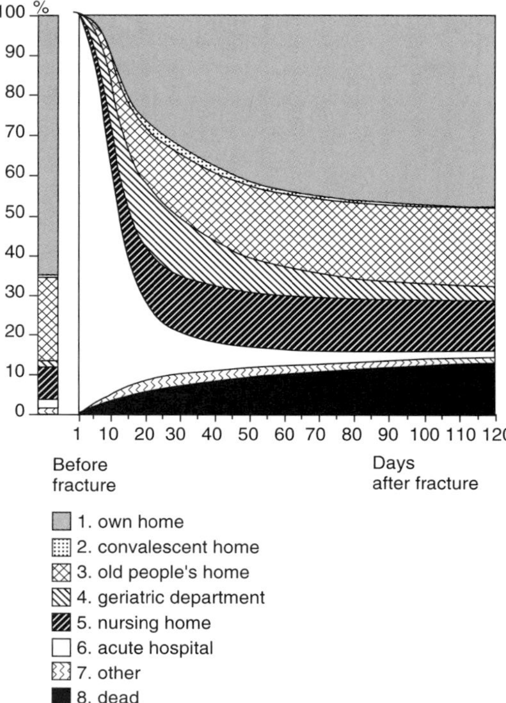

Figure 46.11. Living pattern before and at different time periods after a trochanteric hip fracture in Sweden based on 20,000 patients.

The rest of the patients are rehabilitated through an institution. This is mainly due to other diseases existing before the fracture. Within a month from the fracture the majority of patients from their own home or service house have returned to their previous place of living. After two months a very stable pattern of rehabilitation is apparent from the graph and at four months after the fracture, the majority of patients are back in their pre-fracture way of living.

Irrespective of the philosophy chosen for the treatment of hip fractures it is of the utmost importance to be able to compare the results from the different treatment programs. Different countries have various traditions both socially and in medical treatment, but internationally, comparisons will more rapidly bring out optimized ways of treatment that will be the solution to cope with the increasing amount of hip fractures during the coming decades. Participation in the SAHFE project is a way to accomplish this (for information see www.SAHFE.ort.lu.se).

References

1. Jarnlo G-B, Jakobsson B, Ceder L, Thorngren K-G. Hip fracture incidence in Lund, Sweden, 1966–1986. Acta Orthop Scand 1989;60:278–82.
2. Obrant K, Bengnér U, Johnell O, Nilsson B, Sernbo I. Increasing age-adjusted risk of fragility fractures: a sign of increasing osteoporosis in successive generations? Editorial. Calcif Tissue Int 1989;44:157–67.
3. Wallace WA. The increasing incidence of fractures of the proximal femur: an orthopaedic epidemic. Lancet 1983;I:1413–4.
4. Zetterberg C, Elmersson S, Andersson GBJ. Epidemiology of hip fractures in Göteborg, Sweden 1940–1983. Clin Orthop Rel Res 1984;191:278–82.
5. Melton LJ III. Differing patterns of osteoporosis across the world. In: Chesnut CH III, editor. New Dimensions in Osteoporosis in the 1990s. Hong Kong: Excerpta Medica Asia, 1991;13–18.
6. Johnell O, Gullberg B, Allander E, Kanis JA. The apparent incidence of hip fracture in Europe: a study of national register sources (MEDOS Study Group). Osteoporosis Int 1992;2:298–302.
7. WHO Study Group. Assessment of fracture risk and its application to screening for postmenopausal osteoporosis. WHO Technical Report Series No 843, 1994.
8. Brody JA. Prospects for an ageing population. Nature 1985;315:463–6.
9. Cooper C, Campion G, Melton J III. Hip fractures in the elderly: a worldwide projection. Osteoporosis Int 1992;2:285–9.
10. Parker MJ, Pryor GA, Thorngren K-G. Handbook of Hip Fracture Surgery. Oxford: Butterworth-Heinemann, 1997;1–145.
11. Garden RS. Low-angle fixation in fractures of the femoral neck. J Bone Joint Surg 1961;43-B:647–63.
12. Frandsen PA, Andersen E, Madsen F, Skjödt T. Garden's classification of femoral neck fractures: an assessment of inter-observer variation. J Bone Joint Surg 1988;70-B:588–90.
13. Thorngren K-G. Hip fractures in the geriatric patient. Natural history, therapeutic approach and rehabilitation potential. In: Stein H, editor. Scientific Proceedings SIROT 97. Freund Publishing, 1999:161–70.
14. Koval KJ, Zuckerman JD. Hip fractures. A practical guide to management. New York: Springer, 2000.
15. Raaymakers ELFB, Marti RK. Non-operative treatment of impacted femoral neck fractures: a prospective study of 170 cases. J Bone Joint Surg 1991;73-B:950–4.
16. Cserhati P, Kazár G, Manninger J, Fekete K, Frenyó S. Nonoperative or operative treatment for undisplaced femoral neck fractures: a comparative study of 122 non-operative and 125 operatively treated cases. Injury 1996;27(8):583–8.
17. Strömqvist B, Nilsson LT, Thorngren K-G. Femoral neck fracture fixation with hook-pins, 2-year results and learning curve in 626 prospective cases. Acta Orthop Scand 1992;63(3):282–7.
18. Parker MJ, Pryor GA. Hip Fracture Management. Blackwell Scientific 1993;1–292.

19. Rodrigues J, Herrara A, Canales V, Serrano S. Epidemiologic factors, mortality and morbidity after femoral neck fractures in the elderly – a comparative study: internal fixation vs hemiarthroplasty. Acta Orthop Belg 1987;53:472–9.
20. Rehnberg L, Olerud C. Subchondral screw fixation for femoral neck fractures. J Bone Joint Surg 1989;71-B:178–80.
21. Neander G, Dalén N. Osteosynthesis versus total hip arthroplasty for displaced femoral neck fractures – results after 4 years of a prospective randomized study. Acta Orthop Scand 1999;70 (suppl 287).
22. Rogmark C, Carlsson A, Johnell O, Sernbo I. A prospective randomized trial of internal fixation versus arthroplasty for displaced fractures of the neck of the femur. Functional outcome for 450 patients at two years. J Bone Joint Surg 2002;84-B(2):183–8.
23. Röden M, Schön M, Fredin H. Treatment of displaced femoral neck fractures: a randomized minimum 5-year follow-up study of screws and bipolar hemiprostheses in 100 patients. Acta Orthop Scand 2003;74(1):42–4.
24. Tidermark J, Ponzer S, Svensson O, Söderqvist A, Törnkvist H. Internal fixation compared with total hip replacement for distal femoral neck fractures in the elderly. A randomised, controlled trial. J Bone Joint Surg 2003;85-B(3):380–8.
25. Alberts KA, Isacson J, Sandgren B. Femoral neck fractures – fewer secondary hip arthroplasty procedures with a differentiated treatment protocol. Acta Orthop Scand 1999;70 (suppl 287).
26. Holmberg S, Thorngren K-G. Consumption of hospital resources for femoral neck fracture. Acta Orthop Scand 1988;59:377–81.
27. Thorngren K-G. Optimal treatment of hip fractures. Acta Orthop Scand 1991;62(suppl 241):31–4.
28. Manninger J, Kazar G, Fekete G et al. Avoidance of avascular necrosis of the femoral head following fractures of the femoral neck by early reduction and internal fixation. Injury 1985;16:437–48.
29. Strömqvist B. Femoral head vitality after intracapsular hip fracture: 490 cases studied by intravital tetracycline labeling and Tc-MDP radionuclide imaging. Acta Orthop Scand 1983;Suppl 200.
30. Asnis S, Wanek-Sgaglione L. Intracapsular fractures of femoral neck: results of cannulated screw fixation. J Bone Joint Surg 1994;76A:1793–803.
31. Kyle RF, Cabanela NE, Russel TA, et al. Fractures of the proximal part of the femur. Instruction course lecture. J Bone Joint Surg (Am) 1994;76:924–52.
32. Medoff RJ, Maes K. A new device for the fixation of unstable pertrochanteric fractures of the hip. J Bone Joint Surg 1991;73-A:1192–9.
33. Chinoy MA, Parker MJ. Fixed nail plates versus sliding hip systems for the treatment of trochanteric femoral fractures: a meta analysis of 14 studies. Injury, Int J Care Injured 1999;30:157–63.
34. Parker MJ, Handoll HHG, Bhonsle S, Gilespie WJ. Ender nails compared with nail or screw plate devices for trochanteric femoral fractures: a meta-analysis of randomised trials. Hip Int 1999;9(1):41–8.
35. Thorngren K-G. Fractures in older persons. Disability and Rehabilitation 1994;16:119–26.
36. Berglund-Rödén M, Swierstra BA, Wingstrand H, Thorngren K-G. Prospective comparison of hip fracture treatment, 856 cases followed for 4 months in the Netherlands and Sweden. Acta Orthop Scand 1994;65: 287–94.
37. Borgquist L, Nordell E, Lindelöw G, Wingstrand H, Thorngren K-G. Outcome after hip fracture in different health care districts. Rehabilitation of 837 consecutive patients in primary care 1986–88. Scand J Prim Health Care 1991;244–51.
38. Borgqvist L, Lindelöw G, Thorngren K-G. Costs of hip fractures. Rehabilitation of 180 patients in primary health care. Acta Orthop Scand 1991;62:39–48.
39. Ceder L, Thorngren K-G. Rehabilitation after hip repair (letter). Lancet 1982;2(8307):1097–8.
40. Ceder L, Thorngren K-G, Wallden B. Prognostic indicators and early home rehabilitation in elderly patients with hip fractures. Clin Orthop 1980;152: 173–84.
41. Holmberg S, Thorngren K-G. Rehabilitation after femoral neck fracture: 3,053 patients followed for 6 years. Acta Orthop Scand 1985;56:305–8.
42. Jalovaara P, Berglund-Rödén M, Wingstrand H, Thorngren K-G. Treatment of hip fracture in Finland and Sweden. Prospective comparison of 788 cases in three hospitals. Acta Orthop Scand 1992;63:531–5.
43. Jarnlo G-B, Ceder L, Thorngren K-G. Early rehabilitation at home of elderly patients with hip fractures and consumption of resources in primary care. Scand J Prim Health Care 1984;2:105–12.
44. Snedal J, Thorngren M, Ceder L, Thorngren K-G. Outcome of patients with a nailed hip fracture requiring rehabilitation in hospital for chronic care. Scandinavian Journal of Rehabilitation Medicine 1984;16: 171–6.
45. Thorngren M, Nilsson LT, Thorngren K-G. Prognostic-based rehabilitation of hip fractures. Comprehensive Gerontology 1988;2:12–17.
46. Thorngren K-G. Experience from Sweden. In: Medical Audit. Rationale and Practicalities. Cambridge University Press 1993;365–75.
47. Parker MJ, Currie CT, Mountain JA, Thorngren K-G. Standardised audit of hip fracture in Europe (SAHFE). Hip Int 1998;8:10–15.

47 Normal and Prosthesic Hip Biomechanics

C. Sorbie, R. Zdero, and J. T. Bryant

To assess the diseased human hip, it is necessary to appreciate the motions, forces, and stresses in both normal and abnormal states. Only with this basic understanding can the clinician or engineer bring lasting relief to the patient suffering from hip disease or malfunction. To this end, the following discussion of the biomechanics of the normal and damaged hip is offered.

Normal Hip Biomechanics: Functional Anatomy and Load Transfer

The normal human hip is basically a bony ball-and-socket interface composed of a convex, cartilage-covered femoral head and concave acetabulum. It is surrounded by passive soft tissues and active muscles (Figure 47.1). In shedding light on the workings of the normal human hip during activities of daily living (ADLs) or even traumatic occurrences, it is important to recognize that every anatomical feature has a role to play either in ensuring the necessary biomechanical conditions for the proper completion of ADLs or in preventing injury. In the following, the function of the more important contributing anatomical components and related issues will be discussed.

Ball-and-Socket Congruity

Although the articulating interface of the hip joint is often described as "ball-and-socket" (spheroid), its geometry is in reality non-congruous. The head of the femur forms two-thirds of a sphere [1,2] and is somewhat compressed in the anteroposterior direction [3], whereas the socket, or acetubulum, is not a spheroidal cup but has a horseshoe shape [4]. It might be assumed that non-congruous geometry would be the cause of uneven joint load distributions and, hence, damage to the cartilage-lined joint interface. The opposite, in fact, is true. During the unloaded state, the degree of conformity of the surfaces is unimportant, as they are unengaged and do not facilitate any load transfer [4,5]. However, as loading starts and increases, the acetabular surface geometry changes from being incongruent to congruent, thereby creating the maximum amount of cartilage and bone contact. The joint force is then evenly distributed (Figure 47.2). This surface engagement is the major stabilizing force in hip function. In fact, as shown in Figure 47.3, if the joint was perfectly conforming in the unloaded state, increasing joint force during active loading would lead to a non-conforming articulation, uneven load distribution, diminished contact area, and a predisposition to cartilage trauma [4].

Abnormal congenital ball-and-socket hip joint congruency occurs in Congenital Dislocation of the Hip (CDH), acetabular dysplasia, or subluxation, in which there is inadequate acetabular coverage of the femoral head for proper joint stability and function. Treatments have included the use of the Children's Hospital abduction brace, the Pavlik harness, and the Frejka splint [6]. As shown in Figure 47.4, these methods increase abduction of the hip by applying an abducting treatment force, F, over a period of several months. This increases adductor tension, T, and creates a "tension band" effect. It then follows that the resulting femoral force, R, also rises and progressively deepens penetration of the femur into the acetabulum,

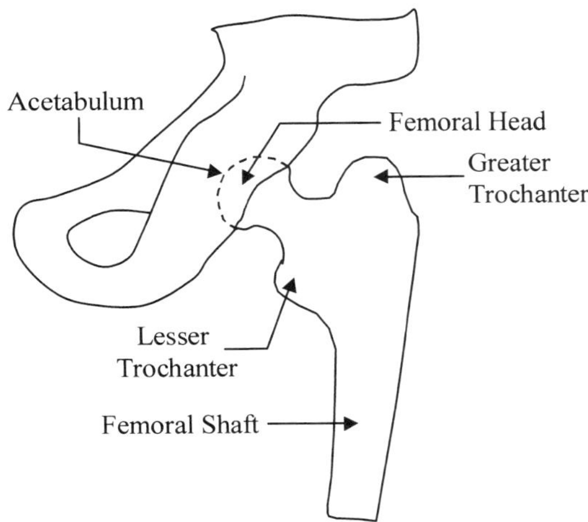

Figure 47.1. Normal hip joint anatomy.

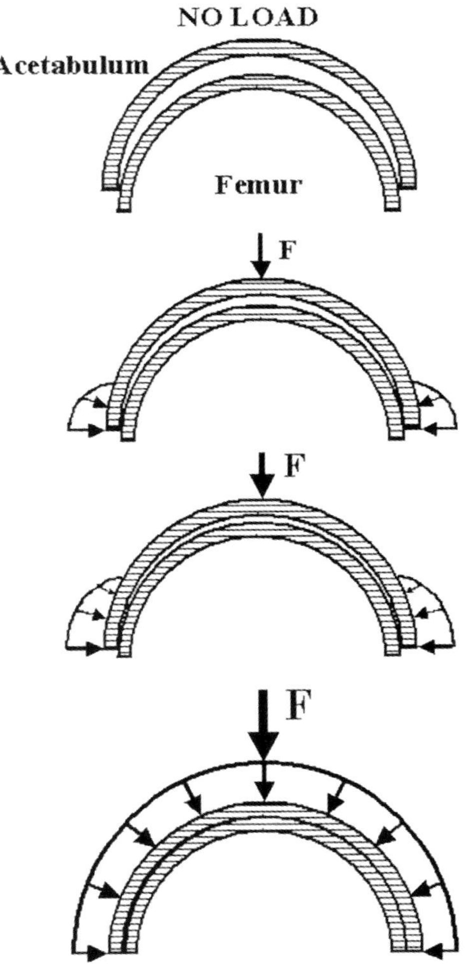

Figure 47.2. The change in congruency of the normal hip joint during loading. The ball-and-socket incongruency at no load ensures that maximum conformity occurs as weight bearing increases. Redrawn with permission from Greenwald [5].

increasing hip joint congruency and stability. Excessive joint reaction force, however, commonly leads to avascular necrosis caused by impairment of blood supply and can result in deformation of the relatively soft femoral head. It is therefore recommended that rigid hip immobilization be avoided, especially in frog-like abduction, unless soft-tissue release is first performed to reduce often-painful adductor tension, T, and the "tension band" effect.

Trabecular Bone

The trabeculae composing the proximal femur are a honeycomb-like network of bony structures whose function it is to maintain the structural integrity of the femur under tensile and compressive forces at the hip joint. As shown in Figure 47.5, the trabeculae are organized into specific zones whose orientations are dictated by the magnitude and direction of the forces they normally encounter [1]. These regions are often classified into three main networks of trabeculae: the arcuate, medial, and lateral systems [1,3].

The arcuate system primarily resists the tensile stresses caused by a bending moment about the neck of the femur. The medial system is a vertically oriented set of trabeculae which sustain compressive loads acting vertically through the femoral head. The lateral system trabeculae are oriented along the trochantric line and resist both compressive and tensile forces caused by hip muscle action.

The areas of greatest strength overall are those where trabecular systems intersect orthogonally due to their ability to resist equally compression and tension in any given direction. However, a region of weakness, termed Ward's triangle, is particularly prone to fractures under excessive loading or during the progression of osteoporosis because, in the triangle, trabeculae do not intersect and are thin. In young adults, loading of the femoral head would need to reach

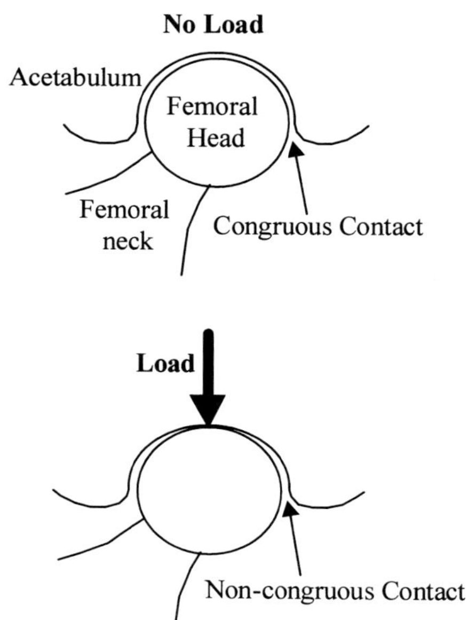

Figure 47.3. Abnormal hip joint loading. If perfect conformity occurred at no load, increased weight bearing would generate non-conforming articulation with a region of cartilage undergoing abnormally high stresses.

or exceed 12–15 times body weight before neck fracture [3]. In the aged, however, it is not uncommon to find complete loss of trabeculae in this triangular region and only marrow fat present [7]; the load to fracture is reduced significantly. This correlates with the observation that the femoral neck is the most common fracture site in the elderly [2].

Femoral Neck Angles

There are two angular relationships of the femoral neck that are particularly important in hip joint motion, namely the neck-shaft angle in the frontal plane (or angle of inclination) and the neck angle in the transverse plane (or angle of torsion), as shown in Figure 47.6.

The *angle of inclination* in an adult hip is 125 ± 5 degrees (Figure 47.6a), with *coxa valga* being the condition when this value exceeds 130 degrees and *coxa vara* when the inclination is less than 120 degrees [1]. The importance of this feature is that the femoral shaft is laterally displaced from the pelvis, thus facilitating freedom for joint motion [2]. If there is significant deviation in angle outside this typical range, the lever arm used to produce motion by the abductor muscles will be either too small or too large [2]. Extreme cases in which the angle is less than 100 degrees may require surgical correction to restore the proper force relationships about the hip [1]. Furthermore, the change in neck-shaft angle over a human life cycle, which steadily decreases from 150 degrees three weeks after birth to 120 degrees in the adult, indicates that there is increasing reliance on the developing musculature to provide joint stability than ball-and-socket engagement, which is the case with infants [1].

The *angle of torsion* is normally anteverted about 12–14 degrees for the adult hip (Figure 47.6b) and facilitates the proper amount of internal–external rotation during gait [2,3]. Anteversion greater than 12–14 degrees may cause a portion of the femoral head to become uncovered, causing a tendency to walk with internal rotation of the hip to keep the femoral head fully engaged with the acetabulum. Conversely, a neck retroverted below 12 degrees will result in a tendency to externally rotate the leg.

Musculature

The ball-and-socket geometry of the hip permits rotational motion in all directions, necessitating a large number of controlling muscles arising from a wide surface area to provide adequate stability [4]. The 22 muscles acting on the hip joint not only provide stability but also the forces required for movement of the femur during activity [8]. These muscles can be classified according to their function as flexors, extensors, adductors, abductors, and external and internal rotators (Figure 47.7).

Flexors and extensors are closely balanced in terms of the forces that each generates [7]. They are primarily responsible for moving the femur in the sagittal plane about the transverse hip joint axis. The two strongest flexors are the iliopsoas, whose tendon inserts at the base of the lesser trochanter, and the rectus femoris. Both perform powerfully during activities such as kicking when the knee is flexed [1]. The gluteus maximus is the principal extensor and extends

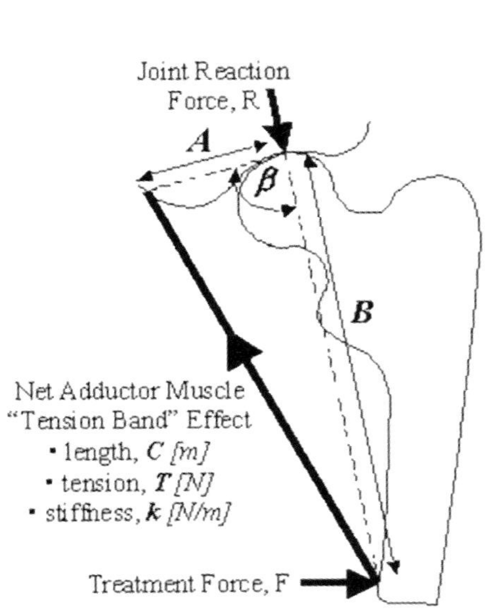

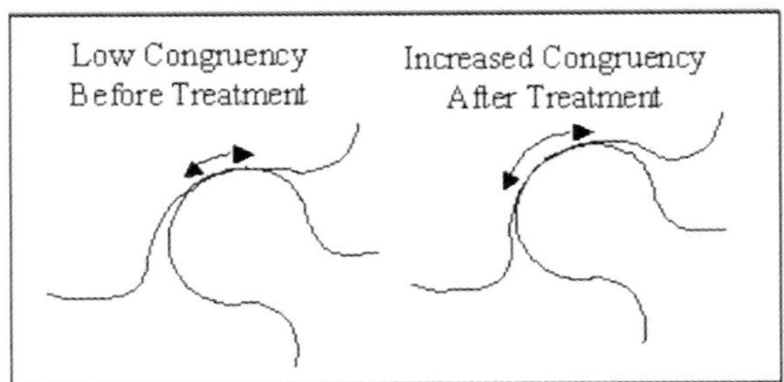

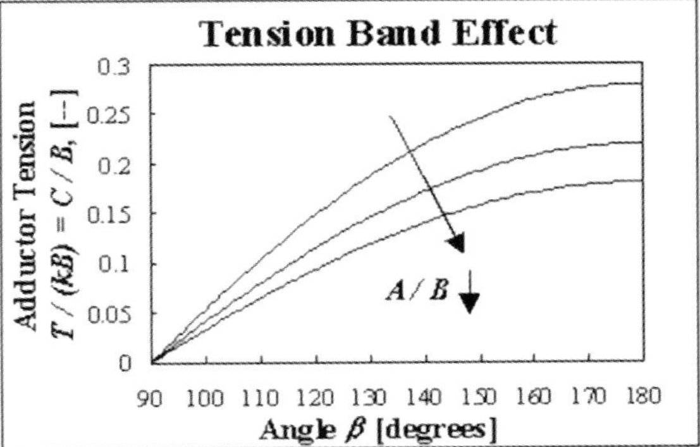

Figure 47.4. Treatment of Congenital Dislocation of the Hip (CDH), dysplasia, or acetabular subluxation intends to increase abduction of the hip by applying a treatment force, F. The femoral force, R, progressively deepens penetration of the femur into the acetabulum, increasing hip joint congruency and stability. Past approaches have included the use of the Children's Hospital abduction brace, the Pavlik harness, or the Frejka splint. The resulting "tension band" effect was modeled mathematically with the simplified assumptions of fixed adductor muscle stiffness, k, and zero initial tension, T, at $\beta = 90$ degrees. The graph illustrates the rise in normalized adductor tension, T/kB, with increasing abduction angle, β, for changes in length ratio, A/B.

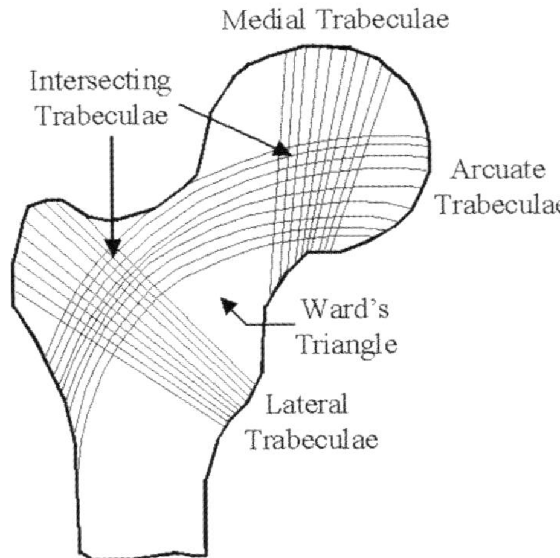

Figure 47.5. The honeycomb-like structure of trabecular bone in the proximal femur. The arcuate, medial, and lateral systems each sustain the bone under different stress states. Regions of trabeculae intersecting orthogonally provide the greatest strength. Ward's triangle is the weakest region, being almost completely osteoporotic.

the hip during activities such as stair climbing or rising from the seated position. The three hamstring muscles, biceps femoris, semimembranosus, and semitendinosus, also assist in hip extension, their contribution increasing as the knee is extended.

Adductors and abductors, which are also equally force balanced [7], are dedicated to rotating the femur in the frontal plane. The adductors form the greatest muscle mass on the medial side of the hip, the primary ones being adductor longus, adductor brevis, and adductor magnus. The main abductor muscle is gluteus medius, a three-part muscle inserting on the greater trochanter. It plays a crucial role in preventing pelvic drop during the single leg stance phase of walking, as does the gluteus minimus, which lies deep within the gluteus medius.

External and internal rotators turn the femur laterally and medially around the long axis of the leg. There are no muscles specifically

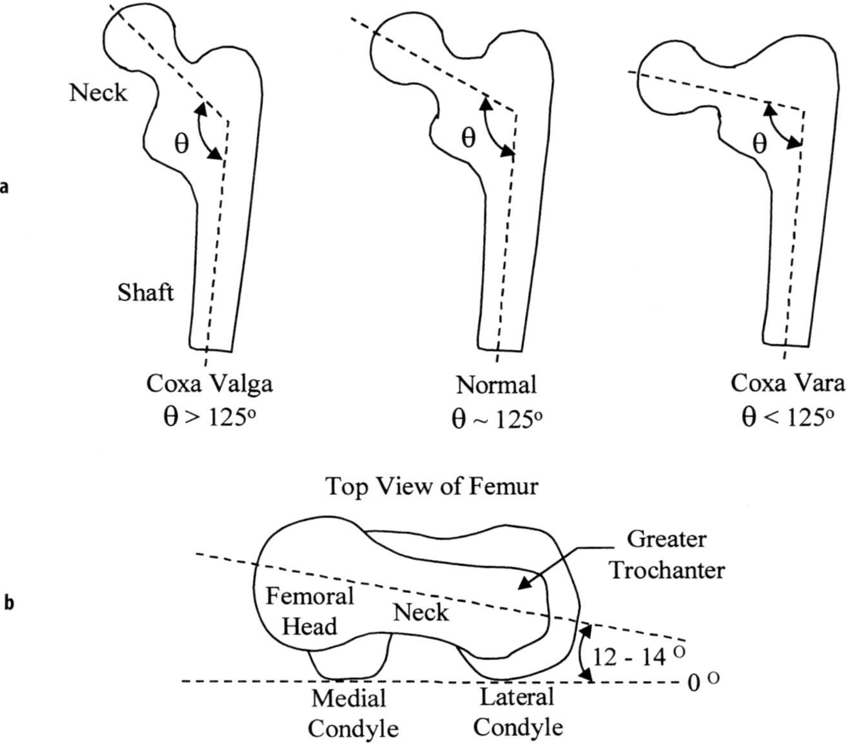

Figure 47.6. Femoral neck angles. **a** The angle of inclination of the femoral neck is approximately 125 degrees in the adult human hip. Significant deviations, as in coxa valga and coxa vara, will alter the biomechanics of the hip. **b** The angle of torsion is anteverted 12–14 degrees in the normal adult.

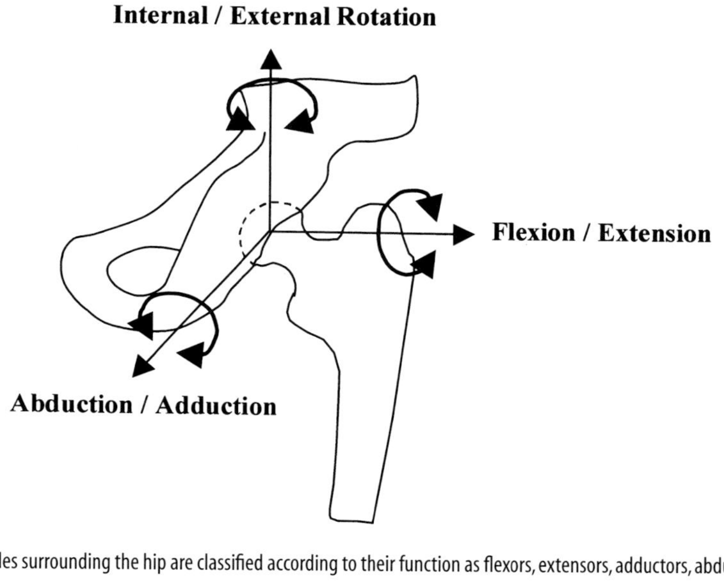

Internal / External Rotation

Flexion / Extension

Abduction / Adduction

Figure 47.7. The muscles surrounding the hip are classified according to their function as flexors, extensors, adductors, abductors, and external and internal rotators.

dedicated to internal rotation, but rather this motion is performed by the secondary action of tensor fascia latae, gluteus medius, and gluteus minimus [1]. However, six muscles which lie normal to the femur and parallel to the neck and head are external rotators, namely obturator internus, obturator externus, gemellus superior, gemellus inferior, quadratus femoris, and piriformis. Their orientation not only facilitates their function as rotators but also provides additional stabilization of the hip joint by compressing the femoral head into the acetabular cavity. The external rotators produce almost three times as much force as the internal ones [7].

As previously stated, the muscles of the hip joint perform several different functions depending on the position of the hip, which is caused by a change in the relationship between a muscle's line of action and the hip's axis of rotation. This is referred to as the "inversion of muscular action" [9] and most commonly manifests as a muscle's secondary function. For example, the gluteus medius and minimus act as abductors when the hip is extended but as internal rotators when the hip is flexed [7]. The adductor longus acts as a flexor at 50 degrees of hip flexion, but as an extensor at 70 degrees [1].

In addition to providing stability and motion for the hip, muscles act to prevent undue bending stresses on the femur [4]. When the femoral shaft undergoes vertical load (Figure 47.8), the lateral and medial sides of the bone experience tensile and compressive stresses, respectively. To resist these potentially harmful stresses, as might occur in the case of an elderly person whose bones have become osteoporotic and susceptible to tensile stress fractures, the tensor fascia lata muscle acts as a lateral tensioning band.

Muscle weakness around the hip is usually compensated by the individual to perform the desired task of walking. An example is the Trendelenburg's sign, noted by the pelvis sagging due to the weakness of the abductor muscles on the weight-bearing leg. This is countered by the individual shifting their center of gravity towards the affected joint by leaning over [5,10,11]. This tilting reduces the force required by the abductors.

Labrum

As shown in Figure 47.9, the labrum is a fibrocartilaginous ring or lip, triangular in cross-section, with densely packed parallel fibers, that completely surrounds the acetabulum and embraces the femoral head in all positions [1,5]. Biomechanically, it has three important functions.

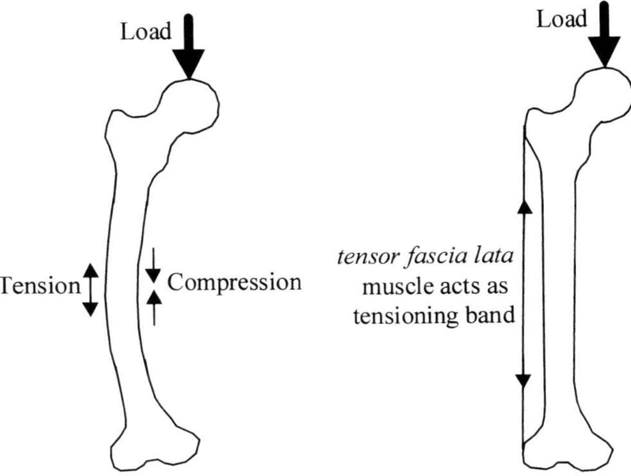

Figure 47.8. Femoral bending under load. The tensor fascia lata muscle acts as a tensioning band to reduce compressive and tensile stresses generated due to bending.

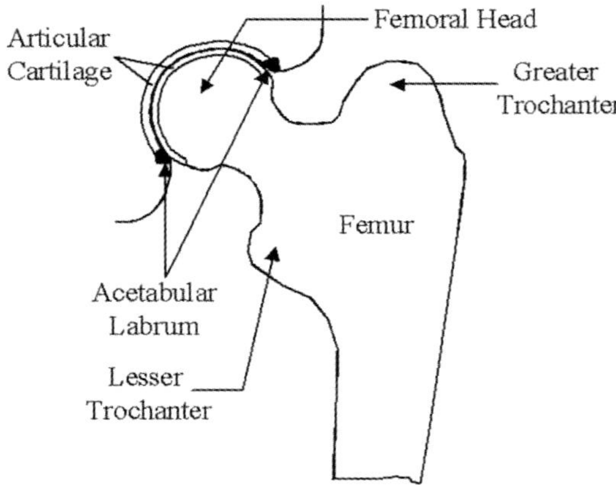

Figure 47.9. The acetabular labrum and other hip joint structures.

First, it provides added joint stability by deepening the acetabulum from less than one-half to greater than one-half of a sphere, thereby allowing for a more secure penetration of the femoral head into the acetabular cavity [1]. The labrum also resists any tendency for femoral head dislocation by the development of large hoop stresses along its fiber-reinforced ring structure [12].

Second, as the free edge of the ring is in close contact with the femoral head in all positions, it isolates the intra-articular joint from the rest of the intracapsular cavity. This creates an effective seal that prevents fluid escaping from the intra-articular joint [12]. It has been shown that, even under a 1,000–1,500 N load, the sealing action of the labrum allows a 0.2–0.6 mm fluid layer to remain between the articulating hip surfaces [13]. Retaining adequate fluid and, hence, maintaining interstitial joint pressure, is necessary for the fluid's role in load carriage, which can be as much as 90–94% of the total load across the hip joint [1,12]. Compromising this load-bearing mechanism would lead to direct or intimate contact between femoral and acetabular cartilage and as much as a 92% increases in contact stress [12]. In addition, such direct contact would cause an increase in compressive, adhesive, and surface shear stresses during movement, which have been linked to cartilage wear [14].

Third, the labrum's compressive material properties may allow it to function as a shock absorber in the hip joint. In extreme positions and movements of the hip, the femoral neck may impinge on the acetabular rim, which in turn could lead to joint dislocation. During such movements, intra-articular fluid flow against the labrum will generate force as the femoral head "swims" through the synovial fluid. The inverse relationship between the labrum's permeability and the drag forces experienced means that viscous dissipation of energy may be enough to prevent dislocation [12].

Cartilage

The articulating surfaces of the hip joint are covered with a layer of porous, wear-resistant and springy articular cartilage. The thickness of the cartilage varies on different parts of the femoral head. It reaches 3 mm at the posterosuperior region and thins down to about 0.5 mm at the periphery [1,7]. Similarly, acetabular cartilage ranges from 2 to 2.5 mm in the posterosuperior areas to 1 mm along the periphery, where it blends in with the labrum [1,7]. The cartilage is composed of a gelatinous matrix in which is immersed a fibrocollagenous framework. The fiber orientation is arranged adequately to resist compressive and tangential stresses [7].

The cartilage, in concert with the synovial fluid it expresses, functions as a vehicle for load transfer across the hip joint. As shown in Figure 47.10, it has been proposed that there are five load-dispersing mechanisms acting synergistically, namely hydrodynamic, squeeze film, boosting, hydrostatic, and boundary layer mechanisms [14,15].

Hydrodynamic lubrication occurs during high-velocity movements of the femur (0.06 m/s for normal walking), by causing the non-Newtonian synovial fluid to increase in viscosity. A "rigid" fluid wedge is developed between the femur and acetabulum. This is a function of gap geometry, sliding speed, and the fluid viscosity. Note that in this case the direction of applied compressive force is normal to the direction of bearing surface motion.

During *squeeze film* lubrication, synovial fluid is squeezed out of the joint space as the femoral head approaches the acetabulum during loading. In this case, the movement of

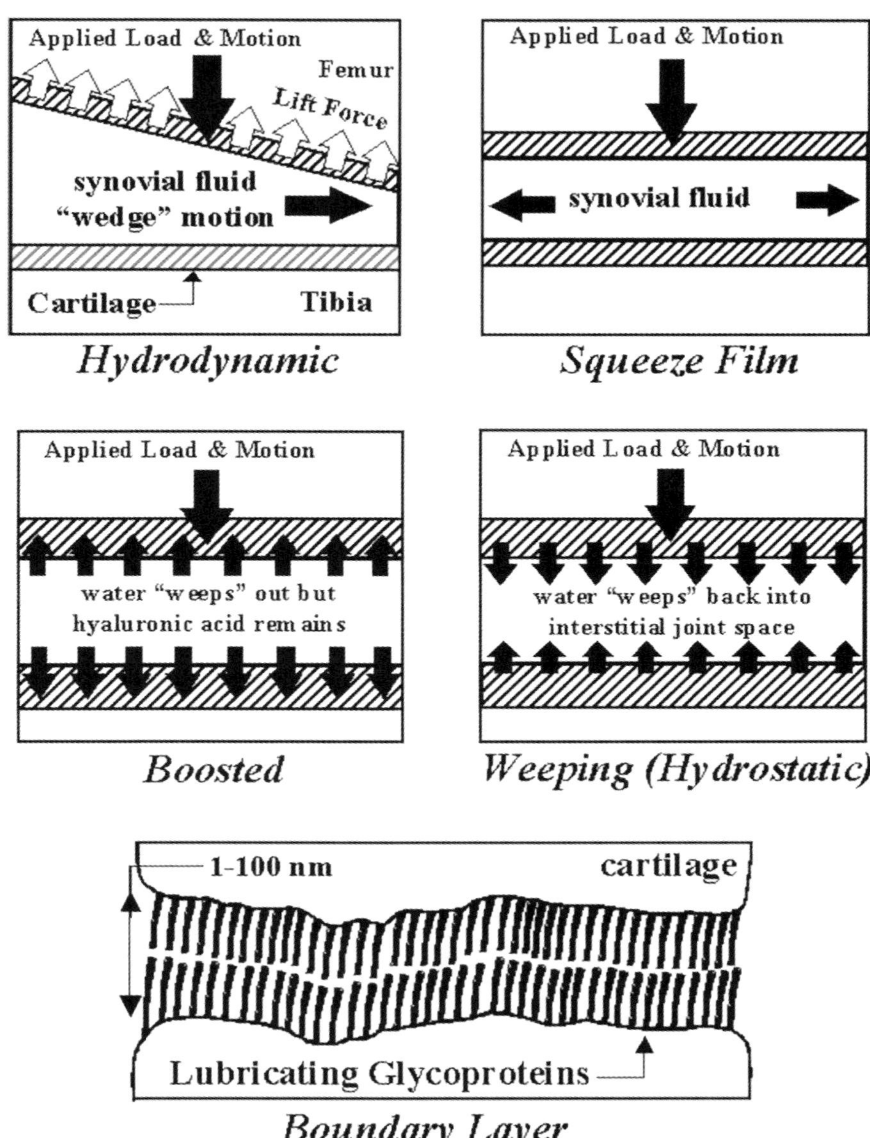

Figure 47.10. Weight bearing and lubrication mechanisms are typically classified as hydrodynamic, squeeze film, boosted, weeping, and boundary layer.

the bearing surfaces is parallel to the direction of applied compressive force. This motion is partially resisted by the viscous synovial fluid. For example, a film thickness of 20 μm can resist several MPa of pressure for several minutes before the fluid layer is depleted.

Boosting occurs when water from the synovial fluid flows into the porous cartilage, leaving the larger hyaluronic acid molecules in the joint space as lubricants.

Hydrostatic, sometimes termed *weeping*, lubrication refers to the expression of water that had been previously absorbed by the porous cartilage back into the joint space.

Finally, *boundary layer* lubrication occurs for very close proximities (1–100 nm) of the articulating surfaces, during which the lubricating glycoproteins (lubricin) on the cartilage surfaces minimize direct femoro-acetabular contact.

The solid matrix of cartilage also protects subchondral bone from damage by acting as a cushion minimizing the direct stress levels. This is often referred to as "stress shielding". In addition, cartilage's time-dependent properties, which are a function of its solid–fluid biphasic nature and its porosity, will resist joint force with a relatively slow rate of deformation, or creep. Furthermore, because of its compliant nature, cartilage is able to conform to the contacting bodies of the hip joint articulation, thereby distributing compressive forces evenly along the joint interface [16].

Contact forces and stress levels across articular cartilage have been measured or estimated using telemetrized hip implants [17,18], endoprostheses [19–21], piezoelectric transducers implanted superficially in articular cartilage [22], subchondral strain gages and sensors [23,24], and gait analysis [25–29]. Forces and stresses are dependent on a number of factors including type of activity, age, joint orientation, joint geometry, cartilage thickness, and material properties of both articular cartilage and subchondral bone [5]. Table 47.1 shows the range of forces and stresses that can act across the in vivo hip during various activities. Forces, for example, can range from 0.3 to 0.5 times bodyweight during passive hip motion to 1.29 to 5.95 times bodyweight in stair descent, with the corresponding average contact pressures reaching 1.7 MPa and 18.2 MPa, respectively [1]. From Fujifilm tests done on a cadaveric hip loaded in the normal anatomical position, Miyanaga et al. [30] demonstrated that loading is not equally distributed over the contact area, there usually being a pressure profile with two local stress peaks aligned anteroposteriorly (Figure 47.11). The location of this peak stress is the site most commonly found to display cartilage loss associated with the progression of osteoarthritis [31].

Although the engaged weight-bearing cartilage area includes the peripheral portion of the femoral head, it is usually concentrated along the superior regions of the articular surface, as shown in Figure 47.12 [5,32]. These anterosuperior areas are thought to be susceptible to cartilage destruction because of long-term exposure to alternating stress fields, a proposal confirmed by the characteristic lesions found in this region among a random sampling of 100

Table 47.1. In vivo hip joint forces and contact pressures (Selected data from Robbins [1])

Activity or Exercise	Force Range (times BW)		Pressure Range (MPa)	
	Min Max	Max Max	Min Max	Max Max
Passive hip ROM	0.3	0.5		1.7
Straight leg raise	0.97	2.0	1.85	5.9
Lifting opposite leg	0.4	2.6		
Isometric quadriceps			2.13	3.44
Isometric hip extension			3.1	6.2
Free speed gait	1.51	4.64	3.69	6.7
Slow gait	1.8	2.93	5.2	7.8
Fast gait	2.7	4.71	4.4	7.7
Crutches or walker (TWB)			2.0	6.5
Crutches or walker (PWB)	1.82		1.39	3.5
Crutches or walker (NWB)			1.08	2.4
Bilateral crutch walking	0.6	3.1	1.39	6.5
Chair raise	0.8	1.6	1.21	18.0
Stairs down	1.29	5.95		18.2
Jogging	4.33	5.84		7.7
Jumping			7.8	16.2
Double-leg stance	0.5	2.42	0.8	1.4
Single-leg stance	2.0	5.4	7.2	9.7
Single-leg stance on opposite leg	0.4	0.94	4.2	9.8

Min Max: Lowest reported maximum. Max Max: Highest reported maximum. TWB: Total weight bearing. PWB: Partial weight bearing. NWB: No weight bearing.

osteoarthritic heads [5,32]. The amount of cartilage area engaged, as with contact forces and stresses, is dependent on numerous factors. In one study, a 2,000 N compressive load, for example, corresponding roughly to the 2.9 times bodyweight force peak for a 70 kg individual during walking, can create a contact area as large as 17 cm^2 [22]. Miyanaga et al. [30] loaded a cadaveric hip in the normal anatomical position for loads between 20 and 2,000 N. Employing the casting method, which utilizes the silicone rubber commonly used in dental casting, they were able to measure the momentary contact area engaged for their range of loads. As shown in Figure 47.13, the weight-bearing area did not increase linearly with load, but rather tended to plateau as compressive force increased, engaging as much as 79% of the total cartilage area available for load transmission. Unlike Strange [33], both Miyanaga et al. [30] and Afoke et al. [31] found that the shape of these areas was kidney-like, being neither symmetric nor uniform even at loads as high as four times bodyweight, indicating the nonconformity of the femur and acetabulum (Figure 47.11). Afoke et al. [31] suggest that these shapes may also arise from a combination of local surface irregularities and local variations in stiffness of cartilage and subchondral bone. The inability of cartilage to adequately distribute the loads evenly may be a factor in the onset of cartilage degeneration.

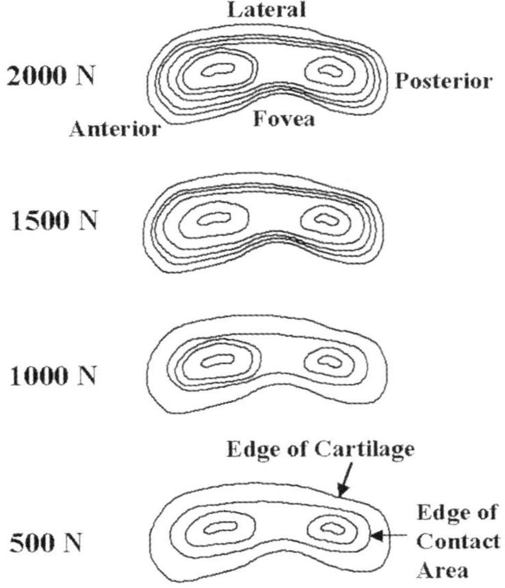

Figure 47.11. Hip joint contact area and pressure distribution for a range of compressive loads. Redrawn with permission from Miyanaga et al. [30].

A recent development reported in the literature by Olson et al. [8,34,35] has been the study of alterations in contact area and stress distribution across cartilage weight-bearing areas in

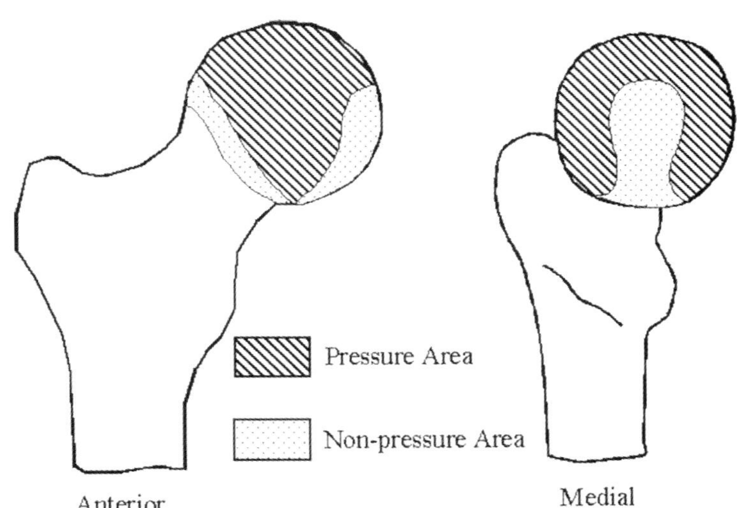

Figure 47.12. Pressure areas (located supero-anteriorly) and non-pressure regions (along the periphery) of the femoral head.

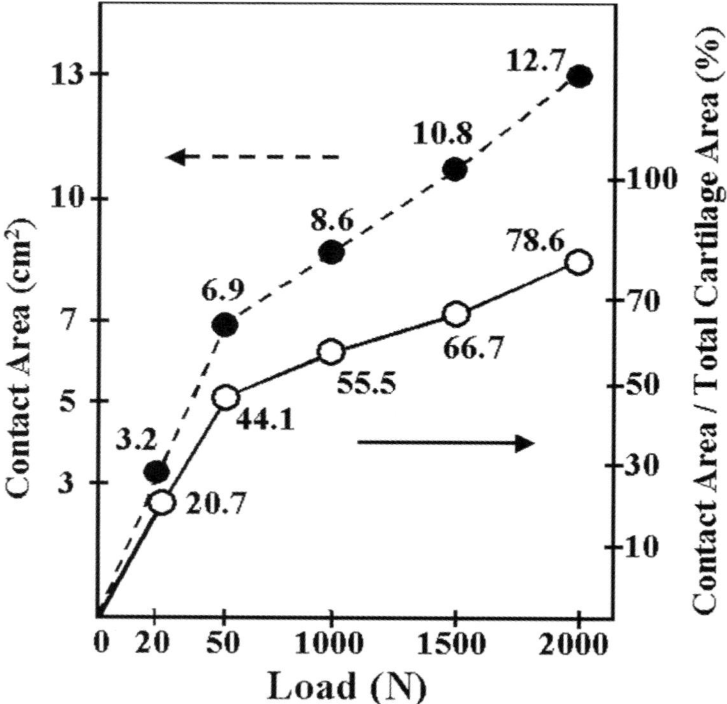

Figure 47.13. Contact area as a function of weight-bearing load. Replotted with permission from Miyanaga et al. [30].

the hip in acetabular fracture. Under normal circumstances, the hip joint demonstrates a reasonably uniform load distribution, with the weight-bearing area being equally distributed among the anterior wall, the superior aspect, and posterior wall of the acetabulum. Olson et al.'s Fujifilm testing has shown that, upon acetabular fracture, this interstitial contact area becomes limited to the periphery and is parallel to the acetabular rim. Transverse pelvic fractures cause a measurable loss of the peripheral weight-bearing area in the hip. These findings, however, may not be significant.

Finally, although the exact sequence of deterioration of articular cartilage is not yet settled, it is thought that once cartilage begins to deteriorate through disease, injury, or age, an imbalance between tissue synthesis and degeneration ensues [36]. This imbalance stimulates the production of enzymes which further cleave the cartilage matrix, making it more susceptible to subsequent mechanical damage. This process leads to osteoarthritis. Osteoarthritis or rheumatoid arthritis to the point of debilitating pain and significant loss in hip function, make an individual a candidate for Total Hip Replacement (THR) or other corrective measures.

Pelvic Deformation

There is evidence from both experimental work and finite element analysis that the pelvis beyond the acetabulum has an important role to play in weight bearing. Investigating the compressive loading of the hip has commonly involved the explantation of the acetabulum from the pelvic ring and fixation of the acetabulum in a pot. The femur and acetabulum are loaded directly to simulate single leg stance. The explanted model, however, does not allow the joint reaction force to be developed indirectly from the effect of muscle action with the hip intact. The hip muscle forces cause a degree of pelvic deformation and displacement [8]. The explanted model overestimates the weight-bearing area and underestimates the corresponding peak contact pressure. The difference in peak contact pressure between intact and explanted models, most significant in the superior region of the acetabulum, can be as high as 28%. In addition, finite element modeling of the entire pelvis has demonstrated large strains at the sacroiliac joints and the pubic symphysis under an array of compressive forces at the acetabulum and tensile forces at muscle origins. This underlines the important role of the entire pelvic structure in supporting hip joint loads [37].

Lever Arm Ratio and Joint Reaction Force

In assessing weight bearing at the hip, it is important to recognize the additional contribution of the abductor muscles to the magnitude of joint reaction force. Consider, for example, a simple, two-dimensional, free body diagram force analysis of the hip joint in the frontal plane for single leg stance (Figure 47.14). The external forces and moments acting at the hip arise from ground reaction forces, the gravitational forces from the weight of body segments, and inertial force effects due to acceleration of body segments. These forces are generated by passive and active tensile forces arising from muscles and ligaments. The resultant of all these forces is counteracted by the hip joint reaction force, J, acting at the femoro–acetabular interface.

An important factor that affects the joint reaction force magnitude is the *lever arm ratio*, which is the ratio between the gravitational force lever arm, b, and the abductor muscle lever arm, c (Figure 47.14). Typical levels for single leg stance are three times bodyweight, corresponding to a lever ratio of 2.5 [38]. However, increased ratios above this can occur in congenital *coxa valga*, in which the short abductor lever arm requires the abductor muscles to generate a greater force from its fixed position, thereby creating an even larger force across the hip. Other changes in lever arm ratio can arise from weak abductors (which cause a reduction in lever arm c and a rise in hip joint force J), leaning towards the hip (which reduces lever arm b and decreases joint load), and the use of a cane in the opposite hand (which can reduce the resultant hip force J by approximately 50%).

Attempts at surgical correction of abnormal load distributions or levels are designed to alter the acetabulum position, the relative location of the hip joint center, the head–neck angle, neck length, anteversion angle, and the muscle lines-of-action. All of these influence the lever arm ratio and, hence, the resulting load at the hip [5,11]. For example, relocating the trochanter to a more lateral position changes both the neck–

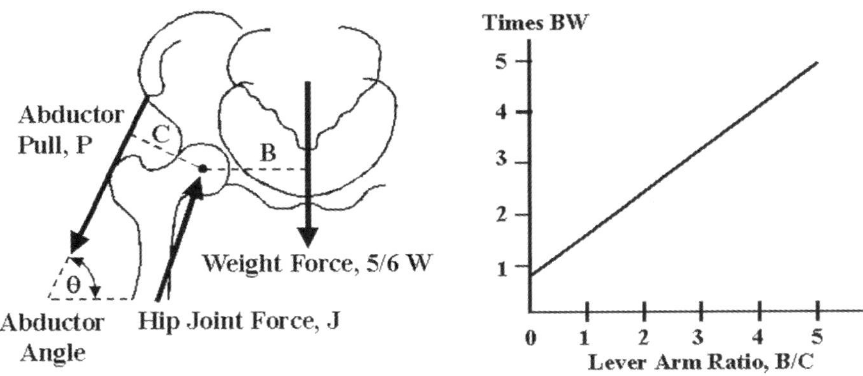

Figure 47.14. Hip joint reaction force as a function of lever arm ratio. Redrawn with permission from Greenwald [5].

shaft and anteversion angles, thereby reducing the abductor force angle, the bending moment acting around the hip joint, and the stresses on the femoro–acetabular interface [39–41].

Prosthetic Hip Biomechanics

Degeneration of articular cartilage, associated with osteoarthritis (OA) and rheumatoid arthritis (RA), leads to excessively high or uneven contact stress distribution in the hip joint [5,11,38]. The progressive changes that follow are loss of joint mobility, instability, deformity, and a variable amount of pain. Surgical treatment has included osteotomy, lateral displacement of the greater trochanter, tenotomy, fusion, and arthroplasty, all of which reduce compressive articular forces to an acceptable level [39]. The most common of these clinical solutions is Total Hip Replacement (THR) or Total Hip Arthroplasty (THA). Between 500,000 and 1 million hip replacements are performed worldwide annually, with a success rate of 93% at 10 years and 85% after 15 years [42]. Success has depended not only on the developments in orthopedic surgical techniques but also on the evolution of design and development of the biomechanical and biomaterial aspects of THRs.

Currently Used THRs

The most common type of THA is total hip replacement (THR) which uses a cobalt–chrome (CoCr) or ceramic femoral head that is modular with a CoCr or titanium alloy (Ti-6Al-4V) intramedullary stem, as shown in Figure 47.15. After the femoral head is excised, the proximal femur medullary cavity is reamed, broached, and the metal stem is placed into the cavity and held there with a grout of bone cement (polymethylmethacrylate, PMMA). It may be fastened by impact and later fixed by the shaft's textured outer surface, which allows bone ingrowth [41,43,44]. This "biologic" fixation may be enhanced by a coating of hydroxyapatite. The femoral head articulates against the surface of an acetabular cup, which is made from medical-grade, ultra-high-molecular-weight polyethylene (UHMWPE), alone or with a shell of CoCr alloy.

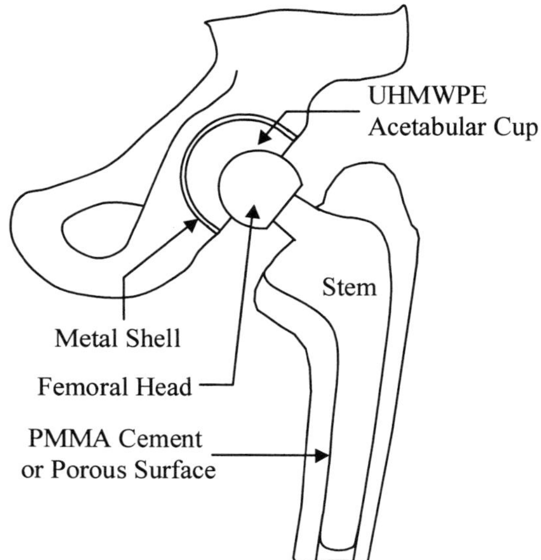

Figure 47.15. Total hip arthroplasty (THA).

Gait Analysis

The basic premise for gait analysis is that, from a knowledge of the ground reaction force acting at the foot during some activity and the corresponding kinematics, i.e., angular positions of the limbs being analyzed, the forces and moments acting at the ankle, knee, and hip joints can be calculated. A two-dimensional, rigid-link segmental mathematical model is used as in Figure 47.16 [11]. Although several methods have been used directly to measure dynamic forces during activity, such as instrumented endoprostheses [19–21], telemetrized THRs [17,18,45], or instrumented nail plates [46–48], the major advantage of gait analysis is its non-invasive nature, making it a much more elegant and patient-friendly measure. As early as the mid 1960s, hip joint load and range of motion were a focus for researchers trying to understand the biomechanics of the normal and diseased hip during dynamic activities [26,28,49–53]. Since the introduction of Charnley-type prostheses in the 1960s as a treatment,

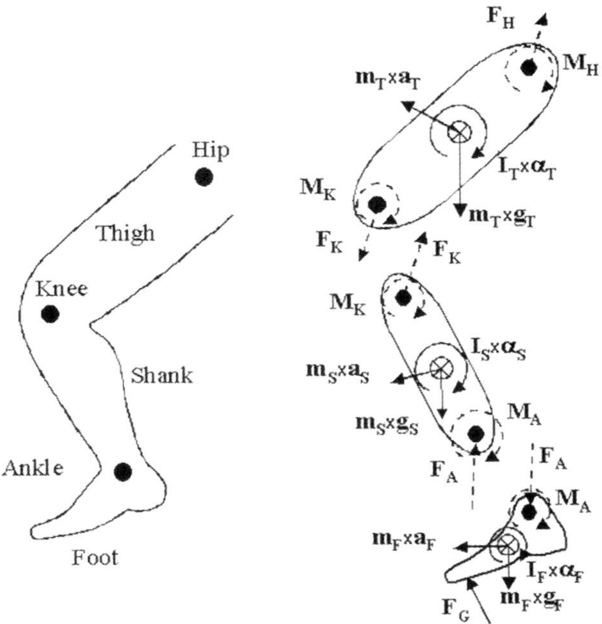

Figure 47.16. Two-dimensional rigid link segment model used in gait analysis. The thigh, shank, and foot are modeled as rigid links, each having their own mass (m), mass moment of inertia (I), and linear (a) and angular acceleration (α). From these inertial parameters and the ground reaction force (F_G), knee, ankle, and hip joint forces (F) and moments (M) may be calculated.

gait analysis has been increasingly used to link pre and postoperative THR biomechanical assessment of patients with their clinical performance [29,54].

In a recent study Tanaka [29] performed force-plate gait analysis on 24 patients with OA of the hip, 85 patients with total hip arthroplasty, and 56 normal control subjects. All subjects were females averaging 60.8 years (41–77 years), those with THA having an average postoperative period of 39.2 months (12–85 months). Temporal factors such as single-stance phase and distance parameters such as step length, gait velocity, and ground reaction force were measured. In addition, using a two-dimensional rigid segment model, Tanaka calculated hip joint moments. Normal subjects had high correlation of age with velocity, load-lifting effect, magnitude of the peak in the acceleration phase, hip abductor muscle force, flexor muscle force, and flexion moment, all of which decreased with age. The parameters unaffected by age were hip flexion–extension angle, extension moment, and abduction moment. Patients with OA of the hip demonstrated lower values for most of the measured parameters compared to the normals tested. Thirty-one of the 85 arthroplasty patients who had no pain nonetheless had a Duchenne lurch, defined as an inclination of the shoulder greater than 5 degrees toward the affected side when walking during stance phase. Although no correlation usually exists between abduction moment and abductor muscle strength, patients with this lurch did show such a correlation. During gait, 13 of the unilateral THA patients showed an insufficient extension of the hip. Hip flexion moment and walking speed were much higher in those patients who had hip extension greater than 0 degrees.

Measuring the outcome of THA using force-plate gait analysis is a more objective measure than the subjective and highly variable clinical assessments usually employed. Although force-plate gait analysis can be used to detect some changes between normals, OA patients, and THA recipients as discussed above [29], several criteria must be clearly met before gait analysis becomes a clinically viable tool [54]. First, it must be evident that gait analysis will produce data that cannot be observed during routine clinical visits. Second, gait parameters must be shown strongly to correlate with the patient's observable functional abnormalities. Third, data must not only be precise and accurate, but sensitive enough to detect small changes in patient function. Fourth, it must be clear that instrumentation and laboratory environment will negligibly affect patient function.

Contact Stresses and Weight-bearing Areas

Across a total hip replacement during articulation, the load transferred through the femoral head creates contact surface stresses on the acetabular polymer component, resulting in the creation of a polymeric weight-bearing contact area. Both stress and contact area are affected by a number of factors including load, geometry, polymer thickness, acetabular metal backing, radial clearance, component orientation, and material properties [42,55–58]. Numerous experimental methods have been employed to quantify these stresses and their areas, including Fujifilm, resistive ink sensors, micro-indentation pads, instrumented pipes, electrical contact resistors, piezoelectric transducers, photoelasticity, and ultrasound [59–61]. However, work on total hip arthroplasty has focused rather on the development of analytical and numerical methods [42,62], perhaps because of the relatively simple geometry of the THR in comparison, for example, to the TKR.

Metal–polymer contact at the joint interface causes polyethylene to deform, resulting in a complex stress distribution pattern across and through the polymer component, with compressive, tensile, and shear components as shown in Figure 47.17. Bartel et al. [56] used an elasticity solution to predict typical maximum stress values for a THR with a 3,000 N compressive load, which approximates 4.3 times body weight for a 70-kg individual. Maximum contact stresses were 13–18 MPa (compressive), shear stresses were 5–7 MPa (compressive), and stresses tangential to the polymer surface ranged from 3 MPa (tensile) to 3 MPa (compressive). All of these stresses occurred at the surface of the polymer. The only stress magnitude of concern, however, is the compressive contact stress, which may exceed the 10–15 MPa plastic yield stress of UHMWPE and contribute to stress risers in the presence of any surface defects.

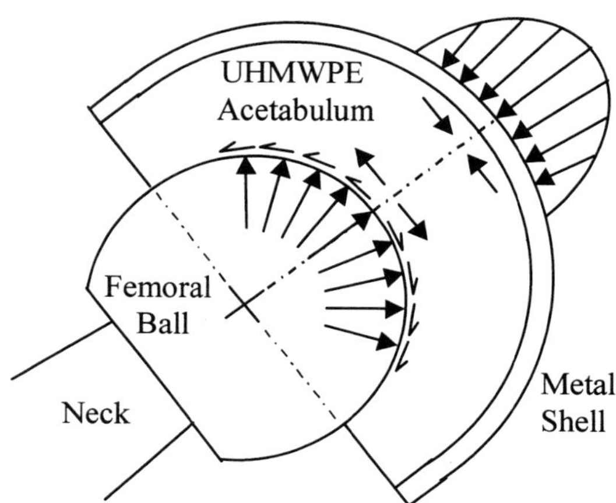

Figure 47.17. Types of stresses experienced by the acetabular cup of a THA.

Pedersen et al. [57] used a two-dimensional axisymmetric finite element model to show that increasing the thickness of the polyethylene acetabular component will reduce peak stress levels at the polymer–cement and cement–bone interfaces, a phenomenon referred to as stress shielding. Similarly, using an elasticity model, Bartel et al. [55] predicted the stress peaks for a highly conforming metal-backed acetabular component undergoing a 2,100 N compressive load. They concluded that, regardless of the clearance between the indenting femur and the acetabulum or the stiffness of the polymer acetabulum, the surface contact stresses decreased rapidly with increased polymer thickness. Peak stresses can reach almost 80 MPa, far beyond the UHMWPE plastic yield stress of about 10–15 MPa [63] for large femoro-acetabular clearances and thin polymer components. The upshot is that a minimum acetabular plastic thickness of 6 mm should be used for THRs in order to ensure that contact stresses remain below 30 MPa.

Typical radial clearances between the femoral head and acetabulum range between 0.05 and

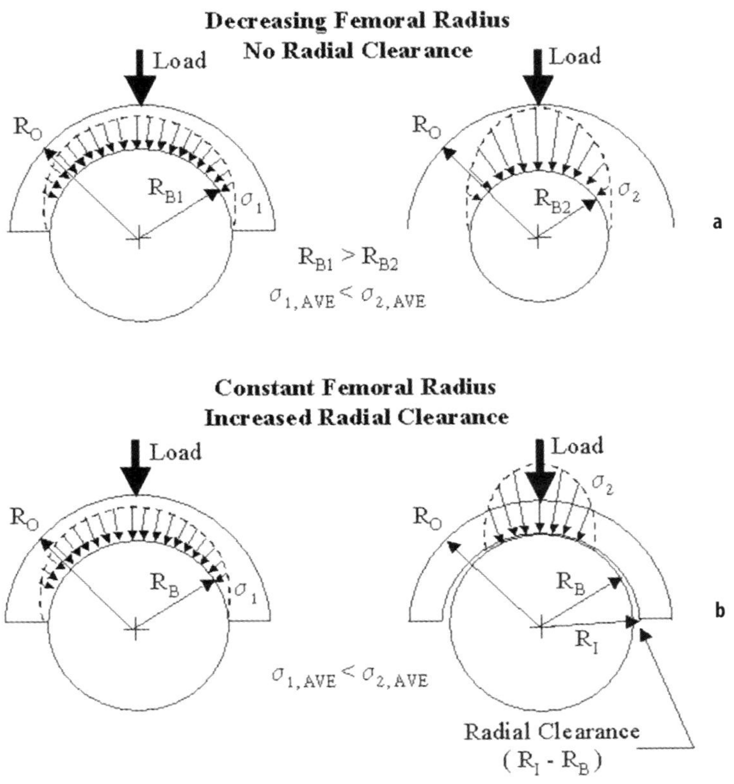

Figure 47.18. The effect of femoral head diameter and radial clearance on THA contact stress at a fixed load. **a** For no radial clearance, greatest contact stress occurs for smallest femoral ball. **b** For a given femoral ball, increasing radial clearance generates a rise in contact stresses.

0.14 mm and will have an effect on peak contact stress levels [41]. Bartel et al. [55] demonstrated this for 11, 14, and 16 mm radius femoral balls indenting a plastic acetabular cup of fixed outer radius. For zero clearance, i.e., complete conformity, the greatest contact stresses occurred with the smallest femoral head because the smallest contact area was engaged (Figure 47.18a). However, for a given femoral ball, radial clearance was increased up to 0.4 mm by decreasing the inner radius of the acetabular cup, causing peak stress to reach 25–29 MPa, moving beyond the plastic yield of the material (Figure 47.18b). The lesson here is simple; the greater the clearance, the greater the contact stress. Stresses above the material plastic yield could in turn initiate a polymer damage sequence leading to eventual implant failure. However, it must be noted that clearances change with time, given the creep of the polyethylene, making it difficult to calculate the ideal femoral head size and to predict the contact stresses of a THR over time [41].

In their three-dimensional computer model of a THR, Robinson et al. [58] examined the relationship between contact area and component position for spherical and truncated femoral head designs. Femoral components were oriented at 0, 10, and 25 degrees of anteversion while the acetabular component was placed in 0, 10, and 25 degrees of anteversion and 30, 40, 45, and 50 degrees of abduction. The resulting 48 combinations of implant positions were tested with five directions of maximum joint motion associated with hip joint dislocation. For both femoral ball designs, the most influential factor on contact area was acetabular abduction, with contact area reaching a maximum of 9.1 cm^2 as acetabular abduction became horizontal. In spherical head designs, however, femoral head anteversion does not affect contact area at all, being the same for a given acetabular orientation.

THR Polymer Wear

One of the most important factors in the long-range survival of THRs is the wear resistance of the polymer acetabular component. Although significant strides have been made recently with optimally cross-linked UHMWPE that is more wear resistant, data need to be forthcoming to assess the long-term physiological survival time of this improved polymer [64–70]. Current THAs last 10 to 20 years before revision surgery is required [42]. The extent to which THA wear occurs will be affected by numerous variables including polymer quality, polymer thickness, manufacturing procedure, method of sterilization, femoral head diameter and material, acetabular metal backing, lubrication, and motion pattern, as well as patient age, gender, weight, and activity level [71–74].

Although UHMWPE wear is a multifactorial problem, the mechanism attributed directly to polymer surface degradation is high levels of femoro–acetabular interfacial friction, the two forms of which are abrasion and adhesion [72,74,75]. *Abrasion* may be defined as a roughening of the polymer component's surface caused by the plowing action of either the asperities of the harder metal ball or by third-body wear, which occurs as extraneous particles of bone or bone cement find their way into the interstitial joint space. *Adhesion*, or burnishing, results in the micropolishing of the bearing surface that exceeds the smoothness of the original surface quality, involving the removal of fine particulate debris from the weaker acetabular surface as well as surface heating. This is unlike the case for total knee implants, which degrade primarily due to a fatigue phenomenon characterized by alternating high contact stresses, inducing the appearance of surface pitting and delamination [72].

The surface degradation of polymer components causes the release of wear particles in very large numbers, most of which are flake-shaped, measuring several micrometers in width and length, but are often less than one micrometer in size [74]. THA in vivo wear studies have indicated that a volume of between 15 and 860 mm^3 of polymer particles per year can be generated [76]. These particles in turn induce an inflammatory foreign-body tissue reaction and the production of damaging enzymes. Osteolysis around the implant components follows due to resorption of periprosthetic bone. This is the leading cause of implant loosening [77]. Although this is the commonly accepted understanding, recent work by Kesteris et al. [78] has challenged the notion that polyethylene debris causes osteolysis by its direct action. These authors hypothesize rather that the particles act indirectly by increasing intracapsular pressure. In their study of 48 cemented total hips at 10-year follow-up, their aim was to discover whether a correlation existed between the thickness of the synovium or synovial contents (i.e., capsular distance), radiographic loosening, and polyethylene wear. Although they found that linear wear, volumetric wear, and capsular distance all correlated positively with the width of radiolucent lines around acetabular components, such was not the case with radiographic signs of femoral loosening. Because a thickened capsule has been demonstrated to elevate peak intracapsular pressures during hip motion [79,80], Kesteris et al. propose that the presence of polyethylene particles is secondary to the rise in fluid pressure in inducing osteolysis [81].

Another important development has been in the understanding of the kinematic conditions of THR contact, i.e., the motion patterns at the femoro–acetabular surfaces, which has helped redefine the previous standard in vitro screen testing procedures of the American Society for Testing and Materials [82]. The accepted practice until recently has been a simple linear metal-pin-on-flat-polymer-disk reciprocation, from which either measurements are taken of linear wear (linear penetration of the metal pin into the polymer disk), volumetric wear, or polymer mass change [74]. Three-dimensional THR computer simulations of Ramamurti et al. [83] and Pedersen et al. [84], however, have shown a series of adjacent elliptical contact paths that cross each other, indicating the bi-directional shear that a point on the polymer surface undergoes. Using a hip simulator,

Bragdon et al. [85] have demonstrated that during normal gait this cross-over pattern creates significantly greater wear (net loss of 24.8 mg/million cycles) than motion patterns with only uni-directional wear (net gain of 2 mg/million cycles). Similarly, McKellop et al. [86] report that in vitro screening tests using linear uni-directional motion yield wear rates 10 to 100 times lower than cross-over motion paths, which actually create the type and amount of wear more closely resembling that found from in vivo studies of THRs [75,83]. To incorporate this new understanding of kinematic contact conditions, pin-on-flat testing machines are being developed that reciprocate in a figure-eight pattern [87], which could become the standard in vitro wear testing configuration since, at this time, no such standard exists.

The Implant–Cement–Bone Interface

By far the regions most susceptible to failure in THAs are those where two surfaces having dissimilar mechanical properties are in direct contact, namely implant–cement, bone–cement, and implant–bone interfaces [41]. The failure of the interface involving PMMA cement (polymethylmethacrylate) is most likely because this requires not only the proper preparation of the cement but also adequate removal of fat and blood from the surface of interest. Back-flow of blood from the bone, for example, has been shown in bench tests to push low-viscosity cement out of a 1 mm diameter hole it had penetrated [88,89]. This situation may consequently cause poor cement penetration, cement cracks, or debonding, leading to an overload of the hip implant and, hence, fracture. Pulsating water lavage and brushing, pressure cementing, precoating with cement, or using an implant with a textured surface, have been shown to minimize these effects and increase bone adhesion and penetration [41,89]. Even so, as Charnley [90] pointed out in his 12–15 year review of his hip operations, although 25% of the cemented sockets demonstrated significant demarcation of the bone–cement interface and even implant migration, the results can be clinically successful.

Static strength bench tests done on bone–cement and implant–cement interfaces have shown a wide range of values (Table 47.2), as they are dependent on differences in test methods, cement curing time, temperature and pressure during joining, surface preparation, mixing techniques of the cement, specimen storage, cement type, implant surface texturing, etc. [41]. It must be remembered, however, that it is the fatigue or cyclic loading that is the real culprit in causing interfacial failure and, thus, reliance on static strength reports should be done cautiously in predicting failure.

Recent Advances and Future Trends in THRs

Although it appears that the shape of THRs has reached a plateau, recent advances and investigations especially in material selection promise to take hip arthroplasty to the next level of development. For example, hip simulator tests

Table 47.2. Interface strength values (selected data from Williams [41])

Interface or Material	Tensile Strength (MPa)	Shear Strength (MPa)	Compressive Strength (MPa)
PMMA	20–26	40–50 (0.001 s^{-1})	95
Cancellous bone	0.7–5.0	7–14 (s^{-1}) 1–2 (0.01 s^{-1})	0.15–16
Old PMMA/New PMMA		23	
PMMA/Stainless steel	6–11	6.2–11.2	
PMMA/Co-Cr-Mo	6.7–9.1	5.3–6.9	
PMMA/Ti-6Al-4V	3.9–8.3	6.3–12.5	
PMMA/UHMWPE		0.2–2.3	
PMMA/Cancellous bone	2–5 (LP) 6–9 (HP)	12–30 (LP) 35–48 (HP)	

LP: Cement under low pressure. HP: Cement under high pressure.

on new, optimally cross-linked UHMWPE, commercially available as Crossfire™ (Howmedica-Osteonics), Durasul™ (Sulzer Medica), Longevity™ (Zimmer), and Marathon™ (DePuy), have demonstrated 90–95% wear reduction in comparison to standard polyethylene [64–66,69,91–93]. Although this new material has shown some promise in clinical studies [68,70], it is questionable whether similar results can be obtained for TKRs due to their more complex geometry and, hence, higher contact stress levels [69].

Increased use of modular ceramic femoral heads and sockets that are modular with metal stems and shells is already occurring and may become more common, primarily because of the lower linear wear rates of ceramic-on-UHMWPE, being 0.05 mm per year in vitro, compared with metal-on-UHMWPE, which is 0.1–0.2 mm per year [94]. Ceramic-on-ceramic, also called alumina-on-alumina, articulation seems even more promising from a tribological perspective, with linear wear rates of 0.005 mm per year [95]. Some researchers are rethinking the use of metal-on-metal articulation due to the wear rate being 100 times less than metal-on-polymer if sufficient tolerance can be maintained [94,96–98]. There is also evidence of polyethylene wear reduction by texturing the femoral ball with concave dimples as in a golf ball [99]. Using a hip simulator, it was found that wear debris was reduced from 23.1 to 7.2 mg when surface dimpling was introduced, the dimples acting as storage crypts for both wear particles and lubricant. These developments are possible because of more sophisticated implant manufacturing technology, which is able to give greater control over surface texture and shape and, hence, the tribological characteristics of implants. Very precise preparation can greatly improve sphericity of the contact surfaces and minimal asperities. The clearance between socket and ball is 10 μm, which may encourage a thin film of tissue joint fluid for hydrodynamic lubrication.

Carbon–carbon composite stems, rather than metal, also seem promising since the material modulus more closely resembles that of bone, thereby reducing stress gradients at the bone–stem interface [43,94,100]. The major advantage of such an "isoelastic" design is that the components are anisotropic, meaning that the material properties will vary depending on the orientation of the stem.

Other advances may provide increased THA longevity, greater patient satisfaction, and provide surgeons with more tools and approaches at their disposal to treat arthritis patients successfully. These include the use of hydroxyapatite (HA) coatings on porous and textured surfaces to increase bony ingrowth, the use of surgical robots, and custom-designed implants [43,94].

References

1. Robbins CE. Anatomy and biomechanics. In: Fagerson TL, editor. The Hip Handbook. Boston: Butterworth-Heinemann, 1998;1–37.
2. Nordin M, Frankel VH. Biomechanics of the hip. In: Nordin M, Frankel VH, editors. Basic Biomechanics of the Musculoskeletal System. Philadelphia: Lea and Febiger, 1989;135–51.
3. Rydell N. Biomechanics of the Hip Joint. Clin Orth Rel Res 1973;92:6–15.
4. Radin EL. Biomechanics of the Human Hip. Clin Orth Rel Res 1980;152:28–34.
5. Greenwald AS. Biomechanics of the hip. In: Steinburg ME, editor. The Hip and Its Disorders. Philadelphia: WB Saunders, 1991;47–55.
6. Chung SMK. Hip Disorders in Infants and Children. Philadelphia: Lea and Febiger, 1981.
7. Harty M. Anatomy. In: Steinburg ME, editor. The Hip and Its Disorders. Philadelphia: WB Saunders, 1991; 27–46.
8. Olson SA, Bay BK, Hamel A. Biomechanics of the hip joint and the effects of fracture of the acetabulum. Clin Orth Rel Res 1997;339:92–104.
9. Kapandji IA. The Physiology of the Joints – Vol. 2: Lower Limb. Edinburgh and London: Churchill Livingstone, 1970;64.
10. Bombelli R. Structure and Function in Normal and Abnormal Hips. New York: Springer-Verlag, 1993.
11. Hurwitz DE, Andriacchi TP. Biomechanics of the Hip. In: Callaghan JJ, Rosenberg AG, Rubash HE, editors. The Adult Hip. Philadelphia: Lippincott-Raven, 1998; 75–85.
12. Ferguson S. Biomechanics of the Acetabular Labrum. 2000. PhD Thesis. Queen's University, Kingston, ON, Canada.
13. Terayama K, Takei T, Nakada K. Joint Space of the Human Knee and Hip Joint under Static Load. Engineering in Medicine 1980;9:67–74.
14. Mow VC, Soslowsky LJ. Lubrication and Wear of Joints. In: Mow VC, Hayes WC, editors. Basic Orthopaedic Biomechanics. New York: Raven Press, 1991;245–92.
15. Mow VC, Foster FJ. Tribology. In: Callaghan JJ, Rosenberg AG, Rubash HE, editors. The Adult Hip. Philadelphia: Lippincott-Raven, 1998;217–29.

16. Armstrong CG, Bahrani AS, Gardner DL. In vitro measurement of articular cartilage deformation in the intact human hip joint under load. JBJS 1979;61A: 744–55.
17. Davy DT, Kotzar GM, Brown RH et al. Telemetric force measurements across the hip joint after total arthroplasty. JBJS 1989;70A:45–50.
18. Bergmann G, Graichen F, Rohlmann A. Hip joint loading during walking and running, measured in two patients. J Biomech 1993;16:969–90.
19. Rushfeldt PD, Mann RW, Harris WH. Improved techniques for measuring in vitro the geometry and pressure distribution in the human acetabulum – 2. Instrumented endoprosthesis measurement of articular surface pressure distribution. J Biomech 1981;14: 315–23.
20. Carlson CE. A proposed method for measuring pressures on the human hip joint. Experimental Mechanics 1971;499–506.
21. Carlson CE, Mann RW, Harris WH. A radio telemetry device for monitoring cartilage surface pressures in the human hip. IEEE Trans Biomed Eng 1974;BME-21: 257–64.
22. Brown TD, Shaw DT. In vitro contact stress distributions in the natural human hip. J Biomech 1983;16: 373–84.
23. Adams D, Kempson GE, Swanson SAV. Direct measurement of local pressures in the cadaveric human hip joint. Med & Biol Eng & Comput 1978;16:113–15.
24. Day WH, Swanson SAV, Freeman MAR. Contact pressures in the loaded human cadaver hip. JBJS 1975; 57B:302–13.
25. Li J, Wyss UP, Costigan PA, Deluzio KJ. An integrated procedure to assess knee-joint kinematics and kinetics during gait using an optoelectric system and standardized X-rays. J Biomed Eng 1993;15:392–400.
26. Paul JP. Bio-engineering studies of the forces transmitted by joints, Part II – Engineering analysis. In: Kenedi R, editor. Biomechanics and Related Bioengineering Topics. Oxford: Pergamon, 1965;369–80.
27. Rohrle H, Scholten R, Sigolotto C et al. Joint forces in the human pelvis–leg skeleton during walking. J Biomech 1984;17:409–24.
28. Sorbie C, Zalter R. Bio-engineering studies of the forces transmitted by joints, Part I – The phasic relationship of the hip muscles in walking. In: Kenedi R, editor. Biomechanics and Related Bioengineering Topics. Oxford: Pergamon, 1965;359–67.
29. Tanaka Y. Gait analysis of patients with osteoarthritis of the hip and those with total arthroplasty. Bio-Med Mater Eng 1998;8:187–96.
30. Miyanaga Y, Fukubayashi T, Kurosawa H. Contact study of the hip joint. Arch Orthop Trauma Surg 1984;103: 13–17.
31. Afoke NYP, Byers PD, Hutton WC. Contact pressures in the human hip joint. JBJS 1987;69B:536–41.
32. Harrison MHM, Schajowitcz F, Trueta J. Osteoarthritis of the hip: a study of the nature and evolution of the disease. JBJS 1953;35B:593
33. Strange C. The Hip. London: W Heinemann, 1969.
34. Olson SA, Bay BK, Chapman MW et al. Biomechanical consequences of fracture and repair of the posterior wall of the acetabulum. JBJS 1995;77A:1184–92.
35. Olson SA, Bay BK, Pollak AN et al. The effect of variable size posterior wall acetabular fractures on contact characteristics of the hip joint. J Orth Trauma 1996; 395–402.
36. Poole AR, Rizkalla G, Reiner A, Ionescu M, Bogoch E. Changes in the extracellular matrix of articular cartilage in human osteoarthritis. In: Hirohata R, Mizuno K, Matsubara T, editors. Trends in Research and Treatment of Joint Diseases. Tokyo: Springer-Verlag, 1992.
37. Dalstra M, Huiskes R. Load transfer across the pelvic bone. J Biomech 1995;28:715–24.
38. Lim L-A, Carmichael SW, Cabanela ME. Biomechanics of total hip arthroplasty. Anat Rec 1999;257:110–16.
39. Maquet PGJ, editor. Biomechanics of the Hip: As applied to Osteoarthritis and Related Conditions. New York: Springer-Verlag, 1985.
40. Pauwels F. Biomechanics of the Normal and Diseased Hip: Theoretical Foundation, Technique and Results of Treatment. Berlin: Springer-Verlag, 1976.
41. Williams JL. Biomechanics of total hip replacement. In: Steinburg ME, editor. The Hip and its Disorders. Philadelphia: WB Saunders, 1991;876–904.
42. Huiskes R, Verdonschot N. Biomechanics of artificial joints: the hip. In: Mow VC, Hayes WC, editors. Basic Orthopaedic Biomechanics. Philadelphia: Lippincott-Raven, 1997;395–460.
43. Siopack JS, Jergesen HE. Total hip arthroplasty. West J Med 1995;162:243–9.
44. Wroblewski M. Cementless versus cemented total hip arthroplasty: a scientific controversy? Orthop Clin North Am 1993;24:591–7.
45. English TA, Kilvington M. In vivo records of hip loads using a femoral implant with telemetric output: a preliminary report. J Biomed Eng 1979;1:111.
46. Frankel VH, Burstein AH, Lygre L, Brown RH. The telltale nail. JBJS 1971;53A:1232.
47. Lygre L. The loads produced on the hip joint by nursing procedures: A telemetrization study. 1970. MS Thesis. Case Western Reserve University.
48. Milde FK. Loads on femoral head during nursing care by a telemetrized nail-plate. 1974. MS Thesis. Case Western Reserve University.
49. Johnson RC, Smidt GL. Measurements of hip-joint motion during walking: evaluation of an electrogoniometric method. JBJS 1969;51A:1083–94.
50. Murray MP. Gait as a normal pattern of movement. Am J Phys Med 1967;46:290–333.
51. Paul JP. Forces at the hip joint. 1967. PhD Thesis. University of Chicago.
52. Rydell N. Forces in the hip joint. Part II – Intravital measurements. In: Kenedi R, editor. Biomechanics and Related Bioengineering Topics. Oxford: Pergamon, 1965;351–7.
53. Rydell N. Forces acting on the femoral head prosthesis. A study on strain gauge pressure prostheses in living persons. Acta Orthop Scand 1966;S88:1–132.
54. Chao EY-S, Kaufman KR, Stauffer RN. Biomechanics. In: Morrey BF, editor. Joint Replacement Arthroplasty. New York: Churchill Livingstone, 1991;529–46.
55. Bartel DL, Burstein AH, Toda MD, Edwards DL. The effect of conformity and plastic thickness on contact stresses in metal-backed plastic implants. J Biomech Eng 1985;107:193–9.
56. Bartel DL, Bicknell VL, Wright TM. The effect of conformity, thickness and material on stresses in ultra-high molecular weight components for total joint replacement. JBJS 1986;68A:1041–51.

57. Pedersen DR, Crowninshield RD, Brand RA, Johnston RC. An axisymmetric model of acetabular components in total hip arthroplasty. J Biomech 1982; 15:305–16.
58. Robinson RP, Simonian PT, Gradisar IM, Ching RP. Joint motion and surface contact area related to component position in total hip arthroplasty. JBJS 1997;79B:140–6.
59. Lewis G. Contact stress at articular surfaces in total joint replacements. Part 1: Experimental methods. Bio-Med Mater Eng 1998;8:91–110.
60. Zdero R. A new diagnostic ultrasound technique for studying TKR contact mechanics. 1999. PhD Thesis. Queen's University, Kingston, ON, Canada.
61. Zdero R, Fenton PV, Rudan J, Bryant JT. Fuji film and ultrasound measurement of total knee arthroplasty contact areas. J Arthroplasty 2001;16(3):367–375.
62. Lewis G. Contact stress at articular surfaces in total joint replacements. Part 2: Analytical and numerical methods. Bio-Med Mater Eng 1998;8:259–78.
63. Stewart T, Shaw D, Auger DD, Stone M, Fisher J. Experimental and theoretical study of the contact mechanics of five total knee replacements. Proc Inst Mech Engrs (Part H) 1995;209:225–31.
64. Clark IC, Good V, Williams P, Oparaugo P, Oonishi H, Fujisawa A. Simulator Wear study of high-dose gamma-irradiated UHMWPE Cups. 23rd Ann Meeting of the Society for Biomaterials, 1997, 71.
65. Jasty M, Bragdon CR, O'Connor DO, Muratoglu O, Permnath V, Merrill E. Marked improvement in the wear resistance of a new form of UHMWPE in a physiologic hip simulator. Trans Soc Biomater 1997;20: 157.
66. McKellop H, Shen F, Yu Y, Lu B, Salovey R, Campbell P. Effect of sterilization method and other modifications on the wear resistance of UHMWPE acetabular cups. In: Anonymous Polyethylene Wear in Orthopaedic Implants Workshop. Minneapolis: Society for Biomaterials, 1997;20–31.
67. McKellop H, Shen F, Salovey R. Extremely low wear of gamma crosslinked/remelted UHMW polyethylene acetabular cups. 44th Ann Meeting of the Orthopedic Research Society (ORS). 1998;97–17.
68. Oonishi H, Saito M, Kadoya Y. Wear of high-dose gamma irradiated polyethylene in total joint replacement – long term radiological evaluation. 44th Ann Meeting of the Orthopedic Research Society (ORS). 1998;97–17.
69. Wang A, Essner A, Polineni V, Sun D, Stark C, Dumbleton J. Joint space of the human knee and hip joint under static load. In: Polyethylene Wear in Orthopaedic Implants Workshop. Minneapolis: Society for Biomaterials, 1997;4–18.
70. Wroblewski BM, et al. Prospective clinical and joint simulator studies of a new total hip arthroplasty using alumina ceramic heads and cross-linked polyethylene cups. JBJS 1996;78B:280–5.
71. Bankston AB, Keating EM, Ranawat C, Faris PM, Ritter MA. Comparison of polyethylene wear in machined versus molded polyethylene. Clin Orth Rel Res 1995; 317:37–43.
72. Cornwall GB, Bryant JT, Hansson CM, Rudan J, Kennedy LA, Cooke TDV. A quantitative technique for reporting the surface degradation patterns of UHMWPe components of retrieved total knee replacements. J Appl Biomater 1995;6:9–18.
73. Schmalzried TP, et al. Quantitative assessment of walking activity after total hip or knee replacement. JBJS 1998;80A:54–59.
74. Schmalzried TP, Callaghan JJ. Wear in total hip and knee replacements. JBJS 1999;81A:115–36.
75. McKellop H, Campbell P, Park SH, et al. The origin of submicron polyethylene wear debris in total hip arthroplasty. Clin Orth Rel Res 1995;311:3–20.
76. Schmalzried TP, Dovey FJ, McKellop H. Commentary: The multifactorial nature of polyethylene wear in vivo. JBJS 1998;80A:1234–42.
77. Lewis G. Design issues in clinical studies of the in vivo volumetric wear rate of polyethylene bearing components. JBJS 2000;82A:281–7.
78. Kesteris U, Jonsson K, Robertsson O, Onnerfalt R, Wingstrand H. Polyethylene wear and synovitis in total hip arthroplasty. J Arthroplasty 1999;14:138–43.
79. Robertsson O, Wingstrand H, Kesteris U et al. Intracapsular pressure and loosening of hip prostheses: preoperative measurements in 18 hips. Acta Orthop Scand 1997;68:231.
80. Wingstrand H, Wingstrand A. Biomechanics of the hip joint capsule – a mathematical model and clinical implications. Clin Biomech 1997;22:273.
81. Schmalzried TP, Akizuki KH, Fedenko AN, Mirra J. The role of joint fluid in periarticular osteolysis. JBJS 1997;79A:447.
82. ASTM. Standard practice for reciprocating pin-on-flat evaluation of friction and wear properties of polymeric materials for use in total joint prostheses (ASTM F732-82). In: American Society for Testing Materials. Philadelphia: ASTM, 1991;262–9.
83. Ramamurti BS, Bragdon CR, O'Connor DO, et al. Loci of movement of selected points on the femoral head during normal gait. J.Arthroplasty 1996;11: 845–52.
84. Pedersen DR, Brown TD, Maxian TA, Callaghan JJ. Temporal and spatial distributions of directional counterface motion at the acetabular bearing surface in total hip arthroplasty. Iowa Orthop J 1998;18:43–53.
85. Bragdon CR, O'Connor DO, Lowenstein JD, Jasty M, Synuita WD. The importance of multidirectional motion on the wear of polyethylene. Proc Inst Mech Eng (Part H) 1996;210:157–65.
86. McKellop H, Clarke I, Markolf KL, Amstutz HA. Friction and wear properties of polymer, metal and ceramic prosthetic joint materials evaluated on a multichannel screening device. J Biomed Mater Res 1981;15:619–53.
87. McConnell AJ, Bryant JT. Differences in surface damage and morphology between conventional and highly cross-linked UHMWPE in multidirectional testing at high contact stresses. Tampa, FL, USA. 28th Ann Meeting Transactions, Society for Biomaterials, 2002, p. 583.
88. Benjamin JB, Gie GA, Lee AJ. Cementing technique and the effects of bleeding. JBJS 1987;69B:620.
89. Markolf KL. Biomechanics of the hip. In: Amstutz HC, editor. Hip Arthroplasty. New York: Churchill Livingstone, 1991;15–23.
90. Charnley J. Low-friction Arthroplasty of the Hip: Theory and Practice. Berlin: Springer, 1979.
91. Howmedica-Osteonics, see web site for more information about Crossfire™ polyethylene: www.osteonics.com/osteonics/hips/crossfire/oscfsplash.htm.

92. Sulzer Medica, see official Sulzer Medica world wide web sites www.durasul.com and www.sulzerorthoeu.ch/technology/tribology/durasul/index.html, Sulzer Orthopedics Ltd., 2000.
93. Zimmer. Longevity Crosslinked Polyethylene: Design Rationale. commercial booklet from Zimmer, Inc., 1999.
94. Bargar WL. New developments and future trends in total hip replacement. In: Steinburg ME, editor. The Hip and Its Disorders. Philadelphia: WB Saunders, 1991;1125–33.
95. Willmann G. Ceramics for total hip replacement – what a surgeon should know. Orthopedics 1998;21: 173–7.
96. Black J. Metal on metal bearings. Clin Orth Rel Res 1996;329S:S244–55.
97. Chan FW, Bobyn JD, Medley JB, Krygier JJ, Yue S, Tanzer M. Engineering issues and wear performance of metal on metal hip implants. Clin Orth Rel Res 1996;333:96–107.
98. Schmalzried TP, Peters PC, Maurer BT, Bragdon CR, Harris WH. Long-duration metal-on-metal total hip arthroplasties with low wear of the articulating surfaces. J Arthroplasty 1996;11:322–31.
99. Ito H, Kaneda K, Yuhta T, Nishimura I, Yasuda K, Matsuno T. Reduction of polyethylene wear by concave dimples on the frictionless surface in artificial hip joints. J Arthroplasty 2000;15:332–8.
100. Christel P, et al. Development of a carbon–carbon hip prosthesis. J.Biomed Mater Res 1987;21:191–218.

48 The Biomechanics of Ligaments

A. A. Amis

Introduction

Ligaments are passive collagenous structures that act primarily as tensile restraints to control the distance between their attachment points. Ligaments normally traverse joints, and so they act to control the relative separation of the bones that they are attached to. Hence, the ligaments control the patterns of movement, or kinematics, of joints, as well as ensuring the stability of joints. In addition to this simple mechanical description of the role of ligaments, they provide more subtle control of joint motion and stability via proprioceptive feedback to the muscles, but this will not be addressed here. This chapter will review the mechanical properties of the ligaments themselves, and then look at how the ligaments act to stabilize joints.

Tensile Properties

Ligament Structure

The structure of ligamentous tissue is well known, with a clearly discernible hierarchical organization. The smallest units are the microscopic collagen protofibrils, themselves consisting of helically arranged long-chain amino acid molecules that group together into fibrils. The fibrils are grouped together into collagen fascicles, and a ligament consists mostly of an array of these fascicles [1]. Different texts describe the microstructure slightly differently, but the overall structure is one where all of the load-bearing collagenous constituents, from the molecular to the macro scale, are aligned closely to the long axis of the structure. The result of this is that ligaments are well adapted to transmit tensile forces. If the structures of individual ligaments are studied in detail, it can be seen that the fascicles may take a helical path, or perhaps fan out to bone attachment areas that are greater than the cross-sectional area of the ligament at its mid part. This has been documented for the cruciate ligaments, for example [2]. As a result of this, there are spaces for ground substance between the fascicles, a mucoid substance with a relatively small contribution to the failure strength. A further consequence of the relatively loose collagen fascicle packing is that ligaments are not as strong as tendons for a given cross-sectional area (i.e., ligaments reach a lower tensile stress before failure), tendons having a denser and more orderly microstructure. Typically, ligaments have an ultimate tensile strength of 36 MPa [3], while tendons may reach 70 MPa [4].

Tensile Test Methods

Since the transmission of tensile load is the main duty of ligaments, it is worth studying their tensile behavior in detail. In order to perform a tensile test on a ligament, it must be prepared for mounting in a tensile test machine. In normal mechanical engineering practice, tensile test specimens are made with a "dog bone" shape that has wide ends for gripping in the vice attachments of the test machine, and a relatively long and narrow central section that is analyzed during the test. This avoids stress concentrations, or the localized effects of the distortions caused by the grips, from affecting the gauge length under study. This approach is not available for ligaments since they do not have sufficient length, and the inter-fibril shear strength is not great. Experience has shown, then, that it is best to test a ligament as a part of a complete bone–ligament–bone complex. This allows either the whole bone to be clamped into the test machine, or for the bone attachments to be isolated and embedded in bone cement or similar potting compound, into holders such as metal tubes that can then be mounted. When this method was used to test the strength of the anterior cruciate ligament [5,6], a greater strength was found than when the ligament fibers were gripped directly [7].

Once mounted in the test machine, the ligament is elongated at a chosen speed (the effects of speed of loading are discussed below), either to a sub-failure load that can be repeated to study cyclic effects, or else until the construct fails. This normally results in a graph of force versus elongation (Figure 48.1a). The measurement of elongation is difficult to do meaningfully on such an irregular specimen geometry as a ligament, and the most common approach is to simply record the movement of the test machine actuator. This, however, assumes that the ligament strength and stiffness are low in comparison to the stiffness of the test machine and the bones in their mountings. It is closer to the truth to monitor the bone–bone separation by means of an extensometer instrument, or else to do a separate stiffness test on the experimental set-up that can correct the tensile test curves.

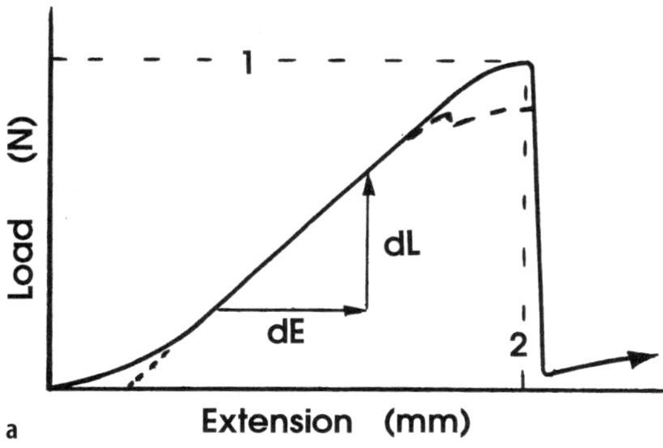

Figure 48.1. a Typical load-extension graph for a bone–ligament–bone preparation that gives structural properties such as failure strength (1) and elongation to failure (2). The interrupted line indicates a premature failure, with sequential fiber ruptures at sub-maximal load. The stiffness is found by dividing the change in load by the change in elongation, so k = dL / dE.

Structural Properties

It is important to note that the force–elongation graph gives details of the *structural properties* of the bone–ligament–bone complex being tested. This means that the properties measured are relevant to that structure, and are not the material properties of the ligament itself. If the ligament had twice the cross-sectional area, and were made of the same material, then the structural property of failure load would be expected to double, yet the material of the ligament remains the same. The structural properties that we can find from the force–elongation graph include the elongation to failure (measured in millimeters, mm), the ultimate strength, or failure load (measured in Newtons, N), and the stiffness of the structure, expressed as the amount of force required to cause a given elongation, so the units are N mm^{-1} (Figure 48.1a).

Material Properties

If we need to know about the properties of the ligament material itself, then the structural properties graph must be further processed to yield the *material properties*. Instead of the failure load, we now divide the load by the cross-sectional area of the ligament to give the stress. The units of stress are derived from force (N) per unit of cross-sectional area (mm^2), or N mm^{-2}. In the S.I. system, the unit of stress (or pressure) is the Pascal (Pa), and 1 Pa is 1 N m^{-2}. This is a tiny pressure (atmospheric pressure being approximately 100,000 Pa), so engineers usually express stress in MegaPascals (MPa), which is the same as N mm^{-2}. Similarly, if the elongation is expressed as a percentage of the original length, then we get the tensile strain as a percentage. Strain is just a ratio, and so it has no units. If we wish to derive a value for the tensile stiffness of the ligament material, the tensile modulus, then we divide a change in stress by the corresponding change in strain (Figure 48.1b). Since stress has units of MPa and strain has no units, it follows that the modulus has the same units as stress, MPa. Thus, the material properties that we obtain include the failure stress, the failure strain, and the tensile modulus (stiffness).

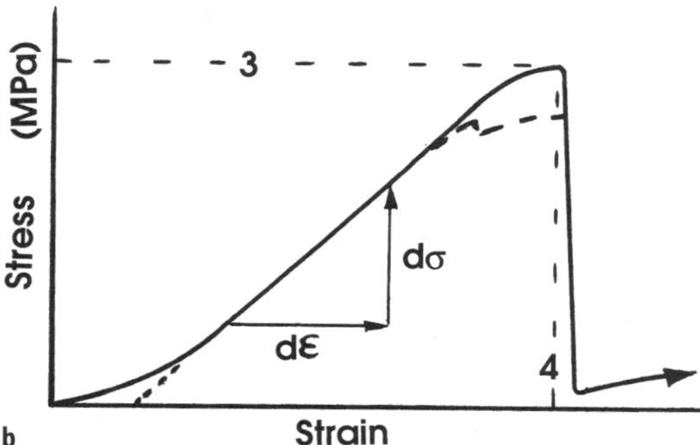

Figure 48.1. **b** Typical stress-strain graph, derived from the load-extension graph by dividing the load by the cross-sectional area to give stress, and elongation by the original length to give strain. These are material properties: failure stress (3) and failure strain (4). The tensile modulus is represented by the steepness of the curve, i.e. change in stress $d\sigma$ divided by change in strain $d\varepsilon$. Strain is often reported as a percentage length change, so 0.04 strain is 4% elongation.

The calculation of tensile stresses requires measurement of the cross-sectional area of the ligament being tested. This cannot be done easily, since the cross-section is always irregular, and the material is very soft, which precludes any means that contacts the ligament. Some papers have used an instrument called an area micrometer [8]. This squeezes the ligament into a regular rectangular recess at a known pressure, but work has shown that this causes fluid to be expelled from the tissue, so that the apparent area diminishes with both time and pressure, causing an underestimation of the area [9]. Another method is the laser micrometer, which illuminates the ligament and estimates the cross-sectional area from an integration of the width of the shadow of the ligament when the specimen is rotated in the laser beam [10]. This also has errors due to the shadow failing to measure any hollows in the shape of the cross-section. A more practical method has been to use silicone rubber to produce a mold of the ligament in situ, and then to measure the area of the cavity in the mold after removing it from the specimen [9].

Fiber Crimp and Recruitment

In contrast to many relatively simple engineering materials that exhibit linear elastic behavior, ligaments and other collagenous tissues show non-linear tensile behavior. The graphs in Figure 48.1 show an initial region of low stiffness (i.e., the graph has a low slope that indicates that elongation requires a small force), followed by a region of strain stiffening (i.e., the slope of the graph gets steeper), before it settles into approximately linear behavior. The initial "toe region" typically occurs over 4% strain and is caused by two factors: progressive tightening of ligament fibers (recruitment), and also the straightening out of fiber crimp. If an unloaded specimen of ligament tissue is viewed under a microscope with polarized light, a clear pattern of alternating color bands is seen, and these show up as a zigzag fiber configuration. The crimp pattern is gradually extinguished as the ligament is extended [11]. It is relatively easy to straighten out the crimp pattern, but after that the fibers must be stretched directly in tension. Hence, strain stiffening occurs. If all the fibers are loaded together, and this is difficult for a non-uniform structure such as a ligament, the point at which the collagen crimp is extinguished corresponds to the transition from the curved "toe" region to linear load-extension behavior [11].

The main factor causing progressive stiffening during ligament extension is fiber recruitment. Since ligaments attach to the bones over areas, and only one point can be at the axis of rotation, it follows that joint motion will cause some fibers to tighten or to slacken. Thus, fibers may need to be tightened to the point of the slack–taut transition before they start to resist tensile loads. This can mean that a uniform extension of a ligament leads to progressive recruitment of fibers across the cross-section, and Figure 48.2 shows how this leads to increasing stiffness. This mechanism is clear during an anterior drawer of the tibia when the knee is flexed. In this posture, the posterior fibers of the anterior cruciate ligament are initially slack, and so the anterior tibial translation is resisted initially by only the anterior fibers of the anterior cruciate ligament. As the tibial displacement increases, so a greater part of the ligament is recruited to oppose the subluxing force.

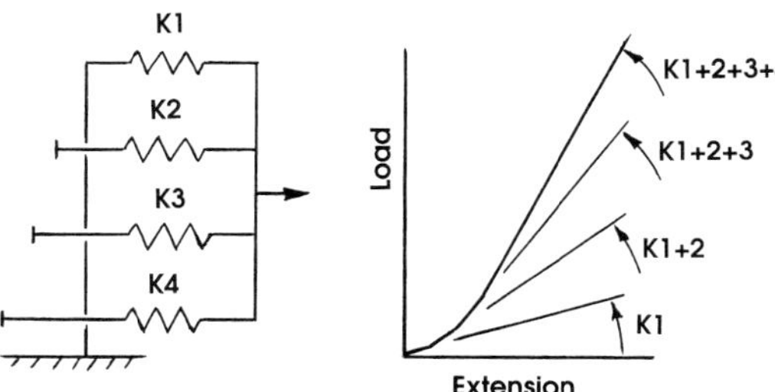

Figure 48.2. Non-uniform initial tightness across the width of a ligament leads to fiber recruitment when it is elongated. This in turn leads to increasing stiffness.

Ligament Failure

After the increasing tensile extension of the ligament reaches a certain limit, failure occurs, and the load recorded by the test machine falls. Some studies have shown a region of "yielding" prior to complete failure [7,12], and the point at which this starts has been called the "linear load" [6,13], as this is also the end of the linear behavior region (Figure 48.1). However, this probably represents sequential fiber failures caused by non-uniform application of strain across the width of the ligament. This has been shown clearly for the posterior cruciate ligament, which has fibers that spread out in widely divergent directions as they pass from the tibia to the femur. If these fibers are separated into two bundles that have approximately parallel fibers, then tensile tests take them up to a sudden complete rupture, and the load drops almost to zero immediately [3]. It is probable that the avoidance of a progressive tearing failure mode allows all of the fibers to bear their failure strength simultaneously, and thus for the ligament to reach a higher force.

The literature includes reports of a wide range of failure strains for the cruciate ligaments, from 18% [3] to 29% [14]. The average lengths of the anterior and posterior cruciate ligaments are reported in the range 30–35 mm [3,15], and so the strains represent elongations of approximately 5–10 mm. However, since a tibial displacement will not usually be in the direction that elongates these ligaments directly, the subluxation to cause a ligament rupture is normally much greater. Thus, Amis and Scammell [16] found that anterior cruciate ligament rupture occurred at a mean of 15 ± 7 mm anterior tibial displacement with the knee at 90° flexion (range 10–28 mm). Similarly, Race and Amis [17] found posterior cruciate ligament rupture at a mean posterior displacement of 15 ± 3 mm, range 12–20 mm, also at 90° knee flexion.

After tensile failure has occurred, further elongation of the bone–ligament–bone construct still requires a small force (Figure 48.1a). This represents the force needed to slide the ligament fibers past each other, presumably causing viscous shearing effects in the interfibrillar ground substance. Although the ligament has failed totally in tension, it may still appear to be intact during this phase of the elongation, until the fibers have slid past each other to the full length of the ligament. This is normally the situation when a slack anterior cruciate ligament is examined arthroscopically.

Effects of Specimen Orientation

The orientation of the specimen at the time of testing can affect the apparent strength and stiffness. Ligaments have distinct patterns of fiber tightening and slackening as the joints move, causing variation of slackness across the cross-section. Similarly, they are adapted to resisting loads in particular physiological configurations. Thus, the bones must be mounted in the test machine in a way that optimizes the ligament strength. The anterior cruciate ligament, for example, has fibers which are approximately parallel and of equal degree of tightness when the knee is extended, and so it is appropriate to test it in this posture, with the fibers of the ligament aligned with the axis of the test machine [18]. If tension is applied with the knee flexed, the tension will fall initially onto the anterior fibers which remain tightest, and they will rupture before the initially slack posterior fibers reach their failure strain. Thus, a sequential failure mode can be anticipated, with the fibers tearing off the attachment at a lower force than with the knee extended. This tendency has been shown by Figgie et al. [19]. Work on animal joints has shown that, even if the knee remains at the same angle of flexion, but the specimen has different orientation to the direction of loading, so different strength and failure mode occur. With the knee at 90° flexion, and the load axis along the tibia, the failure mode was most commonly by bone avulsion: the tibial plateau was lifted. This failure reflects the non-physiological direction of the loading, since injuries will not normally act to distract the tibia axially. If the joint was kept at 90° flexion, but the specimen was loaded so that the load axis was 45° to both the femur and tibia (which meant that the anterior cruciate ligament was close to the load axis), then the specimen reached a significantly higher load before failure, and the failure was then by ligament fiber rupture [20].

Effects of Age

There are progressive changes in the morphology of ligament tissue with maturation and advancing age. The collagen fibers enlarge diametrically and the water content diminishes [21]. These changes combine to cause increased stiffness and reduced elongation to failure. The mechanical effect of age on ligaments has been documented for the anterior cruciate ligament [5] (Figure 48.3). It can be seen that there was a drop of approximately two-thirds reduction in failure strength between specimens aged 20–30 and 70–80 years. Similarly, Noyes and Grood [6] found a significant loss of strength with age: specimens aged 16–26 were an average of 2.4 times stronger than those aged 48–83 years. Not all of this strength loss was attributed to changes in the ligament itself, since the failure mode changed to include many more bone avulsions in the older age group. This is not surprising when the dramatic loss of cancellous bone strength with age is noted. It is accepted that cancellous bone is an order of magnitude weaker in elderly specimens [22]. It is not known if this trend is the same in other ligaments, since they have different surrounding tissues and are subjected to different loads in use.

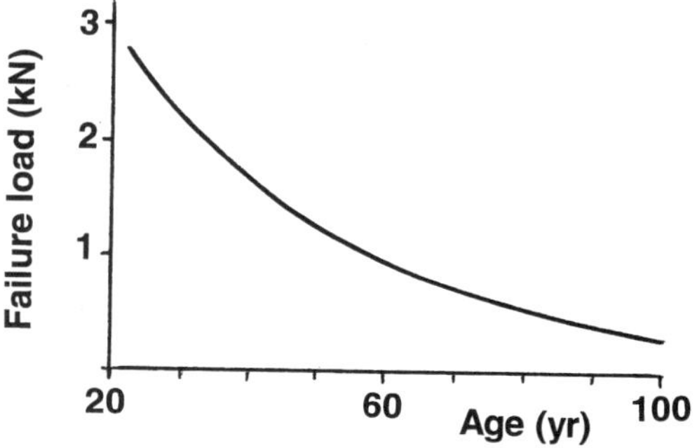

Figure 48.3. Variation of failure strength of bone–anterior cruciate ligament–bone preparations with age. From Woo et al. [5] with permission.

Time-dependent Effects

Ligament tissue exhibits viscoelastic behavior. This means that the tensile behavior is affected by the speed, or strain rate, with which the ligament is stretched. This is also true for other tissues, such as bone and articular cartilage. If a load is hung on a ligament preparation there will be an immediate elastic elongation. If the load is maintained at a sufficient level then the ligament will continue to get slowly longer (Figure 48.4a). This phenomenon is known as creep. If a load that does not cause immediate failure is maintained for long enough then there can be a delayed creep rupture of the ligament [23]. It has also been shown that ligament grafts and healing ligament scars are more likely to creep than the ligament which they have replaced [24], and this may be the explanation for stretch-out of ligament grafts that results in slightly slack ligament reconstructions after rehabilitation.

Another way to look at this phenomenon is to load a ligament as before, but now to hold the preparation at the fixed length to which it has been stretched. This, of course, prevents creep. As time passes, so the stress in the ligament relaxes (Figure 48.4b), and this phenomenon is known as stress relaxation. Static stress relaxation was measured by Viidik [25], who loaded the rabbit anterior cruciate ligament to 100 N and then maintained that extension. The load fell to 84 N after two minutes. A similar effect occurs under cyclic elongations: after ten cycles between 1.6 and 2.4% extension over 160 sec, the peak tensile stress in the canine medial collateral ligament relaxed from 7.8 to 6.5 MPa [26]. Many engineering materials have a direct rela-

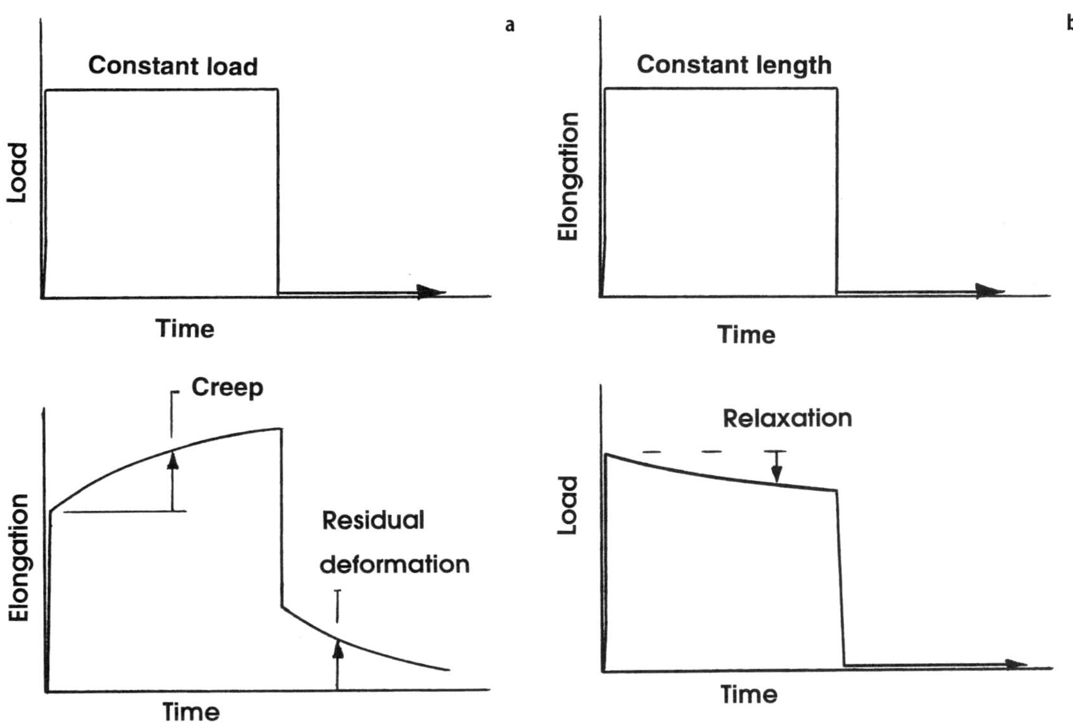

Figure 48.4. **a** When a load is applied to a ligament, there is an immediate elongation. If the load is kept constant the length of the ligament can increase with time due to creep. When the load is released there is an immediate elastic recoil, but this leaves a residual deformation that can recover with time. **b** If the ligament is elongated, and the bones then fixed to keep the ligament at a constant length, the load increases immediately as the load is applied, but then the load relaxes.

tion between stress relaxation behavior and the likelihood of creep.

When the load on a ligament preparation is released, the load drops to zero immediately, but the ligament does not necessarily return immediately to its original length (Figure 48.4a). The amount of residual deformation depends on both the stress that had been imposed and also the time that it had been imposed. With further time after load release, the ligament will gradually recover its initial length. This probably reflects a rearrangement of the collagen fibrils in the viscous ground substance. If the stress had been high enough to cause an irreversible deformation, then clearly the length would not recover its initial value. This probably relates to slippage or ruptures of the collagen fibers as the stress rises closer to ligament failure.

Creep and stress relaxation are usually associated with low-speed, long-time tests. At the other extreme, viscoelastic effects also cause changes in ligament behavior at high strain rates. At a basic level, it is easy to envisage that moving through a viscous medium quickly will require more force than moving slowly. This carries over into tensile test behavior of ligaments, which entails the collagen fibers moving and rearranging themselves amongst the viscous ground substance as the ligament is extended. This means that they exhibit strain-rate sensitivity, and show increasing stiffness at higher speeds of elongation. This is apparent even at relatively low test speeds: a 2.5% extension of the canine medial collateral ligament at 0.5 mm min^{-1} required 56 N, while the same extension at 50 mm min^{-1} required 67 N, so the stiffness increased approximately 20% for a hundred times increase of test speed [26]. This effect continues into higher testing speeds that are representative of impacts in trauma. Pioletti et al. [27] found that more than a half of the stress in human anterior cruciate ligament preparations could be attributed to the strain rate effects when comparing results between 0.1% sec^{-1} with those at 40% sec^{-1}, at strains less than 5%, but found little effect on the linear stiffness beyond the toe region.

As the speed of ligament extension increases, the failure mode changes. It has been observed in laboratory tests in vitro that slow-speed tensile tests of bone–ligament–bone preparations often lead to failure by bone avulsion. This, of course, does not mirror clinical experience, where trauma usually leads to interstitial ligament ruptures. Experiments on the anterior cruciate ligament of primates at two test speeds showed a significant change in failure mode. At 5 mm min^{-1}, there were 29% ligament ruptures and 57% bone avulsions. In contrast, at 500 mm min^{-1} (approximately 40% sec^{-1} strain rate) there were 66% ruptures and 28% avulsions [13]. A further increase in strain rate, to 100% sec^{-1}, led to 97% of failures being by interstitial ligament rupture in tests of the rat anterior cruciate ligament [28]. This change in failure mode can be explained by reference to the changing behavior of the tissues with speed of loading. Although this text has described the strain rate sensitivity of ligaments, it should be noted that bone also exhibits this behavior. The ultimate tensile strength of bovine cortical bone was found to increase by 100% for a hundred times increase in strain rate [29]. The change from bone avulsions seen at low loading rates to ligament failures at high speeds suggests that the strengthening of bone with increasing speed allows the attachment to become stronger in relation to the ligament.

Cyclic Loads and Creep Effects

Cyclic loading can cause an accumulation of the loading/unloading effects described above. If the peak stress of repetitive loading is high enough, and the time for recovery between loading is insufficient, then the deformation induced by each load cycle will not have sufficient time to relax completely. The consequence is that the first load cycle will have left a residual deformation, and so the next load cycle will start from an elongated ligament length. These increments of length can accumulate to give a slack ligament. This has been shown in vitro (Figure 48.5), but the length change shown is exaggerated for the number of load cycles imposed because the peak stress was high, at two thirds of failure stress, and the load cycles were slow, allowing creep to progress [30]. These effects have also been demonstrated in vivo [31] when the ankle was displaced laterally by a force of 32 N and maintained for 150 sec. There was an immediate displacement of 11 mm, then a further 5 mm due to creep. A residual displacement of 4 mm remained after 10 min relaxation, giving behavior similar to that shown in Figure 48.4a. It can be speculated that the process of "warming up" before an athletic event causes these changes in the ligaments, so that the athlete will then be controlling joint stability primarily by muscular actions.

If the peak load is below the level at which irreversible length changes will occur, and the frequency too high to allow significant creep between load cycles, then the ligament length will increase with each load cycle initially. After approximately eight or ten load cycles, the change in length with each load cycle becomes negligible and the behavior settles down with the ligament longer than at the start of the first load cycle (Figure 48.6). This phenomenon is usually known as "conditioning", and most work in vitro imposes a set of conditioning load cycles to a specimen before recording the behavior in a steady state [32]. Figure 48.6 shows that, after the conditioning load cycles are

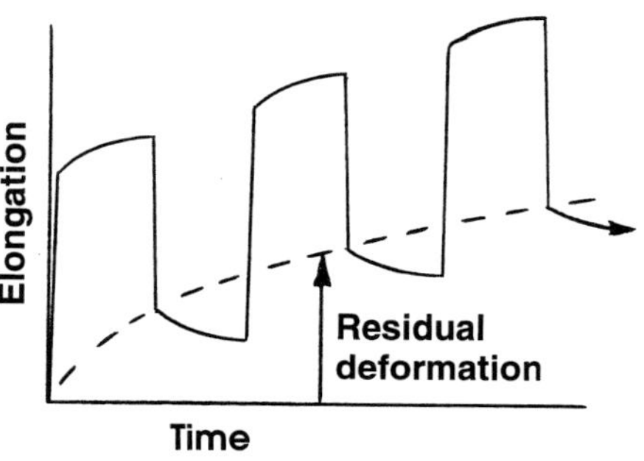

Figure 48.5. Cyclic creep leads to increasing residual deformation, or ligament slackness, if there is insufficient time between the load cycles for the residual deformation to recover completely.

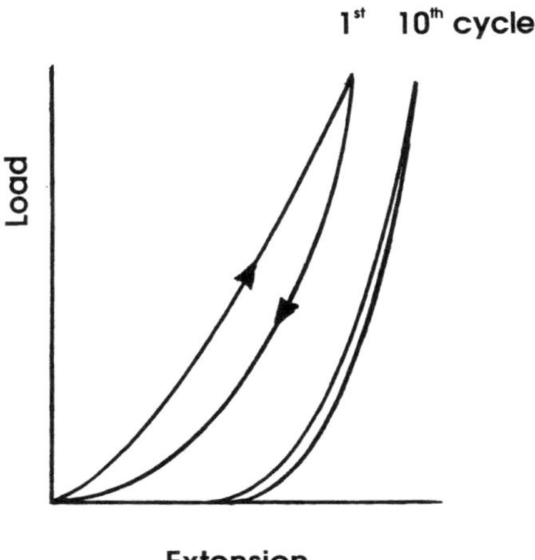

Figure 48.6. Cyclic loading of a bone–ligament–bone preparation causes ligament "conditioning", and the tenth load-unload cycle is seen to exhibit less hysteresis, greater stiffness, and to have residual elongation when compared to the first load cycle.

imposed, there is some slackness before the ligament starts to resist load, the ligament is stiffer, and the loading–unloading curve encloses a smaller area. It has been suggested that these changes result at least partly from loss of water content from the ligament [33]. This load-unload pattern, when the curves enclose an area in an elongated loop, is known as hysteresis. The area enclosed by the loop represents the work, or energy, dissipated in the ligament during the load–unload cycle. This can be understood by noting that work can be expressed as the product of force times distance moved, and each segment of the ligament test graph shows changes in force and length. Clearly, then, more work is done by extending the ligament along the higher curve than is released by relaxing along the lower curve.

Since hysteresis is reduced after imposing a set of conditioning load cycles, the internal rearrangement of the ligament structure during the conditioning load cycles leads it to work more efficiently: it is closer to an elastic spring, with less viscous losses.

Effects of Exercise or Immobilization

Because ligaments are relatively avascular they are slow to respond to alterations in their environment when compared to other tissues. Thus, in general, exercise or immobilization affect the other tissues first, and the ligaments have delayed responses. This is seen most clearly with ligaments that attach to the bone tangentially such as the distal end of the medial collateral ligament. Here, the ligament attaches into the periosteum. A period of immobilization has little immediate effect on the medial collateral ligament itself, but the vascular periosteum responds rapidly, and it soon weakens significantly. This causes a reduction in the failure load, as expected, but also to a change of the failure mechanism: normally, there might be a rupture of the ligament fibers in mid substance; after immobilization, the distal attachment fails by a shear, or sliding, mechanism. This is in contrast to the ligaments that attach at a steep angle, with collagen fibers continuing directly from the ligament into the bone as Sharpey's fibers, such as the anterior cruciate ligament. Here, immobilization affects the bone and leads to bone avulsion failure at reduced force.

Experiments in animals have shown that immobilization does cause a loss of strength of the ligament substance, and that even a long period of remobilization can leave a ligament significantly weaker than normal. Care must be taken when interpreting work on caged animals, however, since they may be suffering from disuse effects prior to any experimental work. This is shown by the frequency of bone avulsions reported in experimental work, which is greater than that seen clinically, where intra substance failure predominates. Laros et al. [34] found that the canine medial collateral ligament had a strength of 3.3 body weight in active animals. This reduced 15% after 9 weeks caged, and 39% after 9 weeks in plaster cast. Although the muscle weight had returned to normal by 12–18 weeks after remobilization, the ligaments were still significantly weaker than normal at 30 weeks. The same trends were shown for the cruciate ligaments by Noyes et al. [35], who immobilized monkeys knees for 8 weeks: the strength fell 39%, and was still 21% below normal after 20 weeks of reconditioning activity. (Figure 48.7). These findings have clear implications for postoperative care and rehabilitation for sports activities.

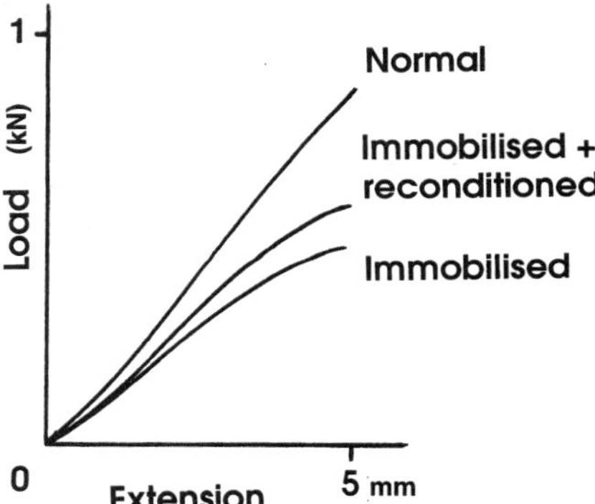

Figure 48.7. Immobilization for eight weeks caused the strength of the anterior cruciate ligament to fall 39%. Twenty weeks reconditioning activity still left a strength deficit of 21%. From Noyes et al. [35], with curves arbitrarily ended at 5 mm elongation.

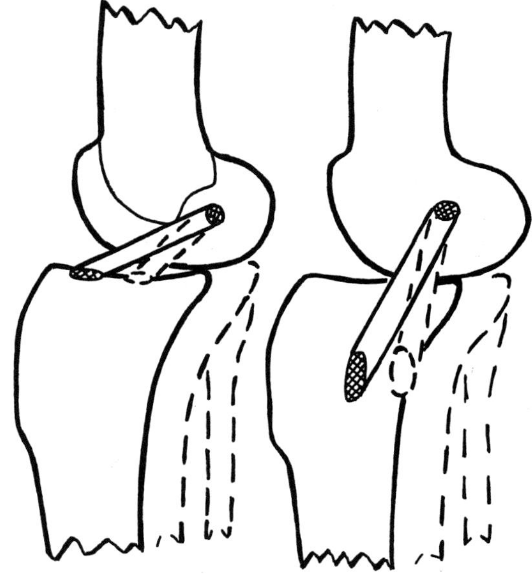

Figure 48.8. Anterior displacement of the tibia has stretched the anterior cruciate ligament more (40%) than the medial collateral ligament (10%) because it is oriented closer to the direction of displacement. Thus, tension rises faster in the anterior cruciate ligament so it is the primary restraint to the subluxation [39]. From Amis [40] with permission.

Joint Stability

Primary and Secondary Restraints

The main function of ligaments is to control the motion of one bone relative to another, thus controlling joint laxity and stability. In this definition, stability refers to the functional symptoms exhibited by patients, when their joints tend to collapse or "give way" during activities such as sports. If the relative movement is excessive, then this can be measured, and this is known as excessive laxity. A common situation is the measurement of excessive tibial anterior translation laxity after damage to the anterior cruciate ligament [36,37].

Most joints are stabilized by several ligaments, and so it can sometimes be difficult to decide which of several possible structures have been injured when examining the joint. The concept of primary and secondary restraints helps to understand both the functions and interactions of ligaments, and hence their relative importance [38]. A primary restraint is the ligament that resists the majority of the force that is tending to displace one bone relative to the other, usually to sublux the joint. Normally, this ligament is oriented approximately in line with the displacement, and so it is stretched directly as one bone attachment moves away from the other. Since ligaments are passive, tensile restraints, their tension depends entirely on how much they are stretched. In contrast, a secondary restraint is usually oriented approximately perpendicular to the bone displacement being induced. This means that it is swung sideways as the bone moves, and so it is not elongated as much as the primary restraint (Figure 48.8). The terminology is probably best known for anterior drawer testing of the knee, when the anterior cruciate ligament is the primary restraint to anterior tibial translation, with the collateral ligaments acting as secondary restraints [39].

A further factor that means that secondary restraints are not efficient at resisting bone displacements is that their tension tends to act in an inefficient direction. The consequence of this is that a secondary restraint must work at a much greater tension than the primary restraint if it is to have the same restraining action on the bone (Figure 48.9). This factor probably explains why secondary restraints tend to stretch if the primary restraint is not repaired or reconstructed after an injury [40].

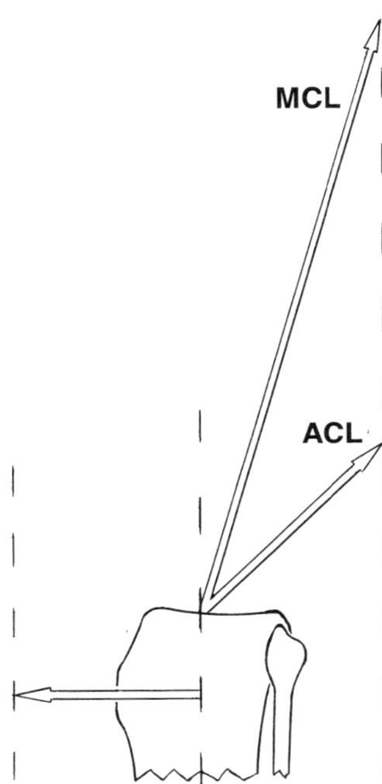

Figure 48.9. The force vectors in the anterior cruciate ligament and the medial collateral ligament that are needed to produce a posteriorly directed component equal and opposite to the anterior drawer force applied to the tibia.

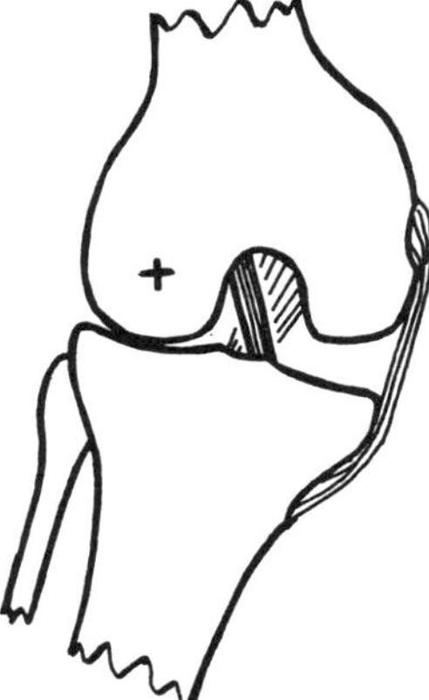

Figure 48.10. During a valgus rotation of the tibia relative to the femur, the axis of rotation is centered in the lateral femoral condyle. This means that the medial collateral ligament is elongated more than the cruciate ligaments and so it is the primary restraint. From Amis [40] with permission.

Since certain bone–bone displacements stretch the primary restraints significantly more than the secondary restraints, it follows that it is the primary restraint that is ruptured by the injury mechanism that causes bone–bone motion akin to that used to define the primary restraints. Thus, for example, it is possible for the anterior cruciate ligament to be ruptured by an anterior tibial translation with the knee at 90° flexion, while the secondary restraints, such as the collateral ligaments, are not damaged [16].

As Figure 48.8 shows, it is usually possible to decide which are likely to be the primary restraints simply by looking at their orientation in relation to the bone–bone displacement being induced. For a relative translation (linear) motion, the primary restraint is likely to be that which is closest to being parallel to the direction of motion, as it is stretched directly. For a rotational displacement, however, the axis of rotation between the bones must be found first. Then, the primary restraint will probably be the one with the greatest moment arm about the axis of rotation. Noting this simple mechanical fact, it is clear why joints are often widened by tuberosities for attachment of the collateral ligaments and have bicondylar articulations. Consideration of varus–valgus rotation of the tibia relative to the femur shows that the axis of rotation is centered within one femoral condyle, while the opposite collateral ligament is stretched. The cruciate ligaments are close to the axis of rotation in this situation, and so they are stretched only a small amount when compared to the elongation imposed onto the collateral ligaments (Figure 48.10). Furthermore, consideration of the orientation of the cruciates in the sagittal plane shows that only a partial component of the tibio-femoral distraction is elongating them directly, since they also are swung sideways as the bones move apart. In contrast, the collaterals are aligned closely to the bone motion, and so they are stretched directly. Thus, although they are weaker, the collateral ligaments are the primary restraints and the cruciates are only secondary restraints [40,41].

If the same argument is used to study tibial internal–external rotation, when the rotation axis is close to the center of the tibial plateau, it is clear that the cruciate ligaments are again too close to the axis of rotation to be the primary restraints – it is the peripheral structures, such as the collateral ligaments, that are the primary restraints in this situation [40]. Noting the observations above, it is apparent that the primary restraint is not necessarily the strongest or stiffest ligament that is capable of resisting the bone–bone motion, but usually the ligament that has the greatest mechanical advantage that is elongated the most.

Length Change Patterns and Isometry

Because ligaments attach to bones over an area, and not just at a single point, it follows that there can only be one ligament fiber which is on the axis of rotation as the joint is moved. Other fibers must be either stretching or slackening depending on which side of the axis they are situated. This effect is important both for understanding the functional anatomy of the ligaments and also for the surgeon when deciding where exactly to place a ligament graft, or how much tension to apply to the graft.

In ligament surgery, the concept of isometry has been popular and instruments have been used to measure isometry of ligament grafts during surgery. The definition of isometry is that motion of the joint, for example, flexion–extension of the knee, does not cause the distance separating the bone attachments to change. The practical importance of isometry is confirmed by several factors: ligaments are elongated irreversibly if the strain cycles exceed a given level, perhaps 7% [42], while tendons that are used in ligament reconstructions can only withstand 4% strain reversibly [43]. It has also been shown that continuous passive motion causes greater slackening of non-isometric reconstructions [44], and that the functional results of anterior cruciate ligament reconstruction diminish if the reconstruction is not isometric [45]. It is, therefore, important to understand the isometric behavior of the natural ligaments that are to be reconstructed.

The anterior cruciate ligament has received the most attention. A review of the literature [42] has shown that most studies have found the antero-proximal corner of the femoral attachment to be closest to isometric, in combination with an antero-medial or central tibial attachment. Measurement of the length change patterns of individual fiber bundles has shown that the more postero-distal fibers slacken significantly as the knee flexes [15,42,46]. Thus, the behavior of the natural anterior cruciate ligament suggests that the antero-medial fiber bundle (defined in terms of the tibial attachment area) is the most important at all angles of knee flexion. This is supported by other work showing that the antero-medial fibers are significantly stronger than the postero-lateral fibers [47] and that this is related to a higher collagen density [48].

It is possible to construct maps of the isometry around the attachment of the anterior cruciate ligament [49], and it has been shown that there is a "transition line" between areas of increasing and decreasing attachment site separation distance over a range of knee flexion [50,51]. This transition line passes through the isometric zone described above. A ligament fiber or graft attached posterior to this line will, overall, slacken with knee flexion. Conversely, a ligament fiber or graft attached anterior to the transition line will be stretched as the knee flexes [52] (Figure 48.11). Some attachments will have a pattern of initial slackening followed by retightening, or of tightening followed by

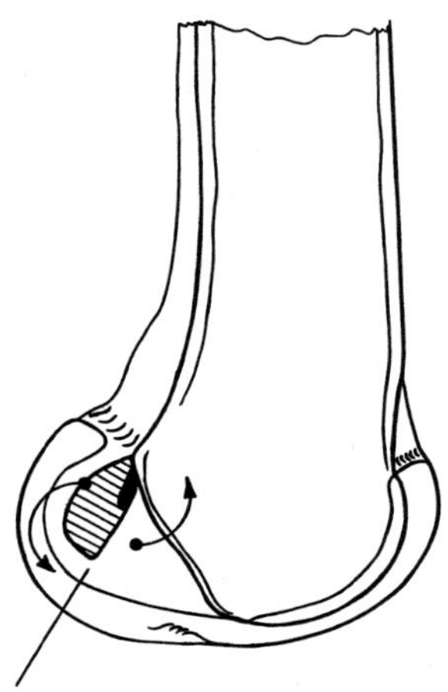

Figure 48.11. The femoral attachment of the anterior cruciate ligament is isometric at its antero-proximal corner. A "transition line" passes through the isometric zone. Fibers that tighten with knee flexion are anterior, and fibers that slacken with knee flexion are posterior to this line. From Amis et al. [52] with permission.

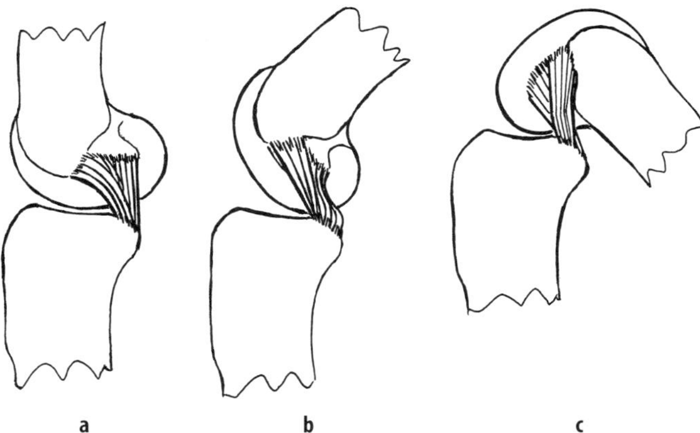

Figure 48.12. This shows a lateral view of a left knee, with the femur split in a midline sagittal plane and the lateral half removed to reveal the posterior cruciate ligament. **a** When the knee is extended, the anterolateral bundle of the posterior cruciate ligament is slack and the postero-medial bundle is tight but not aligned to resist posterior tibial drawer. **b** In mid knee flexion the antero-lateral bundle is tight and the postero-medial bundle is slack. **c** In deep knee flexion the postero-medial bundle is tight and well-aligned to withstand posterior tibial drawer, while the antero-lateral bundle is wrapped against the roof of the femoral intercondylar notch.

slackening [42,53], as they pass through the transition zone between the two main patterns of slackening or tightening as the knee flexes.

The posterior cruciate ligament has more extreme patterns of fiber length change, largely because the femoral attachment covers a large area, typically being 30 mm in anterior–posterior extent. The antero-lateral fiber bundle (defined now as the femoral attachment location) is a large bulk of fibers that joins the posterior tibial attachment to the roof of the femoral intercondylar notch anteriorly. It is clearly seen to be slack when the extended knee is examined by magnetic resonance imaging (Figure 48.12a). As the knee flexes, this fiber bundle tightens and takes a steeper angle from the tibial plateau (Figure 48.12b). In deep knee flexion, this structure rests against the roof of the femoral intercondylar notch. In contrast, the postero-medial fibers of the posterior cruciate ligament are tight in full knee extension, and slacken as the knee flexes. Since these fibers are approximately in a proximal–distal orientation when the knee is extended, they are not aligned efficiently to withstand posterior tibial translation, which is known to be the main role of the posterior cruciate ligament. However, in deep knee flexion, these fibers become tight and well-aligned to control tibial posterior translation (Figure 48.12c) [17]. These observations have been correlated to the results of load testing of knees, with sequential cutting of the fiber bundles. This allows their mechanical role to be found. It confirmed that the antero-lateral fiber bundle was dominant across the mid range of knee flexion, the postero-medial in deep knee flexion, and that other structures, especially postero-lateral ligaments, were the primary restraints to posterior tibial translation when the knee was near full extension [54].

Although most attention has been paid to the cruciate ligaments of the knee, the principles described above also apply to other joints, and ligaments with distinct patterns of tightening and slackening during joint flexion–extension are found frequently, at the ankle, elbow, and metacarpo-phalangeal joints, for example. Although loss of ligament tension or changes of orientation may occur during joint motion that appear to leave the joint less well stabilized, it should be noted that this chapter has considered the passive stabilizing structures in isolation. In many activities of daily life, the joints are stabilized primarily by the muscle actions that both move and stabilize the joints, by compressing the bones together; the ligaments often act only after these active controls have malfunctioned or been overcome. This factor leads to consideration of the proprioceptive role of sensory feedback arising from nerves in or on ligaments that may stimulate the muscles to protect the joint as the limits of motion are approached.

Summary

Ligaments function mechanically as passive restraints to the relative motions of bones, which means that they only act in tension after reaching the transition from slack to taut behavior.

Ligaments have non-linear force–extension behavior, due mostly to the straightening-out of fiber crimp, and then to fiber recruitment.

Ligaments exhibit viscoelastic behavior that is time-dependent, and so can suffer from creep elongation under prolonged loads. This also causes ligaments to stiffen if a rapid elongation is imposed.

Because ligaments are relatively avascular, and so heal poorly, surgical treatment often requires reconstruction using a graft rather than repair. This, in turn, means that the surgeon must have a good working knowledge of factors such as ligament fiber length change patterns as joints move, and therefore an understanding of how to optimize graft placement in relation to the natural ligament attachments and bone anatomy.

Acknowledgements

Work at the author's laboratory is supported by the Arthritis Research Campaign, a charity based in England. Much of the experimental work has been performed by A. M. J. Bull, G. P. C. Dawkins, A. Race, B. E. Scammell, and T. D. Zavras, and the author thanks them for this.

References

1. Amiel D, Billings E, Akeson WH. Ligament structure, chemistry, and physiology. In: Daniel D, Akeson W, O'Connor J editors: Knee Ligaments: Structure, Function, Injury, and Repair. New York: Raven Press, 1990,77–91.
2. Harner CD, Livesay GA, Kashiwaguchi S et al. Comparative study of the size and shape of human anterior and posterior cruciate ligaments. J Orthop Res 1995;13: 429–34.
3. Race A, Amis AA. The mechanical properties of the two bundles of the human posterior cruciate ligament. J Biomech 1994;27:13–24.
4. Pring DJ, Amis AA, Coombs RRH. The mechanical properties of digital flexor tendons related to artificial tendons. J Hand Surg 1985;10:331–6.
5. Woo SLY, Hollis JM, Adams DJ, Lyon RM, Takai S. Tensile properties of the human femur–anterior cruciate ligament–tibia complex: The effects of specimen age and orientation. Am J Sports Med 1991;19:217–25.
6. Noyes FR, Grood ES. The strength of the anterior cruciate ligament in humans and rhesus monkeys, age and species-related changes. J Bone Joint Surg Am 1976;56: 1074–82.
7. Kennedy JC, Hawkins RJ, Willis RB, Danylchuk KD. Tension studies in human knee ligaments, yield point, ultimate failure, and disruption of the cruciate and tibial collateral ligaments. J Bone Joint Surg Am 1976;58: 350–5.
8. Butler DL, Kay MD, Stouffer MD. Comparison of material properties in fascicle-bone units from human patellar tendon and knee ligaments. J Biomech 1986; 19:425–32.
9. Race A, Amis AA. Cross-sectional area measurement of soft tissue. A new casting method. J Biomech 1996;29: 1207–12.
10. Lee TQ, Woo SLY. A new method for determining cross-sectional shape and area of soft tissues. J Biomed Eng 1988;110:110–14.
11. Viidik A. Simultaneous mechanical and light microscopic studies of collagen fibers. Z Anat Entwicklungsgesch 1972;136:204–12.
12. Cabaud HE. Biomechanics of the anterior cruciate ligament. Clin Orthop 1983;182:26–31.
13. Noyes FR, DeLucas JL, Torvik PJ. Biomechanics of anterior cruciate ligament failure: an analysis of strain-rate sensitivity and mechanisms of failure in primates. J Bone Joint Surg Am 1974;56:236–53.
14. Prietto MP, Bain JR, Stonebrook SN, Settlage R. Tensile strength of the human posterior cruciate ligament. Trans 34th Ann Orthop Res Soc 1988;13:195.
15. Amis AA, Dawkins GPC. Functional anatomy of the anterior cruciate ligament: fiber bundle actions related to ligament replacements and injuries. J Bone Joint Surg Br 1991;73:260–7.
16. Amis AA, Scammell BE. Biomechanics of intrarticular and extraarticular reconstructions of the anterior cruciate ligament. J Bone Joint Surg Am 1993;75:812–17.
17. Race A, Amis AA. PCL reconstruction: in vitro biomechanical comparison of 'isometric' versus single and double-bundled 'anatomic' grafts. J Bone Jt Surg Br 1998;80:173–9.
18. Beynnon BD, Amis AA. In vitro testing protocols for the cruciate ligaments and ligament reconstructions. Knee Surg, Sports Traumatol, Arthroscopy 19886;Suppl 1: S70–6.
19. Figgie HE, Bahnuik EH, Heiple KG, Davy DT. The effects of tibial–femoral angle on the failue mechanics of the canine anterior cruciate ligament. Trans 28th Ann Orthop Res Soc 1982;7:309.

20. Amis AA. Biomechanics of ligaments. In: Jenkins DHR, editor. Ligament Injuries and their Treatment. London: Chapman and Hall, 1985;3–28.
21. Tkaczuk H. Tensile properties of human lumbar longitudinal ligaments. Acta Orthop Scand 1968;Suppl 115.
22. Weaver JK, Chalmers J. Cancellous bone: its strength and changes with ageing and an evaluation of some methods for measuring its mineral content. Part 1: Age changes in cancellous bone. Part 2: Osteoporosis. J Bone Jt Surg Am 1966;48:289–308.
23. Smith JW. The elastic properties of the anterior cruciate ligament of the rabbit. J Anat 1954;88:369–81.
24. Thornton GM, Leask GP, Shrive NG, Frank CB. Early medial collateral ligament scars have inferior creep behavior. J Orthop Res 2000;18:238–46.
25. Viidik A. Biomechanics and functional adaptation of tendons and joint ligaments. In: Evans FG, editor. Studies on the Anatomy and Function of Bone and Joints. Berlin: Springer, 1966;17–39.
26. Woo SLY, Gomez MA, Akeson WH. The time and history-dependent viscoelastic properties of the canine medial collateral ligament. J Biomech Eng 1981;103: 293–8.
27. Pioletti DP, Rakotamanana LR, Leyvras PF. Strain rate effect on the mechanical behavior of the anterior cruciate ligament-bone complex. Med Eng Phys 1999; 21:95–100.
28. Cabaud HE, Chatty A, Gildengorin V, Feltman RJ. Exercise effects on the strength of the rat anterior cruciate ligament. Am J Sports Med 1980;8:79–86.
29. Wright TM, Hayes WC. Tensile testing of bone over a wide range of strain rates: effects of strain rate, microstructure and density. Med Biol Eng 1976;14: 671–9.
30. LaBan MM. Collagen tissue: implications of its response to stress in vitro. Arch Phys Med Rehab 1962;43:461–6.
31. Pope MH, Crowninshield R, Miller R, Johnson R. The static and dynamic behavior of the human knee in vivo. J Biomech 1976;9:449–52.
32. Amis AA. Biomechanics of bone, tendon and ligament. In: Hughes SPF, McCarthy I, editors. Sciences Basic to Orthopaedics. London: WB Saunders, 1997;222–39.
33. Chimich D, Frank CB, Shrive NG, Marchuk L, Bray R. Water content alters viscoelastic behavior of the normal adolescent rabbit medial collateral ligament. J Biomech 1992;25:831–7.
34. Laros GS, Tipton CM, Cooper RR. Influence of physical activity on ligament insertions in the knees of dogs. J Bone Jt Surg Am 1971;53:275–86.
35. Noyes FR, Torvik PJ, Hyde WB, DeLucas JL. Biomechanics of ligament failure II. An analysis of immobilization, exercise, and reconditioning effects in primates. J Bone Joint Surg Am 1974;56:1406–18.
36. Daniel DM, Malcolm LL, Losse G, Stone ML. Instrumented measurement of anterior laxity in the knee. J Bone Joint Surg Am 1985;67:720–6.
37. Gurtler RA, Stein R, Torg JS. Lachman test evaluated: quantification of a clinical observation. Clin Orthop 1987;216:141–50.
38. Noyes FR, Grood ES, Butler DL, Paulos LE. Clinical biomechanics of the knee: ligament restraints and functional stability. In: AAOS symposium on the Athlete's Knee. St Louis: CV Mosby, 1980;1–35.
39. Butler DL, Noyes FR, Grood ES. Ligamentous restraints to anterior–posterior drawer in the human knee: a biomechanical study. J Bone Joint Surg Am 1980;62:259–70.
40. Amis AA. The kinematics of knee stability. In: Jakob RP, Fulford P, Horan F., editors. European instructional course lectures vol 4. J Bone Joint Surg, London, 1999; 96–104.
41. Grood ES, Noyes FR, Butler DL, Suntay WJ. Ligamentous and capsular restraints preventing straight medial and lateral laxity in intact human cadaver knees. J Bone Joint Surg Am 1981;63:1257–69.
42. Amis AA, Zavras TD. Isometricity and graft placement during anterior cruciate ligament reconstruction. The Knee 1995;2:5–17.
43. Abrahams M. Mechanical behavior of tendon in vitro. Med Biol Eng 1967;5:433–4.
44. O'Meara PM, O'Brien WR, Henning CE. Anterior cruciate ligament reconstruction stability with continuous passive motion. Clin Orthop 277:201–9.
45. Friederich NF, Muller W. How important is isometric placement of cruciate ligament grafts? Intraoperative measurement versus mid-term clinical follow-up. J Bone Joint Surg Orthop Proc Br 1993;75(Suppl II): 150–1.
46. Sapega AA, Moyer RA, Schneck C, Komalahiranya N. Testing for isometry during reconstruction of the anterior cruciate ligament. J Bone Joint Surg Am 1990;72:259–67.
47. Butler DL, Guan Y, Kay M, Cummings J, Feder S, Levy M. Location-dependent variations in the material properties of the anterior cruciate ligament. J Biomech 1992; 25:511–18.
48. Mommersteeg TJA, Blankevoort L, Kooloos JGM, Hendriks JCM, Kauer JMG, Huiskes R. Nonuniform distribution of collagen density in human knee ligaments. J Orthop Res 1994;12:238–45.
49. Sidles JA, Larson RV, Garbini JL, Downey DJ, Matsen FA. Ligament length relationships in the moving knee. J Orthop Res 1988;6:593–610.
50. Hefzy MS, Grood ES, Noyes FR. Factors affecting the region of most isometric femoral attachments. Part II: the anterior cruciate ligament. Am J Sports Med 1989; 17:208–16.
51. Friederich NF, O'Brien WR. Functional anatomy of the cruciate ligaments. In: Jakob RP, Staubli HU, editors. The knee and the Cruciate Ligaments. Berlin: Springer-Verlag, 1992;78–91.
52. Amis AA, Beynnon B, Blankevoort L, Chambat P, Christel P, Durselen L et al. Proceedings of the ESSKA scientific workshop on reconstruction of the anterior and posterior cruciate ligaments. Knee Surg Sports Traumatol Arthrosc 1994;2:124–32.
53. Bradley J, FitzPatrick D, Daniel D, Shercliff T, O'Connor J. Orientation of the cruciate ligament in the sagittal plane: a method of predicting its length change with flexion. J Bone Joint Surg Br 1988;70:94–9.
54. Race A, Amis AA. Loading of the two bundles of the posterior cruciate ligament: an analysis of bundle function in A–P drawer. J Biomech 1966;29:873–9.

49 Knee Ligamentoplasty: Prosthetic Ligament or Ligament Allograft?

D. G. Poitout and B. Ripoll

Lesions of the central pivot of the knee are responsible for chronic disabling instabilities which, in the long term, lead to irreversible arthrotic destruction of the articular surfaces. Since 1980, we have been using synthetic ligaments and, more recently, since 1986, preserved human ligaments. The principle of ligamentoplasty is quite simply to replace torn ligaments with a prosthesis, aiming to reproduce the anatomical course of the ligament and its functional properties as faithfully as possible.

The main advantages of these techniques are:

shortening of the time under surgery,

the other anatomical structures of the knee are not damaged,

and the possibility of early rehabilitation.

General Biomechanics

Basic Properties

A mechanical evaluation of an artificial ligament is only useful if it is compared with the properties of the corresponding human ligament.

The Parameters

The parameters evolve along the usual tension–elongation curve which determines the behavior of the ligament. Extreme stresses are only encountered in traumatological circumstances.

The rigidity of the ligament directly affects the function of the joint because if it is too rigid, it requires an intense muscular effort to mobilize the knee and risks causing tearing of the ligament or its attachments; if it is too weak, it will prevent the ligament from playing its stabilizing role.

Apart from *rigidity*, the main parameters therefore seem to us to be the values of strength and elastic lengthening which, more than the maximum values, represent the true tolerance limit of the ligament as well as the time it takes for the initial length of the ligament to recover after single stretching.

Measuring Methods

Two main methods can be used:

a specimen involving the ligament and the bone consists of removing knees from cadavers, and dissecting them until only the anterior cruciate ligament connects the two articular surfaces.

another way of testing ligaments consists of attaching them directly to the inside of traction machines, using grips.

The problem with this technique is that with grips there are always phenomena which crush and lacerate the tissue, which can alter its elasticity or its strength, hence the preparation of a cone-shaped device which surrounds the ends of the ligaments fixed in acrylic resin.

Mechanical Properties of the Human Anterior Cruciate Ligament and of Prosthetic Ligaments

As far as the mechanical properties of the anterior cruciate ligament of a cadaver is concerned, it would appear that there is no significant difference between the results obtained on a frozen specimen after being frozen for several months.

On the other hand, the age of the subject displays a difference in the maximum average strength between a group of subjects approximately 60 years of age and a group of subjects approximately 20 years of age with the strength being double.

As far as the rate of elongation is concerned, it seems that the maximum strength of the human anterior cruciate ligament increases with elongation and it is often seen that although a slow rate leads to tearing due to bone avulsion, a rapid rate is more likely to lead to the ligament tearing in its middle part.

The Life of a Ligament

Two series of tests can be used in practice:

cyclical tensioning tests,
cyclical deformation tests.

As far as cyclical tensioning is concerned, the variations in tension experienced by the ligament during flexion–extension of the knee are reproduced, at the end of which each ligament has its residual elongation measured and a maximum elongation test is performed to assess the changes in the strength and rigidity of this ligament, compared with a new ligament.

When a cyclical deformation test is used, the resistance to repeated flexion is analyzed, combining constant, fixed-angle flexion of the ligament and continuous axial rotation of this ligament. This test therefore varies the points where maximum tension and compression forces are applied in a homogeneous manner on the periphery. At the end of this test, the ligament is subjected to a maximum elongation test.

Implantation in vivo in animals is certainly the method most used in research on synthetic ligaments. However, the frequency with which premature tears occur in the implant is regrettable as are the problems associated with applying the observations made in animal studies to man.

Biocompatibility and Rehabilitation

Biocompatibility

The Risks

In the case of ligament prostheses, general tolerance at a distance from the implant can take several forms, whether it be problems associated with general cytotoxicity, allergic reactions, toxicity specific to an organ, or teratogenicity.

As for local tolerance, it has to be dissociated from the local inflammatory reaction connected with the introduction of any kind of foreign body into the system. This reaction does not affect biocompatibility, insofar as it ends after a few days or a few weeks to make way for stabilization of the interface.

Carcinogenicity can be linked either to the chemical structure of the implants, or to their physical structure.

Although most of the polymers used have proved to be carcinogenic in animal studies under certain conditions, these phenomena are very rare in man and are only present in a few rare cases of sarcomas seen near vascular prostheses made of Dacron.

The Tests

Tests in vitro, consisting of placing the material studied in a cell culture and observing the reciprocal interactions between these two elements, are interesting because they are rapid (of the order of a few hours to a few days), reproducible, and specific.

In vivo tests make it possible to take into account the stresses connected with the functioning of the prosthesis and, in particular, involving the immune system of the host through the host's reaction to foreign bodies, the course of which, to a large extent, determines the tolerance of the implant.

Rehabilitation of the Artificial Ligament

Colonization of the prosthesis by the tissues of the host – or rehabilitation – is a phenomenon very much linked to its biotolerance. This idea of the prosthesis being a more or less temporary support for a biological new ligament is very debatable and many papers demonstrate that the mechanical role of this recolonization is extremely limited.

The chemical nature of the implant can become involved in this phenomenon as well as its physical nature and it seems to be an established fact that, far more than the diameter of the fibers or their direction, it is the porosity of the implant which is the decisive factor.

Finally, the mechanical stresses seem to be just as important in maintaining the functional properties of the ligaments. And, conversely, immobilization is responsible for considerable fragilization of the ligament tissue and its insertions.

The Ligaments Used

Artificial Ligaments

Carbon

Carbon ligament prostheses consist of elementary filaments 5 to 10 microns in diameter, grouped into unidirectional bundles or in plaited or twisted strands of several thousand elements. Endurance in traction is theoretically very high. However, when this strength is measured on a traction machine, the figures are often clearly lower than the theoretical value anticipated. Shear strength is poor, which leads to fragmentation of the fibers and, when histological slices are taken, many breaks in the filaments can be seen.

Plaiting the fibers considerably reduces the rigidity of the prostheses. It seems that the best results are obtained for a ligament consisting of 32 strands of 3,000 filaments plaited at an angle of 43°.

As far as biocompatibility is concerned, taking account of the usual fragmentation of carbon fibers, this is seen to turn into a foreign body, rapidly developing into abundant and regular fibrosis, the mechanical properties of which are, however, insufficient to replace a prosthetic ligament. On the other hand, the possibility of lymphatic drainage is proven by the almost constant discovery of carbon in the regional ganglions.

Carbon ligaments also poses problems for the anchorage technique, because the fragility of the fibers makes it difficult to attach them.

To summarize, carbon fiber is a material which, mechanically, has great resistance to traction, very high rigidity, and poor resistance to shearing forces, in spite of the various plaiting and sheathing measures adopted. Used as a cruciate ligament prosthesis, it inevitably breaks with the stresses being relayed progressively to the fibrous new ligament which appears, the mechanical properties of which are inadequate to ensure that this function is effective.

Polyethylene

High-molecular-weight polyethylene is largely used by orthopedists for manufacturing total hip acetabula on account of its excellent resistance to abrasion. Several ligaments have been made since 1974, consisting of strands of plaited filaments, the shape of which varies according to the type of ligament used (polyflex of Cendis).

As far as the basic properties of this ligament are concerned, it appears that:

The elastic strength is very limited,

the recovery time after single tensioning is short,

and experiments performed at different tensions show the slowness with which the ligament returns to its initial length.

As far as its biocompatibility and rehabilitation are concerned, polyethylene displays no cytotoxic effects and is well tolerated by the host tissues.

In summary, high-molecular-weight polyethylene has good biological tolerance. The attachment methods of certain ligaments, such as Cendis, for example, make it easy to use, all the more so that their mechanical properties are wholly satisfactory.

Polypropylene

Polypropylene is a synthetic polymer which has been suggested as a strengthening prosthesis.

An example is Kennedy Lad. This is a plait of twisted polypropylene which is used to strengthen ligamentoplasty and is sutured along the whole course of the transplant.

As far as the basic properties are concerned: the rigidity of the plait is linear, with the tension–elongation curve having a uniform slope up to break point. The maximum elastic strength values are, consequently, difficult to measure.

As far as the in vivo life of the plait itself is concerned, it would appear that the overall solidity of attachment increases with time and stabilizes as from the sixth month; the rigidity reaches values of the order of 100 kg Newton per meter and is therefore similar to that of normal ligaments.

As far as the biocompatibility of the product is concerned, using polypropylene as suture material has never posed major tolerance problems, even in the long term.

Polytetrafluoroethylene

Polytetrafluoroethylene, which is still called Teflon, is used as a biomaterial under the name of Gortex for vascular prostheses or cruciate ligament prostheses. As far as its basic properties are concerned: tearing always occurs suddenly by pulling an attachment eyelet hole, without previously modifying the slope of the curve.

The maximum resistance value is very high on isolated ligaments and its rigidity increases with the degree of elongation, but is close to that of the cruciate ligament for its area of physiological use.

Dacron

Dacron is certainly the most used of the synthetic polymers of the ethylene polyterephthalate family. It is a material with an essentially fibrous structure used in the form of a plaited or knitted strand still called Dacron Velvet. The large mesh of this latter type of ligament is intended to encourage good assimilation of the implant into the host tissue.

As far as its biocompatibility is concerned, tests on cell culture show the absence of any cytotoxic effect by Dacron but an analysis of the implantation phenomena in vivo is interesting because a gradual improvement can be seen in the signs of inflammation, as fibrous encapsulation of the ligament takes place. In the intra-articular position, dacron only seems to produce moderate and transitory synovial reactions where the implant remains intact. On the other hand, major inflammatory reactions with changes in the cartilage and in the synovial membrane can be seen if the prosthesis breaks.

As far as carcinogenicity is concerned; under certain conditions many synthetic polymers, including Dacron, have a carcinogenic effect in animals, almost exclusively in rodents. This phenomenon occurs very rarely in man for the materials currently used. Three cases of sarcoma have been published after a Dacron vascular prosthesis was implanted. This suggests that man is only slightly prone to these phenomena or at least that there are very long latency periods of the order of several decades.

To summarize, Dacron is generally fairly well tolerated biologically. The most worrying problem is not the solidity of the implants but the fairly common occurrence of reactional synovitis which sometimes makes it necessary to remove the synthetic material.

Tendon Allografts

Although ligament or tendon autografts are well tolerated, and widely used, they do, however, mean that a neighboring tendon or ligament has to be sacrificed, the removal of which prolongs the time of the operation and alters the biomechanical conditions of the functioning of the joint.

Xenografts of bovine origin are closer to the ligament prostheses treated with glutaraldehyde for increasing the reticulation of the collagen fibers. They are currently rarely implanted in man because of their poor biocompatibility and their inadequate mechanical properties.

Ligament and tendon allografts make it possible to replace the anterior cruciate ligament as an autograft without sacrificing a neighboring tendon.

The specification for ligament and tendon allografts is as follows:

The allograft has to have a similar morphological structure and mechanical properties to those of the ligament replaced.

It has to be perfectly well tolerated.

In the long term, it has to be able to be recolonized by the cells of the host to which it will serve as directional support.

The way in which it is obtained has to be compatible with the legislation on organ removal.

It has to be able to be preserved and stored in a sterile manner without suffering major decomposition.

Its surgical implantation has to be easy, cause little trauma, and allow the graft to be firmly anchored.

Finally, the properties of the allograft have to be stable over time.

Different Allografts which can be Used

Allografts have to be taken from young donors, so that the force and stress values at rupture are similar to those of a normal ligament (1,725 Newtons). The allograft that would best meet the morphology and structure criteria would, of course, be an allograft of an anterior cruciate ligament, however, the choice of this ligament presents considerable technical problems. The patellar tendon, on the other hand, can be easily removed with two bone insertions on the patella and the tibia; it is sufficiently long and has mechanical properties which are clearly superior to those of the anterior cruciate ligaments, even if it is reduced to its middle third.

Preservation by Cryogenics Seems

In order to avoid transmitting any infectious, bacterial, viral, or mycotic pathologies to the host subject, it is necessary to have a sterile allograft. The allograft is removed in the operating theater from selected donors with all the usual asepsis-related surgical precautions. The graft is then packed in a sterile pack bathed in an antibiotic-containing solution and frozen to −198 °C in the Blood Transfusion Center. Other methods for sterilizing the allograft by irradiation or exposure to ethylene oxide were found not to be satisfactory.

Tolerance in the Host Subject

Although the different cellular and protein components which the ligament allografts removed in isolation contain can trigger an immune response, the apatite contained in the bone as well as in the collagen of the tendon does not trigger a clinically perceptible immune response.

Allografts preserved by freezing do not trigger the appearance of the immune HLA group as has been proved by various immunological studies performed at the Blood Transfusion Center in Marseilles.

Mechanical Studies

The graft consisting of the central third of the patellar tendon together with its insertions has been studied mechanically. Creep tests as well as traction tests right up to rupture show that: freezing does not alter the appearance, the color, or the mechanical properties of the grafts; which is not the case for an irradiated tendon – which takes on a board-like appearance –, or an irradiated and freeze-dried tendon, the fibers of which come apart and have a fibrillary appearance.

The problem with these grafts is the mechanical behavior in situ in the long term during revascularization. Experimental studies are currently being conducted to see what the importance is of this revascularization and its effect on the mechanics of the joint in the months and years following implantation.

Conclusion

Apart from the unexpected gain in solidity, synthetic ligaments often also have the advantage of shortening the operating time and allowing early rehabilitation.

The design of a prosthesis of this kind should, in our view, most definitely be directed towards an implant which is itself capable of bearing all the stresses to which the anterior cruciate ligament is subjected immediately after implantation and in the long term. However, the idea of new ligaments which a more or less biodegradable implant could produce or of a functional unit that combining synthetic material and newly formed tissue would produce, although a satisfying idea, seems to us to be illusory from a practical point of view. We have seen some of the models proposed and have shown the striking correlation which exists between the mechanical properties of the prosthetic ligaments and the quality of the results obtained when they are implanted in man.

As far as the use of preserved ligament allografts are concerned, according to the first clinical results, it would appear that there is a considerable percentage of joint laxity when these are used.

The solution, perhaps, is to combine a preserved allograft and a reinforcing prosthetic ligament, which would prevent stresses being exerted directly on the allograft during its rehabilitation period (2–3 years), with the prosthetic ligament being ruptured when the allograft has regained satisfactory mechanical behavior.

Only the test of time will, of course, be able to confirm whether such a choice is well founded.

50 Correlated Kinematic and Dynamic Studies of the Knee Joint During the Stance Phase of Gait: Biomechanical Disturbances Introduced by a Knee Pathology

E. Berthonnaud, B. Moyen, and J. Dimnet

A specific method for accurate and reproducible in vivo knee studies was developed. The femoral knee axis is represented by a straight line which connects lateral and medial markers fixed on the femoral epicondyles. The tibia is represented by three markers. The relative displacements of the femoral knee axis versus the tibia make it possible to determine a knee local frame and to calculate the angles of abduction and axial rotation in the local frame based upon the tibial plateaus.

In this chapter, curves present the kinematic results and show the variations of the anatomical angles of flexion, abduction, and rotation versus time expressed as a ratio of stance phase duration. The global load (force and moment) between the body and the patient's foot induces loading at the level of the knee joint. Dynamic results correlated with kinematic ones are described by the changes in the moment components projected onto the same local frame. The moment components are expressed as values without dimension in order to make direct comparison between subjects of different height and weight possible.

Two groups of subjects were studied: a group of nine normal subjects and another of nine subjects with anterior cruciate ligament (ACL) deficiency. Numerical treatments of the results obtained from the group of normal subjects give the kinematic and dynamic behaviors of the knee of a "mean normal subject" during the stance phase of gait. Mean curves and corresponding dispersions are presented.

The knees of patients suffering from ACL deficiency show similar biomechanical behavior as regards angle and moment variations. The results obtained from the group of patients are directly compared with those obtained from the normal group. Similarities and differences are analyzed and discussed. The biomechanical anomalies observed in the patients are characteristic of ACL deficiency.

Introduction

Anterior cruciate ligament (ACL) deficiency is frequent. Its short and long-term clinical consequences are well known [1]. Cartilage and meniscus damage are observed as well as episodes of giving way [2], and are attributed to changes in knee kinematics. The absence of ACL has obvious consequences on the stability of the knee under static load conditions [3]. Clinical diagnosis is generally easy and clinical tests are made under static conditions. Cadaver studies show an increase in the laxity of the knee in each of its six degrees of freedom [4,5,6,7]. Giving way during a physical activity is a dynamic symptom. Most of the information about knee displacements are obtained through static studies and tests. During these experiments the knee may be loaded but these loading conditions poorly represent human motion. Several gait analyses focus on the sagittal plane during the entire gait cycle. Paul et al. [8] showed that the stance phase must be specifically analyzed at walk because the knee is loaded during this phase. What are the dynamic consequences of an ACL deficiency during the stance phase at walk? In this paper we suggest a solution through the comparison between the mechanical behavior of normal and ACL-deficient knees in motion.

Three-dimensional in vivo studies of joints imply that each body segment adjacent to the joint be equipped with three reflective markers, the trajectories of which are described with an optoelectronic system. The global kinematics of the joint is directly obtained, using the trajectories of the markers in the laboratory frame. These trajectories are obtained from all the discrete positions of the markers given by the cameras. The relative displacements of the markers that belong to a body segment are calculated in relation to the other body segment assumed to be fixed. Any finite displacement of the joint between two successive positions can be described as a screw displacement, using the following screw parameters: the unit vector of the helical axis, one point along the helical axis, the rotation and the translation around and along the helical axis respectively. Dimnet et al. [9] proposed a vectorial method so as to determine the finite screw displacement parameters from relative finite displacements of markers. Woltring et al. [10] defined a matricial technique and evaluated the effects of the errors in the positioning of markers onto the screw parameters. De Lange et al. [11] generalized this approach.

The three-dimensional position of one body segment in relation to another is given by a sequence of three elementary rotations around anatomical axes (flexion, abduction, axial rotation) and the corresponding translations. A matrix form can represent these three elementary rotations and corresponding translations [12,13]. This sequential representation is equivalent to a global screw displacement. The kinematics of the joint can thus be represented by a set of six curves: three curves of rotation and three curves of translation. Ramakrishnan et al. [14] quantified the errors in the results due to a wrong estimation of the axis of flexion of the knee. Grood et al. [15] assume that the axis of axial rotation of the knee is obtained by connecting the knee and hip centers. Other authors use the direction of the femoral diaphysis as the axis of axial rotation of the knee. The calculation of the three-dimensional angular orientation of a body segment in relation to another one, through a sequence of three anatomical rotations, depends on the sequential order because the axes of rotation move during motion. The same displacement in space can be described differently if the sequential order of description is different. Some teams have chosen the following order: flexion, abduction, and rotation [15,16]. Other teams retained another order: flexion, rotation, and abduction [17]. Whatever the sequential order, the results regarding flexion are similar but the results regarding abduction and axial rotation are very different.

Generally, joint kinematics obtained from the numerical treatment of the trajectories of external markers is supposed to represent the kinematics of bones. Markers, however, are fixed upon the skin and tissues are interposed between bone and skin. Cheze et al. [18] observed two disturbing effects in the relative displacements of external markers in relation to internal bones: the first one is due to skin elasticity. The measurements of the triangle formed by the markers fixed on a body segment (segmental triangle) continuously vary during motion. A solidification procedure can be used to substitute a rigid triangle for each segmental triangle. Laws of solid kinematics can then be applied to the solidified triangles. The second disturbing effect is due to the movements of the soft masses interposed between skin and bone. Cappozzo et al. [19] mentioned the problem. Lafortune et al. [20] excluded the effects of these artifacts in directly implanting tripods equipped with reflective markers in bones. In this chapter another approach is suggested. Two groups of subjects were studied: a group of nine normal subjects and another of nine subjects with ACL deficiency. External markers were placed on osseous landmarks at the level of the knee. We studied the stance phase, which is the phase during which the knee is loaded. During this phase, the relative displacements of the skin in relation to bones are constantly relatively small [21]. Our goal was to obtain accurate and reproducible results regarding abduction and rotation during the stance phase. We therefore defined a new protocol and a new method of numerical treatment. This new protocol and this new treatment were applied to our two groups of subjects. The results appeared very reproducible and significant differences between the two groups were shown.

Materials and Methods

The Two Groups of Subjects

Nine normal subjects (six males and three females) of a mean age of 26 ± 6 years were studied. They had no previous trauma or surgery on their inferior limbs. Only their right side was studied.

Nine patients of a mean age of 26 ± 6 with unilateral anterior cruciate ligament (ACL) deficiency (three females, six males) were selected. They all suffered from knee instability and were scheduled for reconstructive surgery. They all had a positive Lachman test, with a soft end point, and a positive pivot shift test. The passive knee anterior laxity was evaluated using a KT 1000 device. The results showed injured vs. normal side differences of 7 ± 2 mm. In this group, only the injured side was studied.

Experimental Protocol

The retro-reflective markers were fixed on the subjects on the following landmarks: anterior iliac spine, iliac crest, and posterior iliac spine for the pelvic segment, medial and lateral epicondyles and greater trochanter for the thigh segment, anterior tibial tuberosity and medial and lateral ankle malleolli for the tibial segment, posterior aspect of calcaneum, base of third and fifth metatarsal bones for the foot. Great attention was paid to fix the markers on the more prominent parts of the osseous landmarks. The markers were fixed upon a tape which was glued on the skin. The trajectories of the markers in the laboratory frame were computed by the use of a Motion Analysis System. This system is composed of five cameras (the frequency of which ranges from 60 to 200 Hz), a high-resolution software (E.V.A) and a Sun computer. The cameras were positioned so that each marker be observed by at least two cameras at each instant of time. This particular location was justified by the need for a continuous observation of all markers and in particular those which were placed on the medial side and which might have been hidden by the opposite limb during its swing phase. A force plate (OR 6-5 AMTI Advanced Mechanical Technology) was used in addition to the opto-electronic system. The cameras were placed around the force plate so as to determine the kinematics of the knee under load. The opto-electronic frame was the same as that of the force plate. Both opto-electronic and force signals were synchronized.

The first step was the calibration of the three-dimensional space of the laboratory. The measurements of the calibrating object were 50 cm × 75 cm × 80 cm. The calibrating object was placed so that it surrounded the force plate. The linear and angular accuracies after treatment when markers were situated inside the calibration object were 0.5 mm and 0.15° respectively. When outside, the accuracies were 1.8 mm and 0.31° respectively.

Each subject was studied in three static rest positions and during walking. The subject was asked to stand up in three different fixed positions: first feet parallel, second in a free rest position in which the subject felt comfortable, and third the knee axes parallel to the frontal plane. Each static position was recorded during one second. The mean position of each marker during each recording was calculated. Dynamic positions were subsequently studied in relation to these static postures in order to determine the more neutral position. This more neutral position was used as a reference for the studies of the knee in motion. Each patient was then asked to walk at a steady state. He started to walk about 6 m before reaching the force plate and the cameras. This distance represented four complete gait cycles. The patient was asked to follow a straight line drawn on the floor. A great many trials were recorded. Only the trials in which the subject walked along the line drawn on the floor and left his foot in contact with the force plate during the whole stance phase were kept and subsequently treated. Trials were repeated as many times as needed in order to obtain at least 10 correct records. The last step was the recording of a circumduction movement of the inferior limb, knee extended, in order to calculate the location of the internal center of the hip in relation to the external markers [22]. Seven complete circumduction movements were recorded in order to improve the accuracy of the result. The duration of this stance phase varied between two trials and from one subject to another. Time was thus expressed as a ratio of stance phase duration. The vertical component Fz of the ground reaction force was used to determine the beginning and the end of the stance phase. The stance phase corresponded to $Fz > 10\,N$. For the studied group, the stance phase may be considered as a sequence of three sub-phases: heel strike (23% of stance phase duration), foot laid flat on the ground (53%), and lift off (24%). These values are comparable to those given by Biden et al. [23].

A reproducibility test was performed. One normal subject was randomly chosen. He was equipped with the markers and asked to make a series of trials at different moments of the same day. Seven trials were recorded. Two weeks later, the same subject was tested again by a different operator. Four trials were recorded. The goal was to estimate the influence of the positioning of markers on the results. A curve describing knee kinematics was drawn from each trial. The mean curve and standard deviation represent the reproducibility of knee kinematics of the same subject in the same conditions. The trajectories of the markers were numerically treated, so as to obtain different kinds of kinematic results. The trajectories of the markers placed upon a same body segment (segmental markers) make it possible to obtain displacements of this segment in relation to the fixed laboratory frame. The relative displacements of each body segment in relation to the adjacent one, assumed to be fixed, make it possible to obtain the kinematics of the joint placed between the two body segments. The kinematics of all segments and all joints was calculated. In this paper we focus on knee joint kinematics.

Local Knee Kinematics

Usually, knee kinematics is obtained from the calculation of the instantaneous relative positions of the set of markers fixed on the thigh segment in relation to the tibial segmental markers assumed to be fixed. Two kinds of noise disturb the estimation of the internal joint kinematics when it is directly obtained through the trajectories of markers. The first one is due to the elastic deformations of the skin. A solidification procedure [18] must be applied. The second origin of noise is the relative displacements between external markers and internal bone because of interposed smooth masses. A corrective function must be applied to the kinematics obtained through the treatment of external markers so as to determine the joint kinematics, but it requires a previous study for each joint, each subject, and each task.

A new approach is proposed so as to determine knee kinematics. The results obtained are independent of the tested subject. In this case, the results obtained from the treatment of external markers coincide with internal kinematic results. Two markers were located at the level of the medial and lateral epicondyles. A straight line passing through these two markers represents the condyle axis. A frame was affixed to the tibia through the three tibial markers using a usual technique. During the stance phase of gait, the bundle of condyle axes is distributed around a mean plane in the tibial frame. This mean plane is a correct estimation of the mean plane of the tibial plateaus (Figure 50.1). A local tibial frame based on this mean tibial plane was defined. The direction of the condyle axis (which connects the condyle markers) is not affected by the possible changes in the distance between the two markers due to the effect of smooth masses. Each instant location of the condyle axis in relation to the local tibial frame gives the angles of both axial rotation and abduction of the knee. The zero value of these angles was obtained from the treatment of the static posture corresponding to the free rest position. This free rest position appeared to be the more neutral static position and to give the more reproducible dynamic results. The flexion angle was obtained using the 3D kinematic technique. At each instant of time the angular orientation of the knee joint was calculated from the relative position of the femoral frame based on the epicondyles and the hip center in relation to the tibial frame. A sequence of three successive rotations – flexion, abduction, and axial rotation – was applied. The values obtained by the standard method for axial rotation and abduction are significantly different from those obtained through the condyle axis technique described above. The curves of angle

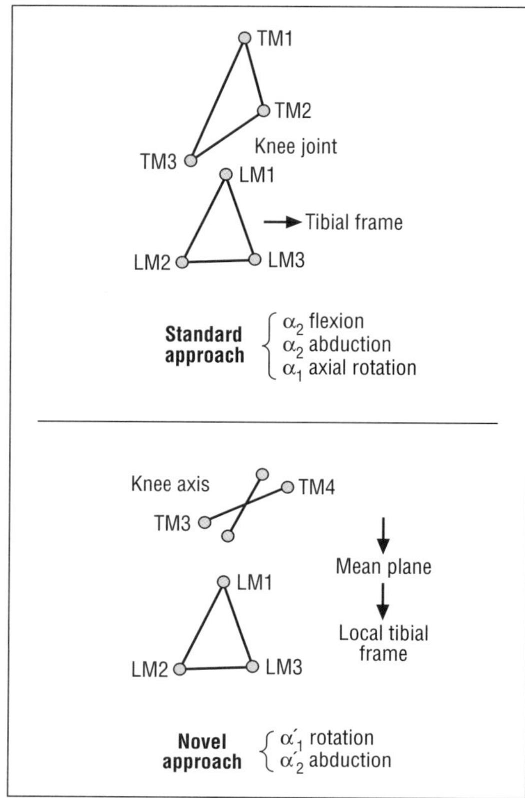

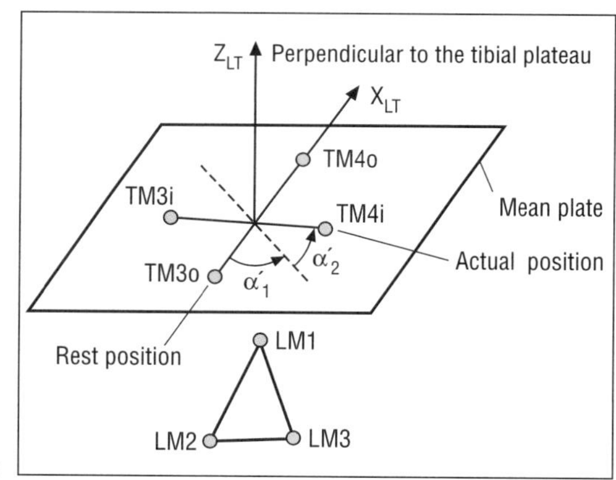

Figure 50.1. Principle of numerical treatments.

variations presented in this paper were obtained by two techniques: calculation of the relative displacements of the femoral frame versus the tibial frame for the angles of flexion (standard method) and calculation of the relative displacements of the condyle axis in relation to the local tibial frame for the angles of rotation and abduction (new method). Flexion, abduction, and internal rotation correspond to positive values. The center of the two epicondyle markers was assumed to be the knee center. It was thus possible to calculate its instant positions in relation to the local tibial frame. The position of this knee center is subject to inaccuracies due to the imprecision of the 3D position of each epicondyle marker. Conversely, the direction of the condyle axis is much more accurate. Because of that, we focused on the shape of the trajectories of the knee center during stance phase rather than on the trajectory itself. This trajectory can be included in a rectangle, the measurements of which were calculated.

Local Knee Dynamics

The signals given by the force plate are available in the form of the instantaneous values of the generalized ground reaction force (force and moment). At each instant of time, three components of force: F_x, F_y, F_z, and three components of moment at the platform center: m_x, m_y, m_z, are measured. The local tibial frame can be located at each instant of time in relation to the laboratory frame. At each instant and for each posture of the lower limb, it was possible to calculate the effects of the ground reaction force at the level of each joint. Force and moment components are presented in normalized form. The force components are expressed as a ratio to the weight of the subject. The moment components are expressed as a ratio to the product weight of subject per height. Kinematic and dynamic results obtained in different subjects can thus be directly compared. Each subject was submitted to previous anthropomorphic measurements.

In this study the weight of the foot and leg segments as well as the forces of inertia were neglected. The generalized local force (local force and local moment) induced by the ground reaction force at the level of the foot was projected onto the local tibial frame. At each instant of time, the kinematic components (flexion, abduction, and rotation angles) can be directly compared with the moment components applied to the joint, as both of them are projected onto the same local frame.

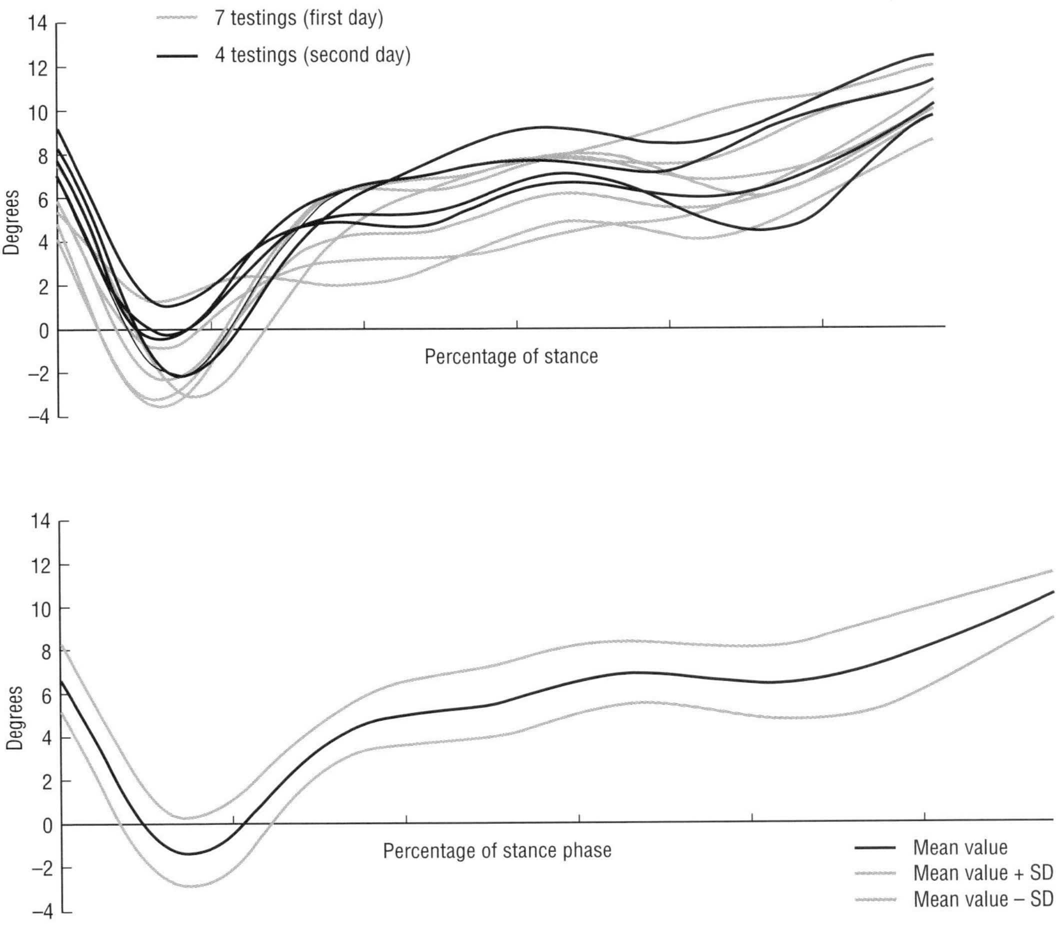

Figure 50.2. Reproducibility of the kinematic results as regards axial rotation in a healthy subject: seven different testings were done on a day (first day) and four other testings were done another day (second day). The markers were all taken off and replaced between first day and second day testings.

Results

Reproducibility Tests

The influence of the positioning of markers and of the moment of the recording is shown for axial rotation of the loaded knee (Figure 50.2) as axial rotation was shown to be the movement that led to the maximum interindividual dispersion.

Figure 50.2 shows the axial rotation obtained in the same subject from two series of tests made on different days, which implied removal and repositioning of markers. Repositioning of markers have no influence on the results, which can thus be considered as reproducible.

The Group of Normal Subjects

The curves describing knee kinematics during the stance phase in the group of normal subjects are presented (Figure 50.3). Kinematic results are obtained as follows: for each instant of time (expressed as a percentage of stance phase duration), for each trial, and for each subject, three angular components were calculated: flexion, abduction, axial rotation. For each angular component and for each instant, 90 values

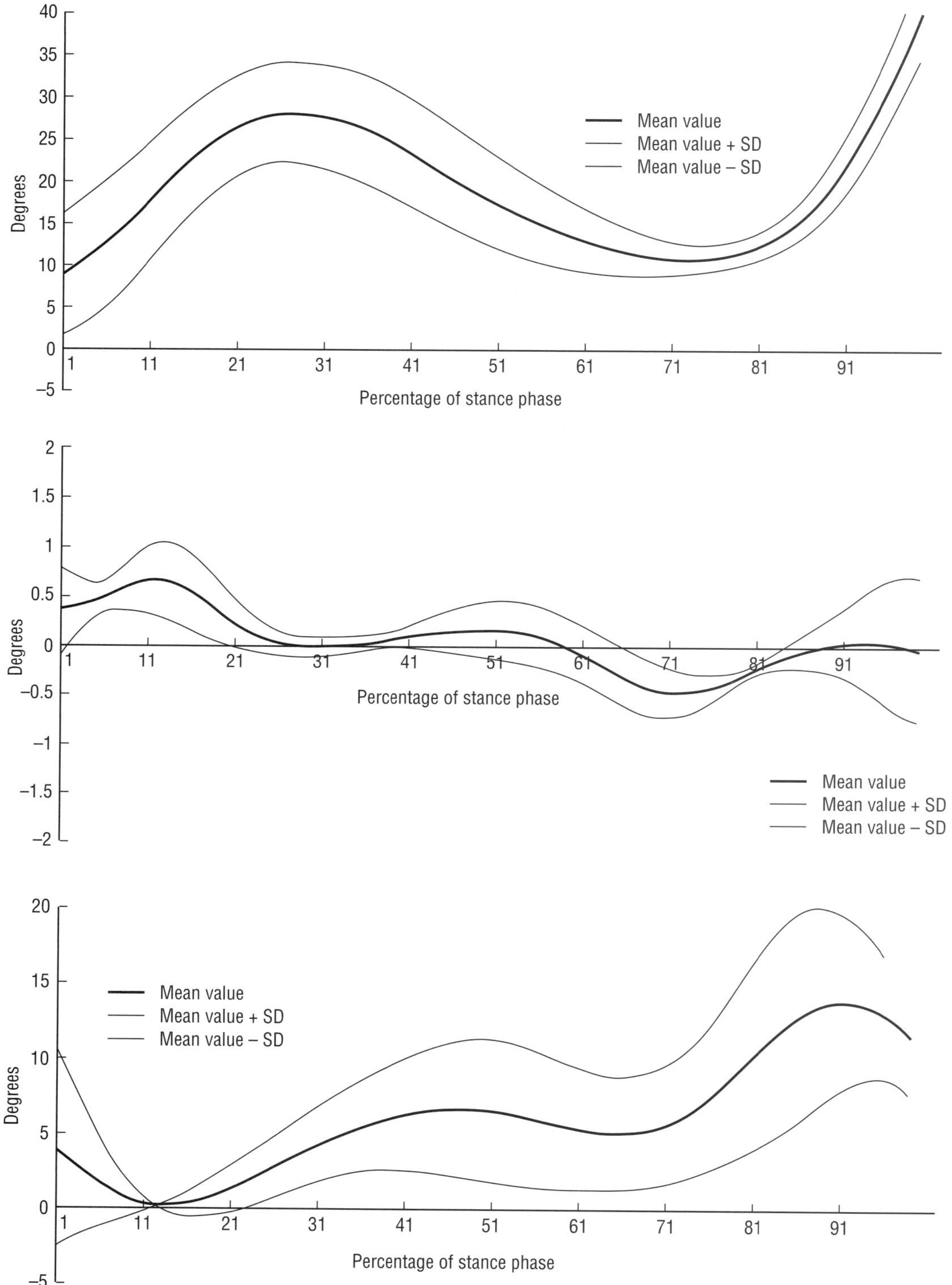

Figure 50.3. Kinematic knee behavior during the stance phase in a "mean normal subject".

corresponding to ten trials per subject and to nine subjects were available. Mean values and corresponding dispersions (± one standard deviation) were then calculated. The same procedure was used for moment components (Figure 50.4). The moment induced by the ground reaction force was reduced at the knee center and then projected onto the same local frame as the one used for kinematic calculations. As time was expressed without dimension (as a percentage of stance phase duration) the curves of variations in angles versus time (Figure 50.3) showed great similarities between normal subjects. Moments were expressed without dimension either, as ratios to weight of subject per height. The corresponding units are called moment units (MU). The curves of variations in moments versus time (Figure 50.4) also showed great similarities between normal subjects. A concept of a "mean normal subject" was thus introduced, which corresponded to the mean values.

The flexion angle of a "mean normal subject" (Figure 50.3a) increases continuously from 9° to 27° during the first subphase, then decreases from 27° to 12° (second subphase) and increases again from 12° to 37°. The dispersion is particularly high during the first subphase (±5°). The mean abduction angle (Figure 50.3b) decreases slightly from 0.4° to −0.4° with a dispersion, the maximum of which is 0.5°. These very small abduction values, which vary around zero, indirectly validate the choice of the mean plane drawn by the bundle of the condyle axes; this mean plane is roughly parallel to the mean plane of the tibial plateaus. The mean angle of axial rotation decreases from 5° to 0° at the beginning of the first subphase and then increases regularly toward the internal side and reaches 14.2° when the foot leaves the ground. Axial rotation behavior displays the greater dispersion. It can be greater than ±5°.

All the curves of variations in moment components (Figure 50.4) show discontinuities at the beginning of the first subphase when the heel strikes the floor. This phenomenon is encountered at the same time. It can be assumed that the moment vector (resultant of the three moment components) is affected by a discontinuity at this time. This discontinuity corresponds to an impact when the heel strikes the floor. This impact lasts for around 20 ms. Curiously enough, it does not appear on kinematic results. This impact apparently has no consequence on movement. The dispersion value is less than ±0.3 MU. In the normal subjects analyzed the greatest interindividual differences observed concern the kinematics of the loaded knee and more particularly the axial rotations. The moment induced at the level of the knee and due to external forces is very similar between subjects.

Group of Patients with ACL Deficiency

For each instant of time, mean values for all trials and for all patients and corresponding dispersions (± one standard deviation) were calculated. The kinematic and dynamic behavior of the ACL-deficient knees are very similar between subjects and thus make it possible to define the mechanical (kinematic and dynamic) behavior of a "mean pathological subject".

Figure 50.5 shows the mean kinematic behavior of a patient with ACL deficiency. The instability previously observed in abduction – adduction in a pathological subject – is found again in all the patients.

Figure 50.6 shows the mean dynamic behavior of the injured knees. The high impact during heel strike is present in all the patients.

Comparison Between the Two Groups (Normal Subjects and Patients with ACL-deficient Knee)

The same figure 50.5 also delivers the kinematic behavior of the loaded knee of a "mean normal subject" during the stance phase of gait and that of a "mean patient" with ACL deficiency. The "mean normal subject" is represented through mean curves and corresponding dispersions. The mean behavior of the group of patients is represented through mean curves only in order to preserve the readability of the diagrams. For flexion angles (Figure 50.5a) the "mean patient" curve is entirely situated below the dispersion zone of the "mean normal subject". There are no significant differences between the curves of variations in abduction angles of the two groups

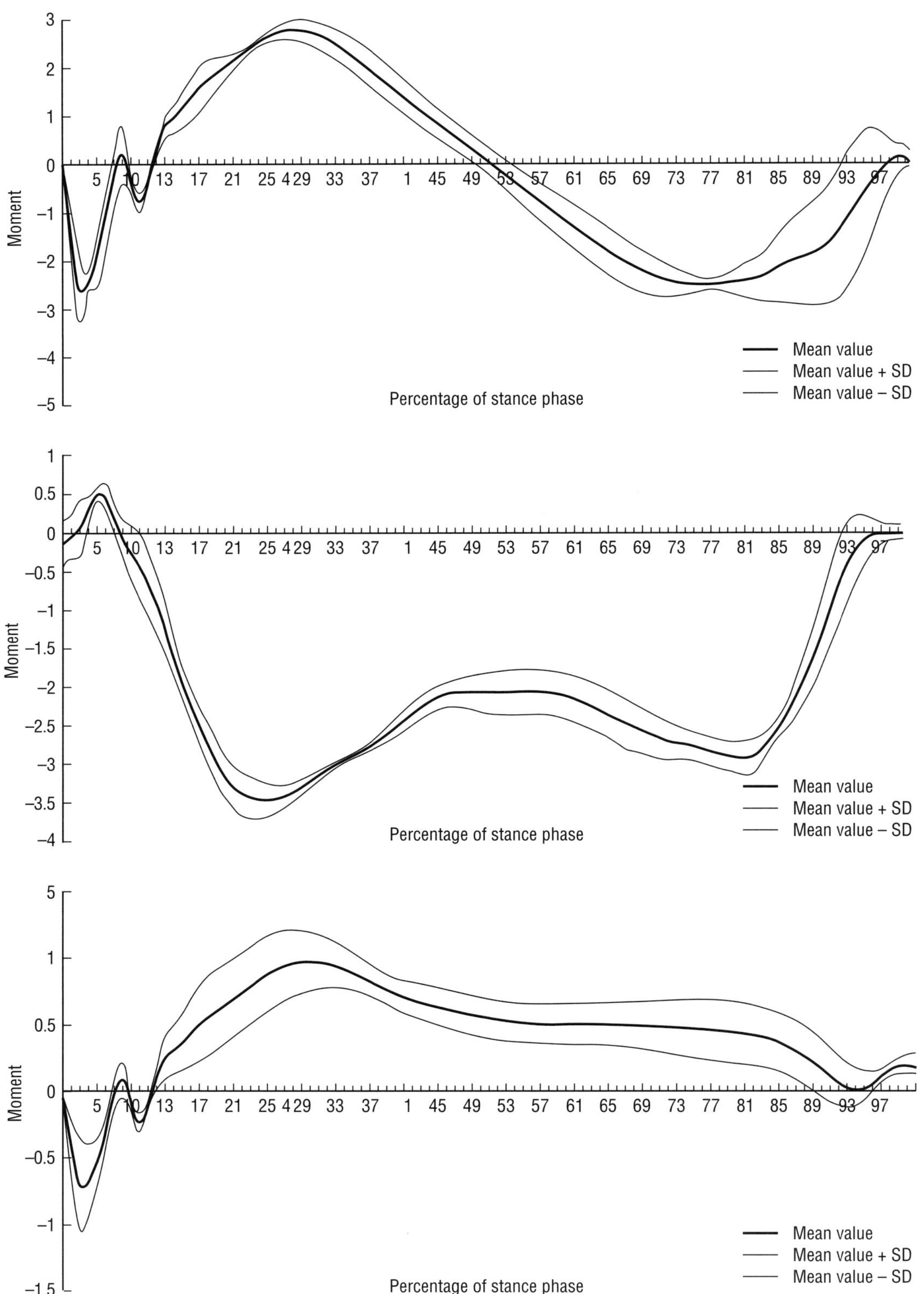

Figure 50.4. Dynamic knee behavior during the stance phase in a "mean normal subject".

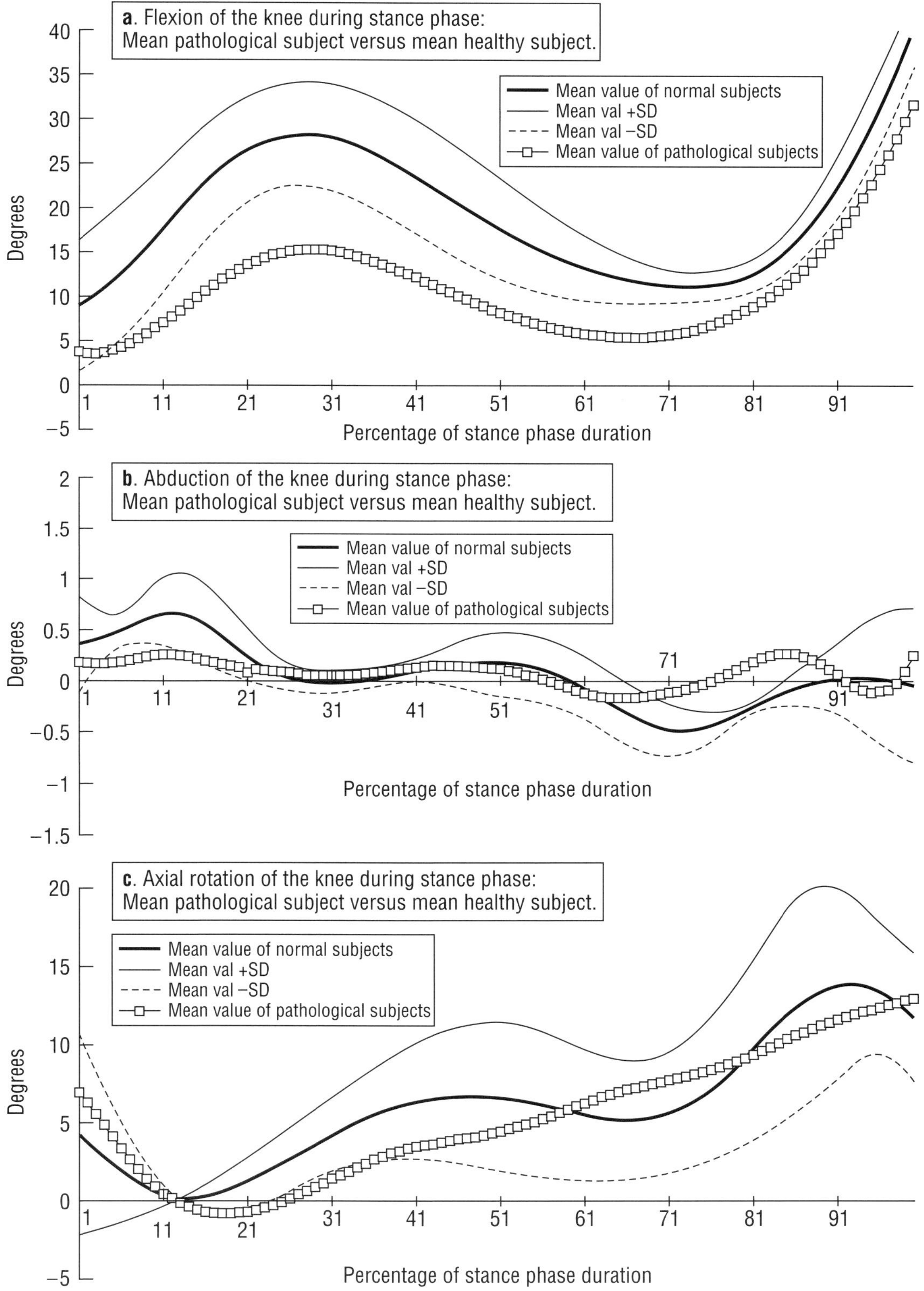

Figure 50.5. Kinematic behavior of a knee joint with ACL deficiency compared with that of a "mean normal subject".

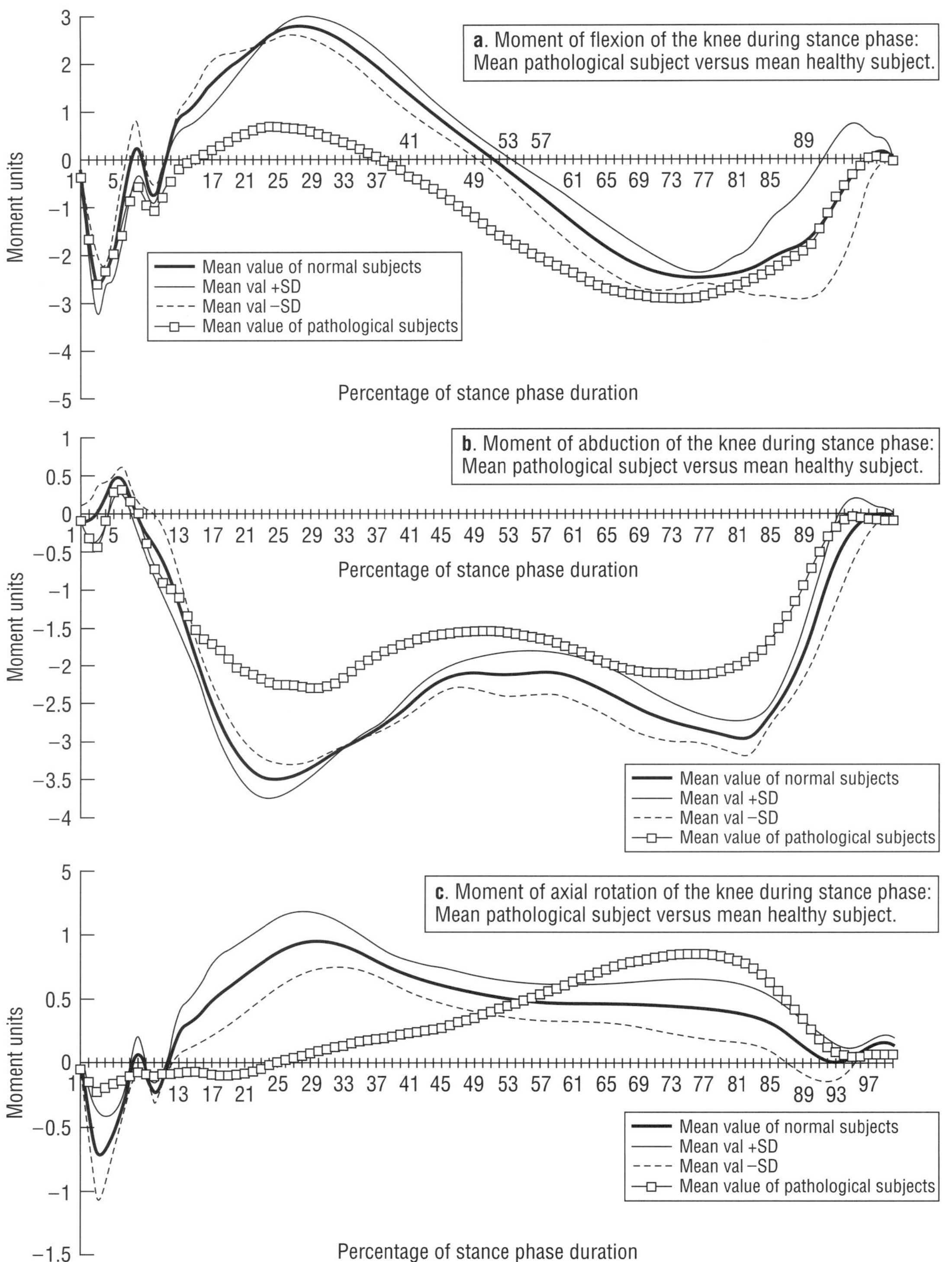

Figure 50.6. Dynamic behavior of a knee joint with ACL deficiency compared with that of a "mean normal subject".

(Figure 50.5b). The variations in the angle of axial rotation of a "mean patient" is entirely included inside the dispersion zone of the "mean normal subject" but the appearance of this curve versus time is very different. The pathological knee continuously rotates with a continuous slope during the last two subphases of the stance phase.

Figure 50.6 shows the dynamic behavior of the knee of a "mean patient" with ACL deficiency and that of a "mean normal subject". The curves of variations in the moment of flexion (Figure 50.6a) and abduction show lower magnitudes in the pathological knee than in the normal one. The moment of axial rotation exerted on a pathological knee is strongly different from that sustained by a normal knee (Figure 50.6c).

The comparison between Figures 50.5 and 51.6 makes it possible to correlate kinematic and dynamic results. In the normal and pathological subjects, kinematic and dynamic behaviors in flexion are similar (same shapes of curves 50.5a and 6a). In the normal and pathological subjects, kinematic and dynamic behaviors in abduction are different (different shapes of curves 50.5b and 6b). In rotation, normal and pathological knees show strongly different mechanical behavior. In the normal knee, the kinematic and dynamic behavior in axial rotation are different (different shapes of curves 50.5c and 6c). In the ACL-deficient knee, the kinematic and dynamic behavior in axial rotation are similar (same shapes of curves 50.5c and 6c).

Influence of the Local Condition on the Kinematic Behavior of Adjacent Joints

An ACL deficiency significantly modifies the kinematics and the dynamics of the knee joint, especially in rotation. What are its effects upon the adjacent joints, i.e., the ankle and the hip joint? Figure 50.7 shows the kinematic behavior of the joints of the pathological lower limb in axial rotation during the stance phase. The results obtained in a pathological patient are compared with those obtained in a "mean normal subject" tested under the same conditions. The results obtained in the patient are presented in the form of a curve while those obtained in a "mean normal subject" are presented as a zone of dispersion.

The kinematic behavior in axial rotation of the hip (Figure 50.7a) and of the ankle (Figure 50.7c) in a patient with ACL deficiency is entirely different from that of the same joints in a normal subject. The adjacent joints compensate for ACL deficiency and for the biomechanical anomalies induced by the condition.

Discussion and Conclusion

In this study, a new protocol based on a different mechanical concept was established and validated. This protocol focuses on the displacements of an axis, the femoral condyle axis, in relation to a local tibial frame based upon the plane of the tibial plateaus. Both kinematic (angular displacements) and dynamic (moment components of external force) results are analyzed in this local tibial frame, which makes comparison between them possible.

The experiment appeared to be reproducible. The positioning of markers does not influence the results significantly.

This new protocol allowed us to obtain results that were comparable between normal subjects, enabling us to define the mechanical (kinematic and dynamic) behavior of a normal knee ("mean normal subject"). Similarly, the results obtained in patients with ACL deficiency were comparable, allowing us to define the mechanical behavior of a knee with ACL deficiency ("mean pathological knee").

The comparison between a "mean normal" and a "mean pathological" subject enabled us to determine mechanical parameters that are characteristic of ACL deficiency. These parameters are:

an instability in abduction–adduction in the ACL-deficient knee at the end of the stance phase (during lift off).

a different pattern of axial rotation in the pathological knee compared with the normal one.

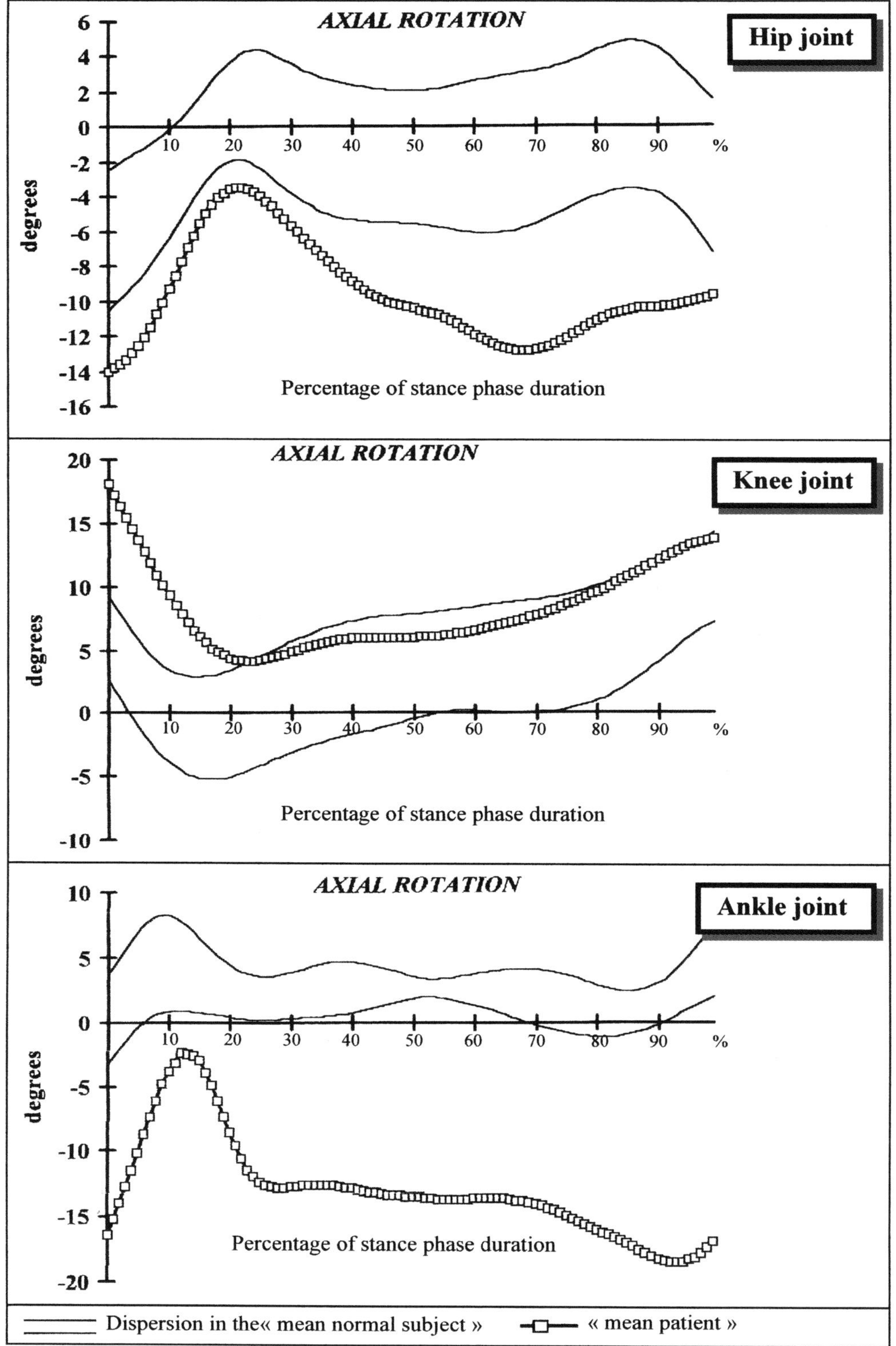

Figure 50.7. Kinematic effects of a knee condition upon the injured and adjacent joints as regards axial rotation.

high moment components (in flexion, abduction, and rotation) during impact (heel strike) in the pathological knee.

a different pattern of the moment of axial rotation in the pathological knee compared with the normal one.

The results present both linear and angular displacements of knee joints during the stance phase of horizontal walking. The linear displacements are strongly influenced by the errors due to the chain of measurement and numerical calculations, contrary to the angular displacements at the level of the knee joint. This is why this paper focuses mainly on the angular results.

The comparison between the kinematic and dynamic results allowed us to look for a physiological link between moments and displacements. Such a link does exist in flexion, both in the normal and in the pathological knee. There is no link in abduction. There is no link in rotation in the normal knee, whereas a link exists in rotation in the pathological knee. ACL deficiency was shown to induce anomalies in the functioning of the adjacent joints (ankle and hip).

The number of subjects studied (nine normal subjects and nine patients with ACL deficiency) is very low. It is not sufficient for clinical applications. The main goal of this study was to test a new experimental protocol and a new numerical treatment so as to obtain clinically significant results and to make sure that ACL deficiency induced biomechanical anomalies. The results of this study incite us to pursue them in the form of clinical applications. The experimental protocol will be simplified so as to take less time, the number of trials will be reduced, but the initial principle which consisted in observing the lateral and medial epicondyles will be kept. A sixth camera will be added to facilitate the experiment.

The software is organized in such a way that the series of normal subjects will be significantly increased. For each new subject, the parameters of the "mean normal subject" are renewed and a new mean curve and corresponding dispersions are calculated. Other tasks will be analyzed: climbing and descending stairs, rotating on a leg, effect of a prosthesis upon the kinematic and dynamic behaviors of the injured knee, the adjacent joints and the controlateral knee.

References

1. Andersson C, Odensten M, Gillquist J. Knee function after surgical or nonsurgical treatment of acute rupture of the anterior cruciate ligament: a randomized study with a long-term follow up period. Clin Orthop 1991;264:255–63.
2. Noyes FR, Dunworth LA, Andriacchi PP, Andews M, Hewett TE. Knee hyperextension gait abnormalities in unstable knees. Am J Sport Med 1996;24:35–44.
3. Grood E, Stowers S, Noyes F. Limits of movements of the human knee. Bone Joint Surg 1988;70A:88–96.
4. Fukubayashi B, Torzilli PA, Sherman MF, Warren RF. An in vitro biomechanical evaluation of anterior–posterior motion of the knee. J Bone Joint Surg 1982;64A:258–64.
5. Gollehon DL, Torzilli PA, Warren RF. The role of the cruciate ligaments in the stability of the human knee. J Bone Joint Surg 1987;69A:233–42.
6. Grood ES, Noyes FF. Diagnosis of knee ligament injuries: biomechanical precepts. In Feagin JA, editor. The Crucial Ligaments. New York: Churchill Livingstone, 1988.
7. Markolf KL, Kocha A, Amstutz HC. Measurement of knee stifness and laxity in patients with documented absence of the anterior cruciate ligament. J Bone Joint Surg 1984;66A:242–53.
8. Paul JP. Mechanics of the knee joint and certain joint replacements. Total Knee Replacement. Springer-Verlag, 1988.
9. Dimnet J, Guingand M. The finite displacement vector method. J Biomech 1984;17:387–94.
10. Woltring HJ, Huiskes R, De Lange A. Finite centroide and helical axis estimation from noisy landmarks measurements in the study of human kinematics. J Biomech 1985;18:379–89.
11. De Lange A, Huiskes R, Kauer GMJ. Measurements errors in roentgen-stereophotogrammetric joint-motion analysis. J Biomech 1990;23(3):259–69.
12. Challis JH. A procedure for determining rigid body transformation parameters. J Biomech 1995;6:703–37.
13. Soderkvist I, Wedin PA. Determining the movements of the skeleton using well-configured markers. J Biomech 1993;12:1473–7.
14. Ramakrishnan KK, Kadaba MP. On the estimation of joint kinematic during gait. J Biomech 1991;24(10): 969–77.
15. Grood ES, Suntay WJ. A joint coordinate system for the clinical description for the three-dimensional motion: applications to the knee. J Biomech Eng 1983;105: 136–44.
16. Chao EY, Laughman RK, Schneider E, Stauffer RN. Normative data of knee joint motion and ground reaction forces in adult level walking. J Biomech 1983;16(3): 219–33.
17. Blankevoort L, Huiskes R, De Lange A. The envelope of the passive knee joint motion. J Biomech 1988;21(9): 705–20.

18. Cheze L, Fregly B, Dimnet J. A solidification procedure to facilitate kinematic analysis based on video system data. J Biomech 1995;28:879–84.
19. Cappozzo A, Gazzani F, Marcellari V. Skin markers artefacts in gait analysis. J Biomech 1990;23:363.
20. Lafortune MA, Cavanagh PR, Sommer HJ, Kalenak A. Three-dimensional kinematics of the human knee during walking. J Biomech 1992;25(4):347–57.
21. Jied A. Etude cinématique et dynamique des groupes articulaires du membre inférieur au cours de la phase d'appui de la marche, à partir de marqueurs externes. Thèse de doctorat en Génie Biologique et Médical n° 15297, Université Claude Bernard Lyon I, 1997.
22. Gutierrez C. Etude des fonctionnalités du membre supérieur en mouvement par l'exploitation de trajectoires de marqueurs externes. Thèse de Mécanique, Université Claude Bernard, Lyon, France, 1996.
23. Biden E, O'Connor J, Collins JJ. Gait analysis. Knee Ligament: Structure, Function and Repair. Raven Press, 1990;291–311.

51 Three-dimensional Kinematic Assessment of Ligament Rupture and Surgery

N. Hagemeister, N. St-Onge, G. Parent, M. Van de Putte, L'H. Yahia, N. Duval, and J. A. de Guise

Introduction

Only a few areas of contemporary orthopedics have raised as many controversies and diverging opinions as optimal reconstruction of the cruciate ligaments of the knee. The annual incidence of ligament tears is 1/3,000 in the United States, adding up to 95,000 ruptures and 50,000 reconstructions per year. It is, therefore, not surprising that there is a large amount of literature on the subject. Over the past 20 years, more than 4,000 papers have been written, including review articles and book chapters [1].

Anterior cruciate ligament (ACL) rupture leads to symptomatic instabilities, thereby increasing the risk of recurring injuries, meniscal tears, and osteoarthrosis [2]. ACL reconstruction is recommended for young, active patients, especially those participating in high-level sports or working in areas involving constant loading of the knees [1].

According to the literature, many methods for reconstructing ligaments seem to give good to excellent outcomes in the short-term, with knee function being quasi-normal in 75–85% of cases [2]. However, long-term results are rarely mentioned. Methods, such as Ellison transfer [3] or reconstruction using the medial meniscus for ACL reconstruction [4], were abandoned without close study. The lack of consensus in the literature on ligamentoplasties is probably due, at least in part, to the difficulty associated with evaluating treatments in a reliable and objective manner. According to Gillquist [2], "*we don't need new ACL operations, we need to know how good or bad the old operations are. This is especially important now with the introduction of artificial ligaments and allografts. A scientific analysis of why things fail is very important. The solution to the problem of why things fail may not be found in a new procedure but in the failure of the old one*".

To evaluate knee injuries affecting joint stability, physicians use subjective questionnaires (KOOS, Tegner), laxity tests which can be manual (Lachmann, pivot shift), or instrumentation (KT1000, Telos). These tools measure displacement of the tibia with respect to the femur in a unidirectional manner (antero-posterior (A-P) translation, tibial rotation) when force is applied to the tibia. Their reliability for diagnostic purposes is generally accepted. However, their use for treatment evaluation has been questioned by several authors [5,6]. Gillquist [2] even suggested that this could be the reason why many reconstruction methods give the same proportion of good and excellent outcomes. Moreover, these tools measure passive knee laxity in a unidirectional manner, while the knee has a three-dimensional (3D) motion. We believe that evaluating 3D kinematics is complementary to measuring passive, unidirectional laxity and that it might allow more objective clinical examination by providing a reliable representation of the functional state of the knee. At the moment, a reliable and repeatable method does not exist for evaluating these two aspects of knee function in vivo.

For the past 10 years, our research group has been working on the problem of precise recording of 3D knee motion. This chapter discusses "proof of the concept" obtained from in vitro studies on human cadaveric knees. In vivo analysis methodology and results concerning accuracy and repeatability of the measurements are also discussed. This methodology is based on the use of an attachment system designed to reduce marker movements relative to the bone, making it possible to record small 3D movements. In addition, a reproducible method of anatomical movement description is presented. Finally, a case study of one subject who suffered an isolated ACL tear is reported, showing that with the proposed methodology, small 3D kinematic variations can be measured non-invasively in vivo.

Evaluating the Outcome of Ligament Reconstruction

In Vitro Study: Confirming the Hypothesis

As already mentioned, knowledge of the failure of reconstruction procedures remains limited. In vitro testing has been widely used to acquire knowledge on knee kinematics, for the development of new surgical reconstruction methods, and for the assessment of old methods. Most protocols employ quasi-static settings, where a force is applied to the tibia or the femur and the resulting displacement is measured [7–21]. These settings mainly reproduce laxity tests performed in clinics. Other authors investigate knee movements in dynamic settings, where the knee is placed in a rig with simulation of every day life movements (descending stairs, rising from a chair, flexion/extension) generated by traction on tendons [22–40]. Relative rotations of the tibia with respect to the femur are then measured. None of these studies have, however, investigated the effect of different types of ligament reconstructions with both biological and synthetic grafts on 3D kinematic parameters. This prompted us to conduct an in vitro study that allowed the experimental comparison of two different ACL reconstruction procedures, using a synthetic ligament and a biological graft.

The following in vitro experimental protocol was performed: electromagnetic Fastrack sensors (Polhemus, Colchester, Vermont, USA) were fixed on the femur and tibia with aluminum screws. Personalized 3D geometric models were first acquired from computer tomography images and reconstructed with a reconstruction software (SliceOmatics, Tomovision, Montreal). The movements of real bones were then matched to the model by means of a calibration procedure. Laboratory-developed software, using imagery and computer graphics coupled with numerical calculation methods, allowed the computation of knee kinematics: flexion/extension, abduction/adduction, internal/external tibial rotation as well as A-P, medio-lateral (M-L), and proximo-distal (P-D) translation [41,42]. The use of personalized geometric models is an original feature of the protocol, since they allow precise documentation of the reference axes used to calculate the 3D kinematic parameters. Kinematics were recorded during flexion/extension movements that were simulated by pulling on the quadriceps tendon. A-P translation and internal/external rotation of the tibia were recorded during manual laxity testing at different fixed knee flexion angles (0°, 30°, 90°) to establish a database for the comparison of our results with those in the literature. The experimental setup is shown in Figure 51.1.

A series of experiments on 10 cadaver specimens was conducted. Two types of reconstructions were tested. The first, called isometric two-tunnel reconstruction, was performed using two different materials: a synthetic Trevira ligament, pretensioned with a tensiometer at 70 N, and a biological bone–patellar tendon–bone autograft, pretensioned manually. The second, modified "over-the-top" (OTT) reconstruction, was performed with a Trevira prosthesis, pretensioned at 70 N. The OTT technique was modified compared to the classical technique described by MacIntosh [43] in that the tibial tunnel entrance was placed dorsomedially (D-M) to the ACL insertion. This method has been proposed by Krudwig [44] to minimize elongation of OTT reconstructions. To the best of our knowledge, ours is the first study to compare the immediate effect of different types of reconstruction methods with a synthetic material versus a biological graft on 3D kinematics.

The results of manual laxity testing are shown in Figure 51.2. ACL dissection created signifi-

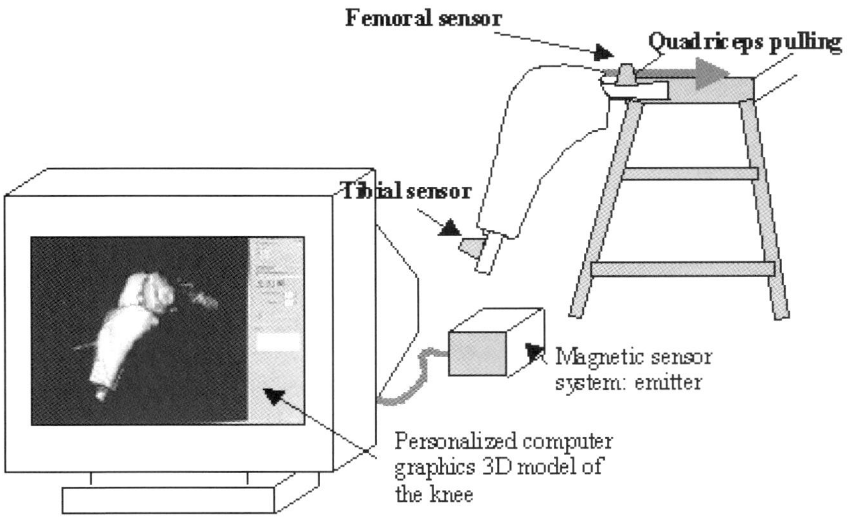

Figure 51.1. Experimental setup.

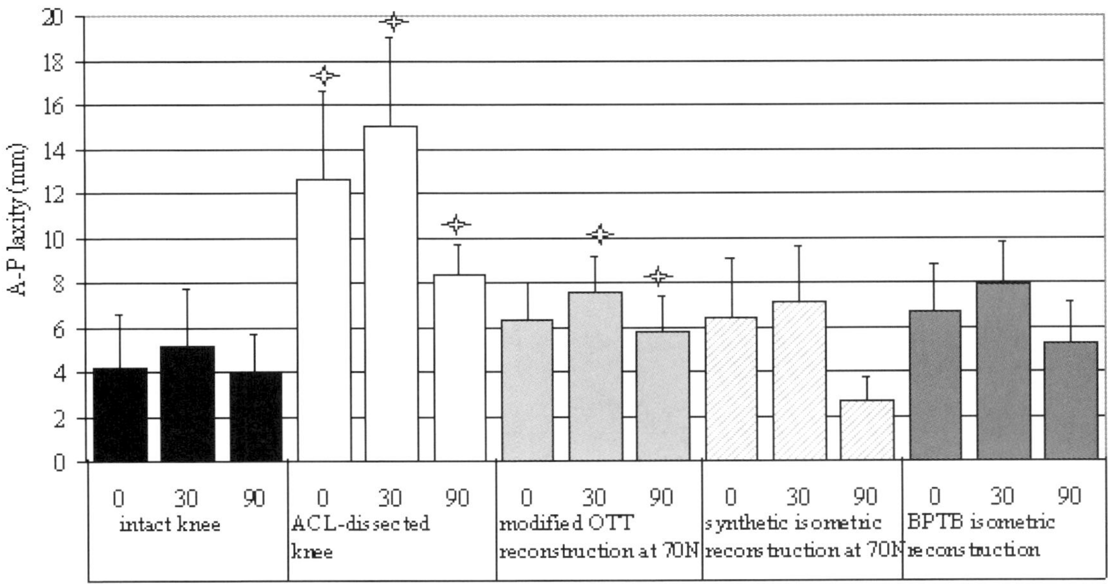

Figure 51.2. A-P laxity at 0°, 30° and 90° of knee flexion. The stars indicate statistical differences from the intact knee (ANOVA, $p < 0.05$; ACL = anterior cruciate ligament, BPTB = bone–patellar tendon–bone, OTT = over-the-top).

cant ($p < 0.05$) A-P translation compared to the intact knee. Modified OTT reconstruction allowed A-P laxity to be restored to a level near that of the intact knee. The two-tunnel reconstructions, performed either with a prosthesis or a biological graft, also adequately restored knee A-P translation. Reconstruction with the synthetic ligament at 70 N pretension achieved the same result. No over-constraint of the knee could be noted at 0 and 30° of flexion, but at 90°

knee laxity was about 1 mm less than for the intact knee. However, this difference was not statistically significant.

The results of flexion/extension movements in intact, dissected, and reconstructed knees are presented in Figure 51.3. Average internal/external rotation is shown for 10 knee specimens. The curves represent the difference between the results obtained with the different methods and the intact knee. ACL dissection created significant internal tibial rotation of the tibia between 60° and 90° of knee flexion. Modified OTT reconstruction did not restore this internal rotation, but increased it. Isometric two-tunnel reconstruction performed with the biological graft increased internal tibial rotation. When it was done with a Trevira prosthesis at 70 N pretension, the isometric reconstruction tended to force the tibia into external rotation. This was observed even between 0° and 60° of knee flexion, where no over-constraining of the knee could be detected by laxity measurements.

In conclusion, this work shows that the measurement of 3D knee kinematics could be essential to assess the quality of surgical treatments of ligament ruptures. It provided complementary information about over- or under-correction generated by reconstruction

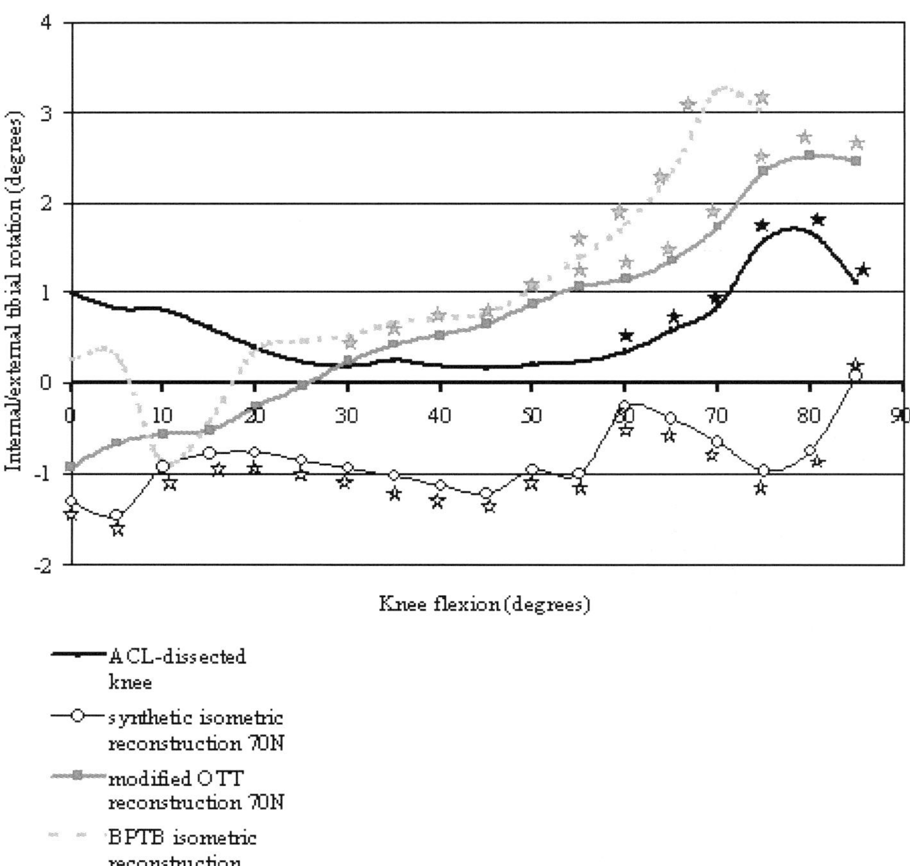

Figure 51.3. Differences in internal/external tibial rotation during flexion on 10 knees compared to the intact knee. External rotation is negative, internal rotation is positive. ACL dissection created significant internal tibial rotation between 60° and 90° of knee flexion; modified OTT reconstruction pretensioned at 70 N significantly increased this internal tibial rotation between 30° and 90° of knee flexion; synthetic isometric (two-tunnel) reconstruction pretensioned at 70 N created significant external tibial rotation. The stars indicate a statistical difference from the intact knee (reference 0 line) (ANOVA, $p < 0.05$).

in vitro, which was not seen by laxity measurements. We also showed that no ACL reconstruction, whether anatomical or functional, using synthetic or biological material, was able to restore all aspects of cadaveric knee function evaluated here, namely, A-P laxity, and 3D kinematics.

In Vivo Study: a Precise and Repeatable Measure of 3D Kinematics

Sophisticated methods exist to measure the functional state of the knee in a dynamic manner. Tools are in use today which can measure step frequency and stride length as well as angles in the sagittal plane (hip, knee, and ankle flexion/extension), angular velocity, and acceleration, using markers positioned on the skin. However, these methods are not precise enough to allow the measurement of small angles such as abduction/adduction, and internal/external tibial rotation. Small displacements are tainted by the noise caused by soft-tissue movements relative to the bones [45,46]. Some authors [47] have dealt with the problem of skin movement by inserting cortical pins into the bones. With this method, they were able to precisely measure angles in the frontal and transverse planes during gait. However, such methods cannot be widely used in a clinical environment.

Reducing Sensor Movements

Sati et al. [48] described an attachment system (Figure 51.4) on which markers can be installed. This system was designed to minimize marker movements relative to the bones. The femoral part of the attachment system consists of an arch that is fixed onto the femur by means of a vertical bar and three orthoplasts. The lateral orthoplast is positioned in the groove between the ilio-tibial band and the biceps femoris tendon, just above the posterior lateral condyle, while the medial orthoplast is positioned in the groove between the medial adductor magnus tendon and anterior to the sartorius tendon, just above the posterior medial condyle. The third orthoplast is positioned on the posterior medial condyle, more specifically on the adductor tubercle. The vertical bar is fixed with a Velcro strap to the proximal part of the thigh. The

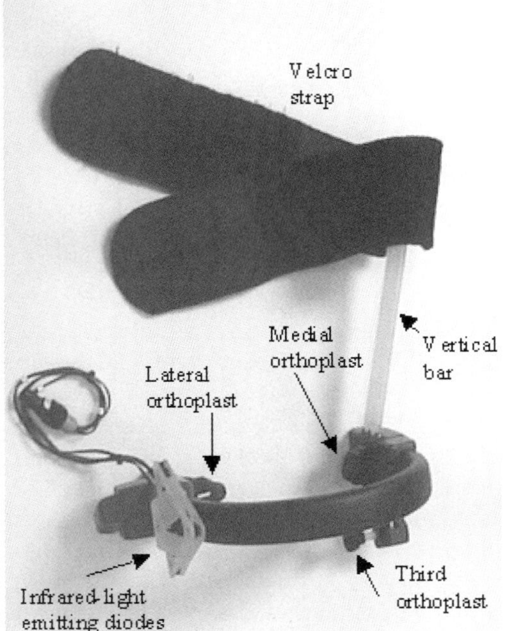

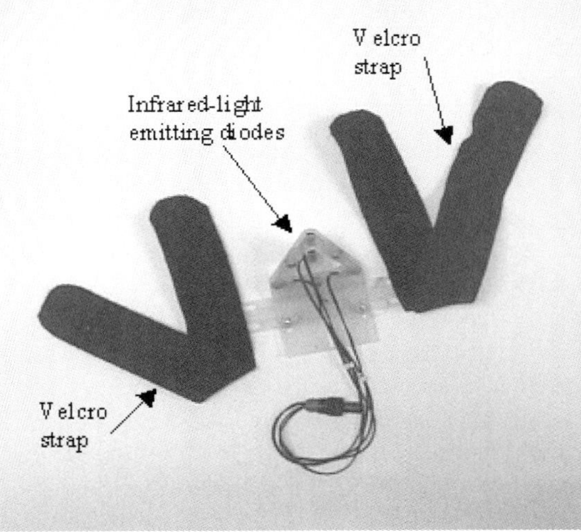

Figure 51.4. Attachment system (developed after the original work of Sati et al. [48] (left: femoral part; right: tibial part) (patent pending).

lateral and medial orthoplasts, which sit on top of the posterior condyles, contain a spring mechanism that allows the absorption of lateral expansion of knee soft tissues during movements. This system improves the subject's comfort and accuracy by preventing deformation of the harness. The tibial part consists of a vertical bar fixed with two Velcro straps on the anterior aspect of the tibia. The whole system is reversible and fits the right or left knee. Recent design improvements minimize the size of the medial side of the femoral attachment system to improve comfort during gait.

A fluoroscopic study was performed to ensure that the femoral part of the attachment system effectively reduces marker movements with respect to the underlying bone. We supposed that the tibial part of the attachment system did not move significantly, because the soft tissue in the area under the vertical bar is basically limited to the skin. The method was initially proposed by Sati et al. [46,48] and then used by Ganjikia et al. [49] with a refined version of the attachment system. It consists of estimating the 3D movement of markers relative to the underlying bone from 2D calibrated projections obtained by fluoroscopy.

Four healthy volunteers participated in the fluoroscopic study [49], and two different settings were considered. The first setting consisted of placing markers directly on the skin, on the lateral and medial sides of the knee. For the second setting, the markers were placed on the attachment system which was installed on the subject's knee. For both settings, each subject went through 3–4 flexion–extension cycles between 0° and approximately 60° of knee flexion while fluoroscopic images were recorded in the sagittal plane. Magnification factors were taken into account by using two radio-opaque rulers positioned on the subject's knee. The geometric parameters of bone angulations were then calculated with the method described by Sati et al. [46,48].

On the lateral side of the knee, movements of the markers in the sagittal plane were reduced from 14.1 ± 5.3 mm when they were placed directly on the skin, to 4.5 ± 1.6 mm when they were placed on the attachment system (mean factor of 3.2; min: 2.2, max: 5.2). Rotations of the markers about the longitudinal axis of the femur decreased from 3.6° ± 1.8° to 1.5° ± 0.7° (mean factor of 2.4; min: 1.7, max: 3.3). Rotations about the axis normal to the frontal plane decreased from 7.0° ± 3.0° to 2.0° ± 0.8° (mean factor of 3.6; min: 2.1, max: 5.7).

On the medial side of the knee, movements of the markers in the sagittal plane were reduced from 10.5° ± 4.8° to 4.0° ± 2.3° (mean factor of 6.9, min: 1.7, max: 26). Rotations of the markers about the longitudinal axis of the femur were decreased from 2.3° ± 0.9° to 1.7° ± 1.1° (mean factor of 5.1; min: 0.8, max: 21). Rotations about the axis normal to the frontal plane decreased from 5.4° ± 3.0° to 1.3° ± 0.9° (mean factor of 7.8; min: 2.4, max: 26.4).

The results of the fluoroscopic study show that the attachment system reduces marker movements relative to the bone, compared to when the markers are positioned directly on the skin. Using the attachment system, we are thus able to record 3D kinematics with a precision of 1.3° in abduction/adduction and 1.7° in internal/external tibial rotation.

Movement Representation

Euler Angles Versus Helical Axis Definition

After a movement has been measured precisely and reliably, it is necessary to represent it in a meaningful way. The knee is not a hinge, and movement about that joint does not occur in a 2D plane. Therefore, it is difficult to represent knee kinematics. Typically, two methods are used to study 3D kinematics of the knee: Euler angles [50], and helical axes [51].

Even though the knee is not gyroscopic, the Euler angle method is the most widely employed. With this method, it is possible to describe a 3D movement as three successive rotations about three different axes defined in space. It is then possible to represent these rotations with respect to each other, for example, abduction/adduction or internal/external tibial rotation versus flexion. These axes can be fixed or floating, and represented locally or globally. The major advantage of this method is that it is easier to interpret the results clinically with anatomical descriptions of movements. With this method, it is also possible to compute A-P, P-D, and M-L translations.

However, the main disadvantage of the Euler angle method is that it is very sensitive to anatomical reference axes definition. Small errors (1–2 mm in the definition of points used to build the coordinate system) cause errors in orientation as well as in kinematic amplitude to the order of 2°. When coordinate systems are built on subjects, errors in landmark definition can be to the order of 30 mm. These large errors make it difficult, even impossible, to compare the results [52,53]. Also, it is not clear if different bone geometries generate various kinematic patterns. For this reason, we do not know the 3D kinematics of the normal knee. Each knee has a kinematic representation associated with a given local coordinate system.

The helical axes method [51] uses the 3D position of each bone to describe the movement of the knee between two moments in time as a unique rotation and a unique translation about a finite rotation axis. Therefore, when employing this method to describe knee kinematics, we need to define the time period during which we want to express the rotation and translation of one bone with respect to the other. The main advantage of this method is that it is independent of an anatomical coordinate system definition. However, it is more difficult to interpret the results in a clinical fashion. Also, the method is sensitive to noise in the measurement and to the time period used for computation of the finite rotation axis.

We chose to represent knee movements with Euler angles to obtain clinical interpretations. We thus needed to define anatomical coordinate systems associated with the femur and tibia. To improve accuracy and repeatability, most investigators use X-rays to build coordinate systems. This involves radiation for the subject and is not practical because of the need to have access to radiological equipment and to a technician.

Landmark Definition

When X-rays are not used, anatomical landmarks can be defined by a pointer or by sticking markers directly on them. This technique generates errors to the order of many mm or even a few cm.

To diminish imprecision when building coordinate systems, we have developed a method with an original feature that uses less anatomical landmarks to define the reference coordinate system than other methods in the literature [54]. To define axes orientation, we employ landmarks that are easy to identify (malleoli, centre of the femoral head).

To build coordinate systems associated with the femur and tibia, the positions of the following anatomical landmarks are needed with respect to markers fixed to the attachment system: center of the femoral head, malleoli, and femoral condyles. For definition of the location of the center of the femoral head, we chose a functional method to avoid having to locate the antero-superior and postero-superior iliac spines, since these anatomical landmarks can potentially generate large errors. This method consists of recording femur movement relative to the pelvis during leg circling. A rigid body with four infrared light-emitting diodes is placed on the pelvis to record its movement. Next, an Optotrak pointer records the position of the other landmarks: lateral and medial malleoli and femoral condyles.

Axes Construction

Employing the position of the femoral condyles, it is possible to compute the 3D position of the knee center. The center is calculated at mean distance on the line joining both centers of the femoral condyles. The femoral and tibial y-axes are defined by the femoral head, knee centre, and inter-malleoli positions.

The z- and x-axes of each bone are localized in the transverse and sagittal planes, respectively. The three anatomical planes are defined by positioning the subject in a standing position with his/her back against a wall. The subject, while leaning against the wall, performs small flexion/extension movements. The instant the knee is in full extension (flexion = 0°) is then defined. At that time, the femoral and tibial y-axes are localized in the sagittal plane. The femoral and tibial x-axes are thus defined as parallel to the sagittal plane and normal to the y-axes. The z-axes are then defined by completing the right-handed coordinate systems. Figure

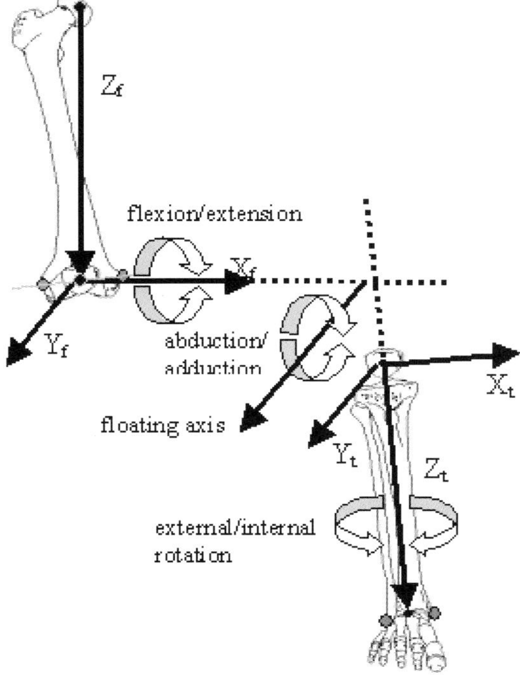

Figure 51.5. Coordinate axes construction using femoral condyles, malleoli (gray dots), and a functional method for definition of the center of the femoral head and the center of the knee (black dots).

51.5 is a representation of the coordinate axes constructed with our method.

Inter- and Intra-tester Variability

To test inter- and intra-tester variability for the calibration procedure, the following protocol was performed. The attachment system was first installed on each subject. Three testers performed the above-described calibration procedure on four subjects. Each tester repeated the procedure five times. Then, the subjects walked on a treadmill at a comfortable speed for three minutes, and finally, 30 gait cycles were recorded. The mean of these 30 cycles was used to compute the kinematic parameters (flexion/extension, abduction/adduction, and internal/external rotation of the tibia as well as A-P translation as a function of percentage of the gait cycle) in association with the 15 calibration procedures. The resulting 15 curves were compared for each subject using an adjusted coefficient of multiple determinations [55]. Table 51.1 shows the results of inter- and intra-tester variability. Figure 51.6 presents an example of the kinematic parameters calculated with five calibration procedures performed by one tester on one subject.

The results show that the calibration method allows the measurement of 3D knee kinematics

Table 51.1. Inter- and intra-tester variability of the calibration procedure

Subject	Kinematic parameter	Tester #1 (intra-tester variability)	Tester #2 (intra-tester variability)	Tester #3 (intra-tester variability)	Testers #1, 2, 3 (inter-tester variability)
Subject #1	Flexion/extension	0.96	0.99	0.99	0.98
	Ab/adduction	0.87	0.96	0.95	0.85
	Int/ext tibial rotation	0.79	0.97	0.89	0.78
	A-P translation	0.86	0.76	0.89	0.73
Subject #2	Flexion/extension	0.99	0.99	0.98	0.99
	Ab/adduction	0.91	0.95	0.88	0.90
	Int/ext tibial rotation	0.82	0.78	0.81	0.81
	A-P translation	0.85	0.94	0.78	0.80
Subject #3	Flexion/extension	0.99	0.99	0.99	0.99
	Ab/adduction	0.42	0.80	0.82	0.63
	Int/ext tibial rotation	0.88	0.88	0.89	0.88
	A-P translation	0.52	0.89	0.67	0.34
Subject #4	Flexion/extension	0.99	0.99	0.99	0.99
	Ab/adduction	0.93	0.92	0.97	0.94
	Int/ext tibial rotation	0.97	0.94	0.96	0.97
	A-P translation	0.77	0.92	0.77	0.58

A-P = antero-posterior. Int/ext = internal/external.

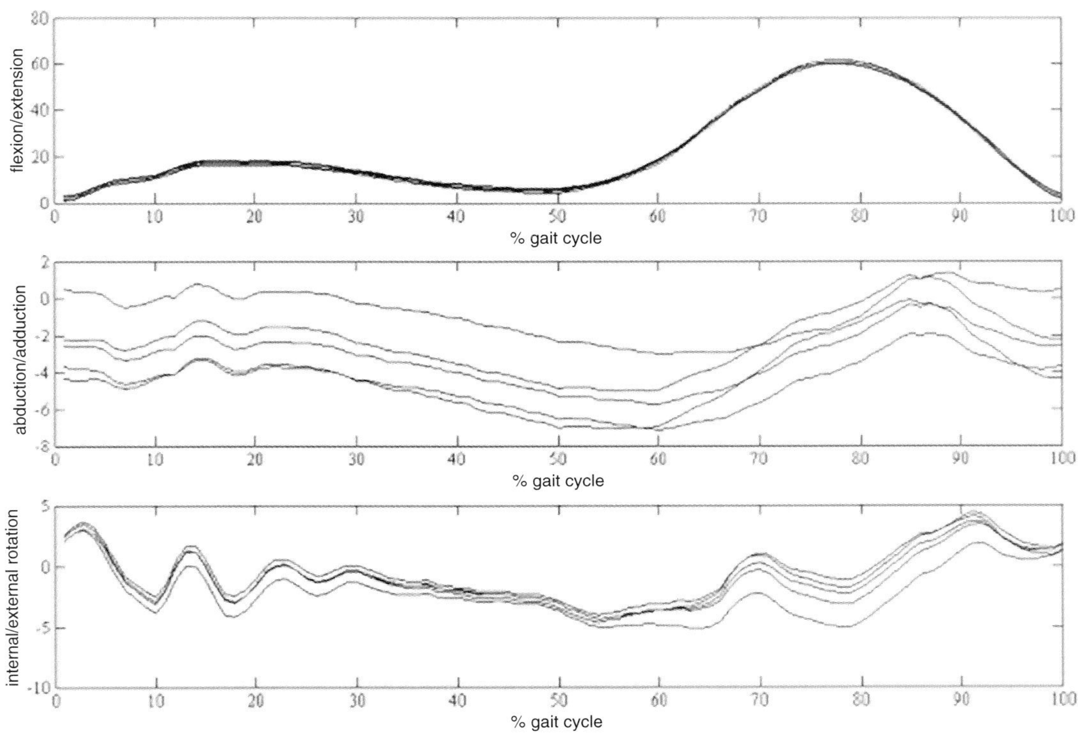

Figure 51.6. Example of kinematic parameters calculated with five calibration procedures performed by one tester on one subject. Abduction is negative and internal rotation is negative. Each curve corresponds to one trial.

with good reproducibility. Mean errors generated by the calibration procedure are 1.1° in flexion/extension, 1.1° in abduction/adduction, 0.8° in internal/external tibial rotation, and 2.6 mm in A-P translation.

Between-day Variability

One of the conventions used in the calibration method described above consists of aligning the femoral and tibial y-axes when the knee is in full extension. When an extension deficit occurs after a knee injury, large errors are introduced in the definition of the knee's position at zero flexion. In that case, another calibration can be used, either from the contralateral knee or from the same knee once it is able to achieve full extension again. However, reinstallation of the system alone also introduces reproducibility errors. This has been estimated in another series of tests, using the same calibration procedure for five different installations on subjects. Figure 51.7 gives an example of the variability of the kinematic parameters for one subject. The standard deviation generated was 4.5° in flexion/extension, 2.6° in adduction/abduction, and 5.3° in internal/external tibial rotation.

Case Study

One of the healthy subjects recorded during the variability protocol unfortunately had a football accident a few weeks after the first assessment of his intact knee. Arthroscopic evaluation diagnosed an ACL rupture associated with intra-articular osteochondral fragments. These fragments were blocking the knee in 20° flexion, putting us in the very situation where a new calibration procedure would be error-prone.

We assessed the knee of this subject after injury and before arthroscopy using the method described above. Since the subject could not extend his knee past 20° flexion, we defined the reference axes after the injury using the calibration method performed at the first testing

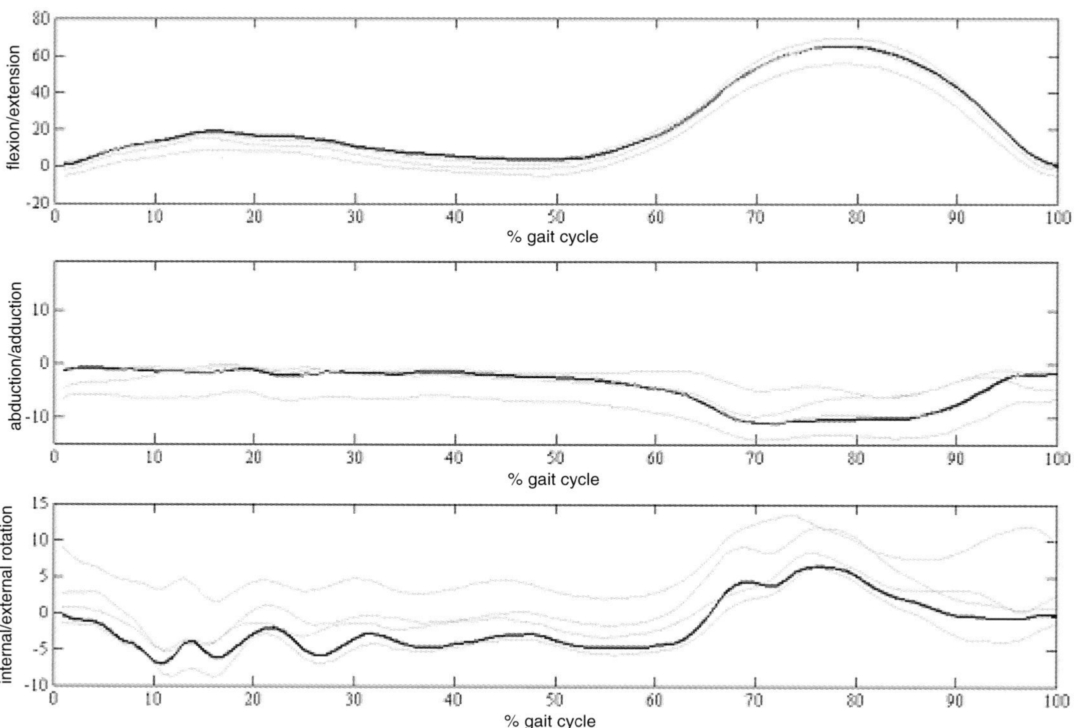

Figure 51.7. Example of variability of the kinematic parameters for five installations of the attachment system with one calibration procedure. Abduction is negative, internal rotation is negative. Black line: Initial installation and calibration. Gray line: individual trials after reinstallation using initial calibration.

when the knee was intact. Testing was repeated once on another day. The results are presented in Figure 51.8. The kinematic curves show a flexion deficit during gait of about 20° compared to the non-injured state of the same knee. Abduction was also affected. Finally, an increase of internal rotation was observed. Interestingly, these changes were also observed in the in vitro experiment, where isolated dissection of the ACL increased internal tibial rotation. All observed changes were greater than the mean standard deviation (σ) generated by re-installation of the attachment system on the same subject.

General Discussion

ACL reconstruction has raised much controversy and has been the subject of a great deal of in vitro and in vivo study. No general consensus has been reached concerning its optimal treatment. The ACL can be reconstructed isometrically or by a non-anatomical OTT route. Presently, the tools of choice for the assessment of such reconstructions, either in vitro or in vivo, are instrumented arthrometers. However, arthrometers do not evaluate knee function, they merely give a measure of knee laxity. Reliable and precise assessment of ligamentous pathologies and treatment is the key for improving surgical reconstructions. However, despite extensive research on this topic, little is known about the effect of ligament surgeries on knee function. This is particularly highlighted by the growing number of procedures arising every year in the literature. The consistently good and excellent results reported over a short period of time, followed by abandonment of these very techniques, indicate the need for a more systematic approach to the development of new procedures.

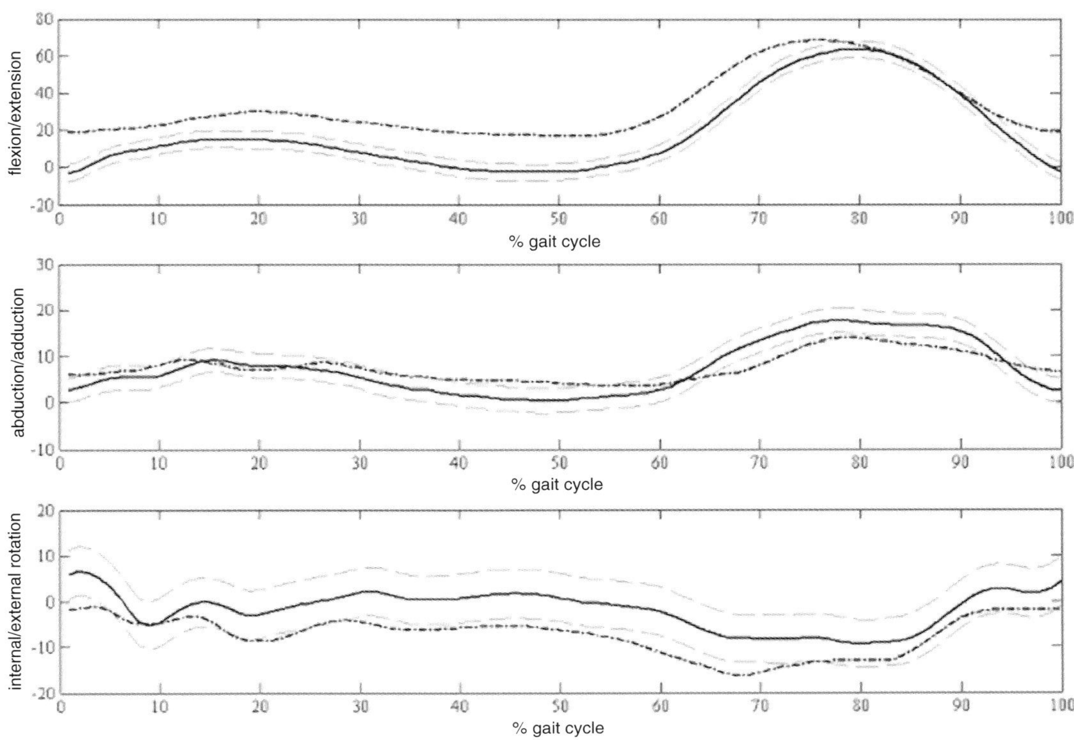

Figure 51.8. Variation of kinematic parameters of a subject before and after ACL rupture. Abduction is negative, internal rotation is negative. Black line: Pre-injured state (Mean). Gray dotted line: Mean curve ± standard deviation (σ). Black dotted line: Post-injury (mean of two trials).

In this chapter, we presented a general methodology for objective and reliable assessment of knee function. Our main hypothesis is that to properly assess knee function, we not only need to measure knee laxity, but also its 3D movement. To confirm this hypothesis, we first conducted an in vitro study. Several authors have done so before, but, to our knowledge, none compared the effect of different reconstruction procedures on laxity and 3D kinematic measurements. Fleming et al. [56] compared knee laxity after isometric and OTT reconstructions, both performed with the same tibial tunnel. Their results indicate that when the knee flexes, OTT reconstruction does not restore initial knee laxity (under-constraining of the knee). Melby et al. [57] have shown, on cadaver knees, that OTT reconstruction, pretensioned at 30° of knee flexion, is as efficient in restoring knee laxity as isometric reconstruction, even for higher knee flexion angles. Nevertheless, another study, conducted by Gertel et al. [14], evaluated the effect of pretensioning on knee laxity and demonstrated that if pretension was applied with the knee in flexion (30°), over-constraining of the knee occurs. It seems, therefore, that classical OTT placement, with a tibial tunnel in the center of anatomical attachment of the ACL, would either under-constrain the knee if pretension is applied in extension, or over-constrain it if pretension is applied in flexion. A cadaveric study on the effect on 3D knee kinematics of isometric or OTT reconstructions of the ACL conducted by Good et al. [28] with synthetic grafts confirmed this fact. For that reason, we evaluated D-M placement of the graft for OTT reconstruction, thereby showing that laxity is restored to nearly normal values without creating non-physiological kinematic curves. It appears, however, that this reconstruction is not able to correct internal tibial rotation created by ACL dissection. Moreover, isometric reconstruction induced non-physiological external tibial rotation, as already noted by other research groups such as Good et al. [28] and Van Heerwaarden [58].

Therefore, it appears not only that no reconstruction procedure seems able to restore laxity and 3D kinematics, but also that 3D knee kinematics could provide useful measurement for better assessment of knee function. Quantitative gait analysis is now recognized as a valuable tool in the objective evaluation of treatment of injuries affecting gait [59]. However, when ligament injuries occur, gait variations are very small, and the recording of rotations in only the sagittal planes might be insufficient. In fact, recording of small rotations in the frontal and transverse plane could provide useful information for the assessment of surgical treatment of ligament injuries.

Nevertheless, in vivo measurement of small rotations of the knee in the frontal and transverse planes is a difficult task. Because of the use of different coordinate systems, it is impossible to compare numerical values among different gait studies. Also, even with recent experimental protocols [60], it is difficult to measure small angles in the frontal and transverse planes (as ab/adduction and internal/external tibial rotation) with good precision and reproducibility. This is due to the fact that bone movements are tainted by the noise of soft tissue movements. Reinschmidt et al. [45] compared knee rotations in subjects whose kinematics were recorded via skin markers and bone pins inserted into the femur and tibia. They showed that average errors during running due to skin movements were about 21% for flexion/extension, 63% for internal/external tibial rotation, and 70% for abduction/adduction. Cappozzo et al. [61] estimated these errors roughly at 10, 50, and 100% respectively. Since bone pins are not a solution that can be used in clinics on a routine basis, we chose an external attachment system aiming at reduction of errors associated with skin movement. The results of the fluoroscopy study show that the attachment system is effective in reducing errors due to skin movements by 3.2 on the lateral side and 6.9 on the medial side of the knee.

However, even if it is now possible to record knee rotations in the frontal and transverse plane with good precision, the representation of these rotations remains another challenge. In fact, as already noted, if a local coordinate system is defined on the bones, small errors in the positioning of these axes generate large errors in rotation calculations [52,58].

Among anatomical landmarks used to define a reference coordinate system on the femur and tibia, the center of the femoral head, used to determine the longitudinal axis of the femur, is one of the most difficult to assess. Basically, two methods estimate the position of the femoral head. Predictive methods [62–67] compute the position of anatomical landmarks and integrate them in regression equations. "Functional" methods estimate the position of the femoral head [54,68] by evaluating the motion of the femur relative to the pelvis during leg circling.

In 1990, Bell et al. [64] proposed a predictive method based on Tylkowski's [67], generating errors of 19.0 mm. They compared this result to those obtained with the functional method for which they found errors of 37.9 mm, and to Andriacchi's method [62,63] for which they noted errors of 36.1 mm. According to us, the large errors occurring with the functional method as tested by Bell et al. [64] are due to the fact that only one marker was installed on the lateral femoral condyle to follow the motion of the thigh, instead of using a four-marker rigid body to get its 3D motion. This generates large acquisition errors. Also, the small error obtained with the predictive method they developed based on Tylkowski's [67] might be due to the fact that they tested it on the same subjects they used to compute the coefficient in the regression equations.

Later, in 1999, Leardini et al. [68] compared two predictive methods with the functional method. They found that the functional method estimates the position of the femoral head within 11.8 mm, while the predictive methods suggested by Bell et al. [64] and Davis et al. [65] provided results within 23.3 mm and 29.1 mm, respectively. For this reason, we preferred a functional definition of the center of the femoral head.

In the literature, the reproducibility of complete calibration methods without the use of X-rays is rarely assessed. One study, performed by Kadaba et al. [59], evaluated reproducibility of a calibration method where they recorded 40 subjects three times in the same day and on three different days, one week apart. They showed that joint angle motion measurements in the frontal and transverse planes display poor between-day repeatability (0.611 and 0.783 for the left and right knees respectively for varus/valgus movement and 0.490 and 0.534 for the left and right knees respectively for tibial rotation movement). The authors imputed these variations to the difficulty in positioning anatomical markers in a reproducible way between days.

We developed a method, presented in this chapter, where as few anatomical landmarks as possible are used to define the coordinate system. The landmarks are generally easy to identify. We demonstrated, by a variability study on four subjects, that this method provides acceptable reproducibility, when performed several times on different subjects by three different testers. In fact, we have evaluated mean inter-observer variability of the whole calibration method on four subjects with five different calibration procedures at 0.98 for the calculation of flexion/extension angle, 0.83 for the ab/adduction angle, 0.86 for the internal/external tibial rotation angle, and 0.61 for A-P translation (adjusted coefficient of multiple determinations). Errors that can be generated by the calibration procedure alone are 1.1° in flexion/extension (2% of the amplitude), 1.1° in abduction (8% of the amplitude), and 0.8° in internal/external rotation (8% of the amplitude). These errors are increased when the patient cannot reach full extension and a previous calibration is used. Maximally, errors can reach 4.5° in flexion/extension (7% of the amplitude), 2.6° in abduction/adduction (26% of the amplitude), and 5.3° in internal/external rotation (35% of the amplitude).

The recording of the same subject's knee before and after ligament injury is generally not possible. Although we are aware of the limitations of experiments on one single patient, the case study presented here suggests, however, that the methodology proposed could allow us to record small gait variations in the frontal and transverse plane after ligament rupture. The absolute values of the kinematic curves in the frontal and transverse planes presented in this chapter cannot directly be compared to those found in the literature, where bone pins were used to record bone movements [47] because of different axes definitions. However, it should be noted that the subject presented here displayed external tibial rotation during the swing phase that appears later in the gait cycle than what was observed for other subjects in our study (Figures 51.6 and 51.7) and by Lafortune et al. [47]. This could have been due in part to an asymmetry of hip flexibility that was clinically assessed on this patient by a physiatrist. It is possible that such asymmetry could have an impact on knee rotation during gait. The recording of a large number of patients will be necessary to define "normal" and pathologic gaits in the frontal and sagittal planes.

The methodology proposed here, combining precise recording of bone movements through a specially designed attachment system with a reproducible method for the definition of an anatomical reference system, provides a precise tool for kinematic evaluation, which can improve the quality of clinical assessment. Once validated on a wider scale, this tool will be helpful for the evaluation of different treatment methods, particularly in the field of ligamentoplasties.

Acknowledgements

The Pathology Department of CHUM-Hôpital Notre-Dame Hospital is acknowledged for technical support. The National Research Council of Canada (NSERC CRD26877-99), the companies BiOp (Joliette, Canada), JK Orthomedic Inc. (Montreal, Canada), and Telos (Marburg, Germany) are acknowledged for grant support of this work. The authors also thank Ovid Da Silva for his editorial work on this manuscript (Research Support Office of the CHUM Research Centre).

References

1. Frank CB, Jackson DW. Current concept review. The science of reconstruction of the anterior cruciate ligament. J Bone Joint Surg Am 1997;79A(10):1556–76.
2. Gillquist J. Repair and reconstruction of the ACL: is it good enough? Arthroscopy 1993;9(1):68–71.
3. Ellison AE. Distal iliotibial band transfer for anterolateral rotatory instability of the knee. J Bone Joint Surg Am 1979;61(3):330–7.
4. Tillberg B. The late repair of torn cruciate ligaments using menisci. J Bone Joint Surg 1977;59:15–23.
5. Forster IW, Warren-Smith CD, Tew M. Is the KT1000 knee ligament arthrometer reliable? J Bone Joint Surg Br 1989;71(5):843–7.
6. Hyder N, Bollen SR, Sefton G, Swann AC. Correlation between arthrometric evaluation of knees using KT1000 and telos stress radiography and functional outcome following ACL reconstruction. Knee 1997;4:121–4.
7. Ahmed AM, Hyder A, Burke DL, Chan KH. In vitro ligament tension pattern in the flexed knee in passive loading. J Orthop Res 1987;7:217–30.
8. Amis AA. Anterior cruciate ligament replacement. Knee stability and the effects of implants. J Bone Joint Surg Br 1989;71(5):819–24.
9. Brantigan V. The mechanism of the ligaments and menisci of the knee joint. J Joint Bone Surg 1941;23: 44–66.
10. Burns WC II, Draganich LF, Pyevich M, Reider B. The effect of femoral tunnel position and graft tensioning technique on posterior laxity of the posterior cruciate ligament-reconstructed knee. Am J Sports Med 1995; 23(4):424–30.
11. Dorlot JM, Christel P, Meunier A, Sedel L, Witvoet J. Analysis of the mechanical function of the cruciate ligaments in antero-posterior knee laxity. An in vitro study. Int Orthop 1983;7(2):91–7.
12. Draganich LF, Reider B, Ling M, Samuelson M. An in vitro study of an intra-articular and extra-articular reconstruction in the anterior cruciate ligament-deficient knee. Am J Sports Med 1990;18(3):262–6.
13. Engebretsen L, Lew WD, Lewis JL, Hunter RE, Benum P. Anterolateral rotatory instability of the knee. Cadaver study of extra-articular patellar-tendon transposition. Acta Orthop Scand 1990;61(3):225–30.
14. Gertel TH, Lew WD, Lewis JL, Stewart NJ, Hunter RE. Effect of anterior cruciate ligament graft tensioning direction, magnitude, and flexion angle on knee biomechanics. Am J Sports Med 1993;21(4):572–81.
15. Hole RL, Lintner DM, Kamaric E, Moseley JB. Increased tibial translation after partial sectioning of the anterior cruciate ligament. The posterolateral bundle. Am J Sports Med 1996;24(4):556–60.
16. Lewis JL, Lew WD, Schmidt J. Description and error evaluation of an in vitro knee joint testing system. J Biomech Eng 1988;110:238–48.
17. Markolf KL, Burchfield DM, Shapiro MM, Cha CW, Finerman GAM, Slauterbeck JL. Biomechanical consequences of replacement of the anterior cruciate ligament with a patellar ligament allograft. Part II: Forces in the graft compared with forces in the intact ligament. J Bone Joint Surg Am 1996;78(11):1728–34.
18. Meystre JL, Trouilloud P. Postero-postero-external instabilities of the knee: experimental study of an extra-articular system to protect reconstructions. Rev Chir Orthop Reparatrice Appar Mot 1994;80(5):420–7.
19. Ogata K, McCarthy JA. Measurements of length and tension patterns during reconstruction of the posterior cruciate ligament. Am J Sports Med 1992;20(3): 351–5.
20. Pearsall AW IV, Pyevich M, Draganich LF, Larkin JJ, Reider B. In vitro study of knee stability after posterior cruciate ligament reconstruction. Clin Orthop 1996;327: 264–71.
21. Shahane SA, Ibbotson C, Strachan R, Bickerstaff DR. The popliteofibular ligament. An anatomical study of the posterolateral corner of the knee. J Bone Joint Surg 1999;Br 81:636–42.
22. Biden E, O'Connor J. Experimental methods used to evaluate knee ligament function. In: Daniel D et al., editors. Knee Ligaments: Structure, Function, Injury and Repair. New York: Raven Press, 1990;135–51.
23. Brower RS, Melby A, Askew MJ, Beringer DC. In vitro comparison of over-the-top and through-the-condyle anterior cruciate ligament reconstructions. Am J Sports Med 1992;20(5):567–74.
24. Claes L, Dürselen L, Kiefer H. Testing of knee ligaments and ligament prostehses by a new computer-controlled knee loading machine. The 12th Annual Meeting of the Society for Biomaterials, Minneapolis, St. Paul, Minnesota, USA, May 29–June, 1986;1:30.
25. Draganich LF, Hsieh Y-F, Ho S, Reider B. Intra-articular ACL graft placement on the average most isometric line on the femur. Does it reproducibly restore knee kinematics? Am J Sports Med 1999;27(3):329–34.
26. Dürselen L, Claes L, Kiefer H. The influence of muscle forces and external loads on cruciate ligament strain. Am J Sports Med 1995;23(1):129–36.
27. Eilerman M, Thomas J, Marsalka D. The effect of harvesting the central one-third of the patellar tendon on patellofemoral contact pressure. Am J Sports Med 1992;20(6):738–41.
28. Good L, Askew MJ, Boom A, Melby A. Kinematic in vitro comparison between the normal knee and two techniques for reconstruction of the anterior cruciate ligament. Clin Biomech 1993;8:243–9.
29. Grood E, Suntay W, Noyes F, Butler D. Biomechanics of the knee-extension exercise. Effect of cutting the anterior cruciate ligament. J Bone Joint Surg Am 1984;66: 725–34.
30. Harding ML, Harding L, Goodfellow JW. A preliminary report of a simple rig to aid the study of the functional anatomy of the cadaver knee joint. J Biomech 1977;10: 517–23.
31. Heegaard J, Leyvraz PF, Van Kampen A, Rakotomanana L, Rubin PJ, Blankevoort L. Influence of soft structures on patellar three-dimensional tracking. Clin Orthop 1994;299:235–43.
32. Hsieh Y-F, Draganich LF, Ho SH, Reider B. The effects of removal and reconstruction of the anterior cruciate ligament on patello-femoral kinematics. Am J Sports Med 1998;26(2):201–9.
33. Landjerit B, Thourot M. The intact, injured and repaired knee: in-vitro experimental biomechanics and level walking. Acta Orthop Belg 1992;58(2):113–21.
34. McLean CA, Ahmed AM. Design and development of an unconstrained dynamic knee simulator. J Biomech Eng 1993;115:144–8.

35. Michel G, Pedros J. Étude in vitro de la cinématique du genou. Technical report. Laboratoire de Biomécanique, ENSAM, Paris, 1997.
36. Miller MD, Olszewski AD. Posterior cruciate ligament injuries. New treatment options. Am J Knee Surg 1995; 8(4):145–54.
37. Ortiz GJ, Schmotzer H, Bernbeck J, Graham S, Tibone JE, Vangsness CT Jr. Isometry of the posterior cruciate ligament. Effects of functional load and muscle force application. Am J Sports Med 1998;26(5):663–8.
38. Perry J, Antonelli D, Ford W. Analysis of knee joint forces during flexed knee stance. J Bone Joint Surg Am 1975;57:961–7.
39. Sakai N, Luo ZP, Rand JA, An KN. The influence of weakness in the vastus medialis oblique muscle on the patellofemoral joint: an in vitro biomechanical study. Clin Biomech 2000;15(5):335–9.
40. Wilson DR, Feikes JD, Zavatsky AB, O'Connor J. The components of passive knee movement are coupled to flexion angle. J Biomech 2000;33:465–73.
41. Hagemeister N, Long R, Yahia L'H, Duval N, Krudwig W, Witzel U et al. Quantitative comparison of three different types of anterior cruciate ligament reconstruction methods: laxity and 3-D kinematic measurements. BioMed Mater Engin 2002;12(1):47–57.
42. Hagemeister N, Duval N, Yahia L'H, Krudwig W, Witzel U, de Guise JA. Computer-based method for the three-dimensional kinematic analysis of combined posterior cruciate ligament and postero-lateral complex reconstructions on cadaver knees. The Knee, 2002;10(3): 249–56.
43. MacIntosh DL. The anterior cruciate ligament: "over-the-top" repair. J Bone Joint Surg Br 1974;56:591.
44. Krudwig WK. Reconstruction of cruciate ligaments using a synthetic ligament of polyethylene terephthalate (Trevira Ligament). In: Yahia L'H, editor. Ligaments and Ligamentoplasties. Heidelberg: Springer-Verlag, 1997; 245–54.
45. Reinschmidt C, van den Bogert AJ, Nigg BM, Lundberg A, Murphy N. Effect of skin movement on the analysis of skeletal knee joint motion during running. J Biomech 1997;30(7):729–32.
46. Sati M, de Guise JA, Drouin G. Quantitative assessment of skin movement at the knee. Knee 1996;3(3): 121–38.
47. Lafortune MA, Cavanagh PR, Sommer HJ, Kalenak A. Three-dimensional kinematics of the human knee during walking. J Biomech 1992;25(4):347–57.
48. Sati M, de Guise JA, Drouin G. Reducing skin movement error in knee kinematic measurements. Knee 1996;3(4): 179–90.
49. Ganjikia S, Duval N, Yahia L, de Guise J. Three-dimensional knee analyzer validation by simple fluoroscopic study. Knee 2000;7(4):221–31.
50. Grood ES, Suntay WJ. A joint coordinate system for the clinical description of three-dimensional motions: application to the knee. J Biomech Eng 1983;105: 136–44.
51. Kinzel GL, Hall AS Jr, Hillberry BM. Measurement of the total motion between two body segments – I. Analytical development. J Biomech 1972;5:93–105.
52. Kadaba MP, Ramakrishnan HK, Wotten ME. Measurement of lower-extremity kinematics during level walking. J Orthop Res 1989;8(3):383–92.
53. Parent G. Contribution au développement d'un système d'analyse de la biocinématique tridimensionnelle du genou, Master thesis, Université de Montréal, Montréal, Canada, 2000.
54. Cappozzo A. Gait analysis methodology. Hum Mov Sci 1984;3:27–50.
55. Winer BJ. Statistical Principles in Experimental Design, 2nd Edition. New York: McGraw-Hill, 1971;261–88.
56. Fleming B, Beynnon B, Howe J, Mcleod W, Pope M. Effect of tension and placement of a prosthetic anterior cruciate ligament on the anteroposterior laxity of the knee. J Orthop Res 1992;10:177–86.
57. Melby A, Noble JS, Askew MJ, Boom AA, Hurst FW. The effects of graft tensioning on the laxity and kinematics of the anterior cruciate ligament reconstructed knee. Arthroscopy 1991;7(3):257–66.
58. Van Heerwaarden RJ. Effect of pretension in reconstructions of the anterior cruciate ligament. Doctoral thesis, Cip Data Koninkljke Biobliotheek, The Hague, 1998.
59. Kadaba MP, Ramakrishnan HK, Wootten ME, Gainey J, Gorton G, Cochran GVB. Repeatability of kinematic, kinetic and electromyographic data in normal adult gait. J Orthop Res 1989;7:849–60.
60. Kaufman KR, Hughes C, Morrey BF, Morrey M, An K-N. Gait characteristics of patients with knee osteoarthritis. J Biomech 2001;34(7):907–15.
61. Cappozzo A, Catani F, Leardini A, Benedetti MG, Croce UD. Position and orientation in space of bones during movement: experimental artefacts. Clin Biomech 1996; 11(2):90–100.
62. Andriacchi TP, Andersson GB, Fermier RW, Stern D, Galante JO. A study of lower-limb mechanics during stair climbing. J Bone Joint Surg Am 1980;62(5):749–57.
63. Andriacchi TP, Galante JO, Fermier RW. The influence of total knee-replacement design on walking and stair climbing. J Bone Joint Surg Am 1982;64(9):1328–35.
64. Bell AL, Pedersen DR, Brand RA. A comparison of the accuracy of several hip center location prediction methods. J Biomech 1990;23(6):617–21.
65. Davis RB III, Ounpuu S, Tyburski D, Gage JR. A gait analysis data collection and reduction technique. Hum Mov Sci 1991;10:575–87.
66. Seidel GK, Marchinda DM, Dijkers M, Soutas-Little RW. Hip joint center location from palpable bony landmarks – a cadaver study. J Biomech 1995;28(8):995–8.
67. Tylkowski CM, Simon SR, Mansour JM. The Frank Stinchfield Award Paper. Internal rotation gait in spastic cerebral palsy. Hip 1982;89–125.
68. Leardini A, Cappozzo A, Catani F, Toksvig-Larsen S, Petitto A, Sforza V et al. Validation of a functional method for the estimation of hip joint centre location. J Biomech 1999;32(1):99–103.

52 Biomaterials for Total Joint Replacements

E. M. Brach del Prever, L. Costa, M. Baricco, C. Piconi, and A. Massé

Introduction

The European Society for Biomaterials defines a *biomaterial* as "a material that interacts with the biological systems to evaluate, treat, reinforce, or replace a tissue, organ, or function of the organism" and the *biocompatibility* as "the ability of a material to perform with an appropriate host response in a specific application" [1]. Biocompatibility of a biomaterial is tested by in vitro screening, in vivo testing, and clinical monitoring; each step evaluates the biological response in different conditions. In vivo, a few seconds after the implantation, the biomaterial is rapidly adsorbed by proteins, whose quantity and organization depend on the characteristics of the biomaterial, such as chemical composition of the bulk and surface, surface geometry, chemical and physical properties, and the properties of the proteins [2]. The host cells contact the protein layer [3]; in total joint replacements, bone cells growing on the prosthetic surface determine an "osseointegration" [4], fibrous cells a "fibrous fixation".

The biological response to wear particles (polyethylene, polymethylmethacrylate, metals, and corrosion products) is defined as bioreactivity; its major determinants are the particle size, concentration, surface chemical composition, surface energy, surface charge, surface roughness, particle shape, and nature of adsorbed proteins [5]; genetics might have an influence in determining the biological response. In periprosthetic bone, wear particles are responsible for osteolysis due to increased bone resorption and reduced bone formation (it was demonstrated that osteoblast cell lines exposed to Ti debris have a down-regulated expression of collagen precursors). Some metal particles, produced by degradation and wear, are able to accumulate in the periprosthetic tissues and enter into the bloodstream, and can be responsible for chromosomal damage and development of cancer. Genotoxicity or mutagenicity, and/or carcinogenicity were demonstrated in experimental studies with CoCr alloys [6]. The carcinogenic ability of Cr is widely described, particularly by epidemiological studies concerning the association of exposure to chromate particles and the incidence of nasal and lung cancer. Nickel is demonstrated to be genotoxic in vitro and carcinogenetic in vivo (lung and ethmoidal bone). The question if such metal biomaterials can be responsible for an increased risk of local and remote neoplasm in patients with joint replacement is open. Up to now, the incidence of local neoplasm at the site of the implant is negligible and no clear answer exists concerning the incidence of remote neoplasm. A slight increase of lymphatic and hemopoietic tumors in patients with a hip prosthesis was demonstrated [7], but an increased risk in patients with hip replacement was never demonstrated [6,8]. Genetic changes in blood lymphocytes were found at the time of revision arthroplasty and were related to metal composition [9]. Further investigations are mandatory to understand how the prosthetic biomaterials work in the body and how neoplasms develop.

The products of metal corrosion or degradation can either act as haptens, bindings to protein carriers, or as adjutants, forming insoluble complexes with the antigens [10], initiating an immune response. Hypersensitivity reactions have been reported to be more frequent with stainless steel or cobalt alloy than with titanium alloy; hypersensitivity to polymethylmethacrylate was found to be 50% in failed total hip implants. It was suggested that in some previously sensitized patients, corrosion products could behave like haptens, and the complex may stimulate memory lymphocytes, initiating an inflammatory process [10].

Ultra-high-molecular-weight Polyethylene (UHMWPE)

A macromolecular chain of PE is represented by the formula $(CH_2\text{-}CH_2)_n$. The word *polyethylene* is used in reference to quite a few polymeric materials, all of which are characterized by the same structural unit, but having chains of different lengths, different space arrangements, and different chain imperfections.

The commercial polyethylenes are: Low-density Polyethylene (LDPE), High-density Polyethylene (HDPE), and in the group of HDPE, High-molecular-weight Polyethylene (HMWPE) with molecular mass in the range 500,000–1,000,000 amu (atomic mass unit) and Ultra-high-molecular-weight Polyethylene (UHMWPE) with a molecular mass of more than 1,000,000 amu (Table 52.1).

In total joint replacements UHMWPE is used owing to its excellent abrasion properties, its very good mechanical properties, and its good biocompatibility. In prosthetic fields, UHMWPE can have different starting characteristics, whether chemical–physical or mechanical: with an exception for the density, there are no limits for the other characteristics. Determination of the characteristics of prosthetic UHMWPE, according to standard ASTM F 648-98 [11], is carried out on the original material, before processing and before sterilization.

UHMWPE powder from the polymerization plant is processed according to two techniques, both using high pressure and controlled heating and cooling cycles: *compression molding* and *ram extrusion*. Compression molding can be carried out in two ways: one directly produces the finished product by sintering the powder into its final shape; the second one compresses the melted powder into large sheets; rectangular sectioned bars are cut out of the sheet and then machined to obtain the finished prosthetic component. Ram extrusion of the melted powder produces circular sectioned bars, which are then mechanically processed to obtain the finished prosthetic component. The very high viscosity of melted UHMWPE may hinder a perfect compaction of the polymer powder: defects or preparation voids may be formed; these defects could have some influence on the tribological behavior of UHMWPE [12].

Both production techniques, i.e., ram extrusion and compression molding, do not significantly modify chemical, physical, and structural characteristics of the starting polymer; therefore prosthetic components, ready to be sterilized, do not differ from the starting material of which they still possess all properties.

The main sterilization processes used nowadays employ ethylene oxide and high-energy radiation, i.e., gamma radiation and, more recently, electron beams. Sterilization by steam

Table 52.1. Properties of different types of PE. The characteristics of UHMWPE, required for use in orthopedics, are those stated in Standard ASTM F 648–98

Properties (Unit)	ASTM standard used	HDPE	LDPE	Orthopedic UHMWPE
Molecular mass (Amu)		50,000–300,000	50,000–200,000	>2,000,000
Crystallinity (%)	From density data	60–75	40–55	50–60
Density (g/cc)	D-792	0.945–0.965	0.91–0.930	0.927–0.944
Melting Point (°C)		128–137	105–115	125–145
Tensile yield strength (MPa)	D 638	15–35	6–11	>19 (19–23)
Elastic modulus (MPa)	D 638	400–1,500	100–500	
Ultimate tensile strength (MPa)	D 638	18–40	7–16	>27
Elongation to breakpoint	D 638	40–1,000	50–800	>300%
Under load deformation	D 621 7 MPa, 24 hrs			2% after 90 min
Shore D hardness	D 2240	60–70		60
Abrasion resistance (mg/cycle)	F-510-81	2–5	10–15	

is not feasible since the required temperatures, about 135 °C, could result in dimensional modifications of the liners. Ethylene oxide (EtO) sterilizes UHMWPE components closed in gas-permeable packages; the treatment is continued for as long as is necessary for the gas to diffuse inside the containers; the packages are then left under vacuum for the time required for complete elimination of the EtO. Prosthetic UHMWPE components sterilized in EtO do not undergo any variation of chemical and physical structure.

Gamma radiations are emitted during decay of a ^{60}Co unstable nucleus. The dose absorbed by the material during sterilization is about 25–30 kGy and depends upon the geometry of the sample and upon its position in relation with the source. The electron beam is produced by thermally exciting a tungsten filament; electrons are accelerated by electric fields up to 10 MeV and then conveyed onto the material to be sterilized. The advantages of this method are the fairly easy control of the apparatus and the very short period of treatment (seconds).

When UHMWPE is exposed to an energy stronger than that which links the atoms of the polymeric chain, e.g., high-energy radiation (ultraviolet and gamma rays, X-rays, electron beam), heat, strong mechanical stress, and so on, some bonds split, and free radicals are formed. If even only one C–C link of the UHMWPE polymer chain breaks and two radicals CH_2 are formed, the length of the chain and, as a consequence, the molecular mass decrease; consequently, some of its chemico-physical characteristics begin to worsen. This process is called degradation; if oxygen is present when the degradation process occurs, it is called oxidative degradation or, more simply, oxidation. Oxygen may be either atmospheric, namely present in the sterilization environment, or absorbed on the surface of the polymer, or even oxygen penetrated by diffusion into the polymer during processing and storage, or during in vivo service.

Oxidation is a series of reactions involving free radicals; once the oxidation process has been started, it cannot be interrupted, and its rate increases continuously (Figure 52.1). The oxidative process, whose extent depends on the

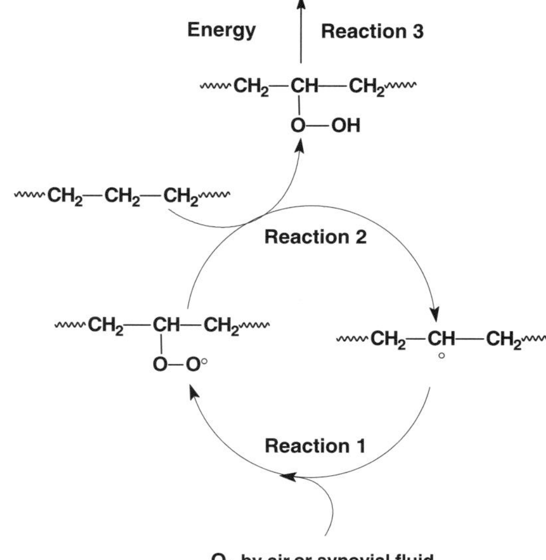

Figure 52.1. Reactions in the process of oxidation.

number of radicals formed during sterilization and on the amount of oxygen in the polymer, can continue during storage and in vivo implantation. The distribution of oxidative products in the prosthetic component is a function of the rate at which gamma radiations have been supplied, of the temperature of the sterilization room, of the amount of oxygen present in the polymer at the time of irradiation, and of the amount of oxygen that can diffuse afterwards [13]. Looking at a section of some new or retrieved UHMWPE prosthetic components, a white halo, known as the *crown effect* or *white band*, can sometimes be seen: it is the macroscopic demonstration of oxidative degradation due to gamma ray sterilization in air. This zone, due to oxidation, has a very consistent molecular mass decreasing with respect to that of the starting material. UHMWPE components with a crown effect have very low mechanical properties, resulting in wear, delamination, and fracture during service in vivo [14].

Whereas polyethylene wear due to sterilization is a problem that has been solved [15], abrasion and production of abraded particles remains a problem: debris is able to initiate an inflammatory reaction and periprosthetic osteolysis [5]. Abrasion is particle loss due to friction caused by the reciprocal movement of the loaded articulate surfaces: for equal mechanical stress, material and interface, it is a function of time. To increase the abrasion resistance, newly cross-linked UHMWPE products are being studied. The cross-linking process links polymer chains by means of a chemical bond; as a consequence, the molecular mass increases (at the end of the process there is one molecule with infinite molecular mass). In comparison to original UHMWPE, cross-linked UHMWPE shows better abrasion resistance and deterioration of some mechanical properties [16], owing to chemical and physical modifications induced by irradiation and heat treatment.

Industrially, for total joint replacements, cross-linking is carried out by irradiation with high-energy radiation, i.e., gamma radiation or electron beam, at a dose between 60 and 100 kGy at room temperature or at molded state; residual radicals are eliminated by heat treatment (at high temperature for a short time or at low temperature for a longer time) [17]. Due to different cross-linking processes, the new cross-linked UHMWPE is very different; whereas normal UHMWPE always has standard properties and maintains them if processed and sterilized in adequate ways, the actual commercial, cross-linked UHMWPE is very different, with highly variable properties. The behavior of these new products is under evaluation.

Recently, it was demonstrated that in UHMWPE prosthetic components, adsorption and deep diffusion of organic molecules present in the synovial liquid occur during service in vivo; the diffused molecules are cholesterol, esters of cholesterol, squalene, β carotene [18]. This diffusion explains the yellowish color in some retrieved components; its relationship with mechanical resistance and behavior in vivo is under evaluation.

Poly(Methylmethacrylate), the Orthopaedic Cement

Orthopedic cement is basically poly(methylmethacrylate) (PMMA) obtained by polymerizing the methyl methacrylate monomer (MMA) [19,20]. It is usually supplied in two separate packages: a brown-colored vial containing about 20 ml of transparent liquid, and one package or two containing 40 g of powder. The liquid contains MMA, usually N,N dimethyl-p-toluidine (DMPT) to accelerate the polymerization process, and traces of hydroquinone to avoid premature polymerization of the monomer. The powder is formed by pre-synthesized PMMA (at times, polymethylmethacrylate-styrene as copolymers are used), benzoyl peroxide (DBP), and barium sulfate (or zirconium dioxide), the latter may be supplied in a separate package. PMMA is in the shape of spherical particles having a variable diameter between 30 and 250 microns; the size of the particles determines the viscosity of the cement. DBP initiates the radical process of polymerization through the effect of heat and polymerization accelerator. Barium sulfate makes the cement radio-opaque. Cements produced by different industrial companies have different chemico-physical characteristics and mechanical properties due various components and their relative concentrations.

Bone cement preparation is characterized by three phases: the wetting phase corresponds to mixing the solid part with the liquid, the setting phase (divided into "dough time" and "working time") corresponds to the initial polymerization process (about 5% of total), and the curing phase corresponds to the final hardening phase and completion of the polymerization process. During mixing, benzoyl peroxide, present on the surface of the PMMA powder, and DMPT, present in the liquid, interact and the polymerization process starts, mainly on the surface of the pre-synthesized poly(methylmethacrylate). Working time starts when a "dough" is obtained which no longer sticks to gloves and temperature increase of the cement is minimal, corresponding to minimal transformation of MMA to PMMA. The final polymerization phase is characterized by the rapid increase of polymerization rate and temperature. The time required for the various phases depends mainly on the temperature in the operating theater: a 10 °C increase causes polymerization to start twice as quickly, cutting mixing times by half. After polymerization, less than 5% of MMA remains free and this percentage may slowly spread into the body. The MMA polymerization reaction is exothermic; the high temperature favors DBP decomposition leading to an increase in radical formation and consequently an increase in polymerization process. Therefore, polymerization speed is initially minimal and gradually increases. Were processing carried out in adiabatic conditions, the bone cement temperature would reach 160 °C. The actual temperature reached by the cement during the surgery depends on the balance between quantity and speed with which the heat is produced, and how easily the heat is dispersed from the surface into surrounding tissues. At the interface with spongy bone, due to vascularization and the trabecular shape of the bone itself, temperatures of 60 °C can be reached, while in the center of the mass of cement the temperature is higher than 100 °C. Schematically, cement produces heat as a function of the amount used, and the temperature at the interface increases with a higher quantity of cement. Based on this assumption, an adequate surgical technique can lower the temperature at the interface by using both an adequate and not too thick layer of cement, and washing liquids in the final polymerization phase. Some cements are declared as "low-temperature polymerization". They are characterized by a smaller quantity of monomer MMA that proportionally lowers the heat developed during transformation of monomer into polymer. High temperature is sought when the cement is used as adjuvant in bone tumors to ensure "sterilization" of a bone surface from which the tumor has been removed; therefore in oncological surgery, standard PMMA is useful.

During the polymerization reaction, a theoretical volumetric shrinking of the PMMA takes place proportional to the amount of MMA used; in orthopedic cement, the volumetric shrinking is 7% of the initial volume. Another characteristic of cement is the porosity due to CO_2 formed during decomposition of the initiator, MMA monomer evaporation, air-bubbles formed during hand preparation of the mixture, and the expansion due to the temperature increase during polymerization. In actual orthopedic cements, the vacuum technique preparation decreases air-bubble formation; other factors can not be eliminated.

Antibiotic-loaded cements are used in order to obtain a greater quantity of local antibiotic and to reduce the systemic quantity, thereby decreasing general toxicity; they are either industrially packaged or prepared in the operating theater according to the antibiotic. The state of the art on how the antibiotic manages to act is the following [21]: the antibiotic dissolves from the surface of PMMA into the tissues; antibiotic molecules of larger size are physically blocked inside the bone cement and, therefore, cannot spread from inside the cement to the surface. The dissolution process depends on the type of antibiotic, on the characteristic of the surface of the cement, and on the way the cement itself is prepared. When the antibiotic is added to the cement during preparation of the cement itself, i.e., in the operating room, only a small part of the antibiotic molecules are on the surface of the cement and will be able to dissolve. This process explains why the actual antibiotic-loaded cements have a limited aseptic action.

Ceramic Biomaterials

Ceramics are solid materials, which have as their essential component inorganic, non-metallic materials. Ceramic biomaterials used in orthopedic devices can be classified as bioinert (alumina and zirconia) [22,23], bioreactive (hydroxyapatite and other calcium phosphate ceramics), and bioresorbable.

Alumina and Zirconia

The development of alumina (aluminum oxide – Al_2O_3) as a biomaterial began in the mid 1960s, the behavior of alumina components (say total hip replacement – THR ball heads) were improved continuously over more than 30 years of clinical use, making alumina one of the better characterized biomaterials [24]. Today more than three million alumina ball heads have been implanted worldwide, coupled either to UHMWPE or to alumina acetabular components. Due to its extremely high hardness and wettability, alumina has shown favorable wear behavior in arthroprostheses. The material used in biomedical applications is α-alumina, known as corundum, one of the most stable oxides, unaffected by corrosion (e.g., absence of ion release from bulk materials and from wear debris) in the most adverse conditions. Alumina shows very good performance in compression, but relatively poor tensile strength, which is improved by higher density and smaller size of grains. The biocompatibility of alumina is a well-established property.

In the first half of the 1980s many researchers focused their efforts on the development of Zirconia-toughened Alumina (ZTA). ZTAs are composite materials consisting of an alumina matrix containing zirconia grains. The mechanical properties of ZTA are significantly improved with respect to alumina, the better performances were obtained using nanometer-sized zirconia grains and other additives, such as, e.g., chromium dioxide, thus obtaining the so-called Alumina Matrix Composite (AMC) ceramics (Table 52.2) [25].

Zirconia (zirconium dioxide – ZrO_2) ceramics were developed and introduced in clinical use in the late 1980s. The early developments were oriented towards Magnesia-partially Stabilized Zirconia (MgPSZ), in which the tetragonal phase is present within large cubic grains (40–50 μm) forming the matrix, a coarse structure that may negatively influence the wear properties of joints. Most of the developments were focused on Yttria-stabilized Tetragonal Zirconia Polycrystal (YTZP), a ceramic consting of tetragonal grains some hundreds of nanometer in size as described elsewhere [26], which is currently a standard material in clinical use for more than 15 years. The excellent mechanical properties of YTZP ceramics (Table 52.2) allow the manufacturing of ball heads with small diameters (22 mm or 26 mm) and long necks. As well as alumina, YTZP also has a wettability superior to metals, as it can bind at its surface a fluid film acting as lubricant. In clinical practice YTZP ball heads are coupled only to UHMWPE sockets. As YTZP may be affected by strength degradation due to phase transformation (a well-known behavior related to temperatures above 100 °C in wet environments), manufacturers have to carefully control several parameters, e.g., density, small and uniform grain size, concentration and distribution of stabilizing oxide Y_2O_3, and sometimes introduction of Al_2O_3 or CeO_2 as additives into the matrix. The biocompatibility of zirconia is well documented, on the basis of in vitro and in vivo tests [26].

Table 52.2. Indicative values of selected properties of biomedical grade Alumina Matrix Composites (AMC) ceramics, Yttria-stabilized Tetragonal Zirconia Polycrystal (YTZP), and Zirconia-toughened Alumina (ZTA) [25]

Properties (Unit)	AMC	YTZP	ZTA
Chemical composition (mol)		$ZrO_2 + 3\% Y_2O_3$	
Density (g/cm^3)	4.36	6.08	5.02
Average Grain size (μm)		0.3	
Bending Strength (MPa)	1,150	1,200	912
Compressive Strength (MPa)	4,700	2,000	
Fracture Toughness (Mpa · m$^{1/2}$)	8.5	9	6.9
Elastic Modulus (GPa)	350	200	285
Hardness (HV)	1,975	1,000–1,300	1,500

The process of formation of dense ceramics is characterized by the selection of raw materials (powders, purity, fine and homogeneous grain size), the processing of powders before firing, and the technology used for powder consolidation. All these parameters influence the microstructural characteristics of the final product. Details on the manufacturing process are described elsewhere [23]. The introduction of Hot Isostatic Pressing (HIP) in bioceramics production has permitted the minimizing of residual stresses within ceramic parts and development of ceramics with a density close to the theoretical one, improving the strength and reliability of ball heads; reliability is improved also by proof testing on all the ball heads and inlays prior to release on the market.

The common method of sterilization of ceramics is by gamma rays; nevertheless, ethylene oxide may also be used. During gamma rays sterilization, ceramics may undergo some changes in color, due to ionization of rare earths that may be present as impurities at parts-per-million levels in the material. Steam sterilization should not be used for zirconia ball heads (MDA SN 9617, June 96).

Almost all THRs make use of a cone taper to join the stem to the ball head (ceramic or metallic). The modularity allows for control of the protrusion of the ball head from the stem neck in several steps, commonly three, allowing for the selection of neck length with the same stem. The drawback of this solution is that the self-locking conical taper transforms the compression stresses in bending stresses on the upper part of the ball head and in tangential (hoop) stresses in the lower part. Careful matching of surface roughness, roundness, and linearity in coupling ceramic tapers with the metallic trunnion plays a relevant role on distribution and intensity of the stresses, which are dependent on, e.g., the cone angle, the extent of the contact, and the friction coefficient among the two surfaces. Mismatches in female-to-male taper – e.g., due to the many angles in clinical use; roundness, roughness, or linearity errors in the taper; material and design of the taper – are among the most likely "technologic" initiators of ceramic ball head failures. Besides the "technologic" failure initiators, when using ceramic ball heads, surgeons have to take care of several technical prescriptions dictated in Safety Notice MDA SN2002-5.

Due to the improvements introduced in manufacturing, fractures of alumina ball heads and inlays currently occur with a very low frequency. Fractures are typically associated with severe trauma, or to technical errors. The problem was relevant in the early years of clinical use, and was characteristic of materials that were withdrawn from the market in the early 1980s. The failure rates observed in materials made after 1980 from different manufacturers are variable, with an average value around 0.05%. Besides the uncertainties present in this kind of evaluations, and bias depending on surgical technique, the scatter of data (some of the series show zero fractures, while others show failure rates up to 2%) appears linked to the size of the study: large series do not show fractures, while the small ones show higher fracture rates. The high fracture rates of YTZP ball heads recently observed by a manufacturer in some batches made between January 1998 and September 1999 is not specific to the material.

Hydroxyapatite (HA)

The use of ceramic coatings to protect THR stems from corrosion began in the 1960s. Following the development of osteoconductive calcium phosphate, ceramic coatings were applied to metallic stems to enhance bony fixation. Long-term follow-ups confirm the results obtained in early works [27]. The rate of bone formation is dependent on the porosity and solubility of the ceramic, as the characteristics of porosity influence cell attachment and spreading, while the solubility, depending on Ca/P ratio, influences the degradation rate of the material. The bioreactivity of the coating and its mechanical stability are dependent on several parameters, e.g., presence of foreign phases, crystallinity, residual porosity, and mechanical properties (shear strength, bond strength, fatigue life) [28].

The most common industrial process to obtain HA coatings is by plasma spray, consisting of injecting powder particles in a plasma gun where they melt. Droplets of melt ceramic are sprayed at high speed onto the surface to be coated, where they splat and cool, thus forming the coating by mechanical adhesion of multiple droplet layers. A critical aspect is the starting Ca/P ratio of the starting powder, as it influence the coating stability. The 100% crystalline HA powders that are used to feed plasma guns experience a severe heating/cooling thermal cycle: formation of amorphous phases and of resorbable calcium phosphate ceramic (CPC) compounds, segregation of CaO, and oxidation reactions may occur during plasma spray. Performing the process at low pressure (Vacuum Plasma Spray – VPS) allows avoidance of undesired oxidation reactions. Sterilization in the presence of HA coating requires similar care as for dense ceramics.

Metallic Materials for Joint Prostheses

For the use of metallic alloys as biomaterials [29], the specific requirement of biocompatibility narrows the range of useful compositions. In fact, the interaction of metallic materials with the biological environment enhances the importance of surface reactivity of alloys, which is also related to composition. As a consequence, the surface of the implant is often chemically modified in order to improve its biocompatibility.

Metallic materials with industrial relevance for joint prostheses belong to three main groups [29–31]: (i) stainless steel; (ii) alloys based on the Co-Cr system; (iii) Ti and its alloys. (i) The stainless steel mainly used for orthopedic implants is austenitic AISI 316. When it is specified as AISI 316L, the carbon content is limited to 0.03 wt% for preserving the good corrosion resistance of this material. (ii) Co-Cr-based alloys have been used for total joint prostheses since the early 1900s and are originating from modifications of the dentistry alloy Vitallium (Haynes Stellite alloy N. 21) [31]. They combine good mechanical properties with a high biocompatibility due to the presence of Cr, which forms a protective oxide layer. The carbon content in the alloy must be carefully controlled, because the formation of carbide phases may be detrimental for mechanical properties. (iii) Ti and Ti-based alloys are widely used as biomaterials for their high biocompatibility, mainly due to a high corrosion resistance related to the formation of a passive oxide layer at the surface. Good mechanical properties and low density constitute an additional benefit for joint prosthesis production. Commercially pure (cP) Ti is used in different grades, as a function of the oxygen content as impurity. Common Ti-based alloys contain aluminum (Al) and vanadium (V), the last often substituted by Niobium (Nb) in order to increase biocompatibility. The main components of most widely used metallic biomaterials for joint prosthesis are collected in Table 52.3 [29].

The industrial production of metallic components for joint prosthesis may be described in different steps. As a first step, raw metals and alloys are processed into stock shapes, such as bars, sheets, rods, plates, tubes, wires, and powders. The second processing step is used to tailor the microstructure of the alloy, which is strongly related to the mechanical properties of the implant, by means of thermo-mechanical treatments. The transformation of stock materials into final products may be obtained by investment casting, machining, forging, and sintering. Techniques used to manufacture various alloys to produce metallic biomaterials for joint prostheses are collected in Table 52.4.

Surface coatings aimed at improving functional properties of implants (i.e., biocompatibility, bone fixation) are often added as a final step. Functionality and duration of implants in a physiological environment are very sensitive to surface properties, which may be considered the most important and selective aspect for joint prosthesis selection. Surface treatments are mainly aimed at increasing hardness and strength of the surface layer, in order to improve the resistance to wear and corrosion. Various techniques for surface modifications have been used for orthopedic alloys: ion implantation,

Table 52.3. Typical composition (maximum amount allowed, wt%), mechanical, and physical properties of metallic biomaterials

Materials	Main comp.	Other comp. (max wt%)	Density (g cm–3)	Yield strength (MPa)	Ultimate Tensile Strength (MPa)	Fatigue Strength (107 cycles) MPa	Fracture Toughness (MPa m1/2)	Elastic modulus (Gpa)	Elongation at fracture (%)
Stainless steels AISI 316	Fe	Ni (14), Cr (19), Mo (2,5), Mn (2)	7.5–8.0	170–790	480–1,000	180–550	75–85	190–200	10–50
Co-Cr based alloys	Co	Cr (30), Ni (37), Mo (10,5), Mn (2.56)	8.2–9.1	250–1,500	650–1,800	300–950	50–60	210–240	8–50
cP-Ti	Ti	Fe (0.5), O (0.4)	4.5	170–485	240–550	200–330	65–75	110	15–25
Ti based alloys	Ti	Al (6.5), V (4.5), Nb (7.5), Fe (3), Mo (15), Zr (6)	4.4–5.3	800–1,050	900–1,100	450–650	50–55	75–115	8–20
Cortical bone				80–150	30	2–12	14–22	0–2	

Table 52.4. Techniques used to produce metallic biomaterials for total joint replacements

Technique	Stainless steels	Co-Cr-based alloys	cP-Ti	Ti-based alloys
Casting	Not used	Investment casting	Difficult	Difficult
Machining	Possible	Difficult	Possible	Possible
Cold working	Rolling	Difficult	Rolling	Difficult
Hot working	Wrought, forged	Wrought, forged	Not used	Wrought, forged
Sintering	Possible	Hot isostatic pressing	Not used	Not used
Thermal treatments	Recrystallization	Precipitation hardening	Recrystallization	Precipitation hardening

chemical and physical vapor deposition, diffusion hardening, and plasma treatment [32]. Typical coatings are constituted by nitrides (CrN or TiN) and, more recently, by diamond-like carbon (DLC) [33]. Porous coatings have been developed to promote bony ingrowth into the implant. Even if metallic biomaterials show good static mechanical properties, they may suffer significantly from fatigue failures. Fatigue strength is defined as the highest periodic stress that does not initiate a failure of the material after a given number of cycles. For hip prostheses, an average of 2×10^6 stress cycles per year can be estimated, so that more than 10^8 cycles may be applied during a lifetime. The applied stress for fatigue failures is in the elastic region of the static loading, so that fatigue strength is significantly lower than ultimate tensile strength. Metallic biomaterials have fatigue strengths in air generally well above the minimum required for joint prosthesis applications [34]. Mechanical properties of most widely used metallic biomaterials for joint prostheses are collected in Table 52.3.

Total joint replacements are subjected to wear and abrasion so that resistance against them is an important criterion for biomaterials [35]. Metal–metal joint coupling seems to be promising [29]. High-carbon Co-Cr-based alloys (F75) significantly improve mechanical properties after working, so that small plastic deformations at the surface significantly increase the hardness of the alloy and, as a consequence, its wear resistance. In addition, the presence of fine, dispersed, hard carbides increases the wear resistance of these alloys. Oxide films formed by passivation at the surface of Cr- and Ti-containing alloys are generally resistant to abrasion. Load required to fracture the oxide surface film is lower for Ti-based alloys with respect to Co-Cr-based alloys [31].

The biological environment can cause corrosive attack. Many types of corrosion may be observed on metallic biomaterials, including general corrosion, pitting and crevice corrosion, stress corrosion, corrosion fatigue, fretting corrosion, and intergranular corrosion [29]. The driving force for oxidation of Cr and Ti is very high, so that orthopedic alloys would corrode rapidly in the absence of the passive oxide film that forms on their surface. These films are very thin (on the order of 5–10 nm) and are very stable in normal conditions, but their composition and structure may change over the lifetime of the prosthesis, resulting in instability of the oxide. Stainless steels appear significantly less resistant to localized corrosion with respect to Ti- and Co-Cr-based alloys.

Bibliography

Brach del Prever EM, Costa L, Baricco M, Piconi C, Massè A. Biomaterials for joint prosthesis. In: EFORT (European Federation of National Associations of Orthpaedics and Traumatology), editor. Surgical Techniques in Orthpaedics and Traumatology. Paris: Elsevier, 2003.

Cales B. Fractures des têtes de prothèses de hanche en zircone. Maitr Orthop 2000;96:26–30.

Li S, Burstein AH. Current concepts review: UHMPWE: the material and its use in total joint implants. J Bone Joint Surg (A) 1994;76-A:1080–9.

References

1. Williams DF. Definition in biomaterials. Proceedings of the Consensus Conference of the European Society for Biomaterials, Chester, England, March 3-5, 1986. Amsterdam: Elsevier, 1987;49–59.

2. Nizard R, Bizot P, Kerboull L, Sedel L. Biomatériaux orthopédiques. Encycl Méd Chir (Elsevier, Paris), Techniques chirurgicales - Orthopédie-Traumatologie, 1996;44-003, 20.
3. Horbett TA, Ratner BD, Schakenraad JM, Schoen FJ. Some background concepts. In: Biomaterial Science. San Diego: Academic Press, 1996:133.
4. Branemark PI, Hansson BO, Adell R, Breine U, Lindstrom J, Hallen O et al. Osseointegrated implants in the treatment of the edentulous jaw. Scand Plast Reconstr Surg 1977;16.
5. Jacobs JJ, Rorebuck KA, Archibeck M, Hallab NJ, Glant TT. Osteolysis: basic science. Clin Orthop Rel Res 2001;393:71-7.
6. Visuri T, Pukkala E, Paavolainen P, Pulkkinen P, Riska EB. Cancer risk after metal on metal and polyethylene on metal total hip arthroplasty. Clin Orthop Rel Res 1996;329S:280-9.
7. Gillespie WJ, Frampton CMA, Henderson RJ, Ryan PM. The incidence of cancer following total hip replacement. J Bone Joint Surg [Br] 1988;70-B:539-42.
8. Nyrén O, McLaughlin JK, Gridley G, Ekbom A, Johnell O, Fraumeni JF Jr et al. Cancer risk after hip replacement with metal implants: a population-based cohort study in Sweden. J Natl Cancer Inst 1995;87(1): 28-33.
9. Doherty AT, Howell RT, Bisbinas I, Learmonth ID, Newson R, Case CP. Increased chromosome translocations and aneuploidy in peripherial blood lymphocytes of patients having revision arthroplasty of the hip. J Bone Joint Surg [Br] 2001;83B(7):1075-81.
10. Baldini N, Cenni E, Granchi D, Ciapetti G, Savarino L, Tigani D, et al. Metal and cement hypersensitivity in patients with arthroplasties. Poster N° 1078. 48th Annual Meeting of the Orthopaedic Research Society, Dallas, 2002.
11. ASTM designation F 648-98 standard specification for UHMWPE powder and fabricated form for surgical implants.
12. Costa L, Jacobson K, Bracco P, Brach del Prever EM. Oxidation on ethylene oxide-sterilized UHMWPE. Biomaterials 2002;23:1613-24.
13. Costa L, Bracco P, Brach del Prever EM, Luda MP. Oxidation in prosthetic UHMWPE. 224th ACS National Meeting, Boston, MA, August 18-22, 2002.
14. Costa L. Brach del Prever EM. UHMWPE for arthroplasty. Torino: Minerva Medica, 2000.
15. Blunn G, Brach del Prever EM, Costa L, Fisher J, Freeman MAR. Ultra-high-molecular-weight polyethylene (UHMWPE) in total knee replacement: fabrication, sterilization and wear. J Bone Joint Surg Br 2002;84:946-9.
16. McKellop H, Shen FW, Lu B, Campbell P, Salovey R. Development of an extremely wear-resistant UHMWPE for total hip replacements. J Orthop Res 1999;17:157-67.
17. Kurtz SM, Muratoglu OK, Evans M, Edidin AA. Advances in the processing, sterilization and crosslinking of the ultra-high-molecular-weight polyethylene for total joint arthroplasty. Biomaterials 1999;20:1659-88.
18. Costa L, Bracco P, Brach del Prever E, Luda MP, Trossarelli L. Analysis of products in vivo diffused in UHMWPE prosthesis components. Biomaterials 2001; 22(4):307-15.
19. Jasty M. Fixation by PMMA. In: Callaghan JJ, Rosenberg AG, Rubash HE, editors. The Adult Hip. Philiadelphia: Lippincott-Raven, 1998;187-200.
20. Wixon RL, Lautenschlager EP. Methyl methacrylate. In: Callaghan JJ, Rosenberg AG, Rubash HE, editors. The Adult Hip. Philiadelphia: Lippincott-Raven, 1998; 187-200.
21. Trippel SB. Antibiotic-impregnated cement in total joint arthroplasty. J Bone Joint Surg (A) 1986;68A: 1297-302.
22. Boutin P, Christel P, Dorlot JM et al. The use of dense alumina-alumina ceramic combination in total hip replacement. J Biomed Mater Res 1988;22:1203-32.
23. Sedel L. Evolution of alumina-on-alumina implants: a review. Clin Orthop 2000;379:113-22.
24. Rieger W. Ceramics in orthopaedics - 30 years of evolution and experience. In: Rieker C, Oberholtzer S, Wyss U, editors. World Tribology Forum in Arthroplasty. Bern, CH: Hans Huber, 2001:309-18.
25. Heimke G, Leyen S, Willmann G. Knee arthroplasty: recently developed ceramics offer new solutions. Biomaterials 2002;23:1539-51.
26. Piconi C, Maccauro G. Zirconia as a ceramic biomaterial. Biomaterials 1999;20:1-25.
27. Geesink RGT. Osteoconductive coatings for total joint arthroplasty. Clin Orthop 2002;395:53-65.
28. Sun L, Berndt C, Gross KA, Kukuc A. Material fundamentals and clinical performances of Plasma spray coatings: a review. J Biomed Mater Res (Appl Biomater) 2001;58:570-92.
29. Helsen JA, Breme HJ, editors. Metals as Biomaterials. Chichester, UK: John Wiley & Sons, 1998.
30. Brunski JB. Metals. In: Biomaterials Science. Academic Press, 1996;37.
31. Gilbert JL. Metals. In: Callaghan JJ, Rosenberg AG, Rubash HE, editors. The Adult Hip. Philiadelphia: Lippincott-Raven, 1998;134.
32. Chu PK, Chen YJ, Wang LP, Huang N. Plasma-surface modification of biomaterials. Mater Sci Eng 2002;R36: 143-206.
33. Bolton J, Hu X. In vitro corrosion testing of PVD coatings applied to a surgical grade Co-Cr-Mo alloy. J Mater Sci Mater in Med 2002;13:567-74.
34. Teoh SH. Fatigue of biomaterials: a review. Int J Fatigue 2000;22:825-37.
35. Ashby MF. Materials Selection in Mechanical Design. Oxford, UK: Butterworth-Heinmann, 1999.

VII

Futures Prospects and New Materials

53 Growth Promoter: General Principles and Experimental Studies on BMP

T. Sam Lindholm

Introduction

Interest in new bone-forming substances to cure bone defects and fractures dates back to the very beginning of the twentieth century. This line of research was especially adopted in the orthopedic field and also in maxillo-facial surgery.

Among relevant historical data on osteoinduction, mention must be made especially of four scientists. G. Levander (1938) was doing research on soluble stimulating agents inducing new bone formation [1] and P. Lacroix (1945) showed that alcoholic extracts of rabbit cartilaginous epiphyses promotes new bone formation. He gave the extract the name “osteogenin” [2]. M. R. Urist (1965) proved that decalcified bone matrix implanted in muscle pouches in the rabbit and rat resulted in cartilage and bone formation [3]. Lastly, a group under J. M. Wozney (1988) published the first results on recombinant bone morphogenetic proteins (BMPs) (BMP-1 through BMP-4) and made the identification of their biochemical and biological characteristics and amino acid sequences [4].

Urist [5], at the very onset of the research on osteoinduction, named the possible agent producing the inductive process the “bone induction principle”. The presentation of scientific papers in the literature especially on DBM (demineralized bone matrix) reached its peak in the year 1989, after which this activity declined. In accord with the osteoinductive activity of DBM the process was called “demineralized bone matrix osteogenesis”.

Reports on chemically extracted native BMP began to appear about 1977 and peaked in 1998. Papers on recombinant BMPs appeared after the discovery of these products, which are now one of the dominating representatives for this kind of research [6]. Today there may be more than 30 different BMPs or agents classified as osteoinductive factors.

Bone Healing and Transplantation

The bone healing process can be classified into two types depending upon the location in the skeleton. Endochondral ossification, e.g., fracture healing of a long bone means that local mesenchymal cell infiltration will take place, followed by differentiation first to chondroblasts and then to osteoblasts and bone tissue. In the process of intramembraneous ossification, e.g., in a defect of the skull, this process proceeds without the cartilaginous phase.

The bone transplantation procedure can be typed according to the origin of the bone material needed. The alternatives are autogenic, allogenic, and xenogenic bone tissue. Autogenic, the patient’s own bone, has the best osteogenic capacity. Today we can, in fact, also use other alternatives in bone transplantation; we can transplant cells and tissues, proteins, molecules, and genes.

BMP

Bone morphogenetic protein is a differentiation factor inducing, e.g., connective tissue cells to produce new bone. BMP also appears to show chemotactic activity and increases the local microcirculation [7]. Some mitogenic activity on primary calvarial derived cells has been shown, but not for other cell types. BMP elicits responses by binding to receptors (serine/threonine kinase) located on the cell surface. Both type I and type II receptors are required for signal transduction. The process also involves other kinds of factors which are currently under research [8].

Native extracted BMP preparation demonstrates an active bone inductive capacity containing a mixture of different growth factors such as TGF-betas, which also are important for new bone formation. Besides TGF-beta 1 and 2, BMP-2, -3, -4, -6, and -7 are present in this protein complex. Additionally there are osteocalcin, osteonectin, albumin, and transferin. The absence of FGF-1 and -2, IGF-1 and -2, PDGF-a and -b has been noted [4,9]. So far no exhaustive study has been made of the complete content of the factors included in this kind of complex extract (Figure 53.1, plate section).

There are also differences between animal species in extracted native BMP in respect of the capacity to induce new bone formation. One report of a comparative test has established that BMP extract from reindeer bone is far superior to the bovine, sheep, and porcine BMP [10].

Extract of Native BMP

There are many modifications of the original chemical extracting technique developed and used by M. R. Urist [11]. The preparation includes pulverization of long bones, demineralization in 0.6 N HCL, extraction with 4 M GuHCL, tangential flow filtration, and dialysis. The end result of this process is called partially purified BMP (Figure 53.2). The purification process can be continued, e.g., by the use of HPLC chromatography, resulting in three major fractions at molecular weights 11 to 40 kDa, 40 to 140 kDa, and 500 to 700 kDa, respectively. The second of the fractions possesses immunogeneic properties and the others are osteoinductive. Further purification is also possible. This kind of extraction process is, however, laborious and time consuming.

Recombinant BMPs

Recombinant BMPs are produced with one or several expression systems such as bacteria, insect, or mammalian cells. BMP-2, -4, -5, -6, and -7 are all osteoinductive. In comparison to these, BMP-5 requires more substance to produce the

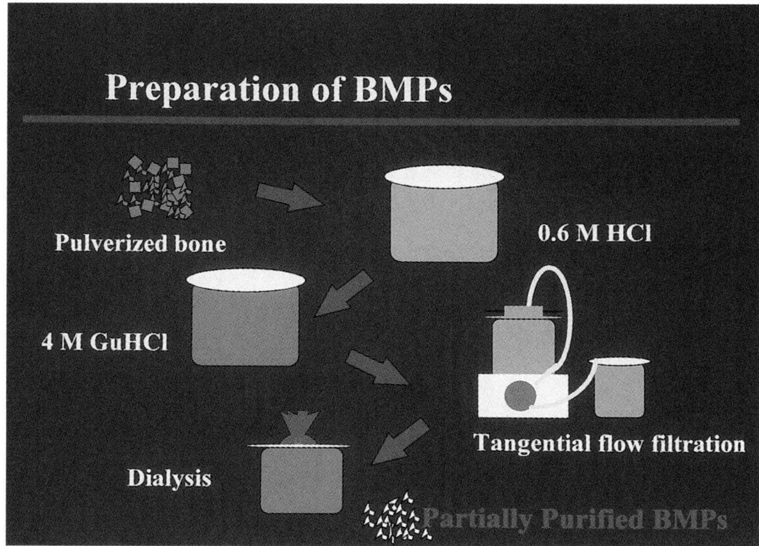

Figure 53.2. Schematic illustration of the extracting technique of native BMP including pulverization of bones, demineralization, extraction, filtration, and dialysis. The end product of this procedure is the partially purified protein complex.

same amount of new bone. The ability of BMP-3 to induce new bone formation is controversial. GDF-5, GDF-6/BMP-13, GDF-7/BMP-12 induce dense connective tissue – tendons and ligaments [12]. The source of the material used for BMP-2 production is human bone tissue and it has been expressed in CHO (Chinese hamster ovarial cells) [13]. Another source for BMP-2 has been the human osteosarcoma U 2 OS cells expressed in *E. coli* [14]. The human BMP-2 expressed in *E. coli*; such a system may be easier and cheaper but involves problems with purification of the remnants according to the bacteria used. Human BMP-4 with a source in murine osteosarcoma cells was expressed in CHO cells [15]. BMP-7 had its source in bovine bone matrix and was expressed in CHO cells [16]. In an as-yet unpublished experiment with different BMPs with an original source in reindeer growing antler anlagen, the site of highly active growth factors, different recombinant BMPs were expressed in *E. coli.* A comparison in amino acid sequences between mature regions of BMPs is shown in Figure 53.3.

Testing of Osteoinductivity

If the aim is to use BMP in experimental work or in clinical testing for bone healing, the most important property of the protein is the capacity for new bone formation. The available techniques for testing new bone formation are as follow: micro-assay (cell culture technique), bioassay (the muscle pouch of, e.g., a mouse), skull defects, segmental long bone defects, and clinical testing with fractures or defects or in bone transplantation procedures.

Micro-assay

In micro-assay many different cell cultures have been used [17]. After application of small amounts of BMP to the culture the ALP activity, 45 Ca incorporation, 3H-thymidine incorporation, or the production of osteocalcin, besides histological staining of cells can be verified [18]. This technique is easy and there is no need for test animals. On the other hand, no comparative results are as yet available between tissue culture results and biological tests in the muscle pouch of rodents. The results obtained so far are not considered definitively comparative [19].

Bioassay

Small amounts of BMP are applied during surgery to a muscle pouch in the hind leg, e.g., of a mouse. A process of cartilage and new bone formation starts and in three weeks a bone ossicle can be verified by X-ray and histologically (Figure 53.4, plate section). The new bone formation correlates to the amount of active BMP incorporated. Combinations of other analyses such as measurement of the content of Ca and the activity of ALP is easily accomplished [20].

Skull Defects

A critical-size skull defect does not have the capacity to heal completely during the lifetime

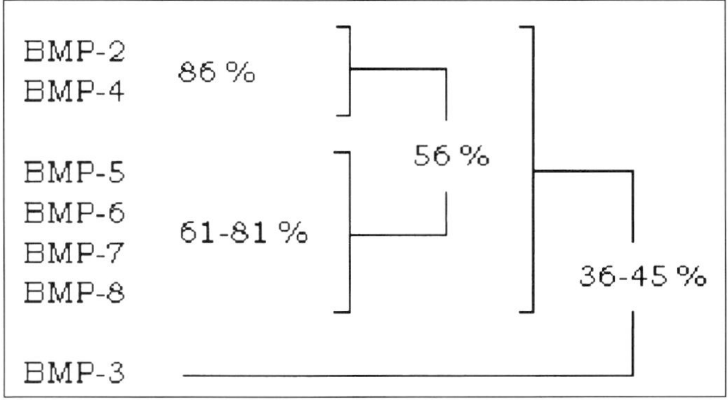

Figure 53.3. Amino acid sequence homologies between mature regions of BMPs.

of the animal. The size of this kind of defect differs depending upon the animal species and even according to the location in the skull area. The process of healing after implantation of BMP in the defect can be analyzed by X-ray and histologically [21].

Segmental Long Bone Defects

The criterion for a segmental defect in a long bone of an animal is that the defect should be twice the breadth of the long bone. Long bone defects have been tested in the femur of rats [22], ulna and radius in rabbits [23], femur in sheep [24], radius in dogs [25], and radius in non-human primates [26]. The mode of fixation of segmental defects in larger animal species is of importance. A long bone defect in a sheep will completely heal in 12 weeks (Figure 53.5).

Maxillo-facial and Dental Reconstruction

The use of osteoinductive implants in the maxillo-facial and dental area may be more important and useful than in orthopedic cases. Pre-clinical studies have been made in mandibular defects in dogs [27], and in non-human primates [28], peri-implantation in non-human primates [29].

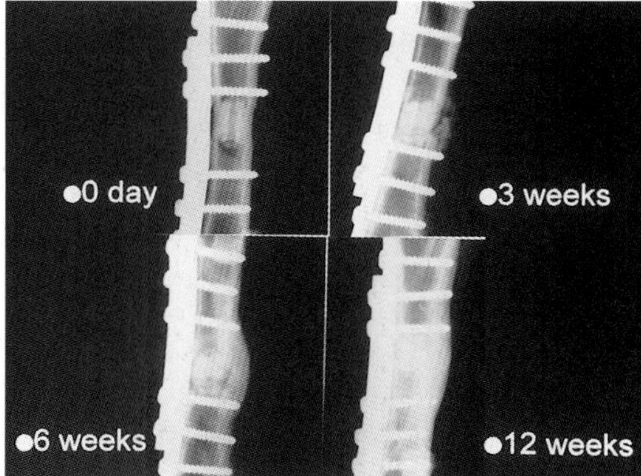

Figure 53.5. A segmental diaphyseal bone defect in a sheep treated with a block of sea-coral and extracted bovine BMP. The lesion was fixed with a plate and screws. There was a complete union of the defect after 12 weeks.

Spinal Fusion

The need for a simple technique to achieve spinal fusion in patients is of crucial importance in view of the increasing numbers of patients to be operated on [30]. Intertransverse process spinal fusion in dogs and the use of inter-body fusion cages in sheep [31] and in non-human primates [32] have been investigated.

Bioimplant

A bioimplant is composed of two parts: the osteoconductive filling or framing material and the carrier used to connect the osteoinductive component or the BMP. The ideal osteoconductive material, sometimes also used as a carrier, must enable vascular and cellular invasion, must be reproducible, non-immunogenic, moldable, and space providing to define the contours of induced bone. The framing material and carrier should be able to start to resorb completely following initiation of bone induction. Different kinds of filling materials used in experimental works have been reported [33]. Sea coral composed of 99% calcium carbonate with 48% pores/volume and resembling trabecular bone has been successfully used in animal and clinical experiments [34]. The BMP can be bound especially to type I collagen with a disulfide or a covalent bonding. On the other hand, methods for assembling BMP with carriers is still restricted to simple mixing, surface coating, substrate soaking, or adsorbing.

The dosage of BMP used in a bioimplant differs according to the location and size of the lesion to be treated. The following recommendations have been made: extract of native human BMP 50–100 mg, extract of bovine BMP 5–30 mg, extract of reindeer BMP 1–10 mg, rhBMP-2 0.15–6.8 mg, rhBMP-7 (OP-1) 2.5–6.8 mg. These values are no more than recommendations according to the literature.

A bioimplant must be sterilized. Here, the following techniques have been used: irradiation, ethylene oxide, and ultrafiltration. Irradiation and ethylene oxide clearly diminish the osteoinductive capacity of BMP. Additionally, ethylene oxide develops poisonous products [35].

Clinical Tests

Tissue engineering usually means reconstruction ex vivo followed by implantation in situ in contrast to the technique of reconstruction of tissues in the host. In a successful reconstruction procedure the regeneration of cells and the extracellular matrix must be in a correct spatial and functional relationship. Clinical trials have been done with DBM, native, and recombinant BMPs.

There are reports in the literature on clinical trials using human, bovine, and reindeer-extracted BMP, and additionally by rhBMP-2 and rhBMP-7. Human BMP extract has been used in more than 100 cases, bovine BMP extract in more than 1,000 clinical cases, and work with reindeer BMP extract has commenced. The indications for the use of osteoinductive bioimplants are orthopedic lesions and uncured fractures, and spinal fusion and partly cases for maxillo-facial and dental surgery.

Extract of Native Human BMP

Native human BMP has been used to compensate for bone loss due to enchondroma, tibial bone defects, femoral non-unions after fractures, and skull defects [36].

Extract of Native Bovine BMP

There are reports on the use of such extracts in maxillo-facial reconstruction and implantology, and humeral, radial, tibial, ulnar, femoral, and phalangeal non-unions [37,38,39]. The product NeO-Osteo, combining the active new-bone-inducing substance GFm (bovine BMP extract) with a collagen carrier and DBM as framing material, yielded excellent results in comparison with autografts in mandibular defect reconstruction and in spinal fusion [40].

A preparation including bovine BMP with type I collagen as a carrier and natural sea-coral as framing material was used to cure previously unsuccessfully treated delayed unions of the humeral, ulnar, tibial diaphyseal bone as in scaphoid non-unions and in sinus lift operations The framing material was dispersed as a block, as granules, or as injectable powder. The success rate was 75% and there were still non-unions in 25% of the long-bone non-unions and pseudoarthrosis. The scaphoid non-unions included in this material showed a failure rate of 66% (Figure 53.6) [41].

Extract of Reindeer BMP

Bioimplants composed of extract of reindeer BMP and type I collagen as carrier and sea-coral as framing material have been preliminarily clinically used in maxillary sinus lift operations with excellent final results and in long-bone pseudoarthroses, primary lower leg fractures, and in bone lengthening [42].

Human Recombinant BMP-2

More than 1,000 patients were included in a testing program using human BMP-2 in hospitals all over the world. Preliminary results were reviewed in cases of tibial fractures, inter-body fusions, maxillary sinus grafting, avascular necrosis of the femoral head, augmentation of localized defects, and preservation of the alveolar ridge [43,44,45]. The success rate was good but not complete, the union of long bone defects and non-unions being 62–75%.

rhBMP-2 does not induce any adverse systemic or toxic effects. The release of BMP-2 from the implant is considered to be relatively slow. No effect has been observed on tumor cell growth.

Human Recombinant BMP-7

A testing program has been started involving 700 patients using OP-1 with type I collagen as carrier; 122 patients at 18 centers in the USA are being tested especially for treatment of tibial non-unions. Projects in the USA, Australia, and Sweden have been launched to cure spondylolisthesis.

Reports have been reviewed in the literature including clavicular pseudoarthrosis, revision of hip prosthesis, long bone non-unions, failed arthrodesis, bone defects, spinal fusions, periprosthetic fractures, sinus floor augmenta-

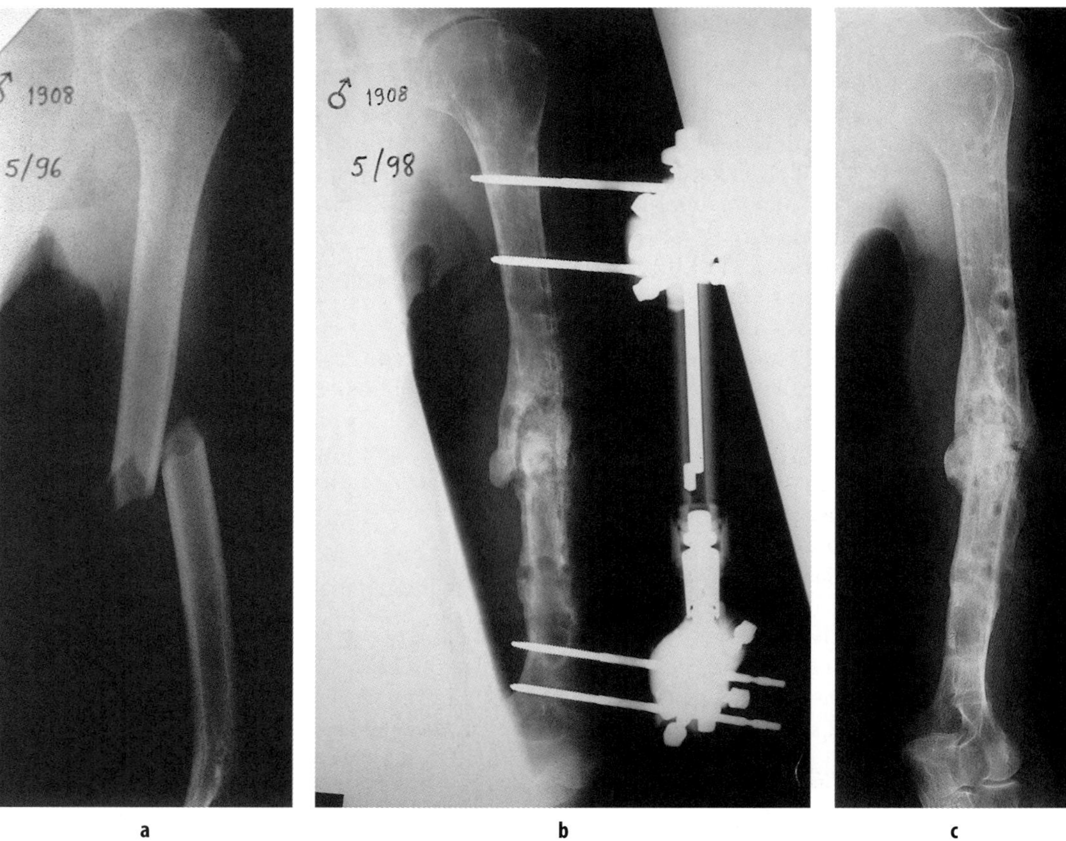

Figure 53.6. A 90-year-old male was surgically treated for a fracture of the humeral shaft (**a**) but developed a non-union. About one year later a re-operation was undertaken together with implantation of an osteinductive implant (sea-coral and native bovine BMP) (**b**). The fixation device was later extracted and the non-union of the fracture completely healed (**c**).

tion, fibular defects, thoracolumbar fractures, and atlanto-axial posterior fusion [46,47,48]. McKee and associates [49] report a randomized trial of previously unsuccessfully autogenous graft-treated atrophic non-unions. The implant consisted of 3.5 mg of OP-1 combined with 1 g of highly purified type I bovine collagen carrier. The success rate was positive in 84% of the patients [49].

Discussion

Today, both rhBMP-2 and rhBMP-7 have been registered as implants for use with special diagnostic treatment groups, the former for spinal fusion and the latter for treatment of lower leg fractures. The economic setting of these osteoinductive bioimplants is so far at a very high level.

Although these preparations constitute effective tools in bone reconstruction, there may be some side effects such as local allergic reactions of the skin, local heterotopic bone formation, and development of antibodies against BMP and collagen in the blood of the patient. The most prominent complication may be infection, probably due to suboptimal internal fixation of the lesion. There may also be problems in respect of the high solubility profile of BMP itself or to the carrier or the framing material. The modes of application of osteoinductive implants are still under investigation and the availability of implants for special surgical applications must be developed. BMP can also augment the action of auto- and allografts. The overall failure rate in treated cases is about 13–32%.

Controversy still prevails as to the difference in action and osteoinductive capacity between native BMP extracts and recombinant BMPs or between rhBMP-2 and rhBMP-7. There may be a better response with a bone-derived (extract) BMP than with a single recombinant [4]. Natural BMP is more active than recombinant, and natural BMPs will retain their values [37].

In view of the failures still observed in BMP treatment the minimum effective dosage of the drug must be found. Combined treatment with different recombinant BMPs or with TGF-betas must be worked out [50]. Other lines of approach may also come to the fore in the very near future, e.g., the development of gene transfer techniques for clinical use of BMP [51].

References

1. Levander G. A study of bone regeneration. Surg Gynecol Obstet 1938;67:705–14.
2. Lacroix P. Recent investigation on the growth of bone. Nature 1945;156:576.
3. Urist MR. Bone: Formation by autoinduction. Science 1965;150:893–9.
4. Wozney JM, Rosen V, Celeste AJ, Mitsock LM, Whitters MJ, Kriz RW et al. Novel regulators of bone formation: Molecular clones and activities. Science 1988;242: 1528–34.
5. Urist MR, Mikulski A, Nakagawa M, Yen K. A bone matrix calcification initiator non-collagenous protein. Am J Physiol 1977;232:115–27.
6. Lindholm TS, Viljanen VV, Mattila M. Thirty years of bone morphogentetic protein research. In: Lindholm TS, editor. Bone Morphogentic Proteins: Biology, Biochemistry and Reconstructive Surgery. San Diego: Academic Press, 1996;3–6.
7. Akioka J, Kusumoto K, Bessho K, Sonobe J, Kaihara S, Wang Y et al. Angiogenesis around induced bone with recombinant human BMP-2 in a latssimus dorsi muscle flap. J Musculoskel Res 2002;6:17–21.
8. Celeste AJ, Iannazi JA, Taylor RC et al. Identification of transforming growth factor beta family members present in bone-inductive protein purified from bovine bone. Proc Natl Acad Sci USA 1990;87:9843.
9. Sampath TK, Maliakal JC , Hauschka PV et al. Recombinant human osteogenic protein-1 (hOP-1) induces new bone formation in vivo with a specific activity combarable with natural bovine osteogenic protein and stimulates osteoblast proliferation and differentiation in vitro. J Biol Chem 1992;267:20352.
10. Jortikka L, Lindholm TS, Marttinen A. Partially purified reindeer (Rangifer tarandus) bone morphogenetic protein has a high bone-forming activity compared with some other arcticodactyles. Clin Orhtop 1993;297: 33.
11. Urist MR, Chang JJ Brownell AG, Huo YK, Lindholm, TS. Native bone morphogenetic protein. Acta Univer Tamp 1992, B;40:27-39.
12. Wolfman NM, Hattersley G, Cox K et al. Ectopic induction of tendon and ligament in rats by growth and differentiation factors 5, 6, and 7 members of the TGF beta family. J Clin Invest 1997;100:321.
13. Israel DI, Nove J, Kerns KM, Moutsatsos IK, Kaufman RJ. Expression and characterization of bone morphogenetic protein-2 in Chinese hamster ovary cells. Growth Factors 1992;7:139.
14. Zhao M, Wang H, Zhou T. Expression of recombinant mature peptide of human morphgenetic protein-2 in Escherichia coli and its activity in bone formation. J Clin Biochem 1994;10:319–24.
15. Takaoka K, Yoshikawa H, Hashimoto J, Ono K, Nakazato H. Transfilter bone induction by Chinese hamster ovary (CHO) cells transfected by DNA encoding bone morphogentic protein (BMP)-4. Clin Orthop 1994;300:269.
16. Özkaynak E, Ruger DC, Drier EA, Corbett C, Ridge RJ, Sampath TK et al. OP-1c DNA encodes an osteogenic protein in the TGF-beta family. J EMBO 1990;9:2085–93.
17. Ulmanen M, Birr E, Hietala O, Lindholm TS. Determination of the biological activity of BMPs in cell cultures is unreliable. In: Lindholm TS, editor. Advances in skeletal reconstruction using bone morphogenetic proteins. New Jersey: World Scientific, 2002;53–61.
18. Jortikka L, Laitinen M, Wiklund J, Lindholm TS, Marttinen A. Rat skeletal muscle myoblasts are target cells for the action of native bone morphogenetic protein. J Musculoskel Res 1997;1:121–9.
19. Atkinson BL, Elton JP, Benedict JJ. Correlation between in vitro and in vivo osteoinduction using a bone morphogenetic protein mixture (Abstract). Int Conf Bone Morphogenetic Prot 2000, Granlibakken, Lake Tahoe, 7–11 June 2000, California, USA.
20. Bessho K, Sinobe J, Kaihara S, Kawai M, Okubo Y, Maeda, J et al. In vivo changes in bone induction by E. coli-derived recombinant human BMP-2. J Musculoskel Res 2002;6:1–7.
21. Lindholm TS, Lindholm TC. The skull defect model in measuring osteoinductivity. J Musculoskel Res 1998; 2:123–39.
22. Yasko AW, Lane JM, Fellinger EJ, Rosen V, Wozney JM, Wang EA. The healing of segmental bone defects induced by recombinant human bone morphogenetc protein (hBMP-2). A radiographic, histological, and biochemical study in rats. J Bone Joint Surg (Am) 1997; 74:659–70.
23. Teixeira JO, Urist MR. Bone morphogenetic protein-induced repair of compartmentalized segmental diaphyseal defects. Arch Orthop Trauma Surg 1998;117: 27–34.
24. Gerhart TN, Kirker-Head CA, Kriz MJ, Holtorp ME, Hennig GE, Hipp J et al. Healing segemental femoral defects in sheep using recombinant human bone morphogentic protein. Clin Orthop 1993;293:317–26.
25. Sciadini MF, Dawson JM, Johnson KD. Bovine-derived protein as a bone graft substitute in a canine segmental defect model. J Orthop Trauma 1997;11:496–508.
26. Cook SD, Wolfe MW, Salkeld SL, Rueger DC. Effect of recombinant human osteogenic protein-1 on healing of segmental defects in non-human primates. J Bone Joint Surg (Am) 1995;77:734–50.

27. Boyne PJ. Animal studies of the application of rhBMP-2 in maxillofacial reconstruction. Bone 1996;19:83S–92S.
28. Hanish O, Tatakis DN, Rohrer MD, Wörhle PS, Wozney JM, Wikesjö UME. Bone formation and osteointegration stimulated by rhBMP-2 following subantral augmentation procedures in nonhuman primates. Int J Oral Maxillofac Implants 1997;12:785–92.
29. Sigurdsson TJ, Fu E, Tatakis DN, Rohrer MD, Wikesjö UME. Bone morphogenetic protein-2 enhances peri-implant bone regeneration and osteointegration. Clin Oral Implants Res 1997;8:367–74.
30. Sandhu HS, Grewal HS, Parvantaneni H. Bone grafting for spinal fusion. Orthop Clin North Am 1999;30: 685–98.
31. Sandhu HS, Kanim LE, Kabo JM, Toth JM, Zegen EN, Liu D et al. Effective doses of recombinant human bone morphogenetic protein-2 in experimental spinal fusion. Spine 1996;21:2115–22.
32. Boden SD, Martin GJ, Morone MA, Ugbo JL, Moskovitz PA. Posterolateral lumbar intertransverse process spine arhrodesis with recombinant human bone morphogenetic protein 2/hydroxyapatite–tricalcium phosphate after laminectomy in the human primate. Spine 1999; 24:1179–85.
33. Urist MR. Experimental delivery systems for bone morphogenetic protein. In: Wise DL, Altobelli DE, Schwartz ER, Gresser JD, Trantolo DJ, Yaszemski M, editors. Handbook of Biomaterials and Applications, Section 3: Orthopedic Biomaterials Applications. Boston: Marcel Dekker, 1995;1093–133.
34. Guillemin G, Patat JL, Fournie J, Chetail M. The use of coral as a bone substitute. J Biomed Mater Res 1987; 21:557–67.
35. Kakiuchi M, Ono K. Preparation of bank bone using defatting, freeze-drying, and sterilization with ethylene oxide gas. Int Orthop 1996;20:147–52.
36. Johnson EE, Urist MR. Human bone morphogenetic protein allografting of resistant femoral nonunions. Clin Orthop Rel Res 2000;371:61–74.
37. Sailer HF, Kolb E. Application of purified bone morphogenetic protein (BMP) in carnio-maxillo-facial surgery. BMP in comprised surgical reconstructions using titanium implants. J Craniomaxillofac Surg 1994; 22:2–11.
38. Bai MH, Liu XY, Ge BF, Yallg C, Chen DA. An implant of a composite of bovine bone morphogenetic protein and plaster of paris for treatment of femoral shaft nonunions. Int Surg 1996;81:390–92.
39. Hu YY. Experimental studies on reconstituted xenograft and its clinical application. Chung Hua Wai Ko Tsa Chih 1993;31:709–13.
40. Camargo PM, Wolinsky LE, Wagner WR, Burgess AV. Use of bovine-derived bone protein complex for treatment of periodontal defects: results of a human feasibility study (abstract). Surfaces in Biomaterials 2000, Aug 30–Sept 2, 2000. Scottsdale Princess Resort, Scottsdale, Arizona, USA, 19-23.
41. Lindholm TS, Hietala O, Birr E, Ulmanen M. Developing a bioimplant of coral and extracted BMP for clinical use. In: Lindholm TS, editor. Advances in Skeletal Reconstruction using Bone Morphogenetic Proteins. New Jersey: World Scientific, 2002;142–56.
42. Lourenco E. BMPs in oral clinical application. In: Lindholm TS, editor. Advances in Skeletal Reconstruction using Bone Morphogenetic Proteins. New Jersey: World Scientific, 2002;290–306.
43. Riedel GE, Valentin-Opran A. Clinical evaluation of rhBMP-2/ACS in orthopedic trauma: A progress report. J Orthop Trauma 1999;22:663–5.
44. Boyne PJ, Marx RE, Nevins M, Triplett G, Lilly L, Alder M et al. A feasibility study evaluating rhBMP-2/absorbable collagen sponge for maxillary sinus floor augmentation. Int J Periodontal Rest Dent 1997;17: 11–25.
45. Boden SD, Zdeblick TA, Sandhu HS, Heim SE. The use of rhBMP-2 in interbody fusion cages. Definitive evidence of osteoinduction in humans: A preliminary report. Spine 2000;25:376–81.
46. Zijderveld SA, Giltaij LR, van den Bergh JPA, ten Bruggenkate CM, Tuinzing DB. Pre-clinical and clinical experiences with BMP-2 and BMP-7 in sinus floor elevation surgery: A comparison. J Musculoskel Res 2002;6:43–54.
47. Giltaij LR. BMP-7 in orthopedic applications: A review. J Musculoskel Res 2002;6:55–62.
48. Shimmin A. Review of 114 challenging orthopaedic cases treated with OP-1. In: Lindholm TS, editor. Advances in Skeletal Reconstruction using Bone Morphogenetic Proteins. New Jersey: World Scientific, 2002;411–27.
49. McKee M, Wild L, Schemitsch E et al. The treatment of atrophic, recalcitrant long-bone nonunion with human recombinant bone morphogenetic protein-7: results of a prospective study. Presented at the Orthopedic Trauma Association 18th Annual Meeting. Oct. 11–13, 2002, Toronto.
50. Duneas N, Crooks J, Ripamonti. Transforming growth factor beta-1: induction of bone morphogentic protein gene expression during endochondral bone formation in the baboon, and synergistic interaction with osteogenic protein-1 (BMP-7). Growth Factors 1998;15: 259–77.
51. Sonobe J, Bessho K, Kaihara S, Okubo Y, Iizuka T. Bone induction by BMP-2 expressing adenoviral vector in rats under treatment with FK 506. J Musculoskel Res 2002;6:23–29.

54 Reconstituted Bone Xenograft as a Novel Approach to Using Xenogeneic Bone

Y. Hu and J. Liu

Introduction

Bone grafting has been widely used in orthopaedic surgery for treatment of nonunions and bony defects. Autograft is the most preferred form of augmenting a fracture healing. However, the bone harvested is always limited and the retrieval procedure may inflict additional morbidity on the patient, especially in the elderly and children. Allografts, which are in common clinical use as an alternative to autograft, are biologically inferior to autografts, are sometimes associated with complications from cross infection of AIDS and hepatitis B and C, and again they have limited sources. Over the years every effort has been made by orthopedists in search of bone grafting materials from other sources.

Xenogeneic Bone Graft

Xenogeneic bone (from animals) is rich in sources, freely available, and proves to be safe for clinical use. But xenogeneic bone, when used in its unprocessed form, generates intense immune rejection which is well documented in the literature and prevents it from being used in clinical seetings [1,2]. The commercially available xenogenous bone products, such as Kiel bone, Oswestry bone, and Bio-oss, all have absent osteoinductivity, serving mainly as an osteoconductive scaffold. These products were once in clinical use with varying degrees of success but have not gained popularity.

It was postulated in the 1960s that the antigenicity of xenograft is primarily localized to the blood elements contained in bone, rather than to the bone cells themselves. Employing frozen sectioning and immunofluorescense assay on undecalcified specimens, Yunyu Hu et al. proceeded with studies of antigenicity of xenografts, and they first reported that the antigenicity of the graft is primarily located in the osteocytes and endothelium of the Haversian canal, while there is hardly any antigenicity detected in the collagen matrix under the circumstances of implantation [3]. They thereby devised a simple, direct-viewing, highly sensitive and reliable method for detecting antigenicity in the bone xenograft, which is useful in the assessment of the varied techniques used to process xenografts [3]. Based on the belief that bone is one of the best carriers for osteoinductive factors, the authors of this study went on to approach processing xenogeneic bone.

Traditional methods for treatment of the xenografts, such as freezing, freeze-drying, decalcifying, calcination, radiation, and deproteinization, all turned out to be unsatisfactory. Some treatments are too weak to eliminate the antigenicity of the xenografts, while others are so strong as to destroy the bioactive agents inherent in the graft. This led the authors to a belief that the key to eliminating the antigenicity of the xenogeneic bone without compromising its osteoinductivity may well lie in the way that the two are treated separately, taking into consideration the fact that the antigenicity and inductivity of the xenograft share the common material base consisting of proteins. With this in mind, the authors of this study developed a xenograft-based biosynthetic material to heal fractures, repair bone defects, and treat several other orthopedic conditions. The experimental results and eight years of clinical trial have provided convincing evidence of its good osteoinductive and osteoconductive properties.

Preparation of Reconstituted Bone Xenograft

Bovine cancellous bone granules 3~5 mm^3 in size were defatted using chloroform and ethanol, deproteinized with hydrogen peroxide, and partially decalcified by immersing them in 0.6 N Hcl at 25 °C for 3 minutes. Scanning electron microscopy (SEM) of the cancellous bone showed a regular porous structure, with the pore size 300~500 μm in diameter and the wall thickness between the pores 60~100 μm (Figure 54.1a). Bone morphogenetic protein (BMP) was obtained according to the method described by Urist et al. with some modifications and the end product is a crude extract of BMP [4]. Then the cancellous bone carrier and BMP were recombined in the following manner: bBMP aggregate was redissolved in 4 M guanidine hydrochloride, and a certain amount of partially decalcified cancellous bone (e.g., with the BMP/carrier ratio being 1:10 or 1:20 by weight) was added, and the air was dispelled from the pores of the cancellous bone framework in a vacuum condition. The resulting composite, designated as reconstituted bone xenograft (RBX), was then dialyzed against distilled water, freeze-dried, and sterilized. SEM of the composite demonstrated a network-like appearance of BMP fraction precipitated in the pores of the cancellous graft (Figure 54.1b).

In view of the less than optimal biomechanical performance of the granular form, especially with regard to some limitations in the reconstruction of segmental bony defect, a massive form of RBX was fabricated as a supplement to RBX products. As starting material, cancellous bone blocks from the proximal part of bovine humerus measuring 20 mm × 15 mm × 10 mm (for use in the canine model) or 5 mm × 5 mm × 15 mm (in the rabbit model) were defatted, partially decalcified, and deproteinized to make of them antigen-extracted massive

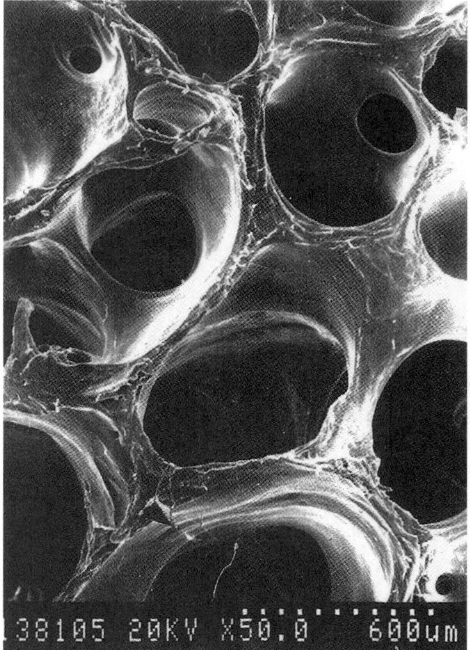

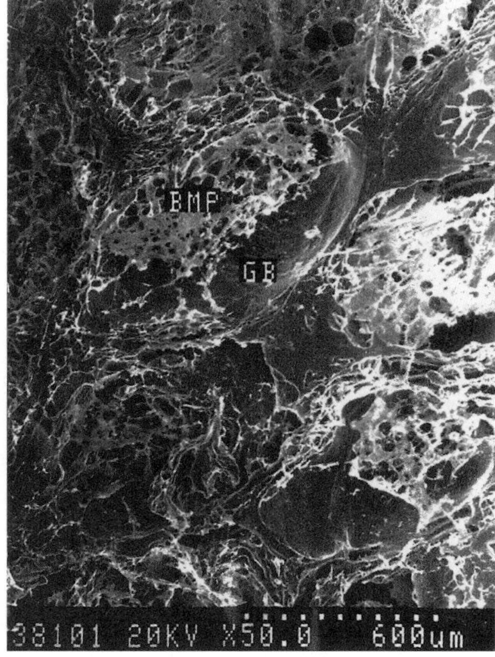

Figure 54.1. a,b SEM showing microstructure of RBX: **a** bovine cancellous bone carrier; **b** cancellous graft with BMP fraction precipitated in the pores.

cancellous bone carriers (MBC). They were then recombined with bBMP in 1:5 ratios (by weight) and dialyzed against water, freeze-dried, and sterilized.

Bioassay of RBX

Twenty milligrams of RBX containing 0.5 mg of bBMP was implanted aseptically into the femoral muscle pouch of the mouse; partially decalcified bone carrier without BMP, which served as a control, was implanted in the same manner. One week after implantation, the grafts of RBX were entrapped and infiltrated with abundant spindle-shaped and hypertrophied cells surrounding islands of cartilage (Figure 54.2); in contrast, the control implants were encased only by fibrous tissue and muscle; in some areas proliferating cells could be seen, but no evidence of cartilage differentiation. At two weeks, the islands of cartilage were enlarged with ossification occurring in the center; the foci of chondroid, osteoid, and woven bone were observed in the pores of the grafts (Figure 54.3). At four weeks, new bone tissue was present in the composite grafts, and some trabecular bone was connected directly to the pore wall of the grafted cancellous bone; in some areas mature lamellar bone and marrow tissue were seen (Figure 54.4). In the control implants of carrier alone, richer granular tissue and even scanty chondrocytes were noted in 2 of 8 samples, while all other controls at this time failed to induce any visible osteochondral differentiation, and more than half the cancellous bone was resorbed and degraded with fibrous tissue encasing the remnants.

For immunological study, blood samples were collected at killing of animals. Undecalcified frozen sections (6 μm) of calf cortical bone were used for immunofluorescent staining, with serum of normal mouse who had received no implantation serving as control. The immunofluorescent assays with the blood sera of the implant recipients showed that all BMP-containing implantations led to positive staining in most of the animals with a very low titer of 1:2, while the animals that had received the implantation of unprocessed xenogeneic cancellous bone gave a titer as high as 1:256. Bright fluorescence in the endothelium of the Haversian canal and osteocytes confirmed the presence of circulating antibodies to the implants in the host [3].

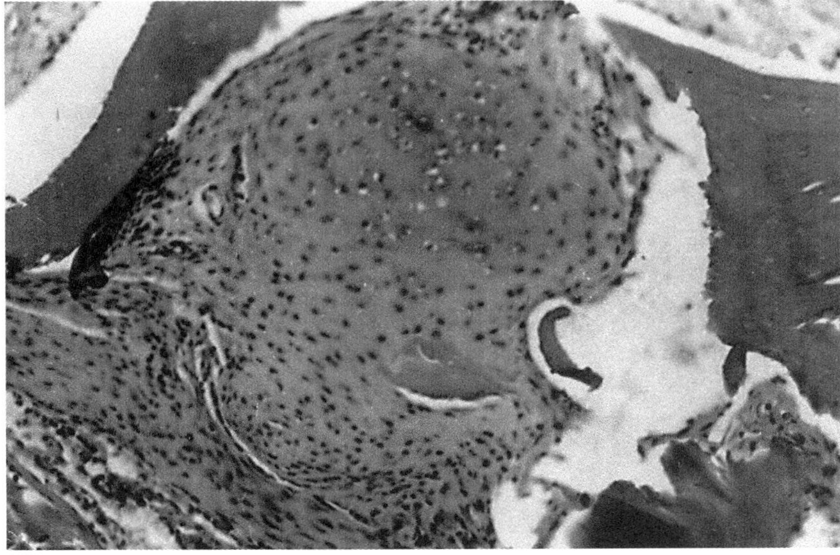

Figure 54.2. One week after implantation, islands of cartilage emerged with spindle-shaped cells surrounding them (H.E. ×80).

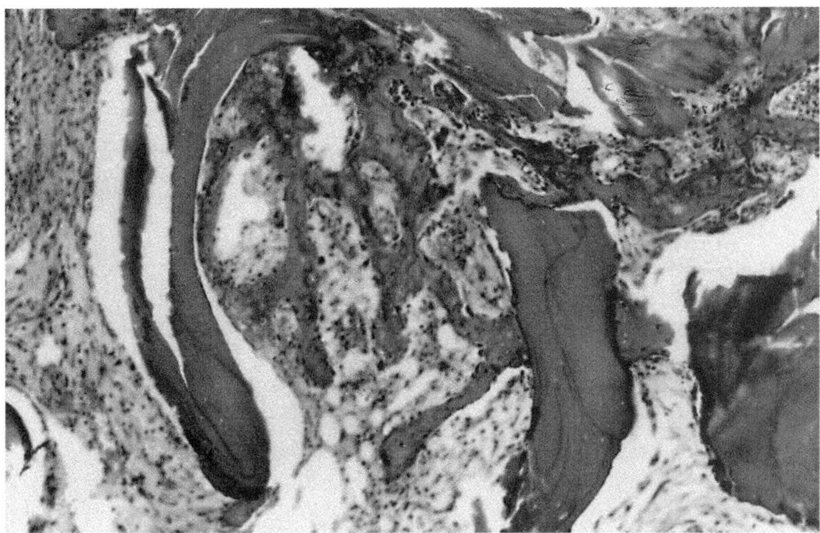

Figure 54.3. At two weeks ossification occurred in the center of cartilage islands, foci of chondroid, osteoid, and woven bone were observed (H.E. ×80).

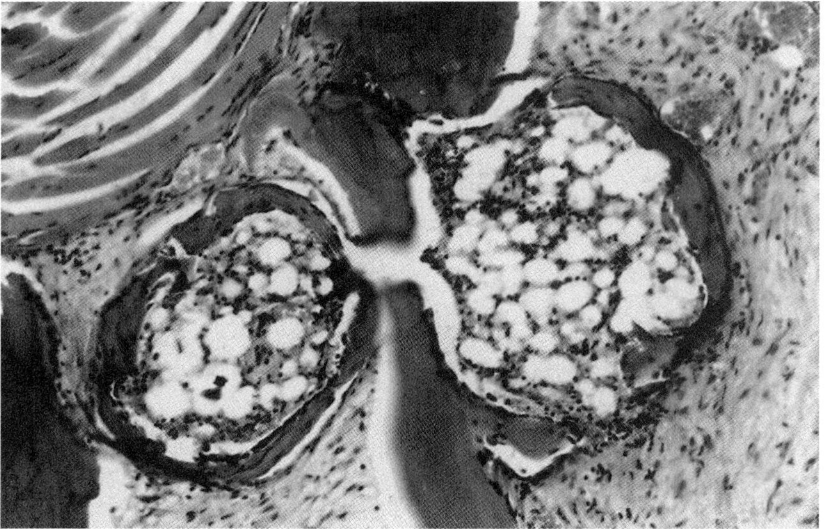

Figure 54.4. At four weeks mature lamellar bone and marrow tissue were seen (H.E . ×80).

Repair of Osseous Defect with Use of Granular Form of RBX

Bone defects 15 mm in size were created in bilateral radii of rabbits, and the defects on the left side were implanted with 80 mg of RBX which contained 4 mg of bBMP with BMP/carrier ratio 1:20, while the defects on the right were left untreated as blank controls. On day 3, spindle-shaped masses against the background of slight swelling could be seen centered over the site of implantation, but there was no redness of the skin, and no exudation either. The masses dwindled in size and increased in consistency in one week, with the wound healed without incident.

On the control side, no conspicuous mass could be discerned and the wound healed by first intention.

Radiologically, considerable callus, dense and irregular in shape, was noted in left radius that had been implanted with RBX, with the implant itself partially resorbed at 4 weeks. At 8 weeks the defects was basically filled up with callus, and conspicuous remodeling could be seen in some of the specimens. At 16 weeks (Figure 54.5), all defects largely had been repaired with formation of lamellar bone and recanalization of the marrow cavity, whereas defect nonunion occurred on the control side.

Histologically, 4 weeks after operation, at the site of RBX implantation, significant chondrogenesis and osteogenesis were identified, with ingrowth of tissue into the graft pores and new bone formation (Figure 54.6). At 8 weeks, increased amounts of new bone tissue was found within and around the graft pores; pronounced creeping substitution was noted in some areas with the graft disorganized, largely resorbed and incorporated with the new bone (Figure 54.7). Periosteum, lamellar bone, and recanalized marrow cavity were identified in most of the specimens at 16 weeks, and the

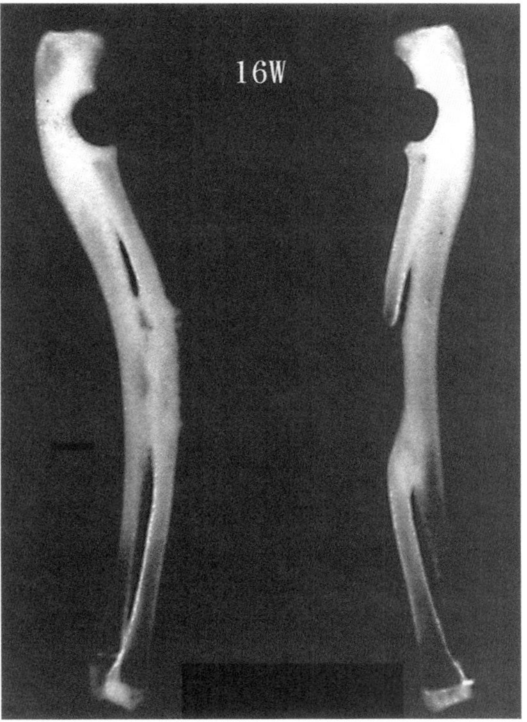

Figure 54.5. Repair of osseous defect with RBX in rabbit model: sixteen weeks postimplantation, the defect in the radius implanted with RBX had been repaired (left); nonunion occurred on the control side (right).

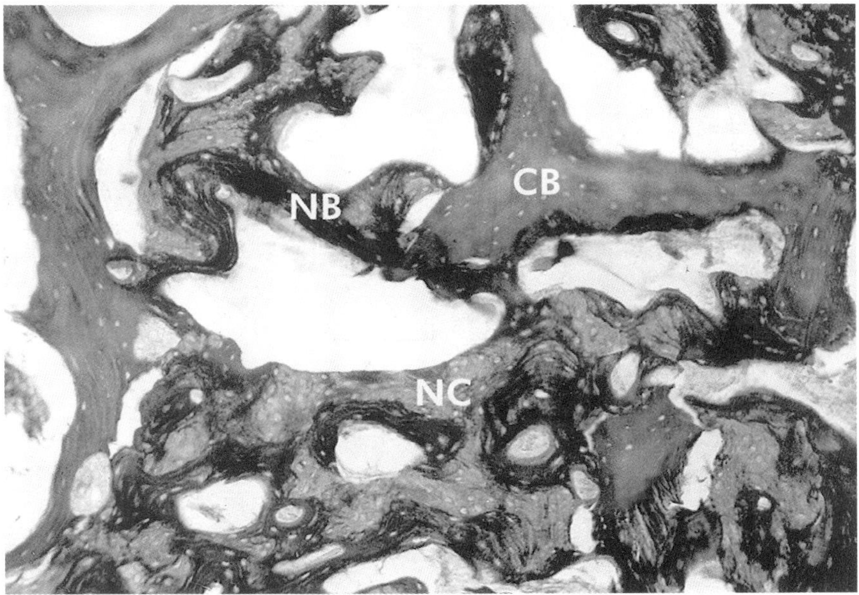

Figure 54.6. Newly formed bone and cartilage were seen surrounding the scaffold of the graft at 4 weeks (NB for new bone; NC for new cartilage; CB for cancellous bone carrier) (×100).

defects were found to have been repaired to an apparent completeness by that time (Figure 54.8). In controls, bony defects were filled up with scar tissue.

Tracing the osteogenesis process using the tetracycline double-labeling technique showed active new bone formation with quite a number of primitive Haversian systems in the early stage after implantation of RBX, while a somewhat sparse distribution of them was noted at the later stage, suggesting that the metabolism of bone tissue had entered a relatively stationary phase and the remodeling process was near completion.

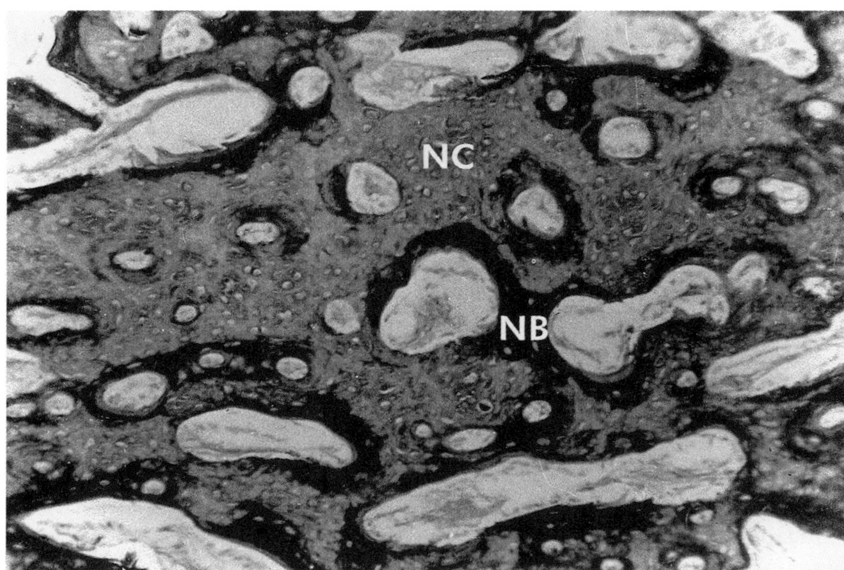

Figure 54.7. At 8 weeks the implants were resorbed and replaced by new cartilage and trabeculae (NB for new bone; NC for new cartilage) (×100).

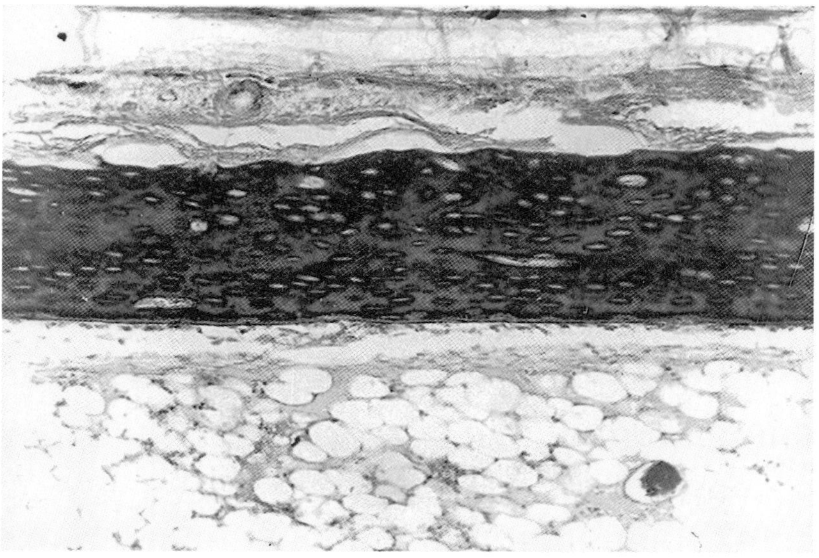

Figure 54.8. At 16 weeks cortical bone was seen with recanalization of the marrow cavity (×50).

Repair of Osseous Defect with Use of the Massive Form of RBX (MRBX)

RBX in its original granular form has some limitations to the reconstruction of segmental bony defect in view of its less than optimal biomechanical performance. The massive form represents an important supplement to RBX products and to some extent has made up for its disadvantages in this regard as evidenced by the findings in this study.

Rabbit Model. Bilateral segmental osseo-periosteal defects 15 mm in length were created in the radii of rabbits. Each animal received MRBX implantation (carrier/bBMP ratio 5:1 by weight) in the left forelimb, and implantation of MBC in the right forelimb. In animals of the study group who had received MRBX implants, fibrous tissue ingrowth was seen in the grafts leading to formation of multiple new cartilaginous and osseous islands at 4 weeks post-implantation; massive new bone formation through endochondral and appositional ossification was identified at 8 weeks; remodeling of the new bone and disorganization and phagocytosis of MRBX was noted at 12 weeks; the contour of a diaphysis with a normal structure which had cortex in continuity and recanalized medullary canal could be discerned at 16 weeks; and finally at 20 weeks the defects were all repaired and no MRBX remnants were seen. Meanwhile, bone union was noted in only 4 of 8 specimens in the MBC group, with induced new bone taking up a much lower proportion in the defect area as compared with its counterpart in the MRBX group. Nonunion occurred in all of the blank controls where the defects had been repaired by fibrous connective tissue. This Suggests a combined effect of osteoinduction and osteoconduction by MRBX in repairing the bony defects.

Canine Model. Adult dogs weighting 20~25 kg were used. Bilateral osseo-periosteal defects 20 mm in length were created in the radii and blocks of MRBX were implanted into the defects on the left side, with the defects on the right side receiving no implantation serving as controls. One month postoperation, some callus of low density was seen at both osteotomy sites, the contour of MRBX grafts becoming obscure and irregular in shape with lower density. Multiple osseous islands were seen scattered at the repair site 2 months later. At 4 months osseous continuity could be defined between the callus at the proximal end of the fracture and the osseous islands in the defects, with the new bone having increased density. The radiodensity of the new bone approximated that of the host bone with well-defined cortical continuity in between and partially recanalized canal by the sixth month (Figure 54.9). Histologically, cartilage and a substantial amount of mature new bone was seen bridging the defect by that time. The bridging reparative tissue assumed an appearance of osseous tissue intermingled with cartilage, with patches of mature lamellar bone that had merged into larger ones at the center of the defect. In contrast to this, the defects that had been left without implantation failed to reach osseous union even by 6 months after operation, there being closed medullary canal in spite of some callus at free bone ends at 3 months. Histologically these defects were found to have been repaired by large amount of fibrous connective tissue.

Significant differences were seen between the limbs in the study group and the controls when evaluated using $^{99}Tc^{m}$ SPECT at 3 months. In the forelimb that had received MRBX implantation there appeared intensified radioisotope imaging, with concurrent image intensification at the defect site, suggesting active bone growth and repair at free bone ends as well as at the site of MRBX implantation. While in the blank controls an intensified image was noted at bone ends only, none was discerned in the defect area.

Masive xenogenous carrier with its integrated structural frame has demonstrated good osteoconductive effects in repairing larger bone defects, so we can look forward to broadening the indications of the use of RBX in the treatment of bone defects and other orthopedic conditions.

Clinical Application of RBX

Clinically, RBX has been used at this institution to treat 198 patients who had nonunions or bone

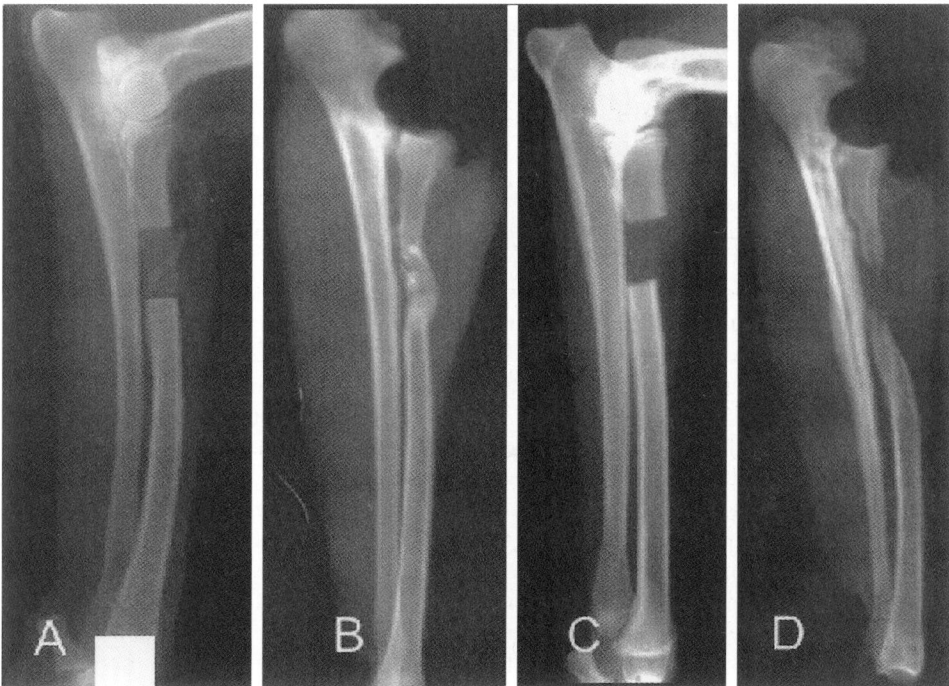

Figure 54.9. **A–D** Repair of osseous defect with MRBX in canine model: (**A**) The framework of MRBX was clearly seen between the cut ends 1 hour postoperation; (**B**) radiodensity of the new bone approximated that of the host bone with well-defined cortical continuity in between and partial recanalization of medullary cavity at 6 months; (**C**) blank control one hour postoperation; (**D**) nonunion persisted in the control even by 6 months after implantation.

defects involving humeri, ulnae, radii, phalanges, femorae and tibiae, caused by trauma or as a result of ablative tumor surgery. Each patient was implanted with 30~60 mg of BMP combined with a certain amount of bovine cancellous bone (bCB) carrier in accordance with the size and location of the defect and other conditions of the patient, with the BMP/carrier ratio being 1:20, 1:50, or 1:100.

The postoperative course was uneventful in all cases treated with RBX implantation, with wounds healed by first intention. A few patients had some measure of local swelling which resolved within one week, necessitating no special care. Also, no abnormalities were detected in relevant immunological testing, including circulating immune complex, soluble interleukin II receptors, and T-lymphocyte subpopulation, etc. One hundred fifty-six patients had a follow-up of three years or longer, with bony union occurring in 91% and the time to bony union ranging from 2 to 6 months. Thus, wound healing, osteogenesis, and reconstruction of bone defect all went unaffected. Postoperatively, SPECT showed that the radionuclide was concentrated at the graft site, pointing to the strong osteoinductive capacity of RBX.

We have also used RBX for the treatment of osteonecrosis of femoral head in a few patients, and the preliminary results are rather encouraging. The results confirmed the potential for RBX to effect more rapid healing and filling up of defects with viable bone. However, further basic and clinical research involving long-term follow-up are needed to establish its role as an effective treatment modality.

Case 1. An eight-year-old boy presented with congenital pseudarthrosis in the lower third of his right tibia, and he had a shortened leg (Figure 54.10). He underwent correction of the deformity and received RBX implantation; in addition he had his proximal tibia lengthened

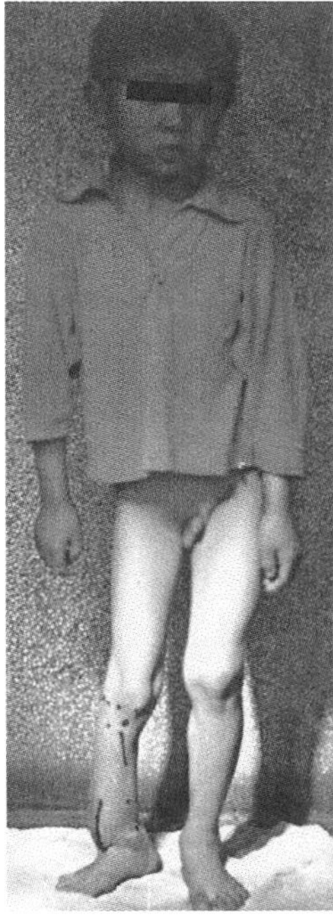
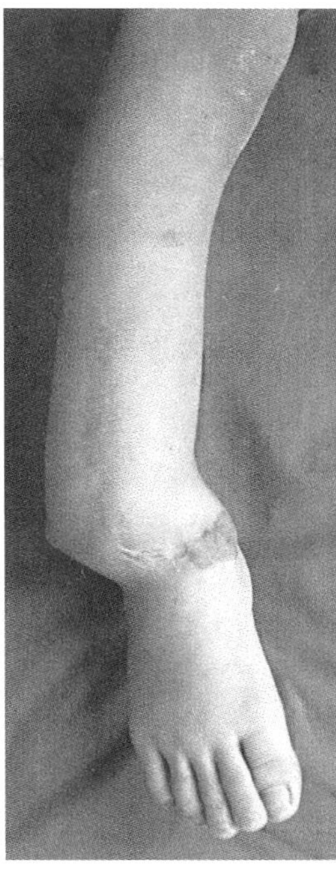

Figure 54.10. The preoperative general view and close-up view of the affected limb in Case 1.

using an external fixator. Two weeks postoperatively, there appeared a small amount of callus, which grew into a massive one to make the fracture line invisible at 2 months; and at 4 months the main findings were remodeling of the graft, recanalization of marrow cavity, and bony union (Figure 54.11). At a follow-up evaluation 3 years later, the patient was found to have equal length of the lower extremities, full weight bearing at the right side and good function of the affected limb (Figure 54.12).

Case 2. A girl of nine was found to have a bone cyst in the lower third of her right humerus 3 years prior to admission, and she had had five operations elsewhere involving autogenous bone grafting and allograft from her parents. All the previous procedures ended with recurrences and pathological fractures. She was treated with curettage and cauterization using 50% zinc chloride of the lesion, in combination with RBX implantation filling up the cavity. Two months postoperatively, massive callus with homogeneous density that was indistinguishable from the surrounding normal bone could be seen; at 4 months recanalization of marrow cavity was

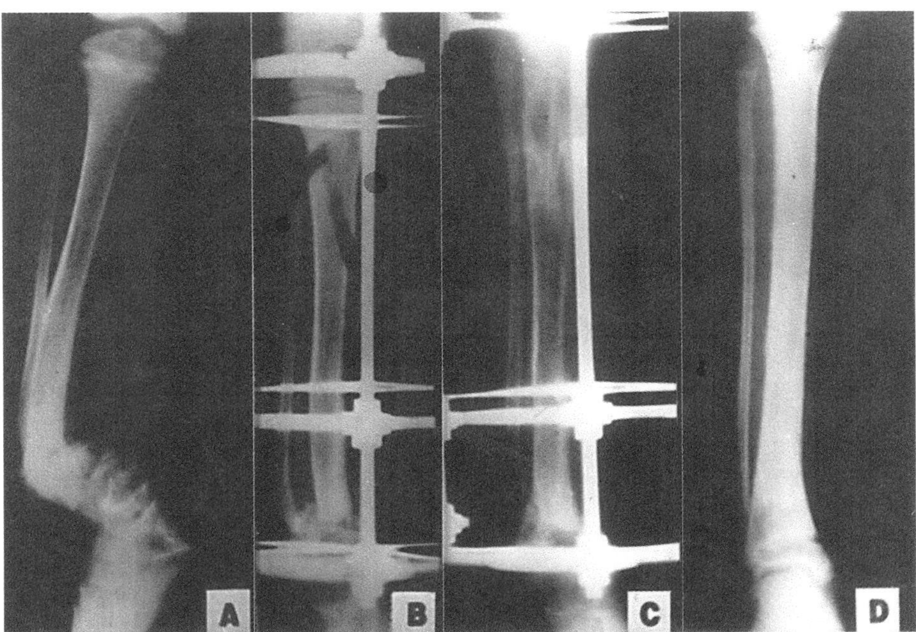

Figure 54.11. A–D Radiographs of the distal tibia of Case 1: (**A**) preoperative X-ray; (**B**) 2 weeks postoperatively; (**C**) 4 months postoperatively; (**D**) 3 years postoperatively.

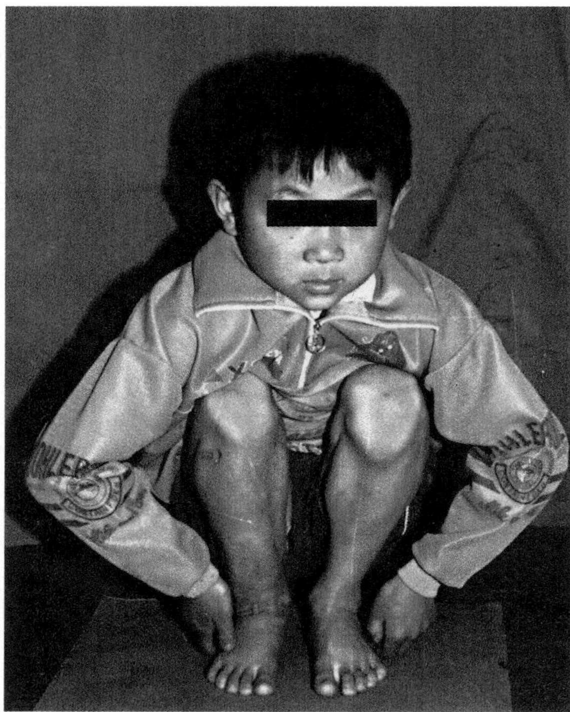

Figure 54.12. At 3 years follow-up the patient in Case 1 had equal length of the lower extremities and good function of the affected limb.

evident. At 3 years postoperation, the patient had no recurrence of the lesion and good limb function, X-ray examination demonstrating graft remodeling (Figure 54.13). Now she has normal function of her right arm.

Case 3. A 24-year-old woman suffered an injury to her left proximal tibia in a traffic accident 3 years earlier. This was a very serious case of open fracture which resulted in nonunion, with bony defects, skin scarring, and stiff knee and ankle (Figure 54.14). She had been advised to amputate her left leg elsewhere. In our hospital, bone lengthening was performed at the distal tibia using an external fixator. Two months later, the fracture ends were brought into contact, then RBX was implanted. At 6 months after RBX implantation bony union was already evident, and at the 3-year-follow-up there appeared well remodeled bone at the implant site, where the diaphysis was considerably thickened (Figure 54.15). The patient was walking with full weight bearing on the affected limb. She was satisfied with the reasonably good function of her leg, and has now married (Figure 54.16).

Immunogenicity of BMP and RBX

RBX is composed of BMP (bBMP) and antigen-extracted bovine cancellous bone (bCB), with a certain amount of collagen matrix retained therein. As shown in our previous studies, low-titer anti-bBMP and anti-RBX antibodies were identified in the sera of the animals who had received the implantation of RBX, while there were absent or minimum specific anti-bCB antibodies in the sera of the recipients [5,6]. In a mouse model, bCB was found to be unable to induce specific antibodies when implanted in the muscle pouch [5]. It was noted that RBX implantation led to significant ectopic new bone formation in all of the cases with the osteogenetic activity in direct correlation with BMP contained in RBX, despite the fact that low-titer antibodies were detected. Low-titer anti-RBX

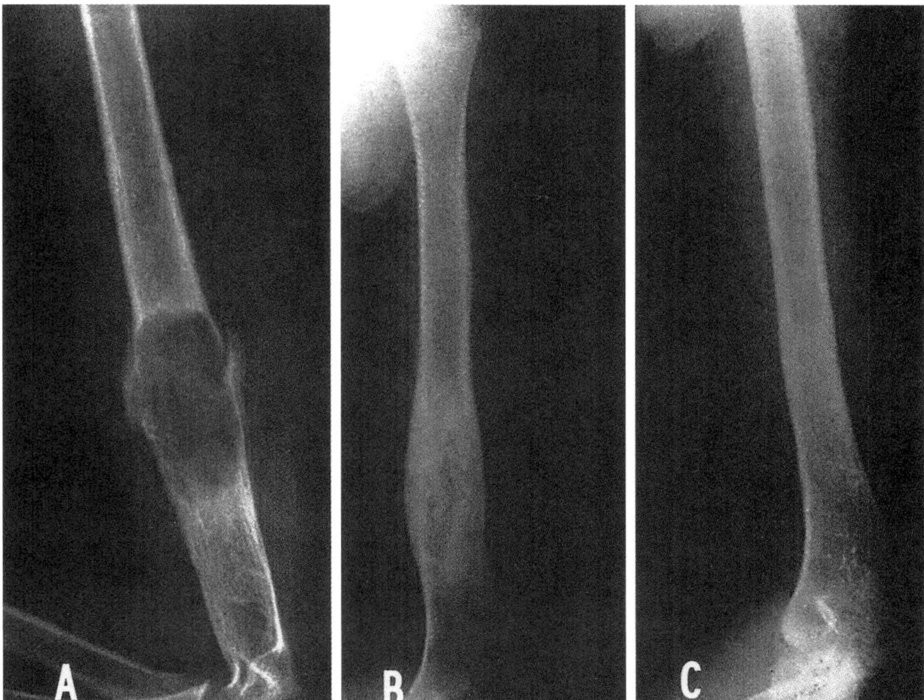

Figure 54.13. A–C The radiographs of Case 2: (**A**) pathological fracture complicating an osteolytic lesion was seen in left humerus; (**B**) four months postoperation bony union and recanalization of marrow cavity were evident; (**C**) at 3 years postoperation, no recurrence of the lesion, but graft remodeling was defined.

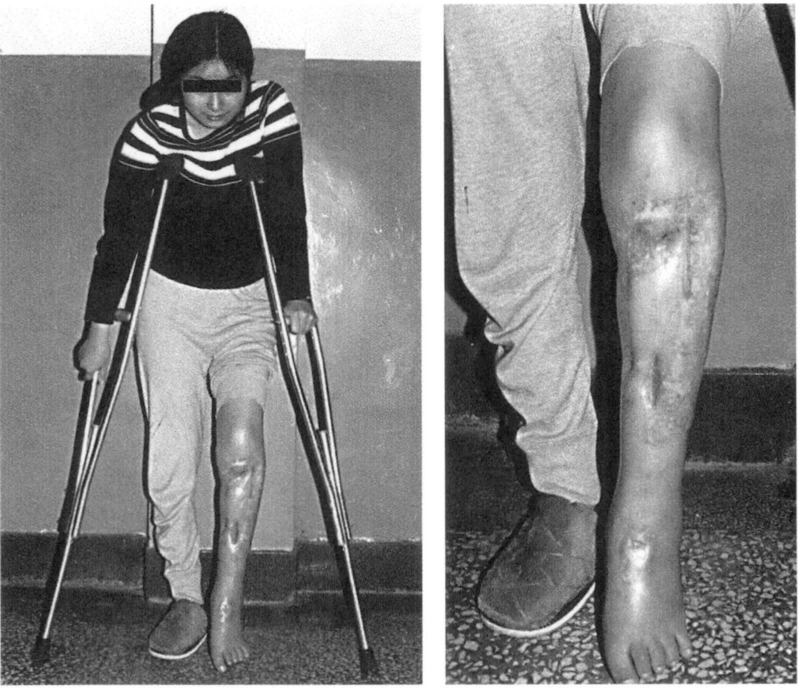

Figure 54.14. The preoperative general view and close-up view of the left lower limb in Case 3.

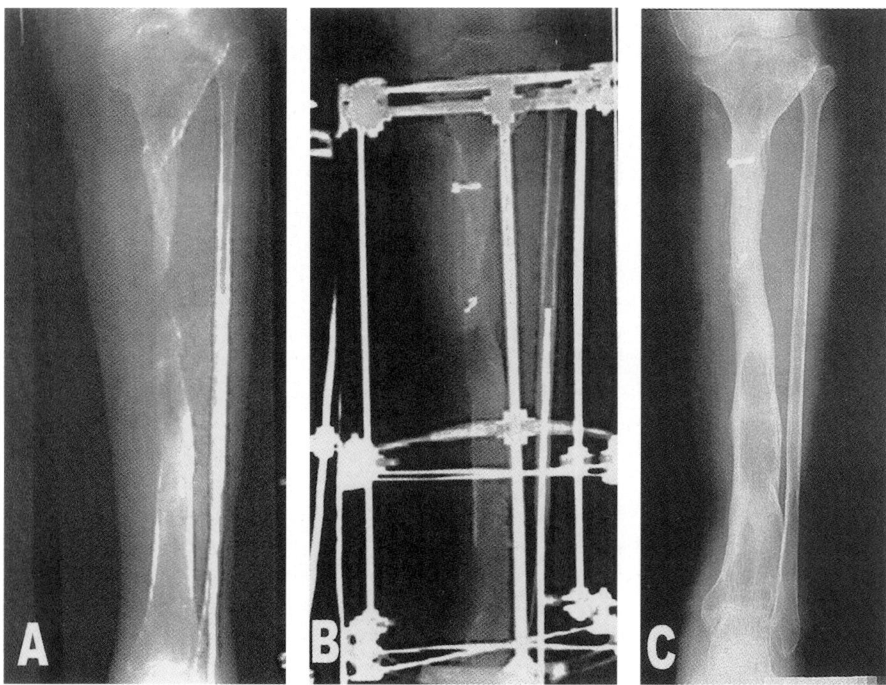

Figure 54.15. Radiographs of left proximal tibia in Case 3. (**A**) preoperative X-ray; (**B**) two months after bone lengthening was performed at the distal tibia using an external fixator, fracture ends were brought into contact. Then RBX was implanted at the fracture site in proximal tibia; (**C**) at the 3-year follow-up bony union was evident with thickened diaphysis and well remodeled bone at the implant site.

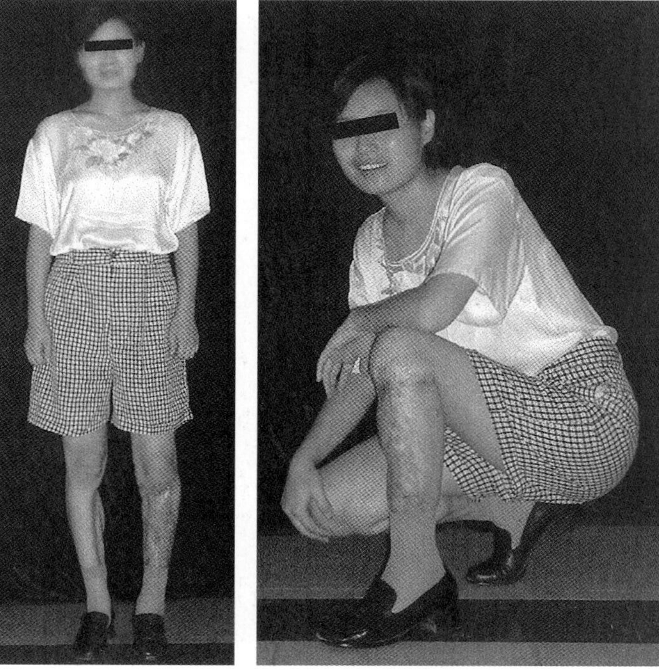

Figure 54.16. The patient in Case 3 was walking with full weight bearing on the affected limb, and she was satisfied with the reasonably good function of her leg.

antibodies were elicited by implantation of RBX products containing different amounts of BMP, with the antibody level in direct correlation to the BMP content. These findings suggest that low-titer anti-RBX antibodies did not interfere with the osteogenetic activity of RBX, and that the antigenicity manifested by RBX was mainly a result of the presence of bBMP. The results with the repair of segmental defect in the radius of the rabbit and canine also showed that there was less of an effect of low-titer antibodies on the osteogenetic activity of RBX. Currently the significance of humoral immunity in graft rejection is uncertain, and it is believed that the humoral aspect of the immune response only is a physiologic reaction to the graft on the part of the host.

BMP is a family of differentiative factors whose principal function is to induce transformation of undifferentiated mesenchymal cells into chondroblasts and osteoblasts. Native and recombinant BMP have been used successfully by many investigators for the treatment of established nonunions and spinal fusion. There have been no reports of any adverse events to date associated with the use of BMP, but the literature on the immunogenicity of BMP is sparse. In the treatment of bone defects in animal models, when the same doses of bBMP or RBX were implanted at 3-week or 6-week intervals, there were significantly higher antibody levels after the second operation than there were after the first one, with significantly decreased osteogenesis induced by bBMP or RBX [7; PDQ Worker et al., unpublished work]. Sensitization of the host by bBMP leading to immunological memory is implicated in this phenomenon, apart from its immunogenicity. Thus the repeated use of RBX or any BMP-containing biosynthetic materials to the same patient should be considered with caution.

It is generally believed that graft rejection is predominantly effected via a T lymphocytes-mediated mechanism. The studies on cellular immunity associated with RBX implantation have provided some evidence that argues for an inhibitory rather than a stimulative effect of RBX on the cellular immunity. Histologically, no conspicuous lymphocyte infiltration or other manifestations typical of graft rejection has been noted, and considerable new bone formation was seen at the graft site [8]. The results with mixed lymphocyte culture showed that RBX exerted an inhibitory effect on lymphocyte proliferation in a dose-dependent manner, varying according to the bBMP content in RBX. This has never been observed with bCB implants, sustaining the assertion that the inhibitory effect that RBX has on T cell proliferation is most likely derived from the bBMP component. In another in vitro study BMP was found to inhibit the polyclonal activator-stimulated T lymphocyte proliferation by suppressing the expression of IL-2R and production of IL-2 by T cells [9]. Furthermore, BMP receptors were found present on the activated T lymphocytes [9]. Taken together, RBX implant can elicit an inhibitory effect on cellular immunity of the host with BMP playing an important dual role as an osteoinductive agent and concurrently an immuno-modulating factor as well.

BMP Carrier

RBX is an additional option for combining BMP with an optimal carrier and a better way of using xenogeneic bone. The biological effect of BMP in repairing a bony defect is dose-dependent, and a strong correlation has been found between the dose of BMP and the response to its implantation in animals of different species [10]. For the BMP aggregates extracted from bone matrix, 1 to 2 mg was considered to be the minimum dose to induce any visible osseous tissue in mouse models [11]. As shown in our previous study using a rabbit model, the BMP fraction in 0.5 mg was rapidly resorbed and did not show any potential for osteogenesis after implantation, despite the fact that it has (at a dose of 2 mg) a definite osteoinductive capacity [5]. However, after recombination with processed cancellous bone this subeffective dose of BMP did induce osteochondral formation, with a typical differentiation course. Also in the rabbit model of bone defect, a mere 4 mg of BMP contained in RBX was capable of inducing chondrogenesis and osteogenesis within and around the graft pores. Incorporation of osteoinductive factors into a biomimetic scaffold can ensure successful bone regeneration; this may be because a proper delivery system can enhence the effect of BMP [12]. The reason for this in our case may be related to the collagenous matrix retained in the graft after limited deproteinization. The collagen component may slow the release of BMP, maintaining an effective concentration of BMP at the implant site; meanwhile, a synergistic effect of BMP and collagen was noted in promoting fracture healing, as has been repeatedly shown in previous studies [13]. Many authors have theorized that collagen may promote osteogenesis by serving as a carrier of BMP, or as a suitable substratum for the attachment of mesenchymal cells [14,15].

On the other hand, as a repair material of bone defects, BMP depends on a suitable framework for mechanical support and release [16]. Many biomaterials have been used as carriers of BMP with varying degrees of success, including hydroxyapatite (HA) [17], β-tricalcium phosphate (β-TCP) [18], polylactide and polyglycolide (PLA/PGA) [19], and plaster of Paris (PLP) [20]. Ideally, a carrier matrix should resorb after the new bone is formed and should not be retained at the graft site for long. Unfortunately, some of the materials are too slow to resorb, and others give rise to deleterious metabolites as they degrade, which makes them suboptimal as carriers. Bone is believed to be one of the best carriers, and the processed cancellous bone used in this study has the following advantages over the others: (1) It has a natural porous structure with varying pore sizes at different sites, which is suitable for ingrowth of tissue as shown by Flatley et al. [21] and our work. (2) It is easy to resorb not only because of its porosity but also because of the homogenicity of bone structures among vertebrate species, which permits host cells to gain easy access to the graft materials, just as in the remodeling process. (3) Biomaterials are superior to any others in that they carry with them information that would enhence the attachment and differentiation of the cells [22]. The collagen matrix contained in the cancellus provides BMP with an optimal combination with and release from the framework, giving full scope to the role of BMP as an osteoinductive agent, which compares very favorably with most artificial materials [15]. (4) Owing to the ready availability of animal tissue, this kind of graft may be economical, convenient, and have unlimited supply.

RBX in Tissue Engineering

In view of the disadvantages inherent to bone grafting as noted above, orthopedic surgery is now in the midst of a transformation from bone graft and the use of bone graft substitutes to bone tissue engineering. Tissue engineering, as an interdisciplinary field that came into being in the early 1980s, is now finding wide applicability in organ and tissue transplantation, repair of bone and cartilage defects in particular [23–25]. Tissue engineering of bone, like most tissues, requires three essential elements, i.e., the scaffolding matrix (carrier), the growth and differentiation factors, and the cellular elements. And there are three general approaches to repairing a bone defect, namely, matrix-based approaches, growth factor-based therapies, and cell-based therapies. RBX, which adopts matrix-based and growth factor-based therapies in combination with cells recruited from the host in response to its use, is in a sense a successful attempt at applying tissue engineering to healing fractures and repairing bone defects. RBX possesses a strong osteoinductive power without evoking an immune rejection, as evidenced by ten years of basic science studies and successful clinical use at the authors' institution. As a newly developed biosynthetic material, it has several advantages over other bone substitutes currently in use. Given its unlimited supply, easy processing and storage as well as its readiness for use, RBX would have a bright prospect for clinical use. And the authors believe that the addition of the growth factors other than BMP or cells with osteogenic potential (from fresh bone marrow or other sources) surely will further potentiate its osteogenic capacity.

References

1. Frank D, Keefe I. Immune response to allogeneic and xenogeneic implants of collagen and collagen derivatives. Clin Orthop 1990;260:263–79.

2. Salama R. Xenogeneic bone grafting in humans. Clin Orthop 1983;174:113–21.
3. Liu Wei, Lu Yu-Pu, Hu Yun-Yu. An immunohistochemical study of the antigenicity in bone exnograft. Clin J Orthop Surg 1989;9:53-4 (in Chinese).
4. Urist MR, Chang JJ, Lietze A, Huo YK, Brownell AG, DeLange RJ. Preparation and bioassay of bone morphogenetic protein and polypeptide fragments. In: Barnes D, Sirbaska DA, editors. Methods in Enzymology. New York. Academic Press 1987;294–312.
5. Liu Wei, Hu Yun-Yu, Lu Yu-Pu. The development of reconstituted bone xenograft and analysis of its bioactivity. Chin Med J 1991;71:378–80 (in Chinese).
6. Zhao Chang-Geng, Hu Yun-Yu, Lu-Rong, Liu Jian. The osteoinductivity and the dose-effect relationship with implantation of reconstituted bone xenograft: An experimental study. Chin J Surg 1998;36:627–9 (in Chinese).
7. Nisson OS, Urist MR. Immune effects on yield of new bone from implants of partially purified bovine BMP in dogs. Acta Orthop Scand 1990;61 suppl 235:47.
8. Luo Zhuo-Jing, Hu Yun-Yu, Wang Qian, Zhang Rong-Qin. The experimantal studies on immune response of antigen-extracted bovine cancellous bone grafting. Clin J Surg 1997;35:690–3 (in Chinese).
9. Huang Chuan-Shu, Jin Bo-Quan, Yang Lian-Jia. Studies on the inhibitory effect of BMP on T lymphocytes proliferation induced by polyclonal activators and its mechanism. Chin J Microbiol Immunol 1991;11:373–7 (in Chinese).
10. Kawai T, Urist MR. Quantitative computation of induced heterotopic bone formation by an image analysis system. Clin Orthop 1998;233:262–7.
11. Mahy PR, Urist MR. Experimental heterotopic bone formation induced by bone morphogenetic protein and recombinant human interleukin-1B. Clin Orthop 1988; 237:236–44.
12. Urist MR, Lietze A, Dawson E. β-tricalcium phosphate delivery system for bone morphogenetic protein. Clin Orthop 1984;187:277–80.
13. Joseph M Lane, Emre Tomin, Mathias PG Bostrom. Biosynthetic bone grafting. Clin Orthop 1999;367S: 107–17.
14. Nakahara H, Takaoka K, Koezuka M, Sugmoto K, Tsuda T, Ono K. Periosteal bone formation elicited by partially purified bone morphogenetic protein. Clin Orthop 1989;239:299–305.
15. Takaoka, Nakahara H, Yoshikawa H. Ectopic bone induction on and in porous hydroxyapatite combined with collagen and bone morpho-genetic protein. Clin Orthop 1988;234:250–4.
16. Jonsson EE, Urist MR, Schmalzried TP. Autogeneic cancellous bone grafts in extensive segmental ulnar defects in dogs. Clin Orthop 1989;243:254–65.
17. Robert WB, Carlton APRT, Holmes RE. Hydroxyapatite and tricalcium phophate bone graft substitutes. Orthop Clin North Am 1987;18:323–34.
18. Hollinger JO, Battistone GC. Biodegradable bone repair materials. Synthetic polymers and ceramics. Clin Orthop 1986;207:290–305.
19. Shmitz JP, Hollinger JO. A preliminary study of the osteogenic potential of a biodegradable alloplastic osteoinductive alloimplant. Clin Orthop 1988;237: 245–55.
20. Yamazaki Y, Oida S, Akimoto Y, Shioda S. Response of mouse femoral muscle to an implant of a composite of bone morphogenetic protein and plaster of Paris. Clin Orthop 1988;234:240–9.
21. Flatly TJ, Lynch KL, Benson M. Tissue response to implants of calcium phosphate ceramics in the rabbit spine. Clin Orthop 1983;179:246–52.
22. Cui Fu-Zhai. Biomimetic and tissue engineering material. In: Cui Fu-Zhai, Feng Qing-Ling, editors. J. Studies of Biomaterials. Beijing: Scientific Works Press, 170–4 (in Chinese).
23. Breibart AS, Grande DA, Kessler R. Tissue engineered bone repair of calvarial defects using cultured periosteal cells. J Plast Reconstr Surg 1998;101:567–74.
24. Chu CR, Coutts RD, Yoshioka M. Articular cartilage repair using allogeneic perichondrocyte-seeded biodegradable porous polylactic acid (PLA): A tissue-engineering study. J Biomed Mater Res 1995;29:1147–54.
25. Klein-Nulend J, Louwerse RT, Heyligers IC. Osteogenic protein 1 (OP-1, BMP-7) stimulates cartilage differentiation of human and goat perichondrium tissue in vitro. J Biomed Mater Res 1998;40:614–20.

55 Biomaterials: European Regulatory and Legal Aspects – a Synthetic Approach

Y. Debacker

Biomaterials include a wide range of different products manufactured with many different materials varying from metallic alloys and synthetic chemicals to biological or human tissue derivatives. It is thus understandable that different regulatory paths will apply to biomaterials and that submission of files may vary from country to country in Europe to gain market clearance. Under the scope of this paper we will discuss mostly the laws governing orthopedic biomaterials with an emphasis on biological derivatives, which are by far the most difficult to understand, with various legislation in many countries, with a specific chapter on bone substitutes and bone grafts.

The Basics

Since June 14 1998, a unique European directive known as the "93/42 directive" applies to all biomaterials with the known exception of human tissue derivatives, which will be specifically discussed later. This directive allows, under certain conditions, the products to be "CE marked". The CE mark theoretically allows the product to be marketed in each country of the EC. I say theoretically because each government has the possibility to deliver a veto to a product which will be considered by experts as dangerous for the safety of the population; thus, even with the CE mark a product or a range of biomaterials can be excluded from a specific market. This is actually the case in France for animal derivatives, which require additional file submissions for virus and microbiological safety.

So, it is obvious that the CE mark does not prevent manufacturers from following local rules to enter the market. Even if a product is well accepted in some European countries with good clinical efficacy it may not be possible to sell it to a neighboring country because of a specific regulation on that particular type of product. Most countries in Europe will accept "biological (animal origin) derivatives" as biomaterials and give market release with the "CE mark" only, but some will ask for special safety submissions either for market release or reimbursement by the national health system. Sometimes, this results in an unusual position, such as in France where, for instance, biological animal derivatives can be sold to hospitals with the CE mark only and not to private clinics requiring a submission to the reimbursement list (TIPS) with a complete safety file.

Classes of Biomaterials in the CE Mark Regulatory System

All biomaterials for surgical application in the CE mark system belong to class IIa, IIb, or III devices; the class will depend on rules edited by the EC based on:

invasiveness,

duration of contact (temporary, medium-term, long-term),

composition of biomaterial and especially existence of biological tissue or derivatives,

ingredients (active substances).

The class of the device will control the safety and security associated with the use of the device: for the highest class, the most safety controls are required to obtain market clearance, and pertinence of the safety risk of biomaterials will be examined in the file submitted to the notified body (the private or public agency in charge of CE agreement by the dossier review).

Somewhat strangely, most products can gain the CE mark approval without completing any clinical study. This is due to the 93/42 EC directive which allows "clinical information" to be obtained from the most pertinent scientific literature of the moment. It is clear that the CE mark ensures the safety minimizes risk of the devices and biomaterials but does not prove the clinical performance of the device.

File Submission by Product Type

Synthetic and Metallic Biomaterials

Most orthopedic devices and prostheses belong to this category. For many years, Europe has undergone a rationale for the homologation of these non-active medical devices. The development of quality insurance management integrated as a system of product development is now fully completed. Manufacturers having a complete in-house quality insurance system and ISO certification submit their files for new product homologation through the multiple "notified bodies" existing in Europe and will get their CE mark after the file review. This process will take only a few months, or even few weeks in certain cases.

Although the CE mark may not always mean clinically tested and verified by good clinical results, this situation (a device authorized without clear clinical evidence of safety or efficacy) may not continue for the long term. For these type of implants the CE mark registration gives market clearance in every country belonging to the EEC organization.

Biological Products

Biological Safety Exemptions

The CE mark procedure allows an individual country to put a "veto" against a product when that local government believes that a risk is associated with the product. In France particularly this safety risk comes with all the internal scandals which have affected the French government in recent years. Some individual countries have organized safety commissions which will examine the product file only with regard to the "biological safety" issue; when the manufacturer has proven the biological safety of the device(s), they will then get market clearance.

Rules

The CE mark is a quality control mark and is quality insurance dependant. It is recommended that manufacturers organize a good internal quality control system prior to applying for a CE mark; the bodies responsible for awarding the CE mark will regard file submissions from an ISO-certified company with a different perspective.

CE Type Review

The CE mark process is a voluntary process from the manufacturer. If the product is biological, safety and viral inactivation will be essential for some markets.

Human Tissues and Derivatives

This category of product is now widely used in orthopedic surgery for grafting when autologous bone is not available in sufficient quantities. At the time of writing this chapter there is no consensus for a unique regulation in Europe, although many multilateral discussions are ongoing. Some European countries have established regulations for donor selection only, others have a complete system from tissue retrieving, donor selection criteria to the tissue banking organization, quality control, and tissue delivery to end user. The most advanced countries in this respect are Belgium, France, Spain, UK, and Germany; some others have few or no regulation such as Italy, Greece, and Finland.

In the very near future (within the next five years) it is expected that a global regulation will be determined for human tissue circulation in Europe, similar to the blood transfusion system. This global system will help tissue circulation in Europe for better graft mapping. Discussions by experts in tissue banking are in progress to implement a proposal for such a directive.

Common Rules in the Existing Banking Systems

Donor selection:

The donation is anonymous.

Donation is free.

Consent is required from living donors, and from the relatives of living patients unable to give consent and of postmortem donors.

Donation:

The medical history is collected.

Serological tests are performed (European standards from the European association in tissue banking) with local exceptions.

Quality insurance of the tissue banks.

Control of grafted tissues.

Inactivation:

Proven methods of viral and bacterial inactivation are applied during the process.

Pooling of donors is banned.

Traceability:

All human grafts or human tissue derivatives need detailed traceability to allow the surgeon to trace the graft from the donor records.

Post implantation survey:

All biomaterials and devices implanted must be notified to the local agency of any malfunction, for example a hip prosthesis failure or breakage, infection, severe life impairment, or death.

These events must be notified by any person who has had access to such malfunctions, not only the surgeon, but also medical staff, pharmacists, GPs, manufacturing team, etc.

All these surveys go for review to a national committee, which will decide on the incidence and gravity of the event, possibly followed by recall of the lot or of all devices after discussion with the manufacturers.

The national committees report to an EC committee.

It is vital to realise that these malfunctions must be notified by any person connected with the case; thus, the responsibility of the person does not declare an adverse effect can be clearly established.

Legal Aspects of New Biomaterials in Orthopedic Practice

Growth Factors

Among the newest developments which will lead to new devices, a number of products include bone growth factors. From a technological aspect these growth factors have two different origins:

DBM and derivatives: growth factors extracted from human demineralized bone tissues, manufactured by acid extraction of the bone proteins including BMP and collagen.

rGF: recombinant growth factors including different fractions of BMP, manufactured by genetic engineering.

DBM will be treated as human tissue derivatives, and will follow the laws according to the country of origin as described above.

rGF are new types of protein belonging to the BMP family, recently discovered and synthesized by genetic engineering. In order to have long term activity on stem cells acting as a growth and transforming factor, these proteins must be introduced into the body by a "carrier" which will protect the growth factor from degradation and will carry them to the site of activity. The longer the carrier plays its role, the better in view of the CE mark legislation the carrier is generally a class III medical device. The combination of growth factor and carrier is also considered to be a medical device as a whole. If the growth factor is considered on its own and injected, for example, it will then become a pharmaceutical and will be classified as such.

Active Implants

In this category are new types of implants including active pharmaceutical ingredients, such as antibiotics, for example, combined with a "traditional" carrier as the bone substitute. In such a situation the 93/42 EC directive considers the final combination as a medical device as long as the primary goal of the final product is to replace the bone. If such a combination claims its antibiotherapy activity first, it would be considered as a pharmaceutical drug. Products on the market now include antibiotic cements and antibiotic mixed bone substitutes.

On a regulatory point of view, active ingredients include:

Pharmaceutical drugs

Recombinant biological substances

Carriers

Cells

Other biological substances

All these categories are reviewed independently by the authorities and the classification of the medical device will depend on the claims that are presented by the manufacturers to these authorities; if the primary claim of a specific product containing an active drug such as an antibiotic (orthopedic cement, bone substitute) is to fill a bone defect, it is accepted that the product will be considered as a medical device and will be commercialized with the CE mark, even if it contains a pharmaceutical substance which act as a helper to the main activity of filling the bone defect.

So, the category in which a product is classified – drug or medical device – depends also on the claims that manufacturers use in the submission, and may lead to some controversy and discussions with the health authorities. Recombinant BMP, for example, is a pharmaceutical substance when used alone, but may also be considered a medical device. Demineralized bone matrix, which is known to contain BMPs, is not a drug (with the exception of Germany where all human tissues are classified under the drug laws).

Cellular Therapy

In the above-mentioned category of "active implants" a special mention has to be made of so-called "cellular therapy" products including autologous chondrocyte cultures, keratinocyte cultures, and fibroblast cultures. The legal situation of these new techniques is not yet uniform in European countries; for example, they are banned in France at the time of writing this article (April, 2001), although a new law on cellular therapy is expected to be implemented; in that country only clinical trials of cellular therapy components is authorized (expect for keratinocyte cultures for life-threatening burns).

In the UK, Germany, Italy, Sweden, and Spain these techniques are authorized. The surgeon wanting to use a new therapy must be aware of the legal situation of the product in his particular governmental environment, and ask for authorization if he is not sure. Surgeons are using biomaterials daily and we have demonstrated that new devices will soon come onto the market with complex compositions, and sometimes legal controversy; the surgeon will then have to manage clinical situations and verify legal situations with his respective local health organization. Because surgeons are also involved in the development of these new biomaterials they must fully understand the regulatory aspects of these biomaterials (Table 55.1).

Table 55.1. Table showing the legal situation of some biomaterials of different classes

Type of biomaterial	Legal situation	Comments
Metallic	Devices, CE mark needed for commercialization	
Ceramics HAP, TCP, biphasic, ionic cements	Devices, CE marked	Hydraulic cements not yet classified for reimbursement in France
Xenomaterials	Devices, CE marked	Plus microbiological safety in some countries
Medical device plus active substance (rBMP, antibiotics, etc.)	Medical device or pharmaceutical	Class depend on claims of activity of active substance: either act as helper or primary component
Human materials and derivatives	Not medical devices in Europe (even class drugs in Germany)	Regulated specifically country by country
Cellular therapy products: Cell cultures, Stem cells	Specific laws Devices in some countries	Regulated country by country; care to be taken as to the exact local situation

Conclusions

Although a great deal of progress has been made with the implementation of the European directive 93/42 for the regulation of medical devices, establishing a unique registration in all European countries for these medical devices, it has been demonstrated in this article that the situation is not yet clear enough. Because some countries have established additional requirements for some classes of products (microbiological safety expert commissions), one must be careful in using products with the CE mark, especially products including different components and biological substances.

Human derivatives are not yet very widely used in Europe (DBM) and are regulated only on a country-by-country basis. A unique regulatory situation is still to be implemented, but some countries such as France, which have in the past experienced "scandals" with contaminated blood, have now set very restrictive laws for the homologation and commercialization of human tissues, and, since January 2001, controls on the production and importation of such biomaterials have been established.

Cellular therapy products are becoming more and more popular, some are in their last developmental stages, some are close to gaining market clearance, and some are already on the market. The regulation of these products is not defined by the European 93/42 directive and, again, the situation has to be examined country by country. In most of the EEC countries, cellular therapy products are authorized for use.

The progress of researchers and clinicians is, as always, quicker than the legal situation of medical devices, but market clearance of new products must become universal in Europe otherwise it may jeopardize the development of new, innovative biomaterials by market restrictions.

Index

A

B

D

Q

R

T